HANDBOOK OF
AIR CONDITIONING
AND REFRIGERATION

Other McGraw-Hill Books of Interest

HANDBOOK OF AIR CONDITIONING AND REFRIGERATION

Shan K. Wang

McGraw-Hill, Inc.

New York   San Francisco   Washington, D.C.   Auckland   Bogotá
Caracas   Lisbon   London   Madrid   Mexico City   Milan
Montreal   New Delhi   San Juan   Singapore
Sydney   Tokyo   Toronto

Library of Congress Cataloging-in-Publication Data

Wang, S. K. (Shan Kuo)
    Handbook of air conditioning and refrigeration / S.K. Wang.
        p.   cm.
    Includes index.
    ISBN 0-07-068138-4 (alk. paper)
    1. Air conditioning.   2. Refrigeration and refrigerating
machinery.   I. Title.
TH7687.W27   1994
697.9'3–dc20   93-1617
                                                     CIP

2 3 4 5 6 7 8 9 0   DOW/DOW   9 9 8 7 6 5 4

ISBN 0-07-068138-4

The sponsoring editor of this book was Robert W. Hauserman, and the production supervisor was Donald Schmidt. It was set in Times Roman by Publication Services.

Printed and bound by R. R. Donnelley & Sons Company.

This book is printed on acid-free paper.

This book is dedicated to my dear wife Joyce
for her encouragement, understanding, and
contributions, and to my daughter Helen
and my sons Roger and David.

CONTENTS

PREFACE

Air conditioning, or more specifically, heating, ventilating, air conditioning, and refrigeration (HVAC&R), was first systematically developed by Dr. Willis H. Carrier in the early 1900s. Because it is closely connected with the comfort and health of the people, air conditioning became one of the most significant factors in national energy consumption. Most commercial buildings in the United States were air conditioned after World War II.

In 1973, the energy crisis stimulated the development of variable-air-volume systems, energy management, and other HVAC&R technology. In 1980s, the introduction of microprocessor-based direct-digital control systems raised the technology of air conditioning and refrigeration to a higher level. Today, the standards of a successful and cost-effective new or retrofit HVAC&R project include maintaining a healthy and comfortable indoor environment with adequate outdoor ventilation air and acceptable indoor air quality with an energy index lower than the Federal and local codes, often using off-air condtioning schemes to reduce energy costs.

The purpose of this book is to provide a useful, practical, and updated technical reference for the design, selection, and operation of air conditioning and refrigeration systems. It is intended to summarize the valuable experience, calculations, and design guidelines from current technical papers, engineering manuals, standards, ASHRAE handbooks and other publications in air conditioning and refrigeration.

It is also intended to emphasize a systemwide approach, especially system operating characteristics at design-load and part-load. It provides a technical background for the proper selection and operation of optimum systems, subsystems, and equipment. This Handbook is a logical combination of practice and theory, system and control, and experience and updated new technologies.

Of the 32 chapters in this Handbook, the first thirty were written by the author, and the last two were written by Walter P. Bishop, P.E., president of Walter P. Bishop, Consulting Engineer, P.C., who has been an HVAC&R Consulting Engineer since 1948. Walter also provided many insightful comments for the other thirty chapters. Another contributor, Herbert P. Becker, P.E., reviewed Chapters 1 through 6 of this Handbook.

ACKNOWLEDGMENTS

The author wishes to express his sincere thanks to McGraw-Hill Senior Editor Robert Hauserman, G.M. Eisensberg, Robert O. Parmley, and Robert A. Parsons for their valuable guidance and kind assistance. Thanks also to ASHRAE, EIA, SMACNA, The Trane Company, Carrier Corporation, Honeywell, Johnson Controls and many others for the use of their published materials. The author also wishes to thank Leslie Kwok, Colin Chan, and Susanna Chang, who assisted in the preparation of the manuscript.

HANDBOOK OF
AIR CONDITIONING
AND REFRIGERATION

CHAPTER
1

INTRODUCTION

1.1 AIR CONDITIONING

Air conditioning is a process that performs many functions simultaneously. It conditions the air, transports it, and introduces it to the conditioned space. It also controls and maintains the temperature, humidity, air movement, air cleanliness, sound level, and pressure differential in a space within predetermined limits for the comfort and health of the occupants of the conditioned space or for the purpose of product processing.

The term HVAC&R is an abbreviation of *heating, ventilating, air conditioning,* and *refrigerating*. The combination of processes in this commonly adopted term is equivalent to the current definition of air conditioning. Because all these component processes were developed prior to the more complete concept of air conditioning, the term HVAC&R is often used by the industry.

1.2 AIR CONDITIONING SYSTEMS

An air conditioning, or HVAC&R, system is composed of components and equipment arranged in sequence to condition the air, to transport it to the conditioned space, and to control the indoor environmental parameters of a specific space within required limits.

Most air conditioning systems have the following functions:

1. Providing the cooling and heating energy required
2. Conditioning the supply air, that is, heating or cooling, humidifying or dehumidifying, cleaning and purifying, and attenuating any objectionable noise produced by the HVAC&R equipment
3. Distributing the conditioned air, containing sufficient outdoor air, to the conditioned space
4. Controlling and maintaining environmental parameters—such as temperature, humidity, cleanliness, air movement, sound level, and pressure differential between the conditioned space and surroundings—within predetermined limits

Parameters such as the size and the occupancy of the conditioned space, the indoor environmental parameters to be controlled, the quality and effectiveness of control, and the cost involved determine the various types and arrangements of components used to provide appropriate characteristics.

Air conditioning systems can be classified according to their applications into (1) comfort air conditioning systems and (2) process air conditioning systems.

Comfort Air Conditioning Systems

Comfort air conditioning systems provide occupants with a comfortable and healthy indoor environment in which to carry out their activities. The various sectors of the economy using comfort air conditioning systems are as follows:

1. The commercial sector includes office buildings, supermarkets, department stores, shopping centers, restaurants, and others. Many high-rise office buildings, including such structures as the World Trade Center in New York and the Sears Tower in Chicago, use very complicated air conditioning systems to satisfy multiple tenant requirements. For shopping malls and restaurants, air conditioning is necessary to attract customers.

2. The public sector includes such applications as indoor stadiums, libraries, museums, cinemas, theaters, concert halls and recreation centers. For example, one of the large indoor stadiums, the Superdome in New Orleans, Louisiana, can seat 78,000 people.

3. The residential and lodging sector consists of hotels, motels, apartment houses, and private homes. Many systems serving the lodging industry and apartment houses are operated continuously, on a 24-hour, seven-day-a-week schedule, since they can be occupied at any time.

4. The healthcare sector encompasses hospitals, nursing homes, and convalescent care facilities. Special air filters are generally used in hospitals to remove bacteria and particulates of submicron size from areas such as operating rooms, nurseries, and intensive care units. The relative humidity in the general clinical area is often maintained at a minimum of 30 percent in winter.

5. The transportation sector includes aircraft, automobiles, railroad cars, and buses. Passengers increasingly demand ease and environmental comfort, especially for long-distance travel. Modern airplanes flying at high altitudes may require a pressure differential of about 5 psi between the cabin and the outside atmosphere.

Table 1.1 lists the area and energy consumption of various types of buildings in the United States. According to the *Nonresidential Building Energy Consumption Survey: Characteristics of Commercial Buildings (1991),* in 1989, among 4,528,000 commercial buildings having 63.184 billion ft^2 of floor area, 81.9 percent of the total floor area was cooled, and 91.6 percent was heated.

Process Air Conditioning Systems

Process air conditioning systems provide needed indoor environmental control for manufacturing, product storage, or other research and development processes.

The following areas are examples of process air conditioning systems:

1. In textile mills, natural and many manufactured fibers are hygroscopic. Proper control of humidity increases the strength of the yarn and fabric

TABLE 1.1
Area and energy consumption of various types of buildings
in the United States

Types of buildings	Area, in billion ft^2		Energy consumption, 1000 Btu/ft^2 · year		Fuels consumption, 10^{15} Btu	
	1979	1983	1979	1983	1979	1983
Assembly	5.35	5.47	83	69	0.44	0.38
Educational	5.97	6.04	86	80	0.51	0.48
Food sales/services	1.82	2.05	200	213	0.36	0.44
Health care	1.96	2.28	240	204	0.47	0.46
Lodging	2.10	2.24	134	163	0.28	0.37
Mercantile/service	10.11	10.35	93	81	0.94	0.84
Office	7.36	8.44	123	123	0.90	1.04
Residential	2.76	2.44	82	73	0.23	0.18
Warehouse	5.96	6.70	110	76	0.65	0.51
Other	2.13	2.74	219	101	0.47	0.28
Vacant	1.14	2.52	81	73	0.09	0.18
Total	46.67	51.28			5.35	5.15

Sources: EIA, *Annual Energy Review,* 1987; and *Nonresidential Building Energy Consumption Survey: Characteristics of Commercial Buildings,* 1988. Reprinted with permission.

during processing. For many textile manufacturing processes, too high a space relative humidity can cause problems in the spinning process. On the other hand, a lower relative humidity may induce static electricity that is harmful to the production processes.

2. Many electronic products require clean rooms for the manufacture of such things as integrated circuits, since their quality is adversely affected by airborne particles. Relative humidity control is also needed to prevent corrosion and condensation and to eliminate static electricity. Temperature control maintains materials and instruments in a stable condition, and is also required for workers who wear dust-free garments.

   For example, in a Class 100 clean room in an electronic factory, the requirements may be temperature of 72°F ± 2°F, relative humidity at 45 percent ± 5 percent, and count of dust particles 0.5 microns diameter or larger not exceeding 100 particles per ft^3.

3. Precision manufacturers always need precise temperature control during their production of precision instruments, tools, and equipment. Bausch & Lomb successfully constructed a constant-temperature control room of 68°F ± 0.1°F to produce light grating products in the 1950s.

4. Pharmaceutical products require temperature, humidity, and air cleanliness control. For instance, liver extracts require a temperature of 75°F and a relative humidity of 35 percent. If the temperature exceeds 80°F, the extracts tend to deteriorate. High-efficiency air filters must be installed for most of the areas in pharmaceutical factories to prevent contamination.

5. Modern refrigerated warehouses not only store commodities in coolers at temperatures of 27°F to 32°F and frozen foods at −10°F to −20°F, but also provide control for perishable foods of relative humidity between 90 percent and 100 percent. Refrigerated storage is used to prevent deterioration. Temperature control can be performed by refrigeration systems only, but the simultaneous control of both temperature and relative humidity in the space can only be performed by process air conditioning systems.

1.3 INDIVIDUAL ROOM AND UNITARY PACKAGED SYSTEMS

Air conditioning systems can also be classified according to their construction and operating characteristics into individual room systems, unitary packaged systems, and central hydronic systems.

Individual Room Air Conditioning Systems

These systems employ either a single, self-contained, packaged room air conditioner installed in a window

(window unit), in an opening through the outside wall (through-the-wall unit) as shown in Fig. 1.1, or separated indoor-outdoor units, connected by refrigerant piping. Unlike a central system, these systems normally use a totally independent unit or units in each room.

The major components in a single packaged room air conditioner include the following:

- *Supply fan.* A device that pressurizes and supplies the conditioned air to the space.
- *Coil.* A heat transfer element made of tubes and fins. When a cooling fluid, the refrigerant, evaporates and expands directly inside the tubes and absorbs the heat energy from the ambient air during the cooling season, the device is called a *direct expansion (DX) coil.* When the hot refrigerant releases heat energy to the conditioned space during heating season, it acts as a heat pump.
- *Air filter.* A device used to remove airborne solid particles from the returned and outdoor air streams. In room air conditioners, air filters are usually made of glass fiber or synthetic material.
- *Compressor.* An apparatus compresses the refrigerant from a lower evaporating pressure to a higher condensing pressure.
- *Condenser.* An air-cooled coil that liquefies refrigerant from the gaseous state and rejects heat during the condensation process.
- *Restrictor.* A tube or valve that reduces the pressure of the refrigerant from the condensing pressure to the evaporating pressure.
- *Grilles.* A supply grille that distributes supply air and a return grille that provides an inlet for the return air.
- *Temperature control device.* A device that senses the space air temperature (sensor) and a controller

that starts or stops the compressor in order to control its cooling and heating capacity.

A *room air conditioner* is usually installed through the window. A packaged room air conditioner in a sleeve through the wall is called a *packaged terminal air conditioner (PTAC).* It becomes part of the building façade.

In separated indoor-outdoor units, the indoor unit houses the supply fan and coil. The outdoor unit houses a compressor and a condenser.

Space air is returned to the room air conditioner through the lower return grille. It is then cooled and dehumidified by the DX-coil during cooling season. The conditioned air is discharged to the space from the top supply grille. Outdoor air is introduced through a small opening before it is mixed with the return air. Some room air conditioners can provide warm air during heating season; this topic will be covered in more detail in Chapter 22.

Individual room air conditioning systems are characterized by the use of a DX-coil for a single room. This is the most simple and direct way of cooling the air. Most of the individual room systems do not employ connecting ductwork. They are available in a capacity less than three tons. A ton of refrigeration is defined as the ability to extract 12000 Btu/h from the conditioned space.

Unitary Packaged Air Conditioning Systems

These systems are also either single, self-contained packaged units or split systems. A single packaged unit contains fan(s), filter(s), DX-coil(s), compressor(s), condenser(s), and other accessories. In the split system, the indoor air-handler is the air system, containing mainly fan(s), filter(s), and DX-coil(s), and the outdoor condensing unit is the refrigeration system, composed of compressor(s) and condenser(s). Figure 1.2 shows a unitary packaged air conditioning system with a rooftop unit.

Unitary packaged systems can be used to serve a single room or multiple rooms. A supply duct is often installed for the distribution of conditioned air and a DX-coil is used to cool it. Heating can be provided by a direct-fired gas furnace, by hot water or steam in coils, or by electric heating. The size and capacity of these air conditioning systems cover a wide range, from 3 to about 220 tons in roof-top units.

Additional components can be added to these systems for operation of a heat pump system, that is, using a centralized system to reject heat during the cooling season and using condensing heat for heating during heating season. Sometimes perimeter base-

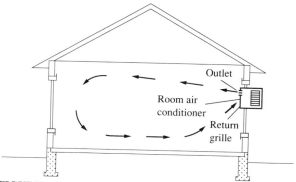

FIGURE 1.1   An individual room air conditioning system.

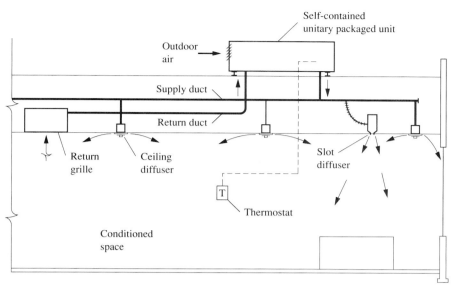

FIGURE 1.2 A unitary packaged air conditioning system.

board heaters or unit heaters, are added as a part of a unitary package system to provide heating required in the perimeter zone.

Although these systems employ a unitary package(s), they are central systems in nature because of the centralized air distributing ductwork or centralized heat rejection systems. Examples of applications are unitary packaged systems with indoor air-handlers, direct-fired gas furnaces and outside condensing units for residences; rooftop unitary packaged systems in supermarkets; and unitary packaged heat pump systems used in many office buildings and apartments.

In the rooftop self-contained unitary packaged system for a store shown in Fig. 1.2, return air from the grilles in the conditioned space enters the ceiling plenum or the return duct and is drawn into the unitary packaged unit. It is then mixed with the outdoor air, and the mixture is cooled and dehumidified at the DX-coil when cooling is required. The conditioned air is then distributed to the conditioned space through the supply duct and ceiling diffusers. During the heating season, the mixture is heated as it passes through a direct-fired gas furnace inside the packaged unit. The heated air is then supplied to the conditioned space. A control system modulates the capacity of the unitary packaged system according to the space or discharge air temperature.

1.4 CENTRAL HYDRONIC AIR CONDITIONING SYSTEMS

In a central hydronic air conditioning system, air is cooled or heated by coils filled with chilled or hot water distributed from a central cooling or heating plant. It is mostly applied to large-area buildings with many zones of conditioned space or to separate buildings.

Water has a far greater heat capacity than air. The following is a comparison of these two media for carrying heat energy at 68°F:

	Air	Water
Specific heat, Btu/lb · °F	0.243	1.0
Density, at 68°F, lb/ft^3	0.075	62.4
Heat capacity of fluid at 68°F, Btu/ft^3 · °F	0.018	62.4

The heat capacity per ft^3 of water is 3466 times greater than air. Transporting heating and cooling energy from a central plant to remote air handling units in fan rooms is far more efficient using water instead of conditioned air in a large air conditioning project. However, an additional water system lowers the evaporating temperature of the refrigerating system and makes a small- or medium-sized project more complicated and expensive.

Applications of such systems include high-rise office buildings, campuses of universities and institutes, hospitals, cultural and recreational centers, large factories, etc. A central hydronic system of a high-rise office building, the NBC Tower in Chicago, is illustrated in Fig. 1.3. A central hydronic air conditioning system consists of an air system, a water system, a central heating/cooling plant, and a control system.

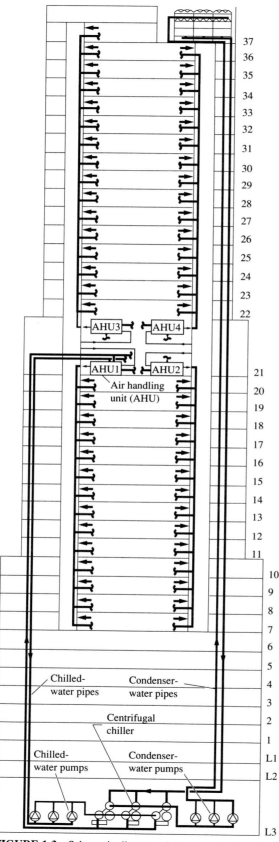

FIGURE 1.3 Schematic diagram of the central hydronic air conditioning system in NBC Tower.

Air System

The function of an air system is to condition, to transport, and to distribute the conditioned air, and to control the indoor environment according to requirements. An air system is sometimes called the air-handling system. The major components in an air system are air-handling units, fan-coil units or fan-powered units, variable-air-volume (VAV) boxes, supply/return ductwork, and space diffusion devices. Fan-coil units, fan-power units, and VAV boxes are called terminals.

A *fan-coil unit* consists of a small fan and a coil. They are usually installed at the floor, under windows, or inside the ceiling plenum directly above the conditioned space. Return air is extracted directly from the space. It is often mixed with outdoor air, then cooled or heated at the coil and supplied to the space.

A *fan-powered terminal unit* employs a small fan with or without a heating coil. It draws return air from the ceiling plenum, mixes it with conditioned air from an air-handling unit, and supplies the mixture to the conditioned space.

A *VAV box* uses a damper to modulate the volume flow rate of supply air supplied to a specific area in the conditioned space.

An *air-handling unit (AHU)* usually consists of supply fan(s), filter(s), a cooling coil, a heating coil, a mixing box, and other accessories. An AHU conditions the mixture of outdoor and recirculating air, supplies the conditioned air to the conditioned space, and extracts the returned air from the space through ductwork and space diffusion devices.

Space diffusion devices include ceiling diffusers mounted in the suspended ceiling; their purpose is to distribute the conditioned air evenly over the entire space according to requirements. They are often connected to the main supply duct by flexible ducts. The return air enters the ceiling plenum through many scattered return slots.

The NBC Tower in Chicago is a 37-story high-rise office complex constructed in the late 1980s. It has a total air conditioned area of about 900,000 ft^2. Of this, 256,840 ft^2 are used by NBC studios and other departments and 626,670 ft^2 are rental offices located on upper floors. Special air conditioning systems are employed for NBC studios and departments at the lower level.

For the rental office floors, four air-handling units are located on the twenty-first floor. Outdoor air either is mixed with the recirculating air or enters directly into the air-handling unit as shown in Fig. 1.4. The mixture flows through the filter and is then cooled and dehumidified at the cooling coil during cooling season. After that, the conditioned air is supplied to the typical floor through the supply fan, the riser, a

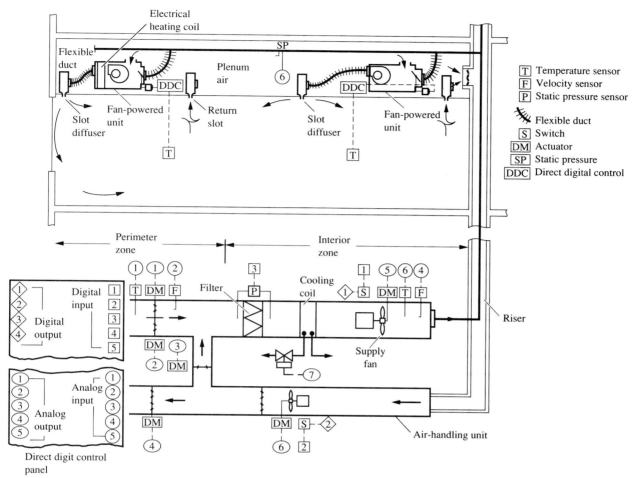

FIGURE 1.4 Schematic drawing of air system for a typical floor of offices in the NBC Tower.

vertical duct, and the main supply duct. For both the perimeter and interior zones of each typical floor, the conditioned supply air from the air-handling unit, often called the primary air, mixes with the plenum recirculating air in fan-powered units. The fan-powered units for the perimeter zone have an electric heating coil that is energized if the space temperature is lower than the predetermined limit. The fan-powered units of the interior zone have no heating coil. Primary air or the mixture is then supplied to the space through slot diffusers. Space air is returned to the air-handling unit through ceiling return slots, the ceiling plenum, the return duct, and a separate return fan.

Water System

Types of water systems include chilled- and hot-water systems, chilled- and hot-water pumps, the condenser-water system, and condenser-water pumps. The purpose of the water system is to transport (1) chilled water and hot water from the central plant

to the air-handling units, fan-coil units, and fan-powered units, and also (2) the condenser water from the cooling tower, well water or other sources of condenser cooling water to the condenser inside the central plant.

In Figs. 1.3 and 1.4, the chilled water is cooled in three centrifugal chillers and then distributed to the cooling coils of various air-handling units located on the twenty-first floor. The temperature of the chilled water leaving the coil increases after absorbing heat from the air stream flowing over the coil. Chilled water is then returned to the centrifugal chillers for recooling through the chilled water pumps.

After the condenser cooling water has been cooled in the cooling tower, it flows back to the condenser of the centrifugal chillers on Lower Level 3. The temperature of the condenser water again rises owing to the absorption of the condensing heat from the refrigerant in the condenser. After that, the condenser water is pumped to the cooling towers by the condenser-water pumps.

Central Plant

The refrigeration system in a central plant is usually in the form of a chiller package. Chiller packages cool the chilled water and act as a cold source in the central hydronic system. The boiler plant, consisting of boilers and accessories, is the heat source. Either hot water is heated or steam is generated in the boilers.

In the NBC Tower, the refrigeration system has three centrifugal chillers located in Lower Level 3 of the basement. Three cooling towers are on the roof of the building. Chilled water cools from 58°F to 42°F in the evaporator when the refrigerant is evaporated. The refrigerant is then compressed to the condensing pressure in the centrifugal compressor and condensed in liquid form in the condenser, ready for re-evaporation in the evaporator.

There is no boiler in the central plant of the NBC Tower. To compensate heat loss in the perimeter zone, heat energy is provided by the warm plenum air and the electric heating coils in the fan-powered units.

Control System

Modern air conditioning control systems for the air and water systems and for the central plant consist of electronic sensors, microprocessor-operated and -controlled modules that can analyze and perform calculations from both digital and analog input signals, that is, in the form of continuous variables. Control systems using digital signals compatible with the microprocessor are called *direct digital control (DDC) systems*. Outputs from the control modules actuate dampers, valves, and relays by means of pneumatic actuators in large buildings and by means of electric actuators for small-size projects.

In the NBC Tower, the HVAC&R system is monitored and controlled by a microprocessor-based DDC system. The DDC panels control the air-handling units, and the DDC controllers control the fan-coil unit, fan-powered unit, and VAV box (also known as the terminals). Both of them communicate with the central operating station through interface modules. In case of emergency, the fire protection system detects alarm conditions. The central operating station gives emergency directions to the occupants, operates the HVAC&R system in a smoke control mode, and actuates the sprinkler water system.

1.5 DISTRIBUTION OF SYSTEMS USAGE IN COMMERCIAL BUILDINGS

According to the survey conducted in 1989 by the Energy Information Administration of the Department of Energy of the United States, the amount of floor space, in million ft^2, in existing commercial buildings using various types of air conditioning systems for cooling and dehumidifying is as follows:

Individual room systems	19,239
Unitary packaged systems	34,753
Central hydronic systems	14,048

The total floor space of cooled buildings is 51,771 million ft^2. Much of the floor space may be included in both individual room and unitary packaged air conditioning systems.

1.6 HISTORICAL DEVELOPMENT

Since the first efforts to air-condition indoor environments in the United States at the beginning of the twentieth century, air conditioning, or HVAC&R, has become a major industry here for the following reasons:

1. Air conditioning is closely related to the demand for indoor comfort in our daily lives, resulting from the rise in living standards.
2. There is also a growing demand for indoor environmental control for industrial manufacturing processes, as well as a recognition of the beneficial effects on health and productivity of a comfortable working environment.
3. Consumption of energy for HVAC&R constitutes a substantial portion of the nation's entire use of energy for all purposes.

The historical development of air conditioning can be summarized briefly.

Central Air Conditioning Systems

As part of a heating system using fans and coils, the first rudimentary ice system, designed by McKin, Mead and White, was installed in New York City's Madison Square Garden in 1880. The system delivered air at openings under the seats. In the 1890s, a leading consulting engineer in New York City, Alfred R. Wolf, used ice at the outside air intake of the heating and ventilating system in Carnegie Hall. Another central ice system in the 1890s was installed in the Auditorium Hotel in Chicago by Buffalo Forge Company of Buffalo, New York. Early central heating and ventilating systems used steam-engine-driven fans. The mixture of outdoor air and return air was discharged into a chamber. In the top part of the chamber were installed pipe coils to heat the mixture with steam. In the bottom part was a bypass passage with damper to mix conditioned air and bypass air according to requirements.

Air conditioning was first systematically developed by Willis H. Carrier, who is recognized as the Father of Air Conditioning. In 1902, Carrier discovered the relationship between temperature and humidity and how to control them. He developed the air washer, a chamber installed with several banks of water sprays for air humidification and cleaning in 1904. His method of temperature and humidity regulation, by controlling the dew point of supply air, is still used in many industrial applications like lithographic printing plants and textile mills.

The U.S. Capitol was air-conditioned by 1929. Conditioned air was supplied from overhead diffusers to maintain a temperature of 75°F and a relative humidity of 40 percent during summer, and 80°F and 50 percent during winter. The volume of supply air was controlled by a pressure regulator to prevent cold drafts in the occupied zone.

Perhaps the first fully air conditioned office building was the Milan Building in San Antonio, Texas, which was designed by George Willis in 1928. This air conditioning system consisted of one centralized plant to serve the lower floors and many small units to serve the top office floors.

During the 1930s, Carrier developed the conduit induction system for multiroom buildings, in which recirculation of space air is induced through a heating/cooling coil by a high-velocity discharging air stream. This system supplies only a limited amount of outdoor air for the occupants.

The variable-air-volume (VAV) systems, which reduce the volume flow rate of supply air at reduced loads instead of varying the supply air temperature as in constant volume systems, were introduced in the early 1950s. These systems gained wide acceptance after the energy crisis of 1973 as a result of their lower energy consumption in comparison with constant volume systems. With many variations, VAV systems are in common use for new high-rise office buildings in the United States today.

Because of the rapid development of space technology after the 1960s, air conditioning systems for clean rooms were developed into sophisticated arrangements with extremely effective air filters.

Unitary Packaged Systems

The first room cooler developed by Frigidaire was installed in 1928 or 1929, and the *Atmospheric Cabinet* developed by the Carrier Engineering Company was first installed in May 1931. The first self-contained room air conditioner was developed by General Electric in 1930. It was a console-type unit with a hermetically sealed motor-compressor (an arrangement in which the motor and compressor are encased to gether to reduce the leakage of refrigerant) and water-cooled condenser, using sulfur dioxide as the refrigerant. Thirty of these self-contained room air conditioners were built and sold in 1931.

Early room air conditioners were rather bulky and heavy. They also required a drainage connection for the municipal water used for condensing. During the post-war period the air-cooled model was developed. It used outdoor air to absorb condensing heat, and the size and weight were greatly reduced. Annual sales of room air conditioners have exceeded 100,000 units since 1950.

Self-contained unitary packages for commercial applications, initially called *store coolers,* were introduced by the Airtemp Division of Chrysler Corporation in 1936. The early models had a refrigeration capacity of three to five tons and used a water-cooled condenser. Air-cooled unitary packages gained wide acceptance in the 1950s, and many of them were split systems incorporating an indoor air-handler and an outdoor condensing unit.

Currently, packaged units enjoy better performance and efficiency with better control of capacity to match the space load. Computerized direct digital control, one important reason for this improvement, places unitary packaged systems in a better position to compete with central hydronic systems.

Refrigeration Systems

In 1844, Dr. John Gorrie designed the first commercial reciprocating refrigerating machine in the United States. The hermetically sealed motor-compressor was first developed by General Electric Company for domestic refrigerators and sold in 1924.

Carrier first introduced the open-type gear-driven centrifugal refrigeration machine, in which the motor is housed separately from the compressor, in 1921; and the hermetic centrifugal chiller, with a hermetically sealed motor-compressor assembly, in 1934. The direct-driven hermetic centrifugal chiller was introduced in 1938 by The Trane Company. Up to 1937, the capacity of centrifugal chillers had increased to 700 tons.

During the 1930s, one of the outstanding developments in refrigeration was the discovery by Midgely and Hene of the nontoxic, nonflammable, fluorinated hydrocarbon refrigerant family called *Freon* in 1931. Refrigerant-11 and Refrigerant-12, the chlorofluorocarbons (CFCs), became widely adopted commercial products in reciprocating and centrifugal compressors. In the 1990s, new refrigerants have been developed by chemical manufacturers like DuPont to replace

CFCs so as to prevent the depletion of the ozone layer.

The first aqueous-ammonia absorption refrigeration system was invented in 1815 in Europe. In 1940, Servel introduced a unit using water as refrigerant and lithium bromide as the absorbing solution. The capacities of these units ranged from 15 to 35 tons. Not until 1945 did Carrier introduce the first large commercial lithium bromide absorption chillers. These units were developed with 100 to 700 tons of capacity, using low-pressure steam as the heat source. Recently, scroll and helical rotary positive displacement compressors with higher efficiency and better capacity control devices have been introduced. They are competitive with reciprocating compressors and with the smaller centrifugal compressors in air conditioning systems. Another trend is the development of more energy efficient centrifugal and absorption chillers for energy conservation.

1.7 SELECTION OF AIR CONDITIONING SYSTEMS

Selection of an air conditioning system depends mainly on the following:

1. Geographic location
2. Climate
3. Design criteria and quality of indoor environmental control
4. Initial investment
5. Local customs
6. Energy consumption and operating cost
7. Building characteristics

Table 1.2 compares three systems of air conditioning. Many factors must be considered when selecting an air conditioning system. Some of them, such as local customs, comfort, reliability, etc., are rather difficult to evaluate. Some items in the listing would not apply to the comparisons shown because of the many variations in design and arrangement of each specific installation in the three categories. These variations are discussed in Chapters 29 and 30.

1.8 AIR CONDITIONING PROJECT DEVELOPMENT

Potential for Air Conditioning

According to data provided by the Energy Information Administration, the total area of commercial buildings in the United States at the end of 1979 was 43.91

billion ft^2 and at the end of 1989 was 63.18 billion ft^2, an annual increase of approximately 3.7 percent from 1979 to 1989. This statistic indicates the need for a considerable number of air conditioning systems in new buildings.

Again, from EIA, at the end of 1989, of the total of 63,184 million ft^2 of floor space in commercial buildings, about 65 percent was built before 1973, that is, before the energy crisis. Of the buildings constructed before 1973, many are still in good condition, but with energy-inefficient systems. Many of these offer abundant opportunity for retrofit and energy conservation. In Korte's paper (1993), according to data compiled by the Air Conditioning and Refrigeration Institute (ARI), the installed value of nonresidential air conditioning at the end of 1991, including new and retrofit projects, was 12 billion dollars.

Basic Steps in Development

The basic steps in the development and use of a large air conditioning system are design, construction, commissioning, operation, and maintenance. Figure 1.5 is a diagram that outlines the relationship between these steps and the parties involved. The owner sets the criteria and the requirements. Design professionals design the air conditioning system and prepare the contract documents. Manufacturers supply the equipment, instruments, and materials. Contractors install and construct the air conditioning system. After construction, the air conditioning system is commissioned by a team and then handed over to the operation and maintenance group of the property management for daily operation.

Principles of Development

During the design, construction, and operation of the air conditioning systems, two principles should always be considered:

1. The system must be effective, that is, it must control and maintain the indoor environmental parameters within predetermined limits to provide satisfactory services.
2. The installation and operating cost of the system must be optimized to be cost effective.

Since the cost of energy in a building is a substantial part of total building operating and maintenance cost and since a considerable amount of energy is consumed by HVAC&R systems, energy conservation should be an important target both in design and operation of air conditioning systems.

TABLE 1.2

Comparison between individual room, unitary packaged, and central hydronic air conditioning systems

Characteristics	Individual room systems	Unitary packaged systems	Central hydronic systems
Area served	Small	Medium to large	Large
Capacity of refrigeration system, ton	Mostly < 2	3–220	Often greater than 100
Amount of outdoor air supply	Influenced by wind direction and velocity	Positive outdoor air supply Providing modulation of volume flow	Positive and guaranteed outdoor air supply Better modulation control
Space temperature control	Greater temperature fluctuation	Temperature may fluctuate as a result of step control of DX control	Better control quality
Air cleanliness	Not satisfactory owing to low-efficiency air filters installed	Satisfactory because of installation of medium-efficiency air filters	Meet any requirement because of the possibility of installing medium- or high-efficiency air filters
Space sound level	Higher space noise criteria NC curve	Lower space NC curve	Lower space NC curve because of better sound attenuation
Load diversity	No diversity	Lower diversity	Lower diversity
System component efficiency	Low	Meduim	Higher
Equipment life	Shorter	Longer	Longest
Maintenance	More maintenance work Maintenance at conditioned space	Less maintenance work Maintenance at the packaged unit	Less maintenance work Maintenance at fan room and plant room
Heat rejection	Easier for perimeter zone	Easier for low-rise building	Easy for any building
Smoke control	Cannot coordinate with smoke control system	May be integrated as part of smoke control system	Can be integrated as part of smoke control system
Operator	No operator	Operator may be required	Operator is required
Characteristics	Individual DX room air conditioner	DX packaged system	Central hydronic system
Initial cost	Lower	Medium	Higher
Operating cost	Lower	Medium	Higher
Plant room	No equipment room is required	Separate equipment room is required	Fan room and central plant room
Flexibility of operation	More flexible	Comparatively flexible	Less flexible
Installation	Simple and fast	Simpler and faster compared with central hydronic systems	Most complex
Energy metering for individual tenants	Simple	Simpler compared with central hydronic systems	Difficult
Future expansion	Flexible	Less flexible compared with individual room systems	Less flexibility
System characteristic	Individual	Central	Central
Cooling	Direct expansion DX coil	Direct expansion DX coil	Water cooling coil

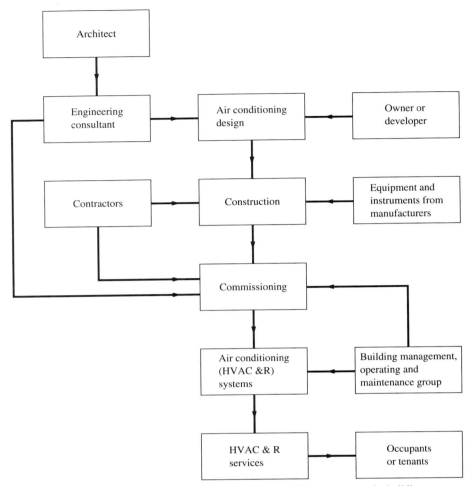

FIGURE 1.5 Steps in the development and use of air conditioning systems in buildings.

Major HVAC&R Problems

In Coad's paper (1985), a study by Wagner-Hohn-Ingles Inc. revealed that many large new buildings constructed in the early 1980s in the United States suffer complaints and major defects. High on the list of these problems are HVAC&R systems. The major problems are these:

1. *Poor indoor air quality (IAQ)—sick building syndrome.* In the 1980s, 346 field investigations conducted by the National Institute of Occupational Safety and Health (NIOSH) found that lack of outdoor air, improper use, and poor operation and maintenance of HVAC&R systems were responsible for more than half of the cases of sick building syndrome. Sick building syndrome will be covered in detail in Chapter 5.

2. *Lack of updated technology.* In recent years, there has been a rapid change in the technology of air conditioning. Various types of VAV systems in-

cluding heat recovery, thermal storage, and DDC controls have become more complicated. Many HVAC&R designers and operators are not properly equipped to apply and to use these systems. Unfortunately these sophisticated systems are constructed and operated under the same budget and schedule constraints as the less sophisticated systems.

3. *Insufficient communication between design professionals, construction groups, and operators.* Effective operation requires the operator to understand the system and to make adjustments if necessary. The operator will operate the system at his or her level of understanding. If adequate operating and maintenance documents are not provided by the designer and the contractor for the operator, the HVAC&R system may not be operated according to the designer's intentions.

4. *Overlooked commissioning.* Commissioning means testing and balancing all systems, func-

tional testing and adjusting of components and the integrated system, and adjusting and tuning the DDC controls. An air conditioning system is different from manufactured products having models and prototypes. All of the defects and errors of the prototypes can be checked and corrected during their individual tests, but the more complicated HVAC&R system, as constructed and installed, is the end product. Therefore, proper commissioning, which permits the system to perform as specified in the design documents, is extremely important. Unfortunately, the specifications seldom clearly designate competent technicians for the responsibility of commissioning the entire integrated system. Commissioning will be covered in detail in Chapter 31.

The Goal—A High-Quality HVAC&R System and a Better Environment

For the 1990s, our design goal is a HVAC&R system that provides high indoor air quality and is, at the same time, energy efficient and cost effective. The followings system features should be part of this goal:

1. Functioning effectively with satisfactory individual zone temperature controls, healthful and comfortable indoor environment, and necessary stairwell pressurization and smoke control systems during a building fire
2. Providing an adequate amount of outdoor air for occupants and an effective maintenance program to maintain a good indoor air quality
3. Energy-efficient refrigeration machinery, fans, pumps, and heating systems
4. DDC controls with intelligent control strategies
5. After-hour access to HVAC&R systems for individual tenant spaces
6. Replacing CFCs with refrigerants that do not deplete the ozone layer and that have less effect on global warming
7. Reducing on-peak operation to minimize operating cost
8. Cost effective equipment and system components

These needs of the owner, property manager, and occupants should be recognized by HVAC&R designers, contractors, and operators.

Design-Bid and Design-Build

There are two types of project development: design-bid and design-build. A design-build project is one in which engineering is done by the installing contractor, whereas a design-bid project separates the design and installation responsibilities. Some reasons for a design-build are that the project is too small to retain an engineering consultant, or that there is insufficient time to follow through the normal design-bid procedures.

1.9 DESIGN FOR AIR CONDITIONING SYSTEM

System design determines the basic characteristics. After an air conditioning system is constructed according to the design, it is difficult and expensive to change the design concept.

Engineering's Responsibilities

The normal procedure in a design-bid project includes the following steps and requirements:

1. Initiation of a construction project by owner or developer
2. Selection of a design team
3. Setting of the design criteria and indoor environmental parameters
4. Selection of conceptual alternatives for systems and subsystems; preparation of schematic layouts of HVAC&R
5. Preparation of contract documents, working drawings, specifications, materials and construction methods, commissioning guidelines
6. Competitive bidding by contractors
7. Evaluation of bids; negotiations and modification of contract documents
8. Advice on awarding of contract
9. Review of shop drawings and commissioning schedule, operating and maintenance manuals
10. Monitoring/supervision/inspection of construction
11. Supervision of commissioning: testing and balancing, functional performance tests
12. Modification of drawings to the as-built condition and the finalization of the operation and maintenance manual
13. Acceptance

Construction work starts at contract award following the bidding and negotiating and ends at the acceptance of the project after commissioning.

It is necessary for the designer to select among the available alternatives for optimum comfort, economics, energy conservation, noise, safety, flexibility, reliability, convenience, and maintainability. Experience, education and judgment all enter into the selection process. If both a complicated system and a simple system result in the same performance, the simpler system is preferred for reliability and operator convenience.

Coordination between Air Conditioning and Other Trades

Air conditioning systems are considered as one of the building services. Coordination between the HVAC&R engineer and the architect, as well as design professionals for the structure and engineers of different building services, becomes important.

Factors requiring input from both the architect and the HVAC&R engineer include the following:

1. Shape and orientation of the building
2. Thermal characteristics of the building envelope, especially the type and size of the windows and the construction of external walls and roofs
3. Location of the ductwork and piping to avoid interference with each other, or with other trades
4. Layout of the diffusers and supply and return grilles
5. Minimum clearance provided between the structural members and the suspended ceiling for the installation of ductwork and piping
6. Location and size of the rooms for central plant, fan rooms, ducts and pipe shafts

If the architect makes a decision that is thermally unsound, the HVAC&R engineer must offset the additional loads by increasing the HVAC&R system capacity. Such a lack of coordination results in greater energy consumption.

HVAC&R and other building services must coordinate on the following:

1. Utilization of daylight and the type of artificial light to be installed
2. The layout of diffusers, grilles, return inlets, and light troffers in the suspended ceiling
3. Integration with fire alarm and smoke control systems
4. Electric power and plumbing requirements for the HVAC&R equipment and lighting for equipment rooms
5. Coordination of the layout of the ductwork, piping, electric cables, etc.

Retrofit, Renovation, and Replacement

Retrofit of a HVAC&R system must be tailored to the existing building and integrated with existing systems. Each retrofit project has a reason, which may be related to safety and health, indoor environment, energy conservation, change of use, etc. For the sake of safety, smoke control systems and stairwell pressurization are items to be considered. For the sake of the occupants' health, an increase in the amount of outdoor air may be the right choice. If energy conservation is the first priority, the following items are among the many that may be considered: efficiency of lighting; energy consumption of the fans, pumps, chillers and boilers; insulation of the building envelope; and the renovation or replacement of fenestration. For improving the thermal comfort of the occupant, an increase in the supply volume flow rate, an increase in cooling and heating capacity, and the installation of appropriate controls may be considered.

Engineer's Quality Control

In 1988, the American Consulting Engineers Council (ACEC) chose "People, Professionalism, Profits: A Focus on Quality" as the theme of the annual convention in New York City. Quality does not mean perfection. In HVAC&R system design, quality means functional effectiveness, for health, comfort, safety, and cost. Quality also implies the best job, using accepted professional practices and talent. Better quality always means fewer complaints and less litigation. Of course, it also requires additional time and higher design cost. This fact must be recognized by the owner and developer.

The use of safety factors allows for the unpredictables in design, installation, and operation. For example, a safety factor of 1.1 for calculated cooling load and of 1.1 to 1.15 for calculated total pressure drop in ductwork is often used to take into account unexpected inferiority in fabrication and installation. A HVAC&R system should not be overdesigned by using a greater safety factor than actually required. The initial cost of an overdesigned HVAC&R system is always higher, and it is energy inefficient.

When a HVAC&R system design is under pressure to reduce initial cost, some avenues to be considered are as follows:

1. Select an optimum safety factor
2. Minimize redundancy, such as stand-by units
3. Conduct a detailed economic analysis for the selection of a better alternative
4. Calculate the space load, the capacity of the system, and the equipment requirements more precisely

5. Adopt optimum diversity factors based on actual experience data observed from similar buildings

A *diversity factor* is defined as the ratio of the simultaneous maximum load of a system to the sum of the individual maximum loads of the subdivisions of a system. It is also called *simultaneous use factor*. For example, in load calculations for a coil, block load is used rather than the sum of zone maximums for sizing coils for AHUs installed with modulating controls for chilled water flow rate.

The needs for a higher-quality design demand that engineers have a better understanding of the basic principles, practical aspects, and updated technology of HVAC&R systems in order to avoid overdesign or underdesign and to produce a satisfactory product.

Design of the Control System

Controlling and maintaining the indoor environmental parameters within predetermined limits depend mainly on adequate equipment capacity and quality of the control system. Energy can be saved when the systems are operated at part-load with the equipment's capacity following the load accurately by means of capacity control.

Because of the recent rapid change of HVAC&R controls from conventional to energy management systems, and then to DDC control with microprocessor intelligence, many designers have not kept pace. In 1982, Haines did a survey on HVAC&R control system design and found that many designers preferred to prepare a conceptual design and a sequence of operations and then to ask the representative of the control manufacturer to design the control system. Only one third of the designers designed the control system themselves and asked the representative of the control manufacturer to comment on it.

HVAC&R system control is a decisive factor in system performance. Many of the troubles with HVAC&R arise from inadequate controls and/or their improper use. The designer should keep pace with the development of new control technology. He or she should be able to prepare the sequence of operations and select the best-fit control sequences for the specific-terminal controllers from a variety of the manufacturers that offer equipment in the HVAC&R field. The designer may not be a specialist in the details of construction or of wiring diagrams of controllers or DDC modules, but he or she should be quite clear about the function and sequence of the desired operation, as well as the criteria for the sensors, controllers, DDC modules, and the controlled devices.

If the HVAC&R system designer does not perform these duties personally, preparation of a systems operation and maintenance manual with clear instructions would be difficult. It would also be difficult for the operator to understand the designer's intention and to operate the HVAC&R system satisfactorily.

Field Experience

It is helpful for the designer to visit similar projects that have been operated for more than two years and talk with the operator before initiating the design process. Such practice has many advantages.

1. The designer can investigate the actual performance and effectiveness of the air conditioning and control systems that he or she intends to design.

2. According to the actual operating records, the designer can judge whether the system is overdesigned, underdesigned, or the exact right choice.

3. Any complaints or problems that can be corrected may be identified.

4. The results of energy conservation measures can be evaluated from actual performance instead of theoretical calculations.

5. The designer can accumulate valuable practical experience from the visit, even from the deficiencies.

Computer-Aided Design and Drafting (CADD)

The use of computer-aided design and drafting (CADD) for air conditioning system design is steadily increasing. Nearly all of the design professionals are more or less involved in CADD. Many large manufacturers of HVAC&R products offer computer-aided load calculation programs. Personal computers provide a low-cost tool for computations and graphics. The use of desktop microcomputers yields a significant economic advantage. Recently developed expert systems for HVAC&R design are a significant advance in CADD. Details of expert systems will be covered in Chapter 18, Section 18.5.

Current CADD in HVAC&R includes the following:

- Space load calculations
- Duct design
- Chilled water and refrigerant piping sizing
- Psychrometric analysis
- Acoustical calculations

- Equipment selection and building operating cost analysis
- Energy estimation

In the future, use of CADD will steadily increase and may become an integral part of a complete computer-aided building design system (CABDS).

Eventually, all computer programs for HVAC&R system design will be developed according to certain basic principles, mathematical models and calculation methods. Many assumptions are made during the preparation of software. Unfortunately, the algorithms of most of the software prepared by private firms are not disclosed to the user, being proprietary in nature; those from government agencies and academic societies are the exception.

Engineering consultants may prefer to develop in-house computer programs in order to perform a comprehensive evaluation of design alternatives or to determine the size and capacity of the equipment. They know the factors and coefficients that are assumed for specific circumstances and improve, modify or expand the software after a given period of application. Computer programs developed in-house leave the user in full control of the application based on personal experience. If the program is complicated, it may be possible to break the program down into separate parts first and then integrate them. In-house computer programs provide effective tools to select systems from up-to-date advanced technology alternatives.

While using one of the many available computer programs, one should verify and calibrate the results against actual measured performance in order to improve the accuracy of the program results when the inputs and affecting factors are varied.

1.10 COST ANALYSIS

Cost analysis is an important tool for choosing between alternatives or evaluating the merits of a proposed retrofit.

Owning Costs and Operating Costs

There are two types of costs: owning costs and operating costs.

OWNING COSTS. Owning costs are those that are incurred even though a facility is unused, including initial costs (such fixed costs as property taxes, rent, and insurance) and salvage value, which reduces the initial cost.

Initial cost (IC), also called first cost, is the sum of the development and construction cost of the system and the part of the building cost allocated to the specific system. Construction cost usually includes equipment, material, and labor costs. Design and system administration expenses should be included as part of the initial cost.

Salvage value (S) is the recoverable value of a system or any component at the end of a depreciation period n', in years; and n' is the time period for which one deducts the depreciation D. For a constant deduction or straight-line depreciation, the annual depreciation D can be evaluated as

$$D = \frac{IC - S}{n'} \qquad (1.1)$$

This depreciation period is established by the United States Internal Revenue Service (IRS) for tax deduction purposes and is not a true measure of actual depreciation.

Administrative expenses include such items as property taxes and insurance. Property tax is often assessed as a percentage of market value of a building, which can be allocated among its various components. Insurance is a periodic premium, paid for benefits to the insured, in case of loss from damage, theft, etc.

OPERATING COSTS. These are expenses incurred from the use or sale of services from a HVAC&R system. They include the costs of energy, salaries and wages, water, materials, maintenance, and other services.

Energy costs for HVAC&R systems may be the largest single operating cost. They include electric and gas utility bills, bills from district heating and cooling plants, and costs for fuel oil from suppliers. Utilities often use rate schedules that differ between peak and off-peak periods. Energy costs vary widely across the nation. Selection of an energy source(s) significantly affects the operating costs of a system.

Maintenance costs include the necessary labor, material supplies, replacements, repairs, cleaning, painting, inspecting, testing, etc. to keep a system operating as desired. Maintenance service can be provided by an in-house staff or an outside firm, or both. Usually, more maintenance will be required during the later years of a system's life.

Life-Cycle Cost Analysis

Life-cycle cost analysis is an economic philosophy that takes into account the selected life period of the HVAC&R system or its components; the time value of money; and the factors of depreciation, inflation, and escalation of costs. This life period, used to make economic comparisons, is called the amortization period n, in years. Selection of this period can be based on *service life* or *useful life*.

SERVICE LIFE. This term reflects the expected life of a system component, as determined from statistics compiled by ASHRAE. In a system with a given amortization, with different component service lives, the life of each component is adjusted to the same amortization period.

USEFUL LIFE. The United States IRS, which lists acceptable useful life periods for various components, established this term. The IRS determines the time period over which the depreciation of the asset can be allocated for income tax purposes. This allowable depreciation, over the entire depreciation period, establishes the allowable income tax deduction each year.

Life cycle cost analysis is more comprehensive and accurate than other simpler methods of cost analysis.

Annualized Costs

ASHRAE Handbook HVAC Application 1991 introduced an economic analysis that takes into account all annualized costs for a particular air conditioning system. Annualized total system owning and operating costs C_y, expressed in constant currency dollars, can be calculated as

$$C_y = - C_{ap} + S_{al} - R_{ep} - C_{en} - M_{ain}$$
$$- P_{ro} + R_{ed} \qquad (1.2)$$

where C_{ap} = capital and interest, dollars
S_{al} = salvage value, dollars
R_{ep} = annual replacements, dollars
C_{en} = energy cost, dollars
M_{ain} = maintenance cost, dollars
P_{ro} = property tax and insurance, dollars
R_{ed} = tax and depreciation deductions, dollars

Capital Recovery Factor and Rate of Return

The *capital recovery factor* (CRF) determines by the following relationship the *uniform annual cost* (UAC) required to pay an expenditure, an initial cost (IC), or a debt:

$$C_{ap} = IC \times CRF \qquad (1.3)$$

For an effective interest rate of i' and an amortization period of n,

$$CRF = \frac{i'(1 + i)^n}{(1 + i')^n - 1} \qquad (1.4)$$

Effective interest:

$$i' = \frac{1 + i_d}{1 + j} - 1 \qquad (1.5)$$

$$i'' = \frac{1 + i_d}{1 + j_e} - 1 \qquad (1.6)$$

where i_d = discount rate
j = inflation rate
j_e = energy inflation rate

If the initial cost of an expenditure, or an investment, yields an *annual saving* (AS), the *rate of return* (ROR) on this investment can be determined from the following relationship by the trial-and-error method:

$$CRF = \frac{AS}{IC} = \frac{ROR(1 + ROR)^n}{(1 + ROR)^n - 1} \qquad (1.7)$$

In Fig. 1.6 one sees the capital recovery factors at various amortization periods n and interest rates i or rates of return ROR. After CRF and n are known, then the ROR or i can be determined from Fig. 1.6.

Refer to *ASHRAE Handbook HVAC Applications 1991* for details.

Simple Payback

Simple payback (SP) analysis gives only the recovery period, in years, of an investment. It overlooks the influences of interest rate and cost escalation, therefore, it is not suitable when there are significant differentials between i and j, or when the initial costs are significantly higher than the estimated first-year savings. Simple payback can be calculated as

$$SP = \frac{IC}{AS} \qquad (1.8)$$

Comparison of Cost Analysis Methods

Although simple payback analysis has its shortcomings in any sophisticated study, it is simple, making it easy to compare alternatives. Hence, it is suitable for comparative evaluations of small projects or for those with such a short payback period that changes in either interest rates or cost escalation factors would not significantly alter the results. It is still one of the most commonly used methods, but if the period of payback is greater than three or four years, a more elaborate cost analysis may be more suitable.

ROR analysis is far more discerning than simple payback, but it fails to account for cost escalation.

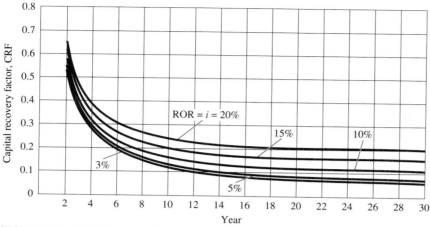

FIGURE 1.6 Capital recovery factors at various amortization periods n and interest rates i or rates of return ROR.

Life-cycle cost analysis is more comprehensive since it not only considers the influence of interest rate and cost escalation, but also can allow for such factors as salvage value, property tax, insurance, and depreciation. For large and more complicated projects, it is often the preferred choice.

Construction Costs of Air Conditioning Systems

Based on the data in *Means Square Foot Costs for Residential, Commercial, Industrial, and Institutional Buildings 1989*, typical U.S. construction costs for residential, commercial and institutional buildings per square foot were as follows:

	$/ft^2
Low-rise, one- or two-story buildings	40–75
High-rise, 8–24 story buildings	60–95

Rough estimates of construction costs for unitary packaged air conditioning systems employing outdoor condensing units for supermarkets, department stores, offices and banks, were the following:

Material, $/ft^2	Installation, $/ft^2	Total, $/ft^2
8–11	4–5.5	12–16.5

Construction costs for some of the more complicated central hydronic systems for modern high-rise office buildings can be still higher. Usually the construction costs of the air conditioning system are about 12 to 20 percent of the entire building cost, including all services. Use a construction cost index for detailed construction cost estimates.

1.11 ENERGY USE AND CONSERVATION

According to data published in the Monthly Energy Review in June 1991, the total energy consumption in 1990 in the United States was 84.2 quad BTU (10^{15} BTU). The United States alone consumed about one fourth of the world's total production. Of the total U.S. energy consumption of 84.2 quads in 1990, the various end-use sectors were as follows:

	Quad BTU	Percent
Residential	10.1	15.9
Commercial	6.6	10.4
Industrial	24.6	38.8
Transportation	22.1	34.9
Total delivered	63.4	100.0
Generation and transmission losses	20.8	

In 1990, the residential/commercial sector consumed slightly more than one fourth of the total energy consumption in the United States. The industrial and transportation sectors each consumed more than one third of the total energy used.

The energy-use by source in 1990 in the United States was

	Quad BTU	Percent
Coal	19.2	22.8
Natural gas and other gas fuels	19.3	22.9
Petroleum	33.6	39.9
Hydro-electric	2.9	3.4
Nuclear	6.2	7.4
Other (including solar)	3.0	3.6
Total	84.2	100.0

Petroleum, coal, and natural gas provided nearly 85 percent of the energy consumed in 1990 in the

United States. Solar energy as an energy source was insignificant, accounting for less than 0.5 percent of the total energy consumption.

Energy use in buildings depends on the use of a building (for example, whether it is an office building or a supermarket, or a factory); configuration of the building; or the type of HVAC&R systems. In Penny and Althof's paper (1992), a breakdown of energy use in commercial buildings in 1990 is as follows:

	Percent
Space heating	31
Air conditioning	10
Ventilation	12
Lighting	25
Others	22
Total	100

The HVAC&R components, combined, consumed approximately 53 percent of the total energy use in commercial buildings.

In the *Annual Energy Review 1987,* the breakdown of household energy consumption in 1984 in the United States was as follows:

	Quad BTU	Percent
Space heating	5.13	56.8
Air conditioning	0.36	4.0
Water heating	1.62	17.9
Appliances	1.92	21.3
Total	9.04	100.0

Considering a heavier energy use in lighting and office equipment in office buildings, if the energy use of HVAC&R systems is taken as about 50 percent of the total energy consumption in commercial buildings, 60 percent in residential buildings, and about 5 to 10 percent in the industrial sector, then—even if the share attributable to the transportation sector is neglected—the national energy use of HVAC&R systems may be around 17 percent, or about one sixth of the total energy consumption in 1990 in the United States. Table 1.1 lists the energy consumption of various types of buildings in 1000 BTU per ft^2 per year according to data published in *Annual Energy Review 1987.*

After the energy crisis in 1973, the United States Congress enacted the Energy Policy and Conservation Act of 1975 and the National Energy Conservation Policy Act of 1978. The enactment of energy conservation legislation by federal and state governments and also the establishment of the Department of Energy (DOE) in 1977 had a definite impact on the implementation of energy conservation.

At the same time, the American Society of Heating, Refrigerating and Air Conditioning Engineers (ASHRAE) published Standard 90-75, Energy Conservation in New Buildings Design, in 1975. Other organizations also offered valuable contributions for conserving energy. All those events started a new era in which energy conservation has become one of the important goals of HVAC&R system design and operation.

In his "Annual Report to Congress" in April 1987, the Secretary of the DOE John S. Herrington wrote: "Thus, energy conservation continues to play an important role in improving the nation's energy and economic efficiency, and, therefore, may be viewed as a significant energy source. Conservation is a prominent component of overall national energy policy and is critical to achieving a balanced and diverse energy resource system. . . . "

The total annual energy consumption in quad BTU and gross national production GNP in billions of 1982 dollars is shown in Fig. 1.7 for the United States between 1949 and 1986. In this figure, the total annual energy consumption can be divided into two stages:

1950 to 1973. Total annual energy consumption increased at an annual rate of about 3.5 percent.

1974 to 1986. The rate of increase of total annual energy consumption hovered around zero, although the GNP increased at an average annual rate of about 2.5 percent. This change from the earlier period has been mainly due to the increased awareness of the need for energy conservation.

1.12 GLOBAL WARMING EFFECT AND CFC PROBLEMS

Global Warming Effect

The surface of the earth is surrounded by a layer of air called *atmosphere.* From the earth's surface to an altitude of about 50 miles, the gases of the atmosphere are relatively well mixed and in stable condition. This lower atmosphere is called homosphere. Clouds exist only in the homosphere. A cloudless homosphere is mainly transparent to short-wave solar radiation but is quite opaque to long-wave infrared rays emitted from the surface of the earth. Carbon dioxide (CO_2), water vapor, and synthetic chlorofluorocarbons (CFCs) play important roles in blocking the direct escape of infrared energy. The term used to describe these phenomena, the transparency to incoming solar radiation and the blanketing of outgoing infrared rays, is the *greenhouse effect* (or, more truly, atmosphere effect) by the climatologists. The greenhouse effect is the main reason why the average temperature of

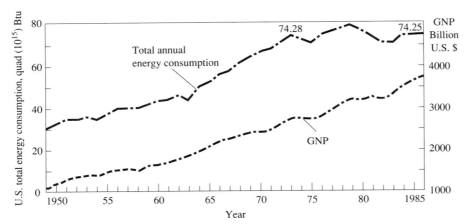

FIGURE 1.7 Total annual energy consumption and GNP of the United States.

the earth's surface is about 60°F higher than it would be if there were no atmosphere.

In addition to CO_2, water vapor, and CFCs, methane (CH_4) and nitrous oxide (N_2O) have an effect in blocking infrared radiation reflected by the earth's surface. The increase in CO_2, water vapor, CFCs, and other gases eventually will result in a rise of air temperature near the earth's surface.

Over the past 100 years, global climate warming has shown an increase of 1°F. For the same period, there was a 25 percent increase of CO_2. During the 1980s, the release of CO_2 to the atmosphere was responsible for about 50 percent of the increase in global warming that was attributable to human activity, and the release of CFCs had a 20 percent share. Some scientists have predicted an accelerated global warming in the coming 50 years because of the increase in the world's annual energy consumption. Further increases in global temperatures may lead to lower rainfalls, drops in soil moisture, more extensive forest fires, more flooding, etc.

Carbon dioxide is the product of many combustion processes. Designers and operators of HVAC&R systems can reduce the production of CO_2 through energy conservation and the replacement of coal power by hydroelectric, nuclear, solar energy, etc. More studies and research are needed to clarify the theory and actual effect of global warming.

Chlorofluorocarbon Problems

Chlorofluorocarbons are used widely as refrigerants in mechanical refrigeration systems, to produce thermal insulation foam and, formerly, to produce aerosol propellants for many household consumer products. Refrigerants CFC-11 (CCl_3F) and CFC-12 (CCl_2F_2) are commonly-used CFCs. Among these, CFC-11 and CFC-12 are very stable, fully halogenated com-

pounds. Fully halogenated means the original hydrogen atoms in a hydrocarbon molecule have all been replaced by chlorine and fluorine atoms. *Halons* are also halogenated hydrocarbons, like BFC-13B1 ($CBrF_3$). If CFCs and halons are leaked or discharged from a refrigeration system during operation and repair to the lower atmosphere, they will migrate to the upper stratosphere and decompose under the action of ultraviolet rays throughout their decades or centuries of life in the atmosphere. The free chlorine atoms will react with the oxygen atoms of the ozone layer in the upper stratosphere and cause a depletion of this layer. The theory of depletion of the ozone layer, the Rowland-Molina theory, and others were proposed in the early to mid-1970s.

The ozone layer filters out harmful ultraviolet rays. Furthermore, changes of the ozone layer may dramatically influence weather patterns.

CFCs and halons influence the atmosphere by increasing the global temperature as well as by causing a depletion of the ozone layer. The United States government and many other countries have taken action to phase out the use of CFCs. More information about CFCs and other refrigerants will be covered in Chapter 4. There is still controversy among scientists about the relationship between CFCs and the ozone layer. Future studies and research are required to clarify the theory of ozone layer depletion.

REFERENCES AND FURTHER READING

ASHRAE, *ASHRAE Handbook HVAC Applications 1991*, ASHRAE Inc., Atlanta, GA, 1991.

Caffrey, R. J., "The Intelligent Building—An ASHRAE Opportunity," *ASHRAE Transactions*, Vol. 94, Part I, pp. 925–933, 1988.

Camejo, P. J., and Hittle, D. C., "An Expert System for the Design of Heating, Ventilating, and Air Conditioning Systems," *ASHRAE Transactions,* Vol. 95, Part I, pp. 379–386, 1989.

Coad, W. J., "Courses and Cures for Building System Defects," *Heating/Piping/Air Conditioning,* pp. 98–100, February 1985.

Coad, W. J., "Safety Factors in HVAC Design," *Heating/Piping/Air Conditioning,* pp. 199–203, January 1985.

Denny, R. J., "The CFC Footprint," *ASHRAE Journal,* pp. 24–28, November 1987.

DiIorio, E., and Jennett, E. J., Jr., "High Rise Opts for High Tech," *Heating/Piping/Air Conditioning,* pp. 83–87, January 1989.

Energy Information Administration (EIA), *Annual Energy Review,* Washington D.C., 1987.

Energy Information Administration, *Monthly Energy Review,* Washington D.C., June 1991.

Energy Information Administration, *Nonresidential Building Energy Consumption Survey: Characteristics of Commercial Buildings,* Washington D.C., 1991.

Faust, F. H., "The Early Development of Self-Contained and Packaged Air Conditioner," *ASHRAE Transactions,* Vol. 92, Part II B, pp. 353–360, 1986.

Grant, W. A., "From '36 to '56—Air Conditioning Comes of Age," *ASHVE Transactions,* Vol. 63, pp. 69–110, 1957.

Guedes, P., *Encyclopedia of Architectural Technology,* Section 4, 1st ed., McGraw-Hill, New York, 1979.

Haines, R. W., "How Control Systems Are Designed," *Heating/Piping/Air Conditioning,* p. 94, February 1982.

Haines, R. W., "Operating HVAC Systems," *Heating/Piping/Air Conditioning,* p. 106, July 1984.

Herrington, J. S., Annual Report to Congress, The Department of Energy of the United States, 1987.

Houghten, F. C., Blackshaw, J. L., Pugh, E. M., and McDermott, P., "Heat Transmission as Influenced by Heat Capacity and Solar Radiation," *ASHVE Transactions,* Vol. 38, pp. 231–284, 1932.

Kohloss, F. H., "The Engineer's Liability in Avoiding Air Conditioning System Overdesign," *ASHRAE Transactions,* Vol. 87, Part I B, pp. 155–162, 1983.

Korte, B., "Existing Buildings: Vast HVAC Resource," *Heating/Piping/Air Conditioning,* pp. 57–63, March 1989.

Korte, B., "The Health of the Industry," Heating/*Piping/Air Conditioning,* pp.139–140, January 1993.

Lewis, L. L., and Stacey, A. E., Jr., "Air Conditioning the Hall of Congress," *ASHVE Transactions,* Vol. 36, pp. 333–346, 1930.

MacCraken, C. D., "The Green House Effect on ASHRAE," *ASHRAE Journal,* pp 52–54, June 1989.

Manley, D. L., Bowlen, K. L., and Cohen, B. M., "Evaluation of Gas Fired Desiccant Based Space Conditioning for Supermarkets," *ASHRAE Transactions,* Part I B, pp. 447–456, 1985.

McClive, J. R., "Early Development in Air Conditioning and Heat Transfer," *ASHRAE Transactions,* Vol. 92, Part II B, pp. 361–365, 1986.

Means, R. S., *Means Square Foot Costs,* R. S. Means, Kingston, MA, 1989.

Miller, A., Thompson, J. C., Peterson, R. E., and Haragan, D. R., *Elements of Meteorology,* 4th ed., Bell & Howell Co., Columbus, OH, 1983.

Montag, G. M., "Life Cycle Cost Analysis Versus Payback for Evaluation Project Alternatives," *Heating/Piping/Air Conditioning,* pp. 75–78, September 1984.

Penny, T., and Althof, J., "Trends in Commercial Buildings," *Heating/Piping/Air Conditioning,* pp. 59–66, September 1992.

Rowland, F. S. "The CFC Controversy: Issues and Answers," *ASHRAE Journal,* pp. 20–27, December 1992.

Simens, J., "A Case for Unitary Systems," *Heating/Piping/Air Conditioning,* pp. 60, 68–73, May 1982.

Whalen, J. M., "An Organized Approach to Energy Management," *Heating/Piping/Air Conditioning,* pp. 95–102, September 1985.

Wilson, L., "A Case for Central Systems," *Heating/Piping/Air Conditioning,* pp. 61–67, May 1982.

PSYCHROMETRICS

2.1 PSYCHROMETRICS

Psychrometrics is the study of the thermodynamic properties of moist air. It is used extensively to illustrate and analyze the characteristics of various air conditioning processes and cycles.

Moist Air

The surface of the earth is surrounded by a layer of air called the *atmosphere* or the *atmospheric air*. From the point of view of psychrometrics, the lower atmosphere, or *homosphere,* is a mixture of dry air

2.1

(including various contaminants) and water vapor, often known as *moist air*.

The composition of dry air is comparatively stable. It varies slightly according to geographic location and from time to time. The approximate composition of dry air by volume percent is the following:

Nitrogen	79.08
Oxygen	20.95
Argon	0.93
Carbon dioxide	0.03
Other gases, like neon, sulfur dioxide, etc.	0.01

The amount of water vapor present in moist air at a temperature range of 0 to 100°F varies from 0.05 to 3 percent by mass. It has a significant influence on the characteristics of the moist air.

Water vapor is lighter than air. A cloud in the sky is composed of microscopic beads of liquid water that are surrounded by thin layers of water vapor. These layers give the cloud the needed buoyancy to float in the air.

Equation of State of an Ideal Gas

The equation of state of an ideal gas indicates the relationship between its thermodynamic properties, or

$$pv = RT_R \qquad (2.1)$$

where p = pressure of the gas, psf (1 psf = 144 psia)
v = specific volume of the gas, ft^3/lb
R = gas constant, ft · lb$_f$/lb$_m$ · °R
T_R = absolute temperature of the gas, °R

Since $v = V/m$, then Eq. (2.1) becomes

$$pv = mRT_R \qquad (2.2)$$

where V = total volume of the gas, ft^3
m = mass of the gas, lb

Using the relationships $m = nM$ and $R = R_o/M$, Eq. (2.2) can be written as

$$pv = nR_oT_R \qquad (2.3)$$

where n = number of moles, mol
M = molecular weight
R_o = universal gas constant, ft · lb$_f$/lb$_m$ · °R

Equation of State of a Real Gas

A modified form of the equation of state for a real gas can be expressed as

$$\frac{pv}{RT_R} = 1 + Ap + Bp^2 + Cp^3 + \cdots = Z \qquad (2.4)$$

where $A, B, C, \ldots$ = virial coefficients
Z = compressibility factor

The compressibility factor Z illustrates the degree of deviation of the behavior of the real gas, moist air, from the ideal gas due to the following:

1. Effect of air dissolved in water
2. Variation of the properties of water attributable to the effect of pressure
3. Effect of intermolecular forces on the properties of water itself

For an ideal gas, $Z = 1$. According to information published by the former National Bureau of Standards of the United States, for dry air at standard atmospheric pressure (29.92 in. Hg) and a temperature of 32 to 100°F, the maximum deviation is about 0.12 percent. For water vapor in moist air under saturated conditions at a temperature of 32 to 100°F, the maximum deviation is about 0.5 percent.

Calculation of the Properties of Moist Air

The most exact calculation of the thermodynamic properties of moist air is based on the formulations developed by Hyland and Wexler of the United States National Bureau of Standards. The psychrometric charts and tables of ASHRAE are constructed and calculated from these formulations.

Calculations based on the ideal gas equations are the simplest and can be easily formulated. According to the analysis of Nelson and Pate, at a temperature between 0 and 100°F, calculations of enthalpy and specific volume using ideal gas equations show a maximum deviation of 0.5 percent from the exact calculations by Hyland and Wexler. Therefore, ideal gas equations will be used in this text for the formulation and calculation of the thermodynamic properties of moist air.

Although air contaminants may seriously affect the health of occupants of the air conditioned space, they have little effect on the thermodynamic properties of moist air since their mass concentration is low. For simplicity, moist air is always considered as a binary mixture of dry air and water vapor during the analysis and the calculation of its properties.

2.2 DALTON'S LAW AND THE GIBBS-DALTON LAW

Dalton's law shows that for a mixture of gases occupying a given volume at a certain temperature the total pressure of the mixture is equal to the sum of the partial pressures of the constituents of the mixture, that is,

$$p_m = p_1 + p_2 + \cdots \qquad (2.5)$$

where p_m = total pressure of the mixture, psia
$p_1, p_2, \ldots$ = partial pressure of constituents 1, 2, ..., psia

The partial pressure exerted by each constituent in the mixture is independent of the existence of other gases in the mixture. In Fig. 2.1 are shown the variation of mass and pressure of dry air and water vapor, at an atmospheric pressure of 14.697 psia and a temperature of 75°F.

The principle of conservation of mass for nonnuclear processes gives the following relationship:

$$m_m = m_a + m_w \qquad (2.6)$$

where m_m = mass of moist air, lb
m_a = mass of dry air, lb
m_w = mass of water vapor, lb

Applying Dalton's law for moist air, we have

$$p_{at} = p_a + p_w \qquad (2.7)$$

where p_{at} = atmospheric pressure or pressure of the moist air, psia
p_a = partial pressure of dry air, psia
p_w = partial pressure of water vapor, psia

Dalton's law is based on experimental results. It is more accurate for gases at low pressures. Dalton's law can be further extended to state the relationship of the internal energy, enthalpy, and entropy of the gases in a mixture as the Gibbs-Dalton law:

$$m_m u_m = m_1 u_1 + m_2 u_2 + \cdots$$
$$m_m h_m = m_1 h_1 + m_2 h_2 + \cdots \qquad (2.8)$$
$$m_m s_m = m_1 s_1 + m_2 s_2 + \cdots$$

where m_m = mass of the gaseous mixture, lb
$m_1, m_2, \ldots$ = mass of the constituents, lb
u_m = specific internal energy of the gaseous mixture, Btu/lb

$u_1, u_2, \ldots$ = specific internal energy of the constituents, Btu/lb
h_m = specific enthalpy of the gaseous mixture, Btu/lb
$h_1, h_2, \ldots$ = specific enthalpy of the constituents, Btu/lb
s_m = specific entropy of the gaseous mixture, Btu/lb · °R
$s_1, s_2, \ldots$ = specific entropy of the constituents, Btu/lb · °R

2.3 AIR TEMPERATURE

Temperature and Temperature Scales

The temperature of a substance is a measure of how hot or cold it is. Two systems are said to have equal temperatures only if there is no change in any of their observable thermal characteristics when they are brought into contact with each other. Various temperature scales commonly used to measure the temperature of various substances are illustrated in Fig. 2.2.

In conventional inch-pound (I-P) units, at a standard atmospheric pressure of 14.697 psia, the Fahrenheit scale has a freezing point of 32°F at the ice point, and a boiling point of 212°F. For the triple point with a pressure of 0.08864 psia, the temperature on the Fahrenheit scale is 32.018°F. There are 180 divisions, or degrees, between the boiling and the freezing point in the Fahrenheit scale. In the International System of Units (SI) units, the Celsius or centigrade scale has a freezing point of 0°C and a boiling point of 100°C. There are 100 divisions between these points. The triple point is 0.01°C. The conversion from Celsius scale to Fahrenheit scale is as follows:

$$°F = 1.8°C + 32 \qquad (2.9)$$

For an ideal gas, at $T_R = 0$ the gas would have a vanishing specific volume. Actually, a real gas has

Thermometer

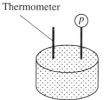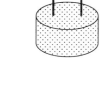

T = 75 °F
m_a = 1 lb
m_w = 0
p_a = 14.482 psia
p_w = 0
p (pressure of mixture) = 14.482 psia

T = 75 °F
m_a = 0
m_w = 0.0092 lb
p_a = 0
p_w = 0.215 psia
p = 0.215 psia

T = 75 °F
m_a = 1 lb
m_w = 0.0092 lb
p_a = 14.482 psia
p_w = 0.215 psia
p_{at} = 14.697 psia

FIGURE 2.1 Mass and pressure of dry air, water vapor, and moist air.

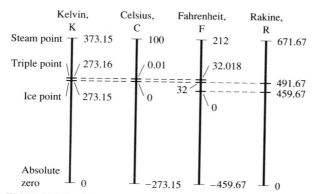

FIGURE 2.2 Commonly used temperature scales.

a negligible *molecular* volume when T_R approaches absolute zero. A temperature scale that includes absolute zero is called an absolute temperature scale. The Kelvin absolute scale has the same boiling-freezing-point division as the centigrade scale. At the freezing point, the Kelvin scale is 273.15 K. Absolute zero on the centigrade scale is $-273.15°C$. The Rankine absolute scale division is equal to that of the Fahrenheit scale. The freezing point is 491.67°R. Similarly, absolute zero is $-459.67°F$ on the Fahrenheit scale.

Conversions between Rankine and Fahrenheit and also between Kelvin and Celsius are

$$R = 459.67 + °F \qquad (2.10)$$

$$K = 273.15 + °C \qquad (2.11)$$

Thermodynamic Temperature Scale

On the basis of the second law of thermodynamics, one can establish a temperature scale that is independent of the working substance and provides an absolute zero of temperature; this is called a *thermodynamic temperature scale*. The thermodynamic temperature T must satisfy the following relationship:

$$T_R/T_{R_o} = Q/Q_o \qquad (2.12)$$

where Q = heat absorbed by a reversible engine, Btu/h

Q_o = heat rejected by a reversible engine, Btu/h

T_R = temperature of the heat source of the reversible engine, °R

T_{R_o} = temperature of the heat sink of the reversible engine, °R

Two of the ASHRAE basic tables, "Thermodynamic Properties of Moist Air" and "Thermodynamic Properties of Water at Saturation," in *ASHRAE Handbook 1989, Fundamentals* are based on the thermodynamic temperature scale.

Temperature Measurements

During the measure of air temperatures, it is important to recognize the meaning of the terms *accuracy, precision,* and *sensitivity.*

1. Accuracy is the ability of an instrument to indicate or to record the true value of the measured quantity. The error indicates the degree of accuracy.
2. Precision is the ability of an instrument to give the same reading repeatedly under the same conditions.
3. Sensitivity is the ability of an instrument to indicate change of the measured quantity.

Liquid-in-glass instruments, such as mercury or alcohol thermometers, were common in the early days for air temperature measurements. In recent years, many liquid-in-glass thermometers have been replaced by remote temperature monitoring and indication systems, made possible by sophisticated control systems.

A typical air temperature indication system includes sensors, amplifiers, and an indicator.

SENSORS. Air temperature sensors needing higher accuracy are usually made from resistance temperature detectors (RTD) made of platinum, palladium, nickel, or copper wires. The electrical resistance of these resistance thermometers characteristically increases when the sensed ambient air temperature is raised, that is, they have a positive temperature coefficient α. In many engineering applications, the relationship between the resistance and temperature can be given by

$$R \approx R_{32}(1 + \alpha T)$$
$$\alpha \approx (R_{212} - R_{32})/(180 \ R_{32}) \qquad (2.13)$$

where R = electric resistance, ohms

R_{32}, R_{212} = electric resistance, at 32° and 212°F, ohms respectively

T = temperature, °F

The mean temperature coefficient α for several types of metal wires often used as RTD are shown below:

	Measuring range, °F	α, ohm/°F
Platinum	−400 to 1350	0.00218
Palladium	−400 to 1100	0.00209
Nickel	−150 to 570	0.0038
Copper	−150 to 400	0.0038

Many air temperature sensors are made from thermistors of sintered metallic oxides, that is, semiconductors. They are available in a large variety of types: beads, discs, washers, rods, etc. Thermistors have a negative temperature coefficient. Their resistance decreases when the sensed air temperature increases. The resistance of a thermistor may drop from approximately 3800 to 3250 ohms when the sensed air temperature increases from 68 to 77°F. Recently developed high-quality thermistors are accurate, stable, and reliable. Within their operating range, commercially available thermistors will match a resistance-temperature curve within approximately 0.1°F. Some manufacturers of thermistors can supply them with a stability of 0.05°F a year.

For direct digital control (DDC) systems, the same sensor is used for both temperature indication, or monitoring, and temperature control. In DDC systems, an RTD with a positive temperature coefficient is generally used.

AMPLIFIER(S). The measured electric signal from the temperature sensor is amplified at the solid-state amplifier to produce an output for indication. The number of amplifiers is matched with the number of the sensors used in the temperature indication system.

INDICATOR. An analog-type indicator, one based on direct measurable quantities, is usually a moving coil instrument. For a digital-type indicator, the signal from the amplifier is compared with an internal reference voltage and converted for indication through an analog-digital transducer.

2.4 HUMIDITY

Humidity Ratio

The *humidity ratio* of moist air w is the ratio of the mass of water vapor m_w to the mass of dry air m_a contained in the mixture of the moist air. The humidity ratio can be calculated as

$$w = \frac{m_w}{m_a} \qquad (2.14)$$

Since dry air and water vapor can occupy the same volume at the same temperature, we can apply the ideal gas equation and Dalton's law for dry air and water vapor. Equation (2.14) can be rewritten as

$$w = \frac{m_w}{m_a} = \frac{p_w V R_a T_R}{p_a V R_w T_R}$$

$$= \frac{R_a}{R_w} \cdot \frac{p_w}{p_{at} - p_w} = \frac{53.352}{85.778} \cdot \frac{p_w}{p_{at} - p_w}$$

$$= 0.62198 \frac{p_w}{p_{at} - p_w} \qquad (2.15)$$

where R_a, R_w = gas constant for dry air and water vapor, ft · lb$_f$/lb$_m$ · °R. Equation (2.15) is expressed in the form of the ratio of pressures, therefore, p_w and p_{at} must have the same units, either psia or psf.

For moist air at saturation, Eq. (2.15) becomes

$$w_s = 0.62198 \frac{p_{ws}}{p_{at} - p_{ws}} \qquad (2.16)$$

where p_{ws} = pressure of water vapor of moist air at saturation, psia or psf.

Relative Humidity

The *relative humidity* φ of moist air, or RH, is defined as the ratio of the mole fraction of the water vapor x_w in a moist air sample to the mole fraction of the water vapor in a saturated moist air sample x_{ws} at the same temperature and pressure. This relationship can be expressed as

$$\varphi = \left. \frac{x_w}{x_{ws}} \right|_{T,p} \qquad (2.17)$$

And also, by definition, the following expressions may be written:

$$x_w = \frac{n_w}{n_a + n_w} \qquad (2.18)$$

$$x_{ws} = \frac{n_{ws}}{n_a + n_{ws}} \qquad (2.19)$$

where n_a = number of moles of dry air, mol
n_w = number of moles of water vapor in a moist air sample, mol
n_{ws} = number of moles of water vapor in a saturated moist air sample, mol

Moist air is a binary mixture of dry air and water vapor; therefore, we find that the sum of the mole fractions of dry air x_a and water vapor x_w is equal to 1, that is,

$$x_a + x_w = 1 \qquad (2.20)$$

If we apply ideal gas equations $p_w V = n_w R_o T_R$ and $p_a V = n_a R_o T_R$, by substituting them into Eq. (2.19), then relative humidity can also be expressed as

$$\varphi = \left. \frac{p_w}{p_{ws}} \right|_{T,p} \qquad (2.21)$$

The water vapor pressure of saturated moist air p_{ws} is a function of temperature T and pressure p, which is slightly different from the saturation pressure of water vapor p_s. Here p_s is a function of temperature T only. Since the difference between p_{ws} and p_s is small, it is usually ignored.

Degree of Saturation

The *degree of saturation* μ is defined as the ratio of the humidity ratio of moist air w to the humidity ratio of saturated moist air w_s at the same temperature and pressure. This relationship can be expressed as

$$\mu = \left.\frac{w}{w_s}\right|_{T,p} \qquad (2.22)$$

Since from Eqs. (2.15), (2.20), and (2.21) $w = 0.62198\, x_w/x_a$ and $w_s = 0.62198\, x_{ws}/x_a$, Eqs. (2.20), (2.21), and (2.22) can be combined, so

$$\varphi = \frac{\mu}{1 - (1-\mu)x_{ws}} = \frac{\mu}{1 - [(1-\mu)(p_{ws}/p_{at})]} \qquad (2.23)$$

In Eq. (2.23), $p_{ws} \ll p_{at}$; therefore, the difference between φ and μ is small. Usually the maximum difference is less than 2 percent.

2.5 PROPERTIES OF MOIST AIR

Enthalpy

The difference in specific enthalpy Δh for an ideal gas, in Btu/lb, at a constant pressure can be defined as

$$\Delta h = c_p (T_2 - T_1) \qquad (2.24)$$

where c_p = specific heat at constant pressure, Btu/lb · °F
T_1, T_2 = temperature of the ideal gas at points 1 and 2, °F

As moist air is approximately a binary mixture of dry air and water vapor, the enthalpy of the moist air can be evaluated as

$$h = h_a + H_w \qquad (2.25)$$

where h_a, H_w = enthalpy of dry air and total enthalpy of water vapor, in Btu/lb. The following assumptions are made for the enthalpy calculations of moist air:

1. Ideal gas equation and the Gibbs-Dalton law are valid.
2. The enthalpy of dry air is equal to zero at 0°F.
3. All water vapor contained in the moist air is vaporized at 0°F.
4. The enthalpy of saturated water vapor at 0°F is 1061 Btu/lb.
5. For convenience in calculation, the enthalpy of moist air is taken to be equal to the enthalpy of a mixture of dry air and water vapor in which the amount of dry air is exactly equal to 1 lb.

Based on the preceding assumptions, the enthalpy h of moist air can be calculated as

$$h = h_a + wh_w \qquad (2.26)$$

where h_w = specific enthalpy of water vapor, Btu/lb.

In a temperature range of 0 to 100°F, the mean value for the specific heat of dry air can be taken as 0.240 Btu/lb · °F. Then the specific enthalpy of dry air h_a is given by

$$
\begin{aligned}
h_a &= c_{pd}T \\
&= 0.240\, T \qquad (2.27)
\end{aligned}
$$

where c_{pd} = specific heat of dry air at constant pressure, Btu/lb · °F
T = temperature of dry air, °F

The specific enthalpy of water vapor h_w at constant pressure can be approximated as

$$h_w = h_{g0} + c_{ps}T \qquad (2.28)$$

where h_{g0} = specific enthalpy of saturated water vapor at 0°F; its value can be taken as 1061 Btu/lb
c_{ps} = specific heat of water vapor at constant pressure, Btu/lb · °F. In a temperature range of 0 to 100°F, its value can be taken as 0.444 Btu/lb · °F

Then, the enthalpy of moist air can be evaluated as

$$
\begin{aligned}
h &= c_{pd}T + w(h_{g0} + c_{ps}T) \\
&= 0.240\, T + w(1061 + 0.444\, T) \qquad (2.29)
\end{aligned}
$$

Here, the unit of h is Btu per lb of dry air. For simplicity, it is often expressed as Btu/lb.

Moist Volume

The moist volume of moist air v, in ft³/lb, is defined as the volume of the mixture of the dry air and water vapor when the mass of the dry air is exactly equal to 1 lb, that is,

$$v = \frac{V}{m_a} \qquad (2.30)$$

where V = total volume of the mixture, ft³
m_a = mass of the dry air, lb

In a moist air sample, the dry air, water vapor, and moist air occupy the same volume. If we apply the ideal gas equation, then

$$v = \frac{V}{m_a} = \frac{R_a T_R}{p_{at} - p_w} \qquad (2.31)$$

where p_{at} and p_w are both in psf. From Eq. (2.15), $p_w = p_{at}w/(w + 0.62198)$. Substituting this expression into Eq. (2.31) gives

$$v = \frac{R_a T_R}{p_{at}} (1 + 1.6078\, w) \qquad (2.32)$$

According to Eq. (2.32), the volume of 1 lb of dry air is always smaller than the volume of the moist air when both of them are at the same temperature and at the same atmospheric pressure.

Density

Since the enthalpy and humidity ratio are always related to a unit mass of dry air, for the sake of consistency, air density ρ_a, in lb/ft^3, should be defined as the ratio of the mass of dry air to the total volume of the mixture, that is, the reciprocal of moist volume, or

$$\rho_a = \frac{m_a}{V} = \frac{1}{v} \qquad (2.33)$$

Sensible Heat and Latent Heat

Sensible heat is that heat energy associated with the change of air temperature between two state points. In Eq. (2.29), the enthalpy of moist air calculated at a datum state 0°F can be divided into two parts,

$$h = (c_{pd} + w c_{ps})T + w h_{g0} \qquad (2.34)$$

The first term on the right-hand side of Eq. (2.34) indicates the sensible heat. It depends on the temperature T above the datum 0°F.

Latent heat (sometimes called h_{fg} or h_{ig}) is the heat energy associated with the change of state of water vapor. Latent heat of vaporization denotes the latent heat required to vaporize liquid water into water vapor. Also, latent heat of condensation indicates the latent heat to be removed in the condensation of water vapor into liquid water. When moisture is added to or removed from a process or a space, a corresponding amount of latent heat is always involved, to vaporize the water, or to condense it.

In Eq. (2.34), the second term on the right-hand side, $w h_{g0}$, denotes latent heat. Both sensible and latent heat are expressed in Btu/lb of dry air.

Specific Heat of Moist Air at Constant Pressure

The specific heat of moist air at constant pressure c_{pa} is defined as the heat required to raise its temperature 1°F at constant pressure. In I-P units, it is expressed as Btu/lb · °F.

In Eq. (2.34), the sensible heat of moist air q_{sen}, in Btu/h, is represented by

$$q_{sen} = \dot{m}_a (c_{pd} + w c_{ps})T = \dot{m}_a c_{pa} T \qquad (2.35)$$

where $\dot{m}_a$ = mass flow rate of the moist air, lb/h.

Apparently,

$$c_{pa} = c_{pd} + w c_{ps} \qquad (2.36)$$

Since c_{pd} and c_{ps} are both a function of temperature, c_{pa} is also a function of temperature and, in addition, a function of humidity ratio.

For a temperature range between 0 and 100°F, c_{pd} can be taken as 0.240 Btu/lb · °F and c_{ps} as 0.444 Btu/lb · °F. Most of the calculations of $c_{pa}(T_2 - T_1)$ have a range of w between 0.005 and 0.010 lb/lb. Taking a mean value of $w = 0.0075$ lb/lb, we find that

$$c_{pa} = 0.240 + (0.0075 \times 0.444)$$
$$= 0.243 \text{ Btu/lb} \cdot °F$$

Dew Point Temperature

The dew point temperature T_{dew} is the temperature of saturated moist air of the same moist air sample, having the same humidity ratio, and at the same atmospheric pressure of the mixture p_{at}. Two moist air samples at the same T_{dew} will have the same humidity ratio w and the same partial pressure of water vapor p_w. The dew point temperature is related to humidity ratio w by the following relationship:

$$w_s(p_{at}, T_{dew}) = w \qquad (2.37)$$

At a specific atmospheric pressure, the dew point temperature determines the humidity ratio w and the water vapor pressure p_w of the moist air.

2.6 THERMODYNAMIC WET BULB TEMPERATURE AND WET BULB TEMPERATURE

Ideal Adiabatic Saturation Process

If moist air at an initial temperature T_1, humidity ratio w_1, enthalpy h_1, and pressure p flows over a water surface of infinite length in a well-insulated chamber as shown in Fig. 2.3, liquid water will evaporate into water vapor and disperse in the air. The humidity ratio of the moist air will gradually increase until the air can absorb no more moisture.

As there is no heat transfer between this insulated chamber and the surroundings, the latent heat required for the evaporation of water will come from the sensible heat released by the moist air. This process results in a drop of temperature of the moist air. At the end of this evaporation process, the moist air is always saturated. Such a process is called an *ideal adiabatic saturation process,* where an adiabatic process is defined as a process without heat transfer to or from the process.

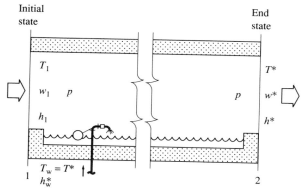

FIGURE 2.3 Ideal adiabatic saturation process.

Thermodynamic Wet Bulb Temperature

For any state of moist air, there exists a *thermodynamic wet bulb temperature* T^* that exactly equals the saturated temperature of the moist air at the end of the ideal adiabatic saturation process at constant pressure.

Applying a steady flow energy equation, we have

$$h_1 + (w_s^* - w_1)h_w^* = h_s^* \qquad (2.38)$$

where h_1, h_s^* = enthalpy of moist air at the initial state and enthalpy of saturated air at the end of the ideal adiabatic saturation process, Btu/lb

w_1, w_s^* = humidity ratio of the moist air at the initial state and humidity ratio of saturated air at the end of the ideal adiabatic saturation process, lb/lb

h_w^* = enthalpy of water as it is added to the chamber at a temperature T^*, Btu/lb

The thermodynamic wet bulb temperature T^* is a unique property of a given moist air sample that depends only on the initial properties of the moist air— $w_1, h_1,$ and p. It is also a fictitious property that only hypothetically exists at the end of an ideal adiabatic saturation process.

Heat Balance of an Ideal Adiabatic Saturation Process

When water is supplied to the insulation chamber at a temperature T^* in an ideal adiabatic saturation process, then the decrease in sensible heat due to the drop of temperature of the moist air is just equal to the latent heat required for the evaporation of water added to the moist air. This relationship is given by

$$c_{pd}(T_1 - T^*) + c_{ps}w_1(T_1 - T^*) = (w_s^* - w_1)h_{fg}^* \qquad (2.39)$$

where T_1 = temperature of moist air at the initial state of the ideal adiabatic saturation process, °F

h_{fg}^* = latent heat of vaporization at the thermodynamic wet bulb temperature, Btu/lb

Since $c_{pa} = c_{pd} + w_1c_{ps}$, we find, by rearranging the terms in Eq. (2.39),

$$\frac{w_s^* - w_1}{T_1 - T^*} = \frac{c_{pa}}{h_{fg}^*} \qquad (2.40)$$

Also

$$T^* = T_1 - \frac{(w_s^* - w_1)h_{fg}^*}{c_{pa}} \qquad (2.41)$$

Psychrometer

A *psychrometer* is an instrument that permits one to determine the relative humidity of a moist air sample by measuring its dry bulb and wet bulb temperatures.

In Fig. 2.4 is shown a psychrometer, which consists of two thermometers. The sensing bulb of one of the thermometers is always kept dry. The temperature reading of the dry bulb is called the *dry bulb temperature*. The sensing bulb of the other thermometer is wrapped with a piece of cotton wick, one end of which dips into a cup of distilled water. The surface of this bulb is always wet; therefore, the temperature this bulb measures is called the *wet bulb temperature*. The dry bulb is separated from the wet bulb by a radiation-shielding plate. Both dry and wet bulbs are cylindrical in shape.

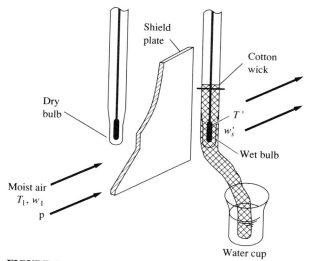

FIGURE 2.4 A psychrometer.

Wet Bulb Temperature

When unsaturated moist air flows over the wet bulb of the psychrometer, liquid water on the surface of the cotton wick evaporates, and as a result the temperature of the cotton wick and the wet bulb drops. This depressed wet bulb reading is called the wet bulb temperature T', and the difference between the dry bulb and the wet bulb temperatures is called the *wet bulb depression*.

Let us neglect the conduction along the thermometer stems to the dry and wet bulbs and also assume that the temperature of the water on the cotton wick is equal to the wet bulb temperature of the moist air. Since the heat transfer from the moist air to the cotton wick is exactly equal to the latent heat required for vaporization, then, at steady state, the heat and mass transfer per unit area of the wet bulb surface can be calculated as

$$h_c(T-T')+h_r(T_{\text{ra}}-T') = h_d(w'_s-w_1)h'_{\text{fg}} \qquad (2.42)$$

where h_c, h_r = mean convective and radiative heat transfer coefficient, Btu/h · ft^2 · °F

$\quad h_d$ = mean convective-mass-transfer coefficient, lb/h · ft^2

$\quad T$ = temperature of the undisturbed moist air at a distance from the wet bulb, °F

$\quad T'$ = wet bulb temperature, °F

$\quad T_{\text{ra}}$ = mean radiant temperature, °F

$\quad w_1, w'_s$ = humidity ratio of the moist air and the saturated air film at the surface of the cotton wick, lb/lb

$\quad h'_{\text{fg}}$ = latent heat of vaporization at the wet bulb temperature, Btu/lb

Based on the correlation of cross-flow forced convective heat transfer for a cylinder, $\text{Nu}_D = C\,\text{Re}^n\,\text{Pr}^{0.333}$, and on the analogy between convective heat transfer and convective mass transfer, the following relationship holds: $h_d = h_c/(c_{\text{pa}}\text{Le}^{0.6667})$. Here, Nu is the Nusselt number, Re the Reynolds number, P_r the Prandtl number, and Le the Lewis number. Also, C is a constant, and n is an exponent.

Substituting this relationship into Eq. (2.42), we have

$$w_1 = w'_s - K'(T - T') \qquad (2.43)$$

In Eq. (2.43), K' represents the wet bulb constant. It can be calculated as

$$K' = \frac{c_{\text{pa}}\,\text{Le}^{0.6667}}{h'_{\text{fg}}}\left[1 + \frac{h_r(T_{\text{ra}} - T')}{h_c(T - T')}\right] \qquad (2.44)$$

The term $(T - T')$ represents the wet bulb depression.

Combining Eqs. (2.43) and (2.44) then gives

$$\frac{w'_s - w_1}{T - T'} = \frac{c_{\text{pa}}\,\text{Le}^{0.6667}}{h'_{\text{fg}}}\left[1 + \frac{h_r(T_{\text{ra}} - T')}{h_c(T - T')}\right] \qquad (2.45)$$

Relationship between Wet Bulb Temperature and Thermodynamic Wet Bulb Temperature

Wet bulb temperature is a function not only of the initial state of moist air, but also of the rate of heat and mass transfer at the wet bulb.

Comparing Eq. (2.40) with Eq. (2.45), we find that the wet bulb temperature measured by using a psychrometer is equal to thermodynamic wet bulb temperature only when the following relationship holds:

$$\text{Le}^{0.6667}\left[1 + \frac{h_r(T_{\text{ra}} - T')}{h_c(T - T')}\right] = 1 \qquad (2.46)$$

2.7 SLING AND ASPIRATION PSYCHROMETERS

Sling and aspiration psychrometers determine the relative humidity through the measuring of the dry and wet bulb temperatures.

A sling psychrometer with two bulbs, one dry and the other wet, is shown in Fig. 2.5a. Both dry and wet bulbs can be rotated around a spindle to produce an air flow over the surfaces of the dry and wet bulbs at an air velocity of 400–600 fpm. Also a shield plate separates the dry and wet bulbs and partly protects the wet bulb against surrounding radiation.

An aspiration psychrometer that uses a small motor-driven fan to produce an air current flowing over the dry and wet bulbs is illustrated in Fig. 2.5b. The air velocity over the bulbs is usually kept at 400–800 fpm. The dry and wet bulbs are located in separate compartments and are entirely shielded from the surrounding radiation.

When the space dry bulb temperature is within a range of 75–80°F and the space wet bulb is between 65–70°F, the following wet bulb constants K' can be used to calculate the humidity ratio of the moist air:

Aspiration psychrometer $K' = 0.000206$ 1/°F
Sling psychrometer $K' = 0.000218$ 1/°F

After the psychrometer has measured the dry and wet bulb temperatures of the moist air, the humidity ratio w can be calculated by Eq. (2.43). Since the saturated water vapor pressure can be found from the psychrometric table, then the relative humidity of moist air can be evaluated through Eq. (2.15).

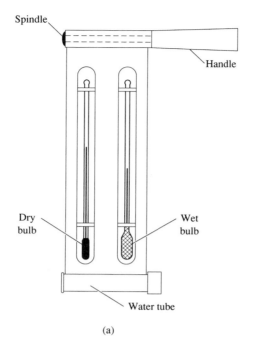

(a)

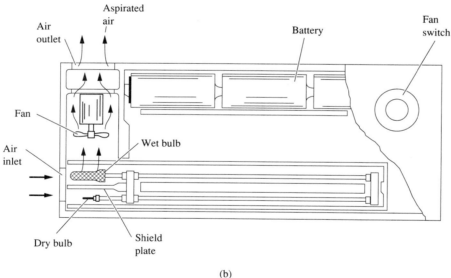

(b)

FIGURE 2.5 Sling and aspiration psychrometers: (a) sling psychrometer; and (b) aspiration psychrometer.

According to the analysis of Threlkeld, for a wet bulb diameter of 0.1 in. and an air velocity flowing over the wet bulb of 400 fpm, if the dry bulb temperature is 90°F and the wet bulb temperature is 70°F, then $(T' - T^*)/(T - T')$ is about 2.5 percent. Under the same conditions, if a sling psychrometer is used, then the deviation may be reduced to about 1 percent. If the air velocity flowing over the wet bulb exceeds 400 fpm, there is no significant reduction in the deviation.

Distilled water must be used to wet the cotton wick for both sling and aspiration psychrometers. Because dusts contaminate them, cotton wicks should be replaced regularly to provide a clean surface for evaporation.

2.8 HUMIDITY MEASUREMENTS

Humidity sensors used in HVAC&R for direct humidity indication or operating controls are in the following categories:

1. *Mechanical hygrometers*
2. *Electronic hygrometers*

Mechanical Hygrometers

Mechanical hygrometers operate on the principle that hygroscopic materials expand when they absorb water vapor or moisture from the ambient air. They contract when they release moisture to the surrounding air. Such hygroscopic materials include human and animal hairs, plastic polymers like nylon ribbon, natural fibers, wood, etc. When these materials are linked to mechanical linkages or electrical transducers that sense the change in size and convert it into electric signals, the results in these devices can be calibrated to yield direct relative humidity measurements of the ambient air.

Electronic Hygrometers

There are three types of electronic hygrometers: Dunmore resistance hygrometers, ion-exchange resistance hygrometers, and capacitance hygrometers.

DUNMORE RESISTANCE HYGROMETER. In 1938, Dunmore of the National Bureau of Standards developed the first lithium chloride resistance electric hygrometer in the United States. This instrument depends on the change in resistance between two electrodes mounted on a hygroscopic material. In Fig. 2.6a is shown a Dunmore resistance sensor. The electrodes could be, for example, a double-threaded winding of noble metal wire mounted on a plastic cylinder coated with hygroscopic material. The wires can also be in a grid type arrangement with a thin film of hygroscopic material bridging the gap between the electrodes.

At a specific temperature, electrical resistance decreases with increasing humidity. Because the response is significantly influenced by temperature, the results are often indicated by a series of isothermal curves. Relative humidity is generally used as the humidity parameter, for it must be controlled in the indoor environment. Also the electrical response is more nearly a function of relative humidity than of the humidity ratio.

The time response to accomplish a 50 percent change in relative humidity varies directly according to the air velocity flowing over the sensor and also is inversely proportional to the saturated vapor pressure. If a sensor has a response time of 10 s at 70°F, it might need a response time of 100 s at 10°F.

Because of the steep variation in resistance corresponding to a change of relative humidity, each of the Dunmore sensors only covers a certain range of relative humidity measurements. A set of several Dunmore sensors is usually needed to measure relative humidity between 1 percent and 100 percent.

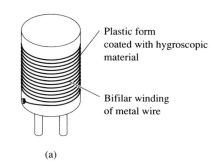

(a)

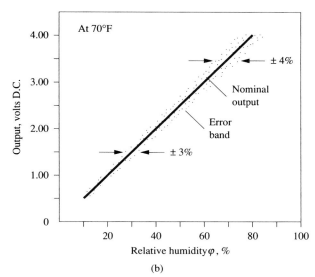

(b)

FIGURE 2.6 Dunmore resistance electronic hygrometer: (a) sensor; and (b) output versus relative humidity φ (Reprinted by permission of Johnson Controls).

A curve for output, in D.C. volts, versus relative humidity is shown in Fig. 2.6b for a typical Dunmore sensor. It covers a measuring range of 10–80 percent, which is usually sufficient for the indication of relative humidity for a comfort air conditioned space. This typical Dunmore sensor has an accuracy of ±3 percent when the relative humidity varies between 10 and 60 percent at a temperature of 70°F (See Fig. 2.6b). Its accuracy reduces to ±4 percent when relative humidity is in a range between 60–80 percent at the same temperature.

In addition to lithium chloride, lithium bromide is also sometimes used as the sensor.

ION-EXCHANGE RESISTANCE HYGROMETER. The sensor of an ion-exchange resistance electric hygrometer is composed of electrodes mounted on a base-plate and a high-polymer resin film, used as a humidity-sensing material, cross-linking the electrodes as shown in Figs. 2.7a and 2.7b. Humidity is measured by the change in resistance between the electrodes.

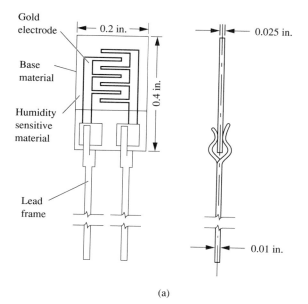

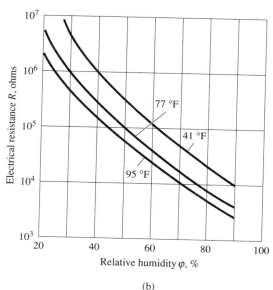

FIGURE 2.7 Ion-exchange resistance electric hygrometer: (a) front and side view of sensor; and (b) characteristic curve of R versus φ (*Source:* General Eastern Instruments. Reprinted by permission).

When the salt contained in the humidity-sensitive material bridging the electrodes becomes ion conductive because of the presence of water vapor in the ambient air, mobile ions in the polymer film are formed. The higher the relative humidity of the ambient air, the greater the ionization becomes, and therefore, the greater the concentration of mobile ions. On the other hand, lower relative humidity reduces the ionization and results in a lower concentration of mobile ions. The resistance of the humidity-sensing material reflects the change of the relative humidity of the ambient air.

In Fig. 2.7b, the characteristic curves of an ion-exchange resistance electric hygrometer show that there is a nonlinear relationship between resistance R and relative humidity φ. These sensors cover a wider range than Dunmore sensors, from 20 to 90 percent relative humidity.

CAPACITANCE HYGROMETER. The commonly used capacitance sensor consists of a thin film plastic foil. A very thin gold coating covers both sides of the film as electrodes, and the film is mounted inside a capsule. The gold electrodes and the dividing plastic foil form a capacitor. Water vapor penetrates the gold layer, which is affected by the vapor pressure of the ambient air and, therefore, ambient relative humidity. The number of water molecules absorbed on the plastic foil determines the capacitance and the resistance between the electrodes.

Comparison of Various Methods

The following table summarizes the sensor characteristics of various methods to be used within a temperature range of 32°F to 120°F and a humidity range of 10 percent to 95 percent RH:

	Operating method	Accuracy, % RH
Psychrometer	Wet bulb depression	±3
Mechanical	Dimensional change	±3 to ±5
Dunmore	Electrical resistance	±1.5
Ion-exchange	Electrical resistance	±2 to ±5
Capacitance	Electrical capacitance and resistance	±3 to ±5

Psychrometers are simple and comparatively low in cost. They suffer no irreversible damage at 100 percent RH, as do the sensors of electronic hygrometers. Unfortunately, complete wet bulb depression readings of psychrometers become difficult when relative humidity drops below 20 percent or when the temperature is below the freezing point. For remote monitoring, keeping sufficient water in the water reservoir is difficult. Therefore, psychrometers are usually used to check the temperature and relative humidity in the air conditioned space manually.

Mechanical hygrometers directly indicate the relative humidity of the moist air. They are also simple and relatively inexpensive. Their main drawbacks are their lack of precision over an extensive period and their lack of accuracy at extremely high and low relative humidities.

Electronic hygrometers, especially the polymer film resistance and the capacitance types are commonly used for remote monitoring and for controls in air conditioning systems.

Both the electronic and the mechanical hygrometers need regular calibration. Initial calibrations are usually performed either with precision humidity generators using two-pressure, two-temperature, and divided-flow systems or with secondary standards during manufacturing (refer to ASHRAE Standard 41.6–1982, Standard Method for Measurement of Moist Air Properties). Regular calibrations can be done with a precision aspiration psychrometer, or with chilled mirror dew-point devices.

Air contamination has significant influence on performance of the sensors of electronic and mechanical hygrometers. This is one of the reasons why they need regular calibration.

2.9 PSYCHROMETRIC CHARTS

Psychrometric charts provide a graphical representation of the thermodynamic properties of moist air, various air conditioning processes, and air conditioning cycles. They also give graphical solutions and help in the solution of complicated problems encountered in air conditioning processes and cycles.

Basic coordinates. The currently used psychrometric charts have two types of coordinates:

1. *h-w chart.* Enthalpy h and humidity ratio w are basic coordinates. The psychrometric charts published by ASHRAE and the Chartered Institution of Building Services Engineering (CIBSE) are *h-w* charts.
2. *T-w chart.* Temperature T and humidity ratio w are basic coordinates. Most of the psychrometric charts published by the large manufacturers in the United States are *T-w* charts.

For an atmospheric pressure of 29.92 in. Hg, an air temperature of 84°F, and a relative humidity of 100 percent, the humidity ratios and enthalpies found from the psychrometric charts published by ASHRAE and Carrier International Corporation are shown below:

	Humidity ratio, lb/lb	Enthalpy, Btu/lb
ASHRAE's chart	0.02560	48.23
Carrier's chart	0.02545	48.20

The last digit for humidity ratios and for enthalpies read from ASHRAE's chart is an approximation. Nevertheless, the differences between the two charts are less than 1 percent, and are considered negligible.

In this handbook, for manual psychrometric calculations and analyses, ASHRAE's chart will be used. If a microcomputer is used for psychrometric graphics

and calculations, then one should adopt a *T-w* chart because of its simplicity for calculations.

Temperature range and barometric pressure. ASHRAE's psychrometric charts are constructed for various temperature ranges and altitudes. In the Appendix only the one for normal temperature, that is, 32–120°F, and a standard barometric pressure at sea level, 29.92 in. Hg, is shown. The skeleton of ASHRAE's chart is shown in Fig. 2.8.

By the aid of microcomputers and software AUTOCAD, a psychrometric chart of normal temperature range and standard barometric pressure at sea level has been constructed. Figure 2.9 is an example of such a microcomputer-constructed psychrometric chart (MCPC). For the sake of simplicity, the skeleton of this chart has only temperature, humidity ratio, relative humidity, and wet bulb temperature lines.

Enthalpy lines. For ASHRAE's chart, the molar enthalpy of moist air is calculated from the formulation recommended by Hyland and Wexler (1983) in their paper "Formulations for the Thermodynamic Properties of Dry Air from 173.5 K to 473.5 K, and of Saturated Moist Air from 173.5 K to 473.5 K, at Pressures to 5 MPa."

For ASHRAE's chart, the enthalpy h-lines incline at an angle of 25 degrees to the horizontal lines. The scale factor for the enthalpy lines C_h, in Btu/lb · ft, is

$$C_h = \frac{h_2 - h_1}{L_h} = \frac{20}{0.4128} = 48.45 \qquad (2.47)$$

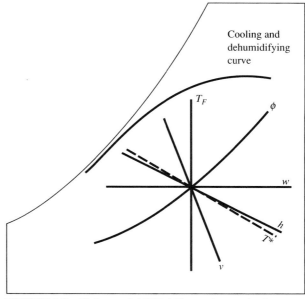

FIGURE 2.8 Skeleton of ASHRAE's psychrometric chart.

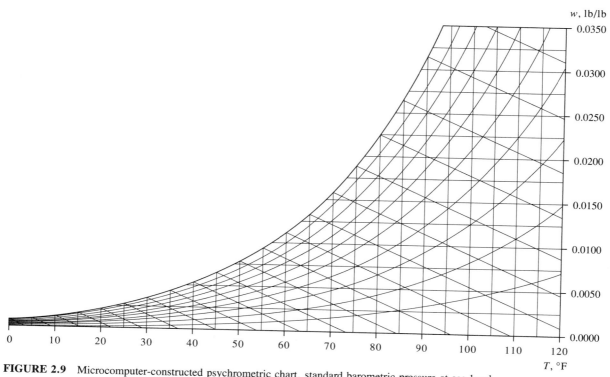

FIGURE 2.9   Microcomputer-constructed psychrometric chart, standard barometric pressure at sea level.

where L_h = shortest distance between enthalpy lines h_2 and h_1, ft. For MCPC, the magnitudes of enthalpy will be calculated by Eq. (2.29) only.

Humidity ratio lines. In both ASHRAE's chart and MCPC, the humidity ratio w-lines are horizontal. They form the ordinate of these psychrometric charts. The scale factor C_w, in lb/lb · ft, for w-lines in ASHRAE's chart is

$$C_w = \frac{w_2 - w_1}{L_w} = \frac{0.020}{0.5} = 0.040 \qquad (2.48)$$

where L_w = vertical distance between w_2 and w_1, ft. For both charts, the humidity ratio w can be calculated by Eq. (2.15). The humidity ratio of saturated moist air w_s in MCPC can also be calculated by Eq. (2.16).

Constant temperature lines. For ASHRAE's chart, since enthalpy is one of the coordinates, only the 120°F constant temperature T-line is a true vertical. All the other constant temperature lines incline slightly to the left at the top.

From Eq. (2.27), $T = h_a/c_{pd}$, therefore, one end of the T-line in ASHRAE'S chart can be determined from the enthalpy scale at $w = 0$. The other end can be determined by locating the saturated humidity ratio w_s on the saturation curve.

For MCPC, T-lines are the abscissa of the psychrometric chart. They are all vertical lines.

Saturation curve. A saturation curve is a locus representing a series of state points of saturated moist air. For ASHRAE's chart, the saturation pressure of water vapor above 32°F is calculated by the formula recommended in the Hyland and Wexler paper.

For MCPC, the humidity ratio of the saturated moist air w_s between 0–100°F can be found by the following polynomial:

$$w_s = a_1 + a_2 T_s + a_3 T_s^2 + a_4 T_s^3 + a_5 T_s^4 \qquad (2.49)$$

where T_s = saturated temperature of moist air, °F
$a_1 = 0.00080264$
$a_2 = 0.000024525$
$a_3 = 2.5420 \times 10^{-6}$
$a_4 = -2.5855 \times 10^{-8}$
$a_5 = 4.038 \times 10^{-10}$

If we use Eq. (2.49) to calculate w_s, the error is most probably less than 0.000043 lb/lb. It is far smaller than the value that can be identified on the psychrometric chart.

Relative humidity lines. For ASHRAE's chart, relative humidity φ-lines, thermodynamic wet bulb T^*-lines, and moist volume v-lines all are calculated and determined based on the formulations in Hyland and Wexler's paper.

For MCPC, the water vapor pressures p_w and p_{ws} at various w and T can be calculated by Eqs. (2.49), (2.15), and (2.16). Then, from the Eq. 2.21, φ is evaluated, and φ-lines can be plotted.

Thermodynamic wet bulb lines. For both ASHRAE's chart and MCPC, only thermodynamic wet bulb T^*-lines are shown. Since the dry bulb and the thermodynamic wet bulb temperatures coincide with each other on the saturation curve, then for MCPC one end of the T^*-line is determined. The other end of the T^*-line can be plotted on the w-line where $w = 0$. Let the state point of the other end of the T^*-line be represented by 1. Then, from Eq. (2.39), at $w_1 = 0$

$$c_{pd}(T_1 - T^*) = w_s^* h_{fg}^*$$

Solving for T^*, we have

$$T^* = \frac{c_{pd}T_1 - w_s^* h_{fg}^*}{c_{pd}} \qquad (2.50)$$

Moist volume lines. For MCPC, moist volume can be calculated by Eq. (2.31). The moist volume of saturated air and the specific volume of dry air form the two ends of the moist volume v-lines.

Cooling and dehumidifying curves. The two cooling and dehumidifying curves plotted on the ASHRAE's chart are based on data on coil performance published in the catalogs of U.S. manufacturers. These curves are very helpful in describing the actual locus of a cooling and dehumidifying process as well as determining the state points of air leaving the cooling coil.

2.10 DETERMINATION OF THERMODYNAMIC PROPERTIES ON PSYCHROMETRIC CHARTS

There are seven thermodynamic properties or property groups of moist air shown on psychrometric charts. They are

1. Enthalpy, h
2. Relative humidity, φ
3. Thermodynamic wet bulb temperature, T^*
4. Barometric or atmospheric pressure, p_{at}
5. Temperature, T, and saturation water vapor pressure, p_{ws}
6. Density, ρ, and moist volume, v
7. Humidity ratio, w; water vapor pressure, p_w; and dew point temperature, T_{dew}

The fifth, sixth, and seventh are thermodynamic property groups.

Usually, atmospheric pressure p_{at} is a known value based on the altitude of the location. Then, in the fifth property group, p_{ws} is a function of temperature T only.

In the sixth property group, according to Eq. (2.33), $\rho_a = 1/v$; that is, air density and moist volume are dependent on each other.

In the seventh property group, for a given value of p_{at}, properties w, p_w, and T_{dew} are all dependent on each other.

When p_{at} is a known value, and if the moist air is not saturated, then any two known independent thermodynamic properties can determine the magnitudes of the remaining unknown properties. If the moist air is saturated, then any independent property will determine the remaining magnitudes.

Example 2.1. The design indoor air temperature and relative humidity of an air conditioned space at sea level is 75°F and 50 percent. Find the humidity ratio, the enthalpy, and the density of the indoor moist air

1. By using the ASHRAE's chart
2. By calculation

Determine also the dew point and the thermodynamic wet bulb temperature of the moist air. The following information is required for the calculations:

Atmospheric pressure at sea level	14.697 psi
Specific heat of dry air	0.240 Btu/lb · °F
Specific heat of water vapor	0.444 Btu/lb · °F
Enthalpy of saturated vapor at 0°F	1061 Btu/lb
Gas constant of dry air	53.352 ft · lb$_f$/ lb$_m$ · °R

Solution

1. Plot the space point r on the ASHRAE's chart by finding first the space temperature $T_r = 75$°F on the abscissa, and then following along the 75°F constant-temperature line up to a relative humidity $\varphi = 50\%$ as shown in Fig. 2.10.

   Draw a horizontal line from the space point r. This line meets the ordinate, humidity ratio, w, at a value of $w_r = 0.00927$ lb/lb. This is the humidity ratio of the indoor space air.

   Draw a line parallel to the enthalpy line from the space point r. This line meets the enthalpy scale line at a value of $h_r = 28.1$ Btu/lb. This is the enthalpy of the indoor space air.

   Draw a horizontal line from the space point r to the left. This line meets the saturation curve at a dew point temperature of 55°F.

   Draw a line parallel to the thermodynamic wet bulb temperature lines through the space point r.

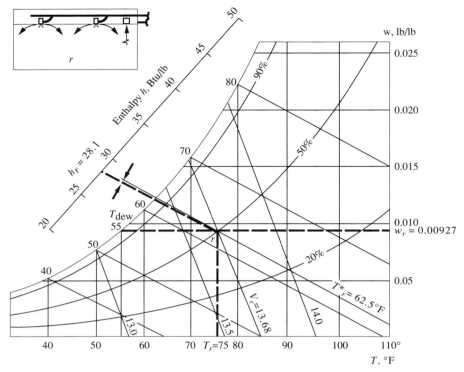

FIGURE 2.10 Thermodynamic properties determined from ASHRAE's psychrometric chart.

The perpendicular scale to this line shows a thermodynamic wet bulb temperature $T_{r*} = 62.5°F$.

Draw a line parallel to the moist volume lines through the space point r. The perpendicular scale to this line shows a moist volume $v_r = 13.68$ ft³/lb.

2. The calculations of the humidity ratio, enthalpy, and moist volume are as follows:

(1) From Eq. (2.49), the humidity ratio of the saturated air at the dry bulb temperature is

$$w_s = 0.00080264 + 0.000024525\,T$$
$$+ 2.542e - 06T^2 - 2.5855e - 08T^3$$
$$+ 4.038e - 10T^4$$
$$= 0.00080264 + 0.0018394 + 0.014299$$
$$- 0.010908 + 0.012776$$
$$= 0.018809 \text{ lb/lb}$$

According to Eq. (2.16), the saturated water vapor pressure of the indoor air is

$$p_{ws} = \frac{w_s\,p_{at}}{w_s + 0.62198}$$

$$= \frac{0.018809 \times 14.697}{0.018809 + 0.62198} = 0.4314 \text{ psia}$$

From Eq. (2.21), the water vapor pressure of indoor air is

$$p_w = \varphi\,p_{ws} = 0.5 \times 0.4314 = 0.2157 \text{ psia}$$

Then, from Eq. (2.15), the humidity ratio of the indoor air is

$$w_r = 0.62198\,\frac{p_w}{p_{at} - p_w}$$

$$= 0.62198\,\frac{0.2157}{14.697 - 0.2157}$$

$$= 0.009264 \text{ lb/lb}$$

(2) From Eq. (2.29), the enthalpy of the indoor moist air is

$$h_r = c_{pd}T + w\,(h_g + c_{ps}T)$$
$$= (0.240 \times 75) + 0.009264 \times$$
$$[1061 + (0.444 \times 75)]$$
$$= 28.14 \text{ Btu/lb}$$

(3) From Eq. (2.32), the moist volume of the indoor air is

$$v_r = \frac{R_a T}{p_{at} - p_w}$$

$$= \frac{53.352 \times 535}{(14.697 - 0.2157)\,144} = 13.688 \text{ ft}^3/\text{lb}$$

Then, from Eq. (2.33), the density of the indoor moist air is

$$\rho_r = 1/v_r = 1/13.688 = 0.07306 \text{ lb/ft}^3$$

3. Comparison between the thermodynamic properties read directly from the ASHRAE's chart and the calculated values is as follows:

	ASHRAE's chart	Calculated values
Humidity ratio, lb/lb	0.00927	0.009264
Enthalpy, Btu/lb	28.1	28.14
Moist volume, ft³/lb	13.68	13.69

Apparently, the differences between the readings from ASHRAE's chart and the calculated values are very small.

Example 2.2. A HVAC&R operator measured the dry and wet bulb temperatures in an air conditioned space as 75°F and 63°F, respectively. Find the relative humidity of this air conditioned space by using ASHRAE's chart and also by calculation. The humidity ratios of saturated air at temperatures of 75°F and 63°F are 0.018809 lb/lb and 0.012355 lb/lb, respectively.

Solution. It is assumed that the difference between the wet bulb temperature as measured by sling or aspiration psychrometer and the thermodynamic wet bulb temperature is negligible.

At a measured dry bulb temperature of 75°F and a wet bulb temperature of 63°F, the relative humidity read directly from the ASHRAE's chart is about 51.8 percent.

From Section 2.6, the wet bulb constant K' for a sling psychrometer is 0.000218 1/°F. Then, from Eq. (2.43), the humidity ratio of the space air can be calculated as

$$w = w'_s - 0.000218(T - T')$$
$$= 0.012355 - 0.000218(75 - 63)$$
$$= 0.009738 \text{ lb/lb}$$

From Eq. (2.16), the vapor pressure of the space air is

$$p_w = w p_{at}/(w + 0.62198)$$
$$= (0.009738 \times 14.697)/(0.009738 + 0.62198)$$
$$= 0.2266 \text{ psia}$$

And from Eq. (2.16), the saturated vapor pressure at a space temperature of 75°F is

$$p_{ws} = w_s p_{at}/(w_s + 0.62198)$$
$$= (0.018809 \times 14.697)/(0.018809 + 0.62198)$$
$$= 0.4314 \text{ psia}$$

Hence, from Eq. (2.21), the calculated relative humidity of the space air is

$$\varphi = p_w/p_{ws} = 0.2266/0.4314$$
$$= 0.5253 \quad \text{or} \quad 52.53\%$$

The difference between the value read directly from the ASHRAE's chart and the calculated one is 52.53% − 51.8% = 0.7%.

REFERENCES AND FURTHER READING

ASHRAE, *ASHRAE Handbook Fundamentals, 1989,* ASHRAE Inc., Atlanta, GA, 1989.

ASHRAE, ANSI/ASHRAE Standard 41.6–1982, *Standard Method for Measurement of Moist Air Properties,* ASHRAE Inc., Atlanta, GA, 1982.

ASHRAE, ASHRAE Standard 41.1–1986, *Standard Measurements Guide: Section on Temperature Measurements,* ASHRAE Inc., Atlanta, GA, 1986.

Aslam, S., Charmchi, M., and Gaggioli, R. A., "Psychrometric Analysis for Arbitrary Dry-Gas Mixtures and Pressures Using Microcomputers," *ASHRAE Transactions,* Part I B, pp. 448–460, 1986.

Hedlin, C. P., "Humidity Measurement with Dunmore Type Sensors," *Symposium at ASHRAE Semiannual Meeting,* ASHRAE Inc., New York, Feb. 1968.

Hyland, R. W., and Wexler, A., "Formulations for the Thermodynamic Properties of Dry Air from 173.15 K to 473.15 K, and of Saturated Moist Air from 173.15 K to 372.15 K, at Pressures to 5 MPa," *ASHRAE Transactions,* Part II A, pp. 520–535, 1983.

Hyland, R. W., and Wexler, A., "Formulations for the Thermodynamic Properties of the Saturated Phases of H_2O from 173.15 K to 473.15 K," *ASHRAE Transactions,* Part II A, pp. 500–519, 1983.

McGee, T. D., *Principles and Methods of Temperature Measurements,* John Wiley, 1st ed., New York, 1988.

Nelson, R. M., and Pate, M., "A Comparison of Three Moist Air Property Formulations for Computer Applications," *ASHRAE Transactions,* Part I B, pp. 435–447, 1986.

Stewart, R. B., Jacobsen, R. T., and Becker, J. H., "Formulations for Thermodynamic Properties of Moist Air at Low Pressures as Used for Construction of New ASHRAE SI Unit Psychrometric Charts," *ASHRAE Transactions,* Part II A, pp. 536–548, 1983.

Threlkeld, J. L., *Thermal Environmental Engineering,* Prentice-Hall, 2nd ed., Englewood Cliffs, NJ, 1970.

The Trane Company, *Psychrometry,* The Trane Company, La Crosse, WI, 1979.

Wang, S. K., *Air Conditioning,* Vol. 1, Hong Kong Polytechnic, 1st ed., Hong Kong, 1987.

AIR CONDITIONING PROCESSES AND CYCLES

3.1 AIR CONDITIONING PROCESSES

An air conditioning process determines the change in thermodynamic properties of moist air between the initial and final states of conditioning and also the corresponding energy and mass transfer between

the moist air and a medium, such as water, refrigerant, or moist air itself during this change.

The energy balance and the conservation of mass in nonnuclear processes are the two principles most often used in the analysis and calculation of the change of thermodynamic properties in air conditioning processes.

In general, for a single air conditioning process, heat transfer or mass transfer is positive. However, for calculations that involve several air conditioning processes, the heat supplied to the moist air is taken to be positive, and the heat rejected from the moist air is taken to be negative.

Sensible Heat Ratio

The *sensible heat ratio* (SHR) of an air conditioning process is defined as the ratio of the absolute value of sensible heat to the absolute value of total heat,

$$\text{SHR} = \frac{|q_{\text{sen}}|}{|q_{\text{total}}|} = \frac{|q_{\text{sen}}|}{|q_{\text{sen}}| + |q_l|} \quad (3.1)$$

In Equation (3.1), total heat (in Btu/h) is given as

$$q_{\text{total}} = |q_{\text{sen}}| + |q_l| \quad (3.2)$$

Figure 3.1 shows an air conditioning process from an initial state s to final state r. The sensible heat change q_{sen} in this process (in Btu/h) can be calculated as

$$q_{\text{sen}} = 60\dot{V}_s \rho_s c_{\text{pa}}(T_r - T_s) = 60\dot{m}_a c_{\text{pa}}(T_r - T_s) \quad (3.3)$$

where $\dot{V}_s$ = volume flow rate of the supply air, cfm
$\dot{m}_a$ = mass flow rate of supply air, lb/min
ρ_s = density of supply air, lb/ft^3
c_{pa} = specific heat of moist air, Btu/lb · °F
T_s, T_r = moist air temperature at initial and final states, °F

Latent heat change q_l in this process (in Btu/h) is given by

$$q_l = 60\dot{V}_s \rho_s (w_r - w_s) h_{\text{fg}}$$

where w_s, w_r = humidity ratio at the initial and final states, lb/lb. Theoretically, h_{fg} in Btu/lb is the latent heat of vaporization of the water at the temperature where vaporization or condensation occurs. But because the sensible heat change of $\dot{m}_a c_{\text{pa}}(w_r - w_s)(T_r - T_s)$ is already included in Eq. (3.3), it is convenient to use $h_{\text{fg.32}} + [c_{\text{ps}}(T - 32)]$ to replace h_{fg}. The error is usually less than 0.03%. Then, the latent heat change q_l can be calculated as

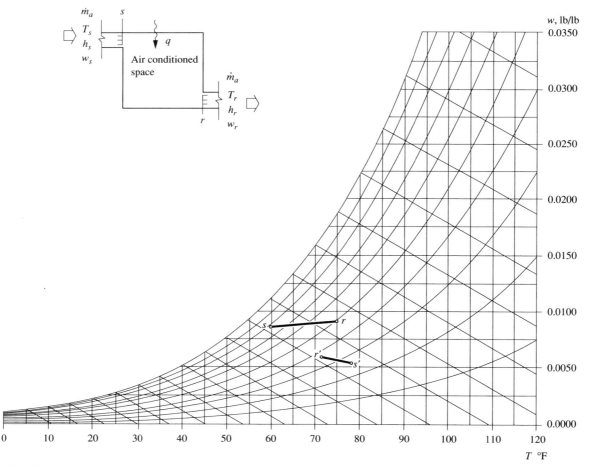

FIGURE 3.1 Space conditioning lines.

$$q_l = 60\dot{V}_s\rho_s(w_r - w_s)h_{\mathrm{fg.32}}$$
$$= 60\dot{V}_s\rho_s(w_r - w_s)(1075)$$
$$= 60\dot{m}_a(w_r - w_s)(1075)$$
$$= 64{,}500\dot{V}_s\rho_s(w_r - w_s) \tag{3.4}$$

In Fig. 3.1, the slope of the air conditioning process *sr* is given by

$$\tan\alpha = \frac{L_w}{L_t} = \frac{C_t(w_r - w_s)}{C_w(T_r - T_s)} \tag{3.5}$$

where α = angle between the air conditioning process and the horizontal line on the psychrometric chart, degrees

L_w, L_t = vertical and horizontal distance between state points r and s, ft

C_w, C_t = scale factor for the humidity ratio lines and the temperature lines, lb/lb · ft and °F/ft

Substituting Eqs. (3.3) and (3.4) in Eq. (3.1), then

$$\mathrm{SHR} = \frac{\dot{m}_a c_{\mathrm{pa}}(T_r - T_s)}{\dot{m}_a c_{\mathrm{pa}}(T_r - T_s) + \dot{m}_a(w_r - w_s)h_{\mathrm{fg.32}}}$$
$$= 1 \Big/ \left[1 + \frac{h_{\mathrm{fg.32}}C_w\tan\alpha}{c_{\mathrm{pa}}C_t}\right] \tag{3.6}$$

In ASHRAE's psychrometric chart as shown in Appendix II.1, additional SHR lines are given by joining the intersection of $T = 78°F$, $\phi = 50\%$, and the outer sensible heat ratio scale. An air conditioning process that is parallel to any one of these SHR lines has the same SHR.

3.2 SPACE CONDITIONING AND SENSIBLE COOLING AND HEATING PROCESSES

Space Conditioning Processes

A space conditioning process is an air conditioning process in which either:

1. Heat and moisture are absorbed by the supply air and removed from the space; or
2. Heat, or sometimes heat and moisture, is supplied to the space to compensate for the transmission and infiltration losses through the building envelope, with moisture being given up by the supply air to the space

The purpose of these processes is to maintain a desirable space temperature and relative humidity.

Figure 3.1 shows two lines, *sr* and *s'r'*, to indicate these two space conditioning processes. These lines are called space conditioning lines. The upper line,

sr, denotes the absorbtion of space heat and moisture during summer, and the bottom line indicates the supply of heat and absorption of moisture during winter.

In space conditioning processes, assuming that the kinetic energy difference between the supply inlet *s* and return exit *r* is negligible, and also that there is no work being done during these processes, the steady flow energy equation can be simplified to

$$60\dot{m}_a h_s + q_{\mathrm{rc}} = 60\dot{m}_a h_r \tag{3.7}$$

where h_r, h_s = enthalpy of space air and supply air, Btu/lb

q_{rc} = heat to be removed from the conditioned space, or the space cooling load, Btu/h

Rearranging the terms, then, the space cooling load can be calculated as

$$q_{\mathrm{rc}} = 60\dot{m}_a(h_r - h_s) = 60\dot{V}_s\rho_s(h_r - h_s) \tag{3.8}$$

where $\dot{V}_s$ = volume flow rate of supply air, cfm

ρ_s = density of supply air, lb/ft^3

Considering the mass balance during the space conditioning process

$$\dot{m}_a w_s + \dot{m}_g = \dot{m}_a w_r$$

where w_r, w_s = humidity ratio at exit r and inlet s respectively, lb/lb

$\dot{m}_g$ = rate of moisture gain in the conditioned space, lb/min

Again, rearranging the terms, the rate of space moisture gain can be calculated as

$$\dot{m}_g = \dot{m}_a(w_r - w_s) = \dot{V}_s\rho_s(w_r - w_s) \tag{3.9}$$

Sensible Heating and Cooling Processes

A *sensible heating process* is a process in which heat is added to the moist air, resulting in an increase in its temperature, while its humidity ratio remains constant. This process is represented by line 12 in Fig. 3.2. The sensible heating process occurs when moist air flows through a heating coil in which heat is transferred from the hot water inside the coil tubes to the moist air, or through a heat exchanger where heat transfer takes place between two fluid streams. The rate of heat transfer from the hot fluid to the cold fluid q_{ch}, in Btu/h, is called the heating coil's load or the heating capacity of the heat exchanger, and can be calculated as

$$q_{\mathrm{ch}} = 60\dot{m}_a(h_2 - h_1) = 60\dot{V}_s\rho_s(h_2 - h_1) \tag{3.10}$$

where h_1, h_2 = enthalpy of moist air entering and leaving the coil (or heat exchanger) respectively, Btu/lb. From Eq. (2.29),

$$h = c_{\mathrm{pd}}T + w(h_{g0} + c_{\mathrm{ps}}T)$$

and also, from Eq. (2.36),

$$c_{pa} = (c_{pd} + wc_{ps})$$

Substituting Eq. (2.29) into Eq. (3.10), assuming that the difference between c_{pd1} and c_{pd2}, and c_{ps1} and c_{ps2} are negligible, and that for a sensible heating process, $w_1 = w_2$,

$$q_{ch} = 60\dot{m}_a c_{pa}(T_2 - T_1) = 60\dot{V}_s \rho_s c_{pa}(T_2 - T_1)$$

$$(3.11)$$

A *sensible cooling process* removes heat from the moist air, resulting in a drop of its temperature while maintaining a constant humidity ratio of the moist air. This process is represented by line $1'2'$ in Fig. 3.2. The sensible cooling process occurs when moist air flows through a cooling coil and there is an indirect heat transfer from the moist air to the chilled water inside the cooling coil; or in an air-to-air heat exchanger where two air streams do not make contact.

In a sensible cooling process, the rate of heat transfer q_{cs}, in Btu/h, from the moist air to the chilled

water inside the cooling coil or from one air stream to another air stream in the heat exchanger is called the sensible cooling coil load or sensible cooling capacity of the heat exchanger. It can be calculated as

$$q_{cs} = 60\dot{m}_a c_{pa}(T_{1'} - T_{2'}) = 60\dot{V}_s \rho_s c_{pa}(T_{1'} - T_{2'})$$

$$(3.12)$$

3.3 HUMIDIFYING PROCESSES

In a *humidifying process,* water vapor is added to moist air, which increases in the humidity ratio if the initial moist air is unsaturated. Large-scale humidification of moist air in a central hydronic system is usually performed by the following methods:

1. Steam injection or submerged heating element
2. Sprayed or wetted media air washers
3. Atomizing device

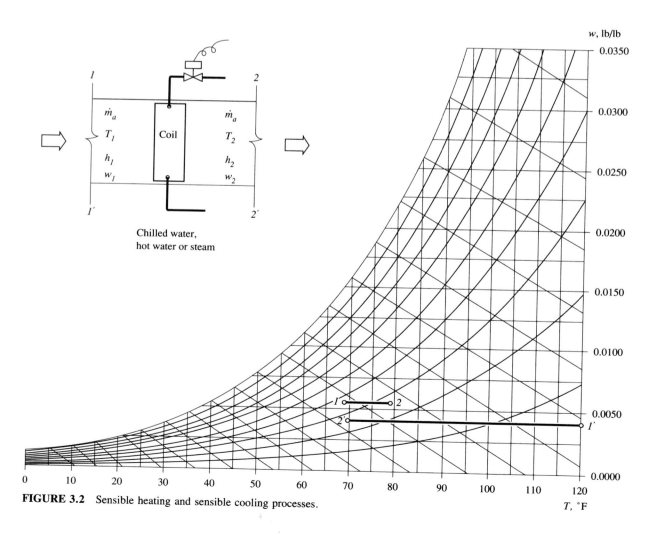

FIGURE 3.2 Sensible heating and sensible cooling processes.

The construction and characteristics of various types of humidifiers will be covered in Chapter 11.

Steam Injection and Heating Element Humidifier

In a humidifying process using steam injection, the steam is often supplied from the main line to a grid type humidifier and then injected directly into the air through small holes, as shown in Fig. 3.3.

The steam injection humidifying process is indicated by line *12*, which is approximately parallel to the constant temperature lines on the psychrometric chart. The slight inclination toward the right-hand side at the top of line *12* is due to the high temperature of the injected steam. The increase of the moist air temperature because of the steam injection can be calculated as follows.

When a mass flow rate of $\dot{m}_s$ lb/min of dry saturated steam at low pressure is injected into a moist air stream of a mass flow rate of $\dot{m}_a$ lb/min, according to the principle of heat balance

$$\dot{m}_a h_1 + \dot{m}_s h_s = \dot{m}_a h_2 \qquad (3.13)$$

where h_1, h_2 = enthalpy of the moist air entering and leaving the steam injection humidifier, Btu/lb

h_s = enthalpy of the injected steam, Btu/lb

The mass flow rate of injecting steam is defined as:

$$\dot{m}_s = \dot{m}_a(w_2 - w_1) \qquad (3.14)$$

where w_1, w_2 = humidity ratio of the moist air entering and leaving the steam injection humidifier, lb/lb.

Let $w_{sm} = \dot{m}_s / \dot{m}_a = (w_2 - w_1)$, then

$$h_1 + w_{sm} h_s = h_2 \qquad (3.15)$$

Assuming that c_{pd} and c_{ps} are constants, as $w c_{ps} \ll c_{pd}$, let $w_{12} = 1/2(w_1 + w_2)$ and replace w_1 and w_2 by w_{12}, then

$$I_5 - T_1 = \frac{w_{sm} c_{ps} T_s}{c_{pd} + w_{12} c_{ps}} \qquad (3.16)$$

where T_s = steam temperature, °F

T_1, T_2 = temperature of moist air entering and leaving the steam injection humidifier, °F.

A heating element humidifier employs a steam coil or electric heating element to provide the latent heat of vaporized water. The saturated water vapor added to the moist air is generally at a temperature higher than the air stream.

Air Washers

An *air washer* is a device that sprays water into air to humidify or cool and dehumidify, and to clean the air.

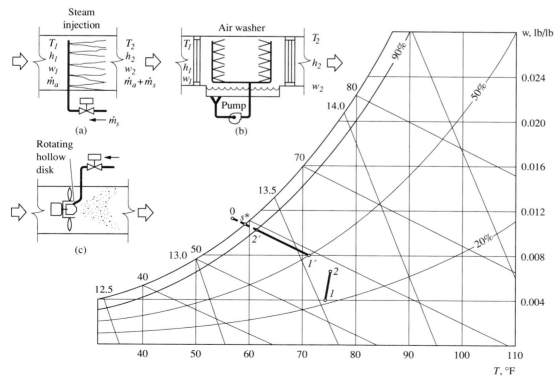

FIGURE 3.3 Humidification processes.

When moist air flows through an air washer sprayed with recirculating water, or water is distributed over a wetted medium as shown in Fig. 3.3b, the moist air is humidified, tending to approach saturation. This actual adiabatic saturation process follows the thermodynamic wet bulb line on the psychrometric chart represented by line $1'2'$ in Fig. 3.3. This process increases the humidity ratio of the air stream while resulting in a reduced air temperature. The cooling effect of the adiabatic saturation process is often called *direct evaporative cooling* and will be covered more in detail in Chapter 13.

A saturated end state is not usually achieved by such a process. The saturation effectiveness η_{sat} of the moist air leaving the air washer can be calculated as

$$\eta_{sat} = \frac{w_2' - w_1'}{w_s^* - w_1} \approx \frac{T_2' - T_1'}{T_s^* - T_1} \qquad (3.17)$$

where T_s^*, w_s^* = temperature and humidity ratio of saturated air at the thermodynamic wet bulb temperature, °F and lb/lb, respectively.

The saturation effectiveness of an actual adiabatic saturation process depends on the contact time and contact surface area between air and water, that is, water-air ratio $\dot{m}_w / \dot{m}_a$ (which indicates the mass flow rate of the spraying water to the mass flow rate of the air flowing through the air washer), the length of the chamber, and the direction of the spray (opposing or following the air flow).

Oversaturation

In water atomization for air humidification, an atomizing device breaks water into fine particles that are injected into the air stream. Whether the atomization is accomplished by centrifugal force, compressed air, or ultrasonic force, the humidification process is an actual adiabatic saturation process. With some atomizing devices, moist air leaving the air washer or pulverizing fan (as shown in Fig. 3.3c) can contain unevaporated water particles w_w great enough to exceed the humidity ratio of saturated air at the thermodynamic wet bulb temperature w_s^*. The excess amount of water particles present in the moist air is called oversaturation. *Oversaturation* is defined as

$$w_o - w_s^* = (w_2' + w_w) - w_s^* \qquad (3.18)$$

The quantity of unevaporated water particles at state point 2, in lb/lb is

$$w_w = w_o - w_2' \qquad (3.19)$$

where w_o = sum of the humidity ratio w_2' and the minute water particles w_w at state point 2, lb/lb

w_2 = humidity ratio at state point $2'$, lb/lb

When adiabatic heat transfer occurs between the air stream and the minute water particles, some evaporation takes place, so the humidity ratio of the moist air will increase. Such a transformation will still follow the thermodynamic wet bulb line $1'O$, as shown in Fig. 3.3.

As moist air flows through an air washer or atomizing device, there can be minute water particles present in the moist air, even if water eliminators are employed. The magnitude of w_w depends mainly on the construction of the humidifying device and water eliminators, and also the air velocity flowing through them. When moist air flows through an air washer, w_w may vary from 0.0002–0.001 lb/lb. If a pulverizing fan is used, w_w may be as high as 0.00135 lb/lb.

3.4 COOLING AND DEHUMIDIFYING PROCESS

In a cooling and dehumidifying process, both the temperature and the humidity ratio of the moist air will drop. This process is represented by curve m cc on the psychrometric chart in Fig. 3.4, where m is the entering mixture temperature of the outside and recirculating air.

There are three types of heat exchangers commonly used in the cooling and dehumidifying process:

1. Water cooling coil with chilled water flowing inside the coil's tubes
2. Direct expansion DX-coil in which refrigerant evaporates directly inside the coil's tubes
3. Air washer where chilled water rather than recirculated water is used for spraying

The temperature of the entering chilled water in a water cooling coil determines whether the process is a sensible cooling process or a cooling and dehumidifying process. If the temperature of entering chilled water T_{we} makes the outer surface temperature of the coil tubes $T_{s.t}$ higher than the dew point of entering air T_{ae}'', this process will be a sensible cooling process. Condensation cannot take place on the outer surface of the coil tubes or associated fins. However, if T_{we} is low so that $T_{s.t} < T_{ae}''$, cooling and dehumidifying occur.

If recirculating water is used in an air washer, the process is an actual adiabatic saturation process. The changes of the properties of the moist air follow the adiabatic saturation line on the psychrometric chart. The temperature of the water spray approaches the thermodynamic wet bulb temperature. When chilled water is used, $T_{s.t} < T_{ae}''$, a cooling and dehumidifying occurs. Temperature, humidity ratio,

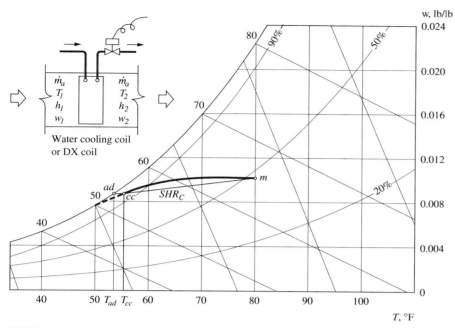

FIGURE 3.4 Cooling and dehumidifying process.

and enthalpy of moist air are changed during a cooling and dehumidifying process.

Based on the principle of heat balance,

total enthalpy of the entering air	=	total enthalpy of the leaving air	+	cooling coil's load (or cooling capacity of the washer)	+	heat energy of the condensate

That is,

$$60\dot{m}_a h_{ae} = 60\dot{m}_a h_{cc} + q_{cc} + 60\dot{m}_w h_w$$

where h_{ae}, h_{cc} = enthalpy of moist air entering and leaving the cooling coil or air washer, Btu/lb

$\dot{m}_w$ = mass flow rate of condensate, lb/min

h_w = enthalpy of the condensate, Btu/lb

The cooling coil's load or the cooling capacity of the air washer q_{cc}, in Btu/h, can be calculated as

$$q_{cc} = 60\dot{m}_a(h_{ae} - h_{cc}) - 60\dot{m}_w h_w$$
$$= 60\dot{V}_s \rho_s(h_{ae} - h_{cc}) - 60\dot{m}_w h_w \quad (3.20)$$

The mass flow rate of the condensate can be evaluated as

$$\dot{m}_w = \dot{m}_a(w_{ae} - w_{cc})$$

where w_{ae}, w_{cc} = humidity ratio of moist air entering and leaving the coil or air washer, lb/lb.

Assuming that the temperature of the condensate $T_w = T_{cc}$, Eq. (3.20) becomes

$$q_{cc} = 60\dot{m}_a[(h_{ae} - h_{cc}) - (w_{ae} - w_{cc})c_{pw}T_{cc}]$$
$$= 60\dot{V}_s\rho_s[(h_{ae} - h_{cc}) - (w_{ae} - w_{cc})c_{pw}T_{cc}]$$

$$(3.21a)$$

where c_{pw} = specific heat of water, Btu/lb · °F. In most cases, $\dot{m}_w h_w$ is less than 0.02 q_{cc}. Because $\dot{m}_w h_w$ is small compared to $m_a(h_{ae} - h_{cc})$, for practical calculations, $\dot{m}_w h_w$ is often ignored, so

$$q_{cc} = 60\dot{m}_a(h_{ae} - h_{cc}) = 60\dot{V}_s\rho_s(h_{ae} - h_{cc}) \quad (3.21b)$$

A straight line has been used to represent a cooling and dehumidifying process, and the intersection of this straight line and the saturation curve is considered the effective surface temperature of the coil, and is called the *apparatus dew point*. This is a misconception.

First, the cooling and dehumidifying process is actually a curve instead of a straight line. Second, the effective surface temperature as a fictitious reference point is affected by many factors, and is difficult to determine in heat and mass transfer calculations.

It is more accurate to define the *apparatus dew point* as the dew point of the moist air leaving the conditioning apparatus, the coil, the air washer, or other heat exchangers. Apparatus dew point is represented by point *ad* on the saturation curve of the psychrometric chart in Fig. 3.4.

However, the sensible heat ratio of the cooling and dehumidifying process SHR_c can be indicated by the slope of the line joining points m and cc as shown in Fig. 3.4. SHR_c can be calculated as the ratio of sensible heat removed during cooling and dehumidifying process q_{cs} to the total heat removed, q_{cc}, both in Btu/h

$$SHR_c = \frac{q_{cs}}{q_{cc}} \qquad (3.22)$$

According to the data published by U.S. manufacturers, a cooling and dehumidifying curve can be plotted for the moist air flowing through a water cooling coil. Figure 3.5 shows such a curve. The operating conditions are as follows:

Entering air   80°F dry bulb and
                            67°F wet bulb

Face air velocity   500 fpm
Entering water temperature 45°F
Water temperature rise   10°F

The number of rows depth, that is, the number of coil tubes along the air flow, varies from 2 to 10 rows, and the fin spacing varies between 6 to 14 fins per inch. Both of these factors affect the outer surface area of the coil.

It should be noted that the relative humidity of the air leaving the coil approaches saturation as the outer surface area of the coil increases. The cooling and dehumidifying process of a DX-coil is similar to that of the water cooling coil. Therefore, for coils with fin spacing of 10 or more fins per inch and for entering air at 80°F dry bulb and 68°F wet bulb, the relative humidity of the conditioned air leaving the cooling coil may be estimated as follows:

4-row coils   90–95 percent
6-row and 8-row coils   96–98 percent

Many factors affect the final state of the conditioned air leaving the coil. A detailed analysis will be provided in Chapter 12.

3.5 DESICCANT DEHUMIDIFICATION AND SENSIBLE COOLING

When moist air flows over a bed of either solid or liquid desiccant, absorption or adsorption occurs at

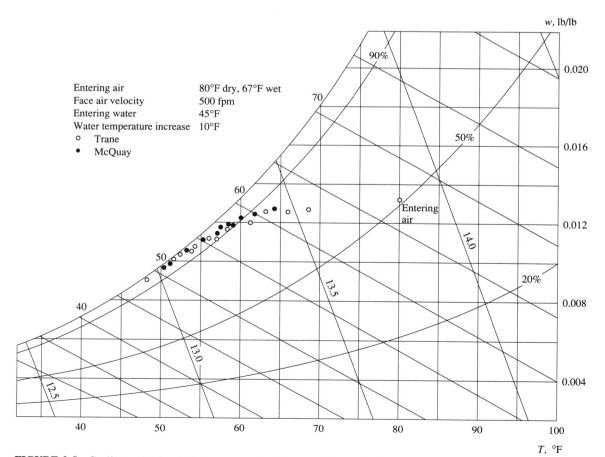

FIGURE 3.5 Cooling and dehumidifying process based on manufacturers' data.

the bed of sorbents. *Absorption* is a moisture sorption process associated with physical or chemical changes, while *adsorption* is one without physical or chemical changes. For liquid absorbants, such as liquid lithium chloride, the absorption of the moisture is mainly caused by the vapor pressure difference between the moist air and the surface of the liquid. Solid adsorbants such as silica gels attract moisture because of the vapor pressure difference and the electric field at the desiccant's surface.

During the process of sorption, heat is released. This released heat, often called the heat of sorption, is the sum of the following:

1. Latent heat due to the condensation of the absorbed water vapor into liquid and

2. Heat of wetting when the surface of the solid sorbent is wetted by the attached water molecules, or the heat of solution when moisture is absorbed by the liquid sorbents

Heat of wetting varies from a relatively large value when new desiccant first absorbs moisture to a very low value when it is nearly saturated.

During the desiccant dehumidification process, the humidity ratio of the leaving moist air is decreased. Although there is a heat loss through the outer casing of the dehumidifier to the ambient air, it is rather small compared with the release of the heat of sorption and any residual heat remaining after the regeneration

process. Regeneration uses heat to drive off the accumulated moisture so that the sorbent can be reused. The temperature of the dehumidified air is considerably increased because of the heat released during sorption. The desiccant dehumidification process line deviates from the thermodynamic wet bulb line and is represented by line *di do* on psychrometric chart in Fig. 3.6.

The decrease of humidity ratio during desiccant dehumidification process is given by

Difference in humidity ratio $= w_{di} - w_{do}$ (3.23)

where w_{di}, w_{do} = humidity ratio at dehumidifier inlet and outlet, lb/lb. The reduction of latent heat q_1, in Btu/h, can be calculated as

$$q_1 = 60\dot{V}_o\rho_o(w_{di} - w_{do})h_{fg.32}$$ (3.24)

where $\dot{V}_o$ = volume flow rate of air entering dehumidifier, cfm
ρ_o = air density of entering air, lb/ft³

After desiccant dehumidification, the hot air is usually sensibly cooled in two stages in order to save energy. This process is represented by line *do sc* on the psychrometric chart in Fig. 3.6.

The first stage uses an indirect evaporative cooling process or an outdoor air stream in an air-to-air heat exchanger. In an indirect evaporative cooling process, the hot air is sensibly cooled by another evaporative-cooled air stream without direct contact.

The first stage sensible cooling process is represented by line segment *do sc1*. The first stage sen-

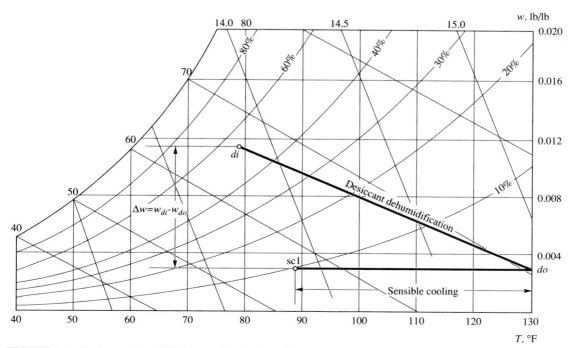

FIGURE 3.6 Desiccant dehumidification and sensible cooling.

sible cooling load q_{sc1}, in Btu/h, can be calculated as

$$q_{\text{sc1}} = 60\dot{V}_o\rho_o c_{\text{pa}}(T_{\text{do}} - T_{\text{sc1}}) \qquad (3.25)$$

where T_{do}, T_{sc1} = temperature of the dehumidified air entering and leaving the indirect evaporative cooler or heat exchanger, °F. The end state of first-stage sensible cooling sc1 can be determined according to local outdoor dry and wet bulb temperature to use fully the indirect evaporative cooling to save mechanical refrigeration in second-stage sensible cooling as described in Section 25.8.

In the second stage, dehumidified air may be mixed with the recirculated air and then enter a water cooling coil or a DX-coil for further sensible cooling.

The second-stage sensible cooling load q_{sc2}, in Btu/h, can be calculated as

$$q_{\text{sc2}} = 60\dot{V}_o\rho_o c_{\text{pa}}(T_{\text{sc1}} - T_{\text{sc2}}) \qquad (3.26)$$

where T_{sc2} = temperature of dehumidified air leaving the coil, °F.

3.6 ADIABATIC MIXING AND BYPASS MIXING PROCESSES

Two-Stream Adiabatic Mixing Process

In a two-stream adiabatic mixing process, two moist airstreams (say, 1 and 2) are mixed together, adiabatically, forming a uniform mixture m in a mixing chamber. Such a mixing process is illustrated by line 1 m1 2 on the psychrometric chart in Fig. 3.7. In actual practice, for the mixing of either recirculating and outdoor airstreams or the conditioned airstream and bypass airstream, the rate of heat transfer between the mixing chamber and the ambient air is small. It is usually ignored.

Based on the principle of heat balance and the continuation of mass

$$\dot{m}_1 h_1 + \dot{m}_2 h_2 = \dot{m}_m h_m \qquad (3.27)$$

$$\dot{m}_1 w_1 + \dot{m}_2 w_2 = \dot{m}_m w_m \qquad (3.28)$$

$$\dot{m}_1 + \dot{m}_2 = \dot{m}_m \qquad (3.29)$$

In Eqs. (3.27), (3.28), and (3.29), subscripts 1, 2, and m represent airstream 1, airstream 2, and the mixture, respectively.

Because $\dot{m}_1 = \dot{m}_m - \dot{m}_2$ and $\dot{m}_2 = \dot{m}_m - \dot{m}_1$, substituting into Eqs. (3.27) and (3.28),

$$\frac{\dot{m}_1}{\dot{m}_m} = \frac{h_2 - h_m}{h_2 - h_1} = \frac{w_2 - w_m}{w_2 - w_1} = K_1$$

and also

$$\frac{\dot{m}_2}{\dot{m}_m} = \frac{h_m - h_1}{h_2 - h_1} = \frac{w_m - w_1}{w_2 - w_1} = K_2 \qquad (3.30)$$

where K_1, K_2 = fraction of airstream 1 and 2 among the mixture.

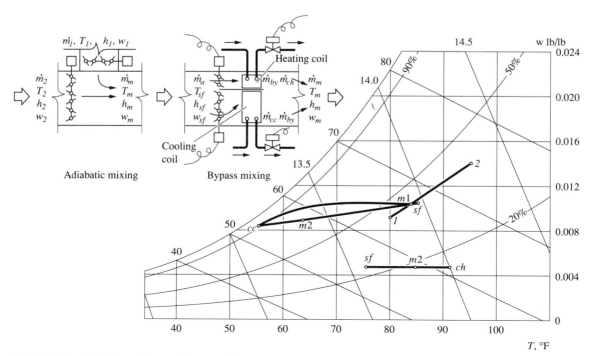

FIGURE 3.7 Adiabatic mixing and bypass mixing processes.

From the above relationships

1. The mixing point m must lie on the line that connects state points 1 and 2
2. The ratio of the mass flow of air stream 1 to the mass flow rate of the mixture is exactly equal to the ratio of the segment m1 2 to segment 12 as shown in Fig. 3.7, that is,

$$\frac{\dot{m}_1}{\dot{m}_m} = \frac{m1\ 2}{12} \qquad (3.31)$$

Similarly, the ratio of the mass flow rate of air stream 2 to the mass flow rate of the mixture is exactly equal to the ratio of segment 1 $m1$ to segment 12, that is,

$$\frac{\dot{m}_2}{\dot{m}_m} = \frac{1\ m1}{12} \qquad (3.32)$$

As the mass flow rate $\dot{m} = \dot{V}\rho$, substituting into Eqs. (3.27), (3.28), and (3.29),

$$\dot{V}_1\rho_1 h_1 + \dot{V}_2\rho_2 h_2 = \dot{V}_m\rho_m h_m$$
$$\dot{V}_1\rho_1 w_1 + \dot{V}_2\rho_2 w_2 = \dot{V}_m\rho_m w_m \qquad (3.33)$$
$$\dot{V}_1\rho_1 + \dot{V}_2\rho_2 = \dot{V}_m\rho_m$$

Because the magnitude of ρ_m always lies between ρ_1 and ρ_2, if the differences between ρ_m and ρ_1, ρ_m and ρ_2, and also the slight differences between the temperature divisions are negligible,

$$\left.\begin{array}{c} \dot{V}_1 h_1 + \dot{V}_2 h_2 = \dot{V}_m h_m \\[4pt] \dot{V}_1 w_1 + \dot{V}_2 w_2 = \dot{V}_m w_m \\[4pt] \dot{V}_1 T_1 + \dot{V}_2 T_2 = \dot{V}_m T_m \\[4pt] \dot{V}_1 + \dot{V}_2 = \dot{V}_m \end{array}\right\} \qquad (3.34)$$

Bypass Mixing Process

In many air handling units, one air stream is divided into an upper hot deck air stream and a lower cold deck air stream as shown in Fig. 3.7. During cooling season, the lower air stream is cooled and dehumidified by the cooling coil, and the hot water supply to the heating coil is shut off. Therefore, the upper air stream becomes a bypass air stream, that is, it bypasses the cooling coil. After the coils, the conditioned air stream is mixed with the bypass air stream and forms a mixture at state point $m2$. This bypass mixing process is shown by line $cc\ m2\ sf$ in Fig. 3.7.

During heating season, the cooling coil is not energized, but the heating coil is. Under such circumstances, the upper conditioned warm airstream and the lower bypass airstream combine to form a mixture m, as indicated by line $ch\ m2\ sf$ in Fig. 3.7.

Similar to the two-stream adiabatic mixing process, the ratio of the mass flow rate of the conditioned stream $\dot{m}_{cc}$ to the mixture $\dot{m}_m$ is

$$\frac{\dot{m}_{cc}}{\dot{m}_m} = \frac{sf\,m2}{sf\,cc} = K_{cc}$$

and

$$\frac{\dot{m}_{ch}}{\dot{m}_m} = \frac{sf\,m2}{sf\ ch} = K_{ch} \qquad (3.35)$$

where K_{cc}, K_{ch} = fraction of conditioning stream flowing through the cooling coil or heating coil. Also

$$\dot{m}_m = \dot{m}_{cc} + \dot{m}_{by} = \dot{m}_{ch} + \dot{m}_{by} \qquad (3.36)$$
$$\dot{m}_m w_m = \dot{m}_{cc} w_{cc} + \dot{m}_{by} w_{sf} \qquad (3.37)$$

Subscript cc represents the condition off the cooling coil, ch the heating coil, by the bypass airstream, which is the same as the state point of air at the supply fan discharge sf. According to the principle of energy balance

$$h_m = (\dot{m}_{cc} h_{cc} + \dot{m}_{by} h_{sf})/\dot{m}_m$$
$$= (\dot{m}_{ch} h_{ch} + \dot{m}_{by} h_{sf})/\dot{m}_m \qquad (3.38)$$
$$w_m = (\dot{m}_{cc} w_{cc} + \dot{m}_{by} w_{sf})/\dot{m}_m \qquad (3.39)$$

Based on the mass flow rate of the mixture, the cooling coils load can be calculated as

$$q_{cc} = K_{cc}\dot{V}_s \rho_s (h_{sf} - h_{cc}) \qquad (3.40)$$

and the heating coil load is

$$q_{ch} = K_{ch}\dot{V}_s \rho_s c_{pa}(T_{ch} - T_{sf}) \qquad (3.41)$$

3.7   BASIC AIR CONDITIONING CYCLE

An air conditioning cycle consists of several air conditioning processes connected in a serial order. Strictly speaking, such processes determine the operating performance of various components of the air side of an air conditioning system including the heat, mass, and other energy transfer, and the changes in property of the moist air.

Different air systems are characterized by different types of air conditioning cycles. The psychrometric analysis of an air conditioning cycle is an important tool to determine the properties of moist air at various state points, the volume flow rates, and the capacities of the major components of the system.

Air conditioning cycles can be divided into two categories: open and closed cycles. In an *open air conditioning cycle,* the moist air at the end state will not resume its original state. An air conditioning cycle of a 100 percent outdoor air make-up system is an open cycle. In a closed air conditioning cycle, moist air at the end state will resume its original state. The cycle of conditioning the mixture of recirculating and outdoor air, supplying it to the space, and recirculating it again is a *closed cycle*.

Based on outdoor weather conditions, the cycles operate in three modes:

1. *Summer mode.* When both outdoor and indoor operating parameters are in summer conditions. Equipment is selected to satisfy the summer outdoor and indoor design conditions.

2. *Winter mode.* When both outdoor and indoor operating parameters are in winter conditions. The same equipment is also selected to satisfy winter outdoor and indoor design conditions.

3. *Spring or fall mode.* When the outdoor air is typical of spring or fall weather and the indoor conditions are between the summer and winter design conditions.

A *basic air system* is an air system with a constant volume of supply air throughout the operating period, serving a single zone with a single supply duct from an air handling unit or a packaged unit. A *single zone* is a conditioned space that is controlled to maintain a unique indoor temperature, relative humidity, cleanliness, and pressure differential. A basic air conditioning cycle is the operating cycle of a basic air system.

3.8 BASIC AIR CONDITIONING CYCLE—SUMMER MODE OPERATION

Figure 3.8 shows a basic air system at summer mode and winter mode operation. In summer mode, return air from the conditioned space may pass through a return fan. Recirculating air enters the air-handling unit (AHU) at state point *ru* and is mixed with outdoor air at point *o* in the mixing chamber. The mixture *m* is then cooled and dehumidified at the cooling coil and leaves it at point *cc*. After that, the conditioned air flows through the supply fan *sf*, and then flows to the conditioned space through the supply outlet at state point *s*. When supply air absorbs the sensible and latent loads from the space, it becomes space air at point *r*. Space air then flows through the upper part of the mall, absorbs heat released from the light fixtures and the roof, and flows through the return duct/return fan *rf*. The recirculated portion enters the air handling unit again at point *ru*.

Summer mode operation consists of the following processes:

1. Sensible heating process *r ru* from the return system heat gain $q_{r.s}$ when return air flows through the ceiling plenum, the return duct, and the return fan. It includes:

- heat gain from the electric lights recessed on the ceiling plenum absorbed by the return air when it flows through the light fixtures and the ceiling plenum q_{rp}
- return fan power heat gain q_{rf}, if there is a return fan
- return duct heat gain q_{rd}

That is,

$$q_{r.s} = q_{rp} + q_{rf} + q_{rd} \qquad (3.42)$$

It is often convenient to use overall temperature increase to denote the system heat gain, such as, $\Delta T_{r.s}$, in °F. Then the return system heat gain is

$$q_{r.s} = \dot{V}_{ret}\rho_{ret}c_{pa}(T_{ru} - T_r) \qquad (3.43)$$

Subscript ret indicates return air and ru the air entering the air-handling unit.

The volume flow rate of return air is often higher than the recirculated air entering the mixing box if there is leakage or exhaust through the exhaust/relief damper.

2. Adiabatic mixing process of recirculating air at state point ru and the outdoor air at state point *o* in the mixing box, which is represented by line *ru m o* on the psychrometric chart in Fig. 3.8. After mixing, the condition of the mixture is *m*, which is also the state point of the moist air entering the cooling coil.

3. Cooling and dehumidifying process, which is indicated by the curve *m cc*. The cooling coils load can be calculated by Eq. (3.21b).

4. Sensible heating process *cc s* from the supply system heat gain $q_{s.s}$ when supply air flows through the supply fan and supply duct. It includes:

- supply fan power heat gain q_{sf} represented by line *cc sf*
- supply duct heat gain q_{sd} indicated by line *sf s*

Supply system heat gain is then given as

$$q_{s.s} = q_{sf} + q_{sd} \qquad (3.44)$$

The supply fan power heat gain and the supply duct heat gain are far greater than the heat gains of the return fan and duct. These differences are caused by the higher fan total pressure in the supply fan and the larger temperature difference between the air inside and outside the supply duct.

5. Space conditioning process, which is represented by line *sr*. The sensible heat removed from the conditioned space or the *space sensible load*, in Btu/h, is

$$q_{rs} = 60\dot{m}_a c_{pa}(T_r - T_s) = 60\dot{V}_s \rho_s c_{pa}(T_r - T_s) \qquad (3.45)$$

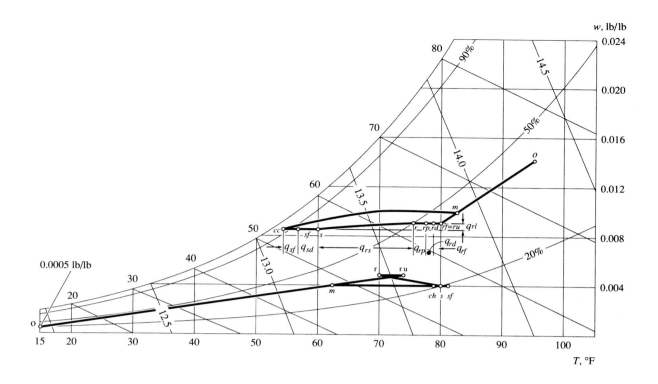

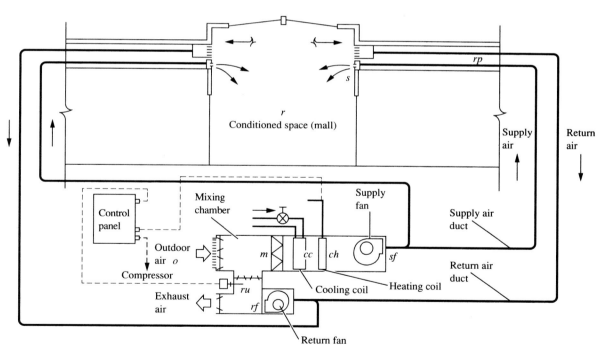

FIGURE 3.8 Basic air conditioning cycle.

The latent heat removed from the conditioned space or the space latent load, in Btu/h, is

$$q_{rl} = 60\dot{m}_a(w_r - w_s)h_{fg.32}$$
$$= 60\dot{V}_s\rho_s(w_r - w_s)h_{fg.32} \qquad (3.46)$$

Therefore, the space cooling load q_{rc}, in Btu/h, is calculated as

$$q_{rc} = q_{rs} + q_{rl} \qquad (3.47)$$

3.9 SUPPLY VOLUME FLOW RATE

Supply air volume flow rate and refrigeration capacity are two primary air conditioning system characteristics.

Based on Space Cooling or Heating Load

For most comfort and many process air conditioning systems, supply air volume flow rate is calculated from the space cooling load. From Eqs. (3.8) and (3.45), the mass flow rate of the supply air, in lb/min, can be calculated as

$$\dot{m}_a = \frac{q_c}{60(h_r - h_s)} = \frac{q_{rs}}{60c_{pa}(T_r - T_s)} \qquad (3.48a)$$

The supply volume flow rate, in cfm, can be calculated as

$$\dot{V}_s = \frac{q_{rc}}{60\rho_a(h_r - h_s)} = \frac{q_{rs}}{60\rho_s c_{pa}(T_r - T_s)} \qquad (3.48b)$$

The intensity of supply air, $\dot{V}_{int}$ in cfm/ft^2, is defined as

$$\dot{V}_{int} = \frac{\dot{V}_s}{A_{fl}} \qquad (3.49)$$

where A_{fl} = floor area of the conditioned space, ft^2. Intensity of supply air is often used in air volume flow estimates.

The following items should be considered during the calculation of the supply air flow rate from Eqs. (3.48a) and (3.48b):

1. *Volumetric vs. mass flow rate.* Supply volume flow rate is widely used in calculations to determine the size of fans, grilles, outlets, and air handling units. However, mass flow rate $\dot{m}$ is sometimes simpler to apply in cooling coil load calculations, or may be more appropriate when the variation of air density affects the result.

2. *Temperature difference vs. enthalpy difference.* Theoretically, in Eqs. (3.48a) and (3.48b), it is more accurate to use enthalpy difference to calculate the supply volume flow rate. This is because of

the elimination of the term c_{pa}, which is actually a variable depending on temperature and humidity ratio. In actual practice, it is far more convenient to use temperature differences on the psychrometric chart than enthalpy differences. The variation of c_{pa} is hardly detectable from the results found on the psychrometric chart.

3. *Summer cooling load vs. winter heating load.* Summer cooling load is usually used to calculate the supply volume flow rate because it is generally greater than using winter heating load. But if the winter weather is very cold, this may require a greater volume flow rate. Based on winter heating load, the supply volume flow rate can be calculated as

$$\dot{V}_s = q_{rh}/\rho_s c_{pa}(T_s - T_r) \qquad (3.50)$$

where q_{rh} = space heating load, Btu/h. Use the greater volume flow rate of the two to select the fans, AHU, and other components.

Based on Requirements Other Than Cooling Load

The supply volume flow rate of many air conditioning systems can also be governed by the following requirements:

1. To dilute the concentration of the air contaminants. Based on the principle of conservation of mass, the mass generated rate of the air contaminants in the conditioned space $\dot{m}_{par}$ is given by

$$\dot{m}_{par} = \dot{V}_s(C_i - C_s)/2118$$

and the supply volume flow rate $\dot{V}_s$, in cfm, can be calculated as

$$\dot{V}_s = 2118\dot{m}_{par}/(C_i - C_s) \qquad (3.51)$$

where C_i, C_s = concentrations of the air contaminants of the space air and supply air, usually in mg/m^3; 1 mg = 0.01543 grain

$\dot{m}_{par}$ = rate of contaminants generated in the space, mg/s

2. To maintain a required relative humidity φ_r as well as a humidity ratio w_r at a specific temperature for the conditioned space. Then, the supply volume flow rate can be calculated as

$$\dot{V}_s = q_{rl}/60\rho_s(w_r - w_s)h_{fg.32} \qquad (3.52)$$

3. To provide a desirable air velocity v_r, in fpm, within the working area in a clean room. The supply volume flow rate is then

$$\dot{V}_s = A_r v_r \qquad (3.53)$$

where A_r = cross sectional area perpendicular to air flow in the working area, ft².

4. To exceed the outdoor air requirement for acceptable air quality for occupants. The supply volume flow rate can then be calculated as

$$\dot{V}_s = n\dot{V}_{oc} \tag{3.54}$$

where n = number of occupants
$\dot{V}_{oc}$ = outdoor air requirement per person, cfm

5. To exceed the sum of the volume flow rates of the exhaust air $\dot{V}_{ex}$ and the space air relief $\dot{V}_{exf}$ through exfiltration to maintain a positive pressure in the conditioned space, both in cfm. Then, the supply volume flow rate is

$$\dot{V}_s = \dot{V}_{ex} + \dot{V}_{exf} \tag{3.55}$$

The supply volume flow rate should be the one that is the greatest from any of the above requirements, the cooling load, or heating load, or outdoor air requirements.

Rated Volume Flow of Supply and Return Fans

The supply volume flow rate $\dot{V}_s$ is determined based on the mass flow rate of supply air $\dot{m}_a$ required to offset the space load at the specific supply condition, with a specific supply air density ρ_s. Because the supply volume flow rate must be provided by the supply fan or the fan's rated volumetric capacity $\dot{V}_{sf.r}$ against a certain fan total pressure Δp_t, it is important to determine the relationship between $\dot{V}_s$ and $\dot{V}_{sf.r}$.

If the supply fan is located downstream from the cooling coil, the humidity ratios of the moist air between state points cc and s are constant, as shown in Fig. 3.8. According to the principle of conservation of mass, the mass flow rate of supply air $\dot{m}_a$ (in lb/min) remains constant, but the volume flow changes as its state changes. The relationships are

$$\dot{m}_a = \dot{V}_{cc}\rho_{cc} = \dot{V}_{sf}\rho_{sf} = \dot{V}_s\rho_s \tag{3.56}$$

Subscript cc represents the state point of conditioned air leaving the cooling coil, sf represents the moist air at the outlet of the supply fan, and s represents the supply air at the outlets.

Fans are rated at standard air condition. Standard air means dry air at a temperature of 70°F and a barometric pressure of 29.92 in. Hg (14.697 psi) with a density of 0.075 lb/ft³. From Eqs. (3.56) and (3.48b), the volume flow rate at the supply fan outlet is calculated as

$$\dot{V}_{sf} = \frac{\dot{V}_s\rho_s}{\rho_{sf}} = \frac{q_{rs}}{60\rho_{sf}c_{pa}(T_r - T_s)} \tag{3.57}$$

For the same fan at a given speed, its volume flow rate remains unchanged against a fixed Δp_t even if the air density varies during rated conditions. Therefore,

$$\dot{V}_{sf.r} = \dot{V}_{sf} \tag{3.58}$$

From Eq. (2.32), the moist volume of 1 lb of moist air v is

$$v = \frac{R_a T_R(1 + 1.6078w)}{p_{at}}$$

For normal summer and winter comfort applications, the humidity ratio varies less than 0.0062 lb/lb. Therefore, $v < 1.01 R_a T_R / p_{at}$, that is, the influence of variation of humidity ratio is smaller than 1 percent. Therefore, it can be ignored. If there is no infiltration or exfiltration at the conditioned space, Eq. (3.56) can be rewritten as

$$\dot{m}_a = \dot{V}_{cc}\rho_{cc} = \dot{V}_{sf}\rho_{sf} = \dot{V}_s\rho_s = \dot{V}_r\rho_r = \dot{V}_{rf}\rho_{rf} \tag{3.59}$$

Subscript rf represents the outlet of the return fan.

Similarly, the rated volume flow rate of the return fan $\dot{V}_{rf.r}$ is given by

$$\dot{V}_{rf.r} = \dot{V}_{rf} = \frac{q_{rs}}{60\rho_{rf}c_{pa}(T_r - T_s)} \tag{3.60}$$

For air leaving the cooling coil T_{cc} at 55°F with a relative humidity of 92 percent, and a T_{sf} of 57°F, the psychrometric chart shows the moist volume v_{sf} = 13.20 ft³/lb. From Eq. (3.57), the rated volume flow rate of the supply fan is

$$\dot{V}_{sf.r} = \frac{13.20 q_{rs}}{60 \times 0.243(T_r - T_s)} = \frac{q_{rs}}{1.105(T_r - T_s)} \tag{3.61}$$

For cold air distribution, which will be covered in detail in Chapters 9 and 25, T_{cc} may equal 40°F with a relative humidity of 98%, and if T_{sf} = 42°F, then v_{sf} = 12.8 ft³/lb, and the rated supply volume flow is

$$\dot{V}_{sf.r} = \dot{V}_{sf} = \frac{12.8 q_{rs}}{60 \times 0.243(T_r - T_s)}$$

$$= \frac{q_{rs}}{1.14(T_r - T_s)} \tag{3.62}$$

For an air system in which the supply fan is upstream from the cooling coil, and if the T_{sf} = 82°F and ϕ_{sf} = 43 percent, then, from the psychrometric chart, v_{sf} = 13.87 ft³/lb, and the supply volume flow rate is

$$\dot{V}_{sf} = \frac{13.87 q_{rs}}{60 \times 0.243(T_r - T_s)} = \frac{q_{rs}}{1.051(T_r - T_s)} \tag{3.63}$$

Thus, an upstream fan requires 9 percent greater fan capacity than a downstream fan with cold air distribution.

3.10 DETERMINATION OF THE SUPPLY AIR CONDITION

The following values are usually known before the summer mode basic air conditioning cycle is developed:

1. Summer outdoor design temperatures: dry bulb T_o and wet bulb T_o'
2. Summer indoor space air temperature T_r and relative humidity φ_r
3. Space sensible and latent load q_{rs} and q_{rl}
4. Outdoor air requirement $\dot{V}_o$
5. Estimated supply system heat gain $q_{s.s}$
6. Estimated return system heat gain $q_{r.s}$

If the state point of the supply air can be determined from the psychrometric chart, the summer mode basic air conditioning cycle can be developed. Eventually, the supply volume flow rate $\dot{V}_s$ and the cooling coil's load q_{cc} can also be determined.

To determine the supply air condition, the following basic concepts need to be considered:

1. For a given design indoor temperature T_r, design space sensible cooling load q_{rs}, and specific supply system heat gain $q_{s.s}$, a lower T_{cc} and T_s always result in a greater supply air differential $\Delta T_s = (T_r - T_s)$ and a lower space relative humidity φ_s. On the other hand, a higher T_{cc} means a smaller ΔT_s and a higher φ_r.
2. A greater ΔT_s reduces the volume flow rate and, therefore, fan size, fan energy use, terminal, and duct sizes. These reduce both construction and operating costs.
3. Temperature of air leaving the cooling coil T_{cc} is closely related to the temperature of the chilled water entering and leaving the coil T_{we} and T_{wl}, or to the evaporating temperature inside the DX-coil T_{ev}. A lower T_{cc} and T_{we} need a correspondingly lower T_{ev} and, therefore, higher power input to the refrigeration compressors.
4. When an air system is used to serve a single zone conditioned space, T_s and T_{cc} should be properly arranged so that the indoor design temperature and relative humidity can be maintained for energy efficient operation.
5. For a central hydronic air conditioning system serving many control zones with many space temperatures and relative humidity requirements, T_{we} and T_{cc} should satisfy the lowest T_{cc} requirement. In actual practice, T_{we} and T_{cc} are often determined by previous experience of similar applications.

Graphical Method

The following graphical method may be used to determine the condition of supply air s for an air system serving a single zone or a dominating zone whose T_r and φ_r must be maintained

1. According to the data given in Section 3.4, the relative humidity of the air leaving a cooling coil (φ_{cc}) with 10 or more fins per inch can be estimated as

   4-row coil φ_{cc} = 93 percent
   6-row coil φ_{cc} = 96 percent
   8-row coil φ_{cc} = 98 percent

2. Draw the space conditioning line sr from the state point r, parallel to the line of known sensible heat ratio SHR_s as shown in Fig. 3.9.
3. State point of supply air s must lie on line sr.
4. The horizontal line cc s represents the magnitude of the supply system heat gain. As segment cc s is moved up and down, there exists a position where cc lies exactly on the selected φ_{cc} curve for the coil, s lies on line sr, and $(T_s - T_{cc})$ exactly matches the temperature increase caused by supply system heat gain.
5. In practice, this required T_{cc} must be reconciled with the selected coil's performance for a given coil configuration T_{we} and air and water velocities.

Influence of Sensible Heat Ratio

For a specific design T_r and φ_r, the sensible heat ratio of the space conditioning line SHR_s has a significant effect on T_s and T_{cc}.

Fig. 3.10 shows such an influence. As SHR_s becomes smaller, for the sake of maintaining the required T_r and φ_r, the temperature of conditioned air leaving the cooling coil T_{cc} must be lower. This requires a lower T_{we} or a lower evaporating temperature in a DX-coil, which eventually increases the energy input to the refrigeration plant, but reduces the air volume flow that must be handled.

For a specific space cooling load q_{rc}, a lower SHR_s always indicates a smaller q_{rs} and a greater ΔT_s. A higher ΔT_s means a lower supply volume flow rate $\dot{V}_s$, smaller equipment size, lower fan energy use, and reduced construction and operating costs.

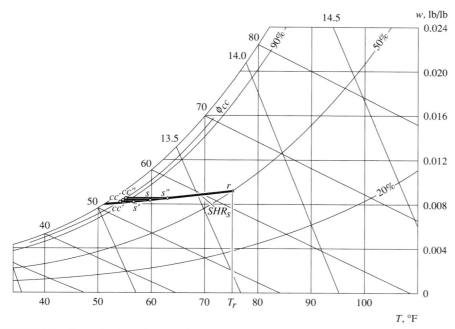

FIGURE 3.9 Determination of state point of supply air.

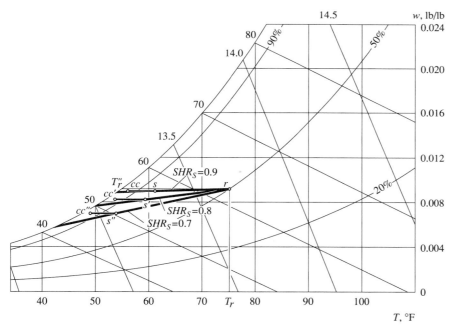

FIGURE 3.10 Effect of sensible heat ratio of space conditioning line on the determination of the condition of supply air.

In Fig. 3.10, when SHR$_s$ = 0.7, the dew point of space air of 75°F and 50 percent relative humidity is 55°F which is higher than the 54°F supply air temperature. Under such circumstances, condensation may occur at the supply outlet when its cold metal frame contacts the space air at a higher dew point level. For cold air distribution, the dew point of space air at 78°F and 35 to 40 percent relative humidity is between 48°F and 51.5°F. If T_s = 44°F, condensation will occur at the surfaces of supply ducts, terminals, and outlets exposed to space air.

Condensation can be avoided by the following measures:

1. Adding an insulation layer, such as fiberglass with vapor barrier, on the surface of ducts, terminals, and outlets with sufficient thickness (typically 1.5 to 2 in.) that are exposed to space air or ambient air whose dew point is higher than the external temperature of the exposed surface;

2. Adopting supply outlet that can induce sufficient space air to mix with low temperature supply air in order to raise T_s, or using a terminal to mix low temperature supply air with plenum air or space air to raise T_s.

For an air system to serve a single-zone conditioned space with specific indoor T_r and φ_r, say 75°F and 50 percent, and a fixed SHR$_s$ and ΔT_s, there will be only one supply air condition that can provide the required indoor T_r and φ_r (see Fig. 3.11). If T_r remains constant at 75°F, an increase of ΔT_s and the reduction

of supply volume flow rate $\dot{V}_s$ at a fixed space sensible load q_{rs} are possible only when the temperature of air leaving the cooling coil T_{cc} and supply temperature are lowered accordingly. The space relative humidity φ_r also will be lower. This again needs a lower chilled water temperature entering the coil T_{we} or a lower evaporating temperature inside the DX-coil T_{ev}.

For cold air distribution, if cold air is directly supplied to the space at 44°F and 84 percent relative humidity with a certain space latent load, and if ΔT_s = 34°F (that is, T_r = 78°F, space relative humidity must be reduced to 35 to 40 percent.)

Example 3.1. A shopping mall uses a basic central hydronic air conditioning system to serve the mall. The summer space sensible cooling load is 350,000 Btu/h and the summer space latent load is 62,000 Btu/h. Other design data are as follows:

Summer outdoor design temperatures:	
dry bulb	95°F
wet bulb	75°F
Summer indoor space conditions:	
air temperature	75°F
relative humidity	50 percent
Temperature rise:	
in the supply fan	2°F
in the supply duct	3°F
because of the heat released from light fixtures and roof at upper part of mall	3°F
in the return duct and return fan	2°F
Relative humidity of air leaving the cooling coil	93 percent
Outdoor air requirement	3200 cfm

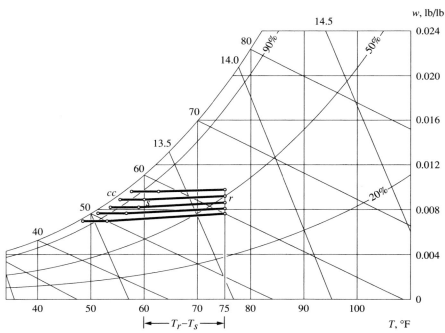

FIGURE 3.11 Influence of supply temperature difference $(T_r - T_s)$ on T_{cc} and ϕ_r.

Determine

1. The condition of supply air at summer design temperatures
2. The rated volume flow rate of supply fan
3. The cooling coil's load

Solution

1. From Eq. (3.1) and the given data, the sensible heat ratio of the space conditioning line can be calculated as

$$\text{SHR}_s = \frac{|q_{rs}|}{|q_{rs}| + |q_{rl}|} = \frac{350,000}{350,000 + 62,000}$$

$$= 0.85$$

On the psychrometric chart in the upper summer mode air conditioning cycle in Fig. 3.8, draw a space conditioning line *sr* from point *r* with $\text{SHR}_s = 0.85$. From the given data, the total temperature rise of the supply system heat gain is: $2 + 3 = 5°\text{F}$. Point *cc* will lie on $\varphi_{cc} = 93$ percent line and point *s* must lie on line *sr*. Let segment $cc\ s = 5°\text{F}$. Move *cc s* until point *cc* is on the 93 percent curve and point *s* is on line sr. The state point of supply air *s* and the condition of air leaving the cooling coil, point *cc*, are then determined. From the psychrometric chart,

$$T_s = 60°\text{F}, \varphi_s = 78 \text{ percent, and}$$

$$w_s = 0.0086 \text{ lb/lb}$$

$$T_{cc} = 55°\text{F}, \varphi_{cc} = 93 \text{ percent,}$$

$$w_{cc} = 0.0086 \text{ lb/lb and } h_{cc} = 22.6 \text{ Btu/lb}$$

2. Because the air temperature at the supply fan outlet $T_{sf} = 55 + 2 = 57°\text{F}$, and $w_{sf} = w_{cc} = 0.0086$ lb/lb, from the psychrometric chart, moist volume $v_{sf} = 13.21$ ft^3/lb. From Eqs. (3.57) and (3.58) the rated supply volume flow rate of the supply fan is

$$\dot{V}_{sf.r} = \dot{V}_{sf} = \frac{q_{rs}}{60\rho_{sf}c_{pa}(T_r - T_s)}$$

$$= 13.21 \times \frac{350,000}{60 \times 0.243(75 - 60)}$$

$$= 21,140 \text{ cfm}$$

3. State point *m* can be graphically determined or calculated. The total temperature rise due to the return system heat gain is $3 + 2 = 5°\text{F}$, therefore, $T_{rf} = T_{ru} = 75 + 5 = 80°\text{F}$. Because line *r ru* is a horizontal line, point *ru* can be plotted on the psychrometric chart.

From the given data, the summer outdoor temperature $T_o = 95°\text{F}$, and the summer design outdoor wet bulb $T'_o = 75°\text{F}$, so the state point of outdoor air *o* can also be plotted.

Connect line *ru o*. Ignore the difference in density between points *ru*, *m*, and *o*. From Eq. (3.32)

$$\frac{ru\ m}{ru\ o} = \frac{3200}{21,140} = 0.15$$

From the psychrometric chart, the length of line section *ru o* is 1.8 in., then, point *m* can be determined as

$$T_m = 82°\text{F}, \varphi_m = 44 \text{ percent,}$$

$$\text{and } h_m = 30.9 \text{ Btu/lb}$$

Mixing point *m* can also be determined by using the analytical method. From the psychrometric chart, the enthalpy of outdoor $h_o = 38.4$ Btu/lb and the enthalpy $h_{ru} = 29.8$ Btu/lb. From Eq. (3.34),

$$h_m = \frac{\dot{V}_o h_o + (\dot{V}_s - \dot{V}_o)h_{ru}}{\dot{V}_m}$$

$$= \frac{(3200 \times 38.4) + (21,140 - 3200)29.8}{21,140}$$

$$= 31.10 \text{ Btu/lb}$$

$$T_m = \frac{\dot{V}_o T_o + (\dot{V}_s - \dot{V}_o)T_{ru}}{\dot{V}_m}$$

$$= \frac{(3200 \times 95) + (21,140 - 3200)80}{21,140}$$

$$= 82.3°\text{F}$$

The differences between the enthalpies and temperatures determined from the graphical and analytical methods are small, less than 1 percent. But the analytical method is more accurate because the line segment lengths are small.

Because $\dot{V}_s \rho_s = \dot{V}_{sf} \rho_{sf}$, from Eq. (3.22), cooling coil load is calculated as

$$q_{cc} = 60\dot{V}_{sf}\rho_{sf}(h_m - h_{cc})$$

$$= 60 \times 21,140\frac{(31.1 - 22.6)}{13.21}$$

$$= 816,154 \text{ Btu/h}$$

3.11 BASIC AIR CONDITIONING CYCLE—WINTER MODE OPERATION

When a basic air system is operated in the winter mode at design outdoor and indoor conditions, the basic air conditioning cycles can be divided into the following four types:

1. Warm air supply without space humidity control
2. Warm air supply with space humidity control
3. Cold air supply without space humidity control
4. Cold air supply with space humidity control

Warm Air Supply Without Space Humidity Control

During winter, if the sum of the heat loss due to transmission through the external windows, walls, and roofs and the heat loss due to infiltration is greater than the sum of the internal heat gains from occupants, electric lights, equipment, and appliances, then a warm air supply is required to maintain a desirable indoor temperature. For many comfort air conditioning systems such as those for office buildings and shopping centers, humidification is usually not included. In winter, warm air supply without space humidity control is often used in these buildings. The lower cycle in Fig. 3.8 also shows the winter mode basic air conditioning cycle using the same basic air system as for the summer mode. In winter, return air at state point r flows through the return grilles, return duct, and return fan. After that, recirculating air at point ru enters the air handling unit and mixes with outdoor air. The mixture, at point m, passes through the cooling coil, which is not energized, and is then heated at the heating coil. Warm air leaves the heating coil at point ch. It leaves the supply fan at point sf after absorbing the fan power heat gain. The warm supply air loses heat in the supply duct and supplies to the conditioned space at point s. At the conditioned space, the supply air absorbs the latent heat from the occupants, supplies heat to compensate for transmission loss through external windows, walls, and roofs, and finally changes to the point r.

The winter cycle in Fig. 3.8 consists of the following processes:

1. Sensible heating process r ru due to the return system heat gain $q_{r.s}$. In winter, the return system heat gain consists of mainly the heat released from the electric lights and the roof q_{rp} and the return fan power heat gain q_{rf}, if any. That is

$$q_{r.s} = q_{rp} + q_{rf} \qquad (3.64)$$

The temperature difference between the air inside and outside the return duct is generally small in winter and can be ignored.

2. Adiabatic mixing process ru m o. The ratio of the volume flow rate of outdoor air to the supply volume flow rate is the same as in summer season in order to provide the required amount of outdoor air for the occupants.

3. Sensible heating process m ch at the heating coil. The heating coil's load can be calculated according to Eq. (3.11) as follows

$$q_{ch} = 60\dot{V}_s \rho_s c_{pa}(T_{ch} - T_m) \qquad (3.65)$$

where T_m, T_{ch} = temperature of mixture and air off the heating coil, °F.

4. Sensible heating process ch sf due to the supply fan power heat gain q_{sf}.

5. Sensible cooling process sf s due to the heat loss from the warm air inside the supply duct through the duct wall to the ambient air q_{sd}.

6. Space conditioning process sr. During winter, heat is supplied to the space to compensate the transmission loss, and moisture is absorbed by the supply air due to the space latent load. Supply air at point s is then changed to r.

The heat supplied to the conditioned space to offset the space heating load q_{rh} can be calculated as

$$q_{rh} = 60\dot{V}_s \rho_s c_{pa}(T_s - T_r) \qquad (3.66a)$$

And the latent heat absorbed by the supply air q_{rl} is

$$q_{rl} = 60\dot{V}_s \rho_s(w_r - w_s)h_{fg.32} \qquad (3.66b)$$

Example 3.2. For the same basic air system as that in Example 3.1, shown in Fig. 3.8, the following winter design conditions are given:

Winter outdoor design conditions:

dry bulb temperature	15°F
relative humidity	30 percent
Winter indoor design temperature	70°F
Outdoor air requirement	3200 cfm
Temperature increase:	
because of heat released from the light fuxtures	2°F
return fan power heat gain	1°F
supply fan	2°F
Temperature drop of the supply duct	1°F
Space heating load	225,000 Btu/h
Space latent load	60,000 Btu/h
Supply volume flow rate	21,140 cfm

Determine

1. The space relative humidity at winter design conditions
2. The temperature of supply air
3. The heating coil's load

Solution

1. To develop a winter mode basic air conditioning cycle, the space humidity ratio w_r must be determined. Because the winter indoor temperature is given, the space relative humidity can then be found from the psychrometric chart.

Assume that the moist volume of the warm supply air $v_s = 13.75$ ft^3/lb. Because $\rho_s = 1/v_s$, from Eq. (3.66b)

$$w_r - w_s = \frac{q_{rl}}{60\dot{V}_s \rho_s h_{fg.32}}$$

$$= \frac{60,000 \times 13.75}{60 \times 21,140 \times 1075}$$

$$= 0.000605 \text{ lb/lb}$$

Because $w_m = w_s$,

$$\frac{w_r - w_s}{w_r - w_o} = \frac{3200}{21,140} = 0.15$$

Therefore,

$$w_r - w_o = \frac{0.000605}{0.15} = 0.0041 \text{ lb/lb}$$

From psychrometric chart, at $T_o = 15°F$, and $\varphi_o = 30$ percent, $w_o = 0.0005$ lb/lb. Then,

$$w_r = w_o + 0.0041 = 0.0005 + 0.0041$$

$$= 0.0046 \text{ lb/lb}$$

Point r can thus be plotted on the psychrometric chart. The relative humidity of the space air at winter design condition is 30 percent.

2. From the given data, $T_{ru} = T_r + 2 + 1 = 70 + 2 + 1 = 73°F$. Because $w_{ru} = w_r$, points ru and o can be plotted from the psychrometric chart.
  Connect $ru\ o$. For an adiabatic mixing process

$$\frac{ru\ m}{ru\ o} = \frac{3200}{21,140} = 0.15$$

Section $ru\ o$ measured from the psychrometric chart is 6.15 in. Then, mixing point m can be determined from the psychrometric chart:

$$T_m = 64.3°F \text{ and } w_m = 0.004 \text{ lb/lb}$$

From Eq. (3.65), the supply air temperature

$$T_s = T_r + \frac{q_{rh}}{60 \dot{V}_s \rho_s c_{pa}}$$

$$= 70 + \frac{225,000 \times 13.80}{60 \times 21,140 \times 0.243}$$

$$= 80.1°F$$

Because $w_m = w_{sf} = 0.004$ lb/lb, point s can be plotted. From the psychrometric chart, $v_s = 13.75$ ft³/lb, which is approximately equal to the assumed value. If the assumed value is too high or too low, the calculation will be repeated with new value of v_s.

3. Warm air temperature off the heating coil at winter mode can be calculated by

$$T_{ch} = T_s + 1 - 2 = 80.1 + 1 - 2 = 79.1°F$$

And from Eq. (3.65), if $\rho_s = (1/13.70)$, the heating coil load is

$$q_{ch} = 60\dot{V}_s \rho_s c_{pa}(T_{ch} - T_m)60 \times 21,140$$

$$\times (1/13.70)\ 0.243\ (79.1 - 64.3)$$

$$= 332,969 \text{ Btu/h}$$

Warm Air Supply with Space Humidity Control

In buildings with special requirements, such as hospitals, certain industrial buildings or buildings located where outdoor air temperature is very low in win-

ter, humidification may have to be incorporated in the warm air supply basic air system in winter.

Figure 3.12 shows a winter mode basic air conditioning cycle with steam injection humidification and the corresponding schematic diagram of the air system. This cycle consists of the following processes:

1. Sensible heating process $r\ ru$ from return system heat gain
2. Mixing process $ru\ m\ o$ of recirculated air and outdoor air
3. Sensible heating process $m\ ch$ at the heating coil
4. Steam injection or heated element humidifying process $ch\ h$
5. Sensible heating process $h\ sf$ from supply fan power heat gain
6. Sensible cooling process $sf\ s$ from duct heat loss
7. Space conditioning process sr

The procedure for constructing this cycle is as follows:

1. Plot the state points of the conditioned space r and outdoor air o.
2. Draw the space conditioning line sr from point r with known sensible heat ratio SHR_s.
3. From Eq. (3.66b), calculate $(w_r - w_s)$. Since w_r, q_{rl}, and the supply volume flow rate from summer mode calculations are known values, w_s can be calculated. Point s must lie on line sr, so the intersection of sr and w_s line determines point s.
4. Draw a horizontal line $r\ ru$ from point r representing the return system sensible heat gain. Plot point ru from its known temperature increase.
5. Connect points ru and o. From the ratio of the volume flow rate of the outdoor air to the supply air $\dot{V}_o/\dot{V}_s$, point m can be found and located on line $ru\ o$.
6. Draw horizontal line $m\ ch$ from point m. Draw also a horizontal line sf $s\ h$ through point s. Temperature difference $(T_{sf} - T_h)$ represents the temperature increase of the supply fan heat gain and $(T_{sf} - T_s)$ is the supply duct heat loss. This defines points sf and h.
7. Draw a vertical line through ch to line $h\ sf$ extended from point h. The intersection of lines $h\ ch$ and $m\ ch$ locates point ch.

Because $w_m = w_{ch}$, the mass flow rate of the injected steam is the mass flow rate of the water vapor $\dot{m}_s$ (in lb/min) added to moist air, and can be calculated from Eq. (3.14)

$$\dot{m}_s = \dot{V}_s \rho_s(w_h - w_m) \qquad (3.67)$$

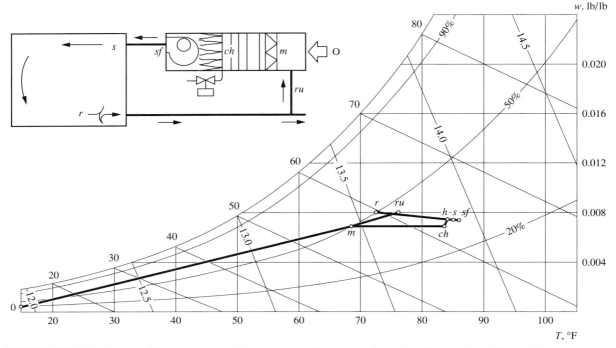

FIGURE 3.12 Basic air conditioning cycle—winter mode operation, warm air supply with space humidity control.

where w_h = humidity ratio of the air leaving humidifier, lb/lb

w_m = humidity ratio of the mixture, lb/lb

Cold Air Supply Without Space Humidity Control

In winter, most central hydronic systems or packaged systems in places like auditoriums, convention halls, and indoor stadiums operate with cold air supply but without space humidity control during occupied periods. Due to high occupancies and the resulting high space latent load, humidification is generally not necessary to maintain indoor humidities around 30 percent at winter design condition in many locations in the United States.

The lower part of the psychrometric chart in Fig. 3.13 shows this free cooling winter cycle with cold outdoor air supply without humidification, and is constructed as follows:

1. From Eq. (3.66b), calculate the humidity ratio difference $(w_r - w_s)$ from known space latent load and supply volume flow rate.

2. The volume flow rate of outdoor air $\dot{V}_o$ should be greater than the occupant requirements, and sufficient to offset the space sensible cooling load during winter without refrigeration. Based on the same supply volume flow rate as in summer, use Eq. (3.3) to calculate the supply air temperature difference $(T_r - T_s)$ from the design winter sensible cooling load, and determine T_s.

3. Temperature of the recirculating air entering the air handling unit or packaged unit T_{ru} can be calculated as

$$T_{ru} = T_r + \text{temperature rise due to return system heat gain}$$

Mixing temperature of outdoor and recirculating air T_m can be determined as

$$T_m = T_s - \text{temperature rise due to supply system heat gain}$$

4. Because $w_r = w_{ru}$ and $w_m = w_s$, the ratio of the difference of the humidity ratio can be evaluated as

$$\frac{w_{ru} - w_m}{w_{ru} - w_o} = \frac{w_r - w_s}{w_r - w_o}$$

$$= \frac{T_{ru} - T_m}{T_{ru} - T_o} \qquad (3.68)$$

From Eq. (3.68), $(w_r - w_o)$ and w_r can be determined.

Example 3.3. An indoor stadium uses a basic central air system. The operating parameters at winter design conditions are as follows:

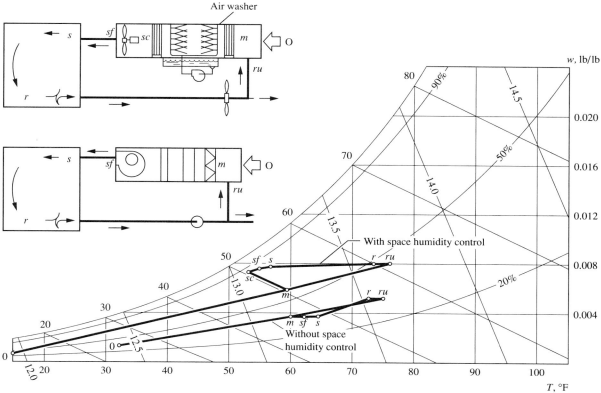

FIGURE 3.13 Basic air conditioning cycle—winter mode operation, cold air supply.

Winter outdoor design conditions:
air temperature   32°F
relative humidity   30 percent
Winter indoor design conditions:
air temperature   72°F
Space sensible cooling load   800,000 Btu/h
Space latent load   600,000 Btu/h
Temperature increase due to supply
system heat gain   4°F
Temperature increase due to return
system heat gain   2°F
Supply volume flow rate   100,000 cfm
Minimum outdoor air requirement   26,250 cfm

The mixture of outdoor air and recirculating air is used as the free cold air supply to offset the space cooling load in winter.

Determine

1. The space relative humidity at winter design conditions
2. The actual amount of outdoor air intake

Solution

1. Assume that the moist volume of the supply air $v_s = 13.3$ ft^3/lb. From Eq. (3.66b) and the given data, the humidity ratio difference between the space and supply air can be calculated as

$$w_r - w_s = \frac{q_{rl}}{60 \dot{V}_s \rho_s h_{fg.32}}$$

$$= \frac{600,000 \times 13.3}{60 \times 100,000 \times 1075}$$

$$= 0.00125 \text{ lb/lb}$$

2. From Eq. (3.3) and the given data, the supply air temperature difference $(T_r - T_s)$ can be calculated as

$$T_r - T_s = \frac{q_{rs}}{60 \dot{V}_s \rho_s c_{pa}}$$

$$= \frac{800,000 \times 13.3}{60 \times 100,000 \times 0.243}$$

$$= 7.30°F$$

3. From the given data

$$T_{ru} = T_r + 2 = 72 + 2 = 74°F$$

and

$$T_m = T_r - (T_r - T_s) - 4$$

$$= 72 - 7.3 - 4 = 60.7°F$$

4. Because $w_{ru} = w_r$ and $w_m = w_s$, from Eq. (3.68),

$$\frac{w_r - w_s}{w_{ru} - w_o} = \frac{T_{ru} - T_m}{T_{ru} - T_o}$$

$$= \frac{74 - 60.7}{74 - 32} = 0.317$$

Then,

$$w_r - w_o = \frac{w_r - w_s}{0.317} = \frac{0.00125}{0.317}$$

$$= 0.00394 \text{ lb/lb}$$

5. From the psychrometric chart, at a temperature of 32°F and a relative humidity of 30 percent, $w_o = 0.0012$, so

$$w_r = w_o + 0.00394 = 0.0012 + 0.00394$$

$$= 0.00514 \text{ lb/lb}$$

At a temperature of 72°F and a $w_r = 0.00514$, the space relative humidity $\varphi_r = 31$ percent as shown in the lower part of the psychrometric chart in Fig. 3.13.

6. The amount of outdoor air intake can be calculated as

$$\dot{V}_o = \frac{\dot{V}_s(T_{ru} - T_m)}{T_{ru} - T_o}$$

$$= 100,000 \times 0.317 = 31,700 \text{ cfm}$$

This amount is greater than the amount of outdoor air required by the occupants (26,250 cfm).

7. At $T_s = 72 - 7.3 = 64.7°F$ and $w_s = 0.00514 - 0.00125 = 0.00389$ lb/lb, from the psychrometric chart, $v_s = 13.3$ ft³/lb, which is equal to the assumed value.

Cold Air Supply with Space Humidity Control

Basic air systems using cold air supply with humidity control during winter season have been adopted in many industrial buildings such as spinning and weaving departments in textile mills. In these workshops, high internal sensible heat loads are released by the processing machines, requiring cold air supply and a specific controlled relative humidity, even in winter.

The upper air conditioning cycle of Fig. 3.13 shows such a winter mode cycle with cold air supply and space humidity control, using a spray type air washer. Air washers have a much greater humidifying capacity than steam humidifiers. They can also provide summer chilled water cooling and clean the air by water spraying.

The following are the air conditioning processes in this cycle and their characteristics:

1. Process *m sc* is an adiabatic saturation process. It proceeds along the thermodynamic wet bulb line on the psychrometric chart.

2. State point *sc* indicates the conditioned air leaving the air washer. Process lines *sc sf*, *sf s*, and *sr* all undergo heating and humidifying processes. Because the airstream leaving the air washer is oversaturated with minute water particles, as the airstream picks up heat from the fan, duct, and machine loads, more evaporation occurs. The increase of the humidity ratios in each of these processes varies from 0.0001 to 0.0003 lb/lb. These processes can be determined from both the increase of the enthalpy and the humidity ratios.

3. Outdoor air is mainly used for the purpose of cooling in quantities that greatly exceed occupancy requirements.

4. Because of the high machine load, the sensible heat ratio of the space conditioning line SHR$_s$ often exceeds 0.90.

REFERENCES AND FURTHER READING

ASHRAE, *ASHRAE Handbook 1987, HVAC Systems and Applications,* ASHRAE Inc., Atlanta, GA, 1987.

ASHRAE, *ASHRAE Handbook 1989, Fundamentals,* ASHRAE Inc., Atlanta, GA, 1989.

Carrier Corporation, *Handbook of Air Conditioning System Design,* McGraw-Hill, New York, 1965.

Manley, D. L., Bowlen, K. L., and Cohen, B. M., "Evaluation of Gas-Fired Desiccant-Based Space Conditioning for Supermarkets," *ASHRAE Transactions,* Part I B, pp. 447–456, 1985.

Meckler, G., "Efficient Integration of Desiccant Cooling in Commercial HVAC Systems," *ASHRAE Transactions,* Part II, pp. 2033–2042, 1988.

REFRIGERATION CYCLES AND REFRIGERANTS

4.1 REFRIGERATION AND REFRIGERATION SYSTEMS

Refrigeration is defined as the process of extracting heat from a lower temperature heat source, substance, or cooling medium, and transferring it to a higher temperature heat sink. Refrigeration maintains the temperature of the heat source below that of its surroundings while transferring the extracted heat, and any required energy input, to a heat sink, atmospheric air, or surface water.

A refrigeration system is a combination of components and equipment connected in a sequential order to produce the refrigeration effect. The refrigeration sytems commonly used for air conditioning can be classified by the type of input energy and the refrigeration process, as follows:

1. *Vapor compression systems.* In *vapor compression systems*, compressors activate the refrigerant by compressing it to a higher pressure and higher temperature level after it has produced its refrigeration effect. The compressed refrigerant transfers its heat to the sink and is condensed into liquid form. This liquid refrigerant is then throttled to a low pressure, low temperature vapor to produce refrigeration effect during evaporation.

2. *Absorption systems.* In an *absorption system*, the refrigeration effect is produced by thermal energy input. After absorbing heat from the cooling medium during evaporation, the vapor refrigerant is absorbed by an absorbent medium. This solution is then heated by direct-fired furnace, waste heat, hot water, or steam. The refrigerant is again vaporized and then condensed into liquid to begin the refrigeration cycle again.

3. *Air or gas expansion systems.* In an *air or gas expansion system*, air or gas is compressed to a high pressure by mechanical energy. It is then cooled and expanded to a low pressure. Because the temperature of air or gas drops during expansion, a refrigeration effect is produced.

4.2 REFRIGERANTS, COOLING MEDIUMS, AND ABSORBENTS

Refrigerants

A *refrigerant* is a primary working fluid used for absorbing and transmitting heat in a refrigeration system. All refrigerants absorb heat at a low temperature and low pressure during evaporation and release heat at a high temperature and pressure during condensation.

Cooling Mediums

A *cooling medium* is the working fluid cooled by the refrigerant to transport the cooling effect between a central plant and remote cooling units and terminals. In a large, centralized system, it is often more economical to use a coolant that can be pumped to remote locations where cooling is required. Chilled water, brine, and glycol are used as cooling mediums in many refrigeration systems. The cooling medium is often called a secondary refrigerant, because it obviates extensive circulation of the primary refrigerant.

Liquid Absorbents

A solution known as a *liquid absorbent* is often used to absorb the vaporized refrigerant (water) after its evaporation in an absorption refrigeration system. This solution, containing the absorbed vapor, is then heated at high pressure. The refrigerant vaporizes and the solution is restored to its original concentration for re-use.

Lithium bromide and ammonia, both in a water solution, are the liquid absorbents used most often in absorption refrigerating systems.

Numbering of Refrigerants

Before the invention of chlorofluorocarbons (CFCs), refrigerants were called by their chemical names. Because of the complexity of these names, especially the CFCs and HCFCs, a numbering system was developed for hydrocarbons and halocarbons, and is used widely in the refrigeration industry.

According to ANSI/ASHRAE Standard 34–1992, the first digit is the number of unsaturated carbon-carbon bonds in the compound. If the number is zero, it is omitted. The second digit (third digit from the right) equals the number of carbon atoms minus one. If the number is zero, it is omitted. The third digit (second digit from the right) gives the number of hydrogen atoms plus one. The last digit denotes the number of fluorine atoms in the compound.

For example, the chemical formula of HCFC–123 is $CHCl_2CF_3$:

No unsaturated C–C bonds, first digit is   0
There are 2 carbon atoms, second digit is $2 - 1 = 1$
There is 1 hydrogen atom, third digit is $1 + 1 = 2$
There are 3 fluorine atoms, last digit is   3

Classification and Prefix of Refrigerants

Before the introduction of chlorofluorocarbons in 1931 (as discussed in Section 1.6), the most commonly used refrigerants were air, ammonia, sulphur dioxide, carbon dioxide, and methyl chloride. Until 1986, nontoxic and nonflammable halogenated hydrocarbons with various ozone depletion potentials were used almost exclusively in vapor-conpression refrigeration systems for air conditioning. Chlorofluorocarbons, halons, hydrochlorofluorocarbons, and hydrofluorocarbons are all *halogenated hydrocarbons* or simply *halocarbons*. The most commonly used refrigerants for air conditioning, most of which are synthetic chemical compounds, can be classified into six groups, as listed in Table 4.1.

1. *Chlorofluorocarbons or fully halogenated chlorofluorocarbons.* CFCs contain only carbon, chlorine, and fluorine atoms, and are designated by the prefix CFC. They are derivatives of hydrocarbons and are obtained by replacing hydrogen atoms with chlorine and fluorine atoms, as shown in Fig. 4.1. CFC–11, CFC–12, CFC–113, CFC–114, and CFC–115 are all CFCs.

    CFCs have an atmospheric lifetime between 60 and 540 years, and cause ozone depletion. Most

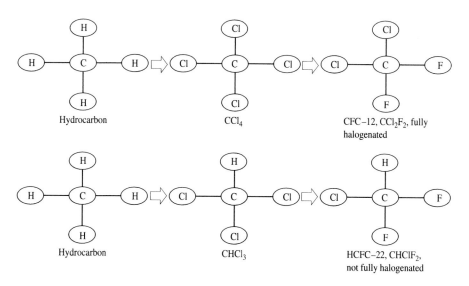

C Carbon atom   H Hydrogen atom   Cl Chlorine atom   F Fluorine atom

FIGURE 4.1 Fully halogenated CFCs and not fully halogenated HCFCs.

TABLE 4.1
Properties of commonly used refrigerants 40°F evaporating and 100°F condensing

	Chemical formula	Molecular mass	Ozone depletion potential† (ODP)	Global warming potential†	Evaporating pressure, psia	Condensing pressure, psia	Compression ratio	Refrigeration effect, Btu/lb
Hydrochlorofluorocarbons HCFCs and hydrofluorocarbons HFCs								
HCFC–22 Chlorodifluoromethane	CHClF$_2$	86.48	0.05	0.40	82.09	201.5	2.46	69.0
HCFC–123* Dichlorotrifluoroethane	CHCl$_2$CF$_3$	153.0	0.02	0.02	5.8	20.8	3.59	
HFC–125§ Pentafluoroethane	CHF$_2$CF$_3$	120.02	0.0	0.84	112.4	276.8	2.46	36.4
HFC–134a* Tetrafluoroethane	CF$_3$CH$_2$F	102.3	0.0	0.26	49.8	138.9	2.79	
Chlorofluorocarbons CFCs and halons BFCs								
CFC–11 Trichlorofluoromethane	CCl$_3$F	137.38	1.00	1.00	6.92	23.06	3.33	68.5
CFC–12 Dichlorodifluoromethane	CCl$_2$F$_2$	120.93	1.00	3.20	50.98	129.19	2.53	50.5
BFC–13B1 Bromotrifluoromethane	CBrF$_3$	148.93	10					
CFC–113 Trichlorotrifluoroethane	CCl$_2$FCClF$_2$	187.39	0.80	4.95	2.64	10.21	3.87	54.1
CFC–114 Dichlorotetrafluoroethane	CCl$_2$FCF$_3$	170.94	1.00	10.60	14.88	45.11	3.03	42.5
Azeotropes or blends								
CFC/HFC–500 CFC–12/HFC152a(73.8/26.2)		99.31			59.87	152.77	2.55	60.5
HCFC/CFC–502 HCFC–22/CFC115(48.8/51.2)		111.63						
Inorganic compounds								
R–717 Ammonia	NH$_3$	17.03	0	0	71.95	206.81	2.87	467.4
R–718 Water	H$_2$O	18.02	0					
R–729 Air		28.97	0					

Source: Adapted with permission from *ASHRAE Handbook Fundamentals* 1989, except

ASHRAE Transactions 1989, Part II

†*Heating, Piping and Air Conditioning*, April 1991

‡*ANSI/ASHRAE Standard* 34-1992

§DuPont HFC–125 Thermodynamic Properties

§§5°F evaporation and 86°F condensation

Reprinted with permission.

TABLE 4.1 (*continued*)
Properties of commonly used refrigerants 40°F evaporating and 100°F condensing

		Chemical formula	Specific volume of vapor ft³/lb	Compressor displacement cfm/ref · ton	Power consumption hp/ref · ton	Critical temperature °F	Discharge temperature§§ °F	Toxicity‡	Flammability	Safety‡
Hydrochlorofluorocarbons HCFCs and hydrofluorocarbons HFCs										
HCFC–22	Chlorodifluoromethane	$CHClF_2$	0.66	1.91	0.696	204.8	127	A	Nonflammable	A1
HCFC–123	Dichlorotrifluoroethane	$CHCl_2CF_3$	5.88					B	Nonflammable	B1
HFC–125	Pentafluoroethane	CHF_2CF_3	0.33			150.9	103	A	Nonflammable	A1
HFC–134a	Tetrafluoroethane	CF_3CH_2F	0.95					A	Nonflammable	A1
Chlorofluorocarbons CFCs and halons BFCs										
CFC–11	Trichlorofluoromethane	CCl_3F	5.43	15.86	0.636	388.4	104	A	Nonflammable	A1
CFC–12	Dichlorodifluoromethane	CCl_2F_2	0.79	3.08	0.689	233.6	100	A	Nonflammable	A1
BFC–13B1	Bromotrifluoromethane	$CBrF_3$	0.21			152.6	103	A	Nonflammable	A1
CFC–113	Trichlorotrifluoroethane	CCl_2FCClF_2	10.71	39.55	0.71	417.4	86	A	Nonflammable	A1
CFC–114	Dichlorotetrafluoroethane	CCl_2FCF_3	2.03	9.57	0.738	294.3	86	A	Nonflammable	A1
Azeotropes or blends										
CFC/HFC–500	CFC–12/HFC152a(73.8/26.2)		0.79	3.62	0.692	221.9	105	A	Nonflammable	A1
HCFC/CFC–502	HCFC–22/CFC115(48.8/51.2)						98	A	Nonflammable	A1
Inorganic compounds										
R–717	Ammonia	NH_3	3.98	1.70	0.653	271.4	207	B	Lower flammability	B2
R–718	Water	H_2O							Nonflammable	
R–729	Air								Nonflammable	

Source: Adapted with permission from *ASHRAE Handbook Fundamentals* 1989, except

*ASHRAE Transactions 1989, Part II

†*Heating, Piping and Air Conditioning,* April 1991

‡*ANSI/ASHRAE Standard 34-1992*

§DuPont HFC-125 Thermodynamic Properties

§§5°F evaporation and 86°F condensation

Reprinted with permission.

CFCs have ozone depletion potentials (ODP) from 0.6 to 1, which will be covered in detail in the next section. CFCs will be phased out worldwide by January 1, 1996 with certain exceptions.

2. *Halons (BFCs).* Halons contain bromine, fluorine, and carbon atoms and are designated by the prefix BFC. BFC–13B1 and BFC–12B1 are halons. They have very high ozone depletion potentials—for example, the ODP for BFC–13B1 is 10. Like CFCs, BFCs will be phased out by January 1, 1996.

3. *Hydrochlorofluorocarbons (HCFCs).* HCFCs contain hydrogen, chlorine, fluorine, and carbon atoms, as shown in Fig. 4.1, and are not fully halogenated. They are designated by the prefix HCFC. HCFCs have much shorter atmospheric lifetimes (2 to 22 years) than CFCs, and cause far less ozone depletion (HCFCs have ODP of 0.02 to 0.1). HCFC–22, HCFC–123, and HCFC–124 are HCFCs. HCFC–22 was the most widely used refrigerant in air conditioning in the early 1990s.

HCFCs will be used as transitional refrigerants in the coming decades. They are scheduled for restriction starting in 2004.

4. *Hydrofluorocarbons (HFCs).* HFCs contain only hydrogen, fluorine, and carbon atoms. They contain no chlorine atoms and therefore cause no ozone depletion. They are designated by the prefix HFC. HFC–134a, HFC–125, and HFC–143a are all HFCs. HFCs may become one of the most widely used refrigerant groups for air conditioning in the next century.

5. *Azeotropes or blends.* An azeotrope is a mixture of two substances that cannot be separated from its constituents by distillation. Azeotropes evaporate and condense as a single substance. Their properties are entirely different from those of their components.

Azeotropes are assigned numbers 500 and up, and are designated by the composite prefixes of their constituents. CFC/HFC–500, HCFC/CFC–501, and HCFC/CFC–502 are all azeotropes. HCFC/CFC–501 is a mixture of 75 percent HCFC–22 and 25 percent CFC–12. Azeotropes containing CFCs cause ozone depletion, and will be phased out by January 1, 1996.

6. *Inorganic compounds.* These compounds include refrigerants used before 1930, such as ammonia (NH_3), water (H_2O), and air. Many of them are still used widely in refrigeration systems because they do not deplete the ozone layer and have other desirable characteristics. Although ammonia is toxic and flammable, it is still used in refrigeration.

Inorganic compounds are assigned numbers 700 and up, and are designated by the prefix R, which stands for *refrigerant*.

4.3 ENVIRONMENTAL CONCERNS

Refrigerant Usage

The use of CFCs and HCFCs is a global concern. Approximately two thirds of all fully halogenated CFCs were used outside the United States in the mid-1980s.

In 1985, the total usage of halocarbons in United States was 611 million pounds. These halocarbons were used in foam insulation, automotive air conditioners, Air Conditioning and Refrigeration Institute (ARI) members in new systems, and other products.

Foam insulation blown by CFCs was the largest user. Automotive air conditioners made up 19 percent of the total usage, and CFCs purchased by ARI members for new systems made up 5 percent of the total usage.

Of the CFCs and HCFCs purchased by ARI members, HCFC–22 made up 77 percent, while CFC–11 and CFC–12 each made up about 10 percent.

Ozone Depletion and Global Warming Potentials

In order to compare the relative ozone depletion caused by various refrigerants, an index called ozone depletion potential (ODP) has been developed. ODP is the ratio of the rate of ozone depletion of one pound of any halocarbon to that of one pound of CFC–11. The ODP of CFC–11 is assigned a value of 1. The following are the ODP values for various refrigerants:

Refrigerant	Chemical formula	ODP value
CFC–11	CCl_3F	1.0
CFC–12	CCl_2F_2	1.0
CFC–13B1	$CBrF_3$	10
CFC–113	CCl_2FCClF_2	0.8
CFC–114	$CClF_2CClF_2$	1.0
CFC–115	$CClF_2CF_3$	0.6
CFC/HFC–500	CFC–12(73.8%) / HFC–152a(26.2%)	0.74
CFC/HCFC–502	HCFC–22(48.8%) / CFC–115(51.2%)	0.33
HCFC–22	$CHClF_2$	0.05
HCFC–123	$CHCl_2CF_3$	0.02
HCFC–124	$CHClFCF_3$	0.02
HCFC–142b	CH_3CClF_2	0.06
HFC–125	CHF_2CF_3	0
HFC–134a	CF_3CH_2F	0
HFC–152a	CH_3CHF_2	0

Like the ODP, global warming potential (GWP) is used to compare the effects of the CFCs, HCFCs, and HFCs on global warming or the greenhouse effect with the effects of CFC–11. The GWPs of various refrigerants are listed in Table 4.1. In addition to the GWP, another global warming index uses CO_2 as a reference gas. For example, 1 lb of HCFC–22 has the same effect on global warming as 4100 lb of CO_2 in the first 20 years after it is released into the atmosphere. Its impact drops to 1500 lb at 100 years. Both ODP and GWP should be considered carefully before selecting a refrigerant.

Phaseout of CFCs, BFCs, and HCFCs

The theory of ozone depletion was proposed in the early and mid-1970s. National Aeronautics and Space Administration (NASA) flights into the stratosphere over the arctic and antarctic circles found CFC residue where the ozone layer was damaged. Approximately the same ozone depletion over the antarctic circle was found in 1987, 1989, 1990, and 1991. By 1988, antarctic ozone levels were 30 percent below those in the mid-1970s. The most severe ozone loss over the antarctic was observed in 1992. Ground monitoring at various locations worldwide in the 1980s showed 5 to 10 percent increases in ultraviolet radiation. The ozone layer may actually be deteriorating at a rapid rate. Although there is controversy about the theory of ozone layer depletion among scientists, action must be taken now before it is too late.

In 1978, the Environmental Protection Agency (EPA) and the Food and Drug Administration (FDA) of the United States issued regulations to phase out the use of fully halogenated CFCs in nonessential aerosol propellants, the major use at that time.

Montreal Protocol and Clean Air Act

On September 16, 1987, the European Economic Community and 24 nations, including the United States, signed the Montreal Protocol. This document is an agreement to phase out the production and consumption of CFCs and BFCs by the year 2000.

The Clean Air Act amendments, signed into law in the United States on November 15, 1990, governed two important issues: the phaseout of CFCs and a ban (effective July 1, 1992) on the deliberate venting of CFCs and HCFCs. Deliberate venting of CFCs and HCFCs must follow the regulations and guidelines of the EPA.

In February of 1992, then-President Bush called for an accelerated phaseout of CFCs in the United States. Production of CFCs will cease at the end of 1995, with limited exceptions for service to certain existing equipment.

In late November, 1992, representatives of 93 nations meeting in Copenhagen also agreed to the complete phaseout of CFCs beginning January 1, 1996, and restrictions on the use of HCFCs beginning in 2004. Consumption levels of HCFCs will be 65 percent of the 1989 level by 2004, 35 percent by 2010, and 10 percent by 2015, and will be completely phased out by 2030.

Action and Measures

The impact of CFCs on the ozone layer may be a serious threat to human survival. The following measures are essential:

- Use substitutes for CFCs, such as HCFC–22 or HCFC–134a to replace CFC–12, and HCFC–123 to replace CFC–11. Other substitutes with ODP between 0.05 and 0 should be used to replace CFCs. It is important to realize that HCFCs themselves may be restricted beginning in 2004. HFCs may be used without limitation. Existing refrigeration machines can be converted to use alternative refrigerants with only minor losses of capacity and efficiency.
- Prevent the deliberate venting of CFCs and HCFCs during manufacturing, installation, operation, service, and disposal of products using CFCs and HCFCs.
- Avoid CFC and HCFC emissions through recovery, recycling, and reclamation. According to ASHRAE Guideline 3–1990, *recovery* is the removal of refrigerant from a system and storage in an external container. *Recycling* involves cleaning the refrigerant for reuse by means of an oil separator and filter drier. In *reclamation*, refrigerant is reprocessed for new product specifications.

4.4 PROPERTIES AND CHARACTERISTICS OF REFRIGERANTS

The first priority for refrigerant selection is the preservation of the ozone layer, then global warming effect. In addition, the following factors should be considered:

- safety requirements
- compressor displacement required per ton of refrigeration capacity
- effectiveness of the refrigeration cycle
- physical properties
- operating characteristics

Safety Requirements

Refrigerant may leak from pipe joints, seals, or component parts during installation, operation, or accident. Therefore, refrigerants must be acceptably safe for humans and manufacturing processes, with little or no toxicity or flammability. Refer to Table 4.1 for these physical properties.

In ANSI/ASHRAE Standard 34–1992, the toxicity of refrigerants is classified as Class A or B. Class A refrigerants are of lower toxicity. A *Class A* refrigerant is one whose toxicity has not been identified when its concentration is less than or equal to 400 ppm, based on threshold limit value–time weighted average or equivalent indices.

Threshold limit value–time weighted average (TLV–TWA) concentration is a concentration to which workers can be exposed over an 8-hour work day and a 40-hour work week without adverse effect. Concentration ppm means parts per million by mass.

Class B refrigerants are of higher toxicity. A *Class B* refrigerant produces evidence of toxicity when workers are exposed to a concentration below 400 ppm based on threshold limit value–time weighted average.

Refrigerants HCFC–22, HFC–134a, CFC–11, CFC–12, and R–718 (water) belong to the Class A (lower toxicity) group. HCFC–123 and R–717 (ammonia) belong to the Class B (higher toxicity) group.

Flammable refrigerants explode when ignited. If a flammable refrigerant is leaked in the area of a fire, the result is an immediate explosion. Soldering and welding for installation or repair cannot be performed near such gases.

ANSI/ASHRAE Standard 34–1992 classifies the flammability of refrigerants into Classes 1, 2, and 3. *Class 1* refrigerants show no flame propagation when tested in air at a pressure of 14.7 psia at 65°F. *Class 2* refrigerants have a lower flammability limit (LFL) of more than 0.00625 lb/ft^3 at 70°F and 14.7 psia, and a heat of combustion less than 8174 Btu/lb. *Class 3* refrigerants are highly flammable, with a LFL less than or equal to 0.00625 lb/ft^3 at 70°F and 14.7 psia or a heat of combustion greater than or equal to 8174 Btu/lb.

Refrigerants HCFC–22, HFC–123, HFC–134a, and most CFCs are nonflammable. Difluoroethane HCFC–152a has a lower flammability limit of 4.1 percent, and ammonia has an explosive limit of 16 to 25 percent by volume. Both of these belong to the lower flammability group.

A refrigerant's safety classification is its combination of toxicity and flammability. According to ANSI/ASHRAE Standard 34–1992, safety groups are classified as follows:

- A1   lower toxicity and no flame propagation
- A2   lower toxicity and lower flammability
- A3   lower toxicity and higher flammability
- B1   higher toxicity and no flame propagation
- B2   higher toxicity and lower flammability
- B3   higher toxicity and higher flammability

HCFC–22 and HFC–134a are in the A1 safety group, HCFC–123 is in the B1 safety group, and R–717 (ammonia) is in the B2 safety group, as listed in Table 4.1. More details regarding refrigerant safety will be presented in Chapter 29.

Compressor Displacement Required per Ton of Refrigeration Capacity

The *compressor displacement* required to produce one ton of refrigeration indicates the ideal volumetric flow rate of vapor refrigerant extracted by the compressor, expressed in cfm/ref.ton, or simply cfm/ton. Compressor displacement determines the size of the compressor for positive displacement compressors, and is one criterion for refrigerant selection.

Compressor displacement depends mainly on the latent heat of vaporization of the refrigerant $h_{fg.r}$ and the specific volume at the suction pressure, or suction volume v_{suc}. The greater the $h_{fg.r}$ and the lower the v_{suc}, the smaller the compressor will be.

Table 4.1 lists the properties and characteristics of commonly used refrigerants. Among these listed in Table 4.1, ammonia requires lowest compressor displacement of 1.70 cfm/ref.ton, HCFC–22 is the next with a value of 1.91 cfm/ref.ton, and CFC–11 has the highest value of 15.86 cfm/ref.ton.

Effectiveness of Refrigeration Cycle

The effectiveness of refrigeration cycles, or *coefficient of performance* (COP), is one parameter that affects the efficiency and energy consumption of the refrigerating system. It will be clearly defined in Section 4.8. The COP of a refrigerant cycle using a specific refrigerant depends mainly upon the isentropic work input to the compressor at a given condensing and evaporating pressure differential, and also the refrigerating effect produced.

In Table 4.1, CFC–11 has the lowest ideal power consumption (in hp/ton). HCFC–123 is less efficient, with about 2 percent more power consumption than CFC–11. HCFC–22 is about 10 percent higher, and CFC–12 is 8 percent higher than CFC–11 in power consumption. The COP of HFC–134a and CFC–12 are similar because HCFC–134a has a higher heat transfer coefficient.

Physical Properties

The physical properties of refrigerants include evaporation and condensation pressure, discharge temperature, dielectric properties, and thermal conductivity.

EVAPORATING AND CONDENSING PRESSURE. It is best to use a refrigerant whose evaporating pressure is higher than that of the atmosphere so that air and other noncondensable gases will not leak into the system and increase condensing pressure. Refrigerants HCFC–123, CFC–11, and CFC–113, and water have subatmospheric evaporating pressures at 40°F.

Condensing pressure should be low because high condensing pressure necessitates heavier construction of the compressor, piping, condenser, and other components. In addition, a higher speed centrifugal compressor may be required to produce a high condensing pressure. HCFC–22 and ammonia both have a condensing pressure higher than 210 psig at a condensing temperature of 100°F.

DISCHARGE TEMPERATURE. Discharge temperature lower than 212°F is preferable because temperatures higher than 300°F may carbonize lubricating oil or damage discharge valves. Ammonia has the highest discharge temperature among the commonly used refrigerants.

DIELECTRIC PROPERTIES. Dielectric properties are important for those refrigerants that will be in direct contact with the windings of the motor (such as refrigerants used to cool the motor windings in a hermetically sealed compressor and motor assembly). The dielectric constants of most CFCs are comparable to those of air.

THERMAL CONDUCTIVITY. The thermal conductivity of a refrigerant is closely related to the efficiency of heat transfer in the evaporator and condenser of a refrigeration system. Refrigerant always has a lower thermal conductivity in its vapor state than in its liquid state. Higher thermal conductivity results in higher heat transfer in heat exchangers. Refrigerant HFC–134a has a significantly higher heat transfer coefficient than CFC–12.

Operating Characteristics

Inertness, oil miscibility, and leakage detection are the major operating characteristics of refrigerants.

INERTNESS. An inert refrigerant does not react chemically with other materials thus avoiding corrosion, erosion, or damage to the components in the refrigerant circuit. Halocarbons are compatible with all containment materials except magnesium alloys. Ammonia, in the presence of moisture, is corrosive to copper and brass.

OIL MISCIBILITY. When a small amount of oil is mixed with refrigerant, the mixture helps to lubricate the pistons, discharge valves, and other moving parts of a compressor. Oil should be returned to the compressor from the condenser, evaporator, and piping, providing continuous lubrication. On the other hand, refrigerant can dilute oil, weakening its lubricating effect, and when the oil adheres to the tubes in the evaporator or condenser, it forms film that reduces the rate of heat transfer.

HCFC–22 is partially miscible, HFC-134a is hardly miscible, and ammonia is immiscible with oil. Oil return measures must be taken into account during the design of refrigeration system, or a shortage of lubricating oil in the compressor may cause shutdown and failure. An appropriate synthetic lubricant should be used when HFC-134a replaces CFC-12.

LEAKAGE DETECTION. Refrigerant leakage should be easily detected. If it is not, gradual capacity reduction and eventual failure to provide the required cooling result. In addition, CFC leakage causes ozone depletion.

Most CFCs, HCFCs, and HFCs are colorless and odorless. Leakage of refrigerant from the refrigerating system can be detected by three different methods:

1. *Halide torch.* This is simple and fast. When air flows over a copper element heated by a methyl alcohol flame, CFC vapor decomposes and changes the color of the flame (green for a small leak, bluish with a reddish top for a large leak).
2. *Electronic detector.* This type of detector reveals a variation of electric current due to ionization of decomposed refrigerant between two oppositely charged electrodes. It is sensitive, but cannot be used where the ambient air contains explosive or flammable vapors.
3. *Bubble detection.* A solution of soap or detergent is brushed over the seals and joints where leakage is suspected, producing bubbles that can be easily detected.

Ammonia leakage can be detected by its objectionable odor, even from small leaks. Other inorganic refrigerant leaks can be detected by burning a sulfur candle, which will form a cloud of sulfuric compound, but ambient air is explosive.

Cost

The cost of refrigerant is such a small percentage of the installation and maintenance cost that it does not affect refrigerant selection.

Most Commonly Used Refrigerants

Between January 1, 1996 and 2030, the most commonly used refrigerants for air conditioning will include the following:

1. HCFCs: HCFC–22 and HCFC–123
2. HFCs: HFC–134a and HFC–125
3. Inorganic compounds: ammonia, water, and air

Use of ammonia is limited to process air conditioning because of its toxicity and flammability.

Storage of Refrigerants

Refrigerants are usually stored in cylinders during transport and on site. During storage, the pressure of the liquid refrigerant must be periodically checked and adjusted. Excessive pressure may cause an explosion.

According to Interstate Commerce Commission (ICC) regulations, liquid refrigerants must not be stored above 130°F, although the containers are designed to withstand up to three times saturated pressure at 130°F. If a container bursts, liquid refrigerant flashes into vapor. Such a sudden expansion in volume could cause a violent explosion inside a building, blasting out windows, walls, and roofs.

Containers should never be located near heat sources without sufficient ventilation. They must also not be put in a car or truck in direct sunlight. The valve of the container is attached by threads only. If the threads are damaged, the force of escaping vapor could propel the container like a rocket.

According to ASHRAE Standard 15–1992, in addition to the refrigerant charge in the system and receiver, refrigerant stored in a machinery room shall not exceed 330 lb. The receiver is a vessel used to store refrigerant after condenser when necessary.

4.5   REFRIGERATION PROCESSES AND REFRIGERATION CYCLES

Refrigeration Processes

A *refrigeration process* indicates the change of thermodynamic properties of the refrigerant and the energy transfer between the refrigerant and the surroundings.

The following refrigeration processes occur during the operation of a refrigerating system:

- Evaporation. In this process, refrigerant evaporates at a lower temperature than its surroundings, absorbing its latent heat of vaporization.
- Superheating. Saturated refrigerant vapor is usually superheated to ensure that liquid refrigerant does not flow into the compressor.
- Compression. Refrigerant is compressed to a higher pressure and temperature for condensation.
- Condensation. Gaseous refrigerant is condensed into liquid form by being desuperheated, then condensed, and finally subcooled, transferring its latent heat of condensation to a coolant.
- Throttling and expansion. The higher pressure liquid refrigerant is throttled to the lower evaporating pressure and is ready for evaporation.

Refrigeration Cycles

When a refrigerant undergoes a series of evaporation, compression, condensation, throttling, and expansion processes, absorbing heat from a lower temperature reservoir and releasing it to a higher temperature reservoir in such a way that its final state is equal in all respects to its initial state, it is said to have undergone a *refrigeration cycle*.

Both vapor compression and air expansion refrigeration cycles will be discussed in this chapter. Absorption refrigeration cycles will be covered in Chapter 24.

Unit of Refrigeration

In Inch–Pound (I–P) units, refrigeration is expressed in British Thermal Units per hour, or simply Btu/h. A *British Thermal Unit* is defined as the amount of heat energy required to raise the temperature of one pound of water one degree Fahrenheit from 59°F to 60°F.

Another unit of refrigeration widely used in the HVAC&R industry is ton of refrigeration, or simply ton. As mentioned before, 1 ton = 12,000 Btu/h of heat removed. This equals the heat absorbed by one ton (2000 lb) of ice melting at a temperature of 32°F over 24 hours.

Because the heat of fusion of ice at 32° is 144 Btu/lb,

$$1 \text{ ton} = \frac{1 \times 2000 \times 144}{24}$$

$$= 12,000 \text{ Btu/h}$$

4.6 GRAPHICAL AND ANALYTICAL EVALUATION OF REFRIGERATION CYCLE PERFORMANCE

Pressure–Enthalpy Diagram

The pressure–enthalpy *p–h* diagram is the most common graphic tool for analysis and calculation of the heat and work transfer and performance of a refrigeration cycle.

A single-stage refrigeration cycle consists of two regions: high-pressure region, high side, and low-pressure region, low side. The change in pressure can be clearly illustrated on the *p–h* diagram. Also, both heat and work transfer of various processes can be calculated as changes of enthalpy, and are easily shown on a *p–h* diagram.

Figure 4.2 is a skeleton *p–h* diagram for refrigerant HCFC–22. Enthalpy *h* (in Btu/lb) is the abscissa and absolute pressure (psia) or gauge pressure (psig), both expressed in logarithmic scale, is the ordinate.

The saturated liquid line separates the subcooled liquid from the two-phase region in which vapor and liquid refrigerants coexist. The saturated vapor line separates this two-phase region from the superheated vapor. In the two-phase region, the mixture of vapor and liquid is subdivided by the constant dryness fraction quality line.

The constant temperature lines are nearly vertical in the subcooled liquid region. At higher temperatures, they are curves near the saturated liquid line. In the two-phase region, the constant temperature lines are horizontal. In the superheated region, the constant temperature lines curve down sharply. Because the constant temperature lines and constant pressure lines in the two-phase region are horizontal, they are closely related to each other. A specific pressure of refrigerant in the two-phase region determines its temperature, and vice versa.

Also in the superheated region, the constant entropy lines incline sharply upward and constant volume lines are flatter. Both are slightly curved.

Temperature–Entropy Diagram

The temperature–entropy *T–s* diagram is often used to analyze the irreversibilities in a refrigeration cycle, as well as in the system, in order to select optimum operating parameters and improve performance of the system. In addition, the *T–s* diagram is more suitable for evaluating the effectiveness of a refrigeration cycle using air as refrigerant.

Figure 4.3 is a skeleton *T–s* diagram for R–729 (air). Entropy *s,* in Btu/lb · °R, is the abscissa of the diagram and temperature *T,* in °R, is the ordinate.

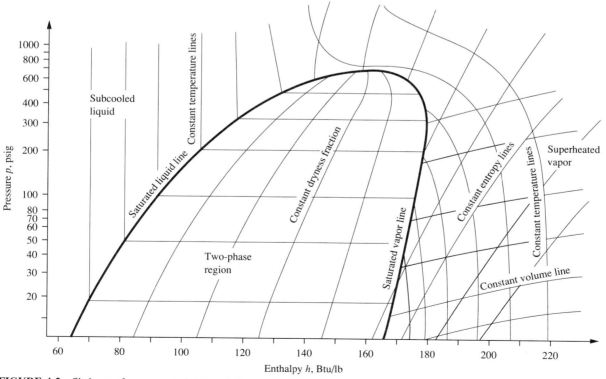

FIGURE 4.2   Skeleton of pressure–enthalpy *p–h* diagram for HCFC–22.

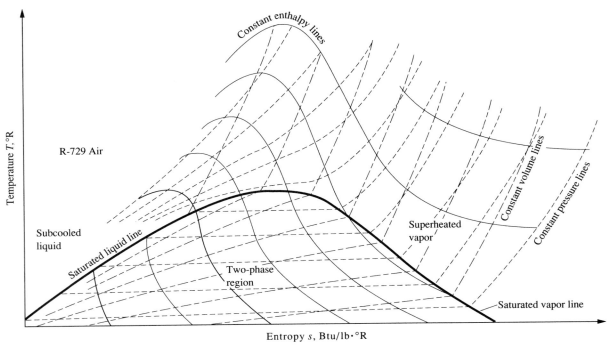

FIGURE 4.3 Skeleton of temperature–entropy T–s diagram for air.

The constant pressure lines are nearly horizontal in the two-phase region and curve upward in the superheated region. The constant volume lines incline slightly upward to the right in the two-phase region and curve upward to the right even more sharply than the constant pressure lines in the superheated region. The constant enthalpy lines look like reverse Ss covering the entire diagram, crossing both the saturated liquid and the saturated vapor lines.

Analytical Evaluation of Cycle Performance

Swers et al. (1972) proposed a thermodynamic analysis of degradation of available energy and irreversibility in a refrigerating system, and Tan et al. (1986) recommended a method of exergy analysis. Exergy of a working substance e, in Btu/lb, is defined as

$$e = h - h_a - [T_{Ra}(s - s_a)] \qquad (4.1)$$

where h, h_a = enthalpy of the working substance and the ambient state, Btu/lb

T_{Ra} = absolute temperature of the ambient state, °R

s, s_a = entropy of the working substance and the ambient state, Btu/lb · °R

Both analyses are effective tools in the selection of optimum design and operating parameters by means of complicated analysis. They require extensive supporting data and information.

For most analyses of refrigeration cycle performance and design and operation of refrigerating systems in actual applications, satisfactory results can be obtained by using the steady-flow energy equation, heat and work transfer, and energy balance principle. If a more precise and elaborate analysis is needed in research or for detailed improvements of refrigeration systems, the references at the end of this chapter can be consulted.

4.7 CARNOT REFRIGERATION CYCLE

The *Carnot refrigeration cycle* is a reverse engine cycle. All processes in a Carnot refrigeration cycle are reversible, so it is the most efficient refrigeration cycle.

Figure 4.4a is a schematic diagram of a Carnot cycle refrigerating system, and Fig. 4.4b shows the Carnot refrigeration cycle using gas as the working substance. This Carnot cycle is composed of four reversible processes:

1. An isothermal process 4–1 in which heat $q_{\#1}$ is extracted at constant temperature T_{R1} per lb of working substance

2. An isentropic compression process 1–2

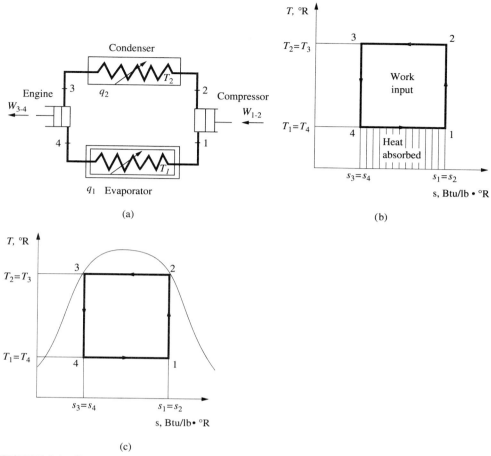

FIGURE 4.4 Carnot refrigeration cycle: (a) schematic diagram, (b) gas cycle, and (c) vapor cycle.

3. An isothermal process 2–3 in which $q_{\#2}$ is rejected at constant temperature T_{R2} per lb of working substance

4. An isentropic expansion process 3–4

Figure 4.4c shows the Carnot refrigeration cycle using vapor as the working substance. Wet vapor is the only working substance of which the heat supply and the heat rejection processes can occur easily at constant temperature. This is because the temperatures of wet vapor remain constant when latent heat is supplied or rejected.

As in the gas cycle, there are two isothermal processes 4–1 and 2–3 absorbing heat at temperatures T_{R1} and rejecting heat at T_{R2} respectively, and two isentropic processes, one for compression 1–2 and another for expansion 3–4.

Performance of Carnot Refrigeration Cycle

According to the First Law of Thermodynamics, often called the Law of Conservation of Energy, when a system undergoes a thermodynamic cycle, the net heat supplied to the system is equal to the net work done, or

$$\text{Heat supply} + \text{Heat rejected} = \text{Net work done} \quad (4.2)$$

Referring to Fig. 4.4a in a Carnot refrigeration cycle,

$$q_{\#1} - q_{\#2} = -W$$

or

$$q_{\#1} = q_{\#2} - W$$

$$q_{\#2} = q_{\#1} + W \quad (4.3)$$

where $q_{\#1}$ = the heat supplied from the surroundings per lb of working substance at temperature T_1; the sign of $q_{\#1}$ is positive

$q_{\#2}$ = the heat rejected to the sink per lb of working substance at temperature T_2; the sign of $q_{\#2}$ is negative

W = the net work done by the system. Its sign is positive, or if it is a work input to the system, the sign is negative.

Heat extracted from the source at temperature T_{R1} by the working substance, that is, the refrigerating effect per lb of working substance, is

$$q_{\#1} = T_{R1}(s_1 - s_4) \qquad (4.4)$$

where s_1, s_4 = entropy at state points 1 and 4 respectively, Btu/lb · °R. Heat rejected to the heat sink at temperature T_{R2} can be calculated as

$$q_{\#2} = -T_{R2}(s_3 - s_2) = T_2(s_2 - s_3) \qquad (4.5)$$

where s_2, s_3 = entropy at state points 2 and 3 respectively, Btu/lb · °R. Because in the isentropic process 1–2, $s_1 = s_2$, and in isentropic process 3–4, $s_3 = s_4$,

$$q_{\#2} = T_{R2}(s_1 - s_4) \qquad (4.6)$$

4.8 COEFFICIENT OF PERFORMANCE OF REFRIGERATION CYCLE

Coefficient of performance (COP) is an index of performance of a thermodynamic cycle or a thermal system. Because the COP can be greater than 1, COP is used instead of thermal efficiency.

Coefficient of performance can be used for the analysis of the following:

- A refrigerator used to produce a refrigeration effect only, that is, COP_{ref}
- A heat pump in which the heating effect is produced by rejected heat, COP_{hp}
- A heat recovery system in which both the refrigeration effect and the heating effect are used at the same time, COP_{hr}

For a refrigerator, COP is defined as the ratio of the refrigeration effect $q_{\#1}$ to the work input W_{in}, both in Btu/lb, that is,

$$COP_{ref} = \frac{\text{Refrigeration effect}}{\text{Work input}}$$

$$= \frac{q_{\#1}}{W_{in}} \qquad (4.7)$$

For Carnot refrigeration cycle, from Eq. (4.3),

$$COP_{ref} = \frac{q_{\#1}}{q_{\#2} - q_{\#1}}$$

$$= \frac{T_{R1}(s_1 - s_4)}{(T_{R2} - T_{R1})(s_1 - s_4)} = \frac{T_{R1}}{T_{R2} - T_{R1}} \qquad (4.8)$$

With a heat pump, the useful effect is the heating effect because of the rejected heat $q_{\#2}$, so COP_{hp} is the ratio of heat rejection to the work input, or

$$COP_{hp} = \frac{q_{\#2}}{W_{in}} \qquad (4.9)$$

For a heat recovery system, the useful effect is $q_{\#1}$ and $q_{\#2}$. In such a condition, COP_{hr} is defined as the ratio of the sum of the absolute values of refrigerating effect and heat rejection to the absolute value of the work input, that is,

$$COP_{hr} = \frac{|q_{\#1}| + |q_{\#2}|}{W_{in}} \qquad (4.10)$$

4.9 IDEAL SINGLE-STAGE VAPOR COMPRESSION CYCLE

The Carnot cycle cannot be achieved in actual practice because liquid slugging would occur during compression of the two-phase refrigerant. In addition, the mixture, mostly liquid, does very little work when it expands after condensation in the heat engine. Therefore, a single-stage ideal vapor compression cycle is used instead of the Carnot cycle.

Figure 4.5 shows an ideal single-stage vapor compression cycle in which compression occurs in the superheated region. A throttling device, such as an expansion valve, is used instead of the heat engine. *Single-stage* means that there is only one stage of compression. An ideal cycle is one in which the compression process is isentropic and the pressure losses in the pipeline, valves, and other components are negligible. All the refrigeration cycles covered in this chapter are ideal cycles except the air expansion refrigeration cycle in Section 4.16.

Vapor compression means that vapor refrigerant is compressed to a higher pressure, and then the condensed liquid is throttled to a lower pressure to produce the refrigerating effect by evaporation. It is different from the absorption or air refrigeration cycle.

Flow Processes

Figure 4.5b and c show the refrigeration cycle on p–h and T–s diagrams. The refrigerant evaporates entirely in the evaporator and produces the refrigerating effect. It is then extracted by the compressor at state point 1, compressor suction, and compressed isentropically from state point 1 to 2. It is then condensed into liquid in the condenser and the latent heat of condensation is rejected to the heat sink. The liquid refrigerant, at state point 3, flows through an expansion valve, which reduces it to the evaporating pressure. In the ideal vapor compression cycle, the throttling process at the expansion valve is the only irreversible process, usually indicated by a dotted line. Some of the liquid flashes into vapor and enters the evaporator at state point 4. The remaining liquid portion evaporates at the evaporating temperature, thus completing the cycle.

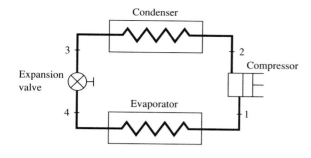

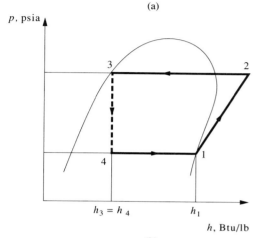

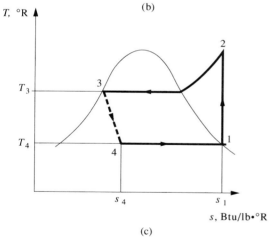

FIGURE 4.5 Single-stage ideal vapor compression cycle: (a) Schematic diagram, (b) p–h diagram, (c) T–s diagram.

Cycle Performance

For the evaporating process between points 4 and 1, according to the steady flow energy equation,

$$h_4 + \frac{v_4^2}{2g_c \times 778} + q_\#$$

$$= h_1 + \frac{v_1^2}{2g_c \times 778} + W \qquad (4.11)$$

where h_1, h_4 = enthalpy of the refrigerant at points 1 and 4 respectively, Btu/lb

v_1, v_4 = velocity of the refrigerant at points 1 and 4 respectively, ft/s

$q_\#$ = heat supplied per lb of working substance during evaporation process, Btu/lb

g_c = dimensional conversion factor, 32 $lb_m \cdot ft/lb_f \cdot s^2$

Because no work is done during evaporation, the change of kinetic energy between points 4 and 1 is small compared with other terms in Eq. (4.11), and is usually ignored. Then,

$$h_4 + q_\# = h_1 + 0$$

The refrigerating effect q_{rf}, in Btu/lb, is

$$q_{rf} = q_\# = (h_1 - h_4) \qquad (4.12)$$

For isentropic compression between points 1 and 2, applying the steady flow energy equation and ignoring the change of kinetic energy,

$$h_1 + 0 = h_2 + W$$

$$-W = (h_2 - h_1)$$

Work input to the compressor W_{in}, in Btu/lb, is given as

$$W_{in} = (h_2 - h_1) \qquad (4.13)$$

Similarly, for condensation between points 2 and 3,

$$h_2 + q_\# = h_3 + 0$$

The heat released by the refrigerant in the condenser $-q_\#$, in Btu/lb, is

$$-q_\# = (h_2 - h_3) \qquad (4.14)$$

For the throttling process between points 3 and 4, assuming that the heat loss is negligible,

$$h_3 + 0 = h_4 + 0$$

or

$$h_3 = h_4 \qquad (4.15)$$

The COP_{ref} of the ideal single-stage vapor compression cycle is

$$COP_{ref} = \frac{\text{Refrigerating effect}}{\text{Work input}}$$

$$= \frac{q_{rf}}{W_{in}} = \frac{h_1 - h_4}{h_2 - h_1} \qquad (4.16)$$

The mass flow rate of refrigerant $\dot{m}_r$ (in lb/min) flowing through the evaporator is

$$\dot{m}_r = q_{rc}/q_{rf} \qquad (4.17)$$

where q_{rc} = refrigerating capacity, Btu/h. From Eq. (4.16), the smaller the difference between the condensing and evaporating pressures, or between condensing and evaporating temperatures, the lower the W_{in} input to the compressor at a specific q_{rc} and, therefore, the higher the COP. A higher evaporating pressure p_{ev} and evaporating temperature T_{ev} or a lower condensing pressure p_{con} and condensing temperature T_{con} will always conserve energy.

Determination of Enthalpy by Polynomials

During the performance analysis of a refrigeration cycle, the enthalpies h of the refrigerant at various points must be determined in order to calculate refrigeration effect, work input, and COP. The enthalpy of refrigerant at saturated liquid and saturated vapor state is a function of saturated temperature or pressure. In other words, saturated temperature T_s and saturated pressure p_s of the refrigerant are dependent upon one another. Therefore, it is more convenient to calculate the enthalpy of refrigerant in terms of saturated temperature T_s within a certain temperature range

$$h = f(T_s) \tag{4.18}$$

The enthalpy differential along the constant entropy line within a narrower temperature range can be calculated as

$$(h_2 - h_1) = f(T_{s2} - T_{s1}) \tag{4.19}$$

where h_1, h_2 = enthalpy of the refrigerant on constant entropy line at point 1 and 2, Btu/lb

T_{s1}, T_{s2} = temperature of saturated refrigerant at point 1 and 2, °F

From the refrigerant tables published by ASHRAE, the following polynomial can be used to calculate the enthalpy of saturated liquid refrigerant h_{lr}, in Btu/lb, from its temperature T_{sl} at a saturated temperature from 20°F to 120°F with acceptable accuracy

$$h_{lr} = a_1 + a_2 T_{sl} + a_3 T_{sl}^2 + a_4 T_{sl}^3 \tag{4.20}$$

where a_1, a_2, a_3, a_4 = coefficients

For HCFC–22,
$a_1 = 10.409$	$a_3 = 0.00014794$
$a_2 = 0.26851$	$a_4 = 5.3429 \times 10^{-7}$

Similarly, the polynomial that determines the enthalpy of saturated vapor refrigerant h_{vr}, in Btu/lb, from its temperature T_{sv} in the same temperature range is

$$h_{vr} = b_1 + b_2 T_{sv} + b_3 T_{sv}^2 + b_4 T_{sv}^3 \tag{4.21}$$

where b_1, b_2, b_3, b_4 = coefficients

For HCFC–22,
$b_1 = 104.465$	$b_3 = -0.0001226$
$b_2 = 0.098445$	$b_4 = -9.861 \times 10^{-7}$

The polynomial that determines the enthalpy changes of refrigerant along the constant entropy line for an isentropic compression process between intial state 1 and final state 2 is

$$(h_2 - h_1) = c_1 + c_2(T_{s2} - T_{s1}) + c_3(T_{s2} - T_{s1})^2 + c_4(T_{s2} - T_{s1})^3 \tag{4.22}$$

where c_1, c_2, c_3, c_4 = coefficients

T_{s1}, T_{s2} = saturated temperature of vapor refrigerant corresponding to its pressure at initial state 1 and final state 2, °F.

For HCFC–22 within a saturated temperature range of 20°F to 100°F:

$c_1 = -0.18165$	$c_3 = -0.0012405$
$c_3 = +0.21502$	$c_4 = +8.1982 \times 10^{-6}$

Computer programs are available that calculate the coefficients based on ASHRAE's refrigerant tables and charts. Equation fitting will be covered in more detail in Section 28.2.

Refrigeration Effect, Refrigeration Load, and Refrigeration Capacity

The *refrigeration effect* q_{rf}, in Btu/lb, is the rate of heat extracted by a unit mass of refrigerant during the evaporating process in the evaporator. It can be calculated as

$$q_{rf} = (h_{lv} - h_{en}) \tag{4.23}$$

where h_{en}, h_{lv} = enthalpy of refrigerant entering and leaving the evaporator, Btu/lb.

Refrigerating load q_{rl}, in Btu/h, is the required rate of heat extraction by the refrigerant in the evaporator. It can be calculated as

$$q_{rl} = 60\dot{m}_r(h_{lv} - h_{en}) \tag{4.24}$$

Refrigerating capacity q_{rc}, in Btu/h, is the actual rate of heat extracted by the refrigerant in the evaporator. In practice, the refrigeration capacity of the equipment selected is often slightly higher than the refrigerating load. This is because the manufacturer's specifications are a series of fixed capacities. Occasionally, equipment can be selected so that its capacity is exactly equal to the refrigeration load required.

Refrigeration capacity can be calculated as

$$q_{rc} = 60\dot{m}_r(h_{alv} - h_{aen}) \tag{4.25}$$

where h_{aen}, h_{alv} = enthalpy of refrigerant actually entering and leaving the evaporator, Btu/lb.

4.10 SUBCOOLING AND SUPERHEATING

Subcooling

Condensed liquid refrigerant is usually subcooled to a temperature lower than the saturated temperature corresponding to the condensing pressure of the refrigerant, as shown in Fig. 4.6a point 3′. This is done to increase the refrigerating effect. The degree of subcooling depends mainly on the temperature of the coolant (atmospheric air, surface water, or well water, for example) during condensation and the construction and the capacity of the condenser.

The enthalpy of subcooled liquid refrigerant h_{sc}, in Btu/lb, can be calculated as

$$h_{sc} = h_{s.con} - c_{pr}(T_{s.con} - T_{sc}) \qquad (4.26)$$

where $h_{s.con}$ = enthalpy of saturated liquid refrigerant at condensing temperature, Btu/lb

where c_{pr} = specific heat of liquid refrigerant at constant pressure, Btu/lb · °F

$T_{s.con}$ = saturated temperature of liquid refrigerant at condensing pressure, °F

T_{sc} = temperature of subcooled liquid refrigerant, °F

Enthalpy h_{sc} is also approximately equal to the enthalpy of the saturated liquid refrigerant at subcooled temperature.

Superheating

As mentioned before, the purpose of superheating is to avoid compressor slugging damage. Superheating is shown in Fig. 4.6b. The degree of superheat depends mainly on the type of refrigerant feed and compressor, as well as the construction of the evaporator. These will be covered in detail in Chapter 21.

Example 4.1. A 500-ton single-stage centrifugal vapor compression system uses HCFC–22 as refrigerant. The vapor refrigerant enters the compressor at dry saturated state. The compression process is assumed to be isentropic. Hot gas is discharged to the condenser and condensed at a temperature of 95°F. The saturated liquid refrigerant then flows through a throttling device and evaporates at a temperature of 35°F.

Calculate

1. The refrigerating effect
2. The work input to the compressor
3. The coefficient of performance of this refrigeration cycle
4. The mass flow rate of the refrigerant

Recalculate the COP and the energy saved in work input if the refrigerant is subcooled to a temperature of 90°F.

Solution

1. From Eq. (4.20), the enthalpy of the saturated liquid refrigerant at a temperature of 95°F, point 3 as shown in Fig. 4.6a, is

$$h_3 = h_4 = 10.409 + 0.26851T_s + 0.0001479T_s^2$$
$$+ 5.3429 \times 10^{-7}T_s^3$$
$$= 10.409 + 0.26851(95) + 0.0001479(95)^2$$
$$+ 5.3429 \times 10^{-7}(95)^3$$
$$= 10.409 + 25.508 + 1.335 + 0.458$$
$$= 37.71 \text{ Btu/lb}$$

From Eq. (4.21), the enthalpy of saturated vapor refrigerant at a temperature of 35°F, point 1, is

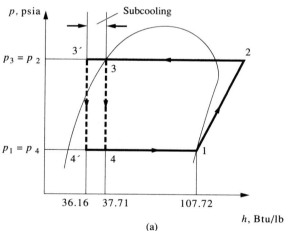

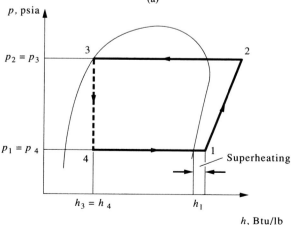

FIGURE 4.6 Subcooling and superheating: (a) subcooling, and (b) superheating.

$$h_1 = 104.465 + 0.098445T_s - 0.0001226T_s^2$$
$$- 9.861 \times 10^{-7}T_s^3$$
$$= 104.47 + 0.098445(35) - 0.0001226(35)^2$$
$$- 9.861 \times 10^{-7}(35)^3$$
$$= 104.47 + 3.445 - 0.150 - 0.042$$
$$= 107.72 \text{ Btu/lb}$$

Then, the refrigerating effect is calculated as

$$q_{rf} = h_1 - h_4 = 107.72 - 37.71 = 70.01 \text{ Btu/lb}$$

2. From Eq. (4.22), the enthalpy differential $(h_2 - h_1)$ on the constant entropy line corresponding to a saturated temperature differential $(T_{s2} - T_{s1}) = (95 - 35) = 60°F$ in the two-phase region is

$$(h_2 - h_1) = -0.18165 + 0.21502(T_{s2} - T_{s1})$$
$$- 0.0012405(T_{s2} - T_{s1})^2$$
$$+ 8.1982 \times 10^{-6}(T_{s2} - T_{s1})^3$$
$$= -0.18165 + 0.21502(60)$$
$$- 0.0012405(60)^2$$
$$+ 8.1982 \times 10^{-6}(60)^3$$
$$= -0.182 + 12.901 - 4.466 + 1.771$$
$$= 10.024 \text{ Btu/lb}$$

That is, the work input $W_{in} = h_2 - h_1 = 10.024$ Btu/lb

3. According to Eq. (4.16), COP_{ref} of the refrigerating system is calculated as

$$\text{COP}_{ref} = \frac{q_{rf}}{W_{in}} = \frac{70.01}{10.024} = 6.98$$

4. From Eq. (4.17), the mass flow rate of the refrigerant can be calculated as

$$\dot{m}_r = q_{rc}/q_{rf} = \frac{500 \times 12,000}{60 \times 70.01} = 1428 \text{ lb/min}$$

5. If the liquid refrigerant is subcooled to a temperature of 90°F, then

$$h_3 = h_4 = 10.409 + 0.26851(90) + 0.0001479(90)^2$$
$$+ 5.3429 \times 10^{-7}(90)^3$$
$$= 10.409 + 24.166 + 1.198 + 0.389$$
$$= 36.16 \text{ Btu/lb}$$

Refrigerating effect is then increased to

$$q_{rf} = 107.72 - 36.16 = 71.56 \text{ Btu/lb}$$

COP_{ref} is also increased to

$$\text{COP}_{ref} = \frac{71.56}{10.024} = 7.14$$

Since 1 ton = 200 Btu/min and 1 hp = 42.41 Btu/min, electric power input to the compressor P_{in} without subcooling is

$$P_{in} = \frac{500 \times 200}{42.41 \times 6.98} = 337.8 \text{ hp}$$

With subcooling,

$$P_{in.s} = \frac{500 \times 200}{42.41 \times 7.14} = 330.2 \text{ hp}$$

Savings in electric energy are calculated as

$$\frac{337.8 - 330.2}{337.8} = 0.022 \text{ or } 2.2\%$$

4.11 MULTISTAGE VAPOR COMPRESSION SYSTEMS

When a refrigeration system uses more than one compression process stage, it is called a *multi-stage system*

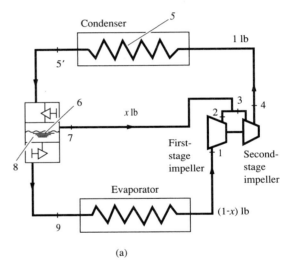

(a)

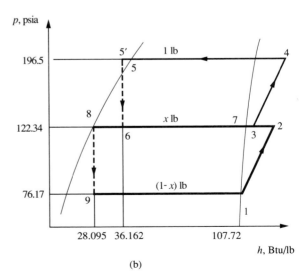

(b)

FIGURE 4.7 Two-stage compound system with a flash cooler: (a) schematic diagram, and (b) refrigeration cycle.

(shown in Figure 4.7), and may include the following:

1. A high-stage compressor and a low-stage compressor
2. Several compressors connected in series
3. Two or more impellers connected internally in series and driven by the same motor or prime mover, as shown in Fig. 4.7a
4. A combination of two separate refrigeration systems

The discharge pressure of the low-stage compressor, which is equal to the suction pressure of the high-stage compressor, is called *interstage pressure*.

The reasons for using a multistage vapor compression system instead of a single-stage system are as follows:

1. The compression ratio R_{com} of each stage in a multistage system is smaller than in a single-stage unit, so compressor efficiency is increased. Compression ratio R_{com} is defined as the ratio of the compressor's discharge pressure p_{dis}, in psia, to the suction pressure at the compressor's inlet p_{suc}, in psia, that is

$$R_{com} = \frac{p_{dis}}{p_{suc}} \qquad (4.27)$$

2. Liquid refrigerant enters the evaporator at a lower enthalpy and increases the refrigeration effect.
3. Discharge gas from the low-stage compressor can be desuperheated at the interstage pressure. This results in a lower discharge temperature from the high-stage compressor than would be produced by a single-stage system at the same pressure differential between condensing and evaporating pressure.
4. Two or three compressors in a multistage system provide much greater flexibility to accommodate the variation of refrigerating loads at various evaporating temperatures during part-load operation.

The drawbacks of the multistage system are higher initial cost and a more complicated system than for a single-stage system.

Compound Systems

Multistage vapor compression systems are classified as compound systems or cascade systems. Cascade systems will be discussed in a later section.

A *compound system* consists of two or more compression stages connected in series. For reciprocating, rotary, screw, or scroll compressors, each compression stage usually requires a separate compressor. In multistage centrifugal compressors, two or more stages may be internally compounded by means of several impellers connected in series.

Interstage Pressure

Interstage pressure is usually set so that the compression ratio at each stage is nearly the same for higher COPs. For a two-stage compound system, interstage pressure p_i, in psia, can be calculated as

$$p_i = \sqrt{p_{con} p_{ev}} \qquad (4.28)$$

where p_{con} = condensing pressure, psia
p_{ev} = evaporating pressure, psia

For a multistage vapor compression system with z stages, the compression ratio R_{com} for each stage can be calculated as

$$R_{com} = \left(\frac{p_{con}}{p_{ev}}\right)^{1/z} \qquad (4.29)$$

Flash Cooler and Intercooler

In compound systems, *flash coolers* are used to subcool liquid refrigerant to the saturated temperature corresponding to the interstage pressure by vaporizing part of the liquid refrigerant. Intercoolers are used to desuperheat the discharge gas from the low-stage compressor and, more often, to subcool also the liquid refrigerant before it enters the evaporator.

4.12 TWO-STAGE COMPOUND SYSTEM WITH A FLASH COOLER
Flow Processes

Figure 4.7a is a schematic diagram of a two-stage compound system with a flash cooler and Fig. 4.7b shows the refrigeration cycle of this system. Vapor refrigerant at point 1 enters the first-stage impeller of the centrifugal compressor at the dry saturated state. It is compressed to the interstage pressure p_i at point 2 and mixes with evaporated vapor refrigerant from the flash cooler, often called an *economizer*. The mixture then enters the second-stage impeller at point 3. Hot gas, compressed to condensing pressure p_{con}, leaves the second-stage impeller at point 4. It is then discharged to the condenser in which the hot gas is desuperheated, condensed, and often subcooled to liquid state at point 5'. After the condensing process, the subcooled liquid refrigerant flows through a throttling device, such as a float valve, at the high-pressure side. A small portion of the liquid refrigerant flashes into vapor in the flash cooler at point 7, and this latent heat of vaporization cools the remaining liquid refrigerant to the saturation temperature corresponding to the interstage pressure at point 8. Inside the flash cooler, the mixture of vapor and liquid refrigerant is at point 6. Liquid refrigerant then flows through another throttling device, a small portion is flashed at point 9, and the liquid and vapor mixture enters the evaporator. The remaining liquid refrigerant is vaporized at point 1 in the evaporator. The vapor then flows to the inlet of the first-stage impeller of the centrifugal compressor and completes the cycle.

Fraction of Evaporated Refrigerant in the Flash Cooler

In the flash cooler, out of 1 lb of refrigerant flowing through the condenser, x lb of it cools down the remaining portion of liquid refrigerant $(1 - x)$ lb to saturated temperature T_8 at interstage pressure p_i.

Because $h_{5'}$ is the enthalpy of the subcooled liquid refrigerant entering the flash cooler, h_6 is the enthalpy of the mixture of vapor and liquid refrigerant after the throttling device, for a throttling process, $h_{5'} = h_6$. Enthalpy h_7 and h_8 are the enthalpies of the saturated vapor and saturated liquid, respectively, at the interstage pressure, and h_9 is the enthalpy of the mixture of vapor and liquid refrigerant leaving the flash cooler after the low-pressure side throttling device. Again, for a throttling process, $h_8 = h_9$.

If the heat loss from the insulated flash cooler to the ambient air is small, it can be ignored. Heat balance of

the refrigerants entering and leaving the flash cooler, as shown in Fig. 4.8a, gives

Sum of the heat energy of the refrigerant entering flash cooler	$=$	Sum of the heat energy of the refrigerant leaving flash cooler

That is

$$h_{5'} = x h_7 + (1 - x)h_8$$

The fraction of liquid refrigerant evaporated in the flash cooler x is given as

$$x = \frac{h_{5'} - h_8}{h_7 - h_8} \qquad (4.30)$$

The fraction, x, also indicates the quality, or dryness fraction, of the vapor and liquid mixture in the flash cooler at the interstage pressure.

Enthalpy of Vapor Mixture Entering Second-Stage Impeller

Ignoring the heat loss from mixing point 3 to the surroundings, the mixing of the gaseous refrigerant discharged from the first-stage impeller at point 2 and the vaporized refrigerant from the flash cooler at point 7 is an adiabatic mixing process. The heat balance at the mixing point before the second-stage impeller, as shown in Fig. 4.8b, is given as

$$h_3 = (1 - x)h_2 + x h_7 \qquad (4.31)$$

where h_2 = enthalpy of the gaseous refrigerant discharged from the first-stage impeller, Btu/lb

h_3 = enthalpy of the mixture at point 3, Btu/lb

h_7 = enthalpy of the saturated vapor refrigerant from the flash cooler at point 7, Btu/lb

Coefficient of Performance

For 1 lb of refrigerant flowing through the condenser, the amount of refrigerant flowing through the evaporator is $(1 - x)$ lb. The refrigeration effect q_{rf} per lb of refrigerant flowing through the condenser can be expressed as

$$q_{rf} = (1 - x)(h_1 - h_9) \qquad (4.32)$$

where h_1 = enthalpy of the saturated vapor leaving the evaporator, Btu/lb

h_9 = enthalpy of the refrigerant entering the evaporator, Btu/lb

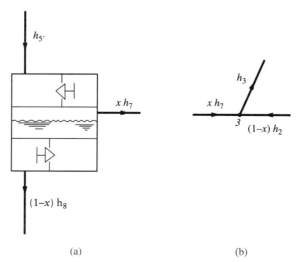

(a)   (b)

FIGURE 4.8 Heat balance of entering and leaving refrigerants in a flash cooler and at the mixing point: (a) in the flash cooler, and (b) at the mixing point 3 before entering the second-stage impeller.

Total work input to the compressor (including the first- and second-stage impeller) W_{in} per lb of refrigerant flowing through the condenser, in Btu/lb, is

$$W_{in} = (1 - x)(h_2 - h_1) + (h_4 - h_3) \quad (4.33)$$

where h_4 = enthalpy of the hot gas discharged from the second stage impeller, Btu/lb. The coefficient of performance of the two-stage compound system with a flash cooler COP_{ref} is

$$COP_{ref} = \frac{q_{rf}}{W_{in}}$$

$$= \frac{(1 - x)(h_1 - h_9)}{(1 - x)(h_2 - h_1) + (h_4 - h_3)} \quad (4.34)$$

The mass flow rate of refrigerant at the condenser $\dot{m}_r$, in lb/min, is

$$\dot{m}_r = \frac{q_{rc}}{q_{rf}} = \frac{q_{rc}}{(1 - x)(h_1 - h_9)} \quad (4.35)$$

Characteristics of Two-Stage Compound System with a Flash Cooler

In a two-stage compound system with a flash cooler, a portion of the liquid refrigerant is flashed into vapor going directly to the second-stage suction inlet, so less refrigerant is compressed in the first-stage impeller. Furthermore, the remaining liquid refrigerant is cooled to the saturated temperature corresponding to the interstage pressure, which is far lower than the subcooled liquid temperature in a single-stage system. The increase in refrigerating effect and the drop in compression work input lead to a higher COP_{ref} than in a single-stage system.

Although the initial cost of a two-stage compound system is higher than for a single-stage system, the two-stage compound system with a flash cooler is widely used in large central hydronic air conditioning systems because of its high COP_{ref}.

Example 4.2. For the same 500-ton centrifugal vapor compression system as in Example 4.1, a two-stage compound system with a flash cooler is used instead of a single-stage centrifugal compressor. Vapor refrigerant enters the first-stage impeller at dry saturated state and the subcooled liquid refrigerant leaves the condenser at a temperature of 90°F. Both compression processes at the first-stage impeller and the second-stage impeller are assumed to be isentropic. Evaporating pressure is 76.17 psia and condensing pressure is 196.5 psia. Other conditions remain the same as in Example 4.1. Calculate:

1. The fraction of liquid refrigerant vaporized in the flash cooler

2. The refrigerating effect per lb of refrigerant flowing through the condenser

3. The total work input to the compressor

4. The coefficient of performance of this refrigerating system

5. The mass flow rate of refrigerant flowing through the condenser

6. The percentage of saving in energy consumption compared with the single stage vapor compression system

Solution

1. Based on the data calculated in Example 4.1, enthalpy of vapor refrigerant entering the first stage impeller $h_1 = 107.72$ Btu/lb and the enthalpy of the subcooled liquid refrigerant leaving the condenser $h_{5'} = 36.162$ Btu/lb, as shown in Fig. 4.7b.

From Eq. (4.28) and the given data, the interstage pressure can be calculated as

$$p_i = \sqrt{(p_{con} p_{ev})}$$
$$= \sqrt{(196.5 \times 76.17)}$$
$$= 122.34 \text{ psia.}$$

From the Table of Properties of Saturated Liquid and Vapor for HCFC–22 in *ASHRAE Handbook 1989, Fundamentals*, for a $p_i = 122.34$ psia, the corresponding interstage saturated temperature T_i in the two-phase region is 63.17°F.

From Eq. (4.20), the enthalpy of liquid refrigerant at saturated temperature 63.17°F is

$$h_8 = h_9 = 10.409 + 0.26851(63.17)$$
$$+ 0.00014794(63.17)^2 + 5.3429(63.17)^3$$
$$= 10.409 + 16.961 + 0.59 + 0.135$$
$$= 28.095 \text{ Btu/lb}$$

Also, from Eq. (4.21), the enthalpy of the saturated vapor refrigerant at a temperature of 63.17°F is

$$h_7 = 104.465 + 0.098445(63.17)$$
$$- 0.0001226(63.17)^2$$
$$- 9.861 \times 10^{-7}(63.17)^3$$
$$= 104.465 + 6.219 - 0.489 - 0.249$$
$$= 109.946 \text{ Btu/lb}$$

Then, from Eq. (4.30), the fraction of vaporized refrigerant in the flash cooler is

$$x = \frac{h_{5'} - h_8}{h_7 - h_8}$$
$$= \frac{36.162 - 28.095}{109.946 - 28.095}$$
$$= 0.09856$$

2. From Eq. (4.32), the refrigerating effect is

$$q_{rf} = (1 - x)(h_1 - h_9)$$
$$= (1 - 0.09856)(107.72 - 28.095)$$
$$= 71.78 \text{ Btu/lb}$$

3. From Eq. (4.22), the enthalpy differential $(h_2 - h_1)$ corresponding to a saturated temperature

differential $(T_{s2} - T_{s1}) = (63.17 - 35) = 28.17°F$ in the two-phase region is:

$$
\begin{aligned}
(h_2 - h_1) &= -0.18165 + 0.21502(28.17) \\
&\quad - 0.0012405(28.17)^2 \\
&\quad + [8.1982 \times 10^{-6}(28.17)^3] \\
&= -0.182 + 6.057 - 0.984 + 0.183 \\
&= 5.074 \text{ Btu/lb}
\end{aligned}
$$

Similarly, from Eq. (4.22), the enthalpy differential $(h_4 - h_3)$ corresponding to a saturated temperature differential

$(T_s - T_i) = (95 - 63.17) = 31.83°F$ is
$$
\begin{aligned}
(h_4 - h_3) &= -0.020 + 0.16352(31.83) \\
&\quad - 0.00035106(31.83^2) \\
&\quad + 1.9177 \times 10^{-6}(31.83)^3 \\
&= -0.020 + 5.205 - 0.356 + 0.062 \\
&= 4.891 \text{ Btu/lb}
\end{aligned}
$$

Then, from Eq. (4.33), the total work input to the compressor is calculated as

$$
\begin{aligned}
W_{in} &= [(1 - x)(h_2 - h_1)] + (h_4 - h_3) \\
&= [(1 - 0.09856)(5.074)] + 4.891 \\
&= 9.465 \text{ Btu/lb}
\end{aligned}
$$

4. The coefficient of performance of this two-stage compound system is

$$
\text{COP}_{ref} = \frac{q_{rf}}{W_{in}} = \frac{71.78}{9.465} = 7.58
$$

5. The mass flow rate of refrigerant flowing through the condenser can be calculated as

$$
\dot{m}_r = \frac{q_{rc}}{q_{rf}} = \frac{500 \times 200}{71.78} = 1393 \text{ lb/min}
$$

6. From the results in Example 4.1, the power input to the single-stage system is 330.2 hp. The power input to the two-stage compound system is

$$
P_{in} = \frac{500 \times 200}{42.41 \times 7.58} = 311.1 \text{ hp}
$$

Energy savings compared with that of the single-stage system is calculated as

$$
\frac{330.2 - 311.1}{330.2} = 0.058 = 5.8\%
$$

4.13   THREE-STAGE COMPOUND SYSTEM WITH A TWO-STAGE FLASH COOLER

To reduce the energy consumption of refrigeration systems in air conditioning, the three-stage compound

system with a two-stage flash cooler was recently introduced to the market as a standard product. Figure 4.9 shows the schematic diagram and refrigeration cycle of this system.

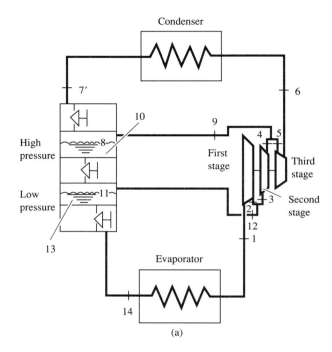

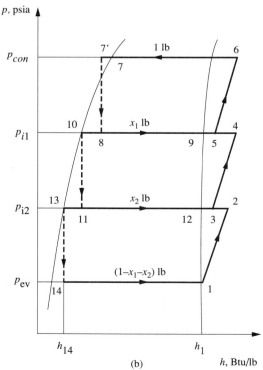

FIGURE 4.9 Three-stage compound system with a two-stage flash cooler (a) schematic diagram and (b) refrigeration cycle.

Flow Processes

In Fig. 4.9, vapor refrigerant enters the first-stage impeller of the centrifugal compressor at dry saturated state, point 1. After the first-stage compression process, at point 2, it mixes with the vaporized refrigerant coming from the second-stage flash cooler at point 12. At point 3, the mixture enters the second-stage impeller. After the second-stage compression process at point 4, it mixes again with vaporized refrigerant from the first-stage flash cooler at point 9. The mixture, at point 5, is then compressed to the condensing pressure in the third-stage impeller. At point 6, the hot gas enters the condenser, condenses into liquid, and subcools to a temperature below the condensing temperature. Subcooled liquid refrigerant leaves the condenser at point 7′ and flows through a two-stage flash cooler and the associated throttling devices in which a small portion of liquid refrigerant is successively flashed into vapor at interstage pressure p_{i1}, point 9, and interstage pressure p_{i2}, point 12. The liquid–vapor mixture enters the evaporator at point 14 and the remaining liquid refrigerant evaporates into vapor completely in the evaporator.

Fraction of Refrigerant Vaporized in the Flash Cooler

Based on the heat balance of the refrigerants entering and leaving the high-pressure flash cooler, as shown in Fig. 4.10a, the fraction of liquid refrigerant vaporized in the high-pressure flash cooler x_1 at interstage pressure p_{i1} can be calculated as

$$h_{7'} = (1 - x_1)h_{10} + x_1 h_9$$

Therefore,

$$x_1 = \frac{h_{7'} - h_{10}}{h_9 - h_{10}} \qquad (4.36)$$

where $h_{7'}, h_9, h_{10}$ = enthalpies of refrigerants at points 7′, 9, and 10 respectively, Btu/lb.

In the same manner, the heat balance of the refrigerant entering and leaving the low-pressure flash cooler, as shown in Fig. 4.10b, may be expressed as

$$(1 - x_1)h_{10} = x_2 h_{12} + (1 - x_1 - x_2)h_{13}$$

where h_{12}, h_{13} = enthalpies of the refrigerant at points 12 and 13, respectively, Btu/lb

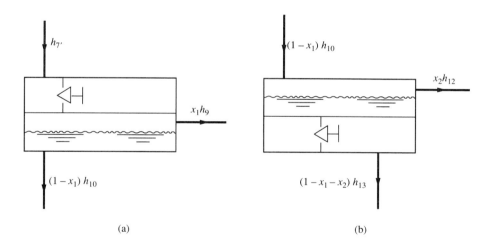

(a)

(b)

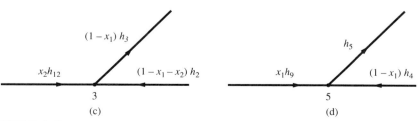

(c)

(d)

FIGURE 4.10 Heat balance of refrigerants entering and leaving the high-pressure, and low-pressure flash cooler and mixing points: (a) high-pressure flash cooler, (b) low-pressure flash cooler, (c) at the mixing point before entering second-stage impeller, (c) at the mixing point before entering the second-stage impeller, (d) at the mixing point before entering the third-stage impeller

The fraction of liquid refrigerant vaporized in the low-pressure flash cooler x_2 at an interstage pressure p_{i2} can be calculated as

$$x_2 = \frac{(1 - x_1)(h_{10} - h_{13})}{h_{12} - h_{13}} \quad (4.37)$$

Coefficient of Performance of Three-Stage Systems

From a heat balance of the refrigerants entering and leaving the mixing point before the inlet of the second-stage impeller, as shown in Fig. 4.10c, the enthalpy of the mixture at point 3, h_3 (in Btu/lb) is given as

$$h_3 = \frac{(1 - x_1 - x_2)h_2 + x_2 h_{12}}{1 - x_1} \quad (4.38)$$

where h_2 = enthalpy of the gaseous refrigerant discharge from the first-stage impeller at point 2, Btu/lb.

As shown in Fig. 4.10d, the enthalpy of the mixture of vapor refrigerants at point 5 h_5, in Btu/lb, can also be calculated as

$$h_5 = (1 - x_1)h_4 + x_1 h_9 \quad (4.39)$$

where h_4 = enthalpy of the gaseous refrigerant discharged from the second-stage impeller at point 4, Btu/lb.

The refrigerating effect q_{rf} in Btu/lb of refrigerant flowing through the condenser is given as

$$q_{rf} = (1 - x_1 - x_2)(h_1 - h_{14}) \quad (4.40)$$

where h_1, h_{14} = enthalpies of the refrigerants leaving the evaporator at point 1 and entering the evaporator at point 14, respectively, Btu/lb. Total work input to the three-stage compressor W_{in}, in Btu/lb of refrigerant flowing through the condenser is given by

$$W_{in} = (1 - x_1 - x_2)(h_2 - h_1) \\ + (1 - x_1)(h_4 - h_3) + (h_6 - h_5) \quad (4.41)$$

where h_6 enthalpy of the hot gas discharged from the third-stage impeller at point 6, in Btu/lb.

From Eqs. (4.40) and (4.41), the coefficient of performance of this three-stage system with a two-stage flash cooler is

$$COP_{ref} = \frac{q_{rf}}{W_{in}} \\ = [(1 - x_1 - x_2)(h_1 - h_{14})]/ \\ [(1 - x_1 - x_2)(h_2 - h_1) \\ + (1 - x_1)(h_4 - h_3) + (h_6 - h_5)] \quad (4.42)$$

A three-stage compound system with a two-stage flash cooler often has a further energy saving of about 2 to 5 percent compared to a two-stage compound system with a flash cooler.

The mass flow rate of the refrigerant flowing through the condenser $\dot{m}_r$, in lb/min, can be calculated as

$$\dot{m}_r = \frac{q_{rc}}{q_{rf}} = \frac{q_{rc}}{(1 - x_1 - x_2)(h_1 - h_{14})} \quad (4.43)$$

4.14 TWO-STAGE COMPOUND SYSTEM WITH INTERCOOLERS

Two-Stage Compound System with a Vertical Coil Intercooler

When an evaporating temperature in the range of $-10°F$ to $-50°F$ is required, a two-stage compound system using reciprocating or other positive displacement compressors is usually applied.

Figure 4.11 shows the schematic diagram and the refrigeration cycle of a two-stage compound system with a vertical coil intercooler. The subcooled liquid refrigerant from the receiver at point 5' is divided into two streams. One stream enters the coil inside the intercooler. The other stream enters its shell after throttling to point 6, the interstage pressure.

In the intercooler shell, some of the liquid refrigerant vaporizes to saturated vapor at point 7, drawing latent heat from the liquid in the coil at point 5', further subcooling it to point 10. This subcooled liquid is throttled by the expansion valve at point 10, and then evaporates to saturated vapor at point 1 in the evaporator.

Vapor refrigerant from the evaporator at point 1 enters the low-stage compressor. The compressed hot gas at point 2 discharges into the intercooler, mixing with the liquid from the receiver at the interstage pressure. The liquid level in the intercooler is controlled by the saturated temperature at the interstage pressure in the intercooler. The saturated mixture, at point 7, enters the high-stage compressor. At point 4, hot gas is discharged from the high stage compressor and then condensed and subcooled to point 5' in the condenser.

In this system, x is the fraction of liquid refrigerant vaporized in the intercooler and h_{10} is the enthalpy of the liquid refrigerant that has been subcooled in the vertical coil. Based on the heat balance of the refrigerants entering and leaving intercooler, as shown in Fig. 4.11a,

$$(1 - x)h_{5'} + x h_{5'} + (1 - x)h_2 = h_7 + (1 - x)h_{10}$$

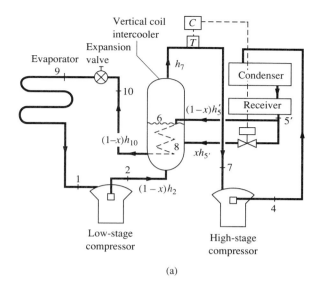

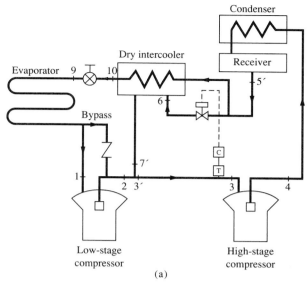

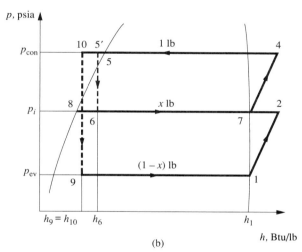

FIGURE 4.11 Two-stage compound system with a vertical coil intercooler: (a) schematic diagram and (b) refrigeration cycle.

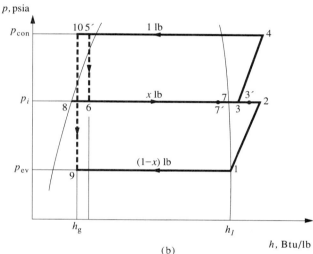

FIGURE 4.12 Two-stage compound system with a dry intercooler: (a) schematic diagram and (b) refrigeration cycle.

Then,

$$x = \frac{(h_2 - h_7) + (h_{5'} - h_{10})}{h_2 - h_{10}} \qquad (4.44)$$

This type of system is often used in low-temperature refrigerated cold storage and other industrial applications. Ammonia is often used as the refrigerant.

Two-Stage Compound System with a Dry Intercooler

The two-stage compound system with a dry intercooler is another type of two-stage system for low-temperature refrigeration systems using positive dis-

placement compressors. Figure 4.12 shows the schematic diagram and refrigeration cycle of such a system.

In Fig. 4.12, subcooled liquid refrigerant leaves the receiver at point 5′ and divides into two streams. One stream enters the coil of the intercooler. The other stream flows through a throttling valve and its pressure is reduced to the interstage pressure at point 6. Part of this liquid stream evaporates in the intercooler, drawing latent heat from the liquid refrigerant inside the coil and cooling it to point 10. It then flows through the expansion valve, enters the evaporator at point 9, and evaporates to saturated vapor in the evaporator. The vapor at point 1 leaves the evaporator, enters the low-stage compressor, and is compressed to the interstage pressure at point 2.

The vapor and the remaining liquid refrigerant leave the intercooler at point 7'. This mixture of vapor and liquid mixes with the hot gas discharged from the low-stage compressor at point 2 and forms another mixture at point 3'. Because of the vaporization of the liquid refrigerant remaining in the mixture and the comparatively low enthalpy of the vapor refrigerant from the intercooler, the mixture at point 3' cools to point 3, and enters the high-stage compressor. It is then compressed to point 4 and enters the condenser where it is desuperheated, condensed, and subcooled. The liquid refrigerant enters the receiver at point 5' from the condenser. The temperature of the desuperheated mixture at point 3 determines the amount of liquid refrigerant fed to the intercooler.

Heat Balance

In this system, x is the sum of the fractions of liquid refrigerant vaporized in the intercooler and desuperheating area per lb of refrigerant flowing through the condenser, and x_i is the fraction of vaporized refrigerant in the intercooler. Then, $(x - x_i)$ represents the fraction of liquid refrigerant vaporized in the desuperheating area. The heat energy of the mixture of vapor and liquid refrigerant leaving the intercooler is

vapor refrigerant	$x_i h_7$
liquid refrigerant	$(x - x_i)h_8$

Based on the heat balance of the refrigerants entering and leaving the intercooler, as shown in Fig. 4.13a,

$$(1 - x)h_{5'} + xh_{5'} = (1 - x)h_{10} + x_i h_7 + (x - x_i)h_8$$

Then,

$$x_i = \frac{x(h_{10} - h_8) + (h_{5'} - h_{10})}{h_7 - h_8} \qquad (4.45)$$

The cooling of the liquid refrigerant inside the coil from point 5' to 10 is

$$(1 - x)(h_{5'} - h_{10})$$

and the cooling of the hot gas from point 2 to 3 is

$$(1 - x)(h_2 - h_3)$$

These coolings are caused by vaporization of the liquid refrigerant in the intercooler and at the desuperheating area

$$x(h_3 - h_{5'})$$

Therefore,

$$(1 - x)(h_{5'} - h_{10}) + (1 - x)(h_2 - h_3) = x(h_3 - h_{5'})$$
$$(4.46)$$

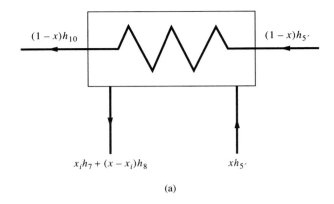

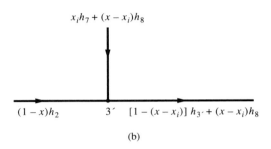

FIGURE 4.13   Heat balance of refrigerants entering and leaving the intercooler, mixing point, and desuperheating area: (a) in the dry intercooler, (b) at the mixing point, (c) at the desuperheating area.

and

$$x = \frac{(h_2 - h_3) + (h_{5'} - h_{10})}{h_2 - h_{10}} \qquad (4.47)$$

For an adiabatic mixing process of the hot gas discharged from the low-stage compressor at point 2 and the mixture of vapor and liquid refrigerant from the intercooler at point 7' as shown in Fig. 4.13b,

$$x_i h_7 + (x - x_i)h_8 + (1 - x)h_2$$
$$= [(1 - (x - x_i)]h_{3'} + (x - x_i)h_8$$

Then,

$$h_{3'} = \frac{x_i h_7 + (1 - x)h_2}{1 - (x - x_i)} \qquad (4.48)$$

After desuperheating, the enthalpy of the vapor refrigerant entering the high-stage compressor at point 3 is

$$h_3 = [1 - (x - x_i)]h_{3'} + (x - x_i)h_8 \qquad (4.49)$$

In Eqs. (4.45) to (4.49), x, x_i, h_3 and $h_{3'}$ should be solved by trial-and-error. Other parameters like q_{rf}, W_{in}, and COP_{ref} can be calculated in the same manner.

Two-stage compound systems with dry intercoolers are often used in low-temperature test rooms or environmental chambers where nontoxic HCFCs or CFCs are used as refrigerants.

Comparison of Flash Coolers, Vertical Coils, and Dry Intercoolers

Hot gas discharged from the low-stage compressor is always desuperheated to a nearly saturated vapor state at the interstage pressure in the vertical coil intercooler. This process is more appropriate for a refrigerant like ammonia, which has a high discharge temperature. In flash coolers and dry intercoolers, desuperheating is caused by the mixing of flashed vapor and hot gas, and will not result in a dry saturated state. Therefore, flash coolers and dry intercoolers are usually used in refrigerating systems using HCFCs or CFCs.

The pressure of liquid refrigerant flowing inside the coils of a vertical coil or dry intercooler can be maintained at a slightly lower pressure than condensing pressure, whereas the pressure of liquid refrigerant in the flash cooler is decreased to the interstage pressure. Some refrigerant may be preflashed before the throttling device, causing a waste of refrigerating capacity. For a flash cooler, the available pressure drop in the throttling device is lower.

4.15   CASCADE SYSTEMS

A *cascade system* consists of two separate single-stage refrigeration systems: a lower system that can better maintain low evaporating temperatures and a higher system that performs better at high evaporating temperatures. These two systems are connected by a cascade condenser in which the condenser of the lower system becomes the evaporator of the higher system as the higher system's evaporator takes on the heat released from the lower system's condenser.

It is often desirable to have a heat exchanger between the liquid refrigerant from the cascade condenser and the vapor refrigerant leaving the evaporator of the lower system. The liquid refrigerant can be subcooled to a lower temperature before entering the evaporator of the lower system shown in Fig. 4.14a. Because the evaporating temperature is low, there is no danger of too high a discharge temperature after the compression process of the lower system.

When a cascade system is shut down while the temperature of the ambient air is 80°F, the saturated vapor pressure of the refrigerant increases. For a lower system using HFC–125 as the refrigerant, this saturated pressure may increase to 209.8 psia. For safety reasons, a relief valve at the cascade condenser connects to an expansion tank, designed to store the refrigerant from the lower system in case it is shut down.

For extremely low evaporating temperatures, a multistage compression system may be used in either the lower or higher system of a cascade system.

Advantages and Disadvantages

The main advantage of a cascade system is that different refrigerants, equipment, and oils can be used for the higher and lower systems. This is especially helpful when the evaporating temperature required in the lower system is less than $-100°F$. The specific volume of the suction vapor v_{suc} is extremely important in such temperature applications. A greater v_{suc} always requires a large compressor and more space. For instance, at $-80°F$, HCFC–22 has a v_{suc} of 9.69 ft^3/lb, whereas HFC–125 v_{suc} is only 4.76 ft^3/lb.

One disadvantage of a cascade system is the overlap of the condensing temperature of the lower system and the evaporating temperature of the higher system for heat transfer in the condenser. The overlap results in higher energy consumption. Also a cascade system is more complicated than a compound system.

Performance of the Cascade System

The performance of the cascade system can be measured in terms of 1 lb of refrigerant in the lower system, for the sake of convenience. In Fig. 4.14b and c, which show the refrigeration cycles on p–h and T–s diagrams, the lower system is indicated by points 1, 2, 3', and 4, and the higher system by points 5, 6, 7, and 8.

The desired low-temperature refrigeration effect of the cascade system can be calculated as

$$q_{rf} = (h_1 - h_4) \qquad (4.50)$$

If the heat transfer between the cascade condenser and the surroundings is ignored, then the heat released by the condenser of the lower system $m_l(h_2 - h_3)$ is equal to the refrigerating load on the evaporator of the higher system $m_h(h_5 - h_8)$, that is

$$\dot{m}_l(h_2 - h_3) = \dot{m}_h(h_5 - h_8)$$

where $\dot{m}_l$, $\dot{m}_h$ mass flow rate of refrigerants in the lower and higher system respectively, lb/min. Therefore, the ratio of the mass flow rate of refrigerant in

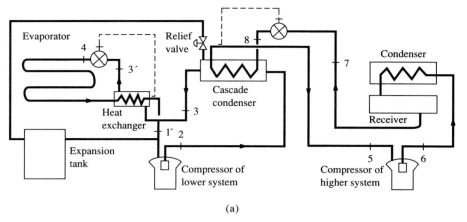

(a)

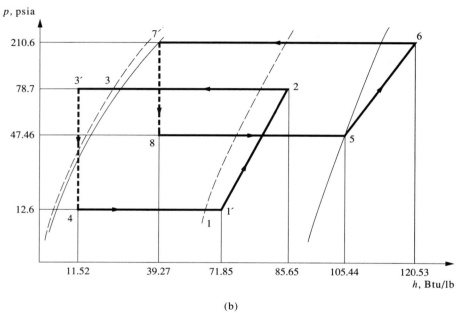

(b)

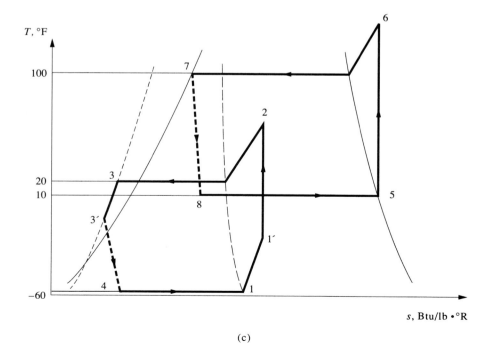

(c)

FIGURE 4.14 Cascade system: (a) schematic diagram, (b) *p–h* diagram, and (c) *T–s* diagram.

the higher system to the mass flow rate in the lower system is

$$\dot{m}_h / \dot{m}_l = \frac{h_2 - h_3}{h_5 - h_8} \qquad (4.51)$$

The mass flow rate of refrigerant in the lower system m_l, in lb/min, is

$$\dot{m}_l = \frac{q_{rc.1}}{60 q_{rf.1}} \qquad (4.52)$$

where $q_{rc.1}$ = refrigeration capacity of lower system, Btu/h

$q_{rf.1}$ = refrigeration effect of lower system, Btu/lb

Total work input to the compressors in both higher and lower systems W_{in}, in Btu/lb of refrigerant in the lower system, can be calculated as

$$W_{in} = (h_2 - h_{1'}) + \frac{\dot{m}_h (h_6 - h_5)}{\dot{m}_l} \qquad (4.53)$$

where h_1' = the enthalpy of vapor refrigerant leaving the heat exchanger, Btu/lb

The coefficient of performance of the cascade system is given as

$$COP_{ref} = \frac{q_{rf}}{W_{in}} = \frac{\dot{m}_l (h_1 - h_4)}{\dot{m}_l (h_2 - h_{1'}) + \dot{m}_h (h_6 - h_5)}$$

$$(4.54)$$

Example 4.3 A cascade system for an environmental chamber is designed to operate at the following conditions during summer:

	Lower system	Higher system
Refrigerant	HFC–125	HCFC–22
Evaporating temperature, °F	–60	10
Evaporating pressure, psia	12.6	47.46
Condensing temperature, °F	20	100
Condensing pressure, psia	78.7	210.6
Suction temperature, °F	–20	
Subcooling after heat exchanger, °F	1	
Refrigerating load, Btu/h	100,000	

Vapor refrigerant enters the compressor of the higher system at dry saturated state and liquid refrigerant leaves the condenser of the higher system without subcooling. Ignore the pressure losses in the pipelines and valves and assume that the compression processes for both the higher and lower systems are isentropic.

Based on data from the refrigerant table of thermodynamic properties for HFC–125, 1991 provided by DuPont Chemicals, the following formulas can be used to calculate the enthalpies of the

refrigerants: Enthalpy of refrigerant HFC–125 at saturated liquid state at temperature T_{sl} between 0 and 50°F h_{sl}, in Btu/lb, is

$$h_{ls} = 11.22 + 0.302 T_{sl}$$

Enthalpy of refrigerant HFC–125 at saturated vapor state at temperature T_{sv} between -80 and $-30°F$ h_{vs}, in Btu/lb, is

$$h_{vs} = 63.27 + 0.143 [T_{sv} - (-80)]$$

Enthalpy difference of vapor refrigerant HFC–125 along the constant entropy line when $T_{con} < 30°F$ and $T_{ev} > -80°F$ is

$$\Delta h_{v.s} = 0.16 (T_{con} - T_{ev})$$

Here T_{con} represents the condensing temperature, and T_{ev} the evaporating temperature, both in °F. The specific heat of saturated vapor of HFC–125 c_{pv} in the temperature range of -80 to $-30°F$ is 0.143 Btu/lb · °F. The specific heat of saturated liquid of HFC–125 c_{pl} in the temperature range of 0 to 50°F is 0.302 Btu/lb · °F.

Calculate

1. The refrigerating effect per lb of refrigerant in the lower system
2. Total work input to the compressors
3. Coefficient of performance of this cascade system
4. Mass flow rates of the refrigerants in the higher and lower systems

Solution

1. From the given values, the enthalpy of saturated vapor refrigerant at $-60°F$ is

$$h_1 = 63.27 + 0.143 [T_{sv} - (-80)]$$

$$63.27 + 0.143 [-60 - (-80)] = 66.13 \text{ Btu/lb}$$

Enthalpy of saturated liquid refrigerant at a condensing temperature of 20°F is:

$$h_3 = 11.22 + 0.302 T_{sl}$$
$$= 11.22 + (0.302 \times 20)$$
$$= 17.26 \text{ Btu/lb}$$

and from Eq. (4.26),
$$h_{3'} = h_4 = h_3 - c_{pl}(T_3 - T_{3'})$$
$$= 17.26 - 0.302(20 - 1) = 11.52 \text{ Btu/lb}$$

Then, the refrigerating effect of the lower system is

$$q_{rf.1} = h_1 - h_4 = 66.13 - 11.52 = 54.61 \text{ Btu/lb}$$

2. Enthalpy of superheated vapor refrigerant after the heat exchanger in the lower system is

$$h_{1'} = h_1 + c_{pv} \text{ (degree of superheat)}$$
$$= 66.13 + 0.143 [-20 - (-60)]$$
$$= 71.85 \text{ Btu/lb}$$

Enthalpy difference $(h_2 - h_{1'})$ along the constant entropy line of the lower system is

$$(h_2 - h_{1'}) = 0.16 (T_{con} - T_{lv})$$

$$= 0.16 [20 - (-60)]$$

$$= 12.80 \text{ Btu/lb}$$

Enthalpy of the hot gas discharged from the compressor of the lower system is

$$h_2 = h_{1'} + (h_2 - h_{1'})$$

$$= 71.85 + 12.80$$

$$= 84.65 \text{ Btu/lb}$$

From Eq. (4.20), the enthalpy of saturated vapor refrigerant HCFC–22 in the higher system is

$$h_5 = 104.465 + 0.098445 T_s - 0.0001226(T_s^2)$$

$$- [9.861 \times 10^{-7}(T_s^3)]$$

$$= 104.465 + 0.098445(10) - 0.0001226(10^2)$$

$$- [9.861 \times 10^{-7}(10^3)]$$

$$= 104.465 + 0.984 - 0.012 - 0.001$$

$$= 105.436 \text{ Btu/lb}$$

From Eq (4.22), the enthalpy difference along the constant entropy line of the higher system is

$$(h_6 - h_5) = -0.18165 + 0.21502 (T_{s2} - T_{s1})$$

$$- 0.0012405 (T_{s2} - T_{s1})^2 + 8.1982$$

$$\times 10^{-6} (T_{s2} - T_{s1})^3$$

$$= -0.18165 + 0.21502(100 - 10)$$

$$- 0.0012405(100 - 10)^2 + 8.1982$$

$$\times 10^{-6}(100 - 10)^3 = 15.09 \text{ Btu/lb}$$

According to Eq. (4.20), the enthalpy of liquid refrigerant HCFC–22 at condensing temperature 100°F in the higher system is

$$h_7 = h_8 = 10.409 + 0.26851 T_s$$

$$+ 0.00014794(T_s^2)$$

$$+ [5.3429 \times 10^{-7}(T_s^3)]$$

$$= 10.409 + 0.26851(100) + 0.00014794(100^2)$$

$$+ [5.3429 \times 10^{-7}(100^3)] = 39.27 \text{ Btu/lb}$$

From Eq. (4.51), the ratio of the mass flow rate of refrigerant in the higher system to that of the lower system is

$$\dot{m}_h / \dot{m}_l = \frac{h_2 - h_3}{h_5 - h_8}$$

$$= \frac{84.65 - 17.26}{105.44 - 39.27}$$

$$= 1.02$$

Then, from Eq. (4.53), the total work input to the compressors is

$$W_{in} = (h_2 - h_{1'}) + \frac{m_h(h_6 - h_5)}{m_l}$$

$$= 12.80 + 1.02(15.09) = 28.19 \text{ Btu/lb}$$

3. Coefficient of the performance of the cascade system is

$$\text{COP}_{ref} = \frac{q_{rf}}{W_{in}} = \frac{54.61}{28.19} = 1.94$$

4. Mass flow rate of refrigerant in the lower system is

$$\dot{m}_l = \frac{q_{rc.l}}{q_{rf.l}} = \frac{100,000}{60 \times 54.61} = 30.52 \text{ lb/min}$$

The mass flow rate of refrigerant in the higher system is

$$\dot{m}_h = 1.02$$

$$\dot{m}_l = 1.02 \times 30.52$$

$$= 31.13 \text{ lb/min}$$

4.16   AIR EXPANSION REFRIGERATION CYCLES

Thermodynamic Principle

According to the steady flow energy equation

$$h_1 + \frac{v_1^2}{(2g_c \times 778)} + q_\# = h_2 + \frac{v_2^2}{(2g_c \times 778)} + W$$

For an adiabatic process, $q_\# = 0$. Assuming that the difference between the kinetic energy $v_1^2/2g_c$ and $v_2^2/2g_c$ is small compared with the enthalpy difference $(h_1 - h_2)$, it can be ignored. For air expanding in a turbine and producing work, the above equation can be rewritten as

$$W = h_1 - h_2$$

Usually air can be considered a perfect gas, so

$$\Delta h = c_p \Delta T_R \qquad (4.55)$$

where c_p = specific heat of air at constant pressure, Btu/lb · °R

T_R = absolute temperature, °R

and

$$W = c_p(T_{R1} - T_{R2}) \qquad (4.56)$$

When the expansion process is a reversible adiabatic process, the temperature–pressure relationship can be expressed as

$$\frac{T_{R1}}{T_{R2}} = \left(\frac{p_1}{p_2}\right)^{(\gamma - 1)/\gamma} \qquad (4.57)$$

where p_1, p_2 = air pressure at the inlet and outlet of the turbine, psia

γ = ratio of specific heat at constant pressure to specific heat at constant volume

Equations (4.56) and (4.57) demonstrate that when air expands in a turbine and produces work, its absolute temperature T_{R1} drops to T_{R2}. The decrease in absolute temperature is proportional to the pressure ratio $(p_1/p_2)^{(\gamma-1)/\gamma}$ and work produced W. The greater the pressure drop in the turbine, the lower the temperature at the outlet of the turbine. This type of refrigeration has been used for air conditioning systems in many aircraft.

Flow Processes of a Basic Air Expansion Refrigeration System for Aircraft

Figure 4.15a is a schematic diagram of a basic air expansion refrigeration system and Figure 4.15b depicts the air expansion refrigeration cycle.

In Fig. 4.15, ambient air that surrounds an aircraft flying at subsonic speed and at high altitude, at point 0 with pressure p_0, is rammed into the engine scoop. Before entering the engine, the pressure of the air at point 1' is increased to p_1. When ram air enters the engine compressor, it is compressed to pressure p_2 and bleeds to a heat exchanger at point 2'. In the heat exchanger, the bleed air is cooled at constant pressure to point 3 by another stream of ram air extracted by a fan. Bleed air is then expanded in a turbine and cooled to point 4'. This cold air is supplied to the cabin and flight deck for air conditioning. Cabin air is then discharged to the atmosphere after absorbing the sensible and latent heat from these conditioned spaces. Air expansion refrigeration cycles for aircraft are open cycles because there is no recirculated air.

An alternative is to add an evaporative cooler between section A-A and B-B, as shown in Fig. 4.15, to cool the supply air a few degrees more. An evaporant such as water or alcohol can be used to extract heat from the supply air by indirect contact.

The stream of ram air, extracted by a fan driven by the cooling turbine to cool the bleed air is also discharged to the atmosphere at the cooling air exit. The work output of the turbine is mainly used to drive the fan.

Air Expansion Refrigeration Cycle

The air expansion refrigeration cycle is different from a vapor compression cycle because the refrigerant, air, remains in gaseous form, is not recirculated, and is used directly as the supply air for cooling the conditioned spaces.

Work input to the engine compressor W_c, in Btu/lb, is given by:

$$W_c = c_p(T_{R2'} - T_{R1'}) \qquad (4.58)$$

where $T_{R1'}$, $T_{R2'}$ = absolute temperature of the ram air and air after the compression process respectively, °R.

The rate of heat transfer in the heat exchanger between the bleed air and the ambient air for cooling q_{ex}, in Btu/h, can be calculated as

$$q_{ex} = 60\,\dot{m}_b c_p(T_{R2'} - T_{R3}) \qquad (4.59)$$

where T_{R3} = absolute temperature of air after heat exchanger, °R

$\dot{m}_b$ = mass flow rate of bleed air, lb/min

Work output from the turbine W_o during the expansion process, in Btu/lb, can be calculated as

$$W_o = c_p(T_{R3} - T_{R4'}) \qquad (4.60)$$

where $T_{R4'}$ = absolute temperature of the cold supply air at the exit of the turbine, °R.

The amount of sensible heat q_{sen} absorbed by the cold supply air in the cabin or in the flight deck, in Btu/h, is

$$q_{sen} = 60\,\dot{m}_b c_p(T_{Rr} - T_{R4'}) \qquad (4.61)$$

where T_{Rr} = absolute temperature of the air cabin, °R.

The relationship between temperature and pressure for an isentropic compression process is

$$\frac{T_{R2}}{T_{R1'}} = \left(\frac{p_2}{p_1}\right)^{(\gamma-1)/\gamma} \qquad (4.62)$$

For an isentropic expansion process, the relationship between temperature and pressure is

$$\frac{T_{R3}}{T_{R4}} = \left(\frac{p_3}{p_4}\right)^{(\gamma-1)/\gamma} \qquad (4.63)$$

where T_{R2}, T_{R4} = absolute temperature of air after isentropic compression and isentropic expansion precesses, °R. The isentropic efficiency for the compressor η_c is

$$\eta_c = \frac{c_p(T_{R2} - T_{R1'})}{c_p(T_{R2'} - T_{R1'})}$$

$$= \frac{(T_{R2} - T_{R1'})}{(T_{R2'} - T_{R1'})} \qquad (4.64)$$

Similarly, isentropic efficiency for the turbine η_t is

$$\eta_t = \frac{T_{R3} - T_{R4'}}{T_{R3} - T_{R4}} \qquad (4.65)$$

If η_c, η_t, $T_{R1'}$, T_{R3}, and pressure ratios p_2/p_1 and p_3/p_4 are known values, $T_{R2'}$ and $T_{R4'}$ can be determined by using Eqs. (4.62) to (4.65).

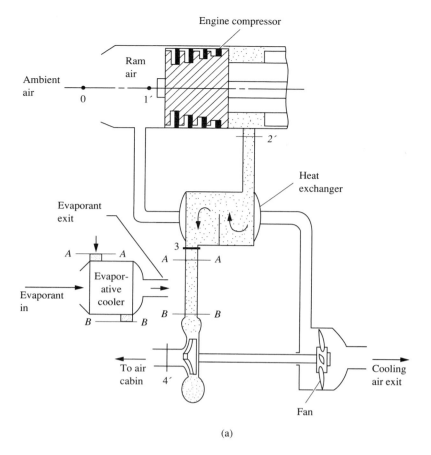

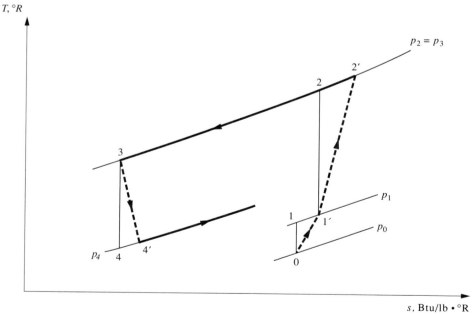

FIGURE 4.15 Air expansion refrigeration system and refrigeration cycle: (a) refrigeration system and (b) air expansion refrigeration cycle.

Effectiveness of the heat exchanger ϵ can be calculated as:

$$
\begin{aligned}
\epsilon &= \frac{c_p \dot{m}_b (T_{R2'} - T_{R3})}{c_p \dot{m}_b (T_{R2'} - T_{Rt1})} \\
&= \frac{T_{R2'} - T_{R3}}{T_{R2'} - T_{Rt1}}
\end{aligned}
\qquad (4.66)
$$

where $T_{R2'}, T_{R3}$ = absolute temperature of bleed air entering and leaving the heat exchanger, °R

T_{Rt1} = absolute total temperature of the cooling ram air entering the heat exchanger, measured at the outer heat transfer surface at rest, °R

From Eq. (4.66), ϵ depends mainly on the construction of the heat exchanger and the ratio $\dot{m}_b / \dot{m}_c$, if the difference of c_p between bleed and cooling ram air is ignored. Here, $\dot{m}_c$ indicates the mass flow rate of cooling ram air, lb/min.

If air at a velocity v with an absolute temperature T_R finally reaches a total temperature T_{Rt} when brought to rest adiabatically, then, based on the steady-flow energy equation,

$$
c_p T_R + \frac{v^2}{2g_c \times 778} = c_p T_{Rt}
$$

or,

$$
T_{Rt} = T_R + \frac{v^2}{(2c_p)(778 g_c)} \qquad (4.67)
$$

When air velocity $v < 200$ ft/s, term $v^2 / [(2c_p)(778 g_c)]$ is small compared with T_R, and can be ignored. Hence, only T_{Rt} is be used for cooling ram air.

Net work input to the air expansion refrigeration cycle is calculated as

$$
W_{net} = W_c - W_t = c_p (T_{R2'} - T_{R1'}) - c_p (T_{R3} - T_{R4'}) \qquad (4.68)
$$

If W_t is used to drive the fan to extract the cooling air, then the coefficient of performance of the basic air expansion refrigeration cycle can be calculated as

$$
\begin{aligned}
COP_{ref} &= \frac{c_p (T_{Rr} - T_{R4'})}{c_p (T_{R2'} - T_{R1'})} \\
&= \frac{(T_{Rr} - T_{R4'})}{(T_{R2'} - T_{R1'})}
\end{aligned}
\qquad (4.69)
$$

The main advantage of an air expansion refrigerating system for an aircraft is its simple construction and light weight. The main drawbacks are the lower COP_{ref} and the need for an additional air conditioning and refrigerating system when the aircraft is on the ground.

REFERENCES AND FURTHER READING

ASHRAE, *ASHRAE Handbook 1988, Equipment,* ASHRAE Inc., Atlanta, GA, 1988.

ASHRAE, *ASHRAE Handbook 1989, Fundamentals,* ASHRAE Inc., Atlanta, GA, 1989.

ASHRAE, *ASHRAE Handbook 1990, Refrigeration Systems and Applications,* ASHRAE Inc., Atlanta, GA, 1990.

ASHRAE, *ASHRAE Standard 15–1992, Safety Code for Mechanical Refrigeration,* ASHRAE Inc., Atlanta, GA, 1992.

ASHRAE, *ANSI/ASHRAE Standard 34–1992, Number Designation and Safety Classification of Refrigerants,* ASHRAE, Inc., Atlanta, GA, 1992.

Denny, R. J., "The CFC Foot Print," *ASHRAE Journal,* pp. 24–28, November 1987.

Dossat, R. J., *Principles of Refrigeration,* John Wiley, 2nd ed., New York, 1978.

DuPont Chemicals, *HFC–125 Thermodynamic Properties,* DuPont Chemicals, Wilmington, Delaware, 1991.

King, G. R., *Modern Refrigeration Practice,* McGraw Hill, 1st ed., New York, 1971.

McGuire, J., "An Industry At Risk, User Concerns," *ASHRAE Journal,* pp. 48–49, August 1988.

Parker, R. W., "Reclaiming Refrigerant in OEM Plants," *ASHRAE Transactions,* Part II, pp. 2139–2144, 1988.

Rowland, S., "The CFC Controversy: Issues and Answers," *ASHRAE Journal,* pp. 20–27, December 1992.

Stamm, R. H., "The CFC Problem: Bigger Than You Think," *Heating/Piping/Air Conditioning,* pp. 51–54, April 1989.

Swers, R., Patel, Y. P., and Stewart, R. B., "Thermodynamics Analysis of Compression Refrigeration Systems," ASHRAE Seminar, New Orleans, LA, January, 1972.

Tan, L. C., and Yin, J. M., "The Exergy–Enthalpy Diagram and Table of R–22 and Their Application," *ASHRAE Transactions,* Part I A, pp. 220–230, 1986.

Wang, S. K., *Principles of Refrigeration,* Vol. I, Hong Kong Polytechnic, 1st ed., Hong Kong, 1984.

Wilson, D. P., and Basu, R. S., "Thermodynamic Properties of a New Stratospherically Safe Working Fluid—Refrigerant 134A," *ASHRAE Transactions,* Part II, pp. 2095–2118, 1988.

INDOOR DESIGN CONDITIONS AND INDOOR AIR QUALITY

5.1 INDOOR PARAMETERS AND DESIGN CONDITIONS

Indoor parameters are those that the air conditioning system influences directly to produce a required conditioned indoor environment in buildings. They include the following:

1. Indoor air temperature and mean radiant temperature
2. Indoor relative humidity
3. Outdoor ventilation rate provided
4. Air cleanliness
5. Air movement
6. Sound level
7. Pressure differential between the space and surroundings

The indoor parameters to be maintained in an air conditioned space are specified in the design document and also are the targets to be achieved during operation.

When specifying the indoor design parameters, the following points need to be considered:

1. Not all of the parameters already mentioned need to be specified in every design project. It is necessary to specify only the parameters that are essential to the particular situation concerned.
2. Even for process air conditioning systems, the thermal comfort of the workers should also be considered. Therefore, indoor design parameters regarding health and thermal comfort for the occupants form the basis of specified design criteria.
3. When one is specifying indoor design parameters, economic strategies of initial investment and energy consumption of the HVAC&R systems must be carefully investigated.

    Design criteria should not be set too high or too low. If the design criteria are too high, the result will be an excessively high investment and energy cost. Design criteria that are too low may produce a poor indoor air quality, resulting in complaints from the occupants, causing low-quality products, and possibly leading to expensive system alterations.
4. Each specified indoor design parameter is usually associated with a tolerance, indicated as a ± sign or as an upper or lower limit. Sometimes, there is a traditional tolerance understood by both the designers and the owners of the building. For instance, although the summer indoor design temperature of a comfort air conditioning system is specified at 75°F or 78°F, in practice a tolerance of ±2–3°F is often considered acceptable.

5. In process air conditioning systems, sometimes a stable indoor environment is more important than the absolute value of the indoor parameter to be maintained. For example, it may not be necessary to maintain a temperature of 68°F (20°C) for all areas in precision machinery manufacturing. More often, a 72°F or an even higher indoor temperature with appropriate tolerance will be more suitable and economical.

5.2 HEAT EXCHANGE BETWEEN THE HUMAN BODY AND THE INDOOR ENVIRONMENT

Two-Node Model of Thermal Interaction

In 1971, Gagge et al. recommended a two-node model of human thermal interaction. In this model, the human body is composed of two compartments: an inner body core including skeleton, muscle and internal organs, and an outer shell of skin surface. The temperatures of the body core and the surface skin are each assumed to be uniform and independent. Metabolic heat production, external mechanical work, and respiratory losses occur only in the body core. Heat exchange between the body core and the skin surface depends on heat conduction from direct contact and the peripheral blood flow of the thermoregulatory mechanism of the human body.

Steady-State Thermal Equilibrium

When the human body is maintained at a steady-state thermal equilibrium, that is, the heat storage at the body core and skin surface is approximately equal to zero, then the heat exchange between the human body and the indoor environment can be expressed by the following heat balance equation

$$M = W + (C + R) + E_{sk} + E_{res} \qquad (5.1)$$

where M = metabolic rate, Btu/h · ft²
 W = mechanical work performed, Btu/h · ft²
 $C + R$ = convective and radiative, or sensible heat loss from the skin surface, Btu/h · ft²
 E_{sk} = evaporative heat loss from the skin surface, Btu/h · ft²
 E_{res} = evaporative heat loss from respiration, Btu/h · ft²

In Eq. (5.1), the ft^2 in the unit $Btu/h · ft^2$ applies to skin surface area. The skin surface area of a naked human body can be approximated by an empirical formula proposed by Dubois in 1916

$$A_D = 0.657 m_b^{0.425} H_b^{0.725} \qquad (5.2)$$

where A_D = Dubois surface area of naked body, ft^2
$\quad\quad m_b$ = mass of the human body, lb
$\quad\quad H_b$ = height of the human body, ft

In an air conditioned space, a steady-state thermal equilibrium is usually maintained between the human body and the indoor environment.

Transient Energy Balance

When there is a *transient energy balance* between the human body and the indoor environment, the thermal interaction of the body core, skin surface, and the indoor environment forms a rate of positive or negative heat storage both in the body core and on the skin surface.

The human body needs energy for physical and mental activity. This energy comes from the oxidation of the food taken into the human body. The heat released from this oxidation process is called *metabolic heat*. It dissipates from the skin surface of the human body into the surroundings.

In a cold environment, the thermoregulatory mechanism reduces the rate of peripheral blood circulation, lowering the temperature of the skin and preventing any greater heat loss from the human body. However, if the heat loss and the mechanical work performed are greater than the rate of the metabolic heat produced, then the temperatures of both the body core and the skin surface fall, and shivering or other spontaneous activities occur to increase the production of heat energy within the human body.

On the other hand, in a hot environment, if a large amount of heat energy needs to be dissipated from the human body, the physiological control mechanism increases the blood flow to the skin surface. This raises the skin temperature. If the heat produced is still greater than the heat actually dissipated and the temperature of the body core increases from its normal temperature of about 97.6°F to about 98.6°F, then liquid water is released from the sweat glands for evaporative cooling.

For a transient state of energy balance between the human body and the indoor environment, the rate of heat storage in the body core S_{cr} and the skin surface S_{sk}, both in Btu/h · ft^2, can be calculated as

$$S_{cr} + S_{sk} = M - W - (C + R) - E_{sk} - E_{res} \quad (5.3)$$

5.3 METABOLIC RATE AND SENSIBLE HEAT LOSSES FROM THE HUMAN BODY

Metabolic Rate

The *metabolic rate M* is the rate of energy release per unit area of skin surface as a result of the oxidative processes in the living cells. Metabolic rate depends mainly on the intensity of the physical activities performed by the human body. The unit of metabolic rate is called *met*. One met is defined as 18.46 Btu/h · ft^2 of metabolic heat produced in the body core. In Table 5.1 are listed the metabolic rates of various activities.

Mechanical Work

Some of the energy released from the oxidative processes within the body core can be partly transformed into external mechanical work through the action of the muscles. Mechanical work W is usually expressed as a fraction of the metabolic rate and can be calculated as

$$W = \mu M \quad (5.4)$$

where μ = mechanical efficiency. For most office work, mechanical efficiency μ is less than 0.05. Only when there is a large amount of physical activity such as bicycling, lifting and carrying, or walking on a slope may μ increase to a value of 0.2 to 0.24.

TABLE 5.1
Metabolic rate for various activities

	Metabolic rate	
Activity level	**Met**	**Btu/h · ft²**
Resting		
Sleeping	0.7	13
Seated, quiet	1.0	18
Office work		
Reading, seated	1.0	18
Typing	1.1	20
Teaching	1.6	30
Domestic work		
Cooking	1.6–2.0	29–37
House cleaning	2.0–3.4	37–63
Walking		
Speed 2mph	2.0	37
4mph	3.8	70
Machine work		
Light	2.0–2.4	37–44
Heavy	4.0	74
Dancing, social	2.4–4.4	44–81
Sports		
Tennis, singles	3.6–4.0	66–74
Basketball	5.0–7.6	90–140
Wrestling	7.0–8.7	130–160

Adapted with permission from *ASHRAE Handbook 1989, Fundamentals.*

Sensible Heat Exchange

Sensible heat loss or, occasionally, sensible heat gain $(R + C)$ represents the heat exchange between the human body and the indoor environment through convective and radiative heat transfer. In Fig. 5.1 is shown the sensible heat exchange between the human body and the environment. The combined convective and radiative heat transfer can be calculated as

$$C + R = f_{\text{cl}} h_c (T_{\text{cl}} - T_a) + f_{\text{cl}} h_r (T_{\text{cl}} - T_{\text{rad}})$$
$$= f_{\text{cl}} h (T_{\text{cl}} - T_o) \tag{5.5}$$

where T_{cl} = mean surface temperature of clothing, °F.

The operative temperature T_o is defined as the weighted average of the mean radiant temperature T_{rad} and indoor air temperature T_a, both in °F, that is,

$$T_o = \frac{h_r T_{\text{rad}} + h_c T_a}{h_r + h_c} \tag{5.6}$$

The surface heat transfer coefficient is defined as

$$h = h_c + h_r \tag{5.7}$$

and the ratio of the clothed surface area to the naked surface area is

$$f_{\text{cl}} = \frac{A_{\text{cl}}}{A_D} \tag{5.8}$$

where h_c, h_r = convective and radiative heat transfer coefficients, Btu/h · ft^2 · °F

A_{cl} = surface area of the clothed body, ft^2

Mean radiant temperature T_{rad} will be discussed in more detail in later sections.

According to Seppanen et al. (1972), the convective heat transfer coefficient h_c for a person standing in moving air, when the air velocity is $30 \leq v \leq 300$ fpm, is

$$h_c = 0.0681 v^{0.69} \tag{5.9}$$

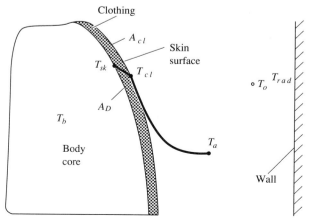

FIGURE 5.1 Sensible heat exchange between the human body and the indoor environment.

When the air velocity $v < 30$ fpm, $h_c = 0.7$ Btu/h · ft^2 · °F.

For typical indoor temperatures and a clothing emissivity nearly equal to unity, the linearized radiative heat transfer coefficient $h_r = 0.83$ Btu/h · ft^2 · °F.

Let R_{cl} be the thermal resistance of clothing, in h · ft^2 · °F/Btu. Then

$$C + R = \frac{T_{\text{sk}} - T_{\text{cl}}}{R_{\text{cl}}} \tag{5.10}$$

where T_{sk} = mean skin surface temperature, °F. If the human body is able to maintain a thermal equilibrium with very little evaporative loss from the skin surface, then the skin temperature T_{sk} will be around 93°F.

Combining Eqs. (5.5) and (5.10), and eliminating T_{cl}, we find

$$C + R = \frac{f_{\text{cl}} h}{R_{\text{cl}} f_{\text{cl}} h + 1} (T_{\text{sk}} - T_o)$$
$$= F_{\text{cl}} f_{\text{cl}} h (T_{\text{sk}} - T_o) \tag{5.11}$$

In this equation the dimensionless clothing efficiency F_{cl} is defined as

$$F_{\text{cl}} = \frac{1}{R_{\text{cl}} f_{\text{cl}} h + 1} = \frac{T_{\text{cl}} - T_o}{T_{\text{sk}} - T_o} \tag{5.12}$$

In Eq. (5.11), if $T_o > T_{\text{sk}}$, then $(C + R)$ could be negative, that is, there could be a sensible heat gain.

Clothing Insulation

Clothing insulation R_{cl} can be determined through measurements on a heated manikin, a model of the human body for laboratory experiments. After $(C + R)$ is measured from the thermal manikin in a controlled indoor environment, R_{cl} can be calculated from Eq. (5.11) since f_{cl}, T_{sk}, T_o and h are also known values. Clothing insulation R_{cl} can be expressed either in h · ft^2 · °F/Btu or in a new unit called *clo*, where 1 *clo* = 0.88 h · ft^2 · °F/Btu.

Clothing insulation (R_{cl}) values and area ratios (f_{cl}) for typical clothing ensembles, taken from McCullough and Jones (1984), are listed in Table 5.2.

5.4 EVAPORATIVE HEAT LOSSES

Evaporative heat loss E is heat loss due to the evaporation of sweat from the skin surface E_{sk} and respiration losses E_{res}. Actually, metabolic heat is mainly dissipated to the indoor air through the evaporation of sweat when the indoor air temperature is nearly equal to the skin temperature.

TABLE 5.2
Insulation values for clothing ensembles*

Ensemble description[†]	R_{cl}, clo	R_{cl}, $\frac{h \cdot ft^2 \cdot °F}{Btu}$	f_{cl}
Walking shorts, short-sleeve shirt	0.41	0.36	1.11
Fitted trousers, short-sleeve shirt	0.50	0.44	1.14
Fitted trousers, long-sleeve shirt	0.62	0.55	1.19
Same as above, plus suit jacket	0.96	0.85	1.23
Loose trousers, long-sleeve shirt, long-sleeve sweater, T-shirt	1.01	0.89	1.28
Sweat pants, sweat shirt	0.77	0.68	1.19
Knee-length skirt, short-sleeve shirt, panty hose (no socks), sandals	0.54	0.48	1.26
Knee-length skirt, long-sleeve shirt, full slip, panty hose (no socks)	0.67	0.59	1.29
Long-sleeve coveralls, T-shirt	0.72	0.63	1.23
Overalls, long-sleeve shirts, long underwear tops and bottoms, flannel long-sleeve shirt	1.00	0.88	1.28

*For mean radiant temperature equal to an air temperature and air velocity less than 40 fpm.

[†]Unless otherwise noted, all ensembles included briefs or panties, shoes, and socks.

Source: Adapted from McCollough and Jones (1984). Reprinted with permission.

Respiration Losses

During respiration, there is convective heat loss C_{res} that results from the temperature of the inhaled air being increased to the exhaled air temperature, or about 93°F. There is also a latent heat loss L_{res} due to evaporation of liquid water into water vapor inside the body core. The amount of respiration losses depends mainly on the metabolic rate. In summer, at an indoor temperature of 75°F and a relative humidity of 50 percent, respiratory losses $E_{res} = (C_{res} + L_{res})$ are approximately equal to 0.09 M. In winter, E_{res} is slightly greater. For simplicity, let $E_{res} = 0.1$ M.

Evaporative Heat Loss from the Skin Surface

Evaporative heat loss from the skin surface E_{sk} consists of (1) the evaporation of sweat as a result of thermoregulatory mechanisms of the human body E_{rsw} and (2) the direct diffusion of liquid water from the skin surface E_{dif}. Evaporative heat loss due to regulatory sweating E_{rsw}, in Btu/h · ft², is directly proportional to the mass of the sweat produced, that is,

$$E_{rsw} = \dot{m}_{rsw} h_{fg} \qquad (5.13)$$

where $\dot{m}_{rsw}$ = mass flow rate of sweat produced, lb/h · ft²
h_{fg} = latent heat of vaporization at 93°F, Btu/lb

Maximum Evaporative Heat Loss Due to Regulatory Sweating

The wetted portion of the human body needed for the evaporation of a given quantity of sweat w_{rsw} is

$$w_{rsw} = E_{rsw} / E_{max} \qquad (5.14)$$

In Eq. (5.14), E_{max} represents the maximum evaporative heat loss due to regulatory sweating when the skin surface of the human body is entirely wet. Its magnitude is directly proportional to the vapor pressure difference between the wetted skin surface and the vapor pressure of the indoor ambient air, and can be calculated as

$$E_{max} = h_{e.c}(p_{sk.s} - p_a) \qquad (5.15)$$

where $p_{sk.s}$ = saturated water vapor pressure at the skin surface temperature, psia
p_a = water vapor pressure of the ambient air, psia
$h_{e.c}$ = overall evaporative heat transfer coefficient of a clothed body, in Btu/h · ft² · psi

Woodcock (1962) proposed the following relationship between $h_{e.c}$ and h_s, the overall sensible heat transfer coefficient, in Btu/h · ft² · psi:

$$i_m LR = h_{e.c} / h_s \qquad (5.16)$$

The *moisture permeability index* i_m denotes the moisture permeability of clothing and is dimensionless. Clothing ensembles worn indoors usually have an $i_m = 0.3 - 0.5$. The moisture permeability indexes i_m of some clothing ensembles are presented in Table 5.3.

TABLE 5.3
Moisture permeability of clothing emsembles*

Ensemble description	i_m
Cotton/polyester long-sleeve shirt, long trousers, street shoes, socks, briefs	0.385
Cotton short-sleeve shirt, long trousers, work boots, socks, briefs, cotton gloves	0.41
Cotton/nylon long-sleeve shirt, cotton/nylon trousers, combat boots, socks, helmet liner (Army Battle Dress Uniform)	0.36

*Measured with $T_{rad} = T_a$, and air velocity = 40 fpm.

Adapted with permission from *ASHRAE Handbook 1989, Fundamentals*.

The Lewis relation LR in Eq. (5.16) relates the evaporative heat transfer coefficient h_e and the convective heat transfer coefficient h_c, both in Btu/h · ft^2 · °F. LR $= f(h_e/h_c)$ has a magnitude of 205°F/psi.

In Eq. (5.16), the overall sensible heat transfer coefficient h_s can be calculated as

$$h_s = \frac{1}{R_t} = \frac{f_{cl}h}{f_{cl}hR_{cl} + 1} \qquad (5.17)$$

where R_t = total resistance to sensible heat transfer between the skin and the indoor environment, h · ft^2 · °F/Btu.

Diffusion Evaporative Heat Loss and Total Skin Wettedness

The minimum level of evaporative heat loss from the skin surface occurs when there is no regulatory sweating and the skin wettedness due to direct diffusion $E_{df.min}$, in Btu/h · ft^2, is approximately equal to 0.06 E_{max} under normal conditions, or

$$E_{df.min} = 0.06E_{max} \qquad (5.18)$$

When there is a heat loss from regulatory sweating E_{rsw}, the diffusion evaporative heat loss E_{dif}, in Btu/h · ft^2, for the portion of skin surface that is not covered with sweat can be calculated as

$$E_{dif} = 0.06\,E_{max}(1 - w_{rsw}) \qquad (5.19)$$

Therefore, the total evaporative heat loss from the skin surface is

$$\begin{aligned}
E_{sk} &= E_{rsw} + E_{dif} \\
&= w_{rsw}E_{max} + 0.06\,E_{max}(1 - w_{rsw}) \\
&= (0.06 + 0.94\,w_{rsw})E_{max} \\
&= w_{sk}E_{max} \qquad (5.20)
\end{aligned}$$

In Eq. (5.20), w_{sk} is called *total skin wettedness*; it is dimensionless, and it can be calculated as

$$w_{sk} = \frac{E_{sk}}{E_{max}} \qquad (5.21)$$

5.5 MEAN RADIANT TEMPERATURE AND EFFECTIVE TEMPERATURE

Mean Radiant Temperature

Mean radiant temperature T_{Rad} is defined as the temperature of a uniform black enclosure in which an occupant would have the same amount of radiative heat exchange as in an actual indoor environment. Mean radiant temperature T_{Rad}, in °R, can be calculated by the expression

$$T_{Rad}^4 = T_{R1}^4 F_{0-1} + T_{R2}^4 F_{0-2} + \cdots + T_{Rn}^4 F_{0-n} \qquad (5.22)$$

where $T_{R1}, T_{R2}, \ldots, T_{Rn}$ = absolute temperature of the surrounding surfaces of the indoor environment, °R

F_{0-1} = shape factor denoting the fraction of the total radiant energy leaving the surface of the occupant's clothing 0 and arriving on the surface 1

F_{0-2} = the fraction of the total radiant energy leaving surface 0 and arriving on surface 2, etc.

Shape factors F_{0-n} depend on the position and orientation of the occupant as well as the dimensions of the enclosure. One can use Figs. 5.2 and 5.3 to estimate the mean value of the shape factor between a seated person and rectangular surfaces. The sum of the shape factors of all the surfaces with respect to the seated occupant in an enclosure is unity.

The temperature measured by a globe thermometer, called *globe temperature*, is often used to estimate the mean radiant temperature. The globe thermometer consists of a hollow copper sphere of 6-in. diameter that is coated with black paint on the outer surface. A precision thermometer or thermocouple is inserted inside the globe with the sensing bulb or the thermojunction located at the center of the sphere.

Because the net radiant heat received at the globe surface is balanced by the convective heat transfer from the globe surface in reaching a thermal equilibrium, such a relationship gives

$$T_{Rad}^4 = T_{Rg}^4 + 1.03 \times 10^8 v^{0.5}(T_{Rg} - T_{Ra}) \qquad (5.22)$$

where T_{Rad} = absolute mean radiant temperature, °R
T_{Rg} = absolute globe temperature, °R
v = ambient air velocity, fpm
T_{Ra} = absolute air temperature, °R

After T_{Rg}, v, and T_{Ra} are measured, mean radiant temperature T_{Rad} can be calculated from Eq. (5.22).

Mean radiant temperature indicates the effect, due to the radiant energy from the surroundings, on radiant exchange between an occupant or any substance and the enclosure. Such an influence may be significant if the mean radiant temperature is several degrees higher than the temperature of the indoor air.

Example 5.1. The dimensions of a private office and the location of a person seated within it are shown in Fig. 5.4. The surface temperatures of the enclosure are as follows:

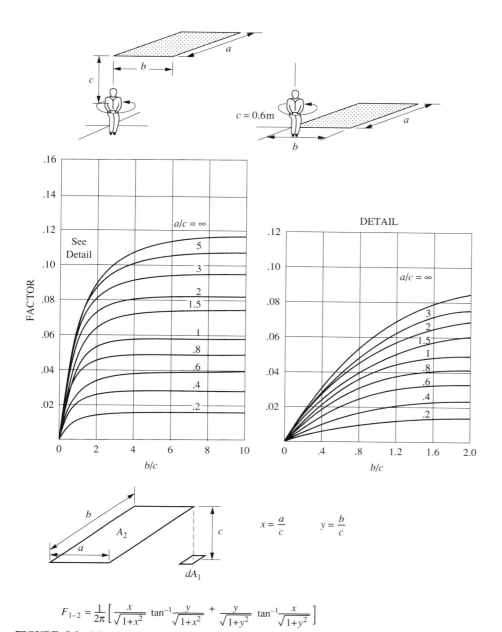

$$F_{1-2} = \frac{1}{2\pi}\left[\frac{x}{\sqrt{1+x^2}} \tan^{-1}\frac{y}{\sqrt{1+x^2}} + \frac{y}{\sqrt{1+y^2}} \tan^{-1}\frac{x}{\sqrt{1+y^2}} \right]$$

FIGURE 5.2 Mean value of shape factor between a sedentary person and a horizontal plane. (*Source*: P. O. Fanger, *Thermal Comfort Analysis and Applications in Environmental Engineering,* 1972. Reprinted with permission.)

West window	88°F
West wall	80°F
North partition wall	75°F
East partition wall	75°F
South partition wall	75°F
Floor	78°F
Ceiling	77°F

Calculate the mean radiant temperature of the enclosure that surrounds this office. The orientation of the seated occupant is unknown.

Solution

1. Regarding the north partition wall, the shape factor denotes the fraction of the total radiant energy that leaves the outer surface of the clothing of the occupant (surface 0) and arrives directly on the north partition wall (surfaces 1, 2, 3, and 4) and is given by

$$F_{0-1,2,3,4} = F_{0-1} + F_{0-2} + F_{0-3} + F_{0-4}$$

Here, F_{0-1} is the shape factor for the radiation from surface 0 to surface 1 of the north partition wall.

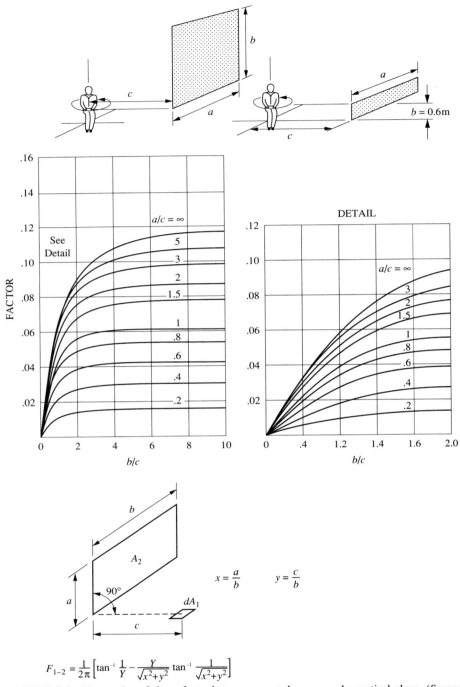

$$F_{1-2} = \frac{1}{2\pi}\left[\tan^{-1}\frac{1}{Y} - \frac{Y}{\sqrt{x^2+y^2}}\tan^{-1}\frac{1}{\sqrt{x^2+y^2}}\right]$$

FIGURE 5.3 Mean value of shape factor between a seated person and a vertical plane. (*Source:* P. O. Fanger, *Thermal Comfort Analysis and Applications in Environmental Engineering*, 1970. Reprinted with permission.)

Based on the curves in Fig. 5.3, for a ratio of $b/L = 1.8/3 = 0.6$ and a ratio of $a/L = 4.5/3 = 1.5$, the shape factor $F_{0-1} = 0.04$. Here, L is the horizontal distance from the occupant to the north partition wall. The shape factors F_{0-2}, F_{0-3}, and F_{0-4} can be calculated in the same manner.

2. To determine the shape factor F_{0-5} for the radiation from the occupant (surface 0) to the floor (surface 5),

we note that the vertical distance L from the center of the seated occupant to the floor is 1.8 ft. From the curves in Fig. 5.2, the ratio $b/L = 3/1.8 = 1.67$, and the ratio $a/L = 4.5/1.8 = 2.5$, thus the shape factor $F_{0-5} = 0.068$.

All the remaining shape factors can be determined in the same manner as listed in Table 5.4.

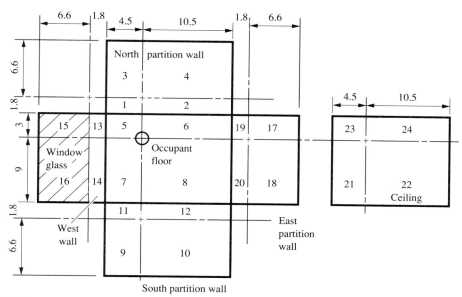

FIGURE 5.4 Dimension, in ft, of a private office.

3. For the north partition wall (surface 1), the product

$$T_{R1}^4 F_{0-1} = (75 + 460)^4 \times 0.04 = 32.77 \times 10^8$$

Other products $T_{Rn}^4 F_{0-n}$ can be similarly calculated, as listed in Table 5.4. The sum of the products $\sum T_{Rn}^4 F_{0-n} = 832.56 \times 10^8$.

Therefore,

$$T_{Rad}^4 = 832.56 \times 10^8$$

That is,

$$T_{Rad} = 537.2°R \quad \text{or} \quad T_{rad} = 77.2°F$$

4. The sum of the shape factors, $\sum F_{0-n} = 0.994$, is nearly equal to 1.

Effective Temperature

Effective temperature ET* is the temperature of an environment that causes the same total heat loss from the skin surface as in an actual environment of an operative temperature equal to ET* and at a relative humidity of 50 percent. ET* can be calculated as

$$\text{ET}^* = T_o + w_{sk} i_m \text{LR } 0.5 p_{ET.s} \qquad (5.23)$$

where $p_{ET.s}$ = saturated water vapor pressure at ET*, psia. The right-hand side of Eq. (5.23) describes the conditions of the indoor air regarding the total heat loss from the human body. The same value of the combination $(T_o + w_{sk} i_m \text{LR } 0.5 p_{ET.s})$ results in the same amount of total heat loss from the skin surface, if other parameters remain the same. Theoretically,

total skin wettedness w_{sk} and clothing permeability index i_m are constants for a specific ET* line.

Because the effective temperature is based on the operative temperature T_o, it is a combined index of T_a, T_{rad}, and p_a. In an indoor air temperature below 77°F, the constant ET* lines are nearly parallel to the skin temperature lines for sedentary occupants with a clothing insulation of 0.6 clo; therefore, ET* values are reliable indexes to indicate thermal sensations at normal indoor air temperatures during low activity levels.

The term *effective temperature* was originally proposed by Houghton and Yaglou in 1923. A new definition of ET* and its mathematical expression were developed by Gagge et al. in 1971. It is the environmental index commonly used in specifying and assessing thermal comfort requirements.

5.6 FACTORS AFFECTING THERMAL COMFORT

Daily experience and many laboratory experiments all show that thermal comfort occurs only under these conditions:

1. There is a steady-state thermal equilibrium between the human body and the environment, that is, heat storage of the body core S_{cr} and the skin surface S_{sk} are both equal to zero.

2. Regulatory sweating is maintained at a low level.

From the heat balance equation at steady-state thermal equilibrium (Eq. 5.1) we have

$$M = W + (C + R) + E_{sk} + E_{res}$$

TABLE 5.4
Values of F_{0-n} and $T_{Rn}^4 F_{0-n}$ in Example 5.1

Surface	Surface temperature, °R	Shape factor	b/L	a/L	F_{0-n}	$T_{Rn}^4 F_{0-n} \times 10^8$
North partition wall	535	F_{0-1}	0.6	1.5	0.04	32.77
		F_{0-2}	0.6	3.5	0.045	36.87
		F_{0-3}	2.2	1.5	0.07	57.35
		F_{0-4}	2.2	3.5	0.087	71.27
East partition wall	535	F_{0-17}	0.63	0.29	0.014	11.47
		F_{0-18}	0.63	0.86	0.03	24.58
		F_{0-19}	0.17	0.29	0.004	3.28
		F_{0-20}	0.17	0.86	0.009	7.37
South partition wall	535	F_{0-9}	0.73	0.5	0.023	18.84
		F_{0-10}	0.73	1.17	0.038	31.13
		F_{0-11}	0.2	0.5	0.008	6.55
		F_{0-12}	0.2	1.17	0.013	10.65
West wall	540	F_{0-13}	0.4	0.67	0.018	15.31
	540	F_{0-14}	0.4	2	0.03	25.51
West window	548	F_{0-15}	1.6	0.67	0.04	36.07
	548	F_{0-16}	1.6	2	0.07	63.13
Floor	538	F_{0-5}	1.67	2.5	0.068	56.97
		F_{0-6}	1.67	5.8	0.073	61.16
		F_{0-7}	5	2.5	0.087	72.89
		F_{0-8}	5	5.8	0.102	85.45
Ceiling	537	F_{0-21}	1.36	0.68	0.033	27.44
		F_{0-22}	1.36	1.59	0.052	43.24
		F_{0-23}	0.45	0.68	0.015	12.47
		F_{0-24}	0.45	1.59	0.025	20.79
				Σ	0.994	$8.32.56 \times 10^8$

Let $E_{\text{res}} = 0.1\,M$ and the mechanical efficiency $\mu = 0.05\,M$. From Eq. (5.11), $(C + R)$ can be determined. And also, from Eqs. (5.14), (5.15), and (5.20), E_{sk} is a known value. If we substitute into Eq. (5.1), the heat balance equation at steady-state thermal equilibrium can be expressed as

$$M(1 - 0.05 - 0.1) = F_{\text{cl}}f_{\text{cl}}h(T_{\text{sk}} - T_o)$$
$$+ w_{\text{sk}}i_m\text{LR}h_s(p_{\text{sk.s}} - p_a)$$

or

$$0.85\,M = F_{\text{cl}}f_{\text{cl}}h(T_{\text{sk}} - T_o) + w_{\text{sk}}i_m\text{LR}h_s(p_{\text{sk.s}} - p_a)$$
$$(5.24)$$

In Eq. (5.24), the physiological and environmental factors that affect the balance of the metabolic rate and the heat losses on the two sides of the equation, are as follows:

1. Metabolic rate M determines the magnitude of the heat energy that must be released from the human body, that is, the left-hand side of the equation.

2. Indoor air temperature T_a is a weighted component of the operating temperature T_o. It also affects the sensible heat loss and the vapor pressure of indoor air p_a in the calculation of the evaporative loss from the skin surface.

3. Mean radiant temperature T_{rad} is another weighted component of the operating temperature T_o. It affects the sensible heat loss from the human body.

4. Relative humidity of the ambient air φ_a is the dominating factor that determines the difference $(p_{\text{sk.s}} - p_a)$ in the evaporative loss from the skin surface. Air relative humidity becomes important when the evaporative heat loss due to regulatory sweating is the dominating heat loss from the human body.

5. Air velocity v_a influences the heat transfer coefficient h and the clothing efficiency F_{cl} in the term in Eq. (5.24) for the sensible heat loss from the human body. It also affects the overall sensible heat transfer coefficient h_s in the evaporative heat loss term and the clothing permeability i_m term in Eq. (5.24).

6. Clothing insulation R_{cl} affects the clothing efficiency F_{cl}, the area ratio f_{cl}, the heat transfer coefficient h, the permeability index i_m, and the overall sensible heat transfer coefficient h_s.

5.7 THERMAL COMFORT

Thermal comfort is defined as the state of mind in which one acknowledges satisfaction with regard to the thermal environment. In terms of sensations, thermal comfort is described as a thermal sensation of being neither too warm nor too cold, defined by the following seven-point thermal sensation scale proposed by ASHRAE:

-3 = cold
-2 = cool
-1 = slightly cool
0 = neutral
$+1$ = slightly warm
$+2$ = warm
$+3$ = hot

Fanger's Comfort Equation

A steady-state energy balance is a necessary condition for thermal comfort, but it is not sufficient by itself to establish thermal comfort. Fanger (1970) calculated the heat losses for a comfortable person, experiencing a neutral thermal sensation, with corresponding skin temperature T_{sk} and regulatory sweating E_{rsw}. The calculated heat losses L are then compared with the metabolic rate M. If $L = M$, the occupant feels comfortable. If $L > M$, then this person feels cool, and if $L < M$, then this person feels warm.

Using the responses of 1396 persons during laboratory experiments at Kansas State University of the United States and Technical University of Denmark, Fanger developed the following equation to calculate the Predicted Mean Vote PMV in the seven-point thermal sensation scale:

$$\text{PMV} = (0.303e^{-0.036M} + 0.276)(M - L) \quad (5.25)$$

In Eq. (5.25), the metabolic rate M and heat losses L are both in Btu/h $\cdot$ ft^2. According to Fanger's analysis, the Predicted Percentage of Dissatisfied vote (PPD) for thermal comfort at a (PMV) = 0 is 5 percent, and at a PMV = ± 1 is about 27 percent.

Tables of PMV and comfort charts including various combinations of operating temperature T_o, air velocity v, metabolic rate M, and clothing insulation R_{cl} have been prepared to determine comfortable conditions conveniently. Fanger's comfort charts also include relative humidity. Six of his comfort charts at various activity levels, wet bulb temperatures, relative humidities, and air velocities are shown in Fig. 5.5. In the first four charts, the air temperatures are equal to the mean radiant temperatures. Three of these four have a clothing insulation of 0.5 clo. The other is for 1 met activity level and 1 clo, because one rarely finds an occupant with such heavy outerwear at an activity level of 2 or 3 mets. In the fifth and sixth charts, the air temperature could be different from the mean radiant temperature with a constant relative humidity of 50 percent. These comfort variations clearly show that all six factors—air temperature T_a, mean radiant temperature T_{rad}, relative humidity φ, air velocity v, metabolic rate M, and clothing insulation R_{cl}—seriously affect thermal comfort.

For example, from Fanger's comfort chart, a sedentary occupant at an activity level of 1.0 met, with a clothing insulation of 0.5 clo, in an air conditioned space at a relative humidity of 50 percent and an air velocity less than 20 fpm, feels comfortable with an air temperature equal to the mean radiant temperature of 78°F. If all values are identical except for a 2 met activity level, the temperature would need to be 67°F for the same level of comfort.

Another factor, the duration of exposure to the indoor thermal environment, should be discussed here. If an indoor environment can provide thermal comfort for the occupant, the duration of the exposure has no significant influence upon the physiological responses of the person's thermoregulatory mechanism. If the indoor environment is uncomfortable, subjecting the subject to a certain degree of heat or cold stress, the time exposure would influence the person's physiological response.

ASHRAE Comfort Zones

Based on results of research conducted at Kansas State University and at other institutions, *ANSI/ASHRAE Standard 55–1981* specified winter and summer comfort zones to provide for the selection of the indoor parameters for thermal comfort (see Fig. 5.6). This chart is based upon an occupant activity level of 1.2 met. For summer, clothing insulation specified is 0.5 clo, that is, light slacks and short-sleeve shirt or comparable ensemble, with an air velocity less than 50 fpm. Standard 55–1981 recommends an optimum effective temperature ET* of 76°F for a summer comfort zone, with boundary effective temperatures of 73 to 79°F. If the clothing insulation is 0.1 clo higher, the boundary temperatures both should be shifted 1°F lower. Rohles et al. (1974) and Spain (1986) suggested that the upper boundary of the summer comfort zone can be extended to 85 or 86°F ET* if the air velocity of the indoor air can be increased to 200 fpm by a ceiling fan or other means.

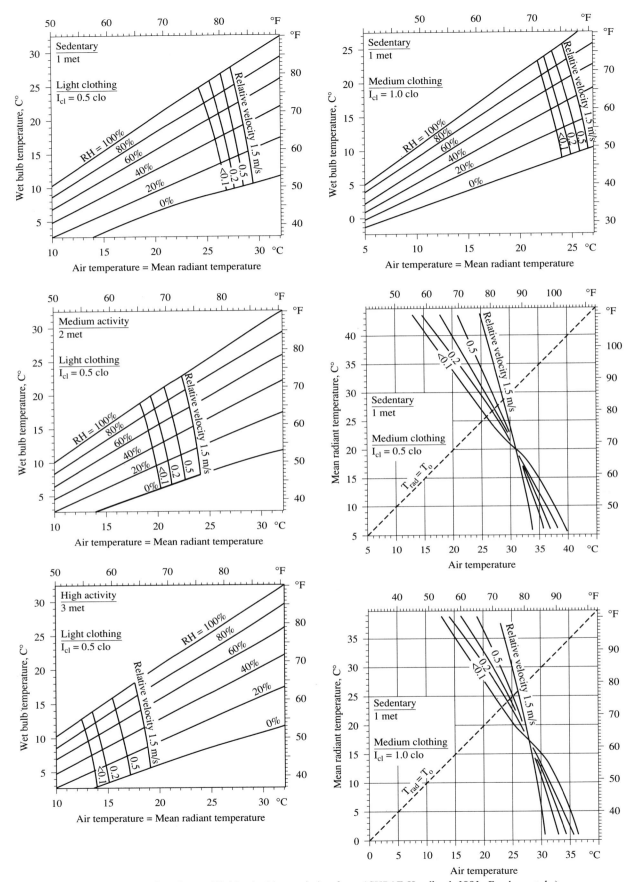

FIGURE 5.5 Fanger's comfort charts. (Abridged with permission from *ASHRAE Handbook 1981, Fundamentals.*)

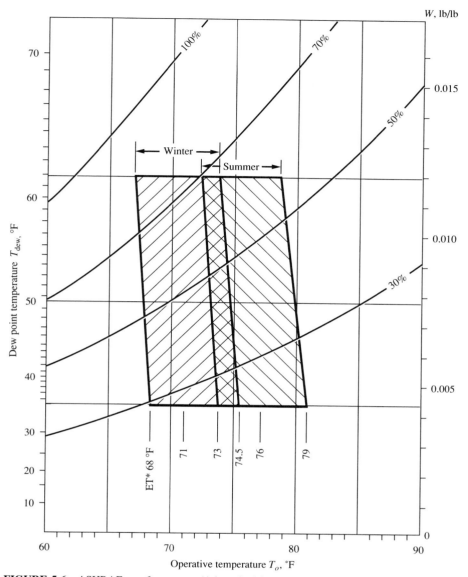

FIGURE 5.6 ASHRAE comfort zones. (Adapted with permission from *ASHRAE Transactions,* Part I B, 1983.)

The winter comfort zone is based upon a 0.9 clo insulation including heavy slacks, long-sleeve shirt, and sweater or jacket at an air velocity of less than 30 fpm. Standard 55–1981 recommended an optimal effective temperature ET* of 71°F for the winter comfort zone with boundary effective temperatures of 68 to 74.5°F.

Indoor air parameters should be fairly uniform in order to avoid local discomfort. According to Holzle et al., 75–89 percent of the subjects tested found the environment within this summer comfort zone to be thermally acceptable.

ASHRAE comfort zones recommend only the optimal and boundary ET* for the determination of the winter and summer indoor parameters. For clothing insulation, activity levels, indoor air velocities close to the values specified in Standard 55–1981, a wide range of indoor design conditions is available.

Comfort-Discomfort Diagrams

A comfort diagram provides a graphical presentation of the total heat loss from the human body at various operative or air temperatures and indoor relative humidities when the activity level, level of clothing insulation, and air velocity are specified. The abscissa of the comfort diagram is the operative T_o or ambient air temperature T_a, whereas the ordinate can be either water vapor pressure p or humidity ratio w. In Fig. 5.7 is presented a comfort diagram with a sedentary activity level, a clothing insulation value of 0.6 clo, and still air conditions, that is, an air velocity $v < 20$

ft/min. The curved lines represent relative humidity and the straight lines represent effective temperature ET*. The short dashed curves diverging from the ET* lines are total skin wettedness w_{sk} lines. The figure is based upon $T_o = T_a$.

Effective temperature ET* lines are calculated according to Eq. (5.23). Because the total skin wettedness w_{sk} is a constant for a specific ET* below 79°F ET*, then ET* lines and w_{sk} lines coincide with one other. At higher ET* values, w_{sk} lines curve to the left at high relative humidities. For low ambient air temperatures, evaporative heat loss from the skin surface E_{sk} is small, therefore, ET* and w_{sk} lines are nearly vertical. As E_{sk} becomes greater and greater, the slope of the ET* and w_{sk} lines decreases accordingly.

The comfort diagram is divided into five zones by the ET* and w_{sk} lines:

1. *Body cooling zone.* For the condition given in Fig. 5.7, if the effective temperature ET* < 73°F, the occupant will feel cold in this zone. Because the heat losses exceed the net metabolic rate, the skin and body core temperatures tend to drop gradually.

2. *Comfort zone.* This is the zone between the lower boundary ET* = 73°F, and w_{sk} = 0.06, and the higher boundary ET* = 86°F, and w_{sk} = 0.25. Steady-state thermal equilibrium is maintained between the occupant and the environment, and regulatory sweating is at a rather low level. The occupant will feel comfortable in this zone, and the heart rate HR is between 76 and 87 beats per minute.

   ASHRAE's winter and summer comfort zones are a part of this zone. The lower boundary of the *ASHRAE* winter comfort zone forms the lower boundary of this comfort zone. The reason that the lower boundary in this diagram is ET* = 73°F whereas ET* = 68°F in ASHRAE's winter comfort zone is that a lower clothing insulation 0.6 clo is used here.

3. *Uncomfortable zone.* In this zone, 86°F < ET* ≤ 95°F and 0.25 < w_{sk} ≤ 0.45. Thermal equilibrium also exists between the occupant and the environment, and the evaporative heat loss due to regulatory sweating dominates. Heart rate shows a range between 87 and 100. The occupant feels uncomfortable, that is, warm or hot, when his or her physiological parameters are in these ranges.

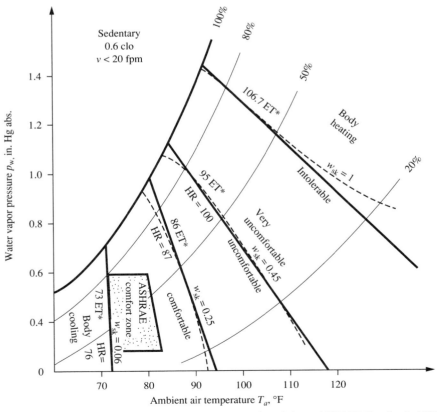

FIGURE 5.7   Comfort and discomfort diagram. (Adapted from *ASHRAE Handbook 1989, Fundamentals.* Reprinted by permission.)

4. *Very uncomfortable zone.* In this zone, 95°F < ET* < 106.5°F, and 0.45 < w_{sk} < 1. Although thermal equilibrium is still maintained with zero heat storage at the skin and the body core, there is a danger of a heat stroke when the ET* > 95°F. Toward the upper boundary of this zone, the skin surface is nearly entirely wet and the heart rate exceeds 120. Under these conditions, the occupant will feel very hot and very uncomfortable.

5. *Body heating zone.* When the ET* = 106.5°F, and w_{sk} = 1, thermoregulation by evaporation fails. At a higher ET* or w_{sk}, the environment is intolerable, and the temperatures of the body core and skin tend to rise gradually.

For an air conditioned space with an occupant at low activity levels, (M < 2 met), the indoor environment is usually maintained within the comfort zone, and the physiological and thermal responses of the occupant are also in the comfort zone. Only at higher activity levels do the thermal responses occasionally fall into the discomfort zone.

5.8 INDOOR AIR TEMPERATURE AND AIR MOVEMENTS

Comfort Air Conditioning Systems

For comfort air conditioning systems, most occupants have a metabolic rate of 1.0–1.5 met. The indoor clothing insulation in summer is usually 0.4–0.6 clo, and in winter it is 0.8–1.2 clo. Relative humidity has a lesser influence on thermal comfort, and will be discussed in the next section, but indoor air temperature and air velocity will be discussed here.

Many researchers have conducted tests to determine the effects of air velocity on the prefered indoor air temperature and the thermal comfort of occupants. The relationship between the prefered indoor air temperatures and various air velocities is presented in Fig. 5.8. Most of the data were taken under these conditions: metabolic rate M = 400 Btu/h, clothing insulation R_{cl} = 0.6 clo, $T_{rad} = T_a$, and relative humidity of the indoor air φ = 50 percent. The one exception is the students in Holzle's experiments, who had 0.54 clo for summer and 0.95 clo for winter.

Examination of Fig. 5.8 shows the following:

1. Higher indoor air temperatures require greater indoor air velocity to provide thermal comfort.

2. Variation of air velocity has a greater influence on preferred indoor air temperature at lower air temperatures.

3. Because the clothing insulation is not the same in summer as in winter, there should be another air

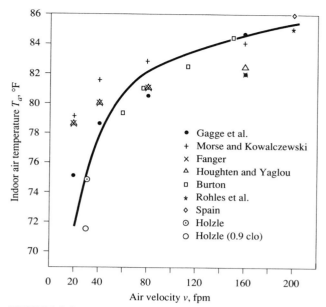

FIGURE 5.8 Preferred indoor air temperatures at various air velocities.

velocity versus indoor air temperature curve for the winter season.

Before specifying T_a for summer conditions, one needs to determine whether occupants are likely to wear suit jackets, such as members of a church congregation or guests in a multipurpose hall. In such cases, a reduction of 4°F of summer optimal ET* may be necessary because of the increase in clothing insulation of about 0.4 clo.

Economic Considerations and Energy Conservation

When one is specifying indoor design conditions, thermal comfort must be provided at optimum cost while using energy effectively. These principles should be considered:

1. To determine the optimum summer and winter indoor design temperatures, consider the local clothing habits and the upper and lower acceptable limits on clothing insulation at various operative temperatures T_o, as shown in Fig. 5.9.

2. It is always more energy efficient to use different indoor design temperatures for summer and winter than a year-round constant value. An unoccupied-period setback during winter always saves energy.

3. For short-term occupancies, or when the metabolic rate is higher than 1.2 met, a strategy of using a low energy-use ceiling fan or a wall-mounted fan to provide higher air velocity may be considered. Thus a higher indoor design temperature within the

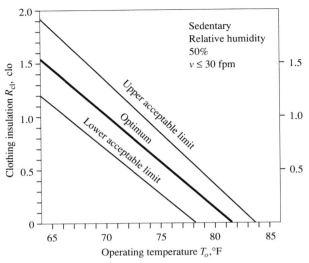

FIGURE 5.9   Relationship between clothing insulation R_{cl} and operating temperature T_o. (*Source: ASHRAE Transactions* 1983, Part I B. Reprinted with permission.)

extended summer comfort zone may be acceptable for occupants, especially in industrial settings.

Indoor Design Temperatures and Air Velocities for Comfort Air Conditioning

According to the *ANSI/ASHRAE Standard 55–1981, Thermal Environmental Conditions for Human Occupancy*, and *ASHRAE/IES Standard 90.1–1989, Energy-Efficient Design of New Buildings Except New Low-Rise Residential Buildings*, the following indoor design temperatures and air velocities apply for comfort air conditioning systems when the activity level is 1.2 met, there is a relative humidity of 50 percent in summer, and $T_a = T_{rad}$:

	Clothing insulation, clo	Indoor design temperature, °F	Air velocity, fpm
Winter	0.8–0.9	69–72	< 30
Summer	0.5–0.6	75–78	< 50

If the space relative humidity can be lowered to 35 to 40 percent in summer, then the upper limit 78°F is often specified.

Indoor badminton and table tennis tournament arenas should have air velocities below 30 fpm. To avoid nonuniformity and to prevent local discomfort, the air temperature difference between 4 in. from the floor and 67 in. above the floor should not exceed 5°F. The radiant temperature asymmetry in the vertical direction should be less than 9°F, and in the horizontal direction less than 18°F.

To determine whether the specified air temperature and air velocity are met, they should be measured at 4-, 24-, and 43-in. levels for sedentary occupants, and 4-, 43-, and 67-in. levels for standing activity. The duration for determining the mean value of the air movement should be 3 minutes or 100 time constants of the measuring instrument, whichever is greater. A *time constant* is the time required for the output of the measuring instrument to complete 63.2 percent of the total rise or decay.

Process Air Conditioning Systems

For process air conditioning systems, indoor design temperature is usually based on previous experiences. For precision manufacturing projects, a basic temperature plus a tolerance, such as 72 ± 2°F for a precision machinery assembling workshop, is often specified. First, the tolerance should be neither too tight nor too loose. Second, either the temperature fluctuation at various times within the working period or the temperature variation within the working space, or both, should be included in this tolerance.

In unidirectional-flow clean rooms, the velocity of the air stream in the working area is usually specified at 90 ± 20 fpm to prevent contamination of the products.

5.9   HUMIDITY

Comfort Air Conditioning Systems

For occupant thermal comfort at low activity levels, such as 1.0 to 1.2 met, and indoor air temperatures between 68 and 79°F, the impact of relative humidity variations between 25 and 80 percent is small. *ANSI/ASHRAE Standard 55–1981* specifies the acceptable dew point temperature boundaries as 62°F and 35°F. Within the winter comfort zone, relative humidity φ varies approximately from 23–86 percent, and within the summer comfort zone, φ varies from 19–72 percent.

The following effects of relative humidity should be taken into consideration in designing and evaluating the performance of comfort air conditioning systems:

1. Many comfort air conditioning systems for public buildings are not installed with humidifiers; in such circumstances, the relative humidity of the indoor air at the winter design conditions is not specified.

2. For comfort air conditioning systems installed with humidifiers, *ASHRAE/IES Standard 90.1– 1989* specifies that space or zone humidity control should prevent "the use of fossil fuel or electric-

ity to produce relative humidity in excess of 30% . . . or to reduce relative humidity below 60% when a humidity control is used for comfort dehumidification."

3. Maintaining a space relative humidity φ_r between 23 and 30 percent in winter prevents or reduces condensation at the inner side of the window glass.

4. The ranges of indoor temperature and humidity specified by the summer and winter comfort zones in *ANSI/ASHRAE Standard 55–1981* include the whole operating period, that is, full-load design conditions and part-load operations. The indoor relative humidity at part-load may be substantially higher than at full-load in some air conditioning systems in summer.

5. Bacterial and viral populations increase at relative humidities above 75 percent. Fungi, a humidity-related problem, grow quickly when the relative humidity is near 80 percent. Therefore, indoor relative humidity should be less than 75 percent.

6. When the indoor relative humidity is below 23 percent, the incidence of respiratory infections increases significantly. If, simultaneously, indoor temperatures are below 70°F, the induced static electricity in carpeted rooms may cause uncomfortable shocks to occupants contacting metal furniture or decorations.

7. In high-occupancy applications, it may be economical to specify summer indoor relative humidities at 55 to 60 percent.

Therefore, for comfort air conditioning systems, the recommended indoor relative humidities are as follows:

	Relative humidity, %	
	Tolerable	Preferred
Summer	30–65	40–50
Winter		
With humidifier		25–30
Without humidifier		Not specified

In surgical rooms and other rooms in hospitals and similar facilities where health and safety may be concerned, the relative humidity should be maintained at 40–60 percent year round.

Process Air Conditioning Systems

Humidity affects the physical properties of many materials and, therefore, their manufacturing processes.

MOISTURE CONTENT. Relative humidity has a marked influence on the moisture content of hygro-scopic materials such as natural textile fibers, paper, wood, leather, and foodstuffs. Moisture content affects the weight of the products and sometimes their strength, appearance, and quality.

DIMENSIONAL VARIATION. Hygroscopic materials often expand at higher relative humidities and contract at lower humidities. A 2 percent increase in moisture content may result in a 0.2 percent increase in dimensions of paper. That is why lithographic printing requires a relative humidity of 45 ± 2 percent.

CORROSION AND RUST. Corrosion is an electro-chemical process. Moisture encourages the formation of electrolytes and, therefore, the corrosion process. A relative humidity over 50 percent may affect the smooth operation of bearings in precision instruments. When indoor relative humidity exceeds 70 percent, rust may be visible on the surface of the machinery and parts made of steel and iron.

STATIC ELECTRICITY. Static electricity may cause minute particles to repel or attract one another, which is detrimental to many manufacturing processes. Static electricity charges minute dust particles in the air, causing them to cling to equipment and work surfaces. Static electricity exists in an indoor environment at normal air temperatures when relative humidity is below about 40 percent.

LOSS OF WATER. Vegetables and fruits lose water vapor through evaporation from their surfaces during storage. Low temperatures and high relative humidities, such as $\varphi = 90$ to 98 percent, may reduce water loss and delay desiccation.

It is important to specify the exact relative humidity required for good quality and cost control. For process air conditioning systems, the specified relative humidity is either a year-round single value or a range. A strict relative humidity requirement always includes a basic value and a tolerance, such as the relative humidity for lithographic printing mentioned before. When temperature and relative humidity controls are both required, they should be specified as a combination. Consider this example:

	Temperature, °F	Relative humidity, percent
Clean room	72 ± 2	45 ± 5

Case Study 5.1. A factory workshop has the following environmental parameters during summer:

Indoor air temperature	79°F
Indoor air relative humidity	50%
Space air velocity	20 fpm
Clothing insulation of the workers	0.5 clo
Activity level	3 met

As a result of comfort complaints by personnel, you are asked to recommend effective and economical corrective measures to improve the indoor environment. The following table includes the information required during analysis:

Area ratio of the clothed body, f_{cl}	1.2
Permeability index of clothing, i_m	0.4
Skin surface temperature, T_{sk}	92.7 °F
Saturation water vapor pressure at 92.7°F	0.764 psia
at 79°F	0.491 psia

Solution

1. When the space air velocity $v = 20$ fpm, the convective heat transfer coefficient can be calculated, from Eq. (5.9), as

$$h_c = 0.0681\, v^{0.69} = 0.0681 \times 20^{0.69}$$
$$= 0.538 \text{ Btu/h} \cdot \text{ft}^2 \cdot °\text{F}$$

Because at normal indoor conditions, the radiative heat transfer coefficient $h_r = 0.83$ Btu/h · ft² · °F, the surface heat transfer coefficient h is

$$h = h_c + h_r = 0.538 + 0.83$$
$$= 1.368 \text{ Btu/h} \cdot \text{ft}^2 \cdot °\text{F}$$

The clothing insulation $R_{cl} = 0.5 \times 0.88 = 0.44$ h · ft² · °F/Btu. From Eq. (5.12), the clothing efficiency is found to be

$$F_{cl} = 1/(R_{cl} f_{cl} h + 1)$$
$$= 1/(0.44 \times 1.2 \times 1.368 + 1)$$
$$= 0.5806$$

Therefore, from Eq. (5.11), we find the sensible heat loss from the skin surface of the worker is

$$(C + R) = F_{cl} f_{cl} h (T_{sk} - T_o)$$
$$= 0.5806 \times 1.2 \times 1.368(92.7 - 79)$$
$$= 13.06 \text{ Btu/h} \cdot \text{ft}^2$$

From Eq. (5.17), the overall sensible heat transfer coefficient h_s is

$$h_s = f_{cl} h / (f_{cl} h R_{cl} + 1)$$
$$= 1.2 \times 1.368/(1.2 \times 1.368 \times 0.44 + 1)$$
$$= 0.9531 \text{ Btu/h} \cdot \text{ft}^2 \cdot °\text{F}$$

From Eq. (5.16), the overall evaporative heat transfer coefficient $h_{e.c}$ is

$$h_{e.c} = h_s i_m \text{LR} = 0.9531 \times 0.4 \times 205$$
$$= 78.16 \text{ Btu/h} \cdot \text{ft}^2 \cdot \text{psi}$$

The maximum evaporative heat loss E_{max} due to regulatory sweating can be calculated from Eq. (5.15) as

$$E_{max} = h_{e.c}(p_{sk.s} - p_a)$$
$$= 78.16[0.764 - (0.5 \times 0.491)]$$
$$= 40.24 \text{ Btu/h} \cdot \text{ft}^2$$

In Eq. (5.24),

$$0.85\, M = 0.85 \times 3 \times 18.46$$
$$= 47.07 \text{ Btu/h} \cdot \text{ft}^2$$

When $\mu = 0.05$ and $E_{res} = 0.1M$, the total evaporative loss from the skin surface of the worker, from Eq. (5.1), is

$$E_{sk} = 0.85\, M - (C + R)$$
$$= 47.07 - 13.06$$
$$= 34.01 \text{ Btu/h} \cdot \text{ft}^2$$

Therefore, the total skin wettedness is

$$w_{sk} = \frac{E_{sk}}{E_{max}} = \frac{34.01}{40.24} = 0.845$$

From Fig. 5.7, when $w_{sk} = 0.845$, a person is very uncomfortable.

2. From Fanger's comfort chart, shown in Fig. 5.5, for a person with an activity level of 3 met and a clothing insulation of 0.5 clo, at a relative humidity of 50 percent and an air velocity of 20 fpm, and with $T_a = T_{rad}$ for neutral thermal sensation, the indoor air temperature should be 56°F. Obviously, this is not economical because too much refrigeration is required.

3. Let us analyze the results if ceiling fans or wall fans are used to increase the space air velocity v to 300 fpm. Then the convective heat transfer coefficient h_c is

$$h_c = 0.681 \times 300^{0.69}$$
$$= 3.486 \text{ Btu/h} \cdot \text{ft}^2 \cdot °\text{F}$$

The surface heat transfer coefficient is

$$h = 3.486 + 0.83$$
$$= 4.316 \text{ Btu/h} \cdot \text{ft}^2 \cdot °\text{F}$$

Also, the clothing efficiency is calculated as

$$F_{cl} = \frac{1}{(0.44 \times 1.2 \times 4.316) + 1}$$
$$= 0.305$$

Then the sensible heat loss is equal to

$$C + R = 0.305 \times 1.2 \times 4.316(92.7 - 79)$$
$$= 21.64 \text{ Btu/h} \cdot \text{ft}^2$$

The overall sensible heat transfer coefficient is

$$h_s = \frac{1.2 \times 4.316}{1.2 \times 4.136 \times 0.44 + 1}$$
$$= 1.58 \text{ Btu/h} \cdot \text{ft}^2 \cdot °\text{F}$$

The overall evaporative heat transfer coefficient $h_{e.c}$ can be shown to be

$$h_{e.c} = 1.58 \times 0.4 \times 205$$
$$= 129.5 \text{ Btu/h} \cdot \text{ft}^2 \cdot \text{psi}$$

Then the maximum evaporative heat loss due to regulatory sweating is

$$E_{\max} = 129.5[0.7604 - (0.5 \times 0.491)]$$
$$= 66.68 \text{ Btu/h} \cdot \text{ft}^2$$

The total skin wettedness is

$$w_{\text{sk}} = \frac{47.07 - 21.64}{66.68} = 0.381$$

That is, w_{sk} has been greatly reduced compared with the value at $v = 20$ fpm. In Fig. 5.7, $w_{\text{sk}} = 0.381$ is in the uncomfortable zone. Workers will feel warm, but the indoor environment has been considerably improved. This may be the most cost-effective solution.

5.10 SOUND LEVEL

Sound and Sound Level

Sound can be defined as a variation in pressure due to vibration in an elastic medium such as air. A vibrating body generates pressure waves, which spread by alternate compression and rarefaction of the molecules within the transmitting medium. Airborne sound is a variation of air pressure, with atmospheric pressure as the mean value.

Because sound is transmitted by compression and expansion of the molecules, it cannot travel in a vacuum. The denser the material, the faster is the traveling speed of a sound wave. The velocity of a sound wave in air is approximately 1130 fps. In water, it is about 4500 fps and in steel 15,000 fps.

Noise is any unwanted sound. In air systems, noise should be compensated for, either by *attenuation*, the process of reducing the amount of sound that reaches the space, or by masking it with other, less objectionable sounds.

Sound Power Level and Sound Pressure Level

Sound power is the ability to radiate power from a sound source excited by an energy input. The intensity of sound power is the power output from a sound source expressed in watts (W). Because of the wide variation of sound output—from the threshold hearing level of 10^{-12} W to a level of 10^8 watts, generated by the launching of a Saturn rocket, a ratio of 10^{20} to 1—it is more appropriate and convenient to use a logarithmic expression to define sound power level, that is,

$$L_w = 10 \log(w/10^{-12} \text{ W})\text{re } 1\text{pW} \qquad (5.26)$$

where L_w = sound power level, dB
w = sound source power output, W

Here 10^{-12} watts, or 1 pW (picowatt), is the international reference base, and re indicates the reference base.

The human ear and microphones are pressure-sensitive. Analogous to the sound power level, the sound pressure level is defined as

$$L_p = 20 \log\left(\frac{p}{2 \times 10^{-5} \text{ Pa}}\right)\text{re } 20\mu\text{Pa} \qquad (5.27)$$

where L_p = sound pressure level, in dB
p = sound pressure, Pa

Here, the reference sound pressure level is 2×10^{-5} Pa (pascal) or 20 μPa, corresponding to the hearing threshold. Because sound power is proportional to the square of the sound pressure, $10 \log p^2 = 20 \log p$. Sound pressure levels of various sources are listed in Table 5.5.

The sound power level of a specific source is a fixed output. It cannot be measured directly and can only be calculated through the measurements of the sound pressure level. On the other hand, sound pressure level is the sound level measured at any one point and is a function of the distance from the source and the characteristics of the surroundings.

Octave Bands

Sound waves, like other waves, are characterized by the relationship between wavelength, speed, and frequency:

$$\text{wavelength} = \frac{\text{speed}}{\text{frequency}} \qquad (5.28)$$

People can hear frequencies from 20 Hz to 20 kHz. To study and analyze sound, we must it break into components. A convenient way is to subdivide the audible range into eight octave bands or sometimes 24 $\frac{1}{3}$-octave bands. An *octave* is a frequency band in which the frequency of the upper band limit is double the frequency of the lower limit. The center frequency of an octave or a $\frac{1}{3}$-octave band, is the geometric mean of its upper and lower band limits. An octave or $\frac{1}{3}$-octave band is represented by its center frequency. The eight octave bands and their center frequencies are listed in Table 5.6.

Addition of Sound Levels

Because sound levels, in dB, are in logarithmic units, two sound levels cannot be added arithmetically. If sound levels A, B, C, ..., in dB, are added together, the combined overall sound level $\sum L$ can be calculated as

$$\Sigma L = 10 \log(10^{0.1A} + 10^{0.1B} + 10^{0.1C} + \cdots) \qquad (5.29)$$

TABLE 5.5
Typical sound pressure levels

Source	Sound pressure, Pa	Sound pressure Level, dB re 20μPa	Subjective reaction
Military jet takeoff at 100 ft	200	140	Extreme
Passenger's ramp at jet airliner (peak)	20	120	Threshold of pain
Platform of subway station (steel wheels)	2	100	
Computer printout room*	0.2	80	
Conversational speech at 3 ft	0.02	60	
Window air conditioner*	0.006	50	Moderate
Quiet residential area*	0.002	40	
Whispered conversation at 6 ft	0.0006	30	
Buzzing insect at 3 ft	0.0002	20	
Threshold of good hearing	0.00006	10	Faint
Threshold of excellent youthful hearing	0.00002	0	Threshold of hearing

*Ambient.

Abridged from *ASHRAE Handbook 1989, Fundamentals.* Reprinted by permission.

Human Response and Design Criteria

The human brain does not respond in the same way to lower frequencies as to higher frequencies. At lower sound pressure levels, it judges a 20 dB sound at 1000 Hz to have the same loudness as a 52 dB sound at 50 Hz. However, at high sound pressure levels, a 100 dB sound at 1000 Hz seems as loud as 110 dB at 50 Hz.

The purpose of noise control in an air conditioned space is to provide a background sound low enough to avoid interference with the acoustical requirements of the occupants. The distribution of the background sound should be balanced over a wide range of frequencies, without whistle, hum, rumble, or audible beats. Three types of criteria for sound control are currently adopted in indoor system design:

1. *A-weighted sound level dBA.* The A-weighted sound level dBA tries to simulate the response of the human ear to sound at low sound pressure levels. An electronic weighing network automatically simulates the lower sensitivity of the human ear to low frequency sounds by subtracting a certain number of dB at various octave bands, such as approximately 27 dB in the first octave band, 16 dB in the second, 8 dB in the third, and 4 dB in the fourth. The A-weighted sound level gives a single value. It is simple and also considers the human judgment of relative loudness at low sound pressure levels. Its main drawback is its failure to cosider the frequency spectrum or the subjective quality of sound.

2. *Noise criteria NC curves. NC curves* represent actual human reactions during tests. The shape of NC curves is similar to the equal loudness contour representing the response of the human ear, as shown in Fig. 5.10. NC curves are intended to indicate the permissible sound pressure level of a broad-band noise at various octave bands by a single NC curve sound level rating. NC curves are practical. They also consider the frequency spec-

TABLE 5.6
Octave bands and their center frequencies

Band number	Band frequency, H2		
	Lower	Center	Upper
	22.4	31.5	45
1	45	63	90
2	90	125	180
3	180	250	355
4	355	500	710
5	710	1000	1400
6	1400	2000	2800
7	2800	4000	5600
8	5600	8000	11200

Abridged with permission from *ASHRAE Handbook 1989, Fundamentals* with permission.

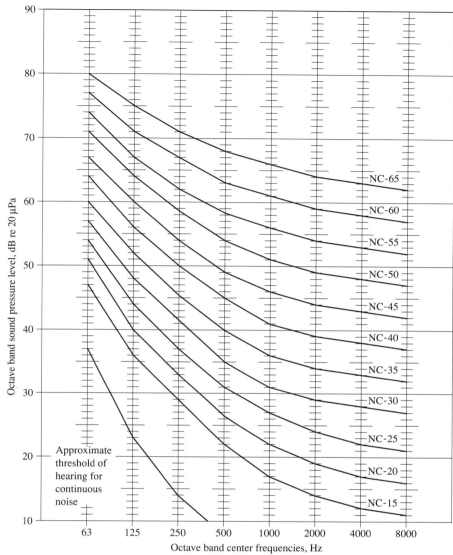

FIGURE 5.10 Noise criteria NC curves. (*Source: ASHRAE Handbook 1989, Fundamentals.* Reprinted with permission.)

trum of the broad-band noise. The main problem with NC curves is that the shape of the curve does not approach a balanced, bland-sounding spectrum that is neither rumbly nor hissy.

3. *Room criteria RC curves. RC curves,* as shown in Fig. 5.11, are similar to NC curves except that the shape of an RC curve is a close approximation of a balanced, bland-sounding spectrum.

ASHRAE recommends the indoor design RC or NC ranges presented in Table 5.7. For sounds con-

taining significant pure tones or impulsive sounds, a 5- to 10-dB lower value should be specified. Noise is always an annoying element and source of complaints in indoor environments. Attenuation to achieve a NC or RC goal below 30 dB for a central hydronic air conditioning system or a packaged system is very expensive. In actual practice, the NC or RC criteria range for private residences and apartments varies greatly because of personal requirements and economic considerations.

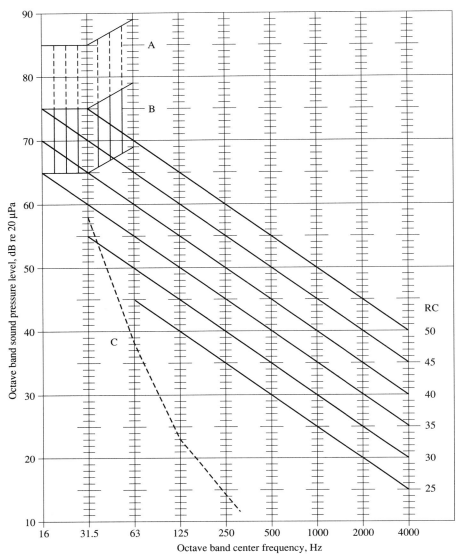

FIGURE 5.11 Room criteria RC curves. (*Source*: *ASHRAE Handbook 1989, Fundamentals.* Reprinted with permission.)

To meet the listed design criteria range, the measured sound pressure levels of at least three of the four octave bands between 250 and 2000 Hz should be within the listed NC or RC range.

In industrial workshops with machinery and equipment, Occupational Safety and Health Administration Standard Part 1910.95, published by the Department of Labor of the United States, specifies: "Feasible administrative or engineering controls shall be utilized when employees are subjected to sound levels exceeding those in Table G-16. If such controls fail to reduce sound levels to below those in Table G-16, personal protective equipment shall be provided and used. Exposure to impulsive and impact noise should not exceed 140 dB peak sound pressure level." Table G-16 is reproduced in this text as Table 5.8.

5.11 SPACE PRESSURE DIFFERENTIAL

Most air conditioning systems strive to maintain a slightly higher indoor pressure than the surroundings. This positive pressure tends to eliminate or reduce infiltration, the entry of untreated air to the space.

TABLE 5.7
Recommended criteria for indoor design RC or NC range

Type of Area	Recommended criteria for RC or NC range
Private residence	25–30
Apartments	25–30
Hotels/motels	
Individual rooms or suites	30–35
Meeting/banquet rooms	25–30
Halls, corridors, lobbies	35–40
Service/support areas	40–45
Offices	
Executive	25–30
Conference	25–30
Private	30–35
Open-plan	30–35
Computer equipment rooms	40–45
Public circulation	40–45
Hospitals and clinics	
Private rooms	25–30
Wards	30–35
Operating rooms	35–40
Corridors	35–40
Public areas	35–40
Churches	25–30
Schools	
Lecture and classrooms	25–30
Open-plan classrooms	30–35
Libraries	35–40
Concert halls	†
Legitimate theaters	†
Recording studios	†
Movie theaters	30–35

Note: These are for unoccupied spaces, with all systems operating.

*Design goals can be increased by 5 db when dictated by budget constraints or when intrusion from other sources represents a limiting condition.

† An acoustical expert should be consulted for guidance on these critical spaces.

Source: *ASHRAE Handbook 1987, HVAC Systems and Applications*. Reprinted with permission.

TABLE 5.8
Occupational noise exposure (Occupational Health and Safety Administration Table G–16)

Duration per day, h	Sound level dBA slow response
8	90
6	92
4	95
3	97
2	100
1 ½	102
1	105
½	110
¼ (or less)	115

Note: If the variations in noise level involve maxima at intervals of 1 second or less, it is to be considered continuous.

In all cases where the sound levels exceed the values shown herein, a continuing, effective hearing conservation program shall be administered.

Source: *Occupational Safety and Health Standard, Part 1910.95*. Reprinted with permission.

For comfort systems, the recommended pressure differential between the indoor and outdoor air is 0.02–0.05 in. WG. The phrase "in. WG" represents the pressure at the bottom of a water column that has a height of the specified number of inches greater than the atmospheric pressure, i.e., the gauge pressure. That is, 1 in. water gauge = 0.03612 psig.

Another type of space pressurization control for the health and comfort of passengers in an aircraft is cabin pressurization control. If an airplane is flying at an altitude of 28,000 ft with an ambient pressure of 4.8 psia, the minimum cabin pressure required is 10.8 psia, and a cabin pressurization control system must be installed to maintain a pressure differential of $10.8 - 4.8 = 6$ psi between the cabin and the ambient air.

Process air conditioning systems, such as for clean rooms, need properly specified space pressurization to prevent contaminated air from entering the clean, uncontaminated, or surrounding semicontaminated areas. According to *Federal Standard No. 209B, Clean Rooms and Work Station Requirements, Controlled Environment,* published by the United States Government in 1973, the minimum positive pressure differential between the clean room and any adjacent area of less clean requirements should be 0.05 in. WG with all entryways closed. When the entryways are open, an outward flow of air is to be maintained to minimize migration of contaminants into the room.

For rooms where toxic, hazardous, contaminated, or objectionable gases or substances are produced, a slightly lower pressure, or negative room pressure, should be maintained to prevent the diffusion of these substances to the surroundings.

The magnitude of the positive or negative pressure to be maintained in the space must be carefully determined. A higher positive pressure always means a greater amount of exfiltrated air, and a higher negative pressure means a greater infiltration. Both create requirements for greater volumes of outdoor air intake, resulting in higher initial and operating costs.

5.12 SICK BUILDING SYNDROME

Sick building syndrome is a kind of building-related illness that received public attention during the 1980s. When more than 20% of a building's occupants complain of a set of discomfort symptoms that include headaches, eye irritation, sore throat, fatigue, dizziness, drowsiness, and nausea for periods of longer than two weeks; when the sources of these complaints are not obvious; and when the occupants obtain rapid relief outside of the building, the building is considered as a kind of sick building syndrome. If there are signs of actual illness, these illnesses are classified as building-related illness.

Meckler's paper reported that the National Institute for Occupational Safety and Health (NIOSH) of the United States classified the reasons for sick building syndrome according to the results of 346 field investigations in the 1980s:

- Improper use or maintenance of HVAC&R systems 51 percent
- Indoor contaminants 17 percent
- Outdoor contaminants 11 percent
- Microbiological contaminants 6 percent
- Building fabric contamination 4 percent
- Unknown sources 11 percent

The HVAC&R problems include lack of outdoor air, poor air distribution, and unsatisfactory indoor temperature and relative humidity. The survey found 70–80 percent of the investigated buildings had no known problems.

In the United States, most people spend about 90 percent of their time indoors. The purpose of specifying the indoor design conditions in the design documents is to provide them with a satisfactory indoor environment at optimum cost.

Effective operation, control, and maintenance of the HVAC&R system will be discussed in later chapters, so only outdoor ventilation rates and air cleanliness will be discussed in the following section.

5.13 OUTDOOR AIR REQUIREMENTS

Outdoor air is required for both comfort and process air conditioning systems. Outdoor air, which contains oxygen, is supplied or infiltrated into the air conditioned space to meet the following requirements:

1. To meet metabolic requirements of the occupants
2. To dilute the air contaminants, odor, and pollutants to maintain required indoor air quality (IAQ)

3. To support any combustion process or replace the amount of exhaust air required in laboratories, manufacturing processes, or restrooms
4. To provide make-up for the amount of exfiltrated air required when a positive pressure is to be maintained in a conditioned space

The rate of outdoor air required is called the outdoor air requirement for ventilation in *ASHRAE Standard 62–1989*, as well as the codes of local governing bodies, in cfm/person or cfm/ft^2 floor area.

The refrigeration capacity required to cool and dehumidify the outdoor air can be a major portion of the total refrigeration requirement during summer, depending upon the occupant density and process and exhaust needs. Therefore, the amount of outdoor ventilation air supplied to the conditioned space should be selected carefully.

The amount of infiltration depends on the wind speed and direction, as well as the outdoor and indoor temperatures and pressure differences and controls, which are variable. Therefore, infiltration is not a reliable source of outdoor air supply. Furthermore, modern buildings can maintain more reliable positive pressures because they are built with much tighter envelopes.

Indoor Air Contaminants

Based on the results of the field investigations of three office buildings by Bayer and Black in 1988, the indoor air contaminants that relate to indoor air quality and the symptoms of the sick building syndrome are the following:

1. *Total particulate concentration*. This parameter includes particulates from building materials, combustion products, and mineral and synthetic fibers. In February 1989, the U.S. Environmental Protection Agency (EPA) specified the allowable indoor concentration level of particulates of 10 μm and less in diameter (which can penetrate deeply into the lungs, becoming hazardous to health) as follows:

    50 μg/m^3(0.000022 grains/ft^3), 1 year

    150 μg/m^3(0.000066 grains/ft^3), 24 hours

    Here "1 year" means "maximum allowable exposure per day over the course of a year," and "24 hours" means "maximum allowable exposure over a 24-hour period."
2. *Nitrogen dioxide*. NO$_2$ is a combustion product from gas stoves and other sources. There is growing evidence that NO$_2$ may cause respiratory disease. The U.S. Occupational Safety and Health

Administration (OSHA) *Code of Federal Regulations, Title 29, Part 1910* specifies a concentration level for the industrial workplace of 5 ppm by volume for NO_2.

3. *Volatile organic compounds.* These include formaldehyde and a variety of aliphatic, aromatic, oxygenated, and chlorinated compounds. Mucous membrane irritation caused by formaldehyde is well established.

4. *Nicotine.* Environmental tobacco smoke is clearly a discomfort factor to many adults who do not smoke. Nicotine and other components of tobacco smoke may also be a health risk for human beings.

5. *Radon.* Radon occurs in soil, rocks, and water by the decay of radium and uranium. It travels through the pores of rock and soil and infiltrates into a house along cracks and other openings in the basement slab or walls. Since the early 1980s, measurements of radon and its derivatives made in residential buildings in United States show that the potential health hazard to occupants in buildings from radon inhalation can be significant. At various locations in the United States, the indoor radon and radon derivative concentrations may vary by as much as a factor of 100. These nuclides may cause cancer. Indoor concentrations of radon progeny can be reduced by increasing the outdoor air supply and also by electrostatic and mechanical filtration. The EPA adopted a concentration limit of 4 picoCuries per liter (pCi/L) for radon progeny.

In addition to the preceding indoor air contaminants, others such as carbon monoxide, sulfur dioxide, and ozone can have significant effects on occupants.

If outdoor air is used to dilute the concentration of indoor contaminants, its quality must meet the *National Primary and Secondary Ambient-Air Quality Standards* provided by the EPA. Some of these time average concentrations are shown below:

Pollutants	Long-term concentration			Short-term concentration		
	$\frac{\mu g}{m^3}$	ppm	Average period of exposure	$\frac{\mu g}{m^3}$	ppm	Average period of exposure
Particulate matter	75		1 yr	260		24 hr
Sulfur oxides	80	0.03	1 yr	365	0.14	24 hr
Carbon monoxide				40,000	35	1 hr
				10,000	9	8 hr
Nitrogen dioxide	100	0.055	1 yr			
Ozone				235	0.12	1 hr
Lead	1.5		3 months			

Only particulate matter is expressed in terms of annual geometric means; for sulfur oxides and carbon monoxide the concentrations are annual arithmetic means. For short term concentrations, the tabular values are not to be exceeded more than once a year.

Outdoor Air Requirements for People

The amount of outdoor air required for metabolic oxidation processes is actually rather small. ASHRAE Standard 62–1989 specifies a minimum requirement of 15 cfm outdoor air per person. This minimum outdoor air ventilation rate should maintain acceptable indoor air quality and dilute odors and contaminants. The concentration of indoor carbon dioxide, $C_{i.CO_2}$, is a common criterion for setting outdoor air rates. An indoor CO_2 concentration of 1000 ppm, or 0.1 percent, under continuous exposure is accepted in this standard as a satisfactory limit. A typical concentration of outdoor CO_2, $C_{o.CO_2}$, is 0.03 percent, and the CO_2 production $\dot{V}_{CO_2}$ of sedentary occupant who is eating a normal diet is 0.0112 cfm. The amount of outdoor air required for each indoor occupant $\dot{V}_{o.oc}$, in cfm, can be calculated as

$$\dot{V}_{o.oc} = \dot{V}_{CO_2}/C_{i.CO_2} - C_o$$
$$= 0.0112/100(0.1 - 0.03)$$
$$= 16 \text{ cfm} \tag{5.30}$$

The calculated 16 cfm is close to the specified value 15 cfm.

Another function of the outdoor air supply in comfort air conditioning systems is to dilute the concentration of smoke and odors from cigarettes. Thayer (1982), based on data from different authors, developed a dilution index that indicates 15 cfm of outdoor air per person will satisfy more than 80 percent of the occupants in the space.

Some of the outdoor air requirements contained in *ASHRAE Standard 62–1989* are indicated in Table 5.9. Usually, the same outdoor air used for the dilution of odors and cigarette smoke is sufficient for metabolic oxygen requirements, for exhausting air from restrooms, and also for replacing exfiltrated air lost from the conditioned space as a result positive pressure.

If the outdoor air supply is used to dilute the concentration of indoor air contaminants other than odors and cigarette smoke, the rate of outdoor air supply $\dot{V}_o$, in cfm, can be calculated as

$$\dot{V}_o = \frac{2118\dot{m}_{par}}{C_i - C_o} \tag{5.31}$$

where $\dot{m}_{par}$ = rate of generation of contaminants in the space, mg/s

TABLE 5.9
Outdoor air requirements recommended by ASHRAE Standard 62–1989

Application	cfm/person
Dining room	20
Bar and cocktail lounges	30
Hotel and conference rooms	20
Office spaces	20
Office conference rooms	20
Retail stores	0.02–0.03*
Beauty shops	25
Ballrooms and discos	25
Spectator areas	15
Theater auditoriums	15
Transportation waiting rooms	15
Class rooms	15
Hospital patient rooms	25
Residences	0.35†
Smoking lounges	60

*cfm/ft^2 floor area

†Air changes/hour

Abridged with permission from *ASHRAE Standard 62–1989.*

$$C_i, C_o = \text{concentrations of air contaminants indoors and outdoors, respectively, mg/m}^3$$

Values of C_o can be found in the EPA National Primary Ambient-Air Quality Standards. The indoor concentration of CO_2 and other contaminants must meet the specified value as stated before.

For laboratories, the outdoor air supply should be greater than each of the following:

1. The rate of outdoor air required to dilute air contaminants odors, or cigarette smoke from the occupants
2. The sum of the exhaust air and the outdoor air required by any combustion or chemical process

For clean rooms, or facilities that control particulates in air, *Federal Standard 209B* specifies the rate of outdoor air, or make-up air, to be 5–20 percent of the supply air. In rooms using a high volume of recirculating air, outdoor air should be specified as a given volume flow rate per person, e.g., 30 cfm per person.

5.14 AIR CLEANLINESS

Indoor air cleanliness is closely related to indoor air quality (IAQ) and is one of the indoor environmental parameters gaining more and more attention for enhanced comfort and health of the occupants.

Indoor air contaminants are discussed in Section 5.13. Both outdoor air and supply air can be used to dilute the concentration of air contaminants. Economic considerations are often the decisive factor. Because volatile organic compounds and various types of odors are composed of extremely small particles, activated carbon filters or solvents are needed to remove them from recirculated air. Because these techniques are comparatively expensive, outdoor air is often used to dilute the objectionable odors and vapors.

To use supply air to dilute the concentration of the air contaminants, one can calculate the required rate of supply air $\dot{V}_s$, in cfm, by Eq. (5.31), except C_s, the particulate concentration in supply air, in mg/m^3, replaces C_o in Eq. (5.31). The magnitude of C_s depends mainly on the air filtration facilities in the air conditioning system.

According to *ASHRAE Handbook 1991, HVAC Applications* and also the recent requirements of still higher classes of air cleanliness in the electronics industry, clean rooms can be classified as follows:

Class 1. Particle count not to exceed 1 particle per ft^3 of a size of 0.5 μm and larger, with no particle exceeding 5 μm.

Class 10. Particle count not to exceed 10 particles per ft^3 of a size of 0.5 μm and larger, with no particle exceeding 5 μm.

Class 100. Particle count not to exceed 100 particles per ft^3 of a size of 0.5 μm and larger.

Class 1000. Particle count not to exceed 1000 particles per ft^3 of a size of 0.5 μm and larger.

Class 10,000. Particle count not to exceed 10,000 particles per ft^3 of a size of 0.5 μm and larger or 65 particles per ft^3 of a size 5.0 μm and larger.

Class 100,000. Particle count not to exceed 100,000 particles per ft^3 of a size of 0.5 μm and larger or 700 particles per ft^3 of a size 5.0 μm and larger.

For other, less critical applications, air cleanliness can be specified as follows:

Better filtered area. This area is served by systems with air filters having atmospheric dust spot efficiency of 85 percent or higher to effectively remove particles causing smudge and stain and to partially remove cigarette smoke.

Average filtered area. This area is served by systems with air filters having atmospheric dust spot efficiency of 35–60 percent to effectively remove pollens and partially remove particles

Rough filtered area. This area is served by systems with air filters of atmospheric dust spot efficiency of 10–20 percent. High-efficiency particulate air (HEPA) filters and atmospheric dust spot efficiency will be discussed in detail in Chapter 15.

Airborne particle counts for sizes 0.5 μm and larger should be taken with instruments using the light-scattering principle, but for particles 5.0 μm and larger, microscopic counting of particles collected on a membrane filter may be used.

REFERENCES AND FURTHER READING

ANSI/ASHRAE, *ANSI/ASHRAE Standard 55–1981, Thermal Environmental Conditions for Human Occupancy,* ASHRAE, Inc., Atlanta, GA, 1981.

ASHRAE, *ASHRAE Standard 62–1989, Ventilation for Acceptable Indoor Air Quality,* ASHRAE Inc., Atlanta, GA, 1989.

ASHRAE/IES, *ASHRAE/IES Standard 90.1–1989, Energy: Efficient Design of New Buildings Except Low-Rise Residential Buildings,* ASHRAE, Inc., Atlanta, GA, 1989.

ASHRAE, *ASHRAE Handbook 1987, HVAC Systems and Applications,* ASHRAE Inc., Atlanta, GA, 1987.

ASHRAE, *ASHRAE Handbook 1989, Fundamentals,* ASHRAE, Inc., Atlanta, GA, 1989.

ASHRAE, *ASHRAE Handbook 1991, HVAC Applications,* ASHRAE Inc., Atlanta, GA, 1991.

Bayer, C. W., and Black, M. S., "IAQ Evaluations of Three Office Buildings," *ASHRAE Journal,* pp. 48–52, July 1988.

Berglund, L., "Mathematical Models for Predicting the Thermal Comfort Response of Building Occupants," *ASHRAE Transactions,* Part I, pp. 735–749, 1978.

Cena, K., Spotila, J. R., and Avery, H. W., "Thermal Comfort of the Elderly Is Affected by Clothing, Activity, and Psychological Adjustment," *ASHRAE Transactions,* Part II A, pp. 329–342, 1986.

EPA, *National Primary and Secondary Ambient Air Quality Standards, Code of Federal Regulations, Title 40 part 50,* 1989.

Fanger, P. O., *Thermal Comfort Analysis and Applications in Environmental Engineering,* McGraw-Hill, New York, 1970.

Federal Supply Service, *Federal Standard No. 209B, Clean Rooms and Work Station Requirements, Controlled Environment,* General Services Administration, Washington, DC, 1973.

Gagge, A. P., Stolwijk, J. A. J., and Nishi, Y., "An Effective Temperature Scale Based on a Simple Model of Human Physiological Regulatory Response," *ASHRAE Transactions,* Part I, p. 247, 1971.

George, A. C., "Measurement of Sources and Air Concentrations of Radon and Radon Daughters in Residential Buildings," *ASHRAE Transactions,* Part II B, pp. 1945–1953, 1985.

Holzle, A. M., Munson, D. M., and McCullough, E. A., "A Validation Study of the ASHRAE Summer Comfort Envelope," *ASHRAE Transactions,* Part I B, pp. 126–138, 1983.

Janssen, J. E., "Ventilation for Acceptable Indoor Air Quality," *ASHRAE Journal,* pp. 40–46, October 1989.

McCullough, E. A., and Jones, B. W., "A Comprehensive Data Base for Estimating Clothing Insulation," *IER Technical Report 84-01,* Kansas State University, 1984.

Meckler, M., "Indoor air Quality—From Commissioning through Building Operation," *IAQ 91, Healthy Buildings,* pp. 77–81, 1992.

OSHA, *Code of Federal Regulations, Title 29, Part 1910,* 1992.

O'Sullivan, P., "Energy and IAQ Can Be Complementary," *Heating/Piping/Air Conditioning,* pp. 37–42, Feb. 1989.

Persily, A., "Ventilation Rates in Office Buildings," *ASHRAE Journal,* pp. 52–54, July 1989.

Rohles, F. H., Jr., Woods, J. E., and Nevins, R. G., "The Effect of Air Movement and Temperature on the Thermal Sensations of Sedentary Man," *ASHRAE Transactions,* Part I, pp. 101–119, 1974.

Seppanen, O., McNall, P. E., Munson, D. M., and Sprague, C. H., "Thermal Insulating Values for Typical Clothing Ensembles," *ASHRAE Transactions,* part I, pp. 120–130, 1972.

Spain, S., "The Upper Limit of Human Comfort from Measured and Calculated PMV Values in a National Bureau of Standards Test House," *ASHRAE Transactions,* Part I B, pp. 27–37, 1986.

Sterling, E. M., Arundel, A., and Sterling, T. D., "Criteria for Human Exposure to Humidity in Occupied Buildings," *ASHRAE Transactions,* Part I B, pp. 611–622, 1985.

Thayer, W. W., "Tobacco Smoke Dilution Recommendations for Comfort Ventilation," *ASHRAE Transactions,* Part II, pp. 291–304, 1982.

Woodcock, A. H., "Moisture Transfer in Textile Systems," *Textile Research Journal,* Vol. 8, pp. 628–633, 1962.

HEAT AND MOISTURE TRANSFER THROUGH BUILDING ENVELOPES

6.1 HEAT TRANSFER FUNDAMENTALS

Heat transfer between two bodies, two materials, or two regions is the result of temperature differences. The science of heat transfer has provided calculations and analyses to predict the rates of heat transfer. The design of an air conditioning system must include estimates of heat transfer between the conditioned space, its contents, and its surroundings to determine cooling and heating loads. Heat transfer analysis can be described in three modes: conduction, convection, and radiation.

Conductive Heat Transfer

Conduction is the mechanism of heat transfer in opaque solid media, such as through walls and roofs. For one-dimensional steady-state heat conduction q_k, in Btu/h, Fourier's law gives the following relationship:

$$q_k = -kA\frac{dT}{dx} \tag{6.1}$$

where k = thermal conductivity, Btu/h · ft · °F
A = cross-sectional area normal to heat flow, ft^2
T = Temperature, °F
x = coordinate dimension along the heat flow, ft

Equation (6.1) shows that the rate of heat transfer is directly proportional to the temperature gradient dT/dx, the thermal conductivity k, and the cross-sectional area A. The minus sign indicates that the heat must flow in the direction of decreasing temperature, that is, if the temperature decreases as x increases, the gradient dT/dx is negative, so heat conduction is a positive quantity.

For steady-state heat conduction through a plane composite wall with perfect thermal contact between each layer, as shown in Fig. 6.1, the rate of heat transfer through each section of the composite wall must be the same. From Fourier's law of conduction,

$$q_k = \frac{k_A A}{L_A}(T_1 - T_2) = \frac{k_B A}{L_B}(T_2 - T_3) = \frac{k_C A}{L_C}(T_3 - T_4) \tag{6.2}$$

where L_A, L_B, L_C = thickness of the layers A, B, and C, respectively, of the composite wall, ft
T_1, T_2, T_3, T_4 = temperature at the surfaces 1, 2, 3, and 4, respectively, °F
k_A, k_B, k_C = thermal conductivity of the layers $A, B,$ and C, respectively, of the composite wall, Btu/h · ft · °F

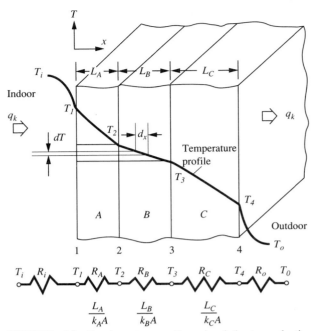

FIGURE 6.1 Steady-state one-dimensional heat conduction through a composite wall.

Eliminating T_2 and T_3, we have

$$q_k = \frac{T_1 - T_4}{(L_A/k_A A) + (L_B/k_B A) + (L_C/k_C A)} \tag{6.3}$$

For a multilayer composite wall of n layers in perfect thermal contact, the rate of conduction heat transfer is given as

$$q_k = \frac{T_1 - T_{n+1}}{(L_1/k_1 A) + (L_2/k_2 A) + \cdots + (L_n/k_3 A)} \tag{6.4}$$

Subscript n indicates the nth layer of the composite wall.

In Eq. (6.2), conduction heat transfer of any of the layers can be written as

$$q_k = \frac{kA}{L}\Delta T = \frac{\Delta T}{R^*}$$
$$R^* = \frac{L}{kA} \tag{6.5}$$

where R^* = thermal resistance, h · °F/Btu. In Eq. (6.5), an analogy can be seen between heat flow and Ohm's law for an electric circuit. Here, the temperature difference $\Delta T = (T_1 - T_2)$ indicates thermal potential, analogous to electric potential. Thermal resistance R is analogous to electric resistance, and heat flow q_k is analogous to electric current.

The total conductive thermal resistance of a composite wall of n layers R_T, in h · °F/Btu, can be calculated as

$$R_T^* = R_1^* + R_2^* + \cdots + R_n^* \tag{6.6}$$

where $R_1^*, R_2^*, \ldots, R_n^*$ = thermal resistances of layers 1, 2,..., n of the composite wall, in $\text{h} \cdot \text{°F/Btu}$. The thermal circuit of a composite wall of three layers is shown in the lower part of Fig. 6.1.

Convective Heat Transfer

Convective heat transfer occurs when a fluid comes in contact with a surface at a different temperature, such as the heat transfer taking place between the air stream in a duct and the duct wall.

Convective heat transfer can be divided into two types: forced convection and natural or free convection. When a fluid is forced to move along the surface by an outside motive force, heat is transferred by *forced convection*. When the motion of the fluid is caused by the density difference of the two streams in the fluid as a product of contacting a surface at a different temperature, the result is called *natural* or *free convection*.

The rate of convective heat transfer q_c, in Btu/h, can be expressed in the form of Newton's law of cooling, as

$$q_c = h_c A(T_s - T_\infty) \tag{6.7}$$

where h_c = average convective heat transfer coefficient, $\text{Btu/h} \cdot \text{ft}^2 \cdot \text{°F}$
T_s = surface temperature, °F
T_∞ = temperature of the fluid away from the surface, °F

In Eq. (6.7), the convective heat transfer coefficient h_c is usually determined empirically. It is related to a dimensionless group of fluid properties, such as the correlation of flat plate forced convection, as shown in the following equation:

$$\text{Nu}_L = \frac{h_c L}{k} = C\text{Re}_L^n \text{Pr}^m \tag{6.8}$$

where C = constant
k = thermal conductivity, $\text{Btu/h} \cdot \text{ft} \cdot \text{°F}$

In Eq. (6.8), Nu_L is the Nusselt number, which is based on a characteristic length L. Characteristic length can be the length of the plate, the diameter of the tube, or the distance between two plates, in ft. The Reynolds number, Re_L, is represented by the following dimensionless group:

$$\text{Re}_L = \frac{\rho v L}{\mu} = \frac{vL}{\nu} \tag{6.9}$$

where ρ = density of the fluid, lb/ft^3
v = velocity of the fluid, ft/s
μ = absolute viscosity of the fluid, $\text{lb/ft} \cdot \text{s}$
ν = kinematic viscosity of the fluid, ft^2/s

The Prandtl number, Pr, is represented by

$$\text{Pr} = \frac{3600 \mu c_p}{k} \tag{6.10}$$

where c_p = specific heat at constant pressure of the fluid, $\text{Btu/lb} \cdot \text{°F}$.

Fluid properties used to calculate dimensionless groups are usually related to the fluid temperature T_f, in °F. That is,

$$T_f = \frac{T_s - T_\infty}{2} \tag{6.11}$$

Convective heat transfer can also be considered analogous to an electric circuit. From Eq. (6.7), the convective thermal resistance R_c^*, in $\text{h} \cdot \text{°F/Btu}$, is given as

$$R_c^* = \frac{1}{h_c A} \tag{6.12}$$

Radiant Heat Transfer

In *radiant heat transfer*, heat is transported in the form of electromagnetic waves traveling at the speed of light. The net rate of radiant transfer q_r, in Btu/h, between a grey body at absolute temperature T_{R1} and a black surrounding enclosure at absolute temperature T_{R2} (for example, the approximate radiation exchange between occupant and surroundings in a conditioned space), can be calculated as

$$q_r = \sigma A_1 \epsilon_1 \left(T_{R1}^4 - T_{R2}^4\right) \tag{6.13}$$

where σ = Stefan-Boltzmann constant = $0.1714 \times 10^{-8} \text{Btu/h} \cdot \text{ft}^2 \cdot \text{°R}^4$
A_1 = area of the grey body, ft^2
ϵ_1 = emissivity of the surface of the grey body
T_{R1}, T_{R2} = absolute temperature of surfaces 1 and 2, °R

If we multiply the right-hand side of Eq. (6.13) by $(T_{R1} - T_{R2})/(T_{R1} - T_{R2})$, and let the thermal resistance

$$R_r^* = \frac{T_{R1} - T_{R2}}{\sigma A_1 \epsilon_1 (T_{R1}^4 - T_{R2}^4)} = \frac{1}{h_r A_1} \tag{6.14}$$

where h_r = radiant heat transfer coefficient, $\text{Btu/h} \cdot \text{ft}^2 \cdot \text{°F}$. We find that

$$q_r = h_r A_1 (T_1 - T_2) \tag{6.15}$$

where T_1, T_2 = temperature of surfaces 1 and 2, °F.

If either of the two surfaces is a black surface or can be approximated as a black surface, as was previously assumed for the conditioned space surroundings, the net rate of radiant heat transfer between surfaces 1 and 2 can be evaluated as

$$q_r = \sigma A_1 F_{1-2}(T_{R1}^4 - T_{R2}^4) \tag{6.16}$$

where F_{1-2} = shape factor for a diffuse emitting area A_1 and a receiving area A_2. The thermal resistance can be similarly calculated.

Overall Heat Transfer

In actual practice, many calculations of heat transfer rates are combinations of conduction, convection, and radiation.

Consider the composite wall shown in Fig. 6.1; in addition to conduction through the wall, convection and radiation also occur at inside and outside surfaces 1 and 4 of the composite wall. At the inside surface of the composite wall, the rate of heat transfer q_i, in Btu/h, consists of convective heat transfer between fluid, the air, and solid surface q_c and the radiant heat transfer q_r, as follows:

$$q_i = q_c + q_r = h_c A_1(T_i - T_1) + h_r A_1(T_i - T_1)$$
$$= h_i A_1(T_i - T_1) \qquad (6.17)$$

where T_i = indoor temperature, °F.

By analogy with Eq. (5.7), the inside surface heat transfer coefficient h_i at the liquid-to-solid interface, in Btu/h · ft² · °F, is

$$h_i = h_c + h_r$$

From Eq. (6.17), the thermal resistance R_i^* of the inner surface to convection and radiation, in h · ft² · °F/Btu, is

$$R_i^* = \frac{1}{h_i A_1} \qquad (6.18)$$

Similarly, at the outside surface of the composite wall, the rate of heat transfer q_o, in Btu/h · ft² · °F, is

$$q_o = h_o A_4(T_4 - T_o) \qquad (6.19)$$

where h_o = outside surface heat transfer coefficient at the fluid-to-solid interface, Btu/h · ft² · °F

A_4 = area of surface 4, ft²

T_4 = temperature of surface 4, °F

T_o = outdoor temperature, °F

The outer thermal resistance R_o^*, in h · °F/Btu, is

$$R_o^* = \frac{1}{h_o A_4}$$

For one-dimensional steady-state heat transfer, the overall heat transfer rate of the composite wall q, in Btu/h, can be calculated as

$$q = q_i = q_k = q_o = UA(T_i - T_o) = \frac{T_i - T_o}{R_T^*}$$
$$(6.20)$$

where U = overall heat transfer coefficient, often called U-value, Btu/h · ft² · °F

A = surface area perpendicular to heat flow, ft²

R_T^* = overall thermal resistance of the composite wall, h · °F/Btu

and

$$R_T^* = R_i^* + R_A^* + R_B^* + R_C^* + R_o^* = \frac{1}{UA} \qquad (6.21)$$

Also, the thermal resistances can be written as

$$R_A^* = \frac{L_A}{k_A A}; \qquad R_B^* = \frac{L_B}{k_B B}; \qquad R_C^* = \frac{L_C}{k_C C}$$

Therefore, the overall heat transfer coefficient U is given as

$$U = 1/[(1/h_i) + (L_A/k_A) + (L_B/k_B)$$
$$+ (L_C/k_C) + (1/h_o)] \qquad (6.22)$$

For plane surfaces, area $A = A_1 = A_2 = A_3 = A_4$. For cylindrical surfaces, because the inside and outside surface areas are different and because $UA = 1/R_T^*$, it must be clarified whether the area is based on the inside surface area A_i, the outside surface area A_o, or any chosen surface area.

For convenient HVAC&R heat transfer calculations, the reciprocal of overall heat transfer coefficient, often called overall R-value R_T, in h · ft² · °F/Btu, is used. R_T can be expressed as

$$R_T = \frac{1}{U} = R_i + R_1 + R_2 + \cdots + R_n + R_o$$
$$(6.23a)$$

where R_i, R_o = R-values of the inside and outside surfaces of the composite wall, h · ft² · °F/Btu

$R_1, R_2, \ldots, R_n$ = R-values of the components 1, 2,..., n, in h · ft² · °F/Btu.

And

$$R_i = 1/h_i, R_1 = L_1/k_1, R_2 = L_2/k_2, \ldots, R_n$$
$$= L_n/k_n \qquad (6.23b)$$

Sometimes, for convenience, the unit of R is often omitted; for example, R-10 means R-value equals 10 h · ft² · °F/Btu.

A *building envelope assembly*, or a building shell assembly, includes the exterior wall assembly (that is, walls, windows, and doors), the roof and ceiling assembly, and the floor assembly. The area-weighted average overall heat transfer coefficient of an envelope assembly U_{av}, in Btu/h · ft² · °F, can be calculated as

$$U_{\text{av}} = \frac{U_1 A_1 + U_2 A_2 + \cdots + U_n A_n}{A_o} \qquad (6.24)$$

where $A_1, A_2, \ldots, A_n$ = area of the individual elements $1, 2, \ldots, n$ of the envelope assembly, ft^2

A_o = gross area of the envelope assembly, ft^2

$U_1, U_2, \ldots, U_n$ = overall heat transfer coefficient of the individual paths $1, 2, \ldots, n$ of the envelope assembly, such as paths through windows, paths through walls, and paths through the roof, $Btu/h \cdot ft^2 \cdot °F$

Heat Capacity

The heat capacity (HC) per ft^2 of an element or component of a building envelope or other structure depends on its mass and specific heat. Heat capacity, in $Btu/ft^2 \cdot °F$, can be calculated as

$$HC = mc$$
$$\frac{HC}{A} = \rho L c \qquad (6.25)$$

where m = mass of building material, lb

c = specific heat of building material, $Btu/lb \cdot °F$

A = area of building material, ft^2

ρ = density of building material, lb/ft^3

L = thickness or height of building material, ft

6.2 HEAT TRANSFER COEFFICIENTS

Determining heat transfer coefficients to be used for load calculations or year-round energy estimates is complicated by the following types of variables:

- Building envelopes, exterior wall, roof, glass, partition wall, ceiling, or floor
- Fluid flow, turbulent flow or laminar flow, forced or free convection
- Heat flow, horizontal heat flow in a vertical surface, or an upward or downward heat flow in a horizontal surface
- Space air diffusion, ceiling or side wall inlet, or others
- Time of operation—summer, winter or other seasons

Among the three modes of heat transfer, convection processes and their related coefficients are the least understood, making analysis difficult.

Coefficient for Radiant Heat Transfer

For a radiant exchange between the inner surface of an exterior wall and the surrounding surfaces (such as the surfaces of partition walls, ceilings, and floors) in an air conditioned room, the sum of the shape factors $\sum F_{1-n}$ can be considered as unity. If all surfaces are assumed to be black, then the radiative heat transfer coefficient h_r, in $Btu/h \cdot ft^2 \cdot °F$, can be calculated as

$$h_r = \frac{\sigma(T_{Ris}^4 - T_{Rrad}^4)}{T_{Ris} - T_{Rrad}} \qquad (6.26)$$

where T_{Ris} = absolute temperature of the inner surface of the exterior wall, roof, or external window glass, °R

T_{Rrad} = absolute mean radiant temperature of the surrounding surfaces, °R

Often T_{Rrad} is approximately equal to the air temperature of the conditioned space T_{Rr} when both are expressed in °R.

Radiant heat transfer coefficients calculated according Eq. (6.26) for various surface temperatures and temperature differences between the inner surface of any building envelope and the surrounded surfaces are presented in Table 6.1. From Table 6.1, h_r depends on the absolute temperature of inner surface T_{Ris} and the temperature difference $(T_{is} - T_{rad})$.

Coefficients for Forced Convection

Before one can select the mean convective heat transfer coefficient h_c in order to calculate the rate of convective heat transfer or to determine the overall heat transfer coefficient U value during cooling load calculations, the type of convection (forced or natural) must be clarified. Forced convection between the surfaces of the building envelope and the conditioned space air occurs when the air-handling system is

TABLE 6.1
Radiative heat transfer coefficients h_r

Temperature difference $(T_{is} - T_{rad})$, °F	Inner surface temperature T_{is}, °F				
	60 (520°R)	70 (530°R)	75 (535°R)	80 (540°R)	85 (545°R)
	$Btu/h \cdot ft^2 \cdot °F$				
1	0.871	0.909	0.935	0.968	0.990
2	0.866	0.910	0.936	0.966	0.990
3	0.862	0.908	0.935	0.963	0.989
5	0.856	0.904	0.930	0.958	0.984
10	0.844	0.892	0.918	0.945	0.971
20	0.819	0.868	0.893	0.919	0.945

Note: Calculations made by assuming that $T_{rad} = T_r$ and the emissivity of inner surface $\epsilon_i = 1$.

operating or there is a wind over the outside surface. In an indoor space, free convection is always assumed when the air-handling system is shut off or there is no forced air motion over the surface involved.

The convective heat transfer coefficient h_c also depends on the air velocity v flowing over the surface, the configuration of the surface, the type of space air diffusion, and the properties of the fluid. According to Kays and Crawford (1980), a linear or nearly linear relationship between h_c and v holds. Even though the air velocity v in the occupied zone may be only 30 fpm or even lower when the air-handling system is operating, the mode of heat transfer is still considered as forced convection and can be expressed as

$$h_c = A + Bv^n \qquad (6.27)$$

where n = exponential index, usually between 0.8 and 1
v = bulk air velocity of the fluid 0.5 to 1 ft from the surface, f/m
A, B = constants

On the basis of test data from Wong (1990), Sato et al. (1972), and Spitler et al. (1991), as well as many widely used energy estimation computer programs, the convective heat transfer coefficient for forced convection h_c, in Btu/h · ft² · °F, for indoor surfaces can be determined as

$$h_c = 1.0 + 0.008\,v \qquad (6.28)$$

For outside surfaces, the surface heat transfer coefficient $h_o = h_c + h_r$, in Btu/h · ft² · °F, can be calculated as

$$h_o = 1.8 + 0.004\,v_{\text{wind}} \qquad (6.29)$$

where v_{wind} = wind speed, fpm.

Coefficients for Natural Convection

For natural convection, the empirical relationship between the dimensionless groups containing the convective heat transfer coefficient h_c can be expressed as follows:

$$\mathrm{Nu}_L = C(\mathrm{Gr}_L\mathrm{Pr})^n = C\mathrm{Ra}_L^n \qquad (6.30)$$

In Eq. (6.30), Gr_L, called the Grashof number, is based on the characteristic length L.

$$\mathrm{Gr}_L = \frac{\beta g \rho^2 (T_s - T_\infty)}{\mu^2} \qquad (6.31)$$

where β = coefficient of volume expansion of the fluid, 1/°R
g = acceleration of gravity, ft/s²
ρ = density of the fluid, lb/ft³
C = constant

Ra is called the Rayleigh number, and Ra = GrPr.

Natural convective heat transfer is difficult to evaluate because of the complexity of the recirculating convective stream of room air that is the result of the temperature distribution of the surfaces and the variation of temperature profile of the stream. Computer programs using numerical techniques have been developed recently. In actual practice, simplified calculations are often used.

Altmayer et al.(1983) in their recent experiments and analyses found that "The ASHRAE free convection correlations provide a fair prediction of the heat flux to the room air from cold and hot surfaces." The errors are mainly caused by the variation of temperature of the convective airstreams after contact with cold or hot surfaces. In simplified calculations, this variation can only be included in the calculation of the mean space air temperature T_∞.

Many rooms have a vertical wall height of about 9 ft. If the temperature difference $(T_s - T_\infty) > 1°F$, the natural convection flow is often turbulent. If $(T_s - T_\infty) \leq 1°F$, as in many cooling load calculations between a partition wall and space air, the flow is laminar.

Based on data published in *ASHRAE Handbook 1989 Fundamentals,* the natural convection coefficients h_c, in Btu/h · ft² · °F, are given as follows:

Vertical plates

Large plates, turbulent flow
$$h_c = 0.19(\Delta T_{\text{sa}})^{0.33} \qquad (6.32)$$

Small plates, laminar flow
$$h_c = 0.29(\Delta T_{\text{sa}}/L)^{0.25} \qquad (6.33)$$

Horizontal plates, facing upward when air is heated or facing downward when air is cooled

Large plates, turbulent flow
$$h_c = 0.22(\Delta T_{\text{sa}})^{0.33} \qquad (6.34)$$

Small plates, laminar flow
$$h_c = 0.27(\Delta T_{\text{sa}}/L)^{0.25} \qquad (6.35)$$

Horizontal plates, facing upward when air is cooled or facing downward when air is heated

Small plates $\quad h_c = 0.12(\Delta T_{\text{sa}}/L)^{0.25} \qquad (6.36)$

Here, ΔT_{sa} indicates the temperature difference between the surface and the air, in °F.

The convective coefficient for natural convection h_c, in Btu/h · ft² · °F, can be determined as follows:

Direction of heat flow	Turbulent flow, $\Delta T_{\text{sa}} = 10°F$	Laminar flow, $\Delta T_{\text{sa}} = 1°F$
Horizontal—upward	0.47	0.215
Vertical—horizontal	0.406	0.231
Horizontal—downward	—	0.095

Surface Heat Transfer Coefficients

The surface heat transfer coefficient h, sometimes called surface conductance, in Btu/h · ft² · °F, is the combination of convective and radiant heat transfer coefficients, that is, $h = h_c + h_r$. Table 6.2 lists h values for various surface types at $\Delta T_{sa} = 10°F$ and $\Delta T_{sa} = 1°F$ during summer and winter design conditions. The values are based upon the following:

- $T_{rad} = T_a$, where T_a is the air temperature, in °F
- ΔT_{sa} indicates the temperature difference between the surface and the air, °F.
- Emissivity of the surface $\epsilon = 0.9$ and $\epsilon = 0.2$.

6.3 MOISTURE TRANSFER

Moisture is water in the vapor, liquid and solid state. Building materials exposed to excess moisture may degrade or deteriorate as a result of physical changes, chemical changes, and biological processes. Moisture accumulated inside the insulating layer also increases the rate of heat transfer through the building envelope. Moisture transfer between the building envelope and the conditioned space air has a significant influence on the cooling load calculations in areas having hot and humid climates.

Sorption Isotherm

Moisture content X, which is dimensionless or else in percentage, is defined as the ratio of the mass of moisture contained in a solid to the mass of the bone-dry solid.

An absorption isotherm is a constant-temperature curve for a material in which moisture content is plotted against an increased ambient relative humidity during an equilibrium state, that is, the rate of condensation of water vapor on the surface of the material is equal to the rate of evaporation of water vapor from the material.

A desorption isotherm is also a constant-temperature curve for a material. It is a plot of moisture content vs a decreased ambient relative humidity during equilibrium state.

Many building materials show different absorption and desorption isotherms. The difference in moisture content at a specific relative humidity between the absorption and desorption isotherms is called *hysteresis*. In Fig. 6.2 are shown absorption and desorption isotherms of a building material.

Temperature also has an influence on the moisture content of many building materials. At a constant relative humidity in ambient air, the moisture content of a building material will be lower at a higher temperature.

When a building material absorbs moisture, heat as heat of sorption is evolved. If liquid water is absorbed by the material, an amount of heat q_1, in Btu per lb of water absorbed, similar to the heat of solution, is released. This heat results from the attractive forces between the water molecules and the molecules of building material. If water vapor is absorbed, the heat released q_v, in Btu/lb, is given by

$$q_v = q_l + h_{fg} \qquad (6.37)$$

where h_{fg} = latent heat of condensation, Btu/lb. Heat of sorption of liquid water q_l varies with equilibrium

TABLE 6.2
Surface heat transfer coefficients h in Btu/h · ft² · °F

Description	Direction of heat flow	Surface emissivity ϵ						
		0.90				0.20		
		Indoor surface				Indoor surface		
		$\Delta T_{sa} = 10°F$				$\Delta T_{sa} = 10°F$		
		Summer	Winter	$\Delta T_{sa} = 1°F$	Outdoor Surface	Summer	Winter	$\Delta T_{sa} = 1°F$
Forced convection								
<30 fpm		2.21	2.11	2.16		1.46	1.43	1.44
50 fpm		2.37	2.27	2.32		1.62	1.59	1.60
660 fpm					4.44			
(7.5 mph)								
1320 fpm					7.08			
(15 mph)								
Free convection								
Horizontal surface	Upward	1.36	1.27	1.17		0.62	0.60	0.44
Vertical surface		1.42	1.33	1.15		0.68	0.66	0.30
Horizontal surface	Downward			1.03				

Note: Assume space temperature $T_r = 74°F$ year-round and $T_r = T_{rad}$; here T_{rad} indicates the mean radiant temperature of the surroundings.

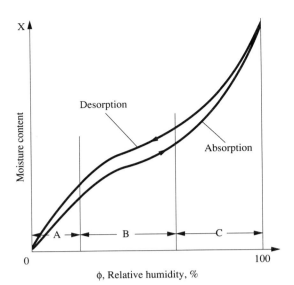

FIGURE 6.2 Sorption isotherms.

moisture content for a given material. The lower the X and the φ of ambient air, the higher will be the value of q_l. For pine, q_l may vary from 450 Btu/lb for nearly bone-dry wood to 20 Btu/lb at a moisture content of 20 percent. Many building materials have very low q_l values compared with h_{fg}, such as a q_l of 40 Btu/lb for sand.

Moisture-Solid Relationship

Many building materials have numerous interstices and microcapillaries of radius less than 4×10^{-6} in. (0.1 μm), which may or may not be interconnected. These interstices and microcapillaries provide large surface areas to absorb water molecules. Moisture can be bound to the solid surfaces by retention in the capillaries and interstices, or by dissolution into the cellular structures of fibrous materials.

When the relative humidity φ of ambient air is less than 20 percent, moisture is tightly bound to individual sites in a monomolecular layer (Region A, Fig. 6.2). Moisture moves by vapor diffusion. The binding energy is affected by the characteristics of the surface, the chemical structure of the material, and the properties of water.

When relative humidity φ is in a range of 20 to 60 percent (region B, Fig. 6.2), moisture is bound to the surface of the material in multimolecular layers and is held in microcapillaries. Moisture begins to migrate in liquid phase and its total pressure is reduced by the presence of moisture in microcapillaries. The binding energy involved is mainly the latent heat of condensation.

When $\varphi > 60$ percent, that is, in region C of Fig. 6.2, moisture is retained in large capillaries. It is relatively free for removal and chemical reactions.

The vapor pressure of the moisture is influenced only moderately because of the moisture absorbed in regions A and B. Moisture moves mainly in a liquid phase.

Because the density of the liquid is much greater than that of water vapor, the moisture content in building materials is mostly liquid.

Unbound moisture can be trapped in interstices having a radius greater than 4×10^{-5} in. (1 μm), without significantly lowering the vapor pressure. Free moisture is the moisture in excess of the equilibrium moisture content in a solid.

For insulating materials in which the interstices are interconnected, air perforates through open cells, moisture can accumulate in the form of condensate and be retained in the large capillaries and pores.

Moisture Migration in Building Materials

Most building envelopes are not constructed with open-cell materials. The airstream and its associated water vapor cannot normally penetrate building materials. Air leakage can only squeeze through the cracks and gaps around windows and joints. However, all building materials are moisture permeable, in other words, moisture can migrate across a building envelope because of differences in moisture content or other driving potentials.

According to Wong and Wang (1990), many theories have been proposed by scientists to predict the migration of moisture in solids. The currently accepted model of moisture flow in solids is based upon the following assumptions:

- Moisture migrates in solids in both liquid and vapor states. Liquid flow is induced by capillary flow and concentration gradients; vapor diffusion is induced by vapor pressure gradients.
- During the transport process, the moisture content, the vapor pressure, and the temperature are always in equilibrium at any location within the building material.
- Heat transfer within the building material is in the conduction mode. It is also affected by latent heat from phase changes.
- Vapor pressure gradients can be determined from moisture contents by means of sorption isotherms.
- Fick's law is applicable.
- Only one-dimensional flow across the building envelope is considered. Building materials are homogeneous.

For simplicity, the mass flux $\dot{m}_w/A$ for one-dimensional flow, in lb/h · ft², can be expressed as

$$\frac{\dot{m}_w}{A} = -\frac{\rho_w D_{lv} dX}{dx} - \frac{D_T dT}{dx} \qquad (6.38)$$

where D_{lv} = mass diffusivity of liquid and vapor, ft^2/h

X = moisture content, lb/lb dry solid

D_T = mass diffusivity due to temperature gradient, $ft^2/h \cdot °F$

ρ_w = density of water, lb/ft^3

x = coordinate dimension along the moisture flow, ft

A = area of the building envelope perpendicular to the moisture flow, ft^2

If the temperature gradient is small, with D_T comparatively smaller than D_{lv}, then $D_T dT/dx$ can be ignored, and Eq. (6.38) is simplified to

$$\frac{\dot{m}_w}{A} = -\frac{\rho_w D_{lv} dX}{dx} \qquad (6.39)$$

Mass diffusivities of some building materials as a function of moisture content are shown in Fig. 6.3.

Moisture Transfer from the Surface of the Building Envelope

Moisture migrating from any part in the building envelope must be balanced by convective moisture transfer from the surface of the building envelope to the ambient air or from the ambient air onto the surface. Such a convective moisture transfer is a part of the latent load. Analogous to the rate of convective heat transfer [Eq. (6.7)], the rate of convective moisture transfer $\dot{m}_w$, in lb/h, can be calculated as

$$\dot{m}_w = h_m A_w (C_{ws} - C_{wr}) \qquad (6.40)$$

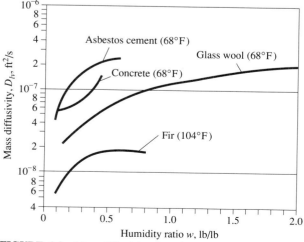

FIGURE 6.3 Mass diffusivity D_{lv} of some building materials. (Abridged with permission from Bruin et al. *Advances in Drying*, Vol. 1, 1979.)

where h_m = convective mass transfer coefficient, ft/h

A_w = contact area between the moisture and the ambient air, ft^2

C_{ws}, C_{wr} = mass concentration of moisture at the surface of the building envelope and of the space air, lb/ft^3

It is more convenient to express the mass concentration difference $(C_{ws} - C_{wr})$ in terms of a humidity ratio difference $(w_s - w_r)$. Here, w_s and w_r represent the humidity ratio at the surface of the building envelope and of the space air, respectively, in lb/lb. In terms of mass concentration we can write

$$C_{ws} = \rho_a w_s; \qquad C_{wr} = \rho_a w_r \qquad (6.41)$$

Then the rate of convective moisture transfer can be rewritten as

$$\dot{m}_w = \rho_a h_m A_w (w_s - w_r) \qquad (6.42)$$

In Eq. (6.42), the surface humidity ratio w_s depends on the moisture content X_s, the temperature T_s, and, therefore, the vapor pressure p_{ws} in the interstices of the surface of the building envelope. From the known X_s, T_s, and the sorption isotherm, the corresponding relative humidity φ_s at the surface can be determined. If the difference between relative humidity φ and the degree of saturation μ is ignored, then

$$w_s = \mu_s w_{ss} \approx \varphi_s w_{ss} \qquad (6.43)$$

where μ_s = degree of saturation at the surface. In Eq. (6.43), w_{ss} represents the humidity ratio of the saturated air. It can be determined from Eq. (2.49) since T_s is a known value.

In Eq. (6.42), the contact area A_w between the liquid water at the surface of the building envelope and the space air is a function of moisture content X_s. A precise calculation of A_w is very complicated, but if the surface area of the building material is A_s, then a rough estimate can be made from

$$A_w = X_s A_s \qquad (6.44)$$

Convective Mass Transfer Coefficients

The Chilton-Colburn analogy relates heat and mass transfer in these forms:

$$j_H = j_D$$

$$\frac{h_c}{\rho_a v c_{pa}} Pr^{2/3} = \frac{h_m}{v} Sc^{2/3} \qquad (6.45)$$

$$\frac{h_c}{\rho_a c_{pa}} \left(\frac{v_a}{\alpha}\right)^{2/3} = h_m \left(\frac{v_w}{D_{aw}}\right)^{2/3}$$

where v = air velocity remote from the surface, ft/s

c_{pa} = specific heat of moist air, Btu/lb $\cdot$ °F

h_m = convective mass transfer coefficient, ft/h

$$\nu = \text{kinematic viscosity, ft}^2/\text{s}$$
$$\alpha = \text{thermal diffusivity of air, ft}^2/\text{s}$$
$$D_{aw} = \text{mass diffusivity for water vapor diffus-}$$
$$\text{ing through air, ft}^2/\text{s}$$
$$\text{Sc} = \text{Schmidt number}$$

Subscript a indicates dry air, and w, water vapor.

For air at 77°F, $c_{pa} = 0.243$ Btu/lb · °F, $\nu_a = 1.74 \times 10^{-4}$ ft²/s, $\alpha = 2.44 \times 10^{-4}$ ft²/s, $D_{aw} = 2.83$ ft²/s, and $\rho_a = 0.0719$ lb/ft³, we can show that

$$\text{Pr} = \frac{\nu_a}{\alpha} = \frac{1.74 \times 10^{-4}}{2.44 \times 10^{-4}} = 0.713$$

and that

$$\text{Sc} = \frac{\nu_a}{D_{aw}} = \frac{1.74 \times 10^{-4}}{2.83 \times 10^{-4}} = 0.614$$

Equation (6.45) is valid for both heat and moisture transfer at a space temperature of 77°F. Therefore,

$$h_m = \frac{h_c \text{Pr}^{2/3}}{\rho_a c_{pa} \text{Sc}^{2/3}}$$
$$= \frac{(0.713)^{2/3}}{0.0719 \times 0.243 \times (0.614)^{2/3}} = 63.2\, h_c$$

Moisture Transfer in Building Envelopes

Moisture transfer in building envelopes can proceed along two paths, as shown in Fig. 6.4:

1. Moisture migrates into the building material mainly in the form of liquid if the relative humidity of the ambient air is over 50 percent. It will be transported to the indoor or outdoor air by convective mass transfer. The driving potentials are the moisture content of the building material, the vapor pressure gradient inside the building material, and the humidity ratio gradient between the surface and the ambient air.

2. Leakage of air and the associated water vapor infiltrates or exfiltrates through the cracks and gaps around the windows, doors, fixtures, outlets, and between the joints. Air and moisture enter the cavities and the air space in the building envelope. If the sheathing is not airtight, the leakage of air and its water vapor penetrate the perforated insulating board and discharge to the atmosphere, as shown in Fig. 6.4. If the sheathing is a closed-cell, airtight insulating board, then the air stream may infiltrate through gaps between the joints of the insulating board or through cracks between the window sill and the external wall and discharge to the atmosphere.

The driving potential for the leakage of air and the associated water vapor is the total pressure differen-

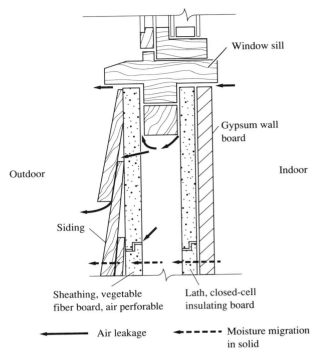

FIGURE 6.4 Moisture transfer in building envelope along two paths.

tial across the building envelope due to wind effects, temperature differences between indoor and outdoor air, or mechanical ventilation.

In leaky buildings, the moisture transfer by means of air leakage is normally far greater than the moisture migration through solids. For better-sealed commercial buildings, moisture migration through the building envelope may be dominant.

6.4 CONDENSATION IN BUILDINGS

When moist air contacts a solid surface whose temperature is lower than the dew point of the moist air, condensation occurs on the surface in the form of liquid water or sometimes frost. Condensation can damage the surface finish; deteriorate the material and cause objectionable odors, stains, corrosion, and mold growth; reduce the quality of the building envelope with dripping water; and fog windows.

Two types of condensation predominate in buildings:

1. Visible surface condensation on the interior surfaces of external window glass, below-grade walls, floor slabs on grade, and cold surfaces of inside equipment and pipes

2. Concealed condensation within the building envelope

Visible Surface Condensation

To avoid visible surface condensation, either the dew point of the indoor air must be reduced to a temperature below that of the surface, or the indoor surface temperature must be raised to a level higher than the dew point of the indoor air.

The indoor surface temperature of the building envelope, such as T_1 of the composite external wall in Fig. 6.1, or any cold surface where condensation may occur can be calculated from Eqs. (6.17) and (6.20) as

$$q = UA(T_i - T_o) = h_i A(T_i - T_1)$$

The inner surface temperature of the building envelope T_1, in °F, is given by

$$T_1 = T_i - \frac{U(T_i - T_o)}{h_i} \qquad (6.46)$$

where q = rate of heat transfer through the building envelope, Btu/h

h_i = inside surface heat transfer coefficient, Btu/h · ft² · °F

Increasing the thermal insulation is always an effective and economical way to prevent condensation because it saves energy and raises the surface temperature of the building envelope and other cold surfaces.

The dew point temperature of the indoor air $T_{r.\text{dew}}$ is a function of humidity ratio w_r of the space air. At a specific space temperature T_r, the lower the relative humidity, the lower the $T_{r.\text{dew}}$. There are several ways to lower the dew point of the indoor space air:

- By blocking and controlling infiltration of hot, humid outside air
- By reducing indoor moisture generation
- By using a vapor retarder, like polyethylene film or asphalt layer, to prevent or decrease the migration of moisture from the soil and the outside wetted siding and sheathing
- By introducing dry outdoor air through mechanical or natural ventilation when doing so is not in conflict with room humidity criteria
- By using dehumidifiers to reduce the humidity ratio of the indoor air

During winter, any interior surface temperature of the wall, roof, or glass is always lower than that of the indoor space temperature. Better insulation and multiple glazing increase the interior surface temperature and, at the same time, reduce heat loss, providing a more comfortable indoor environment than is possible by decreasing the relative humidity.

To reduce excessive indoor humidity in many industrial applications, a local exhausting booth that encloses the moisture-generating source is usually the remedy of first choice.

During summer, the outer surface of chilled water pipes and refrigerant pipes (even the cold supply duct) is at a lower temperature than indoor air temperature. Sufficient pipe and duct insulation is needed to prevent surface condensation. When an air handling system is shut down, the heavy construction mass often remains at a comparatively lower surface temperature as the indoor humidity rises. To avoid indoor surface condensation, the enclosure must be tight enough to prevent the infiltration of hot, humid air from outdoors.

Concealed Condensation within the Building Envelope

Usually, concealed condensation within building materials does not accumulate as a result of moisture migration. Excessive free moisture, usually in the form of liquid or frost, at any location inside the building envelope results in a higher moisture content than in surrounding areas. Therefore, it produces a moisture migration out of the materials rather than into them.

During cold seasons, concealed condensation in building envelopes is mainly caused by the warm indoor air, usually at a dew point temperature of 32°F and above. It leaks outward through the cracks and openings and enters the cavities, the gaps between component layers, or even the penetrable insulating material. It ultimately contacts a surface at a temperature lower than the dew point of the indoor air. Donnelly et al. (1976) discovered that a residential stud wall panel with a poor vapor retarder accumulated about 3 lb/ft² of moisture per ft² of wall area during a period of 31 days in winter. This moisture accumulated in the mineral fiber insulating board adjacent to the cold side of the sheathing and at the interface between the insulating board and the sheathing.

The rate of condensation $\dot{m}_{\text{con}}$, in lb/h, can be calculated as

$$\dot{m}_{\text{con}} = 60 \, \dot{V}_{\text{lk}} \rho_r (w_r - w_{\text{s.con}}) \qquad (6.47)$$

where $\dot{V}_{\text{lk}}$ = volume flow rate of air leakage, cfm

ρ_r = density of the indoor air, lb/ft³

$w_{\text{s.con}}$ = saturated humidity ratio corresponding to the temperature of the surface upon which condensation occurs, lb/lb

Vapor retarders are effective for reducing moisture transfer through the building envelope. They not only decrease the moisture migration in the building material significantly, but also block air leakage effectively if their joints are properly sealed.

Vapor retarders are classified as rigid, flexible, and coating types. Rigid types include reinforced plastic, aluminum, and other metal sheets. Flexible types include metal foils, coated films, and plastic films. Coatings are mastics, paints, or fusible sheets, composed of asphalt, resin, or polymeric layers. The vapor retarder is generally located on the warm side of the insulation layer during winter heating.

6.5 THERMAL INSULATION

Thermal insulation materials, usually in the form of boards, slabs, blocks, films, or blankets, retard the rate of heat transfer in conductive, convective, and radiant transfer modes. They are used within building envelopes or applied over the surfaces of equipment, piping, or ductwork for the following benefits:

- Saving energy by reducing heat loss and heat gain from the surroundings
- Preventing surface condensation by increasing the surface temperature above the dew point of the ambient air
- Reducing the temperature difference between the inside surface and the space air for the thermal comfort of the occupants, when radiant heating or cooling are not desired
- Protecting the occupant from injury due to contact with hot pipes and equipment

Basic Materials and Thermal Properties

Basic materials in the manufacture of thermal insulation for building envelopes or air conditioning systems include

- Fibrous materials such as glass fiber, mineral wool, wood, cane, or other vegetable fibers
- Cellular materials such as cork, foam rubber, polystyrene, and polyurethane
- Metallic reflective membranes

Most insulating materials consist of numerous air spaces, either closed-cells, (i.e., air-tight cellular materials), or open-cells (i.e., fibrous materials, penetrable by air).

Thermal conductivity k, an important physical property of insulating material, indicates the rate of heat transfer by means of a combination of gas and solid conduction, radiation, and convection through an insulating material, expressed in Btu $\cdot$ in./h $\cdot$ ft$^2\cdot$°F or Btu/h $\cdot$ ft $\cdot$ °F. The thermal conductivity of an insulating material depends on its physical structure (such as cell size and shape or diameter of the fibrous materials) density, temperature, and type and amount of binders and additives.

Most of the currently used thermal insulating materials have thermal conductivities within a range of 0.15 to 0.4 Btu $\cdot$ in./h $\cdot$ ft$^2\cdot$°F. Thermal properties of some building and insulating materials, based on data published in the *ASHRAE Handbook 1989, Fundamentals*, are listed in Table 6.3.

The thermal conductivity of many cellular insulating materials made from polymers like polyurethane and extruded polystyrene is not significantly affected by changes in density ρ. Other insulating materials have a density at which the thermal conductivity is minimum. Deviating from this ρ, k increases always, whether ρ is greater or smaller in magnitude.

The thermal conductivity of an insulating material apparently increases as its mean temperature rises. For polystyrene, k increases from 0.14 to 0.32 Btu $\cdot$ in./h$\cdot$ ft$^2\cdot$°F when its mean temperature is raised from 300°F to 570°F.

Moisture Transfer

From the point of view of moisture transfer, penetrability is an important characteristic. Closed-cell airtight board or block cannot have concealed condensation except at the gap between the joints and at the interface of two layers. Only when air and its associated water vapor penetrate an open-cell insulating material can concealed condensation form if they contact the surfaces of the interstices and pores at a temperature lower than the dew point of the penetrating air. Concealed condensation might also accumulate in open-cell insulating materials.

If there is excess free moisture in an insulating material, the conductive thermal resistance may be reduced. A heat transfer increase of 3 to 5 percent has been found by researchers for each percent increase of moisture content.

Economic Thickness

The economic thickness of insulation is the thickness with the lowest owning and operating costs.

Owning cost is the net investment cost of the installed insulation C_{in}, in dollars, less any capital investment that can be made as a result of lower heat loss or gain C_{pt}, in dollars. Theoretically, for a new plant, some small saving might be made because of the reduction of the size of a central plant, but in actual practice, this is seldom considered. For an existing plant where add-on insulation is being considered, C_{pt} is zero because the plant investment has already been made.

Operating cost C_{en} includes the annualized cost of energy over the life of a new plant, or the remaining

TABLE 6.3
Thermal properties of selected materials

	Density, lb/ft^3	Thermal conductivity, Btu/h · ft · °F	Specific heat, Btu/lb · °F		Emissivity
Aluminium (alloy 1100)	171	128	0.214	0.09	
Asbestos: insulation	120	0.092	0.20	0.93	
Asphalt	132	0.43	0.22		
Brick, building	123	0.4	0.2	0.93	
Brass (65% Cu, 35% Zn)	519	69	0.09	0.033	Highly polished
Concrete (stone)	144	0.54	0.156		
Copper (electrolytic)	556	227	0.092	0.072	Shiny
Glass: crown (soda-lime)	154	0.59	0.18	0.94	Smooth
Glass wool	3.25	0.022	0.157		
Gypsum	78	0.25	0.259	0.903	Smooth plate
Ice (32 °F)	57.5	1.3	0.487	0.95	
Iron: cast	450	27.6	0.12	0.435	Freshly turned
Mineral fiber board:					
acoustic tile, wet molded	23	0.035	0.14		
wet felted	21	0.031	0.19		
Paper	58	0.075	0.32	0.92	
Polystyrene, expanded, molded beads	1.25	0.021	0.29		
Polyurethane, cellular	1.5	0.013	0.38		
Plaster, cement and sand	132	0.43		0.91	Rough
Platinum	1340	39.9	0.032	0.054	Polished
Rubber: vulcanized (soft)	68.6	0.08	0.48	0.86	Rough
(hard)	74.3	0.092		0.95	Glossy
Sand	94.6	0.19	0.191		
Steel (mild)	489	26.2	0.12		
Tin	455	37.5	0.056	0.06	Bright
Wood: fir, white	27	0.068	0.33		
oak, white	47	0.102	0.57	0.90	Planed
plywood, Douglas fir	34	0.07	0.29		
Wool: fabric	20.6	0.037			

Adapted with permission from *ASHRAE Handbook 1989, Fundamentals.*

life of an existing plant, in *n* years. It can also be taken as the number of years over which an owner wishes to have a total return of the net investment, considering both the interest and the fuel escalation rate.

Total cost of insulation C_t, in dollars, for any given thickness is

$$C_t = (C_{in} - C_{pt}) + C_{en} \qquad (6.48)$$

When the thickness of the insulating material increases, the quantity $(C_{in} - C_{pt})$ also becomes higher, as shown in Fig. 6.5, and C_{en} decreases. As a result, C_t first decreases, drops to a minimum, and then increases. The optimum economic thickness occurs when C_t drops to a minimum. The closest commercially available thickness is the optimum thickness.

Thermal Resistance of Air Spaces

Thermal resistance of an enclosed air space R_a has a significant effect on the total thermal resistance R_T of the building envelope, especially when the value of R_T is low. Thermal resistance R_a depends on the characteristic of the surface (reflective or nonreflec-

tive), the mean temperature, the temperature difference of the surfaces perpendicular to heat flow, the width across the air space along the heat flow, and the direction of air flow. The R-values of the enclosed air spaces R_a, in h · ft^2 · °F/Btu, abridged from data published in the *ASHRAE Handbook 1989, Fundamentals*, are presented in Table 6.4.

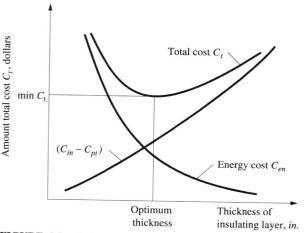

FIGURE 6.5 Optimum thickness of insulating material.

TABLE 6.4
R-Values of enclosed air space $R_a(h \cdot ft^2 \cdot °F/Btu)$

Mean temperature, °F	Temperature difference, °F	Type of surface	Direction of heat flow	Width of air space, in.	Emissivity E*		
					0.05	0.2	0.82
Summer, 90	10	Horizontal	Upward	0.5	2.03	1.51	0.73
				3.5	2.66	1.83	0.80
		Vertical	Horizontal	0.5	2.34	1.67	0.77
				3.5	3.40	2.15	0.85
		Horizontal	Downward	0.5	2.34	1.67	0.77
				3.5	8.19	3.41	1.00
Winter, 50	10	Horizontal	Upward	0.5	2.05	1.60	0.84
				3.5	2.66	1.95	0.93
		Vertical	Horizontal	0.5	2.54	1.88	0.91
				3.5	3.40	2.32	1.01
		Horizontal	Downward	0.5	2.55	1.89	0.92
				3.5	9.27	4.09	1.24

*Emissity $E = 1/(1/e_1 + 1/e_2 - 1)$, where e_1, e_2 indicate the emittances on two sides of the air space.

Abridged with permission from *ASHRAE Handbook 1989, Fundamentals*.

6.6 SOLAR ANGLES

Basic Solar Angles

The basic solar angles between the sun's rays and a specific surface under consideration are shown in Figs. 6.6 and 6.7.

- Solar altitude angle β (Figs. 6.7a, b) is the angle ROQ on a vertical plane between the sun's ray OR and its projection on a horizontal plane on the surface of the earth.
- Solar azimuth φ (Fig. 6.7a) is the angle SOQ on a horizontal plane between the due-south direction line OS and the horizontal projection of the sun's ray OQ.

- Solar declination angle δ (Fig 6.6) is the angle between the earth-sun line and the equatorial plane. Solar declination δ changes with the time of the year. It is shown in Fig. 6.6 on June 21.
- Surface-solar azimuth γ (Figs. 6.7a, c) is the angle POQ on a horizontal plane between the normal to a vertical surface OP and the horizontal projection of the sun's rays OQ.
- Surface azimuth ψ (Fig. 6.7a) is the angle POS on a horizontal plane between OP and the direction line SN.
- Latitude angle L (Fig. 6.6) is the angle $S'O'O$ on the longitudinal plane between the equatorial plane and the line $O'O$ that connects the point of incidence of the sun's rays on the surface of the earth O and the center of the earth O'.

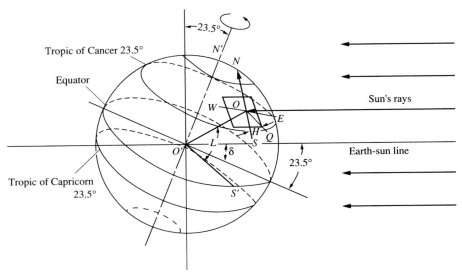

FIGURE 6.6 Basic solar angles and the position of the sun's rays at summer solstice.

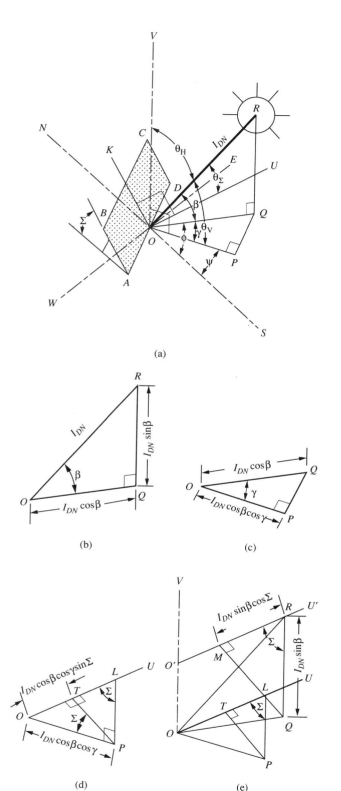

FIGURE 6.7 Solar intensity and angle of incidence.

Hour Angle and Apparent Solar Time

Hour angle H (Fig. 6.6) is the angle SOQ on a horizontal plane between the line OS indicating the noon

of local solar time t_{ls} and the horizontal projection of the sun's ray OQ.

The values of the hour angle H before noon are taken to be positive. At 12 noon, H is equal to 0. After noon, H is negative. Hour angle H can be calculated as

$$H = 0.25 \text{ (number of minutes from local solar noon)}$$
$$(6.49)$$

The relationship between apparent solar time t_{as} as determined by a sundial and expressed in apparent solar time and local standard time t_{st}, both in minutes, is as follows:

$$t_{as} = t_{eq} + t_{st} \pm 4(M - G) \qquad (6.50)$$

where M = local standard time meridian, degree
$\quad\quad G$ = local longitude, degree

In Eq. (6.50), t_{eq}, in minutes, indicates the difference in time between the mean time indicated by a clock running at a uniform rate and the solar time due to the variation of the earth's orbital velocity throughout the year.

Solar Angle Relationships

The relationship among the solar angles is given as

$$\sin \beta = \sin \delta \ \sin L + \cos \delta \ \cos H \ \cos L$$
$$(6.51)$$

and

$$\tan \varphi = \frac{\sin H}{\sin L \ \cos H - \cos L \ \tan \delta} \qquad (6.52)$$

Angle of Incidence and Solar Intensity

The angle of incidence θ (Fig. 6.7a) is the angle between the sun's rays radiating on a surface and the line normal to this surface. For a horizontal surface, the angle of incidence θ_H is ROV; for a vertical surface, the angle of incidence θ_V is ROP; and for a tilted surface, the angle of incidence θ_Σ is ROU. Here, Σ is the angle between the tilted surface $ABCD$ and the horizontal surface.

Let I_{DN} be the intensity of direct normal radiation, or solar intensity, on a surface normal to the sun's ray, in Btu/h · ft². In Fig. 6.7b, I_{DN} is resolved into the vertical component $RQ = I_{DN} \sin \beta$ and the horizontal component $OQ = I_{DN} \cos \beta$.

In Fig. 6.7c, for the right triangle OPQ, angle $POQ = \gamma$, and hence, $OP = I_{DN} \cos \beta \cos \gamma$.

In Fig. 6.7d, OPL is a right triangle. From point P, a line PT can be drawn perpendicular to the line normal OU, and hence, two right triangles are formed: OTP and LTP. In OTP, because angle $OPT = \Sigma$, then the horizontal component of I_{DN}

along the line normal of the tilting surface OU, $OT = I_{DN} \cos \beta \cos \gamma \sin \Sigma$.

In Fig. 6.7e, draw line $O'U'$ parallel to line OU. Again, line QM can be drawn from point Q perpendicular to $O'U'$. Then, the right triangles PTL and QMR are similar. Angles PLT and QRM are both equal to Σ. In the right triangle QMR, the component of RQ that is parallel to the line normal to the tilted surface $MR = I_{DN} \sin \beta \cos \Sigma$.

The intensity of solar rays normal to a tilted surface I_Σ, in Btu/h·ft^2, is the vector sum of the components of the line normal to the tilted surface, that is,

$$I_\Sigma = I_{DN} \cos \theta_\Sigma$$
$$= I_{DN}(\cos \beta \cos \gamma \sin \Sigma + \sin \beta \cos \Sigma)$$
$$(6.53)$$

6.7 SOLAR RADIATION

Solar radiation provides most of the energy required for the earth's occupants, either directly or indirectly. It is the source of indoor daylight and helps to maintain a suitable indoor temperature during the cold seasons. At the same time, its influence on the indoor environment must be reduced and controlled during hot weather.

The sun is located at a mean distance of 92,900,000 miles from the earth, and it has a surface temperature of about 10,800°F. It emits electromagnetic waves at wavelengths between 0.29 and 3.5 μm (micrometers). Visible light has wavelengths between 0.4 and 0.7 μm and is responsible for 38 percent of the total energy striking the earth. The infrared region contains 53 percent. At the outer edge of the atmosphere at a mean earth-sun distance, the solar intensity, called the solar constant I_{sc}, is 434.6 Btu/h·ft^2. The extraterrestrial intensity I_o, in Btu/h·ft^2, varies as the earth-sun distance changes during the earth's orbit.

Based on the data from Miller et al. (1983), the breakdown of solar radiation reaching the earth's sur-

TABLE 6.5
Components of solar radiation that traverse the earth's atmosphere

Components		Breakdowns	
Scattered by air	11%	Reflected to space	6%
		Scattered to earth	5%
Absorbed by water vapor, dust, etc.	16%		
Intercepted by clouds	45%	Reflected to space	20%
		Absorbed by clouds	4%
		Diffused through clouds and absorbed by earth	21%
Traversed through the air	28%	Absorbed by earth	24%
		Reflected by earth	4%

face and absorbed by the earth is listed in Table 6.5. As listed in Table 6.5, only 50 percent of the solar radiation that reaches the outer edge of the earth's atmosphere is absorbed by the clouds and the earth's surface. At any specific location, the absorption, reflection, and scattering of solar radiation depend on the composition of the atmosphere and the path length of the sun's rays through the atmosphere, expressed in terms of the air mass m. When the sun is directly overhead, $m = 1$.

In Table 6.5, the part of solar radiation that gets through the atmosphere and reaches the earth's surface, in a direction that varies with solar angles over time, is called *direct radiation*. The part that is diffused by air molecules and dust, arriving at the earth's surface in all directions, is called *diffuse radiation*.

The magnitude of solar radiation depends on the composition of the atmosphere, especially on the cloudiness of the sky. Therefore, different models are used to calculate the solar radiation reaching the surface of building envelopes.

Solar Radiation for a Clear Sky

ASHRAE recommends use of the following relationships to calculate the solar radiation for clear sky. The solar intensity of direct normal radiation I_{DN}, in Btu/h·ft^2, can be calculated as

$$I_{DN} = \frac{AC_n}{\exp(B/\sin\beta)} \qquad (6.54)$$

where A = apparent solar radiation when air mass $m = 0$; its magnitudes are listed in Table 6.6, in Btu/h·ft^2

B = atmospheric extinction coefficient (Table 6.6), which depends mainly on the amount of water vapor contained in the atmosphere

In Eq. (6.54), the term $1/\sin \beta$ in exponential form denotes the length of the direct radiation path through the atmosphere. Term C_n is the clearness number of the sky; C_n takes into account the dryness of the atmosphere and the dust contained in the air at a geographic location. Estimated C_n values for nonindustrial locations in the United States are shown in Fig. 6.8.

From Eq. (6.53), the direct radiation radiated onto a horizontal surface through a clear sky I_{DH}, in Btu/h·ft^2, is the vertical component of I_{DN}, that is,

$$I_{DH} = I_{DN} \cos \theta_H = I_{DN} \sin \beta \qquad (6.55)$$

and the direct radiation irradiated onto a vertical surface for clear sky I_{DV}, in Btu/h·ft^2, is the horizontal component of I_{DN}, or

TABLE 6.6
Extraterrestrial solar radiation and related data for twenty-first day of each month, base year 1964

	I_0, Btu/h · ft^2	Equation of time, min	Declination degrees	A Btu/h · ft^2	B (Dimensionless ratios)	C
January	448.8	−11.2	−20.0	390	0.142	0.058
Febuary	444.2	−13.9	−10.8	385	0.144	0.060
March	437.7	− 7.5	0.0	376	0.156	0.071
April	429.9	+ 1.1	+11.6	360	0.180	0.097
May	423.6	+ 3.3	+20.0	350	0.196	0.121
June	420.2	− 1.4	+23.45	345	0.205	0.134
July	420.3	− 6.2	+20.6	344	0.207	0.136
August	424.1	− 2.4	+12.3	351	0.201	0.122
September	430.7	+ 7.5	0.0	365	0.177	0.092
October	437.3	+15.4	−10.5	378	0.160	0.073
November	445.3	+13.8	−19.8	387	0.149	0.063
December	449.1	+ 1.6	−23.45	391	0.142	0.057

A: Apparent solar radiation; B: Atmospheric extinction coefficient; C: Diffuse radiation factor.
Source: ASHRAE Handbook 1989, Fundamentals. Reprinted with permission.

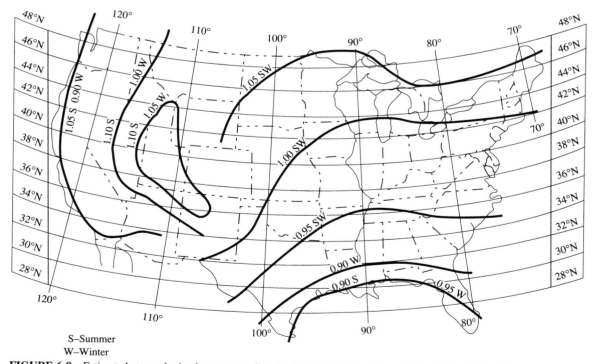

S—Summer
W—Winter

FIGURE 6.8 Estimated atmospheric clearness numbers in the United States for nonindustrial localities. (*Source: ASHRAE Handbook 1989, Fundamentals.* Reprinted with permission.)

$$I_{DV} = I_{DN} \cos \theta_V = I_{DN} \cos \beta \cos \gamma \quad (6.56)$$

From Eq. (6.53), for the direct radiation radiated onto a tilted surface at an angle Σ with the horizontal plane through clear sky I_D, in Btu/h · ft², can be evaluated as

$$I_D = I_\Sigma = I_{DN} \cos \theta_\Sigma$$
$$= I_{DN}(\cos \beta \cos \gamma \sin \Sigma + \sin \beta \cos \Sigma) \quad (6.57)$$

Most of the ultraviolet solar radiation is absorbed by the ozone layer in the upper atmosphere. Direct solar radiation, through an air mass $m = 2$, arrives on the earth's surface at sea level during a clear day with a spectrum of < 3 percent in the ultraviolet, 44 percent in the visible, and 53 percent in the infrared.

The diffuse radiation I_d, in Btu/h · ft², is proportional to I_{DN} on cloudless days and can be approximately calculated as

$$I_d = \frac{C I_{DN} F_{ss}}{C_n^2} \quad (6.58)$$

where C = diffuse radiation factor, as listed in
Table 6.6
C_n = clearness number of sky from Fig. 6.8

In Eq. (6.58), F_{ss} indicates the shape factor between the surface and the sky, or the fraction of short-wave radiation transmitted through the sky that reaches the surface. For a vertical surface, $F_{ss} = 0.5$; for a horizontal surface, $F_{ss} = 1$; and for any tilted surface with an angle Σ,

$$F_{ss} = \frac{1.0 + \cos \Sigma}{2} \quad (6.59)$$

The total or global radiation on a horizontal plane I_G, recorded by the U.S. National Climatic Data Center (NCDC), in Btu/h · ft², can be calculated as

$$I_G = I_D + I_d = I_{DN}\left(\sin \beta + \frac{C}{C_n^2}\right) \quad (6.60)$$

The reflection of solar radiation from any surface I_{ref}, in Btu/h · ft², is given as

$$I_{ref} = \rho_s F_{s-r}(I_D + I_d) \quad (6.61)$$

where ρ_s = reflectance of the surface
F_{s-r} = shape factor between the receiving surface and the reflecting surface

The ground-reflected diffuse radiation falling on any surface I_{sg}, in Btu/h · ft², can be expressed as

$$I_{sg} = \rho_g F_{sg} I_G \quad (6.62)$$

where F_{sg} = shape factor between the surface and the ground
ρ_g = reflectance of the ground.

For concrete, $\rho_g = 0.23$ and for bitumen and gravel, $\rho_g = 0.14$. A mean reflectance $\rho_g = 0.2$ is usually used for ground.

The total intensity of solar radiation I_t, in Btu/h · ft², falling on a surface at a direction normal to the surface on clear days is given by

$$I_t = I_D + I_d + I_{ref}$$
$$= I_{DN} \cos \theta + I_d + I_{ref} \quad (6.63)$$

Solar Radiation for a Cloudy Sky

For cloudy skies, the global horizontal irradiation I_G^*, in Btu/h · ft², usually can be obtained from the NCDC. If it is not available, then it can be predicted from the following relationship:

$$I_G^* = \left(1 + \frac{C_{cc}Q}{P} + \frac{C_{cc}^2 R}{P}\right) I_G \quad (6.64a)$$

Here, C_{cc} indicates the cloud cover on a scale of 0 to 10 and can be calculated by

$$C_{cc} = C_T - 0.5 \sum_{j=1}^{4} C_{cir.j} \quad (6.64b)$$

where C_T = total cloud amount
$C_{cir.j}$ = clouds covered by cirriforms, including cirrostratus, cirrocumulus, and cirrus, in $j = 1$ to 4 layers

Both C_T and $C_{cir.j}$ values can be obtained from the major weather stations. The values of coefficients P, Q, and R, according to Kimura and Stephenson (1969), are listed below:

	P	Q	R
Spring	1.06	0.012	−0.0084
Summer	0.96	0.033	−0.0106
Autumn	0.95	0.030	−0.0108
Winter	1.14	0.003	−0.0082

The direct radiation for a cloudy sky I_D^*, in Btu/h · ft², can be calculated as

$$I_D^* = \frac{I_G^* \sin \beta(1 - C_{cc}/10)}{\sin \beta + C/C_n^2} \quad (6.65)$$

The diffuse radiation for a cloudy sky I_d^*, in Btu/h·ft², is calculated as

$$I_d^* = I_G^* - I_D^* \quad (6.66)$$

6.8   FENESTRATION

Fenestration is the term used for assemblies containing glass or light-transmitting plastic, including appurtenances such as framing, mullions, dividers, and internal, external, and between-glass shading devices, as shown in Fig. 6.12a (see page 22). The purposes of fenestration are to (1) provide visibility of the outside world, (2) permit entry of daylight, (3) admit solar heat as a heating supplement in winter season, (4) act as an emergency exit for single-story buildings, and (5) to add to aesthetics.

Solar radiation admitted through a glass or window pane can be an important heat gain for commercial buildings, with greater energy impact in the Sun Belt. HVAC&R designers are asked to control this solar load while providing the required visibility, daylight, and winter heating benefits as well as fire protection and safety features.

Types of Window Glass (Glazing)

Most window glasses, or *glazing*, are vitreous silicate consisting of silicon dioxide, sodium oxide, calcium oxide, and sodium carbonate. They can be classified as follows:

- *Clear plate or sheet glass or plastic.* Clear plate glass permits good visibility and transmits a greater amount of solar radiation than other types.
- *Tinted heat-absorbing glass.* Tinted heat-absorbing glass is fabricated by adding small amounts of selenium, nickel, iron, or tin oxides. These produce colors from pink to green, including grey or bluish green, all of which absorb infrared solar heat and release a portion of this to the outside atmosphere through outer surface convection and radiation. Heat-absorbing glass also reduces visible light transmission.
- *Insulating glass.* Insulating glass consists of two panes—an outer plate and an inner plate—or three panes separated by metal or rubber spacers around the edges and hermetically sealed in a stainless steel or aluminum alloy structure. The dehydrated space between the glass panes usually has a thickness of 0.125–0.5 in. and is filled with air, argon, or other inert gas. Air- or gas-filled space increases the thermal resistance of the fenestration.
- *Reflective coated glass.* Reflective glass has a microscopically thin layer of metallic or ceramic coating on one surface of the glass, usually the inner surface of a single-pane glazing or the outer surface of the inner plate for an insulating glass. For a single pane, the coating is often protected by a layer of transparent polyester. The chromium and other metallic coatings give excellent reflectivity in the infrared regions with a comparatively lower reduc-

tion of the transmission of visible light than clear plate and heat-absorbing glass.

Reflections from buildings with highly reflective glass may blind drivers, or even kill grass in neighboring yards.

- *Low-emissivity glass coatings (low-E).* Glazing coated with low-emissivity, or low-E, films has been in use since 1978. It is widely used in retrofit applications. A low-emissivity film is usually a vacuum-deposited metal coating, usually aluminum, on a polyester film, at a thickness of about 4×10^{-7} in. Because of the fragility of the metal coating, protection by another polyester film against abrasion and chemical corrosion must be provided. Recently, copper and silver coatings on polyethylene and polypropylene film for protection have been used for better optical transmission.

A low-E film coating reduces the U-value about 25–30 percent for single panes. When combined with other solar control devices, these films can reduce solar heat gain further.

Optical Properties of Sunlit Glazing

When solar radiation impinges on the outer surface of a plate of glass with an intensity of I at an angle of incidence of θ, as shown in Fig. 6.9, a portion of the solar radiation is transmitted, another portion is reflected from both inner and outer surfaces, and the remaining portion is absorbed.

Let r indicate the portion reflected and a the portion absorbed. Also, let θ be the angle of incidence. In Fig. 6.9, it can be seen that the portion of the solar radiation transmitted through the glass is actually the sum of the transmittals after successive multiple reflections from the outer and inner surfaces. The decimal portion of I transmitted through the glass, or τ, represents the transmittance of the window glass. Similarly, the portion of solar radiation reflected from the window glass is the sum of the successive reflections from the outer surfaces after multiple reflections and absorptions, and is identified as ρ, the reflectance. The portion absorbed is the sum of the successive absorptions within the glass α, or its absorptance.

In Fig. 6.10 is illustrated the spectral transmittance of several types of glazing. All are transparent for short-wave solar radiation at a wavelength between 0.3 and 3 μm and opaque to long-wave radiation in the infrared range with a wavelength greater than 3 μm. Most interior furnishings, appliances, and equipment have an outer surface temperature lower than 250°F, emitting almost all long-wave radiation. At such temperatures, glass is opaque to long-wave radiation emitted from inside surfaces, and lets only

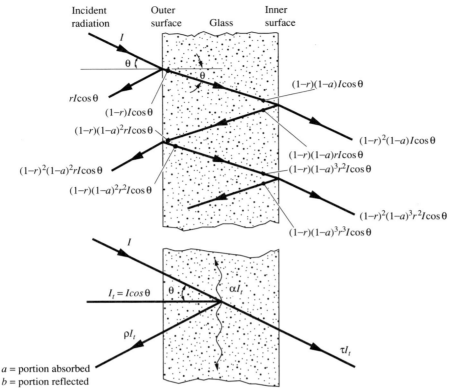

Incident radiation — Outer surface — Glass — Inner surface

I

θ

$rI\cos\theta$

$(1-r)I\cos\theta$

$(1-r)(1-a)^2rI\cos\theta$

$(1-r)^2(1-a)^2rI\cos\theta$

$(1-r)(1-a)^2r^2I\cos\theta$

$(1-r)(1-a)I\cos\theta$

$(1-r)^2(1-a)I\cos\theta$

$(1-r)(1-a)rI\cos\theta$

$(1-r)(1-a)^3r^2I\cos\theta$

$(1-r)^2(1-a)^3r^2I\cos\theta$

$(1-r)(1-a)^3r^3I\cos\theta$

I

$I_t = I\cos\theta$

θ

αI_t

ρI_t

τI_t

a = portion absorbed
b = portion reflected

FIGURE 6.9 Simplified representation of multiple transmissions, reflections, and absorptions of solar radiation at glass surfaces.

short-wave radiation through. This trapping of long-wave radiation indoors is called the *greenhouse effect*.

In Fig. 6.10 one can also see that clear plate glass has a high transmittance, $\tau = 0.87$ for visible light, and that $\tau = 0.8$ for the infrared from 0.7 to 3 μm. Heat-absorbing glass has a lower τ and higher absorp-

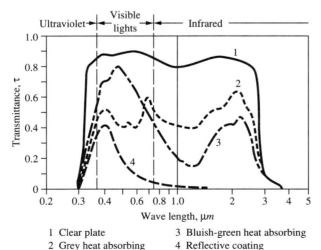

FIGURE 6.10 Spectral transmittance for various types of window glasses. (*Source: ASHRAE Handbook 1989, Fundamentals.* Reprinted by permission.)

1 Clear plate
2 Grey heat absorbing
3 Bluish-green heat absorbing
4 Reflective coating

tance α both for visible light and infrared radiation. Bluish green heat-absorbing glass has a higher τ in the visible range and a lower τ in the infrared range than grey heat-absorbing glass. Some reflective glazing has a high reflectance ρ and a significantly lower τ in the visible light range and is opaque to radiation at wavelengths greater than 2 μm. Such characteristics for heat-absorbing and reflective glasses are effective for reducing the amount of solar heat entering the conditioned space during cooling as well as heating seasons. This fact sometimes presents a dilemma for the designer, who must finally compromise to get the optimum combination of conflicting properties.

Another important property of glass is that τ and α both decrease and ρ increases sharply as the incident angle θ increases from 60° to approaching 90°. At 90°, τ and α are 0 and ρ is equal to 1. That is why the solar radiation transmitted through vertical glass declines sharply at noon during summer with solar altitudes $\beta > 70°$.

For all types of plate glass, the sum of these radiation components is always equal to 1, that is,

$$\tau + \alpha + \rho = 1$$

and

$$I = \tau I + \alpha I + \rho I \qquad (6.67)$$

6.9 HEAT ADMITTED THROUGH WINDOWS

For external glazing without shading, the heat gain admitted into the conditioned space through each ft^2 of sunlit area A_s of window q_{wi}/A_s, in Btu/h·ft^2, can be calculated as follows:

Heat gain through each ft^2 of sunlit window	=	Solar radiation transmitted through the window glass	+	Inward heat flow from the glass inner surface into the conditioned space

That is,

$$\frac{q_{wi}}{A_s} = \tau I_t + \frac{q_{RCi}}{A_s} \qquad (6.68)$$

where q_{RCi} = inward heat flow from the inner surface of an unshaded sunlit window, Btu/h.

Heat Gain for Single Glazing

For an external, sunlit single-glazed window without shading, the inward heat flow from the inner surface of the glass as shown in Fig. 6.11 can be evaluated as

$$q_{RCi} = \text{inward absorbed radiation}$$
$$+ \text{ conductive heat transfer}$$
$$= UA_s\left[\frac{\alpha I_t}{h_o} + (T_o - T_r)\right] \qquad (6.69)$$

where h_o = heat transfer coefficient for outdoor surface of window glass, Btu/h·ft^2·°F
T_o = outdoor air temperature, °F
T_r = space air temperature, °F

Heat admited through a unit area of the single-glazing window glass is

$$\frac{q_{wi}}{A_s} = \tau I_t + U\left[\frac{\alpha I_t}{h_o} + (T_o - T_r)\right]$$
$$= FI_t + U(T_o - T_i) \qquad (6.70)$$

The ratio of the solar heat gain to the incident solar radiation F for a single-glazing window is given as

$$F = \tau + \frac{U\alpha}{h_o} \qquad (6.71)$$

In these last three equations, U indicates the overall heat transfer coefficient of the window, in Btu/h·ft^2·°F, and can be calculated as

$$U = \frac{U_{wg}A_{wg} + U_{eg}A_{eg} + U_fA_f}{A_{wg} + A_{eg} + A_f} \qquad (6.72)$$

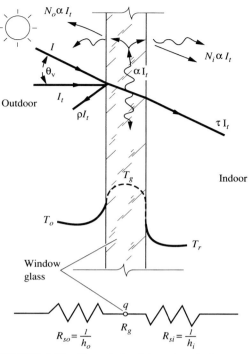

FIGURE 6.11 Heat admitted through a single glazing window glass.

In Eq. (6.72), A represents area, in ft^2; the subscript wg indicates the glass and eg, the edge of the glass including the sealer and spacer of the insulating glass. The edge of glass has a width of about 2.5 in. The subscript f means the frame of the window. Some U-values for various types of windows at winter design conditions are listed in Table 6.7. For summer design conditions at 7.5 mph wind speed, the listed U-values should be multiplied by a factor of 0.92. The U-values of windows depend on the construction of the window, the emissivity of the surfaces of glass or plastic sheets, and the air velocity flowing over the outdoor and indoor surfaces.

As shown in Fig. 6.11, solar radiation having a total intensity I, in Btu/h·ft^2, when it is radiated on the outer surface of a vertical pane with an angle of incidence θ_v, actually consists of

$$I = I_{DN} + I_d + I_{ref.DN} \qquad (6.73)$$

where $I_{ref.DN}$ = component of reflected solar intensity in the direction of the sun's ray, Btu/h·ft^2
$I_t = I\cos\theta_v$

Heat Gain for Double Glazing

For a double-glazed window, the inward heat flow per ft^2 of the inner surface of the glass, as shown in

TABLE 6.7
Overall heat transfer coefficient U-values for windows at winter conditions* with commercial type of frame, in Btu/h·ft² · °F

Type	Gas between glasses	Space between glasses, in.	Emittance† of low-E film	Glass	Edge	Aluminium frame of $U_f = 1.9$	Aluminium frame with thermal break, with $U_f = 1.0$	Wood or vinyl frame of $U_f = 0.41$
				Overall coefficient, in Btu/h · ft² · °F				
Single glass				1.11		1.23	1.10	0.98
Double glass	Air	3/8		0.52	0.62	0.74	0.60	0.51
Double glass	Air	3/8	0.40	0.43	0.55	0.67	0.54	0.45
Double glass	Air	3/8	0.15	0.36	0.51	0.62	0.48	0.39
Double glass	Argon	3/8	0.15	0.30	0.48	0.57	0.43	0.34
Triple glass	Air	3/8		0.34	0.50	0.60	0.46	0.38
Triple glass	Air	3/8	0.40	0.30	0.48	0.57	0.44	0.35
Triple or double glass with polyester film suspended in between	Argon	3/8	0.15	0.17	0.43	0.47	0.34	0.25

**Winter conditions* means 70°F indoor, 0°F outdoor temperature and a wind speed of 15 mph.

†Low-E film can be applied to surface 2 for double glass (see Fig. 6.12b) or surfaces 2, 3, 4, or 5 for triple glass (any surface other than outer and inner surfaces). Abridged with permission from *ASHRAE Handbook, 1989, Fundamentals*.

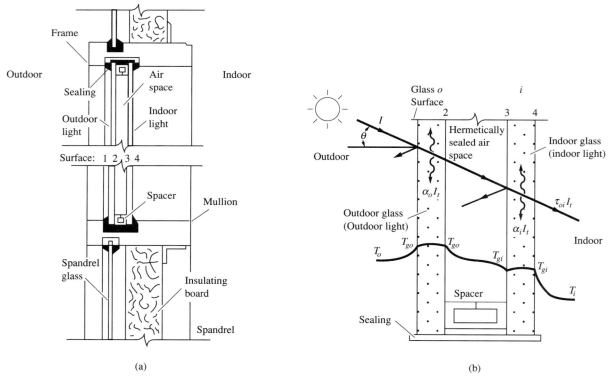

(a) (b)

FIGURE 6.12 Heat flow through an insulating glass (double pane) window: (a) construction of a typical insulating glass; and (b) heat flow and temperature profiles.

Fig. 6.12b, is

$$\frac{q_{RCi}}{A_s} = N_{io}\alpha I_o + N_{ii}\alpha I_i + U(T_o - T_i)$$

$$= U\left[\frac{\alpha I_o}{h_o} + \left(\frac{1}{h_o} + \frac{1}{h_a}\right)\alpha I_i + (T_o - T_i)\right]$$

$$(6.74)$$

where N_{io} = inward fraction of solar radiation absorbed by outdoor glass
N_{ii} = inward fraction of solar radiation absorbed by indoor glass
I_o, I_i = solar intensity irradiated on the outdoor and indoor glass, Btu/h · ft^2

The solar radiation absorbed by the outdoor and indoor panes, in Btu/h · ft^2, is

$$\alpha I_o = \alpha_o I_t; \qquad \alpha I_i = \alpha_i I_t \qquad (6.75)$$

Absorptance of outdoor pane is $\alpha_o = \alpha_1 + \dfrac{\alpha_2 \tau_o \rho_3}{1 - \rho_2 \rho_3}$

Absorptance of outdoor pane is $\alpha_i = \dfrac{\alpha_3 \tau_o}{1 - \rho_2 \rho_3}$

$$(6.77)$$

Subscript o indicates outdoor, i indoor, and 1, 2, 3, and 4 denote the surfaces of the panes as shown in Fig. 6.12b.

The air space heat transfer coefficient h_a, in Btu/h· ft^2·°F, is the R-value of the air space R_a listed in Table 6.4, that is,

$$h_a = \frac{1}{R_a} \qquad (6.78)$$

The transmittance of solar radiation through both outdoor and indoor panes τ_{oi} can be calculated as

$$\tau_{oi} = \frac{\tau_o \tau_i}{1 - \rho_2 \rho_3} \qquad (6.79)$$

Then the heat admitted per ft^2 through a double-glass window $q_{w.oi}/A_s$, in Btu/h · ft^2, can be calculated as

$$\frac{q_{w.oi}}{A_s} = \tau_{oi} I_t + U\left(\frac{\alpha_o}{h_o} + \frac{\alpha_i}{h_o} + \frac{\alpha_i}{h_a}\right) I_t + U(T_o - T_i)$$

$$= F_{oi} I_t + U(T_o - T_i) \qquad (6.80)$$

The solar heat gain ratio for a double-glass window F_{oi} is calculated as

$$F_{oi} = \tau_{oi} + U\left(\frac{\alpha_o}{h_o} + \frac{\alpha_i}{h_o} + \frac{\alpha_i}{h_a}\right) \qquad (6.81)$$

Because a glass plate is usually between 0.125 and 0.25 in. thick, with a thermal conductivity k about 0.5 Btu/h · ft · °F, there is only a small temperature difference between the inner and outer surfaces of the plate when solar radiation is absorbed. For the sake of simplicity, it is assumed that the temperature of the plate T_g is the same in the direction of heat flow.

For a double-glazed window, the glass temperature of the outdoor pane T_{go}, in °F, can be calculated as

$$T_{go} = T_o + \left(\alpha I_o + \alpha I_i - \frac{q_{RCi}}{A_s}\right)\left(\frac{1}{h_o} + \frac{R_{go}}{2}\right) \quad (6.82)$$

The temperature of the indoor pane T_{gi}, in °F, is calculated as

$$T_{gi} = T_r + q_{RCi}\left(\frac{1}{h_i} + \frac{R_{gi}}{2}\right) \qquad (6.83)$$

where R_{go}, R_{gi} = R-value of outdoor and indoor panes, h · ft^2 · °F/Btu
h_i = heat transfer coefficient of the inside Surface 4 (see Fig. 6.12(b)) of the inside plate, Btu/h · ft^2 · °F

Shading Coefficients

Shading coefficient SC is defined as the ratio of solar heat gain of a glazing assembly of specific construction and shading devices at a summer design solar intensity and outdoor and indoor temperatures to the solar heat gain of a reference glass at the same solar intensity and outdoor and indoor temperatures. The reference glass is double-strength sheet glass (DSA) with transmittance $\tau = 0.86$, reflectance $\rho = 0.08$, absorptance $\alpha = 0.06$ and $F_{DSA} = 0.87$ under summer design conditions. The shading coefficient is an indication of the characteristics of a glazing and the associated shading devices, and can be expressed as

$$SC = \frac{\text{Solar heat gain of specific type of window glass}}{\text{Solar heat gain of double-strength sheet glass}}$$

$$= \frac{F_{wi}}{F_{DSA}} = \frac{F_{wi}}{0.87} = 1.15\, F_{wi} \qquad (6.84)$$

where F_{wi} = solar heat gain factor of a specific type of window glass
F_{DSA} = solar heat gain factor of standard reference double-strength sheet glass

Shading coefficients of various types of glazing and shading devices are presented in Table 6.8.

Solar Heat Gain Factors

Solar heat gain factors SHGF, in Btu/h · ft^2, are designated as the average solar heat gain during cloudless days through DSA glass. In the *ASHRAE Handbook 1989, Fundamentals* are tabulated SHGF for various latitudes, solar times, and orientations for load and energy calculations. For calculating the summer cooling peak load, the concept of maximum SHGF has been introduced. This is the maximum value of SHGF on the twenty-first of each month for a specific latitude, as listed in Table 6.9. For high elevations and on very

TABLE 6.8
Shading coefficients for window glass with indoor shading devices

| | | | | Venetian blinds | | Roller shade | | Draperies | |
Type of glass	Thickness of glass, in.	Solar Transmittance	Glass	Med.*	Light†	Opaque, white	Translucent	Med.‡	Light^δ
Clear	3/32	0.87 to 0.79		0.74	0.67	0.39	0.44	0.62	0.52
Heat-absorbing	3/16 or 1/4	0.46		0.57	0.53	0.30	0.36	0.46	0.44
	3/8	0.34		0.54	0.52	0.28	0.32		
		0.24		0.42	0.40	0.28	0.31	0.38	0.36
Reflective-coated			0.30	0.25	0.23				
			0.40	0.33	0.29				
			0.50	0.42	0.38				
Insulating glass		Outer	Inner						
Clear out Clear in	3/32, or 1/8	0.87	.87	0.62	0.58	0.35	0.40		
Heat-absorbing out Clear in	1/4	0.46	.80	0.39	0.36	0.22	0.30		
Reflective glass			0.20	0.19	0.18				
			0.30	0.27	0.26				
			0.40	0.34	0.33				

*Med. indicates medium color

†Light indicates light color

‡Draperies Med. represents draperies of medium color with a fabric openness of 0.10 to 0.25 and yarn reflectance of 0.25 to 0.50

δDraperies Light repesents draperies of light color with a fabric openness below 0.10 and yarn reflectance over 0.50

Adapted with permission from *ASHRAE Handbooks 1989, Fundamentals.*

TABLE 6.9
Maximun solar heat gain factors (max SHGF)

	N (shade)	NNE/NNW	NE/NW	ENE/WNW	E/W	ESE/WSW	SE/SW	SSE/SSW	S	HOR*
				Max SHGF, Btu/h · ft²						
				North latitude, 40 degrees						
Jan.	20	20	20	74	154	205	241	252	254	133
Feb.	24	24	50	129	186	234	246	244	241	180
Mar.	29	29	93	169	218	238	236	216	206	223
Apr.	34	71	140	190	224	223	203	170	154	252
May	37	102	165	202	220	208	175	133	113	265
June	48	113	172	205	216	199	161	116	95	267
July	38	102	163	198	216	203	170	129	109	262
Aug.	35	71	135	185	216	214	196	165	149	247
Sep.	30	30	87	160	203	227	226	209	200	215
Oct.	25	25	49	123	180	225	238	236	234	177
Nov.	20	20	20	73	151	201	237	248	250	132
Dec.	18	18	18	60	135	188	232	249	253	113
				North latitude, 32 degrees						
Jan.	24	24	29	105	175	229	249	250	246	176
Feb.	27	27	65	149	205	242	248	232	221	217
Mar.	32	37	107	183	227	237	227	195	176	252
Apr.	36	80	146	200	227	219	187	141	115	271
May	38	111	170	208	220	199	155	99	74	277
June	44	122	176	208	214	189	139	83	60	276
July	40	111	167	204	215	194	150	96	72	273
Aug.	37	79	141	195	219	210	181	136	111	265
Sep.	33	35	103	173	215	227	218	189	171	244
Oct.	28	28	63	143	195	234	239	225	215	213
Nov.	24	24	29	103	173	225	245	246	243	175
Dec.	22	22	22	84	162	218	246	252	252	158

*Horizontal surface

Abridged with permission from *ASHRAE Handbook 1989, Fundamentals.*

clear days, the actual SHGF may be 15 percent higher than the values listed in Table 6.10. In dusty industrial areas or very humid locations, the actual SHGF may be lower.

TABLE 6.10
Solar altitude (ALT) and solar azimuth (AZ)

	North latitude						
	32 degrees			40 degrees			
	Solar time	Solar position		Solar time	Solar time	Solar position		Solar time
DATE	AM	ALT	AZ	PM	AM	ALT	AZ	PM
DEC	8	10	54	4	8	5	53	4
	9	20	44	3	9	14	42	3
	10	28	31	2	10	21	29	2
	11	33	16	1	11	25	15	1
	12	35	0	12	12	27	0	12
JAN	7	1	65	5	8	8	55	4
+	8	13	56	4	9	17	44	3
NOV	9	22	46	3	10	24	31	2
	10	31	33	2	11	28	16	1
	11	36	18	1	12	30	0	12
	12	38	0	12				
FEB	7	7	73	5	7	4	72	5
+	8	18	64	4	8	15	62	4
OCT	9	29	53	3	9	24	50	3
	10	38	39	2	10	32	35	2
	11	45	21	1	11	37	19	1
	12	47	0	12	12	39	0	12
MAR	7	13	82	5	7	11	80	5
+	8	25	73	4	8	23	70	4
SEP	9	37	62	3	9	33	57	3
	10	47	47	2	10	42	42	2
	11	55	27	1	11	48	23	1
	12	58	0	12	12	50	0	12
APR	6	6	100	6	6	7	99	6
+	7	19	92	5	7	19	89	5
AUG	8	31	84	4	8	30	79	4
	9	44	74	3	9	41	67	3
	10	56	60	2	10	51	51	2
	11	65	37	1	11	59	29	1
	12	70	0	12	12	62	0	12
MAY	6	10	107	6	5	2	115	7
+	7	23	100	5	6	13	106	6
JUL	8	35	93	4	7	24	97	5
	9	48	85	3	8	35	87	4
	10	61	73	2	9	47	76	3
	11	72	52	1	10	57	61	2
	12	78	0	12	11	66	37	1
					12	70	0	12
JUN	5	1	118	7	5	4	117	7
	6	12	110	6	6	15	108	6
	7	24	103	5	7	26	100	5
	8	37	97	4	8	37	91	4
	9	50	89	3	9	49	80	3
	10	62	80	2	10	60	66	2
	11	74	61	1	11	69	42	1
	12	81	0	12	12	73	0	12

Source: ASHRAE Handbook 1981, Fundamentals. Reprinted with permission.

Example 6.1. A double-glass window of a commercial building facing west consists of an outdoor clear plate glass of 0.125 in. and an indoor reflective glass of 0.25 in. thickness with a reflective film on the outer surface of the indoor glass, as shown in Fig. 6.12b. This building is located at 40 degrees north latitude. The detailed optical properties of their surfaces are as follows:

$\tau_0 = 0.80$	$\rho_1 = 0.08$	$\alpha_1 = 0.12$	$e_1 = 0.84$
	$\rho_2 = 0.08$	$\alpha_2 = 0.12$	$e_2 = 0.84$
$\tau_i = 0.16$	$\rho_3 = 0.68$	$\alpha_3 = 0.16$	$e_3 = 0.15$
	$\rho_4 = 0.08$	$\alpha_4 = 0.76$	$e_4 = 0.84$

The R-value of the 0.25 in. thickness indoor glass is $R_g = 0.035$ h · ft^2 · °F/Btu, and of the enclosed air space, $R_a = 1.75$ h · ft^2 · °F/Btu.

At 4 pm on July 21, the outdoor temperature at this location is 100°F, the indoor temperature is 76°F, and the total solar intensity at a direction normal to this west window is 248 Btu/h · ft^2. The outdoor surface heat transfer coefficient $h_o = 4.44$ Btu/h · ft^2 · °F and the heat transfer coefficient of the inner surface of the indoor glass $h_i = 2.21$ Btu/h · ft^2 · °F. Calculate the following:

1. The inward heat flow of this window
2. The temperatures of the outdoor and indoor glasses
3. The shading coefficient of this double-glass window
4. The total heat gain admitted through this window

Solution

1. From Eq. (6.76), the absorption coefficient of the outdoor glass can be calculated as

$$\alpha_o = \alpha_1 + \frac{\alpha_2 \tau_o \rho_3}{1 - \rho_2 \rho_3}$$

$$= 0.12 + \frac{0.12 \times 0.8 \times 0.68}{1 - 0.08 \times 0.68} = 0.189$$

and from Eq. (6.77), the absorption coefficient of the indoor glass is

$$\alpha_i = \frac{\alpha_3 \tau_o}{1 - \rho_2 \rho_3} = \frac{0.16 \times 0.8}{1 - 0.08 \times 0.68} = 0.135$$

From Eq. (6.75), the heat absorbed by the outdoor glass is

$$\alpha I_o = \alpha_o I_t = 0.189 \times 248 = 46.9 \text{ Btu/h} \cdot \text{ft}^2$$

Also, the heat absorbed by the indoor glass is calculated as

$$\alpha I_i = \alpha_i I_t = 0.135 \times 248 = 33.5 \text{ Btu/h} \cdot \text{ft}^2$$

Then, from Eq. (6.25), the overall heat transfer coefficient of this double-glass window is

$$U = \frac{1}{(1/h_i) + R_g + R_a + R_g + (1/h_o)}$$

$$= \frac{1}{(1/2.21) + 0.035 + 1.75 + 0.035 + (1/4.44)}$$

$$= 0.400 \text{ Btu/h} \cdot \text{ft}^2 \cdot {}^{\circ}\text{F}$$

The inward heat flow from the inner surface of the indoor glass, from Eq. (6.74), is given as

$$\frac{q_{RCi}}{A_s} = U\left[\frac{\alpha_o I_t}{h_o} + \alpha_i I_t\left(\frac{1}{h_o} + \frac{1}{h_a}\right) + (T_o - T_r)\right]$$

$$= 0.400\left[\left(\frac{46.9}{4.44}\right) + 33.5\left(\frac{1}{4.44} + 1.75\right)\right.$$

$$\left. + (100 - 76)\right]$$

$$= 40.29 \text{ Btu/h} \cdot \text{ft}^2$$

2. From Eq. (6.82), the temperature of the outdoor glass is

$$T_{go} = T_o + \left(\alpha_o I_t + \alpha_i I_t - \frac{q_{RCi}}{A_s}\right)\left(\frac{1}{h_o}\right) + \frac{R_g}{2}$$

$$= 100 + (46.9 + 33.5 - 40.29)$$

$$\times \left(\frac{1}{4.44} + \frac{0.035}{2}\right)$$

$$= 109.7{}^{\circ}\text{F}$$

and the temperature of the indoor glass is

$$T_{gi} = T_i + \left(\frac{q_{RCi}}{A_s}\right)\left(\frac{1}{h_i}\right) + \frac{R_g}{2}$$

$$= 76 + 40.29\left(\frac{1}{2.21} + \frac{0.035}{2}\right) = 94.9{}^{\circ}\text{F}$$

3. From Eq. (6.79), the transmittance for both panes is

$$\tau_{oi} = \frac{\tau_o \tau_i}{1 - \rho_2 \rho_3} = \frac{0.8 \times 0.16}{1 - (0.08 \times 0.68)} = 0.135$$

and from Eq. (6.81), the solar heat gain ratio for the double-glass window is

$$F_{oi} = \tau_{oi} + \frac{U\alpha_o}{h_o} + \left(\frac{U}{h_o} + \frac{U}{h_a}\right)\alpha_i$$

$$= 0.135 + \frac{0.40 \times 0.189}{4.44}$$

$$+ \left(\frac{0.40}{4.44} + 0.40 \times 1.75\right)0.135$$

$$= 0.259$$

Then, from Eq. (6.84), the shading coefficient of this double-glass window is given by

$$\text{SC} = \frac{F_{oi}}{F_{DSA}} = \frac{0.259}{0.87} = 0.297$$

4. From Eq. (6.80), the total heat gain admitted through this double-glass window per ft^2 of sunlit area is

$$\frac{q_{wi}}{A_s} = F_{oi} I_t + U(T_o - T_i)$$

$$= 0.259 \times 248 + 0.40(100 - 76)$$

$$= 64.2 + 9.6 = 73.8 \text{ Btu/h} \cdot \text{ft}^2$$

Selection of Glazing

During the selection of glazing, visual communication, use of daylight, thermal comfort, summer and winter solar heat gain, street noise attenuation, safety and fire protection, and life-cycle cost analysis should be considered. Energy conservation considerations, including the control of solar heat with the optimum combination of absorbing and reflective glass and various shading devices, will be covered in the next section.

To reduce heat loss through glass during winter, one can install double or triple glazing, storm windows, or low-emission film coating on the surface of the glass.

6.10   SHADING OF GLASS

Shading projected over the surface of glass significantly reduces its sunlit area. Many shading devices increase the reflectance of the incident radiation. There are two types of shading: deliberately installed shading devices, which include indoor and external shading devices, and shading from adjacent buildings.

Indoor Shading Devices

Indoor shading devices not only provide privacy but also are usually effective in reflecting part of the solar radiation back to the outdoors. They also raise the air temperature of the space between the shading device and the window glass, which in turn reduces the conductive heat gain in summer season. Indoor shading devices are easier to operate and to maintain and are more flexible in operation than external shading devices. Three types of indoor shading devices are commonly used: venetian blinds, draperies, and roller shades.

VENETIAN BLINDS. Most horizontal venetian blinds are made of plastic or aluminum slats, spaced 1 to 2 inches apart, and some of rigid woven cloth. The ratio of slat width to slat spacing is generally 1.15 to 1.25. For light-colored metallic or plastic slats at a 45° angle, typical optical properties are $\tau = 0.05$, $\rho = 0.55$, and $\alpha = 0.40$. Vertical venetian blinds of wider slats widely used are in commercial buildings.

Consider a single-glazed window combined with indoor venetian blinds at a slat angle $\psi = 45°$, as shown in Fig. 6.13. Let subscripts g, v, and a repre-

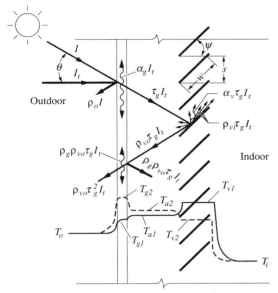

FIGURE 6.13 Heat transfer through a single-glazed window combined with venetian blinds.

sent the glass, the venetian blinds, and the air between the venetian blinds and the glass. Also, let o indicate the outward direction and i the inward direction.

Of the solar radiation transmitted through the glass and radiating on the surface of the slats $\tau_g I_t$,

- A fraction $\rho_{vo}\tau_g I_t$ is reflected outward from the surface of each slat
- Another fraction $\rho_{vi}\tau_g I_t$ is reflected from the slat surface into the conditioned space
- A third fraction absorbed by the slat is either convected away by the space air or reflected from the surface to the indoor surroundings in the form of long-wave radiation

If the glass has a high transmittance, the slat temperature T_{v1} will be higher, as shown by the lower solid temperature curve in Fig. 6.13. If the glass has a high absorptance, its temperature T_{g2} will be higher. In Table 6.8 are listed the shading coefficients of various combinations of venetian blinds and glazing.

According to results of a field survey by Inoue et al. in four office buildings in Japan in 1988, 60 percent of the venetian blinds were not operated during the daytime. The incident angle of the direct solar radiation had more influence on operation of the blinds than the intensity. Automatic control of slat angle and of raising/lowering the blinds is sometimes used and may become more popular in the future.

DRAPERIES. These are fabrics made of cotton, regenerated cellulose (such as rayon), or synthetic fibers. Usually they are loosely hung, wider than the window, and pleated, and can be drawn open or

closed as required. Drapery-glass combinations reduce the solar heat gain in summer and increase the thermal resistance in winter. Reflectance of the fabric is the dominant factor in the reduction of heat gain, and visibility is a function of the openness of the weave.

ROLLER SHADES. These are sheets made of treated fabric or plastic that can be pulled down to cover the window or rolled up. Roller shades have a lower SC than venetian blinds and draperies. When glass is covered with shades, any outdoor visual communication is blocked, and the visible light transmittance is less than for other indoor shading devices.

External Shading Devices

External shading devices include overhangs, side fins, egg-crate louvers, and pattern grilles, as shown in Fig. 6.14. They are effective in reducing the solar heat gain by decreasing the sunlit area. However, the external shading devices do not always fit into the architectural requirements and are less flexible and more difficult to maintain. Pattern grilles impair visibility significantly.

In Fig. 6.15 is shown the shaded area of a glass pane constructed with both overhang and side fin. The profile angle Ω is defined as the angle between a horizontal plane and a tilted plane that includes the sun's rays. We see that $\tan \Omega = UQ/OR = \tan \beta / \cos \gamma$. Let S_W be the width of the shadow and S_H be the height of the shadow projected on the plane of the glass by direct solar radiation, in ft. Also, let W be the width of the glass and H be the height. Then the shadow width on the plane of the glass is

$$S_W = P_V \tan \gamma \qquad (6.85)$$

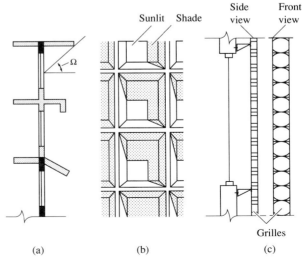

FIGURE 6.14 Types of external shades: (a) overhang; (b) egg-crate louver; and (c) pattern grilles.

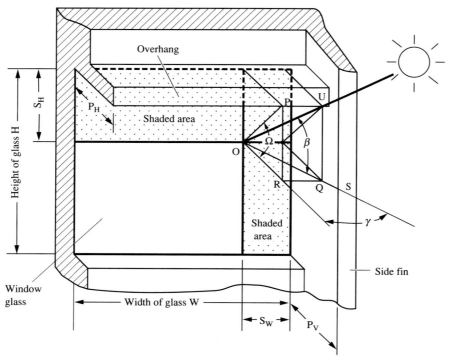

FIGURE 6.15 Shaded area of a window glass constructed with an overhang and sidefin.

and the shadow height is

$$S_H = P_H \tan \Omega = P_H \frac{\tan \beta}{\cos \gamma} \qquad (6.86)$$

where P_V = projection of the side fin plus mullion and reveal, ft
P_H = horizontal projection of the overhang plus transom and reveal, ft

The projection factor due to the external shading device F_{pro} can be calculated as

$$F_{pro} = \frac{P_H}{H} \qquad (6.87)$$

The net sunlit area of the glass A_s, in ft², can then be calculated as

$$A_s = (W - S_W)(H - S_H)$$

and the shaded area of the glass A_{sh}, in ft², is given by

$$A_{sh} = A_g - A_s = WH - (W - S_W)(H - S_H) \qquad (6.88)$$

where A_g = area of the glass, ft².

Shading from Adjacent Buildings

Shadows cast on glass by adjacent buildings significantly reduce the sunlit area of the glass. For example, in Fig. 6.16 we see the area on building A shaded because of the presence of building B. Let two sides of the shaded building A coincide with the X and Y axes on the plan view shown in Fig. 6.16. In the elevation view of Fig. 6.16, the shadow height on the facade of building A is

$$S_H = H_B - L_{AB} \tan \Omega \qquad (6.89)$$

and the shadow width S_W on the facade of building A is

$$S_W = W_{OB} + (W_B - L_{AB} \tan \gamma) \qquad (6.90)$$

where H_B = height of building B, ft
W_B = width of building B, ft
L_{AB} = distance between two buildings along X-axis, ft
W_{OB} = distance between X-axis and building B, ft

Subscript A indicates the shaded building and B the shading building.

Because of the sun's varying position, the solar altitude β and the surface-solar azimuth γ change their values from time to time. Table 6.10 lists the solar attitude β and solar azimuth γ at north latitudes 32° and 40°. In order to evaluate the shaded area of the outer surface of a building, a computer program can determine which of several hundred representative points on this outer surface are sunlit or shaded at specific times of day. Then, a ratio of shaded area to total area can be calculated.

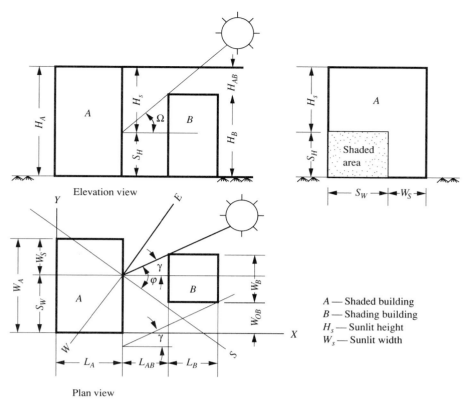

FIGURE 6.16   Shading from adjacent building.

A — Shaded building
B — Shading building
H_s — Sunlit height
W_s — Sunlit width

6.11   HEAT EXCHANGE BETWEEN THE OUTER BUILDING SURFACE AND ITS SURROUNDINGS

Because atmospheric temperature is lower at high altitudes, there is always a radiant heat loss from the outer surface of the building to the sky vault. However, it may be offset or partly offset by reflected solar radiation from the ground on a sunny day. Radiant heat loss from the building must be calculated during nighttime and be included in year-round energy estimation.

In commercial and institutional buildings using glass, concrete, or face brick on the outside surface of the building envelope, the migration of moisture through the glass pane is rather small. Because of the heavy mass of the concrete wall, the influence of the diurnal cyclic variation of the relative humidity of outdoor air on moisture transfer through the building envelope is also small. Therefore, for simplicity, the moisture transfer between the building envelope and the outside air can be ignored.

The heat balance at the outer building surface as shown in Fig. 6.17 can be expressed as follows:

$$q_{sol} + q_{ref} = q_{os} + (q_r - q_{at}) + q_{oi} \qquad (6.91)$$

In Eq. (6.91), q_{sol} represents the solar radiation absorbed by the outer surface of the building envelope,

in Btu/h. It can be calculated as

$$q_{sol} = \alpha_{os}[A_s(I_D + I_d) + A_{sh}I_d] \qquad (6.92)$$

where α_{os} = absorptance of the outer surface of the building envelope. From Eq. (6.61), the reflection of solar radiation from any reflecting surface to the outer surface of the building and absorbed by it, or q_{ref}, in Btu/h, is given by

$$q_{ref} = A\alpha_{os}I_{ref} = A\rho_s\alpha_{os}F_{s-r}(I_D + I_d) \qquad (6.93)$$

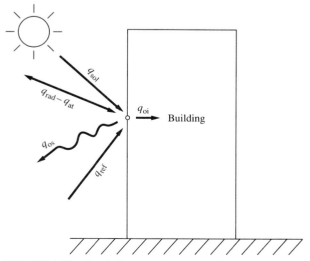

FIGURE 6.17   Heat balance at the outer surface of a building.

where A = total area of the outer surface of the building envelope = $A_s + A_{sh}$, ft^2. The term q_{os} indicates the convective heat transfer from the outer surface of the building outward, in Btu/h. From Eq. (6.7), it can be calculated as

$$q_{os} = h_c A(T_{os} - T_o) \qquad (6.94)$$

where T_{os}, T_o = outer surface temperature of the building and the outdoor air temperature, respectively, °F

h_c = convective heat transfer coefficient of the outer surface, Btu/h · ft^2 · °F

Term q_{oi} denotes the inward heat flow from the outer building surface, in Btu/h. Term $(q_{rad} - q_{at})$ indicates the net heat emitted from the outer surface of the building because of the radiation exchange between the surface and the atmosphere, in Btu/h. Here, q_{rad} represents the long-wave radiation emitted from the surface and q_{at} indicates the atmospheric radiation to the surface.

Kimura (1977) found that atmospheric radiation can be expressed as

$$q_{at} = (1 - C_{cc}K_{cc})\sigma T_o^4 \text{Br} + C_{cc}K_{cc}\sigma T_g^4 \qquad (6.95)$$

where C_{cc} = cloud cover factor, which can be obtained from local climate records, dimensionless

K_{cc} = cloudy reduction factor

Usually, the smaller the value of C_{cc}, the higher the clouds and the smaller the K_{cc} value. For simplicity, it can be calculated as:

$$K_{cc} = 0.83 - 0.4\,C_{cc} \qquad (6.96)$$

Br is actually the emissivity of the atmosphere, and can be expressed by an empirical formula developed by Brunt:

$$\text{Br} = 0.51 + 0.55\sqrt{p_w} \qquad (6.97)$$

where p_w = water vapor pressure, psia. In Eq. (6.95), T_g represents the ground temperature, °F.

Then, radiant heat loss can be written as

$$q_{rad} - q_{at} = \epsilon_{os}A F_{s-at}\sigma$$
$$\times T_{os}^4 - [(1 - C_{cc}K_{cc})T_o^4 \text{Br} + C_{cc}K_{cc}T_g^4]$$
$$(6.98)$$

where ϵ_{os} = emissivity of the outer surface of the building

F_{s-at} = shape factor between the outer surface and the atmosphere

For a sunlit outer surface of a building, if q_{ref} is mainly from ground-reflected solar radiation, it is offset by $(q_{rad} - q_{at})$, and Eq. (6.91) becomes

$$q_{sol} = q_{os} + q_{oi}$$

Let $q_{oi} = h_o A(T_{sol} - T_{os})$. Substituting for q_{oi}, then

$$\alpha_{os}I_t = h_o(T_{os} - T_o) + h_o(T_{sol} - T_{os})$$

and

$$T_{sol} = T_o + \frac{\alpha_{os}I_t}{h_o} \qquad (6.99)$$

In Eq. (6.99), T_{sol} is called *sol-air temperature*. It is a fictitious outdoor temperature that combines the effect of the solar radiation radiated on the outer surface of the building and the inward heat transfer due to the outdoor and indoor temperature difference.

Example 6.2. At midnight of July 21, the outdoor conditions of an office building are as follows:

Outdoor temperature	75°F
Water vapor pressure of outdoor air	0.215 psia
Ground temperature	72°F
Cloud cover factor C_{cc}	0.1
Emissivity of the outer surface	0.90
Shape factor between the surface and sky	0.5

If the outer surface temperature of this building is 76°F and the Stefan-Boltzmann constant $\sigma = 0.1714 \times 10^{-8}$ Btu/h · ft^2·°R^4, find the radiant heat loss from each ft^2 of the vertical outer surface of this building.

Solution. From Eq. (6.97) and given data, we can see

$$\text{Br} = 0.51 + 0.55\sqrt{p_w}$$
$$= 0.51 + 0.55\sqrt{0.215} = 0.765$$

And also, from Eq. (6.96), the cloudy reduction factor is

$$K_{cc} = 0.83 - 0.4\,C_{cc} = 0.83 - 0.4(0.1) = 0.79$$

Then, from Eq. (6.98), the radiant heat loss from the outer surface of the building due to radiant exchange between the surface and the atmosphere is

$$q_{rad} - q_{at} = \epsilon_{os}F_{s-at}\sigma\{T_{os}^4 - [(1 - R_{cc}K_{cc})T_o^4\text{Br}$$
$$+ R_{cc}K_{cc}T_g^4]\} = 0.90 \times 0.5 \times 0.1714$$
$$\times \{(5.36)^4 - [(1 - 0.1 \times 0.79)(5.35)^4$$
$$\times 0.765 + 0.1 \times 0.79 \times (5.32)^4]\}$$
$$= 14.27 \text{ Btu/h} \cdot \text{ft}^2$$

REFERENCES AND FURTHER READING

Altmayer, E. F., Gadgil, A. J., Bauman, F. S. and Kammerud, R. C., "Correlations for Convective Heat Transfer from Room Surfaces," *ASHRAE Transactions*, Part II A, pp. 61–77, 1983.

ASHRAE, *ASHRAE Handbook 1989, Fundamentals*, ASHRAE, Inc., Atlanta, GA, 1989.

ASHRAE, *Procedure for Determining Heating and Cooling Loads for Computerizing Energy Calculations*,

Algorithms for Building Heat Transfer Subroutines, ASHRAE, Inc., New York, 1976.

Bauman, F., Gadgil, A., Kammerud, R., Altmayer, E., and Nansteel, M., "Convective Heat Transfer in Buildings: Recent Research Results," *ASHRAE Transactions,* Part I A, pp. 215–233, 1983.

Bruin, S., and Luyben, K. C. A. M., "Drying of Food Materials: A Review of Recent Development," *Advances in Drying* Vol. 1, Hemisphere Publishing Corp., Washington D. C. 1979.

Chandra, S., and Kerestecioglu, A. A., "Heat Transfer in Naturally Ventilated Rooms: Data from Full Scale Measurements," *ASHRAE Transactions,* Part I B, pp. 211–225, 1984.

Dahlen, R. R., "Low-E Films for Window Energy Control," *ASHRAE Transactions,* Part I, pp. 1517–1524, 1987.

Donnelly, R. G., Tennery, V. J., McElroy, D. L., Godfrey, T. G., and Kolb, J. O., "Industrial Thermal Insulation, An Assessment, *Oak Ridge National Laboratory Report* TM-5283, TM-5515 and TID-27120, 1976.

Fairey, P. W., and Kerestecioglu, A. A., "Dynamic Modeling of Combined Thermal and Moisture Transport in Buildings: Effect on Cooling Loads and Space Conditions," *ASHRAE Transactions,* Part II A, pp. 461–472, 1985.

Galanis, N., and Chatigny, R., "A Critical Review of the ASHRAE Solar Radiation Model," *ASHRAE Transactions,* Part I A, pp. 410–419, 1986.

Glicksman, L. R., and Katsennelenbogen, S., "A Study of Water Vapor Transmission through Insulation under Steady State and Transient Conditions," *ASHRAE Transactions,* Part II A, pp. 483–499, 1983.

Handegrod, G. O. P., "Prediction of the Moisture Performance of Walls," *ASHRAE Transactions,* Part II B, pp. 1501–1509, 1985.

Inoue, T., Kawase, T., Ibamoto, T., Takakusa, S., and Matsuo, Y., "The Development of an Optimal Control System for Window Shading Devices Based on Investigations in Office Buildings," *ASHRAE Transactions,* Part II, pp. 1034–1049, 1988.

Kays, M. M., and Crawford, M. E., *Convective Heat and Mass Transfer,* 2nd ed., McGraw-Hill, New York, 1980.

Kimura, K., *Scientific Basis of Air Conditioning,* Applied Science Publishers, London, 1977.

Kimura, K., and Stephenson, D. G., "Solar Radiation on Cloudy Days," National Research Council Division of Building Research, *Report NRC #418,* 1969.

Miller, A., Thompson, J. C., Peterson, R. E., and Haragan, D. R., *Elements of Meteorology,* 4th ed., Bell and Howell Co., Columbus, Ohio, 1983.

Robertson, D. K., and Christian, J. E., "Comparison of Four Computer Models with Experimental Data from Test Buildings in Northern New Mexico," *ASHRAE Transactions,* Part II B, pp. 591–607, 1985.

Sato, A., Eto, N., Kimura, K., and Oka, J., "Research on the Wind Variation in the Urban Area and Its Effects in Environmental Engineering No. 7 and No. 8 — Study on Convective Heat Transfer on Exterior Surface of Buildings, *Transactions of Architectural Institute of Japan,* No. 191, Jan. 1972.

Spitler, J. D., Pedersen, C. O., and Fisher, D. E., "Interior Convective Heat Transfer in Buildings with Large Ventilative Flow Rates," *ASHRE Transactions,* Part I, pp. 505–514, 1991.

Verschoor, J. D., "Measurement of Water Vapor Migration and Storage in Composite Building Construction," *ASHRAE Transactions,* Part II A, pp. 390–403, 1985.

Wang, S. K., *Air Conditioning,* Vol. 1, Hong Kong Polytechnic, 1987.

Wong, S. P. W., "Simulation of Simultaneous Heat and Moisture Transfer by Using the Finite Difference Method and Verified Tests in a Test Chamber," *ASHRAE Transactions,* Part I, pp. 472–485, 1990.

Wong, S. P. W., and Wang, S. K., "Fundamentals of Simultaneous Heat and Moisture Transfer between the Building Envelope and the Conditioned Space Air," *ASHRAE Transactions,* Part II, pp. 73–83, 1990.

LOAD CALCULATIONS

7.1 OUTDOOR DESIGN CONDITIONS

In principle, the capacity of air conditioning equipment is selected so that indoor design conditions can be maintained when the outdoor weather does not exceed the design values.

The outdoor weather affects the space cooling load and the capacity of the air system to condition the required amount of outdoor air. It is not economical to choose the annual maximum or annual minimum values as the design data. Outdoor design conditions are usually determined according to statistical analysis of the weather data so that only 1 percent, 2.5 percent, or 5 percent of the outdoor design values have been equaled or exceeded. The indoor design conditions can then be attained most of the time. The greater the need for stricter control of indoor environmental parameters, the smaller the percentage occurrence of the total hours that exceed the outdoor design values.

The recommended design values are based on data from the National Climatic Data Center (NCDC), U.S. Air Force, U.S. Navy, Canadian Atmospheric Environment Service, and other appropriate organizations. According to the *ASHRAE Handbook 1989, Fundamentals,* and ASHRAE/IES Standard 90.1–1989, the design conditions recommended for large cities in the United States are listed in Table 7.1.

Summer Outdoor Design Conditions

In Table 7.1 the reader can note the following:

1. Summer outdoor design dry bulb temperature for a specific locality $T_{o.s}$, in °F, is the rounded higher integral number of the statistically determined summer design temperature $T_{o.ss}$ such that the average number of hours of occurrence of outdoor temperatures T_o higher than $T_{o.ss}$ during June, July, August, and September is divided into three grades less than 1 percent, 2.5 percent, or 5 percent of the total number of hours in these summer months, or 2928 hours (122 days × 24 hr/days). The annual occurrence is the average over a period of 15 years.

   For instance, according to data from NCDC, the statistical occurrence of outdoor dry bulb temper-

atures T_o in a period of 15 years in New York City during June, July, August, and September of $15 \times 0.025 \times 2928 = 1098$ hours was between 88°F and 89°F. The 2.5 percent summer outdoor design dry bulb temperature for New York City $T_{o.s}$ is then rounded to 89°F.

   A summer design temperature $T_{o.s}$ having an average annual occurrence of T_o equal to or exceeding $T_{o.s}$ is less than 2.5 percent of 2928 hours, that is, less than 73 hours, is most widely used.

2. The summer outdoor mean coincident design wet bulb temperature $T'_{o.s}$, in °F, is the mean of all wet bulb temperatures occurring at the specific summer outdoor design dry bulb temperature $T_{o.s}$ during the summer.

3. The 2.5 percent summer design wet bulb temperature $T'_{o.s}$ is the design value having an average annual occurrence of $T'_o > T'_{o.s}$ less than 73 hours. $T'_{o.s}$ is used for evaporative cooling systems, which will be covered in Chapter 12.

4. The mean daily range, in °F, indicates the difference between the average daily maximum and average daily minimum temperatures during the warmest month.

5. Regarding the intensity of solar radiation, the *ASHRAE Handbook 1989, Fundamentals* recommended the maximum solar heat gain factors listed in Table 6.10. Data on global horizontal irradiation I_G^* can be obtained from NCDC, as mentioned in Chapter 6.

Also, the annual average daily incident solar energy on an east or west orientation, in Btu/ft² · day, are listed in Table 7.1 for reference.

Winter Outdoor Design Conditions

In Table 7.1, the winter outdoor design dry bulb temperature $T_{o.w}$, in °F, is the rounded lower integral value of the statistically determined winter outdoor design temperature $T_{o.ws}$, such that the average number of hours of occurrence of outdoor temperatures at values $T_o > T_{o.ws}$ should be equaled or exceeded 99 percent or 97.5 percent of the total number of hours in December, January, and February (2160 hours). The annual average number of hours in

TABLE 7.1
Climatic conditions for the United States and Canada

City	State	Latitude, degree	Latitude, min	Elevation, ft	Winter Design dry bulb, 99% °F	Winter Design dry bulb, 97.5% °F	Summer Design dry bulb/mean coincident wet bulb, 1% °F	2.5% °F	5% °F	Mean daily range, °F	Design wet bulb, 2.5% °F	Annual average daily incident solar energy on east or west, Btu/ft²·day	Annual cooling degree days base 65°F, CDD65	Annual heating degree days base 50°F, HDD50	Annual heating degree days base 65°F, HDD65
Albuquerque	New Mexico	35	03	5310	12	16	96/61	94/61	92/61	27	65	1105	1257	1633	4423
Anchorage	Alaska	61	10	90	−23	−18	71/59	68/58	66/56	15	59	538	0	5301	10540
Atlanta	Georgia	33	39	1005	17	22	94/74	92/74	90/73	19	76	807	1566	866	3070
Atlantic City CO	New Jersey	39	23	11	10	13	92/74	89/74	86/72	18	77				
Baltimore	Maryland	39	11	146	10	13	94/75	91/75	89/74	21	77	739	1134	2020	4946
Billings	Montana	45	48	3567	−15	−10	94/64	91/64	88/63	31	66	814	598	3627	7156
Birmingham	Alabama	33	34	610	17	21	96/74	94/75	92/74	21	77	789	1825	765	2882
Bismarck	North Dakota	46	46	1647	−23	−19	95/68	91/68	88/67	27	71	766	496	5196	8992
Boise	Idaho	43	34	2842	3	10	96/65	94/64	91/64	31	66	916	744	2276	5667
Boston	Massachusetts	42	22	15	6	9	91/73	88/71	85/70	16	74	659	695	2416	5775
Bridgeport	Connecticut	41	11	7	6	9	86/73	84/71	81/70	18	74				
Buffalo	New York	42	56	705	2	6	88/71	85/70	83/69	21	73	609	509	3213	6721
Burlington	Vermont	44	28	331	−12	−7	88/72	85/70	82/69	23	72	698	365	4211	7932
Caribou	Maine	46	52	624	−18	−13	84/69	81/67	78/66	21	69	649	121	5297	9483
Casper	Wyoming	42	55	5319	−11	−5	92/58	90/57	87/57	31	61	961	495	3824	7617
Charleston	South Carolina	32	54	41	24	27	93/78	91/78	89/77	18	80	796	2005	435	2194
Charleston	West Virginia	38	22	939	7	11	92/74	90/73	87/72	20	75	667	1008	1816	4587
Charlotte	North Carolina	35	13	735	18	22	95/74	93/74	91/74	20	76	809	1549	1086	3412
Chicago, Midway	Illinois	41	53	610	−5	0	94/74	91/73	88/72	20	75	729	1015	3000	6151
Cincinnati CO	Ohio	39	09	761	1	6	92/73	90/72	88/72	21	75				
Cleveland	Ohio	41	24	777	1	5	91/73	88/72	86/71	22	74				
Concord	New Hampshire	43	12	339	−8	−3	90/72	87/70	84/69	26	73	630	463	3742	7425
Dallas	Texas	32		481	18	22	102/75	100/75	97/75	20	78	875	2448	605	2354
Denver	Colorado	39	45	5283	−5	1	93/59	91/59	89/59	28	63	971	567	2652	6083
Des Moines	Iowa	41	32	948	−10	−5	94/75	91/74	88/73	23	77	788	812	3275	6447
Detroit	Michigan	42	25	633	3	6	91/73	88/72	86/71	20	74	676	922	2799	5997
Honolulu	Hawaii	21	20	7	62	63	87/73	86/73	85/72	12	75	953	4150	0	0
Houston	Texas	29		50	27	32	96/77	94/77	92/77	18	79	805	2891	195	1346
Indianapolis	Indiana	39	44	793	−2	2	92/74	90/74	87/73	22	76	692	951	2624	5620
Jackson	Mississippi	32	19	330	21	25	97/76	95/76	93/76	21	78	833	2330	546	2424
Kansas City	Missouri	39	07	742	2	6	99/75	96/74	93/74	20	77				
Lansing	Michigan	42	47	852	−3	1	90/73	87/72	84/70	24	74				
Las Vegas	Nevada	36	05	2162	25	28	108/66	106/65	104/65	30	70	1136	3043	449	2399
Lincoln CO	Nebraska	40	51	1150	−5	−2	99/75	95/74	92/74	24	77				
Little Rock	Arkansas	34	44	257	15	20	99/78	96/77	94/77	22	79	831	2055	912	3091
Los Angeles CO	California	34	03	312	37	40	93/70	89/70	86/69	20	71	962	472	3	1494
Louisville	Kentucky	38	11	474	5	10	95/74	93/74	90/74	23	77	727	1357	1851	4539
Memphis	Tennessee	35	03	263	13	18	98/77	95/76	93/76	21	79	806	2069	1034	3259
Miami	Florida	25	47	7	44	47	91/77	90/77	89/77	15	79	874	4045	3	185
Milwaukee	Wisconsin	42	57	672	−8	−4	90/74	87/73	84/71	21	74	724	487	3586	7121
Minneapolis/St. Paul	Minnesota	44	53	822	−16	−12	92/75	89/73	86/71	22	75	709	773	4563	8060
New Orleans	Louisiana	29	59	3	29	33	93/78	92/78	90/77	16	80	838	2578	179	1392
New York City, Central Park	New York	40	47	132	11	15	92/74	89/73	87/72	17	75	650	843	1986	5022
Norfolk	Virginia	36	54	26	20	22	93/77	91/76	89/76	18	78	792	648	2773	5907
Pittsburgh CO	Pennsylvania	40	30	1017	3	7	91/72	88/71	86/70	19	73	642	648	2773	5907
Portland CO	Oregon	45	36	57	18	24	90/68	86/67	82/65	21	67	647	272	1151	4577
Providence	Rhode Island	41	44	55	5	9	89/73	86/72	83/70	19	74	677	693	2610	6022
Rapid City	South Dakota	44	03	3165	−11	−7	95/66	92/65	89/65	28	69	819	663	3672	7229

Note: CO designates locations within an urban area. Most of the data are taken from city airport temperature observations. Some semirural data are comparable to airport data.

Abridged with permission from *ASHRAE Handbook 1989, Fundamentals* and ASHRAE/IES Standard 90.1–1989.

TABLE 7.1 (*Continued*)
Climatic conditions for the United States and Canada

City	State	Latitude, degree	min	Elevation, ft	Winter Design dry bulb, 99% °F	97.5% °F	Summer Design dry bulb/mean coincident wet bulb, 1% °F	2.5% °F	5% °F	Mean daily range, °F	Design wet bulb, 2.5% °F	Annual average daily incident solar energy on east or west, Btu/ft^2 · day	Annual cooling degree days base 65°F, CDD65	Annual heating degree days base 50°F, HDD50	Annual heating degree days base 65°F, HDD65
St. Louis CO	Missouri	38	39	465	3	8	98/75	94/75	91/74	18	77	797	1467	2111	4860
Salt Lake City	Utah	40	46	4220	3	8	97/62	95/62	92/61	32	65	975	941	2570	5975
San Antonio	Texas	29		792	25	30	99/72	97/73	96/73	19	76	878	3013	261	1579
San Diego	California	32	44	19	42	44	83/69	80/69	78/68	12	70	950	662	2	1275
San Francisco CO	California	37	37	52	38	40	74/63	71/62	69/61	14	62	941	73	186	3238
Seattle CO	Washington	47	39	14	21	26	85/68	82/66	78/65	19	67	621	106	1382	5281
Shreveport	Louisiana	32	28	252	20	25	99/77	96/76	94/76	20	79	843	2365	447	2265
Spokane	Washington	47	38	2357	−6	2	93/64	90/63	87/62	28	64	758	363	2983	6727
Syracuse	New York	43	07	424	−3	2	90/73	87/71	84/70	20	73	611	513	3448	6856
Tucson	Arizona	32	07	2584	28	32	104/66	102/66	100/66	26	71	1112	2769	187	1601
Tulsa	Oklahoma	36	12	650	8	13	101/74	98/75	95/75	22	78	820	2072	1429	3732
Washington	DC	38	51	14	14	17	93/75	91/74	89/74	18	77	724	1083	2004	4828
Wichita	Kansas	37	39	1321	3	7	101/72	98/73	96/73	23	76				
Wilmington	Deleware	39	40	78	10	14	92/74	89/74	87/73	20	76	726	1078	2133	5084
Canada															
Calgary	Alberta	51	6	3540	−27	−23	84/63	81/61	79/60	25	63				
Montreal	Quebec	45	28	98	−16	−10	88/73	85/72	83/71	17	74				
Regina	Saskatchewan	50	26	1884	−33	−29	91/69	88/68	84/67	26	70				
Toronto	Ontario	43	41	578	−5	−1	90/73	87/72	85/71	20	74				
Vancouver	British Columbia	49	11	16	15	19	79/67	77/66	74/65	17	67				
Winnipeg	Manitoba	49	54	786	−30	−27	89/73	86/71	84/70	22	73				

Note: CO designates locations within an urban area. Most of the data are taken from city airport temperature observations. Some semirural data are comparable to airport data.

Abridged with permission from *ASHRAE Handbook 1989, Fundamentals* and ASHRAE/IES Standard 90.1–1989.

winter when $T_o < 99$ percent $T_{o.ws}$ is 22 hours, and $T_o < 97.5$ percent $T_{o.ws}$ for 54 hours.

A degree day is the product of the difference between a base temperature and the mean daily outdoor air temperature for any one day ($T_{base} - T_{o,m}$), both in °F, and the number of days per annum. Annually, the total number of heating degree days with a base temperature of 65°F, or HDD65, is

$$\text{HDD65} = \sum_{n=1} (65 - T_{o,m})$$

where n = number of days whose $T_{o,m} < 65°F$ per annum. The total number of cooling degree days with a base temperature of 65°F, or CDD65, is

$$\text{CDD65} = \sum_{m=1} (T_{o,m} - 65) \qquad (7.1)$$

where m = number of days whose $T_{o,m} > 65°F$ per annum. Heating and cooling degree days with different base temperatures have been used as climatic parameters to calculate the energy flux through a building envelope, or to determine the U-value and the configuration of the building envelope.

The Use of Outdoor Weather Data in Design

During the design of air conditioning systems, the following factors apply:

1. Outdoor and indoor design conditions are used to calculate the space cooling and heating loads.

2. Summer outdoor dry bulb and coincident wet bulb temperatures are necessary to evaluate the coil's load. The summer outdoor wet bulb temperature is used to determine the capacity of the evaporative coolers and cooling towers.

3. Outdoor weather data presented consecutively for a whole year of 8760 hours, or other simplified form such as a 24-hour diel period in each month, are used for year-round energy estimations.

4. Outdoor climate often has a significant influence on the selection of an air conditioning system and its components.

Outdoor Weather Characteristics and Their Influence

The following are characteristics of outdoor climate in the United States:

1. For clear days during summer, the daily maximum temperature occurs between 2 and 4 P.M. In winter, the daily minimum temperature usually occurs before sunrise, between 6 and 8 A.M.
2. The maximum combined influence of outdoor temperature and solar radiation on load calculations usually occurs in July or August. January is often the coldest month.
3. The mean daily temperature range of the warmest month varies widely between different locations. The smallest mean daily range, 10°F, occurs at Galveston, Texas. The greatest, 45°F, occurs at Reno, Nevada. Many cities have a mean daily range of 20 to 25°F. Usually, coastal areas have smaller mean daily ranges, and continental areas and areas of high elevation have greater values.
4. Extremes in the difference between 2.5 percent design dry bulb temperature in summer and the 97.5 percent design temperature in winter in the United States are
   • A maximum of 111°F at Bemidji, Minnesota
   • A minimum of 18°F at Kaneohe Bay, Hawaii
5. Snelling (1985) studied outdoor design temperatures for different locations all over the country. He found that extremely cold temperatures may last for three to five days. Extremely hot temperatures seldom last more than 24 hours.
6. Among the 62 cities listed in Table 7.1, 19 have a summer 2.5 percent mean coincident wet bulb $T'_{os} \leq 70°F$. For cities with low or very low summer wet bulb design temperatures, using evaporative cooling systems to replace part or all of the refrigerating capacity would be very economical.
7. When commercial buildings are only occupied after 10 A.M. or even after 12 noon, a winter outdoor design temperature higher than 97.5 percent guaranteed value should be considered.

7.2   SPACE LOAD CHARACTERISTICS

Space, Room, and Zone

Space indicates either a volume or a site without a partition or a partitioned room or group of rooms.

A *room* is an enclosed or partitioned space that is usually treated as a single load. A conditioned room usually, but not always, has an individual control system.

A *zone* is a space or several units of space having some sort of coincident loads or similar operating characteristics. It may or may not be an enclosed space, or it may consist of many partitioned rooms. It could be a conditioned space or a space that is not air conditioned. A conditioned zone is always equipped with an individual control system. A control zone is the basic unit of control.

Convective and Radiative Heat

Whether heat enters the conditioned space from an external source or is released to the space from an internal source, the instantaneous heat gains of the conditioned space can be classified into two categories: convective heat and radiative heat, as shown in Fig. 7.1.

When solar radiation strikes the outer surface of a concrete slab, most of its radiative heat is absorbed by the slab; only a small portion is reflected. After absorption, the outer surface temperature of the slab increases. If the slab and the conditioned space are in thermal equilibrium originally, there is then a convective heat and radiative heat transfer from the surface of the slab to the space air and other surfaces. Meanwhile, heat transfer due to conduction takes place from the surface to the inner part of the slab. Heat is then stored inside the slab. The stored heat is released to the space air when the surface temperature falls below the temperature of the inner part of the slab.

To maintain the preset space air temperature, the heat that has been convected or released to the conditioned space should be removed from the space instantaneously.

Space and Equipment Loads

The sensible and latent heat transfer between the space air and the surroundings can be classified as follows:

1. *Space heat gain* q_e, in Btu/h, represents the rate at which heat enters a conditioned space from an external source or is released to the space from an internal source during a given time interval.
2. *Space cooling load,* often simply called cooling load q_{rc}, in Btu/h, is the rate at which heat must be removed from a conditioned space so as to maintain a constant temperature and acceptable relative humidity. The sensible cooling load is equal to the sum of the convective heat transfer from the sur-

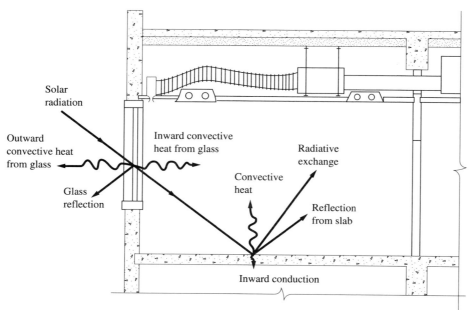

FIGURE 7.1 Convective and radiative heat in a conditioned space.

faces of the building envelope, furnishings, appliances, and equipment.

3. *Space heating load* q_{rh}, in Btu/h, is the rate at which heat must be added to the conditioned space to maintain a constant temperature and sometimes a specified relative humidity.

4. *Space heat extraction rate* q_{ex}, in Btu/h, is the rate at which heat is actually removed from the conditioned space by the air system. The sensible heat extraction rate is equal to the sensible cooling load only when the space air temperature remains constant.

5. *Coil's load* q_c, in Btu/h, is the rate of heat transfer at the coil. The *cooling coil's load* q_{cc}, Btu/h, is the rate at which heat is removed by the chilled water flowing through the coil or absorbed by the refrigerant inside the coil.

The *heating coil load* q_{ch}, in Btu/h, is the rate at which heat is added to the conditioned air from the hot water, steam, or electric heating elements inside the coil.

6. *Refrigerating load* q_{rl}, in Btu/h, is the rate at which heat is absorbed by the refrigerant at the evaporator. For central hydronic systems, the refrigerating load is the sum of the coil's load plus the chilled water piping heat gain, pump power heat gain, and storage tank heat gain. For most water systems in commercial buildings, the water piping and pump power heat gain is only about 5 to 10 percent of the coil's load. In an air conditioning system using DX-coil(s), the refrigerating load is equal to the DX-coil's load.

The instantaneous sensible heat gain of a conditioned space is not equal to the instantaneous sensible cooling load because of storage of part of the radiative heat inside the building structures. Such a phenomenon results in a smaller instantaneous cooling load than that of the heat gain when it is at its maximum value during a diurnal cycle, as shown in Fig. 7.2.

If the space relative humidity is maintained at an approximately constant value, the storage effect of the moisture in the building envelope and furnishings can be ignored. Then the instantaneous space latent heat gain will be the same as the instantaneous space latent cooling load.

Night Shutdown Operating Mode

In many commercial buildings, air systems are often shut down during the nighttime, or unoccupied periods. The operating characteristics of the conditioned space in a 24-hour diel cycle can then be divided into three periods, as shown in Fig. 7.3.

NIGHT SHUTDOWN PERIOD. This period commences when the air system is switched off and ends when the air system is switched on again. When the air-handling unit, package unit, or terminal unit is turned off during summer in a hot and humid area, the infiltrated air (through the elevator shaft, pipe shafts, and window cracks) and any heat transfer from the window glass, external wall, or roofs, will cause a sudden increase in indoor temperature T_r of a few °F.

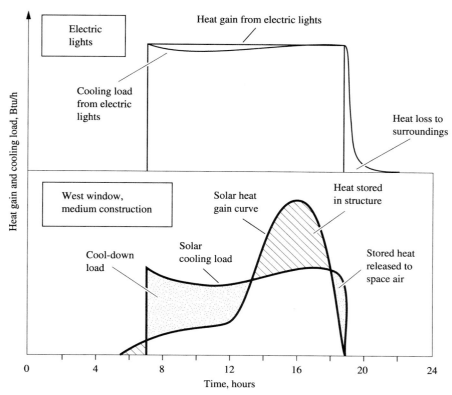

FIGURE 7.2 Solar heat gain and heat gain from electric lights and the corresponding space cooling loads for a night shutdown air system.

After that, T_r will rise further or drop slightly, depending on the difference between outdoor and indoor temperatures.

At the same time, the indoor space relative humidity increases gradually because of the infiltration of hot and humid outdoor air and the moisture transfer from the wetted surfaces in the air system. Higher space relative humidity causes a moisture transfer from the space air to the building envelopes and furnishings.

During winter, after the air system is turned off, the indoor temperature drops because of the heat loss through external windows, walls, and roofs, as well as the infiltration of outdoor cold air. Meanwhile, the space relative humidity increases mainly owing to the fall of indoor space temperature.

COOL-DOWN OR WARM-UP PERIOD. This period commences when the air system begins to operate and ends when the space temperature or other controlled variables has attained predetermined limits.

During summer, the supply of cold and dehumidified air after the air system is turned on causes a sudden drop in space air temperature and relative humidity. Both heat and moisture are transferred from the building envelope to the space air because of the comparatively higher temperature and moisture content of the building envelope. These heat and moisture trans-

fers form the cool-down cooling load. Sometimes, the cool-down load can be the maximum summer design cooling load.

If refrigeration is used for cooling during the cool-down period, then it usually lasts less than one hour, depending mainly on the tightness of the building and the differences of the space temperatures and humidity ratios between the shut-down and cool-down periods. If the cool-down of the space air and the building envelope is by means of the free cooling of outdoor air, the cool-down period may last for several hours.

During winter, the supply of warm air during the warm-up period raises the space temperature and lowers the space relative humidity. Because of the higher space temperature, heat transfer from the space air to the colder building envelope and furnishings, and the heat energy required to raise the temperature of building envelope, forms the warm-up heating load.

CONDITIONING PERIOD. This period commences when the space air temperature has fallen or risen to a value within the predetermined limits. It ends when the air system is shut down.

In summer, cold and dehumidified air is usually supplied to the space to offset the space cooling load and to maintain a required temperature and relative humidity. During winter, warm air is supplied to com-

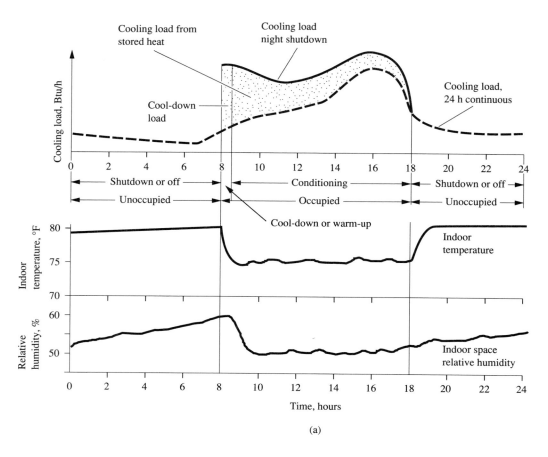

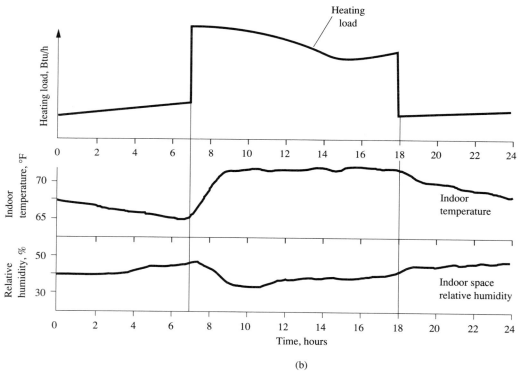

FIGURE 7.3 Operating characteristics of an air conditioned space operated at night shutdown mode: (a) summer cooling mode (hot and humid area); and (b) winter mode.

pensate for heat losses from the conditioned space. Space temperature is controlled and maintained within predetermined limits by control systems.

In commercial buildings located in areas with a cold winter, the air system often operates in duty-cycling mode during nighttime, unoccupied periods in winter, or intermediate seasons to maintain a night setback temperature. In *duty cycling mode,* the fan and heater turn on and off to maintain a desired temperature. Space temperature is set back at night, for example, to 55 or 60°F, to prevent freezing of water pipes and to produce a comparatively smaller temperature lift during a warm-up period.

The operation of the air system in a 24-hour diel cycle in winter is then divided into night setback period, warm-up period, and conditioning period. The required load on the heating coil during a morning warm-up period at winter design conditions is usually the winter design heating load.

Influence of Stored Heat

The curve of solar heat gain from a west window is shown in Fig. 7.2, as well as the cooling load curve due to this solar heat gain for a conditioned space operated at night shutdown mode in summer. The difference between the maximum solar heat gain $q_{hg.m}$ and the maximum cooling load $q_{cl.m}$ during the conditioning period indicates the amount of heat stored inside the building structures. This difference significantly affects the size of air conditioning equipment required. The amount of heat stored depends mainly on the mass of the building envelope (whether it is heavy, medium, or light); the duration of the operating period of the air system within a 24-hour cycle; and the characteristics of heat gain whether radiant heat or convective heat predominates. For solar radiation transmitted through a west window, $q_{cl.m}$ may have a magnitude of only 40 to 60 percent of $q_{hg.m}$.

The *ASHRAE Handbook* divides the mass of the building construction into the following three groups:

Heavy construction. Approximately 130 lb/ft² floor area

Medium construction. Approximately 70 lb/ft² floor area

Light construction. Approximately 30 lb/ft² floor area

7.3 COOLING LOAD AND COIL'S LOAD CALCULATIONS

Components of Cooling Load

Cooling load calculations for air conditioning system design are mainly used to determine the volume flow rate of the air system as well as the coil's and refrigeration load of the equipment. Cooling load usually can be classified into two categories:

External cooling loads. These loads are formed because of heat gains in the conditioned space from external sources through the building envelope or building shell and the partition walls. Sources of external loads include the following cooling loads:

1. Heat gain entering from the exterior walls and roofs
2. Solar heat gain transmitted through the fenestrations
3. Conductive heat gain transferred through the fenestrations
4. Heat gain entering from the partition walls and interior doors
5. Infiltration of outdoor air into the conditioned space

Internal cooling loads. These loads are formed by the release of sensible and latent heat from the heat sources inside the conditioned space. These sources contribute internal cooling loads:

1. People
2. Electric lights
3. Equipment and appliances

If moisture transfers from the building structures and the furnishings are excluded, only infiltrated air, occupants, equipment, and appliances have both sensible and latent cooling loads. The remaining components have only sensible cooling loads.

All sensible heat gains entering the conditioned space represent radiative heat and convective heat except infiltrated air. As in Section 7.2, radiative heat causes heat storage in the building structures, converts part of the heat gain into cooling load, and makes the cooling load calculations more complicated.

Latent heat gains are heat gains from moisture transfer from the occupants, equipment, appliances, or infiltrated air. All of them release heat to the space air instantaneously and, therefore, they are instantaneous cooling loads.

Components of Cooling Coil's Load

If the conductive heat gain from the coil framework and the support is ignored, the cooling coil's load consists of the following components, as shown in Fig. 7.4.

1. Space cooling load, including sensible and latent load
2. Supply system heat gain

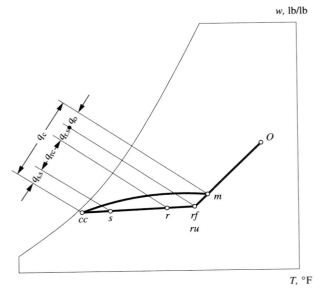

FIGURE 7.4 Difference between cooling load and cooling coil load on psychrometric chart.

3. Return system heat gain
4. Sensible and latent load because of the outdoor ventilation rates to meet the requirements of the occupants and others

Usually, both the supply and return system heat gains are sensible loads. These components are absorbed by the supply air and return air and appear as cooling and dehumidifying loads at the cooling coil.

Difference between the Cooling Load and Cooling Coil's Load

The same basic air conditioning cycle at summer mode operation as shown in Fig. 3.8 appears in Fig. 7.4. Note the following:

1. The space cooling load is represented by q_{rc}, the cooling coil's load is represented by q_c, and

$$q_c = q_{rc} + q_{s.s} + q_{r.s} + q_o \qquad (7.2)$$

where $q_{s.s}, q_{r.s}$ = supply and return system heat gains, Btu/h
q_o = load due to the outdoor air intake, Btu/h

2. The space cooling load is used to determine the supply volume flow rate $\dot{V}_s$ by using Eq. (3.48b), whereas the coil's load is used to determine the size of the cooling coil in an air-handling unit or DX-coil in a packaged unit.

3. A cooling load component influences both $\dot{V}_s$ and the size of the cooling coil, whereas a cooling coil's load component may not affect $\dot{V}_s$.

4. Infiltration is a cooling load. From Fig. 7.4, it is apparent that the load due to the outdoor ventilation air q_o, sometimes called the ventilation load, is a coil's load. If q_o is considered a cooling load, the volume flow rate of the air system will be oversized.

Load Profile

A *load profile* shows the variation of space load within a certain time period, such as a 24-hour operating cycle or an annual operating cycle, as shown in Fig. 7.5. In a load profile, the space load is always plotted against time. For a space cooling load, the magnitude of the curve is positive; for a space heating load, the magnitude is negative. A load profile may be used to illustrate the load variation of an air conditioned space—a room, a zone, a floor, a building, or a project. The shape of a load profile depends on the outdoor climate and, therefore, the latitude, orientation, and structure of the building. It is also affected by the operating characteristics and the variation of the internal loads.

The *load duration curve* is a plot of the number of hours vs. load ratio. *Load ratio* is defined as the ratio of cooling or heating load to the design full load, both in Btu/h, over a certain time period. The time period may be a day, a week, a month, or a year. Load duration curve is helpful in many operating and energy consumption analyses.

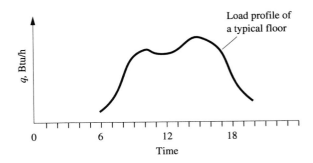

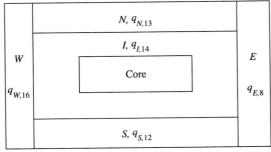

FIGURE 7.5 Load profile, peak load, and block load.

Peak Load and Block Load

The *zone peak load* is the maximum space cooling load in a load profile of a control zone of the same orientation and similar internal load characteristics calculated according to the daily outdoor dry bulb and coincident wet bulb curve containing summer or winter outdoor design conditions.

For a zone cooling load with several components, such as solar load through window glass, heat transfer through roofs, or internal load from electric lights, zone peak load is always the maximum sum of these zone cooling load components at a given time. The *block load* is the sum of the maximum space cooling loads of various load profiles of the control zones of a building floor or a building at a given time.

The *peak cooling load* is the maximum cooling load of a space, room, floor, or building at a given time in a load profile. Space or room peak load may be either zone peak load or block load. Floor or building peak load is always a block load.

The heating load is the peak heating load of a space, room, zone, floor, or building in a load profile. Block load is not needed in heating load calculations because the solar radiation and internal loads are not taken into account. Design heating load depends entirely on indoor-outdoor temperature difference.

Suppose a typical floor of a multistory building has five zones: a north, a south, an east, and a west perimeter zone and an interior zone. For each zone, there is a corresponding daily cooling load profile curve and a zone peak load $q_{N,13}$, $q_{S,12}$, $q_{E,8}$, $q_{W,16}$, and $q_{I,14}$. Here, subscripts N, S, E, W, and I indicate north, south, east, west, and interior, respectively; 8, 12, 13, 14, and 16 represent time, at 8, 12, 13, 14, and 16 hours, respectively.

The block load of this typical floor $q_{max,t}$ is not calculated as the sum of the zone peak loads, $q_{N,13}$, $q_{S,12}$, $q_{E,8}$, $q_{W,16}$ and $q_{I,14}$. Rather, block load is calculated for a specific time. For example, block load $q_{max,14}$, in Btu/h, is the block load at 14 hours, that is,

$$q_{max,14} = q_{N,14} + q_{S,14} + q_{E,14} + q_{W,14} + q_{I,14}$$

Characteristics of Night Shutdown Operating Mode

Compared with the continuous operating mode, night shutdown operating mode has the following characteristics:

1. A greater peak load and, therefore, a higher air system and initial cost
2. A higher maximum heat extraction rate

3. A smaller accumulated heat extraction rate over the 24-hour operating cycle
4. A lower power consumption for the fans, compressors, and pumps

Moisture Transfer from the Building Structures

In hot and humid climates, the frequency of outdoor wet bulb temperatures higher than 73°F exceeds 1750 hours during the six warmest consecutive months annually. At the same time, the air infiltrated through the elevator and pipe shafts during shutdown periods causes a temperature increase of more than 7°F and a relative humidity increase exceeding 10 percent. These result in a significant increase of latent load because of the moisture transfer from the building envelope and furnishings to the space air during the cool-down period in summer. Such an increase in the coil's latent load not only necessitates a greater refrigerating capacity, but also lowers the sensible heat ratio of the space conditioning process.

In industrial applications where a very low relative humidity is maintained in the conditioned space, the latent load due to the moisture transfer from the building envelope should be added to the space cooling load calculations during night shut-down operating mode.

Considerations on Cooling Load Calculations

Before performing cooling load calculations, one should collect indoor design criteria, building characteristics, and information on operating characteristics and internal loads of similar projects.

For peak load calculations, a 2.5 percent, or sometimes 1 percent, outdoor dry bulb and the coincident wet bulb are often used as the summer outdoor design conditions. To calculate the block load, the daily outdoor dry and wet bulb curves containing the foregoing dry and wet bulb values should be used. For winter heating peak load calculations, 99 percent outdoor dry bulb is often used. When humidification is needed in winter, local weather should be analyzed to determine the outdoor design humidity ratio in winter.

Many factors, such as building and operating characteristics, internal loads, outdoor climate, and future development, affect the space cooling load. Calculation of cooling load involves simulating various heat gains of the design project and converting them into cooling loads. The calculated and actual cooling loads will be different. A good designer tends to bring the difference within acceptable limits.

In cooling load calculations, the emphasis should be focused on primary component(s) that have greater weight on total cooling load. The calculations of minor components should be simplified. For unlike projects, different components should be emphasized. In a multistory office building, solar heat gain through the window glass usually is the component that has the greatest influence in the perimeter zone. In an indoor stadium, the presence of occupants is the primary component, and the appliance load is minor. For an industrial application such as textile mill, the machine load has the greatest effect on cooling load, whereas the occupants' load is sometimes negligible.

Cool-down and warm-up load should be considered. In cold climates, air systems may be used to maintain a night setback temperature. In hot climates, sometimes air systems are also operated during nighttime unoccupied periods if the indoor temperature exceeds a certain limit, such as 85°F.

An estimated space cooling load always results in an energy-inefficient air conditioning system and a greater investment.

7.4 HISTORICAL DEVELOPMENT OF COOLING LOAD CALCULATIONS

In the 1930s, Houghton et al. introduced the analysis of heat transmission through the building envelope and discussed the periodic heat flow characteristics of the building envelope. In 1937, the *ASHVE Guide* introduced a systematic method of cooling load calculation involving the division of various load components. In the *ASHVE Guide,* solar radiation factors were introduced and their influence on external walls and roofs were taken into consideration. Both the window crack and the number of air changes methods were used to calculate infiltration.

Mackey and Wright first introduced the concept of sol-air temperature in 1944. In the same paper, they recommended a method of approximating the changes in inside surface temperature of walls and roofs due to periodic heat flow caused by solar radiation and outside temperature with a new decrement factor. In 1952, Mackey and Gay analyzed the difference between the instantaneous cooling load and the heat gain owing to radiant heat incident on the surface of the building envelope.

In 1964, Palmatier introduced the term *thermal storage factor* to indicate the ratio between the rate of instantaneous cooling load in the space and rate of heat gain. One year later, Carrier Corporation published a design handbook in which *heat storage factor* and *equivalent temperature difference* (ETD) were used to indicate the ratio of instantaneous cooling load and heat gain because of the heat storage effect of

the building structure. This cooling load calculation method was widely used by many designers until the current ASHRAE method was adopted.

In 1967, ASHRAE suggested a time-averaging (TA) method to allocate the radiative heat over successive periods of two to three hours or six to eight hours, depending on the construction of the building structure. Heat gains through walls and roofs are tabulated in total equivalent temperature differentials (TETD). In the same year, Stephenson and Mitalas recommended the *thermal response factor*, which includes the heat storage effect, for the calculation of cooling load. Thermal response factor evaluates the system response on one side of the structure according to random temperature excitations on the other side of the structure. This concept had been developed and forms the basis of the *transfer function method* (TFM) and *weighting factors*. In the mid-1970s, ASHRAE and NBS published computerized calculations of heating and cooling loads in energy-estimating programs. In 1977, ASHRAE introduced a single-step cooling load calculating procedure that uses the cooling load factor (CLF) and cooling load temperature difference (CLTD); these are produced from the simplified TFM. Now, both CLF/CLTD and TFM methods are widely used by design engineers in the United States. The TFM method is usually used for more detailed cooling load calculations and analyses.

In 1980, the U.S. Department of Energy sponsored a computer program for energy estimation through hour-by-hour detailed system simulation, called DOE-2, which was published through Los Alamos National Laboratory and Lawrence Berkeley Laboratory. In this program, a custom weighting-factor method is used for heating and cooling load calculations. Many computerized thermal load and energy-calculating computer programs were developed in the 1980s. Some of them use a finite elements method for cooling load calculations and others also contain moisture transfer calculations. At present, the later version of DOE-2, published in the mid-1980s, is the most widely used energy program. It includes hour-by-hour heating and cooling load calculations.

7.5 COOLING LOAD CALCULATIONS BY CLF/CLTD METHOD

General Considerations

For various cooling load calculation methods, the instantaneous heat gain calculations are approximately the same. The differences between the methods are due to conversions from instantaneous heat gain to instantaneous cooling load.

The CLF/CLTD method is a simplified means of calculating the heat gain and the cooling load at the same time. In using the CLF/CLTD method to calculate the cooling load, the following conditions are met during calculation:

1. Consider one-dimensional transient heat flow through building envelopes only.
2. Indoor temperature is considered constant.
3. Building construction is divided into three types: heavy, medium, and light, as defined in Section 7.2.
4. The cooling load is calculated based on the mean value of a time interval of one hour.

Only sensible heat gains and solar radiation transmitted through external walls and roofs and window glass, as well as sensible heat gains from occupants, electric lights, equipment, and appliances need CLF/CLTD to convert heat gains to instantaneous cooling loads. Infiltration and latent heat gains are themselves instantaneous cooling loads.

Transfer Function Coefficients and Weighting Factors

Let us define the following terms before calculating CLF/CLTD.

The transfer function K of an element or a system is the ratio of the Laplace transform of the output Y to the Laplace transform of the input or driving force G, or

$$Y = K \cdot G \qquad (7.3)$$

When a continuous function of time $f(t)$ is represented at regular intervals Δt and its magnitudes are $f(0), f(\Delta), f(2\Delta), \cdots, f(n\Delta)$, the Laplace transform is given as

$$\varphi(z) = f(0) + f(\Delta)z^{-1} + f(2\Delta)z^{-2} + \cdots + f(n\Delta)z^{-n} \qquad (7.4)$$

where Δ = time interval, hour
$z = e^{s\Delta}$

The preceding polynomial in z^{-1} is called the $z-$transform of the continuous function $f(t)$.

In Eq. (7.3), Y, K, and G all can be represented in the form of a $z-$ transform. Then, the sensible cooling load at time t, or $q_{rs,t}$, in Btu/h, can be related to the sensible heat gains and previous cooling loads as

$$q_{rs,t} = \sum_{i=1}(a_o q_{es,t} + a_1 q_{es,t-\Delta} + a_2 q_{es,t-2\Delta} + \cdots)$$
$$- (b_1 q_{rs,t-\Delta} + b_2 q_{rs,t-2\Delta} + \cdots) \qquad (7.5)$$

where $q_{es,t-n\Delta}$ = sensible heat gains at time $t-n\Delta$, Btu/h

$q_{rs,\ t-n\Delta}$ = sensible cooling loads at time $t-n\Delta$, Btu/h
a_n, b_n = transfer function coefficients or weighting factors

Both $q_{es,t-n\Delta}$ and $q_{rs,t-n\Delta}$ are historical values.

Weighting factors are transfer function coefficients presented in the form of z-transform functions. Weighting factors are so-called because they are used to weight the importance of current and historical values of heat gains and cooling loads on currently calculated cooling loads.

Detailed Calculating Procedures

The details of the calculating procedures are as follows:

1. According to the heat balance at the inner surfaces of the building envelope of the conditioned space, a set of equations that equal the number of the inner surfaces is established in the following form:

$$q_{ki}(z) = q_{ci}(z) + \sum_{j=1}^{N} q_{rij}(z) + q_{ei}(z) \qquad (7.6)$$

where q_{ki}, q_{ci} = conductive and convective heat transfer to the ith inner surface, Btu/h
q_{rij} = radiative heat exchange between inner surface i and inner surface j, Btu/h
q_{ei} = solar or other radiative energy from internal loads radiated on inner surface i, Btu/h

2. Because of a unit pulse of radiant energy from solar or internal loads, there is a corresponding change of inner surface temperature $T_{i,t}$. The sensible cooling load, as continuous function of time, can be calculated from the convective heat transfer from the inner surface to the space air.

3. Based on the calculated sensible cooling load, in the form of z-transform, the transfer function coefficients or weighting factors for a specific type of cooling load component at specific building constructional characteristics are calculated.

4. For each type of sensible cooling load component at specific conditions, the magnitudes can be calculated from various sensible heat gains at the same conditions by means of a corresponding set of transfer function coefficients.

5. By dividing the sensible cooling load by the heat gain or the heat gain parameters, CLF or CLTD can be calculated, that is,

$$\text{CLF} = \frac{\text{sensible cooling load}}{\text{sensible heat gain}}$$

$$\text{CLTD} = \frac{\text{sensible cooling load}}{\substack{\text{temperature difference across} \\ \text{the building envelope}}} \quad (7.7)$$

For sensible cooling load due to the solar radiation transmitted through the window glass, the CLF also includes the ratio of the SHGF to the maximum SHGF.

7.6 CALCULATION OF EXTERNAL COOLING LOADS BY CLF/CLTD METHOD

Space Cooling Load Due to Heat Gain Through Exterior Walls and Roofs

This cooling load is mainly due to the heat gain of the combined effect of solar radiation and the heat transfer from the outdoor–indoor temperature difference. According to Eq. (7.6) and the analysis covered in Section 6.11, the inward heat flow from the outer surface of the sunlit exterior wall or roof, that is, the sensible heat gain q_{oi}, is given as

$$q_{oi} = h_o A (T_{sol} - T_{os})$$

For an exterior sunlit wall or roof under the combined effect of solar radiation and the outdoor temperature, this inward heat flow can be represented by the mean value of following expression during a time interval of one hour:

$$q_{oi} = UA(\text{temperature difference}) = UA\,\Delta T \quad (7.8)$$

where U = overall heat transfer coefficient of the exterior wall or roof, Btu/h $\cdot$ ft$^2 \cdot$ °F

A = area of the exterior wall or roof, ft^2

Since the ratio of sensible cooling load to sensible heat gain through the exterior wall or roof $q_{rs.w}/q_{oi}$ = CLTD/ΔT, then the space sensible cooling load $q_{rs.w}$, in Btu/h, can be calculated as

$$q_{rs.w} = U \times A \times \text{CLTD} \quad (7.9)$$

The CLTD recommended by ASHRAE for calculating the space sensible cooling load through flat roofs and sunlit walls of various constructions are listed in Tables 7.2 and 7.3, respectively.

The values in both Tables 7.2 and 7.3 were calculated under the following conditions. In other words, these are the conditions under which the listed data can be applied directly without adjustments:

- Indoor temperature 78°F
- Outdoor maximum temperature 95°F with an outdoor daily mean of 85°F and an outdoor daily range of 21°F

- Solar radiation of 40° north latitude on July 21
- Roof with dark flat surface
- Outer surface R-value R_o = 0.333h $\cdot$ ft$^2 \cdot$ °F/Btu and inner surface R_i = 0.685h $\cdot$ ft$^2 \cdot$ °F/Btu
- No attic fans or return air ducts in suspended ceiling space

The following formula can be used for adjustments when the conditions are different from those just mentioned:

$$\begin{aligned} \text{CLTD}_{corr} = [(\text{CLTD} + \text{LM}) \times K \\ + (78 - T_r) + (T_{om} - 85)] \times f \quad (7.10) \end{aligned}$$

where LM = latitude-month correction, as listed in Table 7.4.

K = color adjustment factor; $K = 1$ for a dark color or any color in an industrial area, and $K = 0.5$ for a permanently light-colored roof in a rural area.

$(78 - T_r)$ = indoor temperature correction; T_r is the indoor temperature, in °F

$(T_{om} - 85)$ = outdoor temperature correction; T_{om} is the outdoor mean daily temperature, in °F

f = factor for attic fans and return ducts; if there is no attic fan or return duct in a suspended ceiling space, $f = 1$. For positive ventilation by attic fans in a suspended ceiling space, $f = 0.75$.

In Table 7.2 the roof terrace system includes the following:

2 in. lightweight concrete

Air space

2 in. insulation (5.7 lb/ft^3)

0.5 in. slag or stone

0.375 in. felt and membrane

For a pitched roof with a suspended ceiling, the area A in Eq. (7.9) should be the area of the suspended ceiling. If a pitched roof has no suspended ceiling under it, then the actual CLTD is slightly higher than the value listed in Table 7.2 because a greater area is exposed to the outdoor air.

Space Cooling Load Due to Solar and Conductive Heat Gain through Fenestration

Although the inward heat flow from solar radiation absorbed by the window glass and the heat flow due

TABLE 7.2
CLTD for calculating sensible cooling loads from flat roofs

Description of construction	Weight, lb/ft²	U-value, Btu/h·ft²·°F	1	2	3	4	5	6	7	8	9	10	11	12	13	14	15	16	17	18	19	20	21	22	23	24	Hour of maximum CLTD	Minimum CLTD	Maximum CLTD	Difference in CLTD
Without suspended ceiling																														
2.5-in. wood with 2-in. insulation	13	0.093	30	26	23	19	16	13	10	9	8	9	13	17	23	29	36	41	46	49	51	50	47	43	39	35	19	8	51	43
4-in. wood with 2-in. insulation	18	0.078	38	36	33	30	28	25	22	20	18	17	16	17	18	21	24	28	32	36	39	41	43	43	42	40	22	16	43	27
With suspended ceiling																														
1-in. wood with 2-in. insulation	10	0.083	25	20	16	13	10	7	5	5	7	12	18	25	33	41	48	53	57	57	56	52	46	40	34	29	18	5	57	52
2.5-in. wood with 1-in. insulation	15	0.096	34	31	29	26	23	21	18	16	15	15	16	18	21	25	30	34	38	41	43	44	44	42	40	37	21	15	44	29
8-in. lightweight concrete	33	0.093	39	36	33	29	26	23	20	18	15	14	14	15	17	20	25	29	34	38	42	45	46	45	44	42	21	14	46	32
4-in. heavyweight concrete with 2-in. insulation	54	0.090	30	29	27	26	24	22	21	20	20	21	22	24	27	29	32	34	36	38	38	37	36	34	33	31	19	20	38	18
2.5-in. wood with 2-in. insulation	15	0.072	35	33	30	28	26	24	22	20	18	18	18	20	22	25	28	32	35	38	40	41	41	40	39	37	21	18	41	23
Roof terrace system	77	0.082	30	29	28	27	26	25	24	23	22	22	22	23	23	25	26	28	29	31	32	33	33	33	33	32	22	33	22	11
6-in. heavyweight concrete with 2-in. insulation	77	0.088	29	28	27	26	25	24	23	22	21	21	22	23	25	26	28	30	32	33	34	34	34	33	32	31	20	21	34	13
4-in. wood with 2-in. insulation	19 20	0.082 0.064	35	34	34	33	32	31	29	27	26	24	23	22	21	22	24	25	27	30	32	34	35	36	37	36	23	21	37	16

Conditions of direct application and adjustments are stated in the text.
Abridged with permission from *ASHRAE Handbook 1989, Fundamentals.*

TABLE 7.3
CLTD for calculating sensible cooling loads from sunlit walls of north latitude

Facing	1	2	3	4	5	6	7	8	9	10	11	12	13	14	15	16	17	18	19	20	21	22	23	24	Hour of maximum CLTD	Minimum CLTD	Maximum CLTD	Difference in CLTD
Group C Walls: typical, outside 1 in. stucco, 2 in. insulation (5.7lb/ft^3, 4 in. concrete, 0.75 in. plaster or gypsum, inside U = 0.119 Btu/h · ft^2 · °F; Mass, 63 lb/ft^2)																												
N	15	14	13	12	11	10	9	8	8	7	7	8	8	9	10	12	13	14	15	16	17	17	17	16	22	7	17	10
NE	19	17	16	14	13	11	10	10	11	13	15	17	19	20	21	22	22	23	23	23	23	22	21	20	20	10	23	13
E	22	21	19	17	15	14	12	12	14	16	19	22	25	27	29	29	30	30	30	29	28	27	26	24	18	12	30	18
SE	22	21	19	17	15	14	12	12	13	16	19	22	24	26	28	29	29	29	29	28	27	26	24		19	12	29	17
S	21	19	18	16	15	13	12	10	9	9	9	10	11	14	17	20	22	24	25	26	25	25	24	22	20	9	26	17
SW	29	17	25	22	20	18	16	15	13	12	11	11	11	13	15	18	22	26	29	32	33	33	32	31	22	11	33	22
W	31	29	27	25	22	20	18	16	14	13	12	12	12	13	14	16	20	24	29	32	35	35	35	33	22	12	35	23
NW	25	23	21	20	18	16	14	13	11	10	10	10	10	11	12	13	15	18	22	25	27	27	27	26	22	10	27	17
Group D walls: typical, outside 1 in. stucco, 4 in. concrete, 1 or 2 in. insulation (2 lb/ft^3), 0.75 in. plaster or gypsum, inside U = 0.119–0.20 Btu/h · ft^2 · °F; Mass, 63 lb/ft^2																												
N	15	13	12	10	9	7	6	6	6	6	6	7	8	10	12	13	15	17	18	19	19	19	18	16	21	6	19	13
NE	17	15	13	11	10	8	7	8	10	14	17	20	22	23	23	24	24	25	25	24	23	22	20	18	19	7	25	18
E	19	17	15	13	11	9	8	9	12	17	22	27	30	32	33	33	32	32	31	30	28	26	24	22	16	8	33	25
SE	20	17	15	13	11	10	8	8	10	13	17	22	26	29	31	32	32	32	31	30	28	26	24	22	17	8	32	24
S	19	17	15	13	11	9	8	7	6	6	7	9	12	16	20	24	27	29	29	29	27	26	24	22	19	6	29	23
SW	28	25	22	19	16	14	12	10	9	8	8	8	10	12	16	21	27	32	36	38	38	37	34	31	21	8	38	30
W	31	27	24	21	18	15	13	11	10	9	9	9	10	11	14	18	24	30	36	40	41	40	38	34	21	9	41	32
NW	25	22	19	17	14	12	10	9	8	7	7	8	9	10	12	14	18	22	27	31	32	32	30	27	22	7	32	25
Group G walls: typical, outside 1 in. stucco, air space, 1, 2, or 3 in. insulation (2 lb/ft^3), 0.75 in. plaster or gypsum, inside U = 0.081–0.78 Btu/h · ft^2 · °F; Mass, 16 lb/ft^2																												
N	3	2	1	0	-1	2	7	8	9	12	15	18	21	23	24	24	25	26	22	15	11	9	7	5	18	-1	26	27
NE	3	2	1	0	-1	9	27	36	39	35	30	26	26	27	27	26	25	22	18	14	11	9	7	5	9	-1	39	40
E	4	2	1	0	-1	11	31	47	54	55	50	40	33	31	30	29	27	24	19	15	12	10	8	6	10	-1	55	56
SE	4	2	1	0	-1	5	18	32	42	49	51	48	42	36	32	30	27	24	19	15	12	10	8	6	11	-1	51	52
S	4	2	1	0	-1	0	1	5	12	22	31	39	45	46	43	37	31	25	20	15	12	10	8	5	14	-1	46	47
SW	5	4	3	1	0	0	2	5	8	12	16	26	38	50	59	63	61	52	37	24	17	13	10	8	16	0	63	63
W	6	5	3	2	1	1	2	5	8	11	15	19	27	41	56	67	72	67	48	29	20	15	11	8	17	1	72	71
NW	5	3	2	1	0	0	2	5	8	11	15	18	21	27	37	47	55	55	41	25	17	13	10	7	18	0	55	55

Direct applications and adjustments are stated in the text.
Abridged with permission from *ASHRAE Handbook 1989, Fundamentals*.

to the outdoor and indoor temperature difference are actually combined, it is simpler and acceptably accurate to separate this composite heat gain into solar heat gain and conductive heat gain.

Space sensible cooling load from solar heat gain flowing through the window facing a specific direction $q_{rs.s}$, in Btu/h, can be calculated as follows:

$$q_{rs.s} = q_{sun} + q_{sh}$$
$$= (A_s \times SHGF_{max} \times CLF_s \times SC)$$
$$+ (A_{sh} \times SHGF_{max.sh} \times CLF_{sh} \times SC)$$
$$(7.11)$$

where

q_{sun} = space cooling load from solar heat gain through sunlit area of the window glass, Btu/h

q_{sh} = space cooling load from solar heat gain through the shaded area of the window glass, Btu/h

A_s, A_{sh} = sunlit and shaded area, ft^2

$SHGF_{max}, SHGF_{max.sh}$ = maximum solar heat gain factor for sunlit and shaded areas, respectively (refer to Table 6.10), Btu/h · ft^2

SC = shading coefficient

CLF_s = cooling load factor for sunlit glass facing the specific direction

TABLE 7.4
CLTD correction for various latitudes and months applied to walls and roofs, in °F

Latitude, North	Month	N	NNE NNW	NE NW	ENE WNW	E W	ESE WSW	SE SW	SSE SSW	S	HOR*
24	Dec	−5	−7	−9	−10	−7	−3	3	9	13	−13
	Jan/Nov	−4	−6	−8	−9	−6	−3	3	9	13	−11
	Feb/Oct	−4	−5	−6	−6	−3	−1	3	7	10	−7
	Mar/Sep	−3	−4	−3	−3	−1	−1	1	2	4	−3
	Apr/Aug	−2	−1	0	−1	−1	−2	−1	−2	−3	0
	May/Jul	1	2	2	0	0	−3	−3	−5	−6	1
	Jun	3	3	3	1	0	−3	−4	−6	−6	1
32	Dec	−5	−7	−10	−11	−8	−5	2	9	12	−17
	Jan/Nov	−5	−7	−9	−11	−8	−4	2	9	12	−15
	Feb/Oct	−4	−6	−7	−8	−4	−2	4	8	11	−10
	Mar/Sep	−3	−4	−4	−4	−2	−1	3	5	7	−5
	Apr/Aug	−2	−2	−1	−2	0	−1	0	1	1	−1
	May/Jul	1	1	1	0	0	−1	−1	−3	−3	1
	Jun	1	2	2	1	0	−2	−2	−4	−4	2
40	Dec	−6	−8	−10	−13	−10	−7	0	7	10	−21
	Jan/Nov	−5	−7	−10	−12	−9	−6	1	8	11	−19
	Feb/Oct	−5	−7	−8	−9	−6	−3	3	8	12	−14
	Mar/Sep	−4	−5	−5	−6	−3	−1	4	7	10	−8
	Apr/Aug	−2	−3	−2	−2	0	0	2	3	4	−3
	May/Jul	0	0	0	0	0	0	0	0	1	1
	Jun	1	1	1	0	1	0	0	−1	−1	2
48	Dec	−6	−8	−11	−14	−13	−10	−3	2	6	−25
	Jan/Nov	−6	−8	−11	−13	−11	−8	−1	5	8	−24
	Feb/Oct	−5	−7	−10	−11	−8	−5	1	8	11	−18
	Mar/Sep	−4	−6	−6	−7	−4	−1	4	8	11	−11
	Apr/Aug	−3	−3	−3	−3	−1	0	4	6	7	−5
	May/Jul	0	−1	0	0	1	1	3	3	4	0
	Jun	1	1	2	1	2	1	2	2	3	2
56	Dec	−7	−9	−12	−16	−16	−14	−9	−5	−3	−28
	Jan/Nov	−6	−8	−11	−15	−14	−12	−6	−1	2	−27
	Feb/Oct	−6	−8	−10	−12	−10	−7	0	6	9	−22
	Mar/Sep	−5	−6	−7	−8	−5	−2	4	8	12	−15
	Apr/Aug	−3	−4	−4	−4	−1	1	5	7	9	−8
	May/Jul	0	0	0	0	2	2	5	6	7	−2
	Jul	2	1	2	1	3	3	4	5	6	1

*Horizontal
Abridged with permission from *ASHRAE Handbook 1989, Fundamentals.*

CLF_{sh} = cooling load factor for shaded area as if the glass is facing north

In Eq. (7.11), at a given time, $A_s + A_{sh} = A_{glass}$. Here A_{glass} indicates the glass area of the window, in ft^2.

In the Northern Hemisphere, the SHGF_{max} of a conditioned space with southern orientation may occur in December instead of June.

Cooling load factors CLF_s and CLF_{sh} can be calculated from Eqs. (7.6) and (7.7). The CLF for window glass with interior shading at north latitudes for all types of room constructions are listed in Table 7.5. Window glass with interior shading is recommended for energy efficiency.

Space sensible cooling load from conduction heat gain transferred through the window $q_{rs.c}$, in Btu/h, is calculated as

$$q_{rs.c} = A_w \times U \times \text{CLTD} \qquad (7.12)$$

where A_w = area of the window including the sash, ft^2

U = overall heat transfer coefficient of the window; its value is listed in Table 6.8, in Btu/h · ft^2 · °F

TABLE 7.5
CLF for window glass with interior shading at north latitudes for all types of room constructions

Window facing	Solar time, h																							
	1	2	3	4	5	6	7	8	9	10	11	12	13	14	15	16	17	18	19	20	21	22	23	24
N	0.08	0.07	0.06	0.06	0.07	0.73	0.66	0.65	0.73	0.80	0.86	0.89	0.89	0.86	0.82	0.75	0.78	0.91	0.24	0.18	0.15	0.13	0.11	0.10
NNE	0.03	0.03	0.02	0.02	0.03	0.64	0.77	0.62	0.42	0.37	0.37	0.37	0.36	0.35	0.32	0.28	0.23	0.17	0.08	0.07	0.06	0.05	0.04	0.04
NE	0.03	0.02	0.02	0.02	0.02	0.56	0.76	0.74	0.58	0.37	0.29	0.27	0.26	0.24	0.22	0.20	0.16	0.12	0.06	0.05	0.04	0.04	0.03	0.03
ENE	0.03	0.02	0.02	0.02	0.02	0.52	0.76	0.80	0.71	0.52	0.31	0.26	0.24	0.22	0.20	0.18	0.15	0.11	0.06	0.05	0.04	0.04	0.03	0.03
E	0.03	0.02	0.02	0.02	0.02	0.47	0.72	0.80	0.76	0.62	0.41	0.27	0.24	0.22	0.20	0.17	0.14	0.11	0.06	0.05	0.05	0.04	0.03	0.03
ESE	0.03	0.03	0.02	0.02	0.02	0.41	0.67	0.79	0.80	0.72	0.54	0.34	0.27	0.24	0.21	0.19	0.15	0.12	0.07	0.06	0.05	0.04	0.04	0.03
SE	0.03	0.03	0.02	0.02	0.02	0.30	0.57	0.74	0.81	0.79	0.68	0.49	0.33	0.28	0.25	0.22	0.18	0.13	0.08	0.07	0.05	0.05	0.04	0.04
SSE	0.04	0.03	0.03	0.03	0.02	0.12	0.31	0.54	0.72	0.81	0.81	0.71	0.54	0.38	0.32	0.27	0.22	0.16	0.09	0.08	0.06	0.06	0.05	0.04
S	0.04	0.04	0.03	0.03	0.03	0.09	0.16	0.23	0.38	0.58	0.75	0.83	0.80	0.68	0.50	0.35	0.27	0.19	0.11	0.09	0.08	0.07	0.06	0.05
SSW	0.05	0.04	0.04	0.03	0.03	0.09	0.14	0.18	0.22	0.27	0.43	0.63	0.78	0.84	0.80	0.66	0.46	0.25	0.13	0.11	0.09	0.08	0.07	0.06
SW	0.05	0.05	0.04	0.04	0.03	0.07	0.11	0.14	0.16	0.19	0.22	0.38	0.59	0.75	0.83	0.81	0.69	0.45	0.16	0.12	0.10	0.09	0.07	0.06
WSW	0.05	0.05	0.04	0.04	0.03	0.07	0.10	0.12	0.14	0.16	0.17	0.23	0.44	0.64	0.78	0.84	0.78	0.55	0.16	0.12	0.10	0.09	0.07	0.06
W	0.05	0.05	0.04	0.04	0.03	0.06	0.09	0.11	0.13	0.15	0.16	0.17	0.31	0.53	0.72	0.82	0.81	0.61	0.16	0.12	0.10	0.08	0.07	0.06
WNW	0.05	0.05	0.04	0.03	0.03	0.07	0.10	0.12	0.14	0.16	0.17	0.18	0.22	0.43	0.65	0.80	0.84	0.66	0.16	0.12	0.10	0.08	0.07	0.06
NW	0.05	0.04	0.04	0.03	0.03	0.07	0.11	0.14	0.17	0.19	0.20	0.21	0.22	0.30	0.52	0.73	0.82	0.69	0.16	0.12	0.10	0.08	0.07	0.06
NNW	0.05	0.05	0.04	0.03	0.03	0.11	0.17	0.22	0.26	0.30	0.32	0.33	0.34	0.34	0.39	0.61	0.82	0.76	0.17	0.12	0.10	0.08	0.07	0.06
HOR*	0.06	0.05	0.04	0.04	0.03	0.12	0.27	0.44	0.59	0.72	0.81	0.85	0.85	0.81	0.71	0.58	0.42	0.25	0.14	0.12	0.10	0.08	0.07	0.06

*Horizontal
Source: ASHRAE Handbook 1989, Fundamentals. Reprinted with permission.

TABLE 7.6
CLTD for calculating space cooling load due to conductive heat gain through windows

Solar Time, h	1	2	3	4	5	6	7	8	9	10	11	12
CLTD, °F	1	0	−1	−2	−2	−2	−2	0	2	4	7	9
Solar Time, h	13	14	15	16	17	18	19	20	21	22	23	24
CLTD, °F	12	13	14	14	13	12	10	8	6	4	3	2

Direct applications: indoor temperature 78°F, outdoor maximum temperature 95°F, and outdoor daily mean temperature 85°F

Adjustments: see Eq. (7.4) in the text.

Source: ASHRAE Handbook 1989, Fundamentals. Reprinted with permission.

In Eq. (7.12), $A_w = A_{glass} + A_{frame}$. Here A_{frame} represents the area of the frame or the sash of the window, in ft^2. Values of CLTD at various hours are calculated from Eqs. (7.6) and (7.7) and listed in Table 7.6. Corrections of CLTDs for conditions different from 78°F indoor design temperature, 95°F, outdoor maximum design temperature, and 85°F outdoor mean daily temperature are made using this equation:

$$CLTD_{corr} = CLTD + (78 - T_r) + (T_{om} - 85) \quad (7.13)$$

Precise calculation of conduction heat gain should increase the CLTDs of the frame or sash because of the radiated solar heat.

Space Cooling Load Due to Heat Gain Through Wall Exposed to Unconditioned Space

When a conditioned space is adjacent to an area that is unconditioned, and if the temperature fluctuation in this area is ignored, then the sensible cooling load from the heat gain q_{rs} transferred through the partitioned walls and interior windows and doors, in Btu/h, can be calculated as

$$q_{rs} = AU(T_{un} - T_r) \quad (7.14)$$

where T_{un}, T_r = daily mean air temperature of the adjacent area that is unconditioned and space temperature, respectively, °F

For an adjacent area that is not air-conditioned and has heat sources inside, such as a kitchen or boiler room, T_{un} may be 15°F higher than the outdoor temperature. For an adjacent area without any heat source other than electric lights, $(T_{un} - T_r)$ may be between 3 and 8°F.

For floors built directly on the ground or located above a basement that is neither ventilated nor conditioned, the space sensible cooling load from the heat gain through the floor can often be ignored.

Infiltration

Infiltration is the uncontrolled inward flow of outdoor air through cracks and openings in the building envelope due to the pressure difference across the envelope. The pressure difference may be caused by any of the following:

1. Wind pressure
2. Stack effect due to the outdoor and indoor temperature difference
3. Mechanical ventilation

A detailed analysis of these causes and the estimation of the volume flow rates of infiltrated air are given in Chapter 19. Modern commercial buildings have their exterior windows well sealed, and if a positive pressure is maintained in the conditioned space when the air system is operating, then normally the infiltration can be considered zero.

When air systems are not in operation, or in hotels, motels, and high-rise residential buildings, ASHRAE/IES Standard 90.1–1989 specifies an infiltration rate of 0.038 cfm/ft² gross area of external wall at all times, 0.15 ach (air change per hour) for the perimeter zone.

Only when exterior windows are not well sealed, the wind velocity is high in winter design conditions, or a door is exposed to outdoor air directly can an infiltration of 0.15 to 0.4 ach be taken into account for the perimeter zone.

As soon as the volume flow rate of infiltrated air $\dot{V}_{if}$, in cfm, is determined, the instantaneous space sensible cooling from infiltration can be calculated as

$$q_{rs.if} = 60 \, \dot{V}_{if} \rho_o c_{pa}(T_o - T_r) \quad (7.15)$$

where ρ_o = the density of outdoor air, lb/ft³. The space latent cooling load from infiltration $q_{rl.if}$, in Btu/h, can be calculated as

$$q_{rl.if} = 60 \dot{V}_{if} \rho_o (w_o - w_r) h_{fg.32} \quad (7.16)$$

where w_o, w_r = humidity ratio of the outdoor and space air, respectively, lb/lb

$h_{\text{fg.32}}$ = latent heat of vaporization at 32°F, Btu/lb

7.7 CALCULATION OF INTERNAL COOLING LOADS BY CLF/CLTD METHOD

People

Human beings release both sensible heat and latent heat to the conditioned space. The radiative portion of the sensible heat gain is about 70 percent when the indoor environment of the conditioned space is maintained within the comfort zone. The latent heat gain from sweating is always an instantaneous cooling load. The space sensible cooling load for occupants staying in a conditioned space can be calculated as

$$q_{\text{rs.}p} = N_p \times \text{SHG}_p \times \text{CLF}_p \qquad (7.17)$$

where N_p = number of occupants in the conditioned space

SHG_p = sensible heat gain of each person, Btu/h

CLF_p = cooling load factor for occupants

Space latent cooling load for occupants staying in a conditioned space $q_{\text{rl.}p}$, in Btu/h, is given as

$$q_{\text{rl.}p} = N_p \times \text{LHG}_p \qquad (7.18)$$

where LHG_p = latent heat gain of each person, Btu/h. In Table 7.7 are listed the heat gains from occupants in conditioned space, as abridged from the

ASHRAE Handbook 1989, Fundamentals. In Table 7.7, total heat is the sum of sensible and latent heat. The adjusted total heat is based on a normally distributed percentage of men, women, and children among the occupants with the stipulations that the heat gain from an adult female is 85 percent of that of an adult male, and that the heat gain from a child is 75 percent of that of an adult male.

In conditioned spaces in which an air system is operated at night in a shutdown mode, during the occupied period when the air system is operating, CLF_p is equal to 1. For a high occupancy density, as in auditoriums and theaters, the radiative heat transfer onto the walls, ceiling, and furnishings is considerably reduced; therefore, CLF_p can also be taken as 1.

Lighting

Electric lighting is one of the primary internal space cooling loads in commercial buildings. It often needs to be calculated more precisely.

INSIDE CONDITIONED SPACE. For electric lights installed inside the conditioned space, such as light fixtures hung below the ceiling, the sensible heat released from the electric lights, the emitting element, and lighting fixtures $q_{\text{s.l}}$ (in Btu/h) depends mainly on the criteria of illumination and the type and efficiency of electric lights. Sensible heat released to the conditioned space $q_{\text{es.l}}$ in Btu/h, is given as

$$q_{\text{es.l}} = 3.413 \, W_{\text{lamp}} \times N_{\text{lamp}} \times F_{\text{use}} \times F_{\text{al}} \qquad (7.19a)$$

TABLE 7.7
Rates of heat gain from occupants of conditioned spaces*

Degree of activity	Typical application	Total heat of adults, male, Btu/h	Total heat adjusted,[†] Btu/h	Sensible heat, Btu/h	Latent heat, Btu/h
Seated at theater	Theater—Matinee	390	330	225	105
Seated at theater	Theater—Evening	390	350	245	105
Seated, very light work	Offices, hotels, apartments	450	400	245	155
Moderately active office work	Offices, hotels, apartments	475	450	250	200
Standing, light work; walking	Department store, retail store	550	450	250	200
Walking; standing	Drug store, bank	550	500	250	250
Light bench work	Factory	800	750	275	475
Moderate dancing	Dance hall	900	850	305	545
Walking 3 miles/h; light machine work	Factory	1000	1000	375	625
Heavy work	Factory	1500	1450	580	870
Heavy machine work; lifting	Factory	1600	1600	635	965
Athletics	Gymnasium	2000	1800	710	1090

*Tabulated values are based on 75°F room dry-bulb temperature. For 80°F room dry-bulb, the total heat remains the same, but the sensible heat values should be decreased by approximately 20 percent and the latent heat values increased accordingly. All values are rounded to nearest 5 Btu/h.

[†]Adjusted heat gain is based on normal percentage of men, women, and children for the application listed, with the postulate that the gain from an adult female is 85 percent of that for an adult male, and that the gain from a child is 75 percent of that for an adult male.

Adapted with permission from *ASHRAE Handbook 1989, Fundamentals.*

The space cooling load from electric lights $q_{rs.1}$, in Btu/h, can be calculated as

$$q_{rs.1} = 3.413\, W_{lamp} \times N_{lamp} \times F_{use} \times F_{al} \times CLF \tag{7.19b}$$

where W_{lamp} = rated input wattage to each lamp, W
N_{lamp} = number of lamps
F_{use} = ratio of wattage in use to the installation wattage

In Eq. (7.19), F_{al} indicates an allowance factor for fixtures, such as ballast losses. For rapid-start 40 W fluorescent fixtures, F_{al} varies from 1.15 to 1.3, with a recommended value of 1.2 (*ASHRAE Handbook 1989, Fundamentals*). For high-efficiency industrial fixtures such as sodium lamps, F_{use} should be determined according to the manufacturers' data.

Sensible heat released by electric lights $q_{s.1}$ can also be expressed as

$$q_{s.1} = 3.413\, W_A A_{fl} \tag{7.20}$$

where A_{fl} = floor area, ft^2
W_A = wattage per ft^2 of floor area

In commercial buildings, W_A usually varies between 1.5 and 2.5 W/ft^2, including the influence of F_{use} and F_{al}.

RECESSED-MOUNTED AND RETURN PLENUM. For situations in which electric lights are recessed-mounted on the ceiling and the ceiling plenum is used as a return plenum, the fraction of the sensible heat gain of electric lights that enters the conditioned space $q_{es.1}$, in Btu/h, is closely related to the type of lighting fixture, the ceiling plenum, and the return system.

If the ceiling plenum is used as a return plenum in a multistory building, as shown in Fig. 7.6, the heat released from recessed fluorescent light fixtures includes the following:

• Radiative and convective heat transfer from the lighting fixture downward directly into the conditioned space q_{ld}, in Btu/h, which is calculated as

$$q_{ld} = q_{s.1} - q_{lp} = (1 - F_{lp})q_{s.1} \tag{7.21}$$

where q_{lp} = heat released by electric lights to return air, Btu/h
F_{lp} = fraction of heat released from the light fixture to plenum air

• Heat carried away by return air from the ceiling plenum q_{ret}, in Btu/h, which is calculated as

$$q_{ret} = 60\, \dot{V}_r \rho_r c_{pa}(T_p - T_r) \tag{7.22}$$

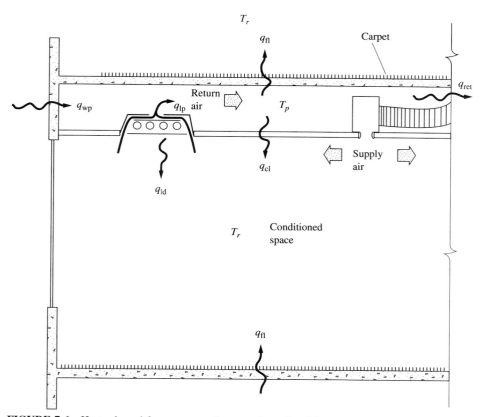

FIGURE 7.6 Heat released from a recessed-mounted ventilated lighting fixture.

where $\dot{V}_r$ = volume flow rate of return air, cfm
ρ_r = density of the return air, lb/ft^3
c_{pa} = specific heat of return air, Btu/lb · °F
T_p, T_r = temperatures of plenum air and space air, °F

If $\rho_r = 0.073$ lb/ft^3 and $c_{pa} = 0.243$ Btu/lb · °F, Eq. (7.22) can be rewritten as

$$q_{\text{ret}} = 60 \times 0.073 \times 0.243\, \dot{V}_r (T_p - T_r)$$
$$= 1.064\, \dot{V}_r (T_p - T_r)$$

- Heat transfer from the plenum air to the conditioned space through the suspended ceiling q_{cl}, and heat transfer from the plenum air to the conditioned space through the composite floor q_{fl}, both in Btu/h, calculated as

$$q_{cl} = U_{cl} A_{cl} (T_p - T_r)$$
$$q_{fl} = U_{fl} A_{fl} (T_p - T_r) \tag{7.23}$$

where U_{cl}, U_{fl} = overall heat transfer coefficient of the ceiling and the composite floor, Btu/h · ft^2 · °F
A_{cl}, A_{fl} = area of the ceiling and composite floor, ft^2

Heat transfer between the outdoor air and the plenum air through the exterior wall of the ceiling plenum q_{wp}, in Btu/h, is calculated as

$$q_{wp} = U_{wp} A_{wp}\, \text{CLTD}_{wp} \tag{7.24}$$

where U_{wp} = overall heat transfer coefficient of the exterior wall, Btu/h · ft^2 · °F
A_{wp} = area of the exterior wall of the plenum, ft^2
CLTD_{wp} = cooling load temperature difference of the exterior wall of the ceiling plenum

For a multistory building where all the floors are air-conditioned, heat gain from the electric lights that enters the conditioned space, heat-to-space, $q_{es.l}$ is given as

$$q_{es.l} = q_{ld} + q_{cl} + q_{fl} \tag{7.25}$$

Of this, about 50 percent is radiative and the rest is convective.

Heat released from the electric lights to the return air including radiative transfer upward, heat-to-plenum, q_{lp} in Btu/h, is given as

$$q_{lp} = q_{\text{ret}} + q_{cl} + q_{fl} \tag{7.26}$$

Of the sensible heat released from the electric lights $q_{s.l}$, the fraction entering the conditioned space directly, $q_{ld} = (1 - F_{lp})q_{s.l}$, depends on the volume flow rate of the return air flowing through the lighting fixture and the type of fixture. Its magnitude should

be obtained from the lighting fixture manufacturer. In Fig. 7.7 is shown the relationship between F_{lp} and the intensity of volume flow rate of return air $\dot{V}_r/A_{fl}$ if the lighting fixture is ventilated. Usually, F_{lp} varies between 0.4 and 0.6 for a ventilated fixture in a return air plenum. For unventilated lighting fixtures recessed-mounted in a return air plenum, F_{lp} varies between 0.15 and 0.5.

The space cooling load due to electric lights recessed-mounted in a return plenum is therefore

$$q_{rs.l} = \text{CLF}\, q_{es.l} \tag{7.27}$$

In a return air plenum, precise calculation of the plenum air temperature T_p is rather complicated. A simplified solution method is to assume a steady-state heat transfer between the plenum air and the conditioned space air and to use a CLTD to calculate the heat transfer through the exterior wall of the plenum. Then, based on the heat balance at the plenum air,

$$q_{lp} + U_{wp} A_{wp} \times \text{CLTD} =$$
$$(60\, \dot{V}_r \rho_r c_{pa} + U_{cl} A_{cl} + U_{fl} A_{fl})(T_p - T_r) \tag{7.28}$$

Plenum air temperature T_p can thus be determined.

SURFACE-MOUNTED UNDER CEILING. If the lighting fixtures are surface-mounted under the suspended ceiling, then the fraction of heat gain downward that enters the conditioned space, or $F_{es.l} = q_{es.l}/q_{s.l}$, varies from approximately 0.8 to 0.95.

The sensible space cooling load from the sensible heat gain entering the conditioned space from the surface-mounted electric lights $q_{rs.l}$, in Btu/h, is

$$q_{rs.l} = F_{es.l} q_{s.l} \times \text{CLF}_l \tag{7.29}$$

where CLF_l = cooling load factor for electric lights.

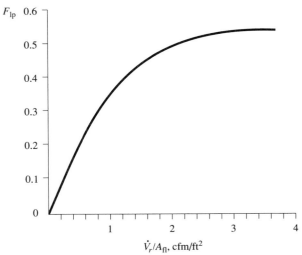

FIGURE 7.7 Relationship between F_{lp} and the intensity of return air $\dot{V}_r/A_{fl}$.

When the air system is operated in night shutdown mode, for electric lights in the conditioned space (recessed or surface-mountred), $CLF_l = 1$. If both the air system and the electric lights are turned on 24 hours a day continuously, $CLF_l = 1$ also.

Example 7.1. A return air plenum in a typical floor of a multistory building has the following constructional and operating characteristics:

Wattage of electric lights	1.5 W/ft^2
Return air volume flow rate	11,800 cfm
Density of return air ρ_r	0.073 lb/ft^3
Fraction of heat to plenum F_{1p}	0.5
U-values: exterior wall	0.2 Btu/h · ft^2 · °F
of the plenum	
suspended ceiling	0.32 Btu/h · ft^2 · °F
floor	0.21 Btu/h · ft^2 · °F
Area of the ceiling and floor	11,900 ft^2
Area of the exterior wall	1920 ft^2
of the plenum	
CLF of electric lights	0.84
CLTD of the exterior wall	24°F
of the plenum	
Space temperature	75°F

Determine the return air plenum temperature and the space cooling load from the electric lights when they are recessed-mounted in the return plenum.

Solution

1. From Eq. (7.20),

$$q_{ld} = (1 - F_{lp})q_{s.1} = 0.5 \times 1.5 \times 12,000 \times 3.413$$

$$= 30,717 \text{ Btu/h}$$

Because $60 \, \dot{V}_r \rho_r c_{pa} = 60 \times 11,800 \times 0.073 \times 0.243 = 12,559$ Btu/h · °F, the return air plenum temperature can be calculated from Eq. (7.28), as

$$T_p = T_r + \frac{q_{lp} + U_{wp}A_{wp}\text{CLTD}}{60 \, \dot{V}_r \rho_r c_{pa} + U_{cl}A_{cl} + U_{fl}A_{fl}} = 75$$

$$+ \frac{30,717 + 0.2 \times 1920 \times 24}{12,559 + 0.32 \times 12,000 + 0.21 \times 12,000}$$

$$= 75 + \frac{39,993}{18919} = 77.11°F$$

2. From Eq. (7.23), heat transfer from the plenum air to conditioned space through the ceiling and floor is calculated as

$$q_{cl} = U_{cl}A_{cl} = (T_p - T_r)$$

$$= 0.32 \times 11,900(77.1 - 75) = 7997 \text{ Btu/h}$$

$$q_{fl} = U_{fl}A_{fl} = (T_p - T_r)$$

$$= 0.21 \times 11,900(77.1 - 75) = 5248 \text{ Btu/h}$$

Then, from Eq. (7.25), heat to space is calculated as

$$q_{es.1} = q_{ld} + q_{cl} + q_{fl} = 30,717 + 7997 + 5248$$

$$= 43,962 \text{ Bth/h}$$

From Eq. (7.27), the sensible space cooling load from electric lights is

$$q_{rs.1} = CLFq_{es.1}$$

$$= 0.84(43,962)$$

$$= 36,928 \text{ Btu/h}$$

Equipment and Appliances

Within equipment and appliances, all energy inputs are converted into heat energy, for example, in the motor windings, combustion chamber, rubbing surfaces, or at the components where mechanical work is performed. A portion of the heat released may be exhausted locally by a mechanical ventilating system.

In many industrial applications, the space sensible cooling load due to the machine load when a motor is located inside the conditioned space $q_{rs.e}$, in Btu/h, can be calculated as

$$q_{rs.e} = \frac{2546 \, P_{hpr} \times F_{load} \times F_{use}(1 - F_{exh})}{\eta_{mo}} \quad (7.30)$$

where P_{hpr} = rated horsepower of the machine, hp
F_{load} = load factor indicating the ratio of actual power required to the rated power
F_{exh} = heat removal factor due to mechanical exhaust system
η_{mo} = motor efficiency

In Table 7.8 are listed the efficiencies of motors from a fraction of a horsepower to 200 hp. The greater the size of the motor, the higher the motor efficiency.

If the motor is located outside the conditioned space, then in Eq. (7.30), $\eta_{mo} = 1$.

In many types of equipment and appliances installed with exhaust hoods in the conditioned space, energy input and F_{exh} are better determined according to the actual performance of similar projects. For example, for food preparation appliances equipped with exhaust hoods, only radiation from the surface of the appliances, which as a fraction of the total heat input is between 0.1 and 0.5, should be counted, or F_{exh} is between 0.9 and 0.5. If there is no mechanical exhaust system, $F_{exh} = 0$.

Because of the widespread installation of microcomputers, display terminals, electric typewriters, printers, copiers, calculators, and fax machines, the heat released from machines in office buildings has increased considerably in recent years. In *ASHRAE/IES Standard 90.1–1989 User's Manual*, the heat gain from the operation of office machines and appliances may have a value between 0.6 and 1.8 W/ft^2. Precise office equipment heat gain can be calculated from manufacturer's data using a load factor F_{load} between 0.3 and 0.5.

TABLE 7.8
Heat gain from typical electric motors

Motor nameplate or rated horsepower	Nominal rpm	Full load motor efficiency, %	Location of motor and driven equipment with respect to conditioned space or airstream		
			Motor in, driven equipment in, Btu/h	Motor out, driven equipment in, Btu/h	Motor in, driven equipment out, Btu/h
Motor type: shaded pole					
0.05	1500	35	360	130	240
0.125	1500	35	900	320	590
Motor type: split phase					
0.25	1750	54	1,180	640	540
0.33	1750	56	1,500	840	660
0.50	1750	60	2,120	1,270	850
Motor type: 3-phase					
0.75	1750	72	2,650	1,900	740
1	1750	75	3,390	2,550	850
1	1750	77	4,960	3,820	1,140
2	1750	79	6,440	5,090	1,350
3	1750	81	9,430	7,640	1,790
5	1750	82	15,500	12,700	2,790
7.5	1750	84	22,700	19,100	3,640
10	1750	85	29,900	24,500	4,490
15	1750	86	44,400	38,200	6,210
20	1750	87	58,500	50,900	7,610
25	1750	88	72,300	63,600	8,680
30	1750	89	85,700	76,300	9,440
40	1750	89	114,000	102,000	12,600
50	1750	89	143,000	127,000	15,700
60	1750	89	172,000	153,000	18,900
75	1750	90	212,000	191,000	21,200
100	1750	90	283,000	255,000	28,300
150	1750	91	420,000	382,000	37,800
200	1750	91	569,000	509,000	50,300
250	1750	91	699,000	636,000	62,900

For motors operating hours more than 750 h/year, it is usually are cost-effective to use a high efficiency motor. Typical efficiency ratings are as follows: 5 hp, 89.5%; 10 hp, 91.0%; 50 hp, 94.1%; 100 hp, 95.1%; 200 hp, 96.2%.

Abriged with permission from *ASHRAE Handbook 1989, Fundamentals.* Reprinted by permission.

The space sensible cooling load due to heat gain from machines and appliances in office buildings $q_{rs.e}$, in Btu/h, can be calculated as

$$q_{rs.e} = 3.413 \, W_{A.e} A_{fl} \times CLF \qquad (7.31)$$

where $W_{A.e}$ = wattage input of the office machines, W/ft^2. For continuous operation, CLF = 1.

The space latent cooling load from the equipment and appliances $q_{rl.e}$, in Btu/h, can be calculated from the mass flow rate of water vapor evaporated:

$$q_{rl.e} = 1075 \, \dot{m}_w \qquad (7.32a)$$

where $\dot{m}_w$ = mass flow rate of water vapor evaporated, lb/h.

Internal load density (ILD) indicates the total internal heat gains of people, light, and equipment, in W/ft^2, and can be calculated as

$$ILD = (SHG_p + LHG_p)/A_{fl} + W_{A.l} + W_{A.e}$$
$$(7.32b)$$

where $W_{A.l}$ = lighting power density, W/ft^2

7.8 SPACE COOLING LOAD OF NIGHT SHUTDOWN OPERATING MODE

In commercial buildings, air systems are often operated in night shutdown mode during unoccupied hours in summer. When an air system is operated in a night shutdown mode, the cooling load factors CLFs for internal loads are equal to 1, as mentioned previously.

Since there are insufficient data to calculate external loads during cool-down and conditioning periods by using the CLF/CLTD methods, the following is a means to estimate the space sensible external loads:

1. Sum up the space external sensible cooling loads of the shutdown period Σq_τ, in Btu, or

$$\Sigma q_\tau = q_\tau + q_{(\tau+1)} + q_{(\tau+2)} + \cdots + q_{(t-2)} + q_{(t-1)}$$
$$(7.33)$$

where τ = the time at which the air system is shut down, h

$\tau+1, \tau+2$ = one and two hours, respectively, after the air system is shut down

t = the time at which the air system is started, h

$t-1, t-2$ = one and two hours, respectively, before the air system is started

In Eq. 7.33,

$$q_\tau = q_{\text{rsw}.\tau} + q_{\text{rsr}.\tau} + q_{\text{rss}.\tau} + q_{\text{rsc}.\tau}$$

$$q_{(\tau+1)} = q_{\text{rsw}.(\tau+1)} + q_{\text{rsr}.(\tau+1)}$$
$$+ q_{\text{rss}.(\tau+1)} + q_{\text{rsc}.(\tau+2)}$$

$$\vdots \qquad\qquad (7.34)$$

In Eq. (7.34), the subscript rsw indicates the space sensible cooling load for the exterior wall, rsr the exterior roof, rss the solar heat gain through fenestration, and rsc the conduction heat gain through the fenestration.

As the space temperature usually rises a few degrees during a shutdown period in hot summer, the space cooling load from the heat gain through partition walls from any area that is not air-conditioned during a shutdown period can be ignored.

2. Take $0.95 \Sigma q_\tau$ and add it to the space sensible cooling loads hour-by-hour during the cool-down and conditioning periods. These are calculated according to the continuous operating mode, that is,

$$q_{\text{rst}.t} = q_{\text{cont}.t} + C_t \Sigma q_\tau \qquad (7.35)$$

where $q_{\text{rst}.t}$ = total space sensible cooling load of night shutdown mode at time t, Btu/h

$q_{\text{cont}.t}$ = total space sensible cooling load at continuous operating mode at time t, Btu/h

C_t = fraction of the sum of the space sensible cooling loads during night shutdown period

Because of the heat loss to the sky vault and surroundings due to a higher space temperature during the night shutdown period, only $0.95 \Sigma q_\tau$ is allocated.

Based on the analysis in Rudoy and Robins (1977) paper and field survey data in offices by Wang, the values of C_t in Table 7.9 are recommended for reference.

Example 7.2. The plan of a typical floor in a multistory office building is shown in Fig. 7.8. The tenants and the partition walls of this floor are unknown during the design stage. The core part contains the restrooms, stairwells, and mechanical and electrical service rooms, and has mechanical exhaust systems only. This building is located at 40° north latitude and has the following constructional characteristics:

Area of the perimeter zone	6300 ft²
Area of the interior zone	5600 ft²
Floor to floor height	13 ft
Floor to ceiling height	9 ft
Window:	
double glazed tinted absorbing glass	0.345
ratio of fenestration to exterior wall	
frame dimensions	36 in. × 56 in.
glass area	33 in. × 53 in.
U-value	0.67 Btu/h · ft² · °F
interior shading, SC	0.36
glasses are recessed-mounted in frame, 2 in.	
Wall: exterior, 4 in. concrete + 2 in. insulation	
U-value	0.12 Btu/h · ft² · °F
partitions, area	1800 ft²
U-value for surface exposed to unconditioned space	0.12 Btu/h · ft² · °F
temperature difference $(T_{\text{un}} - T_x)$	6°F

This typical floor has the following internal loads:

Occupants: 7 persons per 1000 ft² floor area, sensible heat gain for each person is 250 Btu/h and latent heat gain is 150 Btu/h

Lighting: heat-to-space $q_{\text{es.l}}$	1 W/ft² floor area
Appliances and personal computers	1 W/ft² floor area

TABLE 7.9
Recommended C_t values for night shutdown operating mode (10 hours operation)

	Hour				
	8	**9**	**10**	**11**	**12**
C_t	0.15	0.13	0.11	0.10	0.10
	13	**14**	**15**	**16**	**17**
C_t	0.09	0.08	0.07	0.06	0.06

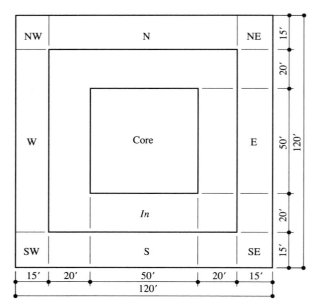

In – interior
All dimensions in ft.
FIGURE 7.8 Typical floor plan of a high-rise office building.

The summer indoor design temperature is 75°F and the maximum summer outdoor design temperature is 89°F with an outdoor daily mean of 80°F. The air-handling system of this typical floor operates from 7:30 A.M. to 5.30 P.M. Let us ignore any infiltrated air to the conditioned space as well as the shading on the window glass from adjacent buildings. At summer design conditions, calculate the following:

1. Zone peak loads
2. The block load of this floor
3. For a supply temperature differential of 20°F, the volume flow per ft^2 conditioned area

Solution

1. Because the locations of the partition walls and the sizes of the rooms are unknown, it is appropriate to divide this typical floor into nine zones:

   - West, east, south, and north perimeter zones each with an area of $90 \times 15 = 1350$ ft^2
   - Southwest, northwest, southeast, and northeast corner zones, each of them $15 \times 15 = 225$ ft^2
   - Interior zone, $4 \times 20 \times (50 + 20) = 5600$ ft^2

2. The window glasses are recessed-mounted 2 in. From Table 6.11, for a north latitude of 40°, on July 21 at 4 P.M., the solar altitude $\beta = 35°$ and the solar azimuth $\varphi = 87°$.

   For the west zone, the surface-solar azimuth $\gamma = 90° - 87° = 3°$ and $\tan \gamma = 0.052$. Therefore,

from Eq. (6.85), the shadow width

$$S_W = P_v \tan \gamma = 2 \times 0.052 = 0.104 \text{ in.}$$

As $\tan \beta = 0.700$ and $\cos \gamma = 0.999$, then, from Eq. 6.86, we see that the shadow height

$$S_H = P_H \frac{\tan \beta}{\cos \gamma} = 2 \times \frac{0.700}{0.999} = 1.40 \text{ in.}$$

From Eq. (6.87), the sunlit area for each frame of glass of 33 in. × 53 in. is

$$
\begin{aligned}
A_s &= (W - S_W)(H - S_H) \\
&= (33 - 0.104)(53 - 1.40) \\
&= 1697.4 \text{ in.}^2
\end{aligned}
$$

The glass area per window frame is $33 \times 53 = 1749$ in.^{2}, or $1749/144 = 12.146$ ft^2. The ratio of sunlit area to glass area $R_{\text{sun}} = 1697.4/1749 = 0.971$.

Because there are 30 frames of window glass in the west zone, the total glass area $A_{\text{glass}} = 12.146 \times 30 = 364.4$ ft^2.

3. From Table 6.9 on July 21, the maximum solar heat gain factor for window glass facing west at 40° north latitude is 216 Btu/h · ft^2, and from Table 7.5, CLF$_s$ for window glass facing west with interior shading at 4 P.M. is 0.82. Then, from Eq. (7.11), the space sensible cooling load from the solar heat gain through the sunlit area of the window glass is

$$
\begin{aligned}
q_{\text{sun}} &= (A_s \times \text{SHGF}_{\text{max}} \times \text{CLF}_s \times \text{SC}) \\
&= 364.4 \times 0.971 \times 216 \times 0.82 \times 0.36 \\
&= 22,562 \text{ Btu/h}
\end{aligned}
$$

Similarly, q_{sun} at 1 P.M. to 3 P.M. and 5 P.M. to 7 P.M. can be calculated and tabulated in Table 7.10.

4. CLF$_s$ at various hours in a diel cycle are calculated based on a 100 percent sunlit area. Actually, sunlit and shaded areas at various sunlit hours are different. For simplicity, the CLF$_s$ at hours other than 1 P.M. to 7 P.M. is multiplied by a mean sunlit area ratio at sunlit hours $R_{\text{m.sun}}$ between 1 P.M. and 7 P.M. that is

$$
\begin{aligned}
R_{\text{m.sun}} &= (0.790 + 0.902 + 0.944 + 0.971 \\
&\quad + 0.976 + 0.974 + 0.970) = 0.932
\end{aligned}
$$

From Eq. (7.11) at 8 P.M. with a CLF$_s = 0.12$, the space cooling load from the solar heat flowing through the sunlit glass during the previous sunlit hours is

$$
\begin{aligned}
q_{\text{sun}} &= 364.4 \times 0.932 \times 216 \times 0.12 \times 0.36 \\
&= 3169 \text{ Btu/h}
\end{aligned}
$$

At 9 P.M. to midnight and at 1 A.M. to 12 noon, q_{sun} values can be similarly calculated. These are listed in Table 7.10.

TABLE 7.10
Space sensible cooling loads for a west zone and a northwest zone (Example 7.2)

West zone	Time, hour												
	1	2	3	4	5	6	7	8	9	10	11	12	13
W-CLF*	0.05	0.05	0.04	0.04	0.03	0.06	0.09	0.11	0.13	0.15	0.16	0.17	0.31
A_s/A_{glass}													0.790
q_{sun}, Btu/h	1321	3121	1056	1056	792	1585	2377	2905	3433	3962	4226	4490	6939
N-CLF†	0.08	0.07	0.06	0.06	0.07	0.73	0.66	0.65	0.73	0.80	0.86	0.89	0.89
q_{sh}, Btu/h	27	24	20	20	24	247	223	221	247	271	292	302	931
q_{sol}, Btu/h	1348	1345	1076	1076	816	1832	2600	3126	3680	4233	4158	4792	7871
Conductive, CLTD	−1	−2	−3	−4	−4	−4	−4	−2	0	2	5	7	10
q_{con}, Btu/h	−281	−563	−844	−1126	−1126	−1126	−1126	−563	0	563	1407	1970	2814
Wall, CLTD	29	25	22	19	16	13	11	9	8	7	7	7	8
q_{wall}, Btu/h	1357	1170	1030	889	749	608	515	421	374	328	328	328	374
C_t								0.15	0.13	0.11	0.10	0.10	0.09
$C_t q$, Btu/h								9127	7910	6693	6085	6085	5476
q_{ext}, Btu/h								12,111	11,964	11,817	12,338	13,175	16,535
Internal loads, Btu/h								11,716	11,716	11,716	11,716	11,716	11,716
Total west zone, Btu/h								23,827	23,680	23,533	24,054	24,891	28,251

Northwest zone													
W-total, Btu/h								3971	3947	3922	4009	4148	4708
N-external, Btu/h								1182	1248	1306	1450	1576	1678
NW-total, Btu/h								5152	5195	5228	5459	5724	6386

West zone	Time, hour											
	14	15	16	17	18	19	20	21	22	23	24	$\sum q_\tau$
W-CLF*	0.53	0.72	0.82	0.81	0.61	0.16	0.12	0.10	0.08	0.07	0.06	
A_s/A_{glass}	0.902	0.944	0.971	0.976	0.974	0.970						Av 0.932
q_{sun}, Btu/h	13,546	19,259	22,562	22,401	16,835	4398	3169	2641	2113	1849	1585	
N-CLF†	0.89	0.82	0.75	0.78	0.91	0.24	0.18	0.15	0.13	0.11	0.10	
q_{sh}, Btu/h	435	229	108	93	118	36	61	51	44	37	34	
q_{sol}, Btu/h	13,981	19,488	22,670	22,494	16,953	4434	3230	2692	2517	1886	1619	43,064
Conductive, CLTD	11	12	12	11	10	8	6	4	2	1	0	
q_{con}, Btu/h	3095	3377	3377	3095	2814	2251	1688	1126	563	281	0	2531
Wall, CLTD	9	12	16	22	28	34	38	39	38	36	32	
q_{wall}, Btu/h	421	562	749	1030	1310	1591	1778	1825	1778	1685	1498	17,784
C_t	0.08	0.07	0.06	0.06								
$C_t q$, Btu/h	4868	4259	3651	3651								60,848
q_{ext}, Btu/h	22,365	27,686	30,447	30,270								
Internal loads, Btu/h	11,716	11,716	11,716	11,716								
Total, West-zone, Btu/h	34,087	39,402	42,163	41,986								

Northwest zone, Btu/h												
W-total, Btu/h	5680	6567	7027	6998								
N-external, Btu/h	1693	1651	1553	1547								
NW-total, Btu/h	7373	8218	8580	8545								

*CLF for window glass facing west.

† CLF for window glass facing north.

5. As the shaded area always is $A_{sh} = A_{glass} - A_{sh}$, the shaded area for a west zone on July 21 at 4 P.M. is

$$A_{sh} = 364.4(1 - 0.971) = 10.568 \text{ ft}^2$$

From Table 6.9, the maximum SHGF for a shaded area at 40° north latitude on July 21 is 38 Btu/h · ft² · °F, and from Table 7.5, the CLF$_{sh}$ for window glass facing north at 4 P.M. is 0.75. Thus, the space cooling load from the solar heat gain through the shaded area of the windows is

$$q_{sh} = A_{sh} \times SHGF_{max.sh} \times CLF_{sh} \times SC$$
$$= 10.568 \times 38 \times 0.75 \times 0.36$$
$$= 108 \text{ Btu/h}$$

For other hours between 1 P.M. and 7 P.M., q_{sh} can be calculated similarly.

For the calculation of the space cooling load between 8 P.M. and noon of the next day caused by the solar heat gain through the shaded area A_{sh} at sunlit hours, A_{sh} should satisfy the relationship at any time: $A_s + A_{sh} = A_{glass}$. Therefore, the mean shaded area at sunlit hours

$$A_{sh} = 364.4(1 - 0.932) = 24.8 \text{ ft}^2$$

For simplicity, at 8 P.M. the space cooling load from solar heat gain through the shaded area during sunlit hours is

$$q_{sh} = 24.8 \times 38 \times 0.18 \times 0.36 = 61 \text{ Btu/h}$$

For other hours between 9 P.M. and noon of next day, q_{sh} can be similarly calculated. These results are listed in Table 7.10.

6. On July 21 at 4 P.M., the space cooling load, from Eq. (7.12), due to the solar heat gain through the window glass

$$q_{rs.s} = q_{sun} + q_{sh} = 22,562 + 108$$
$$= 22,670 \text{ Btu/h}$$

For other hours, $q_{rs.s}$ can be calculated similarly.

7. For an outdoor maximum temperature of 95°F, an outdoor daily mean of 80°F and an indoor space temperature of 75°F, from Eq. (7.13) the correction for the CLTD for conductive heat gain is
vskip -1pt

$$CLTD_{corr} = CLTD + (T_{om} - 85) + (78 - T_r)$$
$$= CLTD + (80 - 85) + (78 - 75)$$
$$= CLTD - 2°F$$

Then from Table 7.6, the CLTD for calculating the space cooling load from conductive heat gain through the window glass on July 21 at 4 P.M. is $14 - 2 = 12°F$. The area of the window glass, including the sash, is $(36 \times 56 \times 30)/144 = 420 \text{ ft}^2$. From Eq. (7.12), the space cooling load from conductive heat gain is

$$q_{rs.c} = A_w \times U \times CLTD = 420 \times 0.67 \times 12$$
$$= 3377 \text{ Btu/h}$$

For other hours, $q_{rs.c}$ can be similarly calculated.

8. At 40° north latitude in July, the correction LM = 0. Because there is no attic fan in the ceiling plenum, $f = 1$. Also, the outside surface of the exterior wall is a dark color, $K = 1$. From Eq. (7.10), the corrections for CLTD to calculate the heat flow through the exterior wall at 4 P.M. are given as

$$CLTD_{corr} = [(CLTD + LM) \times K + (78 - T_r)$$
$$+ (T_{om} - 85)] \times f$$
$$= [(18 + 0) \times 1 + (78 - 75)$$
$$+ (80 - 85)] \times 1 = 16°F$$

From Eq. (7.9), the heat that flows through the exterior wall is

$$q_{rs.w} = U \times A_w \times CLTD$$
$$= 0.12[(9 \times 90) - 420] \times 18$$
$$= 749 \text{ Btu/h}$$

For other hours, $q_{rs.w}$ can be similarly calculated.

9. The external cooling loads during the shutdown period from Eq. (7.34), at 6 P.M., or 18 hours, are

$$q = q_{18} = q_{rsw.\tau} + q_{rss.\tau} + q_{rsc.\tau}$$
$$= 16,953 + 2814 + 1310$$
$$= 21,077 \text{ Btu/h}$$
$$q(\tau + 1) = q_{19} = 4434 + 2251 + 1591$$
$$= 8276 \text{ Btu/h}$$
$$\vdots$$
$$q_{(t-1)} = q_7 = 2600 - 1126 + 515$$
$$= 1989 \text{ Btu/h}$$

The sum of the external cooling loads during the shutdown period is

$$\Sigma q_\tau = 21,077 + 8276 + \cdots + 1989$$
$$= 60,848 \text{ Btu/h}$$

From Eq. (7.35), at 4 P.M. the space cooling load added to the operating period by the external loads in the night shutdown period is calculated as

$$C_t \Sigma q_\tau = 0.06 \times 60,848 = 3651 \text{ Btu/h}$$

Values of $C_t \Sigma q_\tau$ at other hours can be similarly calculated. These results are listed in Table 7.10.

10. On July 21 at 4 P.M., the total of the external sensible cooling loads for a west zone is

$$q_{rs.ex} = 22,670 + 3377 + 749 + 3651$$
$$= 30,447 \text{ Btu/h}$$

The values of other hours are listed in Table 7.10.

11. There are $(7 \times 90 \times 15)/1000 \approx 10$ persons in the west zone. From Eq. (7.17) and Table 7.7, the sensible load from occupants for night shutdown mode operation is

$$q_{rs.p} = N_p \times SHG_p \times CLF_p$$
$$= 10 \times 250 \times 1$$
$$= 2500 \text{ Btu/h}$$

From Eq. (7.27) and given data, the space cooling load from the electric lights is

$$q_{rs.l} = 3.413 \, W_A A_{fl} \times CLF$$
$$= 3.413 \times 1 \times 90 \times 15 \times 1$$
$$= 4608 \text{ Btu/h}$$

From Eq. (7.31), the space sensible cooling load from appliances and computers is

$$q_{rs.e} = 3.413 \, W_{A.e} A_{fl} \times \text{CLF}$$

$$= 3.413 \times 1 \times 90 \times 15 \times 1$$

$$= 4608 \text{ Btu/h}$$

The sum of the internal loads is

$$q_{rs.p} + q_{rs.l} + q_{rs.e} = 2500 + 4608 + 4608$$

$$= 11,716 \text{ Btu/h}$$

12. For a west zone on July 21 at 4 P.M., the total sensible cooling load $q_{rs.t}$, in Btu/h, is

$$q_{rs.t} = 30,447 + 11,716 = 42,163 \text{ Btu/h}$$

The values of $q_{rs.t}$ for other hours during the operating period are listed in Table 7.10. The maximum sensible cooling load of the west zone, 42,163 Btu/h, occurred at 4 P.M.

The total sensible cooling loads for east, south, and north zones can be calculated similarly, and their values are listed in Table 7.11.

13. For the northwest corner zone, there are external loads from walls and windows facing both north and west. Let $q_{rs.w}$ represent the total sensible cooling load of the west zone and $q_{ext.n}$ the total external loads of the north zone, both in Btu/h. On July 21 at 4 P.M., the total sensible load for the northwest zone $q_{rs.nw}$, in Btu/h, is

$$q_{rs.nw} = (q_{rs.w} + q_{ext.n})$$
$$\times \frac{225(\text{area of corner zone})}{1350(\text{area of perimeter zone})}$$

$$= (42,163 + 9321) \times \frac{225}{1350} = 8580 \text{ Btu/h}$$

For the rest of the hours during the operating period, the total sensible cooling loads for the northwest zone and for other corner zones can be calculated similarly, and the results are listed in Table 7.11.

14. In Tables 7.10 and 7.11, the zone maximum sensible cooling load (zone peak loads) are listed. They occur at different hours of the operating period, such as 8 A.M. for the east zone, 4 P.M. for the west and southwest zone.

15. In the interior zone, the sensible cooling load from occupants, electric lights, and appliances is

$$q_{rs} = \frac{5600}{1000} \times 7 \times 250 + (1 + 1) \times 3.413 \times 5600$$

$$= 48,026 \text{ Btu/h}$$

From Eq. (7.14), the sensible cooling load due to the heat gain through the wall exposed to the unconditioned area is

$$q_{rs} = AU(T_{un} - T_r)$$

$$= 50 \times 4 \times 9 \times 0.12 \times 6 = 1296 \text{ Btu/h}$$

and the total sensible cooling load of interior zone is

$$q_{rs,i} = 48,026 + 1296 = 49,322 \text{ Btu/h}$$

16. The total sensible load at 4 P.M. for a typical floor is

$$q_{rs,b} = 42,163 + 24,950 + 21,781 + 21,037$$
$$+ 8705 + 5836 + 8580 + 5711 + 49,322$$

$$= 188,085 \text{ Btu/h}$$

and the total latent load at 4 P.M. for a typical floor is

$$q_{rl} = \left(\frac{6300 + 5600}{1000}\right) \times 7 \times 150$$

$$= 12,495 \text{ Btu/h}$$

From Eq. (3.47), the block load, which indicates the maximum space cooling load occurring at the same time for this typical floor is

$$q_{rc} = q_{rs} + q_{rl}$$

$$= 190,289 + 12,495 = 202,784 \text{ Btu/h}$$

Maximum q_{rc} occurred at 3 P.M. Floor sensible block load is lower than the sum of the zone sensible peak loads, which do not occur at the same time.

17. For an air density of 0.0763 lb/ft^3, a specific heat of moist air of 0.243 Btu/lb · °F, and a supply temperature differential of 20°F, the average load density for this typical floor at 3 P.M. can be calculated as

$$\frac{q_{rc}}{A_{fl}} = \frac{202,784}{6300 + 5600} = 17.04 \text{Btu/h} \cdot \text{ft}^2$$

and the volume flow per ft^2 is

$$\frac{\dot{V}_s}{A_{fl}} = \frac{q_{rs}}{60 A_{fl} \rho_s c_{pa}(T_r - T_s)}$$

$$= \frac{q_{rs}}{60 \times 0.0763 \times 0.243 \times 20}$$

$$= 0.04512 \frac{q_{rs}}{A_{fl}}$$

Maximum load densities and volume flow per ft^2 are calculated and listed in Table 7.11.

7.9 COOLING COIL'S LOAD

In Eq. (3.21), the cooling coil's load is given as

$$q_{cc} = 60 \, \dot{V}_s \rho_s (h_{ae} - h_{cc})$$

Of this, the sensible cooling coil's load q_{cs}, in Btu/h, can be calculated as

$$q_{cs} = 60 \, \dot{V}_s \rho_s c_{pa}(T_{ae} - T_{cc}) \qquad (7.36)$$

TABLE 7.11
Space cooling loads for various zones for a typical floor in a high-rise office building (Example 7.2)

Sensible cooling load in Zone	Hour Btu/h										Zone Peak load Btu/h	Time occurred	Maximum q_{rs}/A_{fl} Btu/ h·ft²	Maximum $\dot{V}_s/A_{fl}$
	8	9	10	11	12	13	14	15	16	17				
Sensible load W	23,827	22,680	23,533	24,054	24,891	28,251	34,081	39,402	42,163	41,986	42,163	16	31.23	1.41
E	42,741	40,651	35,931	30,240	28,392	27,973	27,210	26,375	24,950	23,838	42,741	8	31.66	1.43
S	19,359	20,780	23,312	26,305	28,279	29,268	26,562	24,185	21,781	21,733	28,279	12	20.95	0.95
N	18,807	19,206	19,554	20,414	21,174	21,783	21,876	21,621	21,037	20,998	21,876	14	16.20	0.73
SW	5,245	5,458	5,855	6,441	6,909	7,466	8,154	8,645	8,705	8,674	8,705	16	38.69	1.75
SE	8,397	8,286	7,921	7,472	7,493	7,420	7,009	6,474	5,836	5,649	8,397	8	37.32	1.68
NW	5,152	5,195	5,228	5,459	5,724	6,386	7,373	8,218	8,580	8,545	8,580	16	38.13	1.72
NE	8,305	8,023	7,294	6,490	6,308	6,340	6,228	6,047	5,711	5,520	8,305	8	36.91	1.67
Interior	49,322	49,322	49,322	49,322	49,322	49,322	49,322	49,322	49,322	49,322	49,322		8.81	0.40
Total space sensible cooling load	181,155	180,601	177,950	176,197	178,492	183,209	187,815	190,289	188,085	186,275				
Total space latent load	12,495	12,495	12,495	12,495	12,495	12,495	12,495	12,495	12,495	12,495				

*Based on a supply temperature difference of 20°F.

and the latent coil's load q_{cl}, in Btu/h, is

$$q_{cl} = 60 \, \dot{V}_s \rho_s (w_{ae} - h_{cc}) h_{fg.32} \qquad (7.37)$$

Also,

$$q_{cc} = q_{cs} + q_{cl} \qquad (7.38)$$

Alternatively, the sensible cooling coil's load can be calculated as

$$q_{cs} = q_{rs} + q_{s.s} + q_{r.s} + q_{out.s} \qquad (7.39)$$

where $q_{s.s}$, $q_{r.s}$ = supply and return system heat gain, Btu/h

$q_{out.s}$ = sensible load from the outdoor air intake, Btu/h

and the latent coil's load can be calculated as

$$q_{cl} = q_{rl} + q_{out.l} \qquad (7.40)$$

where $q_{out.l}$ = latent load from outdoor air intake, Btu/h. The components of the supply and return system heat gain are discussed in this section.

Fan Power

In the air duct system, the temperature increase from the heat released to the airstream because of frictional and dynamic losses is nearly compensated by the expansion of air from the pressure drop of the airstream. Therefore, it is usually assumed that there is no significant temperature increase from frictional and dynamic losses when air flows through an air duct system.

Fan power input is almost entirely converted into heat energy within the fan. If the fan motor is located in the supply or return airstream, the temperature increase across the supply (or return) fan ΔT_f, in °F, can be calculated as

$$\Delta T_f = \frac{0.37 \Delta p_t}{\eta_f \eta_m} \qquad (7.41)$$

where Δp_t = total pressure increase, in WG

η_f, η_m = total efficiency of fan and efficiency of motor

If the motor is located outside the airstream, then in Eq. (7.41) $\eta_m = 1$.

The Δp_t of the return fan for central hydronic air conditioning systems in commercial buildings is usually 0.25 to 0.5 of the Δp_t of the supply fan. Therefore, the temperature increase of the return fan is far smaller than that of the supply fan. The temperature increase of the relief fan affects only the relief or exhaust air stream. It is not a part of the supply and return system heat gain. A relief fan is used to re-

lieve excess space pressure when 100% outdoor air is flowing through the supply fan.

Duct Heat Gain

Duct heat gain is the heat transfer caused by the temperature difference between the ambient air and the air flowing inside the air duct. Duct heat gain is affected by this temperature difference, the thickness of the insulation layer, air volume flow rate, and the size and shape of the duct. Detailed calculations are presented in Chapter 8. A rough estimate of the temperature increase of the supply air for an insulated duct is as follows:

Air velocity	Air temperature rise
< 2000 fpm	1°F/100 ft duct length
≥ 2000 fpm	0.75°F/100 ft duct length

Temperature of Plenum Air and Ventilation Load

For a ceiling return plenum, the plenum air temperature can be calculated from Eq. (7.28). The temperature increase of the plenum air, caused by the heat released from the electric lights ($T_p - T_r$), is affected by their power input, type of the lighting fixture, return air volume flow rate, and the construction of the ceiling plenum. The temperature increase of the plenum air ($T_p - T_r$) is usually between 1 and 3°F.

From Eqs. (7.15) and (7.16), the sensible and latent loads q_{os} and q_{ol}, in Btu/h, which are attributable to the outdoor air intake can be similarly calculated, except $\dot{V}_{if}$ in Eqs. (7.15) and (7.16) should be replaced by the volume flow rate of outdoor air $\dot{V}_o$, in cfm.

System heat gains are mainly due to convective heat transfer from the surfaces. For simplification, they are considered instantaneous cooling loads.

7.10 COOLING LOAD CALCULATION BY USING FINITE DIFFERENCE METHOD

Finite Difference Method

Because of the rapid increase in the use of microcomputers in HVAC&R system calculations, it is now possible to use a finite difference method, a numerical approach, to solve transient simultaneous heat and moisture transfer problems in heating and cooling load calculations and energy estimations.

The finite difference method divides building structures into a number of sections. A fictitious node is

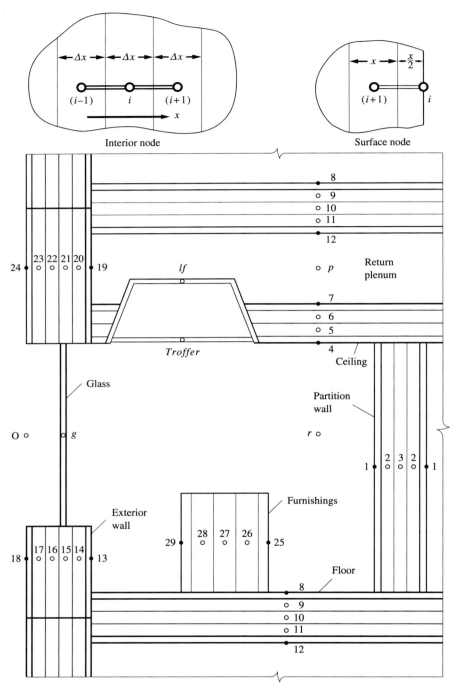

FIGURE. 7.9 Building structures and nodes for a typical room.

located at the center of each section or on the surface, as shown in Fig. 7.9. An energy balance or a mass balance at each node at selected time intervals results in a set of algebraic equations that can be employed to determine the temperature and moisture for each node in terms of neighboring nodal temperatures or moisture contents, nodal geometry, and the thermal and moisture properties of the building structure. The stored heat energy and moisture are expressed as an increase of internal energy and moisture content at the nodes.

Heat conduction can be approximated by using the finite difference form of the Fourier law, as

$$q_{(i+1)\to i} = \frac{kA_i(T^t_{i+1} - T^t_i)}{\Delta x} \qquad (7.42)$$

where k = thermal conductivity, Btu/h · ft · °F

A_i = area of the building structure perpendicular to direction of heat flow, ft²

T = temperature, °F

Δx = spacing between the nodal points, ft

In Eq. (7.42), superscript t denotes time t.

Each nodal equation is solved explicitly in terms of the future temperature of that node. The explicit method is simpler and clearer than the more complex implicit method. The time derivative is then approximated by a forward finite difference in time, or

$$\frac{\partial T_i}{\partial t} \approx \frac{T_i^{t+\Delta t} - T_i^t}{\Delta t} \qquad (7.43)$$

Compared with the transfer function method to calculate the cooling load, the finite difference method has the following benefits:

1. Solves heat and moisture transfer load calculations simultaneously
2. Concept and approach are easily understood
3. Permits custom-made solutions for special problems
4. Allows direct calculation of cooling loads and energy estimates

Its drawbacks are mainly due to the greater number of the computerized calculations and comparatively less computer programs, information, and experience available.

Simplifying Assumptions

When the finite difference method is used to calculate space cooling loads, simplifications are often required to reduce the number of computer calculations and to solve the problem more easily. The errors due to simplification should be within acceptable limits.

For a typical room in the building shown in Fig. 7.9, the following are the simplifying assumptions:

- Heat and moisture flow are one-dimensional.
- Thermal properties of the building materials are homogeneous.
- The properties of the airstream flowing over the surface of the building structures are homogeneous.
- The surface temperature differences between the partition walls, ceiling, and floors are small; therefore, the radiative exchange between these surfaces can be ignored.
- The radiative energy received on the inner surface of the building structures can be estimated as the product of the shape factor and the radiative portion of the heat gains, and the shape factor is approximated by the ratio of the receiving area to the total zone area.

- During the operating period, the heat capacity of the space air is small compared with other heat gains; therefore, it can be ignored.
- When the air system is not operating during night shutdown period, the heat capacity of the space air has a significant influence on the space air temperature; therefore, it should be taken into account.
- Different heat and mass transfer coefficients and analyses are used for operating period and shutdown periods.

Heat and Moisture Transfer at Interior Nodes

Consider an interior node i as shown in the upper part of Fig. 7.9. For a one-dimensional heat flow, if there is no internal energy generation, then according to the principle of heat balance,

Conductive heat from node $(i-1)$	+	Conductive heat from node $(i+1)$	=	Rate of change of internal energy of node i

$$q_{(i-1)\to i} + q_{(i+1)\to i} = \frac{\partial U_i}{\partial t}$$

$$\approx \frac{\rho_b c_b A_i \Delta x (T_i^{t+\Delta t} - T_i^t)}{\Delta t}$$

$$(7.44)$$

where U_i = internal energy of node i, Btu/lb

ρ_b, c_b = density and specific heat of the building material, lb/ft³ and Btu/lb · °F

Δt = selected time interval, seconds or minutes

Substituting Eq. (7.42) in (7.44) and solving for $T_i^{t+\Delta t}$, we have

$$T_i^{t+\Delta t} = \text{Fo}\left[T_{(i-1)}^t + T_{(i+1)}^t\right] + (1 - 2\,\text{Fo})T_i^t \qquad (7.45)$$

In Eq. (7.45), Fo is the Fourier number and is defined as

$$\text{Fo} = \left(\frac{k_b}{\rho_b c_b}\right)\left[\frac{\Delta t}{(\Delta x)^2}\right] \qquad (7.46)$$

Subscript b indicates the building material.

The choice of spacing Δx and the time interval Δt must meet some criterion in order to ensure convergence in the calculations. The criterion is the stability limit, or

$$\text{Fo} \leq 1/2 \qquad (7.47)$$

Similarly, for moisture transfer at the interior nodes,

$$X_i^{t+\Delta t} = \text{Fo}_{\text{mass}}[X_{(i-1)}^t + X_{(i+1)}^t]$$
$$+ (1 - 2\,\text{Fo}_{\text{mass}})X_i^t \qquad (7.48)$$

and

$$\text{Fo}_{\text{mass}} = \frac{D_{1v}\Delta t}{(\Delta x)^2} \le \frac{1}{2} \qquad (7.49)$$

where X = moisture content, dimensionless
D_{1v} = mass diffusivity of liquid and vapor, ft^2/s

Heat and Moisture Transfer at Surface Nodes

For a one-dimensional heat flow, the energy balance at the surface node i of the partition wall as shown in Fig. 7.9 is

Conductive heat from node $(i+1)$	+	Conductive heat transfer from space air	+	Latent heat of moisture transfer from space air
	+	Radiative heat from internal internal loads	=	Rate of change of internal energy of node i

$$\frac{kA_i[T_{(i+1)}^t - T_i^t]}{\Delta x} + h_{\text{ci}}A_i(T_r^t - T_i^t)$$
$$+ \rho_a h_{\text{mi}}A_i X_i^t(w_r^t - w_{\text{is}}^t)h_{\text{fg}}$$
$$+ F_{1\to i}L_r + F_{p\to i}O_r + F_{m\to i}M_r$$
$$= \rho_b c_b A_i \left(\frac{\Delta x}{2}\right)\left[\frac{(T_i^{t+\Delta t} - T_i^t)}{\Delta t}\right]$$

where
h_{ci} = convective heat transfer coefficient of surface i, Btu/h · ft^2 · °F
ρ_a = density of space air, lb/ft^3
h_{mi} = convective mass transfer coefficient of surface i, ft/s
w_{is}^t = humidity ratio corresponding to surface i at time t, lb/lb
h_{fg} = latent heat of vaporization at surface temperature, Btu/lb
$F_{1\to i}, F_{p\to i}, F_{m\to i}$ = shape factor between the surfaces of lights, occupants, appliances, and surface i
L_r, O_r, M_r = radiative portion of heat energy from lights, occupants, and appliances and machines, Btu/h

Solving for $T_i^{t+\Delta t}$,

$$T_i^{t+\Delta t} = 2\,\text{Fo}\left\{ T_{(i+1)}^t + \text{Bi} \right.$$
$$\times \left[T_t^t + \frac{\rho_a h_{\text{mi}} h_{\text{fg}} X_i^t(w_r^t - w_{\text{is}}^t)}{h_{\text{ci}}} \right.$$
$$\left. + \frac{F_{1\to i}L_r}{h_{\text{ci}}A_i} + \frac{F_{p\to i}O_r}{h_{\text{ci}}A_i} + \frac{F_{m\to i}M_r}{h_{\text{ci}}A_i} \right]\right\}$$
$$+ [1 - 2\,\text{Fo}(1 + \text{Bi})]T_i^t \qquad (7.50)$$

In Eq. (7.50), Bi is the Biot number and can be expressed as

$$\text{Bi} = \frac{h_c \Delta x}{k} \qquad (7.51)$$

The stability limit of the surface nodes requires that

$$\text{Fo}(1 + \text{Bi}) \le 1/2 \qquad (7.52)$$

Similarly, according to the principle of conservation of mass, the moisture content at the surface node i is given as

$$X_i^{t+\Delta t} = (1 - 2\,\text{Fo}_{\text{mass}})X_i^t + 2\,\text{Fo}_{\text{mass}}X_{(i+1)}^t$$
$$+ \frac{2\rho_a h_{\text{mi}}\Delta t X_i^t(w_r^t - w_{\text{is}}^t)}{\rho_b \Delta x} \qquad (7.53)$$

The temperature and moisture content at other surface nodes such as floor, ceiling, exterior walls, glass, and Plexiglas of the lighting fixture can be found in the same manner.

According to the *ASHRAE Handbook 1989, Fundamentals*, the radiative and convective portion of the heat gains are as follows:

	Radiative percent	Convective percent
	50%	50%
Flourescent lights	50%	50%
People	33%	67%
External walls and roofs	60%	40%
Appliances and machines	20–80%	80–20%

Space Air Temperature and Cooling Loads

If infiltrated air is ignored, the heat balance on the space air or the plenum air can be described by the following relationship:

Internal energy of supply air	+	Convective heat transfer from building structures	+	Convective heat transfer from internal loads	=	Internal energy of space air

$$60 \, \dot{V}_s \rho_s c_{pa} T_s^{t+\Delta t} + \sum_{k=1}^{n} h_{ck} A_k (T_i^{t+\Delta t} - T_r^{t+\Delta t})$$

$$+ L_c + O_c + M_c = 60 \, \dot{V}_s \rho_s c_{pa} T_r^{t+\Delta t} \qquad (7.54)$$

Solving for $T_r^{t+\Delta t}$, we have

$$T_r^{t+\Delta t} = (60 \, \dot{V}_s \rho_s c_{pa} T_s^{t+\Delta t} + \sum_{k=1}^{n} h_{ck} A_k T_i^{t+\Delta t} + L_c$$

$$+ O_c + M_c) \frac{1}{60 \, \dot{V}_s \rho_s c_{pa} + \sum_{k=1}^{n} h_{ck} A_k} \qquad (7.55)$$

where $\dot{V}_s, T_s$ = volume flow rate and temperature of supply air, cfm and °F

L_c, O_c, M_c = convective heat from lights, occupants, and appliances, Btu/h

The temperature of the return plenum air can be similarly calculated. The space sensible cooling loads, therefore, can be calculated as

$$q_{rs}^{t+\Delta t} = \sum_{i=1}^{n} h_{ci} A_i (T_i^{t+\Delta t} - T_r^{t+\Delta t}) + L_c + O_c + M_c$$

$$(7.56)$$

The latent heat gains are instantaneous latent cooling loads.

7.11 HEATING LOAD

The *design heating load*, or simply heating load, is always the maximum heat energy that might possibly be required to supply to the conditioned space at winter design conditions to maintain the winter indoor design temperature. The maximum heating load usually occurs before sunrise on the coldest days. Solar heat gain and internal loads are not taken into account except for those internal loads q_{in}, in Btu/h, that always release heat to the conditioned space during the operating period.

For a continuously operated heating system, the heating load q_{rh}, in Btu/h, can be calculated as

$$q_{rh} = q_{tran} + q_{if.s} + q_l + q_{mat} - q_{in} \qquad (7.57)$$

where q_{tran} = transmission loss, Btu/h

q_l = heat required to evaporate liquid water, Btu/h

q_{mat} = heat added to the entering colder product or material, Btu/h

As soon as the volume flow rate of infiltrated air $\dot{V}_{if}$, in cfm, is determined, then the sensible heat loss from infiltration $q_{if.s}$, in Btu/h, can be calculated as

$$q_{if.s} = \dot{V}_{if} \rho_o c_{pa} (T_r - T_o) \qquad (7.58)$$

where ρ_o = density of outdoor air, lb/ft³.

Transmission Loss

Transmission loss q_{tran}, in Btu/h, is the heat loss from the conditioned space through the external walls, roof, ceiling, floor, and glass. If the calculation is simplified to a steady-state heat flow, then

$$q_{tran} = \sum AU(T_r - T_o) \qquad (7.59)$$

where A = area of the walls, roof, ceiling, floor, or glass, ft²

U = overall heat transfer coefficient of the walls, roof, ceiling, floor, or glass, Btu/h · ft² · °F

When the winter outdoor design temperature is used for T_o, the heat loss calculated by transient heat transfer will be less than that from Eq. (7.59) because of the cyclic fluctuations of the outdoor temperature and the heat storage in the building structures.

For concrete slab floors on a grade, heat loss q_{fl}, in Btu/h, is mostly through the perimeter instead of through the floor and the ground. It can be estimated as

$$q_{fl} = P C_{fl} (T_r - T_o) \qquad (7.60)$$

where P = length of the perimeter, ft

c_{fl} = heat loss coefficient, Btu/h · ft · °F

For areas having an annual total of HDD65 = 5350 heating degree-days and for a concrete wall with interior insulation in the perimeter having an R-value of 5.4 h · ft² · °F/Btu, C_{fl} = 0.72 Btu/h · °F per ft of perimeter. Refer to the *ASHRAE Handbook 1989 Fundamentals*, Chapter 25, for more details.

For basement walls, the paths of the heat flow below the grade line are approximately concetric circular patterns centered at the intersection of the grade line and the basement wall. The thermal resistance of the soil and the wall depends on the path length through the soil and the construction of the basement wall. A simplified calculation of the heat loss through the basement walls and floor $q_{b.g}$, in Btu/h, is as follows:

$$q_{b.q} = A_{b.g} U_{b.g} (T_{base} - T_o) \qquad (7.61)$$

where $A_{b.g}$ = area of the basement wall or floor below grade, ft²

$U_{b.g}$ = overall heat transfer coefficient of the wall or floor and the soil path, in Btu/h · ft² · °F

The values of $U_{b.g}$ are roughly given as follows:

	0–2 ft below grade	Lower than 2 ft
Uninsulated wall	0.35	0.15
Insulated wall	0.14	0.09
Basement floor	0.03	0.03

Refer to the *ASHRAE Handbook 1989 Fundamentals* for details.

As described in Section 7.6, infiltration can be considered to be 0.15 to 0.4 air changes per hour at winter design conditions only when (1) the exterior window is not well sealed, and (2) there is a high wind velocity. For hotels, motels, and high-rise residential buildings, an infiltration rate of 0.038 cfm/ft² of gross area of external windows is often used for computations for the perimeter zone.

The space heating load in the perimeter zone, including mainly transmission and infiltration losses, is sometimes expressed in a linear density $q_{h.ft}$, in Btu/h per linear foot of external wall, or Btu/h · ft. Many building envelopes in commercial buildings that have been designed in compliance with ASHRAE/IES Standard 90.1–1989 often have a $q_{h.ft} \leq 250$ Btu/h · ft.

Adjacent Unheated Spaces and Others

Heat loss from the heated space to the adjacent unheated space q_{un}, in Btu/h, is usually assumed to be balanced by heat transfer from the unheated space to the outdoor air, and can be calculated approximately by the following formula:

$$q_{un} \sum_{i=1}^{n} A_i U_i (T_r - T_{un})$$

$$= \left(\sum_{j=1}^{m} A_j U_j + \dot{V}_{if} \rho_o C_{pa} \right) (T_{un} - T_o) \quad (7.62)$$

where A_i, U_i = area and overall heat transfer coefficient of the partition surfaces between heated space and unheated space, ft² and Btu/h · ft² · °F

A_j, U_j = area and overall heat transfer coefficient of the building structures exposed to the outdoor air in the unheated space, ft² and Btu/h · ft² · °F

The temperature of the unheated space T_{un}, in °F, can be calculated as

$$T_{un} \sum_{i=1}^{n} A_i U_i T_r$$

$$+ \frac{(60 \dot{V}_{if} \rho_o c_{pa} + \sum_{j=1}^{m} A_j U_j) T_o}{\sum_{i=1}^{n} A_i U_i + 60 \dot{V}_{if} \rho_o c_{pa} + \sum_{j=1}^{m} A_j U_j}$$

$$(7.63)$$

In Eq. (7.57), q_l, in Btu/h, represents the heat required to evaporate the liquid water to raise the relative humidity of the space air or to maintain a specific space relative humidity, that is,

$$q_l = \dot{m}_w h_{fg57}$$
$$= [60 \dot{V}_{oif} \rho_o (w_r - w_o) - \dot{m}_p] h_{fg57} \quad (7.64)$$

where $\dot{m}_v$ = mass flow of water evaporated, lb/h

$\dot{V}_{oif}$ = volume flow rate of outdoor ventilation air and infiltrated air, cfm

$\dot{m}_p$ = mass flow of water evaporated from occupants, lb/h

In Eq. (7.64), h_{fg57} indicates the latent heat of vaporization at a wet bulb of 57°F, that is, 72°F dry bulb and a relative humidity of 40 percent. Its value can be taken as 1061 Btu/lb.

For factories, heat added to the products or materials that enter the heated space within the calculated time interval, q_{mat}, in Btu/h, should be considered part of the heating load, and can be calculated as

$$q_{mat} = \dot{m}_{mat} c_{pm} (T_r - T_o) \quad (7.65)$$

where $\dot{m}_{mat}$ = mass flow rate of cold products and material entering the heated space, lb/h

c_{pm} = specific heat of the product or material, Btu/lb · °F

Setback or Night Shutdown Operation

During a nighttime or unoccupied period, when the space temperature is set back lower than the indoor temperature during the operating period, it is necessary to warm up the conditioned space the next morning before the arrival of the occupants in offices or other buildings. The warm-up or pick-up load depends on the pick-up temperature difference, the outdoor-indoor temperature difference, construction of the building envelope or building shell, and the time required for warm up. There are insufficient data to determine the oversizing factor of the capacity of the heating plant for morning warm-up. According to the tests by Trehan et al. (1989), one can make a rough estimate of the energy required to warm up or raise a space temperature by 12°F for a single- or two-story building with a roof-ceiling U-value of 0.03 Btu/h · ft² · °F and a U-value for external wall of 0.08 Btu/h · ft² · °F as follows:

Outdoor-indoor temperature difference, °F	Warm-up time period, h	Oversizing factor
35	1	40%
55	2	40%
55	1	100%

ASHRAE/IES Standard 90.1–1989 specifies a 30 percent increase in design heating load because of pick-up load during the warm-up period.

7.12 COMPLIANCE WITH ASHRAE/IES STANDARD 90.1–1989 FOR BUILDING ENVELOPES

To design a high-quality energy-efficient cost-effective building and HVAC&R system, it is necessary to select the optimum building envelope and HVAC&R system and calculate the cooling and heating load correctly. An energy-efficient building envelope should meet the requirements and design criteria in *ASHRAE/IES Standard 90.1–1989, Energy Efficient Design of New Buildings Except New Low-Rise Residential Buildings*; the DOE Code of Federal Regulations, Title 10, Part 435, Subpart A, *Performance Standard for New Commercial and Multi-Family High-Rise Residential Buildings*; and local energy codes. The DOE Code is very similar to *Standard 90.1*. The design and selection of the building envelope are generally the responsibility of an architect with the assistance of a mechanical engineer or contractor.

Building envelopes are usually designed, or even constructed, before the HVAC&R system is designed. A speculative building is built of known use and type of occupancy, but the exact tenants are unknown. A shell building is built prior to the use and occupancy is determined.

Compliance with the ASHRAE/IES Standard 90.1–1989 for building envelopes includes several basic requirements. In addition, designers may use the prescriptive method, the system performance method, or the energy cost budget method.

Basic Requirements

The following basic requirements must be met no matter which method of compliance is chosen:

1. Air leakages should not exceed 0.30 cfm/linear ft of sash crack for operable windows, 0.15 cfm/ft² for nonoperable windows, 1.25 cfm/ft² for com-

mercial entrance swinging or revolving doors, and 0.5 cfm/ft² for residential swinging doors.

2. A moisture migration vapor barrier should be installed on the warm, moist side of exterior walls, roofs, or floors to prevent concealed condensation within the building envelope.

3. Maximum skylight area as a percentage of the total roof area may range from 2.3 to 12.3 percent, depending on the following:

- Heating degree days HDD65
- Cooling degree hours based on a temperature 80°F CDH80, which is the number of hours of outdoor temperature exceeding 80° per annum
- Design illumination level (30, 50, or 70 foot-candles)
- Lighting power density (1 to 2.5 W/ft²)
- Visible light transmission (0.75 or 0.5)

Automatic daylight controls must be installed, and the maximum allowable skylight area may be increased by 50 percent if skylight shading devices are used when more than 50 percent of its solar heat gain is reduced.

Prescriptive Method

A set of 38 tables called *Alternate Component Packages (ACP)*, corresponding to 38 climatic areas, is presented in *ASHRAE/IES Standard 90.1–1989*. Each climate area is defined by its own specific combination of the following climatic parameters:

- Heating degree days base 50°F HDD50 and base 65°F HDD65
- Cooling degree days base 65°F CDD65
- Cooling degree hours base 80°F CDH80
- Annual average daily incident solar energy on east or west orientation (VSEW), in Btu/ft² · day

From each table representing a specific climatic area, the design criteria of the following building envelope components can be determined:

1. *Fenestration.* Maximum allowable percent area of fenestration, which is the total area of the fenestration assembly divided by the gross exterior wall area (window-wall ratio), can be selected from the ACP when the following factors are varied:
   - Internal load density (ILD). This varies from 0 to 3.5 W/ft² in three ranges: for most commercial buildings, ILD ranges from 1.51 to 3.0 W/ft²; ILD of warehouses and hotel/motels

ranges from 0 to 1.5 W/ft^2; the ILD of small retail stores ranges from 3.01 to 3.5 W/ft^2.

- Overhang projection factor, which indicates the external shades on glass.
- Shading coefficient, including the effect of internal and external shading devices, and the projection factor.
- No daylighting (the base case) or using perimeter daylighting and installing with automatic daylighting controls.
- Single, double, or triple glazings and their corresponding ranges of U-values of fenestration.

2. *Opaque exterior wall.* The maximum U-value of the opaque wall assembly U_{ow} can be selected from the ACP by using the same ILD range as is used to determine the maximum allowable area of fenestration, in percent. During the selection procedure:

- The heating capacity (HC) may vary from $<$ 5 Btu/ft$^2 \cdot$ °F to 15 Btu/ft$^2 \cdot$ °F. Low-HC lightweight walls are wood-frame, metal structure, and curtain wall construction; mass walls (high-HC) include masonry and concrete construction.
- Window wall ratio.
- Wall insulation installed on the inside surface of wall or integral to the wall mass (INT INS), or wall insulation placed outside the wall (EXT INS).

3. *Other.* U-values for roofs, walls adjacent to unconditioned space, and floors over unconditioned space and R-values for walls below grade and for unheated slabs can be found in the ACP tables.

The prescriptive method provides a simple and easy means of compliance. It allows a certain degree of design flexibility by providing trade-offs between various factors within the fenestration and exterior walls themselves.

System Performance Method

Building envelope compliance using the system performance method allows the designer greater flexibility in selecting variables such as U-values, window-wall ratio, shading coefficients, projection factor, internal load density, and heating capacity of various components. Trade-offs of these design variables between fenestration and the exterior walls/roof are possible.

The procedure for compliance using the system performance method is as follows:

1. Calculate the cumulative annual heat gain criteria WC$_c$ and heat loss criteria WC$_h$ through fenes-

tration and the exterior walls/roof, both in Btu/h·year, for a conditioned space, based on the specific U-values, window-wall ratio, shading coefficient, projection factor, and ILD from the following equations:

$$\text{WC}_c = \sum(q_{\text{es.w}} + q_{\text{es.s}} + q_{\text{es.c}} + q_{\text{int}}) \quad (7.66)$$

$$q_{\text{es.w}} = A_w U_w (T_o - T_r)$$

$$q_{\text{es.s}} = (A_s \times \text{SHGF}_{\text{max}} \times \text{SC})$$
$$+ (A_{\text{sh}} \times \text{SHGF}_{\text{max.sh}} \times \text{SC})$$

$$q_{\text{es.c}} = A_{\text{win}} U_{\text{win}} (T_o - T_r)$$

$$q_{\text{int}} = \text{ILD} \times 3.413 \times A_{\text{fl}} \quad (7.67)$$

where
$q_{\text{es.w}}$ = heat gain of exterior wall or roofs, Btu/h

$q_{\text{es.s}}$ = solar heat gain through fenestration, Btu/h

$q_{\text{es.c}}$ = conduction heat gain through window glass, Btu/h

q_{int} = heat gain from internal loads, Btu/h

$A_{\text{win}}, U_{\text{win}}$ = area and U-values of window glass, ft^2 and Btu/h$\cdot$ft$^2 \cdot$ °F

Similarly, cumulative annual heat loss criteria WC$_h$ can be calculated.

Because T_o is a function of CDD50, CDD65, CDH80, HDD50, and HDD65 of that location, it can be represented in terms of these outdoor climate parameters. Specified U-value, shading coefficient, window-to-wall ratio, and projection factor can be determined by related formulas in *ASHRAE/IES Standard 90.1–1989*.

2. Determine the proposed U-values, shading coefficient, window-to-wall ratio, and projection factor for the exterior walls/roof and windows. Calculate again the cumulative annual heat gain C_{pro} or heat loss H_{pro}, both in Btu/h $\cdot$ year, still using Eqs. (7.66) and (7.67), except the proposed U-values, window-to-wall ratio, and projection factor will replace the specific U-values, window-to-wall ratio, and projection factor.

3. If $C_{\text{pro}} \le \text{WC}_c$ and $H_{\text{pro}} \le \text{WC}_h$, the proposed U-values, window-to-wall ratio, and shading coefficient for the building envelope of a design project are in compliance with the building envelope requirements of *ASHRAE/IES Standard 90.1–1989*. If $C_{\text{pro}} > \text{WC}_c$ or $H_{\text{pro}} > \text{WC}_h$, then change the proposed U-values, window-to-wall ratio, and shading coefficients until $C_{\text{pro}} \le \text{WC}_c$ and $H_{\text{pro}} \le \text{WC}_h$.

A computer program is available to calculate WC$_c$, WC$_h$, C_{pro}, and H_{pro} to find whether the design complies with ASHRAE Standards. Refer to *ASHRAE/IES Standard 90.1–1989* and *User's Manual* for details.

Building Energy Cost Budget Method

This method combines the building envelope, the HVAC&R system, and the equipment together and uses the annual energy cost as the criterion to achieve an energy-efficient building at the specified climatic location.

The building energy cost budget method provides still more flexibility for the designer to achieve an energy-efficient building. It requires an energy estimation program, such as DOE-2, to simulate the energy cost of various buildings and HVAC&R systems. This method is especially suitable for projects with innovative designs or any projects intending to use energy simulation programs to determine alternatives. More details will be discussed in Chapter 28.

For cooling and heating load calculations, *ASHRAE/IES Standard 90.1–1989* recommends the following:

- *Internal heat gains.* For comfort cooling, a sensible load from the occupants of about 0.6 W/ft² can be assumed. Internal heat gains may be ignored during the calculation of heating loads.

- *Safety factor.* Design loads may be increased by up to 10 percent to account for unexpected loads or changes in usage.

- *Pick-up loads.* Warm-up or cool-down transient loads from the off-hour setback or equipment shutdown may be calculated according to the heat capacity of the building, degree of setback, and the desired time of pick-up, or may be assumed to be up to 30 percent for heating and 10 percent for cooling for design loads whose HVAC&R system is continuously operating.

Example 7.3. For the heated and mechanically-cooled multistory office building located at 40° north latitude described in Example 7.2, the climatic parameters are as follows:

Heating degree days base 50°F HDD50	1986
Heating degree days base 65°F HDD65	5022
Cooling degree days base 65°F CDD65	834
Annual average incident solar energy on east or west orientation (VSEW)	650 Btu/ft² · day

Investigate whether the adopted building envelope criteria meet the requirements of the prescriptive criteria of *ASHRAE/IES Standard 90.1–1989*. Recommend possible alternatives if the maximum fenestration is increased to 40 percent.

Solution

1. Table 8A.25 of the Alternate Component Packages (ACP) of *ASHRAE/IES Standard 90.1–1989* can be applied for locations within the following climatic ranges:

HDD50	1751–2600
CDD65	501–1150
VSEW	560–845

2. According to the given data, the sensible internal load density (ILD) for the multistory building in Example 7.2 is

$$\text{ILD} = \left(\frac{250 \times 7}{1000 \times 3.413} \right) + 1 + 1 = 2.5 \text{ W/ft}^2$$

3. In Example 7.2, the projection factor $= \frac{2}{53} = 0.04$, the shade coefficient SC $= 0.36$, and the heat capacity HC of the 4-in. exterior opaque concrete wall with a 2-in. insulation layer external to the concrete is

$$\text{HC} = c_p \rho_{\text{con}} L$$

$$= 0.156 \times 144 \times \frac{4}{12} = 7.5 \text{ Btu/ft}^2 \cdot °F$$

4. For a U-value of the fenestration assembly $U_{\text{of}} = 0.67$ Btu/h · ft² · °F, from ACP Table 8A-25 of Standard 90.1–1989, the required design criteria and the design parameters in Example 7.2 are listed below:

	Required by Standard 90.1–1989	Parameters in Example 7.2
Maximum percent fenestration	34	34.5
U-value of exterior opaque wall	0.12	0.12

Although the maximum percent fenestration in Example 7.2 is 0.5 percent greater than the requirement in the Standard, the shading coefficient SC $= 0.36$ is smaller than the upper limit of 0.379, and the $U_{\text{of}} = 0.67$ Btu/h · ft² · °F is slightly lower than the upper limit of 0.68 Btu/h · ft² · °F in the Standard. Therefore, the building envelope parameters in Example 7.2 meet the requirement of the Standard.

5. If the designer intends to increase the maximum percentage of fenestration from 34 percent to 40 percent, then any one of the following must be changed accordingly:

- The SC of the fenestration assembly must be lower than 0.24

- Use a fenestration with a $U_{\text{of}} < 0.45$ Btu/h · ft² · °F

- Adopt perimeter daylighting with automatic daylighting control

- Use external overhangs with a PF > 0.25

7.13 COST-EFFECTIVE AND ENERGY-EFFICIENT MEASURES

According to the heating and cooling load analyses mentioned before as well as the economic parametric analysis of the thermal design in Johnson et al. (1989), the following are cost-effective and energy-efficient measures for the design of building envelopes for office buildings.

Exterior Wall

An increase of the mass of the exterior wall, that is, its thermal capacitance, reduces only the peak heating and cooling loads. The increase in electrical usage and capital investment often offsets the benefit of the decrease of peak loads. The variation of life-cycle costing is negligible (≤ 1 percent in many instances).

Increasing the insulation to more than 2 in. thick decreases only the natural gas usage for heating. This is cost-effective for areas with very cold winters. In some cases, the increase of the annual cost due to the increase of the capital investments of the insulation may balance the reduction of gas usage. A careful analysis is required.

Windows

The ratio of window area to the gross area of the exterior wall A_w/A_o is the single parameter that influences the building life-cycle costing most among building envelopes. For many office buildings, A_w/A_o lies between 0.2 and 0.3.

Heat-absorbing and -reflective glasses produce a significant building cost savings in areas where solar heat control is important in summer. Double glass, triple glass, and low-emission films are effective in reducing the U-value of the window glass and, therefore, the heating and cooling loads.

Indoor shading devices with window management systems are cost effective. For example, indoor shading can be turned on when solar heat gain exceeds 20 Btu/h · ft². Although overhangs reduce the cooling loads, they may increase the need for electric lighting for daylit buildings. The effect of overhang usage on building life-cycle costing is not significant in many instances.

Infiltration

Infiltration has a significant influence on heating and cooling load. Indoor air quality must be guaranteed by systematic and sufficient outdoor air intake through air systems. It is desirable that windows and cracks in joints be well sealed. In multistory buildings, infiltration through elevator shafts, pipe shafts, and duct shafts should be reduced.

Internal Loads

A daylit building has a lower building cost compared with a nondaylit building.

Localized exhaust systems are always effective in reducing sensible and latent cooling loads from machines and appliances in the conditioned space.

According to EIA Commercial Building Characteristics, as of July 1, 1989, the energy conservation features for commercial buildings of a total area of 63,184 million ft² are as follows:

- Buildings having conservation features   89%
- Roof or ceiling insulation   71%
- Wall insulation   47%
- Storm or multiple glazing   38%
- Tinted, reflective, or shading films   35%
- External or internal shading devices   41%
- Weather stripping and caulking   71%

REFERENCES AND FURTHER READING

ASHRAE, *ASHRAE Handbook 1989, Fundamentals* ASHRAE Inc., Atlanta, GA, 1989.

ASHRAE, *ASHRAE/IES Standard 90.1–1989, Energy Efficient Design of New Buildings Except Low-Rise Residential Buildings,* ASHRAE Inc., Atlanta, GA, 1989.

ASHRAE, *ASHRAE/IES Standard 90.1–1989 User's Manual,* ASHRAE Inc., Atlanta, GA, 1992.

Carrier Air Conditioning Co., *Handbook of Air Conditioning System Design,* 1st ed., McGraw-Hill, New York, 1965.

Deringer, J. J. "An Overview of Standard 90.1: Building Envelope," *ASHRAE Journal,* pp. 30–34, February 1990.

DOE, Code of Federal Regulations, Title 10, Part 435, Subpart A, *Performance Standard for New Commercial and Multifamily High-Rise Residential Buildings,* DOE, Washington D.C., 1989.

EIA, *Commercial Building Characteristics 1989,* EIA, Washington D.C., 1991.

Johnson, C. A., Besent, R.W., and Schoenau, G. J., "An Economic Parametric Analysis of the Thermal Design of a Large Office Building under Different Climatic Zones and Different Billing Schedules," *ASHRAE Transactions,* Part I, pp. 355–369, 1989.

Kerrisk, J. F., Schnurr, N. M., Moore, J. E., and Hunn, B. D., "The Custom Weighting-Factor Method for Thermal Load Calculations in the DOE-2 Computer Program," *ASHRAE Transactions,* Part II, pp. 569–584, 1981.

Kreith, F., and Black, W. Z., *Basic Heat Transfer*, Harper & Row, New York, 1980.

Lawrence Berkeley Laboratory, *DOE-2 User's Guide*, Version 2.1, National Technical Information Service, Springfield, VA, 1980.

Los Alamos National Laboratory, *DOE-2 Reference Manual*, Version 2.1A, National Technical Information Service, Springfield, VA, 1981.

Mackey, C. O., and Gay, N. R., "Cooling Load from Sunlit Glass," *ASHVE Transactions*, pp. 321–330, 1952.

Mackey, C. O., and Wright, L. T., "Periodic Heat Flow—Homogeneous Walls or Roofs," *ASHVE Transactions*, pp. 293–312, 1944.

Palmatier, E. P., "Thermal Characteristics of Structures," *ASHRAE Transactions*, pp. 44–53, 1964.

Persily, A. K., and Norford, L. K., "Simultaneous Measurements of Infiltration and Intake in an Office Building," *ASHRAE Transactions*, Part II, pp. 42–56, 1987.

Rudoy, W., and Duran, F., "Development of an Improved Cooling Load Calculation Method, *ASHRAE Transactions*, Part II, pp. 19–69, 1975.

Rudoy, W., and Robins, L. M., "Pulldown Load Calculations and Thermal Storage during Temperature Drift," *ASHRAE Transactions*, Part I, pp. 51–63, 1977.

Snelling, H. J., "Duration Study for Heating and Air Conditioning Design Temperature, *ASHRAE Transactions*, Part II B, p. 242–249, 1985.

Sowell, E. F., and Chiles, D. C., "Characterization of Zone Dynamic Response for CLF/CLTD Tables," *ASHRAE Transactions*, Part II A, pp. 162–178, 1985.

Sowell, E. F., and Chiles, D. C., "Zone Descriptions and Response Characterization for CLF/CLTD Calculations," *ASHRAE Transactions*, Part II A, pp. 179–200, 1985.

Stephenson, D. G., and Mitalas, G. P., "Cooling Load Calculations by Thermal Response Factor Method," *ASHRAE Transactions*, pp. III 1.1–1.7, 1967.

Sun, T. Y., "Air Conditioning Load Calculation, *Heating/Piping/Air Conditioning*, pp. 103–113, January 1986.

Trehan, A. K., Fortmann, R. C., Koontz, M. D., and Nagda, N. L., "Effect of Furnace Size on Morning Pickup Time," *ASHRAE Transactions*, Part I, pp. 1125–1129, 1989.

Wang, S. K., *Air Conditioning*, Vol. 1, Hong Kong Polytechnic, Hong Kong, 1987.

Williams, G. J., "Fan Heat: Its Source and Significance," *Heating/Piping/Air Conditioning*, pp. 101–112, January 1989.

8.1 BASICS OF AIRFLOW IN DUCTS

Bernoulli Equation

The *Bernoulli equation* relates the mean velocity v, in ft/s, the pressure p, in lb_f/ft^2, and the elevation z, in ft, of a frictionless or ideal fluid at steady state. When a fluid motion is said to be in steady state, the variables of the fluid at any point along the fluid flow do not vary with time. Assuming constant density, the Bernoulli equation can be expressed in the following form:

$$\frac{p}{\rho} + \frac{v^2}{2g_c} + \frac{gz}{g_c} = \text{constant} \qquad (8.1)$$

where p = static pressure, lb_f/ft^2
ρ = fluid density, lb_m/ft^3
g = gravitational acceleration, ft/s^2
g_c = dimensional constant, 32.2 $lb_m \cdot ft/lb_f \cdot s^2$

Steady-Flow Energy Equation

For a real fluid flowing between two cross sections in an air duct, pipe, or a conduit, energy loss is inevitable because of the viscosity of the fluid and the presence of mechanical friction and eddies. The energy used to overcome these losses is usually transformed into heat energy. If we ignore the kinetic energy difference between the value calculated by mean velocity of the cross section and the value calculated according to the velocity distribution of the cross section, the steady-flow energy equation for a unit mass of real fluid is given as

$$\frac{p'_1}{\rho_1} + u_1 J + \frac{v_1^2}{2g_c} + \frac{gz_1}{g_c} + qJ$$
$$= \frac{p'_2}{\rho_2} + u_2 J + \frac{v_2^2}{2g_c} + \frac{gz_2}{g_c} + W \qquad (8.2)$$

where u = internal energy, Btu/lb
J = Joule's equivalent, 778 $ft \cdot lb_f/Btu$
q = heat supplied, Btu/lb
W = work developed, $ft \cdot lb_f/lb_m$

In Eq. (8.2), subscripts 1 and 2 indicate cross sections 1 and 2, respectively, and p'_1 and p'_2 denote the absolute static pressure at cross sections 1 and 2. Signs of q and W follow the convention in thermodynamics, that is, when heat is supplied to the system, q is positive and when heat is released from the system, q is negative. When work is developed by the system, W is positive, and for work input to the system, W is negative.

Multiply both sides of Eq. (8.2) by ρ and rearrange the terms. Then, each term has the unit of pressure, in lb_f/ft^2, or

$$p'_1 + \frac{\rho_1 v_1^2}{2g_c} + \frac{\rho_1 g z_1}{g_c} = p'_2 + \frac{\rho_2 v_2^2}{2g_c}$$
$$+ \frac{\rho_2 g z_2}{g_c} + \rho W + \rho J(u_2 - u_1 - q) \qquad (8.3)$$

For an air duct or a piping work without a fan, compressor, and pump, $\rho W = 0$.

Let the pressure loss from viscosity, friction, and eddies between cross sections 1 and 2 $\Delta p_f = \rho J(u_2 - u_1 - q)$, then, Eq. (8.3) can be rewritten as

$$p'_1 + \frac{\rho_1 v_1^2}{2g_c} + \frac{\rho_1 g z_1}{g_c} = p'_2 + \frac{\rho_2 v_2^2}{2g_c} + \frac{\rho_2 g z_2}{g_c} + \Delta p_f$$
$$(8.4)$$

If both sides of Eq. (8.2) are multiplied by g_c/g, each term of the equation is then expressed in the form of head, in ft or in. of fluid column. That is,

$$\frac{g_c p'_1}{g \rho_1} + \frac{v_1^2}{2g} + z_1 = \frac{g_c p'_2}{g \rho_2} + \frac{v_2^2}{2g}$$
$$+ z_2 + \frac{g_c W}{g} + \frac{g_c \Delta p_f}{\rho g} \qquad (8.5)$$

Static Pressure, Velocity Pressure, and Total Pressure

For an air duct system or a water piping system, pressure is more conveniently expressed as gauge pressure, that is, greater or less than the atmospheric pressure p_{at}.

Consider a supply duct system in a multistory building, as shown in Fig. 8.1. In Eq. (8.4), $p_1' = p_{at1} + p_1$ and $p_2' = p_{at2} + p_2$, where, p_1 and p_2 represent the gauge static pressure and p_{at1} and p_{at2} the atmospheric pressure added on the fluid at cross sections 1 and 2. The relationship of fluid properties between cross sections 1 and 2 can be expressed as

$$(p_{at1} + p_1) + \frac{\rho_1 v_1^2}{2g_c} + \frac{\rho_1 g z_1}{g_c} = (p_{at2} + p_2)$$
$$+ \frac{\rho_2 v_2^2}{2g_c} + \frac{\rho_2 g z_2}{g_c} + \Delta p_f \qquad (8.6)$$

If the air temperature inside the air duct is equal to the ambient air temperature, and if the stack effect because of the difference in air densities between the air columns inside the air duct and the ambient air does not exist, then

$$p_{at1} - p_{at2} = \frac{(\rho_2 z_2 - \rho_1 z_1)g}{g_c}$$

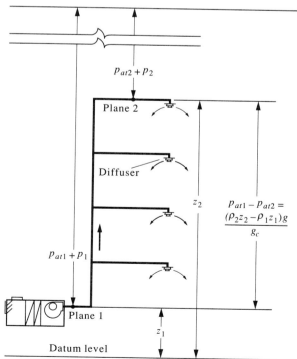

FIGURE 8.1 Pressure characteristics of an air duct system.

Therefore, Eq. (8.6) becomes

$$p_1 + \frac{\rho_1 v_1^2}{2g_c} = p_2 + \frac{\rho_2 v_2^2}{2g_c} + \Delta p_f \qquad (8.7)$$

Equation (8.7) is one of the primary equations used to determine the pressure characteristics of an air duct system that does not contain a fan and in which the stack effect is negligible.

STATIC PRESSURE. In Eq. (8.7), static pressures p_1 and p_2 are often represented by p_s. In air duct systems, its unit can be either Pa (pascals, or newtons per square meter) in SI units, or the height of water column, in inches, for I-P units. All are expressed in gauge pressure. The relationship between the static pressure p_s, in lb_f/ft^2, and the height of a water column H, in ft, is

$$p_s = \frac{\rho_w g H A}{g_c A} = \frac{\rho_w g H}{g_c} \qquad (8.8)$$

where A = cross-sectional area of the water column, ft^2

ρ = density of water, lb/ft^3

When the static pressure is expressed as the height of 1 in. of water column gauge pressure, 1 in. WG, for a density of water $\rho_w = 62.3$ lb/ft^3 and $g = g_c = 32.2$ ft/s^2, we find from Eq. (8.8)

$$p_s \ 1 \ in.WG = \frac{\rho_w g H}{g_c} = \frac{62.3 \times 32.2 \times 1}{32.2 \times 12}$$
$$= 5.192 \ lb_f/ft^2$$

That is, 1 in. WG = 5.192 lb_f/ft^2 gauge pressure. Because 1 lb_f/ft^2 = 47.88 Pa, 1 in. WG = 5.192 × 47.88 = 248.6 Pa.

VELOCITY PRESSURE. In Eq. (8.7), the term $\rho v^2/(2g_c)$ is called *velocity pressure*, or *dynamic pressure*, and is represented by the symbol p_v, that is

$$p_v = \frac{\rho v^2}{2g_c} \qquad (8.9)$$

where ρ = air density, lb/ft^3. For air density $\rho = 0.075$ lb/ft^3, if the velocity pressure p_v is expressed in inches WG and the air velocity in fpm, or ft/min, then, according to Eq. (8.9),

$$\left(\frac{v}{60}\right)^2 = \frac{5.192(2 p_v g_c)}{\rho}$$
$$= \frac{5.192 \times 2 \times 32.2 \times p_v}{0.075} = 4458 \, p_v$$

and

$$v = 4005 \sqrt{p_v}$$

$$p_v = \left(\frac{v}{4005}\right)^2 \qquad (8.10)$$

TOTAL PRESSURE. At any cross-sectional plane perpendicular to the direction of the air flow, the total pressure of the airstream p_t is defined as the sum of the static pressure p_s and the velocity pressure p_v, that is,

$$p_t = p_s + p_v \qquad (8.11a)$$

The unit of p_t must be consistent with p_s and p_v. In I-P units, it is also expressed in inches WG. Substituting Eq. (8.11a) into Eq. (8.7), we see that

$$p_{t1} = p_{t2} + \Delta p_f \qquad (8.11b)$$

Equation (8.11b) is another primary equation that relates the pressure loss from friction and other sources, Δp_f, and the total pressure p_{t1} and p_{t2} at two cross sections of the air duct system.

Stack Effect

When an air duct system has an elevation difference and the air temperature inside the air duct is different from the ambient air temperature, the stack effect exists. It affects air flow at different elevations.

During a hot summer day, when the density of the outdoor air is less than the density of the cold supply air inside the air duct, the pressure exerted by the atmospheric air column on cross section 1 with a height of elevation difference $(z_1 - z_2)$, in ft, as shown in Fig. 8.1, is given as

$$p_{at1} - p_{at2} = \frac{\rho_o g (z_2 - z_1)}{g_c}$$

where ρ_o = mean density of the ambient air, lb/ft^3. Let ρ_i = mean density of the supply air inside the air duct, in lb/ft^3. If the differences between the densities inside the air ducts, $(\rho_1$ and $\rho_i)$ and $(\rho_2$ and $\rho_i)$, are ignored, the pressure exerted by the supply air column in the duct on cross section 1 compared with cross section 2 is

$$\frac{\rho_i g (z_2 - z_1)}{g_c} = \frac{g(\rho_2 z_2 - \rho_1 z_1)}{g_c}$$

Substituting these relationships into Eq. (8.6) yields

$$p_1 + \frac{\rho_1 v_1^2}{2 g_c} + \frac{g(\rho_o - \rho_i)(z_2 - z_1)}{g_c}$$

$$= p_2 + \frac{\rho_2 v_2^2}{2 g_c} + \Delta p_f \qquad (8.12)$$

and

$$p_{t1} + \frac{g(\rho_o - \rho_i)(z_2 - z_1)}{g_c} = p_{t2} + \Delta p_f \qquad (8.13)$$

The second term on the left-hand side of both Eqs. (8.12) and (8.13) is called the stack effect p_{st}, in lb$_f$/ft^2, that is,

$$p_{st} = \frac{g(\rho_o - \rho_i)(z_2 - z_1)}{g_c} \qquad (8.14)$$

If p_{st} is expressed in inches WG (1 lb$_f$/ft^2 = 0.1926 in. WG), then,

$$p_{st} = 0.1926 g (\rho_o - \rho_i)(z_2 - z_1) \qquad (8.15)$$

For an upward supply duct system, $z_2 > z_1$. If cold air is supplied, $\rho_i > \rho_o$, and p_{st} is negative. If warm air is supplied, $\rho_o > \rho_i$, and p_{st} is positive. For a downward supply duct system with a cold air supply, p_{st} is positive; if there is a warm air supply, p_{st} is negative.

When supply air is at a temperature of 60°F and a relative humidity of 80 percent, its density is 0.075 lb/ft^3. Also, if the space air has a temperature of 75°F and a relative humidity of 50 percent, its density is 0.073 lb/ft^3. If $g = g_c$, for a difference of $z_2 - z_1 = 30$ ft, then

$$p_{st} = 0.1926 \times (0.073 - 0.075) \times 30$$

$$= 0.0116 \text{ in. WG}$$

For an air-handling unit or a packaged unit that supplies air to the same floor where it is located, the stack effect is usually ignored.

Laminar Flow and Turbulent Flow

Reynolds identified two types of fluid flow in 1883 by observing the behavior of a stream of dye in a water flow: laminar flow and turbulent flow. He also discovered that the ratio of inertial to viscous forces is the criterion that distinguishes these two types of fluid flow. This dimensionless parameter is now widely known as the *Reynolds number* Re, or

$$Re = \frac{\rho v L}{\mu} = \frac{v L}{\nu} \qquad (8.16)$$

where ρ = density of the fluid, lb/ft^3
v = velocity of fluid, ft/s
L = characteristic length, ft
μ = viscosity or absolute viscosity, lb/ft · s
ν = kinematic viscosity, ft^2/s

Many experiments have shown that laminar flow occurs at Re $\leq$ 2000 in round ducts and pipes. A transition region exists between 2000 < Re $\leq$ 4000. When Re > 4000, the fluid flow is probably a turbulent flow.

At 60°F, the viscosity for air $\mu = 1.21 \times 10^5$ lb/ft · s. For a round duct with a diameter of 1 ft and an air flow through it of 3 ft/s (180 fpm), the Reynolds number is

$$\text{Re} = \frac{\rho v L}{\mu} = \frac{0.075 \times 3 \times 1}{1.21 \times 10^5} = 18,595$$

The Re of such an air duct is far greater than 4000. Therefore, the air flow inside the air duct is usually turbulent except within the boundary layer adjacent to the duct wall.

Velocity Distribution

The *velocity distributions* of turbulent and laminar flow at a specific cross section in a circular duct that result from (1) the mechanical friction between the fluid particles and the duct wall and (2) the shearing stress of the viscous fluid are shown in Fig. 8.2. The difference between these two types of flow is significant.

For a fully developed turbulent flow, the air velocity v at various distances from the wall of a circular duct y, in ft, varies according to Prantdl's *one-seventh power law* as follows:

$$\frac{v}{v_{\max}} = \left(\frac{y}{R}\right)^{1/7} \qquad (8.17)$$

where $v_{\max}$ = maximum air velocity on the centerline of the air duct, ft/s
R = radius of the duct, ft

The mean air velocity v_m lies at a distance about $0.33\,R$ from the duct wall.

Equation of Continuity

For one-dimensional fluid flow at steady-state, the application of the principle of conservation of mass gives the following equation of continuity:

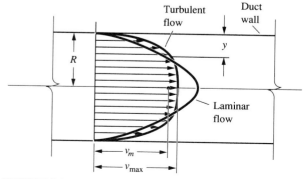

FIGURE 8.2 Velocity distribution in a circular duct.

$$\dot{m} = \rho_1 v_1 A_1 = \rho_2 v_2 A_2 \qquad (8.18)$$

where $\dot{m}$ = mass flow rate, lb/s
A = cross-sectional area perpendicular to fluid flow, ft^2

Subscripts 1 and 2 indicate the cross sections 1 and 2 along the fluid flow.

If the differences in fluid density at various cross sections are negligible, then the equation of continuity becomes

$$\dot{V} = A_1 v_1 = A_2 v_2 \qquad (8.19)$$

where $\dot{V}$ = volume flow rate of the air flow, ft^3/min or cfm
v = mean air velocity at any specific cross section, fpm

Theoretically, the velocity pressure p_{vt} should be calculated as:

$$p_{vt} = \frac{\int (\rho v^2 / 2g_c) v\, dA}{\int v\, dA} \qquad (8.20)$$

Its value is slightly different from p_v calculated from Eq. (8.9), which is based on mean velocity v. For a fully developed turbulent flow, $p_{vt} \approx 1.06\,p_v$. Since most experimental results of pressure loss (indicated in terms of velocity pressure) are calculated by $p_v = \rho v^2 / 2g_c$, for the sake of simplicity p_v will be used here instead of p_{vt}.

8.2 CHARACTERISTICS OF AIR FLOW IN DUCTS

Types of Air Duct

Air ducts can be classified into four types according to their transporting functions:

1. *Supply duct.* Conditioned air is supplied to the conditioned space.
2. *Return duct.* Space air is returned (1) to the fan room where the air-handling unit is installed or (2) to the packaged unit.
3. *Fresh air duct.* Outdoor air is transported to the air-handling unit, to the fan room, or to the space directly.
4. *Exhaust duct.* Space air or contaminated air is exhausted from the space, equipment, fan room, or localized area.

Each of these four types of duct may also subdivide into headers, main ducts, and branch ducts or runouts. Sometimes, a header or main duct is also called a trunk. A header is that part of a duct that connects directly to the supply or exhaust fan before air is supplied to the main ducts in a large duct system. Main ducts have comparatively greater flow rates and size, serve a greater conditioned area, and, therefore, allow higher air velocities. Branch ducts are usually connected to the terminals, hoods, supply outlets, return grilles, and exhaust hoods. A vertical duct is called a *riser*.

Pressure Characteristics of the Air Flow

During the analysis of the pressure characteristics of air flow in an air duct system such as the one in Fig. 8.3, it is assumed that the static pressure of the space air is equal to the static pressure of the atmospheric air, and the velocity pressure of the space air is equal to zero. Also, for convenient measurements and presentation, as previously mentioned, the pressure of the atmospheric air is taken as the datum, that is, $p_{at} = 0$. When $p > p_{at}$, p is positive, and if $p < p_{at}$, then p is negative.

At cross section R_1, as the recirculating air enters the return grille, both the total pressure p_t and static pressure p_s decrease as a result of the total pressure loss of the inlet. The velocity pressure p_v, indicated by the shaded section in Fig. 8.3, will gradually increase until it is equal to the velocity of the branch duct. Both p_t and p_s are negative so that air will flow from the conditioned space at a datum of 0 to a negative pressure. Because velocity pressure

p_v is always positive in the direction of flow, from Eq. (8.11a) $p_t = p_s + p_v$, so p_s is then smaller or more negative than p_t.

When the recirculating air flows through the branch duct section R_1–1_1, elbow 1_1–1_2, branch duct section 1_2–1_3, diffuser 1_3–1_4, and branch duct section 1_4–1_5, both p_t and p_s drop because of the pressure losses. Velocity pressure p_v remains the same between cross sections R_1 and 1_3. It gradually decreases because of the diffuser 1_3–1_4, and remains the same between 1_4 and 1_5.

As the recirculating air flows through the converging tee 1_5–1, this straight-through stream meets with another branch stream of recirculating air from duct section R_2–1, at node or junction 1. The combined air stream then becomes the main stream. Total pressure p_t and static pressure p_s usually both decrease when the straight-through stream flows through the converging tee. It may increase if the velocity of the branch stream is far higher than that of the straight-through stream. However, the sum of the energies of the straight-through and the branch stream at the upstream side of the converging tee is always higher than the energy at the downstream side because of the pressure loss of the converging tee.

In the main duct section 1–2_1, p_t and p_s drop further while p_v increases because of the higher air velocity. This is mainly because of the greater volume flow rate in main duct section 1–2_1. When the recirculating air enters the air-handling unit, its velocity drops sharply to a value between 400 and 600

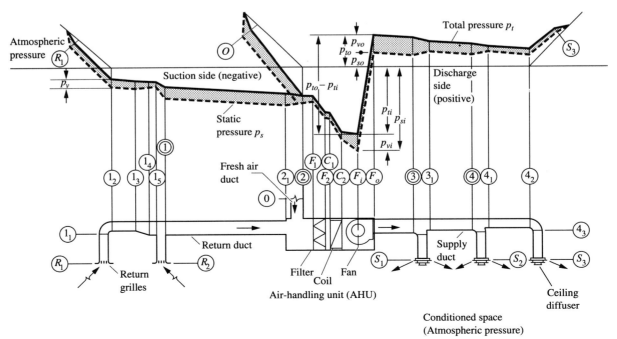

FIGURE 8.3 Pressure characteristics of a fan–duct system.

fpm. The air is then mixed with the outdoor airstream from the fresh air duct at junction 2. After the mixing section, p_t and p_s drop sharply when the mixture of recirculating and outdoor air flows through the filter and the coil. Total pressure p_t and static pressure p_s drop to their minimum value at the inlet of the supply fan F_i.

At the supply fan, p_t is raised to its highest value at the fan outlet F_o. Both p_t and p_s decrease in duct section F_o–3. At junction 3, the airstream diverges into the main stream or straight-through stream and the branch stream. Although p_t drops after the main stream passes through the diverging tee 3–3_1, p_s increases because of the smaller p_v after 3_1. The increase of p_s because of the decrease of p_v is known as the *static regain* $\Delta p_{s.r}$, in inches WG. It can be expressed as

$$\Delta p_{s.r} = \left(\frac{v_3}{4005}\right)^2 - \left(\frac{v_{31}}{4005}\right)^2 \qquad (8.21)$$

where v_3, v_{31} = air velocity at cross section 3 and 3_1, respectively, fpm. In duct sections 3_1–4 and 4–S_3, p_t decreases gradually along the direction of air flow. Finally, p_t, p_s, and p_v all drop to 0 after the supply air is discharged to the conditioned space.

The pressure characteristics along the air flow can be summarized as follows:

- In most sections, p_t of the main airstream decreases along the air flow. However, p_t of the main airstream may increase because of the higher velocity of the combined branch airstream.

- When air flows through the supply fan, mechanical work is done on the air so that p_t and p_s are raised from a minimum negative value at the fan inlet to a maximum positive value at the fan outlet.

- The pressure characteristics between any two cross sections of a fan-duct system are governed by the change of p_t and the pressure loss Δp_f between these two cross sections $p_{t1} = p_{t2} + \Delta p_f$. Static pressure is always calculatd as $p_s = p_t - p_v$.

- In a constant-volume air system, the air flow inside an air duct is considered steady and continuous.

Because the change in p_s in a fan-duct system is small when compared with p_{at}, the air flow is also considered incompressible.

System Pressure Loss

For an air system, system pressure loss Δp_{sy}, in inches WG, is the sum of the total pressure losses of the return air system $\Delta p_{r.s}$, section R_1–2_1 in Fig. 8.3; the air-handling unit Δp_{AHU}, section 2_1–F_o, or the packaged unit Δp_{PU}; and the supply air system

$\Delta p_{s.s}$, section F_o–S_3; all are expressed in inches WG. That is,

$$\Delta p_{sy} = \Delta p_{r.s} + \Delta p_{AHU} + \Delta p_{s.s}$$
$$= \Delta p_{r.s} + \Delta p_{PU} + \Delta p_{s.s} \qquad (8.22)$$

Both $\Delta p_{r.s}$ and $\Delta p_{s.s}$ may include the pressure losses of duct sections; duct fittings such as elbows, diffusers, and converging and diverging tees; components; and equipment.

The sum of $\Delta p_{r.s}$ and $\Delta p_{s.s}$ is called the *external total pressure*, or external pressure $\Delta p_{t.ex}$, as opposed to the pressure loss in the air-handling unit or packaged unit. The external pressure $\Delta p_{t.ex} = \Delta p_{r.s} + \Delta p_{s.s}$, in inches WG.

In commercial buildings, most air systems have a system pressure loss between 2.5 and 6 in. WG. Of this, Δp_{AHU} usually has a value between 1.5 and 3 in. WG, and $\Delta p_{s.s}$ is usually less than 0.6 Δp_{sy} except in large auditoriums and indoor stadiums.

Criteria of Fan Energy Use

ASHRAE/IES Standard 90.1–1989 specifies that for a constant-volume air system, the power required by the motor of the fans, including supply, return, and relief fans in an air system that supply cooling or warm air to the conditioned space, must not exceed 0.8 W/cfm of supply air, or

$$\frac{746 P_{sy}}{\dot{V}_s} \leq 0.8 \qquad (8.23a)$$

where P_{sy} = total power input to the fan motors, hp
$\dot{V}_s$ = volume flow rate of supply air, cfm

In a variable-air-volume (VAV) system, the power input to the motor of the fans in an air system must not exceed 1.25 W/cfm of supply air at design conditions, or

$$\frac{746 P_{sy}}{\dot{V}_{s.d}} \leq 1.25 \qquad (8.23b)$$

where $\dot{V}_{s.d}$ = supply volume flow rate at design conditions, cfm. Power input to the fan motors P_{sy} can be calculated as

$$P_{sy} = \frac{\Delta p_{sy} \dot{V}_s}{6350 \eta_f \eta_m}$$

where η_f, η_m = fan total efficiency and motor efficiency. For a VAV system,

$$P_{sy} = \frac{\Delta p_{sy.d} \dot{V}_{s.d}}{6350 \eta_{f.d} \eta_{m.d}} \qquad (8.24)$$

where $\Delta p_{sy,d}$ = system total pressure loss at design conditions, in WG
$\eta_{f,d}\eta_{m,d}$ = fan total efficiency and motor efficiency at design conditions

Substituting Eq. (8.24) into Eqs. (8.23a) and (8.23b), we find that for constant volume systems, the power consumption per cfm of supply air is

$$\frac{0.1175 \Delta p_{sy}}{\eta_f \eta_m} \le 0.8$$

For VAV systems,

$$\frac{0.1175 \Delta p_{sy}}{\eta_{f,d}\eta_{m,d}} \le 1.25 \qquad (8.25)$$

Example 8.1. Two air systems both have a design supply volume flow rate of 10,000 cfm. One of them is a constant-air-volume (CAV) system and the other is a variable-air-volume (VAV) system. Their operating parameters at design conditions are as follows:

	CAV	VAV
Supply volume flow rate, cfm	10,000	10,000
System total pressure loss, in. WG	4	5
Fan total efficiency η_f	0.75	0.70
Fan motor efficiency η_m	0.85	0.85

Find the fan power consumption per cfm of supply air for these air systems.

Solution
1. For the constant-air-volume system, the design volume flow rate and fan total pressure are the same during operation. According to Eq. (8.25), the power consumption per cfm of supply air is

$$\frac{0.1175 \Delta p_{sy}}{\eta_f \eta_m} = \frac{0.1175 \times 4}{0.75 \times 0.85}$$
$$= 0.737 < 0.8 \text{ W/cfm}$$

2. For the VAV system, the power consumption per cfm of supply air at design conditions is

$$\frac{0.1175 \Delta p_{sy}}{\eta_{f,d}\eta_{m,d}} = \frac{0.1175 \times 5}{0.7 \times 0.85}$$
$$= 0.983 < 1.25 \text{ W/cfm}$$

Both systems meet the requirements specified in ASHRAE/IES Standard 90.1–1989.

8.3 DUCT CONSTRUCTION

Maximum Pressure Difference

Duct systems can be classified according to the maximum pressure difference between the air inside the duct and the ambient air (also called static pressure differential) as ± 0.5 in. WG, ± 1 in. WG, ± 2 in. WG, ± 3 in. WG, ± 4 in. WG, ± 6 in. WG, and ± 10 in. WG. In actual practice, the maximum pressure difference of the supply or return duct system in commercial buildings is usually less than ± 3 in. WG.

In commercial buildings, a low-pressure duct system has a static pressure differential of 2 in. WG or less, and the maximum air velocity inside the air duct is usually 2400 fpm. A medium-pressure duct system has a static pressure differential of 2 to 6 in. WG, with a maximum air velocity about 3500 fpm. In industrial duct systems, including mechanical ventilation, mechanical exhaust, and induStrial air pollution control systems, the pressure difference is often higher. In residential buildings, the static pressure differential of the duct systems is classified as ± 0.5 in. WG or ± 1 in. WG.

Material

Underwriters Laboratory Inc. (UL) classifies duct systems according to the flame spread and smoke development of the duct material during fire as

Class 0. Zero flame spread, zero smoke developed

Class 1. A flame spread rating of not more than 25 without evidence of continued progressive combustion and a smoke-development rating of not more than 50

Class 2. A flame spread of 50 and a smoke development rating of 100

National Fire Protection Association (NFPA) Standard 90A specifies that the material of the ducts must be iron, steel including galvanized sheets, aluminum, concrete, masonry, or clay tile. Ducts fabricated by these materials are listed as Class 0. UL Standard 181 allows Class 1 material to be used for ducts when they do not serve as risers for more than two stories, or are not used in temperatures higher than 250°F. Fibrous glass and many flexible ducts that are factory fabricated are approved by UL as Class 1.

Ducts can be classified according to their shapes into rectangular, round, flat oval, and flexible as shown in Fig. 8.4.

Rectangular Ducts

For the space available between the structural beam and the ceiling in a building, rectangular ducts have the greatest cross sectional area. They are less rigid than round ducts and are more easily fabricated on site. The joints of rectangular ducts have a comparatively greater percentage of air leakage than factory-

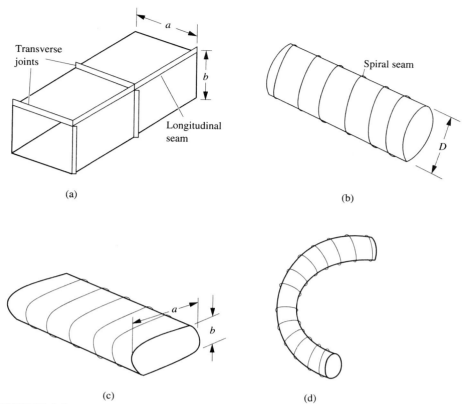

FIGURE 8.4 Various types of air duct: (a) rectangular duct, (b) round duct with spiral seam, (c) flat oval duct, and (d) flexible duct.

fabricated spiral-seamed round ducts and flat oval ducts, as well as fiberglass ducts. Unsealed rectangular ducts may have an air leakage from 15 to 20 percent of the supply volume flow rate. Rectangular ducts are usually used in low-pressure systems.

The ratio of the long side a to the short side b in a rectangular duct is called the *aspect ratio* R_{as}. The greater the R_{as}, the higher is the pressure loss per unit length as well as the heat loss and heat gain per unit volume flow rate transported. In addition, more labor and material are required.

Galvanized sheet or, more precisely, galvanized coated steel sheet, and aluminum sheet are the materials most widely used for rectangular ducts. To prevent vibration of the duct wall by the pulsating air flow, transverse joints and longitudinal seam reinforcements are required in ferrous metal ducts.

The galvanized sheet gauge and thickness for rectangular ducts are listed in Table 8.1. In Table 8.2 are specifications for rectangular ferrous metal duct construction for commercial systems based on the publication of the Sheet Metal and Air Conditioning Contractors' National Association (SMACNA) titled *HVAC Duct Construction Standards—Metal and*

TABLE 8.1

Thickness of galvanized sheet for rectangular ducts

| Gauge | Thickness, in. | | Nominal weight, lb/ft² |
	Nominal	Minimum	
30	0.0157	0.0127	0.656
28	0.0187	0.0157	0.781
26	0.0217	0.0187	0.906
24	0.0276	0.0236	1.156
22	0.0336	0.0296	1.406
20	0.0396	0.0356	1.656
18	0.0516	0.0466	2.156
16	0.0635	0.0575	2.656
14	0.0785	0.0705	3.281
13	0.0934	0.0854	3.906
12	0.1084	0.0994	4.531
11	0.1233	0.1143	5.156
10	0.1382	0.1292	5.781

Note: Minimum thickness is based on thickness tolerances of hot-dip galvanized sheets in cut lengths and coils (per ASTM Standard A525). Tolerance is valid for 48-in. and 60-in. wide sheets.

Source: *ASHRAE Handbook 1988*, Equipment. Reprinted with permission.

TABLE 8.2
Rectangular ferrous metal duct construction for commercial buildings

Minimum galvanized steel thickness, in. (gauge) — Pressure, in. WG

Duct Dimensions, in.	0.0575 (16) ±2	±3	0.0466 (18) ±1	±2	±3	0.0356 (20) ±1	±2	±3	0.0296 (22) ±0.5	±1	±2	±3	0.0236 (24) ±0.5	±1	±2	±3	0.0187 (26) ±0.5	±1	±2
Up through 10																			A-8
12															A-8			A-10	A-5
14												A-8		A-8	A-5			A-8	A-5
16								A-8			A-10	A-8		A-8	A-5			A-8	A-5
18								A-8			A-10	A-8		A-8	A-5	A-10	A-8	A-5	
18								A-8			A-10	A-8		A-8	A-5		A-10	A-8	A-5
20					B-10		B-10	B-8		A-10	B-8	A-5		A-10	A-5	A-5	A-10	A-5	A-5
22				B-10	C-10	A-10	B-10	B-8		A-10	B-8	A-5		A-10	A-5	B-5	A-10	A-5	B-5
24				C-10	C-10	B-10	C-10	B-5		B-10	C-8	B-5		B-10	B-5	B-5	A-10	A-5	B-5
26	C-10	D-10	A-10	C-10	D-10	B-10	C-10	C-5	A-10	B-10	C-8	C-5	A-10	B-8	B-5	C-5	B-8	B-5	B-4
28	C-10	D-10	B-10	D-10	D-10	C-10	C-8	C-5	B-10	C-10	C-5	C-5	B-10	C-8	C-5	C-4	B-8	B-5	C-4
30	D-10	D-10	C-10	D-10	D-8	C-10	D-8	C-5	B-10	C-10	C-5	C-5	B-10	C-8	C-5	C-4	C-5	C-5	
36	E-10	E-8	C-10	E-8	E-5	D-10	D-5	E-5	C-10	D-8	D-5	D-4	C-8	C-5	D-4	D-4	D-5	D-4	—
42	E-8	E-5	D-10	E-5	E-5	D-8	E-5	E-5	D-8	D-5	D-5	E-4	D-8	D-5	E-4	E-3	D-5	D-4	—
48	F-5	G-5	E-10	F-5	G-5	E-5	F-5	F-4	D-8	E-5	E-4	E-3	D-5	E-5	E-3	E-2.5	D-5	—	—
54	G-5	H-5	E-8	G-5	H-5	E-5	F-5	G-3	D-5	E-5	F-3	G-3	D-5	E-4	F-3	E-2.5	E-4	—	—
60	H-5	H-5	F-8	H-5	H-4	F-5	G-4	G-3	E-5	E-5	G-3	G-2.5	E-5	F-4	G-2.5	G-2.5	—	—	—
72	I-5	I-4	G-8	H-4	H-3	G-4	H-3	H-3	F-5	F-5	G-3	H-2.5	F-4	H-2	H-2	H-2	—	—	—
84	J-4	J-3	H-5	J-4	J-3	H-4	I-3	J-2.5	H-5	F-5	H-3		G-4				—	—	—
96	K-4	L-3	I-5	K-3	K-2.5	I-3	J-2.5	J-2	H-4	G-4							—	—	—
Over 96	H-2.5 plus rods	H-2.5 plus rods	H-2.5 plus rods	plus rods	plus rods	—											—	—	—

Note: For a given duct thickness, letters indicate type (rigidity class) of duct reenforcement (see Table 8.3); numbers indicate maximum spacing (ft) between duct reenforcement. Use the same metal thickness on all duct sides.

Adapted with permission from *SMACNA HVAC Duct Construction Standard—Metal and Flexible*. Refer to this standard for complete details.

Flexible. For designing and constructing an economical duct system, it is recommended to select an optimum combination of minimum galvanized sheet thickness, type of transverse joint reinforcement, and its maximum spacing for a specific duct dimension at a specific pressure differential between the air inside the duct and the ambient air.

For rectangular ducts, one uses the same metal thickness for all sides of the duct and evaluates duct reinforcement on each side separately. In Table 8.2,

for a given duct dimension and thickness, letters indicate the type of duct reinforcement (rigidity class) and numbers indicate maximum spacing, in ft. Blanks indicate that reinforcement is not required and dashes denote that such a combination is not allowed.

Transverse joint reinforcements, abridged from SMACNA's publication *HVAC Duct Construction Standard—Metal and Flexible* and *ASHRAE Handbook 1988, Equipment* are presented in Table 8.3. These must be matched with the arrangements in

TABLE 8.3
Transverse joint reinforcement

Pressure, 4 in. WG Maximum*

Minimum rigidity class	Standing drive slip	Standing S		Standing S	Standing S	Standing S (bar-reinforced)	Standing S (angle reinforced)
	$H_S \times T$(min), in.	W, in.	$H_S \times T$(min), in.	$H_S \times T$(min), in.	$H_S \times T$(min), in.	$H_S \times T$(min) plus reinforcement $(H \times T)$, in.	
A	Use Class B	Use Class C		$\frac{1}{2} \times 0.0187$	Use Class D	Use Class F	
B	$1\frac{1}{8} \times 0.0.187$			$\frac{1}{2} \times 0.0296$			
C	$1\frac{1}{8} \times 0.0296$	—	1×0.0187	1×0.0187			
D	—	—	1×0.0236	1×0.0236	$1\frac{1}{8} \times 0.0187$		
E	—	$\frac{3}{16}$	$1\frac{1}{8} \times 0.0356$	—	$1\frac{1}{8} \times 0.0466$		
F	—	$\frac{3}{16}$	$1\frac{5}{8} \times 0.0296$	—	$1\frac{1}{2} \times 0.0236$	$1\frac{1}{2} \times 0.0236$ plus $1\frac{1}{2} \times \frac{1}{8}$ bar	
G	—	$\frac{3}{16}$	$1\frac{5}{8} \times 0.0466$	—	$1\frac{1}{2} \times 0.0466$	$1\frac{1}{2} \times 0.0296$ plus $1\frac{1}{2} \times \frac{1}{8}$ bar	
H	—	—	—	—	—	$1\frac{1}{2} \times 0.0356$ plus $1\frac{1}{2} \times 1\frac{1}{2} \times \frac{3}{16}$ angle	
I	—	—	—	—	—	2×0.0356 plus $2 \times 2 \times \frac{1}{8}$ angle	
J	—	—	—	—	—	2×0.0356 plus $2 \times 2 \times \frac{3}{16}$ angle	

*Acceptable to 36 in. length at 3 in. WG and to 30 in. length at 4 in. WG.

Adapted with permission from *SMACNA HVAC Duct Construction Standards—Metal and Flexible* and *ASHRAE Handbook 1988*, Equipment. Refer to SMACNA Standard for complete details.

Table 8.2. Duct hangers should be installed at right angles to the center-line of the duct. Habjan (1984) recommends the following maximum duct hanger spacing:

Duct area	Maximum spacing, ft
Up to 4 ft^2	8
Between 4 and 10 ft^2	6
Larger than 10 ft^2	4

Round Ducts

For a specified cross-sectional area and mean air velocity, round ducts have less fluid resistance against air flow than rectangular and flat oval ducts. Round ducts also have better rigidity and strength. The spiral- and longitudinal-seamed round ducts used in commercial buildings are usually factory fabricated to improve the quality and the sealing of the ductwork. The pressure losses can be calculated more precisely than for rectangular ducts, and results in a better-balanced system. Air leakage can be maintained at about 3 percent as a result of well-sealed seams and joints. Round ducts have much smaller radiated noise breakout from the duct than rectangular and flat oval ducts.

The main disadvantage of round ducts is the greater space required under the beam for installation.

The standard diameters of round ducts range from 4 to 20 inches in 1-inch increments, from 20 to 36 inches in 2-inch increments, and from 36 to 60 inches in 4-inch increments. The minimum thickness of gal-

vanized sheet and fittings for round ducts in duct systems in commercial buildings is listed in Table 8.4.

Many industrial air pollution control systems often require a velocity around 3000 fpm or higher to transport particulates. Round ducts with thicker metal sheets are usually used in such applications.

Flat Oval Ducts

Flat oval ducts, as shown in Fig. 8.4, have a cross-sectional shape between rectangular and round. They share the advantages of both the round and the rectangular duct with fewer large-scale air turbulences and a small depth of space required during installation. Flat oval ducts are quicker to install and have lower air leakage because of the factory fabrication.

Flat oval ducts are made in either spiral seam or longitudinal seam. The minimum thickness of the galvanized sheet and fittings for flat oval duct systems used in commercial buildings is presented in Table 8.5.

Flexible Ducts

Flexible ducts are often used to connect the main duct or the diffusers to the terminal box. Their flexibility and ease of removal allow allocation and relocation of the terminal devices.

Flexible ducts are usually made of multi-ply polyester film reinforced by a helical steel wire core or corrugated aluminum spiral strips. The duct is of-

TABLE 8.4
Round ferrous metal duct construction for duct systems in commercial buildings

Duct diameter, in.	Minimum galvanized steel thickness, in.						
	Pressure, −2 in. WG			Pressure, +2 in. WG			
	Spiral seam duct	Longitudinal seam duct	Fittings	Spiral seam duct	Longitudinal seam duct	Fittings	Suggested type of joint
Up through 8	0.0157	0.0236	0.0236	0.0157	0.0157	0.0187	Beaded slip
14	0.0187	0.0236	0.0236	0.0157	0.0187	0.0187	Beaded slip
26	0.0236	0.0296	0.0296	0.0187	0.0236	0.0236	Beaded slip
36	0.0296	0.0356	0.0356	0.0236	0.0296	0.0296	Beaded slip
50	0.0356	0.0466	0.0466	0.0296	0.0356	0.0356	Flange
60	0.0466	0.0575	0.0575	0.0356	0.0466	0.0466	Flange
84	0.0575	0.0705	0.0705	0.0466	0.0575	0.0575	Flange

Adapted with permission from SMACNA, *HVAC Duct Construction Standards—Metal and Flexible*. Refer to SMACNA standard for complete details.

TABLE 8.5
Flat-oval duct construction for positive pressure duct systems in commercial buildings

Major axis, in.	Minimum galvanized steel thickness, in.			Suggested type of joint
	Spiral seam duct	Longitudinal seam duct	Fittings	
Up through 24	0.0236	0.0356	0.0356	Beaded slip
36	0.0296	0.0356	0.0356	Beaded slip
48	0.0296	0.0466	0.0466	Flange
60	0.0356	0.0466	0.0466	Flange
70	0.0356	0.0575	0.0575	Flange
Over 70	0.0466	0.0575	0.0575	Flange

Adapted with permission from SMACNA, *HVAC Duct Construction Standards—Metal and Flexible.* Refer to SMACNA Standard for complete details.

ten insulated by a fiberglass blanket 1 or 2 in. thick. The outer surface of the flexible duct is usually covered with aluminum foil or other type of vapor barrier to prevent the permeation of water vapor into the insulation layer.

The inside diameter of flexible ducts may range from 2 to 10 inches in 1-inch increments and from 12 to 20 inches in 2-inch increments. The flexible duct should be as short as possible, and its length should be fully extended to minimize flow resistance.

Fiberglass Ducts

Fiberglass duct boards are usually made in 1-in. thicknesses. They are fabricated into rectangular ducts by closures. Fiberglass duct with a 1.5 in. thickness may be used in the Gulf area of the United States where the climate is hot and humid in summer to minimize duct heat gain. Round molded fiberglass ducts are sometimes used.

Fiberglass ducts have a good thermal performance. For a 1 in. thick duct board, the *U*-value is 0.21 Btu/h · ft^2 · °F at 2000 fpm air velocity, which is better than a galvanized sheet metal duct with a 1 in. inner liner. Fiberglass duct has good sound attenuation characteristics. Its air leakage is usually 5 percent or less, which is far less than a sheet metal rectangular duct that is not well sealed. Another important advantage of fiberglass duct is its lower cost.

The closures, also called taping systems, are tapes used to form rectangular duct sections from duct boards and to join the sections and fittings into an integrated duct system. The improved acrylic pressure-sensitive tapes provide a better bond

than before. Heat-sensitive solid polymer adhesive closures show themselves to be good sealing tapes even if dust, oil, or water is present on the surface of the duct board. Mastic and glass fabric closures are also used in many fiberglass duct systems.

Tests show that the ongoing emission of fiberglass from the duct board was less than that contained in outdoor air. Fiberglass ducts have a slightly higher friction loss than galvanized sheet duct (0.03 in. WG greater for a length of 100 ft). They are also not as strong as metal sheet duct. They must be handled carefully to prevent damage during installation.

More and more applications for fiberglass ducts are being found. They are usually used in duct systems with a pressure differential of ±2 in. WG. Many codes restrict the use of fiberglass in sensitive areas such as operating rooms and maternity wards.

8.4 DUCT HEAT GAIN, HEAT LOSS, AND DUCT INSULATION

Temperature Rise or Drop Due to Duct Heat Gain or Loss

The temperature rise or drop from duct heat gain or loss is one of the parameters that affects the supply air temperature, as well as the supply volume flow rate in the air conditioning system design. Heat gain or loss through the duct wall of a rectangular duct section with a constant volume flow rate q_d, in Btu/h, can be calculated as

$$q_d = UPL\left(T_{am} - \frac{T_{en} + T_{lv}}{2}\right) \qquad (8.26)$$

For a round duct section

$$q_d = U \pi D_d L \left(T_{am} - \frac{T_{en} + T_{lv}}{2} \right) \qquad (8.27)$$

where U = overall heat transfer coefficient of the duct wall, Btu/h · ft² · °F
P, L = perimeter and length of the duct, ft
D_d = diameter of the round duct, ft
T_{am} = temperature of the ambient air, °F
T_{en}, T_{lv} = temperature of the air entering and leaving the duct section, °F

The temperature increase or drop of the air flowing through a duct section is given as

$$T_{lv} - T_{en} = \frac{q_d}{60 \, A_d v \rho_s c_{pa}} \qquad (8.28)$$

where A_d = cross-sectional area of the duct, ft². In Eq. (8.28), the mean air velocity v is expressed in fpm.

Substitute Eq. (8.26) into Eq. (8.28), and for rectangular ducts let

$$y = \frac{120 \, A_d v \rho_s c_{pa}}{UPL} \qquad (8.29)$$

For round duct, let

$$y = \frac{30 \, D_d v \rho_s c_{pa}}{UL} \qquad (8.30)$$

Then the temperature of air leaving the duct section

$$T_{lv} = \frac{2T_{am} + T_{en}(y - 1)}{y + 1} \qquad (8.31)$$

Duct Insulation

Duct insulation is mounted or inner-lined to reduce heat loss and heat gain as well as to prevent condensation on the outer surface of the duct. It is usually in the form of duct wrap, duct liner, or fiberglass duct boards. Duct liner provides both thermal insulation and sound attenuation. The thickness of an insulation layer is based on economic analysis.

ASHRAE/IES Standard 90.1–1989 requires that all ducts and plenums installed as part of an air distribution system inside buildings must be insulated to provide a R-value in h · ft² · °F/Btu, excluding air film resistances, calculated as

$$\begin{array}{lll} TD \le 15°F; & \text{None required} & \\ 40 \ge TD > 15°F; & R = 3.3 & (8.32) \\ TD > 40°F; & R = 5.0 & \end{array}$$

where TD = design temperature differential between the air inside the duct and the ambient air where the duct is located, °F.

For ducts and plenums outside the building, R-values are as follows:

CDD65	HDD65	R-value, h · ft² · °F/Btu
<500	<1500	3.3
500 to 1150	1500-4500	5.0
1151 to 2000	4501-7500	6.5
>2000	>7500	8.0

The preceding requirements do not consider water vapor transmission and surface condensation. The recommended thickness of insulation layer, or duct wrap, for R = 3.3 h · ft² · °F/Btu is 1.5 in. and for R = 5 h · ft² · °F/Btu, 2 to 3 in.

If the temperature of the ambient air is 80°F with a relative humidity of 60 percent, its dew point is 65°F. Only when the outer surface temperature of the duct $T_{sd} < 65°F$ will condensation not occur.

Temperature Rise Curves

Duct heat gain or loss and the temperature rise or drop inside the duct depend on air velocity, duct dimensions, and duct insulation. In Fig 8.5 are shown curves for temperature rise in round ducts. These curves are calculated according to Eqs. (8.28) to (8.31) under these conditions:

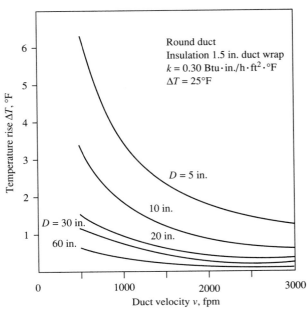

FIGURE 8.5 Temperature rise curves from duct heat gain.

- Thickness of the insulation layer of duct-wrap is 1.5 in. and the thermal conductivity of the insulating material $k = 0.30$ Btu · in./h · ft^2 · °F.
- Heat transfer coefficient of the outer surface of the duct $h = 1.6$ Btu/h · ft^2 · °F.
- Convective heat transfer coefficient of the inside surface h_c, in Btu/h · ft^2 · °F, can be calculated by

$$h_c = 0.023 \frac{k}{D_d} \mathrm{Re}_D^{0.8} \mathrm{Pr}^{0.4} \qquad (8.33)$$

where Re_D = Reynolds number based on the duct diameter as the characteristic length
Pr = Prandtl number

- Air temperature inside the air duct is assumed to be 55°F.
- Temperature difference between the air inside the duct and the ambient air surrounding the duct is 25°F.

If the temperature rise or drop has been determined, duct heat gain or heat loss can either be expressed in percentage of supply temperature differential or calculated by Eqs. (8.26) and (8.27).

8.5 FRICTIONAL LOSSES

In an air duct system, there are two types of resistance against air flow: frictional losses and dynamic losses.

Darcey-Weisbach Equation

Frictional losses result from the existence of shearing stress between the fluid layers of the laminar sublayer, which is adjacent to the surface of the duct wall. Friction also exists when the fluid particles in the turbulent flow bump against protuberances of the duct wall. These lead to the production of eddies and energy loss. Frictional losses occur along the entire length of the air duct.

For a steady, incompressible fluid flow in a round duct or a circular pipe, frictional head loss H_f, in ft of air column, can be calculated by the Darcy-Weisbach equation in the following form:

$$H_f = f \frac{L}{D} \frac{v^2}{2g} \qquad (8.34)$$

If the frictional loss is presented in the form of pressure loss Δp_f, in lbf/ft^2, then

$$\Delta p_f = f \frac{L}{D} \frac{\rho v^2}{2g_c} \qquad (8.35)$$

where f = friction factor
L = length of the duct or the pipe, ft
D = diameter of the duct or pipe, ft
v = mean air velocity in duct, ft/s

If air velocity v is expressed in fpm and Δp_f is expressed in inches WG, then a conversion factor $F_{cv} = (1/5.19) \times (1/60)^2 = 0.0000535$ should be used. That is,

$$\Delta p_f = 0.0000535 f \frac{L}{D} \frac{\rho v^2}{2g_c} \qquad (8.36)$$

In Eqs. (8.34), (8.35), and (8.36), strictly speaking, D should be replaced by hydraulic diameter D_h. But for round ducts and pipes, $D = D_h$.

Friction Factor

Frictional loss is directly proportional to the friction factor f, which is dimensionless. The relationship between f and the parameters that influence the magnitude of the friction factor is shown in Fig. 8.6, which is called a Moody diagram. In Fig. 8.6, the term ϵ represents the absolute roughness of the surface protuberances, expressed in ft, as shown in Fig. 8.7. The ratio ϵ/D indicates the relative roughness of the duct or pipe.

For laminar flow in an air duct when $\mathrm{Re}_D < 2000$, f is affected mainly by the viscous force of the fluid flow and, therefore, is a function of Re_D only. That is,

$$f = \frac{64}{\mathrm{Re}_D} \qquad (8.37)$$

In an ideal smooth tube or duct, that is, $\epsilon/D < 10^{-5}$, if $\mathrm{Re}_D > 4000$, the surface roughness is submerged in the laminar sublayer with a thickness of δ, and the fluid moves smoothly, passing over the protuberances. In this case, f decreases with an increase of Re_D. The relationship between f and Re_D may be expressed by the Blasius empirical formula

$$f = \frac{0.316}{\mathrm{Re}_D^{0.25}} \qquad (8.38)$$

With a further increase of Re_D, the laminar sublayer becomes thinner, even thinner than the height of the irregularities ϵ, that is, $\delta < \epsilon$. The protuberances form the separation of fluid flow, enhance the formation of vortices, and, therefore, increase the pressure loss as well as the value of f at greater Re_D.

If Re_D goes beyond a limit called the *Rouse limit*, f depends mainly on the relative roughness of the duct wall. The Rouse limit line can be determined by $\mathrm{Re}_D = 200/[\sqrt{f}(\epsilon/D)]$, as shown in Fig. 8.6.

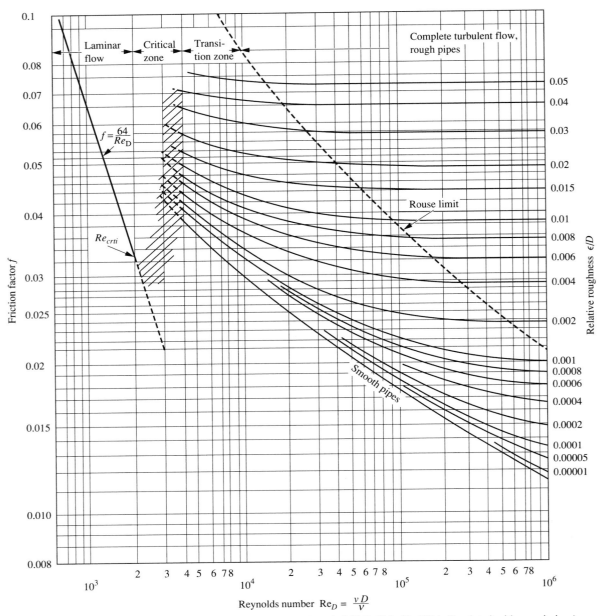

FIGURE 8.6 Moody diagram. (*Source*: Moody, L. F. *Transactions A.S.M.E. Vol. 66*, 1944. Reprinted with permission.)

Duct Friction Chart

In most air ducts, Re_D ranges from 1×10^4 to 2×10^6, and ϵ/D may vary from 0.005 to 0.00015. This covers a transition zone between hydraulic smooth pipes and the Rouse limit line. Within this region, f is a function of both Re_D and ϵ/D. Colebrook (1939) recommended the following empirical formula to relate Re_D, ϵ/D, and f for air ducts:

$$\frac{1}{\sqrt{f}} = -2 \, \log\left(\frac{\epsilon}{3.7D} + \frac{2.51}{\mathrm{Re}_D \sqrt{f}}\right) \qquad (8.39)$$

In Eq. (8.39), ϵ and D must be expressed in the same units. Swamee and Jain suggested the use of an explicit expression that can give nearly the same value of f as Colebrook's formula:

$$f = \frac{0.25}{\left[\log \dfrac{\epsilon}{3.7D} + \dfrac{5.74}{0.9\mathrm{Re}_D}\right]^2} \qquad (8.40)$$

For practical calculations, a friction chart for round ducts, developed by Wright, in the form shown in Fig. 8.8, is widely used. In this chart, air volume flow rate $\dot{V}$, in cfm, and the frictional loss per unit length Δp_f,

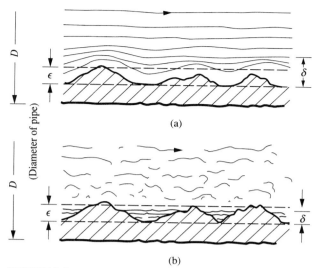

FIGURE 8.7   Modes of air flow when air passes over and around surface protuberances of the duct wall: (a) $\delta > \epsilon$, and (b) $\delta < \epsilon$.

in inches WG per 100 ft, are used as coordinates. The mean air velocity v, in fpm, and the duct diameter are shown as inclined lines in this chart.

The duct friction chart can be applied to the following conditions without corrections:

- A temperature from 41°F to 95°F
- Up to an elevation of 1600 ft above sea level
- Duct pressure of ± 20 in. WG with respect to ambient air pressure
- Duct material of medium roughness

Roughness and Temperature Corrections

The absolute roughness ϵ, in ft, given in *ASHRAE Handbook 1989, Fundamentals*, is listed in Table 8.6. When duct material differs from medium roughness, or when the air duct is installed at an elevation above 1600 ft, or when warm air is supplied at a temperature higher than 95°F, corrections should be made to Δp_f as follows:

$$\Delta p_f = K_{sr} K_T K_{el} \Delta p_{f.c} \qquad (8.41)$$

where $\Delta p_{f.c}$ = friction loss found from the duct friction chart, in. WG. In Eq. (8.41), K_{sr} indicates the

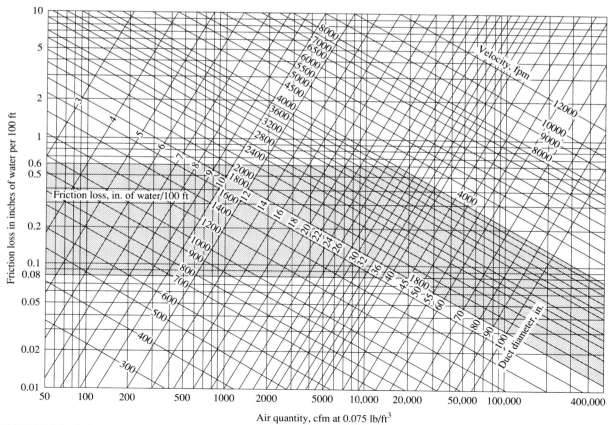

FIGURE 8.8   Friction chart for round ducts. (*Source*: *ASHRAE Handbook 1989, Fundamentals*. Reprinted by permission.)

TABLE 8.6
Duct roughness

Duct material (roughness, ft)	Roughness category	Absolute roughness ϵ, ft
Uncoated carbon steel, clean (0.00015 ft) PVC plastic pipe (0.00003–0.00015 ft) Aluminum (0.00015–0.0002 ft)	Smooth	0.0001

Galvanized steel, longitudinal seams, 4 ft joints
  (0.00016–0.00032 ft)
Galvanized steel, spiral seams, with 1, 2, and 3
  ribs, 12 ft joints (0.00018–0.00038 ft)
	Medium smooth	0.0003

Galvanized steel, longitudinal seams, 2.5 ft joints (0.0005 ft)	Average	0.0005
Fibrous glass duct, rigid Fibrous glass duct liner, air side with facing material (0.0005 ft)	Medium rough	0.003

Fibrous glass duct liner, air side spray-coated
  (0.015 ft)
Flexible duct, metallic (0.004–0.007 ft when fully
  extended)
Flexible duct, all types of fabric and wire
  (0.0035–0.015 ft when fully extended)
Concrete (0.001–0.01 ft)
	Rough	0.01

Source: ASHRAE Handbook 1989, *Fundamentals.* Reprinted with permission.

correction factor for surface roughness, which is dimensionless, and can be calulated as

$$K_{sr} = f_a / f_c \qquad (8.42)$$

Here, f_c denotes the friction factor of duct material of surface roughness that is specified by the duct friction chart, that is, $\epsilon = 0.0005$ ft. The symbol f_a represents the actual friction factor of duct material with surface roughness differing from the f_c. Both f_a and f_c can be calculated from Eq. (8.39) or (8.40).

The term K_T indicates the correction factor for air temperature inside the duct, which affects the density of the air; K_T is dimensionless and can be calculated as

$$K_T = \left(\frac{530}{T_a + 460}\right)^{0.825} \qquad (8.43)$$

where T_a = actual air temperature inside the duct, °F. The term K_{el} indicates the correction factor for elevation, which also affects the density of the air; it is dimensionless. When the elevation is greater than 1600 ft, K_{el} can be calculated as

$$K_{el} = \frac{p_{at}}{29.92} \qquad (8.44)$$

where p_{at} = actual atmospheric or barometric pressure, in. Hg.

Example 8.2. A fabric and wire flexible duct of 8 in. diameter installed in a commercial building at sea level has a surface roughness $\epsilon = 0.12$ in. The mean air velocity inside the air duct is 800 fpm. Find the correction factor of surface roughness. The viscosity of air at 60°F is 1.21×10^{-5} lb/ft · s.

Solution

1. For the galvanized sheet duct specified in the duct friction chart, $\epsilon = 0.0005$ ft, or 0.0060 in. Then

$$Re_D = \frac{\rho v D}{\mu} = \frac{0.075 \times 800 \times 8}{60 \times 12 \times 1.21 \times 10^{-5}}$$

$$= 5.5 \times 10^4$$

From Eq. (8.40),

$$f_c = \frac{0.25}{\left[\log \dfrac{\epsilon}{3.7D} + \dfrac{5.74}{0.9Re_D}\right]^2}$$

$$= \frac{0.25}{\left[\log \dfrac{0.0060}{3.7 \times 8} + \dfrac{5.74}{0.9 \times 5.5 \times 10^4}\right]^2} = 0.0183$$

2. For the fabric and wire flexible duct, Re_D is the same as for the galvanized sheet duct. From the data given, $\epsilon = 0.12$ in., so the actual friction factor is

$$f_a = \frac{0.25}{\left[\log \dfrac{0.12}{3.7 \times 8} + \dfrac{5.74}{0.9 \times 5.5 \times 10_4^2}\right]} = 0.044$$

3. From Eq. (8.42), the correction factor of surface roughness is

$$K_{\text{sr}} = \frac{f_a}{f_c} = \frac{0.044}{0.0190} = 2.32$$

Circular Equivalents

In Eq. (8.35), if D is replaced by D_h, then

$$\Delta p_f = f \frac{L}{D_h} \frac{\rho v^2}{2g_c}$$

Apparently, for circular or noncircular air ducts with different cross-sectional shapes and the same hydraulic diameter, the pressure loss is the same for an equal length of air duct at equal mean air velocities. Circular equivalents are used to convert the dimension of a noncircular duct into an equivalent diameter D_e, in inches, of a round duct when their volume flow rates $\dot{V}$ and frictional losses per unit length $\Delta p_{\text{f.u}}$ are equal. A noncircular duct must be converted to a circular equivalent first before determining its $\Delta p_{\text{f.u}}$ from the duct friction chart. The hydraulic diameter D_h, in inches, is defined as

$$D_h = \frac{4A}{P} \qquad (8.45)$$

where A = area, in.^{2}
$\quad\quad P$ = perimeter, in.

Based on experimental results, Huebscher (1948) recommended the following formula to calculate D_e for rectangular duct at equal $\dot{V}$ and $\Delta p_{\text{f.u}}$:

$$D_e = \frac{1.30(ab)^{0.625}}{(a+b)^{0.25}} \qquad (8.46)$$

where a, b = dimensions of the two sides of the rectangular duct, in.. The circular equivalents of rectangular ducts at various dimensions calculated by Eq. (8.46) are listed in Table 8.7.

For galvanized steel flat oval ducts with spiral seams, Heyt and Diaz (1975) proposed the following formula to calculate the circular equivalent for use of the duct friction chart:

$$D_e = \frac{1.55A^{0.625}}{P^{0.25}} \qquad (8.47)$$

Here A is the cross-sectional area of the flat oval duct, in.^{2}, and is given as

$$A = \frac{\pi b^2}{4} + b(a - b) \qquad (8.48)$$

and the perimeter P, in inches, is calculated as

$$P = \pi b + 2(a - b) \qquad (8.49)$$

The dimensions a and b of flat oval duct are shown in Fig. 8.4c.

8.6 DYNAMIC LOSSES

When air flows through duct fittings, such as elbows, tees, diffusers, contractions, entrances, and exits, or certain equipment, a change in velocity or direction of flow may occur. Such a change leads to flow separation and the formation of eddies and disturbances in that area. The energy loss resulting from these eddies and disturbances is called *dynamic loss* Δp_{dy}, in inches WG. Although a duct fitting is fairly short, the disturbances it produces may persist over a considerable distance downstream.

In addition to the presence of dynamic loss Δp_{dy}, frictional loss Δp_f occurs when an airstream flows through a duct fitting. For convenience in calculation, the length of the duct fitting is usually added to the adjacent duct sections connected with this duct fitting of the same mean air velocity.

When airstreams of the same Reynolds number flow through geometrically similar duct fittings, that is, in dynamic similarity, the dynamic loss is proportional to their velocity pressure p_v. Dynamic loss may be calculated as

$$\Delta p_{\text{dy}} = C_o p_{\text{v}} = \frac{C_o \rho v_o^2}{2g_c \text{cf}} = C_o \left(\frac{v}{4005}\right)^2 \qquad (8.50)$$

where C_o = local loss coefficient or dynamic loss coefficient
$\quad\quad v_o$ = mean air velocity of the airstream at reference cross section o, fpm
$\quad\quad$ cf = conversion factor

Because the mean velocity of the airstream may vary at different ends of a duct fitting, C_o is always specified with respect to a velocity of a reference cross section o in the duct fitting.

Elbows

An *elbow* is a duct fitting in which the air flow changes direction. Elbows are shown in Figs. 8.9 and 8.10.

Consider an elbow that makes a 90° turn in a round duct, as shown in Fig. 8.9a. Because of the change of airstream direction, centrifugal force is created and acts towards the outer wall of the duct. When the airstream flows from the straight part of the duct to the curved part, it is accompanied by an increase in pressure and a decrease in air velocity at the outer wall. At the same time, a decrease in pressure and an increase in air velocity takes place at the inner wall. Therefore, a diffuser effect occurs near the outer wall and

TABLE 8.7
Circular equivalents of rectangular ducts for equal friction and capacity

One side, rectangular duct, in.	Adjacent side, in.																
	4.0	4.5	5.0	5.5	6.0	6.5	7.0	7.5	8.0	9.0	10.0	11.0	12.0	13.0	14.0	15.0	16.0
3.0	3.8	4.0	4.2	4.4	4.6	4.7	4.9	5.1	5.2	5.5	5.7	6.0	6.2	6.4	6.6	6.8	7.0
3.5	4.1	4.3	4.6	4.8	5.0	5.2	5.3	5.5	5.7	6.0	6.3	6.5	6.8	7.0	7.2	7.5	7.7
4.0	4.4	4.6	4.9	5.1	5.3	5.5	5.7	5.9	6.1	6.4	6.7	7.0	7.3	7.6	7.8	8.0	8.3
4.5	4.6	4.9	5.2	5.4	5.7	5.9	6.1	6.3	6.5	6.9	7.2	7.5	7.8	8.1	8.4	8.6	8.8
5.0	4.9	5.2	5.5	5.7	6.0	6.2	6.4	6.7	6.9	7.3	7.6	8.0	8.3	8.6	8.9	9.1	9.4
5.5	5.1	5.4	5.7	6.0	6.3	6.5	6.8	7.0	7.2	7.6	8.0	8.4	8.7	9.0	9.3	9.6	9.9

One side, rectangular duct, in.	6	7	8	9	10	11	12	13	14	15	16	17	18	19	20	22	24	26	28	30	Side rectangular duct
6	6.6																				6
7	7.1	7.7																			7
8	7.6	8.2	8.7																		8
9	8.0	8.7	9.3	9.8																	9
10	8.4	9.1	9.8	10.4	10.9																10
11	8.8	9.5	10.2	10.9	11.5	12.0															11
12	9.1	9.9	10.7	11.3	12.0	12.6	13.1														12
13	9.5	10.3	11.1	11.8	12.4	13.1	13.7	14.2													13
14	9.8	10.7	11.5	12.2	12.9	13.5	14.2	14.7	15.3												14
15	10.1	11.0	11.8	12.6	13.3	14.0	14.6	15.3	15.8	16.4											15
16	10.4	11.3	12.2	13.0	13.7	14.4	15.1	15.7	16.4	16.9	17.5										16
17	10.7	11.6	12.5	13.4	14.1	14.9	15.6	16.2	16.8	17.4	18.0	18.6									17
18	11.0	11.9	12.9	13.7	14.5	15.3	16.0	16.7	17.3	17.9	18.5	19.1	19.7								18
19	11.2	12.2	13.2	14.1	14.9	15.7	16.4	17.1	17.8	18.4	19.0	19.6	20.2	20.8							19
20	11.5	12.5	13.5	14.4	15.2	16.0	16.8	17.5	18.2	18.9	19.5	20.1	20.7	21.3	21.9						20
22	12.0	13.0	14.1	15.0	15.9	16.8	17.6	18.3	19.1	19.8	20.4	21.1	21.7	22.3	22.9	24.0					22
24	12.4	13.5	14.6	15.6	16.5	17.4	18.3	19.1	19.9	20.6	21.3	22.0	22.7	23.3	23.9	25.1	26.2				24
26	12.8	14.0	15.1	16.2	17.1	18.1	19.0	19.8	20.6	21.4	22.1	22.9	23.5	24.2	24.9	26.1	27.3	28.4			26
28	13.2	14.5	15.6	16.7	17.7	18.7	19.6	20.5	21.3	22.1	22.9	23.7	24.4	25.1	25.8	27.1	28.3	29.5	30.6		28
30	13.6	14.9	16.1	17.2	18.3	19.3	20.2	21.1	22.0	22.9	23.7	24.4	25.2	25.9	26.6	28.0	29.3	30.5	31.7	32.8	30
32	14.0	15.3	16.5	17.7	18.8	19.8	20.8	21.8	22.7	23.5	24.4	25.2	26.0	26.7	27.5	28.9	30.2	31.5	32.7	33.9	32
34	14.4	15.7	17.0	18.2	19.3	20.4	21.4	22.4	23.3	24.2	25.1	25.9	26.7	27.5	28.3	29.7	31.0	32.4	33.7	34.9	34
36	14.7	16.1	17.4	18.6	19.8	20.9	21.9	22.9	23.9	24.8	25.7	26.6	27.4	28.2	29.0	30.5	32.0	33.3	34.6	35.9	36
38	15.0	16.5	17.8	19.0	20.2	21.4	22.4	23.5	24.5	25.4	26.4	27.2	28.1	28.9	29.8	31.3	32.8	34.2	35.6	36.8	38
40	15.3	16.8	18.2	19.5	20.7	21.8	22.9	24.0	25.0	26.0	27.0	27.9	28.8	29.6	30.5	32.1	33.6	35.1	36.4	37.8	40
42	15.6	17.1	18.5	19.9	21.1	22.3	23.4	24.5	25.6	26.6	27.6	28.5	29.4	30.3	31.2	32.8	34.4	35.9	37.3	38.7	42
44	15.9	17.5	18.9	20.3	21.5	22.7	23.9	25.0	26.1	27.1	28.1	29.1	30.0	30.9	31.8	33.5	35.1	36.7	38.1	39.5	44
46	16.2	17.8	19.3	20.6	21.9	23.2	24.4	25.5	26.6	27.7	28.7	29.7	30.6	31.6	32.5	34.2	35.9	37.4	38.9	40.4	46
48	16.5	18.1	19.6	21.0	22.3	23.6	24.8	26.0	27.1	28.2	29.2	30.2	31.2	32.2	33.1	34.9	36.6	38.2	39.7	41.2	48
50	16.8	18.4	19.9	21.4	22.7	24.0	25.2	26.4	27.6	28.7	29.8	30.8	31.8	32.8	33.7	35.5	37.2	38.9	40.5	42.0	50
52	17.1	18.7	20.2	21.7	23.1	24.4	25.7	26.9	28.0	29.2	30.3	31.3	32.3	33.3	34.3	36.2	37.9	39.6	41.2	42.8	52
54	17.3	19.0	20.6	22.0	23.5	24.8	26.1	27.3	28.5	29.7	30.8	31.8	32.9	33.9	34.9	36.8	38.6	40.3	41.9	43.5	54
56	17.6	19.3	20.9	22.4	23.8	25.2	26.5	27.7	28.9	30.1	31.2	32.3	33.4	34.4	35.4	37.4	39.2	41.0	42.7	44.3	56
58	17.8	19.5	21.2	22.7	24.2	25.5	26.9	28.2	29.4	30.6	31.7	32.8	33.9	35.0	36.0	38.0	39.8	41.6	43.3	45.0	58
60	18.1	19.8	21.5	23.0	24.5	25.9	27.3	28.6	29.8	31.0	32.2	33.3	34.4	35.5	36.5	38.5	40.4	42.3	44.0	45.7	60
62		20.1	21.7	23.3	24.8	26.3	27.6	28.9	30.2	31.5	32.6	33.8	34.9	36.0	37.1	39.1	41.0	42.9	44.7	46.4	62
64		20.3	22.0	23.6	25.1	26.6	28.0	29.3	30.6	31.9	33.1	34.3	35.4	36.5	37.6	39.6	41.6	43.5	45.3	47.1	64
66		20.6	22.3	23.9	25.5	26.9	28.4	29.7	31.0	32.3	33.5	34.7	35.9	37.0	38.1	40.2	42.2	44.1	46.0	47.7	66
68		20.8	22.6	24.2	25.8	27.3	28.7	30.1	31.4	32.7	33.9	35.2	36.3	37.5	38.6	40.7	42.8	44.7	46.6	48.4	68
70		21.0	22.8	24.5	26.1	27.6	29.1	30.4	31.8	33.1	34.4	35.6	36.8	37.9	39.1	41.2	43.3	45.3	47.2	49.0	70
72			23.1	24.8	26.4	27.9	29.4	30.8	32.2	33.5	34.8	36.0	37.2	38.4	39.5	41.7	43.8	45.8	47.8	49.6	72
74			23.3	25.1	26.7	28.2	29.7	31.2	32.5	33.9	35.2	36.4	37.7	38.8	40.0	42.2	44.4	46.4	48.4	50.3	74
76			23.6	25.3	27.0	28.5	30.0	31.5	32.9	34.3	35.6	36.8	38.1	39.3	40.5	42.7	44.9	47.0	48.9	50.9	76
78			23.8	25.6	27.3	28.8	30.4	31.8	33.3	34.6	36.0	37.2	38.5	39.7	40.9	43.2	45.4	47.5	49.5	51.4	78
80			24.1	25.8	27.5	29.1	30.7	32.2	33.6	35.0	36.3	37.6	38.9	40.2	41.4	43.7	45.9	48.0	50.1	52.0	80
82				26.1	27.8	29.4	31.0	32.5	34.0	35.4	36.1	38.0	39.3	40.6	41.8	44.1	46.4	48.5	50.6	52.6	82
84				26.4	28.1	29.7	31.3	32.8	34.3	35.7	37.1	38.4	39.7	41.0	42.2	44.6	46.9	49.0	51.1	53.2	84
86				26.6	28.3	30.0	31.6	33.1	34.6	36.1	37.4	38.8	40.1	41.4	42.6	45.0	47.3	49.6	51.6	53.7	86
88				26.9	28.6	30.3	31.9	33.4	34.9	36.4	37.8	39.2	40.5	41.8	43.1	45.5	47.8	50.0	52.2	54.3	88
90				27.1	28.9	30.6	32.2	33.8	35.3	36.7	38.2	39.5	40.9	42.2	43.5	45.9	48.3	50.5	52.7	54.8	90
92					29.1	30.8	32.5	34.1	35.6	37.1	38.5	39.9	41.3	42.6	43.9	46.4	48.7	51.0	53.2	55.3	92
96					29.6	31.4	33.0	34.7	36.2	37.7	39.2	40.6	42.0	43.3	44.7	47.2	49.6	52.0	54.2	56.4	96

TABLE 8.7
Circular equivalents of rectangular ducts for equal friction and capacity (*Continued*)

One side, rectangular duct, in.	Adjacent side, in.																				Side rectangular duct
	32	34	36	38	40	42	44	46	48	50	52	56	60	64	68	72	76	80	84	88	
32	35.0																				32
34	36.1	37.2																			34
36	37.1	38.2	39.4																		36
38	38.1	39.3	40.4	41.5																	38
40	39.0	40.3	41.5	42.6	43.7																40
42	40.0	41.2	42.5	43.7	44.8	45.9															42
44	40.9	42.2	43.5	44.7	45.8	47.0	48.1														44
46	41.8	43.1	44.4	45.7	46.9	48.0	49.2	50.3													46
48	42.6	44.0	45.3	46.6	47.9	49.1	50.2	51.4	52.5												48
50	43.6	44.9	46.2	47.5	48.8	50.0	51.2	52.4	53.6	54.7											50
52	44.3	45.7	47.1	48.4	49.7	51.0	52.2	53.4	54.6	55.7	56.8										52
54	45.1	46.5	48.0	49.3	50.7	52.0	53.2	54.4	55.6	56.8	57.9										54
56	45.8	47.3	48.8	50.2	51.6	52.9	54.2	55.4	56.6	57.8	59.0	61.2									56
58	46.6	48.1	49.6	51.0	52.4	53.8	55.1	56.4	57.6	58.8	60.0	62.3									58
60	47.3	48.9	50.4	51.9	53.3	54.7	56.0	57.3	58.6	59.8	61.0	63.4	65.6								60
62	48.0	49.6	51.2	52.7	54.1	55.5	56.9	58.2	59.5	60.8	62.0	64.4	66.7								62
64	48.7	50.4	51.9	53.5	54.9	56.4	57.8	59.1	60.4	61.7	63.0	65.4	67.7	70.0							64
66	49.4	51.1	52.7	54.2	55.7	57.2	58.6	60.0	61.3	62.6	63.9	66.4	68.8	71.0							66
68	50.1	51.8	53.4	55.0	56.6	58.0	59.4	60.8	62.2	63.6	64.9	67.4	69.8	72.1	74.3						68
70	50.8	52.5	54.1	55.7	57.3	58.8	60.3	61.7	63.1	64.4	65.8	68.3	70.8	73.2	75.4						70
72	51.4	53.2	54.8	56.6	58.0	59.6	61.1	62.5	63.9	65.3	66.7	69.3	71.8	74.2	76.5	78.7					72
74	52.1	53.8	55.5	57.2	58.8	60.3	61.9	63.3	64.8	66.2	67.5	70.2	72.7	75.2	77.5	79.8					74
76	52.7	54.5	56.2	57.9	59.5	61.1	62.6	64.1	65.6	67.0	68.4	71.1	73.7	76.2	78.6	80.9	83.1				76
78	53.3	55.1	56.9	58.6	60.2	61.8	63.4	64.9	66.4	67.9	69.3	72.0	74.6	77.1	79.6	81.9	84.2				78
80	53.9	55.8	57.5	59.3	60.9	62.6	64.1	65.7	67.2	68.7	70.1	72.9	75.4	78.1	80.6	82.9	85.2	87.5			80
82	54.5	56.4	58.2	59.9	61.6	63.3	64.9	66.5	68.0	69.5	70.9	73.7	76.4	79.0	81.5	84.0	86.3	88.5			82
84	55.1	57.0	58.8	60.6	62.3	64.0	65.6	67.2	68.7	70.3	71.7	74.6	77.3	80.0	82.5	85.0	87.3	89.6	91.8		84
86	55.7	57.6	59.4	61.2	63.0	64.7	66.3	67.9	69.5	71.0	72.5	75.4	78.2	80.9	83.5	85.9	88.3	90.7	92.9		86
88	56.3	58.2	60.1	61.9	63.6	65.4	67.0	68.7	70.2	71.8	73.3	76.3	79.1	81.8	84.4	86.9	89.3	91.7	94.0	96.2	88
90	56.8	58.8	60.7	62.5	64.3	66.0	67.7	69.4	71.0	72.6	74.1	77.1	79.9	82.7	85.3	87.9	90.3	92.7	95.0	97.3	90
92	57.4	59.3	61.3	63.1	64.9	66.7	68.4	70.1	71.7	73.3	74.9	77.9	80.8	83.5	86.2	88.8	91.3	93.7	96.1	98.4	92
94	57.9	59.9	61.9	63.7	65.6	67.4	69.1	70.8	72.4	74.0	75.6	78.7	81.6	84.4	87.1	89.7	92.3	94.7	97.1	99.4	94
96	58.4	60.5	62.4	64.3	66.2	68.0	69.7	71.5	73.1	74.8	76.3	79.4	82.4	85.3	88.0	90.7	93.2	95.7	98.1	100.5	96

Source: ASHRAE Handbook 1989, Fundamentals. Reprinted with permission.

a bellmouth forms near the inner wall. After turning, the opposite effect takes place as the airstream flows from the curved part to the straight part of the duct. The diffuser effects lead to flow separations from both walls, and eddies and large-scale turbulence form in regions *AB* and *CE* as shown in Fig. 8.9a.

In a rectangular elbow, a radial pressure gradient is also formed by the centrifugal force along its centerline *NR*, as shown in Fig. 8.10c. A secondary circulation is formed along with the main forward airstream.

The magnitude of the local loss coefficient of an elbow is influenced by the following factors:

- Turning angle of the elbow
- Relative radius of curvature R_c/D or R_c/W, where R_c represents the throat radius and W the width of the duct, both in inches
- Installation of splitter vanes, which reduce the eddies and the turbulences in an elbow
- Shape of cross-sectional area of the duct

As the relative radius of curvature becomes greater, the flow resistance of the air stream becomes smaller. However, a greater radius of curvature requires more duct material, a higher labor cost, and a larger

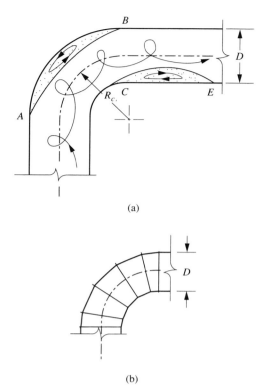

(a)

(b)

FIGURE 8.9 Round elbows: (a) region of eddies and turbulences in a round elbow; and (b) 5-piece 90° round elbow.

allocated space. A value of $R_c/D = 1$ or $R_c/W = 1$ (standard radius elbow) is often used if the space is available.

The installation of splitter vanes, or turning vanes, in rectangular ducts can effectively reduce the pressure loss at the elbow. For a rectangular 90° elbow with $R_c/W = 0.75$ and two splitter vanes as shown in Fig. 8.10a, the local loss coefficient C_o is only 0.04–0.05. The values of R_o, R_1, and R_2 can be found from Table 8.11 on page 8.45. For details, refer to *ASHRAE Handbook 1989, Fundamentals*.

For a mitered rectangular 90° elbow without turning vanes, C_o is about 1.1. If turning vanes are installed, C_o drops to only 0.12 to 0.18. Installation of splitter vanes in mitered elbows is also advantageous for noise control because of the smaller total pressure increase and, therefore, lower fan noise as well as lower air flow noise at the elbow.

Converging and Diverging Tees and Wyes

A branch duct that combines with or diverges from the main duct at an angle of 90° is called a *tee*. However,

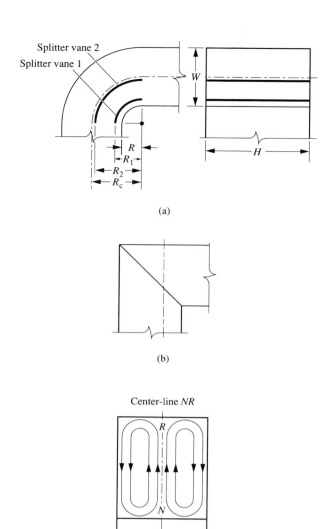

(a)

(b)

Center-line *NR*

(c)

FIGURE 8.10 Rectangular elbows: (a) rectangular elbow, smooth radius, 2 splitter vanes; (b) mitered elbow; and (c) secondary flow in a mitered elbow.

if the angles lie between 15° and 75° it is called a *wye* (Figs. 8.11 and 8.12). Tees and wyes can be round, flat oval, or rectangular in shape. Various types of converging and diverging tees and wyes for round and flat oval ducts appear in Fig. 8.12.

The function of a converging tee or wye is to combine the airstream from the branch duct with the airstream from the main duct. The function of a diverging tee or wye is to diverge part of the air flow from the main duct into the branch take-off.

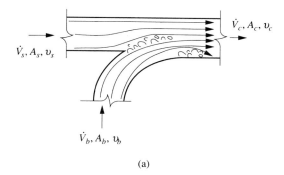

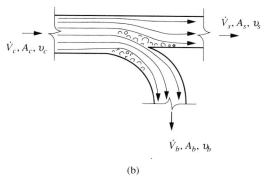

FIGURE 8.11 Converging and diverging wyes: (a) converging wye, rectangular, $\theta = 90°$; and (b) diverging wye, rectangular, $\theta = 90°$.

For airstreams flowing through a converging or diverging tee or wye, the dynamic losses for the main stream can be calculated as

$$\Delta p_{c.s} = \frac{C_{c.s}\rho v_c^2}{2g_c \text{cf}} = C_{c.s}\left(\frac{v_c}{4005}\right)^2$$

$$\Delta p_{s.c} = \frac{C_{s.c}\rho v_c^2}{2g_c \text{cf}} = C_{s.c}\left(\frac{v_c}{4005}\right)^2 \qquad (8.51)$$

For the branch stream,

$$\Delta p_{c.b} = \frac{C_{c.b}\rho v_c^2}{2g_c \text{cf}} = C_{c.b}\left(\frac{v_c}{4005}\right)^2$$

$$\Delta p_{b.c} = \frac{C_{b.c}\rho v_c^2}{2g_c \text{cf}} = C_{b.c}\left(\frac{v_c}{4005}\right)^2 \qquad (8.52)$$

In Eqs. (8.51) and (8.52), subscript c represents the common end, s the straight-through end, and b the branch take-off. The subscript c.s indicates the flow of the main stream from the common end to the straight-through end, and s.c the flow from the straight-through end to the common end. Similarly, c.b denotes the flow of the branch stream from the common end to the branch take-off and b.c the air

flow from the branch duct to the common end. The airstreams flowing through a rectangular converging or diverging tee are shown in Fig. 8.13.

As mentioned in Section 8.2, the total pressure of the main stream may increase when it flows through a converging or diverging wye or tee. This result occurs because energy received from the airstream of higher velocity or the diverging of the slowly moving boundary layer into branch take-off from the main airstream. However, the sum of the energies of the main and branch streams leaving the duct fitting is always smaller than that entering the duct fitting because of the energy losses.

The magnitude of local loss coefficients $C_{c.s}$, $C_{s.c}$, $C_{b.c}$ and $C_{c.b}$ is affected by the shape and construction of the tee or wye, the velocity ratios v_s/v_c and v_b/v_c, the volume flow ratios $\dot{V}_s/\dot{V}_c$ and $\dot{V}_b/\dot{V}_c$, and area ratios A_s/A_c and A_b/A_c.

For rectangular ducts, the converging or diverging tee of the configuration shown in Fig. 8.13 gives the lower C_o and, therefore, less energy loss. For round ducts, the C_o values for various types of diverging tees and wyes with nearly the same outlet direction and the same velocity ratio $v_b/v_c = 0.6$ as shown in Fig. 8.12b are quite different

Tee, diverging, round, with 45° elbow, branch 90° to main	$C_o = 1.60$
Tee, diverging, round, with 90° elbow, branch 90° to main	$C_o = 1.18$
Tee, diverging, round, with 45° elbow, conical branch 90° to main	$C_o = 0.84$
Wye, 45°, diverging, round, with 60° elbow, branch 90° to main	$C_o = 0.68$
Wye, 45°, diverging, round, with 60° elbow, conical branch 90° to main	$C_o = 0.52$

The different C values help to balance the total pressure between different paths of air flow in an air duct system.

In Table 8.11 are listed the local loss coefficients of diverging tees and wye. For details, refer to *ASHRAE Handbook 1989, Fundamentals*.

Entrances, Exits, Enlargements, and Contractions

Entrances and exits are the end openings mounted on a duct or a duct wall, as shown in Fig. 8.14. Because of the change in direction of the streamlines at the entrance, eddies and large-scale turbulences develop along the duct wall when the airstream passes through

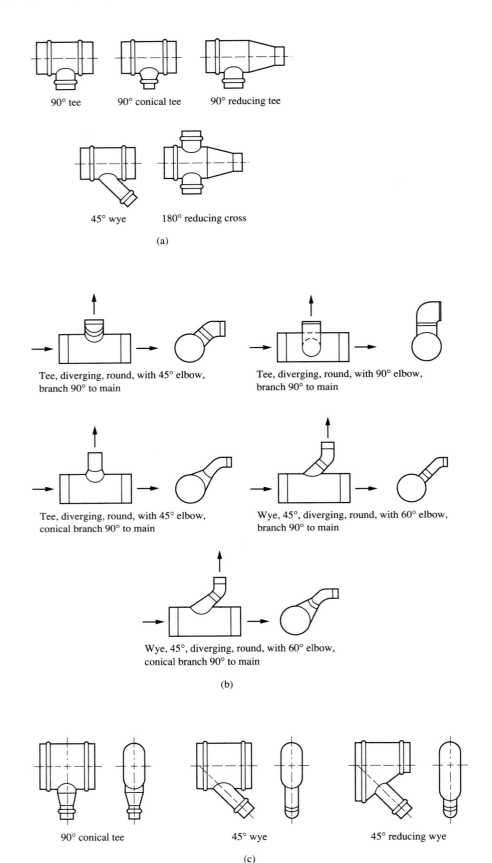

FIGURE 8.12 Round and flat oval tees and wyes: (a) round tees, wyes, and cross; (b) diverging tees and wyes with elbows; and (c) flat oval tees and wyes.

8.24

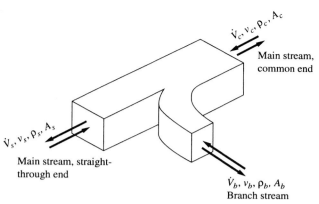

FIGURE 8.13 Air flows through a rectangular converging or a diverging wye.

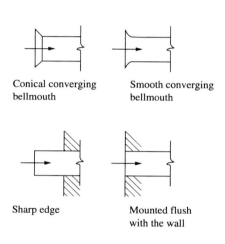

Conical converging
bellmouth

Smooth converging
bellmouth

Sharp edge

Mounted flush
with the wall

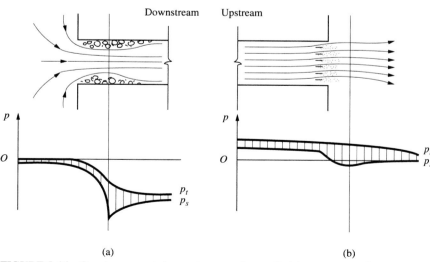

(a)   (b)

FIGURE 8.14 Openings mounted on a duct or a duct wall; (a) entrances; and (b) exits.

the entrance. Generally, the total pressure drop Δp_t of the airstream before it enters the entrance is negligible. A sharp-edged entrance may have a $C_o = 0.9$, whereas for an entrance flush-mounted with the wall, C_o reduces to 0.5. An entrance with a conical converging bellmouth may further reduce C_o to 0.4. If an entrance with a smooth converging bellmouth is installed, C_o could be as low as 0.1.

When the air flows through a wall outlet or an exit, flow separation occurs along the surface of the vanes, wakes are formed downstream, so there is a drop in total pressure. The velocity of the airstream reaches its maximum value at the *vena contracta,* where the cross section of air flow is minimum and the static pressure is negative. The total pressure loss at the outlet always includes the velocity pressure of the discharge airstream.

Various types of return inlets, such as grilles and louvers, and supply outlets, such as diffusers, will be discussed in Chapter 9.

When air flows through an enlargement or a contraction, flow separation occurs and produces eddies and large-scale turbulences after the enlargement, or before and after the contraction. Both cause a total pressure loss Δp_t, as shown in Fig. 8.15. To reduce the energy loss, a gradual expansion, often called a *diffuser,* or converging transition is preferred. An expansion with an including angle of enlargement $\theta = 14°$, as shown in Fig. 8.15a, is ideal. In actual practice, θ may be from 14° to 45° because of limited space. For converging transitions, an including angle of 30° to 60° is usually used.

Reduction of Dynamic Losses

The total of dynamic losses of duct fittings in an air duct system is often greater than the frictional losses. Refer to *ASHRAE Handbook 1989, Fundamentals,* and I. E. Idelchik (1986) for details of local loss coefficients.

To reduce the dynamic losses of the duct path or run that has the highest pressure loss at design flow rates, the following are recommended:

- Maintain an optimum air velocity for air flow through the duct fittings.
- Emphasize the reduction of dynamic losses of duct fittings of higher air velocity; duct fittings that are nearer to the fan outlet often have a higher velocity.
- Use 90° elbows with a R_c/D or R_c/W from 1 to 0.75. If space is not available, use one, two, or three splitter vanes in the elbows with throat radius of $R_c/W = 0.1$; the throat radius should not be smaller than 4 in.
- Set two duct fittings as far apart as allowable; if they are too close together, the eddies and large-scale turbulences of the first duct fitting often affect the velocity distribution in the second duct fitting and considerably increase the pressure loss in the second fitting.

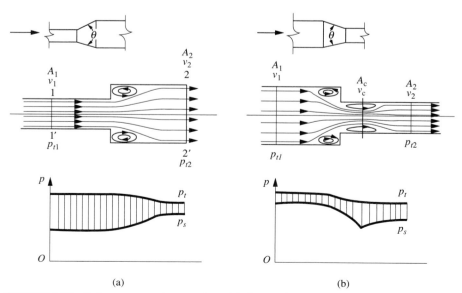

(a)   (b)

FIGURE 8.15   Enlargements and contractions: (a) abrupt enlargement; and (b) sudden contraction.

8.7 PRINCIPLES AND CONSIDERATIONS IN AIR DUCT DESIGN

Optimum Air Duct Design

An optimum air duct system transports the required amount of conditioned, recirculated, or exhaust air to the specific space and meets the following requirements:

- An optimum duct system layout within the allocated space
- A satisfactory system balance, achieved through the pressure balance of various paths by changing duct sizes or using different configurations of duct fittings
- Space sound level lower than the allowable limits
- Optimum energy loss and initial cost
- Installation with only necessary balancing devices such as air dampers and nozzle plates
- National, ASHRAE, and local codes of fire protection, duct construction, and duct insulation met

These requirements result in a design of optimum duct layout, duct size, and total pressure loss of the duct system.

Air duct system design requires comprehensive analysis and computer-aided calculating programs. Different air duct systems have different transport functions and thus have their own characteristics. It is difficult to combine the influences of cost, system balance, and noise together with duct characteristics into one or two representative indices.

Design Velocity

For any air duct system, the nearer the duct section is to the fan outlet or inlet, the greater its volume flow rate. The fan outlet is often the location where maximum air velocity occurs.

For supply air duct systems in high-rise commercial buildings, the maximum air velocity in the supply duct is often determined by the space available between the bottom of the beam and the suspended ceiling, as allocated by the architect, where the main duct traverses under the beam. Because of the impact in recent years of energy-efficient design of HVAC&R systems in commercial buildings (as influenced by ASHRAE standards, local codes, and regulations) the maximum velocity in supply air duct systems usually does not exceed 3000 fpm; however, in special conditions its value may be increased to 3500 fpm. Air flow noise must be checked at dampers, elbows, and branch take-offs to satisfy indoor NC criteria.

Except when there is a surplus pressure, or when a higher branch velocity is required to produce a negative local loss coefficient for the main stream, air velocity in the branch ducts is usually lower than in the main ducts because of the lower volume flow rate and the noise problem.

Higher air velocity results in a higher energy cost, and lower air velocity increases the material and labor costs of the installation. If a commercial building has sufficient headroom in its ceiling plenum, or if an industrial application has enough space at a higher level, an optimization procedure is then possible to reach a compromise between energy and installation costs. This procedure would also include noise reduction measures.

For a particulate-transporting duct system, the air velocity must be higher than a specific value at any section of the duct system to float and transport the particulates.

Design velocities for the components in an air duct system are listed in Table 8.8. The face velocity v_{fc},

TABLE 8.8
Design velocities for air-handling system components

Duct element	Face velocity, fpm
Louvers	
Intake	
1. 7000 cfm and greater	400
2. 2000 to 7000 cfm	250 to 400
Exhaust	
1. 5000 cfm and greater	500
2. Less than 5000 cfm	300 to 400
Filters	
Panel Filters	
1. Viscous impingement	200 to 800
2. Dry-type extended-surface	
Flat (low efficiency)	Duct velocity
Pleated media	up to 750
(medium efficiency)	
HEPA	250
Renewable media filters	
1. Moving-curtain viscous impingement	500
2. Moving-curtain dry-media	200
Electronic Air Cleaners	
Ionizing-type	150 to 350
Heating coils	
Steam and hot water	400 to 600
	(200 min.,
	1500 max.)
Dehumidifying coils	500 to 600
Air washers	
Spray-type	300 to 700
High-velocity-spray-type	1200 to 1800

Adapted with permission from *ASHRAE Handbook 1989, Fundamentals.*

in fpm, is defined as

$$v_{fc} = \frac{\dot{V}}{A_{g.fc}} \qquad (8.53)$$

where $\dot{V}$ = air volume flow rate flowing through the component, cfm

$A_{g.fc}$ = gross face area of the component perpendicular to the air flow, width × height, ft^2

These face velocities are recommended based on the effectiveness in operation of the system component and its optimum pressure loss.

System Balancing

For an air duct system, system balancing means that the volume flow rates from each outlet or flows into each inlet should be (1) equal or nearly equal to the design value for constant-volume systems and (2) equal to the predetermined values at maximum and minimum flow for variable-air-volume systems. System balancing is one of the primary requirements in air duct design. For supply duct systems installed in commercial buildings, using dampers only to provide design air flow often causes additional air leakage, also an increase in installation costs, and in some cases objectionable noise. Therefore, system balancing using dampers only is not recommended.

A typical small supply duct system in which p_t of the conditioned space is equal to zero appears in Fig. 8.16. For such a supply duct system, the pressure and volume flow characteristics are as follows:

- From node 1, the total pressure loss of the airstreams flowing through various paths 1–3, 1–2–4, and 1–2–5 to the conditioned space is always equal, that is,

$$\Delta p_{t,1-3} = \Delta p_{t,1-2-4} = \Delta p_{t,1-2-5} \qquad (8.54)$$

- Volume flow rates $\dot{V}_1$, $\dot{V}_2$, and $\dot{V}_3$ supplied from the branch take-offs 1–3, 2–4, and 2–5 depend on the

size and the configuration of the duct fittings and duct sections in the main duct and in the branch take-offs as well as the characteristics of the supply outlet.

The relationship between the total pressure loss of any duct section Δp_t, the volume flow rate of the duct section $\dot{V}$, and the flow resistance R (discussed in Chapter 19) can be expressed as follows:

$$\Delta p_t = R\dot{V}^2 \qquad (8.55)$$

For duct paths 1–3 and 1–2–45,

$$\Delta p_{t,1-3} = \Delta p_{t,1-2-45}$$
$$= R_{1-3}\dot{V}_1^2 = R_{1-2-45}(\dot{V}_2 + \dot{V}_3)^2$$

Here, path 1–2–45 indicates the air flow having volume flow rate $(\dot{V}_2 + \dot{V}_3)$ and flowing through nodes 1, 2, and a combined parallel path 2–4 and 2–5.

At design conditions, the flow resistance R_{1-2-45} is determined in such a manner that the total pressure loss $\Delta p_{t,1-2-45}$ along path 1–2–45 at a volume flow rate of $(\dot{V}_2 + \dot{V}_3)$ is balanced with the pressure loss $\Delta p_{t,1-3}$.

- If the diverging wye or tee, terminal, duct fittings, and duct section(s) used in path 1–3 are similar to those of paths 2–4 and 2–5, most probably $\dot{V}_1$ will be greater than the required design value as the result of a lower R_{1-3}. In order to have the required $\dot{V}_1$ flowing through path 1–3 at design conditions to provide a system balance, the following means are needed: (1) decrease R_{1-2-45}, including an increase in the size of duct section 1–2; (2) increase R_{1-3}, mainly by using a smaller duct in section 1–3 and a diverging wye or tee of greater local loss coefficient $C_{c.b}$.

- In a duct system, if the flow resistance of each branch duct R_b is great enough, the duct system is more easily adjusted to achieve a system balance. For the branch take-off or connecting duct having a length less than 1 or 2 ft long, it is often difficult to increase R_b by reducing its size. Adding a volume damper directly on the outlet or inlet may alter space air flow patterns. A better remedy is to vary the sizes of the successive main duct sections, with outlets of greater flow resistance, to achieve a better system balance.

- In a duct system of considerable length, a greater R_b for all branch ducts results in a smaller deviation from design flow rates and better system balancing than a smaller R_b.

- A variable-air-volume duct system installed with a VAV box in each branch take-off provides system balancing automatically. However, if half of the modulating capacity of a VAV box is used to

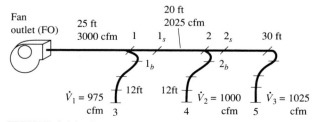

FIGURE 8.16 System balancing and design path of a supply duct system.

provide system balance of a specific branch take-off, the quality of its modulation control at part-load will be impaired.

Design Path

For any air duct system there exists a design path, or duct run, whose total flow resistance $\sum R_{\text{dgn}}$ is maximum compared with other duct paths when the volume flow rates of various duct sections of this system are equal to the design values at design conditions.

A *design path* is usually a duct path with more duct fittings and a comparatively higher volume flow rate; additionally, it may be the longest one. In Fig. 8.16, path FO–1–2–5 may be the design path.

For an energy-efficient air duct design, the local loss coefficients of the duct fittings along the design path should always be minimized, especially at the fan inlet and outlet (see Chapter 19). For other duct paths, if the pressure loss at design flow rate is smaller than the total pressure available at fan outlet, a smaller duct size and duct fittings of greater dynamic loss may provide a better system balance.

Air Leakage

Conditioned air leakage from the joints and seams of the air duct to the space, which is not air conditioned, is always a waste of refrigerating or heating energy as well as fan power. Air leakage depends mainly on the use of sealant on joints and seams, the quality of fabrication, and the shape of the ducts. Heat-sensitive tapes, mastic and glass fabric, and many other materials are used as sealants for the joints and seams.

Duct leakage classifications, based on tests conducted by AISI, SMACNA, ASHRAE, and the Thermal Insulation Manufacturers Association (TIMA), are presented in Table 8.9. Air leakage rate $\dot{V}_L$, in cfm/ft^2 of duct surface area, can be calculated by the following formula:

$$\dot{V}_L = 0.01 C_L \Delta p_{\text{sd}}^{0.65} \qquad (8.56)$$

where C_L = leakage class
Δp_{sd} = static pressure differential between the air inside and outside the duct, in. WG

ASHRAE Standard 90.1–1989 specifies that:

(1) When supply ducts and plenums designed to operate at a static pressure from 0.25 to 2 in. WG are located outside the conditioned space or in ceiling plenums, all duct and plenum joints should be sealed according to Sealing Class C in the SMACNA manuals. Class C requires all transverse joints to be sealed. Pressure-sensitive tape should not be used as the primary sealant when duct static pressure is 1 in. WG or greater.

TABLE 8.9
Duct leakage classification

Type of duct	Predicted leakage class (C_L)	
	Sealed	Unsealed
Metal (flexible excluded)		
Round and oval	3	30
		(6 to 70)
Rectangular		
≤2 in. WG	12	48
(both positive and negative pressures)		(12 to 110)
>2 and ≤ 10 in. WG	6	48
(both positive and negative pressures)		(12 to 110)
Flexible		
Metal, aluminum	8	30
		(12 to 54)
Nonmetal	12	30
		(4 to 54)
Fibrous glass		
Rectangular	6	NA
Round	3	NA

The leakage classes listed in this table are averages based on tests conducted by AISA/SMACNA 1972, ASHRAE/SMACNA/TIMA 1985, and ASHRAE 1988. Leakage classes listed are not necessarily recommendations on allowable leakage. The designer should determine allowable leakage and specify acceptable duct leakage classifications.
Source: ASHRAE Handbook 1989, Fundamentals. Reprinted with permission.

(2) Ducts that are designed to operate at a static pressure exceeding 3 in. WG must meet the following specifications:

- Leak tested according to SMACNA manuals. Testing may be limited to 25% of representative duct sections.
- Leakage rates at design duct pressure must be equal to or less than Leakage Class 6.
- If duct static pressure is 3 in. WG, Class B sealing must be used, that is, all transverse joints and longitudinal seams should be sealed.
- If duct static pressure is 4 in. WG or higher, Class A sealing must be used. That is, in addition to Class B, penetrations of duct walls should be sealed.

Shapes and Materials of Air Ducts

When a designer chooses the shape (round, rectangular or flat oval duct) or material (galvanized sheet, aluminum, fiberglass or other materials) of an air duct, the choices depend mainly on the space available, cost, local customs and union agreements, experience, quality, and the requirements of the project.

In many high-rise commercial buildings, factory-fabricated round ducts and sometimes flat oval ducts with spiral seams are used because they have fewer sound problems, lower air leakage, and many configurations of wye and tees available for easier pressure balance. Round ducts also have the advantage of high breakout transmission loss at low frequencies (see Chapter 16).

For ducts running inside the air conditioned space in industrial applications, metal rectangular ducts are often chosen for their large cross-sectional areas and convenient fabrication. In projects designed for lower cost, adequate duct insulation, and sound attenuation, fiberglass ducts may sometimes be the optimum selection.

Ductwork Installation

Ductwork installation, workmanship, materials, and methods must be monitored at all stages of the design and construction process to assure that they meet the design intent.

Fire Protection

The design of air duct systems must meet the requirements of National Fire Codes *NFPA 90A Standards for the Installation of Air Conditioning and Ventilating Systems,* Warm Air Heating and Air Conditioning Systems, and Blower and Exhaust Systems as well as local codes. Refer to these standards for details. The following are some of the requirements:

- Not only the duct material discussed in Section 8.3 of this text must be made of Class 0 or Class 1 material, but also the duct coverings and linings—including adhesive, insulation, banding, coating, and film covering the outside surface and material lining the inside surface of the duct—must have a flame spread rating not over 25 and a smoke development rating not over 50 except for ducts outside buildings.

  Supply ducts that are completely encased in a concrete floor slab not less than 2 inches thick need not meet the Class 0 or Class 1 requirement.
- Vertical ducts more than two stories high must be constructed of masonry, concrete, or clay tile.
- When ducts pass through the floors of buildings, the vertical openings must be enclosed with partitions and walls with a fire protection rating of not less than 1 hour in buildings less than four stories high and greater than 2 hours in buildings four stories and higher.

- Clearances between the ducts and combustible construction and material must be made as specified in NFPA 90A.
- The opening through a fire wall by the duct system must be protected by (1) a fire damper closing automatically within the fire wall and having a fire protection rating of not less than 3 hours or (2) fire doors on the two sides of the fire wall. A service opening must be provided in ducts adjacent to each fire damper. Many regulatory agencies have very rigid requirements for fire dampers, smoke dampers, fire/smoke dampers in combination, and smoke venting.
- When a duct penetrates through walls, floors, and partitions, the gap between the ducts and the walls, floors, and partitions must be filled with noncombustible material to prevent the spread of flames and smoke.
- Duct systems for transporting products, vapor, or dust in industrial applications must be constructed entirely of metal or noncombustible material. Longitudinal seams must be continuously welded, lapped, and riveted, or spot-welded on maximum centers of 3 in. Transitions must be 5 inches long for every 1 in. change in diameter. Rectangular ducts may be used only when the space is not available and must be made as square as possible.

8.8 AIR DUCT DESIGN PROCEDURE AND DUCT LAYOUT

Design Procedure

Before an air duct system is designed, the supply volume flow rate for each conditioned space, room, or zone should be calculated, and the locations of the supply outlets and return inlets should also be settled according to the requirements of space air diffusion (see Chapter 9). One can use computer-aided programs for more precise calculation and optimum sizing of duct systems.

The design procedure for an air duct system is as follows:

1. Designer should verify local customs, local codes, local union agreements, and material availability restraints before proceeding with duct design.
2. The designer proposes a preliminary duct layout to connect the supply outlets and return inlets with the fan(s) and other system components through the main and branch take-offs. The shape of the air duct is selected. Space available under the beam often determines the shape of the duct and affects the layout in high-rise buildings.

3. The duct layout is divided into consecutive duct sections, which converge and diverge at junctions or nodes. Duct systems should be divided into duct sections at junctions or nodes where flow rates change.

4. Duct fittings are assigned to each duct section if necessary. The local loss coefficients of the duct fittings along the tentative design path should be minimized, especially adjacent to fan inlets and outlets.

5. Duct sizing methods should be selected according to the characteristics of the air duct system. The sizes of various duct sections along the design path are determined. Allocated space under the beam often dictates the maximum velocity in the main duct.

6. The total pressure loss of the air duct system is calculated.

7. The designer sizes the branch ducts and balances the total pressure at each junction of the duct system by varying the duct and component sizes and the configuration of the duct fittings.

8. The supply volume flow rates are adjusted according to the duct heat gain at each supply outlet.

9. The designer resizes the duct sections, calculates again the total pressure loss, and balances the parallel paths from each junction.

10. The airborne and breakout sound levels from various paths should be checked and necessary attenuation added to meet requirements.

Duct System Characteristics

Air duct systems can be classified into the following three categories according to their system characteristics:

1. Supply duct, return duct, or exhaust duct systems with a certain pressure loss in branch take-offs

2. Supply duct, return duct, or exhaust duct systems in which supply outlets or return grilles are either mounted directly on the duct or have only very short connecting ducts between the outlet or inlet and the main duct.

3. Industrial exhaust duct systems to transport dust particles or other particulate products.

Duct Layout

When a designer starts to sketch a preliminary duct layout, the size of the air duct system must be decided first. The size of an air duct system must be consis-

tent with the size of the air system or even the air conditioning system. From the point of view of the air duct system itself, a smaller and shorter system requires less power consumption by the fan and shows a smaller duct heat gain or loss. The air duct system is also comparatively easier to balance and to operate, as discussed in Chapter 30.

If the designer uses a more symmetrical layout, as shown in Fig. 8.17, a more direct and simpler form for the design path can generally be derived. A symmetrical layout usually has a smaller main duct and a shorter design path; it is easier to provide system balance for a symmetrical than a nonsymmetrical layout. A more direct and simpler form of design path always means a lower total pressure loss of the duct system.

For variable-air-volume air duct systems, the next step is to consider connecting the ends of the main ducts to form one or more duct loopings as shown by the dotted lines in Fig. 8.17. Duct looping(s) allow some of the duct sections to be fed from the opposite direction to the main duct. Balance points exist where the total pressure of the opposite flowing airstreams is zero. The positions of the balance points often follow the sun's position and the induced cooling load at various zones during the operating period. Duct looping optimizes transporting capacity and results in a smaller main duct than that without looping.

The designer then compares various alternative layouts and reduces the number of duct fittings, especially the fittings with higher velocity and local loss coefficients along the design path. All these result in a duct system with lower pressure loss.

When duct systems are installed in commercial and public buildings without suspended ceilings, duct runs should be closely matched with the building structures and give a neat and harmonious appearance.

8.9 DUCT SIZING METHODS

Duct sizing determines the dimensions of each duct section in the air duct system. After the duct sections have been sized, the total pressure loss of the air duct system can then be calculated and the fan total pressure can be calculated according to the sum of the total pressure losses of the supply and return duct systems and the pressure loss in the air-handling unit or package unit.

Three duct-sizing methods are currently used:

1. Equal friction method with maximum velocity
2. Constant velocity method
3. Static regain method

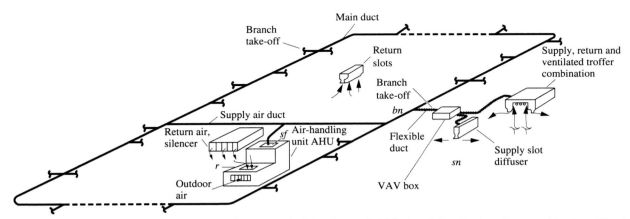

FIGURE 8.17 A typical supply duct system with symmetrical duct layout (bold line) and duct loopings (connected by dotted lines) for a typical floor in a high-rise building.

The *ASHRAE Handbook 1989, Fundamentals* recommended a new duct sizing method, called the *T-method*.

Equal Friction Method

This method sizes the air duct so that the duct frictional loss per unit length $\Delta p_{f.u}$ at various duct sections always remains constant. The final dimensions of sized ducts should be rounded to standard size. The total pressure loss of the duct system Δp_t, in inches WG, should be equal to the sum of the frictional losses and dynamic losses at various duct sections along the design path as follows:

$$\Delta p_t = \Delta p_{f.u}[(L_1 + L_2 + \cdots + L_n) \\ + (L_{e1} + L_{e2} + \cdots + L_{en})] \quad (8.57)$$

where $L_1, L_2, \ldots, L_n$ = length of the duct sections $1, 2, \ldots, n$, ft

$L_{e1}, L_{e2}, \ldots, L_{en}$ = equivalent length of the duct fittings in duct sections $1, 2, \ldots, n$, ft.

If the dynamic loss of a duct fitting is equal to the frictional loss of a duct section of length L_e, in ft, then

$$\frac{C_o \rho v^2}{2g_c} = \frac{f(L_e/D)\rho v^2}{2g_c}$$

and the equivalent length is

$$L_e = \frac{C_o D}{f} \quad (8.58)$$

The selection of $\Delta p_{f.u}$ is usually based on experience, such as 0.1 in. WG per 100 ft for low-pressure systems. A maximum velocity is often used as the upper limit.

The equal friction method does not aim at an optimum cost. Dampers are sometimes necessary for a system balance. Because of its simple calculations, the equal friction method is still used in many hand calculations for low-pressure systems in which airborne noise due to higher air velocity is not a problem or for smaller duct systems.

Constant Velocity Method

The constant velocity method is often used for exhaust systems that convey dust particles in industrial applications. This method first determines the minimum air velocity at various duct sections according to the requirement to float the particles, either by calculation or by experience. On the basis of the determined air velocity, the cross-sectional area and, therefore, the dimension of the duct can be estimated and then rounded to a standard size. The total pressure loss of the duct system Δp_t, in inches WG, along the design path can be calculated as

$$\Delta p_t = [5.35 \times 10^{-5}\rho/(2g_c)]\{[(f_1L_1/D_1) + C_1]v_1^2 \\ + [(f_2L_2/D_2) + C_2]v_2^2 + \cdots \\ + [(f_nL_n/D_n) + C_n]v_n^2\} \quad (8.59)$$

where $v_1, v_2, \ldots, v_n$ = mean air velocity at duct section $1, 2, \ldots, n$ respectively, fpm

Static Regain Method

This method sizes the air duct so that the increase of static pressure (static regain) due to the reduction of air velocity in the supply main duct after each branch take-off exactly offsets the pressure loss of the succeeding duct section along the main duct. As a consequence, the static pressure at the common end of the diverging tee or wye of the sized duct section remains the same as the preceding section.

A rectangular duct section 1–2 between the cross-sectional planes 1 and 2 is illustrated in Fig. 8.18.

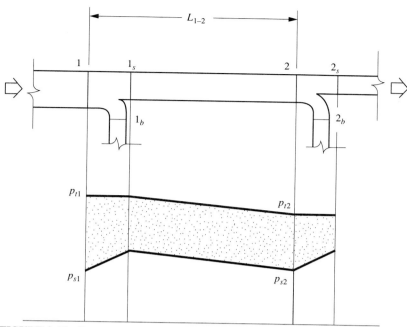

FIGURE 8.18 Pressure characteristics of a main duct section.

The size of this duct section is to be determined. Let v_1 and v_2 be the mean velocities at planes 1 and 2, $\dot{V}_1$ and $\dot{V}_2$ the volume flow rates, and A_1 and A_2 the cross sectional areas. The total pressure loss in duct section 1–2 consists of the duct frictional loss Δp_{f1-2} and the dynamic loss of the main airstream flowing through the diverging tee $\Delta p_{1c.s}$. The relationship between the total pressure at plane 1 and 2 can be expressed as

$$p_{t1} = p_{t2} + \Delta p_{f1-2} + \Delta p_{1c.s} \qquad (8.60)$$

Because $p_t = p_s + p_v$, we can ignore the difference between air densities ρ_1 and ρ_2. Let $\Delta p_{f1-2} = \Delta p_{f.u}L_{1-2}$. Here, L_{1-2} represents the length of the duct section 1–2, in ft. If the static pressures at planes 1 and 2 are equal, that is, $p_{s1} = p_{s2}$, then

$$\frac{\rho_1(v_1^2 - v_2^2)}{2g_c} = \Delta p_{f.u}L_{1-2} + \frac{C_{1c.s}\rho_1 v_1^2}{2g_c}$$

If v is expressed in fpm and $\Delta p_{f.u}$ in inches WG per 100 ft, $\rho = 0.075$ lb/ft³, and $g_c = 32.2$ ft/s² the mean air velocity of the sized duct section is

$$v_2 = \sqrt{(1 - C_{1c.s})v_1^2 - 1.6 \times 10^5 \Delta p_{f.u}L_{1-2}} \qquad (8.61)$$

For any duct section between cross-sectional plane $n - 1$ and n, if the total local loss coefficient of the duct fittings is C_n excluding the local loss coefficient $C_{(n-1)}c.s$, the mean air velocity of the sized duct section is

$$v_n = \sqrt{\frac{(1 - C_{(n-1)c.s})v_{(n-1)}^2 - 1.6 \times 10^5 \Delta p_{f.u}L_n}{1 + C_n}}$$

$$(8.62)$$

Because $v_{(n-1)}$, ρ_n, L_n, and C_n are known values, by using iteration methods, v_n can be determined. The dimension of the duct section and its rounded standard size can also be determined.

The static regain method can be applied only to supply duct systems. It tends to produce a more even static pressure at the common end of each diverging tee or wye leading to the corresponding branch take-off, which is helpful to the system balance. It does not consider cost optimization. The main duct sections remote from the fan discharge often have larger dimensions than those in the equal friction method. Sound level and space required should be checked against determined air velocity and dimension.

When using the static regain method to size air ducts, it is not recommended to allow only part of static regain to be used in the calculation.

T-Method

The *T-method*, first introduced by Tsal et al. (1988), is an optimizing procedure to size air ducts by means of minimizing their life-cycle cost. It is based on the tree-staging idea and is therefore called the *T-method*. The goal of this method is that the ratio between the velocities in all sections of the duct system must be optimum.

The T-method consists of the following procedures:

1. *System condensing*. Condensing various duct sections of a duct system into a single imaginary duct section having the exact same hydraulic characteristics and installation costs as the duct system

2. *Fan selection.* Selecting a fan that provides the optimum system pressure loss

3. *System expansion.* Expanding the imaginary duct section into the original duct system before condensing with the optimum distribution of total pressure loss between various duct sections

During optimization, local loss coefficients are considered constant at various stages of iteration. For details, refer to Tsal et al. (1988).

The T-method can be used for sizing duct systems with certain pressure losses in the branch ducts. However, the local loss coefficients are actually varied at various stages of iteration and should be taken into consideration during optimization.

8.10 DUCT SYSTEMS WITH CERTAIN PRESSURE LOSSES IN BRANCH TAKE-OFFS

Design Characteristics

Supply, return, or exhaust duct systems with certain pressure losses in branch take-offs have the following characteristics:

- Duct is sized based on the optimization of the life-cycle cost of various duct sections of the duct system as well as the space available in the building.

- System balancing is achieved mainly through pressure balancing of various duct paths by changing of duct sizes and the use of various configurations of duct fittings and terminals instead of using dampers or other devices.

- Sound level will be checked and analyzed. Excess pressure at each inlet of VAV box at design conditions should be avoided. Sound attenuation arrangements should be added if necessary.

- Local loss coefficients of the duct fittings and equipment along the design path should be minimized. It may be beneficial to use the surplus total pressure available in the branch take off to produce a higher branch duct velocity and a smaller straight-through local loss coefficient $C_{s.c}$.

- Supply volume flow rates should be adjusted according to the duct heat gain. For VAV systems, use diversity factors to determine the volume flow rate of various duct sections along design path so that the volume flow rate nearly covers the block load at the fan discharge.

Cost Optimization

For any duct section in an air duct system, the total life-cycle cost C_{to}, as shown in Fig 8.19, in dollars, can be calculated as

$$C_{to} = C_e(1/\text{CRF}) + C_{di} \qquad (8.63a)$$

In Eq. (8.63), CRF indicates the capital recovery factor and can be calculated as follows:

$$\text{CRF} = \frac{i(1+i)^n}{[(1+i)^n - 1]} \qquad (8.63b)$$

where i = interest rate
  n = number of years under consideration

The first-year energy cost C_e, in \$, can be calculated as the product of electrical energy consumed at the fan times the unit energy cost E_r, in \$/kWh, times the annual operating hours t_{an}, in h, or

$$C_e = \Delta p_t \left(\frac{0.746 \dot{V} E_r t_{an}}{6350 \eta_f \eta_m} \right)$$

$$= \Delta p_t \left(\frac{1.175 \times 10^{-4} \dot{V} E_r t_{an}}{\eta_f \eta_m} \right)$$

$$= K_e \Delta p_t \qquad (8.64)$$

where Δp_t = total pressure loss of the duct section, in. WG
  $\dot{V}$ = volume flow rate of the duct section, cfm
  η, η_m = total efficiency of the fan and efficiency of the motor

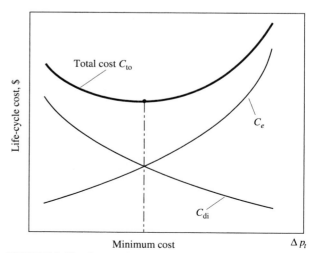

FIGURE 8.19 Cost analysis for a duct system.

Let $z_1 = K_e(1/\text{CRF})$. Then we can write

$$(1/\text{CRF})C_e = z_1 \Delta p_t \qquad (8.65)$$

Installation costs for round duct C_{di}, in \$, can be calculated as

$$C_{di} = \pi D L C_{iu} \qquad (8.66)$$

where D = diameter of the round duct, ft
L = length of the duct section, ft
C_{iu} = unit cost of duct installation, \$/ft^2

The surface area of rectangular duct $A_{rec} = 2(W + H)L$. Here, W is the width of the duct and H is the height of the duct, both in ft. Then for rectangular duct

$$
\begin{aligned}
C_{di} &= 2(W + H)LC_{iu} \\
&= \frac{2(W + H)}{\pi D}\pi D L C_{iu} \\
&= R_{rec}\pi D L C_{iu} \qquad (8.67)
\end{aligned}
$$

where R_{rec} = ratio of surface area of the rectangular duct to the surface area of the round duct.

For any duct section, the total pressure loss Δp_t, in inches WG,

$$\Delta p_t = 5.35 \times 10^{-5}\frac{(fL/D + \sum C)\rho v^2}{2g_c} \qquad (8.68)$$

where $\sum C$ = sum of the local loss coefficients in the duct section, based on the air velocity of the sized duct section
v = mean air velocity in the sized duct section, fpm

For frictional loss $D = f(\Delta p_t^{-0.2})$, and for dynamic loss $D = f(\Delta p_t^{-0.25})$. However, for simplicity, let us take $D = f(\Delta p_t^{-0.22})$ and consider $[(fL/\sqrt{D}) + \sum C \sqrt{D}]$ a constant. Then

$$
\begin{aligned}
D &= 0.1097\left[\frac{(fL/\sqrt{D} + \sum C \sqrt{D})\rho}{g_c}\right]^{0.22} \\
&\quad \times \dot{V}^{0.44}\Delta p_t^{-0.22} \\
&= 0.1097\left(\frac{\rho}{g_c}\right)^{0.22}\frac{K\Delta p_t^{-0.22}}{L}
\end{aligned}
$$

Because $g_c = 32.2$ ft/s^2,

$$D = \{0.0511\left[\left(\frac{fL}{\sqrt{D} + \sum C \sqrt{D}}\right)\rho\right]^{0.22} \dot{V}^{0.44}\}\Delta p_t^{-0.22} \qquad (8.69)$$

where $K = (fL/\sqrt{D} + \sum C \sqrt{D})^{0.22}\dot{V}^{0.44}L$.

Let $z_2 = 0.1097\pi(\rho/g_c)^{0.22}C_{iu}$. By substituting into Eq. (8.67), for a round duct, we have

$$C_{di} = \pi D L C_{iu} = z_2 K \Delta p_t^{-0.22} \qquad (8.70)$$

For a rectangular duct,

$$C_{di} = R_{rec}z_2 K \Delta p_t^{-0.22} \qquad (8.71)$$

Seeking for a minimum cost by taking derivatives of Eq. (8.63) with respect to Δp_t, making it equal to zero, we have

$$dC_{to}/d\Delta p_t = z_1 - 0.22z_2 K\Delta p_t^{-1.22} = 0$$

If $g_c = 32.2$ ft/s^2, then for a round duct

$$
\begin{aligned}
\Delta p_t &= 108\left[\frac{LC_{iu}\eta_f\eta_m}{(1/\text{CRF})E_r t_{an}}\right]^{0.82} \\
&\quad \times \left[\left(\frac{fL}{\sqrt{D}} + \sum C \sqrt{D}\right)\rho\right]^{0.18}\frac{1}{\dot{V}^{0.46}} \qquad (8.72)
\end{aligned}
$$

For a rectangular duct,

$$
\begin{aligned}
\Delta p_t &= 108R_{rec}\left[\frac{LC_{iu}\eta_f\eta_m}{(1/\text{CRF})E_r t_{an}}\right]^{0.82} \\
&\quad \times \left[\left(\frac{fL}{\sqrt{D}} + \sum C \sqrt{D}\right)\rho\right]^{0.18}\frac{1}{\dot{V}^{0.46}} \qquad (8.73)
\end{aligned}
$$

When the total pressure loss of the duct section having the minimum cost is determined from Eq. (8.68), the diameter or the circular equivalent of the duct section can be calculated by iteration.

Condensing Two Duct Sections

According to the T-method, if two duct sections 1 and 2 connected in series are condensed into an imaginary section 1–2, as shown in Fig. 8.20a, then the volume flow rate at each duct section must be equal, that is,

$$\dot{V}_1 = \dot{V}_2 = \dot{V}_{1-2} \qquad (8.74)$$

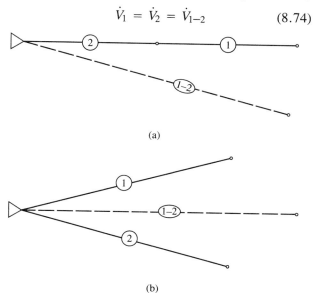

(a)

(b)

FIGURE 8.20 Two duct sections 1 and 2 condensed into an imaginary duct section 1–2: (a) two duct sections 1 and 2 connected in a series; and (b) two duct sections 1 and 2 connected in parallel.

Here, subscripts 1, 2, and 1–2 represent the duct sections 1 and 2 and the imaginary duct section 1–2, respectively.

The total pressure loss of the condensed duct section must equal the sum of the total pressure losses of the duct sections 1 and 2, or

$$\Delta p_{t1} + \Delta p_{t2} = \Delta p_{t1-2} \qquad (8.75)$$

and the installation cost of the condensed section is

$$
\begin{aligned}
C_{di1-2} &= C_{di1} + C_{di2} \\
&= z_2 (K_1 \Delta p_{t1}^{-0.22} + K_2 \Delta p_{t2}^{-0.22}) \\
&= z_2 K_{1-2} \Delta p_{t1-2}^{-0.22} \qquad (8.76)
\end{aligned}
$$

When two duct sections 1 and 2 connected in parallel are condensed into an imaginary duct section 1–2 as shown in Fig. 8.20b, the following relationship holds:

$$\dot{V}_{1-2} = \dot{V}_1 + \dot{V}_2$$

$$\Delta p_{t1-2} = \Delta p_{t1} = \Delta p_{t2}$$

$$C_{di1-2} = C_{di1} + C_{di2} \qquad (8.77)$$

If the whole duct system is condensed into one imaginary duct section, its minimum cost can be found by taking derivatives with respect to Δp_t of the imaginary duct section. In using such a procedure, the minimum cost includes both the main and branch ducts. However, if the local loss coefficients of the terminals and converging and diverging wyes are assumed constant, the benefits of using smaller terminals and different configuration of wyes to balance the branch duct paths are lost, as demonstrated by Dean et al. (1985). A more flexible and often economical alternative is that only the sizes of duct sections along the design path of a duct system should be determined according to the cost optimization procedure—that is, Eqs. (8.72), (8.73), and (8.69)—and rounded to standard size. The sizes of the duct sections of other duct paths, such as branch take offs, should be determined according to the difference between the total pressure at the junction and the space pressure for system or pressure balance by using optimum terminals, and wyes.

Local Loss Coefficients for Diverging Tees and Wyes

For a supply duct system, there are often more diverging tees or wyes than elbows along the design path. Selecting diverging tees or wyes with smaller local loss coefficients $C_{c.s}$ has a definite influence on the design of an optimum duct system.

In Table 8.11 are $C_{c.s}$ values for diverging wyes and tees for round ducts and diverging wyes for rectangular ducts. The following are recommendations for proper selection of wyes and tees:

- Select a diverging wye instead of a diverging tee except when there is surplus total pressure at the branch take offs. A 30° angle between the branch and the main duct has a smaller $C_{c.s}$ value than a 45° or a 60° angle.
- The smaller the difference in air velocity between the common end and the straight-through end, the lower the $C_{c.s}$ value.
- At a greater ratio of branch duct velocity to main duct velocity v_b / v_c, especially when $v_b / v_c > 1.2$, the $C_{c.s}$ value for the diverging wye of a rectangular duct with a smooth elbow take off is negative.

Return or Exhaust Duct Systems

Behls (1978) compared four designs of a bunker ventilation duct system with 10 risers connected to a main exhaust duct. The results of these four designs are as follows:

	Total pressure, in. WG	Riser unbalance, in. WG
Fixed diameter riser, tee/diffuser	−7.39	6.55
Variable diameter riser, tee/diffuser	−5.89	1.77
Fixed diameter riser, conical tee	−5.29	4.06
Variable diameter riser, conical tee	−4.03	1.62

Using a greater branch duct velocity than the main duct, varying the sizes of the risers, and selecting proper duct fittings decrease system total pressure loss and system imbalance.

Example 8.3. For a round supply duct system made of galvanized steel with spiral seams, as shown in Fig 8.16, the operational and constructional characteristics are as follows:

Supply air temperature	60°F
Kinematic viscosity of air	1.59×10^{-4} ft^2/s
Density of supply air	0.075 lb/ft^3
Absolute roughness	0.0003 ft
Local loss coefficients in section FO–1	0.5
Fan total efficiency, average	0.7
Motor efficiency	0.8
Electrical energy cost	$0.08/kWh
Installation cost of duct	$3.5/ft^2
CRF	0.10
Annual operating hours	3000

The total supply volume flow rate at the fan discharge is 3000 cfm, and the adjusted volume flow rates because of duct heat gain for each of the branch take offs are illustrated in Fig. 8.16. First, size this supply duct system. Then, if each branch take off is connected to a VAV box that needs a total pressure of 0.75 in. WG at its inlet, calculate the total pressure loss required for this supply duct system excluding the VAV box and downstream flexible duct and diffuser.

Solution

1. For the first iteration, start with the round duct section 2–5, which may be the last section of the design path. Assume a diameter of 12 in., or 1 ft, and an air velocity of 1200 fpm, or 20 ft/s. Assume also that the local loss coefficient of the straight-through stream of the diverging tee $C_{c.s} = 0.15$, and for the elbow $C_o = 0.22$. The Reynolds number based on the diameter D is

$$\text{Re}_D = \frac{vD}{\nu} = \frac{20 \times 1}{1.59 \times 10^{-4}} = 125{,}786$$

From Eq. (8.40), the friction factor is

$$f = \frac{0.25}{\left[\log \dfrac{\epsilon}{3.7D} + \dfrac{5.74}{0.9\text{Re}_D} \right]^2}$$

$$= \frac{0.25}{\left[\log \dfrac{0.0003}{3.7 \times 1} + \dfrac{5.74}{0.9 \times 125{,}786} \right]^2} = 0.0166$$

From Eq. (8.72), the optimum pressure loss for the round duct section 2–5 is

$$\Delta p_{t2-5} = 108 \left[\frac{L C_{iu} \eta_f \eta_m}{(1/\text{CRF}) E_r t_{an}} \right]^{0.82}$$

$$\times \left[\left(\frac{fL}{\sqrt{D}} + \sum C \sqrt{D} \right) \rho \right]^{0.18} \times \frac{1}{\dot{V}^{0.46}}$$

$$= 108 \left(\frac{30 \times 3.5 \times 0.7 \times 0.8}{10 \times 0.08 \times 3000} \right)^{0.82}$$

$$\times \left[\frac{0.0166 \times 30}{(\sqrt{1} + 0.37\sqrt{1})0.075} \right]^{0.18}$$

$$\times \frac{1}{1025^{0.46}} = 0.13 \text{ in. WG}$$

And from Eq. (8.69), the diameter of the duct section 2–5 when $\Delta p_t = 0.13$ in. WC is

$$D_{2-5} = 0.0511 \left[\left(\frac{fL}{\sqrt{D}} + \sum C \sqrt{D} \right) \rho \right]^{0.22}$$

$$\times \dot{V}^{0.44} \Delta p_t^{-0.22}$$

$$= 0.0511 \left[\left(\frac{0.0166 \times 30}{\sqrt{1}} + 0.37\sqrt{1} \right) 0.075 \right]^{0.22}$$

$$\times 1025^{0.44}(0.13)^{-0.22} = 0.927 \text{ ft or } 11.12 \text{ in.}$$

2. Duct section 1–2 is one of the sections of the design path. For this section, if we assume that $C_{c.s} = 0.05$ and the diameter is 1.2 ft, then the calculated friction factor $f = 0.0155$ and the optimal total pressure loss

$$\Delta p_{t1-2} = 108 \left(\frac{20 \times 3.5 \times 0.7 \times 0.8}{10 \times 0.08 \times 3000} \right)^{0.82}$$

$$\times \left[\left(\frac{0.155 \times 20}{\sqrt{1.2}} + 0.05\sqrt{1.2} \right) 0.075 \right]^{0.18}$$

$$\times \frac{1}{2025^{0.46}} = 0.0575 \text{ in. WG}$$

and the sized diameter is

$$D_{1-2} = 0.051 \left[\left(\frac{0.0155 \times 20}{\sqrt{1.2}} + 0.05\sqrt{1.2} \right) \times 0.075 \right]^{0.22}$$

$$\times 2025^{0.44} 0.0581^{-0.22}$$

$$= 1.229 \text{ ft or } 14.75 \text{ in.}$$

3. Duct section FO–1 is one of the sections of the design path. For this section, $\sum C = 0.5$. Let us assume that the diameter is 1.67 ft and the calculated $f = 0.0147$; then

$$\Delta p_{t\text{FO}-1} = 108 \left(\frac{25 \times 3.5 \times 0.7 \times 0.8}{10 \times 0.08 \times 3000} \right)^{0.82}$$

$$\times \left[\left(\frac{0.147 \times 25}{\sqrt{1.67}} + 0.5\sqrt{1.67} \right) 0.075 \right]^{0.18}$$

$$\times \frac{1}{3000^{0.46}} = 0.0692 \text{ in. WG}$$

And the sized diameter

$$D_{\text{FO}-1} = 0.051 \left[\left(\frac{0.0147 \times 25}{\sqrt{1.67}} + 0.5\sqrt{1.67} \right) \right.$$

$$\left. \times 0.075 \right]^{0.22} 3000^{0.44} 0.0692^{-0.22}$$

$$= 1.734 \text{ ft or } 20.81 \text{ in.}$$

4. Duct section 2–4 is another leg from junction 2. It must have the same total pressure loss as the duct section 2–5. Assume that $C_{c.b} = 1$ and the diameter is 11 in. or 0.917 ft. Then

$$D_{2-4} = 0.051 \left[\left(\frac{0.0166 \times 12}{\sqrt{0.917}} + 1 \right. \right.$$

$$\left. \left. \times \sqrt{0.917} \right) 0.075 \right]^{0.22} 1000^{0.44} 0.13^{-0.22}$$

$$= 0.978 \text{ ft or } 11.78 \text{ in.}$$

5. Duct section 1–3 is another leg from junction 1. It must have the sum of the total pressure loss of duct sections 1–2 and 2–5, that is, $\Delta p_{t1-3} = 0.0575 + 0.13 = 0.1875$ in. WG. Assume that $C_{c.b} = 0.8$ and the diameter is also 11 in. Then

$$D_{1-3} = 0.051 \left[\left(\frac{0.166 \times 12}{\sqrt{0.917}} + 0.8 \times \sqrt{0.917} \right) \right.$$

$$\left. \times 0.075 \right]^{0.22} 975^{0.44} 0.1875^{-0.22}$$

$$= 0.858 \text{ ft or } 10.3 \text{ in.}$$

6. The results of the first iteration are listed in Table 8.10. These results are rounded to standard sizes and provide the information for the selection of the diverging tee and wye and the determination of the local loss coefficient.

7. For branch take offs 2–4 and 1–3, select proper diverging tee and wyes and, therefore, the $C_{c.b}$ values based on the air velocity of the sized duct sections. Vary the size of the duct sections if necessary so that their Δp_t are approximately equal to the Δp_t of section 2–5 for branch duct 2–4 and the Δp_t of the sum of sections 2–5 and 1–2 for branch 1–3. Use Eq. (8.68) to calculate the total pressure loss of sections 2–4 and 1–3 according to the rounded diameters.

8. After the diverging tee or wye is selected, recalculate the optimum total pressure losses and diameters for duct sections 2–5, 1–2, and FO–1 from Eqs (8.72) and (8.69).

9. After two iterations, the final sizes of the duct sections, as listed in Table 8.10, are the following:

Section	2–5	11 in.; wye, 45°, diverging, round, 30° elbow
	1–2	14 in.; tee, diverging, round, 90° elbow
	FO–1	20 in.
	2–4	11 in.
	1–3	12 in.

For duct section 1–2, at a $v_s/v_c = 1.2$, $C_{c.s}$ is about 0.07 according to Idelchik (1986), Diagram 7-17.

The total pressure loss of this supply duct system excluding the VAV box and downstream flexible duct and diffuser, from Table 8.10, is

$$\Delta p_t = 0.116 + 0.0645 + 0.0847$$

$$= 0.2652 \text{ in. WG}$$

8.11 DUCT SYSTEMS WITH NEGLIGIBLE PRESSURE LOSS AT BRANCH DUCTS

Supply Duct Systems

When supply outlets are directly mounted on the main duct without branch take offs, or the connecting duct between the supply outlet and the main duct is about 2 ft or less, the total pressure loss of the branch duct, excluding the supply outlet, is often very small or negligible. In such circumstances, system balancing of the supply duct system depends mainly on the sizes of the successive main duct sections. If volume dampers are installed just before the outlet or in the connecting duct, the damper modulation will also vary the space diffusion air flow pattern.

Pressure Characteristics of Air Flow in Supply Ducts

The rectangular supply duct with transversal slots shown in Fig. 8.21 is an example of a supply duct system with a supply outlet directly mounted on the main duct. The volume flow per ft² of this type of duct system may sometimes exceed 8 cfm/ft². This type of supply duct system is often used in many industrial applications.

Consider two planes, n and $(n + 1)$, with a distance of one unit length, say 1 ft, between them. If the space air is at atmospheric pressure, the total pressure loss

TABLE 8.10
Results of computations of duct sizes and total pressure loss in Example 8.3

Duct section	Volume flow, $\dot{V}$ cfm	First iteration Diameter, ft	First iteration Diameter, in.	Rounded in.	Air velocity fpm	Air velocity fps	Friction factor f	Velocity ratio v_s/v_c	Velocity ratio v_b/v_c	$C_{c.s}$	$C_{c.b}$	C	Diameter	p_t, in. WG
2–5	1025	0.927	11.12	11	1553	25.88	0.0166	0.94		0.011			11	0.116
1–2	2025	1.229	14.75	15	1894	31.97	0.0155	1.20		0.07			14	0.0645
F0–1	3000	1.734	20.81	20	1374	22.92	0.0147					0.5	20	0.0847
2–4	1000	0.978	11.78	11	1515	25.25			0.80		0.65		11	
1–3	975	0.858	10.3	11	1241	20.69			0.90		1.70		12	

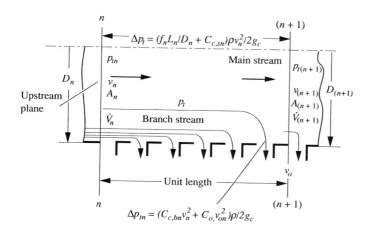

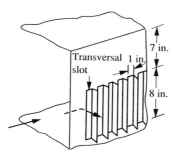

FIGURE 8.21 Rectangular supply duct with transversal slots.

of the supply air that flows from plane n, turns a 90°, and discharges through the slots is

$$\Delta p_{tn} = p_{tn} = \Delta p_{c.bn} + \Delta p_{ton} = \frac{(C_{c.bn} v_n^2 + C_o v_{on}^2)\rho}{2g_c}$$

(8.78)

where $\Delta p_{c.bn}, \Delta p_{ton}$ = dynamic loss of the branch stream when it makes a 90° turn and when it discharges from the slots, lb/ft^2

$C_{c.bn}$ = local loss coefficient of the diverging branch stream with reference to the velocity at plane n

C_o = local loss coefficient of the transversal slots with reference to the velocity at the slot

v_n, v_{on} = air velocity at plane n and at the slot from plane n, ft/s

Similarly, at plane $(n + 1)$ the total pressure loss of the branch stream when it discharges from the slots is

$$p_{t(n+1)} = \frac{\left(C_{c.b(n+1)} v_{(n+1)}^2 + C_o v_{o(n+1)}^2\right)\rho}{2g_c}$$

(8.79)

The relationship between planes n and $(n + 1)$ of the total pressure along the air flow is given as

$$p_{tn} = p_{t(n+1)} + \frac{\left(\dfrac{f_n L_n}{D_n} + C_{c.sn}\right)\rho}{2g_c}$$

(8.80)

In Eq. (8.80), f_n, L_n, D_n, and $C_{s.cn}$ indicate the friction factor, length, circular equivalent, and $C_{c.s}$ value between planes n and $(n + 1)$, respectively. Substituting Eqs. (8.78) and (8.79) into (8.80), Wang et al. (1984) recommended the following formula to calculate $v_{(n+1)}$, in ft/s, based on the balanced total pressure before the transversal slots:

$$v_{(n+1)} = \left\{ \left[\left(v_{on}^2 - v_{o(n+1)}^2 \right) C_o \right. \right.$$
$$\left. + \left(C_{c.bn} - C_{c.sn} - \frac{f_n L_n}{D_n} \right) v_n^2 \right] \times \frac{1}{C_{c.b(n+1)}} \right\}^{0.5}$$

(8.81)

In Eq. (8.81), v_{on} and $v_{o(n+1)}$, in ft/s, are the supply air velocities at the slots. For a cold air supply, usually, $v_{o(n+1)} > v_{on}$ is desirable because of the effect of the duct heat gain.

For a free area ratio of area to duct wall area of 0.5, the C_o for a perforated plate including discharged velocity pressure can be taken as 1.5. The v_n, L_n, and D_n are known values during the calculation of $v_{(n+1)}$. Local loss coefficients $C_{c.bn}$ and $C_{c.sn}$ can be determined from the experimental curves shown in Figs. 8.22 and 8.23. The coefficient $C_{c.b(n+1)}$ can be assumed to have a value similar to $C_{c.bn}$, and f can be calculated from Eq. (8.40). The mean air velocity in the duct at plane $(n+1)$, $v_{(n+1)}$, can then be calculated from Eq. (8.81).

In Figs. 8.22 and 8.23, $\dot{V}_{on}$ indicates the volume flow rate of supply air discharged from the slots per ft length of the duct, in cfm/ft; $\dot{V}_n$ represents the volume flow rate of the supply air inside the duct at plane n.

Sizing of this kind of supply duct system starts from the duct section immediately after the fan discharge. This section can be sized from Eqs. (8.69) and (8.73) based on life-cycle cost optimization. The successive duct sections can then be sized by calculating from Eq. (8.81).

If supply duct systems with negligible pressure at branch ducts are installed with outlets whose $C_{c.bn}$ and $C_{c.sn}$ are not known, the static regain method is recommended for sizing the supply main duct for a large duct system and the equal friction method for a simple or small system.

Example 8.4. A galvanized steel rectangular duct section with transversal slots has the following constructional and operational characteristics at plane n:

Dimensions	4 ft width × 3 ft height
Absolute roughness ϵ	0.0005 ft
Supply air temperature	65°F
Volume flow rate $\dot{V}_n$	23,760 cfm
Volume flow discharged from slots related to plane n	270 cfm/ft
Discharged velocity at the slots related to plane n	7.5 ft/s (450 fpm)
Required discharged velocity at plane $(n + 1)$	7.6 ft/s (456 fpm)

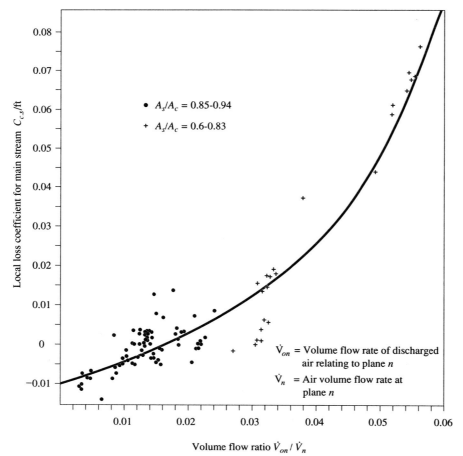

FIGURE 8.22 Local loss coefficient $C_{c.s}$ against volume flow ratio $\dot{V}_{on}/\dot{V}_n$ (Source: *ASHRAE Transactions*, 1984, Part II A. Reprinted with permission.)

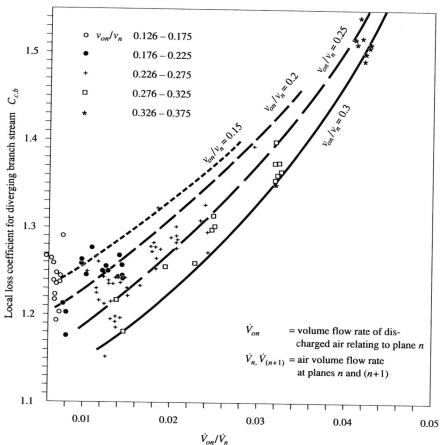

FIGURE 8.23 Local loss coefficiento $C_{c.b}$ against volume flow ratio $\dot{V}_{on}/\dot{V}_n$. (*Source*: *ASHRAE Transactions*, 1984, Part II A Reprinted with permission)

If the height of the rectangular duct remains the same, size the dimension of this duct at plane $(n + 1)$, 10 ft from plane n.

Solution

1. From the information given, the air velocity at plane n

$$v_n = \frac{23,760}{60 \times 4 \times 3} = 33 \text{ ft/s (1980 fpm)}$$

and from Eq. (8.46), the circular equivalent

$$D_e = \frac{1.3(ab)^{0.625}}{(a + b)^{0.25}} = \frac{1.3(3 \times 4)^{0.625}}{(3 + 4)^{0.25}} = 2.9 \text{ ft}$$

The Reynolds number of the supply air

$$\text{Re}_D = \frac{33 \times 2.9}{1.62 \times 10^{-4}} = 596,700$$

The friction factor

$$f = 0.25 \left[\log \left(\frac{\epsilon}{3.7D} + \frac{5.74}{0.9\text{Re}_D} \right) \right]^{-2}$$

$$= 0.25 \left[\log \left(\frac{0.0005}{3.7 \times 2.9} + \frac{5.74}{0.9 \times 596,700} \right) \right]^{-2}$$

$$= 0.0139$$

2. From the information given, the volume flow ratio is

$$\dot{V}_{on}/\dot{V}_n = \frac{270}{23,760} = 0.0113$$

and the velocity ratio is

$$v_{on}/v_n = \frac{7.5}{33} = 0.227$$

From Figs. (8.21) and (8.22), $C_{c.sn} = -0.0025$ per ft and $C_{c.bn} = 1.22$.

3. Assume that $C_{c.b(n+1)} = 1.23$. Then from Eq. (8.81)

$$
\begin{aligned}
V_{(n+1)} = \bigg\{ & \Big[\big(v_{on}^2 - v_{o(n+1)}^2 \big) C_o \\
& + (C_{c.bn} - C_{c.sn} - fL/D) v_n^2 \Big] \frac{1}{C_{c.b(n+1)}} \bigg\}^{0.5} \\
= \bigg\{ & \Big[\big(7.5^2 - 7.6^2 \big) 1.5 \\
& + \Big(1.22 - \Big(\frac{0.0139 \times 10}{2.9} \Big) \\
& - (-0.0025 \times 10) \Big) 33^2 \Big] \frac{1}{1.23} \bigg\}^{0.5} \\
= & \; 32.53 \text{ ft/s (1952 fpm)}
\end{aligned}
$$

4. The required volume flow rate from the slots related to plane $(n + 1)$ is

$$
\dot{V}_{o(n+1)} = \frac{270 \times 7.6}{7.5} = 273.6 \text{ cfm/ft}
$$

Then the volume flow rate of air at plane $(n + 1)$ is

$$
\dot{V}_{(n+1)} = 23{,}760 - (273.6 \times 10) = 21{,}024 \text{ cfm}
$$

The duct area required at plane $(n + 1)$ is

$$
A_{(n+1)} = \frac{21{,}024}{60 \times 32.53} = 10.78 \text{ ft}^2
$$

The width of the duct is

$$
W = \frac{10.78}{3} = 3.59 \text{ ft}
$$

Return or Exhaust Duct Systems

Return duct systems with very short connecting ducts between the return inlet and the main duct are often used in industrial applications. They are often used in large workshops and need nearly equal return volume flow rates for each branch intake. To provide a better system balance, it is necessary to reduce the total pressure difference between the branch inlets at the two ends of the main duct.

For a conditioned space where noise is not a major problem, one effective means of reducing the total pressure loss of a long return main duct is to have a higher branch duct velocity than the velocity in the main duct. Because a total pressure difference between the branch inlets and the main duct at two ends is inevitable, the area of the return grille can be varied along the main duct to provide a more even return volume flow rate. Unlike the supply outlets, the variation of the area of the return grille has a minor effect on the space air diffusion.

A return duct system with short connecting ducts is shown in Fig. 8.24 for a textile mill. Twelve floor grilles are connected to the return main duct at the lower floor. The connecting duct is 2.6 ft. The sizes of the branch ducts vary from 16 in. × 16 in. at the remote end to 12 in. × 8 in. near the fan end. The main duct varies from 16 in. × 32 in. to 60 in. × 32 in. In Fig. 8.24, the variations in total pressure p_t and static pressure p_s along the main duct are shown by the upper and the middle curves, in inches WG, and the velocity in main return duct v, in fpm, is shown

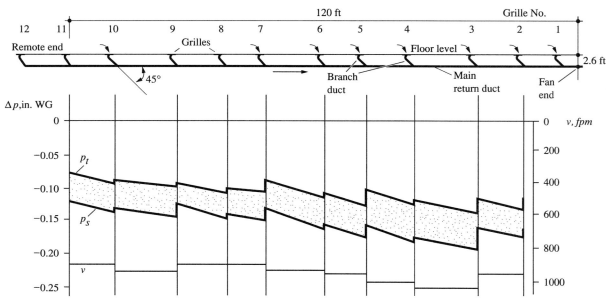

FIGURE 8.24 A return duct system with short connecting ducts for a textile mill.

by the lower curve. The velocity in the branches varies from 700 fpm at the remote end to 2000 fpm at the fan end. Note that p_t at grille 6 is -0.111 in. WG, -0.105 in. WG at grille 5, and -0.114 in. WG at grille 2. Such a satisfactory pressure balance along the return duct is mainly due to the higher branch velocity in the grilles near the fan end.

8.12 REQUIREMENTS OF EXHAUST DUCT SYSTEMS FOR A MINIMUM VELOCITY

Exhaust duct systems to transport dust particles or particulate products contained in the air require a minimum velocity in all duct sections of the system. These systems are used in many industrial applications and usually have air velocity ranging between 2400 and 4000 fpm. In addition to the variation of the sizes of the branch ducts, it is essential to select the proper configuration of duct fittings to provide a better system balance and to reduce the total pressure loss of the system. The following are recommendations for exhaust duct systems designs

- Round ducts usually produce smaller pressure losses; they are more rigid in construction.
- Air velocity inside the duct must not exceed the minimum velocity too much in order not to waste energy.
- Well-sealed joints and seams are important for reducing air leakage at higher pressure differentials.

8.13 PRESSURE AND AIR FLOW MEASUREMENTS

The total pressure of the air flow is the sum of the static pressure, which acts in all directions, and the velocity pressure, which results from the impact and the inertia of the air flow. The Pitot tube and manometers, shown in Figs. 8.25 and 8.26, are widely used to measure total pressure, static pressure, velocity pressure, and, therefore, air flow inside air ducts.

The Pitot tube consists of two concentric tubes having an outside diameter $D = \frac{5}{16}$ in. and an inner tube of diameter of $\frac{1}{8}$ in. as shown in Fig. 8.25. The inner tube opens to the air flow at the nose, and the other end is used as the total pressure tap. Eight equally spaced holes of 0.04 in. diameter allow the air to flow into the hollow space between the outer and inner tubes. The hollow space is connected to the static pressure tap. Because the small holes are located perpendicular to the center-line of the head, they are able to measure the static pressure when the head of the Pitot tube is placed in a position opposite to the direc-

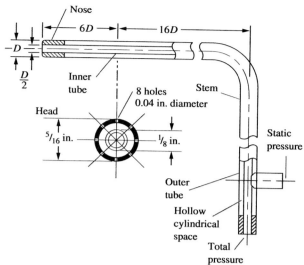

FIGURE 8.25 Pitot tube.

tion of the air flow. The center-lines of the small holes are located at a distance of 8 D from the nose and 16 D from the stem. The negative pressure produced at the nose is nearly balanced by the positive pressure at the stem.

The U-tube shown in Fig. 8.26a is the simplest type of manometer used to measure pressure in air ducts. The vertical difference in the liquid column indicates the pressure reading. For more accurate mea-

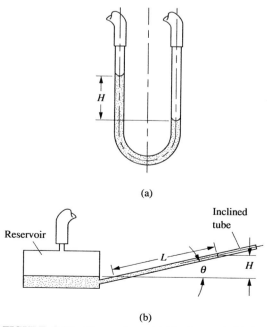

FIGURE 8.26 Manometers: (a) U-tube; and (b) inclined manometer.

surement, an inclined manometer (see Fig. 8.26b) is often used. The relationship between the height of the liquid column H and the length of the magnified inclined scale is

$$H = L \sin \theta \tag{8.82}$$

where L = length of the inclined scale with liquid column, ft
θ = angle of inclination of the inclined tube, degrees

Because the cross-sectional area of the inclined tube is very small compared with that of the reservoir, the change of the liquid level of the reservoir can be ignored.

The following precautions should be observed when one takes pressure measurements in an air duct such as that shown in Fig 8.27:

- The nose of the Pitot tube must always be placed opposite to the direction of air flow whether in the suction side or discharge side of the fan.

- When the total pressure or static pressure is measured, one leg of the U-tube or inclined manometer must be open to the atmospheric air as the reference datum. The smaller pressure is always connected to the open end of the inclined tube of the inclined manometer.

- For a velocity pressure measurement, the total and static pressure taps must be connected to the two ends of the manometer. The total pressure tap is connected to the reservoir and the static pressure tap is connected to the inclined tube. Velocity pressure is always positive.

- Because velocity is usually not uniform across the cross-sectional area, a traverse is often used to determine the average velocity. The cross section should be divided into equal areas. The measuring point is located at the center of each divided area. The number of measuring points should not be less than 16 for rectangular ducts and 20 for round ducts. When the velocity pressure is measured, in inches WG, the air velocity, in fpm, can be calculated from Eq. (8.10).

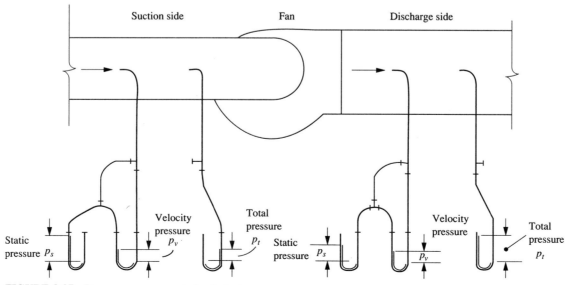

FIGURE 8.27 Pressure measurements in air ducts.

TABLE 8.11
Local loss coefficients for elbows, diverging tees, and diverging wyes

(1) Elbow; 3, 4, and 5-pieces, round

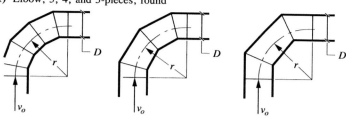

| 5-piece | 4-piece | 3-piece |

Coefficients for 90° elbows C_θ

No.of pieces	R_c/D			
	0.75	**1.0**	**1.5**	**2.0**
5	0.46	0.33	0.24	0.19
4	0.50	0.37	0.27	0.24
3	0.54	0.42	0.34	0.33

Angle correction factors K_θ (Idelchik, 1986; Diagram 6-1)

θ	0	20	30	45	60	75	90	110	130	150	180
K_θ	0	0.31	0.45	0.60	0.78	0.90	1.00	1.13	1.20	1.28	1.40

(2) Elbow, smooth radius with splitter vanes, rectangular

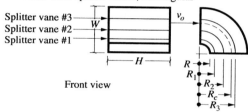

Splitter vane #3
Splitter vane #2
Splitter vane #1

Front view

Side view

Coefficients for elbows with two splitter vanes C_o

R/W	R_c/W	CR	H/W										
			0.25	**0.5**	**1.0**	**1.5**	**2.0**	**3.0**	**4.0**	**5.0**	**6.0**	**7.0**	**8.0**
0.05	0.55	0.362	0.26	0.20	0.22	0.25	0.28	0.33	0.37	0.41	0.45	0.48	0.51
0.10	0.60	0.450	0.17	0.13	0.11	0.12	0.13	0.15	0.16	0.17	0.19	0.20	0.21
0.15	0.65	0.507	0.12	0.09	0.08	0.08	0.08	0.09	0.10	0.10	0.11	0.11	0.11
0.20	0.70	0.550	0.09	0.07	0.06	0.05	0.06	0.06	0.06	0.06	0.07	0.07	0.07
0.25	0.75	0.585	0.08	0.05	0.04	0.04	0.04	0.04	0.05	0.05	0.05	0.05	0.05
0.30	0.80	0.613	0.06	0.04	0.03	0.03	0.03	0.03	0.03	0.03	0.04	0.04	0.04

(3) Wye, 45°, round, with 60° elbow, branch 90° to main

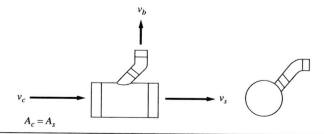

$A_c = A_s$

Branch

v_b/v_c	0	0.2	0.4	0.6	0.8	1.0	1.2	1.4	1.6	1.8	2.0
$C_{c,b}$	1.0	0.88	0.77	0.68	0.65	0.69	0.73	0.88	1.14	1.54	2.2

Main

v_s/v_c	0	0.1	0.2	0.3	0.4	0.5	0.6	0.8	1.0
$C_{c,s}$	0.40	0.32	0.26	0.20	0.14	0.10	0.06	0.02	0

TABLE 8.11(*continued*)

(4) Tee, diverging, round, with 45° elbow, branch 90° to main

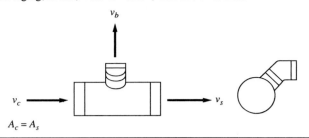

Branch										
V_h/V_c	0	0.2	0.4	0.6	0.8	1.0	1.2	1.4	1.6	1.8	2.0
$C_{c,b}$	1.0	1.32	1.51	1.60	1.65	1.74	1.87	2.0	2.2	2.5	2.7

For main local loss coefficient $C_{c,s}$, see values in (3)

(5) Tee, diverging, rectangular (Idelchik, 1986; Diagram 7-21)

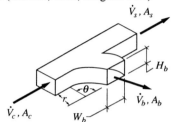

$\theta = 90°$

$r/W_b = 1.0$

Branch, $C_{c,b}$

		$\dot{V}_b/\dot{V}_c$								
A_b/A_s	A_b/A_c	0.1	0.2	0.3	0.4	0.5	0.6	0.7	0.8	0.9
0.25	0.25	0.55	0.50	0.60	0.85	1.2	1.8	3.1	4.4	6.0
0.33	0.25	0.35	0.35	0.50	0.80	1.3	2.0	2.8	3.8	5.0
0.5	0.5	0.62	0.48	0.40	0.40	0.48	0.60	0.78	1.1	1.5
0.67	0.5	0.52	0.40	0.32	0.30	0.34	0.44	0.62	0.92	1.4
1.0	0.5	0.44	0.38	0.38	0.41	0.52	0.68	0.92	1.2	1.6
1.0	1.0	0.67	0.55	0.46	0.37	0.32	0.29	0.29	0.30	0.37
1.33	1.0	0.70	0.60	0.51	0.42	0.34	0.28	0.26	0.26	0.29
2.0	1.0	0.60	0.52	0.43	0.33	0.24	0.17	0.15	0.17	0.21

Main, $C_{c,s}$

		$\dot{V}_b/\dot{V}_c$								
A_b/A_s	A_b/A_c	0.1	0.2	0.3	0.4	0.5	0.6	0.7	0.8	0.9
0.25	0.25	−.01	−.03	−.01	0.05	0.13	0.21	0.29	0.38	0.46
0.33	0.25	0.08	0	−.02	−.01	0.02	0.08	0.16	0.24	0.34
0.5	0.5	−.03	−.06	−.05	0	0.06	0.12	0.19	0.27	0.35
0.67	0.5	0.04	−.02	−.04	−.03	−.01	0.04	0.12	0.23	0.37
1.0	0.5	0.72	0.48	0.28	0.13	0.05	0.04	0.09	0.18	0.30
1.0	1.0	−.02	−.04	−.04	−.01	0.06	0.13	0.22	0.30	0.38
1.33	1.0	0.10	0	0.01	−.03	−.01	0.03	0.10	0.20	0.30
2.0	1.0	0.62	0.38	0.23	0.23	0.08	0.05	0.06	0.10	0.20

Source: ASHRAE Handbook 1989, Fundamentals. Reprinted with permission. For details, refer to *ASHRAE Handbook.*

REFERENCES AND FURTHER READING

ASHRAE, *ASHRAE Handbook 1988, Equipment,* ASHRAE, Inc., Atlanta, GA, 1988.

ASHRAE, *ASHRAE Handbook 1989, Fundamentals,* ASHRAE, Inc., Atlanta, GA, 1989.

ASHRAE/IES Standard 90.1–1989, *Energy Efficient Design of New Buildings Except New Low-Rise Residential Buildings,* ASHRAE, Inc., Atlanta, GA, 1989.

AISI/SMACNA, *Measurement and Analysis of Leakage Rates from Seams and Joints for Air Handling Systems,* Vienna, VA, 1972.

ASHRAE/SMACNA/TIMA, *Investigation of Duct Leakage,* ASHRAE Research Project 308, Atlanta, GA, 1985.

Behls, H., "Balanced Duct Systems," *ASHRAE Transactions,* Part I, pp. 624–646, 1978.

Coe, P. E., "The Economics of VAV Duct Looping," *Heating/Piping/Air Conditioning,* pp. 61–64, Aug. 1983.

Colebrook, C. F., "Turbulent Flow in Pipes with Particular Reference to the Transition Region Between the Smooth and Rough Pipe Laws," *Journal of the Institute of Civil Engineers,* p. 133, February 1939.

Dean, R. H., Dean, F. J., and Ratzenberger, J., "Importance of Duct Design for VAV Systems," *Heating/Piping/Air Conditioning,* pp. 91–104, Aug. 1985.

Habjan, J., "Medium Pressure Duct Sizing and Design," *Heating/Piping/Air Conditioning,* pp. 95–100, Dec. 1984.

Heyt, H. W., and Diaz, M. J., "Pressure Drop in Flat-Oval Spiral Air Duct," *ASHRAE Transactions,* Part II, pp. 221–232, 1975.

Huebscher, R. G., "Friction Equivalents for Round, Square and Rectangular Ducts." *ASHRAE Transactions,* pp. 101–144, 1948.

Idelchik, I. E., *Handbook of Hydraulic Resistance,* 2nd ed., Hemisphere Publishing Corp., Washington, D.C., 1986.

Miller, E. B., and Weaver, R. D., "Computer-Aided Duct Design in a Small Engineering Office," *ASHRAE Transactions,* Part I, pp. 647–664, 1978.

Moody, L. F., "Friction Factors for Pipe Flow," *Transactions A.S.M.E.,* Vol 66, p. 673, 1944.

National Fire Protection Association (NFPA), *NFPA 90A Standard for the Installation of Air Conditioning and Ventilating Systems,* NFPA, Quincy, MA, 1985.

Sheet Metal and Air Conditioning Contractors' National Association, *HVAC Duct Construction Standards—Metal and Flexible,* SMACNA, Chantilly, VA, 1985.

Shitzer, A., and Arkin, H., "Study of Economic and Engineering Parameters Related to the Cost of an Optimal Air Supply Duct System," *ASHRAE Transactions,* Part II, pp. 363–374, 1979.

Stoecker, W. F., and Bertschi, R. I., "Design Duct System for Minimum Life-Cycle Cost," *Conference on Improving Efficiency and Performance of HVAC Equipment and Systems for Commercial and Industrial Buildings,* Purdue University, Vol. I, pp. 200–207, 1976.

Thermal Insulation Manufacturers Association Heating/Piping/Air Conditioning (TIMA/HPAC), "Fiber Glass Duct Systems," *Heating/Piping/Air Conditioning,* Supplement, Oct. 1986.

Tsal, R. J., and Behls, H. F., "Evaluation of Duct Design Methods," *ASHRAE Transactions,* Part I A, pp. 347–361, 1986.

Tsal, R. J., Behls, H. F., and Mangel, R., "T-Method Duct Design, Part I: Optimization Theory," *ASHRAE Transactions,* Part II, pp. 90–111, 1988.

Tsal, R. J., Behls, H. F., and Mangel, R., "T-Method Duct Design, Part II: Calculation Procedure and Economic Analysis," *ASHRAE Transactions,* Part II, pp. 112–150, 1988.

Wang, S. K., *Air Conditioning,* Vol. 2, Hong Kong Polytechnic, Hong Kong, 1987.

Wang, S. K., Leung, K. L., and Wong, W. K., "Sizing a Rectangular Supply Duct with Transversal Slots by Using Optimal Cost and Balanced Total Pressure Principle," *ASHRAE Transactions,* Part II A, pp. 414–429, 1984.

Wright, D. K., Jr., "A New Friction Chart for Round Ducts," *ASHVE Transactions,* pp. 303–316, 1945.

SPACE AIR DIFFUSION

9.1 PRINCIPLES OF SPACE AIR DIFFUSION

Space air diffusion distributes the conditioned air containing outdoor air to the occupied zone (or a given enclosure) in a conditioned space according to the occupants' requirements. An occupied zone is a space with the following dimensions: (width of room − 1 ft) × (depth of room − 1 ft) × a height of 6 ft. A satisfactory space air diffusion evenly distributes the conditioned and outdoor air to provide a healthful, and comfortable indoor environment for the occupants, or the appropriate environment for a specific manufacturing process, at optimum cost.

Because space air diffusion is the last process of air conditioning and takes place entirely within the conditioned space, it directly affects the effectiveness of air conditioning. Because diffused and ambient air are transparent, space air diffusion is also difficult to trace.

Draft and Effective Draft Temperature

One of the most common complaints about air conditioned space is draft. *Draft* is defined as an unwanted local cooling of the human body caused by air movement and lower space air temperature. Recent findings by Fanger et al. (1989) demonstrate that air velocity fluctuations also create drafts. Figure 9.1 depicts such fluctuations in a typical conditioned space. A new dimensionless parameter called *turbulence intensity* (I_{tur}) is introduced and expressed as

$$I_{tur} = \frac{\sigma_v}{v_m} \qquad (9.1)$$

where σ = standard deviation of the fluctuations of air velocity, fpm
v_m = mean air velocity, fpm

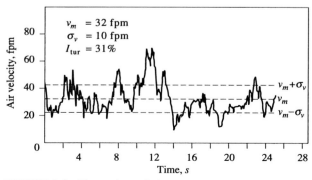

FIGURE 9.1 Fluctuations of air velocity in a typical air conditioned space. (Adapted with permission from ASHRAE journal. April 1989 p. 20.)

Experiments showed that at $T_r = 73.4°F$ and a $v_m = 30$ fpm, the percentage of dissatisfied occupants may increase from about 10 to 15 percent when I_{tur} increases from 0.1 to 0.5.

A parameter called *effective draft temperature* θ, in °F, which combines the effects of uneven space air temperature and air movement, is often used to assess the deviations of local magnitudes from the mean value, and is defined as

$$\theta = (T_x - T_r) - a(v_x - v_{rm}) \qquad (9.2)$$

where T_x, v_x = space air temperature and velocity at a specific location, °F and fpm. In Eq. (9.2), symbol a is a conversion constant to combine the effects of space air temperature and air movements. Its value is 0.07 when T is expressed in °F and v is expressed in fpm.

The desirable mean space air velocity v_{rm}, in fpm, is closely related to the space air temperature T_r to be maintained, the metabolic rate, and the clothing insulation of the occupant. According to Fig. 5.8, for a space relative humidity $\varphi = 50$ percent, $T_{rad} = T_r$, a metabolic rate of 400 Btu/h, and a clothing insulation of 0.6, their relationship may be given as follows:

T_r,°F	70	72	74	76	78	80
v_{rm}, fpm	16	20	25	32	40	55

Air Diffusion Performance Index

To an extent, the magnitude of effective draft temperature θ also reflects the degree of thermal comfort that can be provided by a comfort air conditioning system. During cooling mode in commercial and public buildings, if the space temperature is maintained between 75 to 78°F, $T_{rad} = T_r$, space air velocity $v_r < 55$ fpm, and the space relative humidity is between 30 and 70 percent, most sedentary occupants feel comfortable when $-3°F < \theta < +2°F$. The *Air Diffusion Performance Index* (ADPI), in percent, which evaluates the performance of space air diffusion, is calculated as

$$ADPI = N_\theta \times \frac{100}{N} \qquad (9.3)$$

where N_θ = number of points measured in the occupied zone in which $-3°F < \theta < +2°F$
N = total number of points measured in the occupied zone

The higher the ADPI, the higher the percentage of occupants who feel comfortable. Maximum ADPI approaches 100 percent.

In cooling mode operation, ADPI is an appropriate index to evaluate a space air diffusion system. For heating mode operation, the temperature gradient between two points in the occupied zone may be a better indicator of the thermal comfort of the occupants and the effectiveness of a space air diffusion system. Usually, the temperature gradient is less than 5°F within an occupied zone.

Space Diffusion Effective Factor

The effectiveness of a space air diffusion system can also be assessed by using a *space diffusion effective factor* for air temperature ϵ_T, or for air contamination ϵ_C. Both factors are dimensionless. Effective factor ϵ_T compares temperature differentials and ϵ_C compares contamination differentials as follows

$$\epsilon_T = \frac{T_{re} - T_s}{T_r - T_s} = \frac{T_{ex} - T_s}{T_r - T_s}$$
$$\epsilon_C = \frac{C_{ex} - C_s}{C_r - C_s} \tag{9.4}$$

where T = temperature, °F
C = concentration of air contamination, $\mu g/m^3$ or grains/ft^3

In Eq. (9.4), subscript re represents the recirculating air, ex the exhaust air, r the space air or air at the measuring point, and s the supply air. When $\epsilon \geq 1$, the space air diffusion is considered effective. If $\epsilon < 1$, a portion of supply air has failed to supply the occupied zone and exhausts through the return or exhaust inlets directly. Parameters in both numerator and denominator must be in same units.

Air Exchange Efficiency

Air exchange efficiency ϵ_{ex} is the ratio of the nominal exchange time τ_n, in seconds, to the exchange time for all the air in a room or a conditioned space, in seconds, that is

$$\epsilon_{ex} = \frac{\tau_n}{2\tau_{mean}} \tag{9.5}$$

where τ_{mean} = mean age of all air in the room or conditioned space, s. In Eq. (9.5), the nominal exchange time τ_n is actually the exchange time of the air flow, which depends entirely on the air change rate, or

$$\tau_n = 60 \frac{V_r}{\dot{V}_s} \tag{9.6}$$

where V_r = volume of the room or conditioned space, ft^3
$\dot{V}_s$ = supply volume flow rate, cfm

Air exchange efficiency is closely related to the mean age of the air in the conditioned space, and therefore, the contamination level of the air.

9.2 AIR JETS

An *air jet* is an airstream that discharges from an outlet with a significantly higher velocity than that of the surrounding air, and moves along its central axis until its terminal velocity reduces to a value that equals or approximately equals the velocity of the ambient air. Because of the turbulence of air particles, air jets tend to spread. They also rise or fall depending on the buoyancy of the airstream.

The outer boundary of an air jet where air is moving at a perceptible velocity (such as 150 fpm, 100 fpm, or 50 fpm) is called the *envelope*.

In general, air jets can be classified as free or confined, isothermal or nonisothermal, and axial or radial. A *free air jet* is an ideal air jet whose envelope (outer boundary) is not confined by the enclosure of the conditioned space. A *confined air jet's* envelope is confined by the ceiling, floor, walls, windows, and furniture of the conditioned space.

According to experimental results, the characteristics of an air jet approach those of a free air jet when the relationship $\sqrt{A_r}/D_o > 50$ is satisfied. Here, A_r represents the cross-sectional area of the enclosure perpendicular to the central axis of the air jet, in ft^2, and D_o indicates the diameter or the circular equivalent of the supply outlet, in ft.

An *isothermal air jet* is one whose temperature is equal or nearly equal to the temperature of the ambient air. A *non-isothermal air jet*, is one whose temperature is different from that of the ambient air in the conditioned space. An *axial air jet* projects in one direction, and a *radial air jet* projects radially in all directions.

Free Isothermal Jets

Along the central axis of a free isothermal jet, four zones can be identified, as shown in Fig. 9.2:

1. *Core zone.* In the core zone, the central axis velocity remains unchanged. This zone extends about 4 D_o from the surface of the outlet.
2. *Transition zone.* In the transition zone, the central axis velocity decreases inversely with the square root of the distance from the surface of the outlet. This zone extends about 8 D_o.
3. *Main zone.* In the main zone, the turbulent flow is fully developed and the maximum velocity decreases inversely with the distance from the

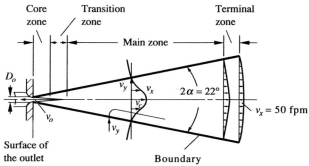

FIGURE 9.2 Four zones of a free isothermal axial air jet.

surface of the outlet. Even when the air jet is discharged from a rectangular outlet, the cross section of the airstream becomes circular in the main zone. This zone extends about 25 to 100 D_o in length.

4. *Terminal zone.* In the terminal zone, the maximum air velocity decreases rapidly to a value less than 50 fpm within a distance of a few outlet diameters.

In core, transition, and main zones, the fluctuating velocity components transport momentum across the boundary. Therefore, ambient air is induced into the air jet and the airstream diverges to a greater spread either vertically or horizontally. The angle of divergence 2α of a free isothermal jet discharged from a nozzle is about 22°. For an air jet discharged from a slot, its angle of divergence 2α perpendicular to the slot is about 33°. If the guide vanes at the outlet are deflected at an angle from the straight position, the spread of the air jet is greater.

The velocity profiles at different cross-sectional planes perpendicular to the air flow in the main zone of a free isothermal jet are similar to each other. At any specific cross-sectional plane, the velocity profile can be approximated by the formula

$$\left(\frac{R}{R_{0.5}}\right)^2 = 3.3 \log \frac{v_c}{v} \qquad (9.7)$$

where v = air velocity at distance R from the centralaxis of the air jet, fpm
$\quad v_c$ = central axis velocity, fpm
$\quad R$ = radial distance from the central axis of the air jet, ft
$\quad R_{0.5}$ = radial distance from the central axis of the air jet to a point where the velocity equals to $0.5\ v_c$, ft

Research shows that the law of conservation of momentum can be assumed in the main zones of a free air jet. The application of this law to deter-mine the central axis velocity of a free air jet gives the following results:

$$\frac{v_c}{v_o} = \frac{KD_o}{x} = \frac{K'\sqrt{A_o}}{x} = \frac{K'H_o}{x} \qquad (9.8)$$

where K, K' = central axis velocity constants, depending mainly on the type of the outlets; normally, $K' = 1.13K$
$\quad H_o$ = width of the radial air jet at the outlet or *vena contracta*, ft
$\quad v_o$ = mean velocity at the *vena contracta* after the outlet, fpm
$\quad x$ = distance from the surface of the outlet to a cross-sectional plane having central axis velocity v_c, ft

In Eq. (9.8), A_o (in ft^2) represents the effective area of the airstream, that is, the minimum area at the *vena contracta*. It can be calculated as

$$v_o = v_{\text{core}} \times C_d \times R_{\text{fa}}$$
$$A_o = A_c \times C_d \times R_{\text{fa}} \qquad (9.9)$$

where A_c = core area of the outlet, that is, the surface area of the opening, ft^2
$\quad v_{\text{core}}$ = face velocity at the core of the outlet, fpm
$\quad C_d$ = discharge coefficient, usually between 0.65 and 0.9
$\quad R_{\text{fa}}$ = ratio of the free area to the gross area; the free area is the net area of the opening through which air can pass

When v_o is between 500 and 1000 fpm, for round free openings, $K = 5$; for rectangular openings and linear slot diffusers, $K' = 4.9$. When v_o is between 2000 and 5000 fpm, for round free openings, $K = 6.2$.

When multiple air jets are discharged into a conditioned space at the same level, each air jet behaves independently until the jets meet. From the point the jets meet, the velocities between the central axes of the air jets increase until they are equal to the central axis velocities of the air jets.

Throw, Entrainment Ratio, and Characteristic Length

Throw, in ft, is defined as the axial distance from the outlet to a cross-sectional plane where the maximum velocity of the airstream at the terminal zone has been reduced to 50 fpm, 100 fpm, or 150 fpm. The throw is indicated by T_v, and the subscript denotes the terminal velocity for which the throw is measured. For instance, T_{50} indicates the throw with a terminal velocity of 50 fpm.

From Eqs. (9.8) and (9.9), throw can be calculated as

$$T_v = \frac{K'\dot{V}_s}{v_{t.max}\sqrt{A_c \times C_d \times R_{fa}}} \qquad (9.10)$$

where $v_{t.max}$ = maximum velocity of the airstream at the terminal zone, fpm. For a specific configuration of supply outlet, throw T_v depends both on supply volume flow $\dot{V}_s$ and supply outlet velocity v_{core}.

The *entrainment ratio* R_{en} is the ratio of the volume flow rate of the total air at a specific cross sectional plane of the air jet $\dot{V}_x$ to the volume flow rate of the supply air $\dot{V}_o$ discharged from the outlet, which is sometimes called *primary air*. Total air is the sum of supply air and induced air.

The entrainment ratio is proportional to the distance or square root of the distance from the outlet. For circular jets in main zone, entrainment ratio can be calculated as

$$R_{en} = \frac{\dot{V}_x}{\dot{V}_o}$$
$$= \frac{2x}{K'\sqrt{A_o}} = \frac{2v_o}{v_c}$$

For a long slot, the entrainment ratio is

$$R_{en} = \frac{\dot{V}_x}{\dot{V}_o} = \sqrt{\frac{2x}{K'H_o}} = \sqrt{2}\frac{v_o}{v_c} \qquad (9.11)$$

The *characteristic length L*, in ft, is either the horizontal distance from the surface of the outlet to the nearest vertical opposite wall, or the horizontal distance from the surface of the outlet to the midplane between two outlets in the direction of air flow, or distance to the closest intersection of air jets. The ratio of throw to characteristic length T_v/L is related to the ADPI of various supply outlets and has been used as a parameter in space diffusion design.

Confined Air Jets

In actual practice, most air jets are confined by the boundary of the room or the conditioned space. For a confined air jet, the total momentum of the fluid elements decreases gradually as distance from the outlet increases because of friction between the airstream and the boundary.

SURFACE EFFECT. When a primary airstream discharged from a supply outlet flows along a surface, the velocity of the primary airstream is significantly higher than that of the ambient air, and a lower pressure region is formed near the surface along the air flow, as shown in Fig. 9.3. Consequently, induced ambient air at a comparatively higher pressure presses the air jet against the surface, even when it is a curved

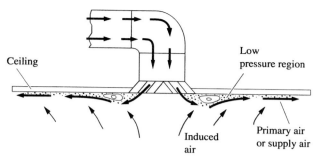

FIGURE 9.3 Surface effect.

surface. Such a phenomenon is called the *surface effect* or the *Conda effect*.

Friction between the air jet and the boundary decreases the central axis velocity of confined air jets, as shown in Fig 9.4. However, because of the surface effect, the throw of a confined air jet is longer, and the drop from the horizontal axis smaller, than those of a free air jet.

AIR FLOW PATTERN. The flow pattern and characteristics of a confined air jet using a side wall outlet were introduced by Russian scientists during the 1950s. In an air conditioned room whose supply outlet is located above the occupied zone and the exhaust opening is on the same side of the supply outlet, the supply air jet clings to the surface of the ceiling and mixes with the room air. An induced reverse airstream, with more even velocity and temperature distribution than that of the air jet, covers the occupied zone.

Figure 9.5 shows the air flow pattern of a typical confined isothermal jet. When supply air is discharged from the circular outlet and moves along the surface of the ceiling, the fluctuating velocity components continue to transport momentum across the boundary of the air jet. Therefore, ambient air is induced into the air jet, and the induced circulating air flow occupies most of the enclosed space.

As the air jet moves forward, its mass flow rate increases and mean air velocity decreases until it arrives at a cross-sectional plane where dimensionless distance $s = 0.22$. Term s is defined as

$$s = \frac{ax}{\sqrt{0.5A_r}} \qquad (9.12)$$

where a = turbulence factor; for a circular nozzle, $a = 0.076$, and for a rectangular outlet without guide vanes, $a \approx 0.15$.

The air jet terminates beyond $s = 0.22$ and the airstream makes a 180°-turn, forming a reverse airstream flowing in the opposite direction. The majority of the reverse airstream turns upward and is induced by the air jet again. Only a portion of it is exhausted outside the room.

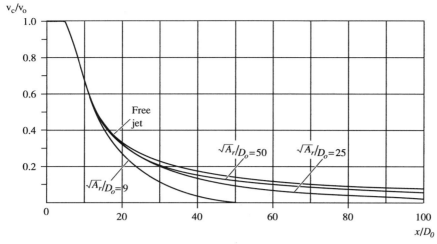

FIGURE 9.4 Central axis velocities of confined air jets.

The angle of divergence, the velocity profile, and the calculation of the entrainment ratio of confined jets are similar to those of free jets. Based on the principle of continuity of mass at steady state, other characteristics of confined air jets can be summarized as follows:

- If there is no infiltration into or exfiltration from the room, the mass flow rate of supply air is exactly equal to the exhaust air.

- If there are no obstructions in the room, the stream lines of the induced air are closed curves.

- If the supply outlet and exhaust inlet are located on the same side of the room, at any cross-sectional plane perpendicular to the horizontal air flow, the mass flow rate of airstreams flowing at opposite directions must be equal.

- The volume flow rate of induced air is equal to or several times greater than the supply air at the cross-sectional plane where $s = 0.22$. The characteristics

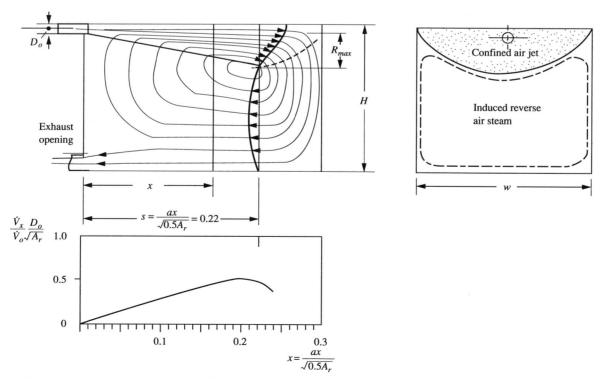

FIGURE 9.5 Air flow pattern of a typical confined air jet.

of airstreams in the occupied zone depend mainly on the induced reverse airstream. The volume flow rate and the air velocity of the reverse airstream are highest at the cross-sectional plane where $s = 0.22$.

Free Nonisothermal Jets

When the temperature of the conditioned air discharged from a supply outlet is different from that of the ambient air in a conditioned space, the buoyancy of the fluid particles cause the trajectory of the air jet to deviate from the axis of the free isothermal jet. A cold air jet will descend, and a warm air jet will ascend. Figure 9.6 shows a free cold air jet discharged horizontally from a nozzle.

In Fig. 9.6, x is the horizontal distance between a cross-sectional plane in a cold air jet and the surface of the outlet and y is the drop of the cold air jet, that is, the vertical distance between the horizontal axis of the nozzle and the center of the specific cross-sectional plane. The velocity profile at the specific cross-sectional plane is indicated by the solid line and the temperature profile is indicated by dotted line.

According to the experimental results by Kostel (1955), the following empirical formula can be used to determine the vertical drop of a cold air jet and the vertical rise of the warm air jet discharged from a nozzle

$$\frac{y}{\sqrt{A_o}} = \frac{x}{\sqrt{A_o}} \tan \alpha + KAr \left\{ \frac{x}{\left(\sqrt{A_o}\right)\cos \alpha} \right\}^3 \tag{9.13}$$

where α = angle between the central axis of the air jet and the horizontal axis of the nozzle, degrees

K = a constant

Ar, known as the Archimedes number, indicates the buoyant forces of the air jet and is given as

$$Ar = \left(g \frac{\sqrt{A_o}}{v_o^2} \right) \frac{\Delta T_o}{T_{Rr}} \tag{9.14}$$

where g = gravitational acceleration, ft/s^2

ΔT_o = temperature difference between the supply air at the outlet and the ambient air in the conditioned space, °F

T_{Rr} = absolute temperature of the space air, °R

For free air jets, Kostel (1955) determined that the constant $K = 0.065$.

For a nonisothermal jet, the relationship between the decay of central axis velocity and the central axis temperature difference ΔT_c, in °F, can be shown as

$$\frac{\Delta T_c}{\Delta T_o} = \frac{(T_c - T_r)}{(T_s - T_r)} = \frac{0.8v_c}{v_o} \tag{9.15}$$

where subscript c represents central axis, o the outlet, and r the space.

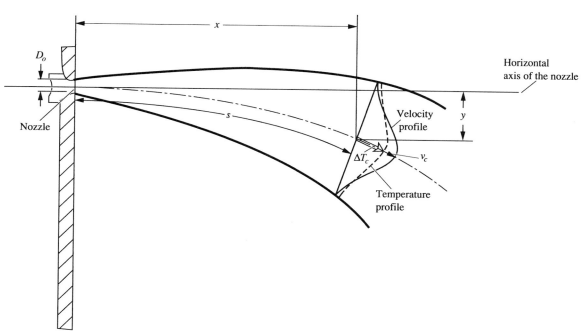

FIGURE 9.6 Path of a cold air jet discharged horizontally from a nozzle.

The characteristics of isothermal radial air jets are similar to those of axial air jets. When determining the central axis velocity, use x/H_o for radial jets instead of $x/\sqrt{A_o}$ in axial jets.

9.3 SUPPLY OUTLETS AND RETURN INLETS

The proper type of supply outlet for a conditioned space is largely determined by the architectural setup of the room, the air flow pattern needed, the indoor environmental requirements, and the load conditions. Five types of supply outlets are currently used: grilles and registers, ceiling diffusers, slot diffusers, light troffer diffusers, and nozzles. A window sill outlet is actually a type of grille mounted at the top of a fan–coil.

Grilles and Registers

A *grille* or grill, as shown in Fig. 9.7a, is an outlet for supply air or an inlet for return air or exhaust air. A *register* is a grille with a volume control damper. Figure 9.7b shows a single deflection grille. It consists of a frame and one set of adjustable vanes. Either vertical or horizontal vanes may be used as face vanes to deflect the airstream.

Figure 9.7c shows a double deflection register. It is able to deflect the airstream both horizontally or vertically. A volume damper is also used to adjust the volume flow through the register. Usually extruded aluminum vanes, aluminum or steel frames, and steel dampers are used. A baked enamel finish gives the grille an attractive appearance.

Grilles have a comparatively lower entrainment ratio, greater drop, longer throw, and higher air velocities in the occupied zone than slot and ceiling diffusers.

In the manufacturer's catalog, the performance of a grille or register is defined by the following parameters:

- Core size or core area A_c, which indicates the total plane area of an opening, ft^2
- Volume flow rate $\dot{V}$, cfm
- Air velocity $v = \dot{V}/A_k$, fpm
- Total pressure loss Δp_t, in. WG
- Throw at terminal velocities of 50 fpm, 100 fpm, and 150 fpm with horizontal vanes deflected at 0°, 22.5°, and 45°
- Noise criteria curve

Here A_k represents the net or corrected area, in ft^2, which is the unobstructed area of a grille, register, or diffuser measured at *vena contracta*.

For Air Diffusion Council-certified performance data, throw is tested on isothermal jets in order to avoid the influence of buoyancy forces at different supply temperature differentials ΔT_s. During testing, the difference between supply and room air temperatures cannot exceed ± 2°F.

Ceiling Diffusers

A *ceiling diffuser* consists of a series of concentric rings or inner cones made up of vanes arranged in fixed directions, and an outer shell or frame as shown in Fig. 9.8a. Ceiling diffusers can be round, square, or rectangular. Supply air is discharged through the concentric air passages or directional passages in one, two, three, or in all directions by using different types of inner cones and vanes. The air discharge pattern of

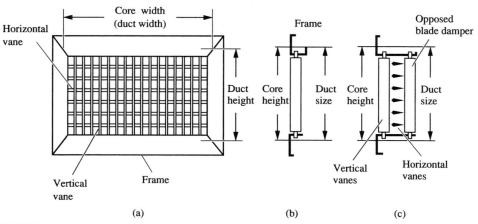

FIGURE 9.7 Supply grille and register: (a) grille, frontview; (b) single deflection, vertical vanes; (c) double deflection register, with vertical and horizontal vanes.

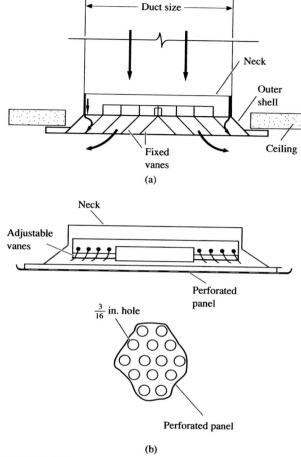

FIGURE 9.8 Ceiling diffusers: (a) square and rectangular ceiling diffusers and (b) perforated ceiling diffuser.

diffusers are designed to discharge air evenly in all directions so that the surrounding ceiling surfaces are less likely to be smudged.

The size of a ceiling diffuser is determined by its neck size, and is therefore closely related to the neck velocity of the diffuser. Ceiling diffusers are made of an outer shell and a removable inner core. The opposed-blade damper with an equalizing device of parallel blades mounted within a frame is often used to adjust the volume flow and to provide a more uniform air flow.

Ceiling diffusers are usually made of aluminum-coated heavy-gauge steel, extruded aluminum, or teak wood.

In addition to high induction ratios, ceiling diffusers also have a good surface effect and a shorter throw, and can supply air in all directions. Horizontally projected ceiling diffusers are suitable for conditioned spaces with low headroom.

The performance data of a ceiling diffuser are defined by the following parameters:

- Neck velocity v_{neck}, that is, mean air velocity at the neck of the ceiling diffuser, fpm
- Volume flow rate, $\dot{V}_s = A_k \times v_s$, cfm; A_k is determined at the testing laboratory, and the average supply velocity v_s is usually measured at a specified position at the outlet by a deflecting vane anemometer, often called a velometer
- Total pressure loss Δp_t, in. WG
- Throw, ft
- Noise criteria curve

Slot Diffusers

A *slot diffuser* consists of a plenum box with single or multiple slots and air deflecting vanes, as shown in Fig. 9.9.

The slots are available in 0.5-in., 0.75-in., 1-in. and 1.5-in. widths. The 0.75-in. and 1-in. slot widths are most commonly used. Slots are typically 2, 3, 4, 5, or 6 ft long. Air discharged from a slot diffuser can be projected horizontally or vertically. With a single slot diffuser, air is always discharged in one direction. With multiple slots, air can be horizontally discharged either left or right, or a combination of both. One slot can discharge vertically while another discharges horizontally. Slot diffusers are often called T-bar diffusers, as they are compatible with standard ceiling T-bars.

Air enters the plenum box from the round inlet at one side. The function of the plenum box is to distribute the air more evenly at the slot. The plenum box is sometimes insulated internally. After passing

many ceiling diffusers may be changed from a horizontal to vertical pattern by means of adjustable inner cones or special deflecting vanes. Ceiling diffusers are often mounted at the center of the conditioned space and discharge air in all directions.

An important characteristic of a ceiling diffuser is its induction effect. *Induction* is the volume flow rate of space air induced into the outlet by the primary airstream. It is also called *aspiration*. The ratio of the volume flow rate of induced air to the volume flow rate of supply air $\dot{V}_{ind}/\dot{V}_s$ (both in cfm) is called the *induction ratio*. A ceiling diffuser with a high induction ratio is beneficial for a higher supply temperature differential ΔT_s for cold air distribution in ice-storage systems, which will be discussed in Chapter 25.

Figure 9.8b shows a perforated ceiling diffuser. A perforated surface increases the induction effect. It is also used to match the appearance of the ceiling tile. The diameter of the holes on the perforated panel is typically $\frac{3}{16}$ in. Some ceiling

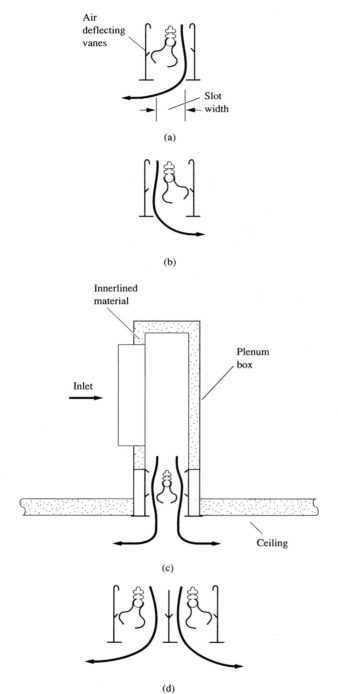

FIGURE 9.9   Slot diffusers: (a) horizontal left, (b) horizontal right, (c) vertical, and (d) two-slot, two-way.

the deflecting vanes, the air is discharged from the slot horizontally (left or right) or vertically according to the required air flow pattern. Figure 9.9 shows the airstreams discharged from slot diffusers in various directions. Different manufacturers use their own patented air deflecting vanes. Deflecting vanes can also be used to block the air flow entirely if necessary.

Extruded aluminum is widely used as the material for the structural member of the slot and a galvanized sheet is used for the plenum box.

The performance of a slot diffuser is defined by the following parameters:

- Width of slot, number of slots, and the length of the slots
- The volume flow rate per ft of slot, or slot intensity, in cfm/ft
- Throw, total pressure loss, and noise criteria curve

Table 9.1 lists the performance data of a typical slot diffuser of 0.75-in. slot width, taken from a manufacturer's catalog.

Nozzles

A *nozzle* is a round supply outlet (see Fig. 9.10). The airstream discharged from a nozzle is contracted just before the outlet, which results in a higher flow velocity and more even distribution. The purpose of using a nozzle instead of other types of supply outlets is to provide a longer throw and a smaller spread. Nozzles are usually used in air diffusion systems with large space volume or in spot cooling/heating, which will be covered in detail in Section 9.8.

Accessories for Supply Outlets

Grilles, registers, and diffusers may require certain accessories to modulate of the volume flow rate of each supply outlet and to distribute the air more evenly at the outlet. Currently used accessories include:

- *Opposed-blade damper.* Usually attached to the register upstream from the vanes, or installed at the outlet collar leading to a ceiling diffuser, as shown in Fig. 9.11a.
- *Split damper.* A piece of sheet metal hinged at one end to the duct wall as shown in Fig. 9.11b. An equalizing grid is often used at the same time to distribute the supply air more evenly to the outlets. A split damper is also often installed at the junction of two main ducts in order to adjust the volume flow rates of the supply air between these ducts.
- *Gang-operated turning vanes.* These vanes are always installed before the take-off to the outlet, as shown in Fig. 9.11c. The tips of the turning vanes through which air enters the take-off remain parallel to the direction of the air flow. Gang-operated turning vanes are especially suitable for adjusting the air flow rate for registers that are mounted on the air duct with a very short outlet collar.

Return and Exhaust Inlets

Various types of return inlets are used to return space air to the air-handling unit or packaged unit in the fan room. Exhaust air inlets are usually connected to the exhaust duct. Before contaminated air is exhausted outside, it is usually cleaned and treated with special air-cleaning equipment.

Return and exhaust inlets can be classified as return or exhaust grilles, return or exhaust registers, return slots, and ventilated light troffers.

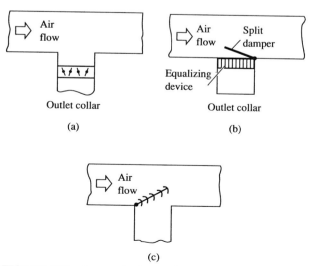

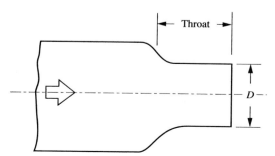

FIGURE 9.10 Round nozzle.

FIGURE 9.11 Accessories for supply outlets: (a) opposed-blade damper, (b) split damper with equalizing device, and (c) gang-operated turning vanes.

TABLE 9.1
Performance data for typical 0.75 in. width slot diffuser with plenum

1 Slot

Capacity, cfm			80	100	120	140	160	180	200	220	240
Projection, ft.		H	4-6-10	5-7-11	6-9-13	7-10-14	8-10-15	9-11-15	9-12-16	10-12-17	10-13-18
		V	3-4-6	3-5-7	4-6-8	5-6-8	5-6-9	5-7-10	6-7-10	6-7-10	6-8-11
Nom. length	Spread ft.	H	6-9-15	7-10-16	9-13-19	10-15-21	12-15-22	13-16-22	13-18-24	15-18-25	15-19-27
		V	4-5-8	4-6-9	5-8-10	6-8-10	6-8-12	6-9-13	8-9-13	8-9-13	8-10-14
36″	TP		.164	.254	.369	.499	.654				
	NC		22	29	35	40	44				
48″	TP		.090	.139	.201	.271	.353	.447	.554	.668	
	NC		—	22	27	32	36	39	43	46	
60″	TP		.058	.093	.133	.186	.244	.307	.377	.458	.539
	NC		—	—	24	29	32	36	40	43	45

2 Slots

Capacity, cfm			130	160	190	220	250	280	310	340	370
Projection, ft.		H	5-7-12	6-9-13	7-10-14	8-11-15	9-11-16	10-12-17	10-13-18	11-13-19	11-14-20
		V	2-4-6	3-4-6	4-5-7	4-5-7	5-6-8	5-6-8	5-6-9	5-6-9	6-7-10
Nom. length	Spread ft.	H	7-10-18	9-13-19	10-15-21	12-16-22	13-16-24	15-18-25	15-19-27	16-19-28	16-21-30
		V	3-5-8	4-5-8	5-6-9	5-6-9	6-8-10	6-8-10	6-8-12	6-8-12	8-9-13
36″	TP		.104	.162	.228	.305	.390	.490	.602		
	NC		24	30	35	40	44	47	50		
48″	TP		.059	.086	.126	.165	.216	.271	.330	.397	.472
	NC		—	21	26	30	34	38	41	43	46
60″	TP		.045	.066	.091	.146	.161	.202	.247	.297	.353
	NC		—	—	24	28	32	36	39	42	44

3 Slots

Capacity, cfm			160	190	220	250	280	310	340	370	400
Projection, ft.		H	5-7-12	6-8-13	6-10-14	7-11-15	8-11-16	9-12-17	10-12-18	10-13-19	11-13-19
		V	2-3-5	3-4-6	3-5-6	3-5-7	4-5-7	4-5-8	5-6-8	5-6-8	5-6-9
Nom. length	Spread ft.	H	7-10-18	9-12-19	9-15-21	10-16-22	12-16-24	13-18-25	15-18-27	15-19-28	16-19-28
		V	3-4-6	4-5-8	4-6-8	4-6-9	5-6-9	5-6-10	6-8-10	6-8-10	6-8-12
36″	TP		.073	.107	.140	.183	.230	.280	.336	.400	.466
	NC		21	26	31	35	38	41	44	46	49
48″	TP		.042	.059	.082	.104	.130	.160	.192	.228	.267
	NC		—	—	23	26	30	33	35	37	40
60″	TP		.023	.037	.046	.060	.078	.092	.110	.133	.156
	NC		—	—	—	21	24	27	30	32	35

Units are tested in accordance with the Air Diffusion Council (ADC) Code 1062R4 and ASHRAE Standard 36-72.

The projection horizontal (H) and vertical (V) are the distances to a terminal velocity of 150 fpm, 100 fpm, and 50 fpm. Spread is the maximum width of the jet defined by the above terminal velocities.

Horizontal projection values based on full-open, one direction.

The NC values are based on a room absorption of 10 dB, re 10^{-12} watts, and one diffuser. The values are 10 dB lower with vertical projection.

Total pressures (TP) are in inches of water.

Source: TITUS. Reprinted with Permission.

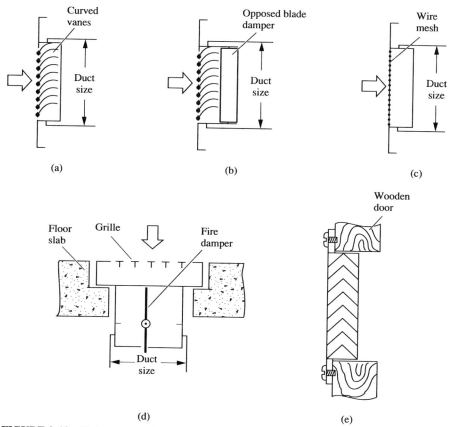

(a)   (b)   (c)

(d)   (e)

FIGURE 9.12 Various types of return grilles and registers: (a) return grille with curved vanes, (b) return register, (c) wire mesh grille, (d) floor type return grille, and (e) door louvers.

Return grilles and registers (see Fig. 9.12) are similar in shape and construction to supply grilles and registers. In Fig. 9.12a, only one set of vanes is required to direct the air into the return duct, so vision-proofed curved vanes are often used instead of the straight vanes. Vanes can be fixed or adjustable. An opposed-blade damper is sometimes installed behind the curved vanes in a return register, as shown in Fig. 9.12b. Wire mesh grilles Fig. 9.12c are simple and inexpensive. They are often installed in places where the appearance of the grille is not important. In many industrial buildings, floor grilles are used when the return duct is below the floor level, as shown in Fig. 9.12d. When room air is required to return to the fan room through a corridor and return duct, a door louver is often used to transport return air through the door. Vision-proofed, right angled louvers are generally used for this purpose, as shown in Fig. 9.12e. Local codes should be consulted before the door louvers are installed.

When a ceiling plenum is used as a return plenum (as in many commercial buildings), return slots are used to draw the return air through the ceiling. Return slots are similar to slot diffusers. They are usually

made with single or double slots, as shown in Fig. 9.13. The width of the slot varies from 0.5 in. to 1.75 in. Slots are usually available in lengths of 2, 2.5, 3, 4, and 5 ft.

The capacity of return slots is also expressed in cfm/ft of slot length at a specific pressure loss, in inches. WG. Return slots are always inner-lined with

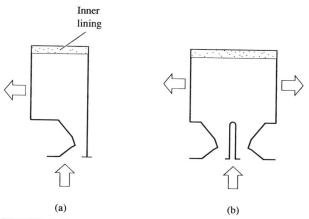

(a)   (b)

FIGURE 9.13 Return slots: (a) single-slot and (b) double-slot.

sound absorption material. This material prevents sound transmission between two rooms via the return slots and the ceiling plenum.

Exhaust inlets are similar to the return inlets but they often have a simpler design.

In return and exhaust inlets, air velocity decreases very sharply as the distance from the surface of the inlet increases. This phenomenon is the major difference in air flow characteristics between a return or exhaust inlet and a supply outlet.

Light Troffer Diffuser and Troffer-Diffuser-Slot

A *light troffer diffuser* combines a fluorescent light troffer and slot diffuser, sometimes in saddle type as shown in Fig. 9.14a. The slot can be used as a supply outlet or return inlet. A troffer-diffuser-slot is a combination of light troffer, slot diffuser, and return slot, as shown in Fig. 9.14b.

Light troffer diffusers and troffer-diffuser-slots serve two main purposes:

1. To maintain a lower air temperature in the light troffer, which increases the luminous efficiency of fluorescent lamps
2. To form an integrated layout of light troffer, diffuser, and return slots on suspended ceilings

In addition to these benefits, a combination of a light troffer and return slots also reduces the space

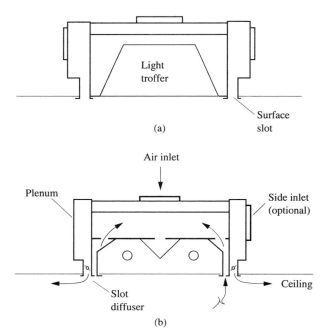

FIGURE 9.14 Light troffer, slot diffuser, and return slot combination: (a) light troffer diffuser slot and (b) troffer-diffuser-slot.

cooling load because return air absorbs part of the heat released by the fluorescent lights.

In a light troffer diffuser, surface slots are often used. Air can be discharged horizontally or vertically. Troffers and slots are available in 2-, 3-, and 4-ft lengths.

To prevent the deposit of dust particles on the surface of the fluorescent tube, the light troffer should be designed so that return air does not come in direct contact with the fluorescent tube. The result is high ventilation efficiency and a neat appearance.

9.4 MIXING FLOW

Air Flow Pattern and Space Air Diffusion

Air flow pattern determines the performance of space air diffusion in the occupied zone of commercial buildings or in the working area of a factory. The optimum air flow pattern for an occupied zone depends mainly on indoor temperature, relative humidity, and indoor air quality requirements; outdoor air supply; and the characteristics of the building.

Four types of air flow patterns are currently used in air conditioned space:

1. Mixing flow
2. Displacement flow
3. Projecting flow
4. Upward flow

Principles and Characteristics of Mixing Flow

The air flow pattern of an air conditioned space is said to be in mixing flow only when the supply air is thoroughly, or nearly thoroughly, mixed with ambient air, and the occupied zone or working area is dominated by the induced recirculating flow.

In many commercial and industrial buildings, cold supply air at 50 to 60°F, and a velocity of 400 to 800 fpm must be thoroughly mixed with the ambient air first. The temperature of the mixture is raised to 72 to 74°F, and its velocity drops to less than 70 fpm before it enters the occupied zone and working area, or drafts occur. Mixing flow is often the best choice for comfort air conditioning systems when the supply temperature differential is greater than 15°F, and the supply air velocity exceeds 300 fpm.

Straub et al. (1956) and Straub and Chen (1957) tested mixing flow patterns by using five types of supply outlets (see Figs. 9.15 to 9.19). These air flow patterns have the following characteristics:

INDUCTION OF SPACE AIR TO THE AIR JET. Mixing of induced air and the air jet reduces the differences in air velocity and temperature between the air jet and the ambient air to acceptable limits.

REVERSE AIRSTREAM IN THE OCCUPIED ZONE. The induced reverse airstream covers the occupied zone with a more even temperature and velocity distribution.

MINIMIZE THE STAGNANT AREA IN THE OCCUPIED ZONE. A *stagnant area* is a zone with natural convective currents. Air velocities in most stagnant zones are lower than 20 ft/min. Air stratifies into layers, with a significant temperature gradient from the bottom to the top of the stagnant zone. If the recirculating flow fills the entire occupied zone, no space remains for stagnant air.

TYPES AND LOCATIONS OF RETURN AND EX-HAUST INLETS. Because the return or exhaust air flow rate is only a fraction of the maximum total air flow of the supply air jet or the maximum air flow of reverse airstream, the location of the return or exhaust inlet does not significantly affect the air flow pattern of mixing flow in the occupied zone.

However, the location of the return and exhaust inlets does affect the effective factor ϵ_T or ϵ_C of space air diffusion. Therefore, for an induced recirculating flow pattern, the return intake should be located near the end of the reverse airstream, in the vicinity of a heat source, or within the stagnant zone.

For a conditioned space in which the space air is contaminated by dust particles, toxic gases, or odors, several evenly distributed return or exhaust grilles should be installed. Collective return grilles, that is, one or two large return inlets instead of many inlets, impair indoor air quality. They are only suitable under the following circumstances:

- When the conditioned space is small
- When it is difficult to provide a return duct
- When lower initial cost is of prime importance

TYPES AND LOCATIONS OF THE SUPPLY OUT-LETS. The type and location of the supply outlets affect the mixing flow pattern and its performance. Currently, mixing air flow incorporates the following types of supply outlets:

- High side outlets
- Ceiling diffusers
- Slot diffusers
- Sill and floor outlets
- Outlets form stratified mixing flow

Mixing Flow Using High Side Outlets

A high side outlet may be a sidewall outlet mounted at the high level of a conditioned space, or an outlet mounted directly on the supply duct. Figure 9.15 shows a mixing flow using a high sidewall outlet. In Fig. 9.15, the shaded black envelope shows the primary air envelope of the air jet, and the dotted envelope the total air envelope.

As the air jet discharges from the high sidewall outlet, the surface effect tends to keep the air jet in contact with the ceiling. During cooling, the air jet induces the ambient air and deflects downwards when

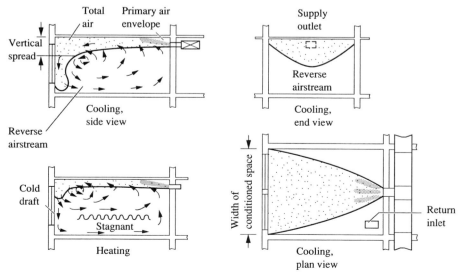

FIGURE 9.15 Mixing flow using high side outlets. (Adapted with permission from *ASHRAE Handbook 1989, Fundamentals.*)

it strikes on the opposite wall. As soon as the maximum air velocity of the air jet drops to about 50 fpm, the air jet terminates as it flows along the opposite wall. The induction of space air from the occupied zone into the air jet forms the reverse airstream and fills the occupied zone.

If the throw of a high sidewall outlet is longer than the sum of the length of the room and the height of the opposite wall, the air jet is deflected by the opposite wall and floor and enters the occupied zone with excessive air velocity. Both the air jet and the reverse airstream fill the occupied zone with a higher air velocity and greater temperature difference.

If the throw is too short, the air jet drops and may enter the occupied zone directly. This causes higher air velocity and lower temperature (cold draft) in the occupied zone.

During heating, warm air tends to rise and results in a shorter throw. As the induced airstream rises, a stagnant zone may form between the floor and the induced airstream. A warm air jet with sufficient velocity and a longer throw can reduce or even eliminate the stagnant zone.

If the airstream at the terminal zone of the warm air jet comes in contact with the cold inner surface of the window, it converts into a cold airstream and forms a cold draft when it flows downward and enters the occupied zone. The installation of a baseboard heater under the windowsill can prevent the formation of this cold draft and reduce the vertical temperature gradient in the occupied zone.

The most suitable location for a return inlet is on the ceiling outside the air jet. Return air flows through the return inlets and often uses the ceiling plenum as a return plenum. Such an arrangement often results in a higher effective factor ϵ_T and a lower installation cost for the return system.

In a room without a suspended ceiling, a cold or warm air jet has a shorter throw because of lack of surface effect. The buoyancy force causes a cold air jet to drop sooner, and it may enter the occupied zone directly. The higher rise of a warm air jet forms a greater stagnant zone above the floor level.

Mixing flow using high side outlets usually gives a longer throw and a higher air velocity in the occupied zone than other supply outlets. Drop must be checked to prevent the air jet from entering the occupied zone directly in a large conditioned space.

Mixing Flow Using Ceiling Diffusers

The air flow pattern of mixing flow using ceiling diffusers is similar to that of high side outlets. Ceiling diffusers produce a better surface effect, a shorter throw, a lower and more even distribution of air velocity, and a more even temperature in the occupied zone.

Figure 9.16 shows mixing flow using ceiling diffusers. During cooling, the cold air jet usually maintains contact with the ceiling, producing a better surface effect than sidewall outlets. The reverse airstream fills the occupied zone. During heating, thinner vertical spread and shorter throw create a larger, higher stagnant zone. The best location for return inlets is on the ceiling within the terminal zone of the air jet.

When two air jets discharge horizontally from ceiling diffusers or other outlets in opposite directions along the ceiling, with a ceiling height of 10 ft, the air velocity where two air jets collide should be less than 120 fpm in order to maintain an air velocity below 50 fpm 5 ft above the floor.

Mixing flow using ceiling diffusers is widely used for conditioned spaces with limited ceiling height. Many ceiling diffusers are designed to have a high induction ratio and to produce a very good surface effect. These ceiling diffusers are often used in variable-air-volume systems.

Mixing Flow Using Slot Diffusers

Slot diffusers are narrower and longer than ceiling diffusers and grilles. When supply air is discharged horizontally from a ceiling slot diffuser, it has a thinner vertical spread and an excellent surface effect. Figure 9.17 shows mixing flow using slot diffusers discharging vertically downward into the perimeter zone and also discharging horizontally along the ceiling in both the perimeter and interior zones of a large office.

Because of its superior surface effect, the horizontally discharged cold air jet remains in contact with the suspended ceiling even when the supply air volume flow rate is reduced in a variable-air-volume (VAV) system. The occupied zone is then filled with the reverse airstream at a more uniform temperature and air velocity.

In order to counteract natural convection along the inner surface of a window in the perimeter zone, the air jet is often projected downward toward the window. A sufficient throw is important in order to produce a downward air jet that offsets the cold draft at the inner surface of the window during winter heating. The occupied zone fills with induced air flow at an air velocity lower than 50 fpm.

For a mixing flow using slot diffusers, the location of the return slot is preferably aligned with the supply slots.

In addition to their excellent surface effect in mixing flow, slot diffusers also have a linear appearance

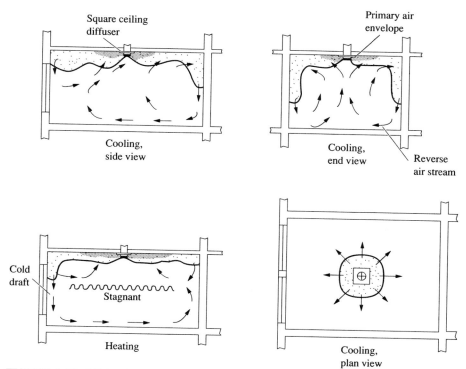

FIGURE 9.16 Mixing flow using ceiling diffusers. (Adapted with permission from *ASHRAE Handbook 1989, Fundamentals.*)

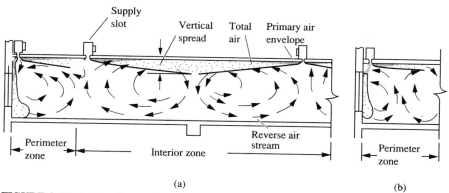

FIGURE 9.17 Mixing flow using slot diffusers: (a) perimeter and interior zone, cooling; and (b) perimeter zone, heating.

that can be easily coordinated with electric lights and ceiling modular arrangements. Therefore, they are widely used in buildings using VAV systems with moderate loads and normal ceiling height.

Mixing Flow Using Sill or Floor Outlets

The purpose of using a sill or floor outlet is to counteract the cold draft flowing downward along the inner surface of the window when the outdoor temperature is below 30°F. Figure 9.18 shows a mixing flow using a sill outlet.

During cooling, a stagnant zone may form above the eddies beyond the terminal zone of the air jet if supply air velocity is not high enough. Meanwhile, the space under the cold air jet is filled with reverse airstreams and cooled. During heating, a stagnant zone may form below the ascending induced airstream.

Mixing flow using sill or floor outlets has been widely used in buildings with large window areas or in raised-floor offices. The direction and amount of air flow from floor diffusers usually can be adjusted according to the requirements of the occupants.

Stratified Mixing Flow

In a building with a high ceiling, it is more economical to stratify the building vertically into two zones (the stratified upper zone and the cooled lower zone)

or three zones (upper, transition, and lower zones) during cooling. The upper boundary of the lower zone is at the level of the supply outlet where the air jet projects horizontally.

Gorton and Sassi (1982) and Bagheri and Gorton (1987) performed a series of model studies and experiments about stratified space air diffusion. Figure 9.19a shows the elevation view of the nuclear reactor facility in which the experiments were conducted. The ceiling height was 41 ft. An air-handling unit was located at floor level and two diffusers were mounted on the supply riser 16 ft above floor level. The supply velocity at the outlet was about 1000 fpm and the supply temperature was around 60°F.

Figure 9.19b shows the space air temperature profile at 4 P.M. at various height levels during cooling. This temperature profile could be divided into:

- An upper zone, from the ceiling down to about 22 ft from the floor level, in which air temperature varied from 82.5 to 79°F
- A transition zone, between 16 and 22 ft from the floor, in which air temperature varied from 79 to 74.5°F
- A lower zone below 16 ft, in which air temperature varied from 72 to 72.5°F

The induced recirculating flow under the cold air jet formed the lower zone. Most of the lower zone was filled with reversed airstreams. The upper zone

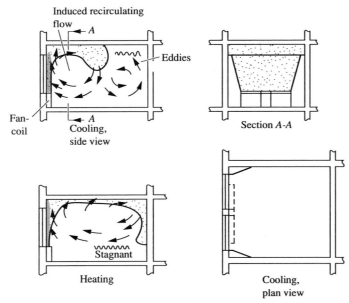

FIGURE 9.18 Mixing flow using a sill outlet. (Adapted with permission from *ASHRAE Handbook 1989, Fundamentals*).

and the transition zone were formed because of the higher cooling loads at the upper level and a weaker recirculating induced flow above the cold air jet. Because of the drop of the cold air jet, a greater amount of recirculating flow was induced into the cold air jet in the lower zone than in the upper zone. Consequently, most of the cooling occurred in the lower zone. The formation of a stagnant layer near the ceiling also retarded air flow and heat transfer between outdoor and indoor air in the upper zone.

Figure 9.19c shows the space temperature profile at various height levels at 4 P.M. during heating. The supply air temperature was about 85°F and the outside temperature 40°F. The air temperature varied from 80 to 81.5°F from a height of 0 to 36 ft from floor level. There was no evidence of significant thermal stratification.

Figure 9.20 shows a stratified mixing flow in a large, high-ceiling indoor stadium using supply nozzles. During summer cooling, cold air jets create two induced recirculating air flows: a cooled lower occupied zone and stratified upper zone. If a higher air velocity is required in the lower occupied zone (the spectator area), the supply nozzles should project at an inclined angle. Otherwise, the cold air jet should project horizontally.

During heating, if the throw of the warm air jet is long enough to arrive at the lowest spectator seat, the whole spectator area is filled with the reversed airstream. No stagnant zone is formed in the lower occupied zone.

Return inlets should be located in the lower cooling zone. They should be evenly distributed under the spectator seats.

Characteristics of the stratified mixing flow during summer cooling mode operation are as follows:

- Convective heat transfer from the hot roof is effectively blocked by the higher temperature or stagnant air layer in the upper zone.

- Cooling loads that occur in the lower zone (windows, walls, lights, occupants, and equipment) should be offset by cold supply air.

- Radiant heat from the roof, upper external walls, and electric lights in the upper zone enters the occupied zone and converts into cooling load.

- Although supply air flow rate and supply temperature affect the throw and the drop of the air jet, the induced recirculating air flow pattern and the stratified upper and lower zones remain the same.

- The height of the supply air jet determines the upper boundary of the lower zone.

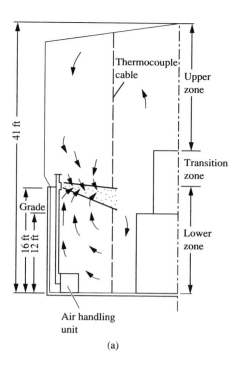

(a)

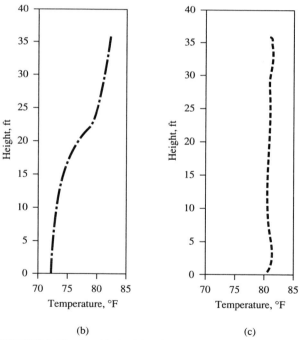

(b) (c)

FIGURE 9.19 Stratified mixing flow in a nuclear reactor facility: (a) elevation view of a nuclear reactor facility, (b) temperature profile during cooling, and (c) temperature profile during heating.

- The location of return inlets influences the cooling load only when they are located in the upper or transition zone. A portion of the exhaust air should be extracted at a higher level.

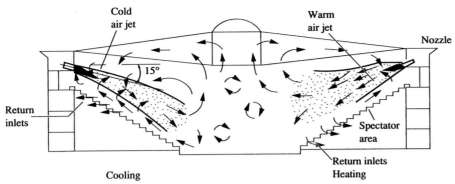

FIGURE 9.20 Stratified mixing flow in a large indoor stadium using supply nozzles.

9.5 DESIGN PROCEDURE OF A MIXING FLOW AIR DIFFUSION SYSTEM

The supply volume flow rates of various control zones at summer and winter design conditions are usually determined before a mixing flow air diffusion system is designed. The design procedure for such an air diffusion system follows.

Select the Type of Supply Outlet

Selection of the supply outlet depends on:

- *Requirements of indoor environmental control.* If the conditioned space needs less air movement or precise air temperature control, a high side outlet is not the right choice.
- *Shape, size, and ceiling height of the building.* For buildings with limited ceiling height, ceiling and slot diffusers are often the best choice. For large buildings with high ceilings, high side outlets mounted at high levels to form stratified induced recirculating flow patterns are recommended. In a perimeter zone, an overhead two-way slot diffuser projected down toward the window and horizontally projected to the room, a ceiling diffuser with a throw to the inner surface of the window glass, or a sill outlet should be used.
- *Surface effect.* A good surface effect is especially important to the VAV system because it allows the supply volume flow rate to be reduced to half or even 30 percent of the design flow.
- *Volume flow per ft² of floor area.* Sidewall outlets are limited to a lower volume flow per ft² of floor area $\dot{V}_s / A_{fl}$ (cfm/ft)² because of the higher air velocity in the occupied zone. The slot diffuser has a narrower slot width and can only project in one or two directions. Therefore, the volume flow per ft² for a slot diffuser is smaller than that of a ceiling diffuser. Table 9.2 lists the volume flow per ft²,

TABLE 9.2
Volume flow per ft² floor area.

Type of outlet	Supply air density cfm/ft² of floor space	Approximate maximum air changes per hour for 10 ft ceiling
Grille	0.6–1.2	7
Slot diffuser	0.8–2.0	12
Perforated panel	0.9–3.0	18
Ceiling diffuser	0.9–5.0	30
Ventilating ceiling	1.0–10.0	60

Source: ASHRAE Handbook 1992, HVAC Systems and Equipment. Reprinted with permission.

volume flow intensity, for various types of supply outlets recommended by *ASHRAE Handbook 1992, HVAC Systems and Equipment.*

- *Appearance.* The shape and configuration of outlets and inlets are closely related to the interior appearance of the building, and should be coordinated with inlets and lighting troffers.
- *Cost.* In many commercial buildings, cost is often an important factor in determining the type of supply outlet.

Cold Air Distribution

Because of the incorporation of ice-storage systems with cold air distribution or low temperature air distribution, supply air temperature can often be reduced from 58°F in the conventional air diffusion system to 42°F to 49°F, typically 44°F. Due to the low dew point of supply air in cold air distribution, space relative humidity is maintained between 35 to 45 percent. It is also possible to maintain a higher space temperature without causing discomfort. Ice-storage systems will be discussed in Chapter 25.

Two types of space diffusion systems are used in cold air distribution:

1. Cold supply air from the air-handling unit (AHU) or packaged unit (PU) is supplied directly to the ceiling diffusers or slot diffusers. In this case, a special high-induction ceiling or slot diffuser should be used to prevent the discharge of a cold air jet into the occupied zone during reduced supply volume flow rates.

2. Using a fan-powered terminal unit to mix low-temperature supply air with return air in the ceiling plenum or directly from the space. The supply air temperature at the diffuser is then increased to about 55°F. Fan-powered terminal units will be discussed in Chapter 27.

In cold air distribution systems, supply air temperature T_s is the air temperature that leaves the supply outlets in the conditioned space. T_s is equal to the temperature discharged from the fan-powered unit during summer cooling mode operation if the duct heat gain between the fan-powered unit and the supply outlet is ignored. Furthermore, T_s is different from the discharged air temperature from the AHU or PU, T_{dis}.

For a $T_s = 44$°F, and a space temperature $T_r = 78$°F, the supply temperature differential for cold air distribution $(T_r - T_s) = 34$°F. Compared with a conventional air distribution system with a $(T_r - T_s) = 20$°F, cold air distribution systems reduce design supply volume flow rate by 40 percent.

To prevent condensation, cold air distribution systems, the AHU or PU, ducts, duct access doors, terminal boxes, flexible ducts, and diffusers should be insulated with a 2-inch layer of fiberglass protected by a vapor barrier. Some manufacturers make the surfaces of high-induction diffusers with insulated material.

Volume Flow Rate per Outlet and Supply Outlet Velocity

For a specific supply volume flow rate in a conditioned space, the volume flow rate per supply outlet $\dot{V}_{s.out}$ in cfm determines the number of outlets in the conditioned space. As defined in Eq. (9.10), both $\dot{V}_{s.out}$ and supply outlet velocity $v_{s.o}$ or v_{core}, in fpm, affect the throw of the supply airstream. However, $\dot{V}_{s.out}$ has a significantly greater influence than $V_{s.o}$, especially for slot diffusers, in which $v_{s.o}$ has only a minor influence.

The volume flow rate per supply outlet depends mainly on the throw required to provide a satisfactory space air diffusion design. Load density, space air diffusion system characteristics, sound control, requirements and cost considerations are also factors that determine $\dot{V}_{s.out}$. In a VAV system in a high-rise office building, avoiding the discharge of cold air into the space directly at a reduced volume flow rate is one of the primary considerations.

For a space air diffusion system using slot diffusers, the supply volume flow per ft length of slot diffuser (the slot intensity) is often an important index especially in the return slots. It is usually between 15 and 40 cfm/ft. All return slots mounted on the same return ceiling plenum must have the same total pressure loss, or for the same type of return slot, the same slot intensity.

In a closed office of an area of 150 ft^2 or less with only one side of external wall, usually one ceiling diffuser is sufficient.

Choose an Optimum Throw and Throw-to-Characteristic-Length Ratio

For most types of supply outlets, the selection of an optimum throw and throw-to-characteristic-length ratio T_{50}/L or T_{100}/L determines the layout and the grid of the supply outlets.

An optimum throw should meet the follow requirements:

- Selected T_{50}/L or T_{100}/L should have an ADPI greater than 70 percent and 80 percent as listed in Table 9.3.
- The spread of the supply air jet covers or almost covers the width of the conditioned space in air flow direction before the air jet enters the occupied zone, as shown in Fig. 9.15.

Miller (1976) and others conducted many experiments and determined the relationship between the ADPI and T_{50}/L or T_{100}/L. They also found that cooling load density is also a factor. Table 9.3 lists the relationship between ADPI and T_{50}/L or T_{100}/L at various load densities. The characteristic length of a sill grill is defined as the length of the room or space in the direction of jet flow. In Table 9.4, T_{50} and T_{100} are throws of isothermal jets given in most of the manufacturers' catalogs for selection convenience.

Generally, high sidewall outlets have a longer throw and therefore a higher T_{50}/L range than ceiling diffusers. Slot diffusers have a wider T_{100}/L range in which ADPI exceeds 80 percent than high sidewall outlets. Perforated ceiling diffusers also have a wider T_{50}/L range than slot diffusers. This characteristic makes them suitable for use at reduced volume flows. Square ceiling diffusers have a T_{50}/L range similar to those of the circular ones. Higher cooling load density usually results in a lower ADPI.

For VAV systems, T_v/L must be selected within the satisfactory range, that is, ADPI > 80 percent, for both maximum and minimum air flow.

TABLE 9.3
Relationship between ADPI and T_{50}/L and T_{100}/L

Terminal device	Load density Btu/h · ft²	T_{50}/L for max. ADPI	Maximum ADPI	For ADPI greater than	Range of T_{50}/L
High sidewall grilles	80	1.8	68	—	—
	60	1.8	72	70	1.5–2.2
	40	1.6	78	70	1.2–2.3
	20	1.5	85	80	1.0–1.9
Circular ceiling diffusers	80	0.8	76	70	0.7–1.3
	60	0.8	83	80	0.7–1.2
	40	0.8	88	80	0.5–1.5
	20	0.8	93	90	0.7–1.3
Sill grille straight vanes	80	1.7	61	60	1.5–1.7
	60	1.7	72	70	1.4–1.7
	40	1.3	86	80	1.2–1.8
	20	0.9	95	90	0.8–1.3
Sill grille spread vanes	80	0.7	94	90	0.8–1.5
	60	0.7	94	80	0.6–1.7
	40	0.7	94	—	—
	20	0.7	94	—	—
Slot diffusers (for T_{100}/L)	80	0.3	85	80	0.3–0.7
	60	0.3	88	80	0.3–0.8
	40	0.3	91	80	0.3–1.1
	20	0.3	92	80	0.3–1.5
Light troffer diffusers	60	2.5	86	80	<3.8
	40	1.0	92	90	<3.0
	20	1.0	95	90	<4.5
Perforated and louvered ceiling diffusers	35–160	2.0	96	90	1.4–2.7
				80	1.0–3.4

Adapted with permission from *ASHRAE Handbook 1989, Fundamentals*. Reprinted with permission.

Determine the Design Characteristics of Slot Diffusers in the Perimeter Zone

In perimeter zones, overhead slot diffusers and ceiling diffusers are widely used. The characteristic length of an overhead slot diffuser in the perimeter zone has not been clearly defined. Based on tests performed by Straub and Cooper (1991), and Lorch and Straub (1983), the design characteristics of overhead slot diffusers in the perimeter zone are as follows:

- Two-slot, two-way overhead slot diffusers should be used. The suitable distance between the overhead slot diffuser and the external wall is about 1 ft. One of the slots near the window is set to blow down, and the other slot blows horizontally into the room. Such an arrangement offsets the cold draft along the window in winter heating.

- During cooling mode operation, ADPI exceeding 80 percent occurs within a wider range of slot intensities, load densities, and spacing between

the diffuser and external wall than during heating mode. Heating mode operation requires careful consideration.

- Greater supply temperature differential $(T_r - T_s)$ should have a smaller slot intensity to avoid a too-short slot in layout.

- According to Straub and Cooper (1991) the criteria for an acceptable space air diffusion in heating mode operation are generally based on a 3 to 4°F temperature gradient from 6-in. to 4-ft levels if there are no excessively higher temperatures above the 4-ft level.

- The supply temperature differential for warm air during heating mode $(T_s - T_r)$ should be less than 20°F to prevent excessive buoyancy effect.

- A terminal velocity of 150 fpm is recommended at the 5 ft level near the window to offset cold drafts in winter.

- An evenly distributed air volume flow rate between the vertically and horizontally discharged slots is recommended.

Select the Specific Supply Outlet from Manufacturer's Catalog

After an optimum T_{50} or T_{100} and characteristic length L are determined from the preliminary layout, the supply outlet can be selected from the manufacturer's catalog from known T_{50} or T_{100} and supply volume flow rates after checking the following parameters:

SOUND LEVEL. The combined sound level of terminal and outlet should be at least 3 dB lower than the recommended NC criteria in the conditioned space.

For optimum noise control, the recommended air velocities at the supply outlet are as follows:

Residences, apartments, churches,   500–750 fpm
    hotel guest rooms, theaters,
    private offices
General offices   500-1250 fpm

The outlet velocity for ceiling diffuser can be calculated by dividing the volume flow by area factor A_k given in manufacturer's catalog.

DROP OF COLD AIR JET. Drop of a cold air jet should be checked if the cold jet enters the occupied zone directly. Figure 9.21a shows the drop and other data of typical sidewall outlets mounted within 1 ft from the ceiling with an adjustable vane deflection of 5° vertical and 0° horizontal. Figure 9.21b shows the drop for sidewall outlets without surface effects and a vane deflection: vertical 15° up, and horizontal, 0°. Both are cold air jets based on a supply temperature differential of 20°F.

TOTAL PRESSURE LOSS OF THE SUPPLY OUTLET. The total pressure loss of supply air when it flows through a slot diffuser with a slot width of

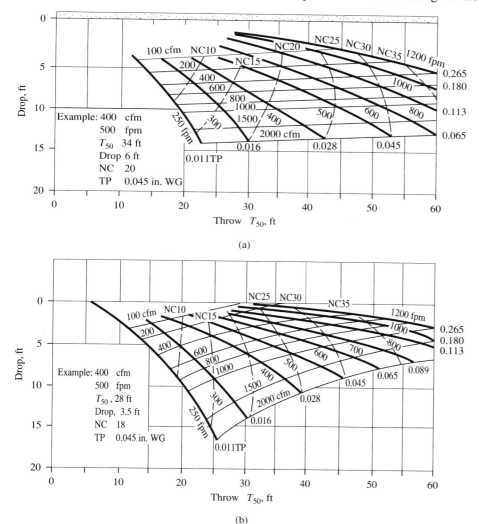

FIGURE 9.21 Performance of cold air jets based on a supply temperature differential of 20°F: (a) performance of cold air jets from typical grilles and registers mounted within 1 ft below ceiling. Deflection: vertical, 5° up, horizontal, 0°; (b) performance of free cold air jets from typical grilles and registers. Deflection: vertical 15° up, horizontal, 0°. (*Source:* Titus Products. Reprinted with permission.)

$\frac{3}{4}$ in. is usually between 0.05 and 0.20 in. WG. For a ceiling diffuser, it is between 0.02 and 0.20 in. WG. A total pressure loss higher than 0.20 in. WG is unsatisfactory.

Determination of Final Layout of Supply Outlets and Return Inlets

The determination of the volume flow rate per outlet, the number of outlets, T_{50}/L, and the selection of specific outlets from the manufacturer's catalog is sometimes an iteration procedure. After that, final layout of supply outlets and return inlets can be determined.

In the perimeter zone, if the outside weather is mild during winter, the high sidewall outlet in a hotel guest room is usually located on the wall opposite from the window. If the weather is cold in winter, the sill outlet is often installed under the window. In the interior zone, ceiling and slot diffusers are usually located directly above the center of the conditioned space.

As mentioned in Section 9.4, if a ceiling plenum is used as a return plenum, the return inlets should be located outside the supply air jet, above the return airstream, or near a concentrated heat source for a better effective factor ϵ.

The recommended face velocities for return inlets are as follows:

Above the occupied zone	800–1000 fpm
Within the occupied zone	400–600 fpm
Door louvers	300–500 fpm

Example 9.1. A large open office has a perimeter length of external wall of 30 ft and a depth of 35 ft, of which 15 ft is perimeter zone, as shown in Fig. 9.22. The supply air volume flow rates and load densities for both perimeter and interior zones are listed below.

	Design supply volume flow rate, cfm
Perimeter zone	900
Interior zone	500
Other design parameters	
Supply temperature differential	20°F
Sound level	NC 40
Cooling load density in perimeter zone	50 Btu/h · ft²
Linear density of heating load	300 Btu/h · ft.

Design a space air diffusion system for this office for year round operation using a VAV system. Its minimum supply volume flow rate is 50 percent of the design volume flow at part-load.

Solution

1. For a large open office using a VAV system, a space air diffusion system of mixing air flow patterns using ceiling slot diffusers is a suitable choice because of its good surface effect and wider T_{100}/L range.

2. For the perimeter zone, the overhead two-way two-slot diffuser should be located on the suspended ceiling parallel to the external wall and window glass. One of the slots should discharge downward toward the window glass. The space between the

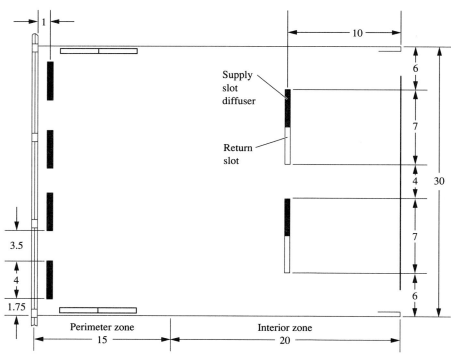

FIGURE 9.22 Layout of slot diffusers for a large open office in Example 9.1. All dimensions in ft.

slot diffusers and the window glass is 1 ft. Another slot discharges horizontally inward from the window glass.

3. In Table 9.3 the range of T_{100}/L for slot diffusers having ADPI > 80 percent at a load density of 50 Btu/h·ft^2 is between 0.3 and 0.95. Considering the reduction of the supply volume flow rate at 50 percent part-load, select a T_{100}/L ratio of 0.6. Because the characteristic length L in the perimeter zone is $L = (15 - 1) = 14$ ft, the required throw for the slot diffuser in perimeter zone is

$$L = 0.6 \times 14 = 8.4 \text{ ft}$$

From Table 9.1, two-way two-slot diffusers with a length of 4 ft and a slot width of 0.75 in. are selected. At a capacity of 110 cfms its performance is as follows:

Throw, horizontal projection, ft	5.5–8–12
vertical projection, ft	3.5–5.5–7.5
Total pressure loss, in WG	0.17
Sound level, NC curve	25

4. If the supply volume flow rate in the perimeter zone is evenly split between horizontal and downward airstreams, the number of slot diffuers is 0.5 (900)/110 = 4.1, so four slot diffusers should be used. For the downward discharged slot, because the ceiling height is 9 ft, the terminal velocity is slightly less than 150 fpm near the window glass at a level from the floor of 5 ft.

5. From Eq. (5.29), the sound level of four two-slot diffusers at 25 dB is

$$\Sigma L = 10 \log(10^{0.1A} + 10^{0.1B} + \ldots)$$
$$= 10 \log(10^{2.5} + 10^{2.5} + 10^{2.5} + 10^{2.5}) = 31 \text{dB}$$

The combined NC curve is still less than 40 dB.

6. Heating load in the perimeter zone is

$$q_{\text{rh}} = 30 \times 300 = 9000 \text{ Btu/h}$$

If the supply volume flow in heating mode operation is still 900 cfm, and if the density of supply air is 0.072 lb/ft^3, the supply differential can then be calculated as

$$(T_s - T_r) = \frac{q_{\text{rh}}}{60\dot{V}_s \rho_s c_{\text{pa}}}$$
$$= \frac{9000}{60 \times 900 \times 0.071 \times 0.243} = 9.7°F$$

This value is far smaller than 20°F.

Although the terminal velocity is slightly less than 150 fpm near the window glass at a level from the floor of 5 ft, because of a far lower $(T_s - T_r)$, cold draft does not occur at the inner glass surface at a heat load linear density of 300 Btu/h·ft.

7. Four 4-ft 0.75-in. wide two-slot return slots with a slot intensity of $900/(2 \times 4) = 28$ cfm/ft are installed in the perimeter zone. Their locations are arranged on two sides of the room in a direction perpendicular to the supply outlets shown in Fig. 9.22 for a better effective factor ϵ.

In Table 9.1, for a capacity of 80 cfm and a length of 3ft, the total pressure loss TP is 0.164 in. WG at a slot intensity of 80/3 = 26.7 cfm/ft. For a capacity of 100 cfm and a length of 3ft, TP is 0.254 in. WG at a slot intensity of 33.3 cfm/ft. TP for selected return slots is 0.17 in. WG.

8. The depth of the interior zone is 20ft. It is better to have two-slot two-way slot diffuser located at its mid-point. If two 4-ft length two-slot two-way slot diffusers with 0.75-in. slot width are selected, throw, T_{100} is still 8ft and the ratio $T_{100}/L = 8/(0.5 \times 20) = 0.8$.

The return inlets in the interior zone must have the same pressure loss TP = 0.17 in. WG, that is, the same slot intensity as those in the perimeter zone. This is because both perimeter and interior zone inlets are installed on the same ceiling and use the ceiling plenum as the return plenum.

For the interior zone, return slots with a slot intensity of 28 cfm/ft must be 500/28 = 18 ft. Select three 0.75 in. wide, 3 ft long, two-slot two-way return slots. The diffusers will be located on the ceiling at a position midway from the depth boundaries of the interior zone, as shown in Fig. 9.22.

9. In the perimeter zone, during cooling mode operation of 50 percent part-load, the throw-to-characteristic-length ratio may reduce to

$$\frac{T_{100}}{L} = \frac{5}{14} = 0.36$$

In the interior zone, it is

$$\frac{T_{100}}{L} = \frac{5}{0.5 \times 20} = 0.5$$

Both are still within the 0.3 to 1.4 range at a load density 25 Btu/h·ft^2.

10. During heating mode part-load operation, the supply air temperature differential and the buoyancy effect will be smaller. This results in smaller vertical temperature gradients than at design-load.

9.6 DISPLACEMENT FLOW AND UNIDIRECTIONAL FLOW

Displacement Flow

Displacement flow is a flow pattern in an air conditioned space in which cold supply air, at a velocity nearly equal to the required velocity in the space, enters the occupied zone or working area and displaces the original space air with a piston-like air flow, without mixing the supply air and the original space air.

Compared with mixing flow, displacement flow provides a better indoor air quality in the occupied zone. If cold air is supplied at a velocity nearly equal to the mean air velocity in the occupied zone with a small supply temperature differential for comfort air conditioning systems, there will be low turbulence intensities and fewer draft problems. On the other hand, displacement flow usually requires a greater supply volume flow rate and a higher construction cost.

There are three types of displacement flow: unidirectional flow, downward uniform flow, and stratified displacement flow.

Unidirectional Flow

In a *unidirectional flow,* the conditioned supply airstream flows in the same direction as a uniform air flow and showers the entire working area or occupied zone. Supply airstreams occupy the entire working area or occupied zone of the conditioned space.

Types of unidirectional flow currently used in clean rooms include downward unidirectional flow and horizontal unidirectional flow, as shown in Fig. 9.23.

Unidirectional flow refers to air flow patterns in which the streamlines of air flow are uniform and move in the same direction. Because of the unidirectional flow, contaminants generated in the space cannot move laterally against the downward air flow, and dust particles will not be carried to higher levels by recirculating flow or large eddies.

Such an air flow pattern in clean rooms was formerly known as *laminar flow.* Laminar flow is often confused with the types of fluid flow distinguished by Reynolds number in fluid mechanics. Almost all forced air flows driven by fans in air conditioned space are turbulent except the boundary laminar layers adjacent to the surface of the building envelope. The Reynolds number of forced air flow, even at very low air velocity, is usually greater than 1×10^4. As of 1991, *ASHRAE Handbook HVAC Applications* refers to this pattern as unidirectional flow.

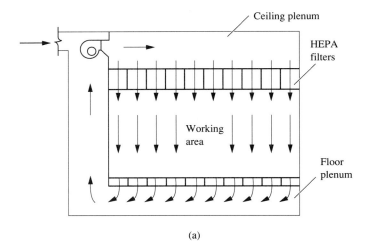

(a)

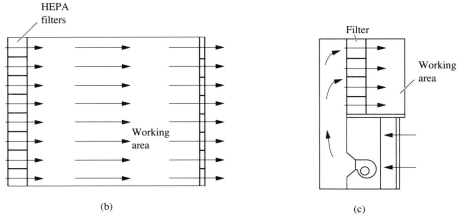

(b)   (c)

FIGURE 9.23 Unidirectional flow for clean rooms: (a) downward unidirectional flow for clean room, (b) Horizontal unidirectional flow for clean room, and (c) unidirectional flow for clean work station.

Unidirectional Flow for Clean Rooms

Figure 9.23a shows a downward unidirectional flow for clean rooms. After flowing through the high-efficiency particulate air (HEPA) filters, ultra-clean air discharges uniformly downward in parallel stream lines and enters the working area. It then flows through the raised floor grating and is returned to the recirculating air unit at the top of the clean room.

Figure 9.23b shows a horizontal unidirectional flow for a clean room. Instead of flowing downward, the clean airstream discharges horizontally from the HEPA filters on one side of the room and flows through the working area. The contamination level near the return inlets of a horizontal unidirectional flow may be higher than those of a downward unidirectional flow.

Figure 9.23c shows a clean work station that provides a small non-contaminated working area for a single worker by means of horizontal or downward unidirectional flow and HEPA filters. Clean work stations can achieve a high degree of contamination control over a limited area for many practical applications.

In order to provide parallel stream lines, air velocity of 60 to 90 fpm is required. Unidirectional flow provides a direct and predictable path of submicron-size dust particles and minimizes the opportunities for these particles to contaminate working parts. It also captures internally generated dust particles and carries them away. Most dust particles in unidirectional flow reestablish their parallel stream lines after the downstream eddies of an obstruction.

The supply temperature differential $(T_r - T_s)$ for unidirectional flow for clean rooms depends mainly on the required space velocity or the cooling loads to be removed within the working area. A case study of a Class 10 clean room will be covered in Chapter 30.

Ventilating Ceiling

A *ventilating ceiling* is sometimes called a perforated ceiling. It creates a downward uniform flow similar to the downward unidirectional flow for clean rooms. In most cases, ventilating ceilings discharge conditioned air through the entire ceiling to form a uniform downward flow, except in the area occupied by light troffers. The primary differences between unidirectional flow for clean rooms and downward uniform flow from ventilating ceilings are as follows:

- Unidirectional flow requires a 60 to 90 fpm air velocity and ultra-clean air in the working area, and ventilating ceilings usually have a mean air velocity of less than 15 fpm of conditioned air.
- There is no mixing of supply and space air in unidirectional flow, whereas just below the perforated ceiling, supply air is mixed with ambient air at a vertical distance of less than 1 ft in downward uniform flow from the ventilating ceiling.

Figure 9.24 shows a downward uniform air flow. In Fig. 9.24, conditioned air is first supplied to the ceiling plenum through supply outlets inside the plenum. It is then squeezed through holes or slots of the ventilating ceiling and discharged to the conditioned space in a downward uniform flow. When parallel airstreams combine together and flow through the working area, they arrive at the floor level, turn to the side return grilles at low levels, and then return to the air-handling unit or package unit in the fan room.

Downward uniform air flow patterns from ventilating ceilings have the following characteristics:

- There is no mixing or induced recirculating flow in the working area or occupied zone, so dust particles contained in the space air at a lower level are not

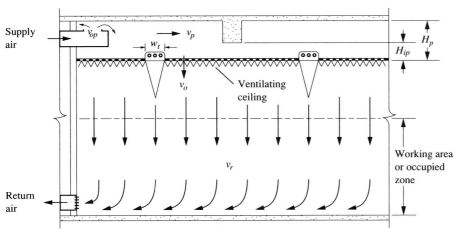

FIGURE 9.24 Ventilating ceiling.

carried to a high level and enter the working area again.

- A very low air velocity can be maintained in the working area or the occupied zone even when the volume flow per ft^2 $\dot{V}/A_{\text{fl}}$, in cfm/ft^2, is very high. Here, $\dot{V}$ represents the supply volume flow rate, in cfm, and A_{fl} represents the floor area, in ft^2.
- When $\dot{V}/A_{\text{fl}} > 3$ cfm/ft^2, the downward uniform flow becomes prominent.

Ceiling Plenum and Supply Air Velocity

In order to create a more uniform supply air velocity v_o in fpm at the perforated holes or slots, the maximum velocity of the airstream crossing the ventilating ceiling inside the ceiling plenum v_p, in fpm, should be low. Usually, $v_p \leq 0.5 v_o$.

Moreover, the clearance between the beam and the ventilating ceiling H_{ip}, in ft, should be always greater than 8 in. to prevent a greater pressure loss against the air flow crossing the ventilating ceiling inside the plenum. The greater the supply volume flow rate, the higher the H_{ip}. If the ceiling plenum is of sufficient height H_p and few small obstructions other than the beams are present, distributing ductwork inside the ceiling plenum is not necessary to provide a uniform air supply.

Supply air discharged from the outlets into the ceiling plenum always moves in an upward direction so that the velocity pressure of the discharged airstream does not affect the supply air velocity at the perforated openings.

The construction of the ceiling plenum must meet the requirements of the National Fire Protection Association (NFPA) and local fire protection codes. Ventilating ceilings are always made of non-combustible material such as metal strips or mineral acoustic tiles. The ceiling plenum must be insulated against heat gain and loss and condensation on its inner surface if it is adjacent to area that is not air conditioned.

Supply air velocity at the perforated openings v_o must be optimal. When v_o exceeds 1000 fpm, objectionable noise is generated. A higher v_o also means higher pressure at the ceiling plenum and, therefore, greater air leakage. On the other hand, a lower v_o may cause unevenness in the supply air flow rate because of a higher v_p/v_o ratio.

Supply air velocity v_o is usually between 200 and 700 fpm. When v_o has been determined, the perforated area A_o, in ft^2, can be calculated as

$$A_o = \frac{\dot{V}_s}{v_o} \qquad (9.16)$$

where $\dot{V}_s$ = supply volume flow rate, cfm.

Generally, the diameter of the circular holes or the slot width is less than $\frac{1}{4}$ in. The number of holes or slots can then be determined accordingly. Based on test results, the local loss coefficient of the perforated holes and slots C_o is about 2.75.

Ventilated ceilings are often used in industrial applications where very low air movements or precise control of space temperature up to $68 \pm 0.1°F$ are needed. They are also used in indoor sport stadiums for badminton and table tennis, where an air velocity of less than 40 fpm should be maintained.

9.7   STRATIFIED DISPLACEMENT FLOW

Two-Zone Stratified Model

Stratified displacement flow is a pistonlike air flow. It supplies conditioned cold air (usually 100 percent outdoor air at lower velocity) at a low-level supply outlet, as shown in Fig. 9.25. The cold supply air, with a volume flow rate of $\dot{V}_s$ (cfm) flows in a thin layer along the floor. Above the heat and contaminant sources, heated air containing contaminants rises upward because of its buoyancy effect. Supply air is then entrained into the upward convective flow with a volume flow rate of $\dot{V}_{\text{conv}}$ (in cfm). When upward convective flow arrives at a height where its volume flow rate is equal to the supply volume flow rate, that is, $\dot{V}_{\text{conv}} = \dot{V}_s$, this height is recognized as the stationary level z_{stat}, in ft. Above z_{stat}, ambient air is induced into the upward convective flow until it reaches the ceiling. Upward convective flow with induced air spreads laterally along the ceiling. An amount nearly equal to the supply volume flow rate exhausts or returns to the fan room through the exhaust or return inlet near the ceiling. The remaining portion containing the contaminant descends to the stationary level to be induced into the upward convective flow and recirculated.

The stratified level divides the room vertically into two zones: an upper zone and a lower zone. In the lower zone, only the supply air is induced into the upward convective flow as its volume flow rate is smaller than the supply volume flow rate, $\dot{V}_{\text{conv}} < \dot{V}_s$. In the upper zone, the portion of air and contaminants that is greater than supply flow rate $\dot{V}_{\text{conv}} > \dot{V}_s$ recirculates. For cold air supply, the mean temperature of the upper zone is usually 1 to 2°F higher than the lower zone.

Operating Characteristics

Stratified displacement flow was introduced to Scandinavian countries in the early 1970s as a means of improving general ventilation in industrial applications.

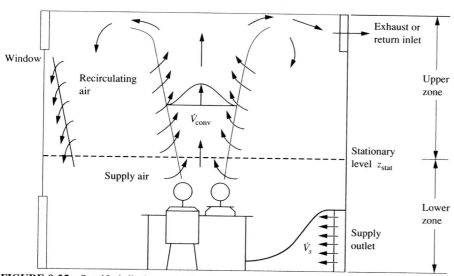

FIGURE 9.25 Stratified displacement flow in a typical room.

For comfort systems in public and office buildings, it was adopted in the early 1980s.

According to Svensson (1989), Mathisen (1989), Sandberg and Blomqvist (1989), and Seppänen et al. (1989), stratified displacement flow has the following characteristics:

- Cold air supply of usually 100 percent outdoor air is used to remove cooling loads in conditioned space. Heating is usually provided by radiating panels or a baseboard heater under the windows or on the walls.

- Air must be supplied at low velocity, generally less than 60 fpm and at a height often less than 1.8 ft from the floor level.

- Cold air is generally supplied at a temperature 5 to 9°F lower than the air in the occupied zone or working area.

- The height of the lower zone, or z_{stst}, generally should be higher than the breath line of a seated occupant (4.5 ft). Stationary level depends mainly on the supply volume flow rate $\dot{V}_s$.

- In the lower zone, all air is supply air, theoretically, except the downward cold drafts. Supply air is supplied to the occupied zone directly without mixing with the ambient air.

- Because of the small supply temperature differential, the maximum cooling load that can be removed from the room is about 13 Btu/h · ft² at a ceiling height of 9 ft. For greater cooling loads, additional cooling panels mounted on the ceiling should be used.

- Return or exhaust inlets are located near the ceiling level.

- According to the required thermal comfort of the occupant, the vertical temperature difference between the 0.3 ft and 5.5 ft level as specified in section 5.8 shall not exceed 5°F.

- The exchange efficiency ϵ_{ex} for complete mixing flow is 0.5, and for stratified displacement flow is from 0.5 to 0.6. For ideal piston-like displacement flow, ϵ_{ex} is 1.

Comparison of Stratified Displacement Flow and Mixing Flow

Compared with mixing flow space diffusion systems, stratified displacement flow has the following advantages:

- Contamination level in the occupied zone is lower and therefore, indoor air quality is better.

- Space diffusion effective factors ϵ_T and ϵ_C are higher.

- Turbulence intensities I_{tur} are lower and therefore, fewer draft problems occur even at higher mean air velocities.

The disadvantages of the stratified displacement flow are as follows:

- Initial cost is significantly higher if cooling load density is greater than 13 Btu/h · ft² and if cooling panels are added.

- Energy cost is comparatively higher.

- Stratified displacement flow is for space cooling only.

9.8 PROJECTING FLOW—SPOT COOLING/HEATING

Benefits of Projecting Flow

In a *projecting flow air pattern,* the cold or warm air jet is deliberately projected into part of the occupied zone or working area, which is often called the *target zone.* The result is control of the environment in a small or localized area, or *micro-environmental control.* This projecting air flow pattern is used in spot cooling/heating, task air conditioning, and personal environments.

Spot cooling/heating using projecting flow has many advantages over conventional space air diffusion:

- Better control of temperature, air cleanliness, and air movement in a micro-environment, occupant's personal environment
- Lower energy consumption
- More direct fresh air supply
- Direct and efficient handling of local loads

The main disadvantages of spot cooling/heating include more complicated supply outlets and duct work, higher cost per unit floor area, and a limited area of environmental control.

Parameters Considered in Design

Air jets in projecting flow are usually free jets with high entrainment ratios because of their large contact area with ambient air. Long-throat round nozzles are often used as supply outlets for spot cooling/heating in large spaces, and small nozzles (3 to 7 in. diameter) are widely used in open offices and other comfort air conditioning systems.

In spot cooling/heating design, the most important parameter is the throw of the air jet or projecting flow. Throw, usually T_{100} in spot cooling/heating, is the distance in feet between the outlet and the center of air jet with a terminal velocity of 100 fpm. T_{100} is largely determined by the supply volume flow rate at the nozzle $\dot{V}_o$, in cfm. The supply air velocity v_{core}, in fpm, has far less influence on T_{100} than $\dot{V}_o$ has. Figure 9.26 shows the relationship between T_{100} and $\dot{V}_o$ for typical small-diameter nozzles. When $\dot{V}_o = 100$ cfm,

T_{100} for a 5 in. diameter nozzle at $v_o = 735$ fpm is 13 ft, whereas T_{100} for a 3.5 in. diameter nozzle at $v_o = 1493$ fpm is only 15 ft.

In spot cooling/heating, it is impossible to have a short T_{100} from a nozzle with high $\dot{V}_o$, or a long T_{100} from a nozzle with low $\dot{V}_o$.

Design parameters other than throw include sound level, pressure drop, and area of the target zone. For a typical 3.5-in. nozzle, if $\dot{V}_o = 120$ cfm, and $v_o = 1790$ fpm, then NC = 27 and total pressure drop at the nozzle is 0.22 in. WG. The size of the target zone can be estimated by T_{100} and known angle of divergence $\alpha = 22°$.

Recommendations in Spot Cooling/Heating Design

Brown (1988) recommends the following for spot cooling/heating design:

- The nozzle should be mounted close to the target zone to provide a greater terminal velocity. This reduces the volume flow rate and induction of room air.
- Optimum terminal velocity provides a better evaporative cooling effect for the comfort of the occupants and a sufficient temperature differential between the air jet in target zone and room air. Optimum terminal velocity depends on the metabolic rate of the occupant, temperature of the target zone, and mean radiant temperature of the surroundings.
- Allow the occupant to adjust both the direction of the air jet and the volume flow rate, if possible, to improve thermal comfort.
- Ideally, the air jet is projected toward the front of the occupant, but side projection is also acceptable.
- Most manufacturers' performance data for spot cooling/heating are for isothermal free jets. If an air jet is projected along a surface, the throw should be 40 percent longer. To estimate the throw of non-isothermal downward warm air jets, consider that throw is reduced 2 percent for each 1°F increase in temperature difference between the supply air and the room air; for downward cold air jets, throw is increased 1 percent for each 1°F increase in temperature difference.

Generally, spot cooling/heating is used where the mean indoor temperature is higher than 82°F during the cooling season and lower than 65°F during the heating season.

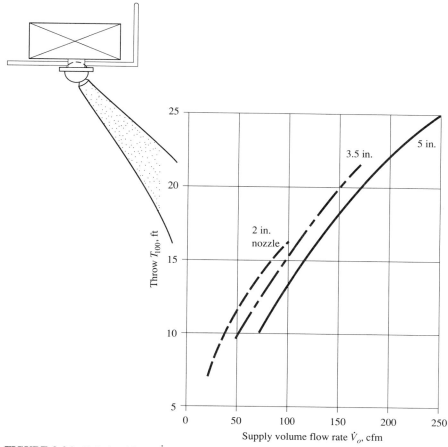

FIGURE 9.26 Relationship of $\dot{V}_o$ and T_{100} for spot cooling/heating. (Adapted with permission from *ASHRAE Transactions 1988* Part 1.)

9.9 UPWARD FLOW—SPACE AIR DIFFUSION FROM A RAISED FLOOR

Upward Flow from a Raised Floor

Upward flow from a raised floor has been successfully used for space air diffusion in computer rooms and other industrial applications with high cooling load density, such as 60 to 300 Btu/h · ft^2 floor area. Figure 9.27a shows a typical upward flow space air diffusion from a raised floor. Conditioned air is supplied to a raised-floor plenum. The floor is usually raised from the structure floor between 12 and 15 in. depending on whether a fan–coil unit is installed inside the floor plenum. The floor plenum can be pressurized at a positive pressure of usually less than 0.08 in. WG. If the plenum is pressurized, air is discharged from the floor outlets directly. If the floor plenum is not pressurized, conditioned air from the floor plenum is extracted by an outlet with a small fan and discharged to the conditioned space. If air is discharged through a desk outlet, either the floor plenum should be pressurized at a higher static pressure or a small fan should be used to overcome the pressure drop of the desk outlet and its connecting pipe as shown in Fig. 9.27b.

Because the floor outlets, fan outlets, and desk outlets are all located in the occupied zone, air supply from these outlets should be at a temperature between 62 and 65°F.

After the space cooling load is absorbed by the supply air, the bouyant space air rises. Return air temperature entering the light troffers is usually between 80 and 82°F. After entering the ceiling plenum, air is returned to the AHU, PU, or terminals.

During winter heating, a perimeter radiant heating system is often used. Sometimes, supply air can be heated at the reheating coil in the zone terminal unit, and is then supplied to the perimeter zone to offset the heating load through floor outlets.

DESK OUTLETS. Many open offices use cubicle partitions to provide privacy. At the same time, localized loads created by personal computers and other electronic equipment have become quite high. Desk outlets supply air directly to the target zone from the floor

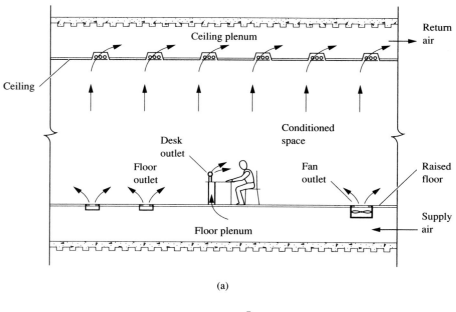

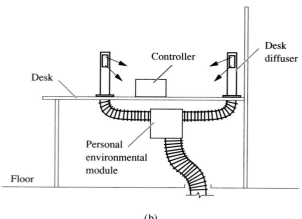

FIGURE 9.27 Upward flow from a raised floor: (a) schematic diagram and (b) desk outlet.

plenum, as shown in Fig. 9.27. In many comfort air conditioning systems (especially in areas with severe winters), desk outlets supply warm air directly to the occupants, with no stagnant layer.

Desk outlets also provide individual control of volume flow and flow direction, which significantly improves the occupants' thermal comfort. The main drawback is their higher cost. Except in applications where high individual control is needed, desk outlets have been replaced by floor outlets in most raised-floor upward flow space air diffusion systems in recent years.

DESIGN CONSIDERATIONS.

- Although supply temperature has been raised to 62 to 65°F, return temperature is also increased to 80 to 82°F, so the supply temperature difference $(T_r - T_s)$ is still approximately the same as that of mixing flow air diffusion systems.

- The diameter of the floor outlet is usually 6 to 8 in. Discharge velocity is about 400 to 800 fpm. For a 6 in. outlet, the maximum supply flow rate $\dot{V}_{max} \leq$ 30 cfm, and for an 8 in. outlet, $\dot{V}_{max} \leq$ 90 cfm.

- Right above the floor outlet, air jet velocity may be 400 fpm at a temperature 7°F lower than the ambient air. Experience shows that a stay longer than a few minutes produces discomfort. Sodec and Craig (1990) recommend that, when considering the layout of the floor outlets, the minimum distance between the work place and a 6-in.-diameter outlet is 2.5 to 3 ft, and 3.5 to 5 ft for an 8-in.-diameter outlet.

- According to Sodec and Craig, attenuation between two floor outlets in adjacent rooms is usually greater than 29 dB. Therefore, conversation transmission through floor outlets is negligible.

APPLICATIONS OF RAISED-FLOOR UPWARD FLOW. Compared with conventional mixing flow air diffusion systems, the primary advantages of upward flow from a raised floor are as follows:

- Supply air is supplied to the occupied zone directly, which results in better indoor air quality.
- When warm air is supplied from floor outlets and returned from the ceiling plenum, there is no stagnant air in the occupied zone.
- Space air is returned at a higher temperature, and therefore, the space diffusion effective factor is higher.

Disadvantages of upward flow from a raised floor include:

- Valuable floor space is occupied by the floor outlets.
- Both floor and ceiling plenums are required.
- The initial investment is higher than in conventional mixing flow systems.

For offices and applications other than computer rooms, raised-floor upward flow for commercial buildings was first developed in Germany, and many raised-floor upward flow systems have recently been adopted in South Africa. Heinemeier et al. (1990) found that in North America, raised-floor upward flow is primarily used in small offices.

Raised-floor upward flow space diffusion systems are still in the developing stage. Where a floor plenum is always needed or where there is a high space cooling load density, a raised-floor upward flow system might be a suitable choice.

Future Development

Task air conditioning for micro-environmental control has great potential because of its lower energy consumption. In the coming century, demand for air conditioning in developing countries in tropical areas is expected to increase dramatically. Energy-efficient systems must be developed. Spot cooling/heating (task air conditioning) or a combination of furniture and air conditioning may be important future technologies.

REFERENCES AND FURTHER READING

ASHRAE, *ASHRAE Handbook 1988, Equipment,* ASHRAE Inc., Atlanta, GA, 1988.

ASHRAE, *ASHRAE Handbook 1989, Fundamentals,* ASHRAE Inc., Atlanta, GA, 1989.

Bagheri, H. M., and Gorton, R. L., "Performance Characteristics of a System Designed for Stratified Cooling Operating During the Heating Season," *ASHRAE Transactions,* Part II, pp. 367–381, 1987.

Bagheri, H. M., and Gorton, R. L., "Verification of Stratified Air-Conditioning Design," *ASHRAE Transactions,* Part II, pp. 211–227, 1987.

Brockmeyer, I. H. P., "Air Flow Pattern and Its Influence on the Economy of Air Conditioning," *ASHRAE Transactions,* Part I, pp. 1127–1142, 1981.

Brown, C. E., "Spot Cooling/Heating and Ventilation Effectiveness," *ASHRAE Transactions,* Part I, pp. 678–684, 1988.

Dorgan, C. E., and Elleson, J. S., "Cold Air Distribution," *ASHRAE Transactions,* Part I, pp. 2008–2025, 1988.

Dorgan, C. E., and Elleson, J. S., "Design of Cold Air Distribution System with Ice Storage," *ASHRAE Transactions,* Part I, pp. 1317–1322, 1989.

Fanger, P. O., Melikow, A. K., Hanzawa, H., and Ring, J., "Turbulence and Draft," *ASHRAE Journal,* pp. 18–25, April 1989.

Genter, R. E., "Air Distribution for Raised Floor Offices," *ASHRAE Transactions,* Part II, pp. 141–146, 1989.

Gorton, R. L., and Sassi, M. M., "Determination of Temperature Profiles and Loads in a Thermally Stratified, Air Conditioning System: Part I—Model Studies," *ASHRAE Transactions,* Part II, pp. 14–32, 1982.

Hanzawa, H., Melikow, A. K., and Fanger, P. O., "Air Flow Characteristics in the Occupied Zone of Ventilated Spaces," *ASHRAE Transactions,* Part I, pp. 524–539, 1987.

Hart, G. H., and Int-Hout, D., "The Performance of a Continuous Linear Diffuser in the Interior Zone of an Open Office Environment," *ASHRAE Transactions,* Part II, pp. 311–320, 1981.

Heinemeier, K. E., Schiller, G. E., and Benton, C. C., "Task Conditioning for the Workplace: Issues and Challenges," *ASHRAE Transactions,* Part II, pp. 678–689, 1990.

Int-Hout, D., and Weed, J. B., "Throw: The Air Distribution Quantifier," *ASHRAE Transactions,* Part I, pp. 667–677, 1988.

Kostel, A., "Path of Horizontally Projected Heated and Chilled Air Jets," *ASHRAE Transactions,* p. 213, 1955.

Lorch, F. A., and Straub, H. E., "Performance of Overhead Slot Diffusers with Simulated Heating and Cooling Conditions," *ASHRAE Transactions,* Part I B, pp. 200–211, 1983.

Mathisen, H. M., "Case Studies of Displacement Ventilation in Public Halls," *ASHRAE Transactions*, Part II, pp. 1018–1027, 1989.

Mayer, E., "Physical Causes for Draft: Some New Findings," *ASHRAE Transactions*, Part I, pp. 540–548, 1987.

Miller Jr., P. L., "Room Air Diffusion Systems: A Re-Evaluation of Design Data," *ASHRAE Transactions*, Part II, pp. 375–384, 1979.

Miller Jr., P. L., "Application Criteria for the Air Diffusion Performance Index (ADPI)," *ASHRAE Transactions*, Part II, pp. 206–218, 1976.

Sandberg, M., and Blomqvist, C., "Displacement Ventilation Systems in Office Rooms," *ASHRAE Transactions*, Part II, pp. 1041–1049, 1989.

Seppänen, O. A., Fisk, W. J., Eto, J., and Grimsrud, D. T., "Comparison of Conventional Mixing and Displacement Air-Conditioning and Ventilating Systems in U.S. Commercial Buildings," *ASHRAE Transactions*, Part II, pp. 1028–1040, 1989.

Sodec, F., and Craig, R., "The Underfloor Air Supply System—The European Experience," *ASHRAE Transactions*, Part II, pp. 690–695, 1990.

Straub, H. E., and Chen, M. M., "Distribution of Air Within a Room for Year-Round Air Conditioning—Part II," University of Illinois Engineering Experiment Station Bulletin No. 442, 1957.

Straub, H. E., and Cooper, J. G., "Space Heating with Ceiling Diffusers," *Heating/Piping/Air Conditioning*, pp. 49–55, May 1991.

Straub, H. E., Gitman, S. F., and Konzo, S., "Distribution of Air Within a Room for Year-Round Air Conditioning—Part I," University of Illinois Engineering Experiment Station Bulletin No. 435, 1956.

Svensson, A. G. L., "Nordic Experiences of Displacement Ventilation Systems," *ASHRAE Transactions*, Part II, pp. 1013–1017, 1989.

Wang, S. K., *Air Conditioning*, Vol. 2, Hong Kong Polytechnic, Hong Kong, 1987.

Wang, S. K., Kwok, K. W., and Watt, S. F., "Characteristics of a Space Diffusion System in an Indoor Sport Stadium," *ASHRAE Transactions*, Part II B, pp. 416–435, 1985.

Wendes, H., "Supply Outlets for VAV Systems," *Heating/Piping/Air Conditioning*, pp. 67–71, February 1989.

CHAPTER

10

FANS, COMPRESSORS, AND PUMPS

10.1 FAN FUNDAMENTALS

Functions and Types of Fans

A *fan* is the prime mover of an air system or ventilating system. It moves the air and provides continuous air flow so that the conditioned air, space air,
exhaust air, or outdoor air can be transported from
one location to another through air ducts or other air
passages.

A fan is also a turbomachine in which air is usually
compressed at a compression ratio R_{com} not greater
than 1.07. The compression ratio, dimensionless, is
defined as

$$R_{com} = \frac{p_{dis}}{p_{suc}} \qquad (10.1)$$

where p_{dis} = discharge pressure at the outlet of the
compressor or fan, lb_f/in^2 abs
or psia
p_{suc} = suction pressure at the inlet of the
compressor or fan, psia

A *blower* is usually an enclosed multiblade rotor
that compresses air to a higher discharge pressure.
There is no clear distinction between a fan and a
blower. Traditionally, blowers do not discharge air at
low pressure, as some fans do.

A fan is driven by a motor directly (direct drive)
or via belt and pulleys (belt drive). Some large industrial fans in power plants are driven by steam or gas
turbines.

Two types of fans are widely used in air conditioning and ventilating systems: centrifugal fans and axial fans (Fig. 10.1). Fans can be mounted individually
as ventilating equipment to provide outdoor air or air
movement inside a building. They can also transport
air containing dust particles or material from one place
to another via air duct systems. In air conditioning
systems, fans are often installed in air-handling units,
packaged units, or other air conditioning equipment.

In both centrifugal and axial fans, the increase of
air static pressure is created by the conversion of
velocity pressure into static pressure. However, in
centrifugal fans, air is radially discharged from the
impeller, also known as the *fan wheel*—air turns 90°
from its inlet to its outlet. In an *axial fan,* the direction of air flow is parallel to the axle of the fan. Additional differences in fan characteristics between these
two types of fans will be discussed in the following
sections.

Crossflow fans force the air flow by means of a
long rotor with many vanes, as shown in Fig. 10.1c.
This type of fan has limited application in small air
conditioning equipment because it is less efficient and
has a higher sound level than centrifugal fans.

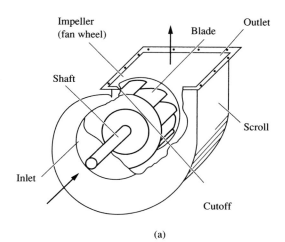

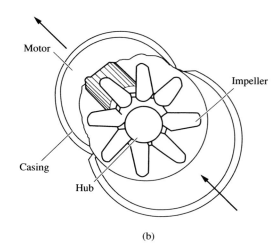

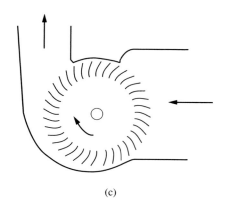

FIGURE 10.1 Types of fans: (a) centrifugal; (b) axial;
and (c) cross-flow.

Fan Capacity or Volume Flow Rate

Fan capacity or the *fan volume flow rate* $\dot{V}_f$, in cfm,
is defined as the rate of volume flow measured at the
inlet of the fan, corresponding to a specific fan total
pressure. It is usually determined by the product of the

duct velocity and the area of the duct connected to the fan inlet, according to the test standard Air Movement and Control Association (AMCA) Standard 210–85 and ASHRAE Standard 51–1985.

Fan volume flow rate is independent of air density ρ. However, fan total pressure is affected by air density. Therefore, the fan volume flow rate is normally rated at standard air conditions, that is, dry air at an atmospheric pressure of 14.696 psia, a temperature of 70°F, and a density of 0.075 lb/ft^3.

Fan Pressure

Fan total pressure Δp_{tf}, in inches WG, is the total pressure increase of a fan, that is, the pressure difference between the total pressure at the fan outlet p_{to} and the total pressure at the fan inlet p_{ti}, or

$$\Delta p_{tf} = p_{to} - p_{ti} \qquad (10.2)$$

Fan velocity pressure p_{vf}, in inches WG, is the pressure calculated according to the mean velocity at the fan outlet v_o, in fpm. From Eq. (8.9), if air density $\rho = 0.075$ lb/ft^3, it can be calculated as

$$p_{vf} = p_{vo} = \frac{\rho v_o^2}{2g_c} = \left(\frac{v}{4005}\right)^2 = \left(\frac{\dot{V}_o}{4005 A_o}\right)^2$$
$$(10.3)$$

where p_{vo} = the velocity pressure at the fan outlet, in. WG

g_c = dimensional constant, 32.2 lb$_m$ · ft/lb$_f$ · s^2

$\dot{V}_o$ = volume flow rate at fan outlet, cfm

A_o = cross-sectional area of the fan outlet, ft^2

Fan static pressure Δp_{sf}, in inches WG, is the difference between fan total pressure and fan velocity pressure, or

$$\Delta p_{sf} = \Delta p_{tf} - p_{vf} = p_{to} - p_{ti} - p_{vo} = p_{so} - p_{ti}$$
$$(10.4)$$

where p_{so} = static pressure at the fan outlet, in. WG.

Fan Power and Fan Efficiency

Air power P_{air}, in hp, is the work done in moving the air along a conduit against a total pressure Δp_{tf}, in inches WG, at a fan volume flow rate of $\dot{V}_f$, in cfm. Because 1 hp = 33,000 ft · lb$_f$/min, and 1 in. WG = 5.192 lb$_f$/ft^2,

$$P_{air} = \Delta p_{tf} \times 5.192 \times \frac{\dot{V}_f}{33,000} = \Delta p_{tf}\frac{\dot{V}_f}{6356} \qquad (10.5)$$

Fan total efficiency η_t is defined as the ratio of air power P_{air} to fan power input P_f, in hp, on the fan shaft

$$\eta_t = \Delta p_{tf}\frac{\dot{V}_f}{6356 P_f} \qquad (10.6)$$

Fan total efficiency is a combined index of aerodynamic, volumetric, and mechanical efficiencies of a fan.

Fan static efficiency η_s is defined as the ratio of the product of the fan static pressure Δp_{sf}, in inches WG, and the fan volume flow rate to the fan power input, that is

$$\eta_s = \Delta p_{sf}\frac{\dot{V}_f}{6356 P_f} \qquad (10.7)$$

In Eqs. (10.6) and (10.7), the fan power input on the fan shaft, often called *brake horse power* (Bhp), can be calculated as

$$P_f = \frac{\Delta p_{tf}\dot{V}_f}{6356\eta_t} = \frac{\Delta p_{sf}\dot{V}_f}{6356\eta_s} \qquad (10.8)$$

Air Temperature Increase Through Fan

If air density $\rho_a = 0.075$ lb/ft^3, the specific heat of air $c_{pa} = 0.243$ Btu/lb · °F, and 1 hp = 42.41 Btu/min, the relationship between fan power input and the air temperature increase when it flows through the fan ΔT_f, in °F, is given as

$$P_f = \dot{V}_f \rho_a c_{pa}\frac{\Delta T_f}{42.41} \qquad (10.9)$$

Combining Eqs. (10.8) and (10.9), then,

$$\Delta T_f = \frac{0.00667\Delta p_{tf}}{\rho_a c_{pa}\eta_t} = \frac{0.37\Delta p_{tf}}{\eta_t} \qquad (10.10)$$

This air temperature increase in a fan is caused by the compression process and energy losses that occur inside the fan. When air flows through the air duct, duct fittings, and equipment, the duct friction loss and dynamic losses cause a temperature increase as mechanical energy is converted into heat energy. However, this temperature increase is offset by a temperature drop caused by the expansion of air due to the reduction of static pressure along the air flow. Therefore, it is more convenient to assume that air temperature increase occurs because of the friction and dynamic losses along the air flow only when air is flowing through the fan.

Fan Performance Curves

Fan characteristics can be described by certain interrelated parameters such as volume flow rate, pressure, power, and efficiency. These characteristics are depicted graphically by fan performance curves in Fig. 10.2.

Fan characteristic curves usually set the volume flow rate $\dot{V}$, in cfm, as the abscissa and either fan total pressure Δp_{tf}, fan static pressure Δp_{sf}, fan power input P_f, or fan total efficiency η_t as the ordinate.

In Fig. 10.2, there are three pressure–volume flow characteristic curves:

1. Total pressure–volume flow Δp_t–$\dot{V}$
2. Static pressure–volume flow Δp_s–$\dot{V}$
3. Velocity pressure–volume flow p_v–$\dot{V}$

The point where the Δp_t–$\dot{V}$ or Δp_s–$\dot{V}$ curve intersects with the ordinate is called the *shut-off point*. At this point, the block-tight or completely shut-off volume flow rate $V_{block} = 0$. Moreover, static pressure is equal to total pressure at this point.

The volume flow rate at the point of intersection between either the Δp_t–$\dot{V}$ and p_v–$\dot{V}$ curves or the Δp_s–$\dot{V}$ and the *x*-axis is called *free delivery*, or the *wide open volume flow rate*. At the point of free delivery, $\Delta p_s = 0$, $\Delta p_t = p_v$, and fan static efficiency $\eta_s = 0$.

In Fig. 10.2, the other curves shown are the fan total efficiency η_t–$\dot{V}$ curve, the static efficiency η_s–$\dot{V}$ curve, and the fan power input P–$\dot{V}$ curve. The efficiency curves show that at shut-off point, both η_t and η_s are zero, and at free delivery, η_s is again zero. It is important that the fan be operated near maximum efficiency as much as possible. Because of friction and dynamic losses, fan power input at the shut-off point is not equal to zero.

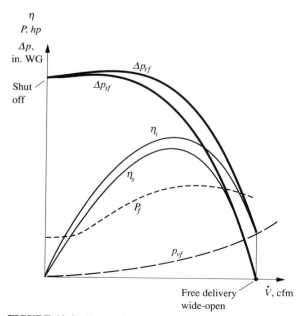

FIGURE 10.2 Fan performance curves.

Influence of Elevation and Temperature

Both elevation and temperature have an influence on air density ρ_a, in lb/ft^3, so they affect fan total pressure. As in Eqs. (8.41), (8.43), and (8.44), the fan total pressure Δp_{tf}, in inches WG, required at high elevations and temperatures can be calculated as

$$\Delta p_{tf} = K_T K_{el} \Delta p_{t,s} \qquad (10.11)$$

where $\Delta p_{t,s}$ = fan total pressure at standard air conditions, in. WG. In Eq. (10.11), K_{el} is an elevation factor and can be calculated as

$$K_{el} = \frac{p_{at}}{29.92} \qquad (10.12)$$

Here, p_{at} represents the atmospheric pressure at high elevation, in inches Hg. Between sea level and an elevation of 5000 ft above sea level, p_{at} can be roughly estimated as

$$p_{at} = 29.92 - 0.001 H_{el} \qquad (10.13)$$

where H_{el} = elevation above sea level, ft.

Temperature factor K_T can be calculated as

$$K_T = \left(\frac{530}{T_a + 460}\right)^{0.825} \qquad (10.14)$$

where T_a = air temperature, °F.

10.2   CENTRIFUGAL FANS

Total Pressure Increase at Fan Impeller

When air flows through the impeller of a centrifugal fan, its total pressure increase is closely related to the peripheral velocities of the impeller as well as the tangential component of the peripheral velocities entering and leaving the impeller. Figure 10.3 shows

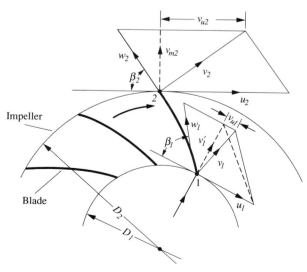

FIGURE 10.3   Velocity triangles at the blade inlet and outlet of a centrifugal fan.

the velocity triangles at the blade inlet and outlet of the impeller of a centrifugal fan.

Let u represent the peripheral velocity of the impeller, w the relative velocity in a direction tangential to the blade profile, and v the absolute velocity of the fluid elements. The angle between the relative velocity w and the peripheral velocity u is called the blade angle β. The tangential component of the absolute velocity is indicated by v_u and the radial component by v_m. Subscripts 1 and 2 denote the velocities at the inlet and outlet of the blade.

Air enters the impeller at the blade inlet with an absolute velocity of v_1. As the fluid elements flow through the blade passage, they accelerate. Moreover, a guiding pressure is exerted on the fluid elements by the blades because of the rotation of the impeller. Air then leaves the blade outlet with an absolute velocity v_2.

If the energy losses due to friction and eddies are ignored, and also if the velocities are expressed in fpm, then the total pressure increase, or fan total pressure Δp_t, in inches WG, of the centrifugal fan when air flows through the impeller is given as

$$
\begin{aligned}
\Delta p_t &= \frac{\rho(v_{u2}u_2 - v_{u1}u_1)}{60 \times 60 \times 5.192 g_c} \\
&= 1.66 \times 10^{-6} \rho(v_{u2}u_2 - v_{u1}u_1) \quad (10.15)
\end{aligned}
$$

where ρ = air density, lb/ft^3

g_c = dimensional constant, 32.2 lb$_m$ · ft/lb$_f$ · s^2

If the impeller is deliberately designed at a specific blade inlet angle β_1 and a radial component v_{m1} so that the tangential component $v_{u1} = 0$, then the fan total pressure becomes

$$
\Delta p_t = 1.66 \times 10^{-6} \rho v_{u2} u_2 \quad (10.16)
$$

From Eqs. (10.15) and (10.16), the fan total pressure of a centrifugal fan depends mainly on the peripheral velocity u_2 and the tangential component v_{u2}, and therefore, the configuration of the blades. The peripheral velocity at the blade outlet u_2, often called *tip speed*, in fpm, can be calculated as

$$
u_2 = \pi D_2 n \quad (10.17)
$$

where D_2 = outside diameter of the fan impeller, ft

n = revolutions per minute of the impeller

For the same type of centrifugal fan, the greater the diameter, the greater the volume flow rate and fan total pressure. For a centrifugal fan of a specific type and size, at each impeller speed, there will be a set of different Δp_t–$\dot{V}$, P–$\dot{V}$, and η_t–$\dot{V}$ curves. The higher the speed, the greater the $\dot{V}$, Δp_t, and P.

Another important parameter that affects the characteristics of centrifugal fans is blast area A_{blast}, in ft^2, as shown in Fig. 10.4. Blast area is the cross-sectional area just above the cutoff point where discharge air is prevented from recirculating. It is always smaller than the outlet area of the centrifugal fan A_o, in ft^2.

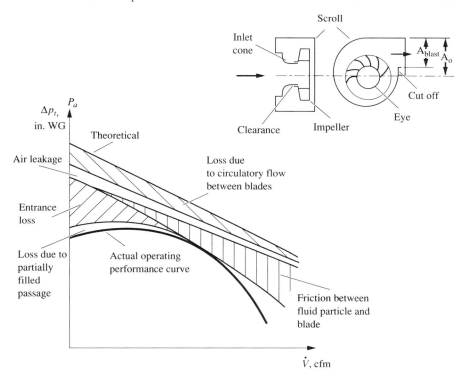

FIGURE 10.4 Operating characteristics for a backward-curved centrifugal fan.

Blast area actually determines the maximum velocity discharged from the fan outlet.

Based on the shape of the blade and the direction of air discharged from the impeller, centrifugal fans can be categorized as backward-curved, radial-bladed, forward-curved, tubular or in-line, and unhoused or cabinet fans, or roof ventilators.

Backward-Curved Fans

In a *backward-curved or backward-inclined centrifugal fan*, the blade tip inclines away from the direction of rotation of the impeller. The β_2 angle of a backward-curved centrifugal fan is smaller than 90°. Figure 10.3 actually shows the velocity triangles of a backward-curved centrifugal fan. The impeller of a backward-curved centrifugal fan usually consists of 8 to 16 blades. For greater efficiency, the shape of the blades is often streamlined to provide minimum flow separation and, therefore, minimum energy losses. Backward-curved centrifugal fans with such blades are called *airfoil fans,* as distinguished from fans with sheet metal blades. The blades in a backward-curved fan are always longer than those of a forward-curved fan. A volute or scroll casing is used. This shape converts some of the velocity pressure into static pressure at the fan outlet.

From the velocity triangles at the outlet, the tangential component $v_{u2} = u_2 - v_{m2} \cot \beta_2$ and the radial component $v_{m2} = \dot{V}/(\pi D_2 b_2)$. Here, b_2 represents the width of the impeller at the blade tip. The total pressure developed is

$$\Delta p_t = C_1 \rho v_{u2} u_2 = C_1 \rho \left(u_2^2 - \frac{u_2 \dot{V} \cot \beta_2}{\pi D_2 b_2} \right)$$

(10.18)

where C_1 = a constant equal to 1.66×10^{-6}. Because $\beta_2 < 90$ degrees, $\cot \beta_2$ is a positive value, and the theoretical Δp_t–$\dot{V}$ curve shown in Eq. (10.18) is a straight line that declines at an increasing volume flow rate, as shown in Fig. 10.4.

In actual operation, however, when the air flows through the centrifugal fan, it encounters the following energy losses:

- Circulatory flow between the blades
- Air leakage at the inlet
- Friction between fluid particles and the blade
- Energy loss at the entrance
- Partially filled passage

These energy losses change the shape of the Δp_t–$\dot{V}$ curve. The actual performance curve of a backward-curved centrifugal fan is a concave curve declining toward the right-hand side, as shown in Fig. 10.4. It can be seen that the maximum total pressure is slightly higher than that at the shut-off condition. After reaching the maximum Δp_t, it drops sharply as the volume flow rate increases.

The fan total efficiency versus volume flow η_t–$\dot{V}$ curves for backward-curved centrifugal fans are concave. All of them have $\eta_t = 0$ at block tight condition as $\dot{V} = 0$. They also have a maximum efficiency between 50 and 65 percent of the wide-open volume flow, as shown in Fig. 10.5. Among these η_t–$\dot{V}$ curves, the airfoil blade backward-curved centrifugal fan has the highest fan total efficiency.

Figure 10.6 shows the fan power–volume flow P–$\dot{V}$ curves for centrifugal fans using impellers of the same size. Centrifugal fans need the minimum amount of power at block-tight condition. This power input is used to offset the aerodynamic and mechanical losses even when the volume flow is zero. For backward-curved centrifugal fans, fan power P tends to increase initially as the volume flow rate increases. Then P reaches a maximum point, after which P declines as the volume flow rate is further increased. The P–$\dot{V}$ curve of a backward-curved centrifugal fan has a shape similar to the Δp_t–$\dot{V}$ curve. If the size of the fan motor is rated at or near the highest fan total efficiency, the power required is often the maximum value and needs no much fan power at the same speed and density at any other operating condition. Such a P–$\dot{V}$ curve is called a *nonoverloading curve.*

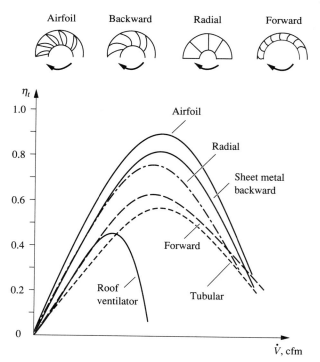

FIGURE 10.5 Total efficiency–volume η_t–$\dot{V}$ performance curves for centrifugal fans.

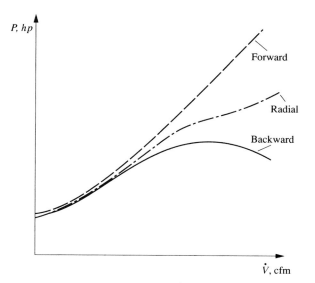

FIGURE 10.6 Power–volume P–$\dot{V}$ performance curves for centrifugal fans with impellers of same diameter.

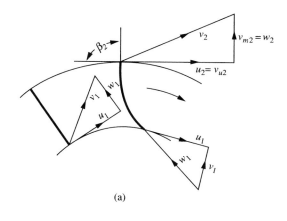

(a)

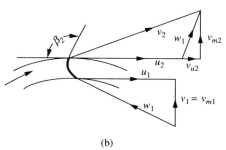

(b)

FIGURE 10.7 Velocity triangles of forward-curved and radial-bladed centrifugal fan: (a) radial-bladed and (b) forward-curved.

Radial-Bladed Fans

The blades in a *radial-bladed centrifugal fan* are either straight or curved at the blade inlet. The blade tip or blade outlet is always radial, that is, $\beta = 90$ degrees as shown in Figure 10.7a. Usually, there are 6 to 10 blades in a radial-bladed impeller. The construction of the radial blades is comparatively simple.

Figure 10.8 shows the Δp_t–$\dot{V}$ curves for centrifugal fans with impellers of the same diameter. The Δp_t–$\dot{V}$ curve of a radial-bladed centrifugal fan is usually steeper than that of a backward-curved fan, with a higher Δp_t and a smaller $\dot{V}$. The fan total efficiency of a radial-bladed fan is lower than that of a backward-curved fan. Furthermore, the fan power input always increases as the volume flow becomes greater. Such an *overloading P–$\dot{V}$ curve* indicates that the motor may be overloaded.

The radial-bladed centrifugal fan is often used in industrial applications to transport particles or products because the spaces between the blades are not easily clogged.

Forward-Curved Fans

In a *forward-curved centrifugal fan,* the blade tip inclines in the direction of rotation of the impeller. At the outlet of the blade, blade angle $\beta_2 > 90$ degrees, as shown in Fig. 10.7b. Because of a greater absolute velocity v_2 at the blade outlet, the blades are shorter and the speed required to produce a specific fan total pressure and volume flow rate is much lower than that for a backward-curved centrifugal fan. The impeller generally has 24 to 64 blades.

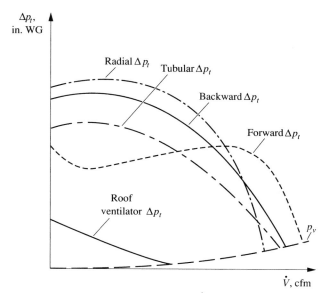

FIGURE 10.8 Pressure–volume Δp_t–$\dot{V}$ performance curves for centrifugal fans with same impeller diameter.

If the diameters of the impellers of a forward-curved fan and a backward-curved fan are the same, then the $\Delta p_t - \dot{V}$ curve of the forward-curved centrifugal fan is flatter than that of the backward-curved fans as shown in Fig. 10.8. Forward-curved fans have a lower fan total efficiency than backward-curved fans and the maximum efficiency of the forward-curved fan occurs between 40 and 50 percent of the wide-open volume flow. The forward-curved fan power curve tends to bend upward more steeply than those of other centrifugal fans, as shown in Fig. 10.6. Therefore, the motor of the forward-curved fan may be overloaded when the volume flow rate is greater than the rated value.

Forward-curved fans are the most compact centrifugal fans available in terms of volume flow rate delivered per unit space occupied. They are also less stable as indicated by the saddle shape of their $\Delta p_t - \dot{V}$ curve. For the same size fan wheel, forward-curved fans have a greater inlet than backward-curved fans.

Tubular or In-Line Fans

A *tubular* or *in-line centrifugal fan* generally consists of an impeller with airfoil blades, an inner cone, a set of fixed vanes, and a cylindrical casing, as shown in Fig. 10.9. Air enters the tubular fan at the inlet and then flows through the blade passages of the impeller. It is discharged radially against the inner surface of the cylindrical casing and is then deflected in a direction parallel to the axle of the fan in order to produce a straight-through flow. The impeller of a tubular centrifugal fan may have 6 to 12 blades. Fixed vanes are used to convert velocity pressure into static pressure to prevent swirls and straighten the air flow.

In Fig. 10.8, it can be seen that the $\Delta p_t - \dot{V}$ curve for a tubular centrifugal fan is similar to that of a backward-curved fan. Because of the 90° turn of the

air flow after the impeller, both total pressure and total efficiency are lower than those of a backward-curved fan, and the sound level is greater. The fan power $P - \dot{V}$ curve of this type of fan is also a non-overloading curve. The main advantage of the tubular fan is its straight-through air flow.

Unhoused or Cabinet Fans

Unhoused centrifugal fans are centrifugal fans with no outer casing or scroll. The impeller is mounted in a cabinet or a plenum with a planned outlet or discharge passage. They are often called *cabinet fans*. Unhoused centrifugal fan are simpler in construction and inlet and outlet configuration. They are especially suitable for spaces where a plenum or cabinet can be used as the outer casing of the impeller, and the two ends of the plenum used as the inlet and outlet of the fan.

Unhoused centrifugal fans with backward-curved impellers have lower fan total efficiency (usually between 58 and 63 percent) than backward-curved centrifugal fans. They are sometimes used in rooftop packaged units or in clean room air systems.

Centrifugal Roof Ventilators

A *centrifugal roof ventilator* is often mounted on a roof to exhaust air from a ventilated space, as shown in Fig. 10.10. Roof ventilators can also be used as intake devices to draw outdoor air through the roof.

In Fig. 10.10, air is extracted from ventilated space and enters the inlet cone directly or through a connecting duct having a total pressure loss usually less than

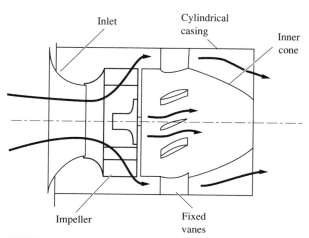

FIGURE 10.9 Tubular centrifugal fan.

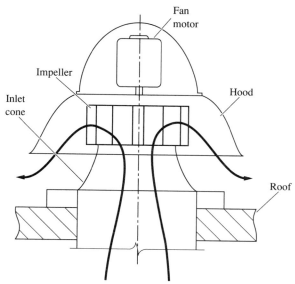

FIGURE 10.10 Centrifugal roof ventilator.

0.2 in. WG. After flowing through a backward-curved centrifugal impeller, air is discharged radially against the inner surface of the hood. The sheet metal or air-foil blades of the centrifugal roof ventilator are made of aluminum alloy, stainless steel, or well-protected structural steel to provide good corrosive resistance above the roof.

The centrifugal roof ventilator usually operates at a low total pressure and a large volume flow rate, as illustrated by a downward inclined Δp_t–$\dot{V}$ curve as shown in Fig. 10.8. The power P–$\dot{V}$ curve is a nonoverloading curve.

10.3   AXIAL FANS

Types of Axial Fans

For an axial fan, a parameter called *hub ratio* is closely related to its characteristics. Hub ratio R_{hub} is defined as the ratio of hub diameter D_{hub}, in ft, to the tip-to-tip blade diameter or diameter of impeller D_{bt}, in ft,

$$R_{hub} = \frac{D_{hub}}{D_{bt}} \qquad (10.19)$$

The higher the hub ratio, the greater the conversion of velocity pressure into static pressure because of the larger difference of the areas: $\pi(D_{bt}^2 - D_{hub}^2)/4$.

The capacity and total pressure increase of axial fans can be increased by raising their rotating speed or through the adjustment of the blade pitch angle to a higher value. This characteristic is important for axial fans that are driven directly by motor, without belts.

Axial fans can be subdivided into the following three types:

PROPELLER FANS. In a *propeller fan,* an impeller having 3 to 6 blades is mounted within a circular ring or an orifice plate as shown in Fig. 10.11a. The blades are generally made of steel or molded plastic, and sometimes may increase in width at the blade tip. If the impeller is mounted inside an orifice plate, the direction of air flow at the blade tip will not be parallel to the axle. Eddies may form at the blade tips. Propeller fans are usually operated at very low static pressure with large volume flow. They often have a hub ratio $R_{hub} < 0.15$.

TUBEAXIAL FANS. The impeller of a *tubeaxial fan* usually has 6 to 9 blades. It is mounted within a cylindrical casing, as shown in Fig. 10.11b. The blades can be airfoil blades or curved sheet metal. Airfoil blades are usually made of cast aluminum or aluminum alloy. The hub ratio R_{hub} is generally less than 0.3 and the clearance between the blade tip and the casing is significantly closer than in propeller fans.

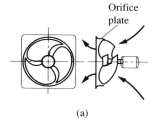

(a)

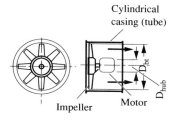

(b)

(c)

FIGURE 10.11   Axial fans: (a) propeller; (b) tubeaxial; and (c) vaneaxial.

In some tubeaxial fans, the blade angle can be adjusted manually when the fan is not in operation. Thus, the volume flow and total pressure increase can be adjusted for tube axial fans driven directly by motor.

VANEAXIAL FANS. The impeller of a *vane axial fan* has 8 to 16 blades, usually airfoil blades. The hub ratio is generally equal to or greater than 0.3 in order to increase total pressure. Another important characteristic of vaneaxial fans is the installation of fixed guide vanes downstream from the impeller, as shown in Figs. 10.11c and 10.12. These curved vanes are designed to remove swirl from the air, straighten the air flow, and convert a portion of the velocity pressure of the rotating air flow into static pressure. Sometimes guide vanes are also installed upstream from the impeller. Automatically controllable pitch (blade angle) for variable-air-volume systems is one of the features of vaneaxial fans.

Velocity Triangles

Consider a blade section of an axial fan as shown in Fig. 10.12. Air enters the impeller with a velocity v_1,

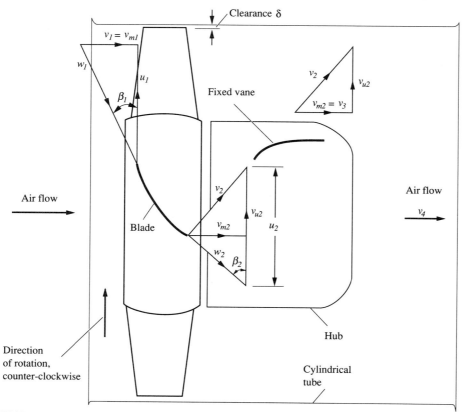

FIGURE 10.12 Velocity triangles for a vaneaxial fan.

which is exactly equal to the axial velocity at the blade inlet v_{m1}. Because of the rotation of the impeller, air leaves the blade at its outlet with a velocity v_2. For a blade section whose diameter at blade inlet equals the diameter at blade outlet, the peripheral velocity $u_1 = u_2$, and the axial velocity $v_{m1} = v_{m2}$. The total pressure increase Δp_t, in inches WG, as air flows through the impeller is given by the same equation as for a centrifugal fan when velocities are expressed in fpm:

$$\Delta p_t = 1.66 \times 10^{-6} \rho v_{u2} u_2$$

To design a blade such that the total pressure developed at various radii of the blade constant, the blade angle near the hub of the same blade must be increased in order to provide a higher v_{u2} when the radius becomes smaller. A greater hub ratio $D_{\text{hub}}/D_{\text{bt}}$ results in a smaller variation of the blade angles of the same blade.

After air leaves the blade and before it enters the downstream fixed guide vanes, there is a rotational component v_{u2} in the direction of rotation. The function of the downstream guide vanes is to convert this

dynamic pressure into static pressure with minimum energy loss.

The static pressure developed in an axial fan is the combined effect of the following:

- The corresponding drop in relative velocity pressure

$$C_2 \rho(w_1^2 - w_2^2)$$

- The conversion of rotating dynamic pressure into static pressure

$$C_2 \rho(v_2^2 - v_{m2}^2)$$

- The conversion of the difference in velocity pressures

$$C_2 \rho(v_3^2 - v_4^2)$$

where v_3, v_4 = mean axial velocities at the cross-sectional areas with and without the hub respectively, fpm
C_2 = constant equal to $1/(60^2 \times 2 \times 32.2 \times 5.192) = 8.3 \times 10^{-7}$

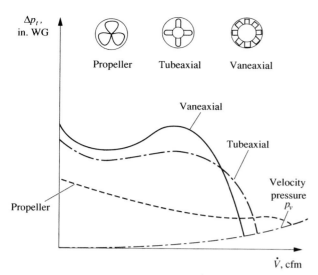

FIGURE 10.13 Pressure–volume Δp_t–$\dot{V}$ for axial fans with the same impeller diameter.

Performance Curves

Figure 10.13 shows the pressure–volume flow Δp_t–$\dot{V}$ curves for various types of axial fans with impellers of the same diameter. Among these fans, the propeller fan has a low fan total pressure Δp_t and a high volume flow rate $\dot{V}$, whereas the tubeaxial fan has a higher Δp_t, and the vaneaxial has the highest Δp_t.

Figure 10.14 shows the total efficiency η_t–$\dot{V}$ curves for axial fans of the same diameters. Because of the installation of airfoil blades and the downstream guide vanes, vane axial fans have the highest η_t of all axial fans. A well-designed vane axial fan may have the same or even higher maximum η_t

than a backward-curved airfoil centrifugal fan. In a vaneaxial fan, it is important to use its velocity pressure, or convert it into static pressure by means of a diffuser. Tubeaxial fans have lower η_t. Propeller fans have the lowest η_t, ranging between 0.4 to 0.55, because of their simple construction.

Figure 10.15 shows the fan power P–$\dot{V}$ curves for axial fans. It is important to see that all axial fans have their maximum fan power input at the shut-off or block-tight condition. In this respect, they are quite different from the centrifugal fans, which have high fan power input at large volume flow rates. For axial fans, the greater the volume flow rate, the lower the P, except in vaneaxial fans, whose P–$\dot{V}$ curves are saddle-shaped. Because the P–$\dot{V}$ curves for vaneaxial and tubeaxial fans are nearly parallel to the Δp_t–$\dot{V}$ curves at the operating range, they are nonoverloading curves. Motors need not be sized according to the maximum fan power input as long as the axial fans will not operate at shut-off condition.

Figure 10.16 shows the performance curves of Δp_t–$\dot{V}$, P–$\dot{V}$, and total efficiency η_t for a typical vaneaxial fan. The impeller is 43 inches in diameter and rotates at a speed of 1750 rpm. From the performance curves, it can be seen that the greater the blade pitch or blade position, the higher the fan total pressure Δp_t and volume flow rate $\dot{V}$.

REVERSE OPERATION. If the rotation of an axial fan is reversed, the direction of air flow is reversed. Propeller and tubeaxial fans without guide vanes deliver about 60 to 70 percent of the volume flow rate in the original forward direction when their rotations

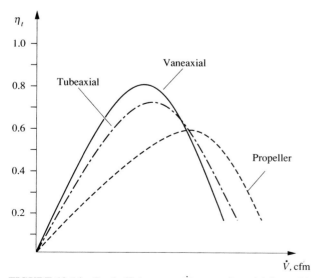

FIGURE 10.14 Total efficiency η_t–$\dot{V}$ curves for axial fans with the same impeller diameter.

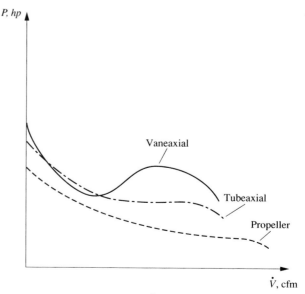

FIGURE 10.15 Fan power P–$\dot{V}$ curves for axial fans with the same impeller of same diameter.

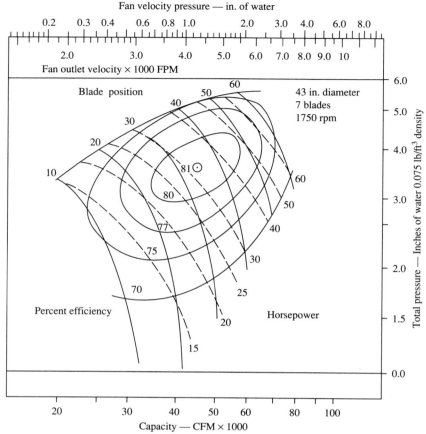

FIGURE 10.16 Controllable pitch vaneaxial fan performance curves. (*Source:* Buffalo Forge Company. Reprinted with permission.)

are reversed. Vaneaxial fans with guide vanes either downstream or upstream from the impeller do not perform efficiently when they rotate in reverse.

TIP CLEARANCE AND NUMBER OF BLADES. For an axial fan, the size of gap between the blade tip and the casing δ as shown in Fig. 10.12, or the ratio of this gap to the outside diameter of the impeller δ/D_{imp}, has a definite effect on fan efficiency and sound power level. Here, D_{imp} indicates the diameter of impeller. Clearance δ and diameter D_{imp} must be expressed in the same units. Generally, δ/D_{imp} should be equal to or less than 0.0025. When the gap becomes greater, the amount of air leakage through the gap increases. Various tests have shown that fan total efficiency may drop from 0.7 to 0.68 if δ/D_{imp} increases from 0.0025 to 0.01.

Theoretically, the higher the number of blades in an axial fan, the better the guidance of air. At the same time, the friction between the fluid elements and the surface of the blade increases. An axial fan with a high number of blades always produces a higher fan total pressure than a fan with fewer blades at the same volume flow rate.

10.4 FAN CAPACITY MODULATION

In a variable-air-volume air system, more than 85 percent of its running hours operate at air flow less than the design volume flow rate. Modulating the fan capacity by providing a new $\Delta p_t - \dot{V}$ curve with lower volume flow and fan total pressure not only corresponds with load reduction in the conditioned space, but also allows significant energy savings at part-load operation.

Four types of fan capacity modulation are currently used in air systems: fan speed, inlet vanes, fan inlet cones, and blade pitch modulation.

Fan Speed Modulation

The peripheral velocity at blade tip u_2 affects both the tangential component v_{u2} and radial component v_{m2} of the air flow, and therefore, the $\dot{V}$ and Δp_t of the fan. The variation of fan speed produces a family of similar $\Delta p_t - \dot{V}$ curves, as shown in Fig. 10.17a.

ADJUSTABLE-FREQUENCY AC DRIVES. *Adjustable-frequency AC drives* modulate the speed of the AC motor by supplying a variable-frequency and variable-

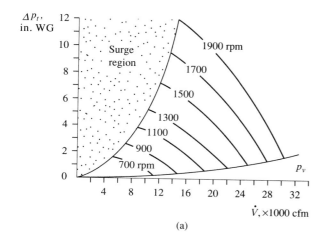

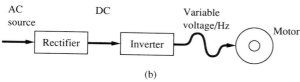

FIGURE 10.17 AC inverter and $\Delta p_t - \dot{V}$ curves of an airfoil fan at various speeds: (a) $\Delta p_t - \dot{V}$ curves at various speeds; and (b) AC-inverter.

voltage power source. The synchronous speed of the motor n_m, in rpm, can be calculated as

$$n_m = 120 \frac{f}{N_p} \qquad (10.20)$$

where f = frequency of the applied power, Hz
N_p = number of poles

If the frequency varies, the motor speed changes accordingly. The voltage must be reduced with the frequency, to maintain a specific voltage-to-frequency ratio, in order to follow the decreasing inductive reactance of the motor.

The incoming three-phase AC power is rectified into DC supply through a rectifier. DC power supply is then shaped into a pseudo-sine wave of predetermined frequency by an inverter as shown in Figure 10.17b. The advantage of an AC inverter is that its speed can be reduced to one tenth (or even less) of its original speed. Through the change of frequency, the speed torque curve of the motor can be adjusted so that it is always operated near its maximum efficiency. The electronics themselves consume little energy. They are usually about 97 to 98 percent efficient. However, because of the AC inverter, motor efficiency degrades 5 to 7 percent, so the total energy loss is about 7 to 8 percent. An AC inverter now costs slightly less than some other variable-speed drives.

There are three types of adjustable-frequency AC drives: *adjustable-voltage inverter (AVI), adjustable-current inverter (ACI)*, and *pulse width modulated inverter (PWM)*. The AVI is a three-phase bridge inverter. Proper sequential firing of the power thyristors produces a pseudo-square waveform with harmonic distortion. The pulsating DC is then filtered by the choke and capacitor and inverts into adjustable-frequency AC drive.

The ACI uses a regulated DC supply, a DC filter choke, and a bridge inverter. Six power-switching devices in the inverter control the current wave form (control is of current instead of voltage).

A PWM inverter operates from a fixed DC bus. Varying the timing on thyristors or transistors produces pulsed output. In large PWM units, sine wave modulation is used to reduce motor loss and increase performance effectiveness at low speeds (less than 6 Hz).

The AVI is simple and reliable. Its speed range is limited above 6 Hz. The ACI is short-circuit-proof and is capable of regenerative operation without an additional converter. It is not suitable for speeds below 6 Hz, and high-voltage spikes are required on motor terminals. The PWM inverter is universally applicable. It operates at a wide range of speeds and multiple motor loads, and generates less noise. The PWM inverter is complicated and needs high-performance thyristors.

Because of the pseudo-square wave form supplied to the induction motor, more heat is produced, which derates the motor capacity (i.e., the motor's rated capacity is decreased). When an AC inverter is used with an induction motor, its capacity is derated to 85 to 90%.

MECHANICAL AND HYDRAULIC VARIABLE SPEED DRIVES. These variable-speed drives vary the effective diameter of the driving pulley of the motor through mechanical or hydraulic mechanisms, thereby changing fan speed. One typical mechanical drive uses a fractional power motor to move the fan motor and pulley a predetermined distance along a steel rail. As the distance between the driving and driven pulleys changes, a spring forces the movable side of the motor pulley to ride up or fall down on the sheave. This changes the diameter of the driving pulley. Mechanical variable-speed drives can modulate fan speed down to 40 percent of design flow.

Inlet Vane Modulation

Inlet vanes are pivoted movable vanes installed at the inlet of the impeller. These vanes are linked mechanically so that they can turn simultaneously around the axis, as shown in Fig. 10.18.

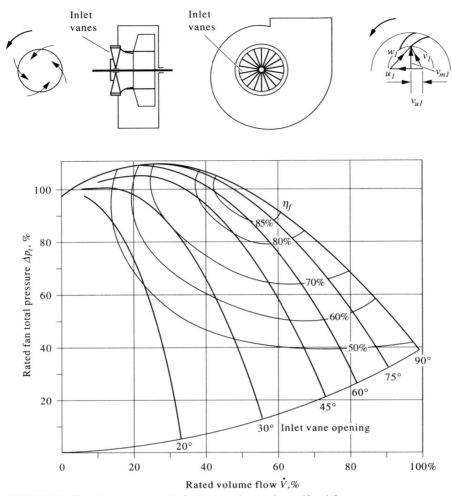

FIGURE 10.18 Inlet vane control of a backward-curved centrifugal fan.

Inlet vanes impart a spin on the airstream before it enters the impeller. From Eq. (10.15),

$$\Delta p_t = 1.66 \times 10^{-6} \rho(v_{u2}u_2 - v_{u1}u_1)$$

If the direction of the spin is the same as the direction of rotation of the impeller, the tangential component v_{u1} has a positive value. The total pressure thus developed is smaller than if air enters the impeller with a radial entry, $v_{u1} = 0$. This produces a new set of Δp_t–$\dot{V}$ curves at various inlet vane opening angles with lower Δp_t and $\dot{V}$.

The fan total efficiency η_t decreases as the inlet vane opening angle decreases, as the air at inlet is further deviated from the shock-free condition and the inlet area is reduced. However, the reduction of shaft power from the decreases of Δp_t and $\dot{V}$ compensates for the effect of the drop in η_t. Energy savings are significant only when the inlet vanes are sufficiently near the eye of the impeller.

Inlet vanes reduce the inlet area. For large fans, this reduced inlet area causes a pressure loss of about 8 percent. For small fans, the additional pressure loss is far greater. If air velocity at the inlet v_i exceeds 5000 fpm, the fan total pressure is significantly reduced. Inlet guide vanes are usually not perferable for small backward-curved centrifugal fans.

A backward-curved centrifugal fan with airfoil blades and inlet vanes has a higher ratio of reduced volume flow power input to design volume flow power input P_{part}/P_{full} than a forward-curved centrifugal fan has. However, a backward-curved airfoil fan has a higher inherent fan total efficiency than a forward-curved fan. Therefore, for a VAV system, actual fan power input should be compared between an airfoil fan and a forward-curved fan at reduced volume flow operation (according to the manufacturer's data) before fan selection.

Use of inlet vanes for fan capacity modulation is still widespread for centrifugal fans because of their

their lower installation cost. Because axial fans require maximum fan power input at shut-off condition, inlet vanes are not suitable for axial fans.

Inlet Cone Modulation

The inlet cone of a backward-curved or airfoil centrifugal fan can be moved so that a portion of the impeller is inactive. Such a modulation results in a lower fan capacity and fan total pressure. Inlet cone modulation has a lower initial cost. Although its fan energy use is higher than that of adjustable-frequency AC drive at part-load conditions, inlet cone modulation does not block the area of the inlet eye. Inlet cone modulation is often used in rooftop packaged units in which airfoil centrifugal fans are installed.

Blade Pitch Modulation

From the velocity triangles shown in Fig. 10.12, for axial fans, the smaller the blade angle β_2, the lower the v_{m2} and the v_{u2}, and therefore, the lower the $\dot{V}$ and Δp_t. Figure 10.16a shows the performance diagram of a typical controllable and adjustable pitch vaneaxial fan in which there is a corresponding Δp_t–$\dot{V}$ curve for each blade angle. For axial fans, blade pitch fan capacity modulation can take place both during the rotation of the fan impeller (controllable pitch) or at rest (adjustable pitch).

Figure 10.19 shows the blade pitch control mechanism of this controllable-pitch vaneaxial fan. The blades are activated internally through lever arms and roller mechanism by pneumatic pressure against a flexible diaphragm. The control mechanism is direct-

HUB DETAIL

1 – Actuator body
2 – Reaction plate
3 – Diaphragm
4 – Actuator cover plate
5 – Rotary union
6 – Cavity (between actuator cover plate and diaphragm)
7 – Base of blade
8 – Lever-arm/roller mechanism
9 – Adjustable stops
10 – Motor or fan shaft

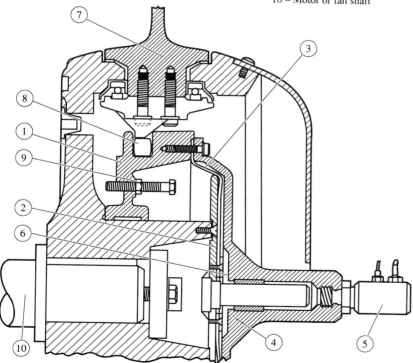

FIGURE 10.19   Controllable pitch vaneaxial fan blade pitch control mechanism. (*Source:* Buffalo Forge Company. Reprinted with permission.)

acting, that is, blade angle increases as the controlled pneumatic air pressure is increased. If the pneumatic pressure fails when the fan is operating, blades return to their minimum blade pitch. A pilot positioner, located outside the fan, ensures that the controlled blade pitch has the minimum deviation from the required one. The pneumatic pressure used to actuate the blade pitch is 25 psig while the pressure used to activate the pilot positioner is 3 to 15 psig.

Of these four fan capacity modulating methods, inlet vane modulation is lower in installation cost and provides significantly lower energy savings than fan speed and blade pitch modulation at reduced air flow. Fan speed and blade pitch modulation also reduce fan noise and increase the fan's life.

10.5 FAN SURGE AND STALL

Fan Surge

Fan surge occurs when the air volume flow through a fan is not sufficient to sustain the static pressure difference between the discharge and suction sides of the fan. Air surges back through the fan impeller and its discharge pressure is reduced momentarily. The surge of air enables the fan to resume its original

static pressure. Fan surge will repeat if the air volume flow is not sufficient to sustain normal operation.

The volume flow and pressure fluctuations caused by fan surge cause noise and vibration. The surge region of a backward-curved centrifugal fan is usually greater than that of a forward-curved fan. Backward-curved or airfoil blade centrifugal fans should not operate in the surge region shown in Fig. 10.17.

Some manufacturers show fan surge region on their fan performance diagrams in their catalogs. If data and curves are not presented in the performance table or diagram, they are probably in or near the fan surge region.

Fan Stall

Stall is an operating phenomenon of an axial fan. For airfoil blades in an axial fan, the angle of attack α is the angle between the motion of the blades and the chord line, as shown in Fig. 10.20a. At normal operation, air flow around the streamlined airfoil blades causes a pressure difference between the top and bottom surfaces of the blades and produces lift. Stall occurs when the smooth air flow suddenly breaks and the pressure difference across the airfoil blades decreases. The axial fan loses its pressure capability drastically.

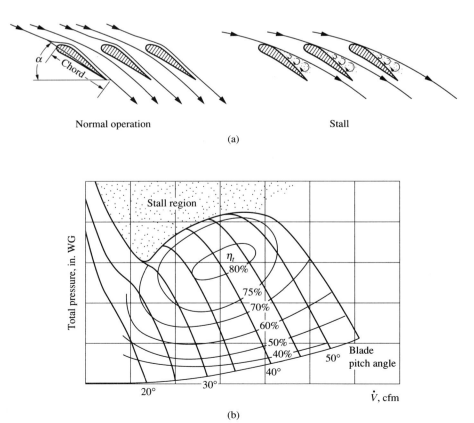

FIGURE 10.20 Stall and stall region of an axial fan: (a) normal operation and stall and (b) stall region of an axial fan.

The stall region of an axial fan is generally located at the upper part of the fan performance diagram beyond the region of highest fan total efficiency as shown in Fig. 10.20b. For each blade pitch angle, there is a corresponding stall region line. Usually, the stall region is also the region beyond which no performance curves are given.

10.6 FAN CONSTRUCTION AND ARRANGEMENTS

Size and Class Standards

The Air Movement Control Association (AMCA) has established standards concerning fan size and classification. The *size standard* specifies the diameter of the impeller for a given fan size. There are 25 sizes, from 12.25 to 132.25 in. diameter. Each size is approximately 10 percent greater in diameter than the next smaller size.

Class standard specifies the construction of the fans as heavy, medium, or light structure. Construction is closely related to the static pressure developed Δp_s (in inches WG) by the fan and the fan outlet velocity v_o (in fpm), as shown in Fig. 10.21. Three classes of fans are widely used in HVAC&R applications: Class I, Class II, and Class III. As shown in Fig. 10.21, a Class I fan must provide 5 in. WG static pressure at an outlet velocity of 2300 fpm and a static pressure of 2.5 in. WG at an outlet velocity of 3200 fpm. A Class II fan must provide 8.5 in. WG static pressure at an outlet velocity of 3000 fpm and 4.25 in. WG static pressure at 4175 fpm.

Fan Width and Inlets

For centrifugal fans, air velocity at the fan inlet v_i should be optimal. A lower v_i means a large inlet. A higher v_i results in a greater energy loss and therefore a lower fan total efficiency. A suitable relationship is $v_i/u_2 = 0.35$ to 0.4.

A *single-width single-inlet (SWSI)* centrifugal fan has a single inlet cone connected to the eye of the impeller, as shown in Fig. 10.22a. A *double-width double-inlet (DWDI)* centrifugal fan has a double-width impeller and provides double inlets from both sides, as shown in Fig. 10.22b. In such an arrangement, v_i can still be maintained at an optimal value and the volume flow rate is approximately doubled. Because of the interaction of the two airstreams in the impeller and the scroll, fan efficiency of a DWDI fan may be 1 to 2 percent lower than that of a SWSI fan.

Drive Arrangements and Direction of Discharge

Fans can be driven by motor directly or through v-belts and sheaves. In a belt drive arrangement, fan speed can be changed by varying the diameter of the sheaves. On the other hand, belt drive requires 3 to 5 percent more energy input than the direct drive.

Drive arrangements include the location of bearing and sometimes the position of motor. For centrifugal fans, there are eight standard drive arrangements, as shown in Fig. 10.23. There are only two drive arrangements, 4 and 9, for axial fans. Arrangements 1, 2, 3, 7, and 8 can include either belt drive or direct drive. Arrangement 4 is for direct drive only. Arrangements 9 and 10 are for belt drive only. However, in Arrangement 9 for centrifugal fans, the motor is located outside the base, whereas in Arrangement 10, the motor is located inside the base. This arrangement is often used with a weatherproof hood for outdoor installation.

Motor location is always specified as W, X, Y, or Z position, facing the fan drive side.

Direction of rotation and discharge position of centrifugal fans are specified from the drive side of the fan, as shown in Fig. 10.24. In single inlet fans, the drive side is always opposite from the fan inlet, regardless of the actual position of the fan drive.

High-Temperature Fans

When fans are operated in high-temperature airstreams, the yield strength of the fan structure may decrease. Operating temperatures are usually classified as −20 to 200°F (normal or standard fan construction), 201 to 300°F, 301 to 400°F, and 401 to 750°F.

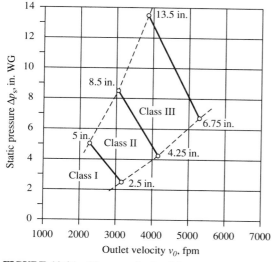

FIGURE 10.21 Class standards of fans.

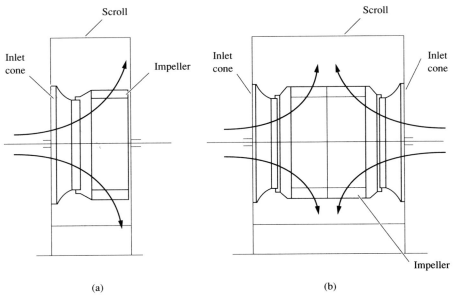

Scroll

Inlet cone

Impeller

(a)

Scroll

Inlet cone

Inlet cone

Impeller

(b)

FIGURE 10.22 Single- and double-width centrifugal fans: (a) single-width single-inlet (SWSI) and (b) double-width double-inlet (DWDI).

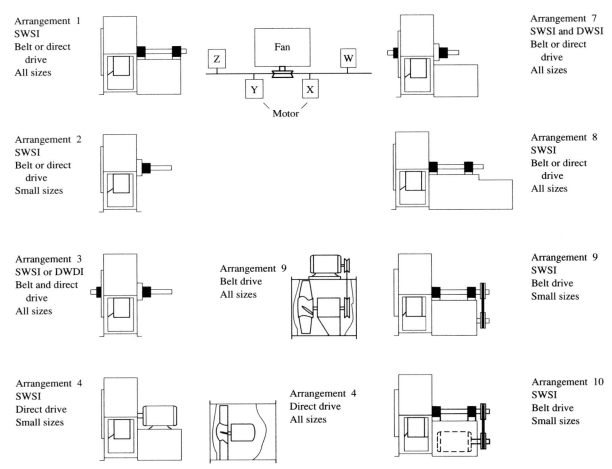

Arrangement 1
SWSI
Belt or direct drive
All sizes

Fan

Z

W

Y

X

Motor

Arrangement 7
SWSI and DWSI
Belt or direct drive
All sizes

Arrangement 2
SWSI
Belt or direct drive
Small sizes

Arrangement 8
SWSI
Belt or direct drive
All sizes

Arrangement 3
SWSI or DWDI
Belt and direct drive
All sizes

Arrangement 9
Belt drive
All sizes

Arrangement 9
SWSI
Belt drive
Small sizes

Arrangement 4
SWSI
Direct drive
Small sizes

Arrangement 4
Direct drive
All sizes

Arrangement 10
SWSI
Belt drive
Small sizes

FIGURE 10.23 Drive arrangements for centrifugal and axial fans and motor positions.

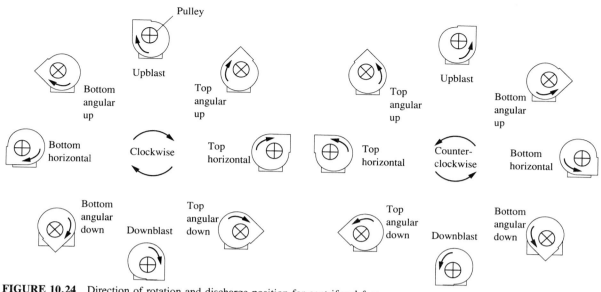

FIGURE 10.24   Direction of rotation and discharge position for centrifugal fans.

The rpm limits on Class I, II, and III must be multiplied by a speed factor of 0.9 when operating temperature is 401 to 500°F. A lower speed factor should be used at higher temperatures. Cooling of the fan bearings is a significant problem. Generally, bearings should be kept out of the airstream except when operating at room temperature. In many anti-friction bearings, grease can be used up to 200°F. Oil bearings or high-temperature grease is required for temperatures higher than 300°F. Special coolants should be applied to the shaft and bearings at higher temperatures. High-temperature aluminum paint can protect surface finishes.

Spark-Resistant Construction

The AMCA specifies three types of spark-resistant fan construction:

Type A construction requires that all parts of the fan in contact with the air be made of nonferrous metal.

Type B construction requires that a fan have a non-ferrous impeller and shaft rings.

Type C construction requires that a shift of the impeller or shaft may not allow the ferrous parts to rub or strike.

The bearings of spark-resistant construction fans should not be placed in the airstream, and all fan parts should be electrically grounded.

Safety Devices

When a fan is installed in a location where it is open to occupants or is accessible during operation, safety devices such as belt guards and protecting screens must be installed.

Even a wire mesh installed at the outlet of a fan with a high air velocity may cause a pressure loss of 0.15 in. WG. Therefore, if the fan is covered by an enclosure or is accessible only when the fan is not in operation (during repair or maintenance), a safety device is not required.

10.7   FAN SELECTION

Selection of a fan for a given type of air system or mechanical ventilating system actually takes place in two stages: selection of fan type and determination of fan size.

Conditions Clarified and Factors Considered

Before the selection, the following conditions must be clarified:

- Setting (in a commercial building to handle clean air at room temperature, or an industrial setting to handle dirty air)
- Special requirements (such as high-temperature operation or spark-resistant construction)
- Function (supply fan or a return fan in an air-handling unit, or supply or exhaust fan in a ventilating system)
- Characteristics of the air system (constant-volume or variable-air-volume)
- Room NC curve
- Approximate annual operating hours
- Unit cost of energy at the specific location

During selection, the following factors should be considered:

- *Pressure–Volume Flow Operating Characteristics.* Selecting a fan to provide the required volume flow rate and total pressure loss for an air system or a ventilating system is of prime importance. An undersized fan results in a uncontrolled indoor environment. An oversized fan wastes energy and money.

- *Fan Capacity Modulation.* A variable-air-volume system operates at a reduced volume flow rate during part-load operation. Effective and economical fan capacity modulation is an important factor that affects the operation of an air system.

- *Fan Efficiency.* Fan efficiency is closely related to the energy consumption of an air system. Fans should be selected so that they can operate at high efficiency during as much of their operating time as possible.

- *Sound Power Level.* Most commercial and public buildings and many industrial applications need a quiet indoor environment. Fan noise is the major source of noise in an air system. Usually, the higher the fan total efficiency, the lower the sound power level of the selected fan. A fan with a low sound power level and sound power at higher frequencies is preferable. High frequency sound is more easily attenuated than low-frequency sound.

- *Air Flow Direction.* In many applications, a straight-through or in-line flow occupies less space and simplifies layout.

- *Initial Cost.* The initial cost of the fan modulation device, sound attenuator(s), and space occupied by a particular type of fan, in addition to the cost of the fan itself, should be considered.

Another important factor in fan selection is the fan inlet and outlet connections. As the effects of these connections are closely related to the characteristics of the air system, they will be covered in Chapter 19.

Fan selection involves other considerations that are beyond the scope of this section. In an air system, a fan is usually operated with a duct system, which may also include other equipment and system components. A comprehensive and detailed analysis of fan operation must consider the characteristics of the duct system.

Estimated Fan Sound Power Level

The estimated sound power level of a fan L_w, in dB, can be calculated as

$$L_w = K_w + 10 \log(\dot{V}/\dot{V_1}) + 20 \log(\Delta p_t / \Delta p_{t_1}) + C$$
$$(10.21)$$

where K_w = specific sound power level of a certain type of fan, dB
$\dot{V}$ = fan volume flow rate or capacity, cfm
$\dot{V_1}$ = 1 cfm
Δp_t = fan total pressure, in. WG
Δp_{t1} = in. WG

In Eq. (10.21), C represents the correction for off-peak operation, that is, fan operation in the region of the performance curve where fan efficiency is not maximum. For a fan operating at 90 to 100% of its maximum static efficiency, $C = 0$; between 85 and 89%, $C = 3$. For each decrease of 10% of fan maximum static efficiency up to 55 percent, the C value increases by 3 dB.

Table 10.1 lists the specific sound power level K_W and blade frequency increments BFI, in dB, of various types of fans. Here, BFI should be added to the sound power level at blade frequency f_{blade}, in Hz. Blade frequency can be calculated as

$$f_{blade} = \frac{\text{fan rpm x number of blades}}{60} \quad (10.22)$$

A comparison of K_w values shows that the backward-curved centrifugal fan with airfoil blades has the lowest K_w value at various octave bands. The differences between K_w values at various octave bands for axial fans are far smaller than those of centrifugal fans. For forward-curved and radial-blade centrifugal fans, K_w values are higher in low frequencies and lower in high frequencies than vaneaxial and tubeaxial fans.

Comparison between Various Types of Fans

Table 10.2 compares the characteristics of various types of fans. The backward-curved centrifugal fan with airfoil blades has the highest fan total efficiency and the lowest specific sound power level, so it is still the most widely used type of fan in commercial, institutional and many industrial applications. The forward-curved centrifugal fan has a compact size, slower speed, and lighter weight per unit volume output. It is generally used in room air conditioners, fan-coils, small air-handling units, and many packaged units. In recent years, vaneaxial fans with controllable pitch, especially used as return fans, have more applications in commercial air systems than before. For exhaust systems that require a large volume flow rate and low total pressure increase, a propeller fan is often the best choice.

Case Study—Selection of Fans

Select the appropriate type of fan for the following applications:

TABLE 10.1
Specific sound power levels K_w for typical fans

Fan type	Octave bands								
	63	125	250	500	1K	2K	4K	8K	BFI
Centrifugal									
AF, BC, or BI wheel diameter									
over 36 in.	40	40	39	34	30	23	19	17	3
under 36 in.	45	45	43	39	34	28	24	19	3
Forward-curved									
All wheel diameters	53	53	43	36	36	31	26	21	2
Radial-bladed									
Low pressure									
(4 to 10 in. of water)	56	47	43	39	37	32	29	26	7
Medium pressure									
(6 to 15 in. of water)	58	54	45	42	38	33	29	26	8
High pressure									
(15 to 60 in. of water)	61	58	53	48	46	44	41	38	8
Axial fans									
Vaneaxial									
Hub ratio 0.3 to 0.4	49	43	43	48	47	45	38	34	6
Hub ratio 0.4 to 0.6	49	43	46	43	41	36	30	28	6
Hub ratio 0.6 to 0.8	53	52	51	51	49	47	43	40	6
Tubeaxial									
Over 40 in.									
wheel diameter	51	46	47	49	47	46	39	37	7
Under 40 in.									
wheel diameter	48	47	49	53	52	51	43	40	7
Propeller									
General ventilation	48	51	58	56	55	52	46	42	5

Note Includes: total sound power level in dB for both inlet and outlet. Values are for fans only—not packaged equipment.
AF - airfoil blades
BC - backward-curved blades
BI - backward-inclined blades
BFI - blade frequency increments
Source: ASHRAE Handbook 1991, HVAC Applications. Reprinted with permission.

1. The weaving room of a textile mill
2. A convenience store
3. A high-rise office building

SOLUTION

1. In the weaving room of a textile mill, the sound power level of the looms is far greater than the sound power level of the fan noise transmitted through supply and return grilles. Also, the fan total pressure required for the air system used in the weaving room is usually less than 3 in. WG. Therefore, a tubeaxial or vaneaxial fan would be suitable.

2. A convenience store usually requires an air system with a volume flow rate less than 5000 cfm. Usually, a small or a medium rooftop packaged unit is used. Forward-curved fans are usually used in this type of packaged unit.

3. A high-rise office building may use a central hydronic VAV system or a unitary packaged VAV system. Design criteria also vary considerably according to the location and design criteria of the building. A more detailed and comprehensive analysis must be made to determine whether backward-curved centrifugal fans with airfoil blades, forward-curved fans, vane axial fans, or a combination of fans should be used.

10.8   REFRIGERATING COMPRESSORS

A refrigerating compressor is the heart of a vapor compression refrigeration system. Its function is to raise the pressure of the refrigerant and provide the primary force to circulate the refrigerant. The refrigerant thus produces the refrigeration effect in the evaporator, condenses into liquid form in the condenser,

TABLE 10.2
Comparison between various types of fans

	Backward, airfoil centrifugal fan	Forward-curved centrifugal fan	Vaneaxial	Propeller fan
Fan total pressure, Δp_{tf}	Higher Δp_t	Comparatively Lower Δp_t	Higher Δp_t	low Δp_t
Flow rate	All flow rates	Larger flow rate	All flow rates	Larger flow rate
Fan power input	Nonoverloading	Overloading	Nonoverloading	Nonoverloading
Fan modulation	Inlet vanes AC inverter	Inlet vanes AC inverter	Controllable pitch AC inverter	
Fan total efficiency	0.7 to 0.86	0.6 to 0.75	0.7 to 0.88	0.45 to 0.6
Sound power level	Lower, higher L_w at low frequencies	Medium, higher L_w at low frequencies	Medium, difference of L_w values is small at various Hz	Higher, higher L_w at high frequencies
Air flow direction	90° turn	90° turn	Paralled to axle	Parallel to axle
Volume and weight	Greater	Less	Greater	Medium volume and lower weight
Initial cost	Higher	Medium	Higher	Low
Applications	Large HVAC&R systems	Lower pressure, small HVAC&R system	Large HVAC&R systems	Low pressure, high volume flow exhaust systems

and throttles to a lower pressure through the throttling device.

Positive Displacement and Centrifugal Compressors

According to the characteristics of the compression process, currently used refrigerating compressors can be classified as positive displacement compressors or centrifugal compressors.

POSITIVE DISPLACEMENT COMPRESSORS. A *positive displacement compressor* increases the pressure of the vapor refrigerant by reducing the internal volume of the compression chamber through mechanical force applied to the compressor. This type of compressor mainly includes reciprocating, screw, rotary, and scroll compressors as shown in Figs. 10.25, 10.26, 10.27, 10.28, and 10.29.

CENTRIFUGAL COMPRESSORS. The only type of non-positive displacement refrigeration compressor widely used in refrigeration systems is the *centrifugal compressor*, shown in Fig. 10.30. In a centrifugal compressor, the increase of the pressure of the vapor refrigerant depends mainly on the conversion of dynamic pressure into static pressure.

Hermetic, Semi-Hermetic, and Open Compressors

HERMETIC COMPRESSOR. This is a compressor in which the motor and the compressor are "sealed" or "welded" in the same housing, as shown in Fig. 10.31a. Hermetic compressors have two advantages: they minimize leakage of refrigerant, and the motor can be cooled by the suction vapor flowing through the motor windings, which results in a smaller and cheaper compressor-motor assembly.

Motor windings in hermetic compressors must be compatible with the refrigerant and lubricating oil, resist the abrasive effect of the suction vapor, and have high dielectric strength. Welded compressors are usually used for small installations from < 1 hp to 24 hp.

SEMIHERMETIC COMPRESSORS. These compressors are also known as accessible hermetic compressors (see Fig. 10.31b). The main advantage semi-hermetic compressors have over hermetic compressors is their accessibility for repair during a compressor failure or for regular maintenance. Other features are similar to those of the hermetic compressor. Most of the medium compressors are semi-hermetic.

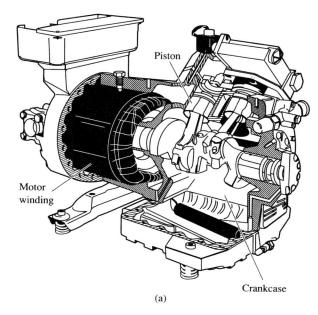

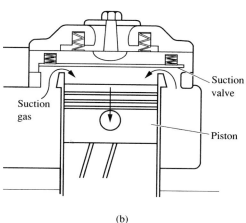

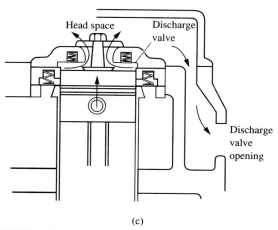

FIGURE 10.25 Semihermetic reciprocating compressor: (a) sectional view; (b) intake stroke; and (c) compression and discharge stroke. (*Source: Trane Reciprocating Refrigeration and Refrigeration Compressors. Reprinted with permission.*)

OPEN COMPRESSORS. In an open compressor, the compressor and the motor are enclosed in two separate housings, as shown in Fig. 10.31c. An open compressor needs shaft seals to minimize refrigerant leakage. In most cases, an enclosed fan is used to cool the motor windings using ambient air. An open compressor does not need to evaporate the liquid refrigerant to cool the hermetic motor windings. Compared with hermetic compressors, open compressors may save 2 to 4 percent of the total power input. Many very large compressors are open compressors.

Direct Drive, Belt Drive, and Gear Drive

Hermetic compressors are driven by motor directly. Both semihermetic and open compressors can be driven directly, driven by motor through v-belts, or driven by gear trains. The purpose of a gear train is to increase the speed of the compressor. Gear drive is compact in size and rotates without slippage. Like belt drive, gear drive needs about 3 percent more power input than direct drive compressors. Some large open compressors may be driven by steam turbine, gas turbine, or diesel engine instead of electric motor.

10.9 PERFORMANCE OF COMPRESSORS

Volumetric Efficiency

The volumetric efficiency η_v of a refrigerating compressor is defined as

$$\eta_v = \frac{V_{a.v}}{V_{dis}} \qquad (10.23)$$

where $V_{a.v}$ = actual induced volume of the suction vapor at suction pressure, ft^3
V_{dis} = theoretical displacement of the compressor, ft^3

Factors that influence the η_v of the compressor are:

- *Clearance volume and compression ratio R_{com}.* Both factors affect the volume of reexpansion gas trapped in clearance volume.
- *Heating effect.* When vapor refrigerant enters the compressor, heat absorbed by the vapor results in a heating effect that increases the specific volume of the refrigerant and, therefore, the $V_{a.v}$ value.
- *Leakage.* Refrigerant leaks through the gap and the clearance across the high- and low-pressure sides of the compressor, such as the clearance between the piston ring and the cylinder in a reciprocating compressor.

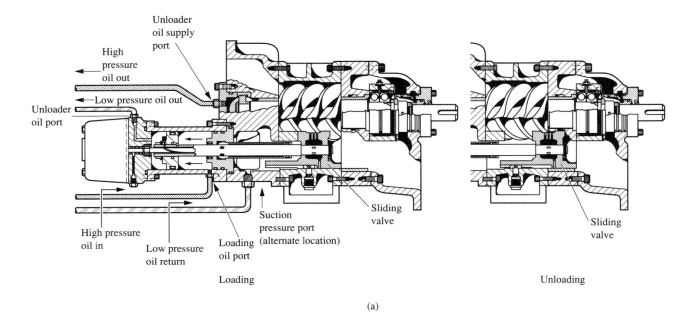

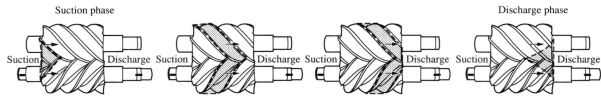

FIGURE 10.26 Twin-screw compressor: (a) loading and unloading of screw compressor and (b) suction-compression-discharge processes. (*Source:* Dunham–Bush Compressors Reprinted by permission.)

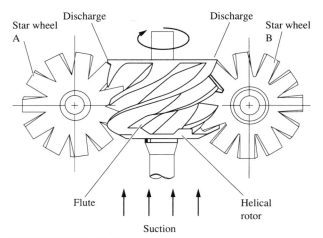

FIGURE 10.27 Single-screw compressor.

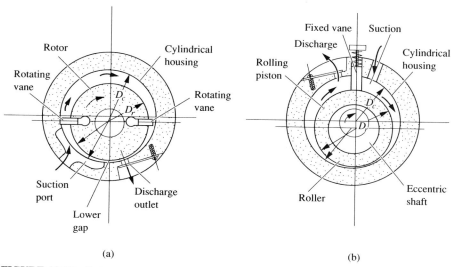

FIGURE 10.28 Rotary compressors: (a) rotating vane; and (b) rolling piston.

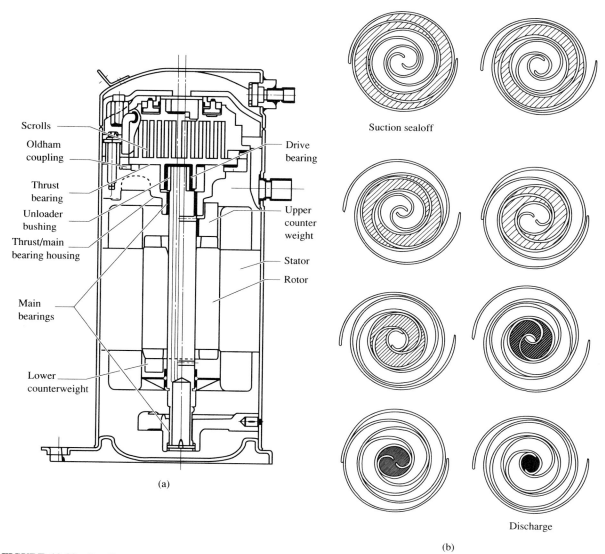

Suction sealoff

Discharge

(a)

(b)

FIGURE 10.29 Scroll compressor and scroll compression process: (a) scroll compressor (Elson et al 1990); and (b) scroll compression process (Purvis 1987). (*Source: ASHRAE Handbook 1992, HVAC Systems and Equipment* Reprinted with permission.)

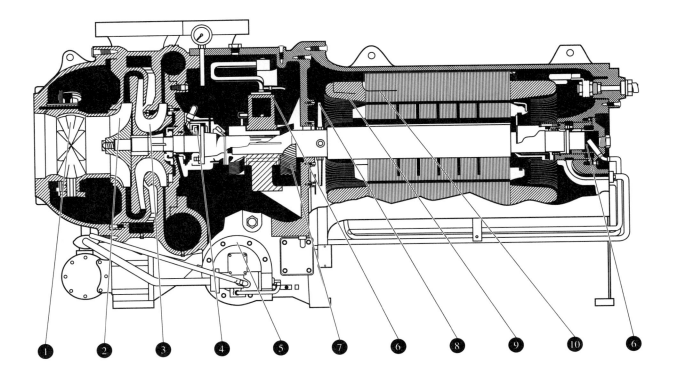

1 – Variable inlet guide vanes
2 – First-stage impeller
3 – Second-stage impeller
4 – Compressor thrust bearing
5 – Lubrication package
6 – Shutdown oil reservoir
7 – Gear transmission
8 – Centrifugal demister system
9 – Internal motor protector
10 – Internal motor protector

FIGURE 10.30   A two-stage hermetic centrifugal compressor. (*Source:* Carries Corporation. Reprinted with permission.)

Motor, Mechanical, and Compression Efficiency

Motor efficiency η_{mo} is defined as

$$\eta_{mo} = \frac{P_{com}}{P_{mo}} \qquad (10.24)$$

where P_{com}, P_{mo} = power input to the shaft of the compressor and to the motor, respectively, hp.

Mechanical efficiency η_{mec} is defined as

$$\eta_{mec} = \frac{W_v}{W_{com}} \qquad (10.25)$$

where W_v, W_{com} = work delivered to the vapor refrigerant and to the compressor shaft, Btu/lb.

Compression efficiency η_{cp} is defined as

$$\eta_{cp} = \frac{W_{isen}}{W_v} \qquad (10.26)$$

where W_{isen} = work required for isentropic compression, Btu/lb. Here, $W_{isen} = (h_2 - h_1)$, where h_1 and h_2 represent the enthalpy of intake vapor refrigerant and of discharged hot gas, respectively, in an isentropic compression process (in Btu/lb).

The difference between work input on the compressor shaft and work delivered to the vapor refrigerant $(W_{com} - W_v)$ is mainly caused by the frictional loss and turbulent loss of refrigerant flow.

Compressor efficiency η_{com} is the product of $\eta_{cp}\eta_{mec}$, that is

$$\eta_{com} = \eta_{cp}\eta_{mec} \qquad (10.27)$$

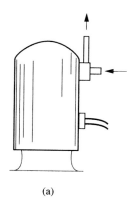

(a)

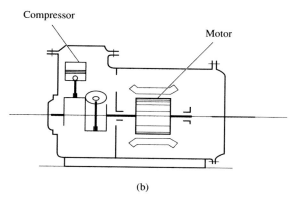

(b)

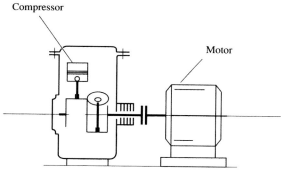

(c)

FIGURE 10.31 Hermetic and open-type compressors: (a) hermetic; (b) semihermetic; and (c) open.

Isentropic and Polytropic Analysis

The isentropic efficiency η_{isen} of a compressor is defined as

$$\eta_{isen} = \frac{h_2 - h_1}{h_2' - h_1} = \eta_{cp}\eta_{mec} \quad (10.28)$$

where h_2' = enthalpy of the discharged hot gas if the compression process is not isentropic, Btu/lb. The difference between h_2' and h_2 implies, first, the deviation of a reversible polytropic process from an isentropic process, and second, the deviation of an irreversible polytropic process from a reversible pro-

cess. Isentropic efficiency is equal to compressor efficiency, or $\eta_{isen} = \eta_{com}$.

The actual power input to the compressor P_{com}, in hp, can be calculated as

$$P_{com} = \frac{\dot{m}_r(h_2' - h_1)}{42.41} = \frac{\dot{m}_r(h_2 - h_1)}{42.41\eta_{isen}}$$
$$= \frac{\dot{m}_r(h_2 - h_1)}{42.41\eta_{com}} \quad (10.29)$$

In Eq. (10.29), $\dot{m}_r$ represents mass flow of refrigerant, in lb/min. For reciprocating compressors, $\dot{m}_r$ can be calculated as

$$\dot{m}_r = \dot{V}_p\eta_v\rho_{suc} \quad (10.30)$$

where $\dot{V}_p$ = piston displacement, cfm (to be covered in Chapter 21)
ρ_{suc} = density of suction vapor, lb/ft^3

Although the actual compression processes for most compressors are irreversible polytropic processes, for simplicity, an isentropic analysis is often used. In other words, actual power input to the compressor P_{com} is usually calculated by Eq. (10.29).

Power Consumption Per Ton Refrigeration

An important index of the energy consumption of a compressor is its electrical power consumption per ton of refrigeration output, or kW/ton.

Power input to a single-stage hermetic compressor P_{com}, in kW, can be calculated as

$$P_{com} = \frac{\dot{m}_r(h_2 - h_1)}{(\eta_{com}\eta_{mo})42.41} \quad (10.31)$$

Here motor efficiency η_{mo} also includes the refrigeration required to cool the motor winding.

Within an operating range of 32 to 120°F, the enthalpy difference along the isentropic process can be expressed as temperature lift, that is, a lift from evaporating temperature T_{ev} to condensing temperature T_{con}. Then

$$h_2 - h_1 = K(T_{con} - T_{ev}) \quad (10.32)$$

For HCFC–123, $K_{123} = 0.16$ Btu/lb · °F. If T_{con} and T_{ev} are expressed in °F, and $\dot{m}_r$ in lb/min, for compressors using HCFC–123 as a refrigerant

$$kW/ton = \frac{(0.16 \times 0.746)(\dot{m}_r/ton)(T_{con} - T_{ev})}{(\eta_{com}\eta_{mo})42.41}$$
$$= \frac{0.0028(\dot{m}_r/ton)(T_{con} - T_{ev})}{\eta_{com}\eta_{mo}}$$
$$(10.33)$$

The kW/ton values obtained from Eq. (10.33) is an estimate. It is better to obtain kW input and refrigeration output from the manufacturer's catalog or from actual measured data.

10.10 CHARACTERISTICS OF REFRIGERATION COMPRESSORS

Reciprocating Compressors

In a *reciprocating compressor,* shown in Fig. 10.25, a single-acting piston in a cylinder is driven by a crankshaft via a connecting rod. At the top of the cylinder are a suction valve and a discharge valve. There are usually two, three, four, or six cylinders in a reciprocating compressor.

Vapor refrigerant is drawn through the suction valve into the cylinder until the piston reaches its lowest position. As the piston is forced upward by the crankshaft, it compresses the vapor refrigerant to a pressure slightly higher than the discharge pressure. Hot gas opens the discharge valve and discharges from the cylinder. The gaseous refrigerant in a reciprocating compressor is compressed by the change of internal volume of the compression chamber caused by the reciprocating motion of the piston in the cylinder.

The refrigeration capacity of a reciprocating compressor ranges from a fraction of a ton to 200 tons. Refrigerant HCFC–22, HFC–134a in comfort and process air conditioning, and R–717 in industrial applications may be the primary refrigerants used in reciprocating compressors by the year 2004 because of their zero or lower ozone depletion factors.

The maximum compression ratio p_{dis}/p_{suc} of a single-stage reciprocating compressor is about 7. Volumetric efficiency of a typical reciprocating compressor η_v decreases from 0.92 to 0.65 when p_{dis}/p_{suc} increases from 1 to 6, and the isentropic or compressor efficiency η_{com} decreases from 0.83 to 0.75 when p_{dis}/p_{suc} increases from 4 to 6.

Methods of capacity control during part-load conditions include on-off, cylinder unloader, and hot bypass controls. In a *cylinder unloader,* the discharge gas is in short circuit and returns to the suction chamber. In a *hot gas bypass control,* hot gas bypasses the expansion valve and the evaporator and mixes with evaporated vapor refrigerant to reduce the refrigeration capacity of the system. *On-off* and *cylinder unloader* are step controls. Hot gas bypass control wastes compression energy. These methods will be covered in Chapter 21.

Reciprocating compressor design is now in its mature stage. There is little room for significant improvement. Although reciprocating compressors have many operating problems, such as liquid slugging, which will be discussed in Chapter 21, they are still a reli-

able and widely used in small and medium refrigeration systems.

Screw Compressors

A *screw compressor* is also called a *helical rotary compressor.* It was designed to incorporate with oil injection and has been used in industrial applications since the 1950s. Screw compressors can be classified as twin-screw compressors or single-screw compressors.

TWIN-SCREW COMPRESSOR. Today, this is the most widely used screw compressor. A typical *twin-screw compressor,* shown in Fig. 10.26, consists of a four-lobe male rotor and a six-lobe female rotor (4 + 6) or five-lobe male rotor and a seven-lobe female rotor (5 + 7), a housing with suction and discharge ports, and a sliding valve for capacity control. Usually, the male rotor is the driver.

When the lobes are separated at the suction port, as shown in Fig. 10.27b, vapor refrigerant is drawn in and the intake process continues until the interlobe space is out of contact with the suction port. The volume trapped in the interlobe space within the meshing point is compressed during successive rotations of the rotor. When the interlobe space makes contact with the discharge port, the compressed gas discharges through the outlet.

The ratio of the volume of vapor trapped during the intake process V_{in} (in ft^3) to the volume of trapped gaseous refrigerant discharged V_{dis} (in ft^3) is called the *built-in volume ratio* V_i. The application of variable V_i instead of fixed V_i eliminates the losses associated with over- or under-compression, especially during part-load operation.

Oil injection plays an important role in twin-screw compression. Injected oil effectively cools the compressor and results in a lower discharge temperature. A small clearance of 0.0005 in. between lobes and the sealing effect of the oil minimize leakage loss. Another important function of the injected oil is lubrication.

The refrigeration capacity of twin-screw compressors ranges from 50 to 1500 tons. The compression ratio of a screw compressor can be as high as 20 : 1. HCFC–22 and HFC–134a (and R–717 for industrial applications), are the most widely used refrigerants for twin-screw compressors.

Screw compressors have no clearance volume. In a typical twin-screw compressor using HCFC–22 as refrigerant, η_v drops from 0.92 to 0.87 and η_{isen} decreases from 0.82 to 0.67 when p_{dis}/p_{suc} increases from 2 to 10.

In a twin-screw compressor, capacity is controlled by moving a sliding valve toward the discharge end, which opens a recirculating passage connected to the suction port, as shown in Fig. 10.26a. This valve allows a portion of trapped gas to leave the interlobe space and return to the suction port through the recirculating passage. This capacity control is a stepless continuous control action.

The twin-screw compressor's low noise and vibration, together with its positive displacement compression, provides reliable operation. Twin-screw compressors are more efficient than reciprocating compressors. Screw compression refrigeration systems will be covered in Chapter 21.

SINGLE-SCREW COMPRESSOR. The *single-screw compressor* was developed in the 1960s and has been used in refrigeration systems since the 1970s. The single-screw compressor has a single helical rotor and two star wheels, as shown in Fig. 10.27.

As the rotor rotates, one of the flutes opens to the suction port and is filled with suction vapor until its suction end meshes with star wheel A. The discharge end of this flute is covered by the rotor casing. When the rotor turns, the meshing of the star wheel A with the flute compresses the trapped gas and raises its pressure. When the discharge end of this flute opens to the discharge port, compression stops and the hot gas is discharged. In a single-screw compressor, compression occurs simultaneously in both the top and bottom of the helical rotor.

The η_v of the single-screw compressor is slightly lower at higher p_{dis}/p_{suc} and the η_{isen} is about 3 to 4 percent lower than in the twin-screw compressor.

Rotary Compressors

Two basic types of rotary compressors are used in refrigeration systems: rotating vane and rolling piston.

In a *rotating vane rotary compressor,* shown in Figure 10.28a, the rotor is concentric with the shaft. Two or more vanes slide in slots engraved on the rotor. The assembly (the rotor and shaft) are eccentric with respect to the cylindrical housing. Vapor refrigerant is drawn through the suction port and enclosed by two sliding vanes. As the rotor rotates, the volume of the gas trapped by the vane and the lower gap is gradually reduced until its pressure reaches the discharge pressure. Hot gas is then squeezed out through the discharge outlet. The rotating vanes are kept in contact with the housing by centrifugal force. Large rotating vane rotary compressors used in industrial applications use more than two vanes.

In a *rolling piston rotary compressor,* a rolling piston mounted on an eccentric shaft (shown in Figure 10.28b) is used instead of rotating vanes. A fixed vane sliding in a slot remains in contact with the roller. As the piston rotates, the vane is in reciprocating motion. Vapor refrigerant enters the compression chamber through the suction inlet and is compressed by the eccentric motion of the roller. When the rolling piston is in contact with the top of the cylindrical housing, the hot gas is squeezed out through the discharge valve.

A rotary compressor has a smaller clearance volume and therefore greater volumetric efficiency than a reciprocating compressor. The mechanical efficiency η_{mec} of a typical rotary compressor operating at a compression ratio of 3.5 is about 0.87. Because of its rotary motion, a rotary compressor makes less noise than a reciprocating compressor.

Small rotary compressors for refrigerators and room air conditioners have a refrigerating capacity up to about 4 tons. HCFC–22 and HCFC–134a are used as refrigerants. Large rotary compressors used in low-temperature refrigeration ($-125°F$ to $-5°F$) may have power inputs from 10 to 600 hp. In addition to HCFC–22 and HFC–134a, R–717 is also used in industrial applications.

Scroll Compressors

The *scroll compressor* was developed in Europe and United States in the 1970s. A scroll compressor consists of two identical spiral scrolls assembled at a phase difference of 180°, as shown in Fig. 10.29. Each scroll is bound on one side to a flat plate. One scroll is stationary and the other moves in an orbit around the shaft center of the motor at an amplitude equal to the orbit radius. The two scrolls make contact at several points and form a series of pockets.

During suction, vapor refrigerant enters the space between the two scrolls through the lateral openings. The lateral openings are then sealed off to form trapped vapor pockets, thus completing the intake process.

During successive revolutions of the motor shaft, the volume of the vapor pockets is reduced and the compression process is complete when the gaseous refrigerant is compressed to its maximum pressure.

During discharge, the trapped gas pockets open to the discharge port. Compressed hot gas is discharged through the small opening. The volume of the gas pockets is reduced to zero and the trapped gaseous refrigerant is squeezed to the discharge line.

The sequence of intake, compression, and discharge occurs simultaneously in the two trapped gas pockets.

In scroll compressors, the components touch each other with sufficient force to create a seal but not sufficient to cause wear. Such a feature is the result of advanced manufacturing technology.

A scroll compressor has a η_v greater than 95 percent at a compression ratio of 4 and an η_{isen} of about 80 percent. The scroll compressor has only half as many parts as the reciprocating compressor, and the same refrigerating capacity. Fewer components means higher reliability and efficiency. Its power input is about 5 to 10 percent less than that of the reciprocating compressor. The scroll compressor also operates more smoothly and quietly.

Currently manufactured scroll compressors have a refrigeration capacity up to 60 tons. They are used in heat pumps and packaged units.

Centrifugal Compressors

A *centrifugal compressor* is a turbomachine. As in a centrifugal fan, the total pressure increase of a centrifugal compressor (usually called *head lift*, in psi) results from the conversion of velocity pressure into static pressure. Although the compression ratio p_{dis}/p_{suc} of a single-stage centrifugal compressor using HCFC–123 and HCFC–22 seldom exceeds 4, the impellers connected in series increase the p_{dis}/p_{suc} and satisfy most compression requirements. For comfort air conditioning systems, a single-stage, two-stage, or sometimes a three-stage centrifugal compressor is used.

Figure 10.30 shows a two-stage centrifugal compressor. Vapor refrigerant is drawn through the inlet vanes and enters the first-stage impeller. The impeller compresses the vapor and discharges it through a diffuser, where it combines with flashed vapor refrigerant from the flash cooler. The mixture then enters the second-stage impeller, where the refrigerant is further compressed. The hot gas is then discharged into the collecting volute through a diffuser. As the gaseous refrigerant flows through the compressor, the area of the flow passage increases. This causes a decrease in gas velocity and, therefore, a conversion of velocity pressure into static pressure.

Because a high head lift is required to raise evaporating pressure to condensing pressure, the hot gas discharge velocity at the exit of the second-stage impeller approaches the acoustic velocity v_{ac} of the saturated vapor. At atmospheric pressure and a temperature of 80°F, for HCFC–123, $\dot{v}_{ac}$ is 420 ft/s, and for HCFC–22, v_{ac} is about 600 ft/s. Centrifugal compressors need a high peripheral velocity and high rotating speeds (from 1800 to 50,000 rpm). The refrigeration capacity of a centrifugal compressor ranges from 100 to 10,000 tons. It is not economical to

manufacture small centrifugal compressors. Because they are turbomachines, centrifugal compressors have higher volume flow than positive displacement compressors. Refrigerants HCFC–123, HCFC–22, and others are used in centrifugal compressors for comfort air conditioning systems, and R–717 is used in industrial applications.

Isentropic efficiency η_{isen} reaches a maximum of about 0.83 for a typical centrifugal compressor operated at design conditions. Its value may decrease to as low as 0.6 during part-load operation. The volumetric efficiency η_v of centrifugal compressors can be estimated at almost 1.

Inlet vanes are widely used as stepless capacity control for centrifugal compressors during part-load operation. AC-inverters, direct steam turbines, and engine drives also provide capacity control by varying the speed of centrifugal compressors. Sometimes, a combination of adjustable-frequency AC drive and inlet vanes is used.

Because centrifugal compressors are reliable, they are the most widely used refrigerating compressors in large central hydronic air conditioning systems.

10.11 SELECTION OF COMPRESSORS

Since the compressor is the heart of a vapor compression refrigeration system, and consumes most of the energy, the selection of a compressor is actually part of the selection of a vapor compression refrigeration system.

The following factors should be considered during the selection of a refrigerating compressor:

- *Size of the project and the number of compressors installed.* If a project needs a refrigeration capacity of 2000 tons, it is not appropriate to install 10 reciprocating chillers. These are expensive and require a large space, even on the roof of a building. It is also not suitable to install only one chiller for a medium-size project. Installation of two or three compressors is recommended to provide backup when a compressor is being repaired.

- *Heat pump operation and air-cooled condensation.* Compressors for heat pump operation or for systems using air-cooled condensers need a greater evaporating and condensing pressure differential than for water-cooled or evaporative-cooled condensers. A greater pressure differential means a higher compression ratio.

- *Efficiency.* Efficiency is closely related to the energy consumption of the compressor and, therefore, the operating cost of the refrigeration system.

- *Capacity control.* A satisfactory capacity control system not only provides smooth, effective system operation and temperature control, but also saves energy.
- *Reliability.* A compressor is said to be reliable if it rarely experiences unexpected failure. Generally, a compressor with fewer parts, especially fewer moving parts, is more reliable than a compressor with many moving parts.
- *Initial cost.* The initial cost of the compressor is also an important factor to consider through payback analysis.

Table 10.3 compares various types of compressors. This table can be used as a reference during selection of refrigerating compressors.

10.12   CENTRIFUGAL PUMPS

Pump Construction and Types of Centrifugal Pumps

Centrifugal pumps are the most widely used pumps for transporting chilled water, hot water, and condenser water in HVAC&R systems for its high efficiency and reliable operation. Like centrifugal fans and compressors, centrifugal pumps accelerate liquid and convert the velocity of the liquid into static head.

A typical centrifugal pump consists of an impeller rotating inside a spiral casing, a shaft, mechanical seals and bearings on both ends of the shaft, suction inlets, and a discharge outlet, as shown in Fig. 10.32. The impeller can be single-stage or multistage. The vanes of the impeller are usually backward-curved.

TABLE 10.3
Comparison of various types of Compressors

	Reciprocating	Twin screw	Rotary	Scroll	Centrifugal
Characteristic of compression process	Positive displacement	Positive displacement	Positive displacement	Positive displacement	Nonpositive displacement
Refrigeration capacity, Ton	< 200	50 to 1500	Commercial < 4 Industrial < 150	Up to 60	100 to 10,000
Refrigerants	HCFC–22 HFC–134a (CFC–12)*	HCFC–22 HFC–134a R–717 (CFC–12)*	HCFC–22 HFC–134a R–717 (CFC–12)*	HCFC–22	HCFC–123 HCFC–22 R–717 (CFC–11)*
Single-stage maximum compression ratio	7	20			4
Volumetric efficiency η_v	0.92 to 0.68 (at R_{com} = 1 to 7)	0.92 to 0.87 (at R_{com} = 2 to 10)	High	High	High
Compressor efficiency η_{com}	0.83 to 0.75 (at R_{com} = 4 to 7)	0.82 to 0.67 (at R_{com} = 4 to 10)	η_{mec} = 0.87 (R_{com} = 3 to 5)	5 to 10% higher than reciprocating	Design 0.83 Part-load 0.6
Capacity control	On-off, cylinder unloader Step control	Sliding valve Stepless			Inlet vanes AC inverters Stepless
Reliability	Reliable	More reliable	Reliable	More reliable	More reliable
Application to refrigerating systems	Medium and small	Large and medium	Commercial: small Industrial: medium	Small	Medium and large central hydronic systems

Note: Because of their high ozone depletion factor, CFC–11 and CFC–12 are being phased out of use before the end of 1995.

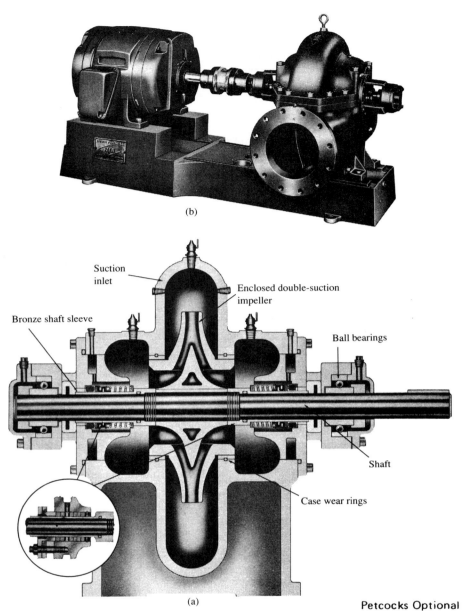

(b)

(a)   Petcocks Optional

FIGURE 10.32 A double-suction, horizontal split case, single-stage centrifugal pump: (a) sectional view; and (b) centrifugal pump and motor assembly. (*Source:* Pacific Pumping Company. Reprinted with permission.)

The pump is usually described as standard-fitted or bronze-fitted. In a standard-fitted construction, the impeller is made of grey iron, and in a bronze-fitted construction, the impeller is made of bronze. For both constructions, the shaft is made of stainless steel or alloy steel, and the casing is made of cast iron.

Three types of centrifugal pumps are often used in water systems: double-suction horizontal split-case, frame-mounted end suction, and vertical in-line pumps, as shown in the upper part of Fig. 10.33. Double-suction horizontal split-case cen-

trifugal pumps are the most widely used pumps in large central hydronic air conditioning systems.

Basic Terminology

Volume flow rate $\dot{V}_p$, in gal/min or gpm, is the capacity handled by a centrifugal pump.

Net static head H_s, in ft, is the head difference between discharge static head H_{dis} and the suction static head H_{suc}, both in ft, as shown in Fig. 10.34.

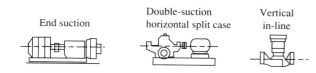

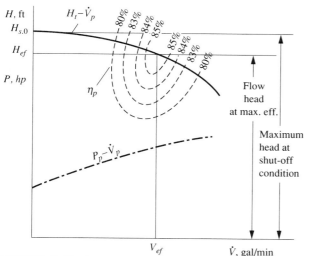

FIGURE 10.33 Performance curves for centrifugal pumps.

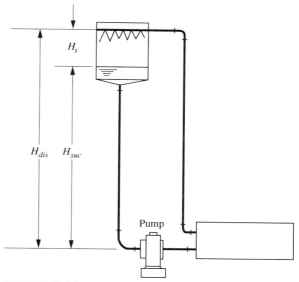

FIGURE 10.34 Net static head.

Velocity head H_v, in ft, can be calculated as

$$H_v = \frac{v_o^2}{2g} \qquad (10.34)$$

where v_o = velocity of water flow at pump outlet, ft/s

g = gravitational acceleration, 32.2 ft/s^2

Total head H_t, in ft, is the sum of net static head and velocity head, that is

$$H_t = H_s + H_v \qquad (10.35)$$

Pump power P_p, in hp, is the power input on the pump shaft, and can be calculated as

$$P_p = \frac{\dot{V}_p H_t g_s}{3960 \eta_p} \qquad (10.36)$$

where g_s = specific gravity, that is, the ratio of the mass of liquid handled by the pump to the mass of water at 39°F

η_p = pump efficiency

Performance Curves

Like centrifugal fan performance, pump performance is often illustrated by a head–capacity H_t–$\dot{V}_p$ curve and a power–capacity P_p–$\dot{V}_p$ curve, as shown in Fig. 10.33.

The head-capacity curve illustrates the performance of a centrifugal pump from maximum volume

flow of the shutoff point. If the total head at shut-off point $H_{s.o}$ is 1.1 to 1.2 times the total head at the point of maximum efficiency H_{ef}, the pump is said to have a flat head–capacity curve. If $H_{s.o} > 1.1$ to 1.2 H_{ef}, it is a steep-curve pump.

Net Positive Suction Head (NPSH)

The lowest absolute water pressure at the suction inlet of the centrifugal pump (shown in Fig. 10.32) must exceed the saturated vapor pressure at the corresponding water temperature. If the absolute pressure is lower than the saturated vapor pressure, the water evaporates and a vapor pocket forms between the vanes in the impeller. As the pressure increases along the water flow, the vapor pocket collapses and may damage the pump. This phenomenon is called *vapor cavitation*.

The sum of the velocity head at the suction inlet and the head loss (due to friction and turbulence) between the suction inlet and the point of lowest pressure inside the impeller is called the *net positive suction head required* (NPSHR), in ft. This factor determined by the pump manufacturer for a given centrifugal pump.

For a specific pump-piping system using a centrifugal pump, the net positive suction head available (NPSHA), in ft, can be calculated as

$$\text{NPSHA} = H_{at} + H_{suc} - H_f - (2.31 \, p_{vap}) \quad (10.37)$$

where H_{at} = atmospheric pressure, usually expressed as 34 ft of water column

H_{suc} = static suction head, ft

H_f = head loss due to friction and dynamic losses of the suction pipework and fittings, ft

p_{vap} = saturated water vapor corresponding to the water temperature at the suction inlet, psia

NPSHA must be greater than NPSHR to prevent cavitation.

Pump Selection

First, the selected pump must satisfy the volume flow and total head requirements, and operate near maximum efficiency most of the time.

Second, for comfort air conditioning systems, quiet operation is an important consideration. A noise generated in a water system is very difficult to isolate and remove. In most cases, the lowest speed pump that can provide required $\dot{V}_p$ and H_t is the quietest and most economical choice.

Third, when a constant speed pump is used to serve a variable flow system that operates with minor changes of head, a flat-curve pump should be selected.

REFERENCES AND FURTHER READING

ASHRAE, *ASHRAE Handbook 1988, Equipment*, ASHRAE Inc., Atlanta, GA, 1988.

ASHRAE, *ASHRAE Handbook 1992, HVAC Systems and Equipment*, ASHRAE Inc., Atlanta, GA, 1992.

Beseler, F., "New Technology and the Helical Rotary Compressor," *Heating/Piping/Air Conditioning*, pp. 127–129, January 1988.

Beseler, F., "Scroll Compressor Technology Comes of Age," *Heating/Piping/Air Conditioning*, pp. 67–70, July 1987.

Branda, M. R., "A Primer on Adjustable Frequency Inverters," *Heating/Piping/Air Conditioning*, pp. 83–87, August 1984.

Coad, W. J., "Centrifugal Pumps: Construction and Application," *Heating/Piping/Air Conditioning*, pp. 124–129, September 1981.

Elson, J., Hundy, G., and Monnier, K., "Scroll Compressor Design and Application Characteristics for Air Conditioning, Heat Pump, and Refrigeration Applications" *Proceedings of the Institute of Refrigeration*, pp. 2.1–2.10, November 1990.

Goldfield, J., "Use Total Pressure When Selecting Fans," *Heating/Piping/Air Conditioning*, pp. 73–82, February 1988.

Grimm, N. R., and Rosaler, R. C., *Handbook of HVAC Design*, McGraw-Hill, 1st ed., New York, 1990.

Gurock, D. R., and Aldworth, D. R., "Fan–Motor Combination Saves Energy," *Heating/Piping/Air Conditioning*, pp. 53–57, November 1980.

Hunt, E., Benson, D. E., and Hopkins, L. G., "Fan Efficiency vs. Unit Efficiency for Cleanroom Application," *ASHRAE Transactions*, Part II, pp. 616–619, 1990.

Lundberg, A., "Design Basics—Screw Compressors," *ASHRAE Transactions*, Part II, pp. 826–836, 1981.

Nader, J. C. and Kanis, T. W., "Industrial vs Commercial Fans," *Heating/Piping/Air Conditioning*, pp. 55–62, February, 1983.

Osborne, W. C., *Fans*, Pergamon Press, 2nd ed., Oxford, England, 1977.

Pillis, J. W., "Design and Application of Variable Volume Ratio Screw Compressor," *ASHRAE Transactions*, Part IA, pp. 289–296, 1985.

Purvis, E., "Scroll Compressor Technology," Heat Pump Conference, New Orleans, 1987.

Rosenbaum, D. P., "Obtaining Proper Fan Performance from Fans Installed in Air Handlers," *ASHRAE Transactions*, Part IB, pp. 790–794, 1983.

Stamm, R. H., "Industrial Refrigeration Compressors," *Heating/Piping/Air Conditioning*, pp. 93–111, July 1984.

The Trane Company, "Adjustable Frequency Fan Speed Control," *Application Engineering Manual*, Trane Company, La Crosse, WI, 1983.

The Trane Company, "Fans and Their Application in Air Conditioning," *Applications Engineering Seminar*, Trane Company, La Crosse, WI, 1971.

The Trane Company, "Helical Rotary Water Chillers," *Air Conditioning Clinic*, American Standard, La Crosse, WI, 1988.

The Trane Company, "The Scroll Compressor," *Air Conditioning Clinic*, American Standard, La Crosse, WI, 1987.

The Trane Company, *Variax Fans*, Trane Company, La Crosse, WI, 1983.

Wilkins, C., "NPSH and Pump Selection: Two Practical Examples," *Heating/Piping/Air Conditioning*, pp. 55–58, October 1988.

Zubair, S. M., and Bahel, V., "Compressor Capacity Modulation Scheme," *Heating/Piping/Air Conditioning*, pp. 135–143, January 1989.

CHAPTER 11

HUMIDIFIERS, AIR WASHERS, AND COOLING TOWERS

11.1 HUMIDIFICATION AND HUMIDIFIERS

Space Relative Humidity

As mentioned in Section 5.9, for comfort air conditioning systems the space relative humidity is tolerrable between 30 and 65 percent in summer, preferably from 40 to 50 percent. During winter, for space served by a comfort air conditioning system that is installed with a humidifier, the space relative humidity should not exceed 30 percent except in hospitals. For air conditioning systems without humidifiers, space relative humidity cannot be specified. For processing air conditioning systems, the space relative humidity should be specified as required by the manufacturing process.

If space temperature is maintained at 72°F and only free convection is provided in the occupied space, the space relative humidity should not exceed 27 percent when the outdoor temperature is 30°F in order to prevent condensation on the inner surface of single-glazed windows during winter. To prevent condensation on the inner surface of double-glazed windows, the space relative humidity should not exceed 33 percent when the outdoor temperature is 0°F.

Humidifiers

A *humidifier* adds moisture to the air. Humidifiers may (1) heat or atomize the liquid water, or force air to flow over a wetted element so that water evaporates and is added to the air as vapor; or (2) inject steam directly into air before it supplies to the conditioned space. All of these increase the humidity ratio of the space air and, therefore, its relative humidity. One important index of a humidifier is its *humidifying capacity* $\dot{m}_{cap}$, in lb of water per hour, or the rate at which water vapor is added to the air.

Humidifying Load

Humidifying load $\dot{m}_{hu}$, in lb of water/h, is the amount of water vapor required to be added to the air by a humidifier so as to maintain a predetermined space relative humidity. The humidifying load of an air system installed with a humidifier can be calculated as shown in Eq. (11.1):

$$\dot{m}_{hu} = 60\dot{V}_s \rho_s(w_{lv} - w_{en}) = 60\dot{V}_s \rho_s(w_s - w_m)$$

(11.1)

where $\dot{V}_s$ = supply volume flow rate of the air-handling unit or packaged unit, cfm
ρ_s = density of supply air, lb/ft³
w_{en}, w_{lv} = humidity ratio of air entering and leaving the humidifier, lb/lb
w_s = humidity ratio of supply air, lb/lb
w_m = humidity ratio of mixture of outdoor air and recirculating air, lb/lb

In Fig. 11.1, s is the state of supply air when the space has a heating load, and s' is the condition when the space has a cooling load $w_s = w'_s$.

The humidifying load can also be calculated from the following relationship:

$$\dot{m}_{hu} = 60(\dot{V}_o + \dot{V}_{if})\rho_o(w_r - w_o) - \dot{m}_{wr}$$ (11.2)

where $\dot{V}_o$ = supply volume flow rate of outdoor air intake, cfm
$\dot{V}_{if}$ = volume flow rate of infiltrated air, cfm
w_r, w_o = humidity ratio of space air and outdoor air, lb/lb

In Eq. (11.2), $\dot{m}_{wr}$ represents the space moisture gains, in lb/h. Space moisture gains include the latent load from the occupants, appliances, equipment, and products. The moisture gains from the building structures are often ignored.

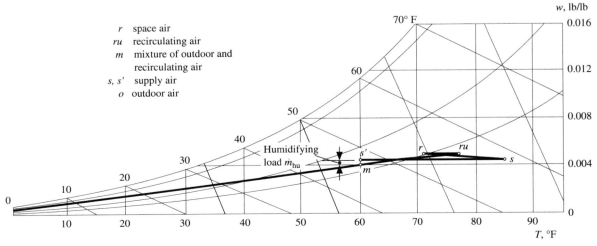

FIGURE 11.1 Humidifying load for a typical air system.

Types of Humidifiers

According to the mechanism used for evaporation of water vapor from water, humidifiers can be classified as steam and heating element humidifiers, atomizing humidifiers, and wetted element humidifiers.

11.2 STEAM AND HEATING ELEMENT HUMIDIFIERS

At a low partial pressure when mixed with dry air, steam is the ready-made water vapor. With proper water treatment at the boiler, steam is free of mineral dust and odor, and it does not support the growth of bacteria that creates sanitary problems.

Two types of steam humidifiers are currently used in air systems: steam grid humidifiers and steam humidifiers with separators.

STEAM GRID HUMIDIFIERS. A steam grid humidifier installed inside a ductwork is shown in Fig. 11.2a. A steam grid humidifier may have a single dis-

tribution manifold or multiple manifolds. Each distribution manifold has an inner steam pipe and an outer jacket. The inner steam tube is connected to a header and controlled by a control valve. Steam is supplied at a pressure less than 10 psig and throttled to a lower pressure after the control valve. It then enters the inner tube and discharges through the small holes to the outer jacket. The dry steam is again discharged through the orifices of the jacket to the ambient airstreams to humidify them, as shown in Fig. 11.2b. The condensate inside the jacket is discharged to a drain pipe and steam traps located at the opposite side. The inner tube and the outer jacket are slightly pitched toward the drain pipe and steam trap.

STEAM HUMIDIFIERS WITH SEPARATORS. A steam humidifer with a separator is illustrated in Fig 11.3. Steam is supplied to a jacketed distribution manifold and then enters a separating chamber with its condensate. It then flows through a control valve, is throttled to slightly above atmospheric pressure, and enters a drying chamber. Because of the lower pressure and temperature in the drying chamber compared with the higher pressure and temperature in the surrounding separating chamber, the steam is superheated. Dry steam is then discharged into the ambient airstream through the orifices of the inner steam discharging tubes. Noise is produced mainly during the throttling of the steam at the control valve. It is attenuated as steam flows through the silencing materials.

Heating Element Humidifiers

A heating element humidifier has a water pan with electric heating elements, steam coils, or hot water coils installed at the bottom of the pan to evaporate the liquid water into water vapor and then add the vapor to the airstream flowing over the water pan, as shown in Fig. 11.4. The humidifying capacity of a heated element humidifier is limited by its heat transfer surface and characteristics of the heating element. This type of humidifier is usually used for small HVAC&R equipment.

Characteristics of Steam and Heating Element Humidifiers

Steam humidifiers are widely used in commercial and industrial applications. Both steam and heating element humidifiers need heat energy from gas, fossil fuel, or electricity to evaporate the liquid water. For a conditioned space that has a cooling load year-round and needs a cold air supply in winter, it is a waste

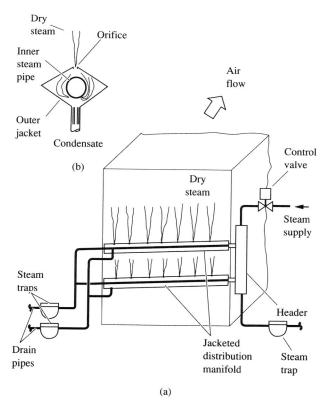

FIGURE 11.2 Steam grid humidifiers: (a) steam grid and piping connections; and (b) cross section of jacketed distribution manifold.

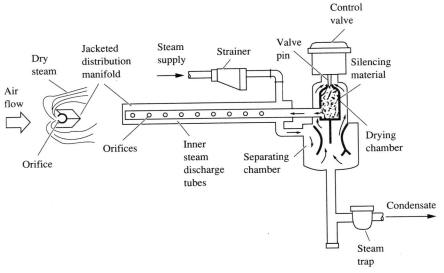

FIGURE 11.3 Steam humidifier with a separator. (Adapted with permission from *The Armstrong Humidification Handbook*, 1991.)

of energy to evaporate the liquid water by using fossil fuel instead of the excess heat gains at the conditioned space.

The capacity of a steam humidifier is often controlled by modulating the valve pin and, therefore, the steam flow rate. The control valve should be integrated with the humidifier and steam jacketed at supply pressure to prevent condensation. An interlocking control should be installed to drain all condensate before the steam humidifier is started.

For electric heating elements, the input wattage can be adjusted when the space relative humidity is too high or too low. For steam and hot water coils, the steam or hot water flow rate is modulated to maintain a preset space relative humidity.

11.3 ATOMIZING AND WETTED ELEMENT HUMIDIFIERS

Humidification Process

Atomizing means producing a fine spray. When liquid water is atomized, the smaller the diameter of the water droplets, the greater the interfacial area between water and air, and thus the higher the rate of evaporation and humidification.

When air flows through a bank of nozzles spraying recirculating water, as described in Section 3.3, the result is an increase in the humidity ratio of air from the addition of evaporated water vapor and a corresponding drop in air temperature. Such a humidification process is an adiabatic saturation process and follows the thermodynamic wet bulb line on the psychrometric chart, as described in Chapter 3.

The performance of atomizing and wetted element humidifiers can also be described by an index called the *saturation effectiveness*, as defined by Eq. (3.17), or

$$\eta_{sat} = \frac{T_2 - T_1}{T_s^* - T_1}$$

Atomizing Humidifiers

Atomizing humidifiers can be classified, according to their configurations and mechanisms of atomizing, as centrifugal atomizing humidifiers, pneumatic atomizing humidifiers, and ultrasonic atomizing humidifiers.

CENTRIFUGAL ATOMIZING HUMIDIFIERS. These humidifiers use the centrifugal force produced by a

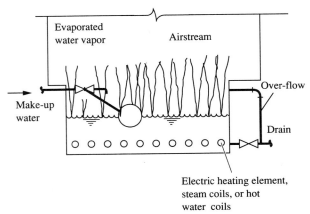

FIGURE 11.4 Heating element humidifier.

rotating device, such as a rotating cone, blades of an axial fan, rotating disc, or rotating drum to break the liquid water film into a fine mist or to fling it into fine water droplets.

When air is forced through an atomizing water spray, produced by a pulverizing fan, water vapor is evaporated and added to air as a result of the mass concentration difference between the saturated air film at the surface of the droplets and the ambient air. The rotating device receives liquid water either from a pressurized water supply or by dipping into the surface of a nonpressurized supply.

A pulverizing fan was shown in Fig. 3.3. The humidifying capacity of a pulverizing fan depends mainly on the volume flow rate of the fan. Because of the oversaturation characteristics of the pulverizing fan, $\dot{m}_{cap}$ may vary from 50 to 150 lb/h.

Many different types of rotating humidifiers with limited humidifying capacities are used directly in the conditioning space in residential and industrial applications. Most of them are portable. One type rotates a suction tube and an impeller with an electric motor. Centrifugal force raises the water in the suction tube to spinning discs, which throw it against an atomizing ring. Water then divides into extremely fine particles of diameters between 0.002 and 0.004 in.

Maintaining a predetermined space relative humidity with a centrifugal atomizing humidifier can be performed by modulating the flow rate of water supplied to the rotating devices.

PNEUMATIC ATOMIZING HUMIDIFIERS. A typical *pneumatic atomizing nozzle* consists of two concentric brass tubes, as shown in Fig. 11.5a. The inner tube has an outer diameter of $\frac{3}{32}$ in. and an inner diameter of $\frac{1}{32}$ in. It is connected to a water tank that can be moved to adjust the difference in water levels between the water tank and the center-line of the nozzle. The conical outer tube has a minimum inner diameter of $\frac{1}{8}$ in. and is connected to a compressed air line with a pressure at 15 psig.

When compressed air is discharged from the annular slot at a very high velocity, it extracts water from the inner tube and breaks the water into very fine mist. A typical pneumatic atomizing nozzle may have a humidifying capacity $\dot{m}_{cap}$ from 6 to 10 lb/h. The magnitude of $\dot{m}_{cap}$ is affected by the configuration of the nozzle, the pressure of the compressed air, and the difference in water levels. In a pneumatic humidifier, $\dot{m}_{cap}$ is usually controlled by adjusting the difference in water levels between the water tank and the center-line of the nozzle.

A pneumatic humidifier produces high-frequency noise. This type of humidifier is usually applied in industries with a certain sound level of machine noise and can be used for direct in-space humidification.

ULTRASONIC ATOMIZING HUMIDIFIER. When a metal plate is electrically powered and vibrated at high frequency, water films in contact with the plate are broken into a fine mist. Air containing this mist can then be blown into the space directly or through a duct system by a fan. Ultrasonic atomizing humidifiers create less equipment noise than other atomizing humidifiers. Ultrasonic atomizing humidifiers with small humidifying capacities have been used directly in residences.

Wetted Element Humidifiers

Wetted element humidifiers include a wetted element, such as an evaporative pad, plastic, or impregnated cellulose, that is dipped with water from the top. Such humidifiers have been installed in air-handling units and packaged units to humidify the air. Characteristics of wetted element humidifiers

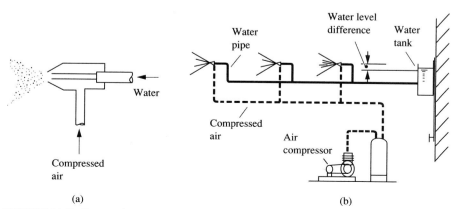

FIGURE 11.5 Pneumatic atomizing humidifier: (a) nozzle; and (b) pneumatic atomizing humidifying system.

are similar to wetted element evaporative coolers (see Chapter 13). Modulation of the water supply to the water dipping device varies the humidifying capacity and maintains desirable space relative humidity.

11.4 AIR WASHERS AND WATER-SPRAYING COILS

The air washer was the first air conditioning equipment developed by Carrier in 1904, and it is still used today to humidify, cool, and clean the air in many factories.

Construction of an Air Washer

An *air washer* has an outer casing, shown in Fig. 11.6a. It is usually made of plastic or a galvanized steel sheet with water-resistant paint for protection.

All joints are well-sealed by water-resistant resin. A water tank either forms a part of the casing at the bottom or is installed separately on the floor to collect or sometimes to mix the recirculating and incoming chilled water. A separately installed water tank is usually made of steel, stainless steel, or reinforced concrete with an insulation layer mounted on the outer surface if chilled water is used for spraying. A bank of guided baffles installed at the entrance provides an evenly distributed air-water contact. Eliminators in the shape of a sinusoidal curve at the exit are installed to remove entrained water droplets from the air. They are preferably made of plastic or stainless steel for convenient cleaning and maintenance. Access doors mounted on the outer casing must be provided for inspection and maintenance. They should be well-sealed to prevent water leakage.

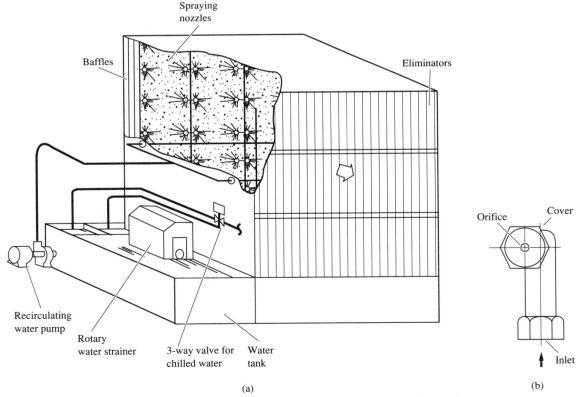

(a)   (b)

FIGURE 11.6 Schematic diagram of a typical air washer: (a) air washer; and (b) spraying nozzle.

In most air washers two banks of spraying nozzles face opposite to each other. Nozzles are often made of brass, plastic, or nylon. The nozzle consists of a cover with an orifice diameter between $\frac{1}{16}$ and $\frac{3}{16}$ in., a discharge chamber, and an inlet with its center-line slightly eccentric from the center-line of the orifice, as shown in Fig. 11.6b. This offset causes the water stream to rotate inside the discharge chamber and break the water into fine droplets. A larger orifice diameter does not clog easily, and a smaller orifice can produce finer water sprays. The distance between two spraying banks is from 3 to 4.5 ft, and the total length of the air washer varies from 4 to 7 ft.

A centrifugal water pump is used to recirculate the spraying water. For air containing much dirt and lint, as in air conditioning systems in the textile industry, an automatic-washing and -collecting water strainer, using copper or brass fine-mesh screen, often precedes the pump. The openings of the fine-mesh screen should be smaller than the diameter of the holes of the spraying nozzles to avoid clogging.

Functions of an Air Washer

Currently, air washers are used to perform one or more of the following functions:

- Cooling and humidification
- Cooling and dehumidification
- Washing and cleaning

As described in Sections 3.3 and 3.4, whether it is a cooling and humidification or cooling and dehumidification process is determined by the temperature of the spraying water. If recirculating water is used as the spraying water, the temperature of water then approaches the wet bulb temperature of the air entering the air washer, and the air is humidified and evaporatively cooled.

If chilled water or a mixture of chilled water and recirculating water is used for spraying water, and its temperature is lower than the dew point of the entering air, the air is then cooled and dehumidified.

Water spraying for the purpose of washing and cleaning air is sometimes practiced when an objectionable gas that is known to be soluble in water is to be removed from the air. For example, exhaust gas from many industrial applications may be removed by water spraying.

For humidification purposes, nozzles must have a smaller orifice and a higher water pressure, such as a $\frac{1}{16}$-in. diameter and a pressure of 40 psig. For cooling and dehumidification, a larger orifice and lower water pressure, such as a $\frac{3}{16}$-in. orifice and 25 psig, are often used.

Performance of an Air Washer

For a humidification process along the thermodynamic wet bulb line, the performance of an air washer can be illustrated by saturation effectiveness η_{sat}, as defined by Eq. (3.17). For a cooling and dehumidification process, the performance of an air washer is better described by a factor called *performance factor F_p* as follows:

$$F_p = \frac{h_{en} - h_{lv}}{h_{en} - h_s^*} \qquad (11.3)$$

where h_{en}, h_{lv} = enthalpy of air entering and leaving the air washer, Btu/lb

$h s^*$ = enthalpy of saturated air at thermodynamic wet bulb of entering air, Btu/lb

The performance of an air washer for a humidification process, indicated by the saturation effectiveness η_{sat}, and for a cooling and dehumidification process, expressed by F_p, depends mainly on the contact time between air and water and the contact surface area, that is, the water-air ratio $\dot{m}_w/\dot{m}_a$, the length of the air washer, and the direction of spray with respect to air flow, whether it is opposing or following the air flow.

An air washer is usually designed at an air velocity between 500 and 800 fpm with respect to its cross-sectional area at the water sprayers. Its greater washer length and the attached water tank result in a bulkier volume than DX-coils and water cooling coils. Total pressure loss of the airstream flowing through an air washer $\Delta p_{t.w}$, in inches WG, depends mainly on the configuration of the eliminators and the air velocity flowing through them. Usually, $\Delta p_{t.w}$ varies from 0.25 to 1 in. WG, and typically, is 0.5 in. WG.

For humidification and evaporative cooling, the water-air ratio $\dot{m}_w/\dot{m}_a$ is usually 0.3 to 0.6. For a 5-ft long air washer with a $\dot{m}_w/\dot{m}_a = 0.45$, that is, 4 gal of water per minute per 1000 ft^3 of air per minute, $\eta_{sat} = 0.85$ to 0.9. For cooling and dehumidification, the water-air ratio is usually between 0.5 and 1.2. For a 5-ft long air washer with a $\dot{m}_w/\dot{m}_a = 0.9$, that is, 8 gal of water per minute per 1000 ft^3 of air per minute, $F_p = 0.75$ to 0.85.

For either the humidification process or the cooling and dehumidification process in an air washer, oversaturation always exists at the exit. As described in Section 3.3, oversaturation $(w_o - w_s^*)$ usually varies between 0.0002 and 0.001 lb/lb. It depends mainly on the construction of the eliminator and also the flow-through air velocity.

Oversaturation is beneficial to a humidification process and interferes with a cooling and dehumidification process. Because of the existence of over-saturation, saturation effectiveness is no longer the dominant index that affects the performance of the air washer during humidification.

Bypass Control

Bypass control is defined as modulating the face damper and the bypass damper of the air washer to vary the proportion of humidifying air and bypass air in order to maintain space relative humidity within predetermined limits. Such a control is often used in an air washer to provide better humidity control than modulating the water pressure or the water flow rate to the spraying nozzles.

Single Stage or Multistage

A group of nozzles that are supplied with water at the same pressure and produce spray at the same temperature are said to be in the same stage. A stage may have one or two spraying banks. A two-stage air washer is illustrated in Fig. 11.7.

For humidification processes or for cooling and dehumidification processes using chilled water from the refrigeration plant, it is more economical to operate a single-stage air washer, usually with two water-

spraying banks. When low-temperature deep well or ground water is used for cooling and dehumidification, it is often advantageous to use two-stage, or even three-stage water spraying to make full use of the cooling capacity of the ground water. The use of ground water is regulated by local and federal environmental protection agencies.

Ground water can be used again as condensing cooling water, or as flushing water in plumbing systems after its discharge from the air washer following necessary water treatment. Used ground water should be drained to another well in the vicinity to prevent subsidence and damage to the soil, which may affect the foundation of the building in the vicinity of the deep well.

Water Spraying Coil

A *water spraying coil* consists of a bank of spraying nozzles, a water-cooling coil, a centrifugal pump for recirculating water, an eliminator, and a water tank including make-up water supply, water strainer, drain, and overflow set-ups, as shown in Fig. 11.8a. During cooling mode operation in summer, recirculating water is sprayed over the water cooling coil. Its temperature approaches the outer surface of the coil so that air is cooled and dehumidified by both the finned coil and the spraying water. Conditioned air leaves the spraying coil at a predetermined dew point temperature and a relative humidity between 95 to 98 percent, as shown in Fig. 11.8b.

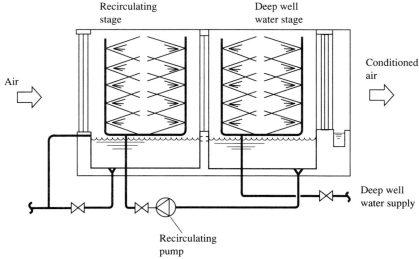

FIGURE 11.7 Two-stage air washer using deep well water.

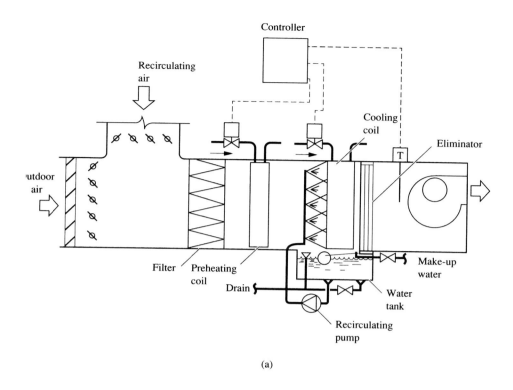

(a)

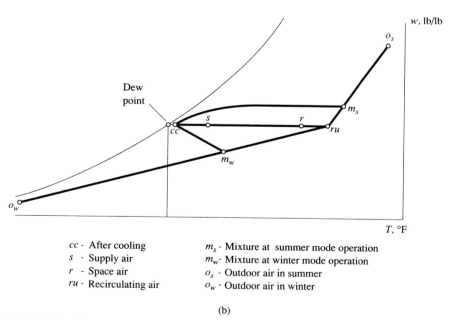

cc - After cooling
s - Supply air
r - Space air
ru - Recirculating air

m_s - Mixture at summer mode operation
m_w - Mixture at winter mode operation
o_s - Outdoor air in summer
o_w - Outdoor air in winter

(b)

FIGURE 11.8 Water-spraying coil: (a) schematic diagram; and (b) dew point control.

During heating mode operation in winter, a mixture of outdoor and recirculating air is heated and then sprayed with recirculating water along the adiabatic saturation line. Air leaves the spraying coil at a preselected dew point temperature and a relative humidity of 80 to 90 percent. A water-spraying coil is often used to provide dew point control, as shown in Fig. 11.8b, for many conditioned spaces need both temperature and humidity control. Dew point control will be discussed in Chapter 18.

Water spraying coils have the capability to humidify, to cool, to cool and dehumidify, and to clean the air, similar to air washers. They can be used to condition either an outdoor and recirculating air mixture or a 100 percent outdoor air make-up system.

A water-spraying coil is shorter than an air washer, making it less bulky. Another advantage is that when a heating coil is installed upstream from the water-spraying coil, the combination can maintain a nearly constant dew point year-round for spaces that need strict temperature and humidity control, such as pharmaceutical and rotary-printing factories, as shown in Fig. 11.8b.

The main disadvantage of a water-spraying coil system is that the dirt and foreign matter contained in air, which have been washed and removed from the air in a water-spraying coil system, still remain in the spray and coil combination. Periodic cleaning and maintenance are essential.

11.5 CHARACTERISTICS OF ATOMIZING AND WETTED ELEMENT HUMIDIFIERS

Atomizing and wetted element humidifiers, including air washers and the water spraying coils for humidification purposes, have the following operating characteristics.

Conditioned air follows the thermodynamic wet bulb line on a psychrometric chart. The humidity ratio of the air increases while it is evaporatively cooled. This characteristic indicates that atomizing and wetted element humidifiers are especially suitable for conditioned spaces that need a cold air supply to offset a space cooling load during or after the humidification process. The heat energy required to evaporate the liquid water, as in steam or heating element humidifiers, has been saved.

For a conditioned space that needs a warm air supply during winter operation, the temperature of evaporatively cooled air after an atomizing or wetted element humidification process should be raised before supplying it to the conditioned space so as to offset the heat transmission loss through the building envelope. The heat energy required is approximately equal to that required to evaporate water in a steam boiler or

in a water pan when a steam humidifier or a heating element humidifier is used.

Oversaturation exists at the exit of the atomizing humidifier even if an eliminator is used. Oversaturation is often advantageous for humidifiers installed in conditioned spaces having excess heat gains because it increases the humidifying capacity.

The size of the water droplets, the local humidifying capacity, and the excessive heat gains determine whether liquid water droplets may fall on the floor, products, or equipment.

According to field measurements, the diameters of water droplets suspended in the airstream discharged from atomizing humidifiers are as follows:

Pneumatic atomizing humidifiers   30 to 80 μm

Centrifugal atomizing humidifiers   50 to 300 μm

When the diameter of the water droplets exceeds 80 μm and the local humidifying capacity is excessive, formation of water droplets may occur. Water dripping must be avoided, even if reducing the humidifying capacity is necessary. It causes product damage and corrosion of appliances, instruments, and equipment.

In the water tank and water pan of atomizing and wetted element humidifiers, algae, bacteria, and other microorganisms may grow. Water spray and the airstream discharged from the atomizing and wetted element humidifiers may distribute the bacteria and sometimes odors. Water treatments, regular cleaning and flushing, and periodic blow-down help to minimize these problems. The designer must include convenient access for such maintenance.

11.6 SELECTION OF HUMIDIFIERS

The following factors should be considered when selecting a humidifier:

- *Humidifying capacity.* Is a large, medium, or small humidifying capacity required? Is oversaturation beneficial in the conditioned space?
- *Quality of humidification.* Humidified air should be clean and free of odor, noise, bacteria, particulate matter, and water droplets.
- *Energy consumption and operating cost.* For a space that needs a cold air supply even in winter, atomizing and wetted element humidifiers save energy and, therefore, have lower operating costs. Among the atomizing humidifiers, pneumatic atomizing humidifiers require compressed air to atomize the liquid water and, therefore, have a higher operating cost than other atomizing humidifiers.

- *Equipment noise.* For in-space humidifiers such as pneumatic humidifiers, equipment noise is a primary concern.
- *Initial cost.* For a steam humidifier, the initial cost should also include the installation costs of the steam boiler and the corresponding steam piping system.
- *Maintenance.* This factor includes the amount and cost of required maintenance work for the humidifiers.
- *Capacity control.* Is the control on-off or is it adjustable, that is, is it a stepless control? Does the control have a fast or slow response?
- *Space occupied.* This is the volume occupied by the humidifier per unit humidifying capacity.

Various types of humidifiers are compared in Table 11.1. For commercial buildings that need a warm air supply during winter and for which a steam supply is available, a steam humidifier is often the most suitable choice. For a small packaged unit serving a healthcare facility, a heating element humidifier may be selected. For a factory in which the space air contains lint and other dirts and for which a cold air supply is needed to offset the cooling load in winter, an air washer is often the most suitable choice. For laboratories with a

cooling load even in winter, a water-spraying coil may be appropriate. Ultrasonic humidifiers have comparatively lower equipment noise. They are often used for in-space humidifiers in residences.

11.7 COOLING TOWERS

A *cooling tower* is a device in which recirculating cooling water from a condenser or cooling coils is evaporatively cooled by contact with atmospheric air.

Most cooling towers used in refrigeration plants for commercial buildings or industrial applications are mechanical draft cooling towers. A mechanical draft cooling tower uses fan(s) to extract atmospheric air. It consists of a fan to extract intake air, a heat transfer medium (or fill), a water basin, a water distribution system, and an outer casing, as shown in Fig. 11.9.

According to the location of the fan corresponding to the fill and to the flow arrangements of air and water, currently widely used mechanical draft cooling towers for HVAC&R can be classified into the following categories:

- Counterflow induced draft
- Crossflow induced draft
- Counterflow forced draft

TABLE 11.1
Comparison of various types of humidifiers

Description	Steam	Heating element	Air washer	Water-spraying coil	Other atomizing humidifiers
Humidifying capacity	Small to large	Small	Large	Medium to large	Small to large
Oversaturation	No	No	Yes	Yes	Yes
Quality of humidification	Good	Good	Good	Good	Acceptable
Dripping of water in conditioned space	No	No	No	No	Possible except with pneumatic atomizing humidifier
Energy consumption and operation cost	Need more energy if cold air supply is needed, so operating cost is higher	Same as steam	Saving energy if cold air supply is needed in winter	Same as air washer	Same as air washer
Initial cost	High	High	Medium	Medium	Low to medium
Maintenance work and maintenance cost	Low	Medium	High	High	High
Capacity control	Modulation	On-off or modulation	Modulation	On-off or modulation	On-off or modulation
Response to control	Fast	Slow	Medium	Medium	Medium
Space occupied	Small	Medium	Large	Medium	Small to medium
Equipment noise	Medium	Low	Medium	Medium	Pneumatic humidifiers, medium Ultrasonic, low

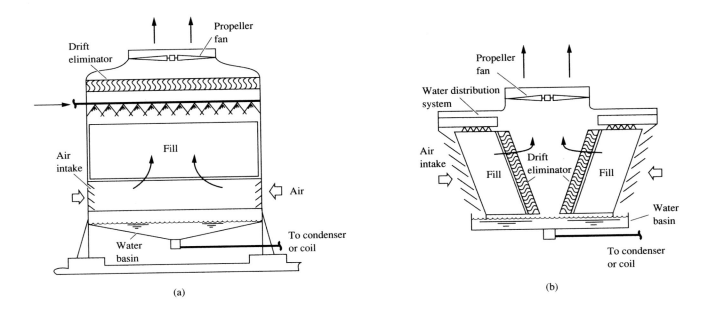

(a)

(b)

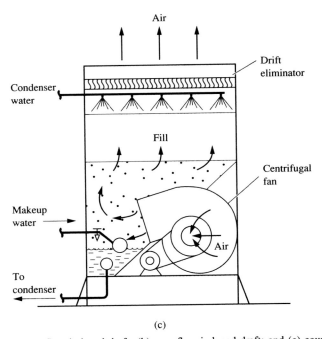

(c)

FIGURE 11.9 Cooling towers: (a) counterflow induced draft, (b) crossflow induced draft; and (c) counterflow forced draft.

Counterflow Induced Draft Cooling Towers

In a counterflow induced draft cooling tower, as shown in Fig. 11.9a, the fan is located downstream from the fill at the air exit. Atmospheric air is drawn by the fan through the intake louver or, more simply, an opening covered by wire mesh. Cooling water from the condenser or recirculating water from the coil, or a mixture of the two is evenly sprayed or distributed over the fill and falls down into the water basin.

Air is extracted across the fill and comes in direct contact with the water film. Because of the evaporation of a small portion of the cooling water, usually about 1 percent of the water flow, the temperature of the water gradually decreases as it falls down through the fill countercurrent to the extracted air. Evaporated water vapor is absorbed by the airstream. Large water droplets entrained in the airstream are collected by the drift eliminators. Finally, the airstream and drift are discharged at the top exit. Drift, or carryover, is the minute water droplets entrained in the airstream discharged out of the tower. The evaporatively cooled water falls into the water basin and flows to the condenser.

In a counterflow induced draft cooling tower, the driest air contacts the coldest water. Such a counterflow arrangement shows a better tower performance than a crossflow arrangement. In addition, air is drawn through the fill more evenly by the induced draft fan and is discharged at a higher velocity from the top fan outlet. Both higher exhaust air velocity and even velocity distribution reduce the possibility of exhaust air recirculation. Compared with the crossflow induced draft cooling tower, the vertical height from the installation level to the inlet of the water spraying nozzles in a counterflow tower is greater and, therefore, requires a higher pump head.

Crossflow Induced Draft Cooling Towers

In a crossflow induced draft cooling tower, as shown in Fig. 11.9b, the fan is also located downstream from the fill at the top exit. The fill is installed at the same level as the air intake. Air enters the tower from the side louvers and moves horizontally through the fill and the drift eliminator. Air is then turned upward and finally discharged at the top exit. Water sprays from the nozzles, falls across the fill, and forms a crossflow arrangement with the airstream.

The crossflow induced draft cooling tower has a greater air intake area. Because of the crossflow arrangement, the tower can be considerably lower than the counterflow tower. However, the risk of recirculation of tower exhaust air increases.

Counterflow Forced Draft Cooling Towers

In a counterflow forced draft cooling tower, as shown in Fig. 11.9c, the fan is positioned at the bottom air intake, that is, on the upstream side of the fill. Cooling water sprays over the fill from the top and falls down to the water basin. Air is forced across the fill and comes in direct contact with the water. Because of the evaporation of the water, its temperature gradually decreases as it flows down along the fill in a counter-flow arrangement with air. In the airstream, large water droplets are intercepted near the air exit by the eliminator. Finally, the airstream containing drift is discharged at the top opening.

Because the fan is located near the ground level, the vibration of the counterflow forced draft tower is small compared with that of the induced draft tower. Also, if the centrifugal fan blow toward the water

surface, there is a better evaporative cooling effect over the water basin. However, the disadvantages of this type of cooling tower are the uneven distribution of air flowing through the fill, which is caused by the forced draft fan. In addition, the high intake velocity may recapture a portion of the warm and humid exhaust air. Counterflow forced draft cooling towers are often used in small and medium installations.

The preceding types of cooling towers, which use the evaporation of water to cool the condenser cooling water, are sometimes called wet towers. There is also a kind of cooling tower called a dry tower. It is essentially a *dry cooler,* a finned-coil and induced fan combination that cools the cooling water flowing inside the tubes. Characteristics of finned coils will be discussed in Chapter 12.

11.8 THERMAL ANALYSIS OF COOLING TOWERS

Technology Terms

The following terms are commonly used when referring to the performance of a cooling tower:

Approach. Temperature difference between the temperature of the cooling water leaving the tower and the wet bulb temperature of the air entering the tower.

Blow-down. Water discharged to the drain periodically in order to avoid buildup of dissolved solids.

Fill. The structure that forms the heat transfer surface within the tower. Cooling water from the condenser or coil is distributed along the flow passages of the fill down to the water basin.

Make-up. Water added to the circulating water to compensate for the loss of water to evaporation, drift, and blow-down.

Range. Temperature difference between the temperature of cooling water entering tower T_{te} and the temperature leaving the cooling tower T_{tl}, both in °F.

Tower coefficient. Because the make-up water in a typical cooling tower is usually about 2 percent of the circulating cooling water, if the heat energy difference between the make-up water is added and the blow-down and drift losses are ignored, and also if the enthalpy increase of the air from the adding of liquid water is ignored, the energy balance between cooling water and air can then be calculated as

$$\dot{m}_w c_{pw}\, dT_w = \dot{m}_a\, dh_a \qquad (11.4)$$

where $\dot{m}_w, \dot{m}_a$ = mass flow rate of water and air, lb/s

c_{pw} = specific heat of water, Btu/lb · °F

T_w = temperature of cooling water, °F
h_a = enthalpy of air, Btu/lb

If the thermal resistance of the saturated air film that separates the cooling water and the airstream is ignored, the combined heat and mass transfer from the air–water interface to the bulk airstream can be evaluated as

$$\dot{m}_a \, dh_a = K_m(h_s - h_a)dA \qquad (11.5)$$

where K_m = mass transfer coefficient, lb/s·ft²
h_s = enthalpy of the saturated air film, Btu/lb
A = surface area at the air–water interface, ft²

In Eq. (11.5), the change of enthalpy, or total heat of air, consists of changes in sensible heat and latent heat. Consider a cooling tower with a fill volume of V and a contact surface area of $A = aV$ ft². Here, a is the surface area of fill per unit volume. Also

let $K = K_m/c_{pw}$. Then, combining Eqs. (11.4) and (11.5),

$$\frac{KaV}{\dot{m}_w} = \int_{T_{w1}}^{T_{w2}} \frac{dT_w}{h_s - h_a} \qquad (11.6)$$

The integrated value of Eq. (11.6) is generally known as the tower coefficient, or the number of transfer units NTU, of the cooling tower.

Heat and Mass Transfer Process for Counterflow Cooling Tower

Figure 11.10 shows the heat and mass transfer processes between cooling water and air in a counterflow cooling tower with a water-air ratio $\dot{m}_w/\dot{m}_a = 1.2$. In Fig. 11.10a, water is cooled from the entering temperature $T_{we} = 85°F$ to the leaving temperature $T_{wl} = 75°F$. The temperature of the saturated air film corresponding to the cooling water temperature drops

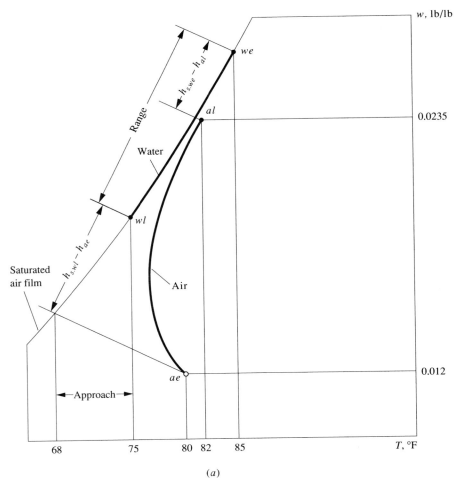

(a)

FIGURE 11.10A Heat and mass transfer process for a counterflow cooling tower: (a) entering air at 80°F dry bulb and 68°F wet bulb.

along the saturation curve and can be represented by the section *we wl*.

Air enters the counterflow tower at a dry bulb $T_{ae} = 80°F$ and a wet bulb $T'_{ae} = 68°F$. Because the temperature of the entering air is higher than the temperature of the leaving water, that is, $T_{ae} > T_{wl}$, air is first evaporatively cooled until the air temperature approaches the water temperature. After that, air is humidified and heated because the water temperature is higher. Air is essentially saturated at the top exit of the tower. The humidifying and heating process of air is illustrated by the curve *ae al* shown in Fig. 11.10a.

The driving potential for heat and mass transfer at the cooling water inlet is indicated by the difference in enthalpy between the saturated air film and the air $(h_{s.we} - h_{al})$. Above the water basin after the fill, it is denoted by $(h_{s.wl} - h_{ae})$.

For humidification with evaporative cooling and heating, the sensible heat transfer from cooling wa-

ter to air is only about 5 percent of the total heat transfer.

In Fig. 11.10b, for the same counterflow cooling tower, air enters at $T_{ae} = 72°F$ and a $T'_{ae} = 70°F$; water may be cooled from 86.5°F to 76.5°F. Because $T_{wl} > T_{ae}$, air is humidified and heated along a curve nearly parallel to the saturation curve. The driving potential is still the difference in enthalpy between a saturated air film and air. Air is also approximately saturated at the top exit and approaches point *al* on the saturation curve. Under this circumstance, the portion of sensible heat transfer is about 23 percent of the total heat transfer. This value indicates that even on rainy days, when air enters the tower in a nearly saturated condition, there is still latent and sensible heat transfer between water and air. The driving potential is mainly due to the higher enthalpy of the saturated air film surrounding the cooling water droplets, which are at a higher temperature than the contacted air.

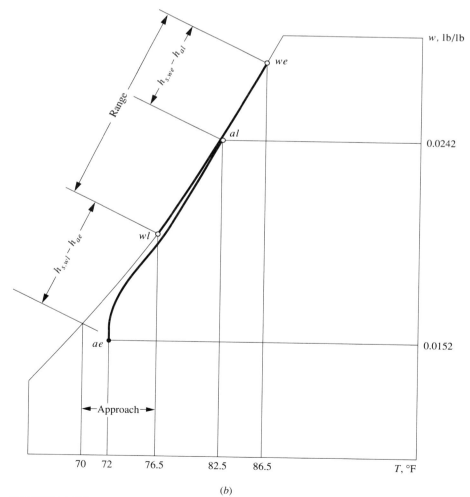

(b)

FIGURE 11.10b Heat and mass transfer process for a counterflow cooling tower; (b) entering air at 72°F dry bulb and 70°F wet bulb

Tower Capacity, Tower Size, and Tower Coefficient

Eq. (11.5) can also be expressed as

$$\dot{m}_w c_{pw}(T_{we} - T_{wl}) = K_m aV(h_s - h_a) \quad (11.7)$$

In Eq. (11.7) the total heat removed from the cooling water, or $\dot{m}_w c_{pw}(T_{we} - T_{wl})$, represents the cooling tower capacity in Btu/h if $\dot{m}_w$ is in lb/h. Tower capacity must meet the required total heat release at the water-cooled condenser.

Tower size is indicated mainly by the volume of the fill V, which includes both the cross-sectional area and the depth of the fill. Strictly speaking, heat and mass transfer also occur in the space between the fill and the water basin as well as between the fill and the spraying water. For simplicity, these can be considered as being included in the volume of fill.

The tower coefficient $K aV / \dot{m}_w$, in lb · °F/Btu actually indicates the heat transfer unit or size of the fill. It is the primary factor that influences the effectiveness of the cooling tower.

An increase of tower capacity may be attributable to a larger tower size or a higher tower coefficient or both. For a fixed water-circulating rate, an increase of tower coefficient may be caused by a larger tower size or a better fill configuration.

By using the numerical integration method, the tower coefficient for a counterflow cooling tower can be calculated as

$$\frac{K aV}{\dot{m}_w} = \frac{T_{w2} - T_{w1}}{h_s - h_a} \quad (11.8)$$

For crossflow cooling towers, the variation in air and water temperature is a function not only of the depth of the fill but also of the transversal position of the fill. A theoretical thermal analysis for crossflow towers is far more complicated than for counterflow towers.

11.9 COOLING TOWER PERFORMANCE

Factors that Affect Cooling Tower Performance

The criteria for selecting a cooling tower for a water-cooled condenser are the effective removal of total heat rejected at the condenser and the minimization of the sum of power consumption in compressors, condenser fans, and condenser water pumps. Proper selection of tower range, water-air ratio, approach, fill configuration, and water distribution system directly affects the performance of a cooling tower.

A cooling tower used for an air conditioning system is rated on the following conditions:

- *Unit of heat release at condenser.* 1 condenser ton = 15,000 Btu/h
- *Water circulating rate.* 3 gpm per condenser ton
- *Entering cooling water temperature.* 95°F
- *Leaving cooling water temperature.* 85°F
- *Outdoor wet bulb temperature.* T_o' 78°F
- *Range.* 10°F
- *Approach.* 7°F

Range and Water-Circulating Rate

Many cooling towers are custom-made, that is, the manufacturer varies the fill configuration, water-circulating rate, and air flow rate to meet the operating characteristics required by the customer. The range $(T_{we} - T_{wl})$, in °F, and the corresponding water-circulating rate $\dot{m}_w$, in gpm, are the primary parameters that should be specified during design.

Range depends on the heat rejected from a water-cooled condenser Q_{rej}, in Btu/h, and the circulating rate of condenser cooling water $\dot{m}_w$, in lb/min. Their relationship can be expressed as

$$T_{te} - T_{tl} = \frac{Q_{rej}}{60 c_{pw} \dot{m}_w} \quad (11.9)$$

The greater the water-circulating rate, the smaller the range. In contrast, a smaller $\dot{m}_w$ results in a greater range. During the selection of a cooling tower at the design stage, Q_{rej}, T_o', and T_{tl} are usually known and fixed. Under such circumstances, a smaller range $(T_{te} - T_{tl})$ and a greater water-circulating rate $\dot{m}_w$ mean the following:

- A lower condensing pressure and temperature
- A greater pumping energy consumption because of the greater $\dot{m}_w$
- A greater tower size because of the smaller $(h_s - h_a)$
- A higher air flow rate for a fixed water-air ratio

A greater range and a smaller $\dot{m}_w$ result in a higher condensing pressure, a lower pumping energy, a smaller tower size, and a lower air flow rate.

Range is usually determined at design conditions, that is, at summer outdoor design wet bulb temperature T_o' and a design Q_{rej}. At a lower T_o' and part-load operation, both the range and T_{tl} are lower. In recent years, there has been a trend to use a range greater than 10°F for air conditioning systems, such as 12°F or even 15°F. A life-cycle cost analysis is recommended to determine the optimum range at specific conditions.

Tower Coefficient and Water-Air Ratio

Cooling tower performance is often determined from field tests and presented in the form of an empirical correlation as

$$\frac{KaV}{\dot{m}_w} = C \left(\frac{\dot{m}_w}{\dot{m}_a}\right)^n z^{-m} \qquad (11.10)$$

where C = constant
  z = depth of the fill, ft

In Eq. (11.10), the average value of exponent n varies from -0.55 to -0.65 and is typically -0.6. According to Webb (1984), the exponent m varies between 0.7 and 1. From the correlation in Eq. (11.10), we see that if all the other conditions remain the same and only the water-air ratio $\dot{m}_w/\dot{m}_a$ varies from 1 to 0.5, the tower coefficient $KaV/\dot{m}_w$ then increases 52 percent.

Face velocity of air that is calculated based on the cross-sectional area of the cooling tower perpendicular to the air flow is usually between 300 to 400 fpm with a total pressure drop across the cooling tower of less than 2 in. WG. This gives a mass flow density of air between 1250 and 1700 lb/h · ft².

When the mass flow density of water drops to 500 lb/h · ft², the water cannot be spread sufficiently over the fill. On the other hand, when the mass flow density of water approaches 3000 lb/h · ft², the fill becomes flooded. For air conditioning systems, the mass flow density of water usually varies between 1000 and 2000 lb/h · ft², and the water-air ratio may vary between 0.7 and 1.5. For a water-circulating rate of 3 gpm per condenser ton, and a $\dot{m}_w/\dot{m}_a = 1$, the air flow rate is about 360 cfm per condenser ton.

Approach

Approach determines the temperature of cooling water leaving the tower T_{tl} corresponding to a local outdoor design wet bulb T_o'. If the range and Q_{rej} are fixed values, a closer approach always means a lower T_{tl} and a lower condensing pressure and temperature.

Approach is closely related to the tower coefficient and the size of the tower. A closer approach always means a larger tower, and a greater approach results in a smaller tower. For cooling towers used for air conditioning systems, approach usually varies from 5 to 12°F. When a cooling tower is connected with a precooling coil forming a water economizer (which will be covered in Chapters 13 and 18), a closer approach, such as 5°F, means a greater percentage of operating hours will be available to replace refrigeration and thereby reduce energy costs.

If other conditions remain the same, when the approach of a typical cooling tower drops from 10°F to 5°F, the tower size may increase around 65 percent. An approach below 5°F requires a very large tower, which is not economical and is, therefore, not recommended.

Outdoor Wet Bulb Temperature

Because heat and mass transfer at the fill is based on the difference in enthalpy between the saturated film and the air, the wet bulb temperature of the outdoor air is the primary parameter that affects the performance of the cooling tower. It is best to use either a 1 percent or 2.5 percent summer outdoor design wet bulb temperature value to guarantee tower performance during most of the operating period in summer, depending on the requirements of the project.

If two different refrigeration systems at different locations have the same range and approach, the one with a higher design outdoor wet bulb temperature will have a smaller tower size than the system with a lower T_o'. This is because the enthalpy of saturated air film is higher, so $(h_s - h_a)$ is greater in the system with a higher design T_o'. However, T_{tl} is lower with a lower T_o'.

Fill Configuration and Water Distribution

Fill configuration has a direct effect on tower coefficient. An efficient fill has a greater surface area per unit volume a, which means more contact surfaces, a longer contact time, and intimate contact between air and water. Currently, two types of fill are widely used in cooling towers for air conditioning systems: splash bars, as shown in Fig. 11.11a, and the more recently developed cellular film, shown in Fig. 11.11b. Splash bars have the advantage of not being easily plugged by dirt contained in the water; the splashing action can also redistribute the water at various levels. However, field practice has shown that when wooden splash bars are replaced by polyvinyl chloride (PVC) cellular film fill, the temperature of condenser water leaving the tower drops 2°F compared with the value before retrofit. Contaminated water must be strained or treated before it is cooled in a PVC cellular film fill.

An even water distribution system is very important to tower performance, especially for counterflow towers. A recently developed large-orifice square plastic nozzle is essentially nonclogging and can provide an even pattern of spraying over the fill. It also has a far greater spraying capacity than small-orifice conical nozzles and, therefore, its maintenance and cleaning are simple.

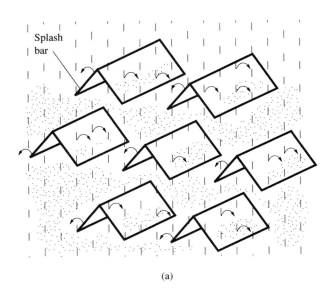

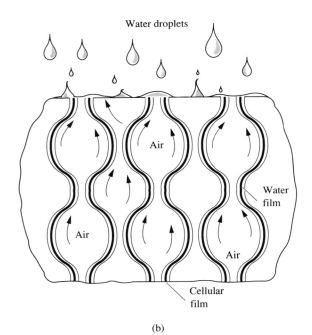

FIGURE 11.11 Two types of fill for cooling towers: (a) splash bars; and (b) PVC cellular film.

11.10 PART-LOAD OPERATION AND OPTIMUM CONTROL

Part-Load Operation

If the water-circulating rate $\dot{m}_w$, the water-air ratio $\dot{m}_w/\dot{m}_a$, and the outdoor wet bulb temperature T_o' remain the same, the performance of a counterflow cooling tower shows the following results when it is operated at part-load (a load ratio less than 1):

• A smaller range due to lower $Q_{\text{rej.}\rho}$

• A closer approach because of a fixed tower coefficient and heat transfer surface and a lower $Q_{\text{rej.}\rho}$

• A lower T_{te} and T_{tl}

The load ratio R_{load} is defined as

$$R_{\text{load}} = \frac{Q_{\text{rej.}\rho}}{Q_{\text{rej.f}}} \qquad (11.11)$$

where $Q_{\text{rej.}\rho}, Q_{\text{rej.f}}$ = heat rejected at condenser at part-load and full-load, respectively, Btu/h. If $\dot{m}_w$, $\dot{m}_w/\dot{m}_a$, and Q_{rej} remain the same, only T_o' falls to a lower temperature than the design condition. Then we note the following:

• A smaller range

• A smaller $(h_s - h_a)$ because of lower enthalpy of saturated air film at lower T_o', and also a lower T_{te} and T_{tl}

• A greater approach because of to a smaller $(h_s - h_a)$

If $\dot{m}_w$ and $\dot{m}_w/\dot{m}_a$ remain the same, both T_o' and $Q_{\text{rej.}\rho}$ drop at part-load. In addition to a smaller range and closer approach, there is a further drop of T_{te} and T_{tl}.

Optimum Control

During part-load operation, the capacity of a cooling tower should be controlled for effective operation of the refrigeration system at minimum energy consumption. In a typical cooling tower, the electric power input to the tower fans is approximately 10 percent of the power input to the centrifugal compressors; and the power input to the circulating pumps for condenser cooling water is about 2 to 5 percent of the power input to the centrifugal compressors.

One common strategy for capacity control of cooling towers during part-load operation is to maintain a fixed temperature T_{tl} for cooling water leaving the tower. This strategy does not minimize the sum of power input to the compressors, tower fans, and circulating pumps when both T_{tl} and the condensing temperature can be lowered. A better strategy for capacity control of cooling towers is to maintain a constant approach, that is, a constant temperature difference between T_{tl} and T_o' by controlling the tower fans until T_{tl} drops to a limit below that at which the refrigeration system cannot be operated properly. For instance, for a centrifugal refrigeration system using multiple orifices as a throttling device, one manufacturer recommends that T_{tl} should not fall below 65°F.

Most tower fans in current use are two- or three-speed fans. The key points for a nearly optimum control of cooling towers according to Braun and Didderich (1990) are as follows:

- Use sequencing of tower fans to maintain a nearly constant approach during part-load to minimize the sum of power input to the compressors and tower fans.

- Heat rejected $Q_{rej.\rho}$ determines the sequencing of tower fans, and $Q_{rej.\rho}$, is measured by the product of the range and the water-circulating rate.

- Establish a simple relationship between the tower capacity and the sequencing of the tower fan.

- Use an open-loop control (see Chapter 17) for a stable operation.

- When an increase of tower capacity is required, the speed of the tower fan operated at the lowest speed should be raised first. Similarly, during a decrease of tower capacity, the highest tower fan speed should be the first one to be reduced.

- If the cooling water temperature leaving the tower T_{tl} drops to a limit below which effective operation of the refrigeration system cannot be provided, the speed of the tower fans should then be reduced in sequence to maintain a T_{tl} not lower than that limit.

Example 11.1. A counterflow induced draft cooling tower is required to cool the condenser cooling water from 95°F to 85°F at a design outdoor wet bulb temperature of 78°F. If the water-air ratio of this tower is 1.2, do the following:

1. Calculate the tower coefficient at this design wet bulb temperature.

2. If the outdoor wet bulb temperature, the water-circulating rate, and air flow rate remain the same as in design conditions, calculate the condenser water entering and leaving the tower if the rate of heat rejection at the water cooled condenser has fallen to 50 percent of its design value. Take the specific heat for water as $c_{pw} = 1$ Btu/lb · °F.

Solution. To find the enthalpy of the saturated air film h_s, the enthalpy of ambient air h_a, and the driving potentials $(h_s - h_a)$, make these calculations:

- Divide the temperature range of condenser cooling water into 1°F divisions, such as 85, 86,..., 93, 94, 95°F. From Appendix 2.2, the enthalpy of the saturated air film that surrounds the water droplets h_s at various temperatures can be determined.

- Find the enthalpies of the air that contacts the cooling water at various temperatures h_a by means of Eq. (11.4), starting with incoming air at a wet bulb of $T'_o = 78$°F in contact with the coldest cooling water at temperature 85°F.

Based on Eq. (11.8), the tower coefficient or fill surface area required to cool the cooling water from 86°F to 85°F and raise the air enthalpy from that at $T'_o = 78$°F to $(h_{a78} + 1.2)$ is $(T_{w2} - T_{w1})/(h_s - h_a)$, or their average, $\frac{1}{2}[1/(h_{s86} - h_{a78}) + 1/(h_{s85} - h_{a78})]$.

The tower coefficient of the cooling tower that cools the cooling water from 95°F to 85°F is the summation or integration of the tower coefficient of 95° to 94°F, ..., 86° to 85°F.

1. From Appendix 2.2, at a wet bulb temperature of 78°F, the enthalpy of saturated air is 41.59 Btu/lb. Because the enthalpy of the saturated air film at a water temperature of 85°F is 49.45 Btu/lb, the enthalpy difference between the saturated air film at 85°F and air at 78°F wet bulb is $(h_s - h_a) = (49.45 - 41.59) = 7.86$ Btu/lb, and $1/(h_s - h_a) = 0.1272$ lb/Btu.

2. From Eq. (11.4), for each degree of increase in cooling water temperature, the corresponding increase of the enthalpy of air is

$$dh_a = \frac{\dot{m}_w}{\dot{m}_a} c_{pw} \, dT_w = (1.2 \times 1) dT_w$$

Then at a cooling water temperature of 86°F, the enthalpy of the saturated air film is 50.68 Btu/lb, and the enthalpy of the corresponding air in contact with this cooling water is

$$h_a = 41.59 + 1.2 = 42.79 \text{ Btu/lb}$$

The enthalpy difference between the saturated air film and air at $T_w = 86$°F is then

$$h_s - h_a = 50.68 - 42.79 = 7.89 \text{ Btu/lb}$$

and

$$\frac{1}{h_s - h_a} = \frac{1}{7.89} = 0.1267 \text{ lb/Btu}$$

3. The average value of $T_w/(h_s - h_a)$ between water temperature 85°F and 86°F is $1(0.1272 + 0.1267)/2 = 0.1270$ lb · °F/Btu.

 Similarly, values of h_a, $(h_s - h_a)$ and $T_w/(h_s - h_a)$ at other water temperatures can be calculated and have been listed in Table 11.2.

4. Using the numerical integration method, we find the tower coefficient is

$$\frac{KaV}{\dot{m}_w} = \frac{\Sigma \Delta T_w}{h_s - h_a} = 0.1272$$

$$+ 0.1262 + 0.1249 + 0.1231$$

$$+ 0.1209 + 0.1182 + 0.1151$$

$$+ 0.1118 + 0.1082 + 0.1045$$

$$= 1.1799 \text{ lb} \cdot \text{°F/Btu}$$

5. At 50 percent part-load operation, the tower coefficient $KaV/\dot{m}_w$, T'_o, the water-circulating rate, and the water-air ratio remain the same. Heat rejection at the condenser $Q_{rej.\rho} = 0.5 Q_{rej.f}$ because

TABLE 11.2
Numerical integration of a counterflow cooling tower for Example 11.1

Cooling water temperature °F	Enthalpy of saturated air film, h_S Btu/lb	Air Enthalpy, h_a Btu/lb	$h_S - h_a$ Btu/lb	$\dfrac{1}{h_S - h_a}$ lb/Btu	Average $\dfrac{\Delta T_w}{h_S - h_a}$ lb/Btu · °F	$\sum\left(\dfrac{\Delta T_w}{h_S - h_A}\right)$ lb/Btu · °F
			Design conditions			
85	49.45	41.59	7.86	0.1272		
					0.1270	0.1270
86	50.68	42.79	7.89	0.1267		
					0.1262	0.2532
87	51.95	43.99	7.96	0.1256		
					0.1249	0.3781
88	53.25	45.19	8.06	0.1241		
					0.1231	0.5012
89	54.58	46.39	8.19	0.1221		
					0.1209	0.6221
90	55.95	47.59	8.36	0.1196		
					0.1182	0.7403
91	57.36	48.79	8.57	0.1167		
					0.1151	0.8554
92	58.79	49.99	8.80	0.1136		
					0.1118	0.9672
93	60.27	51.19	9.08	0.1101		
					0.1082	1.0754
94	61.79	52.39	9.40	0.1064		
					0.1045	1.1799
95	63.34	53.59	9.75	0.1026		
			Part-load operation (50 percent design load)			
82	45.91	41.59	4.32	0.2315		
					0.2328	0.2328
83	47.06	42.79	4.27	0.2341		
					0.2350	0.4678
84	48.23	43.99	4.24	0.2358		
					0.2402	0.7080
85	49.45	45.19	4.26	0.2347		
					0.2338	0.9418
86	50.68	46.39	4.29	0.2331		
					0.2312	1.1730
87	51.95	47.89	4.36	0.2294		

the range has been dropped approximately from 10°F to $0.5 \times 10°F = 5°F$. By trial and error, we find that when the cooling water is cooled from 87°F to 82°F, the calculated tower coefficient is 1.1730 lb · °F/Btu, which is approximately the same as the value at the design condition, or 1.1799 lb · °F/Btu. The results are also listed in Table 11.2.

11.11 OPERATING CONSIDERATIONS FOR COOLING TOWERS

Recirculation and Interference

Most cooling towers are installed outdoors. Location and layout of the cooling towers should be carefully chosen so that recirculation, interference, and insufficient air flow are minimized.

The recapture of a portion of warm and humid exhaust air at the intake by the same cooling tower is called *recirculation*. Recirculation causes an increase in entering wet bulb temperature and, therefore, degrades the tower performance. Recirculation is mainly

caused by insufficient exhaust air velocity, or insufficient vertical distance between the intake and the exit. When a wind encounters a cooling tower, a pressure lower than atmospheric is formed on the leeward side of the tower. Thus, the exhaust air is extracted downward toward the air intake. If the exhaust air velocity is not high enough, recirculation occurs.

If the exhaust air of a cooling tower or warm air from other heat sources located upstream is extracted by another cooling tower downstream, such a phenomenon is called *interference*. Interference can be minimized by considering the summer prevailing wind direction and by properly planning the layout of the cooling towers.

When the space between the intakes of two cooling towers or the space between the wall and the intake is insufficient, or if a barrier is used to shield the view or noise, air flow may be limited, causing an air flow rate lower than the required rate. Insufficient air flow impairs the effectiveness of heat and mass transfer and results in a greater approach and range and therefore, a higher condensing pressure.

Freezing Protection and Fogging

When a cooling tower is operated at locations where outdoor temperatures may drop below 32°F, freezing protection should be provided to avoid the formation of ice on components of the cooling towers (such as intake louvers) which may seriously restrict the air flow.

An electric immersion heater is usually installed in the water basin for freezing protection. In idle towers, fans should not be run and water flow should be completely shut off. Condenser water may bypass the fill and be distributed directly over the inside surface of the intake louvers through a special piping arrangement. This action eliminates the accumulation ice on the surfaces of the louvers.

When warmer and essentially saturated air that is exhausted from the cooling tower is mixed with the cooler ambient air in cold weather, the mixture becomes oversaturated, and the excess water vapor condenses as fog, which is often called *fogging* or *plume*. Fogging limits visibility and may form ice on nearby highways. Fogging should be minimized by locating the cooling towers so that visible plumes and their influence on the nearby environment are not objectionable.

Blow-Down and Legionnaires' Disease

To avoid the concentration of impurities contained in water beyond a certain limit, a small percentage of water in a cooling water system is often purposely drained off or discarded. Such a treatment is called *blow-down* or *bleed-off*. The amount of blown-down is usually 0.8 to 1.0 percent of the total water circulating flow. Because drift in a cooling tower varies from 0.1 to 0.2 percent, if the evaporation loss in a typical location is 1 percent, the amount of make-up water is about $(0.9 + 0.1 + 1.0) = 2$ percent. The practice of blow-down or bleed-off is widely used in condenser cooling water systems.

If simple blow-down is inadequate to control scale formation, chemicals can be added to inhibit corrosion and to limit microbiological growth.

Water treatment to reduce scale formation and corrosion, and to control microbiological growth will be discussed in Chapter 20.

Recently, the possibility that *Legionella* bacteria existed in condenser cooling water systems in cooling towers has received widespread public attention. Many researchers found that regular maintenance of cooling towers and cooling water systems is effective in preventing the growth of *Legionella*. It is important to keep the cooling water clean so as to prevent accumulation of dirt, scale, and impurities, and to control the growth of algae, slime, *Legionella*, and microorganisms through chemicals and other water treatments. Regular inspections, periodic tests, and a planned maintenance schedule are necessary.

The location of a cooling tower should be carefully chosen so that the drift from the cooling tower is not extracted into the outdoor intakes of the air-handling units or packaged units.

Maintenance of Cooling Towers

Maintenance of cooling towers includes the following:

• Inspection of the tower fan, condensing water pump, fill, water basin and the drift carryover to ensure satisfactory operation
• Periodic blow-down, cleaning of wetted surfaces, and effective water treatments to prevent formation of scale and dirt as well as growth of microorganisms
• Periodic lubrication of moving parts

The manufacturer's maintenance schedule should be implemented. Otherwise, a maintenance document must be established.

If noise from the tower fan exceeds the limits specified by local codes, sound attenuators at the fan exit and at the intake louver should be installed. Details of sound control will be discussed in Chapter 16.

REFERENCES AND FURTHER READING

Armstrong Machine Works, *The Armstrong Humidification Handbook*, Armstrong Machine Works, Three Rivers, MI, 1991.
ASHRAE, *ASHRAE Handbook 1988, Equipment*, ASHRAE, Inc., Atlanta, GA.
Braun, J. E., and Diderrich, G. T., "Near-Optimal Control of Cooling Towers for Chilled-Water Systems," *ASHRAE Transactions*, Part II, pp. 806–813, 1990.
Brown, W. K., "Humidification by Evaporation for Control Simplicity and Energy Savings," *ASHRAE Transactions*, Part I, pp. 1265–1272, 1989.
Burger, R., "Cooling Tower Technology," *Heating/Piping/Air Conditioning*, pp. 41–45, August 1987.
Cheremisinoff, N. P., and Cheremisinoff, P. N., *Cooling Towers, Selection, Design and Practice*, Technomic, Lancaster, PA, 1989.
Coad, W. J., "The Spray Coil Option," *Heating/Piping/Air Conditioning*, pp 121–122, November 1988.
Davis, W. J., "Water Spray for Humidification and Air Flow Reduction," *ASHRAE Transactions*, Part II, pp. 351–356, 1989.
Deacon, W. T., "Important Considerations in Computer Room Humidification," *ASHRAE Transactions*, Part I, pp. 1273–1277, 1989.

Facius, T. P., "Certified Cooling Tower Performance," *Heating/Piping/Air Conditioning*, pp. 66–73, August 1990.

Feeley, J. C., and Morris, G. K., "*Legionella*: Impact on Water Systems in Buildings," *ASHRAE Transactions*, Part II, pp. 146–149, 1991.

Galey, M. D., Jr., "Cooling Tower Water Treatment Alternatives," *Heating/Piping/Air Conditioning*, pp. 55–58, July 1990.

Goswami, D., "Computer Room Humidity Control," *Heating/Piping/Air Conditioning*, pp. 123–124, November 1984.

Green, G. H., "The Positive and Negative Effects of Building Humidification," *ASHRAE Transactions*, Part I, pp. 1049–1061, 1982.

Grimm, N. R., and Rosaler, R. C., *Handbook of HVAC Design*, McGraw-Hill, New York, 1990.

Hartman, T., "Humidity Control," *Heating/Piping/Air Conditioning*, pp. 111–114, September 1989.

Hensley, J. C., "Cooling Tower Energy," *Heating/Piping/Air Conditioning*, pp. 51–59, October 1981.

Jourdan, J., "Cooling Tower System Installation," *Heating/Piping/Air Conditioning*, pp. 131–134, January 1988.

Leary, W. M., Jr., "Optimizing Cooling Tower Selections," *Heating/Piping/Air Conditioning*, pp. 67–68, August 1987.

Meitz, A.K., "Clean Cooling Systems Minimize *Legionella* Exposure." *Heating/Piping/Air Conditioning* pp. 99–101, August 1986.

Morton, B. W., *Humidification Handbook*, B. W. Morton and Dri-Steam Humidifying Co., Hopkins, MN, 1981.

Obler, H., "Humidification Alternatives for Air Conditioning," *Heating/Piping/Air Conditioning*, pp. 73–77, December 1982.

Rosa, F., "*Legionella* and Cooling Towers," *Heating/Piping/Air Conditioning*, pp. 75–85, February 1986.

Webb, R. L., "A Unified Theoretical Treatment for Thermal Analysis of Cooling Towers, Evaporative Condensers, and Fluid Coolers," *ASHRAE Transactions*, Part II B, pp. 398–415, 1984.

Wong, S. P. W., and Wang, S. K., "System Simulation of the Performance of a Centrifugal Chiller Using a Shell-and-Tube Type Water-Cooled Condenser and R-11 as Refrigerant," *ASHRAE Transactions*, Part I, pp. 445–454, 1989.

COILS, EVAPORATORS, AND CONDENSERS

12.1   COIL CHARACTERISTICS

Coils are indirect contact heat exchangers that transfer heat between air and medium (such as water, refrigerant, steam, or brine) for the purpose of heating, cooling, dehumidifying, or a combination thereof. Chilled water and refrigerants used to cool and dehumidify moist air are called *coolants*. Coils consist of tubes and external fins arranged in rows along the air flow to provide greater surface contact. Coils are arranged in circuits to create a shorter path length. In a coil, water and refrigerant flow inside the tubes and air flows over the outside surface of the tubes and fins.

Types of Coils

Coils can be classified into four categories according to the medium used.

WATER COOLING COIL. A *water cooling coil* uses chilled water as the coolant inside the tubes. The chilled water cools or cools and dehumidifies the moist air that flows over the external surface of the tubes and fins, as shown in Fig. 12.1a. In order to maintain a high rate of heat transfer, the air and water normally follow a counterflow arrangement, that is, the coldest air meets the coldest water and the warmest air meets the warmest water.

The water tubes are usually copper tubes of $\frac{1}{2}$ to $\frac{5}{8}$ in. diameter with a thickness of 0.01 to 0.02 in. They are spaced at a center-to-center distance of 0.75 to 1.25 in. longitudinally and 1 to 1.5 in. transversely. The tubes may be arranged along the air flows in 2, 3, 4, 6, 8, or 10 rows, either staggered or aligned. The staggered arrangement provides better heat transfer and a higher air pressure drop. Chilled water coils are commonly rated at a pressure of 175 to 300 psig.

DIRECT EXPANSION (DX) COIL. In a *direct expansion coil,* the refrigerant (usually HCFC–22

or HFC–134a) is evaporated and expanded directly inside the tubes to cool and dehumidify the air flowing over it, as shown in Fig. 12.1b. This is why it is called a *DX-coil* or *refrigerant coil*. A DX-coil acts as the evaporator in a refrigerating system.

In a DX-coil, coolant or the refrigerant is fed into a distributor and is evenly distributed to various tube circuits, which are made of copper tubes, typically 0.5 inches in diameter. Refrigerant distribution and loading to various circuits are critical to the performance of a DX-coil. After evaporation, the vapor refrigerant is discharged from the header to the suction line.

WATER HEATING COIL. The water heating coil shown in Fig. 12.1c is similar in construction to the water cooling coil. There are two main differences between them. Hot water, instead of chilled water, is used as the heating medium in a water heating coil. Also, there are fewer rows in the water heating coil than in the water cooling coil. Generally, only 2-, 3-, or 4-row water heating coils are available on the market. Water heating coils are rated at pressures of 175 to 300 psig at temperatures up to 250°F.

Figure 12.2 shows the structure of a water cooling coil and a DX-coil.

STEAM-HEATING COIL. *Steam heating* coils use the latent heat of condensation released by the steam inside the tubes to heat outside air, as shown in Fig. 12.1d. In a standard steam heating coil, steam enters one end of the coil and the condensate comes out the other end. For more even distribution of steam, a baffle plate is often installed just after the inlet.

In a steam heating coil, it is important that the coil core inside the casing expand or contract freely. The coil core is also pitched toward the return connection to facilitate drainage of the condensate. Steam heating coils usually have a rating of 100 to 200 psig at 400°F.

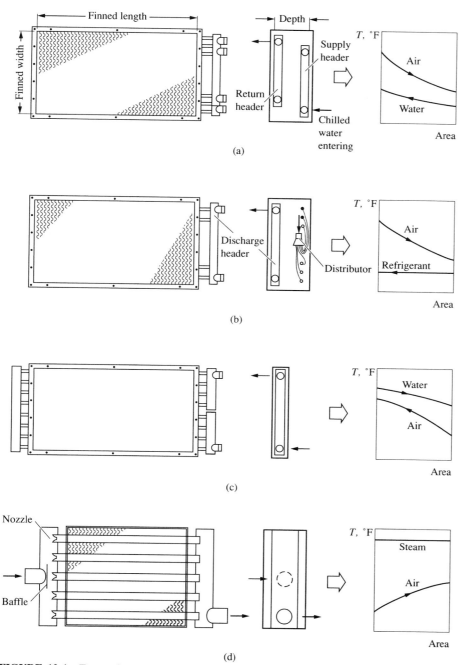

FIGURE 12.1 Types of coils: (a) Water cooling coil (b) Direct-expansion (DX) coil (c) Water heating coil (d) Steam heating coil.

Steam heating coil tubes are usually made of copper, steel, or stainless steel.

Fins

Fins are extended surfaces that are often called the coil's *secondary surface* (the outer surface of the tubes is called the *primary surface* of the coil). Fins are often made of aluminum, with a fin thickness F_t of 0.005 to 0.008 in. They can also be directly extruded, or shaved, from the parent tubes. Copper, steel, or stainless steel fins are sometimes used.

Three types of fins are widely used today:

• *Continuous plate fins* and *corrugated plate fins,* as shown in Fig. 12.3a and b. Plate fins are usually made of aluminum.

• *Smooth spiral* and *crimped spiral fins,* used in many commercial and industrial applications, are shown

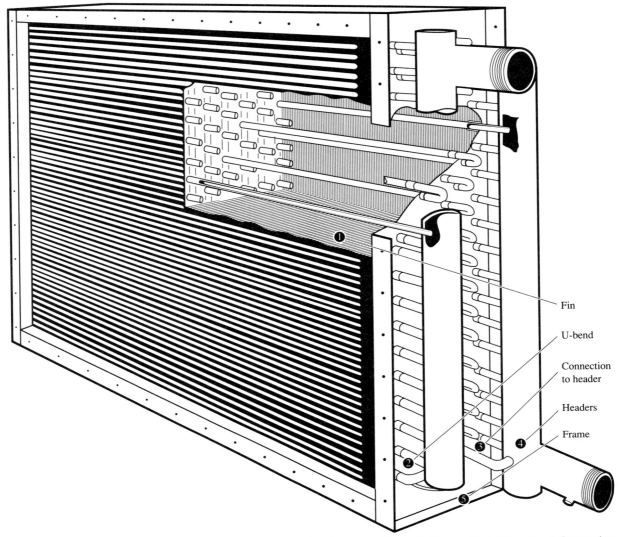

Fin

U-bend

Connection
to header

Headers

Frame

FIGURE 12.2a Structure of a water cooling coil and a DX-coil. (a) water cooling coil. (*Source:* York International Corporation. Reprinted with permission.)

in Fig. 12.3c and d. Smooth spiral fins may be extruded from the parent tubes.

- *Spined pipes,* as shown in Figure 12.3e. Spined pipe is made by shaving the spines and lifting them from the parent tubes.

Corrugated and crimped fins create large turbulences and reduce the thickness of the boundary layer of the airstream. Consequently, their convective heat transfer coefficient is increased significantly. However, corrugated and crimped fins also create a greater air-side pressure drop than plate fins. Various manufacturers have developed rippled corrugated fins to accommodate the higher rate of heat transfer and the allowable air-side pressure drop.

Because spines are part of a spined pipe, they are firmly connected to the primary surface and there is

no contact resistance between the spines and tube. According to Holtzapple and Carranza (1990), heat transfer per expended blower energy of spined pipe is slightly lower than that of plate fins with larger tube spacing S_t, and is slightly better with smaller S_t and small tube diameters.

Fin spacing S_f, in inches, is the distance between two fins. *Fin density* is usually expressed in fins per inch, and varies from 8 to 14 fins/in. for coils used in air conditioning systems. Fin spacing has a definite effect on the extended surface area and, therefore, the capacity and air-side flow resistance of the coil.

For a finned-tube coil, *finned width* indicates the width of the continuous plate fin, as shown in Fig. 12.1a and *fin length* indicates the length of the coil with plate fins.

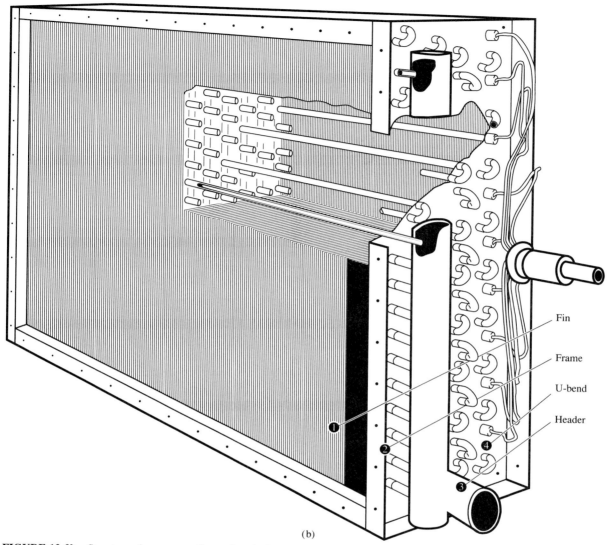

Fin

Frame

U-bend

Header

(b)

FIGURE 12.2b Structure of a water cooling coil and a DX-coil. (b) DX-coil. (*Source:* York International Corporation. Reprinted with permission.)

Contact Conductance

Thermal *contact conductance* between the fins and tube in a coil $h_{c.c}$, in Btu/h · ft^2 · °F, is the steady heat flux across the fin–tube bonding per °F temperature difference. Contact conductance is a function of contact pressure at the fin–tube bonding, fin thickness, fin density, tube diameter, and hardness of the tube. It directly affects the heat transfer capacity of a finned-tube coil.

Contact between the fin and tube is critical to the performance of the coil. Bonding between the fin and tube is usually formed by mechanically expanding the tube from inside using a tool called a *bullet* to produce an interference between the fin collar and the tube, as shown in Figure 12.3f. Interference I, in inches, is the difference between the expanded tube diameter

and the fin collar diameter before expanding. It can be also expressed as

$$I = (D_t + 2W_{tk}) - D_{fc} \qquad (12.1)$$

where D_t = outside diameter of the expanding tool, in.

W_{tk} = tube wall thickness, in.

D_{fc} = inner diameter of the fin collar, in.

According to Wood et al. (1987), the interference of the coils from six manufacturers ranges from 0.003 to 0.0075 in.

Contact conductance is directly proportional to the magnitude of interference. Interference determines the quality of mechanical bonding during manufacturing. A higher interference means a better quality of mechanical bonding.

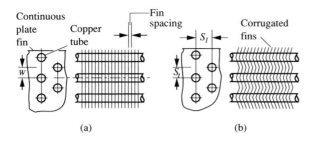

(a)

(b)

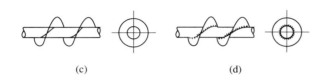

(c)

(d)

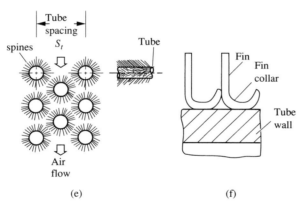

(e)

(f)

FIGURE 12.3 Types of fins: (a) continuous plate fins, (b) corrugated plate fins, (c) smooth spiral fins, (d) crimped spiral fins, (e) spined pipe, and (f) fin collar and tube bondings.

Contact conductance increases as the thickness of the fin increases. It also increases as fin density (fins per inch) increases. Contact conductance decreases as tube diameter increases. It also decreases as the hardness of the copper tube increases in an aluminum fin and copper tube coil.

Water Circuits

In a water cooling or water heating coil, tube feeds or water circuits determine the number of water flow passages. The greater the finned width, the higher the number of tube feeds and thus the higher the number of flow passages. For two finned-tube coils of the same finned width, a difference in the number of tube feeds or water circuits means that the water flow rate (in gpm), number of passes, and pressure drop of the chilled water (in ft WG) inside the two coils are different. One *pass* means that water flows through the coil's finned length once.

The number of serpentines ($\frac{1}{4}$, $\frac{1}{2}$, $\frac{3}{4}$, 1, $1\frac{1}{2}$, or 2) of a water cooling or a water heating coil indicates its water flow arrangement. The greater the number of serpentines, the larger total cross-sectional area of the water circuits and the higher the water volume flow rate. Figure 12.4 shows five water cooling coils, each specified by the number of serpentines, water circuits, passes, and rows, made by the same manufacturer.

In Fig. 12.4, $\frac{1}{2}$ serpentine means that at the first row, there are 8 tubes across the finned-width, but only 4 of them are tube feeds that are connected to the return-header. For a full serpentine coil, all 8 tubes in the first row are tube feeds and connect to the return header.

12.2 SENSIBLE COOLING AND SENSIBLE HEATING COILS—DRY COILS

Heat Transfer in a Sensible Cooling Process

In a sensible cooling process, described in Section 3.2, the humidity ratio w is always constant. A sensible cooling process only exists when the outer surface temperature of the coil T_s, in °F, is equal to or higher than the dew point of the entering air, T''_{ae}, in °F, that is, $T_s \geq T''_{ae}$. A sensible cooling process is indicated by a horizontal line toward the saturation curve on the psychrometric chart.

If heat conduction through the coil framework is negligible, for a sensible cooling process occurring at steady-state within a certain time interval, the rate of sensible heat transfer from the conditioned air q_{cs} (in Btu/h) must be equal to the heat absorbed by the chilled water in the tubes. This relationship can be expressed as

$$q_{cs} = 60\dot{V}_a \rho_a c_{pa}(T_{ae} - T_{al})$$
$$= A_o U_o \Delta T_m = F_s A_a N_r U_o \Delta T_m \quad (12.2)$$
$$= 60\dot{m}_w c_{pw}(T_{wl} - T_{we})$$

where $\dot{V}_a$ = volume flow rate of the conditioned air, cfm

ρ_a = air density, lb/ft^3

T_{ae}, T_{al} = temperature of conditioning air entering and leaving the coil, °F

c_{pa}, c_{pw} = specific heat of the moist air and water, Btu/lb · °F

A_o = total outside surface area of the coil, ft^2

U_o = overall heat transfer coefficient based on the outside surface area of the coil, Btu/h · ft^2 · °F

F_s = coil core surface area parameter

A_a = face area of the coil, ft^2

N_r = number of rows in the coil, rows

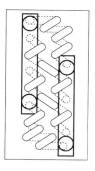

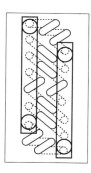

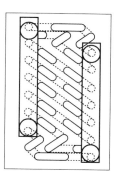

$\frac{1}{2}$ serpentine
4 water circuits
8 passes
4 rows

$\frac{3}{4}$ serpentine
6 water circuits
4 and 6 passes
4 rows

1 serpentine
8 water circuits
6 passes
6 rows

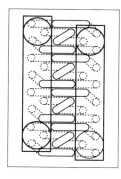

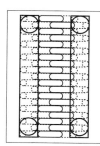

$1\frac{1}{2}$ serpentine
12 water circuits
4 passes
6 rows

2 serpentine
19 water circuits
4 passes
8 rows

FIGURE 12.4 Water circuits for water cooling coils. (Adapted from Catalog of McQuay HAVC. Reprinted with permission of Snyder-General Corp.)

ΔT_m = log-mean temperature difference, °F
$\dot{m}_w$ = mass flow rate of chilled water, lb/min
T_{wl}, T_{we} = temperature of chilled water leaving and entering the coil, °F

Because $\dot{m}_w = \dot{V}_w \rho_w$, here $\dot{V}_w$ indicates the volume flow rate of chilled water, in cfm, and ρ_w = water density, which can be estimated as 62.32 lb/ft^3. Also, 1 gal = 0.1337 ft^3, and $c_{pw} = 1$, so

$$q_{cs} = 60 \dot{V}_w \rho_w c_{pw}(T_{wl} - T_{we})$$
$$= 60 \times 0.1337 \dot{V}_{gal} \rho_w c_{pw}(T_{wl} - T_{we}) \quad (12.3)$$
$$= 500 \dot{V}_{gal}(T_{wl} - T_{we})$$

where $\dot{V}_{gal}$ = volume flow rate of chilled water, gpm.
Coil core surface area F_s is defined as

$$F_s = \frac{A_o}{A_a N_r} \quad (12.4)$$

If the thermal resistance of copper tubes is ignored, the overall heat transfer coefficient based on the out-

side surface area of the coil U_o and the inner surface area U_i (both in Btu/h · ft^2 · °F) can be calculated as

$$U_o = \cfrac{1}{\cfrac{1}{\eta_s h_o} + \cfrac{B}{h_i}}$$
$$U_i = \cfrac{1}{\cfrac{1}{\eta_s h_o B} + \cfrac{1}{h_i}} \quad (12.5)$$

In Eq. (12.5), the finned surface efficiency η_s is given as

$$\eta_s = 1 - \left(\frac{A_f}{A_o}\right)(1 - \eta_f) \quad (12.6)$$

where h_o, h_i = heat transfer coefficient of the outer surface of the coil (including fins and tubes) and the inner surface of the tubes, Btu/h · ft^2 · °F

B = ratio of the outside surface area A_o of the coil to the inner surface of the tubes A_i, both in ft^2, or $B = A_o/A_i$

A_f = area of the fins, ft^2

η_f = fin efficiency

Surface Heat Transfer Coefficients

McQuinston (1981) developed the correlation between Chilton–Colburn j-factors and parameter JP for dry coils as follows

$$j_s = \frac{h_o}{G_c c_{pa}} \Pr^{\frac{2}{3}} = 0.00125 + 0.27 \text{JP} \qquad (12.7)$$

$$\text{JP} = \text{Re}_D^{-0.4}\left(\frac{A_o}{A_p}\right)^{-0.15}; \qquad \text{Re}_D = \frac{G_c D_o}{\mu}$$

$$G_c = 60\rho v_{\min} \qquad (12.8)$$

where G_c = mass velocity at minimum flow area, lb/h · ft^2

$v_{\min}$ = air velocity at minimum flow area, fpm

A_p = primary surface area, or tube outer surface area, ft^2

Pr = Prandtl number, Pr = $\mu c_p/k$, for air at temperature lower than 200°F; its value can be estimated as 0.71

μ = dynamic viscosity of fluid, lb/ft · h

c_p = specific heat of fluid, Btu/lb · °F

k = thermal conductivity of fluid, Btu/h · ft · °F

Re_D = Reynolds number based on diameter D

The outer surface heat transfer coefficient h_o calculated from Eq. (12.7) is for smooth plate fins. For corrugated fins, h_o should be multiplied by a factor F_{cor} between 1.1 and 1.25 to account for the increase in turbulence and rate of heat transfer.

For hot and chilled water at turbulent flow inside the tubes, the inner surface heat transfer coefficient h_i, in Btu/h · ft^2 · °F, can be calculated by Dittus-Boelter equation

$$\frac{h_i D_i}{k_w} = 0.023(\text{Re}_D)^{0.8}\Pr^n \qquad (12.9)$$

where D_i = inner diameter of tubes, ft

k_w = thermal conductivity of water, Btu/h · ft · °F

When the tube wall temperature $T_t > T_{\text{bulk}}$, $n = 0.4$, and when $T_t < T_{\text{bulk}}$, $n = 0.3$. Here, T_{bulk} indicates the bulky water temperature beyond the boundary layer. All fluid properties in Re_D and Pr should be calculated at a mean temperature $T_m = (T_t + T_{\text{bulk}})/2$.

Fin Efficiency

Because fin temperature is not uniform, fin efficiency η_f is used to calculate the rate of heat transfer of a heat exchanger with fins. Fin efficiency is defined as

$$\eta_f = \frac{\text{actual heat transfer}}{\begin{array}{c}\text{heat transfer when fins are}\\\text{at the base temperature}\end{array}} \qquad (12.10)$$

For a finned-tube coil with rectangular plate fins, fin efficiency can be calculated as recommended in *ASHRAE Handbook 1981, Fundamentals*

$$\varphi_{\text{f.max}} = \frac{R_{\text{f.max}} F_t k_f}{W^2} \qquad (12.11)$$

$$R_f = \left(\frac{1}{h_o}\right)\left(\frac{1}{\eta_f} - 1\right) \qquad (12.12)$$

where $\varphi_{\text{f.max}}$ = maximum fin resistance number, dimensionless

$R_f, R_{\text{f.max}}$ = R-value and maximum R-value of fin, h · ft^2 · °F/Btu

F_t = fin thickness, ft

k_f = thermal conductivity of fin material, Btu · h · ft · °F

In Eq. (12.11), W represents a width dimension as shown in Fig. 12.3a, in ft.

Figure 12.5 shows the relationship between $\varphi_{\text{f.max}}$ and ratio W/r_o at various S_L/W values. Here, r_o

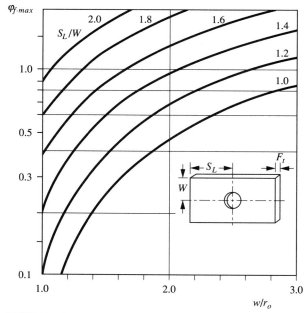

FIGURE 12.5 Maximum fin resistance number $\varphi_{\text{f.max}}$ of rectangular fins. (Adapted with permission from *ASHRAE Handbook 1989, Fundamentals.*)

indicates the outer radius of the tube, and S_L is the longitudinal spacing of the tubes, as shown in Fig. 12.3b.

By assuming that $\varphi_f / \varphi_{f.max} = 0.9$ and then, $R_f / R_{f.max} = 0.9$, fin efficiency η_f can be determined.

Dry Coil Effectiveness

The effectiveness of a heat exchanger ϵ is defined as the ratio of the actual rate of heat transfer between the hot and cold fluids to the maximum possible rate of heat transfer, that is

$$\epsilon = \frac{C_a(T_{ae} - T_{al})}{C_{min}(T_{ae} - T_{we})}$$
$$= \frac{C_w(T_{wl} - T_{we})}{C_{min}(T_{ae} - T_{we})} \qquad (12.13)$$

The heat capacity rate of the hot fluid (moist air) C_a (in Btu/h · °F) in a sensible cooling coil is

$$C_a = 60\dot{m}_a c_{pa} = 60\dot{V}_a \rho_a c_{pa}$$

The heat capacity of the cold fluid (chilled water), C_w (in Btu/h · °F) is

$$C_w = 60\dot{m}_w c_{pw} = 500\dot{V}_{gal} c_{pw} \qquad (12.14)$$

In Eq. (12.13), C_{min} is the smaller value between C_a and C_w, in Btu/h · °F.

In a sensible cooling coil using chilled water as coolant, $\dot{m}_a c_{pa}$ is usually smaller than $\dot{m}_w c_{pw}$. Then,

$$\epsilon = \frac{T_{ae} - T_{al}}{T_{ae} - T_{we}} \qquad (12.15)$$

In Eq. (12.15), T_{al} is usually the required off-coil temperature during the design of a sensible cooling system. Most finned coils are actually a combination of counterflow and crossflow arrangement. Because the effectiveness of a finned coil with a counterflow arrangement is nearly the same as that of a crossflow arrangement, the effectiveness can be calculated as

$$\epsilon = \frac{1 - \exp[-NTU(1 - C)]}{1 - C\exp[-NTU(1 - C)]} \qquad (12.16)$$

where $C = C_{min}/C_{max}$
C_{max} = the greater value between C_a and C_w, Btu/h · °F

Number of transfer units (NTU), dimensionless, actually indicates the size of the coil and can be calculated as

$$NTU = \frac{UA}{C_{min}} = \frac{U_o A_o}{C_a} = \frac{U_i A_i}{C_a} \qquad (12.17)$$

After calculating ϵ of the sensible cooling coil from Eq. (12.16), the sensible cooling coil load can be calculated as

$$q_{cs} = 60\dot{V}_a \rho_a c_{pa}(T_{ae} - T_{we})\epsilon \qquad (12.18)$$

From the energy balances between the moist air and chilled water, the temperature of conditioned air and chilled water leaving the coil T_{al} and T_{wl} (in °F) can be calculated as

$$T_{al} = T_{ae} - \epsilon(T_{ae} - T_{we})$$
$$T_{wl} = T_{we} + C(T_{ae} - T_{al}) \qquad (12.19)$$

Heating Coils

In heating coils, only sensible heat change occurs during the heating process. The humidity ratio w remains constant unless humidifiers are installed before or after the heating coil. For steam heating coils, the steam temperature $T_{steam} \gg T_{ae}$. At the same time, T_{steam} remains constant during the heating process. Therefore, C_{max} equals infinity, and $C_{min}/C_{max} = 0$.

The system performance of both water heating and steam heating coils can also be determined from Eqs. (12.13) through (12.19), in which the hot fluids are hot water or steam and moist air is cold fluid.

Fluid Velocity and Pressure Drop

Air velocity calculated based on the face area of a finned-tube coil v_a is a primary factor that determines the effectiveness of heat transfer, carryover of droplets of condensate in wet coils, air-side pressure drop, and energy consumption of the system. For dry coils, there is no danger of carryover. Their maximum face velocity is usually limited to less than 800 fpm. The maximum air velocity calculated based on the minimum free flow area may be as high as 1400 fpm. For coils used in terminal units such as floor-mounted fan-coil units, face velocity is usually around 200 fpm so that the pressure drop across the coil is low.

In addition to the face velocity v_a, air-side pressure drop Δp_a depends also on fin and tube configuration. For dry coils with a fin spacing of 12 fins/in., at a face velocity $v_a = 600$ fpm, Δp_a may vary from 0.1 to 0.2 in. WG per row depth.

Selection of water velocity and $\dot{V}_{gal}$, in gpm, for a finned-tube coil of a given tube inner diameter and number of water circuits is closely related to the temperature rise or drop. For a finned-tube coil with a face area of 1 ft^2, a full serpentine four-row coil may needs 4.8 gpm of chilled water at a water temperature increase around 10°F, whereas a $\frac{1}{2}$ serpentine coil need only 2.4 gpm and has a temperature increase of 20°F. Heat transfer, pressure drop, erosion, noise, energy, maintenance space, and initial cost should be considered during the selection of the temperature increase or drop and its corresponding water volume flow $\dot{V}_{gal}$ and water velocity. For finned-tube coils, a temperature increase of 10 to 20°F is generally used. Water-side pressure drop is usually limited to 10 psi, about 22.5 ft WG. A water velocity between 2 and 6 ft/s and a pressure drop of 10 ft WG should be maintained for reasonable pump power consumption.

Dry Coil at Part-Load Operation

Because the sensible cooling and heating processes are horizontal lines on the psychrometric chart, if the entering chilled water temperature or hot water temperature remains the same at part-load as in full-load operation, the reduction of water flow rates at part-load tends to shorten the horizontal lines from *el* to *elp* (as shown in Fig. 12.6 on p. 12.11) to maintain the required space temperature at part-load operation.

Coil Construction Parameters

During the analysis and calculation of the performance of a water-cooling, water-heating, or DX-coil, A_o/A_i, fin spacing, and tube spacing are the coil construction parameters that determine coil performance. The following coil construction parameters have been used by many manufacturers to optimize rate of heat transfer, air-side pressure drop, and manufacturing cost:

Longitudinal tube spacing S_L	1.083 in.
Transverse tube spacing S_T	1.25 in.
Outside nominal diameter of copper tube D_o	0.5 in.
Aluminum fin thickness F_t	0.006 in.

Table 12.1 lists the coil construction parameters for smooth fin, based on the data given above. Parameters are given for a face area of 1 ft² and a coil depth of one row, or 1.083 in. If the edge effect, both in longitudinal and traverse direction, is included, then the values of A_o/A_p, A_o/A_i, and F_s may be slightly higher. For corrugated fins, A_o/A_p, A_o/A_i, and F_s should be multiplied by a factor $F_{c.a}$ of 1.1 to 1.20.

The ratio of free flow (or minimum flow) area to the face area A_{min}/A_o is usually between 0.54 and 0.6. For a smaller fin spacing, such as 12 or 14 fins/in., $A_{min}/A_o = 0.58$ to 0.60.

Example 12.1. A counterflow sensible cooling coil using corrugated plate fins has the following construction and operating data:

TABLE 12.1
Finned-tube coil construction parameters

Outside diameter of copper tude D_o	0.528 in.
Inner diameter of copper tude D_i	0.496 in.
Aluminum fin thickness F_t	0.006 in.
Longitudinal tube spacing	1.083 in.
Transverse tube spacing	1.25 in.

Fin spacing S_f						
Fins/in.	in.	$A_o/A_p^\dagger$	A_o/A_i	A_f^*/A_o	F_s	S_f/F_t
8	0.125	7.85	7.95	0.873	9.91	20.8
10	0.100	9.68	9.68	0.896	12.07	16.7
12	0.0833	11.54	11.40	0.913	14.21	13.9
14	0.0714	15.92	13.17	0.925	16.37	11.9

$^*A_f$–Area of the fins
$^\dagger A_p$–Area of the primary surfaces (outside surface of the copper tubes)
Note: For corrugated plate fins, A_o/A_p, A_o/A_i, and F_s should be multiplied by a factor of 1.10 to 1.20

Face area (finned length × finned height)	10 ft²
Fin spacing	12 fins/in.
Number of rows	6 rows
Outside diameter of copper tubes	0.528 in.
Inside diameter of copper tubes	0.496 in.
Entering dry bulb temperature	90°F
Entering chilled water temperature	60°F
Face air velocity	600 fpm
Chilled water flow rate	28 gpm
Specific heat:	
moist air	0.243 Btu/lb · °F
water	1 Btu/lb · °F
Air density	0.075 lb/ft³
Water density	62.4 lb/ft³
At a temperature of 65°F	
air dynamic viscosity μ_a	0.0437 lb/ft · h
water dynamic viscosity μ_w	2.46 lb/ft · h
water thermal conductivity k	0.348 Btu/h · ft · °F

Calculate

1. Outer surface heat transfer coefficient of coil
2. Inner surface heat transfer coefficient of coil
3. Fin surface efficiency
4. Effectiveness of sensible cooling coil
5. Sensible cooling coil's load
6. Off-coil air temperature and temperature of chilled water leaving the cooling coil

Solution

1. For corrugated fins, use a multiplying factor $F_{c.a} = 1.15$. Then, from Table 12.1, parameter $F_s = 14.21$. For a six-row coil with a surface area of 10 ft², the outer surface area is calculated as

$$A_o = 14.21 \times 1.15 \times 6 \times 10 = 981 \text{ ft}^2$$

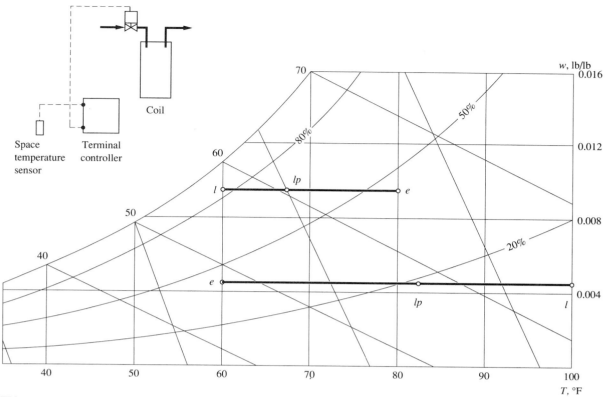

FIGURE 12.6 Dry coils at part-load operation.

From Table 12.1, $A_o/A_i = 11.4 \times 1.15$. The inner surface area of this sensible cooling coil is calculated as

$$A_i = \frac{981}{11.4 \times 1.15} = 74.9 \text{ ft}^2$$

For a fin spacing of 12 fins/in., $A_{\min}/A_a = 0.58$. Therefore, from Eq. (12.8), air mass velocity at minimum flow area

$$G_c = 60\rho v_{\min} = \frac{60 \times 0.075 \times 600}{0.58}$$

$$= 4655 \text{ lb/ft}^2 \cdot \text{h}$$

and the Reynolds number, based on the outside diameter of the tubes, is

$$\text{Re}_D = G_c D_o / \mu = \frac{4655 \times 0.528}{0.0437 \times 12} = 4687$$

Also, parameter JP is calculated as

$$\text{JP} = \text{Re}_D^{-0.4} \left(\frac{A_o}{A_p}\right)^{-0.15}$$

$$= (4687) - 0.4(13.27) - 0.15 = 0.0231$$

From Eq. (12.7), at a JP $= 0.0231$,

$$\left(\frac{h_o}{G_c c_{pa}}\right) \text{Pr}^{\frac{2}{3}} = 0.00125 + 0.27 \times 0.0231$$

$$= 0.00748.$$

For smooth fins,

$$h_{o.s} = 0.00748 \times 4655 \times 0.243 \left(\frac{1}{0.71}\right)$$

$$= 11.92 \text{ Btu/h} \cdot \text{ft}^2 \cdot {}^\circ\text{F}$$

For corrugated fins, a multiplying factor $F_{\text{cor}} = 1.20$ is used to account for the increase in turbulence. Therefore, the outer surface heat transfer coefficient is calculated as

$$h_o = 11.92 \times 1.2 = 14.3 \text{ Btu/h} \cdot \text{ft}^2 \cdot {}^\circ\text{F}$$

2. Assume a water velocity of 4.7 ft/s inside the copper tubes. Then, the Reynolds number of water, based on the inner diameter of copper tube, is

$$\text{Re}_D = \frac{\rho v_w D_i}{\mu_w} = \frac{62.4 \times 4.7 \times 3600 \times 0.496}{2.46 \times 12}$$

$$= 17740$$

And the Prandtl number for water is

$$\text{Pr} = \frac{\mu c_{pw}}{k} = \frac{2.46 \times 1}{0.348} = 7.07$$

From Eq. (12.9), the inner surface heat transfer coefficient can be calculated as

$$h_i = 0.023 \left(\frac{k}{D_i}\right) \text{Re}_D^{0.8} \text{Pr}^{0.4}$$

$$= 0.023 \left(\frac{0.348 \times 12}{0.496}\right) (17740)0.8(7.07)0.4$$

$$= 1061 \text{ Btu/h} \cdot \text{ft}^2 \cdot {}^\circ\text{F}$$

3. For a coil with a longitudinal spacing equal to 1.083 in., $S_T = 1.25$ in., $r_o = 1/2(0.528)$ in., $k_f = 100$ Btu/h · ft · °F, and copper tubes in staggered arrangement, $W = 0.625$ in., and $S_L = 0.542$ in. From Fig. 12.5, at a $W/r_o = 0.625/0.264 = 2.37$, and $S_L/W = 0.87$, $\varphi_{f.max}$ is about 0.5. Then, from Eq. (12.11)

$$R_{f.max} = \frac{\varphi_{f.max}W^2}{F_t k_f} = \frac{0.5 \times (0.625)^2}{0.006 \times 100 \times 12}$$

$$= 0.0271$$

From Eq. (12.12), the maximum possible value of η_f is calculated as

$$\eta_f = \frac{1}{1 + h_o R_{f.max}} = \frac{1}{1 + 14.3 \times 0.0271}$$

$$= 0.721$$

Assuming $\varphi_f / \varphi_{f.max} = 0.9$, then $R_f = 0.9 \times 0.0271 = 0.0244$. Therefore, fin efficiency is calculated as

$$\eta_f = \frac{1}{1 + 14.3 \times 0.0244} = 0.741$$

From Eq. (12.6), fin surface efficiency is calculated as

$$\eta_s = 1 - \left(\frac{A_f}{A_o}\right)(1 - \eta_f)$$

$$= 1 - 0.913(1 - 0.741)$$

$$= 0.763$$

4. From given data, heat capacity rate rates are calculated as

$$C_a = 60 \dot{V}_a \rho_a c_{pa}$$

$$= 60 \times 600 \times 10 \times 0.075 \times 0.243$$

$$= 6561 \text{ Btu/h} \cdot °F$$

$$C_w = 500 \dot{V}_{gal} c_{pw} = 500 \times 30 \times 1$$

$$= 15,000 \text{ Btu/h} \cdot °F$$

Minimum to maximum ratio is

$$C = \frac{C_a}{C_w} = \frac{6561}{15,000} = 0.4374$$

Then, from Eq. (12.5),

$$U_o A_o = \frac{1}{\dfrac{1}{\eta_s h_o A_o} + \dfrac{1}{h_i A_i}}$$

$$= \frac{1}{\dfrac{1}{0.763 \times 14.3 \times 981} + \dfrac{1}{1061 \times 74.9}}$$

$$= 9432 \text{ Btu/h} \cdot °F$$

From Eq. (12.17), the number of transfer units is

$$\text{NTU} = \frac{U_o A_o}{C_a} = \frac{9432}{6561} = 1.438$$

Also,

$$\exp[-\text{NTU}(1 - C)] = \exp[-1.438(1 - 0.4686)]$$

$$= 0.4657$$

From Eq. (12.16), the effectiveness of the sensible cooling coil is

$$\epsilon = \frac{1 - \exp[-\text{NTU}(1 - C)]}{1 - C \exp[-\text{NTU}(1 - C)]}$$

$$= \frac{1 - 0.4657}{1 - (0.4374 \times 0.4657)} = 0.6834$$

5. Then, from Eq. (12.18), the sensible cooling coil's load is

$$q_{cs} = 60 \dot{V}_a \rho_a c_{pa}(T_{ae} - T_{we})\epsilon$$

$$= 60 \times 600 \times 10 \times 0.075 \times 0.243(90 - 60)0.6834$$

$$= 134,522 \text{ Btu/h}$$

6. From Eq. (12.19), the temperature of conditioned air leaving the sensible cooling coil is

$$T_{al} = T_{ae} - \epsilon(T_{ae} - T_{we})$$

$$= 90 - 0.6834(90 - 60)$$

$$= 69.5°F$$

The temperature of chilled water leaving the coil is

$$T_{wl} = T_{we} + C(T_{ae} - T_{al})$$

$$= 60 + 0.4686(90 - 69.5) = 69.6°F$$

12.3   DX-COILS—WET COILS

In a direct-expansion or dry-expansion DX-coil for a comfort air conditioning system, the evaporating temperature T_{ev} (in °F) of refrigerant HCFC–22 or HFC–134a inside the coil tubes is usually between 37 and 52°F. In such a T_{ev}, the surface temperature of the coil is usually lower than the dew point of the air entering the coil. Condensation occurs on the outside surface of the DX-coil, so the coil becomes a *wet coil*. The air conditioning process of a DX-coil is always a cooling and dehumidifying process as well as a heat and mass transfer process.

Construction and Installation

Copper tubes and aluminum fins are widely used for DX-coils using halocarbon refrigerants. The diameter of the copper tubes is usually between $\frac{3}{8}$ and $\frac{5}{8}$ in. Aluminum fins are typically 0.006 in. thick. Corrugated plate fins are often used for additional turbulence and increase in heat transfer. On the inner surface of the copper tubes, micro-fins with a spacing 60 fins per inch and a height of 0.008 in. are widely used to enhance boiling heat transfer.

For even coolant distribution and to reduce pressure drop, flow paths of refrigerant in a DX-coil are always divided into a number of refrigerant circuits, according to the finned height of the coil. Refrigerant is usually supplied through a thermostatic expansion valve. It then flows through a distributor, which distributes the refrigerant evenly, as shown in Fig. 12.7.

Venturi type distributors are usually used. The distributor is connected to various circuits through copper distributor tubes of diameters typically between $\frac{1}{4}$ and $\frac{5}{16}$ in. Equal length of distributor tube and approximately equal circuit lengths ensure even distribution of refrigerant.

To create a counterflow arrangement between air and refrigerant, the suction header should be installed nearest to the air entering side of the coil.

For a DX-coil, copper finned tubes should be reasonably level to ensure proper air venting and condensate drainage. When two or more DX-coils are banked vertically, there must be an individual drain pipe and drain pan for each coil. The drain pipe of the top drain pan should be connected to the lower main drain pan. The lower main drain pan should extend beyond the coil casing a minimum of 10 in. for the overflow of water from the top. The lower main drain pan should be elevated to provide free drainage of condensate, and a trap should be installed to overcome the pressure difference between the coil section and the ambient air.

Two-Region Model

Refrigerant enters the DX-coil at a liquid-vapor two-phase state. It is cooled to evaporating temperature T_{ev} and gradually evaporated along the flow paths until it is completely vaporized. Because of the pressure drop of the refrigerant as it flows through the coil, the evaporating temperature T_{ev} gradually drops along the refrigerant circuit as the evaporating pressure decreases, as shown in Fig. 12.7.

For simplification, the flow path of the refrigerant and its corresponding outer surface in a DX-coil is divided into two regions.

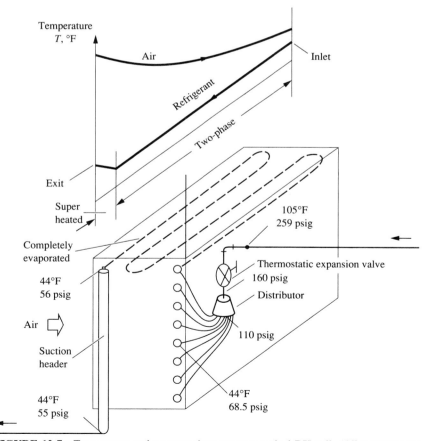

FIGURE 12.7 Temperature and pressure drop across a typical DX-coil. (All pressures are measured at gauge pressure.)

TWO-PHASE REGION. In the two-phase region, the quality of the refrigerant $x_r < 1$. The pressure drop of the liquid refrigerant is negligible and liquid refrigerant is evaporated at a uniform temperature.

In the two-phase region, the boiling heat transfer coefficient h_{boil} of refrigerant HCFC–22 (in Btu/h · ft² · °F) inside a DX-coil is mainly a function of the mass flow rate and quality of the refrigerant. Figure 12.8 shows the boiling heat transfer coefficient of HCFC–22 in a tube with an outside diameter of 0.665 in. at 40°F when the quality of refrigerant is varied.

Schlager et al. (1989) found that naphthenic mineral oil with a viscosity of 150 SSU enhances the surface heat transfer coefficient of refrigerant HCFC–22 inside a microfin tube up to 25 percent at an oil concentration of 3 percent. For 300 SSU oil, the surface heat transfer coefficient inside the micro-fin tube increases slightly and then decreases about 20 percent when the oil concentration increases from 0 to 3 percent.

SUPERHEATED REGION. In the superheated region, the quality of refrigerant $x_r = 1$. The refrigerant-side heat transfer coefficient is low. Because of the relatively high outer surface temperature caused by the superheated refrigerant, the outer surface of the DX-coil corresponding to the superheated region may be dry.

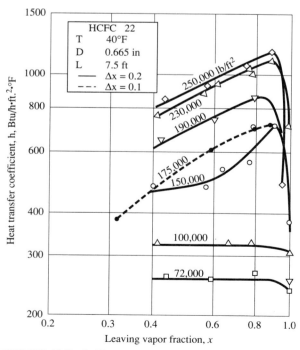

FIGURE 12.8 Boiling heat transfer coefficient of HCFC–22. (*Source: ASHRAE Handbook 1981, Fundamentals.* Reprinted with permission.)

The superheated region is small. For simplification, the rate of heat transfer (including both sensible and latent heat) of the whole DX-coil is often multiplied by a degrading factor F_{super} to account for the existence of the superheated region.

Simultaneous Heat and Mass Transfer

In a DX-coil, the driving potentials for the simultaneous heat and mass transfer during cooling and dehumidifying are the enthalpy difference between the ambient air and the saturated air film at the interface of condensate and air (in Btu/lb) and the temperature difference between the air and the evaporating refrigerant inside the tubes.

This heat and mass transfer $q_{c.wet}$, in Btu/h, is more conveniently calculated as

$$q_{c.wet} = \epsilon_{wet} \dot{V}_a \rho_a (h_{ae} - h_{s.r}) \qquad (12.20)$$

where h_{ae} = enthalpy of entering air, Btu/lb
$h_{s.r}$ = enthalpy of saturated air film at the coil surface corresponding to the evaporating temperature of the refrigerant inside the tubes, Btu/lb

DX-Coil Effectiveness

Because the heat energy of condensate is small compared to that of air and water streams it can be ignored. As in Eq. (12.15), wet coil effectiveness ϵ_{wet} can be calculated as

$$\epsilon_{wet} = \frac{h_{ae} - h_{al}}{h_{ae} - h_{s.r}}$$

or

$$h_{al} = h_{ae} - \epsilon(h_{ae} - h_{s.r}) \qquad (12.21)$$

where h_{al} = enthalpy of air leaving the coil, Btu/lb.

After h_{al} is determined, the dry bulb and wet bulb of the leaving air can be found from the cooling and dehumidifying curve of the psychrometric chart with the same entering air condition.

For convenience, the refrigerant in a DX-coil is assumed to evaporate at a constant temperature, so $C_{min}/C_{max} = 0$. From Eq. (12.16), the wet coil effectiveness is calculated as

$$\epsilon_{wet} = [1 - \exp(-NTU)] \qquad (12.22)$$

The number of transfer units for a DX-coil can also be calculated by Eq. (12.17) as

$$NTU = \frac{U_o A_o}{C_a}$$

Also from Eq. (12.5),

$$U_o A_o = \frac{1}{1/(\eta_s h_o A_o) + 1/(h_i A_i)} \quad (12.23)$$

Face Velocity and Air-Side Pressure Drop of Wet Coil

Selection of face velocity for wet coils must consider the carryover of the water droplets, the influence of coil face velocity on heat transfer coefficients, the pressure drop of the coil, and the size of the coil. If air velocity distribution is even over the entire surface of the DX-coil, the carryover is usually not significant at face velocities lower than 500 fpm for smooth plate fins or lower than 600 fpm for corrugated fins. When the supply fan is located upstream from the DX-coil (a *blow-through* fan and coil arrangement) there may be areas where the face velocity is far higher than in the rest of the coil.

For wet DX-coils, a rough estimate of air-side pressure drop for 12 fins/in. corrugated fins is as follows:

| Face velocity | Fins configuration | |
	Simple	Complicated
500 fpm	0.10 in. WG/row	0.15 in. WG/row
600 fpm	0.24 in. WG/row	0.3 in. WG/row

The air-side pressure drop for a 4-row 12 fins/in. corrugated fin coil from one manufacturer may be 50 percent higher than from another manufacturer, even at the same operating conditions and same unit cooling capacity. It is important to carefully compare the coil performances of different manufacturers.

The lower limit of coil face velocity at design load depends on initial and operating costs. For DX-coils in VAV systems, reduction of face velocity in part-load operation should also be considered. Face velocities between 300 and 450 may be justified in many cases.

DX-Coils at Part-Load Operation

The refrigerant circuits in a DX-coil of capacity greater than 15 tons are usually divided into two or more sections, each with its own expansion valve, distributor, and suction header. The refrigerant circuits are controlled in three ways, as shown in Fig. 12.9.

FACE CONTROL. The refrigerant circuits are divided into upper and lower sections, as shown in Fig. 12.9a. One section can be shut-off by deenergizing its solenoid valve during capacity reduction. Face control reduces the temperature of the conditioned air, and requires thorough mixing of conditioned and bypass airstreams downstream from the DX-coil.

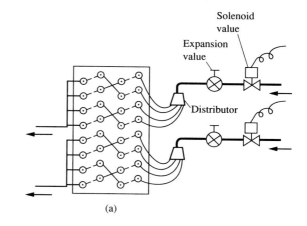

(a)

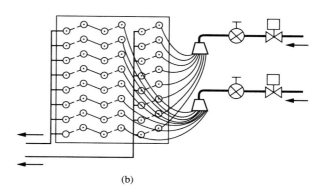

(b)

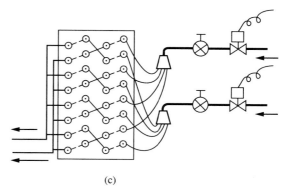

(c)

FIGURE 12.9 Control of DX-coils at part-load operation: (a) face control (b) row control (c) intertwined face control.

ROW-CONTROL. Row control is more effective when applied to a six-row DX-coil, as shown in Fig. 12.9b. One distributor connects to the refrigerant circuits of two rows, another distributor connects to the remaining four rows. The active refrigerant circuits can then be operated at 33, 66 or 100 percent capacity.

INTERTWINED FACE CONTROL. An intertwined face control (shown in Fig. 12.9c) always has a full face active coil even when the refrigerant supply to one of the distributors is cut off by the solenoid valve. When one distributor is inactive, the fin efficiency η_f and the heat capacitance ratio of wet coil C_{wet} are changed, and the capacity of the DX-coil is reduced to about 70 percent.

In the design and selection of a DX-coil for a comfort air conditioning system, it is important to check the minimum refrigerant temperature in the DX-coil during part-load operation. The minimum saturated temperature of the refrigerant $T_{r.min}$, in °F, for a DX-coil at a face velocity greater than 500 fpm must not drop below 26°F. Otherwise, frost will form on the coil surface. Frost formation blocks the air passage and severely reduces the rate of heat transfer. At part-load operation, the evaporating temperature is lowered because less refrigerant is evaporated.

For DX-coils using HCFC–22 or other oil miscible refrigerants, the coil should be selected so that the refrigerant–oil mixture can be returned in a pipe riser at minimum load operation. This will be covered in Chapter 21.

When a DX-coil is installed in a variable-air-volume system, the air volume flow rate and face velocity are decreased during part-load operation. Figure 12.10 shows the relationship between cooling capacity of the DX-coil at a specific face velocity to the cooling capacity at a face velocity of 600 fpm, $(q_c/A_a)/(q_c/A_a)_{600}$. The curve is plotted according to data from the catalogs of three manufacturers at an air entering condition of 80°F dry bulb and 67°F wet bulb, and 45°F suction temperature, or 45°F entering chilled water temperature and 55°F leaving. When the face velocity drops from 600 fpm to 400 fpm, $(q_c/A_a)/(q_c/A_a)$ drops from 1.0 to about 0.8.

For a VAV packaged system using a DX-coil, it is important to have effective multistep capacity unloading and refrigeration circuit cutoff controls during part-load to maintain T_{ev} above 26°F. This will be discussed in Chapters 21 and 29.

Selection of DX-Coils

In the design of an air system, it is often necessary to select a DX-coil at a given coil's load so that the conditioned air leaves the coil at specified conditions.

For DX-coils used in comfort air conditioning systems, many manufacturers are now using 0.5-in. outside diameter copper tubes. The longitudinal tube spacing is usually about 1 in. and transversal spacing 1.25 in. Therefore, parameters F_s, A_o/A_p, and A_o/A_i listed in Table 12.1 can still be applied to DX-coils.

Prior to the selection of a DX-coil, the supply air volume flow rate $\dot{V}_s$ (in cfm) and the coil's load q_c (in Btu/h) at design conditions are usually already calculated. The procedure to select a DX-coil from the manufacturer's catalog according to the required cooling and dehumidifying capacity per unit face area q_c/A_a, in MBtu/h · ft^2 (1000 Btu/h · ft^2), is outlined below. See Table 12.2 for a typical section of this kind of catalog.

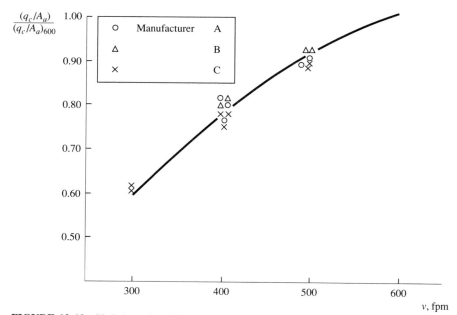

FIGURE 12.10 Variation of cooling capacity vs. face velocity.

TABLE 12.2
DX-coil ratings (R–22, 80°F EDB, 67°F EWB)

Suct Temp, °F	Rows	Fin no.	Face velocity, fpm											
			400			500			600			700		
			MBH	LDB	LWB	MBH	LDB	LWB	MBH	LDB	LWB	MBH	LDB	LWB
35	4	8	20.5	50.9	49.7	23.0	53.2	51.8	25.3	54.9	53.3	27.1	56.4	54.6
	5		23.2	47.5	46.9	26.6	49.7	48.9	29.3	51.7	50.7	31.6	53.3	52.1
	6		25.4	44.9	44.6	29.4	47.1	46.6	32.8	48.9	48.3	35.5	50.6	49.9
	8		28.4	41.4	41.2	33.5	43.2	43.0	38.0	44.8	44.6	41.9	46.4	46.1
	4	10	22.2	48.6	47.9	25.1	51.0	50.1	27.5	52.9	51.9	29.5	54.4	53.3
	5		24.9	45.4	45.0	28.7	47.6	47.2	31.8	49.5	49.0	34.3	51.3	50.6
	6		27.0	43.0	42.8	31.4	45.1	44.8	35.1	47.0	46.7	38.3	48.6	48.3
	8		29.7	39.8	39.6	35.2	41.7	41.5	40.1	43.3	43.1	44.4	44.9	44.7
	4	12	23.6	46.9	46.5	26.8	49.3	48.8	29.2	51.3	50.7	31.4	52.9	52.2
	5		26.1	43.9	43.7	30.2	46.1	45.9	33.5	48.1	47.8	36.4	49.7	49.4
	6		28.1	41.8	41.6	32.8	43.9	43.7	36.9	45.8	45.6	40.3	47.5	47.3
	8		30.5	38.7	38.5	36.4	40.5	40.3	41.6	42.2	42.0	46.2	43.6	43.4
40	4	8	17.6	53.8	52.5	19.9	55.7	54.2	21.6	57.3	55.5	23.1	58.7	56.7
	5		20.0	50.9	50.2	22.9	52.8	51.9	25.3	54.4	53.3	27.2	55.8	54.5
	6		21.9	48.8	48.2	25.3	50.5	49.9	28.2	52.1	51.4	30.7	53.4	52.6
	8		24.7	45.6	45.3	29.0	47.1	46.9	32.8	48.6	48.3	36.1	49.9	49.6
	4	10	19.1	51.8	51.1	21.6	53.8	52.8	23.6	55.4	54.3	25.3	56.8	55.5
	5		21.5	49.0	48.6	24.7	51.0	50.5	27.2	52.7	52.0	29.5	54.0	53.3
	6		23.3	47.0	46.8	27.0	48.8	48.5	30.2	50.4	50.1	32.9	51.8	51.4
	8		25.8	44.3	44.1	30.6	45.8	45.6	34.7	47.2	47.0	38.3	48.6	48.4
	4	12	20.2	50.4	49.9	23.0	52.4	51.6	25.1	54.1	53.3	26.9	55.4	54.6
	5		22.6	47.9	47.5	26.0	49.7	49.4	28.8	51.3	51.0	31.2	52.7	52.3
	6		24.3	46.0	45.8	28.3	47.7	47.5	31.7	49.2	49.0	34.6	50.6	50.4
	8		26.5	43.4	43.2	31.6	44.8	44.6	36.0	46.3	46.1	39.9	47.5	47.3
45	4	8	14.6	56.7	55.3	16.5	58.3	56.6	17.9	59.7	57.7	19.2	60.7	58.6
	5		16.7	54.4	53.4	19.0	55.8	54.8	20.8	57.2	56.0	22.5	58.3	56.9
	6		18.3	52.3	51.8	21.1	54.0	53.2	23.3	55.2	54.4	25.4	56.3	55.4
	8		20.7	49.8	49.6	24.2	51.0	50.8	27.3	52.3	51.9	30.0	53.5	52.9
	4	10	15.9	55.0	54.1	17.9	56.7	55.6	19.5	58.0	56.7	20.9	59.1	57.7
	5		17.9	52.8	52.2	20.5	54.3	53.7	22.5	55.6	54.9	24.3	56.8	55.9
	6		19.5	50.9	50.7	22.5	52.5	52.1	25.1	53.7	53.3	27.3	54.9	54.4
	8		21.7	48.7	48.5	25.6	50.0	49.8	29.0	51.2	51.0	31.9	52.2	51.9
	4	12	16.8	54.0	53.2	19.0	55.4	54.7	20.7	56.8	56.0	22.3	57.9	57.0
	5		18.8	51.6	51.3	21.6	53.3	52.8	23.8	54.5	54.1	25.8	55.6	55.2
	6		20.4	50.1	49.9	23.6	51.5	51.2	26.3	52.9	52.5	28.8	53.8	53.6
	8		22.4	47.9	47.7	26.5	49.1	48.9	30.1	50.3	50.1	33.4	51.4	51.2

Source: York International Corporation. Reprinted with permission.
EDB   Entering Dry bulb;   LDB   Leaving dry bulb;
EWB   Entering wet bulb;   LWB   Leaving wet bulb;
MBH 1000 Btu/h · ft^2 of coil's face area
Suct Temp. Suction temperature of refrigerant HCFC–22
Fin No.   Fins per inch.

1. Choose an optimum face velocity v_a, in fpm, at design load. As mentioned before, for corrugated fins when the supply fan is located downstream from the coil, v_a between 400 and 550 fpm is suitable.

   After v_a has been determined, the face area of DX-coil can be calculated as

$$A_a = \frac{\dot{V}_s}{v_a} \qquad (12.24)$$

2. Calculate the unit coil capacity q_c/A_a, unit sensible coil capacity q_{cs}/A_a (in Btu/h · ft²), and the sensible heat ratio of this cooling and dehumidifying process SHR$_c$ at design load.

3. Select an optimum saturated suction temperature T_{suc}, in °F. A higher T_{suc} means a smaller log-mean temperature difference between the air and refrigerant, a lower unit coil capacity q_c/A_a, a large DX-coil, and a lower power input to the compressor. A lower T_{suc} means a greater log-mean temperature difference, a higher unit coil capacity q_c/A_a, a smaller coil, and greater compressor energy use. For comfort air conditioning systems, T_{suc} should be 35°F or higher to prevent high compressor power input. T_{suc} is usually 2°F lower than T_{ev}. A T_{suc} between 40 and 45°F is often appropriate.

4. Select the optimum number of rows and fins per inch. For example, the air-side pressure drop of four-row, 12 fins/in. coil is about 10 percent lower than for a six-row, 8 fins/in. coil, and the unit coil capacity of the four-row coil is less than two percent lower than that of the six-row coil. Also, the initial cost of a six-row, 8 fins/in. coil is higher than that of a four-row, 12 fins/in. coil. The trend is to use 12 or 14 fins/in. coil.

   It is better to have an even number of coil rows so that the refrigerant distributor and the suction header can be placed on the same side, which is more convenient for operation and maintenance.

   If cooling and dehumidifying processes in comfort air conditioning need a greater latent cooling load at part-load operation, then a six-row coil has a greater primary surface than a four-row coil and provides higher dehumidifying capacity.

5. Select the DX-coil from the manufacturer's catalog according to the required q_c/A_a. If necessary, vary the number of rows and fins/in. of the coil.

6. Calculate the refrigeration tonnage of the DX-coil. According to the selected finned height, determine the number of refrigerant circuits from the manufacturer's catalog. Check the magnitude of refrigeration tonnage per refrigerant circuit. If it falls between 0.8 and 2 tons, it is considered suitable.

7. Determine the air-side pressure drop from the data provided by the manufacturer's catalog.

For detailed coil performance analysis and year-round energy estimates, computer aided DX-coil simulation through simultaneous heat and mass transfer and coil effectiveness computation is sometimes necessary.

Example 12.2. A variable-air-volume air system has a supply volume flow rate of 5500 cfm. Air enters the coil at a dry bulb of 80°F and a wet bulb of 67°F. The conditioned air should leave the coil at a dry bulb of 57°F and a wet bulb of 56°F.

1. Select a DX-coil for the packaged unit of this air system.
2. Calculate the unit cooling capacity and sensible cooling capacity of this DX-coil at full-load, using the following data:
   - A four-row, 12 fins/in. coil with the same constructional parameters as in Example 12.1 except number of rows
   - Dynamic viscosity of air at 55°F is 0.043 lb/ft · h
   - Prandtl number for air is 0.71
   - Boiling heat transfer coefficient for HCFC–22 inside the copper tubes is 700 Btu/h · ft² · °F at $T_{ev} = 45$°F
   - Surface effectiveness for wet fins $\eta_s = 0.76$
3. If it is an intertwined face coil and splits into two sections, calculate the cooling capacity at part-load operation when the face velocity has been reduced to 400 fpm and one section is not active.

Solution
1. From psychrometric chart, the enthalpy of entering air is 31.6 Btu/lb and enthalpy of the leaving air is 23.85 Btu/lb. The DX-coil's cooling and dehumidifying load can then be calculated as

$$q_c = 60\dot{V}_s \rho_s (h_{ae} - h_{al})$$
$$= 60 \times 5500 \times 0.075 \,(31.6 - 23.85)$$
$$= 191,810 \text{ Btu/h}$$

Also, the required sensible cooling coil load is

$$q_{cs} = 60\dot{V}_s \rho_s c_{pa}(T_{ae} - T_{al})$$
$$= 60 \times 5500 \times 0.075 \times 0.243(80 - 57)$$
$$= 138,328 \text{ Btu/h}$$

The SHR$_c$ of this cooling and dehumidifying process is therefore

$$\text{SHR}_c = \frac{q_{sc}}{q_c} = \frac{138,328}{191,810} = 0.72$$

2. For a DX-coil using corrugated plate fins, select a face velocity $v_a = 550$ fpm to provide an efficient

heat transfer and a reasonable air-side pressure drop, and avoiding condensate carryover. Face area A_a, in ft^2, for the DX-coil can then be calculated as

$$A_a = \frac{\dot{V}_s}{v_a} = \frac{5500}{550} = 10 \text{ ft}^2$$

From the given values, the required unit cooling capacity q_c / A_a, in MBtu/h, can be calculated as

$$\frac{q_c}{A_a} = \frac{191,810}{(10 \times 1000)} = 19.18 \text{ MBtu/h} \cdot \text{ft}^2$$

The required unit sensible cooling capacity is

$$\frac{q_{cs}}{A_a} = \frac{138,328}{(10 \times 1000)} = 13.83 \text{ MBtu/h}$$

From Table 12.2, select a 4-row and 12 fins/in. coil at 45°F suction temperature. The listed q_c / A_a can be interpolated as

$$\left(\frac{q_c}{A_a}\right)_{550} = \frac{1}{2(19 + 20.7)} = 19.8 \text{ MBtu/h} \cdot \text{ft}^2$$

The unit sensible cooling capacity of this selected DX-coil is

$$\begin{aligned}\left(\frac{q_{cs}}{A_a}\right)_{550} &= 60 v_a \rho_a c_{pa}(T_{ae} - T_{al}) \\ &= 60 \times 550 \times 0.075 \\ &\quad \times 0.243\left(80 - \frac{55.4 + 56.8}{2}\right) \\ &= 14,374 \text{ Btu/h} \cdot \text{ft}^2\end{aligned}$$

Both of them are higher than the required values.

3. At full-load operation, for a face area of 10 ft^2 and a multiplying factor of 1.15 for the corrugated surface, the outer surface area of this DX-coil is

$$A_o = F_s N_r A_a = 14.21 \times 1.15 \times 4 \times 10 = 654 \text{ ft}^2$$

For corrugated fins, $A_o / A_i = 11.4 \times 1.15$. The inner area is therefore

$$A_i = \frac{654}{11.4 \times 1.15} = 49.9 \text{ ft}^2$$

Also,

$$\frac{A_o}{A_p} = 11.54 \times 1.15 = 13.27$$

Because mass velocity of the airstream is given as

$$\begin{aligned}G_c &= \frac{v_a \rho_a}{A_{\min}} = \frac{550 \times 0.075 \times 60}{0.58} \\ &= 4267 \text{ lb/ft}^2 \cdot \text{h}\end{aligned}$$

The Reynolds number of the airstream can then be calculated as

$$\text{Re}_D = \frac{G_c D_o}{\mu} = \frac{4267 \times 0.528}{0.043 \times 12} = 4366$$

Therefore,

$$\begin{aligned}\text{JP} &= \text{Re}_D^{-0.4}\left(\frac{A_o}{A_p}\right)^{-0.15} = (4366)^{-0.4}(13.27)^{-0.15} \\ &= 0.02374\end{aligned}$$

For smooth plate fins,

$$\begin{aligned}\left(\frac{h_{o.s}}{G_c c_{pa}}\right)\text{Pr}^{2/3} &= 0.00125 + 0.27 \text{ JP} \\ &= 0.00125 + (0.27 \times 0.02374) \\ &= 0.00766\end{aligned}$$

$$\begin{aligned}h_{o.s} &= 4267 \times 0.243 \times \left(\frac{1}{0.71}\right) \times 0.00766 \\ &= 11.19 \text{ Btu/h} \cdot \text{ft}^2 \cdot \text{°F}\end{aligned}$$

For corrugated fins, if $F_{cor} = 1.20$, then

$$h_o = 11.19 \times 1.2 = 13.43 \text{ Btu/h} \cdot \text{ft}^2 \cdot \text{°F}$$

4. As given, for a wet surface $\eta_s = 0.76$. Then, from Eq. (12.23)

$$\begin{aligned}U_o A_o &= \frac{1}{\dfrac{1}{\eta_s h_o A_o} + \dfrac{1}{h_r A_i}} \\ &= \frac{1}{\dfrac{1}{0.76 \times 13.43 \times 654} + \dfrac{1}{700 \times 49.9}} \\ &= 5604 \text{ Btu/h} \cdot \text{°F}\end{aligned}$$

The heat capacity rate for conditioned air is calculated as

$$\begin{aligned}C_a &= 60 \dot{V}_a \rho_a c_{pa} \\ &= 60 \times 550 \times 10 \times 0.075 \times 0.243 \\ &= 6014 \text{ Btu/h} \cdot \text{°F}\end{aligned}$$

The number of transfer units is calculated as

$$\text{NTU} = \frac{U_o A_o}{C_a} = \frac{5604}{6014} = 0.9318$$

From Eq. (12.22), the effectiveness for DX-coil can then be calculated as

$$\begin{aligned}\epsilon_{\text{wet}} &= 1 - \exp(-\text{NTU}) = 1 - e^{-1.023} \\ &= 1 - 0.3938 = 0.606\end{aligned}$$

Considering the influence of superheated region, effectiveness of the DX-coil should be multiplied by a degrading factor $F_{\text{super}} = 0.95$. Then,

$$\epsilon_{\text{DX}} = 0.606 \times 0.95 = 0.576$$

5. From the psychrometric chart, $h_{ae} = 31.6$ Btu/lb and the enthalpy of saturated air film at evaporating

temperature $T_{ev} = 45°F$ is 17.65 Btu/lb. Then, the total cooling capacity of the DX-coil is

$$q_c = 60 v_a A_a \rho_a \epsilon (h_{ae} - h_{s.r})$$
$$= 60 \times 550 \times 10 \times 0.075 \times 0.576$$
$$\times (31.6 - 17.65) = 198,871 \text{ Btu/h}$$

The cooling capacity per ft^2 is

$$\frac{q_c}{A_a} = \frac{198,871}{10 \times 1000} = 19.9 \text{ MBtu/h} \cdot \text{ft}^2$$

which is quite near to the value of 19.8 MBtu/h · ft^2 given in the manufacturer's catalog.

From Eq.(12.21), enthalpy of air leaving the coil is calculated as

$$h_{al} = h_{ae} - \epsilon_{wet}(h_{ae} - h_{s.r})$$
$$= 31.6 - 0.576(31.6 - 17.65)$$
$$= 23.56 \text{ Btu/lb}$$

From the cooling and dehumidifying curve on the psychrometric chart with an entering condition of 80°F dry bulb and 67°F wet bulb, the leaving dry bulb temperature is 56.5°F and the wet bulb is 55.6°F, which are quite near to the data given in the manufacturer's catalog (dry bulb = 56.1°F and wet bulb = 55.4°F).

6. At part-load operation, face velocity has been reduced to 400 fpm. The mass velocity of the airstream becomes

$$G_c = \frac{60 \times 400 \times 0.075}{0.58} = 3103 \text{ lb/ft}^2 \cdot \text{h}$$

and

$$Re_D = \frac{3103 \times 0.044}{0.0437} = 3124$$

When one section is inactive, $A_o/A_p = 23.07$ and

$$JP = (3124)^{-0.4}(23.07)^{-0.15} = 0.02498$$

For smooth plate fins,

$$\left(\frac{h_o}{G_c c_{pa}}\right) Pr^{2/3} = 0.00125 + 0.27 \times 0.02498$$
$$= 0.008$$

$$h_o = 0.008 \times 3124 \times 0.243 \left(\frac{1}{0.71}\right)$$
$$= 8.55 \text{ Btu/h} \cdot \text{ft}^2 \cdot °F$$

For corrugated fins,

$$h_{o.cor} = 8.55 \times 1.2 = 10.26 \text{ Btu/h} \cdot \text{ft}^2 \cdot °F$$

7. If only one section of the coil is active, then $W = 1.25$ in., $W/r_o = 1.25/0.264 = 4.73$, and $S_L/W = 0.542/1.25 = 0.436$. From Fig. 12.5, $\varphi_{f.max}$ may be equal to 0.65. Therefore,

$$R_{f.max} = \frac{0.65 \times (1.25)2}{0.006 \times 100 \times 12}$$
$$= 0.141 \text{ h} \cdot \text{ft}^2 \cdot °F/\text{Btu}$$

and

$$\eta_f = \frac{1}{1 + (14.3 \times 0.141)} = 0.33$$

Let $\varphi_f/\varphi_{f.max} = 0.9$ and $R_f = 0.9 \times 0.141 = 0.1269$ h · ft^2 · °F/Btu. Fin efficiency is then

$$\eta_f = \frac{1}{1 + (14.3 \times 0.1269)} = 0.355$$

and the fin surface efficiency is

$$\eta_s = 1 - 0.913(1 - 0.355) = 0.411$$

Although one of the two sections is inactive, the boiling heat transfer coefficient inside the active refrigerant circuits still can be taken as 700 Btu/h · ft^2 · °F. Therefore, parameter

$$U_o A_o = \frac{1}{\dfrac{1}{0.41 \times 10.26 \times 654} + \dfrac{1}{700 \times 50}}$$
$$= 2377 \text{ Btu/h} \cdot °F$$

8. At part-load,

$$C_a = 60 \times 400 \times 10 \times 0.075 \times 0.243$$
$$= 4374 \text{ Btu/h} \cdot °F$$

and

$$NTU = \frac{2377}{4374} = 0.5434$$

Therefore, effectiveness of the DX-coil at part-load is

$$\epsilon_{wet} = 1 - e^{-0.5434} = 0.4193$$

and the cooling capacity of the DX-coil at part-load is

$$q_{cp} = 60 \times 400 \times 10 \times 0.075 \times 0.4193$$
$$\times (31.6 - 17.65) = 107,550 \text{ Btu/h}$$

Load ratio of the DX-coil is calculated as

$$\text{Load ratio} = \frac{q_{cp}}{q_c} = \frac{107,550}{198,871} = 0.54$$

and the enthalpy of air leaving the DX-coil is

$$h_{al} = 31.6 - 0.4193(31.6 - 17.65)$$
$$= 25.75 \text{ Btu/lb}$$

From the cooling and dehumidifying curve on the psychrometric chart, the leaving air temperature is 61.3°F with a relative humidity of 87 percent.

12.4   WATER COOLING COILS—DRY-WET COILS

As in a DX-coil, the air and water flow arrangement in a water cooling coil is usually a combination of counterflow and crossflow. When chilled water is used as the coolant flowing inside the copper tube of a cooling coil, the outer surface temperature of the coil at the air intake side T_{se} is probably greater than the dew point of the entering air T''_{ae}, that is, $T_{se} > T''_{ae}$. The outer surface temperature of the coil at the air leaving side $T_{sl} < T''_{ae}$. A water cooling coil is then a *dry-wet coil*. It often has a partially dry surface at the air intake side, and partially wet surface at air leaving side.

When the temperature of chilled water entering the coil T_{we} is low, or the dew point temperature of the entering air T''_{ae} is high, and if $T_{se} < T''_{ae}$, the water cooling coil is then a wet coil. The differences between a wet water cooling coil and a wet DX-coil are mainly the lack of a superheated region in a water cooling coil and the variation of chilled water temperature inside a water cooling coil, which is often greater than the variation of evaporating temperature T_{ev} at the two-phase region in a DX-coil.

Dry-Wet Boundary

In a dry-wet coil, there is always a *dry-wet boundary* that divides the coil between the dry surface and the wet surface, as shown in Fig. 12.11. The dry-wet boundary must be determined so that the performance of the dry surface and the wet surface can be calculated separately.

Because the outer surface temperature of the copper tubes at the dry-wet boundary T_{sb} is exactly equal to the dew point of the air entering the coil T''_{ae}, that is, $T_{sb} = T''_{ae}$, the dry-wet boundary can be determined. The dry surface extends from the dry-wet boundary to the air intake side, and the wet surface extends to the air leaving side.

At the dry-wet boundary, if the thermal resistance of the metal tube and the heat energy released by condensate are ignored, then

$$h_{wet} \eta_s A_o (T_{ab} - T_{sb}) = h_i \left(\frac{A_o}{B} \right) (T_{ab} - T_{wb}) \quad (12.25)$$

where T_{ab} = temperature of airstreams at dry-wet boundary, °F

T_{wb} = temperature of chilled water inside the tubes at dry-wet boundary, °F

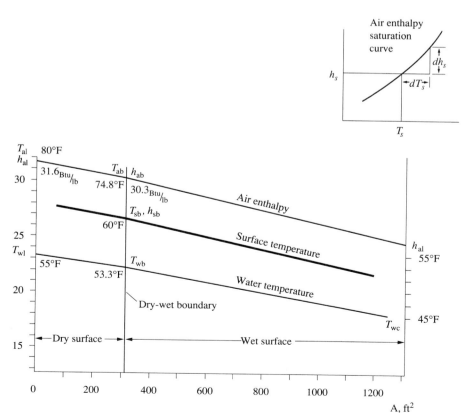

FIGURE 12.11   Dry-wet boundary of a dry-wet coil.

In Eq. (12.25), h_{wet} is the heat transfer coefficient of the wet surface, in Btu/h · ft^2 · °F. According to *ASHRAE Handbook Equipment 1988*,

$$h_{wet} = \frac{m''}{c_{pa}} \cdot \frac{1}{R_{w'}} \qquad (12.26)$$

Here, m'' is the slope of air enthalpy saturation curve $m'' = dh_s/dT_s$, as shown in the upper right-hand corner of Fig. 12.11, where h_s denotes the enthalpy of saturated air film and T_s the saturated temperature.

R_w represents the R-value of the wet film on copper tubes. R_w is nearly equal to $(1/h_o)$. For convenience, their difference is ignored. Then,

$$h_{wet} = \frac{m''}{c_{pa}} h_o \qquad (12.27)$$

In Eq. (12.25), there are two unknowns, T_{sb} and T_{wb}. An additional equation is needed to determine them. As in the dry part of the dry-wet coil, there exists an energy balance between the air and water streams

$$\dot{V}_a \rho_a c_{pa}(T_{ae} - T_{ab}) = 8.33 \dot{V}_{gal} c_{pw}(T_{wl} - T_{wb}) \qquad (12.28)$$

Because the temperature of chilled water leaving the coil T_{wl} is a known value, from Eqs. (12.25) and (12.28), T_{ab} and T_{wb} can then be determined.

Dry Part

The sensible cooling capacity q_{cs} (in Btu/h) of the dry part of the dry-wet coil can be calculated from the left-hand side of Eq. (12.28) as

$$q_{cs} = \dot{V}_a \rho_a c_{pa}(T_{ae} - T_{ab}) \qquad (12.29)$$

The average temperature of the airstream in the dry part $T_{ad.m}$, in °F, is approximately equal to $\frac{1}{2}(T_{ae} + T_{ab})$, and the average temperature of the chilled water corresponding to the dry part of the coil $T_{wd.m}$, in °F, is $\frac{1}{2}(T_{wb} + T_{wl})$. Unit sensible cooling capacity of each ft^2 outer surface area of the dry part of the water-cooling coil, including the tube and fins, can then be calculated as

$$\frac{q_{cs}}{A_{o.d}} = U_o(T_{ad.m} - T_{wd.m})$$

Because U_o can be calculated from Eq. (12.5), the total outer surface area of the dry part of the coil $A_{o.d}$, in ft^2, is

$$A_{o.d} = \frac{q_{cs}}{U_o(T_{ad.m} - T_{wd.m})} \qquad (12.30)$$

Wet Part

As for the DX-coil, the cooling and dehumidifying capacity of the wet part of a dry-wet coil $q_{c.w}$, in Btu/h, can be calculated as

$$q_{c.w} = \epsilon_{wet} \dot{V}_a \rho_a (h_{ab} - h_{s.we}) \qquad (12.31)$$

where $h_{s.we}$ = enthalpy of saturated air film at chilled water entering temperature, °F.

The effectiveness of the wet part of the dry-wet coil ϵ_{wet}, the number of transfer units (NTU), and the heat capacity rates C_a and C_w can be similarly calculated by Eqs. (12.16), (12.17), and (12.14) respectively.

Dry-Wet Coil at Part-Load Operation

When the dry-wet coil of a constant-volume system is at part-load operation, the reduction of the space cooling load results in a drop in space temperature. This causes the control system to modulate the two-way valve at the inlet of the water cooling coil to reduce the mass flow rate of chilled water flowing through the coil, as shown in Fig. 12.12. Because of the reduction of the chilled water flow rate, the following conditions occur:

- The velocity of chilled water inside the coil tubes is reduced and, therefore, the rate of heat transfer between the air and water decreases.
- Although the reduction of the space cooling load tends to lower the temperature of conditioned air leaving the coil T_{al}, T_{al} increases because of the reduction of the chilled water flow rate and has a greater influence than the drop of the space cooling load.
- From the psychrometric analysis shown in Fig. 12.12, the space conditioning line *sp rp* at part-load is shorter because of the reduction of space cooling load.
- The chilled water flow rate is reduced so that the off-coil temperature T_{ccp} and supply temperature T_{sp} match the reduction of space cooling load and maintain a space temperature T_{rp} at part-load.
- The higher dry bulb and dew point temperatures of conditioned air leaving the coil cause an increase in space relative humidity φ_r. Sometimes, φ_r may be raised to 70 percent during part-load operation. This is the main disadvantage in controlling space temperature by means of chilled water flow rate modulation at part-load operation.

Selection of a Dry-Wet Coil

The construction of a water cooling coil, a dry-wet coil, is similar to that of a sensible water cooling

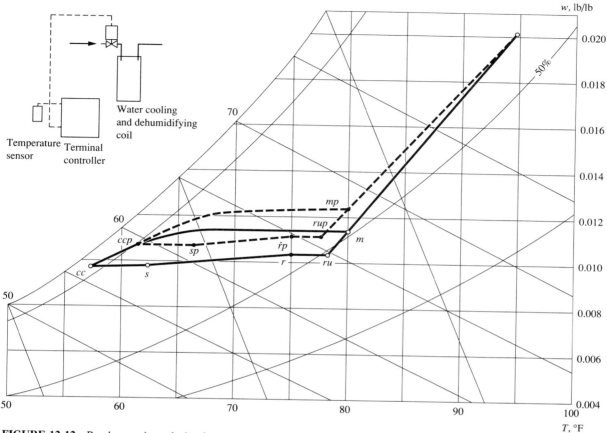

FIGURE 12.12 Psychrometric analysis of part-load operation of a dry-wet coil.

coil except that its face velocity for corrugated fins should not be greater than 550 fpm and condensate drainage must be provided for a dry-wet coil.

As for a DX-coil, the procedure for selecting a water cooling coil according to the manufacturer's data is as follows. See Table 12.3 for a typical listing of catalog data.

1. According to the required supply volume flow rate, choose an optimum face velocity v_a. Calculate the face area A_a of the coil.

2. From the coil's load, calculate the unit cooling and dehumidifying capacity per ft² face area, q_c/A_a.

3. Determine the chilled water temperature entering the coil T_{we}. This is closely related to the required condition of air leaving the coil T_{al}.

4. Select the optimum chilled water temperature rise ΔT_w. A lower ΔT_w means a greater water flow rate, a higher water velocity in the tubes, and a greater heat transfer coefficient. A lower ΔT_w also results in a greater pump power and a higher initial investment in the water system. Water temperature rise ΔT_w also influences air system and refrigeration system performance. It must be

matched with the expected temperature of conditioned air off the coil in the air system as well as the chilled water temperature difference in the chiller.

The current practice is to use a ΔT_w from 12°F to 24°F in order to reduce water flow and save pump power. A detailed analysis will be given in later chapters. When ΔT_w is determined, the $\dot{V}_{gal}$, in gpm, can be calculated accordingly.

5. Check the water velocity inside the tubes v_w, in ft/s, by selecting an appropriate finned width for the water cooling coil. Water velocity between 2 and 6 ft/s (typically between 3 and 5 ft/s at design conditions) is appropriate. As in sensible water cooling coils, the water-side pressure drop should not exceed 10 psi or 22.5 ft WG. Typically, a pressure drop of 10 ft WG across the coil is considered suitable to keep pump power and energy costs reasonable. Many manufacturers provide the formula to calculate v_w at a given finned width and the corresponding tube feeds. Some manufacturers also provide water turbulators, which can be added to the tubes to increase the inner surface heat transfer coefficient if the water velocity is below

TABLE 12.3a
Water cooling coil ratings

45°F Entering water temperature
67°F Entering wet bulb
80°F Entering dry bulb

Water temp. rise °F	FPS	Row	Fin no.	400			500			600			700		
				MBH	LDB	LWB	MBH	LDB	LWB	MBH	LDB	LWB	MBH	LDB	LWB
10	2	4	8	11.9	59.4	57.7	13.3	60.8	58.8	14.5	61.9	59.7	15.5	62.7	60.4
			10	13.0	57.8	56.7	14.6	59.2	57.9	15.9	60.3	58.8	17.0	61.2	59.6
			12	13.9	56.6	55.9	15.6	58.1	57.2	17.0	59.2	58.1	18.2	60.1	58.9
		5	8	13.8	57.3	56.0	15.6	58.5	57.2	17.0	59.6	58.2	18.3	60.6	58.9
			10	15.0	55.7	54.9	16.9	57.1	56.2	18.5	58.2	57.3	20.0	59.1	58.1
			12	15.9	54.7	54.1	18.0	56.0	55.4	19.8	57.1	56.5	21.3	58.1	57.4
		6	8	15.4	55.4	54.6	17.5	56.8	55.8	19.3	57.8	56.8	20.8	58.8	57.7
			10	16.5	54.1	53.5	18.9	55.4	54.8	20.9	56.5	55.9	22.6	57.4	56.8
			12	17.4	53.0	52.8	20.0	54.4	54.0	22.1	55.7	55.2	24.0	56.5	56.0
		8	8	17.9	52.6	52.2	20.6	54.1	53.5	23.0	55.2	54.6	25.1	56.1	55.5
			10	19.0	51.4	51.2	22.0	52.8	52.6	24.6	54.0	53.6	20.9	55.0	54.5
			12	19.8	50.6	50.4	23.0	51.9	51.7	25.8	53.2	52.8	28.3	54.1	53.9
	4	4	8	13.4	58.1	56.4	15.1	59.5	57.6	16.6	60.6	58.5	18.0	61.5	59.2
			10	14.8	56.3	55.1	16.8	57.8	56.4	18.5	58.8	57.3	20.0	59.7	58.1
			12	15.9	55.0	54.1	18.1	56.4	55.4	20.0	57.4	56.4	21.7	58.4	57.2
		5	8	15.3	55.8	54.7	17.5	57.1	55.8	19.4	58.3	56.8	21.0	59.2	57.6
			10	16.7	54.2	53.3	19.2	55.5	54.6	21.3	56.6	55.6	23.2	57.5	56.5
			12	17.8	52.8	52.3	20.6	54.2	53.6	22.9	55.3	54.6	25.0	56.3	55.5
		6	8	16.9	53.9	53.1	19.5	55.3	54.4	21.8	56.4	55.4	23.7	57.3	56.2
			10	18.3	52.3	51.8	21.2	53.8	23.1	23.7	54.9	54.2	26.0	55.8	55.0
			12	19.3	51.2	51.0	22.5	52.5	52.1	25.3	53.7	53.2	27.8	54.6	54.1
		8	8	19.3	51.2	50.8	22.6	52.5	52.0	25.5	53.5	53.0	28.1	54.6	53.9
			10	20.5	49.9	49.7	24.2	51.1	50.9	27.4	52.2	52.0	30.4	53.1	52.7
			12	21.4	49.0	48.8	25.3	50.1	49.9	28.9	51.1	50.9	32.1	52.0	51.8
	6	4	8	14.0	57.6	55.9	16.0	58.9	57.0	17.6	60.0	57.9	19.1	60.9	58.6
			10	15.6	55.7	54.4	17.8	57.0	55.6	19.7	58.1	56.6	21.4	59.0	57.4
			12	16.8	54.0	53.3	19.3	55.5	54.5	21.4	56.6	55.6	23.4	57.5	56.4
		5	8	16.0	55.3	54.0	18.4	56.5	55.2	20.4	57.7	56.2	22.3	58.4	57.0
			10	17.5	53.3	52.6	20.3	54.8	53.8	22.6	55.8	54.8	24.8	56.8	55.7
			12	18.7	51.9	51.4	21.7	53.2	52.7	24.4	54.5	53.7	26.8	55.4	54.6
		6	8	17.5	53.3	52.5	20.4	54.7	53.7	22.9	55.8	54.7	25.1	56.7	55.5
			10	19.0	51.6	51.1	22.2	52.9	52.3	25.0	54.1	53.4	27.6	55.0	54.2
			12	20.1	50.4	50.2	23.6	51.6	51.2	26.7	52.7	52.2	29.6	53.7	53.1
		8	8	19.8	50.7	50.3	23.4	51.8	51.4	26.6	52.9	52.3	29.5	54.0	53.2
			10	21.1	49.3	49.1	25.1	50.4	50.2	28.6	51.4	51.2	31.9	52.3	51.9
			12	21.9	48.4	48.2	26.3	49.3	49.1	30.1	50.3	50.1	33.7	51.1	50.9

4 ft/s. In tubes with water turbulators, the water-side pressure drop is considerably higher than in smooth tubes.

6. Choose the optimum number of rows and fins/in. An even number of rows is preferable because the supply and return header can be placed on the same side.

7. Select the coil from the manufacturer's catalog at given air and water entering and leaving conditions v_a and v_w, and the required q_c/A_a.

8. Determine the air-side and water-side pressure drop.

For detailed analysis and system simulation, calculations by means of dry-wet boundary and wet-coil effectiveness are sometimes necessary.

Example 12.3. A water cooling coil in a constant-air-volume system has a supply volume flow rate of 11,000 cfm. Air enters the coil at a dry bulb of 80°F and a wet bulb of 67°F and leaves the coil at a dry bulb of 57.5°F

TABLE 12.3b
Water cooling coil ratings

45°F   Entering water temperature
68°F   Entering wet bulb
79°F   Entering dry bulb

| Face velocity, fpm | | | | | | | | | | | | Fin no. | Row | FPS | Water temp. rise °F |
| 400 | | | 500 | | | 600 | | | 700 | | | | | | |
MBH	LDB	LWB	MBH	LDB	LWB	MBH	LDB	LWB	MBH	LDB	LWB				
10.8	61.4	59.9	11.9	62.7	60.9	12.9	63.6	61.7	13.8	64.4	62.2	8			
11.8	60.1	59.0	13.2	61.3	60.0	14.3	62.3	60.9	15.3	63.0	61.5	10	4		
12.6	59.0	58.3	14.1	60.4	59.4	15.3	61.3	60.3	16.5	62.0	60.9	12			
12.5	59.4	58.4	14.0	60.6	59.5	15.4	61.7	60.3	16.4	62.6	61.0	8			
13.7	58.1	57.4	15.4	59.3	58.6	16.8	60.5	59.5	18.1	61.2	60.2	10	5		
14.6	57.1	56.6	16.4	58.4	57.9	17.9	59.4	58.8	19.3	60.4	59.6	12		2	
14.1	57.7	57.0	15.9	59.0	58.2	17.4	60.1	59.1	18.9	61.0	59.8	8			
15.3	56.5	56.0	17.3	57.8	57.2	19.0	58.8	58.2	20.5	59.9	59.1	10	6		
16.1	55.6	55.2	18.4	56.9	56.5	20.2	57.9	57.5	21.9	58.8	58.4	12			
16.6	55.2	54.8	19.0	56.5	56.0	21.1	57.6	57.0	22.9	58.5	57.9	8			
17.9	54.0	53.8	20.4	55.4	55.2	22.7	56.5	56.3	24.7	57.4	57.0	10	8		
18.7	53.1	52.9	21.5	54.5	54.3	24.0	55.6	55.4	26.2	56.6	56.4	12			
12.1	60.3	58.8	13.7	61.5	59.8	15.0	62.6	60.6	16.0	63.4	61.3	8			
13.4	58.7	57.6	15.2	59.9	58.7	16.6	60.9	59.6	18.1	61.8	60.3	10	4		
14.5	57.4	56.6	16.4	58.7	57.8	18.1	59.7	58.8	19.6	60.7	59.5	12			
13.9	58.2	57.2	16.0	59.3	58.2	17.6	60.4	59.1	19.1	61.2	59.8	8			
15.4	56.6	55.9	17.5	57.8	57.1	19.4	58.9	58.0	21.0	59.8	58.8	10	5		
16.5	55.4	54.8	18.9	56.6	56.1	21.0	57.7	57.1	22.8	58.6	57.9	12		4	16
15.5	56.4	55.7	17.8	57.7	56.9	19.8	58.7	57.8	21.5	59.6	58.6	8			
17.1	54.8	54.3	19.5	56.2	55.6	21.7	57.3	56.7	23.7	58.1	57.5	10	6		
18.2	53.7	53.3	20.9	55.1	54.6	23.4	56.1	55.6	25.6	57.0	56.6	12			
18.1	53.8	53.3	21.1	55.0	54.5	23.5	56.1	55.6	25.8	57.0	56.4	8			
19.5	52.5	52.0	22.8	53.6	53.2	25.7	54.7	54.3	28.1	55.7	55.3	10	8		
20.5	51.3	51.1	24.1	52.6	52.3	27.3	53.7	53.5	30.0	54.6	54.4	12			
12.7	59.7	58.3	14.4	61.0	59.3	15.9	62.0	60.1	17.2	62.9	60.7	8			
14.2	58.0	57.0	16.1	59.3	58.1	17.9	60.2	58.9	19.4	61.1	59.6	10	4		
15.4	56.6	55.9	17.6	57.9	57.0	19.5	58.9	58.0	21.2	59.8	58.8	12			
14.7	57.5	56.5	16.8	58.8	57.6	18.6	59.8	58.5	20.2	60.7	59.2	8			
16.2	55.8	55.1	18.5	57.1	56.4	20.6	58.2	57.3	22.5	59.1	58.1	10	5		
17.5	54.5	54.0	20.2	55.7	55.2	22.4	56.9	56.3	24.5	57.8	57.1	12		6	
16.3	55.8	55.1	18.7	57.1	56.3	20.9	58.1	57.2	22.8	59.0	58.0	8			
17.8	54.2	53.6	20.7	55.4	54.8	23.0	56.5	55.9	25.2	57.4	56.7	10	6		
19.0	53.0	52.5	22.2	54.2	53.7	24.8	55.3	54.8	27.4	56.1	55.6	12			
18.8	53.3	52.8	21.9	54.4	53.9	24.7	55.4	54.9	27.1	56.4	55.8	8			
20.2	51.6	51.4	23.8	52.9	52.5	26.9	53.9	53.5	29.8	54.8	54.4	10	8		
21.2	50.6	50.4	25.1	51.7	51.5	28.6	52.8	52.4	31.8	53.7	53.3	12			

FPS   ft/s of water velocity

MBH   1000 Btu/h · ft² of coil face area

Source: York International Corporation. Reprinted with permission.

and a wet bulb of 56.5°F. Chilled water enters the coil at 45°F and is expected to leave the coil at 55°F. At an average temperature of 50°F, the fluid properties are as follows:

Thermal conductivity of water k_w	0.339 Btu/h · ft · °F
Dynamic viscosity of water μ_w	3.09 lb/ft · h
Specific heat of water c_{pw}	1.0 Btu/lb · °F

1. Select a water-cooling coil from the manufacturer's catalog.

2. Determine the dry-wet boundary of this dry-wet coil.

3. Calculate the sensible cooling capacity and the outer surface area of the dry part of this dry-wet coil.

4. Determine the cooling and dehumidifying capacity of the wet part as well as the dry-wet coil.

5. At part-load operation, if the water velocity in the copper tubes is reduced to 63 percent of the full-

load value, and the air enters the coil at a wet bulb of 68°F, determine the operating conditions of this dry-wet coil.

Solution

1. For a corrugated coil, select a face velocity $v_a = 550$ fpm. From the given values, the required face area of the coil is

$$A_a = \frac{\dot{V}_a}{v_a} = \frac{11,000}{550} = 20 \text{ ft}^2$$

From the psychrometric chart, the enthalpy of entering air at a dry bulb of 80°F and a wet bulb of 67°F is 31.6 Btu/lb. The enthalpy of leaving air at a dry bulb of 57.5°F and a wet bulb of 56.5°F is 24.2 Btu/lb. From the given conditions, the total cooling and dehumidifying capacity of this dry-wet coil is

$$q_c = 60\dot{V}_a\rho_a(h_{ae} - h_{al})$$
$$= 60 \times 11,000 \times 0.075(31.6 - 24.2)$$
$$= 366,300 \text{ Btu/h}$$

Cooling and dehumidifying capacity per ft² of face area are calculated as

$$\frac{q_c}{A_a} = \frac{366,300}{20}$$
$$= 18,315 \text{ Btu/h or } 18.3 \text{ Mbtu/h}$$

Similarly, the required sensible cooling capacity per ft² of coil face area is

$$\frac{q_{cs}}{A_a} = 60v_a\rho_a c_{pa}(T_{ae} - T_{al})$$
$$= 60 \times 550 \times 0.075 \times 0.243(80 - 57.5)$$
$$= 13,532 \text{ Btu/h} \cdot \text{ft}^2$$

From the given values, the water temperature rise is $(55 - 45) = 10°F$, and the water volume flow rate is therefore

$$\dot{V}_{gal} = \frac{q_c}{500(T_{wl} - T_{we})}$$
$$= \frac{366,300}{500 \times 10}$$
$$= 73.3 \text{ gpm}$$

2. For a face area of 20 ft², choose a coil with a finned width of 30 in. and finned length of 96 in. Assume that the water velocity inside the tubes of this water-cooling coil is 4 ft/s.

   At an air entering condition of 80°F dry bulb and 67°F wet bulb, $v_a = 550$ fpm, $v_w = 4$ ft/s, $T_{we} = 45°F$, $\Delta T_w = 10°F$, and required $q_c/A_a = 18.3$ MBtu/h, a 4-row, 12 fins/in. coil is selected from Table 12.3 that has a cooling and dehumidifying capacity of 19.1 MBtu/h per ft² of face area and a sensible cooling capacity per ft² of

$$\frac{q_{cs}}{A_a} = 60 \times 550 \times 0.075$$
$$\times 0.243\left(80 - \frac{56.4 + 57.4}{2}\right)$$
$$= 13,893 \text{ Btu/h} \cdot \text{ft}^2$$

Both the cooling capacity per ft² of face area and the sensible cooling capacity per ft² face area of this DX-coil are greater than the required values.

From the psychrometric chart, the dew point of the entering air at 80°F dry bulb and 67°F wet bulb is 60°F, that is, $T_{sb} = 60°F$.

The slope of the air saturation curve at 60°F is the difference in saturation enthalpy between 59.5°F and 60.5°F, that is

$$m'' = 28.81 - 26.12 = 0.69 \text{ Btu/lb} \cdot °F$$

Therefore,

$$\frac{m''}{c_{pa}} = \frac{0.69}{0.243} = 2.84$$

Using the same air-side heat transfer coefficient $h_o = 14.3$ Btu/h · ft² · °F as in Example 12.2, from Eq. (12.27), the heat transfer coefficient for the wet surface is

$$h_{wet} = \frac{m''}{c_{pa}}h_o = 2.84 \times 13.4$$
$$= 38.06 \text{ Btu/h} \cdot \text{ft}^2 \cdot °F$$

3. On water side,

$$\text{Re}_D = \frac{\rho v_w D_i}{\mu_w} = \frac{62.4 \times 4 \times 3600 \times 0.496}{3.09 \times 12}$$
$$= 12,020$$
$$\text{Pr} = \frac{\mu_w c_{pw}}{k_w} = \frac{3.09 \times 1}{0.339} = 9.12$$

Then,

$$h_i = 0.023 \text{ Re}_D^{0.8}\text{Pr}^{0.4}$$
$$= 0.023 \times (12,020)^{0.8}(9.12)^{0.4}$$
$$= 838.6 \text{ Btu/h} \cdot \text{ft}^2 \cdot °F$$

Because $B = 11.4 \times 1.15 = 13.1$, from Eq. (12.25), for 1 ft² of outer surface area at the dry-wet boundary,

$$h_{wet}\eta_s(T_{ab} - T_{sb}) = \frac{h_i}{B}(T_{sb} - T_{wb})$$
$$38.06 \times 0.76(T_{ab} - 60) = \frac{838.6}{13.1}(60 - T_{wb})$$

From Eq. (12.28),

$$\dot{V}_a\rho_a c_{pa}(T_{ae} - T_{ab}) = 8.33\dot{V}_{gal}c_{pw}(T_{wl} - T_{wb})$$
$$\times 11,000 \times 0.075 \times 0.243(80 - T_{ab})$$
$$= 8.33 \times 74 \times 1(55 - T_{wb})$$

According to the above simultaneous equations,

$$T_{ab} = 74.8°F \quad \text{and} \quad T_{wb} = 53.3°F$$

The dry-wet boundary can therefore be determined as shown in Fig. 12.11.

4. From Eq. (12.29), the sensible cooling capacity of the dry part of the dry-wet coil is

$$q_{cs} = \dot{V}_a \rho_a c_{pa}(T_{ae} - T_{ab})$$

$$= 11,000 \times 60 \times 0.075 \times 0.243$$

$$\times (80 - 74.8) = 62,548 \text{ Btu/h}$$

The average temperature of the airstream in the dry part is

$$T_{ad.m} = \frac{1}{2(T_{ae} + T_{ab})} = \frac{1}{2(80 + 74.8)} = 77.4°F$$

The average temperature of the water stream corresponding to the dry part of the dry-wet coil is

$$T_{wd.m} = \frac{1}{2(T_{wl} + T_{wb})} = \frac{1}{2(55 + 53.3)}$$

$$= 54.15°F$$

From Eq. (12.5),

$$U_o = \frac{1}{\dfrac{1}{\eta_s h_o} + \dfrac{B}{h_i}} = \frac{1}{\dfrac{1}{0.76 \times 13.42} + \dfrac{13.11}{838.6}}$$

$$= 8.796 \text{ Btu/h} \cdot \text{ft}^2 \cdot °F$$

Then, the outer surface area of the dry part is

$$A_{o.d} = \frac{q_{cs}}{U_o(T_{ad.m} - T_{wd.m})}$$

$$= \frac{62,548}{8.796(77.4 - 54.15)} = 306 \text{ ft}^2$$

5. From Eq.(12.14), the heat capacity rates and their ratios are

$$C_a = \dot{m}_a c_{pa} = 60 \times 11,000 \times 0.075 \times 0.243$$

$$= 12,028 \text{ Btu/h} \cdot °F$$

$$C_w = \dot{m}_w c_{pw} = 74 \times 500 \times 1$$

$$= 37,000 \text{ Btu/h} \cdot °F$$

$$C = \frac{C_{min}}{C_{max}} = \frac{12,028}{37,000} = 0.3251$$

From Eq. (12.4) and Table 12.1, for such a 4-row, 12 fins/in. corrugated plate fin coil, the total outer surface area is

$$A_o = F_s A_a N_r = 14.21 \times 1.15 \times 20 \times 4$$

$$= 1307 \text{ ft}^2$$

The outer surface area of the wet part of the coil is

$$A_{wet} = A_o - A_{o.d} = 1307 - 306 = 1001 \text{ ft}^2$$

Then, from Eq. (12.23),

$$U_o A_o = \frac{1}{\dfrac{1}{\eta_s h_o A_o} + \dfrac{1}{h_i A_i}}$$

$$= \frac{1}{\dfrac{1}{0.76 \times 13.42 \times 1001} + \dfrac{13.11}{838.6 \times 1001}}$$

$$= 8805 \text{ Btu/h} \cdot °F$$

and from Eq. (12.17)

$$\text{NTU} = \frac{U_o A_o}{C_a} = \frac{8805}{12,028} = 0.7321$$

Also,

$$\exp[-\text{NTU}(1 - C)] = \exp[-0.7321(1 - 0.3251)]$$

$$= 0.6101$$

Therefore, from Eq. (12.16), the effectiveness of the wet part of the dry-wet coil is

$$\epsilon_{wet} = \frac{1 - \exp[-\text{NTU}(1 - C)]}{1 - C \exp[-\text{NTU}(1 - C)]}$$

$$= \frac{1 - 0.6101}{1 - 0.3251 \times 0.6101} = 0.4864$$

6. From the psychrometric chart, at a T_{ab} of 74.8°F and a dew point of 60°F, $h_{ab} = 30.3$ Btu/lb. Also, the enthalpy of saturated air film $h_{s.we} = 17.65$ Btu/lb when the temperature T_{we} is 45°F. Then, the cooling and dehumidifying capacity of the wet part of the dry-wet coil is

$$q_{c.w} = \epsilon_{wet} \dot{V}_a \rho_a(h_{ab} - h_{s.we})$$

$$= 0.4864 \times 11,000 \times 60$$

$$\times 0.075(30.3 - 17.65) = 304,571 \text{ Btu/h}$$

The cooling capacity of the dry-wet coil is

$$q_c = q_{cs} + q_{c.w} = 62,548 + 304,571$$

$$= 367,119 \text{ Btu/h}$$

or $367,119/20 = 18.4$ MBtu/h $\cdot$ ft^2 of face area, which is close to the value 19.1 MBtu/h $\cdot$ ft^2 given in the manufacturer's catalog in Table 12.3.

7. If the water velocity inside the copper tubes is reduced to 63 percent of the full-load value $v_w = 0.63 \times 4 = 2.5$ ft/s and the water temperature rise is increased to 16°F, from the performance data shown in Table 12.3, at an entering dry bulb of 79°F, entering wet bulb of 68°F, and an enthalpy of 32.4 Btu/lb, the cooling capacity is reduced from 19.1 to 15.3 MBtu/h $\cdot$ ft^2. Total cooling and dehumidifying capacity of the water cooling coil at part-load is

$$q_c = 15.3 \times 1000 \times 60 \times 20 = 306,000 \text{ Btu/h}$$

The reduction of enthalpy Δh during the cooling and dehumidifying process at part-load is

$$\Delta h = \frac{306,000}{11,000 \times 60 \times 0.075} = 6.2 \text{ Btu/lb}$$

The enthalpy of the air leaving the coil is

$$h_{al} = 32.4 - 6.2 = 26.2 \text{ Btu/lb}$$

At an entering dry bulb of 79°F and wet bulb of 68°F, from the cooling and dehumidifying curve on the psychrometric chart, the condition of air off the coil at part-load is 61.3°F dry bulb and 59.5°F wet bulb.

According to the data in the catalog, the load ratio at part-load = 15.3/19.1 = 0.80.

If the sensible heat ratio and system heat gain at part-load are the same as in full-load operation, the air conditioning cycle at part-load can then be drawn on the psychrometric chart as shown by the dotted line in Fig. 12.12. The space relative humidity has been increased from 55 percent at full-load to 59 percent at 80 percent part-load.

12.5 COIL ACCESSORIES AND FREEZE-UP PROTECTION

Coil Accessories

Coil accessories include air vents, drain valves, isolation valves, flow metering valves, balancing valves, thermometers, pressure gauge taps, condensate drain taps, and even distribution baffles.

AIR VENTS. These devices discharge air from the coils. Air vents are usually located at the highest point of the coil and at the pipe risers. Manual vents are inexpensive and more reliable, whereas automatic vents are more convenient to operate.

DRAIN AND ISOLATING VALVES. Drain valves should be provided for each coil for maintenance and repair. Hose-end type drain valves may be used for remote drains. Isolating valves should be installed at both supply and return piping to cut off the water supply during repair and maintenance.

EVEN DISTRIBUTION BAFFLES. When heating coils are stacked above cooling coils, as in multizone units (to be discussed in Chapter 15), perforated baffles should usually be installed in the path with less flow resistance to balance the difference in pressure losses between two parallel flow paths. Heating coils usually have fewer rows and less flow resistance than cooling coils.

Servicing Problems

Coil cleanliness is important for proper maintenance and operation. If a medium- or high-efficiency air filter is installed upstream from the coil, dirt will not accumulate on the coil. If low-efficiency air filter is used upstream from the coil, dirt accumulation may eventually block the air paths in a coil and significantly increase the pressure drop across the coil. In air systems using low-efficiency air filters, coils should be inspected and cleaned every three months in urban areas. Drain pans should be cleaned monthly to prevent the build-up of bacteria and microorganisms, which can contaminate indoor air.

Improper design of the piping system may cause coil flooding, water hammer, corrosion, and other servicing problems.

Air Stratification and Coil Freeze-Up

One of the most common problems encountered with water and steam coils is *coil freeze-up*. Water inside coil tubes freezes when its temperature is below the freezing point.

Improper mixing of outdoor and recirculating air in the mixing box of an air-handling unit or packaged unit often causes coil freeze-up when outdoor air temperature is below 32°F. A mixture of 30 percent outdoor air at 0°F and 70 percent recirculating air at 70°F theoretically gives an average temperature of 49°F, which is far above the freezing point of water. However, such a temperature is only maintained by thorough mixing. If outdoor air and recirculating air enter the mixing box of an air-handling unit, as shown in Fig. 12.13a, the outdoor air and recirculating air will not mix thoroughly and will form *air stratification*. Low-temperature outdoor air flows over part of the preheat coil, cooling coil, or even the reheat coil, and may freeze part of the coil.

Baffles can be used to direct the airstream to promote thorough mixing, as shown in Fig. 12.13b. The face velocity of the baffles should be between 1000 and 1500 fpm. Temperature sensors, installed at the coil face where lowest air intake temperature may occur, can be used to shut off the fan and the outdoor air damper when the temperature is below freezing. For coils installed downstream from a supply fan, a single inlet fan always provides better mixing than a double inlet fan.

Water Coils and Steam Heating Coils

Freeze-up of idled chilled water coils can be prevented by running the chilled water pump to provide a certain water velocity inside the coil tubes when the outside air temperature is below freezing. If a water velocity

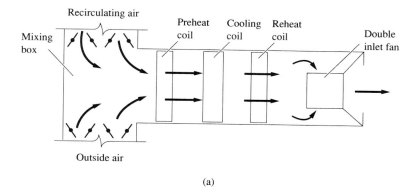

(a)

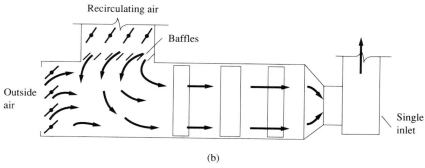

(b)

FIGURE 12.13 Air stratification and freeze-up protection: (a) freeze-up due to air stratification and (b) preventing freeze-up.

of 2.5 ft/s is maintained inside a water cooling coil, the minimum air temperature can drop to 32°F without freezing the coils. Another way to prevent freeze-up in idle coils is to drain the water completely.

For hot water coils, it is better to reset the water temperature rather than to interrupt pump operation during the night. A flow alarm should be used to indicate pump failure. Automatic interlocks should also be provided to close the outdoor air dampers in the event of pump failure or fan shut-down.

If a steam heating coil is installed in a location where the outdoor air temperature may drop below 32°F in winter, as described by Delaney et al. (1984), a coil with inner distributor tubes for freeze-up prevention should be used as shown in Fig. 12.14. Steam condensate can drain from the outer tubes easily without blocking the steam supply.

12.6 EVAPORATORS

Types of Evaporators

The *evaporator* is one of the main components of a refrigerating system, in which refrigerant evaporates for the purpose of extracting heat from the surrounding air, chilled water, or other substances. The evaporator is also an indirect contact heat exchanger.

Evaporators can be classified into three categories, depending on the medium or substance to be cooled:

- An *air cooler* is an evaporator that cools the air directly in a refrigerated space or piece of equipment (such as a packaged unit). Conditioned air is then distributed through air distributing systems. In an air cooler, the refrigerant flows inside the metal tubes or finned tubes while air flows over them.

- In a *liquid cooler,* chilled water is cooled to a lower temperature and is pumped to remote air-handling units, fan–coils, or other terminals for air conditioning or other applications.

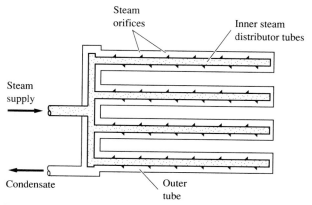

FIGURE 12.14 Steam-heating coil with inner distributor tubes for freeze-up prevention.

- An evaporator can also be used to produce ice directly, as an *ice-maker* in an ice harvester or ice-storage system. These will be discussed in Chapter 25.

The refrigerant feed for air and liquid coolers and ice-makers can be classified mainly into the following three types:

- *Dry expansion or direct expansion (DX)*. In evaporators with dry or direct expansion refrigerant feed, refrigerant flows inside the tubes in an evaporator, and is completely vaporized and superheated to a certain degree before reaching the exit of the evaporator, as shown in Fig. 12.7.
- *Flooded refrigerant feed*. In evaporators with flooded refrigerant feeds, refrigerant vaporizes outside the tubes within a shell. The refrigerant-side surface area is always wetted by the liquid refrigerant, which results in a higher surface heat transfer coefficient.
- *Liquid overfeed*. In liquid overfeed evaporators, liquid refrigerant is fed by a mechanical or gas pump, and is then overfed to each evaporator. The inner surface of the tubes in an overfeed evaporator is also wetted by liquid refrigerant.

The various combinations of cooler types and refrigerant feeds make up the following evaporator groups:

- *DX-coil air cooler with direct expansion refrigerant feed* (see Section 12.4)
- *Flooded shell-and-tube liquid cooler,* or simply *flooded liquid cooler*
- *Shell-and-tube liquid cooler with direct-expansion refrigerant feed*, or simply *direct-expansion liquid cooler*
- *Liquid overfeed cooler*
- *Direct-expansion ice-maker*

Flooded Liquid Cooler

Most medium- and large-size liquid coolers are shell-and-tube flooded liquid coolers. In a flooded liquid cooler, several straight tubes are aligned in a parallel staggered arrangement, usually held in place at both ends by tube sheets, as shown in Fig. 12.15a. Chilled water circulates inside the tubes, which are submerged in a refrigerant-filled shell.

Liquid-vapor refrigerant, usually at a quality x around 0.15 in air conditioning applications, is fed into the bottom of the shell. It is evenly distributed over the entire length of the tubes. As the refriger-

ant boils and bubbles rise, the upper part becomes increasingly bubbly. Vapor refrigerant is discharged from the opening at the top of the cooler. A dropout area, or *eliminator*, is sometimes installed to separate the liquid refrigerant from the vapor. The amount of refrigerant fed to the flooded liquid cooler is controlled by a low-pressure side float-valve, or a multiple-orifice throttling device, which will be discussed in Chapters 21 and 23.

When halocarbons are used as refrigerants, copper tubes are always used because they provide higher thermal conductivity and do not react with halocarbons. Tube diameters usually range from $\frac{1}{2}$ to 1 in. (typically $\frac{5}{8}$ and $\frac{3}{4}$ in.) and the number of tubes inside the shell vary from 50 to several thousand. Integral fins are extruded on the outer surface of the tubes to increase the outer surface area. Fin spacing ranges from 19 to 35 fins/in. at a height of 0.06 in. The surface may also have a porous coating to increase the boiling heat transfer coefficient. The outer to inner surface ratio of integral finned tubes ranges from 2.5 to 3.5. Spiral grooves or other enhancements may be added to the inner surface of the tubes to increase the inner surface heat transfer coefficient. By increasing turbulence, they prevent dirt and suspended solids from settling on the inner surface during operation.

Possible water flow arrangements in flooded shell-and-tube liquid cooler are shown in Fig. 12.15b. A one-pass arrangement has the highest chilled water flow rate, two-pass has a lower rate, and three-pass has the lowest chilled water flow rate. The two-pass water flow arrangement, with the water inlet and outlet on the same side, is the standard arrangement. Performance analyses show that there is no significant difference between a two-pass arrangement with the water inlet and outlet located side-by-side and one with the inlet at the bottom and outlet at the top. Water velocity inside the copper tubes is usually between 4 and 12 ft/s—water velocity higher than 12 ft/s may cause erosion. Water-side pressure drop is usually maintained at or below 10 psi to optimize pump energy consumption. Because of the greater flow passage for refrigerant within the shell, the pressure loss on the refrigerant side is far lower than the pressure loss in DX-coils.

Flooded liquid coolers provide a large surface area and higher heat transfer coefficient, need minimal space and headroom, and are easily maintained. They are most widely used in large refrigeration plants.

Heat Transfer in a Flooded Liquid Cooler

In flooded liquid coolers, the refrigerant is assumed to evaporate at a uniform temperature. As in water cooling coils, the rate of heat transfer between the

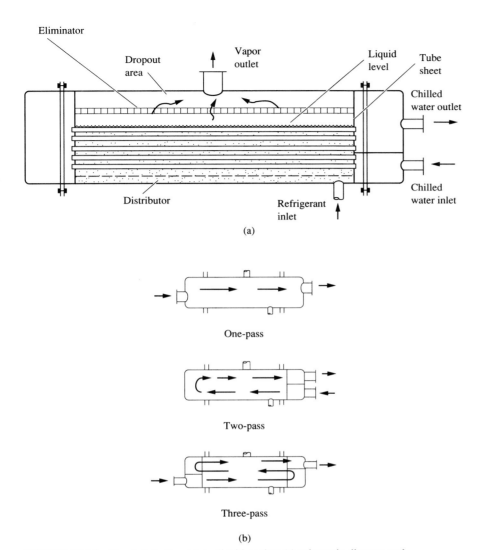

FIGURE 12.15 Flooded shell-and-tube liquid cooler: (a) schematic diagram and (b) passages of water flow.

chilled water and the refrigerant q, (in Btu/h), U_o and h_i (in Btu/h · ft² · °F), ϵ, and NTU can be calculated from Eqs. (12.2), (12.5), (12.9), (12.16), (12.17), and (12.18) as follows:

$$q = 60 \, \dot{m}_w c_{pw}(T_{ee} - T_{el})$$

$$= A_o U_o \Delta T_m$$

$$= 60 \dot{m}_r (h_{rl} - h_{re})$$

$$U_o = \cfrac{1}{\cfrac{1}{h_o} + \cfrac{L_t A_o}{A_i k_t} + \cfrac{A_o R_f}{A_i} + \cfrac{A_o}{A_i h_i}}$$

$$\epsilon = 1 - \exp(-\mathrm{NTU})$$

$$\mathrm{NTU} = \frac{U_o A_o}{c_{pw}} \qquad (12.32)$$

$$q = 60 \dot{m}_w c_{pw}(T_{ee} - T_{ev})\epsilon$$

where h_{re}, h_{rl} = enthalpy of refrigerant entering and leaving the liquid cooler, Btu/lb

T_{ee}, T_{el} = chilled water entering and leaving the liquid cooler, °F

in_r = mass flow rate of refrigerant, lb/min

h_o = average heat transfer coefficient of the refrigerant side surface, Btu/h · ft² · °F

L_t = thickness of the tube, ft

k_t = thermal conductivity of metal tube, Btu/h · ft · °F

T_{ev} = evaporating temperature, °F

For the analysis of heat transfer in a flooded liquid cooler, both the log-mean temperature difference and the NTU method can be used. However, the NTU method is preferred because it clearly describes the effectiveness and characteristics of the liquid cooler.

Fouling Factor

Fouling factor R_f, in $h \cdot ft^2 \cdot °F/Btu$, is the additional resistance caused by the dirty film of scale, rust, silt, or other deposits on the surface of the tube.

The fouling factor may be defined as

$$R_f = \frac{1}{U_{dirty}} - \frac{1}{U_{clean}} \qquad (12.33)$$

where U_{dirty}, U_{clean} = overall heat transfer coefficient through a tube wall with a dirty or clean surface respectively, $Btu/h \cdot ft^2 \cdot °F$.

In 1982, *ASHRAE Journal* published research results on fouling of heat transfer surfaces such as evaporators and condensers in air conditioning. The study showed that with a certain water treatment and in the absence of biological growth and suspended particles, long-term fouling did not exceed 0.0002 $h \cdot ft^2 \cdot °F/Btu$ and short-term fouling did not exceed 0.0001 $h \cdot ft^2 \cdot °F/Btu$.

The new Air Conditioning and Refrigeration Institute (ARI) Standard 550–88 specifies the following:

Field fouling allowance	0.00025 $h \cdot ft^2 \cdot °F/Btu$
ARI Rating Standard (new evaporators and condensers)	0

The ARI Standard recommends specifying 0.00025 $h \cdot ft^2 \cdot °F/Btu$ total fouling for closed-loop liquid coolers and for condensers served by well-maintained cooling towers.

Pool Boiling and Forced Convection

If a surface is in contact with a liquid, and if the temperature of the surface is maintained above the saturation temperature of the liquid, boiling may occur. The rate of heat transfer depends mainly on the temperature difference between the surface giving heat and the saturation temperature of the liquid at a specified pressure.

When the surface is submerged below the free surface of the liquid, the boiling process is known as *pool boiling*. When a mixture of liquid and vapor is forced through a tube or passage and the surface surrounding the mixture is above the saturation temperature of the liquid, *forced-convection boiling* may occur inside the tube or passage.

When the temperature of a surface that releases heat is a few degrees higher than the temperature of the liquid, molecules within the thin layer of superheated liquid adjacent to the surface tend to break away from the surrounding liquid molecules. They form nuclei, which grow into vapor bubbles. As the number and size of these bubbles increase, the bubbles flow upward to the surface of the liquid and

form forced-convection boiling. This process of combined nucleate boiling and forced convection occurs in flooded shell-and-tube liquid coolers.

Chen (1966) developed a forced-convection model that can be applied to flooded shell-and-tube liquid coolers to calculate the boiling heat transfer coefficient on the outer surface of the tube h_b, in $Btu/h \cdot ft^2 \cdot °F$, as follows:

$$h_b = h_{nb} + h_{tp} = S h_{nbl} + F h_{fc} \qquad (12.34)$$

where h_{nbl} = single-tube pool-boiling coefficient, $Btu/h \cdot ft^2 \cdot °F$
   S = suppression factor
   h_{fc} = single-phase forced-convection coefficient for the liquid phase flowing alone in the tube bundle, $Btu/h \cdot ft^2 \cdot °F$
   F = forced convection multiplier

The experimental results of Webb et al. (1990) showed that for CFC–11 and copper tubes with integral fins, the forced-convection portion $h_{nb} = S h_{nbl}$ is 1.6 times larger than the nucleate boiling portion $h_{tp} = F h_{fc}$ at the first row (bottom row) of the tube bundle and 5.3 times larger at the top row. In tubes with high-flux outer boiling surfaces with porous coating, the capacity of the liquid cooler is 36 percent higher than in tubes with integral fins.

Because the average heat transfer coefficient of a tube's outer surface h_o (in $Btu/h \cdot ft^2 \cdot °F$) is a function of the temperature difference between the tube wall and the refrigerant $(T_s - T_r)$, h_o is also a function of the heat flux at the tube's outer surface q_{ev}/A_o, in $Btu/h \cdot ft^2$. Here, q_{ev} represents the rate of heat transfer through the tube surfaces of the liquid cooler or the cooling capacity of a specific cooler, in Btu/h, and A_o the outer surface area of the tubes, in ft^2. Heat transfer coefficient h_o in a flooded liquid cooler can be calculated as

$$h_o = C(A_o/q_{ev})^n \qquad (12.35)$$

Constant C and exponent n are mainly determined by the outer surface conditions of the tubes and the wettability of different refrigerants on tubes.

Parameters That Influence the Performance of Flooded Liquid Coolers

EVAPORATING TEMPERATURE T_{ev}. In comfort air conditioning systems, chilled water usually leaves the liquid cooler at a temperature T_{el} from 42 to 45°F to provide the desirable air off-coil temperature T_{al}. In ice-storage and cold air distribution systems (which will be discussed in Chapter 25), *brine*, a mixture of chilled water and glycol, may leave the liquid cooler at a temperature of 24 to 34°F.

In a conventional chilled-water system, if the value of T_{el} is fixed, the choice of optimum temperature difference $(T_{el} - T_{ev})$ determines the evaporating temperature in the flooded liquid cooler, as shown in Fig. 12.16. A smaller $(T_{el} - T_{ev})$ means a lower log-mean temperature difference ΔT_m between the chilled water and refrigerant. For a specific refrigeration capacity, a smaller ΔT_m needs a larger liquid cooler, a higher T_{ev}, and a lower power input to the compressor. If a larger $(T_{el} - T_{ev})$ is chosen, the liquid cooler can be smaller, but the T_{ev} value will be lower and compressor power input will be higher. A compromise between energy cost and initial investment should be made through life-cycle cost analysis.

Most liquid cooler manufacturers have adapted a value of $(T_{el} - T_{ev})$ between 6 and 10°F. For energy-saving models, $(T_{el} - T_{ev})$ may be as low as 4 to 7°F.

TEMPERATURE DIFFERENCE $(T_{ee} - T_{el})$. For each ton (or 12,000 Btu/h) of refrigeration capacity of a liquid cooler, the water flow rate V_{gal} (in gpm) at various temperature differences of chilled water entering and leaving $(T_{ee} - T_{el})$ can be calculated from Eq. (12.3) as follows:

$(T_{ee} - T_{el})$, °F	$\dot{V}_{gal}$, **gpm**
8	3
10	2.4
15	1.6
20	1.2
24	1.0

Temperature difference $(T_{ee} - T_{el})$ has a direct impact on pump power. In addition, for a given T_{el}, a greater $(T_{ee} - T_{el})$ always means a greater log-mean temperature difference ΔT_m and a greater $(T_{ee} - T_{ev})$ and, therefore, a greater cooler capacity and higher off-coil temperature T_{al}, as shown in Fig. 12.16. Here, log-mean temperature difference is calculated as

$$\Delta T_m = \frac{T_{ee} - T_{el}}{\ln\left(\dfrac{T_{ee} - T_{ev}}{T_{el} - T_{ev}}\right)} \qquad (12.36)$$

At full-load, the chilled-water temperature increase from pump power and piping heat gain is usually negligible. Therefore, temperature difference $(T_{ee} - T_{el})$ is often equal to the temperature difference of chilled water leaving and entering the water-cooling coils $(T_{wl} - T_{we})$ at design full-load, and is closely related to the conditioned air temperature leaving the coil T_{al}. Many liquid coolers use a $(T_{ee} - T_{el})$ between 12°F and 24°F. For a $(T_{ee} - T_{el}) = 20°F$, the water flow rate $\dot{V}_{gal} = 1.2$ gpm/ton refrigeration at full-load.

The minimum chilled-water temperature leaving the liquid cooler should be 40°F and the mass flow rate of chilled water flowing through the flooded liquid cooler should be approximately constant to prevent freezing. If chilled-water temperature is to be lower than 40°F, brine should be used, as in ice-storage systems.

OIL EFFECT. In flooded liquid coolers, vapor velocity above the tube bundle is usually insufficient to return the oil to the compressor. Oil concentration in the refrigerant tends to increase during operation. If oil concentration rises above 5 percent, it may degrade the rate of heat transfer. Generally, oil-rich liquid is taken and returned to the compressor continuously.

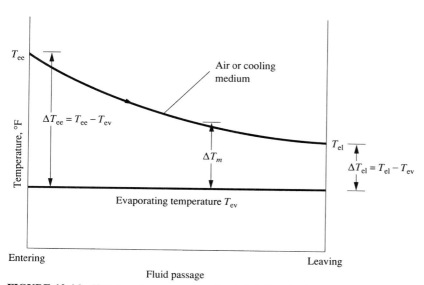

FIGURE 12.16 Heat transfer process in a flooded shell-and-tube cooler.

REFRIGERATION CAPACITY. The refrigeration capacity of a flooded liquid cooler also represents the capacity of the refrigerating system. From Eq. (12.32), for a liquid cooler with specified T_{el}, T_{ee}, and T_{ev}, the refrigeration capacity q_{ev}, in Btu/h, depends on the following factors:

- Tubes' outer surface area A_o and number of transfer units NTU
- Heat flux at the tubes' outer surface q_{ev}/A_o
- Heat transfer coefficients h_o, h_i, and overall heat transfer coefficient U_o
- Heat capacity rate of chilled water

Recent developments in flooded shell-and-tube liquid coolers have emphasized high flux and enhanced boiling coefficient tubes to increase refrigeration capacity.

Part-Load Operation of Flooded Liquid Coolers

When a flooded liquid cooler in a comfort air conditioning system is at part-load operation, the outer surface area A_o is the same as in full-load, so the reduction of its refrigeration capacity is mainly caused by the following:

1. A decrease in heat flux and, therefore, a lower outer surface boiling heat transfer coefficient h_o
2. A smaller difference in $(T_{ee} - T_{ev})$ caused by the drop in T_{ee} and increase in T_{ev}
3. A smaller log-mean temperature difference resulting from a higher T_{ev}
4. A decrease of effectiveness ϵ

Direct-Expansion Liquid Cooler

In a *direct-expansion liquid cooler* (shown in Fig. 12.17), liquid refrigerant evaporates inside the copper tubes while chilled water fills the shell. An expansion valve and sometimes a distributor are used for each group of refrigerant circuits connected to the same suction header and compressor.

In a direct-expansion liquid cooler, various inner surface configurations and enhancements are used to increase the boiling heat transfer. To provide optimum velocity and a higher rate of heat transfer on the water side of the liquid cooler, baffle plates are used to guide the water flow in the shell in multipass arrangements.

Direct-expansion liquid coolers are usually used for refrigerating systems equipped with multiple compressors. In a direct-expansion liquid cooler, refrigerant circuits may be connected to a single header and compressor or two separate headers and multiple compressors.

Liquid Overfeed Cooler

A *liquid overfeed cooler* can be an air cooler or a liquid cooler. Liquid refrigerant is fed to the multiple evaporators either by a mechanical pump or by a gas pump using the high-pressure gas discharged from the compressor. Liquid refrigerant is fed at a mass flow rate two to six times greater than the actual evaporating rate, which causes liquid recirculation, as shown in Fig. 12.18. Liquid recirculation provides a sufficient volume of liquid in the tubes and ensures that the inner surface of the tubes is fully wetted throughout its length.

Compared with DX-coils, liquid overfeed air coolers have the following advantages:

- Higher heat transfer coefficient on the refrigerant side and eliminating the decline of the heat transfer coefficient at the dryout area when the quality of the refrigerant is raised to 0.85 to 0.90 in a DX-coil.
- They allow non-uniform loading between refrigerant circuits or between evaporators, so that none of them are starved, with insufficient refrigerant supply.
- Air-refrigerant temperature difference is lower, which is especially beneficial for low-temperature systems.
- Evaporators can be conveniently defrosted.

Their main disadvantage is their higher initial cost.

During the design of a liquid overfeed cooler, the following choices should be carefully considered:

- *Mechanical Pump or Gas Pump.* Gas pumps generally require more energy than mechanical pumps. According to Cole's (1986) analysis, based on a typical plant of 275 tons of refrigeration capacity and an evaporating temperature of 20°F, the energy requirement of a gas pumping system is about 7 percent of the total plant power, and that of mechanical pump is about 2 percent. However, the difference in energy consumption is a small percentage of the total plant power, so in many circumstances, gas pumps are still used.

   Among mechanical pump systems, centrifugal pumps are the most popular type today. Rotary and gear pumps may be suitable in circumstances where cavitation is critical.

- *Down-Feed or Bottom-Feed.* In down-feed circuiting, refrigerant is fed to the highest tube of the evaporating coil. This arrangement improves oil return and eliminates static head. Up-feed circuiting simplifies liquid distribution. Relative locations of the evaporators and receivers become less important, and the system design and layout may be simplified.

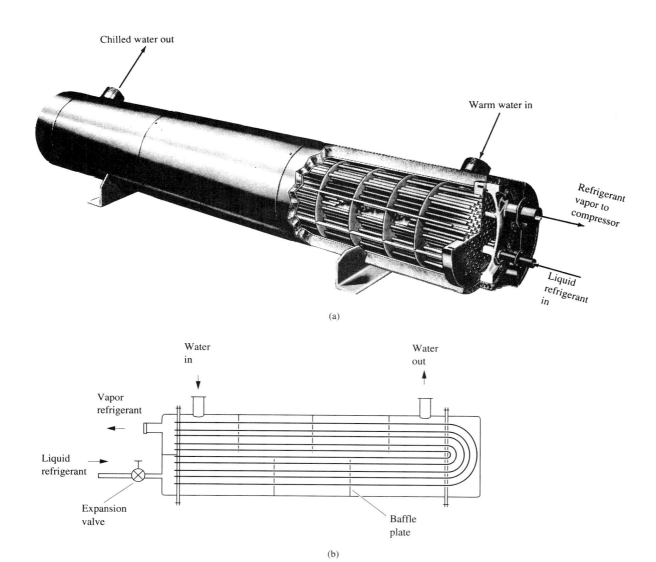

(a)

(b)

FIGURE 12.17   Direct-expansion liquid cooler: (a) cut-away view of a typical DX liquid cooler (*Source:* Trane Company; reprinted with permission); and (b) schematic diagram.

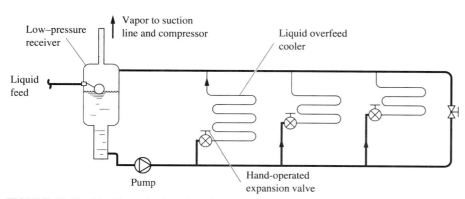

FIGURE 12.18   Liquid overfeed cooler using a mechanical pump.

- *Counterflow or Parallel Flow.* Because of its lack of superheat, as in DX-coils, a parallel flow arrangement between air and refrigerant in a liquid overfeed air cooler has a small performance advantage over the counterflow arrangement. Also, frosting occurs evenly in parallel flow, whereas frosting is heavier in the first and second rows of the finned-tube coil in a counterflow arrangement.

- *Circulating Number.* Circulating number n is defined as

$$n = \dot{m}_{cir} / \dot{m}_{ev} \qquad (12.37)$$

where $\dot{m}_{cir}$ = and $\dot{m}_{ev}$ = mass flow rate of refrigerant, circulated and evaporated, respectively, in lb/min. According to *ASHRAE Handbook 1990, Refrigeration Systems and Applications*, the circulating number for down-feed using ammonia as the refrigerant is 6 to 7. The circulating numbers for up-feed systems are n = 2 to 4 (small-diameter tubes) for ammonia and n = 3 for HCFC–22.

Liquid overfeed air coolers are widely used in low-temperature refrigeration plants for food storage and processing. Recently, overfeed liquid coolers have also been used in ice-storage systems.

12.7   WATER-COOLED CONDENSERS

Condensation Process

When saturated vapor comes into contact with a surface having a temperature below the saturation temperature, condensation occurs on the surface. There are two types of condensation:

1. The condensed liquid, often called the *condensate,* wets the surface and forms a film covering the entire surface. This type of condensation is called *filmwise condensation.*

2. The surface is not totally wetted by the saturated vapor, and the condensate forms liquid droplets that fall from the surface. This type of condensation is called *dropwise condensation.*

Compared with filmwise condensation, dropwise condensation has a greater surface heat transfer coefficient because it has a greater surface area exposed to the saturation vapor. In practice, however, the entire surface of the condenser tubes becomes wet during refrigerant condensation. Filmwise condensate falls in droplets and often disturbs the condensate films of a tube bundle.

The heat transfer process in a refrigerant condenser occurs in three stages:

1. Desuperheating the hot gas
2. Condensing the gas into liquid state and releasing the latent heat

3. Subcooling the liquid refrigerant

Although the surface heat transfer coefficient is lower on the hot gas side during desuperheating, there is a greater temperature difference between the hot gaseous refrigerant and the cooling medium during desuperheating. Subcooling only occupies a small portion of the condenser's surface area. Therefore, for simplification, an average heat transfer coefficient is used for the whole condenser's surface area, and the condensation of refrigerant is assumed to occur at the condensing temperature.

The capacity of a condenser is rated according to its total heat rejection Q_{rej}, in Btu/h. Total heat rejection Q_{rej} is defined as the total heat removed from the condenser during desuperheating, condensation, and subcooling of the refrigerant in the condenser, and is expressed as

$$Q_{rej} = 60\dot{m}_r(h_2 - h_{3'}) \qquad (12.38)$$

where m_r = mass flow rate of the refrigerant in the condenser, lb/min

h_2, h_3 = enthalpy of the hot gas entering the condenser and the enthalpy of the subcooled liquid leaving the condenser, Btu/lb

In refrigerating systems using hermetic compressors, the heat released by the motor is absorbed by the refrigerant. If the heat gains and losses from the ambient air at the evaporator, condenser, suction line, and discharge line are ignored then

$$Q_{rej} = q_{rl} + \frac{2545P_{com}}{\eta_{mot}} \qquad (12.39)$$

where q_{rl} = refrigeration load at the evaporator, Btu/h

P_{com} = power input to the compressor, hp

η_{mot} = efficiency of hermetic compressor motor

For refrigerating systems equipped with open compressors, η_{mot} = 1.

Heat rejection factor F_{rej} is defined as the ratio of the total heat rejection to the refrigeration capacity of the evaporator, and can be calculated as

$$F_{rej} = \frac{q_{rej}}{q_{rl}} = 1 + \frac{P_{com}}{q_{rl}\eta_{mot}} \qquad (12.40)$$

The power input to the compressor $P_{com} = \dot{m}_r W_{in}$. Here, W_{in} represents the work input, in Btu/lb. Therefore, F_{rej} depends mainly on the difference between the condensing and evaporating pressure. F_{rej} is always greater than 1.

Types of Condensers

A condenser is a major system component of a refrigerating system. It is also an indirect contact heat exchanger in which the total heat rejected from the refrigerant is removed by a cooling medium, usually air or water. As a result, the gaseous refrigerant is cooled and condensed into liquid at the condensing pressure. Liquid refrigerant is often subcooled to a temperature up to 15°F below the saturated temperature at condensing pressure to conserve energy.

Based on the cooling medium used, condensers in refrigerating systems can be classified into the following three categories:

1. Water-cooled condensers
2. Air-cooled condensers
3. Evaporative condensers

In a *water-cooled condenser*, cooling water is used to remove condensing heat from the refrigerant. The cooling water is often the recirculating water from the cooling tower. It also could be from a lake, river, or well near the refrigeration plant. Groundwater discharged from the condenser may be treated and used for other purposes. If river or lake water is used as the cooling water, an effective water filter system, or sometimes a heat exchanger, may be necessary to prevent fouling of condenser tubes. Use of river, lake, or groundwater, is a once-through methods, as they cannot be recirculated and used as cooling water again.

If a refrigeration plant is near the seashore, sea water can be used as cooling water if it is more economical than water from the cooling tower. Sea water is corrosive to many metals. Copper-nickel alloy, stainless steel, and titanium are often used for water tubes in sea water condensers. An effective filter system is necessary to prevent dirt and impurities from blocking the water passages. Water treatment is often required to prevent the growth of living organisms. The whole sea water intake system may be very expensive. The use of sea water is also a once-through method.

Two types of water-cooled condensers are widely used for air conditioning and refrigeration purposes: shell-and-tube and double-tube condensers.

A *double-tube condenser* consists of two tubes, one inside the other, as shown in Fig. 12.19. Water is pumped through the inner tube while refrigerant flows in the space between the inner and outer tube in a counterflow arrangement. Because double-tube condensers provide limited condensing area, they are only used in small refrigerating systems such as indoor package units.

Shell-and-Tube Condensers

There are two types of shell-and-tube condensers: horizontal and vertical. *Vertical shell-and-tube condensers* usually incorporate a one-pass water flow arrangement. They consume more cooling water than the horizontal type and are often located outdoors because of their height. Vertical shell-and-tube condensers are sometimes used for ammonia refrigerating systems in industrial applications.

Horizontal shell-and-tube condensers are widely used in both comfort and process air conditioning systems. A horizontal shell-and-tube condenser using halocarbon refrigerant has an outer shell in which copper tubes are fixed in position by tube sheets, as shown in Fig. 12.20. As in the shell-and-tube liquid cooler, the diameter of the copper tubes is typically $\frac{5}{8}$ in. or $\frac{3}{4}$ in. Integral external fins, spaced between 19 and 35 fins/in. and at a height of 0.06 in. are usually used. Surface coating is also used to enhance the condensing coefficient of the tubes's outer surface on the refrigerant side. Spiral grooves or other turbulators are added to the inner surface to increase the water-side heat transfer coefficient.

Hot gas from the compressor enters at the top inlet and is distributed along the two sides of the baffle plate to fill the shell. Cooling water enters the copper tubes at the bottom of the condenser to provide effective subcooling. The water extracts heat from the hot gas and the liquid refrigerant and is discharged near the top of the shell. Water flow arrangements are the same as in shell-and-tube liquid coolers: one-pass, two-pass, and three-pass, with two-pass as the standard arrangement.

After heat has been removed, the hot gas is desuperheated, condensed into liquid, and subcooled to a lower temperature. Because of the high condensing pressure, the liquid refrigerant is forced into the liquid line. In a shell-and-tube condenser, usually about one sixth of the shell's volume is filled with liquid refrigerant for subcooling. This liquid level prevents gas bubbles from entering the liquid line.

Heat Transfer in Shell-and-Tube Condensers

As in shell-and-tube liquid coolers, heat transfer in shell-and-tube condensers can be calculated by using Eq. (12.32) with the following modifications:

- Heat is rejected from the refrigerant in the condenser, whereas heat is absorbed in the liquid cooler.
- Evaporating temperature T_{ev} should be replaced by condensing temperature T_{con}, in °F.

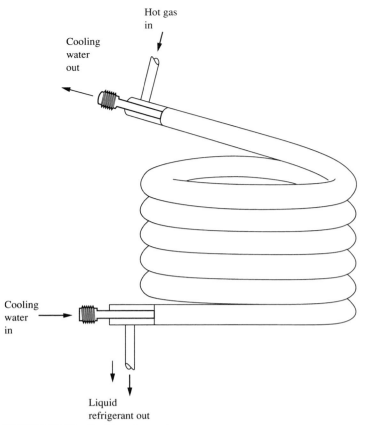

FIGURE 12.19 Double-tube condenser.

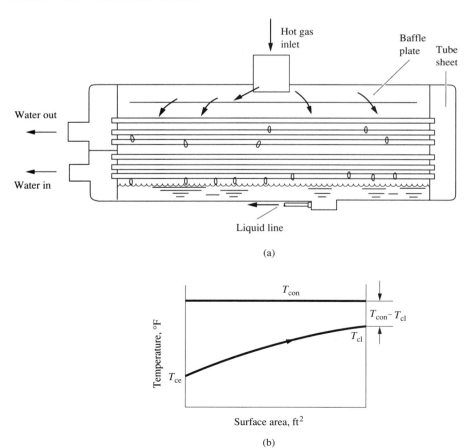

(a)

(b)

FIGURE 12.20 Shell-and-tube condenser with two-pass water flows arrangement: (a) schematic diagram and (b) heat transfer.

- $(T_{ee} - T_{el})$ should be replaced by $(T_{cl} - T_{ce})$. Here, T_{ce} and T_{cl} represent the cooling water entering and leaving the condenser, in °F. $(h_{rl} - h_{re})$ is replaced by $(h_{re} - h_{rl})$ and $(T_{ee} - T_{ev})$ replaced by $(T_{con} - T_{ce})$.

In 1915, Nusselt developed the correlation for laminar film condensation. Actual shell-side condensing coefficient h_{con}, in Btu/h · ft² · °F, is needed to correct for vapor shear, inundation due to tube bundle, surface effects, and enhancements. These factors are usually determined by experiment.

It is important to recognize that the shell side condensing coefficient h_{con} is a function of $1/(T_{con} - T_s)$. Here, T_s is the surface temperature of tube wall, in °F. Therefore, the condensing coefficient can be calculated as

$$h_{con} = h_o = C_{con} \left(\frac{1}{Q_{rej}/A_o} \right)^{1/3} \qquad (12.41)$$

where C_{con} = constant. The difference between the boiling and condensing heat transfer coefficient is that the greater the heat flux q/A_o, the higher the boiling heat transfer coefficient h_b and the lower the condensing coefficient h_{con}.

Parameters That Influence the Performance of Shell-and-Tube Condensers

TEMPERATURE DIFFERENCE $(T_{cl} - T_{ce})$. When the cooling water is from the cooling tower, the temperature difference between the cooling water leaving and entering the shell-and-tube condenser $(T_{cl} - T_{ce})$ has a direct impact on cooling water pump power. The performance of the cooling tower must also be considered. Usually, $(T_{cl} - T_{ce})$ is between 8 and 15°F.

When lake, river, ground, or sea water is used, $(T_{cl} - T_{ce})$ is often determined according to the scarcity of water and water temperature. $(T_{cl} - T_{ce})$ may be between 8 and 15°F, based on economic analysis.

CONDENSING TEMPERATURE T_{con}. In a shell-and-tube condenser, T_{con} is closely related to the temperature of cooling water T_{ce}. If $(T_{cl} - T_{ce})$ has been selected, then T_{cl} is a fixed value and T_{con} is determined by choosing an optimum $(T_{con} - T_{cl})$. This value is directly related to the size of the shell-and-tube condenser, as shown in Fig. 12.20b. As in the shell-and-tube liquid cooler, a smaller $(T_{con} - T_{cl})$ means a lower T_{con} and a larger condensing surface area, whereas a larger $(T_{con} - T_{cl})$ indicates a higher T_{con} and a smaller condenser. Condenser manufacturers usually adopt a $(T_{con} - T_{cl})$ between 6 and 10°F (4 to 7°F for energy-efficient models).

EFFECT OF OIL. If the lubrication oil is miscible with the refrigerant (like many halocarbon refriger-

ants), experiments show that there is no significant effect on condensing coefficient T_{con} when oil concentration is lower than 7 percent. In actual operation, oil concentration is usually lower than 3 percent.

SUBCOOLING. Subcooling increases the refrigeration capacity and coefficient of performance. However, subcooling also uses part of the surface area in the condenser to cool the liquid refrigerant at the bottom of the condenser. The surface heat transfer coefficient on the subcooled liquid refrigerant side is smaller than the condensing coefficient h_{con}. Subcooling also depends on the temperature of entering cooling water T_{ce}. For a shell-and-tube condenser, it may vary between 2 and 8°F.

CONDENSING CAPACITY. Parameters that influence the condensing capacity Q_{rej} of a shell-and-tube condenser are mainly $U_o, A_o, \Delta T_m$, and Q_{rej}/A_o. Size and capacity of the condenser must be matched with the evaporator and compressor. Condensing capacity is always rated at certain operating conditions.

Part-Load Operation

When the heat flux Q_{rej}/A_o reduces at the condensing surface of the shell-and-tube condenser during part-load operation, h_{con} increases. Therefore, the reduction of Q_{rej} at part-load in a shell-and-tube condenser is mainly caused by the drop of condensing temperature T_{con}, and therefore, a smaller ΔT_m between refrigerant and cooling water. At part-load operation, the degree of subcooling is also reduced because of the drop of condensing temperature T_{con}.

12.8 AIR-COOLED CONDENSERS

Construction of Air-Cooled Condensers

Air-cooled condensers use air to extract the latent heat of condensation released by the refrigerant during condensation.

An air-cooled condenser generally consists of a condenser coil with several refrigerant circuits to condense the gaseous refrigerant into liquid, and a subcooling coil at a lower level connected in series with the condensing coil through a liquid accumulator, as shown in Fig 12.21. The condenser coil is usually equipped with copper tubes and aluminum fins when halocarbon is used as the refrigerant. The diameters of the tubes are usually between $\frac{1}{4}$ and $\frac{3}{4}$ in., and the fin spacing is generally 8 to 18 fins/in. On the inner surface of the copper tubes, micro-fins (typically 60 fins/in.) with a height of 0.008 in. are used as in DX-coils. The condenser coil usually has 2 to 3 rows of tubes because of the lower air-side pressure drop provided by the propeller fan.

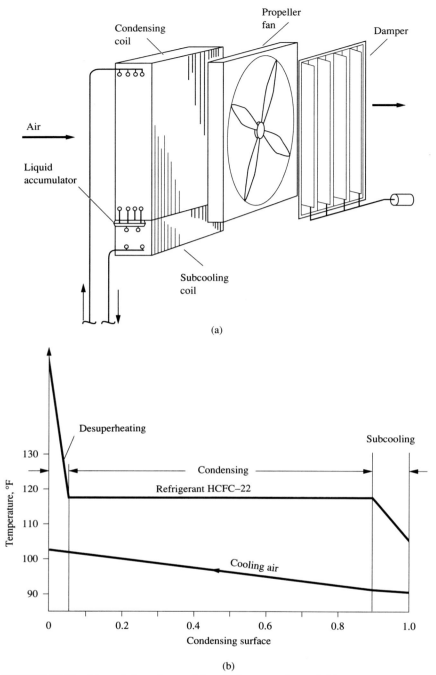

(a)

(b)

FIGURE 12.21 Air-cooled condenser: (a) typical construction and (b) temperature curves of refrigerant HCFC–22 and cooling air.

Hot gas from the compressor enters the refrigerant circuits at the top. This arrangement provides flexibility between the condensing and subcooling areas. As the condensate increases, part of the condensing area can also be used as subcooling for the storage of liquid refrigerant. A receiver is necessary only when all the liquid refrigerant cannot be stored in the condenser coil during the shut-down period of the refrigerant plant in winter.

Cooling air is usually forced through the coil by a propeller fan, as shown in Fig. 12.21. A propeller fan has a lower fan total pressure and large volume flow rate, which makes it more suitable for air-cooled condensers. Fans are usually located downstream from the coils in order to provide an even airstream through the coils. A damper may be installed after or before the fan to modulate the air volume flow rate. In a small air-cooled condenser, the coils, propeller fan, and damper may be installed in-line horizontally. In large air-cooled condensers, condensing and subcooling coils are usually located on two sides, and the propeller fans and dampers are at the top of the unit.

Heat Transfer Processes and Temperature Curves

Heat transfer processes in a typical air-cooled condenser using HCFC–22 as refrigerant are divided into three stages, as shown in Fig. 12.21b: desuperheating of hot gas, condensation of liquid refrigerant, and subcooling. Desuperheating uses about 5 percent of the condensing surface area, condensation uses 85 to 90 percent, and subcooling uses 5 to 10 percent.

For a condensing coil with a depth of 2 to 3 rows, air flow and refrigerant flow are usually in a combined counterflow and crossflow arrangement. Air enters the condensing coil at a temperature equal to the summer outdoor design dry bulb temperature T_o during summer. In a year of average weather, T_o is the temperature that will be exceeded about $0.025 \times 2928 = 75$ hours in summer at the location of the air-cooled condenser. If the air-cooled condenser is installed on the roof; an additional 3 to 5°F should be added to summer outdoor design dry bulb to account for the heating effect of the roof. Let $T_o = 90°F$. For a typical condenser, entering air temperature is gradually raised to T_{al} of about 102°F when it leaves the condensing coil.

Hot HCFC–22 gas may enter the top of the condensing coil at a temperature $T_{gas} = 170°F$ and is desuperheated to a saturated temperature of about 115°F. Gaseous refrigerant condenses into liquid and releases the latent heat of condensation. As the refrigerant flows through the condensing coil, the condensing pressure drops, typically by about 8 psi, before entering the subcooling coil. Because no latent heat is released, the temperature of the subcooled liquid refrigerant drops sharply to 105°F when it leaves the air-cooled condenser.

Equations to calculate heat transfer between refrigerant and air in an air-cooled condenser are similar to those for a water-cooled condenser, except that the temperature of cooling water should be replaced by the temperature of cooling air, and the air-side heat transfer coefficient h_o can be calculated from Eq. (12.7). The condensing coefficient for horizontal tubes can be found in *ASHRAE Handbook 1989, Fundamentals*.

Cooling Air Temperature Increase and Volume Flow

For a specific Q_{rej}, the temperature increase between air entering and leaving air-cooled condenser ($T_{al} - T_o$) depends mainly on the volume flow rate of cooling air per unit of total heat rejection $\dot{V}_{ca}/Q_{u.rej}$. Here, Q_{rej} is often expressed in refrigeration ton capacity at the evaporator, therefore, $\dot{V}_{c.a}/Q_{u.rej}$ is expressed in cfm/ton.

A smaller $\dot{V}_{ca}/Q_{u.rej}$ usually results in a lower air-side heat transfer coefficient h_o, a greater ($T_{al} - T_o$), a higher T_{con}, a lower condenser fan power, and a greater log-mean temperature difference ΔT_m between refrigerant and air. The capacity of the air-cooled condenser is reduced. Conversely, a larger $\dot{V}_{ca}/Q_{u.rej}$ and a smaller ($T_{al} - T_o$) means a lower T_{con}, a smaller ΔT_m, and greater fan power consumption, and probably more noise from the propeller fans. Based on cost analysis, the optimum value of $\dot{V}_{ca}/Q_{u.rej}$ is usually between 600 and 1200 cfm/ton refrigeration capacity at the evaporator. In comfort air conditioning systems, if the heat rejection factor $F_{rej} = 1.25$, air volume flow may be between 40 and 80 cfm per 1 MBtu/h total heat rejection. When $\dot{V}_{ca}/Q_{u.rej} = 900$ cfm/ton refrigeration capacity, ($T_{al} - T_o$) is around 13°F. Fan power consumption in air-cooled condensers is usually between 0.1 and 0.2 hp/ton refrigeration capacity.

Air-Cooled Condenser Temperature Difference (CTD)

The *condenser temperature difference* for an air-cooled condenser is defined as the difference in saturated condensing temperature corresponding to the refrigerant pressure at the inlet and the air intake dry bulb temperature ($T_{con.i} - T_o$). The total heat rejection Q_{rej} of an air-cooled condenser is directly proportional to its CTD. The Q_{rej} of an air-cooled condenser operated at a CTD of 30°F is approximately 50 percent greater than if the same condenser is operated at a CTD of 15°F. On the other hand, for a specific value of Q_{rej}, an air-cooled condenser selected with a CTD of 15°F is larger than a condenser selected with a CTD of 30°F.

Air-cooled condensers are rated at a specific CTD related to the evaporating temperature T_{ev} of the refrigeration system in which the air-cooled condenser is installed. Typical CTD values are as follows:

T_{ev}, °F	CTD, °F
45, for air conditioning	20–30
20	15–20
−20 to −40	10–15

A refrigeration system with a lower T_{ev} is more economical to equip with a large condenser with a small CTD.

Condensing Temperature, Subcooling, and Oil Effect

A high condensing temperature T_{con} means a higher condensing pressure p_{con} and a higher discharge temperature. High T_{con} and p_{con} result in a higher compressor power input and, more often, unsafe operating conditions. Both are undesirable. At a specific location, an air-cooled condenser always has a higher T_{con} than a water-cooled condenser.

A high T_{con} and p_{con} may be caused by an undersized air-cooled condenser or a low $\dot{V}_{ca}/Q_{u.rej}$ value. It may be the result of a high T_o in a specific location or a high roof temperature. It may also be caused by a lack of cooling air or warm air recirculation, which will be discussed below.

In air-cooled condensers, the degree of subcooling usually ranges from 10 to 20°F.

Based on the results of Schlager et al., (1989) for HCFC–22 in micro-fin tubes, the condensing heat transfer coefficient decreased by 15 to 20 percent with a 150 SSU(Saybolt Seconds Universal viscosity) and 300 SSU oil concentration of 2.6 percent. Experiments also showed a larger degradation of heat transfer with the more viscous oil.

Clearance, Warm Air Recirculation, and Dirt Clogging

Air-cooled condensers must provide clearance between the condenser coil and the wall or adequate space between two condenser coils, as specified in the manufacturer's catalog, to provide sufficient cooling air. Sufficient clearance can also prevent recirculation of warm air discharged from air-cooled condensers. Generally, the clearance should not be less than the width of the condenser coil.

Recirculation of discharged warm air from the air-cooled condenser raises the average air temperature entering the condenser coil T_o and, therefore, increases T_{con} and reduces the capacity of the condenser. Recirculation of warm air may be caused by insufficient clearance, insufficient upward discharge air velocity from the condenser, and wind effect.

When an air-cooled condenser is located in industrial areas or in locations where outdoor air has a high concentration of dirt, solid particles, or sometimes pollens and seeds from trees and plants, dirt and particles may clog the condenser coil and considerably reduce the volume flow of cooling air. In these circumstances, the surface of the condenser coil must be inspected and cleaned regularly.

Low Ambient Air Control

When ambient air or outdoor air temperature is low, condensing pressure may drop considerably. If the condensing pressure drops below a certain value because of low outdoor temperature at part-load, a reciprocating vapor compression refrigerating system using an expansion valve as the throttling device will not be operate normally. The supply of liquid refrigerant to the evaporator may be reduced. If refrigerant feed is insufficient, frost may form on the outer surface of the coil. Therefore, various methods of pressure control for air-cooled condensers have been developed to prevent the condensing pressure from falling below a specific value. These include the following:

- Duty cycling, or turning the fans on and off, so that air is not forced through part of the condensing coils. This reduces the cooling air volume flow as necessary.
- Modulating the air dampers to decrease the volume flow of the cooling air
- Reducing the fan speed so that less air passes through the condensing coil.

Low ambient air control should follow the manufacturer's instructions. One manufacturer specifies the following low ambient air control scheme:
Control energized at 65°F ± 2°F;
Control deenergized at 70°F ± 2°F;
Minimum entering outdoor temperature, 25°F.

Selection of Air-Cooled Condensers

Total heat rejection Q_{rej} of air-cooled condensers varies at different cooling air intake temperatures. During selection of an air-cooled condenser, the following steps should be taken:

- Calculate the total heat rejection of the refrigerating system.

- Compare the face area of the condenser coil at the same cooling air intake temperature.

- Choose an air-cooled condenser with the required Q_{rej} at a specific cooling air intake temperature with the appropriate T_{ev}.

- Select a condenser for comfort air conditioning whose $\dot{V}_{ca}/Q_{u.rej}$ value is between 800 and 1000 cfm/ton refrigerating capacity at the evaporator.

- Check the sound power level of the propeller fans. Noise attenuation is always expensive.

12.9 EVAPORATIVE CONDENSERS

Condensation Process

An *evaporative condenser* uses the evaporation of a water spray to remove the latent heat of condensation of the refrigerant during condensation. It is actually a simplified combination of a water-cooled condenser and cooling tower.

An evaporative condenser consists of a condensing coil, a water spray bank, a forced draft or induced draft fan, an eliminator, a circulating pump, a water basin, and a casing, as shown in Fig. 12.22.

Water is sprayed over the outside surface of the condensing coil. Because of the evaporation of water, heat is extracted through the wet surface of the coil. The rest of the spray falls and is collected in the water basin. Air enters from the inlet located just above the water basin. It moves up through the condensing coil, spray nozzles, and water eliminator, is extracted by the fan, and is finally discharged at the top outlet in a counterflow arrangement. Other air flow and water flow arrangements have also been developed.

The condensing coils are usually bare pipes of galvanized steel, copper, or stainless steel. The high heat transfer coefficient of the wetted surface does not need fins to increase its outer surface area. Bare pipes are not easy to foul and are easier to clear.

Heat Transfer in Evaporative Condensation

Heat transfer from the saturated air film to the ambient air and also from the condensing refrigerant through the metal tube to the wetted surface can be similarly calculated from Eq. (12.25) as in the wet part of the water cooling and dehumidifying coils

$$h_{wet}\eta_s A_o(T_{os} - T_o) = h_i A_i(T_{con} - T_{is}) \quad (12.42)$$

where T_{os}, T_{is} = outer and inner surface temperature of the tube, °F. The number of transfer units, rate of heat transfer, coil effectiveness, and parameter $U_o A_o$

can be calculated from Eqs. (12.17), (12.20), (12.22), and (12.23).

In Eq. (12.42), the heat transfer coefficient for wetted surface h_{wet} can be calculated by Eq. (12.27). Because of the higher T_{os} in an evaporative condenser, for instance, $T_{os} = 80°F$, and the ratio $h_{wet}/h_o = m''/c_{pa} = 1.085/0.245 = 4.4$, so h_{wet} is often 4 or 5 times greater than h_o.

Cooling Air and Water Spraying

In evaporative condensers, drift carrying over the water eliminator is about half the amount of water evaporated for extracting the latent heat of condensation. According to *ASHRAE Handbook 1992, HVAC Systems and Equipment*, the total water consumption, including the evaporation and drift carryover, is about 2.4 gal/h per refrigeration ton at the evaporator for air conditioning and 3 gal/h per refrigeration ton for cold storage.

The water circulation rate in an evaporative condenser is far less than in the cooling tower for the same total heat rejection. Usually, it is around 1.6 to 2 gpm/ton refrigeration capacity at the evaporator for air conditioning systems and 3 gpm/ton for cooling towers.

In evaporative condensers, air should not be used for sensible cooling of the outer surface of the condensing coil. An air velocity between 400 and 700 fpm should be maintained when it flows over the condensing coil, in order to provide a higher h_{wet} for the latent heat transfer between the wetted surface and the ambient air. Sufficient air volume flow is necessary to maintain a desirable enthalpy difference between the saturated film and air ($h_s - h_a$).

Site Location and Low Ambient Air Control

Evaporative condensers are usually located outdoors. If they are located indoors, air is ducted outdoors and centrifugal fans are generally used to overcome the greater fan total pressure required.

The evaporative condenser should be located as near to the compressor as possible. If it is located at a distance from the compressor, the discharge pipe should be selected to have a pressure drop of 2°F of saturated temperature difference of the refrigerant between the compressor and the condenser.

If an evaporative condenser is located outdoors and will be operated during winter, its water basin and pump can be located indoors in a heated space. A water heater can be installed in the water basin, and the water should be drained back to the water basin as soon as the pump ceases to operate in order to prevent freezing.

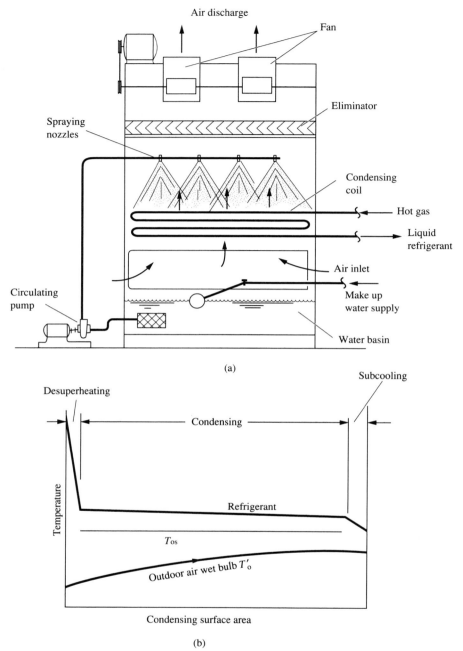

FIGURE 12.22 A typical evaporative condenser: (a) schematic diagram and (b) temperature curves.

In order to prevent the condensing pressure from falling below a certain value at which the expansion valve or other system components are difficult to operate, the capacity of the evaporative condenser is also reduced by means of fan cycling, dampers, and fan speed modulation at low ambient temperatures, as in air-cooled condensers.

Selection and Installation

The procedure for selecting and installing an evaporative cooler is as follows:

1. Before selecting an evaporative condenser, calculate the total heat rejection of the refrigerating system Q_{rej}, in MBtu/h.

2. According to the refrigerant used in the system, select the appropriate material for the condensing coil. Copper tubes should be used for halocarbon refrigerants and stainless steel or galvanized steel should be used for ammonia.

3. Select a condensing coil that can provide a lower condensing temperature—below 95°F, if possible. This will save a lot of energy and reduce scale accumulation on the outer surface of the condensing coil significantly compared with a $T_{con} = 100$ or 105°F.

4. Based on the location of the evaporative condenser and the sound control requirements, it is often best to select a centrifugal fan. Check the volume flow rate and the total pressure increase that can be provided by the fan to overcome the pressure drop of the condensing coil and eliminator or air duct, if any.

5. Check the flow rate of the water spray of the circulating pump.

6. For outdoor installation, select a casing and other system components that are weatherproof and corrosion resistant.

7. Install the evaporative condenser in a location where there is ample space for outdoor air intake and discharge. The influence of nearby exhaust systems should be avoided.

Evaporative condensers have the advantage of lower condensing temperature and lower installation and energy costs. They are widely used in industrial applications and have been adopted more and more frequently in commercial buildings recently. The following reasons may explain why they were not previously popular in commercial and institutional applications:

- Monthly maintenance, inspection, and cleaning are required to prevent buildup of scale on the condensing coil.

- Designers and operators have less experience with evaporative condensers.

- Heat rejected from the evaporative condenser is more difficult to use in winter heating for energy conservation than from a water-cooled condenser.

12.10 HEAT REJECTION SYSTEMS

Types of Heat Rejection Systems

A heat rejection system is a system that extracts condensing heat from refrigerant in the condenser of a refrigerating system, uses some or all of the condensing heat for winter heating or other purposes, if possible, or transfers it to a heat sink, usually atmospheric air or surface water.

A heat rejection system consists of a condenser, heat exchangers, fan(s) or pump, piping work, and duct work if necessary.

Four types of heat rejection systems are commonly used in vapor compression refrigerating systems, as shown in Fig. 12.23:

1. Water-cooled condensers using lake, river, ground, or sea water directly. A water-filtering device is often required to prevent dirt and solid particles from fouling the condenser tubes. Water treatment may be required depending on the quality of the cooling water.

2. Water-cooled condensers using recirculation water from the cooling towers, or a water-cooled condenser and cooling tower combination. This type of heat rejection system is most widely used in large refrigerating plants for commercial buildings.

   Cooling water leaves the water-cooled condenser and enters the top of the cooling tower. It is sprayed over the fills and cooled to a lower temperature by means of evaporation. The cooling water then passes through a strainer and is chemically treated, if necessary. Finally, it is extracted by a water pump and discharged to the water-cooled condenser again.

3. Air-cooled condensers which are widely used in small and medium refrigerating plants or in locations where water is scarce.

4. Evaporative condensers, which extract condensing heat from the coil directly by means of evaporative cooling effect.

Comparison between Various Systems

Systems should be compared at summer design conditions at a specific location with the following outdoor air and water temperatures:

Outdoor dry bulb temperature, 90°F
Outdoor wet bulb temperature, 78°F
Lake, river, or sea water temperature, 78°F

The evaporating temperature and refrigeration capacity are the same for all the heat rejection systems.

During comparison, factors such as condensing temperature, cooling air and water flow rate, maintenance, and initial and energy costs should be considered. The initial cost and energy cost must be measured for the whole refrigerating plant. In addition to the factors mentioned above, designers should also recognize that water is becoming more scarce in many locations in United States.

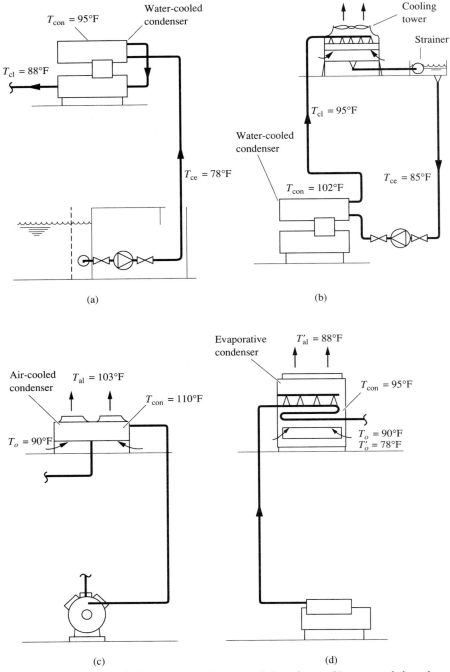

FIGURE 12.23 Heat rejection systems: (a) water-cooled condenser, (b) water-cooled condenser and cooling tower, (c) air-cooled condenser, and (d) evaporative condenser.

is whether the released condensing heat can be easily used for winter heating or the equipment itself can be used for energy conservation. For example, a cooling tower can be used as a part of water economizer to replace part of the refrigeration. Results of this comparison are listed in Table 12.4 for reference.

For small refrigerating systems, air-cooled condensers are still widely used. For large refrigerating plants where condensing heat is not used for winter heating, evaporative condensers should be compared with a water-cooled condenser and cooling tower combination through detailed analysis.

TABLE 12.4
Comparison of various heat rejection systems at summer design conditions

	Water-cooled condenser	Water-cooled condenser and cooling tower	Air-cooled condenser	Evaporative condenser
Condensing temperature, °F	95	102	110	95
Cooling air volume flow rate, cfm/ton*			600–1200	Smaller than air-cooled condenser
Cooling water, gpm/ton*	3	3	3	1.6–2
Make-up water, gal/h · ton‡		2.4		2.4
Maintenance			Periodical cleaning of coil when outdoor air is not clean	Monthly inspection and cleaning of coil
Initial cost (refrigeration plant)	Depending on the initial cost on water intake	Lower than air-cooled condenser in large sizes	Higher for larger refrigeration plant	Lower
Energy consumption (refrigeration plant)	Lower	Lower than air-cooled condenser	Higher at design load	Lower
Condensing heat used for winter heating or equipment itself used for evaporative cooling	Easier	Best applied	Between water-cooled and evaporative condenser	Difficult
Application	Large	Medium and large	Medium and small size, or where water is scarce	Medium and large

Comparison is based on outdoor 90°F dry bulb, 78°F wet bulb; lake, river, or sea water temperature = 78°F.
‡Ton refrigeration capacity at the evaporator.

REFERENCES AND FURTHER READING

ASHRAE, *ASHRAE Handbook 1988, Equipment*, ASHRAE Inc., Atlanta, GA, 1988.

ASHRAE, *ASHRAE Handbook 1989, Fundamentals*, ASHRAE Inc., Atlanta, GA, 1989.

ASHRAE, *ASHRAE Handbook 1992, HVAC Systems and Equipment*, ASHRAE Inc., Atlanta, GA, 1992.

ASHRAE, *ASHRAE Handbook 1990, Refrigeration Systems and Applications*, ASHRAE Inc., Atlanta, GA, 1990.

ASHRAE, "Waterside Fouling Resistance Inside Condenser Tubes, Research Note 31(RP106)" *ASHRAE Journal*, p. 61, June 1982.

Benner, R. L., and Ramsey, J., "Evaporative Condensers," *Heating/Piping/Air Conditioning*, pp. 63–65, August 1987.

Braun, J. E., Klein, S. A., and Mitchell, J. W., "Effective Models for Cooling Towers and Cooling Coils," *ASHRAE Transactions*, Part I, pp. 164–174, 1990.

Braun, R. H., "Problem and Solution to Plugging of a Finned-Tube Cooling Coil in an Air Handler," *ASHRAE Transactions*, Part IB, pp. 385–387, 1986.

Chen, J. C., *Correlation for Boiling Heat Transfer to Saturated Fluids in Connective Flow*, Ind. Eng. Chem. Process Design Develop., Vol. 5, No. 3, 1966.

Cole, R. A., "Avoiding Refrigeration Condenser Problems: I," *Heating/Piping/Air Conditioning*, pp. 97–108, July 1986.

Delaney, T. A., Maiocco, T. M., and Vogel, A. G., "Avoiding Coil Freezeup," *Heating/Piping/Air Conditioning*, pp. 83–85, December 1984.

Denkmann, J. L., "Refrigerant Coils for Air Conditioning and Process Loads," *Heating/Piping/Air Conditioning*, pp. 67–78, December 1986.

HPAC, "ARI Guideline—Fouling Factors," *Heating/Piping/Air Conditioning*, pp. 109–110, February 1988.

Holtzapple, M. T., and Carranza, R. G., "Heat Transfer and Pressure Drop of Spined Pipe in Cross Flow Part III: Air-Side Performance Comparison to Other Heat Exchangers," *ASHRAE Transactions*, Part II, pp. 136–141, 1990.

McQuinston, F. C., "Finned Tube Heat Exchangers: State of the Art for the Air Side," *ASHRAE Transactions*, Part I, pp. 1077–1085, 1981.

Schlager, L. M., Pate, M. B., and Bergles, A. E., "A Comparison of 150 and 300 SSU Oil Effects on Refrigerant Evaporation and Condensation in a Smooth Tube and a Micro-Fin Tube," *ASHRAE Transactions*, Part I, pp. 387–397, 1989.

Sheringer, J. S., and Govan, F., "How To Provide Freeze-Up Protection for Heating and Cooling Coils," *Heating/Piping/Air Conditioning,* pp. 75–84, December 1985.

Stamm, R. H., "Industrial Refrigeration: Condensers," *Heating/Piping/Air Conditioning,* pp. 101–105, November 1984.

Tao, W., and Chyi, D. P., "Coil Design and Selection," *Heating/Piping/Air Conditioning,* pp. 66–73, December 1985.

Wang, S. K., *Air Conditioning* Vol. 3, Hong Kong Polytechnic, Hong Kong, 1987.

Wang, S. K., *Principles of Refrigeration* Vol. 2, Hong Kong Polytechnic, Hong Kong, 1984.

Webb, R. L., "Shell-Side Condensation in Refrigerant Condensers," *ASHRAE Transactions*, Part I, pp. 5–25, 1984.

Webb, R. L., Choi, K. D., and Apparao, T. R., "A Theoretical Model for Prediction of the Heat Load in Flooded Refrigerant Evaporators," *ASHRAE Transactions*, Part I, pp. 326–338, 1990.

Webb, R. L., Apparao, T. R., and Choi, K. D., "Prediction of the Heat Duty in Flooded Refrigerant Evaporators," *ASHRAE Transactions*, Part I, pp. 339–348, 1990.

Wood, R. A., Sheffield, J. W., and Sauer, H. J., Jr., "Thermal Contact Conductance of Finned Tubes: The Effect of Various Parameters," *ASHRAE Transactions* Part II, pp. 798–810, 1987.

EVAPORATIVE COOLING AND EVAPORATIVE COOLERS

13.1 EVAPORATIVE COOLING

Evaporative cooling is an air conditioning process that uses the evaporation of liquid water to cool an airstream directly or indirectly so that the final dry bulb or dry and wet bulb temperatures of the airstream being cooled are lower than before undergoing the evaporative process.

An evaporative cooling system is an air conditioning system in which air is evaporatively cooled.

It consists of evaporative coolers, fans, filters, a mixing box, dampers, and other components. An evaporative cooler is a device in which the evaporative cooling process takes place. An air conditioning system may contain any combination of the following: evaporative coolers, water cooling coil, DX-coil, and desiccant drier.

There are three types of evaporative cooling systems: (1) direct evaporative, (2) indirect evaporative, and (3) indirect-direct evaporative, as shown in Fig. 13.1.

In a direct evaporative cooling system, the airstream to be cooled comes directly into contact with the water spray or wetted medium, as shown in Fig. 13.1a. Air enters the direct evaporative cooler at point 1 and leaves at point 2. The release of the latent heat of evaporation from the directly cooled airstream low-

ers the airstream's temperature; the airstream's humidity ratio increases because of the added water vapor.

In an indirect evaporative cooling system, the primary airstream to be cooled is separated from a wetted surface by a flat plate or a tube wall, as shown in

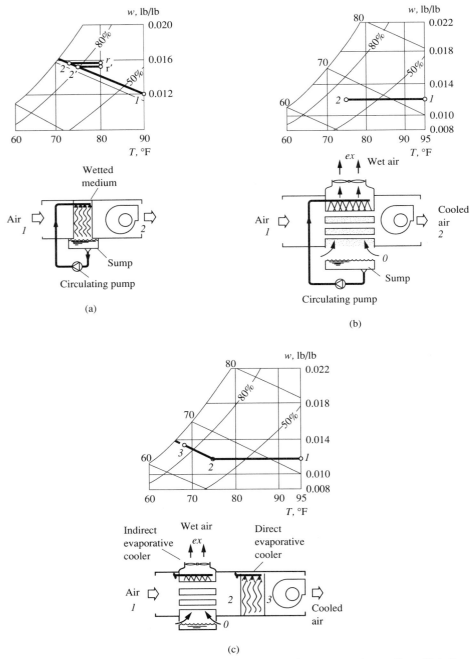

FIGURE 13.1 Types of evaporative cooling systems: (a) direct evaporative cooling; (b) indirect evaporative cooling; and (c) indirect-direct evaporative cooling.

Fig. 13.1b. A secondary airstream flows over the wetted surface so that liquid water will evaporate and extract heat from the primary airstream through the flat plate or tube wall. The cooled air does not directly contact the evaporating liquid.

The purpose of the secondary airstream is to cool the wetted surface, evaporatively approaching the wet bulb temperature, and to absorb the evaporated water vapor. This wet secondary airstream is known as *wet air*.

In an indirect evaporative cooling process, the airstream's humidity ratio remains constant because the air to be cooled does not contact the evaporating liquid. This process is represented by horizontal line *12* on the psychrometric chart shown in Fig. 13.1b.

In an indirect-direct evaporative cooling system, an indirect cooler and a direct cooler are usually connected in series to form a two-stage evaporative cooling system in order to increase their evaporative cooling effect. Figure 13.1c shows a typical indirect-direct evaporative cooling system. The combined evaporative cooling process is illustrated by two connected straight lines (lines *12* and *23*) on the psychrometric chart.

Both direct and indirect evaporative coolers often provide only sensible cooling and humidification. They are unable to dehumidify a mixture of outdoor and recirculating air unless the outdoor wet bulb $T_o' < 60°F$. This is an important difference between evaporative cooling and refrigeration processes. A refrigeration process allows the air to be cooled and dehumidified at any outdoor climate.

Early evaporative cooling systems were of the direct evaporative type. They produced cool and humid air in the hot summers of arid climates. Because of the development of indirect evaporative coolers and multistage evaporative cooling systems, evaporative cooling may partially or entirely replace refrigeration in many applications in which both evaporative and refrigeration systems are present. Evaporative cooling systems combined with desiccant driers are widely used to offset part of the refrigeration load in air conditioning systems, which will be discussed in Chapter 25.

13.2 DIRECT EVAPORATIVE COOLING

Saturation Effectiveness

Saturation effectiveness is an important index used to assess the performance of a direct evaporative cooler. As mentioned in Section 3.3, saturation effectiveness ϵ_{sat} is defined as

$$\epsilon_{sat} = \frac{T_{ae} - T_{al}}{T_{ae} - T_{ae}^*} \qquad (13.1)$$

where T_{ae}, T_{al} = temperature of the air entering and leaving the direct evaporative cooler, °F

T_{ae}^* = thermodynamic wet bulb temperature of the entering air, °F

For a direct evaporative cooler, if T_{ae} and ϵ_{sat} are known, T_{al} can be determined from a psychrometric chart. The value of ϵ_{sat} depends on the following factors:

1. Face velocity v_a of the air flowing through the direct evaporative cooler, in fpm. For a specific cooler (water-spraying or water-dipping) with fixed face area A_a (in ft²) and a given water flow rate $\dot{m}_w$ (in gpm), a higher v_a yields the following results:
   - Higher volume flow rate $\dot{V}_a$ of the cooled air, in cfm
   - Greater evaporative cooling effect $q_{ev.c}$, in Btu/h, which can be calculated as follows:

   $$q_{ev.c} = v_a A_a \rho_a c_{pa}(T_{ae} - T_{al}) \qquad (13.2)$$

   where ρ_a = air density, lb/ft³
   c_{pa} = specific heat of moist air, Btu/lb · °F

   - Lower saturation efficiency η_{sat}, mainly due to a smaller water flow rate for each ft³ of cooled air.

   For most direct evaporative coolers used for comfort cooling, the face velocity should usually be no greater than 600 fpm in order to prevent the carryover of water droplets. Otherwise, a water eliminator should be installed, which significantly increases the air-side pressure drop.

2. Water-air ratio $\dot{m}_w/\dot{m}_a$. This is the ratio of the mass flow rate of spraying water to the mass flow rate of cooled air, both in lb/min. A greater $\dot{m}_w/\dot{m}_a$ indicates a comparatively greater contact area between air and water and, therefore, a higher ϵ_{sat}.

3. Configuration of the wetted surface. Wetted media that provide a greater contact surface and a longer contact time between the air and water yield higher values of ϵ_{sat}.

System Characteristics

When a direct evaporative cooler is used to supply cooled air to maintain a space temperature of 80°F during the summer, it is important to realize that a higher ϵ_{sat} always means a point nearer to the saturation curve, point 2. For a predetermined space temperature $T_r = 80°F$, it means a higher space relative humidity φ_r, as shown in Fig. 13.1a.

In a direct evaporative cooling process, recirculating water is usually used in order to save water, which is often more economical. The temperature of the recirculating water always approaches the wet bulb temperature of the cooled air.

Because the air is sprayed or in contact with dipped water, direct evaporative cooling provides a certain degree of air cleaning. However, if the cooled air contains a great deal of dirt or particulate matter, an additional filter should be used to prevent clogging of the wetted medium or nozzles.

Other parameters that should be considered to assess the performance of a direct evaporative cooler include

- Use of fresh water or make-up water, usually expressed in gal/h per 1000 cfm of cooled air
- Air-side pressure drop, in. WG

Types of Direct Evaporative Coolers

As a self-contained unit that can independently provide cooled air for a conditioned space, a direct evaporative cooler consists of the following: a wetted medium, a fan (which is usually a centrifugal fan to provide the required system total pressure loss and a lower noise level), and a sump at the bottom. For water spraying systems, a circulating pump and piping connection are needed to distribute water evenly. To dip water on the medium from the top (except for rotary evaporative coolers), air filters, dampers, and an outer casing are necessary. Provisions should be made to bleed off the water in order to prevent mineral buildup.

Direct evaporative coolers can be categorized according to the characteristics of the wetted medium as follows:

- *Air washers.* An air washer, or water-spraying chamber, is itself a direct evaporative cooler. The characteristics and performance of air washers were discussed in detail in Chapter 11.
- *Evaporative pads.* These media are generally made of 2-in. thick aspen wood fibers with the necessary chemical treatment and additives to increase wettability and to prevent the growth of microorganisms, as shown in Fig. 13.2a. Evaporative pads are mounted in removable galvanized steel or plastic frames. Because evaporative pads require comparatively lower face velocities, in a self-contained direct evaporative cooler integrated with a centrifugal fan, three sides of the fan cabinet are often mounted with evaporative pads to increase its surface area.
- *Rigid media.* These are sheets of rigid and corrugated material made of plastic, impregnated cellulose, or fiberglass, as shown in Fig. 13.2b. Air and water typically flow in a crossflow arrangement so

that horizontal channels for air flow and vertical channels for water flow meet between two corrugated sheets. The depth of the rigid medium is typically 12 in. in the direction of airflow but may vary from 8 to 16 in. Rigid media need no supporting frame. They have lower air pressure drops and can easily be cleaned by water flushing.

- *Rotary wheel.* A wetted medium in the shape of a rotary wheel is made of corrosion-resistant materials such as plastic, impregnated cellulose, fiberglass, or copper alloy, as shown in Fig. 13.2c. The depth of the wheel along the airstream direction is from 6 to 10 in. The rotary wheel is often driven by a motor and gear box and rotates slowly at a speed of 1 to 2 rpm. The bottom of the wheel is submerged in a water tank. Air flows through various channels of the medium in the direction along the depth of the rotary wheel.

Operating Characteristics

Table 13.1 lists the operating characteristics of various direct evaporative coolers. The evaporative pad is the traditional type of direct evaporative cooler, widely used in residential and small commercial buildings. It has a low initial cost and is easy to operate and maintain.

Rigid media such as impregnated cellulose need no support structure, do not emit debris, and have a service life as long as that of aspen pads. They withstand a comparatively higher face velocity, provide a lower air pressure drop, and have a slightly greater saturation efficiency than aspen pads.

The rotary wheel has a more complicated structure. However, it has no water-recirculating system. It is easier to connect in series with other refrigerative coolers or desiccant driers in an air conditioning systems for energy-efficient and cost-effective operation.

An air washer is a large-capacity, bulky, and expensive direct evaporative cooler. It is usually used for both humidification and evaporative cooling in industrial applications. The saturation effectiveness values for direct evaporative coolers usually range from 0.75 to 0.95.

13.3 INDIRECT EVAPORATIVE COOLING

Indirect Evaporative Coolers and the Cooling Process

Fig. 13.3 shows a typical indirect evaporative cooler made by an Australian manufacturer. The main components of this cooler are a plate heat exchanger, a water spray and recirculating system, an outdoor air intake with filters, a supply fan and exhaust fan con-

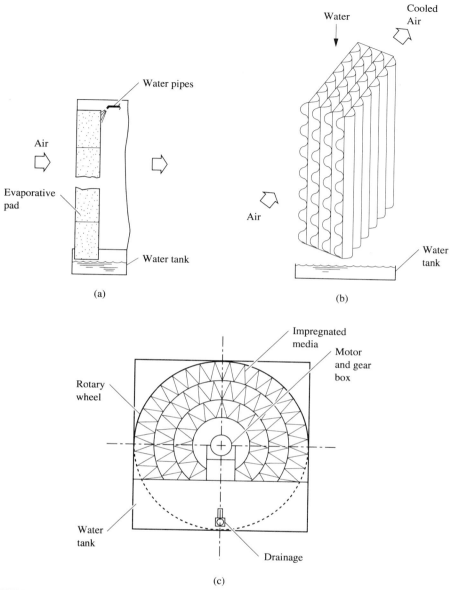

FIGURE 13.2 Wetted media for direct evaporative coolers: (a) evaporative pad; (b) rigid media; and (c) rotary wheel.

TABLE 13.1
Operating characteristics of various direct evaporative coolers

Type of cooler	Saturation efficiency ϵ_{sat}	Face velocity, fpm	Air-side pressure drop, in. WG	Water-air ratio $\dot{m}_w/\dot{m}_a$	Water usage, gal/h · 1000 cfm	Remarks
Air washer	0.80–0.90	400–800	0.2–0.5	0.1–0.4		
Evaporative pad	0.80	100–300	0.1		1.3	Pad thickness of 2 in.
Rigid media	0.75–0.95	200–400	0.05–0.1			Thickness of 8–12 in.
Rotary wheel		100–600	0.5			

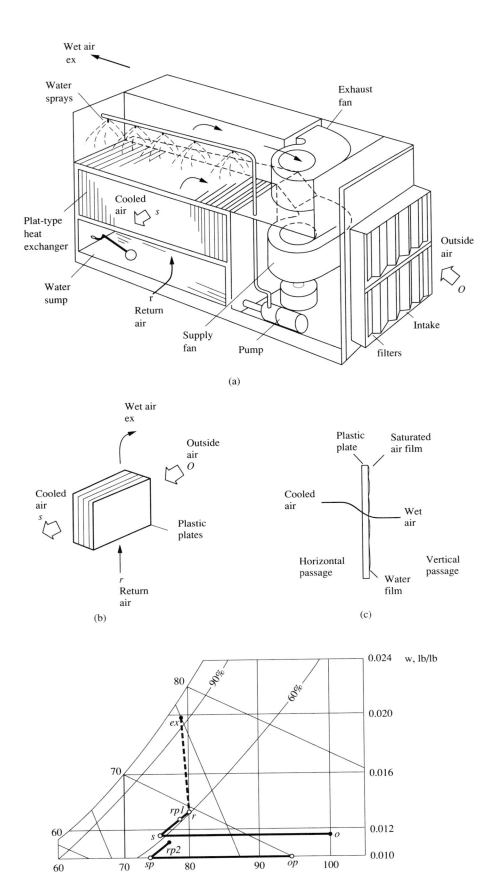

FIGURE 13.3 A typical indirect evaporative cooler: (a) schematic diagram; (b) airstream flowing through the passages; (c) heat transfer through plastic plates; and (d) processes on the psychrometric chart.

nected by the same vertical shaft, and a fiberglass or stainless steel casing to prevent corrosion.

The core part of this indirect evaporative cooler is the plate heat exchanger. The plates are made of dimpled, thin polyvinyl chloride plastic. These plates are spaced from 0.08 to 0.12 in. apart and form alternate horizontal and vertical passages (i.e., the air to be cooled flows horizontally, and the air that is sprayed flows vertically). Because the plates are only about 0.01 in. thick, the thermal resistance of each plastic plate is very small, although the thermal conductivity of the plastic is low.

Hot, dry outdoor air at point o enters the intake and filters and is extracted by the supply fan. It then enters the back of the exchanger and is forced through the horizontal passages, in which it releases its heat through the plastic plates to the adjacent wetted surfaces of the vertical passages. The cooled air at point s flows out the front to the conditioned space, as shown in Fig. 13.3a. After it absorbs the space cooling load, its state point changes from s to r.

Water sprays over the vertical passages at the top of the plate heat exchanger, and forms both wetted surfaces and water droplets. Evaporation from these wetted surfaces and droplets absorbs heat released from the air flowing horizontally through the plastic plates. Excess water drops to a sump, which recirculates it to spraying nozzles by means of a pump. Make-up water is supplied from the city water supply to account for the evaporation and carryover. Water is periodically bled off to prevent the buildup of solid matter.

Return air from the conditioned space at point r is drawn through the vertical passages between the plastic plates. It absorbs the evaporated water vapor, and its humidity ratio increases. The higher the velocity of the wet airstream, the greater the wet surface heat transfer coefficient h_{wet}, the larger the enthalpy difference $\Delta h_{s \cdot w}$ between the saturated air film at the wetted surface and the wet airstream, and the higher the pressure drop of the wet airstream. Wet air is then forced through the exhaust fan and discharged to the outdoor atmosphere at point ex.

Other types of indirect evaporative coolers may use absorbant-lined vertical passages to dip water from the top through distributing troughs; instead of using water sprays. Propeller fans may be used instead of centrifugal fans for wet airstream exhaust. Dampers may be used to extract outdoor air or return air from the conditioned space.

Heat Transfer Process

There are three fluid streams in a plate heat exchanger: cooled air, wet air, and water films along the vertical passages. Because the temperature of the saturated air film above the wetted surface is nearly equal to the wet bulb temperature of the wet airstream flowing over the surface, the heat from the airstream to be cooled on the other side of the plastic plate is transferred to the wetted surface to evaporate liquid water. The heat transfer process in an indirect evaporative cooler takes place mainly between the cooled and wet airstreams.

On the cooled-air side, the amount of water vapor that permeates the plastic plate is very small and can be ignored; therefore, the entering air is sensibly cooled from point o to point s at a constant humidity ratio, i.e., along a horizontal line os on the psychrometric chart shown in Fig. 13.3d. According to Dowdy et al. (1987), the heat transfer coefficient h_{air} on the air side, in Btu/h $\cdot$ ft^2 $\cdot$ °F, is a function $h_{air} = f(\text{Re}_o^{0.8}\text{Pr}^{0.33})$ and can therefore be calculated as

$$h_{air} = 0.023\left(\frac{k_a}{D_h}\right)\text{Re}_D^{0.8}\text{Pr}^{0.3} \qquad (13.3)$$

where k_a = thermal conductivity of air, Btu/h $\cdot$ ft $\cdot$ °F
D_h = hydraulic diameter of the cooled-air passage, ft
L_a = length of the air passage, ft

In Eq. (13.3) the hydraulic diameter D_h is given as

$$D_h = \frac{4A_{ca}}{P_{ca}} \qquad (13.4)$$

where A_{ca} = area of the cooled-air passage, ft^2
P_{ca} = wetted perimeter of the cooled-air passage, ft

On the wet-air side, water is sprayed onto the wet airstream; however, because of the heat transfer from the cooled airstream through the plastic plate to the wet airstream, the saturation process is no longer adiabatic. Return air from the conditioned space (which becomes the wet airstream) is humidified from point r to ex as shown in Fig. 13.3d. According to Wu and Yellott (1987), the relative humidity of the air exhaust from the indirect evaporative cooler is about 95 percent, and the change in the dry bulb temperature is rather small. Consequently, there is an increase in wet bulb and air enthalpy due to the increase in latent heat.

In a plate heat exchanger, the cooled air and wet air are in a crossflow arrangement. The temperature of the saturated air film on the wet-air side depends on the wet bulb of the local wet airstream and the wet bulb of the wet air gradually increases during the humidifying process. The increase in the wet bulb of the wet air can be determined from its enthalpy increase Δh_{wet}, in Btu/lb, and can be calculated as

$$\Delta h_{wet} = (h_{ex} - h_r) = \frac{\dot{V}_{ca}\rho_{ca}c_{pa}(T_o - T_s)}{\dot{V}_{wet}\rho_{wet}} \qquad (13.5)$$

where h_{ex}, h_r = enthalpy of wet air at points ex and r, Btu/lb

T_o, T_s = temperature of the cooled air at points o and s, °F

$\dot{V}_{ca}, \dot{V}_{wet}$ = volume flow rate of the cooled air and wet air, cfm

ρ_{ca}, ρ_{wet} = cooled air and wet air density, lb/ft³

c_{pa} = specific heat of moist air, Btu/lb · °F

The average temperature $T_{s.a}$ of the saturated film on the wet-air side, in °F, is approximately equal to the average water temperature $T_{w.s}$ in the water sump, in °F. According to actual observation, $T_{w.s}$ is about 3°F higher than the wet bulb temperature of the return air for this indirect evaporative cooler.

As with Eq. (12.27), the surface heat transfer coefficient on the wet-air side, h_{wet}, can be calculated as

$$h_{wet} = \frac{m''}{c_{pa}} h_{dry} \qquad (13.6)$$

In Eq. (13.6), h_{dry} indicates the sensible heat transfer coefficient from the wetted surface when it is dry; it can be calculated as in Eq. (13.3):

$$h_{dry} = 0.023 \left(\frac{k_{wet}}{D_{h.w}} \right) \text{Re}^{0.8} \text{Pr}^{0.4} \qquad (13.7)$$

where k_{wet} = thermal conductivity of the wet air, Btu/h · ft · °F

$D_{h.w}$ = hydraulic diameter of the wet-air passage, ft

L_{wet} = length of the wet-air passage, ft

Cooler Effectiveness and Performance Factor

The performance of an indirect evaporative cooler can be determined by its effectiveness value. Usually, the cooled airstream has a smaller heat capacity rate than the wet-air stream. Therefore, the indirect evaporative cooler effectiveness, sometimes called the *performance factor* (PF), is defined as follows:

$$\epsilon_{in} = \frac{T_{ca.e} - T_{ca.l}}{T_{ca.e} - T_{s.a}} \qquad (13.8)$$

where $T_{ca.e}$ = temperature of the air to be cooled entering the indirect evaporative cooler, °F

$T_{ca.l}$ = temperature of the cooled air leaving the indirect evaporative cooler, °F

$T_{s.a}$ = temperature of the saturated air film on the wet-air side (about 3°F higher than the wet bulb temperature of the entering wet air), °F

For a crossflow plate heat exchanger in which cooled and wet airstreams are not mixed, the cooler effectiveness can be calculated as

$$\epsilon_{in} = 1 - \exp\left\{ \left[\frac{1}{\text{NTU}^{-0.22}C} \right] \times [\exp(-\text{NTU}^{0.78}C) - 1] \right\} \qquad (13.9)$$

From Eqs. (12.17) and (12.14),

$$\text{NTU} = \frac{AU}{C_{min}} = \frac{AU}{C_{ca}} \qquad (13.10)$$

and from Eq. (12.16),

$$C = \frac{C_{min}}{C_{max}} = \frac{C_{ca}}{C_{wet}}$$

$$C_{ca} = 60\dot{V}_{ca}\rho_{ca}c_{pa}$$

$$C_{wet} = 60\dot{V}_{wet}\rho_{wet}c_{sat} \qquad (13.11)$$

where C_{ca}, C_{wet} = heat capacity rate of cooled air and wet air, Btu/h · °F

$\dot{V}_{ca}, \dot{V}_{wet}$ = volume flow rate of cooled air and wet air, cfm

ρ_{ca}, ρ_{wet} = density of cooled air and wet air, lb/ft³

c_{pa} = specific heat of moist air, Btu/lb · °F

In Eq. (13.11), c_{sat} indicates the saturation specific heat per degree wet bulb temperature of the wet air at constant pressure, in Btu/lb · °F; it is given as

$$c_{sat} = \frac{dh_s}{dT'} \qquad (13.12)$$

where h_s = enthalpy difference along the saturation curve, Btu/lb. The total cooling capacity of the indirect evaporative cooler q_c, in Btu/h, can then be calculated as

$$q_c = \dot{V}_{ca}\rho_{ca}c_{pa}\epsilon_{in}(T_{ca.e} - T_{s.a}) \qquad (13.13)$$

The actual operating effectiveness of the indirect evaporative cooler shown in Fig. 13.3 during a hot summer in Phoenix, Arizona could be as high as 0.85 at rating conditions.

Operating Characteristics

Self-contained indirect evaporative coolers are made in sizes to handle volume flow rates of 1060, 2600, and 3200 cfm. The size of the indirect evaporative cooler mentioned above is 2600 cfm. Based on observation, its maximum power consumption on hot summer days is 1.68 hp (1250 watts), and the fan total pressure of the centrifugal fan is from 1 to 1.3 in. WG.

In indirect evaporative coolers, either outside air or return air from the conditioned space can be used as the air to be cooled or the wet airstream. It depends on which has a lower wet bulb and which can provide better indirect evaporative cooling results. In Phoenix, a 2.5 percent summer design wet bulb is 75°F. For a summer space temperature of 80°F and a relative humidity of 50 percent, the wet bulb of return air is only 66.5°F. It is more beneficial to use return air for the wet airstream in this case. Only when the wet bulb of outdoor air drops below 66.5°F should outdoor air be used for both cooled and wet airstreams.

The *energy efficiency ratio* (EER) is defined as the ratio of the net cooling capacity of a device, in Btu/h, to the electric power input to that device, in W, under designated operating conditions. The actual maximum EER of the typical indirect evaporative cooler mentioned here, operating under hot summer conditions, is about 50, compared to an average of 8.5 to 11.5 for self-contained packaged units.

The operating characteristics of indirect evaporative coolers are affected by the flow rates and the pressure drops on the cooled-air and wet-air sides. For a specific cooler, the greater the volume flow, the greater the heat transfer coefficients, the pressure drop, and the air velocity flowing through the passages in the plate heat exchanger.

The air velocity of the cooled airstream flowing through the passages is usually from 400 to 1000 fpm. It is important to limit the air velocity of the wet airstream in order to prevent carryover of the water droplets. The indirect evaporative cooler effectiveness usually ranges from 0.6 at an air-side pressure drop of about 0.2 in. WG to an effectiveness of 0.8 at a pressure drop of about 1 in. WG.

Usually, the cooled-air-side pressure drop of indirect evaporative coolers ranges from 0.2 to 1.5 in. WG, depending on the air velocities in the heat exchanger and the distributing duct. The wet-air-side pressure drop varies from 0.5 to 1 in. WG. The volume flow ratio of wet air to cooled air changes from 0.6 to 1.2. This ratio affects the heat capacity ratio C and, therefore, the cooler's effectiveness. The altitude of the unit's location also has a significant effect on its air density and, therefore, its performance.

Part-Load Operation and Control

For a constant airflow unit, if the conditions of the outdoor air and the sensible heat ratio of the space conditioning line both remain constant when there is a reduction of the space cooling load, the space conditioning line *sr*, as shown in Figure 13.3d, tends to extend a shorter distance from point *s*. The space temperature T_{rp1} drops and the space relative humidity

φ_{rp1} is slightly higher at part-load as shown by point rp1 on the psychometric chart.

If the space cooling load remains the same and the condition of the outdoor air changes from point *o* to point *op*, with a lower outdoor dry bulb and wet bulb, then the supply temperature T_{sp}, space air temperature T_{rp2}, and space relative humidity φ_{rp2} will all be lower at part-load.

In a small, self-contained indirect evaporative cooler, the fans automatically cycle on-and-off by means of a control system according to the space temperature at part-load operation. In large coolers, a multispeed fan motor can often be modulated at part-load when the space temperature drops below a predetermined limit.

Example 13.1. An indirect evaporative cooler has the following constructional and operating characteristics:

Volume flow rate: cooled air	2600 cfm
wet air	2400 cfm
Cooled-air velocity between plates	800 fpm
Wet-air velocity between plates	740 fpm
Length of cooled-air passage	1.5 ft
Area of each cooled-air passage	0.00872 ft^2
Wetted perimeter	2.68 ft
Outside air dry bulb temperature	100°F

At an air temperature of 80°F, the fluid properties are as follows:

Dynamic viscosity	0.0448 lb/ft · h
Thermal conductivity	0.0151 Btu/h · ft ·°F
Air density	0.0725 lb/ft^3
Prandtl number	0.71

The return air (wet air) enters the vertical passages at a wet bulb of 68°F and leaves at 76°F. Calculate the effectiveness of this indirect evaporative cooler and its cooling capacity. Also calculate the temperature of cooled air leaving the cooler. The density of the air at the supply fan outlet is $\rho_{fo} = 0.07$ lb/ft^3.

Solution

1. From Eq. (13.4), the hydraulic diameter of the cooled-air passage is

$$D_h = \frac{4A_{ca}}{P_{ca}} = \frac{4(0.00872)}{2.68} = 0.013 \text{ ft}$$

and the Reynolds number for cooled air is

$$\text{Re}_{Dh} = \frac{\rho v D_h}{\mu} = \frac{800(60)(0.0725)(0.013)}{0.0448} = 988$$

For wet air,

$$\text{Re}_{Dh} = \frac{740(60)(0.071)(0.013)}{0.048} = 912$$

2. On the cooled air side, the surface heat transfer coefficient h_{air} can be calculated from Eq. (13.3) as follows:

$$h_{air} = 0.023 \frac{k_a}{D_h} Re_D^{0.8} Pr^{0.3}$$

$$= 0.023 \left(\frac{0.0151}{0.013} \right) (988)^{0.8} (0.71)^{0.3}$$

$$= 5.96 \text{ Btu/h} \cdot \text{ft}^2 \cdot {}^\circ F$$

On the wet-air side, if the surface is dry, the heat transfer coefficient is

$$h_{dry} = 0.023 \frac{0.0151}{0.013} (912)^{0.8} (0.71)^{0.4}$$

$$= 5.43 \text{ Btu/h} \cdot \text{ft}^2 \cdot {}^\circ F$$

3. From the psychrometric table (Appendix II.2), the enthalpy of saturated air at 68°F is 32.42 Btu/lb and at 76°F is 39.57 Btu/lb. Thus, the saturation specific heat per degree of wet bulb temperature is

$$c_{sat} = \frac{39.57 - 32.42}{76 - 68} = 0.893 \text{ Btu/lb} \cdot {}^\circ F$$

From Eq. (13.6), the heat transfer coefficient on the wet-air side is

$$h_{wet} = \frac{m''}{c_{pa}} h_{dry} = \frac{0.893}{0.243} 5.43$$

$$= 20.0 \text{ Btu/h} \cdot \text{ft}^2 \cdot {}^\circ F$$

4. From Eq. (12.5), the parameter UA is given as

$$UA = \frac{1}{1/(h_{ca}A) + 1/(h_{wet}A)}$$

$$= \frac{1}{1/(5.96 \times 1430) + 1/(20.0 \times 1430)}$$

$$= 6567 \text{ Btu/h} \cdot {}^\circ F$$

and the heat capacity rates are

$$C_{ca} = \dot{V}_{ca} \rho_{ca} c_{pa}$$

$$= 2600(60)(0.0725)(0.243)$$

$$= 2748 \text{ Btu/h} \cdot {}^\circ F$$

$$C_{wet} = \dot{V}_{wet} \rho_{wet} c_{sat}$$

$$= 2400(60)(0.71)(0.893)$$

$$= 9130 \text{ Btu/h} \cdot {}^\circ F$$

$$C = \frac{C_{ca}}{C_{wet}} = \frac{2748}{9130} = 0.3$$

From Eq. (13.10),

$$\text{NTU} = \frac{AU}{C_{ca}} = \frac{6567}{2748} = 2.39$$

and

$$\text{NTU}^{-0.22} = \frac{1}{\text{NTU}^{0.22}} = 0.826$$

5. From Eq. (13.9), the cooler's effectiveness is

$$\epsilon_{in} = 1 - \exp\left\{ \left(\frac{1}{\text{NTU}^{-0.22}C} \right) \right.$$

$$\left. \times \left[\exp(-\text{NTU}^{0.78}C) - 1 \right] \right\}$$

$$= 1 - \exp\left\{ \left(\frac{1}{0.3 \times 0.826} \right) \right.$$

$$\left. \times \left[\exp(-(2.39)^{0.78} \times 0.3) - 1 \right] \right\}$$

$$= 0.83$$

Because $T_{s.a} = 68 + 3 = 71°F$, the cooling capacity of the indirect evaporative cooler is then

$$q_c = 60 \dot{V}_{ca} \rho_{ca} c_{pa} \epsilon_{in} (T_{ca.e} - T_{s.a})$$

$$= 2600(60)(0.0725)(0.243)(0.83)(100 - 71)$$

$$= 66,152 \text{ Btu/h}$$

6. The temperature of cooled air leaving the cooler is

$$T_s = T_o - \frac{q_c}{60 \dot{V}_s \rho_{fo} c_{pa}}$$

$$= 100 - \frac{69,739}{60(2600)(0.07)(0.243)}$$

$$= 75.1°F$$

13.4 INDIRECT-DIRECT EVAPORATIVE COOLING

When cooled air leaves an indirect evaporative cooler during a hot summer, its dry bulb may still be above 70°F with a relative humidity between 60 and 80 percent. It is beneficial in such cases to add a direct evaporative cooling process so that the temperature of the cooled air will drop further with an increase in relative humidity. A higher relative humidity of the supply air is often acceptable when the cooled supply air absorbs the space sensible load and maintains a desirable space relative humidity during a hot summer. In a two-stage indirect-direct cooler, a direct evaporative cooler is always connected in series after an indirect evaporative cooler. A two-stage indirect-direct cooler can provide cooled air with both dry and wet bulb temperatures below 58°F when the outside wet bulb is below 65°F during a hot summer. Indirect and direct evaporative cooling processes connected in series are illustrated in Figure 13.1c along with a psychrometric chart.

For a two-stage indirect-direct cooler at part-load operation, one stage, either direct or indirect, can be turned off based on the indoor and outdoor conditions.

Structure and Performance

In two-stage indirect-direct coolers, plate heat exchangers are widely used in the indirect cooler. On the wet-air side, either a water spray or a wetted surface treated with water-absorbent material is used. In the direct cooler, a rigid medium with impregnated cellulose is generally used. The following describes two applications of two-stage indirect-direct coolers and discusses their performances:

SCOTTSDALE, ARIZONA. The first-stage indirect cooler uses a plate heat exchanger as described in Wu (1989). Outdoor air serves as both the air to be cooled and the wet air. The wet air flows vertically along the absorbant-lined wetted surfaces and discharges at the top by means of a propeller fan. In the second-stage direct cooler, rigid-medium wetted surfaces provide direct evaporative cooling.

During July 1987, the performance of this two-stage cooler was monitored. At an average outdoor dry bulb of 92.4°F and an average wet bulb of 63.9°F, the dry bulb of the cooled air dropped to about 77.3°F and the wet bulb dropped to 61.8°F after the first-stage indirect cooler. The average effectiveness of the first-stage indirect cooler was 0.53. In the second-stage direct cooler at a saturation effectiveness of 0.9, the dry bulb dropped an additional 10 to 12°F.

This two-stage indirect-direct cooler has a volume flow of 3900 cfm. The rated power input to the supply fan is 1 hp and to the exhaust fan 1/5 hp.

DENVER, COLORADO. In this application, as described by Scofield et al. (1984), the indirect-direct evaporative cooler is used for a variable-air-volume system. The first-stage indirect cooling is provided by a plate heat exchanger. Outdoor air or return air from the space can be cooled in the plate exchanger or used as the wet airstream.

Outdoor air at a dry bulb of 93°F and a wet bulb of 59°F enters the first-stage indirect cooler and is evaporatively cooled to 67.5°F dry bulb and 49.8°F wet bulb with a cooler effectiveness of 0.76. In the second-stage direct cooler, the supply air is further cooled to a dry bulb of 53.3°F and a wet bulb of 49.8°F at a saturation effectiveness of 0.8.

The pressure drop for the indirect cooler heat exchanger is 0.81 in. WG, and the fan static pressure for the direct cooler is 0.25 in. WG. The volume flow rates of the cooled and wet airstreams in the indirect cooler are nearly equal. The supply fan is equipped with a two-speed motor for capacity control.

13.5 INCORPORATING EVAPORATIVE COOLING WITH OTHER COOLERS

Incorporating Evaporative Cooling with Refrigeration

In locations where outdoor air has a higher wet bulb, evaporative cooling alone cannot provide the required supply air conditioning to maintain a desirable indoor environment. In many applications, incorporating evaporative cooling with refrigeration, as shown in Fig. 13.4a, is often more economical than using refigeration alone.

Anderson (1986) has compared an air conditioning system using an indirect-direct cooler and a DX-coil with a system using only a DX-coil. The operating parameters, installation costs, and electric power consumption of these systems are as follows:

Outdoor conditions	100°F dry, 70°F wet
Supply air	57.5°F dry, 56.5°F wet
Space temperature	78°F
Space relative humidity	50 percent
System pressure drop	
DX-coil only	2 in. WG
Evaporative cooling and DX-coil	2.75 in. WG
Wet air	0.85 in. WG
Fan efficiency	0.6
Indirect cooler effectiveness	0.7
Direct cooler saturation effectiveness	0.9
EER refrigeration	9 Btu/W
Outdoor air (make-up air)	15 percent
Ratio of installation cost of evaporative coolers and DX-coil to DX-coil only	2.25
Electric power consumption (both air and refrigeration sides)	
Evaporative coolers and DX-coil	1.10 kW/ton
DX-coil only	1.79 kW/ton

In the evaporative cooler and DX-coil combination, return air is used for the wet airstream, so outdoor air is used only for the cooled airstream. In a cooling system equipped with only a DX-coil, the return air absorbs lighting heat in the plenum, mixes with the outdoor air, and is then cooled and dehumidified in the DX-coil. For cooling capacity control of such a three-stage system, components are usually energized in the following sequence: indirect cooler first, then direct cooler, and finally the DX-coil.

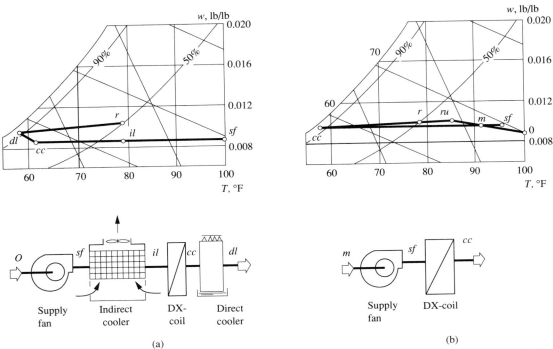

FIGURE 13.4 Evaporative cooling incorporating DX-coil vs DX-coil only: (a) evaporative cooling incorporating DX-coil and (b) DX-coil.

If the electric power rate is $0.05/kWh, the number of payback hours for an evaporative cooling system incorporating refrigeration is 7948; if the rate is $0.075/kWh, the number of payback hours is 5295; and if the rate is $0.1/kWh, the number of payback hours is 3974. If the annual number of operating hours is 1000, the number of payback years is between 4 and 8.

As evaporative cooling becomes more and more popular, the ratio of installation costs for evaporative cooling incorporating a DX-coil to the cost of a DX-coil alone will decrease, as will the number of payback years.

Tower–Coil and Rotary Wheel Combination

When a cooling tower is connected to a water cooling/heating coil, as shown in Fig. 13.5a, the tower–coil becomes an indirect evaporative cooler. During the cooling season, the cooling water from the tower is forced through the coil to cool the air flowing over it. Such a cooling coil is often used as a precooling coil, because there may be another cooling coil downstream. In the heating season, hot water from the condenser may flow through the coil and heat the air.

Using water from the cooling tower to cool the air by means of a precooling coil (to replace all or part

of the refrigeration) is often called a *water economizing process*. The combination of a cooling tower and the connected water cooling coil is called a *water economizer* and is shown in Fig. 13.5a.

If the tower–coil is in series with a direct evaporative cooler using a rotary wheel, the resulting indirect-direct cooler combination can maintain space conditions in summer similar to those achieved with refrigeration for areas where the outdoor wet bulb is below 65°F.

Water returns from the precooling coil, typically at a temperature of 78°F, enters the cooling tower, and is evaporatively cooled to about 70°F. Water is then drawn through the precooling coil, where it absorbs the heat from the air flowing over the coil. It is then pumped back to the tower at a temperature of about 78°F to be evaporatively cooled again.

Outdoor air at a dry bulb of 100°F and a wet bulb of 65°F, is drawn through the precooling coil by the supply fan and is sensibly cooled to 75°F. It then flows through a rotary-wheel-type direct cooler and is evaporatively cooled, typically to 57.5°F dry bulb and 56°F wet bulb. After that, air is supplied to the conditioned space. Recirculating air may be used instead of outdoor air when its enthalpy is lower than that of the outdoor air.

In an effective tower–coil combination, the approach of the cooling tower should be around 5°F.

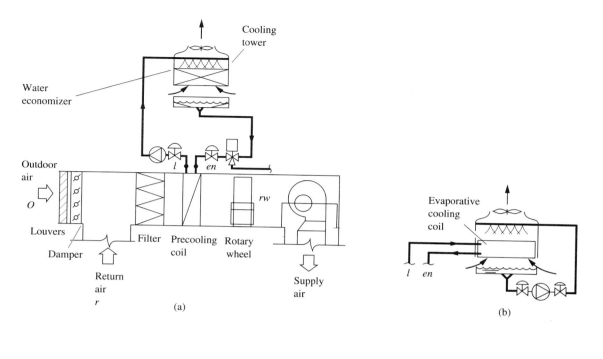

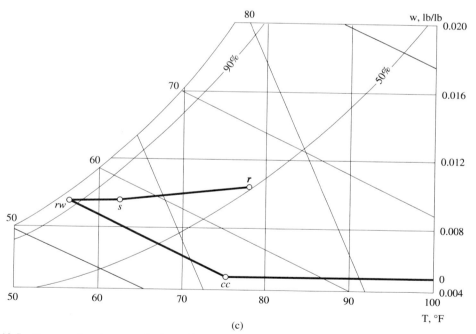

FIGURE 13.5 Tower–coil and rotary wheel combination and evaporative coil–coil and rotary wheel combination: (a) tower–coil and rotary wheel combination, (b) evaporative coil–coil and rotary wheel combination, and (c) evaporative cooling processes.

In order to have such an approach, the cooling tower must be 60 percent larger than a tower with a 10°F approach. The row depth and fin spacing of the pre-cooling coil should be selected to sensibly cool the outdoor air from 100°F to 75°F at an entering water temperature of 70°F.

Field experience and tests have shown that a low face velocity through the evaporating medium of the rotary wheel results in a saturation effectiveness above 0.90 and prevents carryover. In these tests, the air velocity was about 700 fpm with a pressure drop of 0.25 in.WG. The rotary wheel revolved at a rate of approximately 1.5 rpm.

In an evaporative cooling system, the horsepower and number of operating hours of the fan always far exceeds those of the water pump. To save energy and reduce operating costs for a large unit, a two-speed (or even three-speed fan) is often economical. The efficiency of the cooling tower, cooling coil, and rotary wheel cooler depends largely on the conditions of the fills in the tower, the inner surface of the cooling coil, and the evaporative medium in the rotary wheel cooler. If the wetted surfaces are clogged with dirt and scale, the efficiency will decreases proportionally to the resulting drop in evaporation and air flow. Periodic bleed-off and other necessary water treatments are essential for good performance.

Another type of evaporative coil–rotary-wheel combination is shown in Fig. 13.5b. When an evaporative cooling coil is connected to a cooling coil, water that has been evaporatively cooled in the former can be pumped to the cooling coil to absorb heat from the ambient air. Such an evaporative-coil–coil combination is actually an indirect cooler. If this indirect cooler is connected with a rotary wheel or other type of direct cooler, the resulting combination has a system performance similar to that of a tower–coil and rotary wheel combination. The heat transfer between two airstreams by means of an evaporative-coil–coil is always higher than in a coil-to-coil arrangement, which is commonly located inside the outdoor air and exhaust airstreams or between make-up air and exhaust airstreams, and is often called a *run-round system*.

Tower, Plate-and-Frame Heat Exchanger, and Coil Combination

When the outdoor wet bulb drops below a certain value, (about 40°F), it is possible to use the cooling water from the cooling tower to cool the air and thus replace all or part of the chilled water from the refrigeration plant, as shown in Fig. 13.6. In order to prevent the dirt and solid matter contained in the open-circuit tower cooling water from scaling the inner surface of the chilled water coil, a plate-and-frame heat exchanger is often used so that the tower cooling water does not enter the coil directly.

In Fig. 13.6, the cooling water from the tower at 45°F enters the plate-and-frame heat exchanger. It cools all or part of the chilled water returning from the cooling coil to 50°F. Chilled water from the plate-and- frame heat exchanger is then mixed with the water from the centrifugal chiller and enters the chilled-water cooling coil at 45°F.

In order to cool the chilled water at the heat exchanger when the outdoor wet bulb is 40°F, the approach of the cooling tower should be no greater than 5°F and the temperature difference at the heat exchanger between the cooling water from the tower and the chilled water from the coil also should not exceed 5°F. Cost analyses are required to determine the optimum size of the cooling tower and heat exchanger and the flow rates of the cooling water and chilled water.

Plate-and-Frame Heat Exchanger

A *plate-and-frame heat exchanger*, as shown in Fig. 13.7, is a liquid-to-liquid heat exchanger. It consists of a number of corrugated metal plates that are usually made of stainless steel, titanium, or aluminum-brass alloy. Corrugated plates and gaskets form alternate passages that separate two different fluids. Gaskets are usually made of elastomers. Warm fluid flows downward on one side of the corrugated plate, and cold fluid flows upward on the other side in a counterflow arrangement. The corrugated plates are compressed together by a fixed end frame and a movable end frame with clamping bolts. Proper selection of the gasket material and proper operating conditions are important to prevent fluid leakage. The plates and movable end frame are suspended from an upper carrying bar and lower guide bar. Fluid connections are located in the fixed end frame.

The gap between two adjacent plates is rather small, usually ranging from 0.1 to 0.2 in. Because the plates are corrugated, a high degree of turbulence is produced, which results in a high heat transfer coefficient. The following equation may be used for the calculation of the heat transfer:

$$h = 0.2536 \frac{k}{L_g} \mathrm{Re}_L^{0.65} \mathrm{Pr}^{0.4} \qquad (13.14)$$

where k = thermal conductivity of the fluids, Btu/h · ft · °F

L_g = the spacing of the gap, ft

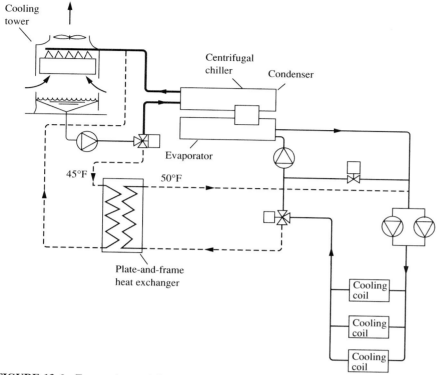

FIGURE 13.6 Tower, plate-and-frame heat exchanger, and coil combination.

The water velocity is usually between 60 and 200 fpm. The higher the fluid velocity, the higher the heat transfer coefficient and the greater the pressure drop of the liquid flowing through the heat exchanger. For a typical plate-and-frame heat exchanger, the overall heat transfer coefficient may be 740 Btu/h · ft^2 · °F through the corrugated plate between the warm and cold liquids at a pressure drop of about 14.7 psi. The maximum working pressure for a high-pressure model can be as high as 350 psig.

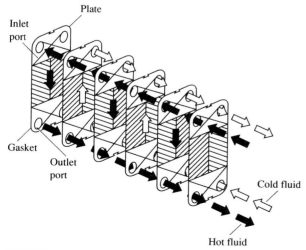

FIGURE 13.7 Plate-and-frame heat exchanger.

The plate-and-frame heat exchanger has a high heat transfer coefficient on the surfaces of its corrugated plates, a smaller temperature difference between the fluids on the two sides of each plate, and a compact size. It is easily dismantled for cleaning and routine maintenance.

13.6 DESIGN CONSIDERATIONS

Scope of Applications

In locations where the outdoor wet bulb $T'_o \leq 60°F$, a stand-alone direct evaporative cooler can normally provide a comfortable indoor environment for the occupants at a space temperature of 78°F, a relative humidity of about 60 percent, and a space air velocity smaller than 100 fpm.

In locations where $60°F \leq T'_o \leq 68°F$, a self-contained indirect-direct cooler also can maintain a comfortable indoor environment for the occupants.

In locations where $T'_o \geq 72°F$, or when dehumidification of the supply air is required, refrigeration may be required. A combined indirect-direct cooler and DX-coil or chilled-water cooling coil is often economical. However, because the installation cost of indirect-direct coolers is higher than that of refrigerating systems, life-cycle cost analysis is necessary, especially for small and medium systems or systems with a low annual number of operating hours during the cooling season. Evaporative cooling is best suited to applications where both winter humidification and summer cooling are required.

Summer Outdoor Design Wet Bulb Temperature

The performance of an evaporative cooling system is closely related to the outdoor wet bulb temperature T_o'. The value of T_o' not only determines the temperature of cooled air leaving the direct cooler but also has a decisive effect on the lowest possible wetted surface temperature in an indirect cooler. Outdoor air at a high dry bulb T_o and low wet bulb T_o' can easily be cooled in a direct cooler. Outdoor air with a high wet bulb may need refrigeration.

In the arid southwestern areas of the United States, a higher outdoor dry bulb always corresponds to a lower coincident wet bulb. The 2.5 percent summerdesign wet bulb is often several degrees higher than the mean coincident wet bulb at 1 percent design dry bulb and the mean coincident wet bulb. For example, in Denver the 2.5 percent wet bulb is 63°F and the 1 percent mean coincident wet bulb is 59°F (a difference of 4°F). If the 1 percent design dry bulb and mean coincident wet bulb are used as the design wet bulb temperature, there may be more than 150 hours (or 50 days) in the summer season between 1 P.M. and 4 P.M. in which the indoor temperature exceeds the design value.

For applications whose summer indoor space design temperature is 78°F or lower and a space relative humidity of $\varphi_r \leq 60$ percent, a 2.5 percent summer design wet bulb and 2.5 percent design dry bulb are recommended. For an indoor design temperature of 80°F, a 1 percent design wet bulb and 2.5 percent design dry bulb are recommended.

Serial Combination of Evaporative Coolers

When two indirect evaporative coolers or two direct evaporative coolers are connected in series to form a two-stage system, the dry and wet bulb drop of the second-stage evaporative cooler is always far less than in the first-stage cooler. Usually, it is not be economical when two direct evaporative coolers or two indirect evaporative coolers are connected in series. However, an indirect cooler is often connected in series with a direct cooler, as mentioned before.

Suppose two indirect coolers are connected in series, forming an indirect-indirect combination. If outdoor air at 100°F dry bulb and 67°F wet bulb is used for both cooled air and wet air in the first stage,

if outdoor air is used for the cooled air in the second stage, return air at 78°F dry bulb and 66°F wet bulb is used for wet air in the second stage, and if the cooler effectiveness values for both the first- and second-stage indirect coolers are 0.75, then, from Eq. (13.8), the drop in dry bulb temperature in the first stage is

$$
\begin{aligned}
T_{ca.e} - T_{ca.l} &= \epsilon_{in}(T_{ca.e} - T_{s.a}) \\
&= 0.75[100 - (67 + 3)] \\
&= 22.5°F
\end{aligned}
$$

From the psychrometric chart, the wet bulb temperature of the cooled air leaving the first-stage cooler is 62.5°F and the drop of the wet bulb in the first-stage cooler is 7.5°F.

In the second-stage indirect cooler, the drop in the dry bulb temperature is

$$
T_{ca.e} - T_{ca.l} = 0.75[(100 - 22.5) - (66 + 3)] = 6.4°F
$$

From the psychrometric chart, the drop in wet bulb temperature in the second-stage indirect cooler is about 2.3°F.

Cost Analysis

In an indirect evaporative cooler, a higher air velocity through the plate heat exchanger improves cooler effectiveness, raises the air-side pressure drop, and raises the energy cost.

In an indirect cooler with an air-side pressure drop of 0.2 in. WG, a cooler effectiveness usually ranges between 0.6 and 0.7. A pressure drop of 1 in. WG. usually corresponds to a cooler effectiveness of 0.75 to 0.80. Life-cycle analysis should be performed to determine optimum cooler effectiveness and air-side pressure drop.

Water Scaling

Water bleed-offs and necessary water treatments are essential in order to maintain a clean and efficient evaporative cooling system. The quantity of bleed-off should be no greater than the rate of evaporation and usually equals half the evaporation rate. These considerations should be taken in to account during system design.

REFERENCES AND FURTHER READING

Anderson, W. M., "Three-Stage Evaporative Air Conditioning versus Conventional Mechanical Refrigeration," *ASHRAE Transactions*, Part I B, pp. 358–370, 1986.

ASHRAE, *ASHRAE Handbook 1988, Equipment*, ASHRAE Inc., Atlanta, GA.

Brown, W. K., "Fundamental Concepts Integrating Evaporative Techniques in HVAC Systems," *ASHRAE Transactions*, Part I, pp. 1227–1235, 1990.

Dombroski, L., and Nelson, W. I., "Two-Stage Evaporative Cooling," *Heating/Piping/Air Conditioning*, pp. 87–92, May 1984.

Dowdy, J. A., and Karabash, N. S., "Experimental Determination of Heat and Mass Transfer Coefficients in Rigid Impregnated Cellulose Evaporative Media," *ASHRAE Transactions*, Part II, pp. 382–395, 1987.

Dowdy, J. A., Reid, R. L., and Handy, E. T., "Experimental Determination of Mass-Transfer Coefficients in Aspen Pads," *ASHRAE Transactions*, Part II A, pp. 60–70, 1986.

McClellan, C. H., "Estimated Temperature Performance for Evaporative Cooling Systems in Five Locations in the United States," *ASHRAE Transactions*, Part II, pp. 1071–1090, 1988.

McDonald, G. W., Turietta, M. H., and Foster, R. E., "Modeling Evaporative Cooling Systems with DOE-2.1D," *ASHRAE Transactions*, Part I, pp. 1236–1240, 1990.

Meyer, J. R., "Evaporative Cooling for Energy Conservation," *Heating/Piping/Air Conditioning*, pp. 111–118, September 1983.

Mumma, S. A., Cheng, C., and Hamilton, F., "A Design Procedure to Optimize the Selection of the Water-Side Free Cooling Components," *ASHRAE Transactions*, Part I, pp. 1250–1254, 1990.

Scofield, C. M., and DesChamps, N. H., "Indirect Evaporative Cooling Using Plate-Type Heat Exchangers," *ASHRAE Transactions*, Part I B, pp. 148–153, 1984.

Sun, T. Y., "Design Experience with Indirect Evaporative Cooling," *Heating/Piping/Air Conditioning*, pp. 149–155, January 1988.

Supple, R. G., and Broughton, D. R., "Indirect Evaporative Cooling—Mechanical Cooling Design," *ASHRAE Transactions*, Part I B, pp. 319–328, 1985.

Watt, J. R., "Nationwide Evaporative Cooling is Here!" *ASHRAE Transactions*, Part I, pp. 1237–1251, 1987.

Wu, H., "Performance Monitoring of a Two-Stage Evaporative Cooler," *ASHRAE Transactions*, Part I, pp. 718–725, 1989.

Wu, H., and Yellott, J. I., "Investigation of a Plate-Type Indirect Evaporative Cooling System For Residences in Hot and Arid Climates," *ASHRAE Transactions*, Part I, pp. 1252–1260, 1987.

Yellott, J. I., and Gamero, J., "Indirect Evaporative Air Coolers for Hot, Dry Climates," *ASHRAE Transactions*, Part I B, pp. 139–147, 1984.

HEATING SYSTEMS, FURNACES, AND BOILERS

14.1 HEATING SYSTEMS

According to the data in *Characteristics of Commercial Buildings* (1989), prepared by the Energy Information Administration (EIA), for the 57.8 billion ft^2 of heated commercial buildings in the United States in 1989, the percentage of use of the various types of heating systems was as follows:

Ducted warm air heating system using warm air furnaces	27 percent
Ducted warm air systems heating by coils in AHUs, packaged units, and fan-coils	34 percent
Radiators, baseboards, an space heaters	39 percent

According to Modera, approximately 50 percent of U.S. houses use ducted forced-air heating systems with warm air furnaces. In commercial buildings in 1989, the use of a boiler as a primary heat source accounted for about 34 percent of the supply of hot water or steam to heating coils, radiators, baseboard heaters, and heating panels. The heating systems in new commercial projects that use boilers are mainly hot water systems. In newly constructed residences, the heat sources are mainly gas-fired warm air furnaces.

Heating systems that use finned coils to heat the air, including water heating coils, electric heating coils, steam coils, condensing coils of heat pumps in air-handling units, packaged units, and fan-coil units are *convective* or *warm air (all air) heating systems*. Heating systems that use high-temperature infrared heaters, radiant panels, or radiators in which emitted radiant heat exceeds the released convective heat are *radiant heating systems*.

Selecting a Heating System

Several factors must be considered before selecting a suitable heating system, including the following:

- Whether it is a separate heating system or part of the air conditioning system
- An open or enclosed space, or a space with high infiltration
- Size of heating system (small, medium, or large)
- Available existing heat source, such as hot water or steam
- Cost of gas, oil, or electricity
- Design criteria
- Local customs

In both commercial and residential buildings, a warm air furnace is generally linked with unitary packaged air conditioning systems and, therefore, packaged units. In buildings with rooftop, split, or indoor packaged units, a warm air heating system using a warm air furnace is often the most direct, economical, and suitable choice. For buildings with central hydronic air conditioning systems, a hot water heating system using a boiler is often a suitable choice because the central primary plant and boilers are usually far away from the air handling units and conditioned space.

ASHRAE/IES Standard 90.1–1989 specifies that for areas with high ceilings, scattered work spaces using spot heating, or space heating in areas where infiltration load exceeds 2 ach at design heating conditions (such as open areas for loading docks or stadiums), radiant heating systems should be considered instead of convective or warm air heating systems.

14.2 WARM AIR FURNACES

Types of Warm Air Furnaces

A *warm air furnace* is a combustion and heating device in which gas or oil is directly fired to heat the air through a heat exchanger, or air is directly heated by electric resistance elements, in order to supply warm air to the conditioned space.

Warm air furnaces can be classified into various categories according to

- Fuel type: natural gas, liquefied petroleum gas (LPG), oil, electric energy, or wood
- Air flow direction: upflow, horizontal, or downflow
- Application: residential, commercial, or industrial

Natural gas is the primary fuel used in warm air furnaces. Warm air furnaces used for residences usually have a heating capacity up to 175,000 Btu/h. Upflow models are the most popular models in residences. For commercial applications, heating capacities are usually greater than 150,000 Btu/h. Horizontal models mounted inside a rooftop packaged unit or packaged heat pump for supplementary heating are widely used.

Upflow Gas-Fired Furnaces

An *upflow natural gas-fired warm air furnace* consists of one or more gas burners, a heat exchanger, a forced-draft fan, a venting system, a filter, and an outer casing, as shown in Fig. 14.1.

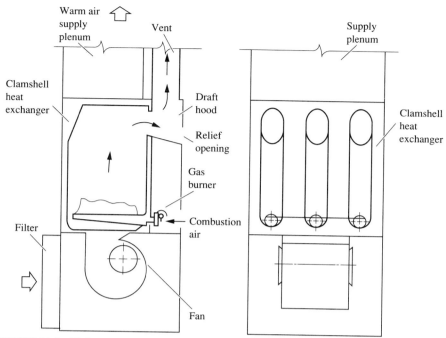

FIGURE 14.1   Upflow gas-fired furnace.

GAS BURNERS. Gas burners in small furnaces in residences are often atmospheric burners. An *atmospheric burner* consists of an air shutter, a gas orifice, and outlet ports and is usually die-formed and made of aluminum-painted, heavy-gauge steel or aluminized steel (or sometimes stainless steel). Atmospheric burners are either in-shot or up-shot and are installed in single or multiple ports. In-shot burners are installed horizontally and are often used for Scotch Marine boilers. Up-shot burners are placed vertically and are suitable for vertical fire-tube boilers. Both Scotch Marine and fire-tube boilers will be discussed in later sections. Atmospheric burners are simple, require only a minimal draft of air, and maintain sufficient gas pressure for normal functioning.

A power burner that uses a fan to supply and control combustion air is often used for large furnaces. Conversion burners are integrated with furnaces for safety and efficiency. Older gas furnaces often use conversion burners.

Recently, a new approach to fuel burning called *pulse combustion* has been developed. Pulse combustion intermittently draws gas and combustion air without the use of a fan. The pulse pressure waves are strong enough to pull air through the inlet pipe. Air and gas are extracted to the burner in very small amounts many times per second. The valves are set to provide the minimum excess air for combustion. Pulse combustion allows for a higher efficiency than conventional burners.

IGNITION. The ignition device is often a standing pilot ignition. These pilots are small. The pilot's flame is monitored by a sensor that shuts off the gas supply if the flame is extinguished. Another type of ignition is called spark ignition. It ignites intermittently only when required. Spark ignition saves more gas fuel than a standing pilot if the furnace is not operating.

HEAT EXCHANGER. This is in the shape of a clamshell that completely surrounds the burner. Air to be heated flows over the outer surface of the heat exchanger. A heat exchanger is usually made of aluminum-painted heavy gauge steel, aluminized steel, or sometimes stainless steel.

FAN OR BLOWER. A fan or blower is always installed in a warm air furnace to force the air to flow over the heat exchanger and to distribute it to the conditioned space, except when the warm air furnace is a part of an air-handling unit or packaged unit, in which a supply fan is always provided. A centrifugal fan with forward-curved blades and a double-inlet intake is usually used for packaged units.

FILTER. A disposable filter, which will be discussed in the next chapter, is often used to remove dust from the recirculating air. It is often located upstream from the fan.

VENTING ARRANGEMENTS. In a *natural-vent* warm air furnace, a draft hood is employed to connect the flue gas exit at the top of the heat exchanger section to a vent pipe or chimney. A relief air opening is also used to guarantee that the pressure at the flue gas exit is always atmospheric. A draft hood also diverts the backdraft from the chimney, bypassing the burner without affecting the combustion operation. A *direct-vent* warm air furnace does not have a draft hood. If the vent pipe or chimney is blocked, a control system shuts down the warm air furnace. A *power-vent* furnace uses a fan to force flue gas to flow through the gas burner and the heat exchanger.

CASING. The outer casing is usually made of heavy-gauge steel with removable access panels.

FURNACE OPERATION. Gas is generally brought from the main to the pressure regulator. The regulator reduces the gas pressure to 3.5 in. WG. Gas then flows through a gas valve controlled by a room thermostat. When a solenoid gas valve opens, the gas flows to the burners and mixes with the necessary outside primary air for combustion. The primary air-gas mixture then discharges from the port slots, mixes with ambient secondary air, and is burned. The combustion products flow through the heat exchanger and the draft diverter and discharge outdoors through the vent pipe or chimney.

A mixture of space recirculating air and outdoor air is pulled by the fan from the return ducts and enters the bottom inlet of the warm air furnace. This air is then forced through the heat exchanger. For a heating-only furnace, air temperature is raised by between 50 and 80°F. Warm air discharges from the top outlet and is distributed to various conditioned spaces through the supply duct and outlets.

An upflow gas-fired furnace is usually installed indoors and is often installed within a vented closet or vented basement in residential buildings.

Most natural-gas furnaces can use liquefied petroleum gas (LPG). The main difference is their gas pressure. Natural gas usually needs a pressure of 3 to 4 in. WG at the manifold, whereas LPG needs a higher pressure of about 10 in. WG and more primary air for gas burners.

Horizontal Gas-Fired Furnaces

A typical *horizontal gas-fired furnace* in a rooftop packaged unit is shown in Fig. 14.2. It mainly consists of multiple gas burners, a heat exchanger, a combustion blower, and an ignition device. The supply fan used to force air to flow through the gas-fired furnace is the same supply fan used to force air through the filters and DX-coil in the rooftop unit.

A power burner is often used in a horizontal gas-fired furnace. This type of burner provides better combustion and higher efficiency than atmospheric burners. A power draft centrifugal blower may be added to extract the combustion products and discharge them to a vent or chimney. The gas supply to the burner is controlled by a pressure regulator and a gas valve for the purpose of controlling the firing rate. In a premix power burner, gas and primary air are mixed first; the mixture is then forced to mix with secondary air in the combustion zone. A power burner

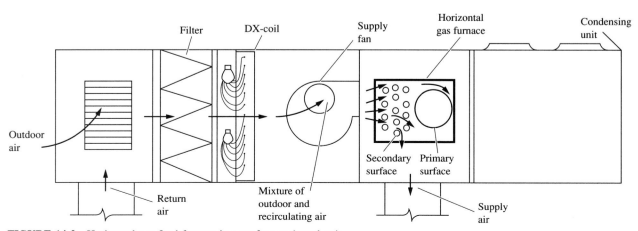

FIGURE 14.2 Horizontal gas-fired furnace in a rooftop packaged unit.

usually has a higher gas pressure than the atmospheric burners used in residences.

The heat exchanger usually has a tubular two-pass arrangement, typically with 16-gauge stainless steel for primary surfaces and 18-gauge stainless steel for secondary surfaces, as shown in Fig. 14.2. The primary surface is the heat transfer surface of the combustion chamber. The secondary surface is the surface of the tubes through which flue gas flows after the combustion chamber. A cone-shaped flame is injected into a tubular or drum-shaped combustion chamber. A centrifugal blower is used to provide secondary air for forced combustion. Another small centrifugal blower may be used to induce the flue gas at the exit of the heat exchanger, the power vent, to maintain a negative pressure at the heat exchanger section so as to prevent the mixing of any leaked flue gas with the heated air. The mixture of outdoor and recirculating air is heated when it flows over the primary and secondary heating surfaces.

Furnace Performance Factors

The performance of a gas-fired furnace is usually measured by the following parameters:

- *Thermal efficiency* E_t, in percent, is the ratio of the energy output of the fluid (air or water) to the fuel input energy. Input and output energy should be expressed in the same units. The value of E_t can be calculated as

$$E_t = 100 \frac{\text{Fluid energy output}}{\text{Fuel energy input}} \qquad (14.1)$$

- Annual fuel utilization efficiency (AFUE) is the ratio of annual output energy from air or water to the annual input energy expressed in the same units:

$$\text{AFUE} = 100 \frac{\text{Annual output energy}}{\text{Annual input energy}} \qquad (14.2)$$

AFUE also includes nonheating-season standing pilot input energy loss. AFUE is similar to thermal efficiency E_t, except that AFUE is the ratio of annual energy output to energy input, whereas E_t is the ratio of energy output to energy input at specific test periods and conditions.

- The *steady-state efficiency* (SSE) is the efficiency of a given furnace according to an ANSI test procedure, and is calculated as

$$\text{SSE} = 100 \frac{\text{Fuel input} - \text{Fuel Loss}}{\text{Fuel input}} \qquad (14.3)$$

The steady-state efficiency of gas-fired furnaces varies from 65 to 95 percent.

Test data in Jakob et al. (1986) and Locklin et al. (1987) for ASHRAE Special Project SP43, based on a nighttime setback period of 8 hours and a setback temperature of 10°F, gave the following performance factors for gas-fired furnaces of two test houses with different construction characteristics:

Construction characteristics	AFUE, percent	SSE, percent
Natural vent		
Pilot ignition	64.5	77
Intermittent ignition	69	77
Intermittent ignition + vent damper	78	77
Power vent		
Noncondensing	81.5	82.5
Condensing	92.5	93

ASHRAE/IES Standard 90.1–1989 specifies a minimum AFUE of 78 percent for both gas-fired and oil-fired warm air furnaces of heating capacity $<$ 225,000 Btu/h according to the DOE Test Procedure, *Code of Federal Regulations (CFR), Title 10, Part 30.*

For gas-fired warm air furnaces $\geq$ 225,000 Btu/h at maximum rating capacity (steady state), minimum thermal efficiency E_t is 80 percent, and at minimum rating capacity, E_t is 78 percent. For oil-fired furnaces $\geq$ 225,000 Btu/h, at both maximum and minimum rating capacity, minimum E_t is 81 percent as of January 1, 1992.

Energy Savings and Performance

Based on the above results, factors that affect the energy savings of the fuel and improve the furnace performance are as follows:

- *Condensing or noncondensing.* When the water vapor in the flue gas is condensed by indirect contact with recirculating air, part of the latent heat released during condensation is absorbed by the recirculating air, which increases furnace efficiency and saves fuel. The difference in the values of AFUE between condensing and noncondensing power vent furnaces may be around 10 percent.
- *Preheated outdoor combustion air.* If combustion air is taken from the outdoors and preheated through an annular pipe that surrounds the flue pipe, part of the heat in the flue gas will be saved.
- *Automatic vent damper.* An automatic thermal or electric vent damper located in the vent pipe closes

the vent when the furnace is not in operation. This decreases exfiltration from the house and restricts the amount of heat escaping from the heat exchanger.

- *Power vent and forced-combustion air supply.* Such a setup usually has a high flow resistance, which blocks the combustion airflow when the combustion blower is off.
- *Intermittent ignition.* Intermittent ignition saves more energy than standing pilot ignition, especially during times when the furnace is not operating.

Control and Operation

CAPACITY. The heating capacity of a gas-fired furnace is controlled by a gas valve and ignition device. For small furnaces the gas valve is usually controlled by a room thermostat that has on-off control. Either standing pilot or intermittent ignition may be used. For large furnaces a two-stage gas valve, controlled by a two-stage thermostat, may be operated with a full gas supply, or at a reduced rate when the outdoor weather is mild. Energy or fuel savings may not be significant unless both the gas and the combustion air supply are controlled.

NIGHTTIME SETBACK. ASHRAE research project SP43 found that a nighttime thermostat setback of 10°F lower for eight hours improved E_t slightly (only a 0.4 percent increase). However, there is a 10 to 16 percent annual savings in energy input compared to those furnaces without nighttime setbacks.

Oversizing the furnace shortens the morning pickup time caused by nighttime setback. Use of an oversized furnace has a significant effect on the swing of space air temperature when an on-off control is used for the gas valve. Project SP43's results showed that when the furnace size corresponded to 1.4 times the design heating load (DHL), the furnace had a space temperature swing of 4.9°F and a morning pickup time of about 1 hour. If the furnace size was based on 1.7 times the DHL, the space temperature swing increased to 5.9°F, and the pickup time reduced to about 0.5 hour. A furnace size based on 1.4 times the DHL is more suitable for a nighttime setback period of 8 hours and a setback temperature of 10°F.

FAN OPERATION. In the past, continuous fan operation in small upflow gas-fired burners was said to offer the benefits of better air circulation, reduced noise (because the fan didn't start and stop), and an even temperature distribution. Project SP43 showed that continuous fan operation resulted in a higher furnace efficiency. However, continuous operation consumes more electricity than intermittent operation in which the fan shuts off and the supply temperature drops below 90°F. In many locations with a high electricity-to-fossil-fuel cost ratio, an energy cost analysis based on SEUF may determine whether continuous or intermittent operation is more efficient.

The fan often starts about one minute after the burner starts. Such a delay allows the heat exchanger to warm up and prevents a flow of cold air. The fan will shut down two to three minutes after the burner is shut off. The supply of residual heat from the heat exchanger also improves the performance of the furnace.

SAFETY. For safety, the vicinity of the furnace should be free of combustible gas, vapor, and material. Any passage to provide combustion air must be carefully planned. Gas and vent pipes should be installed according to local and federal codes.

14.3 HOT WATER BOILERS

A *hot water boiler* for space heating is an enclosed pressure vessel in which water is heated to a required temperature and pressure without evaporation. Hot water boilers are manufactured according to American Society of Mechanical Engineers (ASME) boiler and pressure vessel codes. Boilers are usually rated based on their gross output heat capacity, that is, the rate of heat delivered at the hot water outlet of the boiler, in MBtu/h (thousands of Btu/h). Hot water boilers are available in standard sizes for up to 50,000 MBtu/h.

Fuel Selection

Natural gas, oil, coal, and electricity are energy sources that can be used in hot water boilers. It is necessary to provide for an adequate supply during normal and emergency conditions and to take into account the limitations imposed by any building and boiler codes for certain types of equipment due to safety and environmental concerns. In addition, storage facilities and cost should be considered before a fuel is selected.

Whereas natural gas and electricity are supplied by a utility, LPG, oil, and coal all need space for storage within and outside the boiler plant.

Costs includes energy costs, initial costs, and maintenance costs. A gas-fired boiler plant requires the lowest initial costs and maintenance costs, oil-fired boiler plants are moderately higher, and coal-fired boiler plants are significantly higher (although their energy cost is lowest). An electric boiler is simple to operate and maintain. In addition, it does not require a combustion process, chimney, or fuel storage. In locations where electricity costs are low, electric boilers become increasingly more attractive.

According to the data in the EIA's *Characteristics of Commercial Buildings* (1991), the percentages of floor area in all commercial buildings served by different kinds of fuel used in hot water and steam boilers in 1989 in the United States are as follows:

Gas-fired boilers	69 percent
Oil-fired boilers	19 percent
Electric boilers	7 percent
Others	5 percent

Types of Hot Water Boilers

According to their working temperature and pressure, hot water boilers can be classified as follows:

1. *Low-pressure boilers.* These hot water boilers are limited to a working pressure of 160 psig and a working temperature of 250°F.
2. *Medium- and high-pressure boilers.* These boilers are designed to operate at a working pressure above 160 psig and a temperature above 250°F.

A low-pressure hot water boiler is generally used for a low-temperature water (LTW) heating system in a single building, regardless of the building's size. Medium- and high-pressure boilers are often used in medium-temperature water (MTW) and high-temperature water (HTW) heating systems for campuses or a building complexes in which hot water temperature may range from 300 to 400°F.

Based on their construction and materials, hot water boilers can also be classified as fire tube boilers, water tube boilers, cast-iron sectional boilers, and electric boilers. Water tube boilers, mainly used to generate steam, will not be discussed here. Electric boilers will be discussed in the next section.

Fire Tube Boilers

A *fire tube boiler's* combustion chamber and flue gas passages are in tubes, which are all enclosed in a shell filled with water. Heat released from the combustion process and the flue gases is absorbed by the sur-rounding water, the temperature of which is increased to a required value.

Many kinds of fire tube boilers have been developed. One of the more recently developed models is known as the modified Scotch Marine boiler, which is a compact and efficient design originally used on ships. The Scotch Marine boiler is probably the most popular hot water boiler manufactured today.

SCOTCH MARINE PACKAGE BOILER. A *Scotch Marine boiler* consists mainly of a gas, oil, or gas/oil burner; a mechanical draft system; a combustion chamber; fire tubes; and a flue vent. A *packaged boiler* is a one-piece, integrated, factory-assembled boiler that includes a burner, outer steel shell, fire tubes, draft system, external insulation, controls, interconnecting piping, and wiring. A schematic diagram of a typical Scotch Marine packaged boiler is shown in Fig. 14.3a and a cutaway illustration in Fig. 14.3b.

Flow processes. Gas or oil and air are taken into the burner in measured quantities. The burner then injects a combustible air-fuel mixture containing the necessary combustion air into the combustion chamber, where it is initially ignited by an ignition device. The mixture burns, and the combustion process sustains itself once the heat it generates is greater than the heat it transfers to the water, i.e., once a high enough temperature in the combustion chamber is attained. The injected air–fuel mixture is then burned spontaneously.

The direct radiation from the flame and the high temperature in the combustion chamber both conduct heat through the wall of the chamber—the primary surface adjacent to the water that fills the shell. The combustion product from the combustion chamber, flue gas, is directed into fire tubes by headers. As the flue gas at a higher temperature flows through the fire tubes, it transfers heat to the surrounding water through the pipewall of the tubes—the secondary surface. The flue gas temperature drops and its volume contracts. Because the number of fire tubes continually decreases in the second, third, and fourth pass of fire tubes, which matches the volume contraction of flue gas, the gas velocity is maintained in a more uniform manner. Flue gas leaves after the fourth pass and is vented into the stack.

Return water enters the side of the boiler, sinks to the bottom, is heated, and rises again. Hot water is finally supplied at the top outlet. Such an arrangement prevents cold return water from surrounding the combustion chamber and producing thermal shock. It also promotes good water circulation.

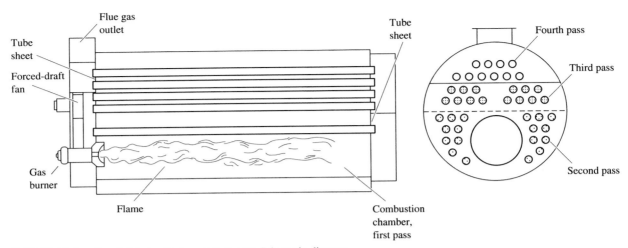

FIGURE 14.3a Scotch Marine Packaged Boiler (a) Schematic diagram.

Constructional characteristics. The constructional characteristics of Scotch Marine packaged boilers are as follows:

- *Four flue gas passes.* As with the water in a shell-and-tube evaporator and condenser (as discussed in Chapter 12) each pass means a horizontal run of the fluid flow passage. A four-pass flow arrangement for the flue gas, with a gradual reduction in gas flow area as the flue gas becomes colder, helps maintain a higher gas velocity and, therefore, a clean surface and a higher rate of heat transfer.

- *Sufficient heat transfer surface area.* In a packaged Scotch Marine boiler, the combustion chamber, fire tubes, and tube sheets at both ends are all heating surfaces. The amount of heating surface area directly affects a boiler's output. Three to five ft² of heating surface area for each boiler hp, or 33,475 Btu/h, of output is the key value for providing sufficient capacity and a long-lasting boiler. A smaller heating surface area for each boiler hp of output often results in a higher heat flux, probably a higher surface temperature, and therefore a shorter life for the boiler.

- *Forced-draft arrangement.* In a forced-draft arrangement, the fan is located adjacent to the burner. The fan forces air into the combustion chamber. It also forces the flue gas to flow through the fire tubes and to discharge from the vent or chimney. A forced-draft fan supplies a controllable quantity of combustion air to the combustion chamber. The force-draft fan is located upstream from the combustion chamber, so it handles dense, clean, boiler-room air at a comparatively lower temperature and lower volume flow rather than the hot, dirty, high-temperature, expanded flue gas downstream. The temperature fluctuation for the room air is smaller than for the exhausted flue gas, so more accurate control of combustion air is possible than with an induced draft design.

Cast-Iron Sectional Boilers

A cast-iron sectional boiler consists of many vertical cast-iron hollow sections in the shape of an inverted U filled with water. When the sections are linked together by bolts, the center part of the inverted U forms the furnace or the combustion chamber.

Because of its thick, heavy sections, a cast-iron sectional boiler has a large heat storage capacity and thus is slow to heat up. This reduces temperature swings when heat demand varies. Thick, heavy sections are also helpful in extending the boiler's service life by preventing corrosion.

The heating capacity of a cast-iron sectional boiler depends on the number of sections connected together.

FIGURE 14.3b Scotch Marine Packaged Boiler (b) A cutaway photograph of a typical product.

Because of its relatively large combustion chamber and lower flow resistance, such a boiler is able to use atmospheric gas burners with a lower chimney height.

Cast-iron sectional boilers are low-pressure boilers used in residences and in small and medium commercial buildings. They can be field-assembled and fitted in existing buildings.

Gas and Oil Burners

When natural gas is used as the fuel in boilers, as in gas-fired furnaces, atmospheric and power gas burners are usually used.

Oil burners used in boilers differ from gas burners in that oil requires atomization and vaporization prior to combustion. Combustion air is supplied either by natural drafts or forced drafts. Ignition is usually provided by an electric spark or a standing pilot using gas or oil.

Oil burners used in residences are usually pressure atomizing-gun burners. This type of burner either directly atomizes oil at 100 to 300 psig high pressure or uses combustion air to atomize the oil. Combustion

air is supplied by a blower, and electric spark ignition is generally used.

Oil burners for boilers in commercial buildings inject oil and atomize it into fine sprays. Burners also force combustion air to mix with atomized oil and ignite the mixture, often with an electric spark. Complete combustion can be sustained with as little as 20 percent excess air.

Sometimes a boiler can be designed to use either gas or oil. A combination of a ring-type gas burner with an oil burner at the center is often used. Combination gas–oil burners are more often used than single-fuel burners in large boilers.

Condensing and Noncondensing Boilers

In condensing boilers, the water vapor in the flue gas is condensed and drained by means of a heat exchanger, as in gas furnaces. Thus, the latent heat of condensation can be recovered. In hot water boilers, the cooling medium is usually the return water from the conditioned space. The lower the temperature of the return water, the higher the amount of heat recovered. If the dew point of the flue gas is 130°F, return water or service water at a temperature below 125°F may be used as the condensing cooling medium. Corrosion in the heat exchanger and flue gas passage caused by the condensate should be avoided. Noncondensing boilers have no way to condense the water vapor contained in the flue gas.

Boiler Efficiency

Boilers can be assessed by two efficiency values. The combustion efficiency E_c, in percent, is the ratio of heat output from the hot water or steam q_{out} to the heat content rate of the fuel consumed q_{fuel}, that is,

$$E_c = \frac{100 q_{out}}{q_{fuel}} \qquad (14.4)$$

Both q_{out} and q_{fuel} are expressed in Btu/h. Annual fuel utilization efficiency AFUE is also an efficiency value used for hot water boilers. For noncondensing boilers, E_c varies from 80 to 85 percent. For condensing boilers, E_c varies from 85 to 90 percent.

ASHRAE/IES Standard 90.1–1989 specifies that gas-fired or oil-fired hot water boilers with a heating capacity < 300,000 Btu/h should have a minimum AFUE of 80 percent as of January 1, 1992. For gas-fired hot water boilers with a heating capacity ≥ 300,000 Btu/h, minimum AFUE is 80 percent at both maximum and minimum rated capacity (steady state). For oil-fired hot water boilers ≥ 300,000 Btu/h, minimum AFUE is 83 percent as of January 1, 1992.

The chimney, or stack, is the vertical pipe or structure used to discharge flue gas. The breeching, or lateral, is the part of the horizontal duct that connects the flue gas outlet of the boiler and the vertical chimney. Breeching is commonly fabricated from 10-gauge steel covered with a high-temperature insulation layer. Flue gas discharge from hot water boilers that burn gas or oil usually has a temperature rise of 300 to 400°F over the ambient temperature.

Chimneys for burning gas or oil should be extended to a certain height above adjacent buildings according to the local codes and topographical conditions. In high-rise buildings, locating the boiler room at the basement level may be very expensive because of the space occupied by the vertical stack and insulation inside the building.

Operation and Safety Controls

HEATING CAPACITY CONTROL. For hot water boilers, heating capacity control during periods of reduced heat demand is achieved by sensing the return water temperature and controlling the firing rate of the gas and oil burners. The firing rate of burners can be controlled in on-off, off-low-high, and modulating modes.

A on-off control mode is usually used for small boilers in which hot water temperature control is not critical. A solenoid valve is used to open or close the fuel supply line to the burner.

An off-low-high control mode offers better output control than an on-off control mode. If the demand for heat is low, the boiler starts to fire at approximately half of its full capacity (the low fire setting). Upon further increase in demand, the burner fires at its maximum capacity (the high fire setting). A two-stage control valve and damper are used to control the supply of fuel and combustion air to the burners.

With modulating control, the boiler starts at a low fire of about one-third of the full load capacity. As the demand for heat increases, an increasing amount fuel and combustion air are supplied to the burner through the modulation of either a butterfly valve in a gas burner or an orifice of variable size in an oil burner. Both types of burner are linked to a damper to vary the supply of combustion air.

For gas burners, two pressure sensors are often provided to maintain the gas pressure within a narrow range for proper operation. If one of the sensors detects an improper pressure, the gas supply is cut off and the boiler shuts down.

In order to maintain an optimum air-fuel ratio, the amount of O_2 in the stack flue gas can be maintained at 1 to 5 percent by modulating the air damper of combustion air to provide higher combustion efficiency. Usually, CO_2 is measured and monitored in small boilers, and O_2 is measured and monitored in large boilers.

SAFETY CONTROL. One or more pressure or temperature relief valves should be equipped for each boiler. These are mechanical devices that open when the boiler pressure or temperature exceeds the rated value. An additional limit control is installed to open the switch and shut down the boiler as soon as the sensed water pressure or temperature exceeds the predetermined limit. When the pressure or temperature falls below the limit, the switch closes and the firing starts again.

A flame detector is used to monitor the flame. When the flame is extinguished, the controller closes the fuel valve and shuts down the boiler. An airflow sensor is also used to verify a continous supply of combustion air. Once the combustion air is not present, the fuel valve closes before a dangerous fuel-air ratio can form.

Modern packaged boilers are often include a totally enclosed and factory-assembled control cabinet that offers the latest microprocessor-based programmable direct digital controls, including flame safeguard and control systems.

14.4 ELECTRIC FURNACES, HEATERS, AND BOILERS

Electric Heating Fundamentals

Except for hot water electric boilers using electrodes, most electric furnaces, electric heaters, and electric heating coils used in HVAC&R installations are of the resistance type. When an electric current flows through a resistor under electric potential, heat is released to the ambient air or water. Electric resistance heaters of small wattage (e.g., an electric furnace in a residence with a heating capacity of less than 8 kW) usually use a single-phase 120/240-V supply. Large-capacity electric resistance heaters for commercial and industrial applications usually use a three-phase 240/480-V supply.

For a single-phase electric heater, the electric power input P, in W, which is equal to the heat released from the resistor, and can be calculated as

$$P = EI = I^2 R \qquad (14.5)$$

where E = electric voltage, V
I = electric current, A
R = electric resistance, Ω

For a three-phase electric heater, the electric power input is given as

$$P = 1.73 E_l I_l = 1.73 I_l^2 R \qquad (14.6)$$

where E_l = line voltage, V
I_l = line current, A

When a three-phase electric heater is in a delta connection, as shown in Fig. 14.4b, then

$$E_p = E_l \quad \text{and} \quad I_p = \frac{I_l}{1.73} \qquad (14.7)$$

where E_p = phase voltage, V
I_p = phase current, A

When a three-phase electric heater is in a wye connection, then

$$E_p = \frac{E_l}{1.73} \quad \text{and} \quad I_p = I_l \qquad (14.8)$$

Several kinds of overload protecting devices are used for electric heaters to open the circuit if the electric current reaches a value that will cause a dangerous temperature:

- *Fuses* are alloys in the form of links to be inserted in the electric circuit. A fuse melts at a preset temperature opening the circuit, when an overload current passes through.
- *Circuit breakers* are mechanical devices that open their contacts, by means of a warping action of a bimetallic strip or disc, when an excessive load passes through and heats them.

There are also thermal overload current sensing devices and magnetic overload relays, both of which open a circuit when the electric circuit is overloaded.

National Electric Codes (NEC) and local codes must be followed during the design, installation, and operation of electric heaters.

Electric Furnaces, Electric Heaters, and Duct Heaters

Electric furnaces installed in rooftop packaged units, electric heaters equipped in air-handling units and packaged units, electric duct heaters mounted in air systems, and electric unit heaters installed directly above the conditioned space are all electric resistance warm air heaters. Air is forced through the electric heating element by the supply fan in a rooftop packaged unit, or by the separately installed propeller fan in an electric unit heater.

Because the mixture of outdoor air and recirculating air or outdoor air alone is clean and nonharzadous after passing through the air filter, most electric resistance heaters use heating elements made of bare wire of various alloys, such as 80 percent nickel and 20 percent chromium wire. Bare wire is supported by bushes and brackets of insulating materials such as ceramics. Sheathed elements are also used for better protection.

The face velocity of air flowing through warm air electric heaters in rooftop packaged units is approximately the same as the face velocity of air at the cooling coil in AHUs. For a face velocity of 500 to 600 fpm, the temperature rise of warm air is often between 30 and 60°F.

The total pressure loss of the warm air flowing through an electric resistance heater in AHU or packaged units is usually less than 0.3 in. WG. For electric duct heaters, the air velocity and pressure loss may be higher.

Figure 14.4a shows a typical wiring diagram for an electric heater using a single-phase electric power supply. Figure 14.4b shows a wiring diagram for a 3φ electric power supply. In the design and installation of an electric duct heater, the following requirements should be fulfilled:

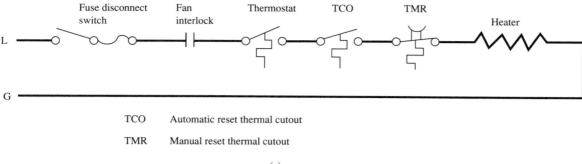

TCO   Automatic reset thermal cutout

TMR   Manual reset thermal cutout

(a)

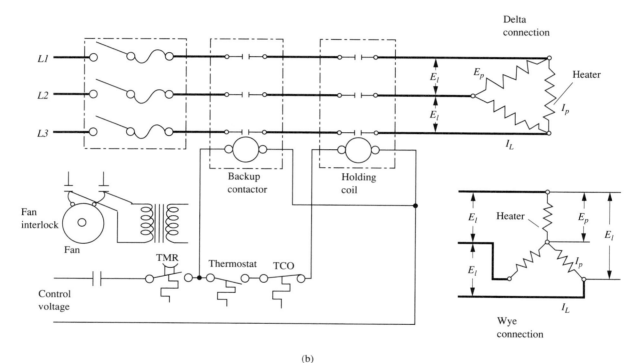

(b)

FIGURE 14.4 Typical electric heater wiring diagram (a) Single-phase line voltage control with thermal cutout and fan interlock, and (b) three-phase line voltage control with fan interlock and backup contactor.

- An electric duct heater should be installed at least 4 ft downstream of the fan outlet, elbow, or other obstructions. Otherwise, devices should be installed to provide for an even distribution of air flow over the face of the heater.

- Do not install two electric duct heaters in series. Install electric duct heaters at least 4 ft from the heat pumps and air conditioners. Following these suggestions prevents excessively high temperatures and the accumulation of moisture in the heater.

- A fan interlock circuit must be provided so that the electric heater does not operate unless the fan is on.

- An automatic-reset overload-cutout safety control should be placed in series with the thermostat. A manual-reset overload-cutout safety control is used in the backup safety control circuit. Limit controls may be used instead of a manual-reset cutout control. Disconnecting devices should be installed within the sight of the electric heater.

- On each electric circuit, the maximum load of the electric heater is 48 A. The maximum fuse size is

60 A. Divide larger electric heater into smaller heaters with smaller loads. The loading of an electric heater is often divided into stages for step control. Small heaters may be divided into two, or three stages; large electric heater may have as many as 16 stages.

Electric Hot Water Boilers

There are two types of electric hot water boilers: *resistance* and *electrode*. Most electric boilers in comfort air conditioning systems are resistance boilers. Resistance electric boilers use metal sheathed heating elements submerged in hot water. Such safety valves as a relief valve, thermal overload cutout, and a high-limit switch should be provided. As with electric heaters, the electric load is divided into stages. The heating capacity is controlled in steps by sensing the return water temperature. Resistance boilers are available in sizes with outputs up to about 2500 kW. Electrode boilers use water as a resistor. They require high-quality water and may operate at high voltages.

14.5 LOW-PRESSURE DUCTED FORCED-AIR HEATING SYSTEMS

System Characteristics

A low-pressure ducted forced-air heating system has the following system characteristics:

- It is often equipped with an upflow gas-fired furnace, oil-fired furnace, or electric heater, and a centrifugal fan to force air through the furnace or electric heater.
- The external static pressure loss for the supply and return duct system is usually no greater than 0.5 in. WG.
- It uses a furnace heating capacity Q_f to air flow $\dot{V}_a$ ratio ranging from 50 to 70 Btu/h per cfm and a temperature rise immediately after the furnace between 50 and 70°F.
- Because of the heat storage capacity of the supply duct system, the supply temperature differential $(T_s - T_r)$ is often between 20 and 35°F. Here, T_s represents the supply air temperature at the supply register, in °F, and T_r indicates the space temperature.
- The heating system may be integrated with cooling system to form a heating/cooling air conditioning system.
- The heating system usually has a capacity no greater than 100,000 Btu/h.

Low-pressure ducted forced-air heating systems are usually used in residences and sometimes in small commercial buildings.

Types of Low-Pressure Ducted Forced-Air Heating Systems

Low-pressure ducted forced-air heating systems can be divided into two categories according to their duct systems:

- Supply and return duct system
- Supply duct and return plenum

Figure 14.5a shows a typical low-pressure ducted forced-air heating system with a supply and return duct system for a single family house. Recirculating air from the living room, dining room, and bedrooms flows through return ducts and enters the lower part of the upflow gas furnace. It mixes with outdoor air. The mixture is forced through a gas furnace and heated to a required supply temperature at the supply plenum. Warm air is then supplied to various rooms through supply ducts and registers.

Figure 14.5b shows a typical low-pressure ducted forced-air heating system with supply ducts and a return plenum. There are no return ducts. Warm air is supplied to various rooms through supply ducts. Recirculating air flows back to the return plenum by means of undercut or louvered doorways and corridors. Once an interior door between rooms is closed, air squeezes through the door undercuts and louvers to the return plenum as well as other cracks and gaps to the attic or outdoors. The room is therefore maintained at a positive pressure. This causes a reduction in the supply flow rate and an increase in the infiltration rate of the whole house.

Field tests performed by Cummings and Tooley (1989) on five Florida houses showed that, for a low-pressure ducted forced-air heating system with supply ducts and a return plenum, the average whole-house infiltration was 0.31 ach (air changes/hour) when all the interior doors between rooms were open. If the interior doors were closed, the average whole house infiltration increased to 0.91 ach. The typical pressure difference between rooms separated by a closed door was 0.032 in. WG (8 Pa), for a door undercut of 0.5 in.

Heat Supplied to Conditioned Space

Heat supplied to the conditioned space by a low-pressure ducted forced-air heating system, q_{sh} (in Btu/h) can be calculated as

$$q_{sh} = 60 \dot{V}_s \rho_s c_{pa}(T_s - T_r) \qquad (14.9)$$

where $\dot{V}_s$ = volume flow rate of supply air, cfm
T_s, T_r = temperature of supply and space air, °F

In Eq. (14.9), ρ_s represents the density of supply air. For a supply air temperature of 100°F and a relative

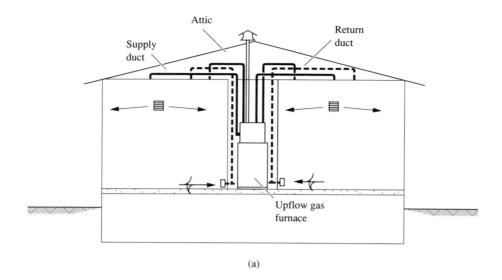

(a)

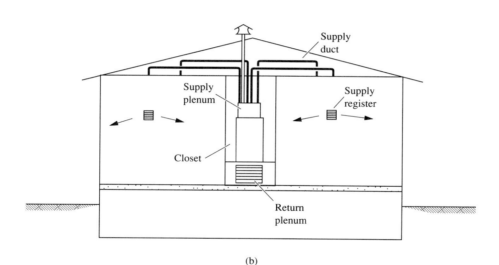

(b)

FIGURE 14.5 Typical low-pressure ducted forced-air heating system, (a) Supply and return duct, and (b) Supply duct and return plenum.

humidity around 20 percent, ρ_s can be taken as 0.07 lb/ft^3. The term c_{pa} indicates the specific heat of moist air. For simplicity, its value can still be taken as 0.243 Btu/lb · °F.

Duct Efficiency and System Efficiency for Heating

The duct efficiency for heating, $\eta_{du.h}$, can be calculated by dividing the heat energy output from the supply and return duct system by the heat energy input to it, that is,

$$\eta_{du.h} = \frac{q_{ho.s} + q_{ho.r}}{q_{hi.s} + q_{hi.r}}$$

$$= 1 - \frac{q_{hl.s} + q_{hl.r}}{q_{hi.s} + q_{hi.r}} \qquad (14.10)$$

where $q_{ho.s}$, $q_{ho.r}$ = output heat energy from the supply ducts and return ducts, Btu/h

$q_{hi.s}$, $q_{hi.r}$ = input heat energy to the supply ducts and return ducts, Btu/h

$q_{hl.s}$, $q_{hl.r}$ = heat loss from the supply ducts and return ducts, Btu/h

and the supply duct efficiency for heating is

$$\eta_{sd.h} = \frac{q_{ho.s}}{q_{hi.s}} = 1 - \frac{q_{hl.s}}{q_{hi.s}} \qquad (14.11)$$

The system efficiency for heating, $\eta_{sy.h}$, in percent, can be calculated as

$$\eta_{sy.h} = \frac{q_{ho.s} + Q_{f.h}}{Q_{f.in}} = E_t \eta_{du.h} + \frac{Q_{f.h}}{Q_{f.in}} \qquad (14.12)$$

where $Q_{f.h}$ = jacket loss and equipment losses from the furnace and releases to the conditioned space, Btu/h

$Q_{f.in}$ = total energy input to the furnace, including auxiliary energy input, Btu/h

Locations of Furnace and Duct Insulation

The location of the furnace has a significant effect on the system heating efficiency. In Locklin et al. (1987), if the gas furnace is installed in a closet, the supply duct is mounted inside the conditioned space, and the equipment losses of the furnace become the direct heat gains of the conditioned space, $\eta_{sy.h}$ might then be 20 percent higher than for those installations where the furnace and supply ducts are in the attic or basement.

Jakob et al. (1986) also showed that if ducts had an exterior or interior insulation of R5, that is, an insulation layer with an R-value of 5 h · ft^2 · °F/Btu, the duct efficiency increased from 61 percent (without duct insulation) to about 78 percent. The system efficiency had a smaller increase because a portion of the heat loss from the duct without insulation had been used to heat the attic or basement, which in turn reduced the heat loss from the conditioned space.

Duct Leakage

Supply and return ducts can be designed and sized in accordance with the procedure covered in Chapter 8. Registers are used to adjust the supply flow rate to each room manually during testing and balancing. As mentioned in Chapter 8, ASHRAE Standard 90.1–1989 specifies that HVAC duct leakage should be less than or equal to leakage class 6, that is, less than about 6 percent for commercial buildings. Field tests in many houses have shown that actual duct leakage in many low-pressure ducted forced-air heating systems is considerably higher.

Lambert and Robinson (1989) analyzed the duct leakage, whole-house leakage, and heat energy use of 800 electric-heated houses built since 1980 in the Pacific Northwest. Tested houses were divided into two groups: highly energy-efficient Model Conservation Standard (MCS) houses and a control group built to current regional practice or standards (CP). Ducted heating systems in CP houses had 26 percent more air leakage than unducted systems and used 40 percent more heating energy. Ducted heating systems in MCS houses had 22 percent more air leakage and used 13 percent more heating energy than unducted MCS houses.

Gammage et al. (1986) studied the ducted forced-air systems of 31 Tennessee houses. They found that the air infiltration rate was 0.44 ach when the forced-air systems were off and 0.78 ach when the forced-air systems were on.

Field tests have also shown that leakages are greater for the return duct than for the supply duct because of the greater importance of the supply air. Also, the return plenum is often not carefully sealed. In five Florida homes with low-pressure ducted forced-air systems with supply and return plenums, Cummings and Tooley (1986) found that when repairs were made to seal the return plenums, the infiltration in these five homes dropped from an average of 1.42 ach to 0.31 ach.

When a ducted forced-air heating system is operating, supply duct leakage in such nonconditioned spaces as the attic or basement raises the space pressure to a positive value and promotes exfiltration. Return duct leakage extracts the space air, lowers the space pressure to a negative value, and promotes infiltration. Both types of leakage increase the whole-house infiltration.

Suggested remedies to reduce duct leakage and energy usage are as follows:

- Externally seal the duct with tapes; seal the ducts internally if possible.
- Seal the return plenum and equipment if there is leakage. Seal the duct, pipes, and cable penetrations through the structures.
- Avoid locating ducts in unconditioned spaces. Provide insulation for ducts running in unconditioned spaces.
- For a low-pressure ducted forced-air system without return ducts, an adequate door undercut or door louver should be provided.

Thermal Stratification

A low-pressure ducted forced-air heating system with a gas-fired furnace always has a high supply temperature differential ΔT_s. If ΔT_s exceeds 30°F, or if there is a high ceiling, thermal stratification may form in the conditioned space. The vertical temperature difference may be greater than 5°F and cause discomfort, as mentioned in Section 5.8. In addition, a higher temperature near the ceiling may increase heat transfer through the ceiling, attic, and roof.

A greater supply-air volume flow rate, lower ΔT_s, higher downward air jet velocity, and suitable location for the supply grille are remedial measures that reduce thermal stratification and vertical temperature differences.

Part-Load Operation and Control

For low-pressure ducted forced warm air heating systems, a thermostat or temperature sensor is usually installed in a representative space to control the gas valve of a furnace operating under an on-off or two-stage step control mode to maintain the required space temperature. The proportion of on- and off-time in each operating on-off cycle varies to meet the varying space heating load. Figure 14.6 shows a typical on-off control for a gas furnace, resulting in a temperature variation for a point 20 ft downstream from the supply plenum and at the supply register.

When the space temperature increases above the upper limit as the space heating load falls, the controller shuts off the gas valve so that the heat supply to the space is cut off. When the space temperature falls below the lower limit, the gas valve will open again and raise the space temperature. The time period for an on-off operating cycle is generally between 5 and 15 minutes. Too short of a cycle may result in unstable operation—a condition known as *hunting*.

14.6 DUCTED FORCED-AIR HEATING SYSTEMS

Ducted forced-air heating systems are heating systems that are part of the air-handling units, packaged units, packaged heat pumps, or fan–coil units that use water heating coils, gas furnaces, electric heaters, or indoor coils in heat pumps to heat the air. Warm air is then supplied to the conditioned space through the duct system.

A ducted forced-air system has the following characteristics:

- It is often integrated with a cooling system, forming a heating/cooling system.
- Its external static pressure is usually between 0.5 and 2 in. WG, except fan-coil units.
- The heating capacity is controlled by modulating the water flow rate to the water heating coil, by modulating the gas valve, or by controlling the refrigeration flow in an on-off or step-control mode in the heat pumps.
- Heating systems often have a capacity greater than 100,000 Btu/h.

Because the perimeter zones of office buildings need strict control of the vertical temperature difference in their conditioned space, a smaller supply air temperature differential, such as $\Delta T_s = 15$ to $20°F$, is usually used. The performance and design of these types of heating and cooling systems will be discussed in later chapters.

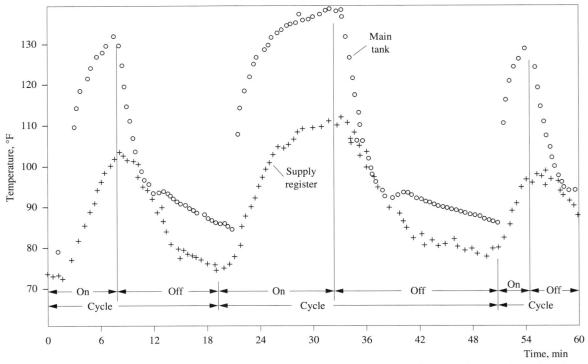

FIGURE 14.6 Typical on-off control of a gas furnace and the variation of temperature of supply plenum and supply register. (*Source: ASHRAE Transactions Part II B, 1986,* Jakob et al. Reprinted by permission.)

14.7   HOT WATER HEATING SYSTEMS USING FINNED-TUBE HEATERS

Types of Hot Water Heating Systems

Like hot water boilers, hot water heating systems can be classified according to their operating temperature into two groups:

- *Low-temperature water systems.* These operate at a temperature not exceeding 250°F, typically 190°F supply and 150°F return, and a maximum working pressure not exceeding 150 psig, usually less than 30 psig.
- Medium- and high-temperature water systems. In medium-temperature water systems, the operating temperature is 350°F or less and the operating pressure is less than 150 psig. In high-temperature water systems, the maximum operating temperature can be from 400 to 450°F and the maximum operating pressure is 300 psig. In both medium- and high-temperature systems, the hot water supplied to and returned from the heating coils and space heat are typically 190°F and 150°F, respectively.

Low-temperature hot water heating systems are widely used for space heating in residential and commercial buildings. Medium- and high-temperature hot water heating systems are often used in central heating plants for university campuses or groups of buildings, or in industrial applications for process heating. Only low-temperature hot water systems will be discussed in this section.

Two-Pipe Individual Loop Systems

Most current low-temperature hot water heating systems that use finned-tube heaters are equipped with zone control facilities. Without such a control system, a building's rooms that face south may be overheated and rooms that face north may be underheated in northern latitudes because of the effects of solar radiation.

The typical piping arrangement for a low-temperature hot water heating system is the *two-pipe individual-loop system,* as shown in Fig. 14.7. On a cold winter day, water in a typical individual-loop system returns from various finned-tube baseboard heaters through different individual return loops at a temperature between 150 and 155°F. It is heated at the hot water boiler to a temperature of 190°F, extracted by the on-line circulating pump on each individual loop, and distributed to the finned-tube baseboard heaters by means of supply and branch pipes. Because of the higher average temperature of hot water in the baseboard heaters, heat is released to the space air to offset the heating load by means of radiation and convection in various control zones, such as north, south, east and west zones. The temperature of hot water then drops to between 150 and

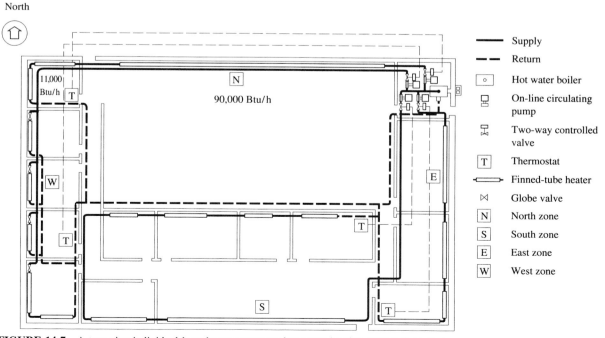

FIGURE 14.7   A two-pipe individual-loop low-temperature hot water heating system for a factory.

155°F and returns to the hot water boiler again via a return main.

In Fig. 14.7, there are several individual loops, or control zones. In each individual loop, several finned-tube baseboard heaters in a large room can be connected in series. Finned-tube baseboard heaters in small rooms belonging to the same individual loop can be connected in a reverse return arrangement. In such an arrangement, the length of pipeline that hot water travels is nearly the same for each baseboard heater for a better system balance. Details will be discussed in Chapter 20.

The design procedure for a hot water heating system involves the following:

- Calculating the space heating load, as discussed in Chapter 7
- Selecting suitable finned-tube heaters
- Dividing the heating space into various control zones, or individual loops
- Planning a piping layout, containing branches and necessary piping fittings
- Locating the boiler, circulating pumps, and expansion tank in the mechanical equipment room and determining their capacities
- Specifying the control functions of the capacity and safety control systems, including the sequence of operation

Finned-Tube Heaters

A *finned-tube heater* is a terminal unit installed directly inside the conditioned space to deliver heat energy to the space by means of radiation and convection. A finned-tube heater consists of a finned-tube element and an outer enclosure, as shown in Fig. 14.8. The tubes are often made of copper and steel. Copper tubes are usually of 0.75-in., 1-in., and 1.25-in. diameters, and steel tubes of 1.25-in. and 2-in. diameters. For copper tubes, fins are often made of aluminum. For steel tubes, the fins are made of steel. Fin spacing varies from 24 to 60 fins per ft. A finned-tube heater may have a length up to 12 ft.

Although various configurations of the outer enclosure have been designed and manufactured, each enclosure must have a bottom inlet and a top outlet for better convection. The enclosure is usually made of 18-gauge steel with corrosion-resistant coating to provide protection and improve appearance. To allow a higher heating capacity, two finned-tube elements can be set in a two-tier (two-row) arrangement.

The most widely used finned-tube heater, is the baseboard heater, which is often installed at a level of 7 to 10 in. above the floor. It is usually 3 in. deep and has only one tier of finned-tube elements. Baseboard heaters are usually mounted on cold walls and release heat nearly at the floor level. They also interfere less with indoor decor than other heaters.

A wall finned-tube heater has a greater height and is available in various shapes to meet the architectural interior design. A convector has a cabinet-type enclosure with one or two tiers of finned-tube heating elements installed under the windowsill.

The rated heating capacity of a finned-tube heater depends mainly on its length, the fin spacing, the average water temperature flowing through the heating element, and the temperature of air entering the heater. Table 14.1 lists the rated heating capacity of typical finned-tube heaters for entering air at 65°F and different average hot water temperatures.

Design Considerations

In the older low-temperature hot water heating system design, a water temperature drop of $\Delta T_w = 20°F$ at the finned-tube element was usually used. The current trend is to use a greater temperature drop, such as 20 to 50°F. A detailed cost analysis should be performed to determine the optimum temperature drop in a finned-tube heater.

The sizing of low-temperature hot water pipes is usually based on a pressure drop of 0.5 to 1.5 psi (1 to 3 ft) per 100 ft of pipe length. Friction charts for hot water steel pipes and the equivalent length of pipe fittings will be discussed in Chapter 20.

For small low-temperature hot water heating systems, an open expansion tank is usually used. For

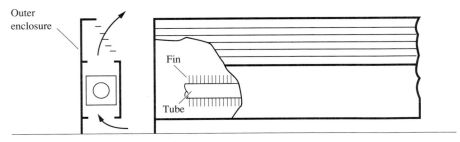

FIGURE 14.8 Baseboard finned tube-heater.

TABLE 14.1
Heating capacity of finned-tube elements for an entering air temperature of 65°F, Btu/h · ft

Type of finned-tube heater	No. of rows	Average water temperature, °F					
		220	210	200	190	180	170
Steel tube: 1.25-in. dia., steel fin	1	1260	1140	1030	940	830	730
	2	2050	1850	1680	1520	1350	1190
Copper tube: 1-in. dia., aluminum fin	1	1000	900	820	740	660	580
	2	1480	1340	1210	1100	970	860

Source: Adapted with permission from *Handbook of HVAC Design 1990.*

medium and large systems, a diaphragm tank may be more suitable. Circulating water pumps are often on-line pumps with low pump head. Both expansion tanks and circulating pumps will be discussed in Chapter 20.

Part-Load Operation and Control

In a low-temperature hot water heating system, one of the basic part-load controls is the variation of the hot water supply temperature from the boiler in response to a variation in outdoor temperature. For instance, in the Midwest, a low-temperature hot water heating system has a winter-design outdoor temperature $T_o = 0°F$. At the winter-design outdoor temperature, the hot water supply temperature T_{ws} is 190°F. When the outdoor temperature drops, T_{ws} is reset as follows:

Outdoor temperature T_o, °F	Supply temperature T_{ws}, °F
0	190
32	135
60	85

A hot water temperature sensor located at the hot water exit of the boiler, whose set point is reset by an outdoor temperature sensor, is used to control the firing rate of the boiler by means of a controller. Zone control can be performed better by sensing the hot water temperature that returns from each individual loop and then modulating the control valve to vary the mass flow rate of hot water supplied to that zone through an on-line circulating pump (see Fig. 14.7).

For a low-temperature hot water heating system installed with multiple boilers, the control strategy is to decide when to turn a boiler on or off. This optimum operation control can be accomplished by using a microprocessor-based controller that fires a standby boiler according to a preprogrammed software instruction. This strategy depends not only on the increase or reduction in heating demand but also on how the

operating cost of such a hot water heating system can be minimized.

Example 14.1. A two-pipe individual-loop low-temperature hot water heating system is used to heat a factory that has a layout as shown in Fig. 14.7. At winter-design conditions, hot water is supplied to the heated space at a temperature of 190°F and returns from the baseboard finned-tube heaters at a temperature of 150°F.

1. If the space heating load for the largest room (facing north) is 90,000 Btu/h and for the northwest corner room is 11,000 Btu/h, and if steel tubes and fins are used, determine the number of feet of finned tubing required for each of these two rooms.

2. If a pressure drop of 1 ft/100 ft of pipe is used and the hot water system is equipped with an open expansion tank, determine the diameter of the hot water supply main for these two rooms.

3. Divide this hot water system into appropriate control zones, or individual loops.

Solution

1. For the largest room, if a two-row finned-tube heater is used, then, from Table 14.1, for an average hot water temperature of $(190+150)/2 = 170°F$, the heat output of each foot of two-row finned tube is 1190 Btu/h. The number of feet required is therefore

$$\frac{90,000}{1190} = 75.63 \text{ ft (76 ft)}$$

For the northwest corner room, a single-row finned-tube heater is used. The number of feet required is

$$\frac{11,000}{730} = 15.06 \text{ ft (15 ft)}$$

Along the north side's external wall, an 8-ft finned tube is mounted. On the west side's external wall, a 7-ft finned tube is mounted.

2. Because the density of water at 170°F is 60.8 lb/ft³, from Eq. (12.3), the flow rate of hot water supplied

to the largest room and the northwest corner room, in gpm, can be calculated as

$$V_{gal} = \frac{q}{60\,(0.1337)\rho_w c_{pw}(T_{we} - T_{wl})}$$

$$= \frac{90,000 + 11,000}{60\,(0.1337)(60.8)(1)(190 - 150)}$$

$$= 5.18 \text{ gpm}$$

Then, from the water friction chart for open water systems in Chapter 20, for a pressure drop of 1 ft/100 ft, the pipe diameter for the hot water supply main to the largest room and northwest corner room is 1.25 in.

3. This two-pipe individual-loop low-temperature hot water heating system should be divided into four zones: north, south, east, and west. Each zone, or individual loop, includes the rooms whose outer walls face that direction, (except the south zone, which also includes interior rooms without external walls). Such a setup can offset the variation in solar radiation and the operating conditions for each zone to prevent overheating of rooms facing south and underheating of rooms facing north. For each individual loop, or control zone, a thermostat senses the return hot water temperature and modulates the control valve and the flow rate of hot water supplied to that loop in order to meet the variations in zone heating load.

14.8 INFRARED HEATING

Basics

Infrared heating uses radiant heat transfer from a gas-fired or electrically-heated high-temperature tube or panel to provide heating to a localized area for the comfort of the occupant or the maintenance of a suitable environment for a manufacturing process.

Heat radiates from an infrared heater in the form of electromagnetic waves in all directions. Most infrared heaters have reflectors to focus the radiation on a localized target; hence, they are often known as *beam radiant heaters*.

Infrared heating is widely used in factories, warehouses, garages, gymnasiums, skating rinks, outdoor loading docks, and racetrack stands.

In environments with a low indoor ambient air temperature during the winter, the warmth that an occupant feels depends on the radiant energy absorbed by the occupant from all sources that have temperatures higher than that of the indoor ambient air. This warmth is called the effective radiant field I_{rad}, or ERF, in W/ft^2. The effective radiant field can be calculated as

$$I_{rad} = h_r(T_{rad} - T_a)$$

$$= \frac{\alpha_s f_{eff} F_p I_{hr}}{L^2} \qquad (14.13)$$

where h_r = radiative heat transfer coefficient, W/ft$^2 \cdot$ °F

T_{rad} = mean radiant temperature, °F

T_a = indoor ambient air temperature, °F

α_s = absorptance of skin-cloth surface, which usually varies from 0.85 to 0.95

I_{hr} = irradiance from infrared beam heater, W/sr

L = distance between the heater and the center of the occupant, ft

In Eq. (14.13), f_{eff} indicates the effective radiation area factor and is the ratio of the projecting area of the occupant at a plane normal to the beam radiation of the infrared heater, A_p, to the DuBois surface area of the human body, A_D, defined by Eq. (5.2), that is, $f_{eff} = A_p/A_D$. An average value of $f_{eff} = 0.71$ is used for standing or seated person.

In Eq. (14.13), F_p indicates the radiation area factor. When infrared beam radiation is at an altitude angle $\beta = 45°$ and it radiates on the side of the human body, $F_p = 0.18$; and if $\beta = 60°$ (also for side irradiation), $F_p = 0.15$.

In Eq. (14.13), $(T_{rad} - T_a)$ indicates the amount T_{rad} is above the indoor ambient air T_a; it is important because it influences the ERF as well as the infrared radiant heat transfer to the occupant. There are two types of infrared heaters: gas infrared heaters and electric infrared heaters.

Gas Infrared Heaters

Gas infrared heaters can be divided into *indirect infrared heaters* and *porous matrix infrared heaters*. An indirect infrared heater consists of a burner, a radiating tube, and a reflector. Combustion takes place within the radiating tube at a temperature up to 1200°F. In a porous matrix infrared radiation heater, a gas-air mixture is supplied to an enclosure and distributed evenly through a porous ceramic, stainless steel, or metallic screen that is exposed at the other end. Combustion takes place at the exposed surface and has a maximum temperature of about 1600°F.

Gas infrared heaters are usually vented and have a small conversion efficiency. Only 10 to 20% of the energy output of an open combustion gas infrared heater is in the form of infrared radiant energy. Gas infrared heaters should not be operated under conditions in which the ambient air contains ignitable gas or materials that may decompose to hazardous or toxic gases or vapors.

Adequate combustion air must be provided. Venting is preferred in order to prevent a buildup of combustion products. Usually, 4 cfm of make-up air is required for 1000 Btu/h gas input.

If unvented infrared heaters are used, humidity and condensation control should be provided to account for the accumulation of water vapor that forms during combustion.

A thermostat usually controls the supply of gas by means of a gas valve in on-off mode. For standing pilot ignition, a sensing element and controller are also used to cut off the gas supply when the pilot flame extinguishes. If the combustion air is blocked, the gas supply is also cut off for safety.

Electric Infrared Heaters

Electric infrared heaters are usually made of nickel-chromium wire or tungsten filaments inside an electrically insulated metal tube or quartz tube with or without an inert gas. The heaters also contain a reflector, which directs the radiant beam to the area that needs heating. Nickel-chromium wires often operate at a temperature between 1200 and 1800°F. The tungsten filament can stand a temperature as high as 4000°F while the outer envelope is at a temperature of about 1000°F.

Electric infrared heaters also use a thermostat to switch on or cut off the electric current in order to control the input to the nickel-chromium electric heater. Some input controllers can preset the on-off time period. For a quartz tube, the output can be controlled by varying its voltage supply.

As with gas infrared heaters, electric infrared heaters should not be used where there is a danger of igniting flammable dust or vapors or decomposing contaminated matter into toxic gases.

Electric infrared heaters have a far higher infrared radiant energy conversion efficiency than gas infrared heaters. They are clean and more easily managed. Although the cost of electric energy is several times higher than that of natural gas, a comprehensive analysis should be conducted before selecting an electric or gas infrared heater.

Design and Layout

DETERMINATION OF WATTS DENSITY. According to tests, an acceptable temperature increase $T_{rad} - T_a$ for normal clothed occupants in an indoor environment using infrared heating is often between 20 and 25°F. The effective radiant field, or required watts density, provided by the infrared heaters has been determined from field experiments, with the data listed in Table 14.2 for different operating conditions.

SINGLE OR MULTIPLE HEATERS. Based on the required coverage area and the width and length of the floor area that can be covered with the specified watts density listed in Table 14.3, it is possible to determine whether a single heater or multiple heaters should be installed. It is also necessary to select the coverage pattern: three-fourths overlap or full overlap. If there are many occupants within the coverage area or if the occupant will stay for a long period, a full overlap pattern is preferable.

MOUNTING HEIGHT AND CLEARANCES. The mounting height of infrared heaters should not be below 10 ft; otherwise, the occupant may feel discomfort from the radiant beam overhead. Because of the spread of the radiant beam, a higher mounting height results in a smaller watts density and a greater coverage, and allows a greater spacing between heaters. An optimum mounting height should be selected based on the ceiling height of the building or the outdoor structures.

Adequate clearance (recommended by the manufacturer) between the infrared heater and any combustible material, especially between the heater and the roof must be maintained to prevent fire.

OTHER CONSIDERATIONS. When infrared heaters are used for total indoor space heating, the following must be taken into account:

TABLE 14.2
Required watt density, W/ft^2

Temperature increase $T_{rad} - T_a$ q_c	Tight uninsulated building	Drafty indoors or large glass area	Loading area, one end open	Outdoor shielded, less than 5 mph wind speed	Outdoor unshielded, less than 10 mph wind speed
20	30	40	50	55	60
25	37	50	62	70	75
30	45	60	75	85	90
40	60	80	100	115	120
50	75	100	125	145	150

Source: Abridged with permission from *Handbook of HVAC Design 1990,* Chapter 7 (Lehr Associates).

TABLE 14.3
Watt density and coverage

Two asymmetric heaters					
	Three-fourths overlap			Full overlap	
Mounting height, ft	**Watt density, W/ft^2**	**W × L, ft**	**Spacing s, ft**	**Watt density, W/ft^2**	**W × L, ft**	**Spacing s, ft**
10	33	13 × 12	8.5	38	11 × 12	6
11	27	14 × 13	9	33	12 × 13	6.5
12	24	15 × 14	10	29	12 × 14	7
13	21	16 × 15	10.5	25	13 × 15	7.5
14	18	17 × 16	11	22	14 × 16	8
15	16	18 × 17	12	20	15 × 17	8.5
16	14	20 × 18	12.5	18	16 × 18	9
18	11	22 × 20	14	14	18 × 20	10
20	9.6	24 × 22	15.5	11	20 × 22	11
22	8.0	26 × 24	17	9.6	22 × 24	12
25	6.4	30 × 27	19	8	24 × 27	13.5

Single heater					
	90°		60°		60° Asymetric
Mounting height, ft	**Watt density, W/ft^2**	**W × L, ft**	**Watt density, W/ft^2**	**W × L, ft**	**Watt density, W/ft^2**	**W × L, ft**
10	18	12 × 12	25	8.5 × 12	19	11 × 12
11	15	13 × 13	22	9 × 13	16	12 × 13
12	13	14 × 14	18	10 × 14	14	12 × 14
13	11	15 × 15	16	10 × 15	13	13 × 15
14	10	16 × 16	14	11 × 16	11	14 × 16
15	9	17 × 17	13	12 × 17	9.6	15 × 17
16	8	18 × 18	11	12 × 18	8.8	16 × 18
18	6.4	20 × 20	8.8	14 × 20	7.2	18 × 20
20	5.2	22 × 22	7.6	15 × 22	5.8	20 × 22
22	4.4	24 × 24	6.2	17 × 24	4.8	24 × 24
25	3.5	27 × 27	5.0	19 × 27	3.9	24 × 27

W = width, L = length, both in ft.

Source: Abridged with permission from *Handbook of HVAC Design 1990*, Chapter 7 (Lehr Associates).

- The space heating load can be lower because of a lower space temperature.
- The area near the external walls needs a greater intensity of infrared heating.
- Heaters are often mounted around the perimeter at a suitable height and are usually arranged 15 to 20 ft from the corners of the building.
- The key to successful design is to supply a proper amount of heat evenly to the occupying zone.

REFERENCES AND FURTHER READING

Adams, C. W., "Performance Results of a Pulse-Combustion Furnace Field Trial," *ASHRAE Transactions*, Part I B, pp. 693–699, 1983.

The American Boiler Manufacturer Association, "Why Packaged Firetube Boilers?" *Heating/Piping/Air Conditioning*, pp. 79–83, November 1990.

Andrews, J. W., "Impact of Reduced Firing Rate on Furnace and Boiler Efficiency," *ASHRAE Transactions*, Part I A, pp. 246–262, 1986.

ASHRAE, *ASHRAE Handbook 1988, Equipment* ASHRAE Inc., Atlanta, GA.

ASHRAE, *ASHRAE Handbook 1987, HVAC Systems and Applications* ASHRAE Inc., Atlanta, GA.

ASHRAE, *ASHRAE/IES Standard 90.1–1989 Energy Efficient Design of New Buildings Except New Low-Rise Residential Buildings*, ASHRAE Inc., Atlanta, GA, 1989.

Axtman, W. H., "Boiler Types and General Selection Requirements," *Heating/Piping/Air Conditioning*, pp. 5–9, November 1987.

Cummings, J. B., and Tooley, J. J., "Infiltration and Pressure Differences Induced by Forced Air Systems in Florida Residences," *ASHRAE Transactions*, Part II, pp. 551–560, 1989.

DOE, *Code of Federal Regulations, Title 10, Part 430, Appendix N*, DOE, Washington D.C., 1992.

EIA, *Characteristics of Commercial Buildings*, EIA Washington D.C., 1991.

Fischer, R. D., Jacob, F. E., Flanigan, L. J., and Locklin, D.W., "Dynamic Performance of Residential Warm-Air Heating Systems—Status of ASHRAE Project SP43," *ASHRAE Transactions*, Part II B, pp. 573–590, 1984.

Gammage, R. B., Hawthrone, A. R., and White, D. A., "Parameters Affecting Air Infiltration and Air Tightness in Thirty-One East Tennessee Homes," *Measured Air Leakage in Buildings*, ASTM STP 904, American Society of Testing Materials, Philadelphia, 1986.

Grimm, N. R. and Rosaler, R. C., *Handbook of HVAC Design 1990*, McGraw-Hill, New York, 1990.

Int-Hout, D., "Analysis of Three Perimeter Heating Systems by Air-Diffusion Methods," *ASHRAE Transactions*, Part I B, pp. 101–112, 1983.

Jakob, F. E., Fischer, R. D., and Flanigan, L. J., "Experimental Validation of the Duct Submodel for the SP43

Simulation Model," *ASHRAE Transactions*, Part I, pp. 1499–1514, 1987.

Jakob, F. E., Fischer, R. D., Flanigan, L. J., Locklin, D. W., Herold, K. E., and Cudmik, R.A. "Validation of the ASHRAE SP43 Dynamic Simulation Model for Residential Forced-Warm-Air Systems," *ASHRAE Transactions*, Part II B, pp. 623–643, 1986.

Jakob, F. E., Locklin, D. W., Fisher, R. D., Flanigan, L. G., and Cudnik, R. A., "SP43 Evaluation of System Options for Residential Forced-Air Heating," *ASHRAE Transactions*, Part II B, pp. 644–673, 1986.

Kesselring, J. P., Blatt, M. H., and Hough, R. E., "New Option in Commercial Heating," *Heating/Piping/Air Conditioning*, pp. 41–43, July, 1990.

Lambert, L. A., and Robison, D. H., "Effects of Ducted Forced-Air Heating Systems on Residential Air Leakage and Heating Energy Use," *ASHRAE Transactions*, Part II, pp. 534–541, 1989.

Locklin, D. W., Herold, K. E., Fischer, R. D., Jakob, F. E., and Cudnik, R.A., "Supplemental Information from SP43 Evaluation of System Options for Residential Forced-Air Heating," *ASHRAE Transactions*, Part II. pp. 1934–1958, 1987.

Modera, M. P., "Residential Duct System Leakage: Magnitude, Impacts, and Potential for Reduction," *ASHRAE Transactions*, Part II, pp. 561–569, 1989.

Parker, D.S., "Evidence of Increased Levels of Space Heat Consumption and Air Leakage Associated with Forced Air Heating Systems in Houses in the Pacific Northwest," *ASHRAE Transactions*, Part II, pp. 527–533, 1989.

Palm, Jr., R. B., "Pulse Combustion: A New Approach," *Heating/Piping/Air Conditioning*, pp. 148–150, January 1989.

Patani, A., and Bonne, U., "Operating Cost of Gas-Fired Furnace Heating Systems with Add-on Heat Pumps," *ASHRAE Transactions*, Part I B, pp. 319–329, 1983.

Robison, D. H., and Lambert, L. A., "Field Investigation of Residential Infiltration and Heating Duct Leakage," *ASHRAE Transactions*, Part II, pp. 542–550, 1989.

Spolek, G. A., Herriott, D. W., and Low, D. M., "Air Flow in Rooms with Baseboard Heat: Flow Visualization Studies," *ASHRAE Transactions*, Part II A, pp. 528–536, 1986.

Tao, W., "Modern Boiler Plant Design," *Heating/Piping/Air Conditioning*, pp. 69–82, November 1984.

Trehan, A. K., Fortmann, R. C., Koontz, M. D., and Nagda, N. L., "Effect of Furnace Size on Morning Picking Time," *ASHRAE Transactions*, Part I, pp. 1125–1129, 1989.

The Trane Company, *Basics of Heating with Electricity*, La Crosse, WI, 1973.

The Trane Company, *Boilers for Steam and Hot Water*, La Crosse, WI, 1971.

Trewin, R. R., Langdon, F. W., Nelson, R. M., and Pate, M. B., "An Experimental Study of A Multipurpose Commercial Building with Three Different Heating Systems," *ASHRAE Transactions*, Part I, pp. 467–481, 1987.

AIR FILTRATION, AIR-HANDLING UNITS, AND PACKAGED UNITS

15.1 AIR CLEANING

Air Filtration and Industrial Air Cleaning

Air cleaning is the process of removing airborne particles present in the air. It can be classified into two categories:

- Air filtration
- Industrial air cleaning

Air filtration involves the removal of airborne particles present in outdoor air as well as recirculated air from a given space. Most airborne particles removed by air filtration are < 1 micrometer in size, and the concentration of these particles in the airstream seldom exceeds 2 mg/m^3. The purpose of air filtration is to benefit the health and comfort of the occupants in the conditioned space. Air filtration is closely related to the indoor air quality of the conditioned space.

Industrial air cleaning mainly involves the removal of dusts and gaseous contaminants from industrial manufacturing processes, and provides pollution control to the exhaust gas and flue gas. Air contaminants that are discharged to the outdoor environment are governed by the U.S. Environmental Protection Agency (EPA) and those that are exhausted to the indoor working space are regulated by the Occupational Safety and Health Administration (OSHA). The size of airborne particles may range from < 1 micrometer to several hundred micrometers. The amount of airborne particles present in the airstream often varies from several mg to 40,000 mg/m^3.

In this chapter, only air filtration will be discussed.

Atmospheric Dust and Space Air Contaminants

Any materials other than the oxygen, nitrogen, carbon dioxide, water vapor, and rare gases present in air are called *air contaminants*. They are a mixture of dusts, smoke, fumes, fogs, mists, and living organisms. Dusts are defined as solid granular particles and fibers of sizes less than 100 μm (micrometer, 1 μm = 0.00004 in.) derived from natural and mechanical processes. Smoke is a solid, a liquid, or a mixture of solid and liquid particles of sizes averaging from 0.1 to 0.3 μm. It is the product of incomplete combustion. Fumes are solid particles smaller than 1 μm formed in the condensation of vapors. Fogs are very small liquid particles, ranging from 2 to 60 μm in size, formed by the condensation of vapors, and mists are relatively large liquid particles, from 60 to 200 μm, produced by atomizing and spraying processes. Living organisms include viruses, from 0.003 to 0.06 μm, and bacteria, usually from 0.4 to 5 μm. Both of these attach themselves to a particle of larger size for transportation. Fungal spores and pollens, which derive from living organisms, range in size from 10 to 100 μm. The mixture of granular particles, fibers, smoke, fumes, and mists are often collectively called *dust* or particulates, and a mixture of air and dust is called an *aerosol*.

Airborne particles that are smaller than 0.1 μm remain suspended and moving owing to Brownian movement, whereas particles that range from 0.1 to 1 μm in size have a negligible settling velocity. Particles between 1 and 10 μm have a constant and significant settling velocity and tend to drop onto a solid surface. Particles that are larger than 10 μm settle on a surface even more rapidly.

Atmospheric dust denotes dust contained in the outdoor air, or outdoor atmosphere. According to Section 5.13, concentration of particulate matter in atmo-

spheric air should meet the EPA Ambient Air Quality Standard of 260 μg/m^3 within a period of 24 hours. In atmospheric dust, (1) 99 percent of the dust particles are smaller than 0.3 μm and they constitute only 10 percent of the total weight; and (2) 0.1 percent of the dust particles are larger than 1 μm and make up 70 percent of the total weight.

The dust concentration of an indoor conditioned space may vary considerably according to the requirements of the space, the sources releasing dust particles, the type of equipment and devices used in removing the dust particles, and the type of building enclosing the conditioned space. The concentration of total airborne particles of a conditioned space usually varies, from the very low particle count in a clean room to about 0.2 mg/m^3.

15.2   RATING AND ASSESSMENTS

An *air cleaner* is a device that removes dust particles from an airstream passing through it. An *air filter* is an air cleaner that removes dust particles from a conditioned airstream by means of fibrous, metallic, or other filter media. *Filter media* are fabricated materials that perform air filtration.

The operating characteristics of an air filter depend on the size, shape, specific gravity, concentration, and electrical properties of the dust particles as well as the filter medium. However, the most important factor for effective removal is the particle size, or size distributions in a dust sample. For air filtration in air conditioning systems, the dust particles are small and their concentrations are comparatively low.

The rating and assessment of air filters are mainly based on the following characteristics:

- *Efficiency.* The average efficiency of an air filter during the operating period shows the effectiveness of the dust removal process and, therefore, is the primary rating index. The efficiency rating of a specific filter is affected by the size of the dust particles, or the test methods for various sizes of dust particles.

- *Pressure drop.* Total pressure drop of an air filter Δp_t, in inches WG, can be calculated as

$$\Delta p_t = C \left(\frac{v_f}{4005}\right)^2 = C_a \left(\frac{v_a}{4005}\right)^2 \quad (15.1)$$

where C, C_a = local loss coefficients of the air filter
v_f = air velocity flowing through the filter media, in fpm
v_a = face velocity of the air filter, in fpm

The pressure drop of an air filter is affected by v_f or v_a, as well as by the amount of dust held inside the air filter, that is, its dust holding capacity m_{dust}, in grains/ft^2. The greater the v_f, v_a, and m_{dust}, the higher the Δp_t.

At the rated volume flow rate, the pressure drop of a clean air filter when it first starts to collect dust particles is called the *initial pressure drop*.

• *Service life*. The pressure drop of an air filter at which the air filter is about to be replaced or cleaned is called its *final pressure drop*. The operating period between the initial and final pressure drop is called the *service life* of an air filter. When the pressure drop of an air filter is at its final pressure drop, the dust-holding capacity of the air filter is usually the maximum dust holding capacity $m_{max.d}$.

The efficiency of an air filter is significantly affected by the service life, or dust-holding capacity of an air filter. The rating and assessment of a specific air filter at a specific volume flow rate require complete data on its efficiency and initial and final pressure drops.

15.3   TEST METHODS FOR AIR FILTERS

Different test methods to assess air filters use different types of test dust and procedures. The efficiency rating for the same air filter may vary considerably if different test methods are used. When an air filter is said to have a certain efficiency in removing airborne dust particles, the test method must also be specified.

The performance of an air filter is usually tested in a test unit as shown in Fig. 15.1. The test unit consists of a fan, a test duct, a tested filter, two samplers, two flow meters, a vacuum pump, a flow nozzle, two sets of Pitot tubes and inclined manometers, and other necessary instruments and accessories. Synthetic dust or atmospheric dust is fed into the test unit from one end. The draw-through fan mounted at the other end of the unit extracts the airstream containing the dust particles through the tested filter. A precision balance, an optical instrument, or sometimes a particle counter is used to determine the degree of dust concentration in the airstream before and after the filter to be tested. Three different test methods, described below, are used for testing low-, medium-, and high-efficiency air filters.

Weight Arrestance Test

Standard synthetic dust particles, which are considerably coarser than atmospheric dust, are fed into the air filter test unit. The size distribution by weight for typical standard dust particles used in the weight arrestance test is <5 μm, 39 percent; 5 to 10 μm, 18 percent; 10 to 20 μm, 16 percent; 20 to 40 μm, 18 percent; and 40 to 80 μm, 9 percent. The dust concentration in the airstream before the tested air filter can be controlled by the rate of dust feed. Within the samplers, the dust particles are held by the high-efficiency membrane filter. Only air is extracted through the membrane by using a vacuum pump.

By measuring the weight of dust fed and the weight gain due to dust particles on the membrane filter after the tested air filter, both for the same time in-

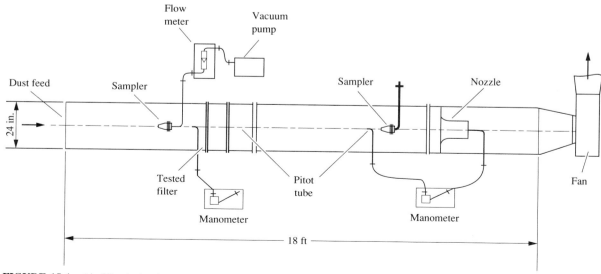

FIGURE 15.1   Air filter test unit.

terval, one can calculate the arrestance, in percent, as follows:

$$\text{Arrestance} = 100 \left(1 - \frac{W_{mf}}{W_{df}}\right) \quad (15.2)$$

where W_{mf} = weight gain of the membrane filter, grains
W_{df} = weight of dust fed, grains

The weight arrestance test primarily determines the capacity of an air filter to remove large dust particles present in the airstream. This test is suitable to assess the performance of low-efficiency air filters.

Dust Spot Efficiency Test

This test method is an integral part of ASHRAE Standard 52–76. The discoloration of white filter-paper targets in the sampler of this method simulates the objectionable discoloration of building interiors.

Untreated atmospheric dusts are fed into the test unit. Air samples taken before and after the tested air filter are drawn through identical fiber filter-paper targets at approximately equal volume flow rates. The downstream air sample is drawn continuously, whereas the upstream sample is drawn intermittently. The time ratio of the off-period to the continuously drawn period should approximately equal the efficiency of the air filter.

By measuring the change in light transmission of these discolored white filter papers, the efficiency of the air filter E, in percent, can be calculated as

$$E = 100 \left[1 - \frac{\dot{V}_1}{\dot{V}_2} \frac{(T_{20} - T_{21})}{(T_{10} - T_{11})} \frac{(T_{10})}{(T_{20})}\right] \quad (15.3)$$

where $\dot{V}_1, \dot{V}_2$ = volume flow rate of air drawn through the upstream and downstream papers, cfm
T_{10}, T_{11} = initial and final light transmission of the upstream filter paper
T_{20}, T_{21} = initial and final light transmission of the downstream filter paper

The dust spot efficiency test is widely used for medium-efficiency air filters. It measures the ability of the air filters to remove dust particles of finer size, especially to reduce smudging and discoloration of building interiors.

DOP Penetration and Efficiency Test

According to U.S. Military Standard MIL-STD-282 (1956), a smoke cloud of uniform *di-octyl phthalate* (DOP) droplets 0.18 μm in diameter, as measured by a laser spectrometer, is generated from the condensa-

tion of DOP vapor and fed into the test unit. DOP vapor produced by thermal generation is actually oily liquid particles, and is usually called *hot DOP*. It may penetrate the tested air filter or leak through the cracks of its gasket. By measuring the concentration of these particles in the air sample upstream and downstream from the tested air filter with an electronic particle counter or a laser spectrometer, the penetration of the air filter P, in percent, can be calculated as

$$P = 100 \frac{\text{downstream concentration}}{\text{upstream concentration}} \quad (15.4)$$

and the efficiency of the air filter, in percent, is given as

$$E = 100 \left(1 - \frac{\text{penetration}}{100}\right) \quad (15.5)$$

The DOP penetration and efficiency test is mainly used to assess high-efficiency air filters, that is, the ability of air filters to remove dust particles of submicron size from the airstream.

New filter testing methods have been developed during recent years. Compressed-air-generated polydispersed *cold DOP* particles in sizes of less than 0.4 to 3 μm and polydispersed particles of ethylene glycol (using atomizers and electrostatic classifiers to generate monodispersed aerosols of sizes from 0.01 to 0.4 μm) have been developed. Particles are then detected by a condensation nucleus counter. This counter can measure concentrations from < 0.01 to 1×10^7 particles/cm^3 and determine particle size distribution within various ranges or *channels*.

15.4 AIR FILTERS

Filtration Mechanisms

The removal or collection of dust particles in air filtration is performed by various combinations of the following mechanisms.

INERTIAL IMPACTION. When an airstream containing dust particles makes a sudden change in direction as it flows around the fibrous media, the inertia of the dust particles causes a collision with the fibers. The greater the sizes of dust particles, the larger the effect of impingement and impaction.

STRAINING. If the spaces between the fibers or other filter media through which the flow passes are smaller than the size of the dust particles, the particles are trapped and collected.

DIFFUSION. For very fine dust particles < 0.4 μm in size, Brownian movement causes the dust particles to settle and deposit on fibrous media.

DIRECT INTERCEPTION. When dust particles follow the airstream, they contact the fibrous media and remain there.

Classification of Air Filters

Air filters can be classified into various types according to the following properties:

- *Shape*. Filters can be panel filters, extended-surface filters, rotary filters, or automatic renewable rolling filters, as shown in Fig. 15.2.

- *Coating*. *Viscous filters* are coated with adhesives; *dry filters* are not.
- *Operating Life*. Filters can be disposable or renewable.
- *Efficiency or grade*. Filters are classified as low-, medium-, or high-efficiency.

Because the efficiency or grade clearly indicates the primary operating characteristic of an air filter and is closely related to the design criteria of the conditioned space and the selection of the air filters, it is the primary classification criterion of air filters.

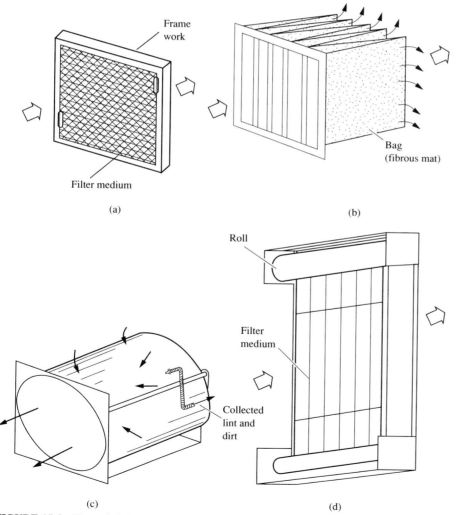

FIGURE 15.2 Type of air filters: (a) panel fitter; (b) extended surface (bag type) filter; (c) rotary filter; and (d) automatic renewable rolling filter.

Low-Efficiency Air Filters

The ASHRAE weight arrestance for low-efficiency air filters is between 60 and 95 percent, and the ASHRAE dust spot efficiency for low-efficiency filters is usually less than 20 percent. These filters are generally panel filters. The dimensions of the framework of these filters are typically 20 in. × 20 in. or 24 in. × 24 in.. The depth of the filter may vary from 1 to 4 in.

For low-efficiency filters, the filter media remove dusts mainly by the straining and impingement mechanisms. Filter media for low-efficiency filters are divided into three categories:

- *Viscous and reusable.* Some examples are corrugated wire mesh and screen strips. They are usually coated with oil, which acts as an adhesive, to increase their dust removal ability. Detergents may be used to wash off the dust when the filter media are to be cleaned and reused.
- *Dry and reusable.* Certain materials such as synthetic fibers (nylon, terylene) and polyurethane foam can be cleaned or washed if reuse is required.

- *Dry and disposable.* Glass fiber mats, with most of the glass fibers greater than 10 μm in diameter belong to this category. The air filter is discarded as soon as the final pressure drop is reached.

The face velocity of panel filters usually lies between 300 and 600 fpm. The initial pressure drop usually varies from 0.05 to 0.25 in. WG, and the final pressure drop varies from 0.2 to 0.5 in. WG.

The characteristic curves of a disposable panel filter are shown in Fig. 15.3. The dimensions of this filter are 24 in. × 24 in. × 2 in. It is filled with a glass fiber filter medium and the weave becomes progressively tighter along the air flow.

In Fig. 15.3, the arrestance or efficiency and pressure drop are affected by both the air velocity flowing through the filter and the amount of dust fed or the amount of dust held. At first, the dust fed reduces the flow passage and raises its arrestance. When the amount of dust held exceeds a specific limit, dust in the airstream may strike an already held dust particles and release them again to the airstream. Therefore,

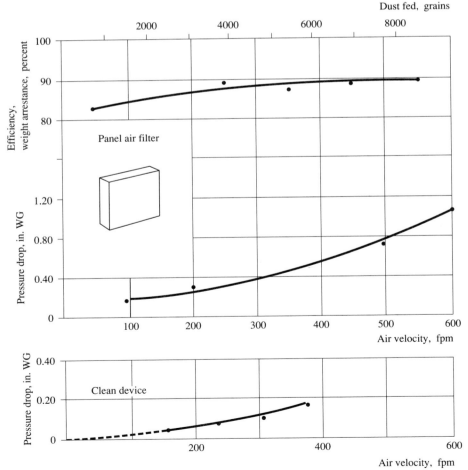

FIGURE 15.3 Characteristic curves of a disposal panel filter. (*Source:* American Air Filter. Reprinted by permission.)

as the amount of dust held further increases, the efficiency curve approaches a horizontal line.

Low-efficiency air filters are widely used in fan–coil units, room air conditioners, packaged units, and sometimes in air-handling units.

Medium-Efficiency Air Filters

Medium-efficiency air filters operate at a range between 0.20 and 0.95 ASHRAE dust spot efficiency. Filter media for medium-efficiency air filters are often made of glass fibers with diameters between 1 and 10 μm in the form of a mat or paper using nylon fibers to join them together. They are usually dry and disposable. Their main characteristics are as follows:

- As the dust spot efficiency increases, the diameter of the glass fibers is reduced and they are spaced comparatively closer together.
- Extended surfaces such as pleated mats or bags are commonly used to increase the surface area of the filter medium and to reduce the air velocity flowing through it to a range of 6 to 90 fpm. A lower air velocity will eventually increase the diffusive effect of the small particles and decrease the pressure drop across the filter medium.
- Most medium-efficiency air filters have a face velocity around 500 fpm, to match the face velocity of the coil in the air-handling unit or packaged units.
- The initial pressure drop of medium-efficiency filters varies from 0.20 to 0.60 in. WG. It may be more effective and economical to renew a medium-efficiency air filter when the final pressure drop is 0.70 to 1.20 in. WG.

Recent trends are to use medium-efficiency air filters in air-handling units and packaged units to provide better indoor air quality and prevent smudging and discoloration of the building interiors.

High-Efficiency Particulate Air (HEPA) Filters and Ultra Low-Penetration Air (ULPA) Filters

High-efficiency particulate air filters have a DOP efficiency of 99.97 percent for dust particles ≥ 0.3 μm in diameter. Ultra low-penetration air filters have a DOP efficiency of 99.999 percent for dust particles ≥ 0.12 μm. The filter media for HEPA and ULPA filters are dry and disposable.

In Fig. 15.4 is a typical HEPA filter. The dimensions of this filter are 24 in. $\times$ 24 in. $\times$ 11.5 in. The filter media are made of glass fibers of submicrometer diameter that are formed into pleated paper mats. Some of the larger fibers act as the carrier of the web. The performance of the filter media is often assessed by an index called the *alpha value* α, which can be calculated as follows:

$$\alpha = [2 - (\log P)]\frac{100}{25.4 \, \Delta p_t} \qquad (15.6)$$

where Δp_t = pressure drop of the filter media, in. WG. Both penetration P, in percent, and pressure drop Δp_t are measured at an air velocity flowing through the filter media of 10.5 fpm. The α value is usually between 10 and 11. Because of the development of new media with lower pressure drops, an α value of 13 or even higher can now be achieved. The surface area of the filter media may be 50 times the face area of the HEPA filter, and the rated face velocity may vary from 190 to 390 fpm for regular HEPA filters at a pressure drop of 0.65 to 1.35 in. WG for clean filters. The face velocity of high-capacity HEPA filters can be raised to 500 fpm.

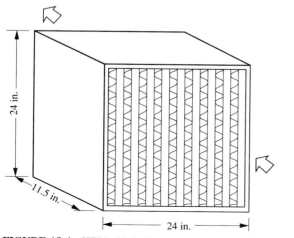
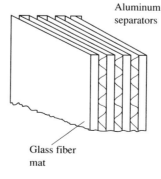

FIGURE 15.4 HEPA (high-efficiency particulate air) filter.

The filter media themselves have efficiencies higher than the mounted filter. Sealing of the filter pack within its frame, and sealing between the frame and gasket are critical factors that affect HEPA filter penetration and efficiency. Penetration of dusts represents aerosol passing through the medium and through pinholes in the filter medium, as will as leaks between the pack and the frame and between the frame and the gasket.

In the microelectronics industry, the latest generation of devices, called *very large scale integrated (VLSI) chips*, have clearances as close as 1 μm, or 3.94×10^{-5} in. Consequently, the maximum allowable size of contamination during the processing of VLSI chips is 0.1 μm. The development of new test methods, described previously, and the use of the laser spectrometer make it possible to count airborne particles as small as 0.12 μm and to measure the concentration to the nearest 0.001 percent.

A ULPA filter contains filter media similar to those in the HEPA filter. Both its filter media and sealing are more efficient than the HEPA filter.

To extend the service life of an HEPA filter or a ULPA filter, both should be protected by either a medium-efficiency filter or two filters: a low-efficiency filter and a medium-efficiency filter located just upstream from the HEPA filter or ULPA filter. The removal of large particles in the prefilter reduces the dust load and prolongs the life of the HEPA or ULPA filter.

HEPA filters and ULPA filters are widely used in clean rooms and clean spaces for the microelectronics industry, the pharmaceutical industry, precision manufacturing, and operating theaters in hospitals.

Selection of Air Filters

During the selection of air filters, the following factors should be carefully considered:

- The degree of air cleanliness required in the conditioned space must be identified, especially the design criteria for clean spaces or clean rooms.
- The size and the concentration of the dust particles to be removed must be specified.
- The efficiency rating and the corresponding test method must be indicated.
- The average pressure drop during the operating period, which affects the energy consumption of the filter as well as the air system, must be determined.
- Service life of the air filter, which influences the installation cost, and its pressure drop are closely related to the energy consumption. Together they determine the minimum cost.

- The inclusion of two or three low-efficiency filters connected in series does not improve their efficiency to collect submicron-sized atmospheric dust.
- HEPA or ULPA filters must be protected by low- and medium-efficiency prefilters to extend the service life of the HEPA or ULPA filter.
- Monitoring the pressure drop of the air filters and periodic maintenance of washable low-efficiency air filters have a direct impact on filter's performance.

15.5 ELECTRONIC AIR CLEANERS

An electronic air cleaner uses the attraction between particles with opposite charges. Dust particles charged within the cleaner attract and agglomerate to greater sizes at the collecting plates. They are therefore easily removed from the airstream.

A typical electronic air cleaner is shown in Fig. 15.5. A high DC potential of 1200 V is supplied to the ionizing field. The positive ions generated from the ionizer wire charge the dust particles. Right after the ionizing section, the dust particles come to a collecting section, which consists of several plates that are alternatively grounded and insulated. A strong electric field is produced by supplying a DC potential of 6000 V to these plates. The positively charged dust particles are attracted by the grounded plates of the opposite charge, and attach themselves to the plates.

Because of the numerous points of contact, the bond between particles held together by intermolecular forces is greater than that between the particles and plates. Therefore, the dust particles agglomerate and grow to such sizes that they are blown off and carried away by the airstream. The agglomerates are then collected by a medium-efficiency air filter located downstream from the collecting section.

Electronic air cleaners are efficient for removing small particles such as atmospheric dust and cigarette smoke. The dust spot efficiency of an electronic cleaner may be up to 90 percent. The pressure drop across the ionizer section and collecting section is low and ranges from 0.15 to 0.25 in. WG against an air velocity of 300 to 500 fpm. Safety measures must be provided for protection against the high DC potential. If the positive charges are not removed at the collecting section, the accumulation of these positively charged dust particles may build up a space charge. This space charge drives the charged particles to the walls and other building envelope surfaces in the conditioned space, causing them to be smudged.

15.6 ACTIVATED CARBON FILTERS

Activated carbon filters are most widely used to remove objectionable odors and irritating vapors of

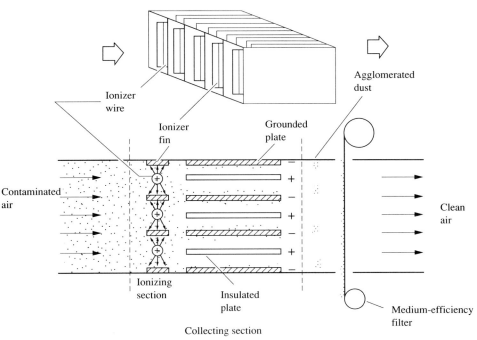

FIGURE 15.5   An electronic air cleaner.

gaseous airborne particulates, typically 0.003 to 0.006 μm in size, from the airstream by adsorption. *Adsorption* is the physical condensation of a gas or vapor on to an activated substance. Activated substances are highly porous. One pound of extremely porous carbon contains more than 5,000,000 ft^2 of internal surface. Gas molecules diffuse to the micropores or macropores of activated carbon, bond to these surfaces, and come in contact with the carbon granules. One pound of activated carbon may adsorb 0.2 to 0.5 pounds of odorous gases.

Activated carbon in the form of granules or pellets is made from coal, coconut shells, or petroleum residues. They are heated in steam and carbon dioxide to remove foreign matters such as hydrocarbons and to produce internal porosity. Activated carbon is placed in special trays, which slide easily into position, to form activated carbon beds that are sealed into the cell housing by face plates, as shown in Fig. 15.6. A typical carbon tray, which is 23 in. × 23 in. × $\frac{5}{8}$ in. thick, weighs 12 lb. Low-efficiency air filters are used as prefilters for protection. When air flows through a typical assembly with a face velocity of 375 to 500 fpm, the corresponding pressure drops are between 0.2 and 0.3 in. WG. Activated carbon can also be mounted in fixed frames with perforated sheets in continuous pleats.

Adsorption capacity is defined as the amount of carbon tetrachloride adsorbed by a given weight of

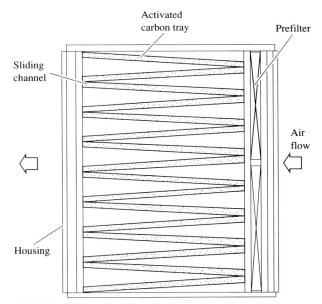

FIGURE 15.6   Activated carbon filter assembly.

activated carbon. For various odors, the adsorption capacity is also affected by the operating temperature and humidity. In general, a higher humidity or higher operating temperature usually decreases the adsorption capacity of the activated carbon. The maximum operating temperature is 100°F.

The adsorption efficiency of the carbon bed remains relatively constant during its service life. When activated carbon has reached its maximum adsorption capacity, that is, when the activity of the carbon has been used up, it must be replaced either by reactivation or regeneration. *Reactivation* is the process of removing spent carbon and replacing it with fresh carbon. *Regeneration* is the process by which the spent carbon is converted to fresh carbon. Regeneration can only be performed by the activated carbon manufacturer. The simplest way to investigate whether the carbon needs to be replaced is to detect downstream the odor that the carbon filter is supposed to remove. A sample of carbon can also be tested to determine its remaining capacity.

Gaseous airborne contaminants can also be removed by means of chemisorption media. During chemisorption, gas molecules bond to the surface of the chemisorption media through chemical action instead of physical adsorption. Chemisorption media are effective gaseous airborne contaminants remover because they are effective over wide particle size distributions.

15.7 AIR-HANDLING UNITS

Functions of Air-Handling Units

An air-handling unit (AHU) is the primary equipment in an air system of a central hydronic system; it handles and conditions the air and distributes it to various conditioned spaces. In an AHU, the required amounts of outdoor air and recirculating air are mixed and conditioned. The temperature of the discharge air is then maintained within predetermined limits by means of control systems. After that, the conditioned supply air is provided with motive force and is distributed to various conditioned spaces through ductwork and space diffusion devices.

The basic components of an air-handling unit are a supply fan, a fan motor, a water cooling coil, filters, a mixing box, dampers, control systems, and a casing. A return or relief fan, heating or preheating coils, or both, precooling coils, and a humidifier may also be included, depending on the application. Supply volume flow rates of AHUs vary from 2000 to 63,000 cfm.

Whether a return fan or a relief fan should be added to an air system depends on the construction and operating characteristics of the air system and the total pressure loss of the return system (see Chapter 19). The purpose of a preheating coil is to prevent coil freeze-up. A preheating coil is always located before the cooling and heating coils. A heating coil is mainly used in the air-handling unit that serves the perimeter zone, or for morning warm-up in heating season. The use of a precooling coil, drawing cooling water from the cooling tower, as a water economizer will be discussed in Chapter 18. Humidifiers are employed for process air conditioning where space humidity must be controlled, for comfort air conditioning in which the space relative humidity may drop below 30 percent, or in healthcare facilities.

Classifications of Air-Handling Units

Air-handling units may be classified according to their structure, location, and conditioning characteristics.

HORIZONTAL OR VERTICAL UNIT. In a horizontal unit, the supply fan, coils, and filters are all installed at the same level, as shown in Fig. 15.7a. Horizontal units need more floor space for installation, and they are mainly used as large-size AHUs. Most horizontal units are installed inside the fan room. Occasionally, small horizontal units may be hung from the ceiling inside the ceiling plenum. In such circumstances, fan noise and vibration must be carefully controlled if the unit is adjacent to the conditioned space.

In a vertical unit, the supply fan is not installed at the same level as the coils and filters but is often at a level higher, as shown in Fig. 15.7b. Vertical units require less floor space. They are usually smaller, so that the height of the coil section plus the fan section, and the height of the ductwork that crosses over the AHU under the ceiling is less than the headroom (the height from the floor to the ceiling or the beam of the fan room). A fan room is the room used to house AHUs and other mechanical equipment.

DRAW-THROUGH UNIT OR BLOW-THROUGH UNIT. In a draw-through unit, the supply fan is located downstream from the cooling coil section and the air is drawn through the coil section, as shown in Fig. 15.7a and Fig. 15.7b. In a draw-through unit, conditioned air is evenly distributed over the entire surface of the coil section. Also, the discharge air from the AHU can be easily connected to a supply duct of similar high velocity. Draw-through units are the most widely used AHUs.

In a blow-through unit, the supply fan is located upstream from the coil section, and the air blows through the coil section, as shown in Fig. 15.7c. A blow-through unit is often used as a multizone air-handling unit. In a multizone AHU, the coil section is divided into the hot deck and the cold deck. The heating coil is installed in the hot deck just above the cold deck, where the cooling coil is located. The hot deck is connected to ductwork that supplies warm air to the

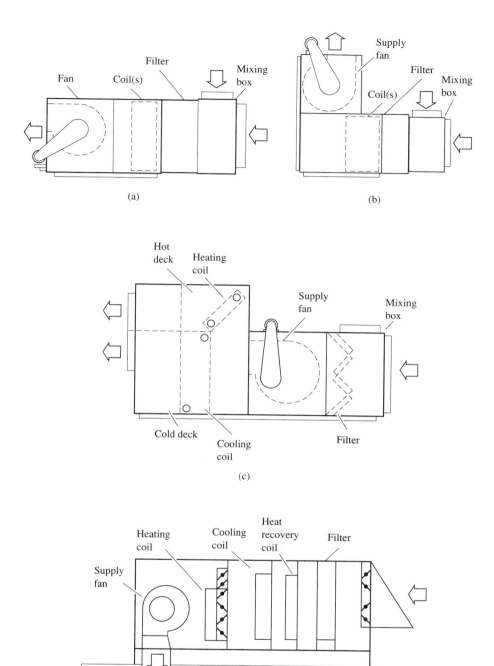

FIGURE 15.7 Type of air handling units (AHUs): (a) horizontal draw-through unit; (b) vertical draw-through unit; (c) blow-through unit, multizone AHU; and (d) make-up air AHU, custom-built, rooftop unit.

perimeter zone through the warm duct. The cold deck is connected to a cold duct that supplies cold air to both the perimeter and interior zones.

A blow-through unit also has the advantage of treating the supply fan heat gain as a part of the coil load and, thus, reduces the supply system heat gain.

MAKE-UP AIR AHUS AND GENERAL-PURPOSE AHUS. Most general-purpose AHUs can be used to condition either outdoor air only, or a mixture of outdoor air and recirculating air, whereas a make-up air AHU is used only to condition 100 percent outdoor air, as shown in Fig. 15.7d.

A make-up air AHU is a once-through unit; there is no return air and mixing box. It is often a constant-volume system because the required outdoor air quantity usually does not vary even at part-load operation. In a make-up air AHU, the cooling coil is usually a 6- to 10-row depth coil because of the greater enthalpy difference during cooling and dehumidification in summer. A steam heating coil or other freezing protections are necessary in locations where outdoor temperatures may be below 32°F in winter. A heat recovery coil or a water economizer precooling coil is often installed in make-up air AHUs for energy savings.

FACTORY-FABRICATED AHU OR FIELD BUILT-UP AHU, CUSTOM-BUILT OR STANDARD FABRICATION. One important reason to use factory-fabricated AHUs or standard fabrications is their lower cost and higher quality. Factory labor and controlled manufacturing techniques provide more efficient and better quality construction than field labor and assembly.

Custom-built and field built-up AHUs provide more flexibility in structure, system component arrangements, dimensions, and specialized functions than standard fabricated products. Custom-built and field built-up AHUs also need more comprehensive, detailed specifications. Standard fabricated products are usually less expensive and can be delivered in a shorter time.

ROOFTOP AHU OR INDOOR AHU. A rooftop AHU is an outdoor penthouse, as shown in Fig. 15.7d. It is usually curb-mounted on the roof and should be completely weathertight. The outside casing is usually made of heavy-gauge galvanized steel with sealant at the joints, both inside and outside. The fan motor, water valves, damper actuator linkages, and controls are all installed inside the casing. Access doors are necessary for service and maintenance of fans, coils, and filters. An indoor AHU is usually located in the fan room. Small AHUs are sometimes ceiling-mounted.

Main Components

CASING. To provide a good thermal insulation, a double-skin structure is often used, that is, a 1- or 2-in. thickness of insulating material such as mineral wool or fiberglass is sandwiched between two steel panels with a U-value between 0.12 and 0.25 Btu/h · ft^2 · °F.

FAN(S). A double-inlet airfoil, backward-inclined centrifugal fan is often used in large AHUs for its higher efficiency and lower noise. Vane-axial fans

with carefully designed sound absorptive housings, sound attenuators at inlet and outlet, and other treatments can now provide a sound rating of NC 55 in the fan room. Although a forward-curved centrifugal fan has a lower efficiency at full-load, its part-load operating characteristics are better than those of a backward-curved fan. It is often used in small AHUs. An axial relief fan, or a centrifugal or axial return fan, may be added as an optional component.

As mentioned in Chapter 10, an adjustable-frequency AC drive saves more energy than inlet vanes for VAV systems during part-load operation. However, it is more expensive. Inlet vanes are not suitable for small backward-curved centrifugal fans because they block the air passage at the fan inlet cone. Operating characteristics and manufacturing information should be carefully analyzed through life-cycle cost analysis or simple payback years before inlet vanes or an AC inverter is used for fan capacity control.

COILS. In AHUs, the following types of coil are used: water heating coils, steam heating coils, water economizer precooling coils, and water cooling coils. The construction and characteristics of these coils were discussed in Chapter 12.

Sometimes, electric heating coils are also used in AHUs. Electric heating coils are made with nickel-chromium wire as the heating element in an open coil. Ceramic bushes float the heating elements and vertical brackets prevent the elements from sagging. In a finned tubular element sheathed construction, the electric heating coil is usually made with a spiral fin brazed to a steel sheath. An electric heating coil is usually divided into several stages for capacity control.

When an electric heating coil is installed inside an AHU, the manufacturer should have the assembly tested by the Underwriters' Laboratory (UL) to ensure that its requirements are met. Otherwise, the heater may only be installed outside the AHU as a duct heater, with a minimum distance of 4 ft from the AHU. Various safety cut-offs and controls must be provided according to the National Electrical Codes and other related codes.

FILTERS. Both medium-efficiency bag and pleated-mat air filters and low-efficiency panel air filters are used in AHUs. More and more AHUs now include medium efficiency air filters with a dust spot efficiency between 40 and 95 percent to improve the indoor air quality and to prevent smudging and discoloration of building interiors. A low-efficiency prefilter may be used to protect the bag filter for a dust spot efficiency greater than 65 percent. Low-efficiency filters may be considered only for applications where smoking is not allowed and occupants will stay for short periods of time.

HUMIDIFIERS. Steam grid or electric heating element humidifiers are widely used in AHUs where a warm air supply and humidity control are needed in the winter. For industrial applications where a cold air supply is needed in the heating season, an air washer is often used to humidify the air.

Component Layout

In a typical horizontal, draw-through AHU, the layout of the components in serial order is usually as follows:

1. (Relief or return fan, exhaust air passage and damper, optional)
2. Mixing box with outdoor air and recirculating air dampers
3. Filters: prefilter and medium-efficiency filter
4. Preheating coil
5. Precooling coil
6. Cooling coil
7. (Reheating coil, optional)
8. (Steam grid humidifier, optional)
9. Supply fan
10. (HEPA filter, optional)

The relief fan is always located in the relief or exhaust passage.

If there is a return fan, it should be located upstream from the mixing box. The volume flow and fan total pressure of the return fan should be carefully determined so that at the location before the exhaust damper, the pressure is positive; inside the mixing box, the pressure must be negative (see Chapter 20).

In an AHU, conditioning the mixture is more economical than conditioning the outdoor air and recirculating air separately. Therefore, the mixing box should be located before the filters and coils.

Prefilters and medium-efficiency filters should be located before the coils in order to protect the coil's outer surface from fouling and to prevent the clogging of the air passage. Dirty coils and condensate pans significantly degrade the indoor air quality.

Preheating coils should always be located before the water-heating and -cooling coils for the sake of freezing protection. A precooling coil is always located before the cooling coil for a greater temperature

difference between the air and water. A reheating coil, if any, is located after the cooling coil to offset the space loads during part-load operation when the AHU is used to serve a single-zone precision manufacturing process.

A steam grid humidifier is usually located after the heating coil because humidification is more effective at a higher air temperature.

If there are HEPA filters, they should always be located as near to the clean room or clean space as possible to prevent pollution with dust particles from the ductwork or elsewhere after the HEPA filter.

Coil Face Velocity

The size, or more accurately, the width and height of a horizontal AHU is mainly determined by the face velocity of the coil. A higher coil face velocity results in a smaller coil, a higher heat transfer coefficient, a greater pressure drop across the coil and filter, and a smaller fan room, which directly affects the space required. On the contrary, a lower coil face velocity has a larger coil, a lower heat transfer coefficient, and a smaller pressure drop.

The maximum face velocity is usually determined according to the value required to prevent carryover of water droplets due to condensate from a cooling coil. As mentioned in Chapter 12, cooling coil face velocity should not exceed 500 fpm for a smooth fin coil and 600 fpm for corrugated fins. The lower limit of the coil face velocity depends mainly on the initial and energy cost analysis, including cost of the AHU, fan room size, the total number of operating hours annually, and the unit rate of electric power. A lower limit of face velocities between 400 and 450 fpm at design conditions may be considered appropriate in many circumstances.

For a fan room of adequate headroom, it is preferable to use a high coil to reduce the face velocity and the pressure drop of coils and filters. Use of a high coil has little influence on the floor area of the fan room.

Mixing of Outdoor and Recirculating Air

Poor mixing of the outdoor and recirculating airstreams in the mixing box causes stratification of the mixture entering the coil and is one of the primary causes of coil freezing in locations where the outdoor air temperature is below 32°F.

Parallel airstreams result in poor mixing. For good mixing, airstreams should meet at a 90° angle or opposite each other, as shown in Fig. 15.8.

Selection of AHUs

In Table 15.1 are listed general data of the supply fan and coil of a typical horizontal draw-through AHU; *Fan B* means a Class II fan. In Table 15.2 is presented volume flow–fan static pressure performance of this horizontal draw-through modular AHU (unit size 30) with inlet vanes. A backward-inclined centrifugal fan of $22\frac{1}{4}$ in. diameter is used. In Table 15.2, the following items are listed:

- Supply volume flow rate $\dot{V}_s$, in cfm of standard air
- Air velocity at fan outlet, fpm
- Fan static pressure, in inches WG; fan total pressure can be obtained by adding fan velocity pressure to the fan static pressure; velocity pressure p_v, in inches WG, can be calculated by $p_v = (v_{out}/4005)^2$; here v_{out} indicates the fan outlet velocity, in fpm
- Revolutions per minute (rpm) of fan impeller
- Brake horsepower (bhp) input to fan shaft.

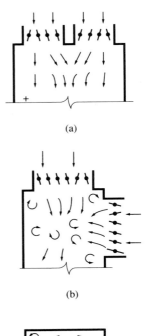

(a)

(b)

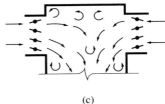

(c)

FIGURE 15.8 Mixing of outdoor and recirculating airstreams: (a) parallel air streams (poor mixing), (b) air streams at 90° (good mixing), and (c) opposite airstreams (good mixing).

The following are recommendations for selection of AHUs from a manufacturer's catalog:

- The size of the AHU is selected so that the face velocity of the cooling coil v_{coil} is optimum. For corrugated fins, the maximum v_{coil} should not exceed 600 fpm. The optimum v_{coil} depends on the configuration of the AHU and the local energy cost, as mentioned in Chapter 12. It may be between 550 and 600 fpm.

- For large AHUs, choose a backward centrifugal fan with airfoil blades or a backward-inclined fan for higher efficiency. Select the rpm of the supply fan or supply and return fans in order to meet the required system total pressure loss, that is, external total pressure plus the pressure loss of the AHU.

  For VAV systems, an adjustable-frequency AC drive fan speed control should be compared with inlet vanes through life-cycle cost analysis. Energy consumption also should be analyzed at part-load operation. For a large AHU, an adjustable-frequency AC drive may be cost effective.

- The required coil's load is met through the variation of the number of rows of coil and the fin spacing.

- To improve indoor air quality and prevent smudging and discoloration of building interiors in buildings where smoking is allowed, medium-efficiency filters of dust spot efficiency greater than 65 percent are preferable. A prefilter can extend the service life of the bag filter.

- Use an air or water economizer to save energy. Use indirect and direct evaporative coolers to replace part of the refrigeration if they are applicable and economical.

- In locations where outdoor air temperatures go below 32°F, protection of the coil against freezing by means of a preheat coil, and improvement of the mixing effect in the mixing box should be considered.

15.8   PACKAGED UNITS

Types of Packaged Units

A packaged unit (PU) is a self-contained air conditioner. It is also the primary equipment of a unitary packaged system. A packaged unit not only conditions the air and provides the motive force to supply the conditioned air to the space, but also provides gas heating, electric heating, and refrigeration from its own gas-fired furnace and refrigeration equipment or from its own heat pump. These features are why it is called a self-contained or a packaged unit.

TABLE 15.1
General data on supply fans and coils of a typical horizontal draw-through AHU

Description		3	6	8	10	12	14	17	21	25	30
						Unit Size Number					
Fan Data											
Fan	Size	—	—	—	$13\frac{1}{2}$	15	$16\frac{1}{2}$	$18\frac{1}{4}$	20	$22\frac{1}{4}$	$22\frac{1}{4}$
BI	Shaft size (in.)	—	—	—	$1\frac{11}{16}$	$1\frac{15}{16}$	$1\frac{15}{16}$	$2\frac{3}{16}$	$2\frac{7}{16}$	$2\frac{11}{16}$	$2\frac{11}{16}$
	Outlet area (sq. ft.)	—	1.41	1.90	2.31	2.79	3.39	4.14	5.05	6.30	6.30
	Max. RPM/static pressure	—	—	—	3536/8	3183/8	2894/8	2616/8	2387/8	2164/8	2164/8
	Motor hp range	—	—	—	1–10	1–15	1–15	1–20	1–25	1–25	1–30
Coil Data											
Unit coils ($\frac{1}{2}$ in. Tube)											
Cooling	Sq. ft.	3.32	5.86	7.54	9.64	12.3	14.2	16.8	20.8	24.4	29.0
	Dimensions (in.)	21 × 23	23 × 36	27 × 40	27 × 51	32 × 55	35 × 59	37 × 65	45 × 67	51 × 69	51 × 82
Heating	Sq. ft.	2.34	4.31	5.49	7.01	9.46	10.2	12.3	15.0	17.8	21.2
	Dimensions (in.)	15 × 23	17 × 36	20 × 40	20 × 51	25 × 55	25 × 59	27 × 65	32 × 67	37 × 69	37 × 82
$\frac{5}{8}$ in. or 1 in. tube coils											
	Sq. ft.	2.75	4.38	6.50	8.33	11.3	12.1	14.7	16.5	19.8	23.6
	Dimensions (in.)	18 × 22	18 × 35	24 × 39	24 × 50	30 × 54	30 × 58	33 × 64	2–18 × 66*	1–18 × 68*	1–18 × 81*
										1–24 × 68	1–24 × 81

BI: Backward-inclined centrifugal fan

Source: The Tramne Company. Reprinted with permission.

* Two coils are used for this unit.

TABLE 15.2
Volume flow and fan static pressure of a typical horizontal draw-through modular AHU (size 30) with inlet vanes (Backward-inclined centrifugal fan of 22.25 in. diameter)

CFM std. air	Outlet velocity	3.50 rpm	3.50 bhp	3.75 rpm	3.75 bhp	4.00 rpm	4.00 bhp	4.25 rpm	4.25 bhp	4.50 rpm	4.50 bhp	4.75 rpm	4.75 bhp	5.00 rpm	5.00 bhp
							Fan static pressure, in. WG								
10,000	1587	1524	10.39	1561	10.91	1597	11.40	1632	11.87	1667	12.32	1701	12.77	1734	13.23
11,000	1746	1570	11.81	1607	12.54	1643	13.23	1678	13.86	1712	14.44	1745	14.99	1777	15.51
12,000	1905	1620	13.21	1656	14.00	1691	14.81	1725	15.62	1759	16.41	1792	17.16	1824	17.85
13,000	2063	1678	14.90	1710	15.67	1742	16.46	1774	17.28	1806	18.14	1838	19.02	1870	19.90
14,000	2222	1741	16.82	1771	17.61	1800	18.41	1830	19.23	1860	20.08	1889	20.94	1919	21.84
15,000	2381	1809	18.98	1836	19.80	1864	20.63	1891	21.47	1919	22.32	1946	23.20	1974	24.09
16,000	2540	1880	21.36	1906	22.23	1931	23.10	1957	23.97	1982	24.85	2008	25.75	2034	26.65
17,000	2698	1953	23.97	1978	24.88	2002	25.80	2026	26.72	2050	27.64	2074	28.57	2098	29.71
18,000	2857	2028	26.83	2052	27.78	2075	28.74	2098	29.71						
19,000	3016	2104	29.96												

CFM std. air	Outlet velocity	5.25 rpm	5.25 bhp	5.50 rpm	5.50 bhp	5.75 rpm	5.75 bhp	6.00 rpm	6.00 bhp	6.25 rpm	6.25 bhp	6.50 rpm	6.50 bhp
10,000	1587	1766	13.70	1798	14.19	1830	14.71	1861	15.26	1892	15.82	1923	16.41
11,000	1746	1809	16.01	1840	16.51	1871	17.00	1901	17.51	1931	18.02	1960	18.55
12,000	1905	1854	18.48	1885	19.09	1915	19.68	1944	20.24	1973	20.79	2002	21.33
13,000	2063	1901	20.76	1932	21.58	1962	22.35	1991	23.07	2020	23.76	2048	24.41
14,000	2222	1949	22.77	1979	23.71	2008	24.66	2037	25.60	2066	26.50	2094	27.36
15,000	2381	2002	25.00	2030	25.94	2057	26.90	2085	27.9	2113	28.90	2140	29.92
16,000	2540	2060	27.57	2086	28.51	2112	29.46						

Outlet vel—outlet velocity, fpm

rpm—revolutions/minute

bhp—brake horsepower

Source: The Trane Company. Reprinted with permission.

A packaged unit is always equipped with DX-coil(s) for cooling. This characteristic is the primary difference between a packaged unit and an air-handling unit. The portion that handles conditioned air in a packaged unit is called an air handler, to distinguish it from an air-handling unit. Refrigerants HCFC–22 and HFC–134a are now used in packaged units. Most PUs are factory-built standard fabrication units.

A packaged unit can be either enclosed in a single package or split into two units: an indoor air handler and an outdoor condensing unit.

A packaged unit can also be a packaged heat pump. In a packaged heat pump, in addition to the fan, DX-coil, filters, compressors, condensers, expansion valves, and controls, there are also four-way reversing valves to reverse the refrigerant flow when cooling mode operation is changed to heating mode operation.

Packaged units can be classified according to their place of installation as rooftop packaged units, indoor packaged units, and split packaged units.

Rooftop Packaged Units

A rooftop packaged unit is mounted on the roof of the conditioned space, as shown in Fig. 15.9. It is usually enclosed in a weatherproof outer casing. Outdoor air and recirculating air are conditioned in the rooftop packaged unit and supplied to the conditioned space on the floors below. Based on the types of the heating and cooling sources, rooftop packaged units can be subdivided into the following:

- *Gas/electric rooftop packaged unit.* In this unit, heating is provided by a gas-fired heater, and cooling is provided by electric power-driven reciprocating or other compressors.
- *Electric/electric rooftop packaged unit.* In this unit, heating is provided by an electric heating coil and cooling by reciprocating or other compressors.
- *Rooftop packaged heat pump.* In this unit, heating and cooling are provided by the heat pump, with auxiliary electric heating if necessary.

A rooftop packaged unit is a single packaged unit. Its cooling capacity may vary from 3 to 220 tons, and its supply volume flow rate may vary from 1200 to 80,000 cfm. A typical rooftop packaged unit consists mainly of filters; DX-coil; indoor fan; gas-fired heater; compressor(s); air-, water-, or evaporative-cooled condensers; expansion valves; and controls. A relief or return fan, humidifiers, and four-way reversing valves are included if they are required.

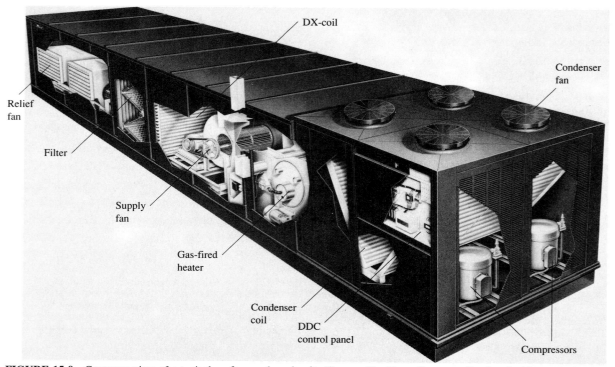

FIGURE 15.9 Cutaway view of a typical rooftop packaged unit. (*Source:* The Trane Company. Reprinted with permission.)

CASING. The outer casing is made from heavy-gauge galvanized steel panel, phosphatized and coated with corrosion-resistant finishes such as epoxy primer and enamel. A ≥ 1-in. thick thermal insulation is provided for all the interior surfaces that contact with airstreams.

FILTERS. In a rooftop packaged unit, air filters are generally low- or medium-efficiency. Sometimes, a low-efficiency prefilter is located before the bag filter for protection.

DX-COIL. A DX-coil with two separate refrigerant circuits, associated expansion valves, and distributors is usually used for better capacity control except in small units with on-off control.

FANS. In small rooftop packaged units, a forward-curved centrifugal supply fan is often used. In large rooftop packaged units, a higher-efficiency backward-curved or airfoil centrifugal supply fan is often used. An axial relief fan may be located in the relief or exhaust passage, or a backward-curved centrifugal return fan may be located before the mixing box. Inlet vanes could be added to the supply or return fan for VAV systems. A maximum of 2.5 in. WG of external total pressure is usually provided by the supply fan and up to 1.5 in. WG external total pressure by the return fan to overcome the pressure loss of the ductwork and terminals. The ratio of the supply volume flow to the refrigeration capacity is from 350 to 450 cfm per refrigeration ton.

GAS-FIRED FURNACE AND ELECTRIC HEATING COIL. The gas-fired furnace is composed of a gas burner, fire tubes, and flue tubes and is sometimes called a gas-fired heater, as discussed in Chapter 14. An electric heating coil is often divided into multiple stages for capacity control.

COMPRESSORS. Semihermetic, hermetic reciprocating, scroll, or screw compressors are usually used. For medium and large rooftop packaged units, two or three compressors of equal or sometimes unequal horsepower input are preferable for better capacity control in steps.

AIR-COOLED CONDENSER. Multirow, $\frac{3}{8}$ in. copper tubes and aluminum fin air-cooled condensing coils connected with subcooling coils are used. Condensing coils may cover the two sides of the condenser or be in a V-shape at the middle. The ratio of the face area of the condensing and subcooling coils to the refrigeration capacity is about 1.5 to 2 ft²/refrigeration ton (ref. ton). Propeller fans are used for the induced

draft condenser fan. The ratio of volume flow of the cooling air to the refrigeration capacity is between 600 and 900 cfm/ref. ton. Evaporative condensers or shell-and-tube water-cooled condensers are also used.

CONTROLS. A DDC with a programmable or factory-preprogrammed microprocessor is used for discharge air, minimum outdoor air, economizer cycle, and other controls.

HEAT PUMP. A rooftop packaged heat pump is a packaged unit installed with four-way reversing valves to change the refrigerant flow after the compressor. In an air-to-air heat pump, the DX-coil is often called an indoor coil, and the air-cooled condensing coil the outdoor coil. During cooling mode operation, hot gas from the compressor is first discharged to the outdoor coil for condensing and subcooling. The liquid refrigerant then enters the expansion valve and indoor coil to produce refrigeration. During heating mode of operation, the reversing valve changes its connections and the direction of refrigerant flow. Hot gas from the compressor now enters the indoor coil to release its condensing heat first. The liquid refrigerant is then discharged to the outdoor coil to absorb heat from the ambient air for evaporation. Details of operation and system performance will be discussed in Chapter 22.

Indoor Packaged Units

An indoor packaged unit is also a single packaged, factory-built unit. It is usually installed indoors inside a fan room or a machine room, as described in Chapter 29. A small or medium indoor packaged unit may sometimes be floor-mounted directly inside the conditioned space with or without connecting ductwork, such as the indoor packaged unit in computer rooms as shown in Fig. 15.10. The cooling capacity of the indoor packaged unit may vary from 3 to 100 tons, and its supply volume flow rate from 1200 to 40,000 cfm.

Indoor packaged units can be classified as follows:

- *Indoor packaged cooling units.* Only cooling is provided by the DX-coil.
- *Indoor packaged heating/cooling units.* These units not only provide cooling from the DX-coil, but also provide heating from a hot water coil, steam-heating coil, or electric heater.
- *Indoor packaged heat pump.* When an indoor packaged unit is connected to an air-cooled condenser and equipped with reversing valves, the change of refrigerant flow also causes the change of cool-

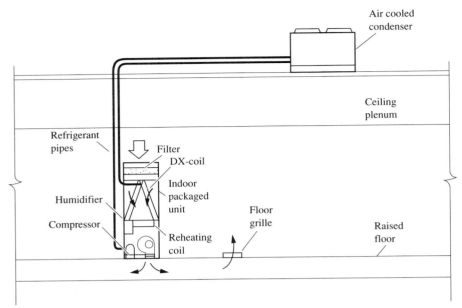

FIGURE 15.10   A typical indoor packaged unit.

ing mode and heating mode operation and provides heating and cooling as required.

In indoor packaged units, usually only a supply fan is installed in small units. Because of the compact size of the units, generally a forward-curved centrifugal fan is used.

For computer room indoor packaged units, return air entering the unit at a high level and supply air discharged at a low level are common. Low-efficiency filters or medium-efficiency filters with a dust spot efficiency between 40 and 85 percent are usually used. A steam humidifier or other type of humidifier is always an integral part of computer room units so as to maintain a required space relative humidity in winter and prevent static electricity. Multiple semihermetic, hermetic reciprocating compressors, or scroll or screw compressors—all with dual refrigerant circuits—are used in medium and large units for capacity control. The current trend is to use microprocessor-based DDC controls.

An indoor packaged unit differs from a rooftop packaged unit in condensing arrangements. Usually, there are two alternatives in indoor packaged units:

- With an air-cooled condenser, hot gas from the compressor is discharged to an air-cooled condenser through the discharge line located on the rooftop. Liquid refrigerant is returned to the DX-coil from the air-cooled condenser through the liquid line.

- A shell-and-tube or a double tube water-cooled condenser is installed inside the unit, and the condenser cooling water is supplied from the cooling tower or from other sources.

An economic analysis based on the local conditions should be made to determine whether an air-cooled, water-cooled, or even evaporative condenser should be installed. If a water-cooled condenser using cooling water from cooling tower is used, a precooling coil is sometimes installed in the indoor packaged unit as a component of the water economizer.

Split Packaged Units

A split packaged unit, sometimes called a split system, splits the packaged unit into an indoor air handler and an outdoor condensing unit, which is most probably mounted outdoors, on the rooftop, a podium, or some other adjacent place, as shown in Fig. 15.11. A condensing unit is a combination of refrigeration compressors, condensers, and other accessories. Indoor and outdoor units are connected by refrigerant pipes.

An air handler is similar to an AHU except that a DX-coil is installed in an air handler instead of the water-cooling coil installed in an AHU. An air handler is composed mainly of a centrifugal supply fan, low- and medium-efficiency filters, a mixing box, and DDC controls. A relief fan, water heating coils, steam heating coils, electric heaters, and humidifiers are optional. Large air handlers are usually installed

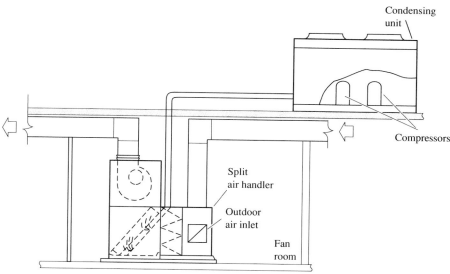

FIGURE 15.11 A typical split packaged unit.

inside the fan rooms, whereas the small air handler may be hung under the ceiling. An air handler for split packaged units usually has a supply volume flow rate between 1200 and 30,000 cfm, a cooling capacity from 3 to 75 tons, and a maximum external pressure of 2.5 in. WG.

An outdoor condensing unit can be either air cooled or water cooled. Most often, an air-cooled condensing unit is used in split packaged units. If a water-cooled condenser using cooling water from a cooling tower is used as in the indoor packaged units, a water economizer precooling coil may be installed in the air handler. The compressors and air-cooled and water-cooled condensers are similar to those in rooftop packaged units.

A split packaged unit always has its compressors in its outdoor condensing unit, whereas an indoor packaged unit has its compressor indoors. A split packaged heat pump is also a kind of split packaged unit. In such a unit, additional reversing valves or changeover arrangements force the refrigerant flow from the compressor to the outdoor air-cooled condensing coil during the cooling mode operation, and change to the indoor DX-coil during the heating mode operation. The cooling capacity of specific models of split packaged heat pumps varies from 10 to 30 tons and the heating capacity from 100,000 to 400,000 Btu/h at rated conditions.

Minimum Performance

In order to reduce energy use in the packaged units, ASHRAE/IES Standard 90.1–1989 specifies minimum equipment performance for the PUs. The minimum performance of various PUs (including packaged heat pumps) except packaged terminal air conditioners (PTAC) and room air conditioners are listed in Table 15.3. PTACs are also a kind of air conditioner for individual rooms, as mentioned in Chapter 1. Almost all PUs are driven and operated electrically.

In Table 15.3, SEER indicates the seasonal energy efficiency ratio. This is the total cooling capacity of a PU during its normal annual usage period for cooling, in Btu, divided by the total electric energy input during the same period, in watt-hours.

IPLV indicates the integrated part-load value. This is a single index of merit that is based on part-load EER or COP. It expresses part-load efficiency for PUs and packaged heat pumps based on the weighted operation at various load capacities.

HSPF indicates the heating seasonal performance factor. This is the total heating output of a heat pump during its normal annual usage period for heating, in Btu, divided by the total electric energy input to the equipment during the same period, in watt-hours.

Selection of Packaged Units

The procedure for selection of PUs is different from that for an AHU because the number of rows and fin spacing of a DX-coil are fixed for a specific model and size of PU. The size of a PU is determined primarily by the required cooling capacity of the DX-coil at various operating conditions. The cooling capacity of a typical rooftop PU is listed in Table 15.4, and in Table 15.5 is present the supply fan performance of this typical rooftop PU.

The selection procedure for PUs, based on data from manufacturers' catalogs can be outlined as follows:

TABLE 15.3
Minimum performance for packaged units (including packaged heat pumps), electrically operated except packaged terminal air conditioner and room air conditioner

Category	Cooling capacity q_c, Btu/h	Indoor temperature, °F	Outdoor temperature, °F	Efficiency rating*	Rating beginning Jan. 1, 1992	Reference standards
		Rating condition				
Air-cooled,						
Single package, cooling mode†	< 65,000			SEER	9.7	ARI 210-81
Split system, cooling mode†	< 65,000			SEER	10.0	ARI 210/240-84
Single and split, cooling mode‡	< 65,000		95	EER	9.5	
Single and split, cooling mode‡	< 65,000		80	IPLV	8.5	
cooling mode	65,000 ≤ q_c < 135,000		95	EER	8.9	
cooling mode	65,000 ≤ q_c < 135,000		80	IPLV	8.3	
Single package, heating mode (heat pumps†)	< 65,000	Seasonal		HSPF	6.6	
Split system, heating mode (heat pumps†)	< 65,000	Seasonal		HSPF	6.8	
Single and split, heating mode	65,000 ≤ q_c < 135,000		47 db/43 wb	COP	3.0	
Single and split, heating mode	65,000 ≤ q_c < 135,000		17 db/15 wb	COP	2.0	
Water-cooled,						
Single package	< 65,000	80 db/67 wb	85	EER	9.3	ARI 210-81
Single package	< 65,000		75	IPLV	8.3	ARI 210/240-84
Single package	65,000 ≤ q_c < 135,000	80 db/67 wb	85	EER	10.5	CTI 201(86)
Evaporatively cooled	< 65,000	80 db/67 wb	95 db/75 wb	EER	9.3	
	< 65,000			IPLV	8.5	
	65,000 ≤ q_c < 135,000	80 db/67 wb	95 db/75 wb	EER	10.5	
	65,000 ≤ q_c < 135,000			IPLV	9.7	
Air-cooled	135,000 ≤ q_c < 760,000			EER	8.5	ARI 380-86
	> 760,000			EER	8.2	
	> 135,000			IPLV	7.5	
Water/evaporatively-cooled	> 135,000			EER	9.6	
	> 135,000			IPLV	9.0	
Packaged heat pumps, air-cooled, cooling mode	135,000 < q_c ≤ 760,000			EER	8.5	ARI 340-86
	≥ 760,000			EER	8.7	
	> 135,000			IPLV	7.5	
heating mode	> 135,000		47	COP	2.9	
	> 135,000		17	COP	2.0	
Condensing unit, air-cooled	> 135,000			EER	9.9	ARI 365-87
	> 135,000			IPLV	11.0	CTI 201-86
Condensing unit, water/evaporatively-cooled	> 135,000			EER	12.9	
	> 135,000			IPLV	12.9	

Water/evap-cooled: Condenser is water cooled or evaporatively-cooled.

*SEER: seasonal energy efficiency ratio. EER: energy efficiency ratio. COP: coefficient of performance. HSPF: heating seasonal performance factor. IPLV: integrated part-load factor.

Adapted from *ASHRAE/IES Standard 90.1–1989*. Reprinted with permission.

†Single-phase AC current

‡3-phase AC current

TABLE 15.4
Cooling capacity of a typical rooftop PU, nominal capacity 55 ton
Volume flow rate, standard air, of 19,250 cfm

Entering dry bulb, °F	Ambient temperature, °F											
	85						95					
	Entering wet bulb, °F											
	61		67		73		61		67		73	
	MBH	SHR	MBH	SHR	MBH	SHR	MBH	SHR	MBH	SHR	MBH	SHR
75	502	81	559	55	621	33	473	83	528	56	587	33
80	506	95	559	69	620	46	479	97	528	71	586	47
85	530	100	559	83	620	59	505	100	528	85	586	60
90	558	100	564	95	619	71	533	100	536	96	585	73

Entering dry bulb, °F	105						115					
	61		67		73		61		67		73	
	MBH	SHR	MBH	SHR	MBH	SHR	MBH	SHR	MBH	SHR	MBH	SHR
75	443	86	496	57	552	33	413	89	463	58	516	33
80	453	98	495	73	551	47	427	100	462	75	516	48
85	480	100	496	88	551	61	454	100	464	91	515	63
90	508	100	507	98	550	75	481	100	481	100	515	78

MBH: 1000 Btu/h

SHR: Sensible heat ratio

Source: The Trane Company. Reprinted with permission.

TABLE 15.5
Supply fan performance of a typical rooftop PU, nominal capacity 55 tons

cfm standard air	Total static pressure, in. WG															
	2.250		2.500		2.750		3.000		3.250		3.500		3.750		4.000	
	rpm	bhp	rpm	bhp	rpm	bhp	rpm	bhp	rpm	bhp	rpm	bhp	rpm	bhp	rpm	bhp
10,000	842	8.09	887	9.18	927	10.28	965	11.37	1000	12.47	1034	13.59	1067	14.74	1099	15.90
12,000	843	9.17	887	10.31	931	11.50	972	12.75	1012	14.03	1050	15.34	1085	16.65	1118	17.96
14,000	857	10.72	897	11.85	936	13.01	974	14.25	1012	15.54	1050	16.90	1087	18.31	1123	19.76
16,000	876	12.59	914	13.81	952	15.05	988	16.31	1023	17.60	1057	18.93	1091	20.30	1124	21.74
18,000	900	14.80	936	16.08	972	17.40	1006	18.76	1040	20.14	1073	21.54	1105	22.96	1137	24.40
20,000	926	17.31	962	18.75	996	20.14	1029	21.56	1061	23.01	1093	24.50	1124	26.01	1154	27.54
22,000	946	19.85	986	21.57	1022	23.22	1055	24.79	1086	26.33	1116	27.88	1146	29.46		
22,500	950	20.50	991	22.28	1027	23.97	1061	25.64	1093	27.24	1123	28.84				
24,000	963	22.44	1004	24.40												

rpm   Revolutions per minute

bhp   Brake horsepower

Source: The Trane Company. Reprinted with permission.

1. Calculate the coil's load and coil sensible load (refer to Chapter 7) of the conditioned space that is served by the PU. For a unitary packaged system using DX-coils, the coil's load is equal to the refrigeration load of the refrigeration system.

   Select the model and size of the PU based on the cooling capacity that is equal to or greater than the required coil load and coil sensible load at the specified design conditions. These include the dry and wet bulb temperatures of air entering the DX-coil and the outdoor air entering the air-cooled condenser or other type of condenser. ASHRAE/IES Standard 90.1–1989 specifies that equipment capacity may exceed the design load only when the equipment selected is the smallest size needed to meet the load.

   If the required supply air volume flow rate deviates from the nominal rated value, in percent as shown below, then the cooling capacity and sensible cooling capacity can be roughly multiplied by a multiplier as follows:

	−20%	−10%	0%	+10%	+20%
Cooling capacity	0.965	0.985	1.0	1.015	1.025
Sensible cooling	0.94	0.97	1.0	1.03	1.06

2. Calculate the heating coil's load during the conditioning or warm-up period. Determine the capacity of the gas-fired heater, electric heater, water- or steam-heating coil. For a packaged heat pump, calculate the supplementary heating capacity of the gas-fired or electric heater.

3. Evaluate the external total pressure. Determine the speed of the supply fan, relief fan, or return fans such that the volume flow and the fan total pressure of the supply fan, supply fan/return fan combination is equal to or greater than the sum of the external total pressure and the total pressure loss in the PU.

   For fans for which only a fan static pressure is given, roughly a 0.4 in. WG of velocity pressure can be added to the fan static pressure for purposes of rough estimation of fan total pressure.

4. For small PUs, a medium-efficiency air filter of lower pressure drop, such as a final pressure drop ≤ 0.3 in. WG, should be selected.

5. For a large PU, a programmable DDC panel is preferable. For a small PU, a DDC controller may be appropriate. DDC panels and controllers will be discussed in Chapter 18.

Example 15.1. Select an AHU or a rooftop PU for a typical floor in a commercial building with the following operating characteristics:

Supply volume flow rate	16,000 cfm
Cooling coil's load	520,000 Btu/h or 43 ton
Coil sensible cooling load	364,000 Btu/h
Outdoor air temperature	95°F
Entering coil dry bulb	80°F
Entering coil wet bulb	67°F
External total pressure	2.0 in. WG
Total pressure loss of AHU or PU	2.25 in. WG

Solution

1. Divide the supply volume flow rate by 500 fpm, that is, $16,000/500 = 32$. From Table 15.1 and 15.2, select an AHU of size 30, which gives a maximum static pressure up to 8 in. WG and a face area of cooling coil of 29 ft^2. The actual face velocity of the cooling coil is

$$v_{\text{coil}} = \frac{16,000}{29} = 551 \text{ fpm}$$

There will be no carryover of condensate water droplets; therefore, it is acceptable.

2. From Table 15.2, select a size-30 horizontal draw-through unit with a backward-inclined centrifugal fan at an impeller diameter of $22\frac{1}{4}$ in., and a fan speed of 1931 rpm. The fan static pressure is then 4.00 in. WG at a volume flow of 16,000 cfm.

   From Table 15.2, because the fan outlet velocity is 2540 fpm, its velocity pressure is then

$$p_v = \left(\frac{2540}{4005}\right)^2 = 0.40 \text{ in. WG}$$

and the total pressure increase provided by this fan is

$$4.00 + 0.40 = 4.40 \text{ in. WG}$$

This value is greater than the required fan total pressure, or

$$2 + 2.25 = 4.25 \text{ in. WG}$$

From the manufacturers' catalogs, the height of the fan and coil modules are both 4 ft $6\frac{3}{4}$ in. For such a height, the headroom of the fan room is usually sufficient.

3. As in Example 12.3, the cooling and dehumidifying capacity per ft^2 of the coil face area of the water cooling coil is 19.1 MBtu/h · ft^2, and the sensible cooling capacity is 13.89 MBtu/h · ft^2. For a coil face area of 29 ft^2, the total cooling and dehumidifying capacity is

$$q_c = 19,100 \times 29 = 553,900 \text{ Btu/h}$$

and the sensible cooling capacity is

$$q_{cs} = 13,890 \times 29 = 402,810 \text{ Btu/h}$$

These capacities both are greater than the required cooling capacity of 520,000 Btu/h and the sensible capacity of 364,000 Btu/h.

4. From Table 15.4, select a rooftop PU with a cooling capacity of 55 tons and a coil sensible heat ratio $SHR_c = 0.69$. When the supply volume flow rate of this PU is 19,250 cfm, at a dry bulb of 80°F and a wet bulb of 67°F and an outdoor air temperature of 95°F, the cooling capacity is 528 MBtu/h, and the sensible cooling capacity is $0.71 \times 528 = 375$ Mbtu/h. Both of them are greater than the required values. The selected 55-ton PU is suitable.

If the volume flow is reduced to 17,625 cfm $(19,250 - 17,625)/16,000 = 0.10$, or 10 percent greater than the required volume flow rate of 16,000 cfm, then the cooling capacity is reduced to

$$528,000 \times 0.985 = 520,080 \text{ Btu/h}$$

and the sensible cooling capacity will be reduced to

$$375,000 \times 0.97 = 363,750 \text{ Btu/h}$$

Only the sensible cooling capacity is slightly less than the required value 364,000 Btu/h.

5. Assume that the supply fan outlet velocity of the rooftop PU is 2000 fpm, that is, the velocity pressure at fan outlet is

$$p_v = \left(\frac{2000}{4005}\right)^2 = 0.25 \text{ in. WG}$$

From Table 15.5, we see that at a supply volume flow of 17,625 cfm and a fan speed of about 1135 rpm, this rooftop PU can provide a fan total pressure of

$$\Delta p_t = 4.0 + 0.25 = 4.25 \text{ in. WG}$$

This fan total pressure can meet the required value 4.25 in. WG.

15.9 FAN ROOM

Types of Fan Rooms

A fan room is an enclosure in which an AHU, indoor PU, air handler, and other equipment are located. In low-rise buildings of three stories and less, a fan room may be used to serve up to three floors, whereas in high-rise buildings of three stories and less, a fan room may be used to serve one or more floors, depending on the characteristics of the air system, its initial cost, and its operating cost. The optimum size of air systems will be discussed in Chapter 30. Fan rooms can be classified according to their pressure characteristics as open or isolated.

OPEN FAN ROOM. An open fan room is open to the filter end of the AHU or PU, as shown in Fig. 15.12. The return ceiling plenum is directly connected to the fan room through an inner-lined return duct or a return duct with a sound attenuator.

Outdoor air is often forced to the fan room by an outdoor air fan or a make-up AHU. The fan room becomes the mixing box of the AHU or PU. Its pressure is always lower than the static pressure in the return plenum and the outdoor atmosphere. Return air is then extracted to the fan room through the return duct. An exhaust fan is often installed on the external wall of the fan room to maintain this pressure difference.

The advantages of this kind of fan room are a positive outdoor air supply and less ductwork in the fan room. The disadvantages include the following:

- There may be infiltration of outdoor air if the fan room is not airtight.
- The fan room is entirely exposed to outdoor air.
- It may be more difficult to provide proper mixing of outdoor and recirculating air.

ISOLATED FAN ROOM. In this kind of fan room, shown in Fig. 15.13, the AHU or PU is not open to the fan room. The outdoor air, the return air, and the exhaust air are all isolated from the fan room air because of the ductwork. The static pressure in the fan room depends mainly on the air passage connecting the fan room and the outdoor atmosphere, and on the air leakage from or to the AHU or air handler. This kind of fan room is most widely used in commercial buildings.

Fan rooms located at the perimeter of the building often have direct access to the outside walls. They are more conveniently situated with respect to the outdoor air intake and exhaust. For fan rooms located in the interior core, large outdoor and exhaust air risers are required when an air economizer cycle is used.

In Fig. 15.14 is shown the plan and sectional views of an isolated fan room for an indoor packaged unit located in the interior core of the building. The unit is also equipped with a coil module for water cooling and heating coils and water economizer precooling coils.

Layout Considerations

A satisfactory fan room layout should meet the following requirements:

- It should be compact yet provide sufficient space for the maintenance workers to pull out fan shafts, coils, and filters. Piping connections on the same side of the AHU or PU are preferable.
- Outdoor intakes and exhaust outlets are located either on outside walls that are perpendicular to each other, or on different walls with a certain distance between them.

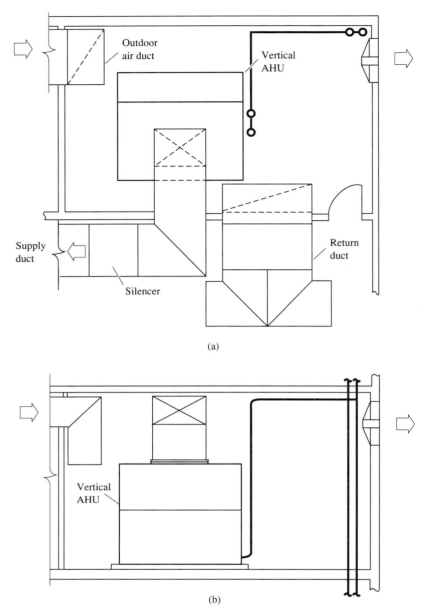

(a)

(b)

FIGURE 15.12 Open fan room: (a) plan view; and (b) sectional view.

- Fire dampers should be installed to separate the fan room and the fire compartment according to the fire codes. Fan room ventilation and exhaust should also be provided to meet the requirements of the codes.
- Inner lined square elbows or elbows with 2-in. or 3-in. thick duct liners are used for better sound attenuation at low frequencies. Sound attenuating devices are required for both the supply and return side of the fan.
- A vertical unit occupies less floor space than a horizontal unit; therefore, it is often the first choice if the headroom is sufficient.

- If a return fan is used, an axial fan with adequate sound attenuation may be the suitable choice to meet the fan characteristic of a considerably lower fan total pressure than the supply fan.

In Fig. 15.13a, at point rf after the return fan, the pressure is positive. At point m in the mixing box, the pressure must be negative in order to extract outdoor air. There must be a damper and pressure drop between point rf and m to guarantee such a positive to negative pressure conversion. This will be discussed in Chapter 19.

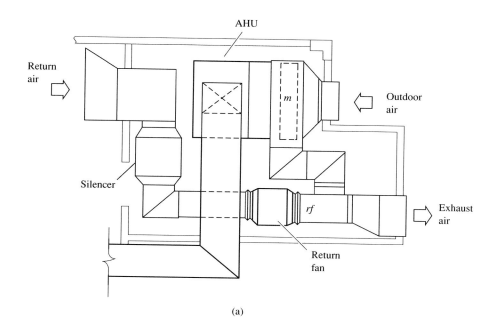

(a)

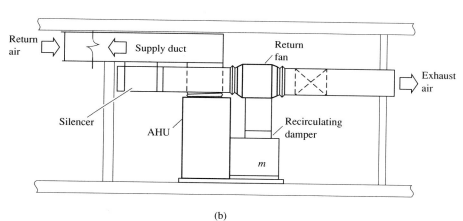

(b)

FIGURE 15.13 Isolated fan room: (a) plan view; and (b) sectional view.

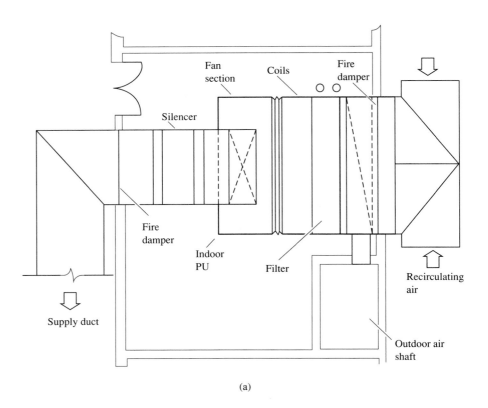

(a)

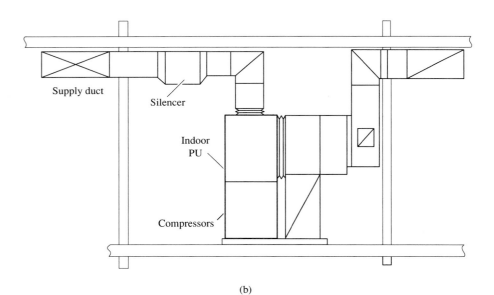

(b)

FIGURE 15.14 Interior core fan room: (a) plan view; and (b) sectional view.

REFERENCES AND FURTHER READING

ASHRAE, *ASHRAE Handbook 1988, Equipment*, ASHRAE, Inc., Atlanta, GA.

ASHRAE, *ASHRAE Standard 52–76, Method of testing Air Cleaning Devices Used in General Ventilation for Removing Particulate Matter*, ASHRAE Inc., Atlanta, GA., 1976.

ASHRAE/IES *Standard 90.1–1989, Energy Efficient Design of New Buildings Except New Low-Rise Residential Buildings*, ASHRAE Inc., Atlanta, GA, 1989.

Avery, R. H., "Selection and Uses of HEPA and ULPA Filters," *Heating/Piping/Air Conditioning*, pp. 119–123, January 1986.

Bierwirth, H. C., "Packaged Heat Pump Primer," *Heating/Piping/Air Conditioning*, pp. 55–59, July 1982.

Brasch, J. F., "Electric Duct Heater Principles," *Heating/Piping/Air Conditioning*, pp. 115–130, March 1984.

Haines, R. W., "Stratification," *Heating/Piping/Air Conditioning*, pp. 70–71, November 1980.

McGuire, A. B., "Custom Built HVAC Units," *Heating/Piping/Air Conditioning*, pp. 115–122, January 1987.

Remiarz, R. J., Johnson, B. R., and Agarwal, J. K., "Filter Testing with Submicrometer Aerosols," *ASHRAE Transactions*, Part II, pp. 1850–1858, 1988.

Riticher, J. J., "Low Face Velocity Air-Handling Units," *Heating/Piping/Air Conditioning*, pp. 73–75, December 1987.

Rivers, R. D., "Interpretation and Use of Air Filter Particle-Size-Efficiency Data for General-Ventilation Applications," *ASHRAE Transactions*, Part II, pp. 1835–1849, 1988.

Scolaro, J. F., and Halm, P. E., "Application of Combined VAV Air Handlers and DX Cooling HVAC Packages," *Heating/Piping/Air Conditioning*, pp. 71–82, July 1986.

Waller, B., "Economics of Face Velocities in Air-Handling Unit Selection," *Heating/Piping/Air Conditioning*, pp. 93–94, March 1987.

Wang, S. K., *Air Conditioning*, Vol. 3, Hong Kong Polytechnic, Hong Kong, 1987.

Weisgerber, J., "Custom Built HVAC Penthouses," *Heating/Piping/Air Conditioning*, pp. 115–117, November 1986.

16.1 SOUND CONTROL AND SOUND PATHS

Sound Control

Sound control for an air conditioned space or, more accurately, an occupied zone is provided to attenuate the HVAC&R equipment sound to an acceptable background level and to provide a suitable acoustic environment for people in the occupied zone. A suitable acoustic environment is as important as a comfortable thermal environment to the occupants. Noise, or any unwanted sound, is always annoying.

The objective of sound control should also be to provide an unobtrusive background sound at a level low enough that it does not interfere with human activities. Sound quality depends on the relative intensities of the sound levels in various octave bands of the audible spectrum. Unobtrusive sound quality means the following: a balanced distribution over a wide frequency range; no tonal characteristics such as hiss, whistle, or rumble; and a steady sound level.

The background sound level to be provided must be optimized. An NC or RC design criterion that is too low always means an unnecessarily high cost. The recommended indoor design RC or NC criteria ranges is listed in Table 5.7.

Sound Paths

When HVAC&R equipment or a component generates sound, it is often called a *sound generator* or *sound source*. Sound created at the source is received by the occupant through various sound transmission paths. HVAC&R equipment that acts as noise generators include fans, compressors, pumps, and dampers as well as other air flow noise in ducts. Of these, the fan is the major noise source in HVAC&R equipment to be controlled and attenuated because fans are widely used and have more sound transmission paths than compressors and pumps.

Sound transmits from the source to the receiver via the following paths:

- *Duct-borne paths*. These are noise transmission paths through ducts, duct fittings, and other air system components, either at the supply air side or the return air side, as shown in Figure 16.1. The noise thus transmitted is often called *system noise*.

- *Radiated sound path*. This is the path along which noise radiates through the duct walls or casings into the ceiling plenum or into an adjacent area and then transmits to the receiver.

- *Airborne path*. This path includes equipment noise transmitted to adjacent areas through the air.

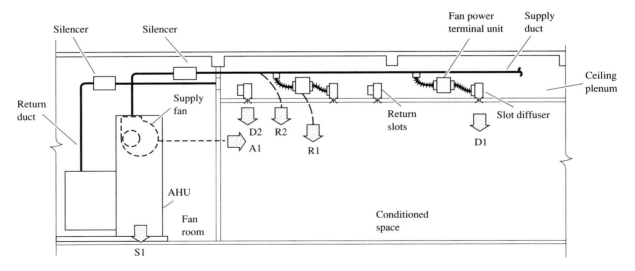

Duct–borne	D1	Fan → Silencer → Supply duct → Elbows → Branch power division → Flexible duct → Space effect
	D2	Fan → Silencer → Return duct → Elbow → Ceiling plenum → Space effect
Radiated sound	R1	Terminal unit casing → Ceiling plenum → Space effect
	R2	Supply duct breakout → Ceiling plenum → Space effect
Airborne	A1	Fan → Fan room → Wall → Space effect
Structure-borne	S1	Fan → Building structure

FIGURE 16.1 Fan room and sound paths.

- *Structure-borne path.* These are the paths along which equipment vibration and noise are transmitted through the building structure.

Structure-borne transmission is actually a combination of sound and vibration effects. This problem is usually solved through the combined effort of the structural and HVAC&R designers.

Controlled at Design Stage

The evaluation and control of all noise transmission paths should take place at the design stage and they should be a cooperative effort of the architect and the structural and mechanical engineers. If an excessive noise level occurs due to improper design—the designer fails to analyze the potential noise problem, there are mistakes in analysis, or data are used improperly—remedial measures to reduce the noise level are usually expensive and are often less effective.

One effective measure to reduce noise in buildings is to locate the fan room or other sound source away from critical areas, such as conference rooms and executive offices. For purposes of noise control the fan room may be located near restrooms, stairwells, or copy rooms.

For noise control, theoretical estimation and prediction schemes are mainly guidelines. Previous experience and field performance are important. Especially for projects with strict acoustic requirements, field/laboratory tests and checkouts are often necessary.

16.2 RECOMMENDED PROCEDURE FOR NOISE CONTROL

Before the analysis and evaluation of sound levels, the indoor design NC or RC criteria are usually determined, and the necessary data to evaluate sound levels at various points of interest are collected and investigated. The recommended procedure for noise control uses the following concept:

Source → Path (attenuation) → Receiver

The basic procedure is as follows:

1. Determine the sources of noise from the fan, compressor, and pump and the sound power level generated. Use certified manufacturer's data.
2. Carefully analyze all the possible sound paths that can transmit noise from the source to the occupied zone. An overlooked sound path may affect the final results.

3. Calculate all the sound attenuation and transmission losses in each sound path during transmission.
4. Investigate the difference, in db re 10^{-12} W, between air flow noise $L_{a.f}$ in duct-borne paths due to dampers, elbows, or junctions and the attenuated fan noise near the damper or turning vane $L_{at.fan}$. Because both $L_{at.fan}$ and $L_{a.f}$ are in dB, if $L_{at.fan} - L_{a.f} > 8$ dB, the addition of $L_{a.f}$ to $L_{at.fan}$ does not significantly increase the magnitude of $L_{at.fan}$; thus, $L_{a.f}$ can be ignored. Only when $L_{at.fan} - L_{a.f} \leq 8$ db should $L_{a.f}$ be added to $L_{at.fan}$ for duct-borne transmission calculations.
5. For duct-borne path or system noise, determine the attenuated sound power level of fan noise in dB re 10^{-12} W at the supply outlet (or room source L_{wr}) that affects the receiver. Convert L_{wr} into a sound pressure level received by the receiver at a chosen point in a room L_{pr} or at a plane 5 ft from the floor L_{pt}, both in dB re 20 μPa.
6. For radiated sound transmissions or airborne transmissions, the sound power level of the sound radiated into the receiving room should be converted to a room sound pressure level.
7. The resultant sound pressure level $L_{\Sigma p}$ at the center frequency of each octave band, in dB re 20 μPa, that is perceived by the receiver at a chosen point or at a plane 5 ft from the floor level is the sum of the sound pressure levels from all sound paths. Determine the NC or RC $L_{\Sigma.NC}$, in dB re 20 μPa, from the calculated $L_{\Sigma p}$ in each octave band.
8. Compare $L_{\Sigma.NC}$ with the design criteria NC or RC $L_{p.NC}$. If $L_{\Sigma.NC} > L_{p.NC}$, then one should add a silencer into the duct-borne sound path, change the configuration of the duct system, or build structures to bring $L_{\Sigma.NC}$ down to or slightly below $L_{p.NC}$.
9. Check this noise control prediction scheme with the actual field results of projects of similar noise control arrangements if possible.

16.3 FAN, COMPRESSOR, PUMP, AND AIR FLOW NOISE

Fan Noise

Use the fan noise data provided by the manufacturer. If manufacturer's data are not available, estimate the sound power level for the fan by Eq. (10.21). The specific sound power levels at full-load operation are listed in Table 10.1.

Fans should be selected for maximum efficiency so as to produce less fan noise. An oversized fan causes a higher velocity to flow through the wheel and housing and, therefore, greater noise. In undersized fan, flow separation at the blades produces noise. When a fan is operated at off-peak conditions, according to *ASHRAE Handbook 1991 HVAC Applications,* a correction factor C, in dB, should be added to the full load value as shown below:

Static efficiency, % of peak	Correction factor C, dB
99 – 100	0
85 – 89	3
75 – 84	6
65 – 74	9
55 – 64	12
50 – 54	15

It is therefore important to analyze the sound power level at various operating conditions to make the best choice for lower fan noise.

Noise from Chillers and Pumps

If manufacturer's data are not available, according to *ASHRAE Handbook 1991, HVAC Applications,* the sound pressure level L_{pA}, in dBA for centrifugal chillers at a distance 3.3 ft from the chiller can be calculated as

$$L_{pA} = 60 + 11 \log (\text{TR}) \qquad (16.1)$$

where TR = refrigeration capacity, in tons

For reciprocating chillers, L_{pA} in dBA at a distance of 3.3 ft is

$$L_{pA} = 71 + 9 \log (\text{TR}) \qquad (16.2)$$

The sound pressure level L_p, in dB, at the center frequency of various octave bands can be obtained by adding the following values at each octave band to the calculated L_{pA}:

	63	125	250	500	1000	2000	4000
Centrifugal chiller	−8	−5	−6	−7	−8	−5	−8
Reciprocating chiller	−19	−11	−7	−1	−4	−9	−14

The centrifugal chiller previously mentioned is constructed hermetically, has internal gears, and operates at medium or full load. At light load operation, L_p may increase 10 to 13 db representing the shaft frequency in the band, 8 to 10 dB in the blade pass frequency, and 5 dB for the remaining octave bands.

For circulating pumps, L_{pA} at a distance of 3.3 ft is

$$L_{pA} = 77 + 10 \log (\text{hp}) \qquad (16.3)$$

where hp = power input to the pump, hp.

Air Flow Noise

Air flow noise is an aerodynamic noise generated in flow passages or at duct fittings; it is sometimes called *regenerated noise.* Air flow noise is the result of vortices passing around obstacles that cause local acceleration and deformation and thereby produce local dynamic compression of the air. Air flow noise can also be generated when vortices pass through solid discontinuities.

Air flow noise is mainly determined by the velocity of the air flowing through the air passage and the constrictions due to duct fittings. The geometry of the duct fittings also affects the air flow turbulence and vortices and, therefore, the air flow noise.

According to *ASHRAE Handbook 1991, HVAC Applications,* air flow noise can be classified into the following categories based on its nature and influence:

STRAIGHT TURBULENT FLOW. Noise generated by straight air flow distributed across the duct is usually broadband in nature. Such a noise has no significant influence upon the sound power level in a straight duct section if the air velocity $v \le 2000$ fpm.

DUCT FITTINGS SUCH AS DAMPERS, ELBOWS, AND BRANCH TAKE-OFFS AND OBSTRUCTIONS. Such a noise is also generally broadband and limited to a small frequency range. Damper and obstruction noise should be prevented at the design stage. The sound power level of air flow noise $L_{a.f}$, in dB re 10^{-12} W, for dampers and elbows and at branch take-offs can be calculated as

$$L_{a.f} = K + 10 \log f + 50 \log v_{con} \\ + 10 \log S + 10 \log D + K_s \qquad (16.4)$$

where K = characteristic spectrum of the duct fitting
K_s = special parameter of the duct fitting
v_{con} = air velocity at the constricted part of the duct fitting or at the branch takeoff, fpm
S = cross-section area of the duct where damper is installed, or cross-sectional area of the elbow or branch duct, ft^2
D = duct height normal to the damper axis, or height of the elbow, ft
F = octave band center frequency, Hz

In Eq. (16.4), the special parameter $K_s = -107$ for a damper. For elbows with turning vanes,

$$K_s = 10 \log n - 107 \qquad (16.5)$$

where n = number of turning vanes
For elbows without turning vanes or for branch takeoff junctions,

$$K_s = -107 + \Delta L_{ro} + \Delta L_{tur} \qquad (16.6)$$

If the ratio of the radius r of the branch takeoff to the branch duct diameter d_{br}, or r/d_{br}, equals zero, then the takeoff rounding correction $\Delta L_{ro} = 4$ to 6 dB; and if $r/d_{br} = 0.15$, $\Delta L_{ro} = 0$.

The turbulence correction ΔL_{tur}, in dB, is necessary only if there is another damper, elbow, or takeoff upstream within 5 duct diameters. In such circumstances, $\Delta L_{tur} = 1$ to 5 dB, depending on the degree of turbulence in the upstream airstream.

Air velocity in the constricted part of the damper or duct fitting can be calculated as

$$v_{con} = SF_{block}/\dot{V} \qquad (16.7)$$

where $\dot{V}$ = air volume flow rate, cfm
For a multiblade damper, the blockage factor $F_{block} = 1$ if the pressure loss coefficient $C_{pre} = 1$.

If C_{pre} does not equal 1,

$$F_{block} = \frac{\sqrt{C_{pre}} - 1}{C_{pre} - 1} \qquad (16.8)$$

Here, the pressure loss coefficient, dimensionless, is given as

$$C_{pre} = \frac{15.9 \times 10^6 \Delta p_t}{(\dot{V}/S)^2} \qquad (16.9)$$

where Δp_t = total pressure loss across the duct fitting, in. WG.
For a single-blade damper, if $C_{pre} \leq 4$, Eq. (16.8) holds true. If $C_{pre} > 4$, then

$$F_{block} = 0.68\, C_{pre}^{-0.15} - 0.22 \qquad (16.10)$$

The characteristic spectrum K depends on the magnitude of the Strouhal number St, which can be calculated as

$$St = \frac{60fD}{v_{con}} = \frac{60fD}{v_{br}} \qquad (16.11)$$

where v_{br} = air velocity at the branch duct, fpm. When $St \leq 25$, K for dampers can be calculated as

$$K = -36.3 - 10.7 \log St \qquad (16.12)$$

If $St \geq 25$, then K for dampers is given as

$$K = -1.1 - 35.9 \log St \qquad (16.13)$$

For elbows with turning vanes, K can be calculated as

$$K = -47.5 - 7.96\, (\log St)^{2.5} \qquad (16.14)$$

The relationship between K and the St number for branch takeoffs is shown by curves in Fig. 16.2, where v_{ma} indicates the air velocity at the main duct, in fpm.

Example 16.1. A supply main duct is connected to several rectangular branch takeoffs. The first branch takeoff has a volume flow rate of 2200 cfm and dimensions of 16-in. width × 10 in. height. A single-blade damper is installed in the first branch takeoff and has a total pressure loss of 0.5 in. WG across this damper. Determine the sound power level of air flow noise from this single-blade damper.

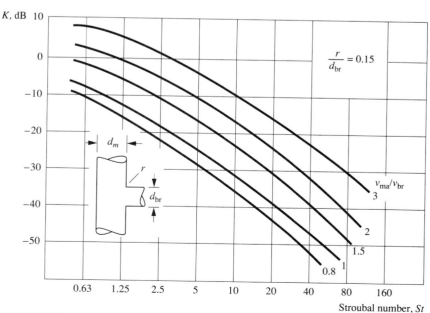

FIGURE 16.2 Relationship between K and St number for branch takeoffs. (Adapted from *ASHRAE Handbook HVAC Applications 1991*. Reprinted with permission.)

Solution

1. From the data given, the cross-sectional area of the branch duct is (16 in. × 10 in.)/144 = 1.1 ft². From Eq. (16.9),

$$C_{\text{pre}} = \frac{15.9 \times 10^6 \Delta p_t}{(\dot{V}/S)^2}$$

$$= \frac{15.9 \times 10^6 \times 0.5}{(2200/1.1)^2} = 1.98$$

As $C_{\text{pre}} < 4$, the blockage factor

$$F_{\text{block}} = \frac{\sqrt{C_{\text{pre}}} - 1}{C_{\text{pre}} - 1} = \frac{\sqrt{1.98} - 1}{1.98 - 1} = 0.418$$

From Eq. (16.7), the air velocity at the constricted part of the damper is

$$v_{\text{con}} = \frac{\dot{V}}{S F_{\text{block}}} = \frac{2200}{1.1 \times 0.418} = 4741 \text{ fpm}$$

From Eq. (16.11), the Strouhal number is

$$\text{St} = \frac{60 f D}{v_{con}} = \frac{60 \times 10}{12 \times 4741} f = 0.01054 \, f.$$

When the octave band center frequency $f = 63$ Hz,

$$\text{St} = 0.01054 \times 63 = 0.66$$

If $f = 4000$ Hz,

$$\text{St} = 0.01054 \times 4000 = 42.2$$

2. Because for dampers, $K_s = -107$, let

$$Z = K_s + 50 \log v_{\text{con}} + 10 \log S + 10 \log D$$

$$= -107 + 50 \log 4741 + 10 \log 1.1 + 10 \log (10/12)$$

$$= -107 + 183.8 + 0.45 - 0.79 = 76.5 \text{ dB}$$

At an octave band center frequency $f = 63$,

$$10 \log f = 10 \log 63 = 18$$

Also, when $\text{St} \leq 25$ and $f = 63$ Hz, the characteristic spectrum of the duct fitting is

$$K = -36.3 - 10.7 \log \text{St}$$

$$= -36.3 - 10.7 \log S_t$$

$$= -34.3 \text{ dB}$$

When $\text{St} > 25$ and $f = 4000$ Hz,

$$K = -1.1 - 35.9 \log \text{St}$$

$$= -1.1 - 35.9 \log 42.2 = -59.4 \text{ dB}$$

3. From Eq. (16.4), sound power level of air flow noise for this single blade damper is:

$$L_{\text{a.f}} = K + 10 \log f + 50 \log v_{\text{con}}$$

$$+ 10 \log S + 10 \log D + K_s$$

and can be listed as

	63	125	250	500
St	0.66	1.32	2.63	5.27
K	−34.3	−37.6	−40.8	−44.0
$10 \log f$	18	21	24	27
Z	76.5	76.5	76.5	76.5
$L_{\text{a.f}}$	60.2	59.9	59.7	59.5

	1000	2000	4000	8000
St	10.54	21.08	42.2	84.3
K	−47.2	−50.6	−59.4	−70.2
$10 \log f$	30	33	36	39
Z	76.5	76.5	76.5	76.5
$L_{\text{a.f}}$	59.3	58.9	53.1	45.3

From the calculations in Example 16.1, the key factor that influences the sound power level of the air flow noise at the damper is the velocity at the constricted part of the damper, v_{con}. Both the duct velocity at the branch duct v_{br} and the total pressure loss Δp_t at the damper (to balance the air flow at the branch takeoffs) affect v_{con}. If Δp_t is reduced from 0.5 in. WG to 0.25 WG, F_{block} then increases from 0.418 to 0.5, and v_{con} drops from 4741 fpm to 4000 fpm. The estimated $L_{\text{a.f}}$ values at various octave band center frequencies decrease about 4 to 7 dB.

POOR FAN ENTRY AND DISCHARGE CONDITIONS. Noise is generated because of the abrupt air passage constrictions and sudden changes in air flow direction. Both cause flow turbulence and separation. Such low-frequency noise results in a duct rumble, and is very difficult to attenuate. The designer should carefully design the fan intake and discharge connections. Doing so is the best way to control this kind of air flow noise.

Grilles and Diffusers. Grille and diffuser noise is primarily a function of their face or neck velocity. Most manufacturers' catalogs list reliable test NC data for grilles and diffusers. However, the following conditions must be taken into consideration:

- The manufacturer's test data are based on a normal velocity distribution before the air enters the grilles and diffusers. In actual setup, grilles may be installed after an abrupt turn and generate higher noise levels.

- If a designer chooses a grille with double deflection vanes and opposed-blade dampers, additional noise will be generated by the vanes and blades, the setting of the rear deflection vanes, and the reduction of the effective area or free flow area. Not all of these may be included in the manufacturers' data.

- Imbalance of volume flow between grilles from a main duct may create a greater face velocity than rated values.
- The actual condition of the room may be different from the value assumed in the catalog.

Air flow noise generated at the grilles or diffusers at the end of duct-borne path is difficult to attenuate except by reducing their face or neck velocities.

16.4 SOUND ATTENUATION ALONG DUCT-BORNE PATHS

Sound Attenuation in Ducts

Sound attenuation is the reduction of the intensity of sound, expressed in watts per unit area, as it travels along a sound transmission path from a source to a receiver. Sound attenuation is achieved by (1) the absorption of sound energy by the absorptive material, (2) spherical spreading and scattering, and (3) reflection of sound waves incident upon a surface.

Sound attenuation in duct sections, duct fittings, silencers, and other sound-reducing elements can be indicated by insertion loss. *Insertion loss* (IL) at a specific frequency is the reduction of sound power level, in dB re 10^{-12} W, measured at the receiver when a sound attenuation element is inserted in the transmission path between the sound source and the receiver. Noise reduction is the difference of sound pressure levels between any two points along the sound transmission path.

Sound absorptivity is the ability of a material to absorb sound energy. When a sound wave impinges on the surface of a porous sound-absorbing material, air vibrates within the small pores. The flow resistance of air and its vibration convert a portion of the absorbed sound energy into heat. The fraction of the incident sound power that is absorbed is called the *sound absorption coefficient* α. Most sound-absorbing materials have a low α at low frequencies and a high α at high frequencies. For a typical sound-absorbing material, α is equal to 0.15 in the octave band whose center frequency is 63 Hz, and $\alpha = 0.9$ at 1000 Hz.

Ducts can be unlined or inner-lined with fiberglass or other sound-absorbing materials, usually at a thickness of 1 in., (sometimes 2 in. or 3 in.) to absorb low-frequency sound. Inner-lined ducts have a better sound attenuation than unlined ducts, and the inner-lined sound-absorbing layer can also serve as thermal insulation.

In unlined round ducts with or without external thermal insulation, the sound attenuation, called *nat-*

TABLE 16.1

Approximate natural attenuation of sound in unlined rectangular sheet metal ducts

P/A ratio, in./in.²	Octave band center frequency, Hz		
	63	125	250 and over
	Attenuation,† dB/ft		
Over 0.31	0	0.3	0.1
0.31 to 0.13	0.3	0.1	0.1
Under 0.13	0.1	0.1	0.1

*P/A ratio indicates perimeter/area ratio.
†Double these values if the duct is externally insulated.
Source: ASHRAE Handbook 1991, HVAC Applications. Reprinted with permission.

ural attenuation, is about 0.03 dB/ft of duct length when the frequency is below 1000 Hz; it increases to 0.1 dB/ft irregularly at frequencies higher than 1000 Hz. Natural attenuation of unlined rectangular ducts is listed in Table 16.1.

For inner-lined ducts, the sound attenuation, or insertion loss, of a given length of duct section depends on the cross-sectional area and the properties of the absorptive material. Sound attenuation in inner-lined round ducts with a 1-in. thick of duct liner is listed in Table 16.2.

According to *ASHRAE Handbook 1991, HVAC Applications*, for inner-lined rectangular ducts, the following equation can be used to calculate the total insertion loss IL_{low} at frequencies below 800 Hz, in dB at any duct length:

$$IL_{low} = \frac{t^{1.08}h^{0.356}(P/A)Lf^{(1.17+0.19d)}}{1190d^{2.3}} \quad (16.15)$$

where t = thickness of duct liner, in.
h = smaller inside dimension of lined duct, in.
P = inside duct perimeter, in.
A = inside duct area, in.²
L = length of duct section, ft
f = frequency, Hz
d = density of the lining material, lb/ft³

At frequencies of 800 Hz and above, the total insertion loss of inner-lined rectangular duct IL_{hi}, in dB, can be calculated as

$$IL_{hi} = \frac{K(P/A)L_{rec}f^{[1.53-1.61\log(P/A)]}}{w^{2.5}h^{2.7}} \quad (16.16)$$

where $K = 2.11 \times 10^9$
w = greater inside dimension of lined duct, in.

TABLE 16.2
Round duct attenuation

Duct diameter, in.	Octave band center frequency, Hz						
	63	125	250	500	1000	2000	4000
	Approximate attenuation for 1 in. duct liner, dB/ft						
6	0.38	0.59	0.93	1.53	2.17	2.31	2.04
12	0.23	0.46	0.81	1.45	2.18	1.91	1.48
24	0.07	0.25	0.57	1.28	1.71	1.24	0.85
48	0	0	0.18	0.63	0.26	0.34	0.45

High-frequency (800Hz and over) attenuation values are applicable to a maximum of 10 ft.
Abridged with permission from *ASHRAE Handbook 1991, HVAC Applications*.

TABLE 16.3
Sound attenuation in straight lined sheet metal rectangular ducts lining thickness, 1 in.

Internal cross-sectional dimensions, in.	Octave and center frequency, Hz						
	63	125	250	500	1000	2000	4000
	Attenuation, dB/ft						
8 × 8	.10	.28	.77	2.12	5.82	6.08	2.95
8 × 16	.08	.21	.58	1.59	4.37	3.89	2.17
12 × 12	.08	.22	.60	1.64	4.48	4.52	2.67
12 × 24	.06	.16	.45	1.23	3.36	2.89	1.97
18 × 18	.06	.17	.46	1.26	3.45	3.37	2.42
18 × 36	.05	.13	.34	.94	2.59	2.15	1.78
24 × 24	.05	.14	.38	1.05	2.87	2.73	2.26
24 × 48	.04	.10	.29	.78	1.90	1.75	1.66
36 × 36	.04	.11	.29	.81	2.01	2.03	2.04
36 × 72	.03	.08	.22	.60	1.02	1.30	1.50
48 × 48	.03	.09	.24	.67	1.30	1.65	1.90
48 × 96	.02	.07	.18	.50	.66	1.05	1.40

1. Based on measurements of surface-coated duct liners of 1.5 lb/ft^3 density. For the specific materials tested, liner density had a minor effect over the nominal range of 1.5 to 3 lb/ft^3.

2. Add natural attenuation (Table 16.1) to obtain total attenuation.

Source: Abridged with permission from *ASHRAE Handbook 1991, HVAC Applications*.

In Eq. (16.16), L_{rec} represents the length of the duct section, in ft, when $L_{rec} \leq 10$ ft. If $L_{rec} \geq 10$ ft, assume $L_{rec} = 10$ ft. To avoid excessive IL values for rectangular duct, IL for any straight, lined duct section should not exceed 40 dB at any frequency. Sound attenuation (insertion loss) for 1 in.-thick inner-lined straight rectangular ducts is listed in Table 16.3.

The insertion loss of lined round, flat-oval, and rectangular ducts is a function of their inside diameter (or the smaller inside dimension of rectangular and flat-oval ducts). Generally, as the inside diameter increases, the insertion loss (IL) decreases. Surprisingly, a 3:1 aspect ratio flat-oval duct has approxi-

mately the same IL as a round duct with the same diameter as the minor axis.

The thickness of the duct liner has a definite effect on the IL of the lined ducts. When the thickness of duct liner is increased to 2 in., the sound attenuation is nearly double that of a 1-in. duct liner for octave band center frequencies $f \leq 500$. For example, for a 24 in. × 48 in. rectangular duct, at an octave band center frequency $f = 250$ Hz, a 1-in. thick duct liner has an IL = 0.29 dB/ft, whereas for a 2-in. thick duct liner, IL = 0.61 dB/ft. Two-inch or three-in. thick duct liners can be used effectively to control low-frequency noise if the ceiling space is available.

According to Bodley (1981), the density of the duct liner material has a minor effect on the IL of inner-lined ducts. The IL is affected by the absorption coefficient of the lining material. Lining material is protected by a perforated sheet, often made of galvanized steel or aluminum. The purpose of having a perforated inner surface is to protect the sound-absorbing material against air flow and to retain its acoustical properties. Neither the size of hole nor the percent open area of the perforated surface has an appreciable effect on the absorption coefficient of the lining material. A perforated inner surface liner of 22 percent open area is generally accepted as the standard type of perforated inner liner.

Sound Attenuation at Elbows and Branch Takeoffs

Sound is reflected back by elbows. Lined or unlined elbows provide sound attenuation to reduce the duct-borne noise. Sound attenuation depends on the size of duct, whether the elbow has a duct lining before and/or after the elbow, and whether turning vanes are provided. The estimated sound attenuation for unlined round elbows is indicated in Table 16.4.

A mitered 90° elbow, often called a square elbow, provides effective noise attenuation by means of repletion loss in the octave bands whose center frequencies equal or exceed 125 Hz. Rumble noise is difficult to attenuate even for a square elbow. In Table 16.5 are listed the insertion losses for elbows that are followed or preceeded by duct lining. Square elbows with turning vanes are slightly less effective in sound attenuation.

There are no sound attenuation data available for branch takeoffs. However, for any branch takeoff, an acoustic energy distribution takes place between the main duct and the branch duct. That part of the branch power division $L_{w.b}$, in dB re 10^{-12} W, is proportional

TABLE 16.4
Insertion loss of unlined round elbows

	Insertion Loss, dB
$fw < 1.9$	0
$1.9 < fw < 3.8$	1
$3.8 < fw < 7.5$	2
$fw > 7.5$	3

$fw = f \times w$

f = frequency, kHz

w = diameter for round duct, in.

Source: ASHRAE Handbook 1991, HVAC Applications. Reprinted with permission.

TABLE 16.5
Insertion loss of square elbows

	Insertion Loss, dB	
	Unlined	**Lined**
Without turning vanes		
$fw < 1.9$	0	0
$1.9 < fw < 3.8$	1	1
$3.8 < fw < 7.5$	5	6
$7.5 < fw < 15$	8	11
$15 < fw < 30$	4	10
$fw > 30$	3	10
With turning vanes		
$fw < 1.9$	0	0
$1.9 < fw < 3.8$	1	1
$3.8 < fw < 7.5$	4	4
$7.5 < fw < 15$	6	7
$f_w > 15$	4	7

$fw = f \times w$

f = frequency, kHz

w = duct width or depth of square duct or diameter for round duct, in.

to the ratio of the area of branch A_{br}, in ft², to the sum of the area of branch A_{br} and the straight-through end of the main duct after the takeoff A_{st}, in ft², and can be calculated as

$$L_{w.b} = 10 \log \frac{A_{br}}{A_{br} + A_{st}} \qquad (16.17)$$

The branch power division $L_{w.b}$ should be subtracted from the sound power level of duct-borne noise at the common end of the diverging wye in the main duct just before the branch takeoff L_{wc}, in dB re 10^{-12} W, to obtain the sound power level of the straight-through end in the main duct just after the branch takeoff L_{ws}, in dB re 10^{-12} W.

End Reflection Loss

When a sound wave propagates through the end opening of a duct section into a room, the sudden enlargement of the room space causes a certain amount of sound to reflect back to the duct. This effect significantly reduces the low-frequency sound that reaches the receiver in the occupied zone. According to *ASHRAE Handbook 1991, HVAC Applications*, the end reflection loss should not be taken into account if (1) the diffusers are tapped directly on the plenum, (2) the diffuser is not preceded by a straight duct section of less than 3 duct diameters and (3) the diffusers are connected to a flexible duct.

Examples of configurations that have end reflection losses include (1) air discharged from a nozzle mounted at the end of a branch duct of sufficient

TABLE 16.6
End reflection loss, in dB

Mean duct width, in.	Octave band center frequency, Hz				
	63	125	250	500	1000
6	18	12	8	4	1
8	16	11	6	2	0
10	14	9	5	1	0
12	13	8	4	1	0
16	11	6	2	0	0
20	9	5	1	0	0
24	8	4	1	0	0
28	7	3	1	0	0
32	6	2	0	0	0
36	5	1	0	0	0
48	4	1	0	0	0
72	1	0	0	0	0

Do not apply for liner diffusers or diffusers tapped directly into primary ductwork. If duct terminates in a diffuser, deduct at least 6 dB.
Source: ASHRAE Handbook 1991, HVAC Applications. Reprinted with permission.

length, (2) air discharged from an outlet downward to a suspended plaque with a connecting duct that has a length of at least one duct diameter, or (3) plain return opening or bar return grille with a length of connecting duct 3 to 5 duct diameters. End reflection losses are listed in Table 16.6.

Duct-borne Crosstalk

Duct-borne crosstalk or *plenum crosstalk* is the sound transmitted between two rooms through the duct systems or ceiling plenum. Usually, the sound attenuation in the duct system or in the ceiling plenum should be 5 dB greater than the transmission loss of the architectural partition between two rooms. Transmission loss will be discussed in the following section.

Duct lining, devious duct runs, flexible ducts, inner lining in the slot diffusers, and return slots are often used to attenuate the sound transmitted along duct-borne or plenum-connected paths to avoid crosstalk. A well-sealed architectural partition in the ceiling plenum with an inner-lined return air passage effectively prevents crosstalk through the ceiling plenum for applications with strict NC or RC criteria in the occupied zone.

Attenuation Along Duct-borne Paths

For a typical VAV system with a supply fan installed in the AHU inside a fan room, as shown in Fig. 16.1, the sound attenuation along the duct-borne paths—

the supply duct side and the return duct side—are as follows:

Supply ductborne path, D1	Return duct borne path, D2
Supply side silencer	Return side silencer
Supply duct attenuation	Return duct attenuation
Elbow attenuation	Elbow attenuation
Branch power division	Ceiling plenum attenuation or branch power division—
Flexible duct attenuation	return slot attenuation
Space effect	Space effect

On the return air side, an air filter made of a fiberglass mat may have a certain degree of sound attenuation. This could be considered an added safety factor in the return duct-borne path. Ceiling plenum attenuation will be discussed in the following section.

All sound attenuations should be subtracted from the sound power level of the supply fan $L_{w.f}$, in dB re 10^{-12} W, and converted to the sound pressure level at the receiver $L_{p.r}$, in dB re 20 μPa, through space effect.

16.5 RADIATED NOISE AND TRANSMISSION LOSSES

Breakout and Break-in

When a duct section enters the ceiling plenum or directly passes over occupied spaces, carrying duct-borne fan noise that is not well-attenuated, the noise radiates through the duct wall, raises the sound power level in the ceiling plenum or the occupied zone, and is then transmitted to the receiver. Noise that radiates through the duct wall and causes the duct wall to vibrate, caused either by internal sound waves or by airflow turbulence, is called *breakout*. Noise can also be transmitted into a duct and then travel along the duct-borne path, either discharging into a space through the duct opening or breaking out into ambient air, where it is transmited to the receiver. The transmission of external noise into a duct section through the duct wall is called *break-in*.

Breakout fan noise from the discharge duct directly under a rooftop packaged unit into the ceiling plenum may cause a serious acoustic problem in the underlying conditioned space.

Transmission Loss (TL)

Transmission loss is the reduction of sound power when sound is transmitted through a wall, a partition, or a barrier. Transmission loss (TL), in dB re

10^{-12} W, is defined as ten times the logarithmic ratio of incident sound power W_{in}, in W, to the transmitted sound power W_{tr}, in W, that is,

$$TL = 10 \log \frac{W_{in}}{W_{tr}} \qquad (16.18)$$

The relationship between transmission loss and mass of the material indicates that the sound transmission loss of a homogenous solid partition is a function of its mass and the frequency of sound f transmitted through it. This relationship can be expressed in the following form:

$$TL = F(\log m_s, \log f) \qquad (16.19)$$

where m_s = surface density of the homogeneous partition, lb/ft^2

Concrete walls and galvanized sheet ducts show that for each doubling of the mass partition, TL increases dB about 2 to 3 for low-frequency noise (less than 800Hz), and TL increases about 5 to 6 dB for high-frequency sound (800 Hz and over).

The TL and mass relationship is valid only when sound is incident on the surface of the partition in a normal direction and is an approximation. Actual measured data may be show a considerable deviation from the predicted values. Factors that cause deviations are nonhomogeneity, cracks, stiffness, and resonance.

Cracks and gaps around doors, windows, duct and piping sleeves, or other openings on the partition or wall may considerably reduce TL. The TL of a well-sealed partition of 100 ft^2 surface area may drop from 40 dB to 20 dB because of a total of 1 ft^2 of openings on that partition.

According to Cummings (1985), the model of breakout transmission loss (TL$_{out}$), in dB, can be described as

$$TL_{out} = 10 \log \frac{W_i A_o}{W_r A_i} \qquad (16.20)$$

where W_i = inlet sound power in the duct, dB
W_r = sound power radiated through the duct wall, dB
A_o = outer surface area of the duct section, ft^2
A_i = inner cross-sectional area of the duct section, ft^2

Such a breakout transmission model is shown in Fig. 16.3.

Break-in transmission loss for rectangular ducts TL$_{in}$, in dB, can be expressed as

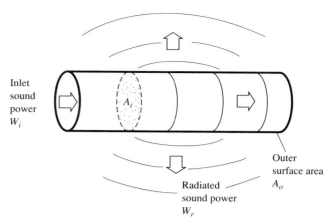

FIGURE 16.3   Breakout transmission loss TL$_{out}$.

$$TL_{in} = 10 \log \frac{W_{in}}{2W_t} \qquad (16.21)$$

where W_{in} = total incident sound power, dB
W_t = total transmitted sound power propagated inside the duct in a specific direction, dB

Transmission Loss for Rectangular Ducts

For rectangular ducts, Cummings recommended estimating the radiated noise through the duct wall based on a line source embodying axial traveling waves propagating at the speed of sound in the duct section. At higher frequencies, the duct wall response to an internal plane wave is essentially given by the TL and mass relationship.

According to *ASHRAE Handbook 1991, HVAC and Applications*, let the limiting frequency $f_L = 24,134/(ab)^{0.5}$, in Hz. When sound frequency $f < f_L$, when plane mode transmission dominates, TL$_{out}$ for rectangular ducts can be estimated from the following formula:

$$TL_{out} = 10 \log[fm^2/(a+b) + 17 \qquad (16.22)$$

where m = mass per unit area of the duct wall, lb/ft^2
a, b = longer and shorter dimensions of the rectangular duct, respectively, in.

If $f \geq f_L$, then TL$_{out}$ due to multimode sound waves in the duct section can be estimated as

$$TL_out = 20 \log mf - 31 \qquad (16.23)$$

In Tables 16.7 and 16.8 are listed the TL$_{out}$ and TL$_{in}$ for rectangular ducts at various octave band center frequencies.

TABLE 16.7
TL_{out} **for rectangular ducts at various octave bands, in dB**

Duct size		Octave band center frequency, Hz							
Inches	Gauge	63	125	250	500	1000	2000	4000	8000
12 × 12	24	21	24	27	30	33	36	41	45
12 × 24	24	19	22	25	28	31	35	41	45
12 × 48	22	19	22	25	28	31	37	43	45
24 × 24	22	20	23	26	29	32	37	43	45
24 × 48	20	20	23	26	29	31	39	45	45
48 × 48	18	21	24	27	30	35	41	45	45
48 × 96	18	19	22	25	29	35	41	45	45

The data are tests on 20-ft long ducts, but the TL values are for ducts of the cross section shown regardless of length.
Adapted with permission from *ASHRAE Handbook 1991, HVAC Applications.*

TABLE 16.8
TL_{in} **for rectangular ducts at various octave bands, in dB**

Duct size		Octave band center frequency, Hz							
Inches	Gauge	63	125	250	500	1000	2000	4000	8000
12 × 12	24	16	16	16	25	30	33	38	42
12 × 24	24	15	15	17	25	28	32	38	42
12 × 48	22	14	14	22	25	28	34	40	42
24 × 24	22	13	13	21	26	29	34	40	42
24 × 48	20	12	15	23	26	28	36	42	42
48 × 48	18	10	19	24	27	32	38	42	42
48 × 96	18	11	19	22	26	32	38	42	42

The data are tests on 20-ft long ducts, but the TL values are for ducts of the cross section shown regardless of length.
Source: Adapted with permission from *ASHRAE Handbook 1991, HVAC Application.*

Transmission Loss for Round and Flat Oval Ducts

Transmission loss characteristics for round ducts are quite different from rectangular ducts. At low frequencies and for internal propagation of a plane wave, the wall of an ideal round duct shows extremely high resistance to motion. TL falls with an increase in frequencies.

Measured values of TL for round ducts are much lower than the theoretical predictions. It is difficult to predict accurately the TL for round ducts. The TL_{out} values for round ducts at various octave band center frequencies are presented in Table 16.9.

The cross-sectional geometry of a flat-oval duct as well as its actual measured TL is between that for rectangular and round ducts. In Table 16.10 are listed the TL_{out} for flat-oval ducts at various octave bands center frequencies.

Comparison of TL Between Round, Rectangular, and Flat-Oval Ducts

Round ducts have the highest TL_{out} and TL_{in}, flat-oval ducts the second-highest TL_{out} and TL_{in}, and rectangular ducts the lowest TL_{out} and TL_{in} among these three kinds of ducts. For projects in which acoustic requirements are critical, round ducts in most cases are the first choice. Two or three round ducts can be run in parallel to reduce the diameter of the round duct and, therefore, save plenum height.

The static pressure differential between the inside and outside air across the duct wall, elbows, duct supports, and other crossbreakings does not appear to have a dramatic effect on TL.

External lagging for rectangular ducts is expensive to apply and needs additional duct supports. A better method to reduce TL is to use flat-oval or round ducts to replace rectangular ducts. If this is impossible, a

TABLE 16.9
TL$_\text{out}$ **for round ducts at various octave bands, in dB**

Duct size and type	Octave band center frequency, Hz							
	63	**125**	**250**	**500**	**1000**	**2000**	**4000**	**8000**
8 in. dia., [26 ga (0.022 in,)] long seam, length = 15 ft	> 45	(53)	55	52	44	35	34	26
14 in. dia., [24 ga (0.028 in.)] long seam, length = 15 ft	> 50	60	54	36	34	31	25	38
22 in. dia., [22 ga (0.034 in.)] long seam, length = 15 ft	> 47	53	37	33	33	27	25	43
32 in. dia., [22 ga (0.034 in.)] long seam, length = 15 ft	(51)	46	26	26	24	22	38	43
8 in. dia., [26 ga (0.022 in.)] spiral wound, length = 10 ft	> 48	> 64	> 75	> 72	56	56	46	29
14 in. dia., [26 ga (0.022 in.)] spiral wound, length = 10 ft	> 43	> 53	55	33	34	35	25	40
26 in. dia., [24 ga (0.028 in.)] spiral wound, length = 10 ft	> 45	50	26	26	25	22	36	43
26 in. dia., [16 ga (0.028 in.)] spiral wound, length = 10 ft	> 48	> 53	36	32	32	28	41	36
32 in. dia., [22 ga (0.034 in.)] spiral wound, length = 10 ft	> 43	42	28	25	26	24	40	45
14 in. dia., [24 ga (0.028 in.)] long seam, with two 90° elbows, length = 15 ft plus elbows	> 50	54	52	34	33	28	22	34

In cases where background noise swamped the noise radiated from the duct walls, a lower limit on the TL is indicated by a > sign, Parentheses indicate measurements in which background noise produced a greater uncertainty than usual in the data.
Adapted with permission from *ASHRAE Handbook 1991, HVAC Applications.*

TABLE 16.10
TL$_\text{out}$ **for flat-oval ducts at various octave bands, in dB**

Duct size (a × b)		Octave band center frequency, Hz							
Inches	**Gauge**	**63**	**125**	**250**	**500**	**1000**	**2000**	**4000**	**8000**
12 × 6	24	31	34	37	40	43	—	—	—
24 × 6	24	24	27	30	33	36	—	—	—
24 × 12	24	28	31	34	37	—	—	—	—
48 × 12	22	23	26	29	32	—	—	—	—
48 × 24	22	27	30	33	—	—	—	—	—
96 × 24	20	22	25	28	—	—	—	—	—
96 × 48	18	28	31	—	—	—	—	—	—

The data are tests on 20-ft long ducts, but the TL values are for ducts of the cross section shown regardless of length.
Adapted with permission from *ASHRAE Handbook 1991, HVAC Applications.*

TABLE 16.11
Transmission loss of some building structures

Building structures	Octave band center frequency, Hz						
	63	125	250	500	1000	2000	4000
4 in. dense concrete or solid concrete block, 48 lb/ft^2	32	34	35	37	42	49	55
4 in. hollow core dense aggregate concrete block, 28 lb/ft^2	29	32	33	34	37	42	49
8 in. hollow core dense aggregate concrete block	31	33	35	36	41	48	54
Standard dry-wall partition, $\frac{5}{8}$ in. gypsum board on both sides of 2 in. × 4 in. wood studs	12	17	34	35	42	38	44
Standard dry-wall partition, two layers of $\frac{5}{8}$ in. gypsum board on each side of 3 $\frac{5}{8}$ in. metal studs	25	36	43	50	50	44	55
$\frac{1}{2}$ in. plate glass	11	16	23	27	32	28	32
Double glazing, two $\frac{1}{2}$ in. panes, $\frac{1}{2}$ in. air space	12	16	23	27	32	30	35
Typical mineral fiber lay-in acoustic ceiling (insertion loss)	1	2	4	8	9	9	14
Typical gypsum board ceiling (insertion loss)	9	15	20	25	31	33	27
Roof construction, 6 in. thick, 20 gauge (0.0396 in) steel deck with 4 in. lightweight concrete topping, $\frac{5}{8}$ in. gypsum board ceiling on resilient hangers	25	41	47	56	65	68	69
*Plenum/ceiling cavity effect: lay-in fiberglass tile $\frac{1}{2}$ in., 6 lb/ft^3	4	8	8	8	10	10	14
*Plenum/ceiling cavity effect: lay-in mineral fiber tile $\frac{5}{8}$ in., 35 lb/ft^3		5	9	10	12	14	15
*Plenum/ceiling cavity effect: finished Sheetrock, $\frac{5}{8}$ in.	10	15	21	25	27	26	27
Acoustic equipment housing, 20 gauge steel outer shell, 2 in. thick acoustic insulation, 22 gauge (0.0336 in.) perforated inner shell	15	18	21	39	38	49	55
Solid-core wood door, normally closed†		23	27	29	27	26	29

Source: ASHRAE Hanbook 1991, HVAC Applications, except
*ASHRAE Transactions 1989, Part I
†Handbook of HVAC Design, 1990.

direct and inexpensive way is to double the duct wall thickness and, therefore, the mass per unit area which gives an increase in TL$_{out}$ for rectangular ducts of about 2 to 6 dB at various frequencies.

Transmission Loss for Selected Building Structures

The transmission loss for selected building structures—including walls, partitions, window glass, ceiling and plenum, and acoustic equipment housing—are listed in Table 16.11. These are mainly abridged from *ASHRAE Handbook 1991, HVAC Applications.* In Table 16.11, the combined plenum/ceiling cavity effect indicates the combined effect of the plenum sound absorption and the transmission through the ceiling material. These values are based on data from several manufacturers' laboratory and mock-up spaces.

16.6 SILENCERS

The purpose of a silencer is to reduce the sound power level of a fan, an air flow noise, or other sound source transmitted along a duct-borne path or airborne path to a required level.

Types of Silencers

The type of silencer is determined by its configuration.

RECTANGULAR SPLITTER SILENCER. A typical rectangular splitter silencer is shown in Fig. 16.4a. Inside the rectangular casing are a number of flat splitters, depending on the width of the silencer. These splitters direct the air flow into small sound-attenuating passages. Usually, the splitter is made from a strong, perforated steel envelope containing sound-absorbing materials with protected noneroding

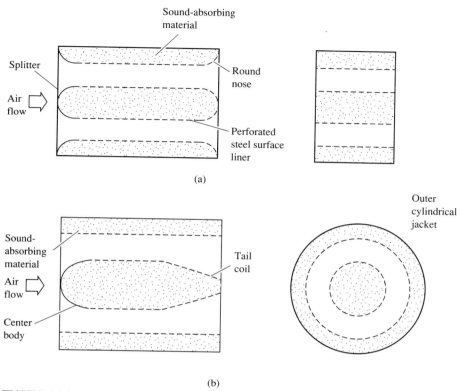

FIGURE 16.4 Rectangular and cylindrical silencers: (a) rectangular silencer; and (b) cylindrical silencer.

facing material inside the envelope. Splitters often have a round instead of a flat nose in order to reduce their air flow resistance. A rectangular silencer is often connected with rectangular ducts or sometimes with rectangular fan intakes and discharges.

CYLINDRICAL SILENCER. A cylindrical silencer has an outer cylindrical jacket and an inner concentric center body, as shown in Fig. 16.4b. The surfaces of the center body that contact the air flow are covered with perforated steel sheet and noneroding facing material. Sound-absorbing material lines the inside of the facing material. A cylindrical silencer is often used in conjunction with vane-axial fans and in round duct systems.

SOUND PLENUM. A sound plenum consists of several splitters in the form of two, three, or four successive square elbows, as shown in Fig. 16.5. Splitters are covered with noneroding facing material and sometimes also perforated galvanized steel sheet. The inner part of the splitter is filled with sound-absorbing material. A sound plenum effectively attenuates low-frequency noise and is often located between the collective return grille and the return air side of a pack-

aged unit. A low air velocity is important in a sound plenum because of the high local loss coefficient of square elbows.

Characteristics of Silencers

The acoustic and aerodynamic characteristics of a silencer are indicated by four parameters.

INSERTION LOSS (IL). This is the capability of a silencer to reduce the sound power level of fan noise or other sound source at various octave band center frequencies.

The IL of a silencer is affected by the air flow direction, especially when air velocity is greater than 2000 fpm. A sound wave that propagates in the same direction as the air flow is said to be in forward flow. A sound wave that propagates opposite to the air flow is said to be in reverse flow.

At low frequencies, reverse flow has a longer contact time and, therefore, a higher IL in a silencer. At higher frequencies, sound waves tend to refract toward the absorptive surface in a silencer under forward flow, and away from the absorptive surface under reverse flow. Therefore, these effects increase the

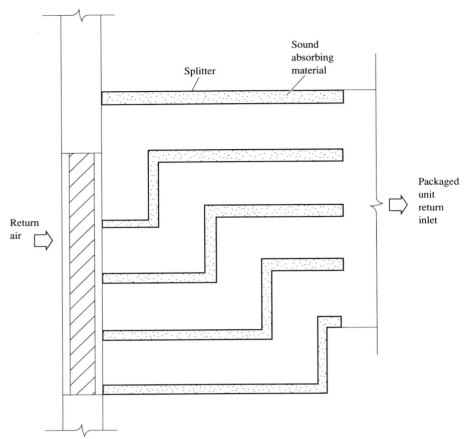

FIGURE 16.5   Sound plenum.

high-frequency attenuation in the forward flow and decrease the attenuation in reverse flow. The difference in IL between forward and reverse flow of a rectangular silencer at 2000 fpm face velocity is presented in Fig. 16.6.

OPERATING PARAMETERS. The operating parameters for a silencer are as follows:

- Volume flow rate of air through the silencer $\dot{V}_{sil}$, in cfm
- Cross-sectional area of the silencer, including the free area and the cross-sectional area of the absorptive material A_{sil}, ft^2
- Free area ratio R_{free}, which is defined as the ratio of free area A_{free}, that is, area of passages through which air flows, in ft^2 to the cross-sectional area A_{sil}, in ft^2
- Face velocity v_{sil}, which is defined as the ratio of the silencer's volume flow rate to its cross-sectional area, $\dot{V}_{sil}/A_{sil}$, in fpm

Many silencers have a free area ratio between 0.3 and 0.8. The actual mean air velocity inside the air flow passages in a silencer v_{free}, in fpm, may be 1.25 to 3.3 times the face velocity v_{sil}. Because of the configuration of square elbows of the splitters in a sound plenum, silencer face velocity should be in a range between 500 and 1000 fpm.

SELF NOISE. This is the sound power level, in dB, generated by a silencer at a given face velocity v_{sil} and a stated free area ratio R_{free}. The self noise of a silencer is mainly a function of its v_{free}. Self noise also indicates the lower limits of sound power level that a specific silencer can approach at various octave band center frequencies.

PRESSURE DROP Δp_{sil}, IN. WG. This is the pressure drop of air when it flows through a silencer. The pressure drop of a silencer is a function of its face velocity, free area ratio, length, and the configuration of the splitter or the center body. The pressure drop of a silencer usually does not exceed

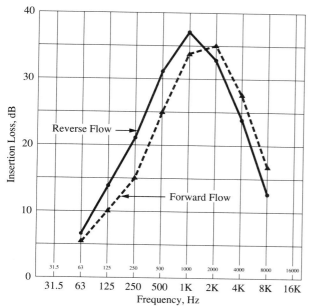

FIGURE 16.6 Difference in insertion loss of a rectangular silencer between forward and reverse flow (at 2000 fpm face velocity). *Source: Handbook of HVAC Design 1990.* Reprinted by permission.

0.5 in. WG. A cylindrical silencer with an attached tail cone may reduce the pressure drop up to 30 percent.

Location of Silencers

The IL and Δp_{sil} of a silencer are tested in a straight section of ductwork with a minimum of 2.5 D (diameter) of duct length upstream and 5 D of duct length downstream, according to ASTM test codes. A length of 2.5 D upstream ensures a uniform approach velocity, and 5 D downstream provides sufficient length to allow static regain to occur.

For maximum acoustical performance and minimum pressure drop, the optimum distance of the silencer from the fan discharge of both centrifual and axial fans is one duct diameter for every 1000 fpm average duct velocity. The actual pressure drop of the silencer $\Delta p_{\text{si,a}} = \Delta p_{\text{tt}}$ (the laboratory-tested pressure drop, both in inches WG).

The location of silencer from the fan intake for both centrifugal and axial fans should be 0.75 duct diameter for every 1000 fpm average duct velocity. When a cooling tower is equipped with rectangular silencers directly at its discharge outlet and intake, $\Delta p_{\text{si,a}} = 2\Delta p_{\text{tt}}$.

Selection of Silencers

A silencer should not be under- or oversized. An undersized silencer results in higher NC criteria in

conditioned space, which are not desirable. An oversized silencer always means a higher initial cost and waste of energy due to excessive pressure drop.

For a specific type of silencer, variations in its length and free area ratios often meet most of the requirements for IL and pressure drops. Silencers are usually made in lengths of 3, 5, 7, and 10 ft. The IL and Δp_{sil} of a silencer are proportional to its length. When the length of a typical rectangular silencer increases from 3 ft to 10 ft, its IL at 63 Hz increases from 3 dB to 13 dB, and the IL at 125 Hz increases from 15 dB to 42 dB. For sound attenuation at lower frequencies, a thicker splitter or cylindrical jacket is desirable.

A lower face velocity and greater free area ratio for a specific silencer at a given length always result in a lower pressure drop and, therefore, lower energy consumption. If space is allowed, a lower v_{sil} is beneficial for both IL and Δp_{sil}.

Active Duct Sound Attenuators

An active duct sound attenuator generates interfering sound waves from a sound source 180° out of phase in order to cancel the fan noise or other low-frequency noise. A computer is used to analyze and invert the unwanted noise so that it is 180° out of phase. According to *ASHRAE Handbook 1991, HVAC Applications*, attenuations of 20 to 38 dB have been reported at 40 to 400 Hz by active duct sound attenuating systems.

16.7 RELATIONSHIP BETWEEN ROOM SOUND POWER LEVEL AND ROOM SOUND PRESSURE LEVEL

Single or Multiple Sound Sources

Based on field measurements, Schultz (1985) recommended the following empirical formula to estimate the room sound pressure level L_{pr}, in dB re 20 μPa, from the room sound power level of single or multiple sound sources at a chosen point in a room:

$$L_{\text{pr}} = L_{\text{wr}} - 5 \log V - 3 \log f - 10 \log r + 25 \quad (16.24)$$

where L_{wr} = sound power level of the room source, dB re 10^{-12} W

V = volume of the room, ft^3

f = octave band center frequency, Hz

r = distance from the source to the receiver or reference point, ft

For a single sound source in the room, Eq. (16.24) can be applied directly. If there are multiple sound sources, estimate L_{wr} for each sound source and add their contributions together to obtain the total sound pressure level at the receiver due to the multiple sound sources.

Ceiling Diffusers

Offices and large rooms in commercial buildings often have several diffusers mounted on the ceiling. If the number of ceiling diffusers is greater than or equal to four and the spacing of the ceiling diffusers is on the order of the ceiling height, the resulting room sound pressure level at a level 5 ft above the floor level L_{p5}, in dB, can then be estimated as a function of the number of diffusers installed on the ceiling,

$$L_{p5} = L_{ws} - 5\log X - 28\log h + 1.3\log N - 3\log f + 31$$

$$(16.25)$$

where L_{ws} = sound power level of single outlet or a diffuser, dB re 10^{-12} W

h = ceiling height, ft

N = number of ceiling outlets in the room (N should be ≥ 4)

In Eq. (16.25),

$$X = \frac{A_{out}}{h^2} \qquad (16.26)$$

A_{out} = floor area served by each outlet, ft^2.

For slot or linear diffusers installed on the ceiling, each slot or linear diffuser of a certain length, or section of a slot or linear diffuser, can be considered one outlet.

The difference between the room sound pressure and the room sound power level $L_p - L_w$ is often called the *space effect*.

16.8 NOISE CONTROL FOR A TYPICAL AIR SYSTEM

In many commercial buildings using a variable-air-volume air system employing VAV boxes, there are two noise sources, as shown in Fig. 16.1:

- Fan noise due to the supply fan of the air-handling unit usually installed in a fan room
- Air-flow noise due to the VAV box with a single-blade damper installed inside the ceiling plenum

Sound is transmitted from these two sources to the receiver in the conditioned space via the following paths:

- Duct-borne path, supply side
- Duct-borne path, return side
- Radiated noise breakout from the supply duct in the ceiling plenum

- Radiated sound breakout from the VAV box casing in the ceiling plenum
- Airborne path through fan room walls to adjacent area

Combination of Supply Fan Noise and Terminal Noise

Variable-air-volume (VAV) boxes are tested for radiated sound power level and discharge-side sound power level in accordance with ADC/ARI Industry Standard 880–83 and the results are listed in the manufacturers' catalogs. The attenuated duct-borne sound power level of the supply fan from the air-handling unit at the terminal unit L_{AHUi} should be added to the discharge sound power level from the VAV box L_{ter} to form a resultant sound power level if their difference $L_{ter} - L_{AHUi} \leq 8$ dB. If their difference is greater than 8 dB, the smaller of these two sound power levels can be ignored.

Estimate Sound Pressure Level for Spaces Served by Terminal Units

Recently, the ADC and ARI jointly developed Industry Standard 885 to estimate the space sound pressure levels when terminal units are installed inside the ceiling plenum. Consider a terminal unit connected to two slot diffusers through flexible ducts to serve a conditioned space, as shown in Fig. 16.7, where C represents the terminal casing, D the duct-borne or discharge side, D_1 and D_2 the discharge outlets 1 and 2, and O_1 the supply outlet 1. Another outlet, O_2, is not shown in Fig. 16.7. Along with the arrows, B represents the transmission loss of the breakout from duct and casing, P the sound attenuation due to the plenum ceiling, and S the space effect.

There are altogether seven sound paths by which the receiver can hear fan, terminal, and slot diffuser noises.

Path 1. Radiated sound power level of the terminal unit $\longrightarrow$ plenum ceiling $\longrightarrow$ space effect $\longrightarrow$ receiver

Path 2. Discharge sound power level $\longrightarrow$ breakout $\longrightarrow$ plenum ceiling $\longrightarrow$ space effect $\longrightarrow$ receiver

Path 3. Ductborne sound power level at the outlet of slot diffuser $\longrightarrow$ space effect $\longrightarrow$ receiver

Path 4. Sound power level of the slot diffuser $\longrightarrow$ space effect $\longrightarrow$ receiver

Sound path 3 transmits radiated sound breakout from another flexible duct after the terminal unit, and

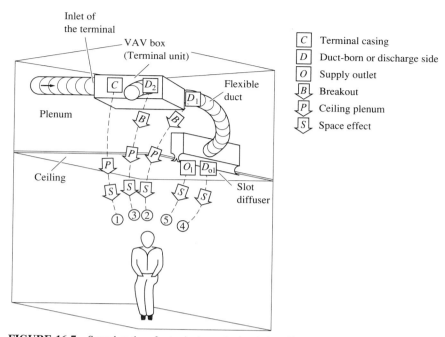

FIGURE 16.7   Sound paths of a typical terminal unit installed in the ceiling plenum.

Legend:
- C — Terminal casing
- D — Duct-born or discharge side
- O — Supply outlet
- B — Breakout
- P — Ceiling plenum
- S — Space effect

is similar to path 2. The duct-borne noise is transmitted along sound path 6 at another slot diffuser and

is similar to sound path 4. The air flow noise of another slot diffuser transmits along sound path 7 and is similar to sound path 5.

The radiated sound power level that breaks out at the inlet of the terminal unit has the same kind of sound path as paths 2 and 3.

The sound power level of duct-borne noise at D_{O1} equals the sound power level at D_1 or D_2 minus the sound attenuation of the flexible duct and the slot diffuser.

Environmental Adjustment Factor

According to Industry Standard (IS) 880, sound power levels of the terminal unit as listed in the manufacturers' catalogs are based on free field (outdoor) calibration of the reference sound source. At low frequencies, actual rooms are highly reverberant. The adjustment factor that takes into account the difference in sound power levels of a commonly used reference sound source measured in a free field and a reverberant field is called the *environmental adjustment factor* E_f, in dB. IS 885 recommends the following E_f values:

	Octave center band frequency, Hz							
	63	125	250	500	1000	2000	4000	8000
E_f, dB, re 10^{-12} W	+7	+3	+2	+1	+1	+1	+1	+1

Plenum Ceiling Effect

Sound attenuation of the plenum ceiling combination not only includes the transmission loss of the ceiling, but also considers the effect of the plenum. For a $\frac{5}{8}$-in. thick mineral fiber acoustic tile ceiling with a density of 35 lb/ft^3, IS 885 recommends the following values, which are the same as those in Table 16.11:

Octave center band frequency, Hz	125	250	500	1000	2000	4000
Plenum ceiling effect, dB		-5	-9	-10	-12	-14	-15

Example 16.2. A variable-air-volume box (terminal unit) receives conditioned air from an air handling unit through main and branch ducts. The VAV box is installed inside a ceiling plenum and is connected to two slot diffusers by 10-in. diameter flexible ducts, each of which is

5 ft long. Information is available from manufacturers' catalogs on the duct-borne fan noise at the inlet of the VAV box, the radiated sound power level $L_{w.rad}$, the discharge sound power level $L_{w.dis}$ of the VAV box, and the sound power level of the slot diffuser $L_{w.s}$. From measurements the transmission loss of the breakout noise through flexible duct wall and the insertion loss of the flexible duct (IL) are known. These data are summarized as follows:

	Octave center band frequency, Hz		
	125	250	500
Fan noise at inlet, dB	65	63	61
$L_{w.rad}$, dB	60.5	55	50.5
$L_{w.dis}$, dB	63	60.5	47
$L_{w.s}$, dB	40	39	35
Flexible, TL_{out}, dB	−8	−11	−14
Flexible, IL, dB/ft	0.50	0.85	1.48
	1000	**2000**	**4000**
Fan noise at inlet, dB	52	46	45
$L_{w.rad}$, dB	46.5	41.5	37.5
$L_{w.dis}$, dB	47.5	54.5	52
$L_{w.s}$, dB	33	31	24
Flexible, TL_{out}, dB	−17	−20	−25
Flexible, IL, dB/ft	2.2	2.04	1.64

If the volume of this room is 5040 ft³, estimate the total sound pressure level in the occupied zone with an average distance from the receiver of 8 ft. Ignore the fan noise transmitted through the return duct system.

Solution

1. From Eq. (16.24), the space effect at the octave band of center frequency of 125 Hz of this occupied zone is

$$L_{pr} - L_{wr} = -5 \log V - 3 \log f - 10 \log r + 25$$
$$= -5 \log 5040 - 3 \log 125 - 10 \log 8 + 25$$
$$= -8.8 \text{dB}$$

The space effect of other octave bands can be estimated similarly as follows:

	Octave center band frequency, Hz					
	125	250	500	1000	2000	4000
Space effect, dB	−9	−9.5	−11	−11.5	−12.5	−13.5

2. For path 1, after considering the influence of the environmental adjustment factor E_f, one finds that the radiated sound power levels of the VAV box $L_{rad.r} = L_{w.rad} - E_f$, in dB, are as follows:

	Octave center band frequency, Hz					
	125	250	500	1000	2000	4000
Reverberant field, dB	57.5	53	49.5	45.5	40.5	36.5

Because the transmission loss of the casing of the VAV box is rather high, the breakout of duct-borne fan noise through the casing is far smaller than $L_{rad.r}$, and is ignored.

The sound pressure level in the occupied zone due to the radiated sound power level from the VAV box can be calculated by subtracting the plenum ceiling effect from the $L_{rad.r}$ and then converting to the sound pressure level through the space effect, or

	Octave center band frequency, Hz		
	125	250	500
$L_{rad.r}$, dB	57.5	53	49.5
Ceiling-plenum effect, dB	−5	−9	−10
Space effect, dB	−9	−9.5	−11
Total sound pressure level, dB	43.5	34.5	28.5
	1000	**2000**	**4000**
$L_{rad.r}$, dB	45.5	40.5	36.5
Ceiling-plenum effect, dB	−12	−14	−15
Space effect, dB	−11.5	−12.5	−13.5
Total sound pressure level, dB	22	14	8

3. For path 2, the discharge sound power level of the VAV box changes from the free field $L_{w.dis}$ to the reverberant field $L_{dis.r}$ and the duct-borne fan noise at the terminal unit $L_{w.f}$. Their combination is as follows:

	Octave center band frequency, Hz					
	125	250	500	1000	2000	4000
$L_{dis.r}$, dB	60	58.5	46	46.5	53.5	51
Fan noise $L_{w.f}$, dB	65	63	61	52	46	45
$L_{dis.r} + L_{w.f}$, dB	66.5	64.5	61.5	53.5	54.5	52

The sound pressure level in the occupied zone due to the combination ($L_{dis.r} + L_{w.f}$) transmitted along sound path 2 can be estimated by subtracting the transmission loss of the flexible duct wall, the plenum ceiling effect, and the space effect from ($L_{dis.r} + L_{w.f}$). That is,

	Octave center band frequency, Hz		
	125	250	500
$L_{dis.r} + L_{w.f}$, dB	66.5	64.5	61.5
Flexible, TL_{out}, dB	−8	−11	−14
Ceiling-plenum, dB	−5	−9	−10
Space effect, dB	−9	−9.5	−11
Sound pressure level, dB	44.5	35	26.5
	1000	2000	4000
$L_{dis.r} + L_{w.f}$, dB	53.5	54.5	52
Flexible, TL_{out}, dB	−17	−20	−25
Ceiling-plenum, dB	−12	−14	−15
Space effect, dB	−11.5	−12.5	−13.5
Sound pressure level, dB	13	8	< 0

4. For path 4, there are two flexible ducts and slot diffusers connected to the terminal unit, therefore, the branch power division is 3 dB. The sound pressure level in the occupied zone due to duct borne noise transmitted along sound path 4 can be evaluated by subtracting the branch power division $L_{w.b}$, the insertion loss of the flexible duct IL_{flex}, and the space effect from the combination $(L_{dis.r} + L_{w.f})$.

	Octave center band frequency, Hz		
	125	250	500
$(L_{dis.r} + L_{w.f})$, dB	66.5	64.5	61.5
$L_{w.b}$, dB	−3	−3	−3
IL_{flex}, dB	−2.5	−4	−7
Space effect, dB	−9	−9.5	−11
Sound pressure level, dB	52	48	40.5
	1000	2000	4000
$(L_{dis.r}, + L_{w.f})$ dB	53.5	54.5	52
$L_{w.b}$, dB	−3	−3	−3
IL_{flex}, dB	−11	−10	−8
Space effect, dB	−11.5	−12.5	−13.5
Sound pressure level, dB	28	29	27.5

5. For path 5, change the sound power level of the slot diffuser from free field $L_{w.s}$ to reverberant field $L_{s.r}$. Then the sound pressure level in the occupied zone due to the slot diffuser can be estimated by subtracting the space effect from $L_{s.r}$.

	Octave center band frequency, Hz		
	125	250	500
$L_{s.r}$, dB	37	37	34
Space effect, dB	−9	−9.5	−11
Sound pressure level, dB	28	27.5	23
	1000	2000	4000
$L_{s.r}$, dB	32	30	23
Space effect, dB	−11.5	−12.5	−13.5
Sound pressure level, dB	20.5	17.5	9.5

6. The resultant estimated sound pressure level in the occupied space $L_{p.t}$, in dB re 20 μPa, is the sum of the contributions of all sound paths and can be calculated from Eq. (5.29):

	Octave center band frequency, Hz					
	125	250	500	1000	2000	4000
Path 1	43.5	34.5	28.5	22	14	8
Path 2	44.5	35	26.5	13	8	0
Path 3	44.5	35	26.5	13	8	0
Path 4	52	48	40.5	28	29	27.5
Path 5	52	48	40.5	28	29	27.5
Path 6	28	27.5	23	20.5	17.5	9.5
Path 7	28	27.5	23	20.5	17.5	9.5
$L_{p.t}$, dB	56	51.5	44	32.5	32.5	31

These resultant sound pressure levels at the center frequencies of various octave bands are all equal to or below the NC 40 curve.

16.9 ROOFTOP PACKAGED UNITS

Rooftop packaged units are widely used in low rise and sometimes even high-rise commercial buildings. Many rooftop packaged units are curb-mounted on lightweight roof deck construction. The distances between the fan and ceiling plenum, and between the fan and the conditioned space are sometimes shorter than the distance between the fan in the AHU and space. They may cause unique noise control problems.

Locating the rooftop packaged unit away from areas with critical sound control requirements, such as conference rooms and executive offices, is the most effective and economical method of noise control.

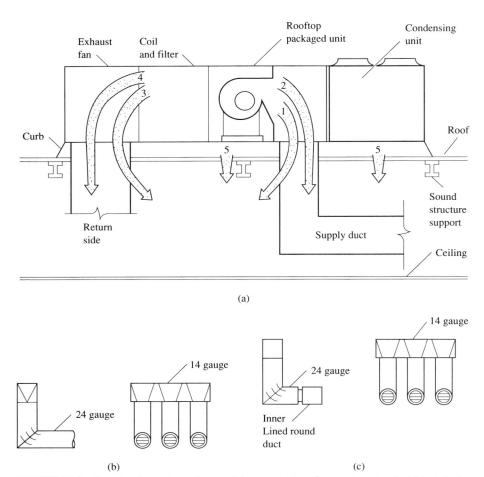

(a)

(b)   (c)

FIGURE 16.8 Sound paths and sound control for a typical rooftop packaged unit: (a) installation of rooftop unit with rectangular supply duct; (b) rectangular to multiple drop: round mitered elbows with turning vanes; and (c) rectangular to multiple drop: round mitered elbows with turning vanes, inner lined round duct after mitered elbow.

Sound Sources and Paths

In a typical rooftop packaged unit, the paths by which fan and compressor noise is tranmitted from the rooftop packaged units to the receiver in the occupied space are shown in Fig. 16.8a:

1. Radiated fan noise breaks out from the supply duct in the ceiling plenum $\longrightarrow$ plenum ceiling effect $\longrightarrow$ space effect $\longrightarrow$ receiver.
2. Duct-borne fan noise on the discharge side $\longrightarrow$ attenuation of ducts, elbows, and silencer $\longrightarrow$ space effect $\longrightarrow$ receiver.
3. Radiated fan noise breaks out from the return air duct in the ceiling plenum $\longrightarrow$ plenum ceiling effect $\longrightarrow$ space effect $\longrightarrow$ receiver.
4. Duct-borne fan noise on the return side $\longrightarrow$ attenuation of ducts, elbows, and silencer $\longrightarrow$ end reflection $\longrightarrow$ space effect $\longrightarrow$ receiver.

5. Structure-borne fan or compressor noises $\longrightarrow$ roof construction $\longrightarrow$ plenum ceiling effect $\longrightarrow$ receiver.

Discharge Side Duct Breakout

Experience has shown that low-frequency noise breakout from the supply duct inside the ceiling plenum (path 1) and subsequent transmissions to the receiver through the ceiling is the critical noise problem in most rooftop packaged units.

Round ducts have a far greater TL_{out} than rectangular ducts. Based on laboratory tests, Beatty (1987) recommended using rectangular to multiple round ducts drops with round mitered elbows and turning vanes, as shown in Fig. 16.8b and 16.8c to contain low-frequency rumble noise. Compared with rectangular ducts without turning vanes, these configurations have an additional noise reduction of 17.5 dB in octave band 1, 11.5 dB in octave band 2, and 13 dB in octave band 3 when the configuration is the same as in Fig. 16.8b. If the multiple drops of round duct have the configuration shown in Fig. 16.8c, there will be an additional noise reduction of 18 dB in octave band 1, 13 dB in octave band 2, and 16 dB in octave band 3 compared with rectangular ducts without turning vanes.

Careful design of the fan outlet configurations will prevent sharp turning of airstreams. Use heavier-gauge sheet metal to fabricate the rectangular plenum or duct before the round duct drops to reduce the breakout noise. Inner lining of the ductwork and sometimes a silencer are necessary at the downstream side for more effective noise control.

Sound Path on the Return Side

On the return side, a rectangular return duct may be used if breakout noise is not critical. Select a lower air velocity, preferably lower than 1000 fpm. A 2-in. thick inner lining in the return side of the packaged unit and ducts affords better sound attenuation. Use tees with a minimum leg length of three to five equivalent diameters to increase end reflection loss.

Structure-Borne Noise

Structure-borne noise can be controlled by using suitable vibration isolators. Special curb-mounting bases should be used to provide better sound and vibration control. Sound structural support should be provided for rooftop packaged units.

REFERENCES AND FURTHER READING

ADC/ARI, *ADC/ARI Industry Standard 880–83 for VAV Terminals*, ADC, Chicago, IL 1983.

ADC/ARI, *ADC/ARI Industry Standard 885, Method of Predicting Sound Pressure Levels*, ADC, Chicago, IL, 1989.

ASHRAE, *ASHRAE Handbook 1991, HVAC Applications*, ASHRAE Inc., Atlanta, GA.

ASTM, *ASTM E477 Standard Method of Testing Duct Liner Materials and Prefabricated Silencers for Acoustical and Airflow Performance*, Amercian Society of Testing Material, Philadelphia, PA 1973.

Beatty, J., "Discharge Duct Configurations to Control Rooftop Sound," *Heating/Piping/Air Conditioning*, pp. 53–58, July 1987.

Blazier, W. E., "Noise Rating of Variable-Air-Volume Terminal Devices," *ASHRAE Transactions*, Part I, pp. 140–152, 1981.

Bodley, J. D., "An Analysis of Acoustically Lined Duct and Fittings and the State of the Art of Their Use," *ASHRAE Transactions*, Part I, pp. 658–671, 1981.

Cummings, A., "Acoustic Noise Transmission through Duct Walls," *ASHRAE Transactions*, Part II A, pp. 48–61, 1985.

Ebbing, C., and Waeldner, W. J., "Industry Standard 885: An Overview, Estimating Space Sound Levels for Air Terminal Devices," *ASHRAE Transactions*, Part I, pp. 529–533, 1989.

Ebbing, C. E., Fragnito, D., and Inglis, S., "Control of Low Frequency Duct-Generated Noise in Building Air Distribution Systems," *ASHRAE Transactions*, Part II, pp. 191–203, 1978.

Grimm, N. R., and Rosaler, R. C., *Handbook of HVAC Design*, McGraw-Hill, New York, 1990.

Hirschorn, M., "Acoustic and Aerodynamic Characteristics of Duct Silencers for Air Handling Systems," *ASHRAE Transactions*, Part I, pp. 625–646, 1981.

Reese, J., "The Case for Certified Ratings," *Heating/Piping/Air Conditioning*, pp. 85–87, August 1986.

Reynolds, D. D., and Bledsoe, J. M., "Sound Transmission Through Mechanical Equipment Room Walls, Floor, or Ceiling," *ASHRAE Transactions*, Part I, pp. 83–89, 1989.

Schultz, T. J., "Relationship between Sound Power Level and Sound Pressure Level in Dwellings and Offices," *ASHRAE Transactions*, Part I A, pp. 124–153, 1985.

Sessler, S. M., and Angevine, E. N., "HVAC System Noise Calculation Procedure—An Update," *ASHRAE Transactions*, Part II B, pp. 697–709, 1983.

Smith, M. C., "Industry Standard 885 Acoustical Level Estimation Procedure Compared to Actual Acoustic Levels in an Air Distribution Mock-Up," *ASHRAE Transactions*, Part I, pp. 543–548, 1989.

Smith, M. C., and Int-Hout, D., "Using Manufacturers' Catalog Data to Predict Ambient Sound Levels," *ASHRAE Transactions*, Part II B, pp. 85–96, 1984.

The Trane Company, *Acoustics in Air Conditioning*, American Standard Inc., La Crosse, WI, 1986.

Ver, I. L., "Noise Generation and Noise Attenuation of Duct Fittings—A Review: Part II," *ASHRAE Transactions*, Part II A, pp. 383–390, 1984.

/

CHAPTER 17

BASICS OF AIR CONDITIONING CONTROL SYSTEMS

17.1 CONTROL SYSTEMS AND HISTORICAL DEVELOPMENT

Control System

The primary function of an air conditioning control system, or automatic control system, is to adjust the air conditioning equipment so as to maintain predetermined levels of certain parameters within an enclosure or control zone, so that the fluid leaving the equipment will meet the conditioning load for any climate changes with optimum energy consumption and safety. The predetermined parameter or variable to be controlled is called the *controlled variable*. In air conditioning or HVAC&R, the controlled variable can be temperature, relative humidity, pressure, enthalpy, fluid flow, or contaminant concentration.

Figure 17.1 shows a control diagram of an electronic control system for a single-zone, variable-air-volume system. *Single-zone* means that the load characteristics of the whole conditioned space are similar and will be monitored and controlled by one controller to maintain a predetermined controlled variable.

Control Loops

The basic element of a control system is a *control loop*. A control loop consists of a *sensor* (or sometimes two or more sensors) (such as temperature sensor T2 in Fig. 17.1) which senses and measures the controlled variable of recirculating air; a *controller* (C3 in Fig. 17.1), which usually compares the sensed input signal with the predetermined condition (the *set point*) and produces an output signal to actuate the third element, a *controlled device* or *control element* (such as a damper or a valve in Fig. 17.1). Modulation—minute adjustments or on-off control of such devices as dampers and valves—will change the controlled device's position or operating status, which affects the controlled variable by changing the air, water, and steam flow, electric power supply, and so on. The controlled variable is thus varied toward the predetermined value, the set point. Air, water, and steam are *control mediums* or *control agents*. Air flow, water flow, and steam flow are *manipulated variables*. The equipment that varies the output capacity by changing the opening position of the valves and dampers is called the *process plant*.

There are two types of control loops: *open* and *closed*. An open-loop system assumes a fixed relationship between the controlled variable and the input signal being received. The sensed variable is not the controlled variable, so there is no feedback. An example of an open-loop system would be a ventilating fan that turns on when the outdoor temperature exceeds a specified set point. The sensed variable is the outdoor temperature, and the controlled variable is the state of the fan (on or off).

A closed-loop system depends on sensing the controlled variable to vary the controller output and modulate the controlled device. In Fig. 17.1, temperature sensor T2 senses the controlled variable, the recirculating air temperature T_{ru} entering the air-handling unit; controller C3 receives this sensed signal input and produces an output to modulate the damper actuator DM4 and thus the position of the final controlled device, the inlet vanes at the supply fan. As the inlet vanes open and close, the supply-air volume flow rate

varies and the space temperature T_r and recirculating temperature T_{ru} change accordingly. This change in T_{ru} is sensed again by T2 and fed back to the controller for further modulation of the inlet vanes to maintain values of T_{ru} that approach their set points. These components form a closed loop system.

Figure 17.2 shows a block diagram of this closed-loop system. It shows a secondary input to the controller, such as the outdoor air temperature T_{ru}, which may reset the set point in the controller to provide a better and a more economical control. The disturbances that affect the controlled variable are load variations and changes in the outdoor weather. After the controller senses the signal feedback, it sends a corrective signal to the controlled device based on the difference between the sensed controlled variable and the set point. Thus, T_{ru} is under continuous comparison and correction.

A control system—or its component, the control subsystem—used to control the controlled variable(s) in a conditioned space or within a mechanical device or equipment, may contain only one control loop, or it may contain two or more loops.

Sequence of Operation

The *sequence of operation* is a description of the sequential order of the functional operations that a control system is supposed to perform. For the single-zone VAV system shown in Fig. 17.1, when cold air supply is required during full occupancy, even in the winter, the sequence of operation is as follows:

1. The supply fan is started and stopped according to a time schedule. Manual override is possible. The smoke detector in the return air or the low-temperature limit sensor of the mixed air will stop the supply fan if necessary. The supply fan status (on or off) is determined by the pressure differential switch across the fan. After the fan is turned on, the control system is energized.

2. When the recirculating air temperature T_{ru} drops below the set point, the opening of the inlet vanes of the supply fan becomes smaller, resulting in a lower supply volume flow rate. If the sensed T_{ru} is higher than the set point, the opening of the inlet vane becomes larger, resulting in a higher supply volume flow rate.

3. Outdoor air, recirculating air, and relief air dampers are controlled according to an air economizer cycle, which is a kind of optimum application of low-temperature outdoor air for free cooling that works as follows:

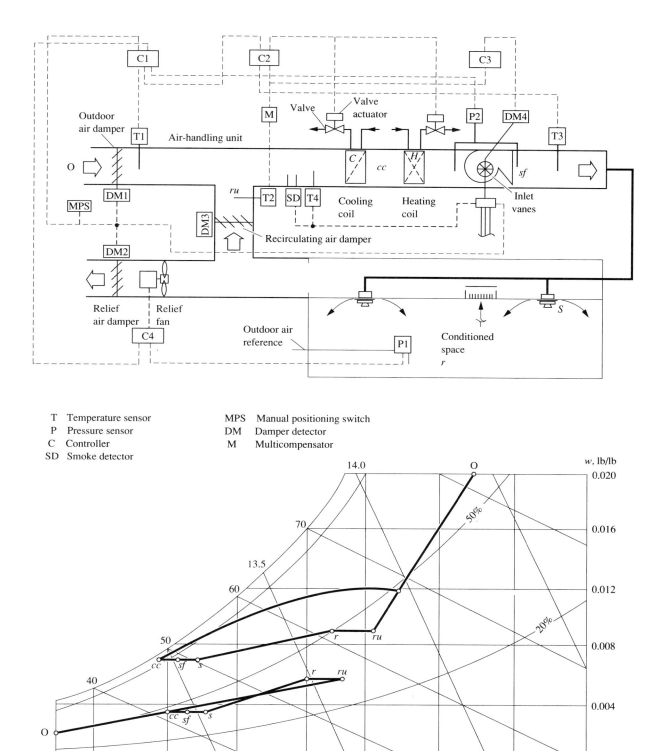

T   Temperature sensor
P   Pressure sensor
C   Controller
SD Smoke detector

MPS Manual positioning switch
DM   Damper detector
M   Multicompensator

FIGURE 17.1 Schematic diagram and psychrometric analysis of an electronic control system for a single zone, VAV system.

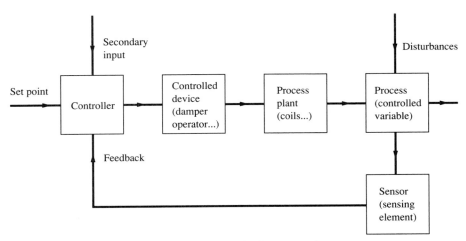

FIGURE 17.2 Block diagram for a closed-loop feedback control system.

- When the supply fan is switched off, the outdoor and relief air dampers are closed and the recirculating air damper is opened.
- The recirculating air temperature T_{ru}, sensed by the temperature sensor T2, is compared with the outdoor air temperature T_o, sensed by T1. If $T_o > T_{ru}$, the outdoor air damper opens to a minimum position, that is, outdoor air will be 25 percent of the total supply air. A manual positioning switch (MPS) determines the minimum outdoor and relief air damper positions.
- If $T_o \leq T_{ru}$, the outdoor and relief air dampers will be fully opened and the recirculating air damper will be closed. One hundred percent of outdoor air is extracted to minimize the refrigeration capacity, and an all-outdoor air supply is admitted to the conditioned space.

4. The control valve of the cooling coil, the outdoor air and recirculating air dampers, and the control valve of the heating coil are modulated in sequence to maintain a preset discharged temperature T_{dis}. The set point of the discharge temperature T_{dis} and recirculating air temperature T_{ru} will be reset by the outdoor air temperature T$_o$ according to the following schedule:

T_o	T_{ru}	T_{dis}
$\geq 75°F$	80°F	50°F
$70°F < T_o < 75°F$	77.5°F	52.5°F
$\leq 70°F$	75°F	55°F

5. Only when the outdoor air damper is opened and in a position greater than the minimum position is the relief fan turned on and the relief damper fully opened. If the static pressure difference between the space air and the outdoor air is greater than the set point 0.05 in. WG, the speed of the relief fan and its exhaust volume flow rate will be increased by means of an AC inverter to maintain the predetermined static pressure difference in the space. If the static pressure difference between the space air and the outdoor air is less than or equal to 0.05 in. WG, the relief fan is shut off.

In the design and operation of control systems for air conditioning, the necessary documentation includes the sequence of operation, control diagrams, specifications, instructions, and the operation and maintenance manual.

Historical Development

Early controls for comfort air conditioning systems used in the Capitol building since 1928 were pneumatic controls that included high- and low-limit thermostats in the supply-air discharge of the air-handling unit. These controls were used to maintain the desired supply air temperature. Dew point control of the supply air maintained room humidity within a desirable range. A thermostat in the return-air passage was used to control the air volume flow rate passing through the cooling coil, and a sixteen-points recorder connected with resistance thermometers was used to record the room temperature at various locations.

During the 1940s, improvements were made to the sensitivity and appearance of thermostats and controllers as well as the modulating characteristics of control valves and dampers. In the 1950s, sensitivity, accuracy, and simplicity were improved through the introduction of electronic sensing and control.

The concept of a building automation system (BAS), or building management system (BMS), which centralizes the monitoring, operations, and management of building services so as to provide greater efficiency, was devised in the early 1950s. These were hardwired systems. All of the monitored data and control points were wired directly to a centralized control center. An operator could start or stop equipment; monitor the temperature, relative humidity, and operating status of the system; and adjust the control point settings. Multiplexed wiring was the most important improvement in BMSs in the early 1960s; it selectively connected the centrally located selection switches and control wires by means of relays. Multiplexing also reduced the number of wires required for centralized control and monitoring operations.

During the late 1960s and early 1970s, a major development for BMSs was the use of solid-state electronics. Solid-state electronic components improved the scanning process and serial transmission, and reduced many wires in the trunk wiring to a single pair.

After the energy crisis in 1973, many energy management systems (EMSs) were installed for HVAC&R systems in buildings, primarily for the purpose of saving energy. An EMS optimized the operation to maintain the required indoor environmental parameter while minimizing the energy consumption of the HVAC&R system. An EMS is part of a BMS. The introduction of the minicomputer and then the microcomputer, or personal computer (PC), greatly increased the ability to monitor, control, optimize, and operate HVAC&R systems. The microcomputer widened computer-based EMS applications in commercial buildings. The development of microprocessor-based direct digital control (DDC) and controllers in the early 1980s changed the basic EMS structure from one of central supervision to a more capable, flexible, and reliable control system.

17.2 CONTROL METHODS

The term *control methods* represents the various types of control signals and different kinds of energy used to operate a control system.

Analog or Digital

There are two types of control signals: *analog* and *digital*. An analog signal is in the form of a continuous variable, e.g., using the magnitude of the pneumatic pressure to represent the air temperature. a digital signal is a series of on and off pulses used to transmit information.

A conventional analog controller receives a continuous analog signal, such as a voltage or a pneumatic signal, that is proportional to the magnitude of the sensed variable. The controller compares the signal received from the sensor to the desired value (i.e., the set point) and sends a signal to the actuator in proportion to the difference between the sensed value and the set point. Figure 17.1 shows an electronic control system using analog control.

A digital controller, or microprocessor-based controller, receives an electronic signal from a sensor. It converts the electronic signal to digital pulses of different time intervals to represent the signal's values. The microprocessor of the digital controller performs mathematical operations on these values. The output from the microprocessor can either be in digital form to actuate relays or converted into an analog signal (say, a voltage or a pneumatic pressure) to operate the actuator.

Direct digital control (DDC) is a means of control that involves using a microprocessor-based digital controller to perform mathematical operations according to predetermined control algorithms or computer programs. The key element of DDC control in comparing with the analog control, is the digital processing of data, that is, the digital computation performed in the microprocessor-based digital controller.

Figure 1.4 shows a DDC control system for an air-handling unit in a typical floor of the NBC Tower.

Pneumatic, Electric, and Electronic Control

According to the type of energy used to transmit the signal and operate the system, control systems can be classified as either *pneumatic, electric,* or *electronic*.

PNEUMATIC CONTROL SYSTEMS. This kind of control system uses compressed air to operate the sensors, controllers, and actuators and transmit the signals. A pneumatic control system consists of a compressed air supply and distribution system, sensors, controllers, and actuators. Figure 17.3 shows a typical pneumatic control system. In Fig. 17.3, a filter is used to remove dust particles, including submicron particles, contained in the air. The function of the pressure-reducing valve is to reduce the pressure of compressed air discharged from the compressor to the required value in the main supply line. The discharged compressed air is usually at a gage pressure of 18 to 25 psig, and the pressure signal required to actuate the valve or damper actuator is between 3 and 13 psig.

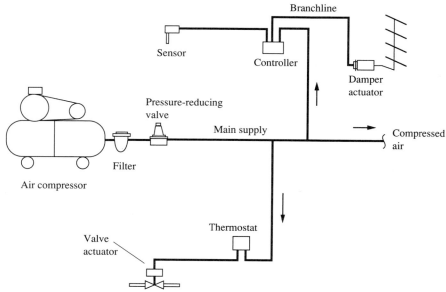

FIGURE 17.3 A typical pneumatical control system.

The advantages of pneumatic control systems are as follows:

- The compressed air itself is inherently a proportional control signal (which will be discussed later).
- The cost of modulating actuators is low, especially for large valves and dampers.
- They require less maintenance and have fewer problems.
- They are explosion-proof.

The disadvantages stem mainly from comparatively fewer control functions, the higher cost of sophisticated pneumatic controllers, and the comparatively higher first cost of a clean and dry compressed air supply for small projects.

ELECTRIC OR ELECTRONIC CONTROL SYSTEMS. Both of these types of control systems use electric energy as the energy source for the controllers. Relays, contactors, and electromechanical or solid-state actuators are used as control devices. Electric control systems often offer two-position on-off control. They are generally used for low-cost, small, and simpler control systems. Electronic control systems have accurate sensors, and solid-state controllers with sophisticated functions and can easily interface with the building management system. They are higher in cost and need skillful personnel for maintenance and trouble-shooting.

Comparison of Control Methods

Because of increasing demand for more complicated controllers to satisfy the needs of better indoor environmental control, improved energy savings, lower cost, and greater reliability, the recent trend is to use direct digital control (DDC). Current DDC controls consist of electronic sensors, a microprocessor-based controller, an analog-to-digital (A/D) or digital-to-analog (D/A) signal converter, and a pneumatic actuator (especially for large operators) or electric actuators. Electronic sensors and actuators can both be used for DDC control systems. It is often more convenient, effective, and cheaper to use analog output signals from a DDC controller to operate a pneumatic actuator, rather than using electric or electronic actuators in a large DDC system.

DDC offers the following advantages:

1. Flexibility of application and the ability to coordinate multiple functions directly from complicated software programs (it even mimics a human expert to a limited extent)

2. More precise and accurate control actions provided by the microprocessor-based controller

3. The possibility of using high-level self-checking and self-tuning system components, which increases the system reliability

4. Competitive first cost in comparison to other control methods such as pneumatic or electronic controls, that is, a lower cost per control function for DDC control

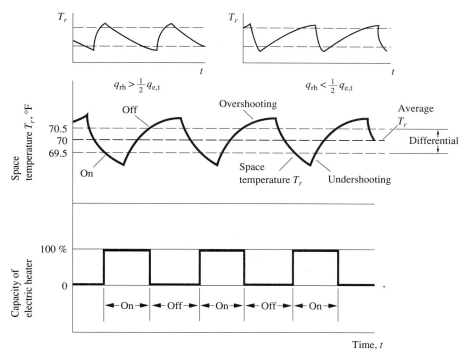

FIGURE 17.4 A typical two-position control mode.

One of the main disadvantages of DDC control is that is still so new. Nonetheless, many room air conditioners today use microprocessor chips to provide more sophisticated control functions. In the 1990s, the trend in HVAC&R control for new and retrofit projects is to use DDC systems. As more and more DDC control systems are installed, DDC will become familiar to us. The cost of microprocessor-based controllers and other DDC system components will drop further in the future.

17.3 CONTROL MODES

Control modes describe how the corrective action of the controller takes place as well as its effect on the controlled variable. For applications in HVAC&R, control modes can be classified as two-position, step, floating, proportional, proportional-integral, or proportional-integral-derivative.

Before discussing control modes in detail, the term *lag,* or *time lag,* should be introduced. According to ASHRAE terminology, lag is (1) the time delay required for the sensing element to reach equilibrium with the controlled variable or (2) any retardation of an output with respect to the causal input, including the delay, because of the transport of material or the propagation of a signal.

Two-Position Control

In *two-position control,* the controller controls the final control element at one of two positions: maximum and minimum (except during the short period when it changes position). Examples of two-position control include starting and stopping the motor of a fan, pump, or compressor; and turnng on or off an electric heater. Sometimes, it is also called *on-off control.*

Figure 17.4 shows a two-position control mode for an electric heater installed in a branch duct. In the middle diagram, the ordinate indicates the controlled variable, the space temperature T_r; in the lower diagram, the ordinate denotes the output capacity of the final control element, the electric heater. If the controller turns on the electric heater when the sensor senses a space temperature 69.5°F and turns off at 70.5°F, the result is a cyclic operation of the electric heater and the rise and fall of T_r toward the two positions 69.5°F and 70.5°F. The thermal storage effects of the electric heater, the branch duct, the building envelope that surrounds the space, and the sensor itself will all have a time lag effect on T_r. The rise in T_r is a convex curve with an overshoot higher than 70.5°F, and the drop of T_r is a concave curve with a undershoot lower than 69.5°F. The difference between the two points, on and off, is called the *differential.*

If the heating capacity of the electric heater that results from an on-and-off cyclic operation is represented by $q_{e,t}$, and if the actual space heating load $q_{rh} < \frac{1}{2}q_{e,t}$, the slope of the rising T_r curve will then be greater than the slope of the falling T_r curve. As a consequence, the "on" period will be shorter than the "off" period. Such a cyclic T_r curve is shown in the upper left corner of Fig. 17.4. If $q_{rh} > \frac{1}{2}q_{e,t}$, the condition will be reversed (see the upper right corner of Fig. 17.4). If $q_{rh} = \frac{1}{2}q_{e,t}$, the "on" period is equal to the "off" period. The two-position control system varies the ratio of on and off periods to meet any variation in the space heating load.

In order to reduce the overshoot and undershoot of the controlled variable in a two-position control mode, a modification called *timed two-position control* has been developed. A small heating element attached to the temperature sensor is energized during "on" periods. This additional heating effect on the sensor shortens the "on" timing. During "off" periods, the heating element is deenergized.

The differential of two-position control, the overshoot, and the undershoot all result in a fluctuation of the controlled variable. A suitable differential is always desirable in two-position control in order to prevent very short cycling, which causes *hunting*, a phenomenon of short cycling of the controlled variable. Two-position control is not suitable for precise control of the controlled variable, but it is often used for status control, such as opening or closing a damper, turning equipment on or off, etc., and for lower-cost control systems.

Step Control

In *step control,* the controller operates the relays or switches in sequence to vary the output capacity of the process plant element in steps or stages. The greater the deviation of the controlled variable from the set point, the higher the output capacity of the process plant.

Figure 17.5 depicts automatic staging of three reciprocating chillers by means of a simpler step control. The controlled variable is the entering chilled water temperature T_{ee}, and the set point is 75°F. As soon as T_{ee} exceeds 75.5°F, chiller No. 1 is turned on. If T_{ee} increases further to 76°F, chiller No. 2 will turn on. Chiller No. 3 will turn on when T_{ee} exceeds 76.5°F. If T_{ee} drops below 74.5°F, chiller No. 3 will switch off, and chillers No. 2 and No. 1 will be turned off at 74°F and 73.5°F, respectively.

The temperature difference between the "on" and "off" points for a particular chiller is the differential, and the temperature difference between the "on" point of the third chiller and the "off" point of the last chiller is the *throttling range,* which will be introduced in detail later in this chapter. The interval between the "on" points of two chillers has a direct influence on the magnitude of the throttling range. A smaller interval requires more precise control of the "on" and "off" points. Usually the interval ranges between 0.35 and 1°F.

A more elaborate step-control mode using DDC is now available. This will be covered in detail in Chapter 29. Step control has been widely used

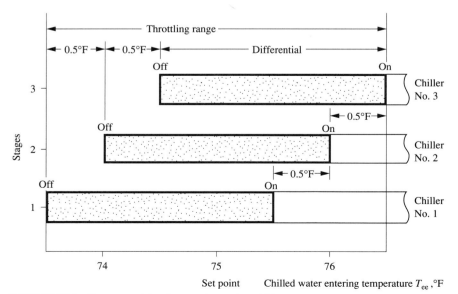

FIGURE 17.5 Step control for three reciprocating chillers.

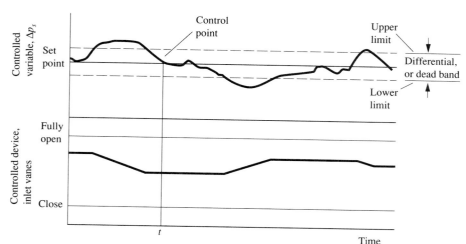

FIGURE 17.6 Floating control mode.

for multiple units and multiple-stage electric heaters operating in sequence.

Floating Control

In a *floating control* mode, the controller moves the controlled device by means of the actuator toward the set point only when the *control point* is out of the differential, or dead band, as shown in Fig. 17.6. A control point is the actual value of the controlled variable at a certain time instant.

Figure 17.6 presents a duct static pressure control system using a floating control mode by opening and closing the inlet vanes of a supply fan. The controlled devices, inlet vanes, can be moved in either an opening or closing direction depending on whether the control point over the upper limit of the differential or under the lower limit. In Fig. 17.6, when the controlled variable, the duct static pressure Δp_s, is outside the dead band above the upper limit of the differential, the controller then closes the inlet vanes. If Δp_s is outside the dead band below the lower limit of the differential, the controller opens the inlet vanes.

A floating control mode is more suitable for control systems with a minimal lag between the sensor and the control medium. A control medium is the medium in which the controlled variable exists. For example, floating control could also be used in a discharge air temperature control system in which the sensor is located immediately after the final controlled device or process plant (damper, coil, etc.) in the discharge air.

Proportional Control

In a *proportional control* mode, the controller moves the controlled device to a position such that the change in its output capacity is proportional to the deviation of the controlled variable from the set point. The position of the controlled device is linearly proportional to the magnitude of the controlled variable.

Figure 17.7 represents the control of the space temperature through the modulation of a two-way valve of a cooling coil using proportional control. In Fig. 17.7, the controlled variable is the space temperature T_r, and the controlled device is the valve.

In a proportional control mode, the *throttling range* is the change in the controlled variable when the controller moves the controlled device from the position of maximum output to the position of minimum output. The controlled variable's range of values that will move the proportional controller through its operating range is called the *proportional band*. In a proportional control system, the throttling range is equal to the proportional band.

The set point is the desired value of the controlled variable, or the desired control point that the controller seeks to achieve. The difference between the control point and the set point is called the *offset, or deviation*. In proportional control, when the controlled variable is at the bottom line of the throttling range, the controller will position the actuator at the closed position. At the set point, the actuator will be at 50 percent of the open position. When the controlled variable is at the top of the throttling range, the actuator will be at 100 percent of the open position.

In a proportional control mode, since the output signal V of the controller is proportional to the deviation of the control point from the set point, their relationship can be expressed as

$$V = K_p e + M \qquad (17.1)$$

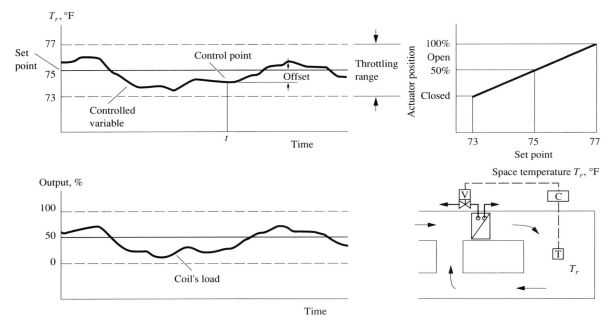

FIGURE 17.7 Proportional control mode.

where K_p = proportional gain, proportional to $1/(\text{throttling range})$

e = error signal, that is, the deviation, or offset

M = output value when the deviation is zero (usually, the output value at the middle of the output range of the controller)

For space and discharge air temperature control using proportional control, the offset is directly proportional to the space load, and the coil's load—the space or discharge temperature T, in °F—can be calculated as

$$T = R_{\text{load}}T_{\text{t,r}} + T_{\min} \qquad (17.2)$$

where $T_{\text{t,r}}$ = throttling range, expressed in terms of the space or discharge temperature, °F

$T_{\min}$ = space and discharge temperature when the space load or coil's load is zero, °F

In Eq. (17.2), R_{load} represents the load ratio of the space load or coil's load, which is dimensionless and can be evaluated as

$$R_{\text{load}} = \frac{\text{Actual load}}{\text{Design load}} \qquad (17.3)$$

The actual load and design load must be in same units, such as Btu/h.

For a proportional control mode, a certain degree of offset, or deviation, is inherent. Only when offset exists will the controller position the actuator and

the valve or damper at a position greater or smaller than 50 percent open. The throttling range or proportional band is primarily determined by the HVAC&R systems characteristics and cannot be changed after the system is designed. A proportional control mode is suitable for an HVAC&R system that has a large thermal capacitance, resulting in a slow response and stable system. For such a condition, the throttling range as well as the offset will be smaller. Proportional control that is too fast-acting or has a too small proportional band may resemble a two-position control mode and cause instability and hunting.

Proportional-Integral (PI) Control

In a proportional-integral control mode, a second component is added to the proportional action in order to eliminate the offset. The output of the controller can thus be expressed as

$$V = K_p e + K_i \int e\ dt + M \qquad (17.4)$$

where K_i = integral gain

t = time

In Eq. (17.4), the second term on the right-hand side (the integral term) indicates that (1) the error, or offset, is measured at regular time intervals and (2) the product of the sum of these measurements and K_i is added to the output of the controller to eliminate the offset. The longer the offset exists, the greater the response of the controller. Such a control action is

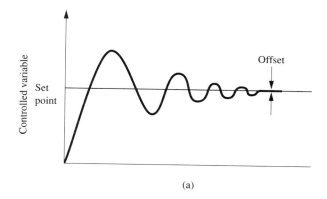

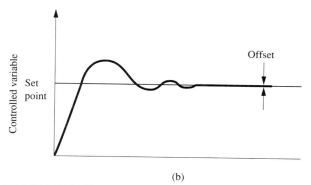

FIGURE 17.8 Proportional-integral and proportional-integral-derivative control mode: (a) proportional-integral (PI) and (b) proportional-integral-derivative (PID).

equivalent to resetting the set point in order to increase the controller output to eliminate the offset. As such, PI control is sometimes called as *proportional plus reset control*. Figure 17.8a shows the variations of a controlled variable for a proportional-integral control mode.

For PI control, the controlled variable does not have any offset once it has achieved a stable condition, except due to any inaccuracy in the instrumentation. Proper selection of the proportional gain K_p and integral gain K_i is important for system stability and control accuracy.

PI control may be applied to fast-acting control systems with a greater throttling range setting at the controller for better system stability—for example, discharge air temperature, discharge chilled water temperature, and duct static pressure control systems.

Proportional-Integral-Derivative (PID) Control

A PID control mode has additional control action added to the PI controller: a derivative function that opposes any change and is proportional to the rate of change. The output of such a controller can be described by the following equation:

$$V = K_p e + K_i \int e \ dt + K \frac{de}{dt} + M \qquad (17.5)$$

where K is the derivative gain.

The effect of adding the derivative function $K(de/dt)$ is that the quicker the control point changes, the greater the corrective action provided by the PID controller. Figure 17.8b shows the variation of the controlled variable for a PID control mode. As with a PI control mode, PID control also has no offset once the controlled variable has reached a stable condition except due to instrument inaccuracy.

Compared to a PI control mode, PID control exhibits faster corrective action and a smaller overshoot and undershoot following a change in the controller variable. The controlled variable is brought to the required set point in a shorter time. However, it is more difficult to determine properly the three constants, or gains (K_p, K_i, and K) for a PID controller.

Compensation Control, or Reset

Compensation control, or reset, is a type of control mode in which a compensation sensor is generally used to reset a main sensor to compensate for a variable change sensed by the compensation sensor. The purpose is to achieve operation that is more effective, energy-efficient, or both.

In the design of a reset mode, the first things to decide are the control point at which the main sensor will be reset and the variable to be sensed by the compensation sensor. The main sensor, which senses the mixed air temperature in the air-handling unit, is usually reset by a compensation sensor that senses the outdoor air temperature to compensate for a change in outdoor temperature, as shown in Fig. 17.9.

Another thing to decide is the relationship between the variables sensed by the main and compensation sensors—the *reset schedule*. Many reset schedules have a different relationship between these two sensors at various stages. For example, in Fig. 17.9, when the outdoor temperature T_o is less than 30°F, it is within the range of stage I. No matter what the magnitude of T_o, the set point of the mixed temperature is 65°F. When 30°F $\leq T_o \leq$ 95°F, it is in stage II. The linear relationship between the temperatures sensed by the main and compensation sensors in Fig. 17.9 can be expressed by the following equation:

$$T_m = 65 - \frac{65 - 55}{95 - 30}(T_o - 30) \qquad (17.6)$$

When $T_o > 95$°F, the set point of the mixed temperature is always 55°F.

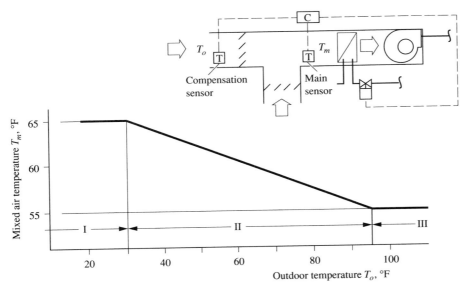

FIGURE 17.9 Reset control and schedule.

Compensation control modes have been widely adopted to reset space temperatures, discharged air temperatures from air-handling units, or water discharge temperatures from central plants.

Applications of Various Control Modes

Selection of a suitable control mode depends on

- operating characteristics
- process or system characteristics, such as whether the thermal capacitance should be taken into consideration
- characteristics of load changes

If a simpler control mode can meet the requirements (say, two-position on-off control versus PID control), the simpler control mode is always the best choice. It should be also noted that there is a significant difference between a proportional control mode with an inherent offset and the offset-free PI and PID controls.

For HVAC&R processes, PI control can satisfy most of the requirements. PID control mode is more appropriate for duct static pressure control and airflow control. In modern DDC systems, PI control and two-position control are widely used. When a PID control mode is used for air static pressure or airflow controls, it is recommended to set the controller with a large proportional band so as to provide control system stability, a slow reset to eliminate deviation, and a derivative action to provide a quick response.

17.4 SENSORS AND TRANSDUCERS

A sensor is a device that acts as a component in a control system to detect and measure the controlled variable and send a signal to the controller. A sensor consists of a sensing element and accessories, as shown in Fig. 17.10a (see page 17.14). The term *sensing element* often refers to that part of the sensor that actually senses the controlled variable. In HVAC&R systems, the most widely used sensors are temperature sensors, humidity sensors, pressure sensors, and flow sensors. Electronic sensors that send electric signals to electronic controllers can also be used for the DDC controllers. The current trend is to use solid-state miniature sensing elements. In the selection of sensors, accuracy, sensitivity of response, reliability, maintainability, and especially the possibility of contamination by dust particles due to contact with the control medium should be considered.

Temperature Sensors

Temperature sensors fall into two categories: those that produce mechanical signals and those that give electric signals. Bimetal and rod-and-tube sensors that use sensing elements to produce a mechanical displacement during a sensed temperature change either open or close an electric circuit in an electric control system or adjust the throttling pressure by means of a bleeding nozzle in a pneumatic control system. For a sealed bellows sensor, a change in temperature causes a change in the pressure of a liquid in a remote bulb. The expansion and contraction of vapor then moves the mechanism of the controller.

Temperature sensors that produce electric signals as shown in Fig. 17.10a are the same as the sensors for temperature measurement and indication mentioned in Section 2.3.

In addition to resistance temperature detectors (RTD) and thermistors, sensors also sometimes use thermocouples. A thermocouple uses wires of two dissimilar metals, such as copper and constantan or iron and nickel, connected at two junctions, to generate an electromotive force between the junctions that is directly proportional to the temperature difference between them. One of the junction is kept at constant temperature and is called the *cold junction*. Various system have been developed to maintain the cold junction at a constant temperature. This task makes the use of thermocouples more complicated and expensive. The electromotive force produced between the two junctions can be used as the signal input to a controller.

Bimetal and rod-and-tube temperature sensors are simple and reliable. However, at the same time, they cannot also be used as an electronic sensor in a system. Among electronic sensors, thermocouples are generally used for precision control and high-temperature industrial applications. Platinum and nickel RTDs are stable, reliable, and accurate. They need calibration to compensate for the effects of having external wires and are widely adopted in DDC. High-quality, drift-free, and interchangeable thermistors can provide stable, reliable, and cost-effective temperature sensors, too, and are also used in many applications.

Humidity Sensors

As mentioned in Section 2.8, humidity sensors fall into two categories: mechanical and electronic. When the same humidity sensor is used for both monitoring and control, electronic sensors, especially of the Dunmore resistance type and the capacitance type, are often adopted.

Pressure Sensors

A pressure sensor usually senses the difference in pressure between the controlled medium (air or water) and a reference pressure; or the pressure differential across two points, such as the pressure differential across a filter. The reference pressure may be an absolute vacuum, atmospheric pressure, or the pressure at any adjacent point. Output signals from pressure sensors may be pneumatic, electric analog, or on-off.

Pressure sensors used for HVAC&R can be divided into high-pressure and low-pressure sensors. High-pressure sensors measure in psi, and low-

pressure sensors measure in inches WG. The sensing elements for high-pressure sensors are usually Bourdon tubes, bellows, etc. For low-pressure sensors, large diaphragms or flexible metal bellows are usually used.

Flow Sensors

Flow sensors usually sense the rate of air and water flow in cfm for air and gpm for water. For airflow sensors, the average-velocity pressure p_v, in inches WG, is usually sensed and measured. The average air velocity v_a, fpm is then calculated according to Eq. (8.10) as

$$v_a = 4005K \sqrt{p_v} \qquad (17.7)$$

where K is the flow coefficient, which depends on the type of pitot tube array used and the dimensions of the round or rectangular duct. After that, the volume flow rate can easily be determined as the product of $v_a A$. Here, A represents the cross sectional area of the air passage, perpendicular to the airflow, in ft^2.

Various forms of pitot tube arrays have been developed and tested to determine the average velocity pressure of a rectangular or circular duct section by measuring the difference between the total and static pressures of the airstream. A typical pitot tube flow-measuring station with flow straighteners to provide more even airflow is shown in Fig. 17.10.

Electronic air velocity sensors such as hot-wire anemometers and thermistors have also been widely adopted to measure airflow, especially for VAV boxes. They need frequent calibration.

Water flow sensors of a differential pressure type, such as orifice plates, flowing nozzles, and pitot tubes have only a limited measurement range. Turbine or magnet-type flow meters can apply to a wider range, but they are more expensive and need periodic calibration.

Transducers (or Transmitters)

A transducer, or transmitter, is a device that converts energy from one form to another or amplifies an input or output signal. In HVAC&R control systems, a transducer may be used to convert an electric signal to a pneumatic signal (E/P transducer), for example, a pneumatic proportional relay that varies its branch air pressure from 3 to 15 psig in direct proportion to changes in the electrical input from 2 to 10 V. E/P transducers are also used between microprocessor-based or electronic controllers and pneumatic actuators. On the other hand, a pneumatic signal can be converted to an electric signal in a P/E transducer. For example, a P/E relay closes a contact when the

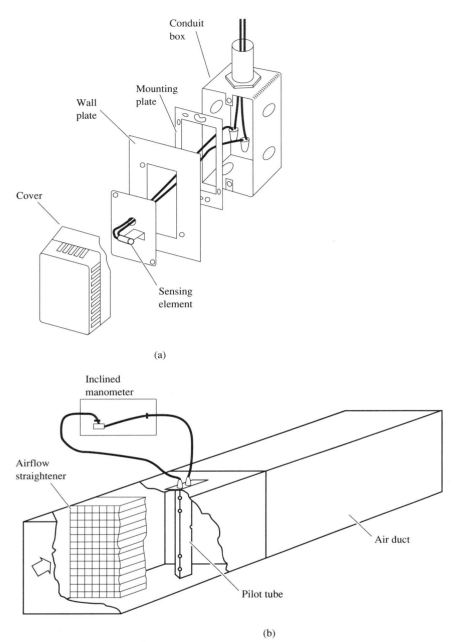

FIGURE 17.10 A temperature sensor (a) and a typical pitot tube flow-measuring station (b).

air pressure falls and opens the contact when the air pressure rises above a predetermined value.

A transducer or a transmitter may also be used to sense the concentration of carbon dioxide (CO_2) and then send an electric signal to the controller.

17.5 CONTROLLERS

A controller is a device that analyzes the inputs received from sensors or other sources, compares the controlled variable with the set point, and performs mathematical and other operations in accordance with a software program or a setup. It then sends an output signal to modulate the actuator so as to move the controlled variable toward the predetermined set point or to direct the controlled device to attain a required operating status.

A thermostat is a combination of a temperature sensor and a temperature controller, whereas a humidistat is a combination of a humidity sensor and a humidity controller.

Direct-Acting or Reverse-Acting

A *direct-acting* (DA) controller increases its output signal upon an increase in the sensed controlled variable, and it decreases its output signal upon a decrease in the sensed controlled variable. Conversely, a *reverse-acting* (RA) controller decreases its output signal upon an increase in the sensed controlled variable and increases its output signal upon a decrease in the controlled variable.

Normally Closed or Normally Open

A controlled device, a valve or damper, that is said to be in the *normally closed* (NC) position indicates that when the input signal to the controller falls to zero or below a critical value, the controlled device will be in the closed position. The closing of the valve or damper is most likely due to the action of a spring or supplementary power supply. A valve or a damper that is said to be *normally open* (NO) indicates that when the input signal to the controller fall to zero or below a critical value, the valve, damper, or associated process plant will be in the open position.

Pneumatic Controllers

In a pneumatic controller, the basic mechanism used to control the air pressure in the branch line supplied to the actuators is a nozzle-flapper assembly plus a restrictor. Figure 17.11 shows a pneumatic controller with such a mechanism. Compressed air sup-

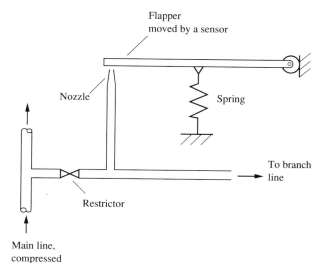

FIGURE 17.11 A pneumatic controller with a nozzle-flapper assembly and a restrictor.

plied from the main line flows through a restrictor and discharges at the opening between the nozzle and the flapper, which has a spring pulling it downward. The nozzle and restrictor are sized in such a manner that when the flapper moves away from the nozzle, all the air escapes from the nozzle and the branch-line pressure is zero. When the flapper covers the nozzle or if there is no airflow in the branch line, the pressure of the branch line is equal to that in the main line.

If a sensor, such as a bimetal sensor, moves the flapper upward according to the magnitude of the controlled variable sensed by the sensor, the input signal from the sensor determines the opening between the nozzle and flapper and hence, the compressed air pressure in the branch line.

For a direct-acting pneumatic temperature controller used to control the space temperature during summer, the branch-line pressure may change from 9 to 13 psig when the space temperature increases from 73°F to 77°F. This nozzle-flapper-assembly pneumatic controller operates in proportional control mode. Many other more complicated pneumatic controllers have been developed to perform other control modes and additional functions.

Electric and Electronic Controllers

An electric controller uses switches, relays, and bridge circuits formed by potentiometers to position the actuators in on-off, floating, and proportional control modes according to the input signal from the sensor and the predetermined set point. An electronic controller can provide far more functions than electric controllers. It may receive input signals from both the main sensor and compensation sensor with amplification and combination.

In the control circuit, an electronic controller basically provides proportional or proportional-integral control modes. The output signal from the controller can be used to position an actuator or to provide the sequencing of actuators, or to change to two-position, floating, or even PID control modes in conjunction with additional circuits. DDC controllers will be covered in detail in Chapter 18.

17.6 WATER CONTROL VALVES AND VALVE ACTUATORS

Water valves are used to regulate or stop water flow in a pipe either manually or by means of automatic control systems. Water control valves adopted in water systems can modulate water flow rates by means of automatic control systems.

Valve Actuators

An actuator, sometimes called an operator, is a device that receives a pneumatic or electric analog control signal from the controller, either directly or through a digital/analog converter. It then closes or opens a valve or damper, modulating the associated process plant, and causes the controlled variable to change toward its set point.

Valve actuators are used to position control valves. They are mainly of the following types:

SOLENOID ACTUATORS. These use a magnetic coil to move a movable plunger connected with the valve stem. Most solenoid valve actuators operate at two positions (on and off). They are used mainly for small valves.

ELECTRIC MOTORS. These move the valve stem by means of a gear train and linkage. Different electric-motor valve actuators can be classified based on the mode they use:

1. *On-off mode.* For this type of actuator, the motor moves the valve in one direction, and when the electric circuit breaks, the spring returns the valve stem to the top position (either open or closed, depending on whether it is a normally open or closed valve).
2. *Modulating mode.* The motor can rotate in both directions, with spring-return when the electric circuit breaks.
3. *Modulating mode with supplementary power supply.* The motor rotates in two directions and without a spring-return arrangement. When the power is cut off, a bypass signal is usually sent to the electric motor to drive the valve to its open or closed position, depending on whether it is a normally open or closed valve.

Unfortunately, it may take minutes to fully open a large valve using an electric-motor valve actuator.

PNEUMATIC ACTUATORS. A pneumatic valve actuator consists of an actuator chamber whose bottom is made of a flexible diaphragm or bellows connected with the valve stem. When the air pressure in the actuator chamber increases, the downward force overcomes the spring compression and pushes the diaphragm downward, closing the valve. As the air pressure in the actuator chamber decreases, the spring compresses the diaphragm, moving the valve stem and valve upward. A pneumatic valve actuator is powerful, simple, and reliable. It is widely used to actuate large valves.

Types of Control Valves

Water control valves consist mainly of a valve body, one or two valve discs or plugs, one or two valve seats, a valve stem, and a seal packing.

Based on their structure, water control valves can be classified into the following types:

1. *Single-seated.* A single-seated valve has only a single valve disc and seat, as shown in Figs. 17.12b and c. It is usually used for water systems that need a tight shutoff.
2. *Double-seated.* A double-seated valve has two valve discs connected to the same valve stem and is designed so that the fluid pressure exerted on the valve discs is always balanced. Consequently, less force is required for the operation of a double-seated valve, as shown in Fig. 17.12a.
3. *Butterfly.* A butterfly valve consists of a cylindrical body, a shaft, and a disc that rotates around an axis, as shown in Fig. 17.12d. When the valve closes, it seats against a ring inside the body. A butterfly valve exhibits low flow resistance when it is fully opened. It is compact and is usually used in large water pipes.

According to the pattern of the water flow, water control valves can again be classified as two-way valves or three-way valves. A two-way valve has one inlet port and one outlet port. Water flows straight through the two-way valve along a single passage as shown in Fig. 17.12a.

In a three-way valve, there are three ports: two inlet ports and one common outlet port for a three-way mixing valve, as shown in Fig. 17.12b, and one common inlet port and two outlet ports for a three-way diverting valve, as shown in Fig. 17.12c. In a three-way mixing valve, the main water stream flows through the coil or boiler, and the bypass stream mixes with the main stream in the common mixing outlet port. In a three-way diverting valve, the supply water stream divides into two streams in the common inlet port. The main water stream flows through the coil, and the bypass stream mixes with the main water stream after the coil.

A three-way mixing valve is always located downstream of the coil. On the other hand, a diverting valve is always located upstream of the coil.

A diverting valve should never be used as a mixing valve. The unbalanced pressure difference between the two inlet ports and the outlet port at a closed position may cause disc bouncing and valve wear when the valve disc travels between the two extremes.

Valve Characteristics and Ratings

The different types of control valves and the characteristics that are important during the selection of these valves are as follows:

1. *Equal-percentage valve:* A control valve that changes the water flow rate by a certain percentage for that same percentage of lift in the valve stem when the upstream versus downstream water pressure difference across the valve (its *pressure drop*) is constant.

2. *Linear valve:* A control valve that shows a directly proportional relationship between the flow rate and the lifting of the valve stem for a constant pressure drop

3. *Quick-opening valve:* A control valve that gives the maximum possible flow rate when the valve disc or plug is just lifted from its seat

Rangeability is defined as the ratio of the maximum flow rate to the minimum flow rate under control. An equal-percentage valve may have a very good range-

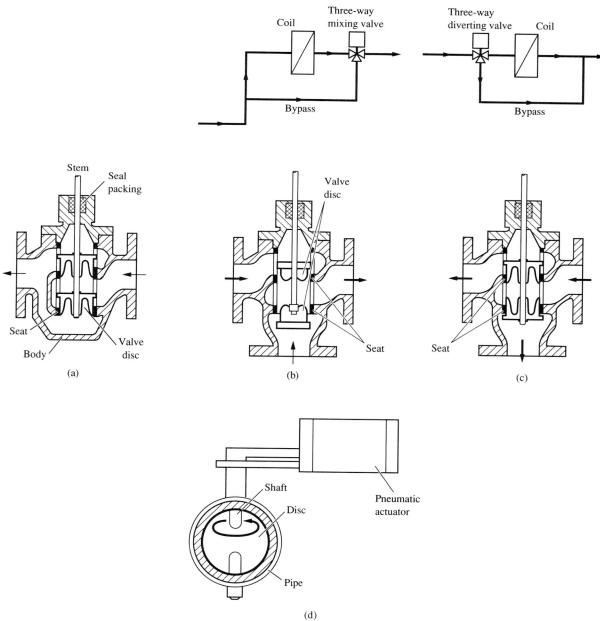

FIGURE 17.12 Various types of control valves: (a) double-seated two-way valve, (b) single-seated three-way mixing valve, (c) single-seated three-way diverting valve, and (d) butterfly valve.

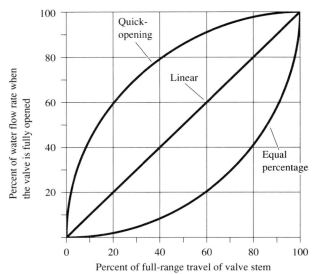

FIGURE 17.13   Typical flow characteristics of various types of control valve.

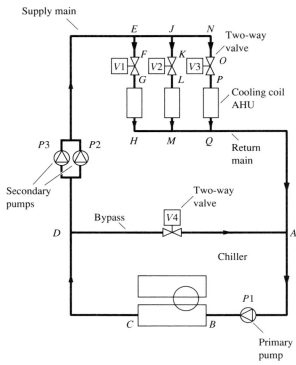

FIGURE 17.14   A typical chilled water system.

ability of 50 : 1. A linear valve may have a rangeability of 30 : 1.

The flow characteristics of equal-percentage, linear, and quick-opening valves are shown in Fig. 17.13.

The following control valve ratings should be considered during the selection and sizing of a valve:

1. *Body rating.* The *nominal* body rating of the valve is the theoretical rating of the valve body only, in psig. The *actual* body rating is the permissible safe water pressure for the valve body, in psig, at a specific water temperature.

2. *Close-off rating.* That is the maximum pressure difference between the inlet and outlet ports that a valve can withstand without leakage when the valve is fully closed, in psi.

3. *Maximum pressure and temperature.* These are the maximum pressure and temperature of water that the whole valve, including body, disc, seat, packing, etc., can withstand.

Valve Selection

Proper selection of water control valves depends on water system performance, load variations, pipe size, control modes, etc.

Ideally, a linear relationship between a change in the controlled variable and the amount of travel of the valve stem, or a linear system control characteristic over the operating range, is most desirable. Hence, a linear valve is often used for water systems for which

the controlled variable has a linear relationship with the water flow or valve opening; or for applications that do not have wide load variations.

When a control valve is used to modulate the water flow rate of a hot or chilled water coil, a large reduction in the flow rate causes only a small reduction in the heating or cooling output of the coil. Given such circumstances, the nonlinear behavior of an equal-percentage valve combined with the nonlinear output performance of hot or chilled water coil will provide more linear system behavior.

When three-way valves are being used, the water flow rate before or after the common port is approximately constant, no matter how wide the openings of the various ports in the three-way valves. As such, three-way valves are used in constant or approximately constant water flow rate systems for which the coil's load changes.

As a two-way valve closes, the flow rate of the water system decreases and its static pressure increases. A two-way valve is thus used for water systems that have variable volume flow during a variation in system load, as shown in Fig. 17.14.

Valve Sizing

The size of a control valve is closely related to its water flow rate $\dot{V}$, in gpm, and the pressure drop across the valve Δp_{vv}, in psi. Their relationship can be expressed as

$$\dot{V} = C_v \sqrt{\Delta p_{vv}}$$

$$C_v = \frac{\dot{V}}{\sqrt{\Delta p_{vv}}} \qquad (17.8)$$

where C_v is the flow coefficient for a flow rate of 1 gpm at a pressure drop of 1 psi. The flow coefficients of control valves can be found in manufacturers' catalogs.

For a water system that has several cooling coils between the supply and return mains, as shown in Fig. 17.14, modulating the water flow is effective only when opening and closing the control valve has a significant effect on the change in the pressure drops Δp_{EH}, Δp_{JM}, or Δp_{NQ} across piping sections *EH, JM,* or *NQ* between the supply and return mains of Fig. 17.14. As Δp_{EH} decreases, if there is no concurrent change in flow resistance for piping sections *JM* and *NQ*, a greater water flow will then pass through piping section *EH*. If Δp_{EH} increases, less water will flow through *EH*. If Δp_{EH}, Δp_{JM}, and Δp_{NQ} all increase, based on the characteristic curve of the centrifugal pump, the water flow rate through pump P2 or P3 will decrease for an increase in the system head. If Δp_{EH}, Δp_{JM}, and Δp_{NQ} all drop, the water flow rate through pump P2 or P3 will increase.

For a piping loop such as *ABCDA* in Fig. 17.14 and a change in the pressure drop across the two-way valve V4, the water flow rates flowing through the coils and the bypass line will be changed accordingly. According to the previous analysis:

1. For any piping loop, the pressure drop Δp_{vv} for the control valve should be greater than or equal to the pressure drop through the coil (or other heat exchanger) plus the pressure drop of all piping and fittings within this piping loop at design operating conditions.

2. For a section of piping loop between the supply and return main pipes, the pressure drop Δp_{vv} for the control valve should be greater than or equal to the pressure drop through the coil plus all the pressure drops of the piping and fittings connecting the coil to the supply and return mains in that section at design operating conditions.

3. For a two-way valve mounted on a bypass line, the size of the valve should be determined according to the value of Δp_{vv} when the water flow through the bypass line is at a maximum.

Example 17.1. A water system supplies chilled water to the cooling coil of three air-handling units (AHUs),

as shown in Fig. 17.14. At design flow conditions, the pressure drops of the cooling coil, pipeline, and fittings (excluding the pressure drop of valve V1 between the supply main pipe at point *E* and the return main pipe at point *H*) are as follows:

Cooling coil	5 ft WG
Pipe fittings	8 ft WG
Pipe line	10 ft WG

For the chiller loop *ABCDA*, the pressure drops are as follows:

Chiller	20 ft WG
Pipe fittings	10 ft WG
Pipe line	5 ft WG

If the flow rate of the chilled water flowing through two-way valve V1 is 150 gpm and the rate through both P1 and P2 is 600 gpm (P3 is a standby pump), select and size valves V1 and V4, respectively.

Solution

1. Because the relationship between the coil's load and the water flow rate at the cooling coil is nonlinear, in order to achieve a linear system control, an equal-percentage valve is selected for valve V1.

2. Between point *E* of the supply main pipe and point *H* of the return main pipe, the total pressure loss is $5 + 8 + 10 = 23$ ft, and the pressure drop expressed is $23 \times 0.433 = 9.96$ psi. Let the pressure drop across V1 be equal to this total pressure loss. Therefore, from Eq. (17.8), the flow coefficient is

$$C_v = \frac{\dot{V}}{\sqrt{\Delta p_{vv}}} = \frac{150}{\sqrt{9.96}} = 48$$

An equal-percentage valve from a manufacturer's catalog can be sized with $C_v = 48$ for V1.

3. For valve V4 in the bypass line, the relationship between the valve stem travel of V4 and the controlled variable, the pressure differential between *D* and *A*, is directly proportional within a certain operating range. Select a linear valve for valve V4.

4. When valves V1, V2, and V3 are all closed, all the chilled water from the chiller will flow through valve V4 on the bypass line. The total pressure drop for the chiller loop is $20 + 10 + 5 = 35$ ft WG, or $35 \times 0.433 = 15.15$ psi. Then, the flow coefficient for valve V4 is

$$C_v = \frac{600}{\sqrt{15.15}} = 154$$

A linear valve V4 can be sized from a manufacturer's catalog with a $C_v = 154$.

In actual operation, when valves V1, V2, and V3 all are closed, the chiller and the plant pump P1 should be turned off.

17.7 DAMPERS AND DAMPER ACTUATORS

A dampers is a device that controls the airflow in an air system or ventilating systems by changing the angle of its blades and therefore the area of its flow passage.

In HVAC&R systems, dampers can be divided into volume-control dampers and fire dampers. Fire dampers were covered in Chapter 8. In this section, only volume-control dampers will be discussed.

Types of Volume-Control Dampers

Volume-control dampers can be classified as *single-blade* dampers or *multiblade* dampers according to their construction. Various types of volume control dampers are shown in Fig. 17.15.

BUTTERFLY DAMPERS. A butterfly damper is a single-blade damper. A butterfly damper is made from either a rectangular sheet mounted inside a rectangular duct or a round disc placed in a round duct, as shown in Fig. 17.15*a*. It rotates about an axle and is able to modulate the air volume flow rate of the duct system by varying the size of the opening of the flow passage for the airflow.

GATE DAMPERS. A gate damper is a single-blade damper. It also may be rectangular or round in shape.

It slides in and out of a slot in order to shut off or open up a flow passage, as shown in Fig. 17.15b. Gate dampers are mainly used in industrial exhaust systems with high static pressure.

SPLIT DAMPERS. A split damper is also a single-blade damper. It is a piece of movable sheet metal that is usually installed at the Y-connection of a rectangular duct system, as shown in Figure 17.15c. The movement of the split damper from one end to the other modulates the volume flow of the air flowing into the two legs or branches. A split damper is usually modulated only during air balancing after installation or during periodic air balancing.

OPPOSED-BLADE DAMPERS. An opposed-blade damper is a type of multiblade damper that is often rectangular in shape, as shown in Figure 17.15d. It is usually used for a flow passage of large cross-sectional area. The damper blades may be made from galvanized steel, aluminum alloy, or stainless steel sheets, usually not exceeding 10 in. in width. Rubber or spring seals can be provided at the fully closed position to control the air leakage rating, which usually does not exceed 50 cfm/ft^2 at a static pressure drop across the damper of 3 in. WG. The bearing used to support the blade axle should be made of a corrosion-resistant material such as copper alloy or nylon. Teflon-coated bearings may also be used

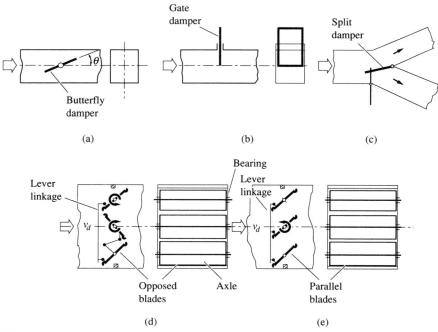

FIGURE 17.15 Various types of volume control dampers: (a) butterfly damper, (b) gate damper, (c) split damper, (d) opposed-blade damper, and (e) parallel-blade damper.

to ensure a smooth operation of the damper. Lever linkages are used to open and close the damper blades. The characteristics of opposed-blade dampers will be covered later in this section.

The maximum static pressure drop across closed opposed-blade dampers is 6 in. WG for a 36-in.-long damper (the length of the damper blade) and 4 in. WG for of 48-in.-long damper.

PARALLEL-BLADE DAMPERS. A parallel-blade damper is also a type of multiblade damper used mainly for large cross-sectional areas, as shown in Fig. 17.15e. The blade material and the requirements for the seals and bearings are the same as those for opposed-blade dampers.

Damper Actuators

Damper actuators are used to position dampers according to a signal from the controller. As with valve actuators, damper actuators can be classified as either electric or pneumatic:

1. *Electric damper actuators.* These are either driven by electric motors in reversible directions or are unidirectional and spring-returned. A reversible electric actuator is used more often for more precise control. It has two sets of motor windings. When one set is energized, the actuator's shaft turns in clockwise direction, and when the other set is energized, the actuator's shaft turns in a counterclockwise direction. If neither motor winding is energized, the shaft remains in its current position. Such an electric actuator can provide the simplest floating control mode, as well as other modes if required.

2. *Pneumatic damper actuators.* Their construction is similar to that of pneumatic valve actuators, but the stroke of a pneumatic damper actuator is longer. They also have lever linkages and crank arms to open and close the dampers.

Volume Flow Control between Various Airflow Paths

For air conditioning control systems, most of the dampers are often installed in parallel-connected airflow paths to control their volume flow, as shown in Fig. 17.16. The types of airflow volume control are as follows:

1. *Mixing air control.* In Fig. 17.16a, there are two parallel airflow paths: the recirculating path *um*, in which a recirculating air damper is installed, and the exhaust and intake path *uom*, in which exhaust and outdoor air dampers are installed. The outdoor

air and recirculating air are mixed together before entering the coil. Both the outdoor damper and the recirculating damper are located just before the mixing box are often called *mixing dampers*.

2. *Bypass control.* In the flow circuit for bypass control, as shown in Fig. 17.16b, the entering air at the common junction m_1 is divided into two parallel air flow paths: the *bypass path*, in which a bypass damper is installed, and the *conditioned path*, in which the coil face damper is installed in series with the coil, or the washer damper with the air washer. The bypass and conditioned airstreams are then mixed together at the common junction m_2.

3. *Branch flow control.* In a supply main duct that has many branch take-offs, as shown in Fig. 17.16c, there are many parallel airflow path combinations: paths *b1s1* and *b1b2s2*, *b2s2*, *b2b3s3*, etc. In each branch flow path, there is a damper in the VAV box, and *s1*, *s2*, *s3*, etc., are the status points of the supply air.

Parallel airflow paths such as those shown in Fig. 17.16 have the following characteristics:

1. The total pressure losses of two airflow paths that connect the same endpoints are always equal; for example, $\Delta p_{um} = \Delta p_{uom}$, $\Delta p_{m1\ by\ m2} = \Delta p_{m1\ con\ m2}$, etc.

2. If the total pressure loss of one parallel path increases, the airflow through this path will reduce and the airflow in other parallel paths will increase until the pressure losses of these flow paths are again equal to each other.

3. The total pressure loss of an airflow path between two common junctions determines the volume flow rate of air passing through that path.

4. In most parallel airflow paths installed with damper(s), the duct friction loss is often small and is usually negligible, except for the pressure losses of flexible ducts and duct fittings.

Flow Characteristics of Opposed- and Parallel-Blade Dampers

A parallel-blade or an opposed-blade damper that is installed in a single airflow path to modulate its airflow is often called a throttling damper or a volume-control damper. For throttling dampers, a linear relationship between the percentage of the damper opening and the percentage of full flow is desirable. (Full flow is the air volume flow rate when the damper is fully opened at design conditions.) The actual relationship is given by the installed characteristic curves of parallel-blade and opposed-blade dampers shown

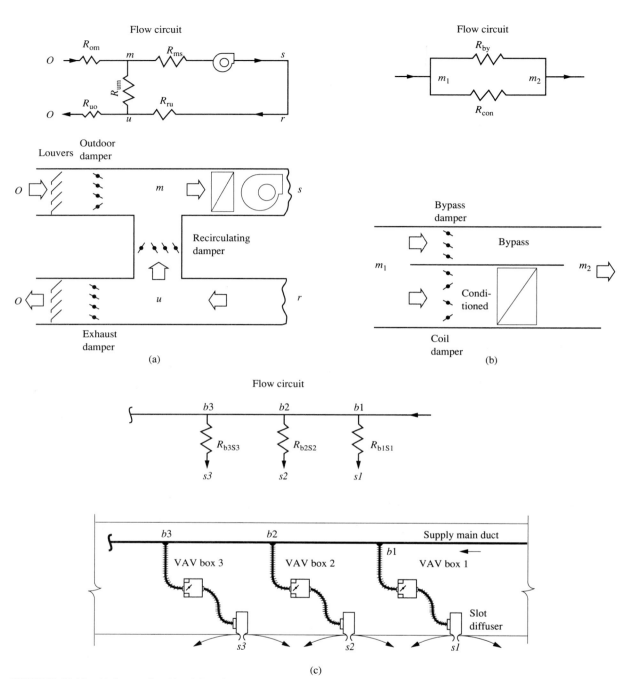

FIGURE 17.16 Airflow paths: (a) mixing air control, (b) bypass control, and (c) Branch flow control.

in Figs. 17.17a and b. For the sake of energy savings, it is also preferable to have a lower pressure drop when air flows through the damper at the fully open condition.

In Fig. 17.17, α is called the *damper characteristic ratio* and may be calculated as:

$$\alpha = \frac{\Delta p_{path} - \Delta p_{od}}{\Delta p_{od}} = \frac{\Delta p_{path}}{\Delta p_{od}} - 1$$

$$= \frac{\Delta p_{p\text{-}od}}{\Delta p_{od}} \qquad (17.9)$$

where Δp_{path} = total pressure loss of the airflow path, in. WG

Δp_{od} = total pressure loss of the damper when it is fully opened, in. WG

$\Delta p_{p\text{-}od}$ = total pressure loss of the airflow path excluding the damper, in. WG

For parallel-blade dampers, in order to have a linear relationship between the percentage opening and the percentage of full flow, $\alpha = 2$. For opposed blade dampers, $\alpha = 10$.

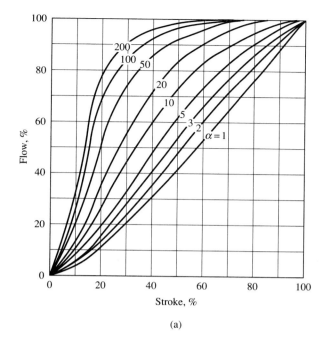

(a)

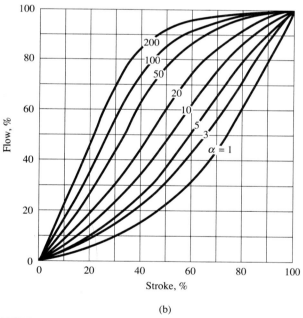

(b)

FIGURE 17.17 Flow characteristic curves for dampers: (a) parallel-blade and (b) opposed blade.

Flow Characteristics of Mixing and Bypass Dampers

When outdoor air is used for free cooling in an air economizer, the outdoor and recirculating dampers are always interlocked and operated in opposition to each other. When one of them is fully opened, the other will be fully closed, and vice versa.

For a parallel-blade damper, when the stroke to open the damper blades is at its midpoint, there is an open area between blades typically equal to 60 to 70 percent of the area when the dampers are wide open. If parallel-blade dampers are used for both outdoor air dampers and recirculating dampers in mixing air control, the open area at midpoint for both dampers is 120 to 140 percent of the open area when one of them is wide open and the other is fully closed. The result is a smaller pressure drop across the mixing dampers and an increased volume flow rate when the stroke is at midpoint.

For an opposed-blade damper at midpoint, its open area is typically only 20 to 30 percent of its fully open area. If opposed-blade dampers are used for outdoor damper and recirculating damper in mixing air control, the results are a greater pressure drop across the dampers at the mid point opening and a smaller volume flow rate.

Tests conducted by a damper manufacturer showed that the control characteristics of outdoor and recirculating dampers in mixing air control can be changed by mixing parallel-acting and opposed-acting blades in the same damper, which results in a parallel-opposed blade damper. Figure 17.18 shows the pressure drops for a parallel-opposed blade damper at various degrees of opening of the blades. The fraction of opposed blades F_{blade} can be calculated as

$$F_{blade} = \frac{\text{Number of opposed blades}}{\text{Total number of blades}} \quad (17.10)$$

Based on the test results, the pressure drop across the dampers remains approximately constant for various degrees of blade opening when $F_{blade} = 0.2$ to 0.25.

Based on the recommendations made by Alley (1988), when the face area A_{dam} of the damper, in ft^2, is smaller than the ducted area A_d of the airflow passage, in ft^2, the local loss coefficient C_{dam} of the damper for different setups can be determined from Fig. 17.19 (see page 17.25).

Instead of using a parallel-opposed blade damper, a nearly constant pressure drop across the mixing dampers for various degrees of blade opening can also be achieved in mixing air control by using an all-parallel bladed damper with additional specially shaped baffle-type components.

For face and bypass dampers like those shown in Fig. 17.16b, the pressure drop across the dampers will be more even if opposed blades are used for coil face dampers and parallel blades are adopted for use as bypass dampers. The system pressure drop across the face damper will shift from the coil to the opposed-blade face damper when the conditioned airflow is reduced. Such an arrangement also provides more linear system control characteristics.

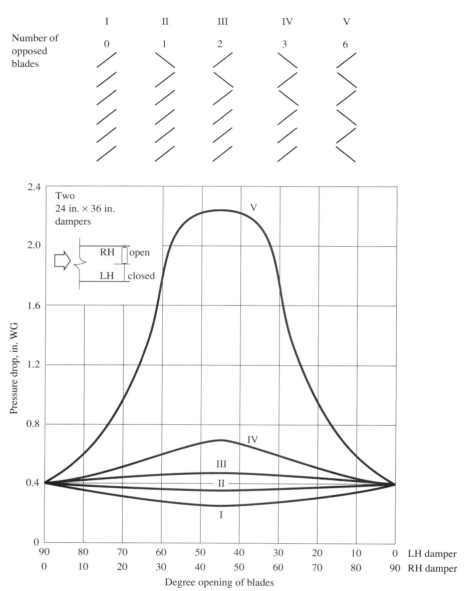

FIGURE 17.18   Flow characteristics of parallel-opposed blade dampers. (Adapted from *ASHRAE Transactions 1988, Part I*. Reprinted by permission.)

Damper Selection

For proper selection of dampers, consider the following:

1. Butterfly dampers are usually used in ducts of small cross-sectional area or in places like VAV boxes.

2. For throttling dampers (that is, volume control dampers in a single airflow path) in order to have a linear relationship between the percentage of damper opening and the percentage of full flow,

an opposed blade damper is recommended if many dynamic losses other than the damper itself (such as a coil or air washer, heat exchanger, louvers, etc.) exist in the airflow paths. If the damper is the primary pressure drop in the airflow path, a parallel-blade damper is recommended.

3. For mixing dampers, in order to have a nearly constant pressure drop across the dampers when the interlocked outdoor air and recirculating dampers are open to varying degrees, parallel-opposed blade dampers are recommended, with the fraction of opposed blades F_{blade} from 0.2 to 0.25.

Damper positions:

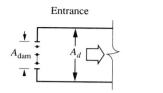
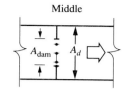

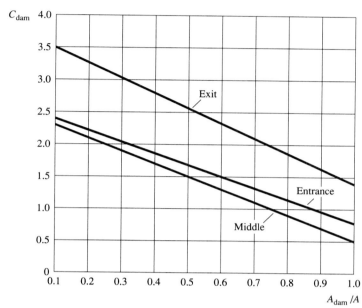

FIGURE 17.19 Local loss coefficient C_{dam} of air damper. (*Source: ASHRAE Transactions 1988, Part I.* Reprinted by permission.)

4. For face and bypass dampers, an opposed-blade coil face damper and a parallel-blade bypass damper will give better control quality.

5. For two-position control dampers, a parallel-blade damper is always used for the sake of lower energy consumption.

Damper Sizing

The damper sizing procedure for a volume control damper is as follows:

1. Calculate the total pressure loss of the airflow path $\Delta p_{\text{p-od}}$ according to Eq. (8.50). In order to achieve a linear relationship, the total pressure loss across the damper Δp_{od} should be determined as follows:

$$\Delta p_{\text{od}} = \frac{\Delta p_{\text{p-od}}}{2} \quad \text{for a parallel-blade damper}$$

$$\Delta p_{\text{od}} = \frac{\Delta p_{\text{p-od}}}{10} \quad \text{for an opposed-blade damper} \quad (17.11)$$

2. The pressure drop across the damper when the damper is fully opened Δ/p_{od}, in inches WG$_O$ can be calculated as

$$\Delta p_{\text{od}} = C_{\text{dam}} \left(\frac{v_{\text{dam}}}{4005}\right)^2 \quad (17.12)$$

and the face velocity of the damper v_{dam}, in fpm, is given as

$$v_{\text{dam}} = \frac{\dot{V}_{\text{dam}}}{A_{\text{dam}}} \quad (17.13)$$

where $\dot{V}_{\text{dam}}$ is the air volume flow rate flowing through the damper, in cfm.

3. Depending on whether the damper is at the entrance, exit, or middle of the flow path, and on the area ratio A_{dam}/A_d, the local loss coefficient C_{dam} can be found from Fig. 17.19. The face area of the damper can then be calculated as

$$A_{\text{dam}} = 0.00025 \, \dot{V}_{\text{dam}} \left(\frac{C_{\text{dam}}}{\Delta p_{\text{od}}}\right)^{0.5} \quad (17.14)$$

The area ratio A_{dam}/A_d is usually between 0.5 and 0.9.

For mixing dampers, if the face velocity of the damper v_{dam} can be determined beforehand, the face area of the outdoor and recirculating air dampers can be calculated from Eq. (17.13) as $A_{dam} = \dot{V}_{dam}/v_{dam}$. Once the area ratio A_{dam}/A_d is selected, C_{dam} can be obtained from Fig. 17.19 and the pressure loss Δp_{od} across the damper can be calculated from Eq. (17.12).

The face area for a coil face damper will be approximately the same as the cross-sectional area of the coil. For the bypass damper, the pressure loss across the fully opened damper when the coil face damper is fully closed should be nearly equal to the pressure loss when the face damper is fully opened and the bypass damper is fully closed. The face velocity of the bypass damper is often several times higher than than that of the face damper.

Example 17.2 For the mixing air control system shown in Fig. 17.16a, the volume flow rates are as follows:

	Volume flow, ft^3/min
Outdoor air flow path	1500–10,000
Exhaust air flow path	1500–10,000
Recirculating flow path	8500–0

Select the type of damper best suited for these flow paths. Also calculate the damper areas of the outdoor, exhaust, and recirculating dampers.

Solution

1. For mixing dampers, in order to have a nearly constant pressure drop Δp_{od} across the dampers when the flow rates of the outdoor and recirculating air are varied, parallel-opposed blade dampers should be selected.

2. Let the face velocity of the outdoor air damper be 1500 fpm. The face area of this damper based on the maximum volume flow rate is

$$A_{dam} = \frac{\dot{V}_{dam}}{V_{dam}} = \frac{10,000}{1500} = 6.67 \text{ ft}^2, \text{ or } 960 \text{ in.}^2$$

If the size of this damper is selected to be 24 in. × 40 in., then for a blade width of 6 in., there will be 4 blades, one of which is opposed. The fraction of opposed blades will be $F_{blade} = 1/4 = 0.25$.

3. For factory-built air-handling units, the outdoor air damper is located at the exit of the outdoor air duct. If the area ratio $A_{dam}/A_d = 0.7$, Fig. 17.19, yields $C_{dam} = 2.1$. Then, from Eq. (17.12), the total pressure loss across the outdoor air damper is

$$\Delta p_{od} = C_{dam}\left(\frac{v_{dam}}{4005}\right)^2 = 2.1\left(\frac{1500}{4005}\right)^2$$
$$= 0.295 \text{ in. WG}$$

4. The exhaust air damper is usually located at the middle of the exhaust duct. If the size of the exhaust damper and the area ratio A_{dam}/A_d are the same as for the outdoor damper, then Fig. 17.19 yields $C_{dam} = 1.1$. The pressure loss across the exhaust damper is then

$$\Delta p_{od} = C_{dam}\left(\frac{v_{dam}}{4005}\right)^2 = 1.1\left(\frac{1500}{4005}\right)^2$$
$$= 0.154 \text{ in. WG}$$

5. The recirculating damper is often at the exit of the recirculating airflow path. When the fully opened recirculating air damper has a maximum volume flow rate of 8500 cfm, the outdoor air damper has a volume flow rate of 1500 cfm. For mixing dampers, the pressure drop across the recirculating damper should be equal to that across the outdoor damper, 0.295 in. WG.

If the area ratio A_{dam}/A_d for the recirculating damper is 0.6, then Fig. 17.19 yields $C_{dam} = 2.3$. Then, the face velocity v_{dam} of the recirculating damper is

$$v_{dam} = 4005\left(\frac{\Delta p_{od}}{C_{dam}}\right)^{0.5} = 4005\left(\frac{0.295}{2.3}\right)^{0.5}$$
$$= 1434 \text{ cfm}$$

and the face area A_{dam} of the recirculating damper is

$$A_{dam} = \frac{\dot{V}_{dam}}{v_{dam}} = \frac{8500}{1434} = 5.93 \text{ ft}^2, \text{ or } 854 \text{ in.}^2$$

Thus, the size of the recirculating damper is determined to be: 24 in. × 36 in. It has four blades, each of which has a width of 6 in. One of the blades is an opposed blade.

REFERENCES AND FURTHER READING

Alley, R. L., "Selecting and Sizing Outside and Return Air Dampers for VAV Economizer Systems," *ASHRAE Transactions,* Part I, pp. 1457–1466, 1988.

Asbill, C. M., "Direct Digital vs Pneumatic Controls," *Heating/Piping/Air Conditioning,* pp. 111–116, November 1984.

ASHRAE Inc., *ASHRAE Handbook 1991, HVAC Applications,* Atlanta, GA.

Grimm, N. R., and Rosaler, R. C., *Handbook of HVAC Design 1990,* McGraw-Hill, New York.

Haines, R. W., "Proportional Plus Integral Control," *Heating/Piping/Air Conditioning,* pp. 131–132, January 1984.

Haines, R. W., "Reset Schedules," *Heating/Piping/Air Conditioning,* pp. 142–146, September 1985.

Honeywell Inc., *Engineering Manual of Automatic Control for Commercial Buildings,* Minneapolis, MN, 1988.

Petze, J., "Understanding Temperature Sensing Methods and Myths," *Heating/Piping/Air Conditioning,* pp. 193–208, November 1986.

CHAPTER 18

ENERGY MANAGEMENT SYSTEMS WITH DIRECT DIGITAL CONTROL

18.1 SYSTEM ARCHITECTURE

Building Automation Systems and Energy Management Systems

A building automation system (BAS), as mentioned in Section 17.1, is a centralized monitoring, operation, and management system for HVAC&R, lighting, fire protection, security, and other building services that is used to achieve more efficient operations.

An energy management system (EMS) optimizes the operations, conditioning processes, and indoor environmental parameters of HVAC&R systems in order to maintain a satisfactory indoor environment at minimum energy use. An energy management system is a part of a building automation system.

ASHRAE/IES Standard 90.1–1989 and the DOE Standard specify that an EMS should be considered for buildings larger than 20,000 ft^2 gross area and should provide the following features:

- Gas and electricity individually metered, monitored, and recorded except for those buildings less than 5000 ft^2 gross area
- For HVAC&R equipment with energy input exceeding 20 kVA or 60,000 Btu/h, a facility to measure their input and output performace individually
- Means to read, to record, and to summarize their energy consumption
- Optimum starting and stopping of the HVAC&R systems, and indication of their operating time schedule
- Monitoring and resetting of HVAC&R control system to ensure proper operation
- Off-hour readily accessible means and energy-efficient operation schedule

An HVAC&R EMS also includes the following additional functions. The first four functions are also required by ASHRAE/IES Standard 90.1–1989.

- Free cooling economizer cycle
- Unoccupied period setback
- Dead-band control
- Supply air and space air temperature reset
- Optimal preoccupancy warm-up and cool-down
- Duty cycling
- Chiller optimization
- Boiler optimization
- Ability to investigate worn or faulty equipment or component
- Load limiting and shifting, kW demand control
- Submetering of gas and electricity end-use for building tenants

Architecture of a Typical EMS with a Direct Digital Control System

In 1988, *Heating/Piping/Air Conditioning* sponsored a survey among 16 manufacturers of EMSs for HVAC&R. The average sales volume percentages in current lines of HVAC&R control devices and systems are as follows: 72 percent DDC-related, 15 percent on-off, and 13 percent pneumatic. Nearly 80 percent of the manufacturers will still produce pneumatic actuators after pneumatic controls have been phased out. Some of them will produce only large pneumatic actuators. DDC is the trend in HVAC&R EMS control in medium and large buildings for now and the coming future.

Due to the rapid development of microprocessor technology, there have been significant changes in DDC systems during the past few years. Figure 18.1 shows the system architecture of a current typical EMS incorporating DDC. Such a DDC system has three main levels:

1. *Building level.* This is where an operator may interface with the EMS through the use of a keyboard/mouse, monitor, printer, and central personal computer (PC). The powerful PC, which is the operator workstation, is the predominant tool. The operator interface is where people can interface with the DDC panel and terminal DDC controllers as well as the HVAC&R, lighting, fire protection, and security control systems. The basic functions of the central computer are (1) to upload the status of all points from region and zone levels, (2) to download set points, and (3) to override (correct) commands, time schedules, and control software if necessary. Many other functions, such as alarm handling, color graphics displays, and remote access, are also provided.

2. *The region,* or *floor(s), level.* This level corresponds to specific DDC panels. A *DDC panel* is a combination of many controllers. It also has an on-board capacity, programmed by an operator, to execute complicated HVAC&R and other control programs. A DDC panel interfaces with numerous field sensors/transducers by means of input/ouput (I/O) connections. Sometimes, terminal controllers are configured on a separate trunk and connected to the DDC panels. DDC panels can coordinate communications among themselves. Because a DDC panel can execute software programming and implement control functions all by itself, it is often called a *stand-alone panel.* It is the brain of modern DDC systems.

   A *slave panel* is a DDC panel that does not have the on-board computer capacity to execute

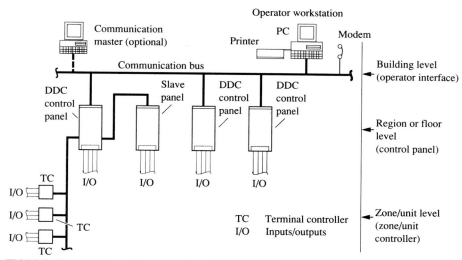

FIGURE 18.1   System architecture of a typical EMS with DDC.

software and the corresponding control functions. It is only an expansion of the I/O point connections of a DDC panel and an expansion of that DDC panel's software executing capacity.

A DDC panel is generally used for the control of chillers, boilers, air-handling units (AHUs), packaged units (PUS), especially built-up units, and any other equipment that need custom-made control programs. The use of multiple DDC panels also reduces the effect on a DDC system when one of them fails to operate.

3. *Zone/unit level.* This level is controlled by a DDC terminal controller, or simply DDC controller. A DDC controller is a small and specialized controller whose software is mostly preprogrammed. Only the time schedules, set points, and tuning constants can be changed by the user. Some of the most recently developed DDC controllers are also programmable to a limited extent. For any DDC controller, the manufacturer provides a variety of preprogrammed control sequences, and designers can specify the required control sequence that best fits their designs.

DDC controllers are often used in factory-made small AHUs and PUs, fan-coils, VAV terminal boxes, etc.

An architecture of EMS system incorporating DDC such as that in Fig. 18.1 adheres to a distributed processing model; that is, most of its control is at the terminal DDC controller level. The zone/unit level is separate from the numerous communication and data processing functions, which greatly simplifies system operation. A DDC controller can be installed in the equipment at the factory with the necessary preprogramming, only set points need to be entered at the field location. This approach is more cost-effective than field installation.

Data Communication Protocols

A *protocol* refers to the rules by which two or more devices communicate with each other. Two or more devices that transmit data to each other must obey the same rules.

Data communication between multiple DDC panels can be divided into two categories: *poll-response* and *peer-to-peer*. A poll-response, or master-slave, configuration uses a master communication computer that periodically polls status, alarm, and other required data from the DDC panels. Each DDC panel has an assigned address and responds when its address appears on the bus. The advantage of poll-response data communication is that it requires less memory and a lower processing capacity. Its main disadvantage is its dependence on the communication master, since all communication must flow through the master.

In peer-to-peer data communication, each panel can communicate with any other panel on the peer-to-peer level. Each DDC panel has a bus-connecting device connected to a communication bus, as shown in Fig. 18.1. A direct message containing the address of the specific bus-connecting device issued by one of the DDC panels will lead the message to that addressed DDC panel. The main advantage of peer-to-peer data communication is that it is more reliable because it is not dependent on the single master.

The cabling used in the network is either shielded twisted-pair or coaxial cable. Shielded twisted-pair is the economical choice. Copper conductors of 16- to 24-gauge are commonly used in single buildings. Bus lengths up to 4000 ft are common. Coaxial cable is more expensive and is generally used for greater distances and a larger capacity.

18.2 DDC SYSTEM COMPONENTS

Sensors

Resistance temperature devices (RTDs) with positive temperature coefficients are commonly used as temperature sensors in DDC systems. Platinum, silicon, and other materials are used for the sensing elements. A typical space air temperature sensor might be as follows:

Sensing element	Platinum film element, 3000 ohms Positive temperature coefficient, 4.8 ohms/°F
Set point range	60 to 90°F

For a 50-ft lead wire used to connect such an RTD temperature sensor to a DDC panel or zone/unit controller, the error due to the additional resistance of the lead wire is about 0.07°F.

A solid-state humidity-sensing element measures the relative humidity and sends a proportional voltage to the DDC panel or DDC controller. A typical space air humidity sensor might be as follows:

Power source	12 VDC ± 0.5 VDC
Humidity range	10 to 80 percent
Output range	1.14 to 1.93 VDC
Accuracy (70°F)	±3 percent at 10 to 60 percent range ±4 percent at 60 to 80 percent range
Speed of response	90 percent response time at 8 minutes
Mounting	Wall mount

Pressure sensors used in DDC systems are usually transmitters that convert a pressure differential change to an electric signal, such as a voltage, to be used as an analog input to the DDC panel or DDC controller. A change in the pressure differential compresses or stretches a flexible diaphragm as well as a strain gauge. The change in the total resistance of the gauge is detected and amplified, and this signal is sent to the controller. A typical differential pressure transmitter might be as follows:

Differential pressure range	0 to 1.33 in. WG
Supply voltage	5 VDC
Output signal	1 to 2 VDC
Maximum pressure at pressure connections	1 psig
Diaphragm material	Rubber

Space air sensors should be located so that the sensing element will be well exposed to the air and the air can flow through the sensor without obstructions. Air sensors should not be affected directly by supply air jet. Sensors should be shielded from radiation and mounted firmly on a structure member or duct wall, free from vibration.

Structure of DDC Panels and Controllers

HARDWARE. Inside a DDC panel, there are usually two printed circuit boards. One of them is called the *control board*, which contains one or more microprocessors and the necessary memory to execute control programs. Another board is called the *input/output (I/O) board*, which locates the I/O connections for field services. A slave DDC panel, or a DDC controller usually contains only one board.

MEMORY. Many control boards use a 16-bit microprocessor and have 1 megabyte of battery-backed memory for use in storing control programs and user databases. The types of memory included in a DDC panel and DDC controller are as follows:

- Read-only memory (ROM), which stores the software provided by the manufacturer and should not be modified by the user. This type of memory is nonvolatile.

- Random-access memory (RAM), which stores custom control software developed during installation or prepared by the user. This type of memory is volatile (i.e., it can be read from and written to) and requires battery backup.

- Electrically erasable programmable read-only memory (EEPROM), which stores custom control software and is volatile. The advantage of EEPROM over RAM is that EEPROM does not need battery backup. However, EEPROM cannot be used for as many writing/erasing/rewriting cycles as RAM.

INPUT/OUTPUT (I/O). There are four kinds of I/O points: *analog input* (AI); *binary*, or *digital, input* (BI); *analog output* (AO); and *binary*, or *pulsed, output* (BO). In the latest DDC panels, universal points, that is, points that can handle any kind of signal, are also provided. Typical analog input signals are 0 to 10 VDC or 4 to 20 mA. In a two-board DDC panel,

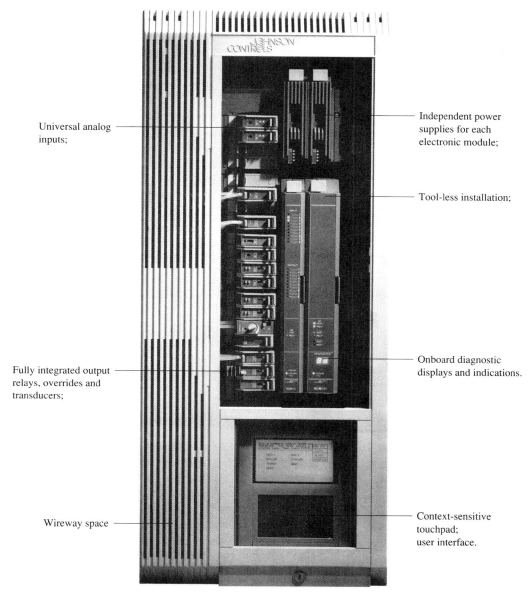

Universal analog inputs;

Independent power supplies for each electronic module;

Tool-less installation;

Onboard diagnostic displays and indications.

Fully integrated output relays, overrides and transducers;

Wireway space

Context-sensitive touchpad; user interface.

FIGURE 18.2   A typical DDC panel. (Source: Johnson Controls. Reprinted by permission.)

there are usually 20 to 40 I/O points, which can be extended to 100 points if necessary. A DDC controller has a limited number of I/O points (typically 8 inputs and 8 outputs or less). A typical DDC panel has the following I/O points capacity:

18   analog/digital inputs
 6   universal analog/digital input/outputs
12   digital outputs
 4   totalizer inputs, that is, pulsed inputs

Figure 18.2 shows a DDC panel.

DDC CONTROLLERS. The new generation of DDC controllers have greater memory to handle complicated control software and provide time and calendar scheduling, data storage, and other functions. In a DDC controller, there may be 4 to 20 input/outputs altogether. Inputs can take the following forms:

Analog inputs	0 to 10 VDC
Digital	switch, relay, or transistor
Microbridge sensor	0 to 3 in. WG pressure differential

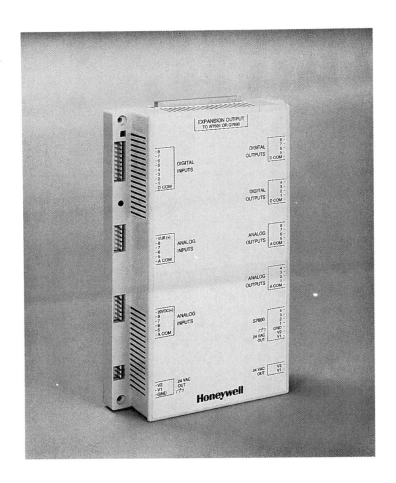

FIGURE 18.3 A typical DDC controller. (Source: Honeywell Inc. Reprinted by permission.)

Outputs may be in the following forms:

Analog	0 to 10 VDC, up to 20 mA
Digital	30 VAC
Pneumatic	3 to 15 psig
Triac	on-off output for an electric heater, fan motor, etc.

Figure 18.3 shows a DDC controller.

TUNING DDC PANEL AND CONTROLLERS. Most DDC panels and DDC controllers use proportional-integral (PI) control modes. The controller's proportional gain and integral gain should be properly selected for optimum control. Tuning DDC panels and DDC controllers typically includes selecting the appropriate proportional and integral gains and time interval to assess controller performance as well as selecting other compensation terms. A high gain may result in unstable control, whereas a low gain will produce a slow response. Unfortunately, the proportional and integral gains selected at design full-load operation are not necessarily correct at part-load operation. Computer software for self-tuning of DDC panels and DDC controllers has been developed and will be covered in Section 18.5.

Valve and Damper Actuators

These components are the same as those covered in Chapter 17.

18.3 PROGRAMMING FOR DDC SYSTEMS

Programming Language

Once an HVAC&R system is designed and the control sequence of operation has been determined by the designer, a specification is written for bidding. The successful control contractor would submit control drawings that show each of the control devices connected to the control system. For a DDC system, computer software also will be prepared by a control engineer working as the manufacturer's representative or by the control contractor. After installation, the HVAC&R and the control system will be commissioned and the DDC panel or DDC controller will be tuned. It will then be handed over to the operator for normal operation.

During operation, in addition to modification of the set points and time scheduling, the custom-made software may sometimes require modification to make improvements. The control engineer should keep this in mind while preparing the custom-made software, and the operator need to understand the programming to some degree for the sake of improving system performance.

In DDC panels, a programming language is provided in the erasable programmable read-only memory (EPROM). The language most commonly provided is BASIC, though PASCAL is also sometimes used. The majority of the public-domain computer programs available for HVAC&R control are written in BASIC, which is a language known to some degree by many potential users. Of course, other programming languages can be specified instead of BASIC if the designer feels that doing so is necessary. Simplicity and ease of preparation and understanding for users and installers are always important factors in selecting a programming language.

Computer software or modifications can be downloaded from an operator workstation or loaded to a DDC panel or DDC controller by means of additional portable computers or portable programming terminals.

Flow Charting and the Graphic Programming Language

Yaeger (1986) recommends an I/O summary that includes hardware inputs and outputs and a flow chart which defines the functional details of the control sequence of operation as well as the step-by-step order of control algorithms. Since DDC control functions are far more complicated than pneumatic control, a DDC flow chart incorporating symbols that represent

DDC operators will be extremely helpful to control engineers and programmers during the preparation of computer software.

A leading U.S. control product manufacturer has recommended a kind of simplified programming method known as the *graphics programming language* (GPL). Programming using GPL proceeds as follows:

1. A more complicated system in need of control, such as an air system, chilled water system, or boiler, is divided into many basic control processes. Each control process, such as mixing air control, duct static pressure control, or cooling and dehumidifying process control, represents a control function and may contain one or more control loops.

2. On a computer screen, the basic control loop of the control process is selected and illustrated. Required engineering input data are specified by the control engineer or operator.

3. With the help of presetting programs in the computer, additional lines and symbols are added to the basic control loop to represent signal selection, control logic, and mathematical operations. Programming for the control process is thus complete, as illustrated by the completed logic diagram on the screen.

4. Each control process is represented by one of the component control logic diagrams or one unit on the screen. The control logic diagrams and their relationships, as shown by the connecting lines, determine the computer programs for the system as well as their operation.

The detail programs and mathematical operations are predetermined by the manufacturer, but a control engineer using GPL can clearly see, for a given set of conditions, how a selection is made and which factors will influence the final output, and then determine the value of that output. GPL is also helpful during troubleshooting.

18.4 HVAC&R PROCESSES AND BUILDING ENVELOPE

The characteristics of an air conditioning or HVAC&R process are of vital importance to the optimum control and maintenance of a required indoor environment in buildings. The following process characteristics should be considered during DDC systems design and operation.

Load

The term *load* refers to the magnitude of the space load, coil's load, refrigeration load, or boiler load that determines the amount of supply air, chilled water, or hot water needed to control and maintain the controlled variable at the desirable value(s). Load variation affects the controlled variable in three circumstances: part-load operation, intermittent operation, and disturbances.

PART-LOAD OPERATION. The sizes of an air conditioning system and its components are always selected based on the magnitude of the load for the design condition, which is often called the *design load* or *full load*. In actual operation, most air conditioning systems and their respective components operate at part-load operation most of the time. Many air systems spend 85 to 90 percent of their annual operating hours at part-load operation. The load can drop below the design load because of

1. Changes in the outdoor climate
2. Changes in the internal loads at the time of operation

For part-load operation, the output of an air conditioning system must be appropriately reduced to maintain a desirable indoor environment; the DDC system should control the air conditioning system in the most efficient way possible.

INTERMITTENT OPERATION. Many air systems do not operate continuously. Some operate only few hours within a diurnal cycle. Some operate only in the daytime and shut down at night when the building is unoccupied. During the warm-up and cool-down periods, space loads vary a a great deal from the design load as well as from a continuous part-load. They represent transient loads of a dynamic model. Optimum starting and stopping of intermittently operated air systems is important for better indoor environmental control and energy savings.

DISTURBANCES. These are load changes that affect the controlled variable. A sudden disturbance is a sudden load change that occurs within a short time period, say, one hour. The offset of a controlled variable resulting from a disturbance can be eliminated by a DDC system with a proportional-integral control mode. For sudden disturbances resulting from a sudden climate change that affects a 100 percent outdoor air-handling unit, or from a sudden switching on of the spot lights in a conditioned space, a DDC system that incorporates proportional-integral-derivative control mode may be more suitable.

Thermal Capacitance and Lag

Thermal capacitance, which is sometimes called thermal inertia, is related to the mass of the system components and the building envelope. It is usually calculated as the product of the specific heat of the material and the mass of the material. The *lag* is the time delay that occurs before the control system can provide a corresponding and complete response to a change in the controlled variable. The thermal capacitance of the HVAC&R process and of the building envelope influence the performance of a control system as follows:

First, the high thermal capacitance of the building envelope, the equipment, chilled and hot water, etc., reduce the effect a disturbance has in varying the controlled variable. For example, consider a sudden increase in the lighting load in a conditioned space. As the space air temperature increases because of the heat from the electric lights, a large amount of heat will at the same time be transferred to the building envelope because of the additional temperature difference between the space air and the building structure. The increase in the space temperature resulting from the disturbance is thus considerably reduced.

Second, during the warm-up and cool-down periods, because of the high thermal capacitance of the building envelope, heat exchanger, ducts, and pipes, a greater system capacity and longer period of time are needed to create a certain change in a controlled variable such as the space air temperature or discharge temperature of air and water.

Third, the influence of the weather and external load changes on conditions in the indoor space air through the building envelope is rather slow, often needing several hours to fully take effect. Furthermore, the final variation in the space air temperature and relative humidity are reduced because of the high thermal capacitance of the building envelope. However, any influence on the indoor space air because of internal load changes or changes in the supply air temperature occurs considerably more quickly.

Finally, because of the thermal capacitance of the sensing element and the motion of the fluid flow, there is always a time lag between when the change in the controlled variable occurs, when that change is sensed by the sensing element, when the controlled device is modulated, when the capacity of the process plant varies, etc.

18.5 CONTROL STRATEGIES

Because of the use of microprocessor-based controllers, many new control strategies have been developed through the use of computer software for EMSs that incorporate DDC systems. Among these strategies, the following should be considered for better control and energy conservation.

Predictive Control

The concept of *predictive* or *dynamic control* for an HVAC&R EMS with DDC was introduced by Hartman (1988). Predictive control predicts the upcoming weather, utilizes the thermal capacitance of the building envelope, and ensures that all HVAC&R system components work in unison to provide a comfortable indoor environment with energy-efficient operation.

Predictive control has the following main characteristics:

1. It employs free heating energy sources, such as recirculating air that has already been heated and free cooling energy sources, such as outdoor air with an enthalpy or temperature lower than that of the recirculating air.

   For an air system that conditions a space including perimeter zones, let the heating balance temperature $T_{h,bal}$, in °F, denote the outdoor temperature at which the space internal load balances both the heat transmission loss through the building envelope and the amount of heat required to be added to the minimum outdoor air intake $\dot{V}_{o,min}$, in cfm (see Fig. 18.4). Also, let the cooling balance temperature $T_{c,bal}$, in °F, denote the outdoor temperature at which the space internal load balances the heat absorbed by the maximum outdoor air intake $\dot{V}_{o,max}$, in cfm. When $T_o > T_{h,bal}$, free heating is available and when $T_o < T_{c,bal}$, free cooling is available. In Fig. 18.4, q_{int} indicates the space

internal load, q_{tran} the transmission loss, and q_o the heat absorbed by the minimum outdoor air intake; all of them are in Btu/h. The term ρ represents the air density, in lb/ft^3.

The use of free cooling and free heating should be maximized, and the use of refrigeration and primary heat energy, such as energy from gas, oil, and electric heaters, should be minimized.

2. Predictive control establishes the basic operating mode of the air system, cooling mode or heating mode, for that operating day. It also predicts an upcoming day's maximum outdoor temperature $T_{o,max}$ and a day's minimum outdoor temperature $T_{o,min}$.

   In cooling mode operation, precooling should be used before the normal conditioning period so as to precool the mass temperature of the building envelope T_{en} to a temperature that depends on $T_{o,max}$. The higher the value of $T_{o,max}$, the lower the value of T_{en}, and vice versa. The thermal capacitance of the building envelope at a lower temperature would delay the use of refrigeration. In heating mode operation, the recirculating air should be used to elevate T_{en} in order to delay the use of primary heating energy.

3. Predictive control allows the conditioned space average temperature to drift between acceptable limits that still feel comfortable to most occupants (for example, T_r between 74 and 76°F for cooling mode operation, and T_r between 70 and 72°F for heating mode operation.)

Predictive control is more effective in areas where the outdoor temperature in winter is low and free

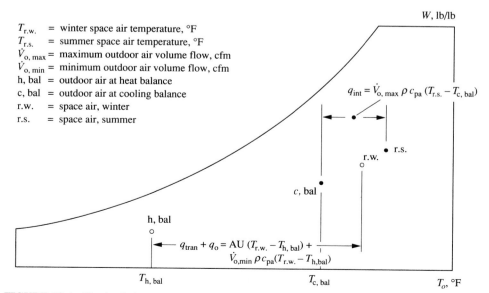

$T_{r.w.}$ = winter space air temperature, °F
$T_{r.s.}$ = summer space air temperature, °F
$\dot{V}_{o,max}$ = maximum outdoor air volume flow, cfm
$\dot{V}_{o,min}$ = minimum outdoor air volume flow, cfm
h, bal = outdoor air at heat balance
c, bal = outdoor air at cooling balance
r.w. = space air, winter
r.s. = space air, summer

W, lb/lb

$q_{int} = \dot{V}_{o,max}\, \rho\, c_{pa}\, (T_{r.s.} - T_{c,bal})$

r.s.

r.w.

c, bal

h, bal

$q_{tran} + q_o = AU\,(T_{r.w.} - T_{h,bal}) + \dot{V}_{o,min}\, \rho\, c_{pa}(T_{r.w.} - T_{h,bal})$

$T_{h,bal}$ $T_{c,bal}$ T_o, °F

FIGURE 18.4 Heating balance temperature $T_{h,bal}$ and cooling balance temperature $T_{c,bal}$.

cooling is available for a longer operating period annually.

Adaptive Control

Adaptive control is a type of control strategy in which the system learns from experience with system operation and adapts its control parameters by means of programmed software. In HVAC&R control systems, many such control parameters vary as the system load changes.

Adaptive control provides the following functions:

- Self-tuning of proportional and integral gains
- Resetting of set points, proportional band, and time scheduling
- Energy management control (for example, optimum starting and stopping)

With adaptive control, the DDC system will provide optimal control for various system loads and air and water flow conditions.

Expert Systems

An *expert system* is a computer program that mimics human expertise in a given knowledge domain, asks questions to obtain desirable data, uses mathematical operations to perform calculations, and takes advantage of chaining of knowledge, data, and logic to reach its best solution to a problem.

For an expert system, Van Horn (1986) lists the following characteristics:

- All the decision rules and all the data used to solve the problem are not necessarily represented in terms of numbers and algebraic equations.
- For any set of data, there may be more than one solution considered.
- Missing data do not cease execution of the computer program; a solution is still reached even if the user doesn't have all the needed data.
- The computer program assigns solutions in order of certainty; if much data are missing, the expert system will provide a solution of low certainty.

In an expert system, if a type of software structure known as an *expert system shell* is adopted, then adding to and changing the existing knowledge domain do not require changing the entire expert system program. An expert system shell consists mainly of a rules editor, an inference engine, a knowledge base, and a user interface. Figure 18.5 shows the structure of an expert system for an EMS with DDC (the dotted lines represent HVAC&R system design concerns). The components of an expert system shell have the following functions:

1. The rules editor is used to develop the rules for the knowledge base.

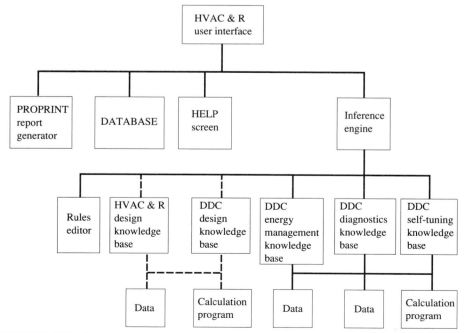

FIGURE 18.5 Structure of an expert system for an EMS with DDC and for HVAC&R system design.

2. The inference engine executes reasoning algorithms by applying the knowledge from the knowledge base, asks for inputs through the user interface, and makes conclusions based on the rules. The *mode of reasoning* reflects the way in which the program is organized. There are two kinds of reasoning: forward and backward. In most engineering design problems, forward reasoning is adopted. For instance, knowledge of the building envelope leads to load calculations, which leads next to determining the equipment capacity, and so on. In troubleshooting, backward reasoning is used. For instance, collecting facts leads to determining the cause of the problem.

The *mode of chaining* describes the method that the inference engine uses to prove or disprove the knowledge-base rules. Backward chaining tries to minimize the number of questions asked during a test, whereas forward chaining tries to minimize the number of irrelevant solutions tested.

3. The knowledge base is the most important part of an expert system. It contains all of the knowledge of the expert system. This knowledge is in the form of If-Then-Else rules. For an office building, for example if the local summer outdoor design wet bulb $T_o \le 60°F$, direct evaporative cooling is always the choice for summer cooling. Every rule in the knowledge base is based on either experience, published handbooks, engineering manuals, field surveys, etc. Data files are called and made available to use in accordance with a program contained in the rules of the knowledge base.

The structure of the knowledge base must be able to accommodate growth, because knowledge acquisition is a continuous accumulation process.

4. The user interface allows the user to interact with the expert system shell. Use of either a natural language for input and output or a menu-type interface is acceptable.

In addition to the expert system shell, there are also auxiliary programs:

- PROPRINT is mainly a text processor. It reads data files and prints the information in a desirable format.

- DATABASE is a database editor for random-access files.

- HELP is an on-line help screen that contains instructions on how to use the programs of the expert system and how to use the shell program, as well as information about the knowledge base.

An expert system can be used for energy management, energy audits, DDC system diagnostics, and self-tuning, as well as DDC and HVAC&R systems

design. An expert system offers great potential. It will not replace human experts, but it is still a very helpful tool for its users.

18.6 SINGLE-ZONE ZONE TEMPERATURE DDC SYSTEMS

Single-Zone Zone Temperature Control

For a single-zone zone temperature DDC system, the control system applies to either a large, single-zone room or one individual zone in a multizone conditioned space. ASHRAE/IES Standard 90.1–1989 requires that the supply of heating or cooling to each zone should be controlled by an individual control system. It consists mainly of a sensor, a DDC controller, controlled device(s) for fan-coils, fan-power terminal units, VAV boxes, reheating VAV boxes, etc., as well as an air-handling unit or packaged unit in the case of a large, single-zone conditioned space such as auditorium, arena, or indoor stadium.

For the sake of energy saving, ASHRAE/IES Standard 90.1–1989 also requires the following for zone temperature control:

1. *Setpoint.* The zone setpoint for space heating should be capable of being set down to 55°F or lower, and for space cooling it should be capable of being set up to 85°F or higher.

2. *Dead band control.* The dead band is the range of the controlled variable within which a variation in the sensed input to the controller does not initiate any noticeable change in the output. In other words, the process plant will be neither heating nor cooling, or the cooling or heating will be minimized when the controlled variable is within the dead band.

For both comfort heating and cooling, zone temperature control must be capable of providing a dead band of at least 5°F except for special occupancy and usage.

3. *Prevention of simultaneous heating and cooling.* The zone temperature or humidity control system must be capable of supplying cooling and heating in sequence and to prevent:

- Reheating
- Recooling
- Mixing of simultaneous supply of air cooled by refrigeration and air heated by gas, oil, or electricity.

Exceptions for prevention of simultaneous heating and cooling include a variable-air-volume system, or a zone peak supply flow rate of 300 cfm or less, or zone have specified humidified level, or 75% of heating energy is recovered.

Terminals

A *terminal unit*, or simply a *terminal*, is a piece of equipment that is installed directly in the conditioned space or above the space. A terminal unit may heat, cool, clean, mix supply air with recirculating air, or modulate the volume flow of supply air in order to maintain the zone temperature within preset limits. Fan-coil units, fan-power units, and variable-air-volume (VAV) boxes are all terminal units.

Fan-Coil Units

Consider a single-zone zone temperature DDC system that uses a fan-coil unit as a terminal, as shown in Fig. 18.6. A fan-coil unit typically consists of a small centrifugal fan, a two-row cooling coil, and a one-row heating coil. For a control zone located in the perimeter of a typical floor, year-round operation can be divided into three modes: summer cooling, dead band, and winter heating.

The system heat gain created by the fan power of the centrifugal fan is about 0.5°F, and the supply-duct heat gain after the fan-coil unit is small and thus negligible.

When the space air temperature $T_r \geq 75$°F, the fan-coil operates in summer cooling mode. If the outdoor weather is at summer design conditions, since T_r is higher than 75°F, it opens the modulating two-way valve on the chilled-water supply line V1. The chilled water then enters the cooling coil. At the same time, the modulating two-way valve on the hot-water supply line V2 remains closed. The mixture of the conditioned primary air and the recirculated air approaches the coil at state point m. The mixture is cooled and dehumidified at the cooling coil and leaves the coil at point cc. It is then supplied to the conditioned space, where it absorbs heat and moisture along the space conditioning line $cc\ r$, and reaches the space air condition r, as shown by a solid line in Fig. 18.6.

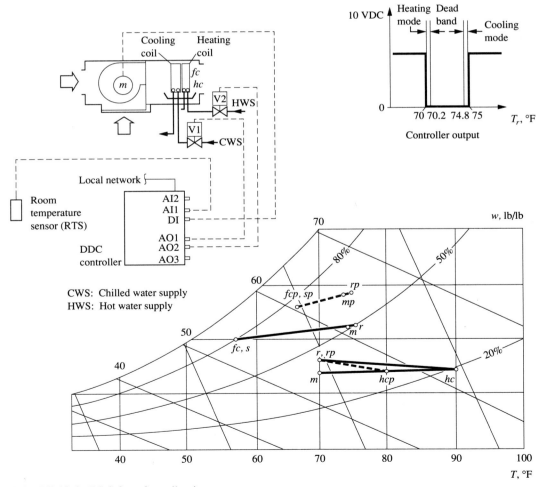

CWS: Chilled water supply
HWS: Hot water supply

FIGURE 18.6 DDC for a fan-coil unit.

At part-load operation, the space sensible cooling load reduces. The drop in the space temperature at part-load T_{rp} due to the reduction of the space cooling load is sensed by the room temperature sensor (RTS), and a signal is sent to the DDC controller at the analog input AI. The DDC controller compares this signal with the set point, say 75°F, and sends an analog output, a reduced DCV (direct-current voltage) from AO1 to the valve actuator of the two-way valve V1 for the cooling coil. The valve is then closed to a smaller opening. This action will reduce the chilled water mass flow rate flowing through the coil as well as reduce the rate of heat transfer from the coil to the airflow over the coil. The result is an increase in the temperature of chilled water leaving the coil and thus a rise in air off-coil temperature or the supply-air temperature at part-load T_{sp}. For a PI control mode, T_{sp} increases until the zone temperature at part-load is again maintained at $T_{rp} = 75$°F.

An increase in the temperature of the chilled water leaving the cooling coil at part-load also raises the humidity ratio w_{fcp} of the conditioned air off the cooling coil. Therefore, conditioned air will be supplied to the space at a higher temperature and higher humidity ratio, point fcp, and the result will be a higher value of φ_{rp}, the zone relative humidity at part-load. Usually, at part-load operation, the sensible heat ratio (SHR$_p$) of the space conditioning line $fcp\,rp$ is smaller than at the design condition. Here the subscript p indicates the state point at part-load. This will further increase φ_{rp}. If φ_r at the design condition is 50 percent, the value φ_{rp} will then be greater than 50 percent. According to the analysis of Howell et al., for a ratio of sensible cooling load to design sensible load equal to 0.3, φ_{rp} could be higher than 75 percent. The lower the part-load, the higher the value of φ_{rp}. Although modulating the chilled water flow rate in a cooling coil at part-load can maintain the required zone temperature, the resulting zone relative humidity will be significantly higher.

Even when the water velocity inside the tubes of the cooling coil, v_w, has been reduced to a value between 1 and 2 ft/s, both the relative humidity of the conditioned air off the cooling coil, φ_{al}, and the temperature difference between the air and water leaving the cooling coil, $\Delta T_{aw} = (T_{al} - T_{wl})$ (in °F), depend on the area ratio of the coil $F_s N_r = A_o/A_a$, the heat transfer and mass transfer coefficients, and the log-mean temperature difference between the air and water. Here, F_s indicates the coil core surface-area parameter, N_r the number of rows of the coil, A_o the total outer surface area of the tube and fins (in ft^2), and A_a the face area (in ft^2).

Determination of the coil performance for cooling-mode part-load operation is the same as that covered in the Section 12.4.

When the zone temperature drops to a value 70.2°F $\leq T_r \leq$ 74.8°F, the fan-coil unit is operating in the dead-band mode. Only when $T_r \geq 75$°F will the fan-coil leave dead-band mode and reenter cooling mode. In dead-band mode operation, both the chilled water and hot water supplies to the fan-coil are cut off. The space temperature T_r may drift between 70.2 and 74.8°F. Only the centrifugal fan in fan-coil is operating.

If $T_r \leq 70$°F, the fan-coil is in winter heating mode operation. Once the fan-coil has dropped below the dead-band mode, it will reenter only when $T_r \geq 70.2$°F. When the RTS senses that T_r has dropped below 70°F, it sends a signal to the DDC controller. The controller then actuates modulation of the two-way valve on the hot-water supply line V2 and starts to open its valve port. Hot water enters the coil. In heating-mode operation, if the space temperature T_r increases above a limit, say 70°F, the DDC controller sends a signal to position the actuator so as to close the hot-water valve to a smaller opening. The hot water flow rate is then reduced, and the air temperature leaving the coil is also lowered accordingly. As a result, the zone temperature is maintained at 70°F. In heating-mode operation, the DDC should be reverse-acting; that is, the higher the space air temperature, the smaller the output (which is the reverse of direct-acting control in cooling-mode operation). A reverse action element is usually placed in the valve actuator to produce such a reverse-acting effect.

In heating-mode operation, regardless of whether the system is at the winter-design condition or at a part-load condition, the state of air leaving the coil always moves along the horizontal line $m\,hc$. If the moisture gain and the supply volume flow rate of the fan coil at part-load operation are the same as for the design condition, the space air temperature and relative humidity at part-load will be nearly equal to the winter-design values.

If the humidity ratio of the mixed air entering the coil for part-load heating-mode operation is higher than for the winter-design condition because of a change in the outdoor weather, the humidity ratio of supply air at part-load, w_{hcp}, will then be correspondingly higher. Although the zone temperature at part-load will still be maintained the same as for the design condition, the zone relative humidity φ_{rp} will be higher than the winter-design value that is more comfortable to the occupants.

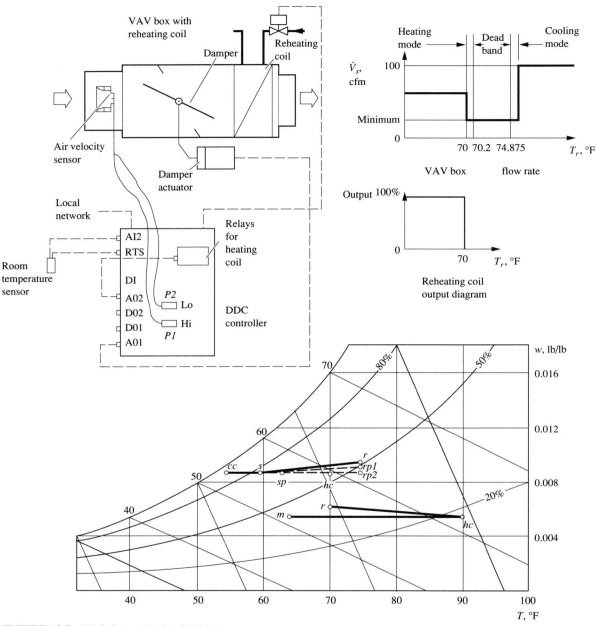

FIGURE 18.7 DDC for a reheating VAV box.

VAV Boxes

Variable-air-volume (VAV) boxes are the type of terminals most widely adopted to maintain a desirable space air temperature in commercial buildings. A typical DDC-controlled VAV box with a reheating coil, the reheating VAV box, is shown in Fig. 18.7. This VAV box and the incorporated DDC controller include a zone temperature sensor, a butterfly damper and an associated electric actuator, a two-way

valve-modulated hot water reheating coil, a DDC controller with a microprocessor chip to compute the operating parameter and operate the controller, and an electrically erasable programmable read-only memory (EEPROM) chip. This is a pressure-dependent VAV box. The volume flow of a pressure-dependent VAV box is modulated according to the signal from the zone temperature sensor to the controller. Usually, a PI control mode is adopted by such a DDC controller. For a pressure-independent VAV box, there is an additional sensitive air velocity sensor at the inlet as shown in Fig. 18.7.

The volume flow of a pressure-independent VAV box is affected by variations in the duct static pressure. Pressure-dependent and pressure-independent VAV boxes will be covered in Chapter 27 in detail.

Year-round control of the zone temperature served by a VAV box can be divided into cooling mode, dead-band mode, and heating mode.

During cooling-mode operation, if the zone temperature at part-load operation, T_{rp}, drops below 75°F because of a reduction in the space sensible cooling load q_{rsp}, the DDC controller receives a corresponding analog input from the RTS. In accordance with pre-programmed software instructions, this signal is compared with the set point 75°F. If $T_{rp} < 75$°F, the DDC controller will send an output to the damper actuator and start to reduce the opening of the flow passage of the VAV box. The total pressure loss across the terminal is then increased, and so the volume flow rate of the conditioned air supplied to the conditioned space, $\dot{V}_{sp}$, is reduced. If the reduced $\dot{V}_{sp}$ is balanced with the space sensible load, T_{rp} will remain at 75°F. Further drops in q_{rsp} will cause the damper to close more and more until it reaches the minimum opening necessary to supply the required amount of outdoor air for this zone. If these drops in q_{rsp} cause T_{rp} to drop below 74.8°F, the VAV box will operate in dead-band mode.

On the psychrometric chart at summer design conditions (see Fig. 18.7), point *cc* represents the conditioned air coming off the cooling coil. Solid line *cc s* is the sensible heating process due to the fan power and duct heat gain up to the supply air condition, point *s*. The supply air then absorbs the space sensible and latent heat along the space conditioning line *sr* and reaches state point *r*. At part-load operation, both the volume flow rate and the fan power heat gains drop, resulting in nearly the same temperature rise as in the design condition. Since the duct heat gain is only slightly reduced, the reduction in the supply volume flow rate causes the supply air temperature T_{sp} to increase to point *sp*. If the SHR_{sp} at part-load operation drops slightly, the space relative humidity φ_{rp} will be slightly lower, as shown by *rpl* in Fig. 18.7, when the space temperature at part-load T_{rp} is maintained the same as for the design load.

When the space temperature drops within the range 70.2°F $\leq T_{rp} \leq$ 74.8°F, the VAV box operates in dead-band mode. In dead-band mode,

- The damper closes to the minimum opening.
- The minimum $\dot{V}_s$ may be at a value of 20 to 50 percent of the design volume flow rate $\dot{V}_{s.des}$, typically at 30 percent.

- The reheating coil is deenergized.
- The space relative humidity will be slightly lower than at the design condition.

As with a fan-coil, when a VAV box is in dead-band mode operation, it will reenter cooling-mode operation only if $T_r \geq$ 70.2°F. The damper remains in minimum opening.

When $T_r \leq$ 70.0°F, the VAV box and the zone area that it serves enters winter heating-mode operation. Modulation of the hot water flow to the reheating coil may proceed according to the following schedule:

1. When in winter heating mode, the DDC controller will send a signal to a reverse-action relay so that the direct-acting mode will be changed to reverse-acting mode.
2. $\dot{V}_s$ may be maintained at a value equal to 50 percent to 100 percent of $\dot{V}_{s.des}$. $\dot{V}_s$ will remain constant during winter heating-mode operation.
3. During winter heating-mode operation, the reheating coil is energized. If $T_r \leq$ 70.0°F, the DDC controller will modulate the two-way valve to admit hot water to the reheating coil in order to maintain $T_r =$ 70.0°F.

At winter design conditions, the supply air temperature may be as high as 90°F; the zone temperature will be maintained at 70.0°F; and the zone air relative humidity φ_r may be between 25 and 50 percent depending on the outdoor air humidity ratio, percentage of outdoor air intake at the air-handling unit, and the internal latent load. If the outdoor weather is warmer than the design conditions, the space relative humidity will be higher than the value at winter design conditions.

Fan-Powered Units

A fan-powered unit is considered a terminal as well as a system component in a VAV system. Figure 18.8 shows a parallel-flow fan-powered unit. It mainly consists of a small centrifugal fan, a reheat coil, a VAV modulating damper, an air velocity sensor, a DDC controller, and an outer casing.

For a single-zone area in the perimeter zone of a typical floor served by a fan-powered unit, year-round operation can be divided into cooling-mode, dead-band mode, and heating-mode operations.

For summer cooling mode and dead-band mode, both the centrifugal fan and the reheat coil are deenergized. The cooled conditioned air from the air-handling unit or package unit, the primary air, flows through the modulating damper and is supplied to the

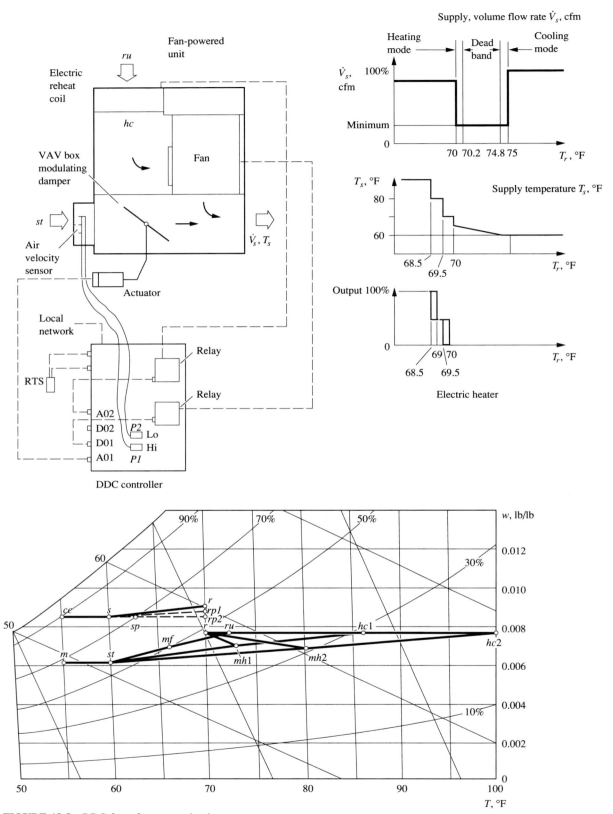

FIGURE 18.8   DDC for a fan-powered unit.

conditioned space. Operation of the DDC controller and trends in the zone temperature for cooling mode and dead-band mode are the same as for a VAV box.

When $T_r \leq 70°F$, the fan-powered unit operates in winter heating mode. Winter heating mode can be divided into the *fan operating stage* and the *heater operating stage*.

When $69.5°F < T_r \leq 70°F$, the fan-powered unit is in the fan operating stage. Only the centrifugal fan is energized. Ceiling plenum air at point *ru* is extracted through the deenergized reheat coil and mixed with primary air from the air-handling unit or packaged unit at point *st*. Supply air from the fan-powered unit, at point *mf*, is supplied to the conditioned space. The DDC controller positions the damper actuator of the VAV box and modulates the damper based on input from the zone temperature sensor in order to maintain $T_r = 70°F$. Because the centrifugal fan is operating, the supply volume flow rate from the fan-powered terminal unit, including the primary air and the extracted plenum air, is about 80 to 100 percent of the design volume flow rate of the cooling mode. The zone temperature must be at a value $T_r \geq 70.2°F$, or the fan-powered terminal will reenter dead-band mode again.

When $T_r < 69.5°F$, the fan-powered unit is in the heater operating stage. Both the fan and the first-stage heater are energized. The plenum air leaves the reheat coil at point *hc* and is mixed with the primary air at point *st*. The mixture is then supplied to the conditioned space at point *mh*. The DDC controller modulates the damper of the VAV box according to input from the zone temperature sensor in order to maintain a zone temperature of 69.5°F.

If $T_r < 68.5°F$, the fan and the first- and second-stage fan heaters all are energized to maintain a space temperature $T_r = 68.5°F$. When the space temperature rises from $T_r = 68.5°F$, the second-stage heater will be deenergized at 69.0°F, and the first-stage heater gap will be deenergized at 70°F. A 0.5°F gap between on/off switching of the first and second-stage heaters is needed to prevent hunting.

During winter heating-mode operation, the zone temperature T_r is maintained at a value between 68.5 and 70°F; the zone relative humidity is usually higher than for a fan-coil unit and VAV box because of the nature of *double mixing*—the mixing of outdoor air and recirculating air in the air-handling unit and the mixing of primary air and plenum air in the fan-powered unit. The actual amount of outdoor air supply to the conditioned space must be carefully checked against the required amount when primary air is supplied at the minimum level. Because the centrifugal fan is on during heating-mode operation, the space air

movement of a VAV system with a fan-powered unit is greater than that when only a VAV box is in use.

Example 18.1 A designer is considering whether to use fan-coil units, VAV boxes, or fan-powered units as the terminal units for a control zone in a high-rise office building. The operating characteristics of these terminal units at summer full-load and at 50 percent part-load (space sensible cooling load) are as follows:

	Full-load	50 percent part-load
Space temperature, °F	75	75
Space relative humidity, percent	50	
Space sensible cooling load, Btu/h	10,000	5000
Sensible heat ratio of space conditioning line	0.85	0.8
Supply air temperature, °F	60	63
Supply air relative humidity, percent		
Fan-coil	80	82
Others	80	73
Supply system heat gain, °F		
Fan-coil unit (fan heat gain and duct heat gain after fan-coil)	0	0
VAV box	5	8
Fan-powered unit	5	8

If the air density is taken as 0.075 lb/ft³ and the specific heat of moist air is 0.243 Btu/lb · °F, determine the zone relative humidity, supply volume flow rates, and system heat gain for these terminals at 50 percent part-load during cooling-mode operation. Both the volume flow through the VAV box and the primary air supplied to the fan-powered unit can be reduced to 30 percent of the full-load volume flow.

Solution

1. From Eq. (3.48b), the supply volume flow rate for summer full-load conditions is

$$\dot{V}_s = \frac{q_{rs}}{60\rho_s c_{ps}(T_r - T_s)}$$

$$= \frac{10,000}{60(0.075)(0.243)(75 - 60)} = 601 \text{ cfm}$$

2. For a fan-coil unit at 50 percent of the sensible cooling load, the supply volume flow rate is still 601 cfm. The supply air temperature difference is

$$T_{rp} - T_{sp} = \frac{5000}{60(601)(0.075)(0.243)} = 7.5°F$$

On the psychrometric chart as shown in Fig. 18.6, draw a line *sp rp* from point *sp* at a temperature of $T_{sp} = 75 - 7.5 = 67.5°F$ and a relative humidity of 82 percent. Line *sp rp* intersects the 75°F temperature line at point *rp*. From the psychrometric chart, the zone relative humidity at 50 percent part-load for a fan-coil unit is 67 percent.

3. For a VAV box or fan-powered unit, since the supply system heat gain is 8°F at part-load operation and is 5°F at design load, a sensible heat line *s sp* may be drawn from point *s* with a temperature increment of (8 − 5) = 3°F; from the chart as shown in Fig. 18.7, T_{sp} = 63°F and φ_{sp} = 72 percent.

From point *sp*, draw a line *sp rp1* with a SHR_p = 0.8. Line *sp rp1* intersects the 75°F line at *rp1*. The zone relative humidity at 50 percent part-load using a VAV box or fan-powered unit is 49 percent.

4. For both a VAV box and fan-powered unit, the supply temperature difference is 75 − 63 = 12°F. Therefore, the supply volume flow rate has been reduced to

$$\dot{V}_{sp} = \frac{5000}{60(0.075)(0.243)(12)} = 381 \text{ cfm}$$

and the system heat gain at 50 percent part-load is

$$q_{sys} = 60(381)(0.075)(0.243)(8) = 3333 \text{ Btu/h}$$

Comparison of Various Control Methods at Part-Load Operation

Various methods of control have been used to maintain a desirable zone temperature T_r at part-load operation, including the following:

Fan-coil unit	Water flow rate control
VAV box	Variable-air-volume control
	Reheat control
Fan-powered unit	Variable-air-volume control
	Primary air and plenum air mixing control
	Reheat control

If reheat control is used to maintain a 75°F zone temperature at part-load operation, as shown in Fig. 18.7, supply air must be reheated to point *sp* along the sensible heating line *s sp*. The reheating capacity q_{reh}, in Btu/h, is approximately equal to the reduction in the space sensible cooling load.

All these control methods can maintain the desirable space temperature at part-load operation. Their

influence on space relative humidity, supply system heat gain, and energy consumption during cooling-mode operation are listed in Table 18.1.

During summer cooling-mode operation, a VAV box used for reheat control can maintain the same or an even slightly lower space relative humidity than for summer design conditions at part-load. A fan-coil unit with chilled-water flow rate control has a significantly higher space relative humidity at part-load. During winter heating-mode operation, if there is no humidifying device, the smaller the percentage of outdoor air used in the air system, the higher the space relative humidity in a range from 25 to 45 percent.

Reheat control consumes far more energy than other types of control, and a fan-coil has the smallest supply system heat gain and thus the lowest energy consumption. However, a more comprehensive comparison of a fan-coil system and a VAV system should take into account indoor air quality, noise level, etc. These issues will be covered in Chapter 26.

18.7   YEAR-ROUND OPERATION OF A VAV AIR SYSTEM

Operating Parameters

Consider a typical single-duct VAV system with an associated refrigerating plant or evaporative coolers used to serve the interior zone of a typical floor in a commercial building. During part-load operation, the supply volume flow rate will be controlled by modulating the dampers in the VAV boxes to maintain the required space temperature.

During winter heating mode, a cold air supply is still required to maintain a year-round space temperature of 75°F. It is also assumed that the building is in a location where no humidification is required to maintain a space relative humidity no lower than 30 percent during the winter season.

During summer cooling-mode operation, the required off-coil temperature is 55°F with an off-coil

TABLE 18.1
Comparison of various control methods used during summer cooling-mode part-load operation (space sensible cooling load 5000 Btu/h)

	Space air		Supply volume flow rate, cfm	Supply system heat gain, Btu/h	Amount of reheat, Btu/h
	Temperature, °F	Relative humidity, percent			
Fan-coil water flow rate control	75	67	601	Far less than VAV control	
Reheat control	75	Slightly less than 50 percent	601	Less than VAV control	5000
Variable-air-volume control	75	49	381	3333	

relative humidity of 95 percent. The recirculating air enters the mixing box at 78°F, as shown on the psychrometric chart of Fig. 18.9a. It is mixed with an amount of outdoor air approximately equal to 20 percent of the total design supply volume flow rate and is cooled and dehumidified in the cooling coil. During winter heating-mode operation, outdoor air and recirculating air are mixed together to provide a mixture at a temperature of 55°F.

Year-Round Operating Modes

In order to achieve effective and energy-efficient operation when the outside weather changes from hot summer to cold winter, the year-round operation of the air system described above can be divided into the following operating modes, corresponding to the various outside weather regions on the psychrometric chart shown in Fig. 18.9a:

REGION I: REFRIGERATION AND EVAPORA-TIVE COOLING. The lower left boundary line of region I is the (dashed) enthalpy line that passes through the state point ru of recirculating air entering the mixing box. When the outside weather at point o is inside region I, the enthalpy of the outside air is higher than that of the recirculating air, that is, $h_o \geq h_{ru}$. In these instances, it is more energy-efficient to condition the recirculating air than the outside air. As such, the quantity of outdoor air that will be mixed with the recirculating air will be the minimum required to meet the outdoor air requirement in the conditioned space.

In region I, refrigeration, or evaporative cooling, or both are required to cool and dehumidify or to cool and humidify the mixture of outdoor and recirculating air to point cc in order to maintain the predetermined space condition r.

For region I, the operating characteristics of the VAV system include the following:

- The outdoor damper is opened to such a degree that only the minimum required amount of outdoor air is extracted.
- The recirculating damper is fully opened.
- Supply air is cooled and dehumidified, or cooled and humidified to cc.
- A refrigerating plant, evaporative cooler(s), or both are operating.

REGION II: FREE COOLING AND REFRIGERA-TION/EVAPORATIVE COOLING. Region II is enclosed by four boundary lines: the enthalpy line through point ru at the upper right, part of the sat-uration curve, the 55°F temperature line, and cc ru line.

In region II, the enthalpy of the outdoor air is lower than that of the recirculating air, that is, $h_o < h_{ru}$. Less energy will be used in conditioning if 100 percent outside air is cooled and dehumidified instead of a mixture of outdoor air and recirculating air. Both free cooling by the outdoor air and refrigeration cooling are required to bring the 100 percent outside air down to 55°F with a relative humidity of 95 percent.

The operating characteristics of this VAV system when outdoor air is inside region II include the following:

- The outdoor damper and the exhaust damper or relief damper both are fully opened.
- The recirculating damper is closed.
- Outdoor air is cooled and dehumidified to point cc. Point cc is at a temperature of 55°F and a humidity ratio of 0.0088 lb/lb.
- A refrigerating plant is operating.

REGION III: FREE COOLING AND EVAPORA-TIVE COOLING. Region III is bounded by the line cc ru, 55°F temperature lines, and by enthalpy lines passing through points ru and cc' for $w_{cc'} = 0.005$ lb/lb, as shown in Fig. 18.9a. For outdoor air inside this region, $h_o < h_{ru}$, the humidity ratio of outdoor air w_o is approximately less than 0.009 lb/lb. Indirect and direct evaporative cooling are required to maintain a space temperature of 75°F and a space relative humidity greater than or equal to 30 percent.

When outdoor temperature $T_o \leq 78°F$, it is energy-efficient to extract 100 percent outdoor air (to make use of free cooling) and also to use direct and indirect evaporative cooling to cool and humidify the outdoor air to a temperature of 55°F with a humidity ratio $w_{cc'} \geq 0.005$ lb/lb.

When $T_o > 78°F$, it is energy efficient to extract minimum outdoor and mix it with recirculated air. The mixture is then evaporative cooled indirectly-directly to $T_{cc} = 55°F$ and $w_{cc} \geq 0.005$ lb/lb.

The operating characteristics of the VAV system when outdoor air is inside region III and when 100 percent outdoor air is extracted are as follows:

- The outdoor air damper and the exhaust or relief air damper both are fully opened.
- The recirculating damper is closed.
- Evaporative cooler(s) are used for evaporative cooling and humidifying.
- Refrigeration is not required.

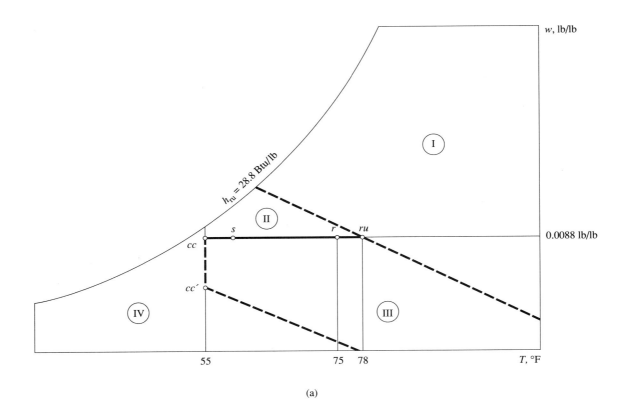

(a)

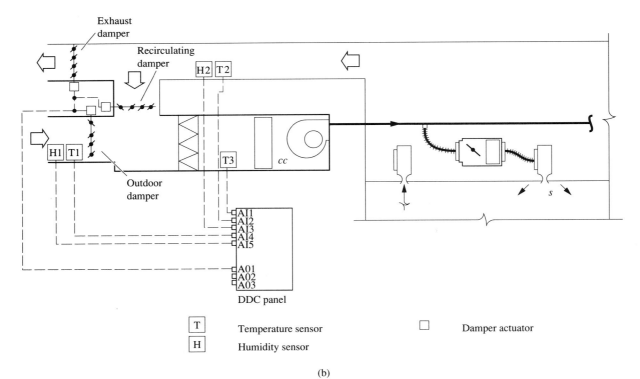

(b)

FIGURE 18.9 Enthalpy air economizer control: (*a*) outside weather regions and (*b*) enthalpy air economizer.

When only a minimum of outdoor air is mixed with recirculating air, then:

- The outdoor air damper is opened to minimum.
- The recirculating damper is fully opened.
- Evaporative cooler(s) are used for evaporative cooling and humidifying.
- Refrigeration is not required.

REGION IV: FREE COOLING, HUMIDIFICATION, AND HEATING. Region IV is enclosed by part of the saturation curve, the 55°F temperature line, and the enthalpy line passing through point cc' for $w_{cc'} = 0.005$ lb/lb, as shown in Fig. 18.9a. For outdoor air in region IV, $T_o \leq 55°F$ and $h_o \leq h_{cc'}$ when $w_{cc'} = 0.005$ lb/lb.

Modulation of outdoor and recirculating dampers is required to provide a mixed air temperature of 55°F or a mixed air enthalpy around 22.8 Btu/lb. Free cooling will be utilized. Necessary optimum humidifying methods should be adopted to maintain a space relative humidity of not less than 25 percent.

The operating characteristics of the VAV system include the following:

- Outdoor, exhaust and recirculating dampers are modulated.
- Humidification or cooling and humidification may be required.
- The refrigeration plant is not operating.

Year-round operation of a perimeter zone in a building is similar to that for an interior zone with the exception that heating may be required in region IV.

18.8 ECONOMIZER CYCLE AND CONTROL

Economizer Cycle and Economizers

An economizer cycle is an air conditioning cycle that utilizes the free cooling capacity of either outdoor air or cooling water from the cooling tower, instead of refrigeration, to provide cooling/dehumidification so as to maintain a required space temperature.

The components and devices used in the operation of an economizer cycle are collectively called an *economizer*, and the type of control used to operate the economizer cycle effectively and energy-efficiently is called *economizer control*.

There are two types of economizers:

1. *Air or air-side economizers.* These bring in up to 100 percent outdoor air to utilize its free cooling

capacity, rather than using refrigeration or evaporative cooling, to offset the space cooling load when the enthalpy of the outdoor air is lower than the enthalpy of the recirculating air, or when the outdoor air temperature is lower than the recirculating air temperature.

An air economizer consists of outdoor, exhaust, relief, and recirculating ducts and dampers in the AHU or PU, as well as a control system to operate them (see Fig. 18.9b). Air economizer control can be subdivided into enthalpy *(enthalpy-based)* air economizer control and temperature *(temperature-based)* air economizer control.

2. *Water or water-side economizers.* As described in Chapter 11, these use the free cooling capacity of the cooling water from the cooling tower to cool the supply air when the cooling water temperature is lower than the mixing temperature of outdoor and recirculating air. A water economizer consists mainly of a cooling tower, a water-cooling coil in the AHU or PU, a circulating pump to circulate the cooling water, and the associated control system.

Enthalpy Air Economizer Control

The operation of an enthalpy air economizer control system is based on the comparison of air enthalpies. The sequence of operation of a typical enthalpy air economizer control system is as follows:

1. The outdoor air temperature T_o is sensed by sensor T1, and the relative humidity φ_o of outdoor air is sensed by humidity sensor H1. Both signals are sent to the DDC panel, where calculations are performed to combine them into an outdoor air enthalpy h_o signal. This signal is compared with the recirculating air enthalpy h_{ru}, which is the combined signal from temperature sensor T2 and humidity sensor H2, at the DDC panel.

2. If $h_o > h_{ru}$, DDC panel sends signals to actuate the damper actuators so that the outdoor air and exhaust air dampers will be at their minimum opening positions as determined by the manual positioning switch. Meanwhile, the recirculating damper is fully opened.

3. If $h_o \leq h_{ru}$, the DDC panel sends signals to actuate the damper actuators to fully open the outdoor air and exhaust air dampers; at the same time, the recirculating damper will be closed.

4. Temperature sensor T3 senses the mixed air temperature of outdoor and recirculating air T_m for airflow leaving the mixing box. This signal is com-

pared with the set point $T_{m.s} = 55°F$ in the DDC panel. When the outdoor temperature $T_o \leq 55°F$, the DDC panel then actuates the damper actuators to modulate the openings of the outdoor, exhaust, and recirculating air dampers to maintain a predetermined mixed air temperature of $T_m = 55°F$.

When $T_o \leq 55°F$, the refrigeration compressors or chillers that serve this air system can be shut down.

Temperature Air Economizer Control

Theoretically speaking, it is more energy-efficient to compare the enthalpy of outdoor air with that of recirculating air during air economizer control. However, comparing enthalpies necessitates having both temperature and humidity sensors or temperature and dew-point sensors. In actual practice, humidity sensors may demonstrate considerable errors (sometimes up to 10 percent). Dew-point sensors are delicate, expensive, and cause many maintenance difficulties. Therefore, it is simpler and more convenient to use only temperature sensors and to compare the outdoor air temperature T_o with the recirculating temperature T_{ru} (or a predetermined set point) instead of sensing and comparing enthalpies. This method of control is called *temperature air economizer control.* Outside weather regions using temperature air economizer control are shown in Fig. 18.10a and are similar to the outside weather regions for enthalpy air economizer control.

The sequence of operation of a temperature air economizer control system is as follows:

1. The outdoor air temperature T_o is sensed by temperature sensor T1, as shown in Fig. 18.10b. T_o is then compared with the recirculating air temperature T_{ru} sensed by temperature sensor T2. If $T_o > T_{ru}$, the DDC panel sends signals to damper actuators so that the outdoor and exhaust air dampers will be at their minimum opening, as determined by the manual positioning switch. The recirculating damper should be fully opened.

2. If $T_o \leq T_{ru}$, the outdoor and exhaust air dampers should be fully opened and the recirculating damper should be closed.

3. Temperature sensor T3 senses the mixed air temperature T_m after the mixing box. When the outdoor temperature $T_o \leq 55°F$, the DDC panel modulates the outdoor, exhaust, and recirculating air dampers to maintain a nearly constant mixed air temperature of 55°F.

4. When $T_o \leq 55°F$, the refrigeration plant used to serve this air system is shut off.

In temperature air economizer control, instead of comparing T_o with T_{ru}, a single dry bulb temperature set point T_{db}, in °F, is sometimes used as an indicator of when to change between using the minimum amount of outdoor air and using 100 percent outdoor air. The changeover is determined from a comparison of T_o with T_{db} instead of T_o with T_{ru}. The proper magnitude of T_{db} depends on the outside climate and operating experience of the air system. T_{db} varies from 65 to 78°F between various locations. For hot and humid climates, T_{db} should be lower. Using a single T_{db} set point is a simpler approach, but, according to Wacker (1989), comparing T_o and T_{ru} to determine the changeover saves more energy. Instead of a fixed T_{db} set point, a fixed enthalpy h_{fix} is sometimes used as a set point for comparison.

Comparison of Enthalpy and Temperature Control

In addition to the instrumentation and maintenance problems of enthalpy control, the main difference between enthalpy and temperature air economizer cycles depends on whether the status points of outside weather fall in the dotted area A or shaded area B shown in Fig. 18.10a. When the outside weather is inside the dotted area A, an enthalpy air economizer cycle consumes less energy to cool and dehumidify the air, whereas when the outside weather falls in the shaded area B, a temperature air economizer cycle usually consumes less energy.

Therefore, consider the following:

1. For locations having hot and humid climates, enthalpy air economizer control is recommended.

2. For most locations having moderate outdoor humidity ratios, the energy savings offered by using enthalpy economizer control are small. According to Spitler et al. (1987), the two types of control differ by only about 10 to 20 percent. Because of the maintenance difficulties of humidity sensors, temperature air economizer control is recommended.

3. For locations having dry climates, temperature economizer control is recommended because it offers nearly the same energy-saving capacity as enthalpy economizer control.

4. More precise and reliable temperature and enthalpy sensors are preferable.

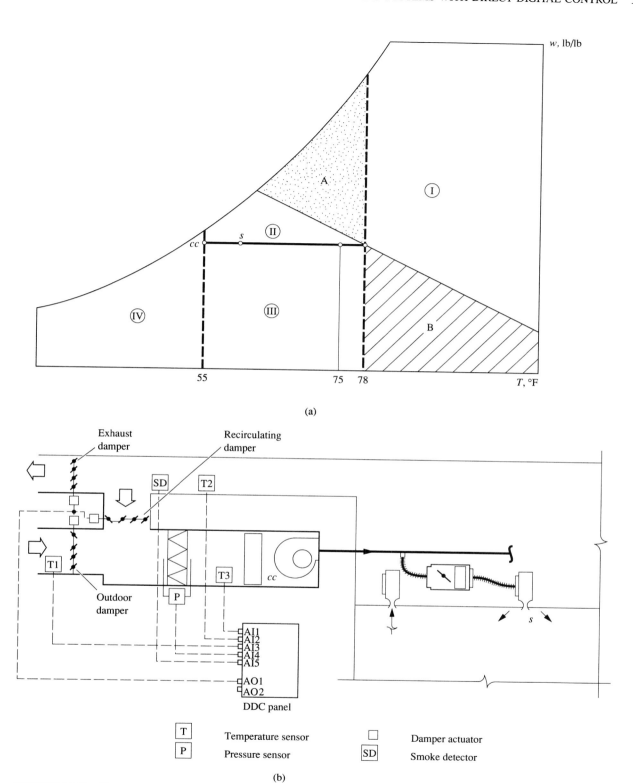

FIGURE 18.10 Temperature air economizer control (*a*) outside weather regions and (*b*) temperature air economizer.

Water Economizer Control

Consider a typical water economizer that uses a pre-cooling coil in either an AHU or a PU, as shown in Figure 18.11b, and has the following operating characteristics:

- The cooling tower approach is assumed to be 5°F.

- The mixing air temperature of the outdoor air and recirculating air, T_m, is 78°F.

- The minimum temperature difference between T_m and the cooling water temperature $T_{con.w}$ is 5°F.

- When the outdoor wet bulb $T_o' \leq 40°F$, the temperature of the cooling water used to replace the entire refrigeration through a heat exchanger is equal to or less than 47°F.

- The outdoor wet bulb $T_{o.we}'$ below which the water economizer can be operated effectively is

$$T_{o.we}' = 78 - 5 - 5 = 68°F$$

During the operation of a water economizer, the outside weather can be divided into three zones based on the outdoor wet bulb, as shown in Fig. 18.11a:

Region I: The water-side economizer is not operating.

Region II: The water-side economizer is operating and replaces part of the mechanical refrigeration.

Region III: The water-side economizer replaces entirely the mechanical refrigeration.

The dividing line between regions I and II is the wet bulb $T_{o.we}'$ line. The dividing line between regions II and III is the wet bulb $T_{r.of}'$ line. $T_{r.of}'$ represents the outdoor wet bulb below which the refrigeration plant can be shut down.

The sequence of operation for a water-side economizer using a precooling coil is as follows:

1. Temperature sensor T2 senses the temperature of the cooling water discharged from the cooling tower and sends a signal to the DDC panel. This signal is compared to a high-limit set point $T_{o.we}'$, such as 68°F. If the value T2 senses is less than 68°F, the DDC panel actuates the three-way valves V1 and V2 so that all the cooling water discharged from the cooling tower will flow through the precooling coil before going to the water-cooled condenser. Cooling water discharged from the condenser will return to the cooling tower.

2. When the discharged air temperature T_{dis} sensed by T1 is lower than a predetermined set point, the DDC panel then actuates the three-way valves V1 and V2 to reduce the water flow to the precooling coil in order to maintain a required discharge temperature.

3. When the value sensed by T2 is less than 50°F, the DDC panel will stop the refrigeration plant serving this air system.

4. When the discharge air temperature T_{dis} sensed by T1 is lower than its set point, and the cooling water temperature sensed by the sensor T2 drops below a lower limit, say 45°F, the DDC panel actuates the three-way valve V3 so that part of the cooling water returned from the condenser will bypass the cooling tower to maintain a required discharge air temperature.

5. If the value sensed by T2 is less than 35°F, the DDC panel will shut down the water economizer to provide freezing protection.

Comparison of Air and Water Economizers

The economizer cycle is a popular means of saving energy in the year-round operation of air conditioning systems. The amount of energy savings, compared with air systems for which the economizer cycle is not adopted, ranges from 15 to 40 percent of the total energy use in various locations.

The advantages of the air economizers over water economizers include the following:

1. Because of the intake of a large quantity of outdoor air, the indoor air quality is improved.

2. The energy input to an air conditioning system that uses an air economizer is approximately 20 percent less than that for a water economizer.

3. The cooling tower requires less makeup water, fewer water treatments, and reduced maintenance. Less maintenance work is required for the precooling water system.

However, the disadvantages of air economizers are as follows:

1. For an AHU or PU located in the interior core fan room and used to serve several floors in a high-rise building, the large vertical shafts used to transport the outdoor air and exhaust air often occupy a lot of valuable rental space.

2. When outdoor weather suddenly changes, the space air may also produce corresponding fluctuations. Building pressurization is not as stable as with a water economizer.

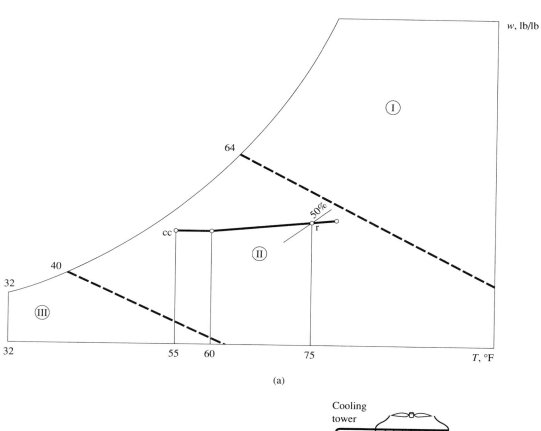

(a)

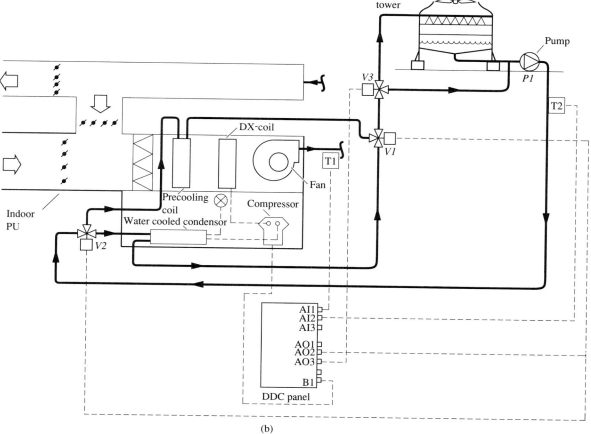

(b)

FIGURE 18.11   Water economizer control: (*a*) outside weather regions and (*b*) water economizer of an indoor PU.

3. During the winter season, the intake of drier outdoor air always results in a lower space relative humidity than when using recirculating air. Humidification may be required.

4. For packaged units, the first cost of an air economizer is about 25 to 40 percent of the installation cost of the packaged unit. For a water economizer, the first cost is only about 10 percent of the installation cost of the PU.

5. Coil freezing protection is more critical in locations where the outdoor air temperature drops below 32°F.

For the use of free cooling to reduce refrigeration, ASHRAE/IES Standard 90.1–1989 specified that each air system should include either of the following:

1. A temperature or enthalpy air economizer to provide 85 percent of design supply flow for outdoor air intake and recirculating air.

2. A water economizer using direct-indirect evaporative cooling that is capable of replacing refrigeration entirely when outdoor air is 50°F dry bulb and 45°F wet bulb and below.

Exceptions include: (1) small systems whose capacities are less than 3000 cfm or 90,000 Btu/h cooling capacity, and the sum of these exceptions in a building do not exceed 10 percent of total cooling capacity or 600,000 Btu/h, and (2) systems for which an economizer is shown to be energy-inefficient or cost-ineffective.

18.9 DISCHARGE AIR TEMPERATURE CONTROL

Subsystems in AHU and PU Control

The purpose of a control system in an air-handling unit (AHU) or a packaged unit (PU) for a multizone floor in a commercial building is to provide an optimum discharge air temperature and humidity ratio at predetermined supply volume flow rates to various terminals in an energy-efficient manner. The control system in an AHU or a PU, including the control of its components by subsystems, comprises of most of the controls in an air system aside from the zone temperature control in terminals. AHU or PU control consists of the following main subsystems:

- Discharge air temperature control
- Minimum outdoor air control
- Supply, return, and relief fan volume flow control
- Economizer control

- Humidity control
- Dew-point control
- Optimum start and stop control
- Fire alarm control
- Filter pressure drop indication

Economizer controls were covered in the last section. In this section, discharge air temperature, minimum outdoor air, and dew point control will be covered.

Mutual interactions often occur among these control subsystems. Interactions can be divided into the following categories:

1. *Sequence control.* For this type of control action, several individual control loops operate in sequence in order to maintain a particular controlled variable between predetermined dead band(s).

2. *Override.* If two control loops control the same controlled variable, such as the opening of the outdoor air damper for both minimum outdoor air control and mixed air control, there must exist a priority between them to actuate the damper. The predetermined precedence of one control signal over another in the DDC panel or controller is called *override.*

3. *Chain effect.* Modulating the controlled device of one of the control loops affects the controlled variable of another control loop, resulting in a chain effect of control actions.

Discharge Air Temperature Control

The discharge air temperature T_{dis}, in °F, is the temperature of air coming off the coil, T_{cc}, plus the temperature rise across the supply fan, ΔT_f, typically $55 + 2 = 57$°F. The discharge air temperature control subsystem is the basic control system of an AHU or a PU. The difference between the supply temperature from the terminals and the discharge temperature from the AHU or PU, $T_s - T_{\text{dis}}$, in °F, is the rise or drop in temperature resulting from the duct heat gain or heat loss ΔT_{du}. For a specific air system, $T_s - T_{\text{dis}}$ is usually a function of the supply volume flow. The relationship between the discharge temperature and the supply temperature can often be expressed in fixed terms:

$$T_s = T_{\text{dis}} \pm \Delta T_{\text{du}}$$

A typical discharge air temperature control subsystem for a single-duct VAV system supplying cold air during the winter occupying period and in a location having a cold winter consists of the following control loops, as shown in Fig. 18.12a:

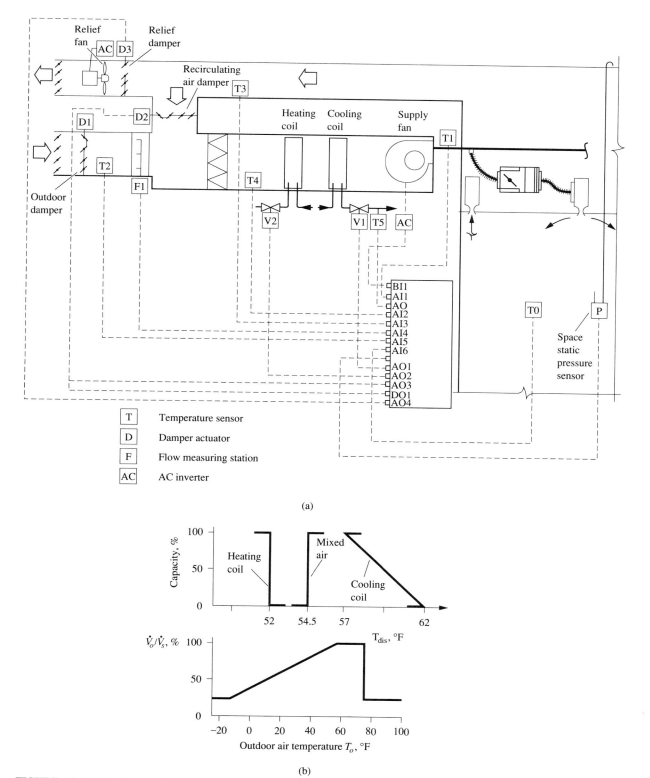

FIGURE 18.12 Discharge air temperature control for a typical AHU: (*a*) control diagram, (*b*) output diagram, and (*c*) air conditioning cycles.

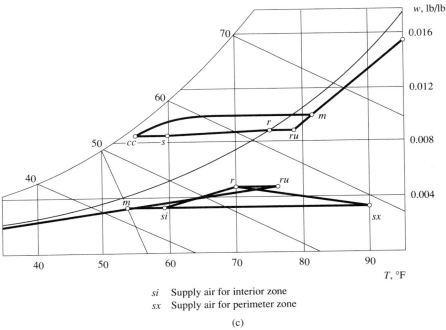

si Supply air for interior zone
sx Supply air for perimeter zone

(c)

FIGURE 18.12C

- Cooling coil control: discharge air temperature sensor T1, DDC panel, cooling-coil valve actuator V1
- Mixing air control: discharge air temperature sensor T1, DDC panel, and outdoor air damper actuator D1 and recirculating air damper actuator D2
- Heating coil control: discharge air temperature sensor T1, DDC panel, and heating-coil valve actuator V2
- Economizer control: outdoor air temperature sensor T2, recirculating air temperature sensor T3, DDC panel, and damper actuators D1 and D2
- Low-limit control (coil freezing protection): temperature sensor T4, DDC panel, outdoor air damper actuator D1.

Cooling coil, mixing air, and heating coil control loops are usually operated in sequence so as to maintain a predetermined discharge air temperature while preventing simultaneous cooling and heating. For a typical air-handling unit, the set point of the discharge air temperature for a cooling coil control loop may be 57°F. In order to avoid sudden repeated changes in operating mode among the cooling coil, mixing air, and heating coil control loops, resulting in unstable operation, the set point for mixing air control is better set at 54.5°F and for heating coil control at 52°F, as shown in Fig. 18.12b.

The modulation of valve V1 in the cooling coil control loop of the discharge air temperature subsystem can be performed in parallel with the control operations in economizer control loops: minimum opening of the outdoor air damper and maximum

opening of the recirculating damper or 100 percent opening of the outdoor air damper and closing of the recirculating damper. Although mixing air control is one of the functions of economizer control, it is often a part of sequence control for a discharge air temperature control subsystem. Mixing air control functions only when $T_o < 54.5°F$.

Sequence of Operation for Discharge Air Temperature Control

The sequence of operation for discharge air temperature control is as follows:

1. When the supply fan is turned on, the control system is energized.
2. If the discharge air temperature T_{dis} sensed by the temperature sensor T1 is higher than 57°F, the DDC panel sends a signal (in a PI control mode) to open and modulate the cooling-coil valve V1 until $T_{dis} = 57°F$.
3. In cooling-mode operation, sensor T5 of chilled water temperature T_{wl} leaving the cooling coil will reset the discharge-air set point $T_{d.s}$ according to a linear relationship between $T_{d.s}$ and T_{wl}: when $T_{wl} = 55°F$, $T_{d.s} = 57°F$; when $T_{wl} = 50°F$, $T_{d.s} = 59.5°F$; when $T_{wl} = 47°F$, $T_{d.s} = 61°F$; and so on, as shown Fig. 18.12b.
4. The outdoor air temperature T_o is sensed by T2 and the recirculating air temperature T_{ru} by T3. These values are compared at the DDC panel. If $T_o > T_{ru}$, the DDC panel positions the outdoor/relief

dampers to the minimum opening position, as determined by the manual positioning switch; the recirculating damper is fully opened (see Figs. 18.12a and b). If $T_o \le T_{ru}$, the outdoor/relief dampers are fully opened and the recirculating damper is closed.

5. When valve V1 closes to its minimum position, T_{dis} further drops until it is less than 54.5°F. The DDC panel then modulates the outdoor and recirculating dampers, until the outdoor damper is at its minimum opening, to maintain T_{dis} at 54.5°F.

6. When the outdoor air damper is at its minimum opening, if T_{dis} drops below 52°F, the DDC panel starts to open and modulate the heating-coil valve V2 to maintain T_{dis} at 52°F. The outdoor air damper remains at its minimum opening.

7. Low-limit temperature sensor T4 senses the mixing air temperature after the mixing box, T_m, to provide coil freezing protection. If T_m drops below a certain limit, say 40°F, the DDC panel will close the outdoor air damper and stop the supply fan.

ASHRAE/IES Standard 90.1–1989 requires that when an air system supplies cold air or warm air to more than one zone, it must provide control to reset discharge air temperature at low loads, except for VAV systems that have minimum settings at low loads or air systems that are not reheated or recooled. Discharge air temperature reset is more energy-efficient for upward resets during cooling-mode operation and for perimeter zones having a greater load variation.

Control of Coils

For water-cooling and water-heating coils, two-way valves are usually used instead of three-way valves because they offer pump power savings and lower cost. Two-way valves should be closed when the supply fan is turned off. For direct-expansion coils, the refrigerant flow is usually controlled, using two-position control by the solenoid valves located at the liquid lines. As mentioned in Section 12.3, having two separate refrigerant circuits with two-stage or three-stage row-control or intertwined face-control is preferable in DX-coil capacity controls. DX-coil capacity control should be coordinated with compressor capacity control. A coil face-bypass damper may cause frost formation when the air flowing through the DX-coil has been cooled below a certain limit.

Minimum Outdoor Air Control

The minimum outdoor air is the quantity of outdoor air that must be supplied to the conditioned space at any time according to local codes as well as ASHRAE Standard 62-1989. This air is needed to provide the outdoor air requirements for the occupants and to maintain an acceptable indoor air quality.

For a VAV system, although the outdoor air supply is typically well over the outdoor air requirement specified by the ASHRAE Standard, say 20 cfm per person at design conditions, the outdoor air supply may be inadequate when the supply volume flow rate has been reduced to the minimum value at part-load operation, such as at 40 percent of the design flow rate.

A typical minimum outdoor air control subsystem is shown in Fig. 18.12a. A flow sensor F1 is located in the outdoor air passage after the outdoor air damper to measure the outdoor air velocity and thus the outdoor-air volume flow rate. This signal is sent to the DDC panel and compared with a minimum set point. If the measured value is smaller than the set point, the DDC panel sends a signal to override the signal from the economizer or mixing air control system. It opens the outdoor air damper and closes the recirculating damper until the minimum volume flow requirement is met. A sufficient distance between the outdoor damper and the pitot-tube flow-measuring station, as well as a flow straightener, must be provided for a more accurate velocity pressure measurement. It is helpful if the outdoor damper is divided into two sections: (1) a minimum opening and (2) a modulating section up to 100 percent outdoor air intake.

Dew-Point Control

For a conditioned space that needed both close control of the space temperature and relative humidity and a sensible heat ratio of the space conditioning line SHR$_s$ above 0.95, dew-point control is often used.

Dew-point control involves using a water-cooling coil, DX-coil, water-spraying coil, or air washer to cause a specific amount of conditioned air to cool and dehumidify to point cc as that on Fig. 11.8. Because of the high relative humidity of the departing air, the temperature T_{cc} is approximately equal to the dew point on the saturation curve.

When the indoor space temperature is to be maintained at 75°F, with a relative humidity of 50 percent, if $T_{cc} = 55°F$, $\phi_{cc} = 97$ percent and SHR$_s >$ 0.95, the state point of the conditioned space r must move approximately along the horizontal dew-point line. Either the sensible cooling load is absorbed by the supply air or reheating is provided to the supply air by a reheating coil to compensate the sensible load at part-load, if the space temperature is maintained at 75°F and the relative humidity at 50 percent.

Dew-point control is simple and effective. Its main drawback is its need for reheating, which wastes energy. Dew-point control is no longer a popular control method except for industrial applications.

18.10 FAN WARM-UP, COOL DOWN, AND HUMIDIFIER CONTROL

Supply Volume Flow Control

In a VAV air system, if more than half of the VAV boxes close their dampers at the same time to reduce the volume flow supplied to various zones at part-load operation, the static pressure of the supply duct will be raised. Such a high static pressure may cause objectionable noise and volume flow in some of the terminals and result in unstable operation. At part-load, the volume flow and the fan total pressure rise of the supply fan should be reduced to achieve smooth and balanced system operation.

There are two methods of reducing the supply volume flow rate and the buildup of pressure in a VAV system at part-load operation: (1) duct static pressure control, and (2) terminal-regulated air volume.

DUCT STATIC PRESSURE CONTROL. The purpose of supply-duct static pressure control is to maintain a nearly constant static pressure at a specific loca-

tion in the main supply duct. This is the method most widely adopted in VAV systems. A duct static pressure control system consists mainly of a static pressure sensor, a DDC panel or a DDC controller, and a fan speed modulator (AC inverter) or inlet vanes actuator, as shown in Fig. 18.13a. When the static pressure at a predetermined fixed point in the supply duct is higher than a set point, the DDC panel reduces the fan speed or positions the inlet vanes to reduce the fan volume flow and fan total pressure, so as to maintain a nearly constant static pressure at that point.

The following points need to be considered when duct static pressure control is selected:

1. Duct static pressure shows a small capacitance lag, or time constant. Therefore, duct static pressure responds quickly to changes in the supply fan output. Proportional-integral- derivative (PID) control action and a wider proportional band are more appropriate.

2. The set point of static pressure should be the minimum static pressure for all the branch take-offs and VAV boxes under all operating conditions. For most VAV systems, the pressure sensor is usually located in the main duct between $0.3L_{main}$ from the farthest branch take-off and near the farthest branch take-off in order to provide the required volume flow for all VAV boxes and the greatest

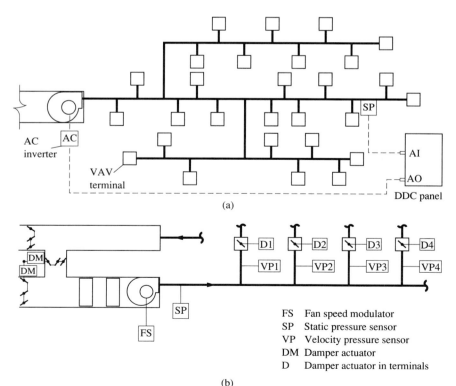

FS   Fan speed modulator
SP   Static pressure sensor
VP   Velocity pressure sensor
DM   Damper actuator
D   Damper actuator in terminals

FIGURE 18.13 Supply-fan volume flow control: (*a*) supply-duct static pressure control and (*b*) Terminal-regulated air volume control.

energy savings. Here L_{main} indicates the length of the main duct.

3. Modulating the fan speed by means of an AC inverter saves more energy than modulating of the inlet vanes. However, the initial cost of inlet vanes is considerably lower than that of an AC inverter.

4. A high-limit pressure sensor should be installed after the fan outlet to prevent damaging duct system by means of excessive pressure due to a malfunction, a closing of the fire or smoke dampers, etc. When these circumstances arise, the excess static pressure buildup due to a blockage of the air passage can be sensed by the high-limit pressure sensor and the DDC panel will send a signal to turn off the fan.

TERMINAL-REGULATED AIR VOLUME (TRAV).
Terminal-regulated air volume uses computer software to control the supply fan volume flow by means of fan speed modulation based on real-time measured airflow at the VAV boxes. This is possible because of the employment of a DDC system like the one shown in Fig. 18.13b.

The primary benefit of using TRAV over duct static control stems from its lower fan power consumption at volume flows less than the design flow. Another benefit is the even distribution of air flow in various VAV boxes achieved by resetting the set point at each box when the total volume flow is significantly lower than the required flow during cool-down or warm-up periods. The third benefit is a compensation in fan static pressure when this pressure is affected by a change in the positions of outdoor and recirculating dampers during mixing air control.

TRAV with a high static pressure is sometimes required to provide a high volume flow rate to a specific area, such as a computer room, when the air system is operating under a low load. The ductwork should be carefully designed to meet all possible requirements at part-load operation.

Relief Fan and Return Fan Control

If a VAV system is installed with a supply fan and a relief fan (as shown in Fig. 18.12) or a supply fan and a return fan, the relief fan or return fan is usually controlled to maintain the required space pressurization.

A direct space pressurization control consists of a static pressure sensor located in an open area of the conditioned space away from the doors. The outdoor reference static pressure pick-up is better at a level above the rooftop shielded from the influence of wind and rain. A DDC panel or controller and

an AC inverter are used to modulate the fan speed. When the outdoor air damper is at the minimum opening, the relief fan is usually turned off and the relief damper is closed. When the outdoor air damper is fully opened, or for mixing air control, the relief air damper is opened and the space static pressure sensor actuates the relief fan speed modulator to maintain a nearly constant space static pressure. Direct space pressurization control is also used for the return fan by varying its volume flow to maintain a preset space pressure.

Flow tracking control has been used by designers to maintain preset volume flow relationships between supply and return fans at reduced volume flow rates. In a flow tracking control system, flow measuring stations measure the volume flow rates of the supply air $\dot{V}_s$ and return air $\dot{V}_{ret}$, and these rates are fed to a DDC controller for comparison. The return fan volume flow rate $\dot{V}_{ret}$, in cfm, is then modulated by varying its fan speed according to the following relationship:

$$\dot{V}_{ret} = \dot{V}_s - (\dot{V}_{ex} + \dot{V}_{ef}) \qquad (18.1)$$

where $\dot{V}_{ex}$ = exhaust air, cfm
$\dot{V}_{ef}$ = exfiltrated air, cfm

Return fan flow tracking control cannot provide minimum outdoor air control unless outdoor and recirculating dampers are modulated at the same time. Many return fan flow tracking control systems have been abandoned in favor of direct space pressurization control. The main difficulty is due to the variation of pressure characteristics of the supply and return fans during part-load operation.

Warm-Up and Cool-Down Control

The purpose of warm-up control is to recover the set-back in space heating beneath the set point that developed during an unoccupied period. The purpose of cool-down control is to cool and dehumidify the space air down to predetermined limits following an increase in the space air temperature and humidity when the air system is turned off during an unoccupied period.

When the supply fan of the AHU or PU is turned on and the difference between the space temperature sensed by the temperature sensor and the set point exceeds a certain limit, such as 5°F, the air system will enter a warm-up or cool-down operating mode, whichever is appropriate. Both warm-up and cool-down control close the outdoor and relief/exhaust dampers and open fully the recirculating damper. In addition, warm-up control opens and modulates the two-way valve of the heating coil, and cool-down control opens and modulate the two-way valve of the cooling coil, until the space temperature is within predetermined limits.

One of the problems in warm-up or cool-down control is determining the optimum start time. The optimum start is affected by the following factors:

- Space temperature and relative humidity just prior to the warm-up or cool-down period
- Structure of the building envelope
- Capacity of the heating/cooling equipment
- Internal loads
- Outdoor temperature

An adaptive control strategy should be used to determine the optimum start or stop times. There are also software packages available on the market to find the optimum start time. A simple and effective way that is worthwhile to try is as follows: record the outdoor air temperature, space temperature, relative humidity, internal load, and equipment and damper operating conditions at suitable time intervals, say 10 minutes to 1 hour, during the warm-up or cool-down period; analyze and determine the optimum start time once sufficient data have been accumulated.

Steam Humidifier Control

A steam humidifier is the most widely used type of humidifier in an AHU or PU. A humidity sensor is located in the supply duct just after the fan discharge to sense the relative humidity of the discharge air. The DDC panel receives the sensed signal and positions the two-way valve to modulate the steam flow rate entering the steam jet humidifier until the relative humidity of the discharge air meet the required set point.

When the fan is turned on, the humidifier control is energized. The steam valve closes when the fan is turned off.

Off-Hour, Smoke Detection, and Pressure Drop Indication

ASHRAE/IES Standard 90.1–1989 requires that an HVAC&R system shall be equipped with automatic controls to reduce energy use through setback set points and equipment shutdown during unoccupied period off-hours. Small system load demand of 2 kW(6826 Btu/h) or less may be controlled by manual off-hour controls.

Other types of control in an AHU or PU include fire alarm control and filter pressure drop indication. A smoke detector is usually located in the return air passage. Once the smoke detector detects smoke in the return airstream, it sends a signal to the DDC

panel and energizes the fire alarm system (see Fig. 18.10b).

A pressure sensor is usually used to sense the pressure drop across the filter. As soon as the pressure drop exceeds a predetermined limit, an alarm signal appears at the control panel to indicate the need for replacement or maintenance.

18.11   REQUIREMENTS FOR EFFECTIVE CONTROL

The following are requirements for effective control:

1. Equipment, control valves, and air dampers should be properly selected and sized. Oversized equipment and components degrade the control quality.

2. Proportional-integral (PI) control mode is preferable for most of the control loops. PI control eliminates the offset, resulting in a better indoor environment and a lower energy consumption. A wider proportional band and properly selected proportional and integral gains for controller provides a stable and effective operation.

3. A sensor should be located at a location where the value it senses represents the controlled variable of the whole zone to be controlled. A sensor should be exposed and located in a place that has sufficient air movements. However, it should not be directly under a supply air jet, in which the air temperature and relative humidity are quite different from the occupied zone. It also should not be located in a stratified airstream.

   To measure the duct static pressure, if a long section of straight duct is available (say, a length of duct section greater than or equal to 10 duct diameters), a single point, pitot-tube-type duct static pressure sensor can be used. Otherwise, a multipoint pitot-tube array or flow measuring station with airflow straighteners should be used for velocity pressure or static pressure measurement. The small holes used to measure static pressure in a pitot tube should never be directly opposite to an airstream with a velocity pressure that can affect the reading.

   A reference pressure should be picked up at a point with low air velocity outside the duct, at a point served by the same air system, or in the ceiling plenum.

   A space static pressure sensor should be located in an open area of the conditioned space where the air velocity is less than 40 fpm and where its reading is not affected by the opening of doors. The reference pressure pickup is best located at the rooftop, at a level 10 ft above the building and placed to avoid the influence of wind.

4. If a simpler control system can do the same job as a more complex system, the simpler system is always the first choice.

5. A sufficient, clear, and well-followed operation manual and a well-implemented maintenance schedule are key factors for an effective control system.

REFERENCES AND FURTHER READING

Alley, R. L., "Selecting and Sizing Outside and Return Air Dampers for VAV Economizer Systems," *ASHRAE Transactions*, Part I, pp. 1457–1466, 1988.

Anderson, D., Graves, L., Reinert, W., Kreider, J. F., Dow, J., and Wubbena, H., "A Quasi-Real-Time Expert System for Commercial Building HVAC Diagnostics," *ASHRAE Transactions*, Part II, pp. 954–960, 1989.

ASHRAE Inc., *ASHRAE Handbook 1991, HVAC Applications*, Atlanta, GA.

ASHRAE/IES, *Standard 90.1–1989 Energy Efficient Design of New Buildings Except New Low-Rise Residential Buildings*, AGHRAE Inc., Atlanta, GA, 1989

ASHRAE, Inc., *ASHRAE/IES Standard 90.1–1989 User's Manual*, ASHRAE, Atlanta, GA, 1992.

Barker, K. A., "Programming for DDC Control," *Heating/Piping/Air Conditioning*, pp. 83–87, April 1989.

Brothers, P. W., "Knowledge Engineering for HVAC Expert Systems," *ASHRAE Transactions*, Part I, pp. 1063–1073, 1988.

Burt, W. T., "Adaptive Control," *Heating/Piping/Air Conditioning*, pp. 89–104, November 1987.

Camejo, P. J., and Hittle, D. C., "An Expert System for the Design of Heating, Ventilating, and Air-Conditioning Systems," *ASHRAE Transactions*, Part I, pp. 379–386, 1989.

Coad, W. J., "The Air System in Perspective," *Heating/Piping/Air Conditioning*, pp. 124–128, October 1989.

Coad, W. J., "DX Control Problems with VAV," *Heating/Piping/Air Conditioning*, pp. 134–139, January 1984.

Coggan, D., "Mixed Air Control with DDC," *Heating/Piping/Air Conditioning*, pp. 113–115, May 1986.

DOE, *Code of Federal Regulations (CFR), Title 10, Part 435, Subpart A, Performance Standards for New Commercial and Multifamily High-Rise Residential Buildings*, DOE, 1989

Goldschmidt, I., "State-of-the-Art Building Automation," *Heating/Piping/Air Conditioning*, pp. 41–50, May 1990.

Goswami, D., "VAV Fan Static Pressure Control with DDC," *Heating/Piping/Air Conditioning*, pp. 113–117, December 1986.

Grimm, N. R., and Rosaler, R. C., *Handbook of HVAC Design 1990*, McGraw-Hill, New York.

Hahn, W. G., "Is DDC Inevitable?" *Heating/Piping/Air Conditioning*, pp. 81–95, November 1988.

Haines, R. W., "Outside Air Volume Control in A VAV System," *Heating/Piping/Air Conditioning*, pp. 130–132, October 1986.

Haines, R. W., "Control Strategies for VAV Systems," *Heating/Piping/Air Conditioning*, pp. 147–148, September 1984.

Hartman, T. B., "Stand-Alone Panels and Terminal Controllers," *Heating/Piping/Air Conditioning*, pp. 41–47, November 1990.

Hartman, T. B., "TRAV — A New HVAC Concept," *Heating/Piping/Air Conditioning*, pp. 69–73, July 1989.

Hartman, T. B., "Dynamic Control: Fundamentals and Considerations," *ASHRAE Transactions*, Part I, pp. 599–609, 1988.

Honeywell Inc., *Engineering Manual of Automatic Control for Commercial Buildings*, Minneapolis, MN, 1988.

Howel, R. H., Ganesh, R., and Sauer, H. J., Jr., "Comparison of Two Control Strategies to Simulate Part-Load Performance of a Simple Air Conditioning System," *ASHRAE Transactions*, Part II, pp. 1768–1780, 1987.

Madsen, T. L., Schmidt, T. R., and Helk, U.,"How Important Is the Location of the Room Thermostat," *ASHRAE Transactions*, Part I, pp. 847–852, 1990.

Muxen, S. A., and Chapman, W. F., "Versatile Application-Specific Controllers For Hotel Guest Rooms," *ASHRAE Transactions*, Part I, pp. 1530–1538, 1988.

Nordeen, H., "Ventilation Air Control in HVAC Systems," *Heating/Piping/Air Conditioning*, pp. 71–80, November 1983.

Paoluccio, J. P., and Burfield, J. A., "University Evaluates Dead Band Control Strategy," *ASHRAE Transactions*, Part I, pp. 864–870, 1981.

Petze, J., "Open Protocol for Terminal Unit Controllers," *Heating/Piping/Air Conditioning*, pp. 129–131, September 1989.

Seem, J. E., Armstrong, P. R., and Hancock, C. E., "Algorithms for Predicting Recovery Time from Night Setback," *ASHRAE Transactions*, Part II, pp. 439–446, 1989.

Sosoka, J. R., and Peterson, K. W., "Building Control Systems from the Bottom Up Using Application-Specific Controllers," *ASHRAE Transactions*, Part I, pp. 1521–1529, 1988.

Spitler, J. D., Hittle, D. C., Johnson, D. L., and Pederson, C. O., "A Comparative Study of the Performance of Temperature-Based and Enthalpy-Based Economy Cycles," *ASHRAE Transactions*, Part II, pp. 13–22, 1987.

Spratt, D. K., Sadler, G. W., and Moodie, K. R., "Dynamic Control — A Case Study," *ASHRAE Transactions*, Part II, pp. 193–200, 1989.

Van Horn, M., *Understanding Expert Systems*, Toronto Bantam Books, 1986.

Wacker, P. C., "Economizer Saving Study," *ASHRAE Transactions*, Part I, pp. 47–51, 1989.

Williams, V. A., "VAV System Interactive Controls," *ASHRAE Transactions*, Part I, pp. 1493–1499, 1988.

Yaeger, G. A., "Flow Charting and Custom Programming," *ASHRAE Transactions*, Part I B, pp. 167–174, 1986.

AIR SYSTEM BASICS AND SPACE PRESSURIZATION

19.1   FLOW RESISTANCE

Flow resistance is a property of fluid flow that measures the characteristics of a flow passage that resist the fluid flow in that passage with a corresponding total fluid pressure loss at a specific volume flow rate.

Consider a round duct section, between two cross-sectional planes 1 and 2, with a diameter D and length L, in ft. The fluid flow has a velocity of v, in fpm, and a volume flow rate of $\dot{V}$, in cfm. Figure 19.1 shows the flow resistance R of such a round duct section at a total pressure loss Δp_t, in inches WG.

From Eq. (8.36), the total pressure loss of the fluid flow Δp_t, in inches WG, between planes 1 and 2 can be calculated as

$$\Delta p_t = C_v f \left(\frac{L}{D} \right) \left(\frac{\rho v^2}{2g_c} \right) = \left(\frac{8 C_v f L \rho}{g_c \pi^2 D^5} \right) \dot{V}^2 \quad (19.1)$$

where f = friction factor
  ρ = density of the fluid, lb/ft3
  v = mean air velocity in duct, fpm
  g_c = dimensional constant, 32.2 $\text{lb}_m \cdot \text{ft/lb}_f \cdot \text{s}^2$
  C_v = conversion constant

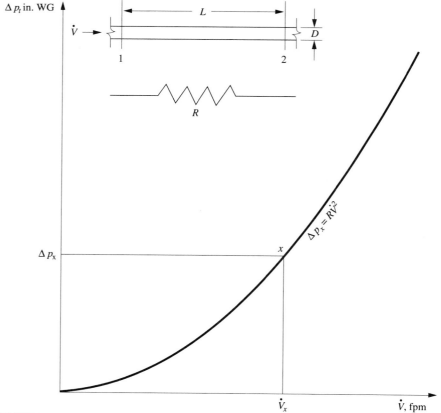

FIGURE 19.1   Total pressure loss Δp_t and flow resistance R of a round duct section.

Let $R = 8C_v fL\rho/(g_c\pi^2 D^5)$, so the total pressure loss becomes

$$\Delta p_t = R\dot{V}^2$$

$$R = \frac{\Delta p_t}{\dot{V}^2} \qquad (19.2)$$

In Eq. (19.2), R is defined as flow resistance, which indicates the resistance to fluid flow of this duct section, and is characterized by its specific total pressure loss and volume flow rate. Flow resistance in I-P units is expressed as in. WG/(cfm)^2.

For a given duct section, D and L are constants. In addition, the difference in the mean values of f and ρ at different volume flow rates is small, so R can be considered a constant. The relationship between Δp_t and $\dot{V}$ of this section can be represented by a parabola whose vertex coincides with the origin.

Differentiating Eq. (19.2) with respect to $\dot{V}$, then,

$$\frac{d\Delta p_t}{d\dot{V}} = 2R\dot{V} \qquad (19.3)$$

This means that the slope of any segment of the curve $\Delta p_t = R\dot{V}^2$ is $2R\dot{V}$. Consequently, the greater the flow resistance R, the steeper the slope of the characteristic curve of this duct section.

Because the flow resistance R of a specific duct section is related to $\dot{V}$ by means of Eq. (19.2) the total pressure loss of the fluid flow in this duct section at any $\dot{V}$ can then be calculated. Therefore, the total pressure loss Δp_t curve of this duct section can be plotted at different $\dot{V}$ values, and forms the Δp_t–$\dot{V}$ diagram. On the Δp_t–$\dot{V}$ diagram, each characteristic curve represents a unique value of flow resistance R.

If a duct section contains additional dynamic losses $C_v C_{dy}\rho v^2/2g_c$ as mentioned in Chapter 8, the relationship between Δp_t, R, and $\dot{V}$ defined by Eq. (19.1) still holds, but the flow resistance R is changed to

$$R = \frac{8C_v\rho[(fL/D) + C_{dy}]}{g_c\pi^2 D^4} \qquad (19.4)$$

where C_{dy} = local loss coefficient.

Flow Resistances Connected in Series

Consider a ductsystem that consists of several duct sections connected in series with air flowing inside the duct, as shown in Fig. 19.2a.

When these duct sections are connected in series, the volume flow rate of air that flows through each section must be the same. The total pressure loss be-

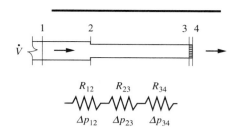

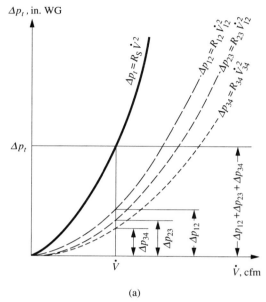

(a)

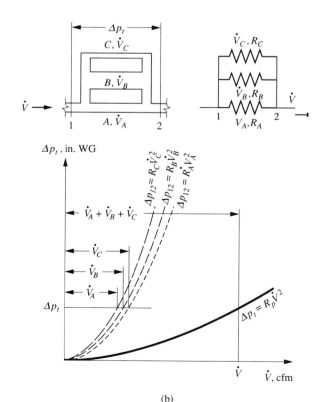

(b)

FIGURE 19.2 Combination of flow resistances: (a) connected in series and (b) connected in parallel.

tween planes 1 and 4 is the sum of the total pressure losses of each duct section, that is

$$\Delta p_t = \Delta p_{12} + \Delta p_{23} + \ldots + \Delta p_n$$
$$= (R_{12} + R_{23} + \ldots + R_n)\dot{V}^2 = R_s \dot{V}^2 \quad (19.5)$$

where $R_{12}, R_{23}, \ldots R_n$ = flow resistances of duct sections 1, 2, ... n, in. WG · (cfm)^2

$\Delta p_{12}, \Delta p_{23} \ldots \Delta p_n$ = total pressure losses across flow resistances $R_{12}, R_{23}, \ldots R_n$, in. WG

$\dot{V}_{12}, \dot{V}_{23}, \ldots \dot{V}_n$ = volume flow rates of air flowing through flow resistances $R_{12}, R_{23}, \ldots R_n$, cfm

The flow resistance of the duct system R_s is equal to the sum of the individual flow resistances of segments connected in series, or

$$R_s = R_{12} + R_{23} + \ldots + R_n \quad (19.6)$$

If the characteristic curves of each individual duct section with flow resistances of $R_{12}, R_{23}, \ldots R_n$ are plotted on a Δp_t–$\dot{V}$ diagram, the characteristic curve of the entire duct can then be plotted either by using the relationship $R_s = \Delta p_t / \dot{V}^2$, or by graphical method. Plotting by graphical method is done by drawing a constant $\dot{V}$ line. The points of Δp_t on the characteristic curve of the duct can be found by summing the individual total pressure losses $\Delta p_{12}, \Delta p_{23}, \ldots \Delta p_n$ of individual duct sections 1, 2, ... n, as shown in Fig. 19.2a.

Flow Resistances Connected in Parallel

When a ductsystem has several segments connected in parallel, as shown in Fig. 19.2b, the total volume flow rate of this duct $\dot{V}$ that flows through cross-sectional planes 1 and 2 is the sum of the individual volume flow rates $\dot{V}_A, \dot{V}_B, \ldots \dot{V}_n$ of each duct section, that is,

$$\dot{V} = \dot{V}_A + \dot{V}_B + \ldots + \dot{V}_n$$
$$= \sqrt{\frac{\Delta p_{12}}{R_A}} + \sqrt{\frac{\Delta p_{12}}{R_B}} + \ldots + \sqrt{\frac{\Delta p_{12}}{R_n}} \quad (19.7)$$

where Δp_{12} = total pressure loss between planes 1 and 2, in. WG

$R_A, R_B, \ldots R_n$ = flow resistance of the duct section A, B, ... n, in. WG · (cfm)^2

As for the entire duct, $\dot{V} = \sqrt{\Delta p_{12}/R_p}$. Here, R_p represents the flow resistance of the entire duct whose sections are in parallel connection. Then,

$$\frac{1}{\sqrt{R_p}} = \frac{1}{\sqrt{R_A}} + \frac{1}{\sqrt{R_B}} + \ldots + \frac{1}{\sqrt{R_n}} \quad (19.8)$$

If duct sections are connected in parallel, the total pressure loss across the common junction 1 and 2 must be the same, or

$$\Delta p_t = \Delta p_{12} \quad (19.9)$$

For two segments with flow resistances R_A and R_B connected in parallel, the flow resistance of the combination is then

$$R_p = \frac{R_A R_B}{R_A + R_B + 2\sqrt{R_A R_B}} \quad (19.10)$$

As in series connection, the characteristic curve of the duct with segments connected in parallel can be plotted on Δp_t–$\dot{V}$ diagram either by the relationship $\Delta p_t = R_p \dot{V}^2$, or by graphical method. In graphical method, draw a constant Δp_t line; the total volume flow rate on the characteristic curve of the duct can be found by adding the sum of the volume flow rates of the individual sections, as shown in Fig. 19.2b.

Flow Resistance of a Y-Connection

When a round duct consists of a main duct section and two branches as shown in Fig. 19.3, a *Y-connection flow circuit* is formed. In a Y-connection flow circuit, the volume flow of the main duct section 01, $\dot{V}_{01}$ is the sum of the volume flow rates at branches *11'* and *12'*, that is, $\dot{V}_{01} = \dot{V}_{11'} + \dot{V}_{12'}$. If the ambient air that surrounds the duct is at atmospheric pressure, then $\Delta p_{11'} = \Delta p_{12'}$ and $\Delta p_{01'} = \Delta p_{02'}$.

Because $\Delta p_{11'} = R_{11'}\dot{V}_{11'}^2 = \Delta p_{12'} = R_{12'}\dot{V}_{12'}^2$,

$$\dot{V}_{11'} = \frac{1}{1 + \sqrt{R_{11'}/R_{12'}}} \dot{V}_{01} = K_{11'}\dot{V}_{01}$$
$$\dot{V}_{12'} = \frac{1}{1 + \sqrt{R_{12'}/R_{11'}}} \dot{V}_{01} = K_{12'}\dot{V}_{01}$$
$$(19.11)$$

Therefore, total pressure loss at ductwork 01' and 02' is calculated as

$$\Delta p_{01'} = \Delta p_{01} + \Delta p_{11'}$$
$$= R_{01}\dot{V}_{01}^2 + R_{11'}\dot{V}_{11'}^2 = (R_{01} + R_{11'}K_{11'}^2)\dot{V}_{01}^2$$
$$= R_{01'}\dot{V}_{01}^2 \quad (19.12)$$

Similarly,

$$\Delta p_{02'} = R_{02'}\dot{V}_{01}^2 \quad (19.13)$$

For a given duct, $C_{dy}, L_{01}, L_{11'}, L_{12'}, D_{01}, D_{11'}$, and $D_{12'}$ are constants. Air density ρ can be considered a constant. Although the friction factor f is a function of air velocity v and the volume flow rate $\dot{V}$, at high

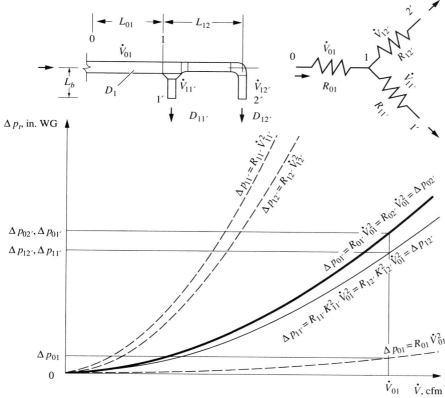

FIGURE 19.3 Flow resistance for a Y-connection.

Reynolds numbers f is least affected, and can also be taken as a constant. Therefore, $R_{01'}$ and $R_{02'}$ can be considered constants.

Figure 19.3 also shows the characteristic curve of a Y-connection plotted on a Δp_t–$\dot{V}$ diagram.

Flow Resistance of a Duct System

Supply and return duct systems are composed of Y-connections. If the ambient air is at atmospheric pressure, the total pressure loss of a supply duct system and its flow characteristics as shown in Fig. 19.4a are given as

$$\Delta p_{0n'} = \Delta p_{01} + \Delta p_{12} + \ldots + \Delta p_{(n-1)n'}$$
$$= R_{01} \dot{V}_{01}^2 + R_{12} K_{12}^2 \dot{V}_{01}^2 + \ldots$$
$$+ R_{(n-1)n'} K_{(n-1)n'}^2 \dot{V}_{01}^2$$
$$= R_{0n'} \dot{V}_{01}^2 \qquad (19.14)$$

and

$$R_{0n'} = R_{01} + R_{12} K_{12}^2 + \ldots + R_{(n-1)n'} K_{(n-1)n'}^2 \quad (19.15)$$

The characteristic curve of a duct system on a Δp_t–$\dot{V}$ diagram is called the *system curve*. It is shown in Fig. 19.4b.

19.2 FAN–DUCT SYSTEMS

A fan–duct system is a system in which a fan or several fans are connected to a duct or to a duct and equipment. More concisely, a fan–duct system is a combination of fan and flow resistance. A fan–duct system is illustrated in Fig. 19.5.

System Operating Point

A system operating point indicates the operating condition and characteristics of a fan–duct system. Because the operating point of the fan must lie on the fan performance curve and the operating point of the duct system must lie on the system curve, the system operating point of a fan–duct system must be the intersection point P of the fan performance curve and the system curve, as shown in Fig. 19.5. At the operating point of the fan–duct system, its volume flow rate is V_P and its total pressure loss is Δp_P.

If the flow resistance of the duct system remains the same while fan performance changes, and its characteristics are represented by a new fan performance curve, the operating point of the fan–duct system moves to point Q in Fig. 19.5. On the other hand, if the fan performance curve remains the same while the

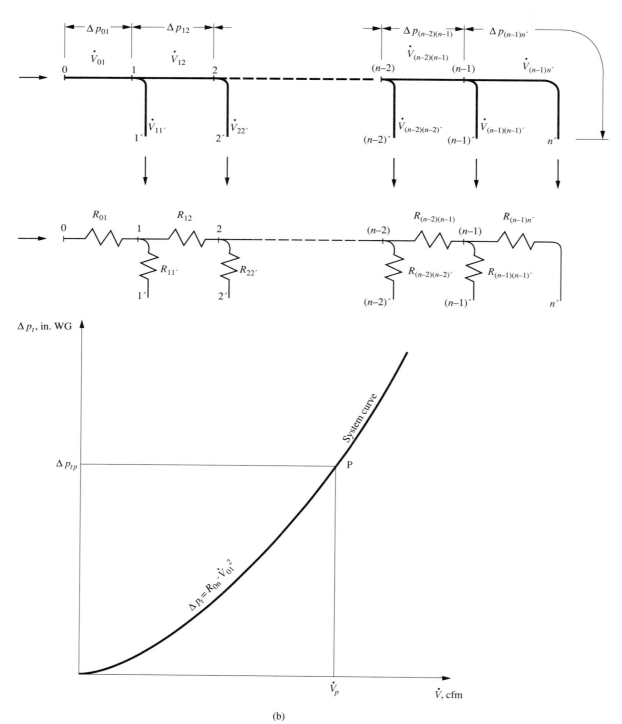

(b)

FIGURE 19.4 Flow characteristics of a supply duct system: (a) schematic diagram and (b) system curve.

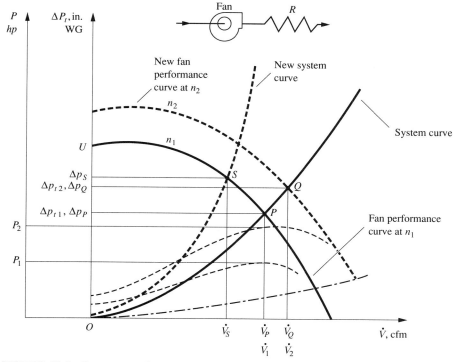

FIGURE 19.5 System operating point.

flow resistance of the duct system is represented by a new system curve, the operating point of the fan–duct system moves from P to S.

The fan in a constant-volume air system should be selected with a $\dot{V}$ and Δp_t so that the system operating point P is near the point of maximum fan total efficiency η_t. The actual operating point of a constant volume system is usually within a range of $\pm$ 10 percent of the wide-open volume from the point of maximum η_t.

Buckingham π Method and Fan Laws

The relationship between various interdependent variables that represent fan characteristics for dynamically similar systems can be determined by using *Buckingham π method.* For two systems to be dynamically similar, they must be similar in shape and construction, that is, geometrically similar. Their velocity distribution or profile of fluid flow should also be similar. The Buckingham π method gives the relationship as

$$\pi = F D^a n^b \rho^c \qquad (19.16)$$

where F = dependent variable such as $\dot{V}$, Δp_t, or power

$\quad\quad D$ = diameter of the impeller, ft

n = fan speed, rpm
ρ = air density, lb/ft^3

In Eq. (19.16), *a, b,* and *c* are indices to be determined so that the π group as a whole is dimensionless.

Variables corresponding to fan–duct systems and their dimensions in terms of mass M, length L, and time t are as follows:

Variables	Units	Indices		
		M	**L**	**t**
Volume flow, $\dot{V}$	cfm	0	3	−1
Total pressure, Δp_t	lb$_f$/ft^2	1	−1	−2
Power, P	ft · lb$_f$/min	1	2	−3
Impeller diameter, D	ft	0	1	0
Speed, n	rpm	0	0	1
Air density, ρ	lb$_m$/ft^3	1	−3	0

For the dimensionless group π_1 including dependent variable volume flow $\dot{V}$ and independent variables D, n, and ρ, their relationship can be found by determining the indices as follows:

$$\pi_1 = \dot{V} D^a n^b \rho^c$$
$$0 = (L^3 t^{-1})(L)^a (t^{-1})^b (ML^{-3})^c$$

Setting up equations in M, L, and t,

$$M \quad 0 = c$$
$$L \quad 0 = 3 + a - 3c, \quad a = -3$$
$$t \quad 0 = -1 - b, \quad b = -1$$

That is,

$$\pi_1 = \dot{V} D^{-3} n^{-1} = \frac{\dot{V}}{D^3 n}$$

Or, $\dot{V}/(D^3 n) = $ constant, then

$$\dot{V} = K_v D^3 n \qquad (19.17)$$

where $K_v = $ volume flow constant.
Similarly,

$$\Delta p_t = K_p D^2 n^2$$
$$P = K_P D^5 n^3 \rho \qquad (19.18)$$

where $K_p = $ pressure constant
$\quad\quad K_P = $ power constant

Eqs. (19.17) and (19.18) can be used to determine the performance of a fan–duct system under the following two conditions:

• *Condition I.* For the same fan–duct system operating at different speeds, $D_1 = D_2$, if the difference in air density is negligible, the volume flow rate ratio is proportional to the speed ratio, or

$$\frac{\dot{V}_2}{\dot{V}_1} = \frac{K_v D_2^3 n_2}{K_v D_1^3 n_1} = \frac{n_2}{n_1} \qquad (19.19)$$

In Eq. (19.19), subscripts 1 and 2 indicate the original and changed operating conditions, as shown in Fig. 19.5.

The total pressure increase ratio is equal to the square of the speed ratio

$$\frac{\Delta p_{t2}}{\Delta p_{t1}} = \frac{K_p D_2^2 n_2^2 \rho_2}{K_p D_1^2 n_1^2 \rho_1} = \frac{n_2^2}{n_1^2} \qquad (19.20)$$

The fan power ratio is equal to the cube of the speed ratio

$$\frac{P_2}{P_1} = \frac{K_P D_2^5 n_2^3 \rho_2}{K_P D_1^5 n_1^3 \rho_1} = \frac{n_2^3}{n_1^3} \qquad (19.21)$$

• *Condition II.* For fan–duct systems with geometrically and dynamically similar fans and the same duct system, that is, the same type of fans having same K_v, K_p, and K_P, operating at high Reynolds numbers and installed in the same duct system, the volume flow ratio, total pressure increase ratio, and power ratio can be expressed as

$$\frac{\dot{V}_2}{\dot{V}_1} = \frac{K_v D_2^3 n_2}{K_v D_1^3 n_1} = \frac{D_2^3 n_2}{D_1^3 n_1}$$

$$\frac{\Delta p_{t2}}{\Delta p_{t1}} = \frac{K_p D_2^2 n_2^2 \rho_2}{K_p D_1^2 n_1^2 \rho_1} = \frac{D_2^2 n_2^2}{D_1^2 n_1^2} \qquad (19.22)$$

$$\frac{P_2}{P_1} = \frac{K_P D_2^5 n_2^3 \rho_2}{K_P D_1^5 n_1^3 \rho_1} = \frac{D_2^5 n_2^3}{D_1^5 n_1^3}$$

Eqs. (19.19) through (19.22) relate the parameters of fan–duct system at various operating conditions. They are often called the *fan laws*.

19.3 SYSTEM EFFECT

Mechanism of System Effect

The *system effect* of a fan–duct system describes the loss of fan performance caused by uneven and nonuniform velocity profiles at the inlet and after the fan outlet using actual operating inlet and outlet connections, as compared with the performance of that fan test unit during laboratory ratings.

Individual fans are tested based on ANSI/ASHRAE Standard 51–1985 and ANSI/AMCA Standard 210–85. The test configuration is shown in Fig. 19.6b. There are no inlet connections. There is also a minimum length of 10 duct diameters of straight duct connected to the fan outlet.

The system effect of a fan–duct system consists of *inlet system effect* and *outlet system effect*.

Inlet System Effect

Consider a 90° elbow that is connected to the fan inlet with a very short connecting duct between the elbow and the fan inlet. Eddies, large-scale turbulences, and a nonuniform velocity profile form after the elbow and are not smoothed to a uniform velocity distribution because of the short connecting duct. Air enters the fan with a nonuniform velocity profile that is quite different from the fan inlet conditions in the standard laboratory test unit. This results in a fan performance curve with lower fan total pressure and volume flow than the catalog fan performance curve, as shown in Fig. 19.6a.

The difference between the fan total pressure in the fan–duct system with a nonuniform velocity profile at fan inlet and the fan total pressure during standard rating at the laboratory is called the inlet system effect. Fans must be selected at a fan total pressure that compensates for the inlet system effect, so that the actual fan performance curve can meet the required fan total pressure at design volume flow at operating point P_d, as shown in Fig. 19.6a.

If a fan–duct system design does not consider the pressure loss caused by the inlet system effect, it will actually operate at point P_a with a lower volume flow

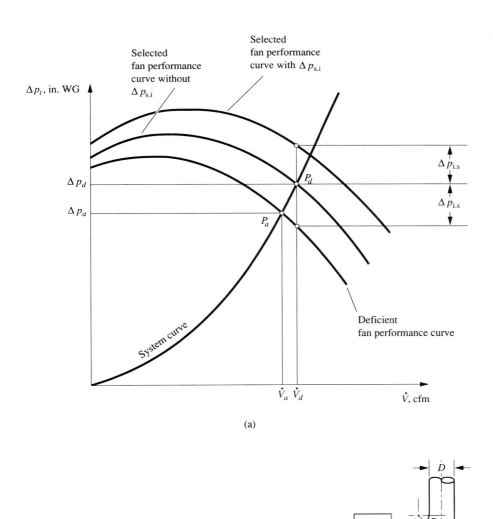

FIGURE 19.6 Inlet system effect: (a) inlet system effect on Δp_t-$\dot{V}$ diagram, (b) fan test unit at laboratory, and (c) round elbow before fan inlet.

and fan total pressure than the calculated values, even though the calculated design operating point is at P_d.

Inlet System Effect Loss

Inlet system effect loss $\Delta p_{s.i}$ (in inches WG) is the difference in pressure loss between actual and catalog performance curves caused by the inlet system effect, and can be calculated as dynamic loss, as defined in Eq. (8.50)

$$\Delta p_{s.i} = C_{s.i}\left(\frac{v_{fi}}{4005}\right)^2 \qquad (19.23)$$

where $C_{s.i}$ = inlet system effect loss coefficient.

In Eq. (19.23), v_{fi} is the air velocity at the fan inlet based on the area of the inlet collar, fpm.

Inlet system effect loss should be added to the system total pressure loss calculated at the design volume flow rate, as described in Eq. (8.22). The selected catalog fan performance curve should meet the system total pressure loss, including inlet system effect loss.

Inlet System Effect Loss Coefficient

According to the Air Movement and Control Association (AMCA) *Fan and Systems Publication 201* (1973), if an elbow is connected by a round duct to the inlet of a centrifugal fan and the length of the connecting duct is equal to 2 duct diameters, the inlet system effect loss coefficients are as follows:

R/D	$C_{s.i}$
0.75	0.8
1.0	0.66
2.0	0.53

Here, R is the radius of the elbow, and D is the duct diameter, both in ft.

If the length of the connecting duct is $5D$, $C_{s.i}$ is approximately half the values listed above. If there is no connecting duct, then $C_{s.i}$ is about twice the values listed above.

According to AMCA *Fan and Systems* (1973), for a square elbow (with or without turning vanes) with an inlet transition and a connecting duct of length $2D$, as shown in Fig. 19.7, the inlet system effect loss coefficients are as follows:

R/H of square elbow	$c_{s.i.}$ Without turning vanes	With turning vanes
0.5	1.6	0.47
0.75	1.2	–
1.0	0.66	0.33
2.0	0.47	0.22

Here, H indicates the dimension of either side of the square duct, in ft.

For a square elbow before a fan inlet with a connecting duct of $5D$ and with turning vanes and an inlet transition, $C_{s.i}$ is approximately 0.6 of the values listed above.

If a square elbow before a fan inlet has turning vanes and an inlet transition with no connecting duct, for an $R/H = 1$, $C_{s.i} = 0.53$; and for an $R/H = 0.5$, $C_{s.i} = 0.8$.

For double-width double-inlet fan installed in a plenum or in a cabinet, the distance between the fan inlet and the wall of the plenum or cabinet should not be less than 0.4 diameter of the fan inlet. If it is only 0.2 diameter of the fan inlet, an inlet system effect loss of $C_{s.i} = 0.08$ should be added to the system total pressure loss at design volume flow rate. For details, refer to AMCA *Fan and Systems* (1973).

The fans in most air-handling units and packaged units installed according to ASHRAE and ARI standards are rated and tested as complete units. The condition of fan inlet of the actual operating fan–duct system is usually the same as the rated condition. There is no inlet system effect loss in such a unit.

Outlet System Effect

If a duct fitting is connected to a fan outlet with a short connecting duct, its dynamic loss Δp_{dy} and local loss coefficient C_{dy} increase because of the nonuniform velocity profile of the airstream discharged from the fan outlet. The difference between the dynamic losses of a duct fitting connected to a fan outlet with a nonuniform and uniform velocity profile at the duct fitting inlet is known as the *outlet system effect loss* $\Delta p_{s.o}$, in inches WG. Outlet system effect results in an increase of flow resistance of the duct system, as shown in Fig. 19.8a.

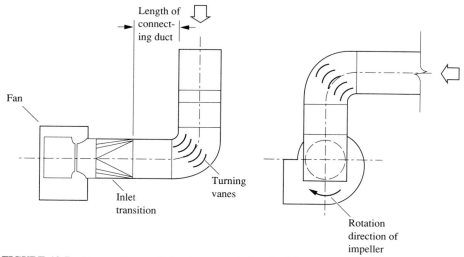

FIGURE 19.7   Square elbow with turning vanes before fan inlet.

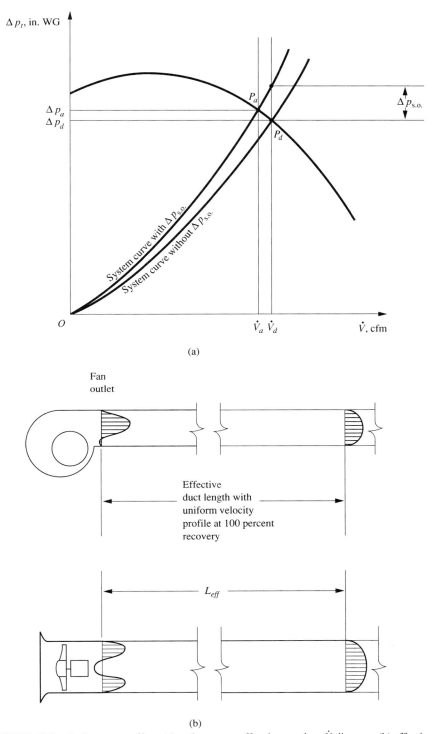

FIGURE 19.8   Outlet system effect: (a) outlet system effect loss on Δp_t-$\dot{V}$ diagram, (b) effective duct length, and (c) orientation of elbow connected to duct.

As in Eq. (19.23), outlet system effect loss at design volume flow rate can be calculated as

$$\Delta p_{s.o} = C_{s.o} \left(\frac{v_{fo}}{4005}\right)^2 \qquad (19.24)$$

where $C_{s.o}$ = outlet system effect loss coefficient
v_{fo} = fan outlet velocity based on fan outlet area, fpm

Outlet system effect loss at design volume flow should be added to the calculated total system pressure loss.

Outlet System Effect Loss Coefficient

Outlet system effect loss $\Delta p_{s.o}$ and outlet system effect loss coefficient $C_{s.o}$ both depend on the effective duct length ratio L/L_{eff}, the air velocity at the fan outlet v_{fo}, and the configuration of the duct fitting with respect to the fan outlet. Here, L indicates the length of duct section between fan outlet and the duct fitting in duct diameters.

Effective duct length L_{eff} is the length of duct between the fan outlet and a cross-sectional plane at which the uniform velocity profile has been completely recovered, in duct diameters or equivalent diameters D. L and L_{eff} must be expressed in the same units. Effective duct length can be calculated as

$$L_{eff} = 2.5 + \frac{v_{fo} - 2500}{1000} \qquad (19.25)$$

For a single-width single-inlet (SWSI) centrifugal fan, the configurations of a 90° elbow with respect to the fan outlet at the end of the connecting duct can be classified as: 90° turn right or 90° turn down, 90° turn left or 90° turn up, parallel with inlet, and opposite to inlet, as shown in Fig. 19.9.

According to AMCA *Fan and Systems,* Publication 201 (1973), the magnitudes of $C_{s.0.8}$ at various L/L_{eff} and elbow orientations with a blast area to fan outlet area ratio $A_{blast}/A_{out} = 0.8$ are as follows:

L/L_{eff}	0.12	0.25	0.5	1.0
90° turn right or down	0.64	0.44	0.20	0
90° turn left or up, parallel with inlet	1.16	0.76	0.36	0
90° turn left or up, opposite to inlet	0.80	0.52	0.24	0

For a SWSI centrifugal fan with $A_{blast}/A_{out} = 0.7$, a multiplier $K_{blast} = 1.4$ should be applied to the above values. If $A_{blast}/A_{out} = 0.6$, $K_{blast} = 2$, and if $A_{blast}/A_{out} = 0.9$, $K_{blast} = 0.75$.

For double-width double-inlet (DWDI) centrifugal fans, an additional multiplier K_{DI} should be applied to the $C_{s.0.8}$ values. For a 90° elbow connected to the duct at horizontal turning orientation, $K_{DI} = 1.25$.

The outlet system effect coefficient $C_{s.o}$ can then be determined as

$$C_{s.o} = K_{DI}K_{blast}C_{s.0.8} \qquad (19.26)$$

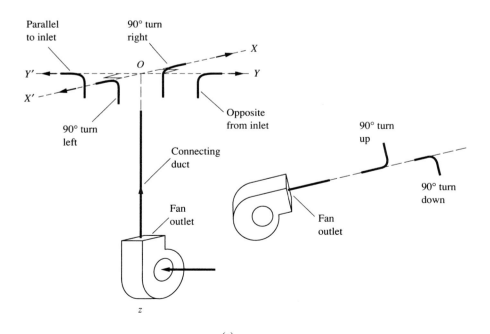

(c)

FIGURE 19.8 Outlet system effect: (a) outlet system effect loss on $\Delta p_t - V$ diagram, (b) effective duct length, and (c) orientation of elbow connected to duct.

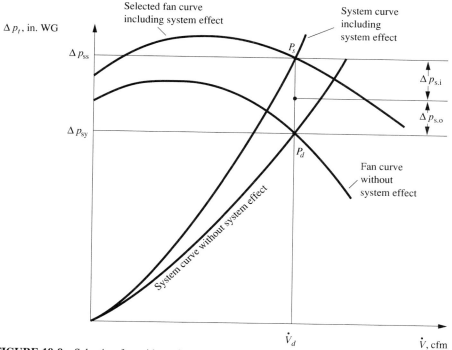

FIGURE 19.9   Selecting fan with performance including system effect loss.

For details, refer to AMCA *Fan and Systems* (1973).

For fans installed in air-handling units and packaged units and rated as complete units, the ratings are based on a straight duct length of two duct diameters from the unit outlet. The outlet system effect loss is not included in the laboratory-rated fan performance curve.

Selecting Fans Considering System Effect Losses

If the calculated volume flow rate of the design operating point P_d (as shown in Fig. 19.9) is $\dot{V}_d$ (in cfm) and the calculated system total pressure loss is Δp_{sy} (in inches WG), the selected fan that compensates for inlet and outlet system effect losses must operate on a performance curve that intersects with point P_s, which has a volume flow rate of $\dot{V}_d$ and a total pressure increase of Δp_{ss} including system effect losses (in inches WG). That is,

$$\Delta p_{ss} = \Delta p_{sy} + \Delta p_{s.i} + \Delta p_{s.o} \qquad (19.27)$$

System effect directly affects the performance of the air system as well as the function of the air conditioning system. Because the air velocity at the fan inlet and outlet is usually between 1500 and 3000 fpm, system effect losses may amount to 5 to 20 percent of the system pressure loss of the air system.

Inlet and outlet system effect losses must be calculated correctly. Moreover, systems must be designed to reduce system effect losses to increase energy efficiency.

19.4   COMBINATION OF FAN–DUCT SYSTEMS

The purposes of combining fan–duct systems are as follows:

- To increase the volume flow rate of the combined system
- To provide the required pressure characteristics for the combined fan–duct system
- To balance the volume flow of the combined fan–duct system
- To maintain a required space pressure characteristic

Two Fan–Duct Systems Connected in Series

Figure 19.10 shows two fan–duct systems connected in series equipped with fans F_1 and F_2 and the corresponding two duct systems represented by flow resistances R_1 and R_2.

Flow resistance R_1 may be the combination of two flow resistances R_{1A} and R_{1B} connected in series, and flow resistance R_2 may be the combination of two flow resistances R_{2A} and R_{2B}. The volume flow

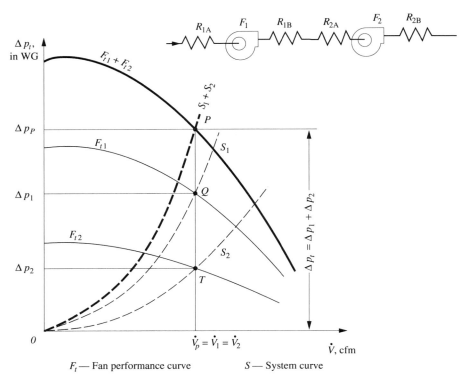

FIGURE 19.10 Two fan–duct systems connected in series.

rate of the combined fan–duct system $\dot{V}_P$, in cfm, must be the same as the volume flow rates $\dot{V}_1$ or $\dot{V}_2$ of fan–duct system 1 or 2 and their components fan F_1, fan F_2, or

$$\dot{V}_P = \dot{V}_1 = \dot{V}_2 \qquad (19.28)$$

The total pressure loss across flow resistance R_1 has a magnitude Δp_1, and the relationship $\Delta p_1 = R_1 \dot{V}_1^2$ holds. This relationship also determines the system curve S_1. Here, the total pressure loss in the duct system includes system effect loss. Fan performance curve F_1 can be obtained from the data given in the manufacturer's catalog.

Total pressure loss Δp_1 and volume flow $\dot{V}_1$ also determine the operating point of fan–duct system 1 at point Q, which is graphically the intersection of fan performance curve F_{t1} and system curve S_1. For fan–duct system 2, Δp_2, $\dot{V}_2$, R_2, fan performance curve F_{t2}, system curve S_2, and system operating point T can be similarly determined.

The total pressure loss of the combined fan–duct system Δp_t, in inches WG, is

$$\Delta p_t = \Delta p_1 + \Delta p_2 \qquad (19.29)$$

Graphically, the combination of the curves of two fans connected in series on a $\Delta p_t - \dot{V}$ diagram can be

performed by drawing several constant volume flow rate $\dot{V}$ lines. For each constant volume line, the fan total pressure of the combined fan curve ($F_{t1} + F_{t2}$) always equals the sum of the sections of fan total pressure of fan curve F_{t1} and fan curve F_{t2}, represented by Δp_1 and Δp_2 as shown in Fig. 19.10. The purpose of connecting two fan–duct systems in series is to increase the fan total pressure that can be provided by the combined system.

Figure 19.11 shows the system pressure characteristics of a combined fan–duct system with fans F_1 and F_2 and a single fan–duct system that uses only one fan F. It can be seen that the static pressure across the duct wall is considerably higher in a fan–duct system with a single fan than in the combined fan–duct system with two fans. Of course, a single-fan system has a lower initial cost.

Two fan–duct systems are connected in series, the volume flow rates of the two fans should be similar. If a large fan is connected in series with a small fan, at large volume flow rates, the fan total pressure of the combined system may be less than if the large fan is operated alone. This loss of efficiency results from the influence of negative total pressure of the small fan when it is operated at a volume flow rate greater than its wide-open volume flow.

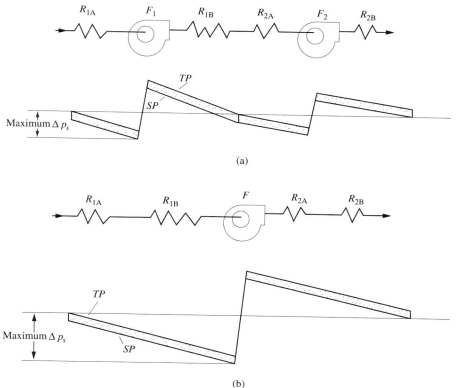

FIGURE 19.11   Comparison of pressure characteristics: (a) combined fan–duct system and (b) single fan–duct system.

Fans Combined in Parallel and Connected in Series with a Duct System

When two fans F_1 and F_2 combined in parallel are connected to a duct system represented by flow resistance R, as shown in Fig. 19.12, the volume flow rate of the parallel combined fans $\dot V_P$ is the sum of the volume flow rates of individual fans $\dot V_1$ and $\dot V_2$.

The purpose of such a combination is to increase the volume flow rate of the system. Usually, the fans in parallel combination are identical fans, that is, fans of the same model with the same rated volume flow and fan total pressure increase. As in fan–duct systems connected in series, a large fan should not be combined in parallel with a small fan because of the possible negative effect of the small fan.

The procedure for determining the combined fan performance curve $(F_{t1} + F_{t2})$ of the two identical fans connected in parallel, is as follows:

1. Plot the fan performance curve F_{t1} according to the manufacturer's data.

2. Draw the constant total pressure increase Δp_t lines (horizontal lines). Points on the combined curve along the constant Δp_t lines have a volume flow

of $2\dot V_1$. Here, $\dot V_1$ indicates the volume flow of each identical fan along the constant Δp_t lines.

3. Plot the combined curve $(F_{t1} + F_{t2})$ by connecting points having $\dot V = 2\dot V_1$ on the constant Δp_t lines.

The system curve can be plotted based on the relationship $\Delta p_t = R\dot V^2$. The intersection of the combined curve $(F_{t1} + F_{t2})$ and the system curve is the system operating point P. The volume flow at the system operating point $\dot V_P$, in cfm, is the volume flow rate flowing through the duct system at a flow resistance R. The combined system total pressure at the system operating point Δp_P, in inches WG, is the fan total pressure of the identical fans or the system total pressure loss of the duct system.

Example 19.1.   An air conditioning system uses two identical fans combined in parallel and connected in series with a duct system, as shown in Fig. 19.12. The pressure–volume and power–volume characteristics of each of the identical fans are listed below:

$\dot V$, cfm	2500	5000	7500	10,000	12,500	15,000
Δp_t, in. WG	5.00	4.88	4.54	4.00	3.13	1.75
Power P_f, hp	3.93	5.90	7.14	7.87	8.20	7.60

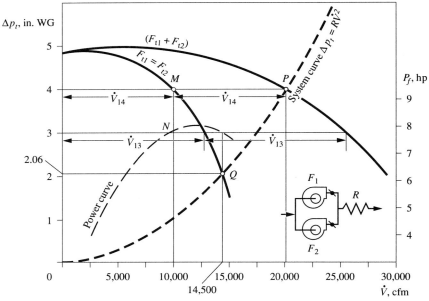

FIGURE 19.12 Two fans combined in parallel and connected in series with a duct system.

If the total pressure loss of this air conditioning system is known to be 1 in. WG at an air flow of 10,000 cfm, determine the following:

1. The system operating point, the corresponding air volume flow rate, and the system total pressure when both fans are operating

2. The operating point, the corresponding air flow rate, and the fan total pressure when only one fan is running, assuming that the air damper after the other fan is closed

3. The fan total efficiency when two fans or only one fan is running

Solution

1. From the given data, the flow resistance of this air conditioning system is

$$R = \frac{\Delta p_t}{\dot{V}^2}$$

$$= \frac{1}{10,000^2} = 1 \times 10^{-8} \text{ in. WG/(cfm)}^2$$

The system pressure loss at various air flow rates are as follows:

$\dot{V}$, cfm	5000	10,000	15,000	20,000	25,000
Δp_t, in. WG	0.25	1.0	2.25	4.0	6.25

After plotting the single-fan performance curve, the combined curve of the two fans in parallel can be drawn. Plot the system curve from the data given above. The volume flow rate and the total pressure loss at operating point P, which is the intersection of the combined curve and the system curve when two

fans are running (as shown in Fig. 19.12), are found to be

$$\dot{V}_P = 20,000 \text{ cfm} \qquad \Delta p_P = 4.0 \text{ in. WG}$$

2. When only one fan is operating and the damper after the other fan is closed, the intersection of the single-fan curve and the system curve is at point Q (see Fig. 19.12). At point Q

$$\dot{V}_Q = 14,500 \text{ cfm} \qquad \Delta p_Q = 2.06 \text{ in. WG}$$

3. When two fans are running, each fan has a volume flow of $20,000/2 = 10,000$ cfm and a fan total pressure of 4 in. WG. From the power curve, the corresponding fan power input is 7.75 hp. From Eq. (10.6), the fan total efficiency is then

$$\eta_{\text{two}} = \frac{\Delta p_t \dot{V}}{6356 P_f} = \frac{4 \times 10,000}{6356 \times 7.75}$$

$$= 0.81 \text{ or } 81 \text{ percent}$$

When only one fan is running, the corresponding fan power input is also 7.75 hp. The fan total efficiency is

$$\eta_{\text{one}} = \frac{2.06 \times 14,500}{6356 \times 7.75}$$

$$= 0.61 \text{ or } 61 \text{ percent}$$

Two Parallel Fan–Duct Systems Connected with Another Duct System

When two parallel fan–duct systems are connected in series to a third duct system represented by flow

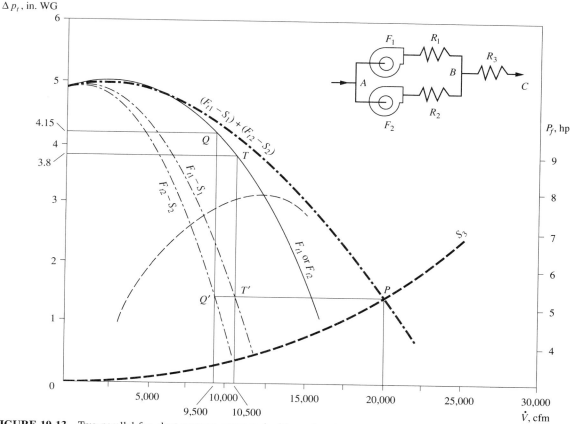

FIGURE 19.13 Two parallel fan–duct systems connected with another duct system.

resistance R_3, as shown in Fig. 19.13, the volume flow rate $\dot{V}_3$ flowing through R_3 is the sum of volume flow rates flowing through the parallel combined fan–duct systems $\dot{V}_1$ and $\dot{V}_2$:

$$\dot{V}_{BC} = \dot{V}_3 = \dot{V}_1 + \dot{V}_2.$$

Although the total pressure loss Δp_{AB} between points A and B of parallel combined fan–duct systems 1 and 2 are the same, flow resistances R_1 and R_2 may be different from each other, and therefore, at their own operating points, the fan total pressure of fan F_1, Δp_{t1} may also be different from the fan total pressure of fan F_2, Δp_{t2}, even though fans F_1 and F_2 are identical.

Through the concept of residual pressure of a fan–duct system, the performance of two parallel combined fan–duct systems can be determined.

When fan F_1 is connected in series with flow resistance R_1, the total pressure loss across R_1 is $\Delta p_{R1} = R_1 \dot{V}^2$. If the volume flow rate of air flowing through R_1 is of a given value $\dot{V}_1$, the residual pressure after flowing through the flow resistance R_1, Δp_{res1}, in inches WG, is

$$\Delta p_{res1} = \Delta p_{F1} - \Delta p_{R1} = \Delta p_{F1} - R_1 \dot{V}_1^2 = F_{t1} - S_1$$

$$(19.30)$$

where $\Delta p_{F1}, F_{t1}$ = fan total pressure of fan F_1, in. WG

$\Delta p_{R1}, S_1$ = total pressure loss across flow resistance R_1 or total pressure loss of duct system 1, in. WG

For fan–duct system 2, residual pressure can be similarly calculated.

For two fan–duct systems connected in parallel, the residual pressure of fan–duct system 1 after R_1, Δp_{res1} must be equal to the residual pressure of fan–duct system 2 after R_2, Δp_{res2} because there is only one unique total pressure p_{tB} at point B.

If two parallel combined fan–duct systems are connected in series with another duct system having flow resistance R_3, the residual pressure provided after fan–duct systems 1 and 2 can be used to overcome the total pressure loss across flow resistance R_3, as follows:

$$\Delta p_{res1} = \Delta p_{res2} = R_3 \dot{V}_3^2 \qquad (19.31)$$

Because the pressure at points A and C is equal to atmospheric pressure (zero), the performance of two parallel combined fan–duct systems connected in series with another duct system can be determined as follows:

1. Plot two fan curves F_{t1} and F_{t2}, the residual pressure curves $(F_{t1} - S_1)$ and $(F_{t2} - S_2)$, and the combined residual pressure curve $(F_{t1} - S_1) + (F_{t2} - S_2)$. The residual pressure $(F_t - S)$ curve can be plotted as follows:

   - At a volume flow of $\dot{V}_A$, find a corresponding fan total pressure $\Delta p_{tfA} = F_{tA}$ from the fan curve given by manufacturer.
   - Calculate the pressure drop across the duct system, the flow resistance R at a volume flow $\dot{V}_A$, $S_A = \Delta p_{tRA} = R\dot{V}_A^2$.
   - Calculate residual pressure $\Delta p_{resA} = F_{tA} - S_A$ at a volume flow of $\dot{V}_A$.
   - Plot point A on the residual pressure $(F_t - S)$ curve at a volume flow of $\dot{V}_A$ and a residual pressure of Δp_{resA}.
   - Similarly, plot points B, C, ... on the residual pressure curve.
   - Residual pressure curve $(F_t - S)$ can be drawn by joining points A, B, C, etc.

   At each point on a residual curve, there is a corresponding residual pressure $\Delta p_{res} = (F_t - R\dot{V}^2)$ that can be provided at a corresponding volume flow of $\dot{V}$ to overcome the flow resistance R of a connected series.

   Combining residual pressure curves is performed in the same way as combining fan curves.

2. Calculate R_3 from the required volume flow rate and total pressure loss in that duct system represented by R_3. From the relationship $\Delta p_t = R_3 \dot{V}_3^2$, draw system curve S_3.

3. Determine the intersection point P of the combined residual pressure curve $(F_{t1} - S_1) + (F_{t2} - S_2)$ and the system curve S_3, which denotes the operating condition at R_3. The residual pressure at point P is the pressure that can be provided by fan–duct systems 1 and 2 to overcome the total pressure loss across R_3 at that volume flow rate $\dot{V}_3 = \dot{V}_1 + \dot{V}_2$.

   In fan–duct system design, if the volume flow rate through R_3 thus found is greater or smaller than the required volume flow rate, select a larger or smaller fan to make them approximately the same.

4. Draw a horizontal line PQ' from point P, which intersects with residual pressure curves at points T' and Q', and draw vertical lines QQ' and TT' from points Q' and T'. Points Q and T on fan curve F_1 or F_2 indicate the volume flow, fan total pressure of the fan–duct systems 1 and 2, and fan power input to fan F_1 and fan F_2.

Example 19.2. An air conditioning system is equipped with two parallel combined fan–duct systems connected in series with another duct system, as shown in Fig. 19.13. The fans in the fan–duct systems are identical and have the same pressure–volume and power-volume characteristics as listed in Example 19.1.

Flow resistances R_1, R_2, and R_3 have the following pressure–volume characteristics:

- R_1 has a Δp_t loss of 2.29 in. WG at a $\dot{V}$ of 10,500 cfm.
- R_2 has a Δp_t loss of 2.82 in. WG at a $\dot{V}$ of 9500 cfm.
- R_3 has a Δp_t loss of 1.50 in. WG at a $\dot{V}$ of 20,000 cfm.

Determine

1. Air volume flow rate flowing through the duct system having flow resistance R_3
2. Operating point of fan–duct systems 1 and 2
3. Fan total efficiency of fans 1 and 2

Solution

1. From the given data, the flow resistances can be calculated as

$$R_1 = \frac{\Delta p_t}{\dot{V}^2} = \frac{2.29}{(10,500)^2}$$

$$= 2.08 \times 10^{-8} \text{ in. WG/(cfm)}^2$$

$$R_2 = \frac{2.82}{9500^2} = 3.12 \times 10^{-8} \text{ in. WG/(cfm)}^2$$

$$R_3 = \frac{1.5}{20,000^2} = 3.75 \times 10^{-9} \text{ in. WG/(cfm)}^2$$

2. From Eq. (19.30), the residual pressure of fan–duct systems 1 and 2 after R_1 and R_2 at a volume flow rate of 2500 cfm can be calculated as

$$\Delta p_{res1} = \Delta p_{F1} - R_1 \dot{V}_1^2$$

$$= 5.0 - [2.08 \times 10^{-8} \times (10,500)^2]$$

$$= 4.87 \text{ in. WG}$$

$$\Delta p_{res2} = 5.0 - [3.12 \times 10^{-8} \times (9500)^2]$$

$$= 4.80 \text{ in. WG}$$

Similarly, the residual pressure of other volume flow rates can be calculated as listed below:

$\dot{V}$, cfm	2500	5000	7500	10,000	12,500
F_t, in. WG	5.0	4.88	4.54	4.0	3.13
R_1, in. WG/(cfm)^2	0.13	0.52	1.17	2.08	3.25
R_1, in. WG/(cfm)^2	0.20	0.78	1.76	3.12	4.88
$F_{t1} - S_1$, in. WG	4.87	4.36	3.37	1.92	
$F_{t2} - S_2$, in. WG	4.80	4.10	2.78	0.88	

3. At various volume flow rates flowing through the duct system with flow resistance R_3, the total pressure loss $\Delta p_t = R_3 \dot{V}_3^2$ can be calculated as follows:

$\dot{V}$, cfm	5000	10,000	15,000	20,000	25,000
$S_3 = R_3 \dot{V}_3^2$, in. WG	0.09	0.38	0.84	1.50	2.34

4. Plot the single fan curve F_t, the residual pressure curves $(F_{t1} - S_1)$ and $(F_{t2} - S_2)$, the combined residual pressure curve $(F_{t1} - S_1) + (F_{t2} - S_2)$, and the system curve S_3 on $\Delta p_t - \dot{V}$ diagram as shown in Fig. 19.13. The intersection point P of the combined residual pressure curve $(F_{t1} - S_1) + (F_{t2} - S_2)$ and the system curve S_3 gives the volume flow rate in the duct system with flow resistance R_3. From the diagram, $\dot{V}_P = 20,000$ cfm.

5. Draw horizontal line PQ' from point P to intersect the $(F_{t2} - S_2)$ curve at Q' and the $(F_{t1} - S_1)$ curve at point T'. Again, draw vertical lines $Q'Q$ and $T'T$ from points Q' and T'. The lines intersect fan curve F_t at points Q and T. For fan–duct system 1, operating point T gives

$$\dot{V}_1 = 10,500 \text{ cfm} \quad \text{and} \quad \Delta p_{R1} = 3.8 \text{ in. WG}$$

For fan–duct system 2, the operating point Q gives

$$\dot{V}_2 = 9500 \text{ cfm} \quad \text{and} \quad \Delta p_{R2} = 4.15 \text{ in. WG}$$

6. From the plotted power–volume curve in Fig. 19.13, for operating point T, the fan power input $P_{f1} = 7.8$ hp, and for operating point Q, the fan power input $P_{f2} = 7.6$ hp. Fan total efficiency of fan 1 is therefore

$$\eta_1 = \frac{\dot{V}_1 \Delta p_{R1}}{6356 P_{f1}} = \frac{10,500 \times 3.8}{6356 \times 7.8} = 0.80 \text{ or } 80 \text{ percent}$$

Fan total efficiency for fan 2 is

$$\eta_2 = \frac{9500 \times 4.15}{6356 \times 7.6} = 0.82 \text{ or } 82 \text{ percent}$$

19.5 MODULATION OF THE FAN–DUCT SYSTEM

Modulation Curve

The *modulation curve* of a fan–duct system or, practically, an air system, is its operating curve when its volume flow rate is modulated at part-load operation. A modulation curve is also the locus of the system operating points at reduced system loads and volume flow rates.

For a variable-air-volume VAV system with duct static pressure control, the pressure loss of the air system can be divided into fixed part Δp_{fix} and variable part Δp_{var} when the volume flow rate varies. This will be described in detail in the Section 19.8.

Figure 19.14 shows a typical modulation curve of a VAV system with duct static pressure control using inlet vanes to modulate fan capacity. This VAV system has many fan curves F_t at various volume flow rates and a system curve for the variable pressure loss S_{var}.

During reduction of the volume flow rate of a VAV system, the dampers in the VAV boxes close partially. The variable part of the pressure loss of the air system Δp_{var} from the AHU or PU up to the pressure sensor of duct static pressure control varies as volume flow is reduced. Its pressure–volume characteristics

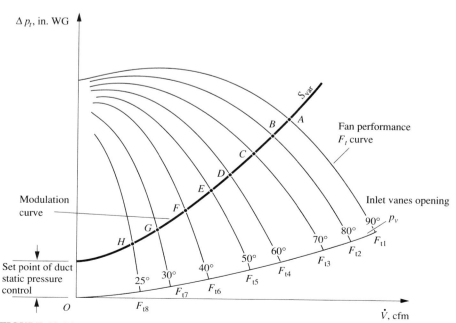

FIGURE 19.14 Modulation curve of a VAV system installed with duct static pressure control.

are indicated by the system curve S_{var}. However, the pressure loss of the duct system, branch take-offs, and VAV boxes beyond the duct static pressure sensor remains constant, and is equal to the set point of duct static control Δp_{fix}. The modulation curve is actually the system curve of the part of VAV air system with variable pressure loss S_{var}.

At the same time, the inlet vanes at the fan inlet also reduce their opening in response to the sensed higher duct static pressure. Therefore, a new fan curve F_{t2} is formed. The intersection of F_{t2} and system curve S_{var} at the reduced volume flow rate is the system operating point B. Similarly, at a further reduction of volume flow rate, the system operating points are C, D, E, etc.

As the volume flow rate of the air system approaches zero, theoretically, the total pressure of the air system will equal the fixed part of pressure loss Δp_{fix}, which is equal to the set point of duct static pressure control, as shown in Fig. 19.14.

The modulation curve of a fan–duct system should not enter the fan surge area or stall region, or unstable operation will result. However, a forward-curved

centrifugal fan may be operated in the surge area if system pressure is below 1.5 in. WG. Refer to the manufacturer's data for detailed analysis.

Modulation of Fan–Duct Systems

The performance of a fan–duct system can be modulated either by changing the fan characteristics or by varying the flow resistance of the duct system. This can be achieved by any of the following methods.

MODULATION USING DAMPERS. Use of a damper is the simplest method of modulation. A multiblade damper is usually installed inside the main duct after the centrifugal supply fan.

When the damper closes, the system flow resistance increases and the system operating point moves from point A to B on a different system curve. This method is known as "riding on the curve" as shown in Fig. 19.15a. If the fan F_t curve is flatter, then the increase in Δp_t during reduction of the volume flow rate of the fan–duct system is small.

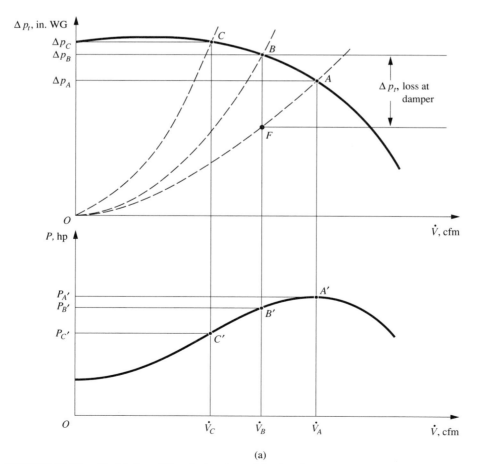

(a)

FIGURE 19.15a Modulation of fan–duct systems: (a) using dampers.

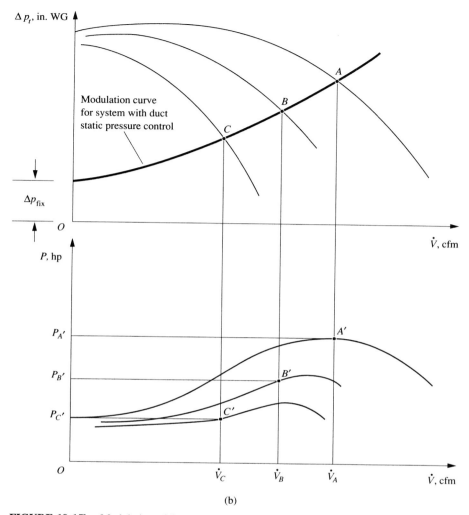

FIGURE 19.15b   Modulation of fan–duct systems: (b) using inlet vanes.

Because of the considerable total pressure loss across the damper, power input to the fan is only slightly decreased by this method. Modulation dampers have the lowest installation cost and only small energy savings.

MODULATION USING INLET VANES. Modulation by varying the opening of inlet vanes at the centrifugal fan inlet gives different fan performance curves and therefore, different system operating points. It is widely used in many VAV systems. The surge area of a fan with inlet vanes is smaller than that of a fan without inlet vanes.

For a VAV system with duct static pressure control, a modulation curve and the corresponding fan power inputs are shown by points A', B', and C' on the power–volume curves as shown in Fig. 19.15b.

Compared to modulation by damper, modulation using inlet vanes has a moderately high installation cost and considerable energy savings.

MODULATION USING INLET CONES. Moving the inlet cone of a backward-curved or airfoil centrifugal fan also gives different fan performance curves so that its modulation curve is similar to that of inlet vanes.

Modulation by inlet cone for backward-curved and airfoil centrifugal fans has a comparatively low installation cost and considerable energy savings.

MODULATION BY BLADE PITCH VARIATION OF AXIAL FAN. Modulation of the volume flow of a fan–duct system by blade pitch (blade angle) variation of an axial fan changes the fan characteristics and, therefore, the system operating points.

Modulation by automatic blade pitch variation significantly lowers the fan energy consumption at reduced volume flow rate. Although automatic blade pitch variation is rather expensive, it has become more popular in recent years.

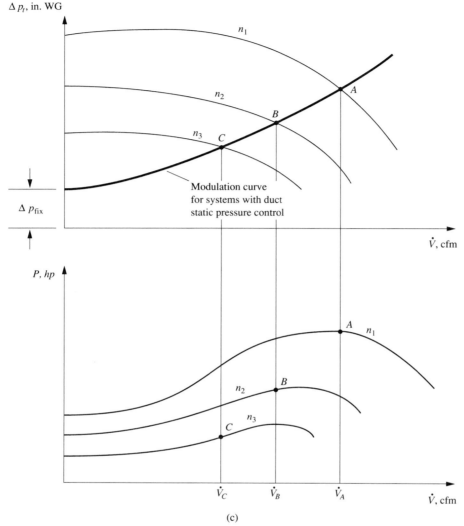

Δp_t, in. WG

n_1

n_2

n_3 C

B

A

Modulation curve
for systems with duct
static pressure control

Δp_{fix}

$\dot{V}$, cfm

P, hp

A n_1

B

n_2

C n_3

$\dot{V}_C$ $\dot{V}_B$ $\dot{V}_A$ $\dot{V}$, cfm

(c)

FIGURE 19.15c Modulation of fan–duct systems: (c) using an AC inverter to vary fan speed.

MODULATION BY VARYING FAN SPEED. Modulation of the fan–duct system by varying the fan speed renders new fan performance curves N_2 and N_3 and, therefore, new system operating points B and C at reduced volume flow rates as shown in Fig. 19.15c. The modulation curve of a VAV system with duct static pressure control can be obtained in the same way as for inlet vanes.

Fan power input for fan speed modulation is lower than for damper or inlet vanes modulation. Because of the fan speed reduction at reduced $\dot{V}$, the fan sound power level is reduced accordingly.

Fan speed variation using adjustable-frequency AC drives in a VAV system has become more popular. An adjustable-frequency AC inverter and motor consume only 7 to 8 percent more energy input to the fan motor. The high initial cost of adjustable-frequency AC drives is often cost-effective for large centrifugal fans. Initial cost may drop further in the future as more and more AC inverters are installed.

Because of the differences in fan characteristics, inlet-vane and AC-drive fan speed modulation are widely used in centrifugal fan–duct systems, and blade-pitch and AC-drive modulation are used for axial fan–duct systems.

Example 19.3. A variable-air-volume VAV system equipped with a backward-curved centrifugal fan with airfoil blades operated at 1700 rpm has the following pressure–volume and power–volume characteristics:

$\dot{V}$, cfm	5000	10,000	15,000	20,000	25,000
Δp_t, in. WG	6.0	6.0	5.8	4.65	2.70
Power, hp	10.0	15.2	18.7	21.0	20.0

At design conditions, this system has a volume flow rate of 20,000 cfm and a system total pressure loss of 4.65 in. WG. When it is operated at 50 percent of its design flow rate and its flow rate is modulated by inlet vanes, its fan performance curves have the following characteristics:

$\dot{V}$, cfm	5000	7500	10,000	12,500	15,000
Δp_t, in. WG	4.88	4.55	4.0	3.12	1.87
P_f, hp	9.6	11.2	12.0	12.3	12.0

Determine the fan power input if the volume flow rate is reduced to 50 percent of its design flow rate by the following methods:

1. Volume control damper

2. Inlet vanes, where system total pressure loss at 50 percent design flow is 4 in. WG

3. Fan speed variation by means of an adjustable-frequency AC inverter

Solution

1. From the given data, the flow resistance of this VAV system at design condition is

$$R_d = \frac{\Delta p_t}{\dot{V}^2} = \frac{4.65}{(20,000)^2}$$

$$= 1.16 \times 10^{-8} \text{ in. WG/(cfm)}^2$$

The system total pressure loss at various volume flow rates can then be calculated as

$$S = \Delta p_t = R_d \dot{V}^2 = 1.16 \times 10^{-8}\dot{V}^2.$$

Plot the fan performance curve F_t and the system curve S. The intersection of these curves is the system operating point P at design conditions, as shown in Fig. 19.16a.

2. When the volume flow rate is reduced by a damper to 50 percent of its design volume flow, that is, $0.5 \times 20,000 = 10,000$ cfm, its operating point must lie on the fan performance curve with a $\dot{V} = 10,000$ cfm, point Q. The fan power input at point Q is 15.2 hp, as shown in Fig. 19.16a.

3. When volume flow is reduced by modulation of inlet vanes to 10,000 cfm, the system total pressure loss is 4 in. WG, so the flow resistance of this VAV system at reduced flow using inlet vanes is

$$R_{iv} = \frac{4}{(10,000)^2} = 4 \times 10^{-8} \text{ in. WG/(cfm)}^2$$

From the given data, plot the fan performance curve using inlet vanes F_{iv} and the system curve $S_{iv} = R_{iv}\dot{V}^2$, as shown in Fig. 19.16b. Also plot the power–volume curve. The intersection point P_{iv} of F_{iv} and S_{iv} curves is the system operating point at 10,000 cfm volume flow, and its fan power input is 12 hp, as shown in Fig. 19.16b.

4. For the same fan–duct system, from the fan law, the fan speed required at reduced volume flow of 10,000 cfm is

$$n_2 = n_1\left(\frac{\dot{V}_2}{\dot{V}_1}\right) = 1700\left(\frac{10,000}{20,000}\right) = 850 \text{ rpm}$$

Fan power input at a fan speed $n_2 = 850$ rpm is

$$P_2 = P_1\left(\frac{n_2}{n_1}\right)^3 = 21\left(\frac{10,000}{20,000}\right)^3 = 2.63 \text{ hp}$$

The difference in fan power inputs between using damper and AC-drive fan speed variation is the power consumption at the damper: $15.2 - 2.63 = 12.5$ hp. At reduced fan speeds, fan total efficiency may be reduced. Because of the degradation of motor efficiency and AC inverter loss at part-load, actual power input at 50 percent part-load for AC inverter fan speed modulation is far higher than 2.63 hp.

19.6 SPACE PRESSURIZATION
Space Pressure Characteristics

Space pressurization is a process that maintains a static pressure difference between space air and the air in the surrounding area, either outdoor air at atmospheric pressure or air in adjacent rooms, for improved indoor environmental control.

As mentioned in Chapter 5, normally, slightly positive space pressure should be maintained if the required level of contamination and environmental control of the conditioned space is higher than that of surrounding area. In laboratories with toxic gas exhaust systems, negative space pressure should be maintained. The pressure differential between space air p_o and outdoor air p_o, $\Delta p_{r.o}$, is usually between ± 0.02 and ± 0.1 in. WG, typically ± 0.05 in. WG. A space positive pressure of $+0.1$ in. WG exerts a force of 8 lb_f on a door with an area of 16 ft^2. An excessive pressure differential may make opening and closing doors difficult.

A positive space pressure means that $p_r > p_o$ or $p_r > p_{sur}$. Here p_{sur} denotes static pressure in the surrounding area, in inches WG. When positive pressure is maintained in the conditioned space, in addition to the air returned and exhausted from the space, a certain amount of air is squeezed out or exfiltrated from the space through openings and cracks in the building shell. Unconditioned outdoor air or contaminated air from the surrounding area is prevented from infiltrating the conditioned space because of this positive space pressure. A negative space pressure means that $p_r < p_o$ or $p_r < p_{sur}$, which induces infiltration.

Except in air jets, the velocity pressure in the conditioned space is small and is usually ignored in space pressurization analysis.

The space pressure difference between the conditioned space and the surrounding area is affected by the stack effect, wind effect, and forced ventilation of air systems or other mechanical ventilation.

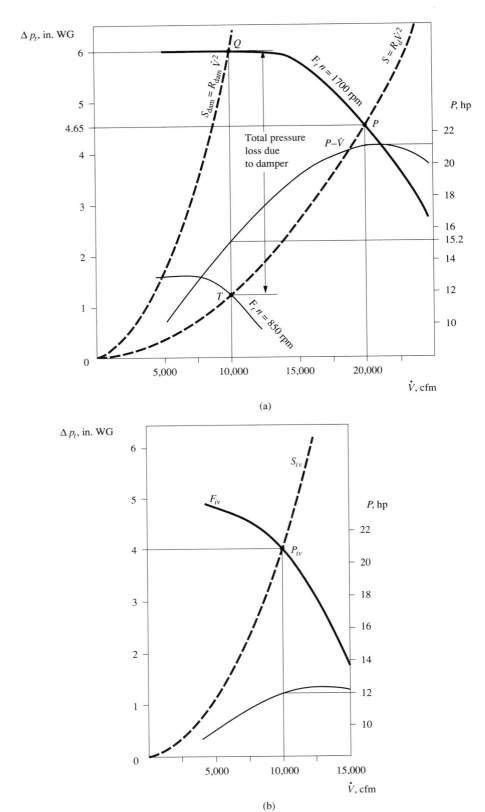

FIGURE 19.16   Modulation of a VAV system using dampers, inlet vanes, and fan speed variation: (a) using dampers and fan speed variation and (b) using inlet vanes.

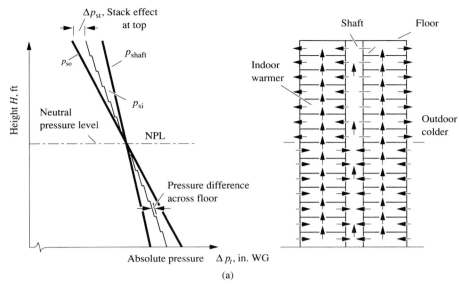

FIGURE 19.17a Space pressure characteristics and air flow around buildings: (a) Space pressure characteristics of a high-rise building under stack effect and $T_{Ri} > T_{Ro}$.

Stack Effect

The temperature difference between the cold outdoor and warm indoor air columns causes a density difference between these air columns that, in turn, creates a pressure difference between the cold and the warm air columns. This phenomenon is called *stack effect*.

In any building under stack effect alone, there is a level at which the indoor and outdoor pressures are equal. This level is called the *neutral pressure level (NPL)*, as shown in Fig. 19.17a. During the heating season, outdoor cold air pressure is greater than indoor warm air pressure below the NPL. This pressure difference causes the outdoor colder air to enter the building through the lower inlets, and the indoor warmer air to flow upward and discharge from the top openings. During the cooling season, the reverse effect occurs but the stack effect is smaller because of the smaller temperature difference between outdoor and indoor air.

The pressure difference due to the stack effect between outdoor and indoor air Δp_{st} (in inches WG) at height H (in ft), from a reference level, is given as

$$\Delta p_{st} = p_{so} - p_{si} = 0.15926(\rho_o - \rho_i)(H - H_{NPL})$$

$$(19.32)$$

where p_{so}, p_{si} = outdoor and indoor absolute static pressure, in. WG
ρ_o, ρ_i = outdoor and indoor air density, lb/ft^3
H_{NPL} = vertical distance between the NPL and reference level, ft

For a high-rise building with two openings on the external wall, H_{NPL} (in ft) measured from lower opening can be calculated as

$$H_{NPL} = \frac{H_o}{1 + (A_1/A_2)2(T_{Ri}/T_{Ro})} \qquad (19.33)$$

where H_o = vertical distance between two openings, ft
A_1, A_2 = area of the lower and higher openings, ft^2
T_{Ri}, T_{Ro} = absolute temperatures of indoor and outdoor air, °R

In Eq. (19.33), $T_{Ri} > T_{Ro}$. If the indoor air is cooler, that is, $T_{Ro} > T_{Ri}$, T_{Ri}/T_{Ro} should be changed to T_{Ro}/T_{Ri}. For conditioned spaces with mechanical ventilation systems, when $T_{Ri} > T_{Ro}$, the chimney and exhaust systems raise the NPL level and the outdoor air supply system lowers the NPL level.

In Fig. 19.17a, it can be seen that the absolute static pressure of outdoor air p_{so} is greater at the lower level than at the higher level. This is because the air column is taller at the lower level.

Stack effect becomes the dominant factor that influences indoor space pressure characteristics only when the air system is not operating, usually when the space is not occupied. Under these circumstances, maintaining a specific indoor static pressure is not as important as when the space is occupied.

Additionally, in an ideal single-cell building with no inside partitions, with a vertical distance of 50 ft between the lower air inlet and upper air outlets and an indoor and outdoor air temperature difference of

50°F, the static pressure difference Δp_{st} across the inlet 25 ft from the NPL is about 0.04 in. WG.

Most buildings have inside partitions, that is, they are multicell buildings. They have doors at the lower entering inlet, windows at the upper discharge outlets, and doors to the stairwells. Usually doors and windows are closed. When outdoor cold air flows through these doors, windows, and partitions, its velocity and flow rate at the cracks and gaps is significantly reduced because of these dynamic losses.

Although stack effect sometimes has a significant effect on space pressure in high-rise buildings, its influence is generally ignored for simplicity during space pressurization analysis.

Wind Effect

As wind flows over a building, it creates positive pressure on the windward side of the building and negative pressure on the leeward side, as shown in Fig. 19.17b. Wind pressure on the building surface relative to the approaching wind speed p_w, in inches WG, can be calculated as

$$p_w = C_p p_{vw} = C_p \left(\frac{v_{wa}}{4005}\right)^2 \qquad (19.34)$$

where C_p = surface pressure coefficient
p_{vw} = velocity pressure at approaching wind speed, in. WG
v_{wa} = approaching wind speed at wall height on windward side, fpm

If the wind direction is normal to the windward surface of a high-rise building with a width to depth ratio of 4, the average C_p value is about 0.60 on the windward side, about −0.5 on the leeward side and

on the flat roof, and about − 0.25 on the other two sides.

Wind speed measured from a meteorological station v_{met}, in fpm, is usually measured at a height of 33 ft. The approaching wind speed at wall height v_{wa} should be corrected for height and terrain roughness, as follows:

$$v_{wa} = A_o v_{met} \left(\frac{H_w}{H_{met}}\right)^n \qquad (19.35)$$

where H_w = height of the wall where wind pressure is exerted, ft
H_{met} = height at which wind speed is measured at meteorological station, ft

In Eq. (19.35), A_o is the correction factor for terrain roughness and n is the velocity profile exponent. For suburban areas, $A_o = 0.6$ and $n = 0.28$; for urban areas, $A_o = 0.35$ and $n = 0.4$; for airports, $A_0 = 1.0$ and $n = 0.15$.

The mean wind speeds measured at meteorological stations v_{met} in most U.S. cities in winter are below 1200 fpm. Wind pressure on a wall at a height of 33 ft in suburban areas with a $v_{met} = 1200$ fpm (13.6 mph) is about 0.054 in. WG.

The influence of wind effect on space and building pressure characteristics is as follows:

1. Rooms on the windward side of the building are usually at a positive pressure and on the leeward side a negative pressure relative to the corridor pressure. It is best to build clean spaces, such as conference rooms, on the windward side of the prevailing wind, and laboratories with toxic gas exhaust systems on the leeward side of the building.

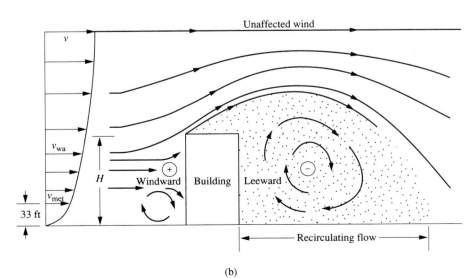

(b)

FIGURE 19.17b   Space pressure characteristics and air flow around buildings: (b) air flow around buildings.

2. Outdoor air intake should be located on the side with a positive surface pressure coefficient C_p in the prevailing wind. Exhaust outlets should be located where C_p is negative, preferably on the rooftop.

3. Sufficient total pressure must be provided by the supply fan to overcome the negative pressure at the outdoor intake and by the exhaust fan to overcome the positive pressure at the exhaust outlet. Alternatives should be provided to allow outdoor intake and outlet when the wind direction is different from that of the prevailing wind.

Because wind speed may vary from zero to its maximum speed and wind pressure is often significantly smaller than the pressure provided by air systems and exhaust systems, for simplicity, wind effect is generally ignored during the analysis of space pressurization.

Air Systems or Mechanical Ventilation Systems

If a space is served by air systems or mechanical ventilating systems, including supply, return, and exhaust systems, the space pressure characteristics are dominated by these air systems or mechanical ventilation systems during most of the operating time.

An *air system or an air-handling system* is a system that conditions and distributes air. An air system is a part of an air conditioning system.

A *mechanical ventilation system* is a system that supplies air to the indoors or exhausts air from the indoors using a fan–duct system without conditioning (except heating and filtration). It mainly consists of fans, supply and exhaust duct systems, filters, heater, and grilles.

The following are the types of mechanical ventilation systems currently used:

- *Make-up air systems,* which supply outdoor air, often heated and filtered
- *Pressurizing mechanical ventilation systems,* which pressurize enclosed spaces such as stairwells for escape routes in high-rise buildings during building fires
- *Ventilation cooling and contamination control systems,* which supply outdoor air to a space to cool an enclosure or dilute air contamination to acceptable levels, as in garages and transformer rooms
- *Exhaust systems,* which discharge indoor air or exhaust toxic gas from localized sources

Air is usually not cooled, humidified, or dehumidified in mechanical ventilation systems. The performance of a mechanical ventilation system is similar to that of an air system that does not cool, dehumidify, or humidify.

19.7 INFILTRATION AND EXFILTRATION

Infiltration is the uncontrolled inward flow of outdoor air through cracks and openings in the building shell when the pressure of the outdoor air is higher than the pressure of the indoor air at the same level. *Exfiltration* is the uncontrolled outward flow of air through cracks and openings in the building shell when the pressure of indoor air is higher than that of the outdoor air.

Infiltration and exfiltration can be induced by stack effect, wind effect, air systems or mechanical ventilating systems, or combination thereof. The volume flow rate of infiltration $\dot{V}_{if}$ or exfiltration $\dot{V}_{ef}$, in cfm, can be calculated as

$$\dot{V}_{if} = \dot{V}_{ef} = C_d A_{e.1} \sqrt{\frac{2\Delta p}{\rho}} = C_d A_{e.1} 4005 \sqrt{\Delta p}$$

(19.36)

where $A_{e.1}$ = effective leakage area on building shell, ft^2

C_d = discharge coefficient; its value is 1 for effective leakage area

ρ = air density, lb/ft^3

Δp = pressure difference between outdoor and indoor air at same level, in. WG

Infiltration is often expressed in air changes per hour (ach), where ach = $60\dot{V}_{if}/V_r$. Here, V_r indicates the space volume of room, in ft^3.

Effective Leakage Area

The *effective leakage area* of a building is the amount of open wall area that allows the same volume flow rate of air flowing through it as flows through the actual building when the pressure difference across the open wall area and the shell of the actual building are the same. The $A_{e.1}$ value is usually determined by a fan-pressurization method, in which a blower installed at the door of a building is used to maintain a pressure difference across the building shell. The air flow rate can then be measured, and the $A_{e.1}$ value is calculated from Eq. (19.36).

According to field measurements of air leakage from office buildings in Persily and Linteris (1983), the air leakage of external walls in U.S. office buildings are as follows:

	$A_{e.1}$, ft^2/ft^2	$\dot{V}_{if}$ at $\Delta p = 0.3$ in. WG, cfm/ft^2
Tight wall	0.00005	0.1
Average	0.00014	0.3
Leaky wall	0.00027	0.6

Here, the unit ft^2/ft^2 means ft^2 of effective leakage area per ft^2 of external wall area.

Volume Flow Rate of Infiltration

Determining the volume flow rate of infiltration is complicated because actual buildings are usually multicell structures with inside partitions, doors, and corridors. Stack effect, wind effect, and the air systems must also be considered. As discussed in Chapter 7, infiltration is considered zero at summer cooling load design calculations because slightly positive pressure is maintained in the conditioned space and wind speed is often lower during summer.

According to field measurements, the following infiltration rates, in ach, should be considered in winter heating load calculations for rooms with windows on one side of a commercial building that are not well-sealed:

Winter mean wind speed v_{met}, fpm	Infiltration rate, ach
1000	0.2
1200	0.3

For rooms with windows on two sides of the building, the above values should be multiplied by a factor of 1.5.

19.8 AIR SYSTEMS AND OPERATING MODES

Mass Flow Balance

If the principle of continuity of mass is applied to a steady flow process, the total mass flow rate of the air entering an enclosed space must equal the total mass flow rate of the air leaving the space (see Fig. 19.18), or

$$\dot{m}_s + \dot{m}_{if} = \dot{m}_{rt} + \dot{m}_{ex} + \dot{m}_{ef} \qquad (19.37)$$

$$A_s \dot{m}_s = \dot{m}_o + \dot{m}_{ru} \quad \text{and} \quad \dot{m}_{rt} = \dot{m}_{eu} + \dot{m}_{ru}, \text{ so}$$

$$\dot{m}_o + \dot{m}_{if} = \dot{m}_{eu} + \dot{m}_{ex} + \dot{m}_{ef} \qquad (19.38)$$

where $\dot{m}$ = mass flow rate, lb/min

$\dot{m}_s$ = mass flow rate of air supplied to the conditioned space

$\dot{m}_{rt}$ = mass flow rate of air returned from the space to the fan room

$\dot{m}_{ru}$ = mass flow rate of air recirculated to the AHU or PU

$\dot{m}_o$ = mass flow rate of outdoor air

$\dot{m}_{ex}$ = mass flow rate of air exhausted from the space to the outdoors directly

$\dot{m}_{eu}$ = mass flow rate of air returned to the AHU or PU and exhausted to the outdoors

$\dot{m}_{if}$ = mass flow rate of air infiltrated into the space

$\dot{m}_{ex}$ = mass flow rate of air exfiltrated from the space

Because $\dot{m} = \dot{V}\rho$, where the difference in air density ρ is ignored, then

$$\dot{V}_s + \dot{V}_{if} = \dot{V}_{eu} + \dot{V}_{ex} + \dot{V}_{ef} \qquad (19.39)$$

where $\dot{V}$ = volume flow rate, cfm.

For a specific conditioned space, if the space static pressure is maintained at a positive value p_r, $\dot{V}_{ef}$ is approximately constant, even if $\dot{V}_s$ varies. The exhaust volume flow rate from places such as restrooms and kitchens $\dot{V}_{ex}$ is usually also constant.

Figure 19.18b shows a flow circuit of an air system. Flow circuits are helpful in analyzing the characteristics of more complicated air systems.

Air Systems of Various Types of Fan Combination

In air systems used for commercial, institutional, and industrial applications, the various fan combinations can be classified into the following three main categories:

- Supply fan and constant volume exhaust fan combination
- Supply fan and relief fan combination
- Supply fan and return fan combination

Supply, exhaust, relief, and return fans can either be a single fan or multiple fans. Air systems of various fan combinations are installed to meet the following requirements:

- To provide required outdoor ventilation air for the occupants and space according to ASHRAE Standard 62–1989
- To supply conditioned air to the space
- To recirculate space air for energy conservation
- To operate in an air economizer cycle
- To exhaust objectionable toxic gases from the space
- To maintain a proper space pressure

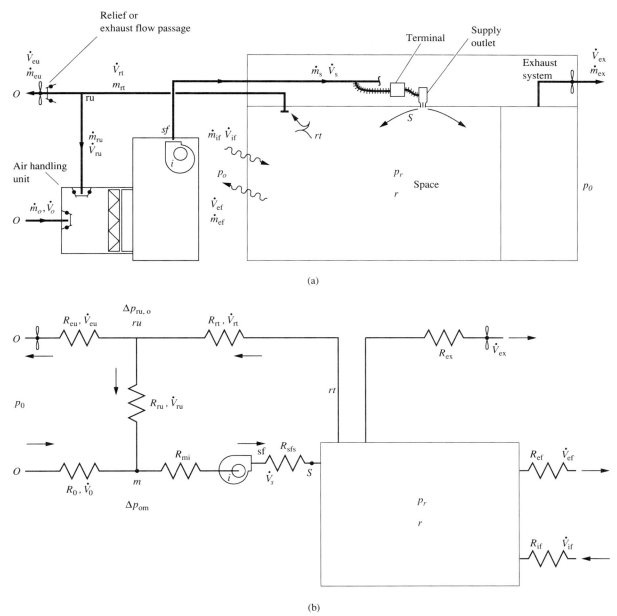

FIGURE 19.18 Mass flow balance of an enclosed space: (a) mass flow balance and (b) flow circuit.

Operating Modes

In this chapter, only volume and pressure characteristics of the air systems are discussed. A more comprehensive analysis, including thermal behavior, is described in Chapters 26 and 27.

Air systems that use the cooling capacity of outdoor air to save energy may be operated in the following modes:

- *Minimum outdoor air recirculation mode,* in which only minimal outdoor air is supplied to the space for the occupants and dilution of contaminated air

- *Air economizer mode,* in which only the outdoor air or a mixture of outdoor and recirculating air is supplied to the space to replace the part or all of the refrigeration

- *Warm-up and cool-down mode,* in which space air temperature is warmed or cooled to predetermined limits after the air system is started.

In analyzing the pressure–volume characteristics of an air system at various space loads, it is assumed that the supply volume flow may be at 100 or 50 percent of design volume flow during full-load or part-load operation. For each operating mode, $\dot{V}_s$, $\dot{V}_o$, and total pressure p_t at key points of the air system should be analyzed.

When outdoor air temperature T_o is below a specified value, although the interlocked outdoor and recirculating dampers modulate the outdoor and recirculating air in order to maintain a specified mixed air temperature (as described in Sections 18.8 and 18.9), the volume flow rates of outdoor air, exhaust air, and recirculating air depend on not only the size of the damper openings, but also the pressure loss across the dampers, such as Δp_{om}, $\Delta p_{ru.o}$, and $\Delta p_{ru.m}$, as shown in Fig. 19.18.

As described in Chapter 17, the interlocked outdoor and recirculating damper, or mixing damper, can maintain a nearly constant pressure drop across the dampers at various opening sizes by using parallel-opposed blade dampers.

System Pressure Loss

For a VAV air system using duct static pressure control, system pressure loss can be divided into two parts: variable part Δp_{var} and fixed part Δp_{fix}. The pressure loss between the outdoor intake of the AHU

and the static pressure sensor of the duct static pressure control is expressed as Δp_{var}, and the pressure loss between the static pressure sensor and the ceiling diffuser including the VAV box is expressed as Δp_{fix}. As the supply volume flow rate in a VAV system varies, Δp_{var} varies accordingly. The fixed part pressure loss Δp_{fix} is independent from the variation of the supply flow rate in VAV system, as shown in Fig 19.19.

System pressure loss Δp_{sys} (in inches WG) of a VAV system can be calculated as

$$\Delta p_{sys} = \Delta p_{var} + \Delta p_{fix} = R_{var} \dot{V}_{var}^2 + \Delta p_{fix}$$
$$= (\Delta p_{AHU} + \Delta p_{d.var}) + \Delta p_{fix} \quad (19.40)$$

where Δp_{AHU} = pressure drop across AHU or PU, in. WG

$\Delta p_{d.var}$ = pressure loss of supply duct system before the duct static pressure control pressure sensor, in. WG

R_{var} = flow resistance representing the variable part of the air system, in. WG $\cdot$ (cfm)^2

$\dot{V}_{var}$ = volume flow rate of the variable part of the air system, cfm

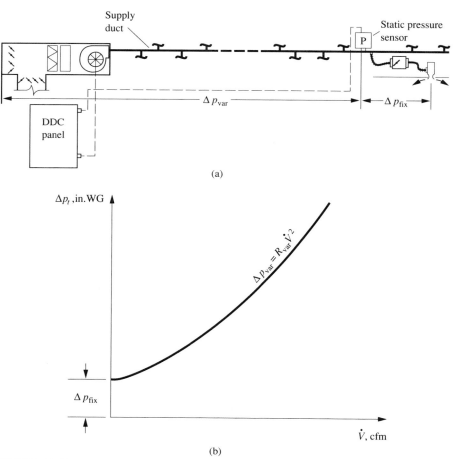

(a)

(b)

FIGURE 19.19   Duct static pressure control and system pressure loss: (a) duct static pressure control and (b) system pressure loss.

If the pressure sensor is installed closer to the supply fan, the fan power savings at reduced volume flow decrease because of the increase of Δp_{fix}. The set point of the static pressure sensor may vary from 0.5 to 1.5 in. WG.

In a VAV air system serving a single floor in a multistory building, if the maximum design air velocity in main duct v_{max} is 3000 fpm, and if the length of the main duct is 150 ft, the design total pressure loss $\Delta p_{d.var}$ may vary from 0.75 to 1.5 in. WG, typically 1 in. WG, depending on the duct fittings in the main duct. If v_{max} increases to 4000 fpm, $\Delta p_{d.var}$ increases from 1 in. WG to 1.78 in. WG, and if v_{max} increases to 5000 fpm, $\Delta p_{d.var}$ then increases to 2.78 in. WG.

19.9 SUPPLY FAN AND CONSTANT-VOLUME EXHAUST FAN COMBINATION

System Characteristics

Air systems equipped with a single supply fan and a constant-volume exhaust fan are often used in buildings in which the total pressure loss of the return duct system is low. A constant volume of air is usually exhausted from rooms such as restrooms. A barometric relief damper is often installed in the conditioned space or in the return plenum to avoid excessively high space pressure. When the space positive pressure on the damper is greater than the weight of the damper, the damper opens and the space pressure is relieved and is thus maintained below a predetermined value. Such an air system may be equipped with a water economizer.

In order to analyze the pressure characteristics of the airstream at two cross-sectional planes along the air flow in an air system, total pressure is determined from Eq. (8.11b), $p_{t1} = p_{t2} + \Delta p_f$.

To determine whether air is flowing from one enclosed space to another, or from an enclosed space to the outdoors, the static pressure difference between these two spaces $\Delta p_s = p_{s1} - p_{s2}$ should be calculated. However, the static pressure inside an enclosure of uniform pressure, or the mean static pressure on a cross-sectional plane along air flow, must be determined from Eq. (8.11a), $p_s = p_t - p_v$. Here, p_v represents velocity pressure.

Because atmospheric pressure $p_o = 0$, static pressure at the mixing chamber p_m must be negative in order to extract outdoor air. The total pressure difference between the outdoor air and the mixing chamber $\Delta p_{om} = p_o - p_m$ consists mainly of the pressure loss of the outdoor dampers and louver.

If a space is maintained at a positive pressure, the volume flow rate of exfiltration from the space $\dot{V}_{ef}$, in cfm, can be calculated from Eq.

(19.36). From Eq. (19.36), the pressure loss across the building shell Δp_{ro}, in inches WG, can be calculated as

$$\Delta p_{ro} = p_r - p_o = \left(\frac{1}{4005A_{e.1}}\right)^2 \dot{V}_{ef}^2 = R_{ef}\dot{V}_{ef}^2$$
(19.41)

The flow resistance of exfiltrated air across building shell, R_{ef}, in inches WG · (cfm)^2 is

$$R_{ef} = \left(\frac{1}{4005A_{e.1}}\right)^2$$

Because the volume flow rates of the supply air and recirculating air are different, the flow resistance of the supply system and return system as well as the flow resistance of the fixed part and variable part of the air system pressure loss should be calculated separately.

Operating Characteristics

Consider a VAV rooftop packaged system with a supply fan and a constant-volume exhaust fan serving a typical floor in a commercial building, as shown in Fig. 19.20. This air system has the following operating characteristics at design conditions:

Suppy volume flow rate	20,000 fpm
Total pressure losses:	
Across recirculating damper at design flow	0.4 in. WG
Filters and coils,	2.5 in. WG
Supply main duct, $\Delta p_{d.var}$	0.85 in. WG
Recirculating system	0.15 in. WG
VAV box, branch duct, and diffuser	0.75 in. WG
Effective leakage area on the building shell	0.4 ft^2
Constant volume exhaust fan	3000 cfm
Minimum outdoor ventilation air required	3358 cfm
Space pressure	+0.05 in. WG

In minimum outdoor air recirculating mode, the pressure losses of the outdoor louver can be ignored. A barometric relief damper is mounted in the conditioned space and is opened when the space positive pressure is greater than 0.2 in. WG. The supply fan has the same Δp_t–$\dot{V}$ characteristics as in Example 19.3.

The following control systems are installed to maintain the required operating parameters:

- Space temperature control
- Discharge temperature control
- Duct static pressure control

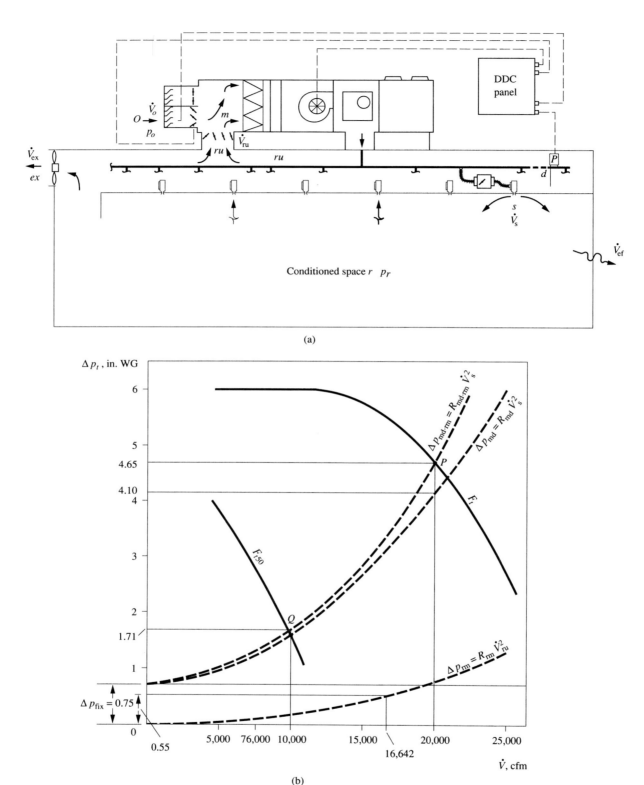

(a)

(b)

FIGURE 19.20   Air system of simple supply fan and constant-volume exhaust fan combination: (a) schematic diagram and (b) system characteristics on $\Delta p_t - \dot{V}$ diagram.

- Minimum outdoor air control
- High pressure limit control

Only the operation of duct static pressure and outdoor minimum air control are be described here.

This air system can be operated in recirculating mode, air economizer mode, or warm-up or cool-down mode.

Recirculating Mode, Design Volume Flow

Minimum outdoor air is extracted through the outdoor intake and mixed with recirculating air in the mixing chamber m. The mixture is conditioned in the rooftop packaged unit. After conditioning, the supply air then flows through the supply fan and the supply duct, and is discharged to the conditioned space. In the conditioned space, a constant volume flow rate is exhausted through an exhaust fan and another small portion is exfiltrated through the openings on the building shell. The major portion of the space air is returned to the rooftop unit, where it is mixed with outdoor air for recirculation.

When the air system of a single supply fan and constant-volume exhaust combination is operated in recirculating mode with design volume flow, the inlet vanes at the supply fan inlet are fully open, and the outdoor air damper is at minimum opening position. The minimum outdoor ventilation air required is $\dot{V}_o = 3358$ cfm.

The fixed part of pressure loss of such an air system is $\Delta p_{fix} = 0.75$ in. WG. The variable part of pressure loss of the air system consists of the following two sections:

- Mixing chamber point m to the static pressure sensor, point d, at a supply volume flow rate of 20,000 cfm:

$$\Delta p_{md} = 2.5 + 0.85 = 3.35 \text{ in. WG}$$

Its corresponding flow resistance can be calculated as

$$R_{md} = \frac{\Delta p_{md}}{\dot{V}_s^2} = \frac{3.35}{(20,000)^2}$$
$$= 8.38 \times 10^{-9} \text{ in. WG/(cfm)}^2$$

- Recirculating system from point r to m:

$$\Delta p_{rm} = R_{rm}\dot{V}_{ru}^2 = 0.15 + 0.4 = 0.55 \text{ in. WG}$$

Because of a positive space pressure of $+0.05$ in. WG and an effective leakage area $A_{e.l}$ of 0.4 ft^2, the volume flow rate of exfiltration from the conditioned space $\dot{V}_{ef}$ is calculated as

$$\dot{V}_{ef} = 4005 A_{e.l} \sqrt{\Delta p_{ro}}$$
$$= 4005 \times 0.4 \sqrt{0.05} = 358 \text{ cfm}$$

The flow resistance of the opening in the building shell is

$$R_{ef} = \frac{\Delta p_{ro}}{\dot{V}_{ef}^2} = \frac{0.05}{(358)^2}$$
$$= 3.9 \times 10^{-7} \text{ in. WG/(cfm)}^2$$

The recirculating volume flow rate $\dot{V}_{ru}$ through the recirculating system is then equal to

$$\dot{V}_{ru} = \dot{V}_s - (\dot{V}_{ef} + \dot{V}_{ex}) = 20,000 - (358 + 3000)$$
$$= 16,642 \text{ cfm}$$

The flow resistance of the recirculating system based on pressure loss between points r and m is

$$R_{rm} = \frac{\Delta p_{rm}}{\dot{V}_{rm}^2} = \frac{0.55}{(16,642)^2}$$
$$= 1.99 \times 10^{-9} \text{ in. WG/(cfm)}^2$$

The fan total pressure of the supply fan at design volume flow rate is

$$\Delta p_{sf} = 0.75 + 3.35 + 0.15 + 0.4$$
$$= 4.65 \text{ in. WG}$$

The system operating point of this supply and constant-volume exhaust fan combination P is plotted in Fig. 19.20b. The pressure–volume characteristics for the key points in this system are shown below:

Point	O	m	r	ru
p, in. WG	0	-0.5	$+0.05$	-0.1
V, cfm	3358	20,000	20,000	16,642

From Eqs. (19.14) and (19.15), for the combined supply and recirulating system,

$$\Delta p_{md.rm} = \Delta p_{md} + \Delta p_{rm}$$
$$= R_{md}V_s^2 + R_{rm}K_{rm}^2 V_s^2$$
$$= R_{md.rm}V_s^2 \qquad (19.42)$$

and the flow resistance is

$$R_{md.rm} = R_{md} + R_{rm}K_{rm}^2$$
$$= 8.38 \times 10^{-9} + 1.99 \times 10^{-9}\left(\frac{16,642}{20,000}\right)^2$$
$$= 9.75 \times 10^{-9} \text{ in. WG/(cfm)}^2$$

Recirculating Mode, 50 Percent Design Flow Rate

When the space sensible cooling load is reduced at part-load operation, the supply volume flow rate may be reduced to 50 percent of the design volume flow rate, that is, $0.5 \times 20,000 = 10,000$ cfm.

As the dampers in the VAV boxes close to smaller openings than at design volume flow, the static pressure in the supply main duct rises. The sensor senses the increase, and the direct digital control (DDC) panel instructs the inlet vanes to close to a smaller opening, until the duct static pressure is maintained approximately at its set point at the location where pressure sensor is mounted. The duct static pressure is equal to the pressure loss of the VAV box and the last branch takeoff plus the space static pressure: $(0.75 + 0.05) = 0.8$ in. WG. The fan performance curve is now a new fan curve F_{t50}, as shown in Fig. 19.20b.

The variable part pressure loss between the mixing box and point d is now decreased to

$$\Delta p_{md50} = R_{md} \dot{V}_{s50}^2$$
$$= 8.38 \times 10^{-9}(10{,}000)^2 = 0.84 \text{ in. WG}$$

Assume that the space pressure at 50 percent design flow is -0.14 in. WG. The volume flow rate of infiltrated air into the space is

$$\dot{V}_{if} = 4005 A_{e.1} \sqrt{\Delta p}$$
$$= 4005 \times 0.4 \sqrt{0.14} = 600 \text{ cfm}$$

The recirculating volume flow rate is then

$$\dot{V}_{ru50} = \dot{V}_{s50} - \dot{V}_{ex} + \dot{V}_{if50}$$
$$= 10{,}000 - 3000 + 600 = 7600 \text{ cfm}$$

The pressure loss between the space (point r) and the mixing box (point m) becomes

$$\Delta p_{rm} = R_{rm} \dot{V}_{rm50}^2$$
$$= 1.99 \times 10^{-9}(7600)^2 = 0.12 \text{ in. WG}$$

Therefore, the total pressure in the mixing box is

$$p_m = -0.14 - 0.12 = -0.26 \text{ in. WG}$$

The flow resistance of the outdoor passage is

$$R_{om} = \frac{\Delta p_{om}}{\dot{V}_o^2} = \frac{0.5}{(3358)^2}$$
$$= 4.43 \times 10^{-8} \text{ in. WG/(cfm)}^2$$

Because of a pressure difference $\Delta p_{om} = -0.26$ in. WG, outdoor intake $\dot{V}_o$ is therefore

$$\dot{V}_o = \sqrt{\frac{\Delta p_{om}}{R_{om}}} = \sqrt{\frac{0.26}{4.43 \times 10^{-8}}} = 2422 \text{ cfm}$$

If the difference in air density is ignored, the volume flow balance of incoming and outward outdoor air in this air system becomes

$$\dot{V}_o + \dot{V}_{if} - \dot{V}_{ex} = 2422 + 600 - 3000 = 22 \text{ cfm}$$

Volume flow balance of $\dot{V}_o$ is roughly constant.

From the above analysis, the system operating point at 50 percent design volume flow rate, point Q, has a supply volume flow of 10,000 cfm and a fan total pressure of

$$\Delta p_{t50} = 0.75 + 0.84 + 0.12 = 1.71 \text{ in. WG}$$

During minimum outdoor air recirculating mode at 50 percent design volume flow, the outdoor air volume flow rate through outdoor intake passage decreases to 2422 cfm. Space is maintained at a negative pressure of -0.14 in. WG. An infiltration rate of 600 cfm enters the space through openings and cracks in the building shell.

The p_t–$\dot{V}$ characteristics at key points during 50 percent design volume flow are:

Points	O	m	r	ru
p, in. WG	0	-0.26	-0.14	-0.17
$\dot{V}$, cfm	2422	10,000	10,000	7600

Air Economizer Cycle

When an air economizer cycle is operated at a design flow rate of 100 percent outdoor air in an air system with a single supply fan and a constant-volume exhaust flow combination, the recirculating damper is closed, and the outdoor damper is fully open. Outdoor air is extracted to the mixing chamber through the filters and coils, and supplied to the conditioned space. Because only a small portion of supply air is exhausted and exfiltrated, the space positive pressure increases to about 0.2 in. WG. The barometric relief damper then opens. Most of the supply air is discharged through the relief damper.

When the supply and constant-volume exhaust fan combination is operated at 50 percent design volume flow in the air economizer cycle, space pressure is still limited to 0.2 in. WG and air is exhausted from the space as in the design volume flow.

Warm-Up and Cool-Down Mode

During warm-up and cool-down mode, the outdoor damper is closed and the recirculating damper is fully open, while the constant-volume exhaust fan is turned off. Recirculated air from the space is conditioned in the packaged unit and is then supplied to the space. Because of the negative pressure in the mixing chamber, outdoor air is leaked into the chamber through the closed outdoor damper and exfiltrated from the space because of the higher positive space pressure.

Pressure Variation at the Mixing Chamber

During the recirculating mode, when the supply volume flow rate is reduced from design flow to 50 percent of design flow, the decrease of the recirculating volume flow rate causes a corresponding drop in the pressure loss of the recirculating system. Therefore, the pressure at the mixing chamber p_m increases from -0.5 in. WG to -0.26 in. WG. The greater the pressure loss of the return system at design flow rate, the higher the fluctuation of Δp_{om}. In order to balance the outdoor air flow, space pressure must be changed from $+0.05$ in. WG to -0.14 in. WG, and the exfiltration changes to infiltration. The pressure loss at the outdoor air intake passage (excluding the damper), like intake louver and duct friction loss Δp_{lou}, should be taken into account only in the outdoor air economizer cycle. During minimum outdoor air recirculating mode, Δp_{lou} should be negligible, or the very high Δp_{lou} value in the air economizer cycle will reduce energy efficiency.

Pressure fluctuation Δp_{om} at design flow and at 50 percent design flow causes insufficient outdoor air intake. This pressure fluctuation can be reduced or eliminated by using minimum outdoor air control or by adding a minimum outdoor fan.

19.10 SUPPLY FAN AND RELIEF FAN COMBINATION

Consider an air system with a supply fan, relief fan, and constant-volume exhaust fan combination, as shown in Fig. 19.21a with the same operating characteristics as mentioned in Section 19.9. A *relief fan* is a fan that is installed in the relief flow passage, just after the recirculating air enters the mixing box, to relieve the undesirably high positive space pressure. A relief fan is different from an *exhaust fan,* which is often used to exhaust fumes or toxic gases and to maintain a negative pressure in the space or in a localized enclosure. An axial fan is often used as a relief fan because of its large volume flow and smaller fan total pressure.

The sub-control systems used in such a combination are similar to those of a supply fan and constant-volume exhaust combination, except that a relief fan control is added. The relief fan has the following pressure–volume characteristics:

$\dot{V}$, cfm	10,000	15,000	16,640	17,250
Δp_t, in. WG	4.0	2.10	1.08	0.43

Recirculating Mode

During minimum outdoor air recirculating mode operation, the relief fan is not operating and the relief damper is closed. Outdoor air drawn into the air system is balanced by the constant-volume exhaust fan and the exfiltration through the openings in the building shell. In this case, the system becomes a supply fan and constant-volume combination, and its operating characteristics are the same as those described and analyzed in Section 19.9.

Air Economizer Mode, Design Volume Flow Rate

During air economizer mode with 100 percent outdoor air at design volume flow rate, the outdoor damper is fully open and the recirculating damper is closed. The relief fan relieves the space air pressure and maintains it at a predetermined value. Outdoor air is extracted through the intake louver and the outdoor damper, and flows through the filter and coils. The conditioned air is then supplied to the space. A small part of the space air is exhausted, and another part is exfiltrated. The majority of space air is relieved by the relief fan.

At design volume flow, the duct friction loss of the outdoor intake passage is negligible, and the pressure loss of the intake louver is 0.2 in. WG. The variable part of the pressure loss of the air system is given as

$$\Delta p_{var} = 0.2 + 0.4 + 2.5 + 0.85$$
$$= 3.95 \text{ in. WG}$$

The fixed part of pressure loss of the air system is $0.75 + 0.05 = 0.80$ in. WG. Here, 0.05 in. WG is the pressure required to maintain space positive pressure. The fan total pressure of the supply fan is then equal to

$$0.80 + 3.95 = 4.75 \text{ in. WG}$$

The flow resistance of the variable part of the air system can be calculated as

$$R_{or} = \frac{\Delta p_{or}}{\dot{V}_s^2} = \frac{3.95}{20,000^2}$$
$$= 9.88 \times 10^{-9} \text{ in. WG/(cfm)}^2$$

The fan total pressure of the relief fan is used to overcome the pressure loss of the return system, the relief damper, and the louver in the relief flow passage. Usually, the duct friction loss of the relief passage is ignored. At design volume flow rate, for a space pressure of $+0.05$ in. WG, an exfiltration of 358 cfm, and a constant exhaust volume flow rate of 3000 cfm, the volume flow rate of relief air is $20,000 - 3000 - 358 = 16,642$ cfm.

As in the supply fan and exhaust fan combination, from the given data, the pressure loss between points r and ru is

$$\Delta p_{r.ru} = R_{r.ru} \dot{V}_{ru}^2 = 0.15 \text{ in. WG}$$

and flow resistance is

$$R_{r.ru} = \frac{\Delta p_{r.ru}}{\dot{V}_{rc}^2}$$
$$= 5.42 \times 10^{-10} \text{ in. WG/(cfm)}^2$$

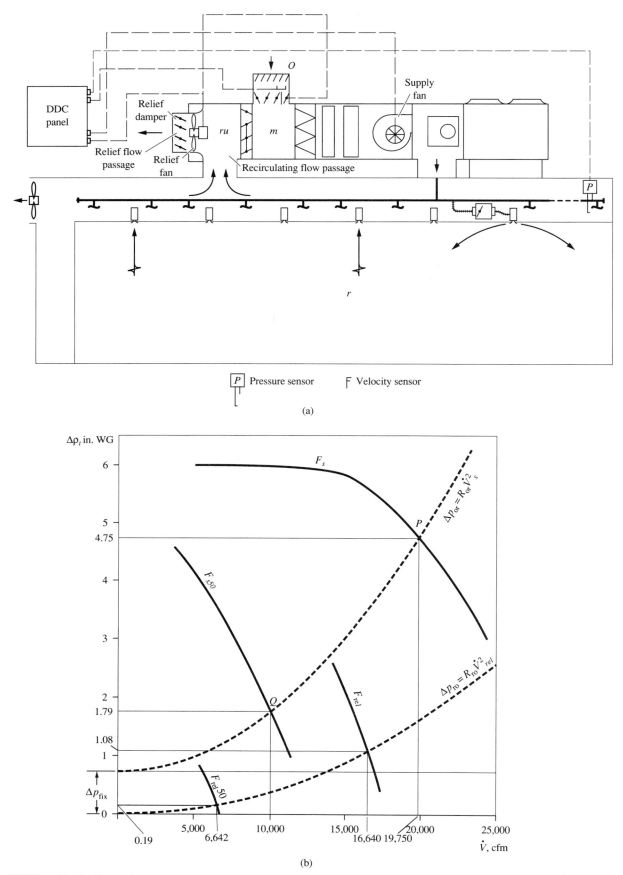

FIGURE 19.21 Supply fan and relief fan combination: (a) schematic diagram and (b) system characteristics on $\Delta p_t - \dot{V}$ diagram, air economizer mode.

If the pressure loss of the relief louver at a volume flow of 16,642 cfm is 0.2 in. WG, the pressure loss of the relief damper is 0.4 in. WG, and the velocity pressure of the axial fan is 0.38 in. WG, the flow resistance of the relief system is

$$R_{rel} = \frac{0.2 + 0.4 + 0.38}{(16,642)^2}$$

$$= 3.54 \times 10^{-9} \text{ in. WG/(cfm)}^2$$

The sum of flow resistance of the return and relief systems is

$$R_{ro} = R_{r.ru} + R_{rel} = 5.42 \times 10^{-10} + 3.54 \times 10^{-9}$$

$$= 4.08 \times 10^{-9} \text{ in. WG/(cfm)}^2$$

Therefore, the fan total pressure of the relief fan is calculated as

$$\Delta p_{rel} = 0.2 + 0.4 + 0.38 + (0.15 - 0.05)$$

$$= 1.08 \text{ in. WG}$$

Here, 0.05 in. WG is the total pressure provided by the space positive pressure.

The supply fan curve F_s, the relief fan curve F_{rel}, the system curve of the supply system $\Delta p_t = R_{or} \dot{V}^2$, and the system curve of the return and relief system $\Delta p_t = R_{ro} \dot{V}^2$ can then be plotted. At design flow of an air economizer cycle, the system operating point for supply system P has a $\dot{V} = 19,750$ cfm and a $\Delta p_t = 4.75$ in. WG. The system operating point of the relief system has a $\dot{V} = 16,400$ cfm and a $\Delta p_t = 1.08$ in. WG.

The pressure–volume characteristics at key points for air economizer mode at design volume flow are as follows:

Points	O	m	r	ru
p, in. WG	0	−0.6	+0.05	−0.1
$\dot{V}$, cfm	19,750	19,750	19,750	16,400

Air Economizer Mode, 50 Percent Design Flow

At 50 percent design volume flow rate in air economizer mode, the variable part of the pressure loss of the air system when inlet vanes are partly closed is

$$\Delta p_{var50} = 9.88 \times 10^{-9}(10,000)^2 = 0.99 \text{ in. WG}$$

and the fan total pressure of the supply fan at 50 percent design flow is

$$\Delta p_{sf50} = 0.99 + 0.75 + 0.05 = 1.79 \text{ in. WG}$$

The total pressure loss of the return and relief systems is then

$$\Delta p_{r.ru50} = 4.08 \times 10^{-9}(6642)^2 = 0.18 \text{ in. WG}$$

At 50 percent design flow, the velocity pressure of the relief fan is

$$p_{v.rf} = \frac{6642}{(16,642)^2} \times 0.38 = 0.06$$

The fan total pressure of the relief fan at 50 percent design flow is therefore

$$\Delta p_{ref50} = (0.18 - 0.05) + 0.06 = 0.19 \text{ in. WG}$$

Pressure–volume characteristics at key points in air economizer mode at 50 percent design volume flow are

Points	O	m	r	ru
p, in. WG	0	−0.15	+0.05	+0.03
$\dot{V}$, cfm	10,000	10,000	10,000	6642

Warm-Up and Cool-Down Mode

During warm-up and cool-down mode, the relief fan is turned off and the relief damper is closed. As in the supply fan and constant-volume exhaust fan combination, the constant-volume exhaust fan is also turned off, the outdoor damper is closed, and the recirculating damper is fully open. Air is recirculated from the space to the AHU or PU and to the space again.

Controls

The operation of recirculating mode or air economizer mode is actuated by the air economizer control, which is either temperature-based or enthalpy-based, as described in Section 18.8. Warm-up and cool-down mode is activated when the sensed space temperature deviates from the set point after the AHU or PU is turned on, as described in Section 18.10.

Zone supply volume is controlled by modulation of dampers at the VAV boxes. If the space needs 10,000 cfm to offset the sensible cooling load, the zone temperature sensors modulate the dampers by means of DDC controllers to supply approximately 10,000 cfm of cold air to the appropriate zones.

At minimum outdoor air recirculating mode when the relief fan is not operating, space positive pressure is maintained because of the air balance of outdoor air. Suppose that 20,000 cfm of air is supplied to the space. If the space pressure is zero, as 3000 cfm is exhausted from space by the exhaust fan, the recirculating air is increased to 17,000 cfm. As the recirculating volume flow extracted by the supply fan is 16,642 cfm, the greater volume flow of supply air raises the space pressure until the 20,000 cfm supply air is balanced with the sum of 16,642 cfm recirculating air, 3000 cfm exhaust air, and 358 cfm exfiltrated air at + 0.05 in. WG space positive pressure.

The volume flow rate of the relief fan is usually controlled by varying the fan speed. The relief fan is often controlled by a space pressure sensor and a DDC panel to maintain a predetermined space positive pressure, as described in Section 18.10.

In addition to space pressurization, volume flow of the relief fan is sometimes controlled by the combined effect of the position of the outdoor damper and inlet vanes of the supply fan. The greater the opening of the outdoor damper and inlet vanes, the higher the relief fan speed. Control of relief volume flow by means of outdoor dampers and inlet vanes is a kind of open loop control. It often has a high tolerance, but a low installation cost.

The main drawback of the supply and relief fan combination is the increase in pressure at the mixing chamber m when the supply volume flow rate of a VAV system is reduced during minimum outdoor air recirculating mode. The higher the pressure loss of the return system, the greater the increase of p_m. Higher pressure in the mixing chamber and, therefore, a lower Δp_{om} causes a deficiency of outdoor air. The operation of a relief fan reduces both mixing chamber and space pressure. Avery (1984) recommends using a separate outdoor air fan and fan speed controller to maintain proper outdoor air intake. Measuring the outdoor air flow and opening the outdoor damper wider is another method often used to maintain required outdoor air flow at a reduced supply volume flow rate.

Design Considerations

During the design of a VAV system with a supply and relief fan combination, the following measures are recommended

- If possible, the pressure drop of the recirculating air passage between points ru and m, including the recirculating damper, should be nearly equal to the pressure drop of the outdoor air passage, including the damper, during minimum outdoor air recirculating mode.
- Use parallel opposed-blade dampers for mixing dampers, outdoor dampers, and recirculating dampers. Use parallel-blade dampers as relief dampers.

- The outdoor and recirculating dampers should have a maximum velocity of about 1500 fpm.
- Use a proportional-integral-derivative (PID) controller to control the volume flow of the relief fan.

19.11 SUPPLY AND RETURN FAN COMBINATION

Supply fan and return fan combinations are widely used, especially in air systems that serve large conditioned areas.

A return fan is located upstream from the junction of the recirculating flow passage and the exhaust air passage, point ru, as shown in Fig. 19.22a. Consider an AHU or PU with a supply fan and return fan combination with a constant-volume exhaust system. Its operating characteristics are the same as in Section 19.9, except the following:

- A return fan is installed with the following pressure–volume characteristics:

$\dot{V}$, cfm	10,000	15,000	20,000
Δp_t, in. WG	1.47	1.23	1.10

- The pressure loss of the exhaust damper after point ru in the exhaust passage is 0.4 in. WG and is 0.2 in. WG for the air louver at maximum exhaust volume flow rate. At minimum outdoor air flow, damper and louver pressure losses can be ignored.

Recirculating Mode

During minimum outdoor air recirculating mode, outdoor air is drawn through the outdoor air louver, duct, and damper and mixed with recirculating air. The mixture then flows through the filters and coils, and is supplied to the conditioned space. At design volume flow, a small part is exhausted, another part is exfiltrated, and the remaining portion is extracted by the return fan flowing through the return system. At point ru, most of the return air is recirculated through the recirculating damper, and only a small portion of return air is discharged to the outdoor air through the exhaust passage.

During minimum outdoor air recirculating mode, it is similar to a supply and constant-volume exhaust fan combination except for the following characteristics:

- At design volume flow rate, if the return fan is not operating, the fan total pressure of the supply fan is then higher than 4.65 in. WG because of the air flowing through the turned-off return fan.

If the return fan is operating, its fan total pressure is only 0.10 in. WG if the pressure at point ru

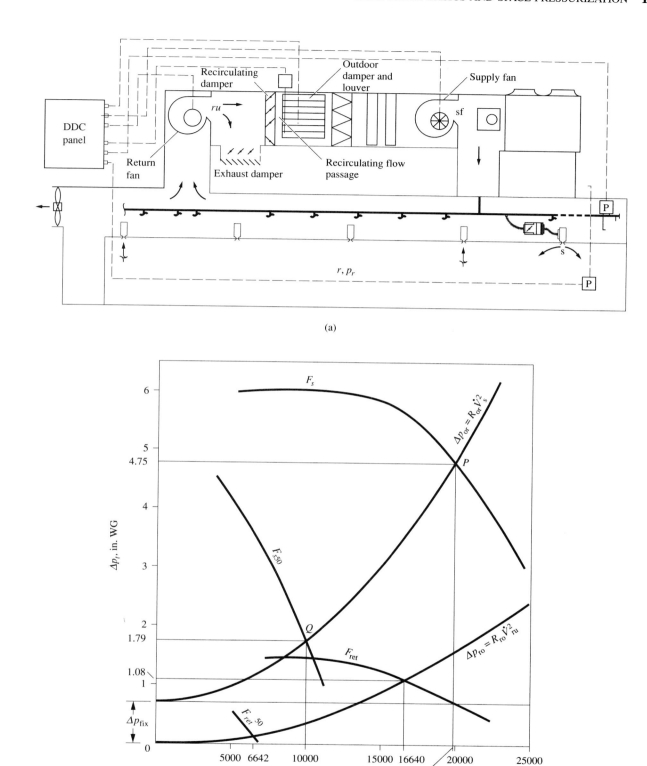

(a)

(b)

FIGURE 19.22 Supply fan and return fan combination: (a) schematic diagram and (b) system characteristics on $\Delta p_t - \dot{V}$ diagram, air economizer mode.

is 0 and is -0.40 in. WG in the mixing box, point m. If the pressure at point ru is raised higher than 0 to overcome the pressure loss in the exhaust passage, more energy is wasted across the recirculating damper.

- At 50 percent of design flow rate, if the return fan is turned off and the exhaust damper after point ru is closed, the fan total pressure of the supply fan will be higher than 1.71 in. WG, as in the supply fan and constant-volume exhaust fan combination. Space pressure p_r drops to -0.14 in. WG. Total pressure at point ru may be lower than -0.17 in. WG, and at point m lower than -0.26 in. WG because of the pressure loss of air flowing through the turned-off return fan.

If the return fan is operating and the exhaust damper is closed, the total pressure at points ru and m increases, and space pressure p_r decreases further because of the balance of outdoor air. The fan total pressure of the supply fan decreases because of the effects of the return fan.

Air Economizer Mode

During air economizer mode operation at design volume flow rate, the supply and exhaust volume flow rate and the pressure losses of various sections of the air system are the same as in a supply and relief fan combination. Because of the effects of the return fan total pressure at point ru is raised to $+0.98$ in. WG to overcome the pressure loss of the exhaust flow passage, and the discharge velocity pressure at the outlet is raised to 0.38 in. WG. The fan total pressure of the return fan is then

$$\Delta p_{\text{ret.f}} = 0.60 + (0.15 - 0.05) + 0.38$$
$$= 1.08 \text{ in. WG}$$

Pressure–volume characteristics at the key points in air economizer mode at design volume flow are as follows:

Points	O	m	r	ru
p, in. WG	0	-0.6	$+0.05$	$+0.98$
V,/cfm	20,000	20,000	20,000	16,642

During air economizer mode at 50 percent of design volume flow, the return fan raises the total pressure at point ru to $+0.16$ in. WG, including a discharge velocity pressure of 0.06 in. WG. The fan total pressure of the return fan is

$$\Delta p_{\text{ret50}} = 0.16 - (0.05 - 0.02) = 0.13 \text{ in. WG}$$

The pressure–volume characteristics at the key points are as follows:

Points	O	m	r	ru
p, in. WG	0	-0.15	$+0.05$	$+0.16$
V, cfm	10,000	10,000	10,000	6642

Controls

For a VAV system using a supply and return fan combination, the purpose of return fan control is to modulate its volume flow rate and fan total pressure in order to maintain the required space pressure and conserve return fan energy.

In air economizer mode, outdoor air supply is not a problem. In recirculating mode at a reduced supply volume flow rate, modulation of the inlet vane openings or the return fan speed during flow tracking control does not alleviate the deficiency of outdoor air intake, and may even result in a lower space pressure.

If the pressure drop of the return system between the space and point ru is small (less than 0.25 in. WG), it may be best to turn off the return fan in recirculating mode. Use minimum outdoor air control or an additional minimum outdoor fan with a speed controller to alleviate outdoor air intake deficiency at a reduced supply volume flow rate. When the return fan is not operating, the exhaust damper must be closed and a return fan bypass may be used to conserve energy.

A DDC panel or controller using PID control mode is recommended for return fan control. In air economizer mode, the exhaust damper should be fully open. Other controls are similar to those of a supply and relief fan combination.

19.12   COMPARISON OF THE THREE AIR SYSTEMS

An air system with a supply fan and constant-volume exhaust combination is simpler and less expensive. Such a system is not suitable for operation in air economizer mode. In a supply fan and constant-volume exhaust fan combination, a comparatively low pressure drop in the return system is necessary to prevent an unacceptably low pressure in the mixing box ($p_m < -1$ in. WG). Such low mixing box pressure induces air leakage and may impede minimum outdoor air control.

The operating characteristics of the supply and relief fan combination and the supply and return fan combination can be compared as follows:

- In a supply and relief fan combination in recirculating mode, total pressure at point ru may be negative; in a supply and return fan combination, total pressure at ru must be positive in order to overcome the pressure loss of the exhaust passage.
- With a relief fan, the mixing chamber pressure p_m is always negative. With a return fan, p_m may be positive if the return fan is too large. A positive p_m means that there will be no outdoor air intake in the AHU or PU.
- If the pressure drop of the return system is considerable, the variation in negative pressure in the mixing box point m during recirculating mode is greater when using a relief fan than when using a return fan as the supply volume flow varies.
- When negative space pressure must be maintained, a relief fan causes a greater decrease in mixing box pressure than a return fan during minimum outdoor air recirculating mode.
- When there is a considerable pressure drop in the exhaust or relief flow passage, a return fan requires high positive pressure at point ru, so energy is wasted across the recirculating damper.
- A supply and relief fan combination is more energy-efficient than a supply and return fan combination during recirculating mode operation for the following reasons: (1) the pressure drop across the recirculating damper is smaller; (2) the relief fan heat gain does not increase coil load; and (3) if there is no bypass passage, there is an additional pressure drop across the idle return fan.
- An axial relief fan is often simpler in layout and installation, and is therefore less expensive to install than a centrifugal return fan.

Therefore, for those air systems with a smaller pressure drop in the return duct system (normally less than 0.6 in. WG) or a considerable pressure drop in the exhaust or relief flow passage, a supply and relief fan combination is recommended.

For air systems with a greater pressure drop in the return system (greater than 1 in. WG, for example) or those that require a negative space pressure, a supply and return fan combination may be more appropriate.

REFERENCES AND FURTHER READING

Alcorn, L. H., and Huber, P. J., "Decoupling Supply and Return Fans for Increased Stability of VAV Systems," *ASHRAE Transactions,* Part I, pp. 1484–1492, 1988.

Alley, R. L., "Selecting and Sizing Outside and Return Air Dampers for VAV Economizer Systems," *ASHRAE Transactions,* Part I, pp. 1457–1466, 1988.

AMCA, *Fan and Systems Publication 201,* AMCA, Arlington Heights, IL, 1973.

Anderson, S. A., "Control Techniques for Zoned Pressurization," *ASHRAE Transactions,* Part II, pp. 1123–1139, 1987.

ASHRAE, *ASHRAE Handbook 1987, HVAC Systems and Applications,* ASHRAE Inc., Atlanta, GA, 1987.

Avery, G., "VAV Economizer Cycle: Don't Use a Return Fan," *Heating/Piping/Air Conditioning,* pp. 91–94, August 1984.

Aynsley, R. M., "The Estimation of Wind Pressures At Ventilation Inlets and Outlets on Buildings," *ASHRAE Transactions,* Part II, pp. 707–721, 1989.

Bentsen, L. J., "Zoned Airflow Control," *ASHRAE Transactions,* Part II, pp. 1140–1145, 1987.

Coward, C. W., Jr., "A Summary of Pressure Loss Values for Various Fan Inlet and Outlet Duct Fittings," *ASHRAE Transactions,* Part I B, pp. 781–789, 1983.

Driscoll, D. J., "System Effect—The Balancer's Dilemma?" *ASHRAE Transactions,* Part I B, pp. 795–801, 1983.

Grimm, N. R., and Rosaler, R. C., *Handbook of HVAC Design,* McGraw-Hill, New York, 1990.

Haines, R. W., "Supply Fan Volume Control in a VAV System," *Heating/Piping/Air Conditioning,* pp. 107–111, August 1983.

Holness, G. R., "Pressurization Control: Facts and Fallacies," *Heating/Piping/Air Conditioning,* pp. 47–51, February 1989.

Kalasinsky, C. C., "The Economics of Relief Fans vs Return Fans in Variable Volume Systems with Economizer Cycles," *ASHRAE Transactions,* Part I, pp. 1467–1476, 1988.

Kettler, J. P., "Field Problems Associated with Return Fans on VAV Systems," *ASHRAE Transactions,* Part I, pp. 1477–1483, 1988.

Mumma, S. A., and Wong, Y. M., "Analytical Evaluation of Outdoor Airflow Rate Variation vs Supply Airflow Rate Variation in Variable-Air-Volume Systems When the Outdoor Air Damper Is Fixed," *ASHRAE Transactions,* Part I, pp. 1197–1208, 1990.

O'Connor, J. F., "The System Effect and How It Changes Fan Performance," *ASHRAE Transactions,* Part I B, pp. 771–775, 1983.

Persily, A. K., and Linteris, G. T., "A Comparison of Measured and Predicted Infiltration Rates," *ASHRAE Transactions,* Part II B, pp. 183–200, 1983.

Smith, R. B., "Importance of Flow Transmitter Selection for Return Fan Control in VAV Systems," *ASHARE Transactions,* Part I, pp. 1218–1223, 1990.

The Trane Company, *Air Conditioning Fans,* The Trane Company, La Crosse, WI, 1985.

The Trane Company, *Building Pressurization Control,* The Trane Company, La Crosse, WI, 1982.

Wang, S. K., *Air Conditioning,* Vols 2 and 3, Hong Kong Polytechnic, Hong Kong, 1987.

Winter, S., "Building Pressurization Control with Rooftop Air Conditioners," *Heating/Piping/Air Conditioning,* pp. 89–94, October 1982.

CHAPTER
20

WATER
SYSTEMS

20.1 FUNDAMENTALS

Types of Water Systems

Water systems that are part of an air conditioning system and that link the central plant, chiller/boiler, air-handling units (AHUs), and terminals may be classified into the following categories according to their use:

CHILLED WATER SYSTEMS. In a chilled water system, water is first cooled in the water chiller—the evaporator of a reciprocating, screw, or centrifugal refrigeration system located in a centralized plant—to a temperature between 40 and 50°F. It is then pumped to the water cooling coils in AHUs and terminals, in which air is cooled and dehumidified. After flowing through the coils, the chilled water increases in temperature from 50°F to 60°F and then returns to the chiller.

Chilled water is widely used as a cooling medium in central hydronic air conditioning systems. When the operating temperature is below 40°F, inhibited glycols, such as ethylene glycol or propylene glycol, may be added to water to create an aqueous solution with a lower freezing point.

EVAPORATIVE-COOLED WATER SYSTEM. In arid southwestern parts of the United States, evaporative-cooled water is often produced by an evaporative cooler to cool the air.

HOT WATER SYSTEMS. These systems were discussed in Chapter 14.

DUAL-TEMPERATURE WATER SYSTEMS. In a dual-temperature water system, chilled water or hot water is supplied to the coils in AHUs and terminals and returned to the water chiller or boiler mainly through the following two distribution systems:

- Use supply and return main and branch pipes separately
- Use the common supply and return mains, branch pipe, and coil for hot and chilled water supply and return.

The changeover from chilled water to hot water and vice versa in a building or a system depends mainly on the space requirements and the temperature of the outdoor air.

Hot water is often produced by a boiler; sometimes it comes from a water source heat pump or heat recovery system, which will be discussed in later chapters.

COOLING WATER SYSTEMS. In a cooling water system, latent heat of condensation is removed from the refrigerant in the condenser by the cooling water. This cooling water is either from the cooling tower or is surface water taken from a lake, a river, the sea, or a well. For an absorption refrigeration system, heat is also removed from the solution by cooling water in the absorber. The temperature of the cooling water depends mainly on the local climate.

Water systems can also be classified according to their operating characteristics into the following categories:

Closed systems. In a closed system, chilled or hot water flowing through the coils, heaters, chillers, boilers, or other heat exchangers forms a closed recirculating loop, as shown in Fig. 20.1a. In a closed system, water is not exposed to the atmosphere during its flow process. The purpose of recirculation is to save water and energy.

Open Systems. In an open system, the water is exposed to the atmosphere, as shown in Fig. 20.1b. For example, chilled water comes directly into contact with the cooled and dehumidified air in the air washer, and cooling water is exposed to atmospheric air in the cooling tower. Recirculation of water is used to save energy.

Open systems need more water treatments than closed systems because dust and impurities in the air may be transmitted to the water in open systems. A greater quantity of make-up water is required in open systems to compensate for evaporation, drift carry-over, or blow-down operation.

Once-through systems. In a once-through system, water flows through the heat exchanger only once and does not recirculate, as shown in Fig. 20.1c. Lake, river, well, or sea water used as condenser cooling water represents a once-through system. Although the water cannot recirculate to the condenser because of its rise in temperature after absorbing the heat of condensation, it can still be used for other purposes, such as flushing water in a plumbing system after the necessary water treatments, to conserve water. In many locations, the law requires that well water be pumped back into the ground.

Volume Flow and Temperature Difference

The heating and cooling capacity of water when it flows through a heat exchanger, q_w, in Btu/h, can be

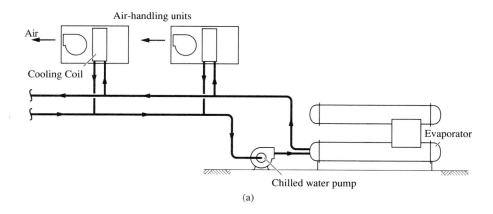

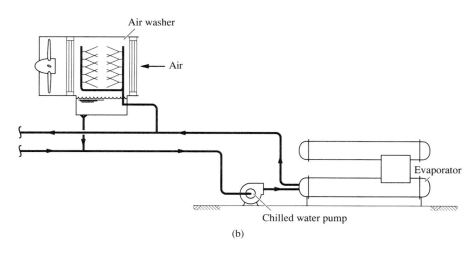

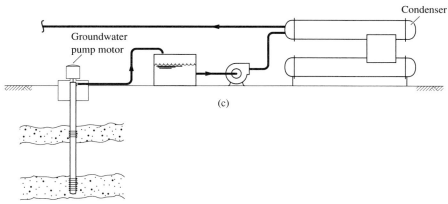

FIGURE 20.1   Types of water systems: (a) closed system, (b) open system, and (c) once-through system.

calculated by Eq. (12.3) as

$$q_w = 500\dot{V}_{gal}(T_{we} - T_{wl}) = 500\dot{V}_{gal}\Delta T_w$$

where $\dot{V}_{gal}$ = volume flow rate of water, gpm
T_{we}, T_{wl} = temperature of water entering and leaving the heat exchanger, °F

ΔT_w = temperature drop or temperature rise of water after flowing through the heat exchanger, °F

This equation also gives the relationship between ΔT_w and $\dot{V}_{gal}$ during the heat transfer process.

The temperature of water leaving the water chiller should normally be no lower than 40°F to prevent freezing. If the chilled water temperature is lower than 40°F, brine, ethylene glycol, or propylene glycol should be used. Brine will be discussed in Chapter 25. For a dual-temperature water system, the hot water temperature leaving the boiler often ranges from 100 to 150°F and returns at a ΔT_w between 20 and 40°F.

For most dual-temperature water systems, the value of $\dot{V}_{gal}$ and the pipe sizing are determined based on the cooling capacity requirement for the coils and water coolers. This is because chilled water has a comparatively smaller ΔT_w than hot water. Furthermore, the system cooling load is often higher than the system heating load. As described in Section 12.6, for a chilled water system to transport each refrigeration ton of cooling capacity, a ΔT_w of 8°F requires a $\dot{V}_{gal}$ of 3 gpm, whereas for a ΔT_w of 24°F, $\dot{V}_{gal}$ is only 1.0 gpm.

The temperature of water entering the coil, T_{we}, the temperature of water leaving the coil, T_{wl}, and the difference between them, $\Delta T_w = (T_{wl} - T_{we})$, are closely related to the performance of a chilled water system, air system, and refrigeration system:

- T_{we} directly affects the power consumption in the compressor.
- The temperature differential ΔT_w is closely related to the volume flow of chilled water, $\dot{V}_{gal}$, and thus the size of the water pipes and pumping power.
- Both T_{we} and ΔT_w influence the temperature and humidity ratio of air leaving the coil.

If the chilled water temperature leaving the water chiller and entering the coil is between 40 and 45°F, the off-coil temperature in the air system is usually around 55°F for conventional comfort air conditioning systems. In low-temperature cold air distribution systems, chilled water leaving the chiller may be as low as 34°F and the off-coil temperature is often between 42 and 47°F (typically 44°F).

The greater the valve of ΔT_w for chilled water, the lower the amount of water flowing through the coil. Current practice is to use a value of ΔT_w between 12 and 20°F, or sometimes 24°F. For chilled water systems in a campus-type central plant, a value of ΔT_w between 16 and 24°F is often used.

Water Velocity and Pressure Drop

The maximum water velocity in pipes is governed mainly by pipe erosion, noise, and water hammer.

Erosion of water pipes is the result of the impingement of rapidly moving water containing air bubbles and impurities on the inner surface of the pipes and fittings. Solden and Siegel (1964) increased their feedwater velocity gradually from 8 ft/s to an average of 35.6 ft/s. After three years, they found no evidence of erosion in the pipe or a connected check valve. Erosion occurs only if solid matter is contained in water flowing at high velocity.

Velocity-dependent noise in pipes results from flow turbulence, cavitation, release of entrained air, and *water hammer* which results from the transient pressure impact on a suddenly closed valve. Ball and Webster (1976) performed a series of tests on $\frac{3}{8}$-in. copper tubes with elbows. At a water velocity of 16.4 ft/s, the noise level was below 53 dBA. Tests also showed that cold water at speeds up to 21 ft/s did not cause cavitation. In copper and steel pipes, water hammer at a water velocity of 15 ft/s exerted a pressure on 2-in. diameter pipes less than 50 percent of their design pressure.

Given the above results, excluding the energy cost for the pump power, the maximum water velocity in certain short sections of a water systems may be raised to an upper limit of 12 ft/s for a special purpose, such as enhancing the heat transfer coefficients.

Normally, water flow in coils and heat exchangers becomes laminar and seriously impairs the heat transfer characteristics only when its velocity drops to a value less than 1 ft/s and its corresponding Reynolds number is reduced to below 4000.

When pipes are being sized, the optimum pressure drop ΔH_f, commonly expressed in ft of head loss of water per 100 ft of pipe length, is a compromise between energy costs and investment costs. At the same time, the aged-corrosion of pipes should also be considered.

Generally, the pressure drop for water pipes inside buildings, ΔH_f, is in a range of 0.75 ft/100 ft to 4 ft/100 ft, with a mean of 2.5 ft/100 ft used most often. Because of a comparatively lower increase in installation costs for smaller-diameter pipes, it may be best to use a pressure drop lower than 2.5 ft/100 ft when the pipe diameter is 2 in. or less.

Aged-corrosion results in an increase in the friction factor and a decrease in the effective diameter. The factors that contribute to aged-corrosion are slimming, caking of calcareous salts, and corrosion. Many scientists recommend an increase in friction loss of 15 to 20 percent, resulting in a design pressure drop of 2 ft/100 ft, for closed water systems; and a 75 to 90 percent increase in friction loss, or a design pressure drop of 1.35 ft/100 ft, for open water systems.

ASHRAE/IES Standard 90.1–1989 specifies that water piping systems should be designed at a pressure loss rate of no more than 4.0 ft/100 ft of pipe.

Figures 20.2, 20.3, and 20.4 show the pressure drop charts for steel, copper, and plastic pipes, respectively, for closed water systems. Each chart shows the volume flow $\dot{V}_{gal}$, in gpm, pressure drop ΔH_f, in ft/100 ft, water velocity v_w, in ft/s, and water pipe diameter of D, in inches. Given any two of these parameters, the other two can be determined. For instance, for a steel water pipe that has a water volume flow of 1000 gpm, if the pressure drop is 2 ft/100 ft, the diameter should be 8 in. and the corresponding velocity is about 8 ft/s.

It is a common practice to limit the water velocity to no more than 4 ft/s for water pipes 2 in. or less in diameter in order to prevent an excessive ΔH_f. The pressure drop should not exceed 4 ft/100 ft for water pipes of over 2-in. diameter.

20.2 WATER PIPING

Piping Material

For water systems, the piping materials most widely used are steel, both black (plain) and galvanized (zinc-coated) in the form of either welded-seam steel pipe or seamless steel pipe; ductile iron and cast iron; hard copper; and polyvinyl chloride (PVC). The piping materials for various purposes are shown below:

Chilled water	Black and galvanized steel
Hot water	Black steel, hard copper
Cooling water and drains	Black steel, galvanized ductile iron, PVC

Copper, galvanized steel, galvanized ductile iron, and PVC pipes have better corrosion resistance than black steel pipes.

Technical requirements, as well as local customs, determine the selection of piping materials.

Piping Dimensions

The steel pipe wall thicknesses currently used were standardized in 1930. The thicknesses range from Schedule 10, light wall, to Schedule 160, very heavy wall. Schedule 40 is the standard thickness for a pipe with a diameter up to 10 in. For instance, a 2-in. standard pipe has an outside diameter of 2.375 in. and an inside diameter of 2.067 in. The nominal pipe size is only an approximate indication of pipe size, especially for pipes of small diameter. Table 20.1 list the dimensions of commonly used steel pipes.

The outside diameter of extruded copper is standardized so that the outside diameter of copper tub-

ing is $\frac{1}{8}$ in. larger than the nominal size used for soldered or brazed socket joints. As is the case with steel pipes, the result is that the inside diameters of copper tubes seldom equal the nominal sizes. Types K, L, M, and DWV designate the wall thickness of copper tubes: type K is the heaviest, and DWV is the lightest. Type L is generally used as the standard for pressure copper tubing. Type DWV is used for drainage at atmospheric pressure.

Copper tubes are also categorized as *hard* and *soft* copper. Soft pipes should be used in applications for which the pipes will be bent in the field. Table 20.2 list the dimensions of copper tubes.

Thermoplastic plastic pipes are the most widely used plastic pipes in air conditioning. They are manufactured with dimensions that match steel pipe dimensions. The advantages of plastic pipes include resistance to corrosion and scaling and resistance to the growth of algae and fungi. They have smooth surfaces and negligible aged-allowance. *Aged-allowance* is the allowance for corrosion and sealing for plastic pipes during their service life. Most plastic pipes are low in cost, especially compared with corrosion-resistant metal tubes.

The disadvantages of plastic pipes include the fact that their pressure ratings decrease rapidly when the water temperature rises above 100°F. PVC pipes are weaker than metal pipes and must usually be thicker than steel pipes if same working pressure is to be maintained. Plastic pipes may experience expansion and contraction during temperature changes that is four times greater than that of steel. Precautions must be taken to protect plastic pipes from external damage and to account for its behavior during fire. Some local codes do not permit the use of some or all plastic pipes. It is necessary to check with local authorities.

Pipe Joints

Steel pipes of small diameter (2 in. or less) threaded through cast-iron fittings are the most widely used type of pipe joint. For steel pipes of diameter 2 in. and over, welded joints, bolted flanges, and grooved ductile iron joined fittings are often used. Galvanized steel pipes are threaded together by galvanized cast-iron or ductile iron fittings.

Copper pipes are usually joined by soldering and brazing socket end fittings. Plastic pipes are often joined by solvent welding, fusion welding, screw joints, or bolted flanges.

Vibrations from pumps, chillers, or cooling towers can be isolated or dampened by means of flexible pipe couplings. Arch connectors are usually constructed of nylon, dacron, or polyester and neoprene.

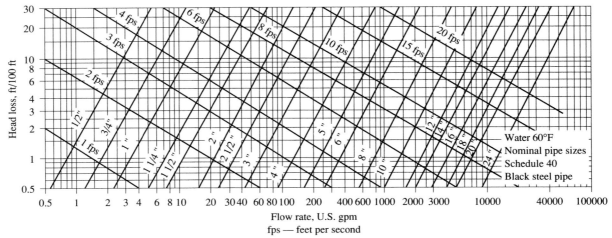

FIGURE 20.2 Friction chart for water in steel pipes (schedule 40). (*Source: ASHRAE Handbook 1989 Fundamentals*. Reprinted with permission.)

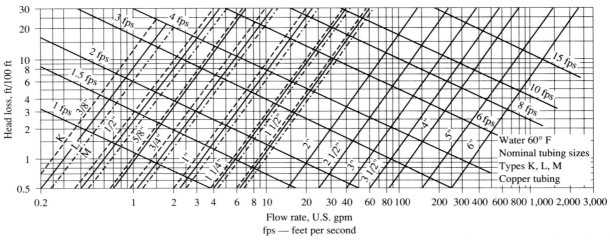

FIGURE 20.3 Friction chart for water in copper tubing (type K, L, M). (*Source: ASHRAE Handbook Fundamentals*. Reprinted with permission.)

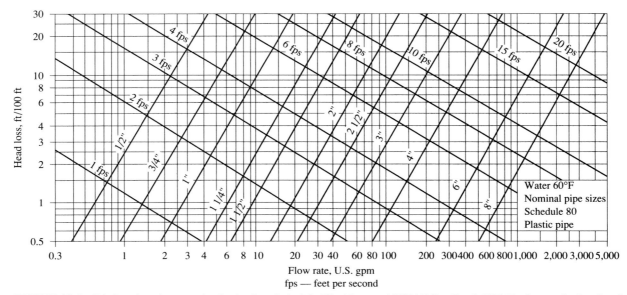

FIGURE 20.4 Friction chart for water in plastic pipes (schedule 80). (*Source: ASHRAE Handbook 1989 Fundamentals*. Reprinted with permission.)

20.6

TABLE 20.1
Dimensions of commonly used steel pipes

Nominal size and pipe O.D. D, in.	Schedule number or weight	Wall thickness t, in.	Inside diameter d, in.	Surface area Outside, ft²/ft	Surface area Inside, ft²/ft	Cross-sectional Metal area in²	Cross-sectional Flow area in²	Weight of Pipe lb/ft	Weight of Water lb/ft	Mfr. process	Joint type	psig
$\frac{1}{4}$ D = 0.540	40 ST	0.088	0.364	0.141	0.095	0.125	0.104	0.424	0.045	CW	Thrd	188
	80 XS	0.119	0.302	0.141	0.079	0.157	0.072	0.535	0.031	CW	Thrd	871
$\frac{3}{8}$ D = 0.675	40 ST	0.091	0.493	0.177	0.129	0.167	0.191	0.567	0.083	CW	Thrd	203
	80 XS	0.126	0.423	0.177	0.111	0.217	0.141	0.738	0.061	CW	Thrd	820
$\frac{1}{2}$ D = 0.840	40 ST	0.109	0.622	0.220	0.163	0.250	0.304	0.850	0.131	CW	Thrd	214
	80 XS	0.147	0.546	0.220	0.143	0.320	0.234	1.087	0.101	CW	Thrd	753
$\frac{3}{4}$ D = 1.050	40 ST	0.113	0.824	0.275	0.216	0.333	0.533	1.13	0.231	CW	Thrd	217
	80 XS	0.154	0.742	0.275	0.194	0.433	0.432	1.47	0.187	CW	Thrd	681
1 D = 1.315	40 ST	0.133	1.049	0.344	0.275	0.494	0.864	1.68	0.374	CW	Thrd	226
	80 XS	0.179	0.957	0.344	0.251	0.639	0.719	2.17	0.311	CW	Thrd	642
$1\frac{1}{4}$ D = 1.660	40 ST	0.140	1.380	0.435	0.361	0.669	1.50	2.27	0.647	CW	Thrd	229
	80 XS	0.191	1.278	0.435	0.335	0.881	1.28	2.99	0.555	CW	Thrd	594
$1\frac{1}{2}$ D = 1.900	40 ST	0.145	1.610	0.497	0.421	0.799	2.04	2.72	0.881	CW	Thrd	231
	80 XS	0.200	1.500	0.497	0.393	1.068	1.77	3.63	0.765	CW	Thrd	576
2 D = 2.375	40 ST	0.154	2.067	0.622	0.541	1.07	3.36	3.65	1.45	CW	Thrd	230
	80 XS	0.218	1.939	0.622	0.508	1.48	2.95	5.02	1.28	CW	Thrd	551
$2\frac{1}{2}$ D = 2.875	40 ST	0.203	2.469	0.753	0.646	1.70	4.79	5.79	2.07	CW	Weld	533
	80 XS	0.276	2.323	0.753	0.608	2.25	4.24	7.66	1.83	CW	Weld	835
3 D = 3.500	40 ST	0.216	3.068	0.916	0.803	2.23	7.39	7.57	3.20	CW	Weld	482
	80 XS	0.300	2.900	0.916	0.759	3.02	6.60	10.25	2.86	CW	Weld	767
4 D = 4.500	40 ST	0.237	4.026	1.178	1.054	3.17	12.73	10.78	5.51	CW	Weld	430
	80 XS	0.337	3.826	1.178	1.002	4.41	11.50	14.97	4.98	CW	Weld	695
6 D = 6.625	40 ST	0.280	6.065	1.734	1.588	5.58	28.89	18.96	12.50	ERW	Weld	696
	80 XS	0.432	5.761	1.734	1.508	8.40	26.07	28.55	11.28	ERW	Weld	1209
8 D = 8.625	30	0.277	8.071	2.258	2.113	7.26	51.16	24.68	22.14	ERW	Weld	526
	40 ST	0.322	7.981	2.258	2.089	8.40	50.03	28.53	21.65	ERW	Weld	643
	80 XS	0.500	7.625	2.258	1.996	12.76	45.66	43.35	19.76	ERW	Weld	1106
10 D = 10.75	30	0.307	10.136	2.814	2.654	10.07	80.69	34.21	34.92	ERW	Weld	485
	40 ST	0.365	10.020	2.814	2.623	11.91	78.85	40.45	34.12	ERW	Weld	606
	XS	0.500	9.750	2.814	2.552	16.10	74.66	54.69	32.31	ERW	Weld	887
	80	0.593	9.564	2.814	2.504	18.92	71.84	64.28	31.09	ERW	Weld	1081
12 D = 12.75	30	0.330	12.090	3.338	3.165	12.88	114.8	43.74	49.68	ERW	Weld	449
	ST	0.375	12.000	3.338	3.141	14.58	113.1	49.52	48.94	ERW	Weld	528
	40	0.406	11.938	3.338	3.125	15.74	111.9	53.48	48.44	ERW	Weld	583
	XS	0.500	11.750	3.338	3.076	19.24	108.4	65.37	46.92	ERW	Weld	748
	80	0.687	11.376	3.338	2.978	26.03	101.6	88.44	43.98	ERW	Weld	1076
14 D = 14.00	30 ST	0.375	13.250	3.665	3.469	16.05	137.9	54.53	59.67	ERW	Weld	481
	40	0.437	13.126	3.665	3.436	18.62	135.3	63.25	58.56	ERW	Weld	580
	XS	0.500	13.000	3.665	3.403	21.21	132.7	72.04	57.44	ERW	Weld	681
	80	0.750	12.500	3.665	3.272	31.22	122.7	106.05	53.11	ERW	Weld	1081
16 D = 16.00	30 ST	0.375	15.250	4.189	3.992	18.41	182.6	62.53	79.04	ERW	Weld	421
	40 XS	0.500	15.000	4.189	3.927	24.35	176.7	82.71	76.47	ERW	Weld	596
18 D = 18.00	ST	0.375	17.250	4.712	4.516	20.76	233.7	70.54	101.13	ERW	Weld	374
	30	0.437	17.126	4.712	4.483	24.11	230.3	81.91	99.68	ERW	Weld	451
	XS	0.500	17.000	4.712	4.450	27.49	227.0	93.38	98.22	ERW	Weld	530
	40	0.562	16.876	4.712	4.418	30.79	223.7	104.59	96.80	ERW	Weld	607
20 D = 20.00	ST	0.375	19.250	5.236	5.039	23.12	291.0	78.54	125.94	ERW	Weld	337
	30 XS	0.500	19.000	5.236	4.974	30.63	283.5	104.05	122.69	ERW	Weld	477
	40	0.593	18.814	5.236	4.925	36.15	278.0	122.82	120.30	ERW	Weld	581

*Numbers are schedule number per ASTM B36.10; ST = standard weight; XS = extra strong.

†Working pressures have been calculated per ASME/ANSI B31.9 using furnace butt weld (continuous weld, CW) pipe through 4 in. and electric resistance weld (ERW) thereafter. The allowance A has been taken as: (a) 12.5 percent of t for mill tolerance on pipe wall thickness, plus (b) An arbitrary corrosion allowance of 0.025 in. for pipe sizes through NPS 2 and 0.065 in. from NPS $2\frac{1}{2}$ through 20 plus (c) A thread cutting allowance for sizes through NPS 2.

Because the pipe wall thickness of threaded standard weight pipe is so small after deducting the allowance A, the mechanical strength of the pipe is impaired. It is good practice to limit standard-weight threaded pipe pressures to 90 psig for steam and 125 psig for water.

Source: ASHRAE Handbook 1992, HVAC Systems and Equipment. Reprinted with permission.

TABLE 20.2
Dimensions of copper tubes

Nominal diameter, in.	Type	Wall thickness t, in.	Outside diameter D, in.	Inside diameter d, in.	Surface area Outside, ft²/ft	Surface area Inside, ft²/ft	Cross-sectional Metal area, in²	Cross-sectional Flow area in²	Weight of Tube, lb/ft	Weight of Water, lb/ft	Working pressure* ASTMB 888 to 250°F Annealed, psig	Working pressure* ASTMB 888 to 250°F Drawn, psig
$\frac{1}{4}$	K	0.035	0.375	0.305	0.098	0.080	0.037	0.073	0.145	0.032	851	1596
	L	0.030	0.375	0.315	0.098	0.082	0.033	0.078	0.126	0.034	730	1368
$\frac{3}{8}$	K	0.049	0.500	0.402	0.131	0.105	0.069	0.127	0.269	0.055	894	1676
	L	0.035	0.500	0.430	0.131	0.113	0.051	0.145	0.198	0.063	638	1197
	M	0.025	0.500	0.450	0.131	0.008	0.037	0.159	0.145	0.069	456	855
$\frac{1}{2}$	K	0.049	0.625	0.527	0.164	0.138	0.089	0.218	0.344	0.094	715	1341
	L	0.040	0.625	0.545	0.164	0.143	0.074	0.233	0.285	0.101	584	1094
	M	0.028	0.625	0.569	0.164	0.149	0.053	0.254	0.203	0.110	409	766
$\frac{5}{8}$	K	0.049	0.750	0.652	0.196	0.171	0.108	0.334	0.418	0.144	596	1117
	L	0.042	0.750	0.666	0.196	0.174	0.093	0.348	0.362	0.151	511	958
$\frac{3}{4}$	K	0.065	0.875	0.745	0.229	0.195	0.165	0.436	0.641	0.189	677	1270
	L	0.045	0.875	0.785	0.229	0.206	0.117	0.484	0.455	0.209	469	879
	M	0.032	0.875	0.811	0.229	0.212	0.085	0.517	0.328	0.224	334	625
1	K	0.065	1.125	0.995	0.295	0.260	0.216	0.778	0.839	0.336	527	988
	L	0.050	1.125	1.025	0.295	0.268	0.169	0.825	0.654	0.357	405	760
	M	0.035	1.125	1.055	0.295	0.276	0.120	0.874	0.464	0.378	284	532
$1\frac{1}{4}$	K	0.065	1.375	1.245	0.360	0.326	0.268	1.217	1.037	0.527	431	808
	L	0.055	1.375	1.265	0.360	0.331	0.228	1.257	0.884	0.544	365	684
	M	0.042	1.375	1.291	0.360	0.338	0.176	1.309	0.682	0.566	279	522
	DMV	0.040	1.375	1.295	0.360	0.339	0.168	1.317	0.650	0.570	265	497
$1\frac{1}{2}$	K	0.072	1.625	1.481	0.425	0.388	0.351	1.723	1.361	0.745	404	758
	L	0.060	1.625	1.505	0.425	0.394	0.295	1.779	1.143	0.770	337	631
	M	0.049	1.625	1.527	0.425	0.400	0.243	1.831	0.940	0.792	275	516
	DWV	0.042	1.625	1.541	0.425	0.403	0.209	1.865	0.809	0.807	236	442
2	K	0.083	2.125	1.959	0.556	0.513	0.532	3.014	2.063	1.304	356	668
	L	0.070	2.125	1.985	0.556	0.520	0.452	3.095	1.751	1.339	300	573
	M	0.058	2.125	2.009	0.556	0.526	0.377	3.170	1.459	1.372	249	467
	DWV	0.042	2.125	2.041	0.556	0.534	0.275	3.272	1.065	1.416	180	338
$2\frac{1}{2}$	K	0.095	2.625	2.435	0.687	0.637	0.755	4.657	2.926	2.015	330	619
	L	0.080	2.625	2.465	0.687	0.645	0.640	4.772	2.479	2.065	278	521
	M	0.065	2.625	2.495	0.687	0.653	0.523	4.889	2.026	2.116	226	423
3	K	0.109	3.125	2.907	0.818	0.761	1.033	6.637	4.002	2.872	318	596
	L	0.090	3.125	2.945	0.818	0.771	0.858	6.812	3.325	2.947	263	492
	M	0.072	3.125	2.981	0.818	0.780	0.691	6.979	2.676	3.020	210	394
	DWV	0.045	3.125	3.035	0.818	0.795	0.435	7.234	1.687	3.130	131	246
$3\frac{1}{2}$	K	0.120	3.625	3.385	0.949	0.886	1.321	8.999	5.120	3.894	302	566
	L	0.100	3.625	3.425	0.949	0.897	1.107	9.213	4.291	3.987	252	472
	M	0.083	3.625	3.459	0.949	0.906	0.924	9.397	3.579	4.066	209	392
4	K	0.134	4.125	3.857	1.080	1.010	1.680	11.684	6.510	5.056	296	555
	L	0.110	4.125	3.905	1.080	1.022	1.387	11.977	5.377	5.182	243	456
	M	0.095	4.125	3.935	1.080	1.030	1.203	12.161	4.661	5.262	210	394
	DWV	0.058	4.125	4.009	1.080	1.050	0.741	12.623	2.872	5.462	128	240
5	K	0.160	5.125	4.805	1.342	1.258	2.496	18.133	9.671	7.846	285	534
	L	0.125	5.125	4.875	1.342	1.276	1.963	18.665	7.609	8.077	222	417
	M	0.109	5.125	4.907	1.342	1.285	1.718	18.911	6.656	8.183	194	364
	DWV	0.072	5.125	4.981	1.342	1.304	1.143	19.486	4.429	8.432	128	240
6	K	0.192	6.125	5.741	1.603	1.503	3.579	25.886	13.867	11.201	286	536
	L	0.140	6.125	5.845	1.603	1.530	2.632	26.832	10.200	11.610	208	391
	M	0.122	6.125	5.881	1.603	1.540	2.301	27.164	8.916	11.754	182	341
	DWV	0.083	6.125	5.959	1.603	1.560	1.575	27.889	6.105	12.068	124	232
8	K	0.271	8.125	7.583	2.127	1.985	6.687	45.162	25.911	19.542	304	570
	L	0.200	8.125	7.725	2.127	2.022	4.979	46.869	19.295	20.280	224	421
	M	0.170	8.125	7.785	2.127	2.038	4.249	47.600	16.463	20.597	191	358
	DWV	0.109	8.125	7.907	2.127	2.070	2.745	49.104	10.637	21.247	122	229
10	K	0.338	10.125	9.449	2.651	2.474	10.392	70.123	40.271	30.342	304	571
	L	0.250	10.125	9.625	2.651	2.520	7.756	72.760	30.054	31.483	225	422
	M	0.212	10.125	9.701	2.651	2.540	6.602	73.913	25.584	31.982	191	358
12	K	0.405	12.125	11.315	3.174	2.962	14.912	100.554	57.784	43.510	305	571
	L	0.280	12.125	11.565	3.174	3.028	10.419	105.046	40.375	45.454	211	395
	M	0.254	12.125	11.617	3.174	3.041	9.473	105.993	36.706	45.863	191	358

*When using soldered or brazed fittings, the joint determines the limiting pressure.

Working pressures calculated using ASME B31.9 allowable stresses. A 5 percent mill tolerance has been used on the wall thickness. Higher tube ratings can be calculated using allowable stress for lower temperatures.

If soldered or brazed fittings are used on hard drawn tubing, use the annealed ratings. Full-tube allowable pressures can be used with suitably rated flare or compression-type fittings.

Source: ASHRAE Handbook 1992, HVAC Systems and Equipment. Reprinted with permission.

TABLE 20.3
Maximum allowable pressures at corresponding temperatures

Application	Pipe material	Weight	Joint type	Fitting		System	
				Class	Material	Temperature, °F	Maximum allowable pressure at temperature, psig
Recirculating water							
2 in. and smaller	Steel (CW)	Standard	Thread	125	Cast iron	250	125
	Copper, hard	Type L	95-5 solder	—	Wrought copper	250	150
	PVC	Sch 80	Solvent	Sch 80	PVC	75	350
	CPVC	Sch 80	Solvent	Sch 80	CPVC	150	150
	PB	SDR-11	Heat fusion	—	PB	160	115
			Insert crimp	—	Metal	160	115
2.5–12 in.	A53 B ERW steel	Standard	Weld	Standard	Wrought steel	250	400
			Flange	150	Wrought steel	250	250
			Flange	125	Cast iron	250	175
			Flange	250	Cast iron	250	400
			Groove	—	MI or ductile iron	230	300
	PB	SDR-11	Heat fusion		PB	160	115
Refrigerant							
	Copper, hard	Type L or K	Braze	—	Wrought copper	—	—
	A53 B SML steel	Standard	Weld		Wrought steel	—	—

Note: Maximum allowable working pressures have been de-rated in this table. Higher system pressures can be used for lower temperatures and smaller pipe sizes. Pipe, fittings, joints, and valves must all be considered.

Note: A53   ASTM Standard A53
PVC   Polyvinyl chloride
CPVC   Chlorinated polyvinyl chloride
PB   Polybutylene

Abridged with permission from *ASHRAE Handbook 1992, HVAC Systems and Equipment.*

They can accommodate deflections or dampen vibrations in all directions. Restraining rods and plates are required to prevent excessive stretching. A flexible metal hose connector includes a corrugated inner core with a braided cover. It is available with flanged or grooved end joints.

Working Pressure and Temperature

In a water system, the maximum allowable working pressure and temperature are not limited to the pipes only; joints or pipe fittings, especially valves, may often be the weak links. Table 20.3 lists types of pipes, joints, and fittings and their maximum allowable working pressures for specified temperatures.

Expansion and Contraction

During temperature changes, all pipes expand and contract. The design of water pipes must take into consideration this expansion and contraction. Not only should temperature change during the operating pe-

riod be considered; the possible temperature change between the operating and shut-down periods should also be considered. For chilled and cooling water, which have a possible temperature change of 40 to 100°F, expansion and contraction cause considerable movement in a long run of piping. Unexpected expansion and contraction cause excess stress and possible failure of the pipe, pipe support, pipe joints, and fittings.

Expansion and contraction of hot and chilled water pipes can be better accommodated by using loops and bends. The commonly used bends are U-bends, Z-bends, and L-bends, as shown in Fig. 20.5. Anchors are the points where the pipe is fixed so that it will expand or contract between them. Reaction forces at these anchors should be considered when the support is being designed. The *ASHRAE Handbook 1992, HVAC Systems and Equipment* gives the required calculations and data for determining these stresses. Guides are used so that the pipes expand laterally.

Empirical formulas are often used instead of detailed stress analyses to determine the dimension of the offset leg L_o, in ft. Waller (1990) recommended

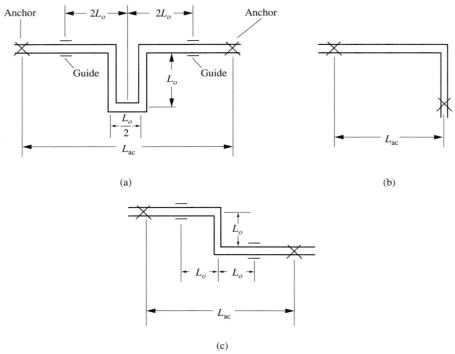

FIGURE 20.5 Expansion loops (a) U-bends, (b) L-bends, and (c) Z-bends.

the following formulas:

U-bends: $\quad L_o = 0.041D^{0.48}L_{ac}^{0.46}\Delta T$

Z-bends: $\quad L_o = (0.13DL_{ac}\Delta T)^{0.5}$ $\qquad$ (20.1)

L-bends: $\quad L_o = (0.314DL_{ac}\Delta T)^{0.5}$

where D = diameter of pipe, in.
$\quad L_{ac}$ = distance between anchors, hundreds of ft
$\quad \Delta T$ = temperature difference, °F

If there is no room to accommodate U-, Z-, or L-bends (such as in high-rise buildings or tunnels), mechanical expansion joints are used to compensate for movement during expansion. Packed expansion joints allow the pipes to slide to accommodate movement during expansion. Various types of packing are used to seal the sliding surfaces in order to prevent leakage. Another type of mechanical joint uses bellows or flexible metal hose to accommodate movement. These types of joints should be carefully installed to avoid distortion.

Piping Supports

Types of piping support include hangers, which hang the pipe from above; supports, which usually use brackets to support the pipes from below; anchors to control the movement of the piping; and guides to guide the axial movement of the piping.

Table 20.4 lists the recommended spacing of pipe hangers. Piping support members should be constructed based on the stress at their point of con-

TABLE 20.4
Recommended pipe hanger spacing, ft

Nominal pipe diameter, in.	Standard-weight steel pipe (water)	Copper tube (water)	Rod size, in.
1/2	7	5	1/4
3/4	7	5	1/4
1	7	6	1/4
1 1/2	9	8	3/8
2	10	8	3/8
2 1/2	11	9	3/8
3	12	10	3/8
4	14	12	1/2
6	17	14	1/2
8	19	16	5/8
10	20	18	3/4
12	23	19	7/8
14	25		1
16	27		1
18	28		1 1/4
20	30		1 1/4

Note: Spacing does not apply where concentrated loads are placed between supports such as flanges, valves, and specialties.
Source: ASHRAE Handbook 1992, HVAC Systems and Equipment. Reprinted with permission.

nection to the pipe as well as on the characteristics of the structural system. Pipe supports must have sufficient strength to support the pipe, including the water inside. Except for the anchors, they should also allow expansion movement.

Pipes should be supported around the connections to the equipment so that the pipe's weight and expansion or contraction do not affect the equipment. For insulated pipes, heavy-gauge sheet-metal half-sleeves are used between the hangers and the insulation. Corrosion protection should also be carefully considered.

Piping Insulation

External pipe insulation should be provided when the inside water temperature $105°F < T_w < 55°F$ for the sake of energy savings, surface condensation, and high-temperature safety protection. The optimum thickness of the pipe insulation depends mainly on the operating temperature of the inside water, the pipe diameter, and the types of service. As with duct insulation, there is a compromise between initial cost and energy cost. ASHRAE Standard 90.1–1989 specifies the minimum pipe insulation for water systems, as listed in Table 20.5.

20.3 VALVES, PIPE FITTINGS, AND ACCESSORIES

Types of Valves

As described in Section 17.6, valves are used to regulate or stop the water flow in pipes, either manually or by means of automatic control systems. Valves used in automatic control systems are called control valves, which were discussed in Chapter 17. In this section, only manually operated valves, or simply valves, will be discussed.

Hand-operated valves are used to stop or isolate flow, to regulate flow, to prevent reverse flow, and to regulate water pressure. The basic construction of a valve consists of the following (see Fig. 20.6): a disc to open or close the water flow; a valve body to seat the disc and provide the flow passage; a stem to lift or rotate the disc, with a handle wheel or a handle and corresponding mechanism to make the task easier; and a bonnet to enclose the valve from the top.

Based on the shape of the valve disc, the valve body, or its function, commonly used valves can be classified into the following types:

GATE VALVES. The disc of a gate valve is in the shape of a "gate" or wedge, as shown in Fig. 20.6a. When the wedge is raised at the open position, a gate valve does not add much flow resistance. The wedge can be either a solid wedge, which is most commonly used, or a split wedge, in which two disc halves being forced outward fit tightly against the body seat. Gate valves are used either fully opened or closed, an on-off arrangement. They are often used as isolating valves for pieces of equipment or key components, such as the control valve, for service during maintenance and repair.

GLOBE VALVES. These are so named because of the globular shape of the valve body, as shown in

TABLE 20.5
Minimum pipe insulation, in.

Fluid design operating temperature range, °F	Insulation conductivity		Nominal pipe diameter					
	Conductivity range, Btu·in./h·ft²·°F	Mean rating temperature, °F	Runouts* up to 2 in.	1 in. and less	$1\frac{1}{4}$ to 2 in.	$2\frac{1}{2}$ to 4 in.	5 and 6 in.	8 in. and up
Heating systems (steam, steam condensate, and hot water)								
Above 350	0.32–0.34	250	1.5	2.5	2.5	3.0	3.5	3.5
251–350	0.29–0.31	200	1.5	2.0	2.5	2.5	3.5	3.5
201–250	0.27–0.30	150	1.0	1.5	1.5	2.0	3.5	3.5
141–200	0.25–0.29	125	0.5	1.5	1.5	1.5	2.0	3.5
105–140	0.24–0.28	100	0.5	1.0	1.0	1.0	1.5	1.5
Cooling systems (chilled water, brine, and refrigerant)†								
40–55	0.23–0.27	75	0.5	0.5	0.75	1.0	1.0	1.0
Below 40	0.23–0.27	75	1.0	1.0	1.5	1.5	1.5	1.5

*Runouts (branch pipes) to individual terminal units not exceeding 12 ft in length.

†The required minimum thicknesses do not consider water vapor transmission and condensation. Additional insulation, vapor retarders, or both may be required to limit water vapor transmission and condensation.

Source: Abridged with permission from *ASHRAE Standard 90.1–1989.*

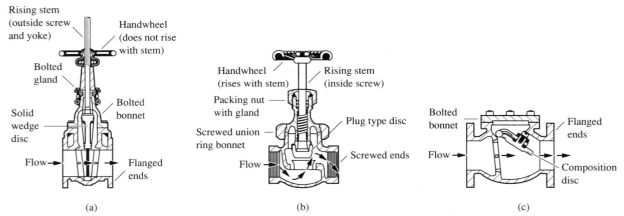

FIGURE 20.6 Types of valves: (a) gate valve, (b) globe valve, and (c) check valve, swing check. (*Source: Carrier's Handbook of Air Conditioning System Design*. Reprinted with permission.)

Fig. 20.6b. Globe valves have a round disc or plug-type disc seated against a round port. Water flow enters under the disc. Globe valves have high flow resistances. They can be opened or closed substantially faster than gate valves. Angle valves are similar to globe valves in their seats and operation. Their basic difference is that the valve body of an angle valve can also be used as a 90° elbow at that location.

Globe valves are used to throttle and to regulate the flow. They are sometimes called *balancing valves*. They are deliberately designed to restrict fluid flow, so they should not be used in applications for which full and unobstructed flow is often required.

CHECK VALVES. Check valves, as their name suggests, are valves used to prevent, or "check," reverse flow. There are basically two types of check valves: *swing check* and *lift check*.

A swing check valve has a hinged disc, as shown in Fig. 20.6c. When the water flow reverses, water pressure pushes the disc and closes the valve. In a lift check valve, upward regular flow raises the disc and opens the valve and reverse flow pushes the disc down to its seat and stops the backflow. A swing check valve has a lower flow resistance than a lift check valve.

PLUG VALVES. These valves use a tapered cylindrical plug disc to fit the seat. They vary from fully open to fully closed positions with a quarter-turn. Plug valves may be used for throttling control during the balancing of a water system.

BALL VALVES. These valves use a ball as the valve disc to open or close the valve. As with plug valves, they vary from fully open to fully closed positions with a quarter-turn. As with gate valves, ball valves

are usually used for open-close service. They are less expensive than gate valves.

BUTTERFLY VALVES. A butterfly valve has a thin rotating disc. Like a ball or plug valve, it varies with a quarter-turn from fully open to fully closed. The flow characteristics of a butterfly valve are the same as described in Section 17.6. The difference between a butterfly valve used for control purposes and a hand-operated butterfly valve is that the former has an actuator and can be operated automatically. Butterfly valves are lightweight, easy to operate and install, and lower in cost than gate valves. They are primarily used as fully open or fully closed, but they may be used for throttling purposes. Butterfly valves are gaining in popularity, especially in large pipes.

PRESSURE RELIEF VALVES. These valves are safety valves used to prevent a system that is over-pressurized from exceeding a predetermined limit. A pressure relief valve is held closed by a spring or rupture member and is automatically opened to relieve the water pressure when it rises above the system design working pressure.

Valve Connections and Ratings

The type of connection used between a valve and the pipes is usually consistent with the type of joint used in the pipe system. A water piping system with flanged joints requires a valve with flanged ends. The most commonly used types of valve connections are as follows:

- *Threaded ends*. These connections are mainly used for small pipes with diameters from $\frac{1}{4}$ in. to 2 in. Threaded-end valves are usually inexpensive and simple to install.

- *Flanged ends.* These connections are commonly used for larger pipes (2-½ in. and above). Flanged ends are more easily separated when necessary.
- *Welded ends.* Steel valves, when used at higher pressure and temperature, are often connected with welded ends. Welded ends exhibit the fewest instances of leakage.
- *Grooved ends.* These connections use circumferential grooves in which a rubber gasket fits and are enclosed by iron couplings. Butterfly valves are often connected with grooved ends.
- *Soldered ends.* Bronze valves in copper piping systems use soldered ends. Tin alloy soldering is the type of soldering commonly used. Lead soldering cannot be used in a potable water system because it will contaminate the water.

Valves are usually rated based on their ability to withstand pressure at a specified temperature. Metal valves have two different ratings, one for steam (working steam pressure—WSP), which should correspond to its operating temperature, and the other for cold water, oil, or gas (WOG). The following are the commonly used ratings:

- 125 psig WSP, 250 psig WOG
- 150 psig WSP, 300 psig WOG
- 300 psig WSP, 600 psig WOG

Here, psig represents gauge pressure in pounds per square inch.

As listed in Table 20.3, a wrought steel valve flange joint with a 150 psig rating can be used for a hot or chilled water system with a maximum allowable pressure of 250 psig at a temperature below 250°F, for pipes of diameters between 2.5 and 12 in.

Valve Materials

Valve materials are selected according to their ability to withstand working pressure and temperature, their resistance to corrosion, and their relative cost. The most commonly used materials for valves are as follows:

- *Bronze.* It has a good corrosion resistance and is easily machined, cast, or forged. Bronze is widely used for water valves only up to a size of 3 in. because of its high cost. For valves above 3 in., bronze is still often used for sealing elements and stems because it is machinable and corrosion resistant.
- *Cast iron and ductile iron.* These materials are used for pressure-containing parts, flanges, and glands in valve sizes 2 in. and larger. Ductile iron has a higher tensile strength than cast iron.

- *Steel.* Forged or cast steel provides a higher tensile strength as well as toughness in the form of resistance to shock and vibration than bronze, cast iron, and ductile iron. Steel is used in applications that require higher strength and toughness than bronze and ductile iron can provide.
- *Trim materials.* These include the elements and components that are easily worn as well as those parts that need to be resistant to corrosion, such as the disc, seating elements, and stem. Stainless steel, stellite (a kind of cobalt-chromium-tungsten alloy), and chromium-molybdenum steel are often used for trim material in valves.

Pipe Fittings and Water System Accessories

Water pipe fittings include elbows, tees, and valves. Water pipe elbows and tees are often made of cast iron, ductile iron, or steel. Pressure losses due to the water pipe fittings are usually expressed in term of an equivalent length of straight pipe, for the sake of convenience. Table 20.6 allows calculation of the pressure losses for various types of pipe fittings, in terms of an equivalent length of straight pipe. The equivalent length for a fitting can be estimated by multiplying the elbow equivalent to that fitting by the equivalent length for a 90° elbow.

Water system accessories include drains, strainers, and air vents. Drains should be equipped at all low points of the system. Arrangements should be made so that a part of the system or individual components can be drained rather than draining the entire system.

A condensate drain pipe is always required for cooling and dehumidifying coils. Galvanized steel is often used for this purpose. It is usually piped to a plumbing drain or other suitable location. A condensate drain pipe should be insulated so as to avoid surface condensation.

Water strainers are often installed before the pumps, control valves, or other components to protect them from dirt and impurities. Air vents will be discussed in the next section.

20.4 WATER SYSTEM PRESSURIZATION AND AIR CONTROL

Water System Pressurization Control

For an open water system, the maximum operating gauge pressure is the pressure at a specific point in the system where the positive pressure exerted by the water pumps to overcome the pressure drops across the equipment, components, fittings, and pipes plus the static head due to the vertical distance between the highest water level and that point, is at a maximum.

TABLE 20.6
Pressure losses for pipe fittings and valves, expressed in terms of an equivalent length (in ft) of straight pipe

Velocity, ft/s	\%	\%	1	1¼	1½	2	2½	3	3½	4	5	6	8	10	12
	½	¾	1	1¼	1½	2	2½	3	3½	4	5	6	8	10	12
1	1.2	1.7	2.2	3.0	3.5	4.5	5.4	6.7	7.7	8.6	10.5	12.2	15.4	18.7	22.2
2	1.4	1.9	2.5	3.3	3.9	5.1	6.0	7.5	8.6	9.5	11.7	13.7	17.3	20.8	24.8
3	1.5	2.0	2.7	3.6	4.2	5.4	6.4	8.0	9.2	10.2	12.5	14.6	18.4	22.3	26.5
4	1.5	2.1	2.8	3.7	4.4	5.6	6.7	8.3	9.6	10.6	13.1	15.2	19.2	23.2	27.6
5	1.6	2.2	2.9	3.9	4.5	5.9	7.0	8.7	10.0	11.1	13.6	15.8	19.8	24.2	28.8
6	1.7	2.3	3.0	4.0	4.7	6.0	7.2	8.9	10.3	11.4	14.0	16.3	20.5	24.9	29.6
7	1.7	2.3	3.0	4.1	4.8	6.2	7.4	9.1	10.5	11.7	14.3	16.7	21.0	25.5	30.3
8	1.7	2.4	3.1	4.2	4.9	6.3	7.5	9.3	10.8	11.9	14.6	17.1	21.5	26.1	31.0
9	1.8	2.4	3.2	4.3	5.0	6.4	7.7	9.5	11.0	12.2	14.9	17.4	21.9	26.6	31.6
10	1.8	2.5	3.2	4.3	5.1	6.5	7.8	9.7	11.2	12.4	15.2	17.7	22.2	27.0	32.0

The header row reads: **Equivalent length, in ft of pipe, for 90° elbows**, **Pipe size, in.**

Iron and copper elbow equivalents

Fitting	Iron Pipe	Copper Tubing
Elbow, 90°	1.0	1.0
Elbow, 45°	0.7	0.7
Elbow, 90° long turn	0.5	0.5
Elbow, 90° welded	0.5	0.5
Reduced coupling	0.4	0.4
Open return bend	1.0	1.0
Angle radiator valve	2.0	3.0
Radiator or convector	3.0	4.0
Boiler or heater	3.0	4.0
Open gate valve	0.5	0.7
Open globe valve	12.0	17.0

Source: ASHRAE Handbook 1989, Fundamentals. Reprinted with permission.

In a closed chilled or hot water system, a variation of the water temperature will cause an expansion of water that may raise the water pressure above the maximum allowable pressure. The purposes of system pressurization control for a closed water system are as follows:

- To limit the pressure of the water system to below its allowable working pressure
- To maintain a pressure higher than the minimum water pressure required to vent air
- To assist in providing a pressure higher than the net positive suction head (NPSH) at the pump suction to prevent cavitation
- To provide a point of known pressure in the system

Expansion tanks, pressure relief valves, pressure-reducing valves for make-up water, and corresponding controls are used to achieve water system pressurization control. These are two types of expansion tanks for closed water systems: open and closed.

Open Expansion Tanks

An expansion tank is a device that allows for the expansion and contraction of water contained in a closed water system when the water temperature changes between two predetermined limits. Another function of an expansion tank is to provide a point of known pressure in a water system.

An open expansion tank is vented to the atmosphere and is located at least 3 ft above the highest point of the water system, as shown in Fig. 20.7. Make-up water is supplied through a float valve, and an internal overflow drain is always installed. A *float valve* is a globe or ball valve connected with a float ball to regulate the make-up water flow according to the liquid level in the tank. An open expansion tank is often connected to the suction side of the water pump to prevent the water pressure in the system from dropping below the atmospheric pressure. The pressure of the liquid level in the open tank is equal to the atmospheric pressure, which thus provides a reference point of known pressure to determine the water pres-

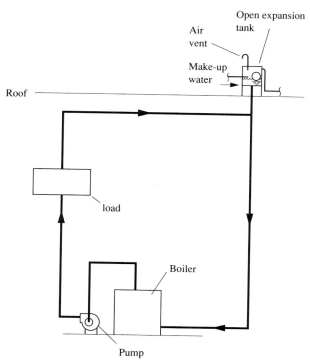

Air vent
Open expansion tank
Make-up water
Roof
load
Boiler
Pump

FIGURE 20.7   Open expansion tank.

the closed expansion tank to separate the filled air and the water permanently. Such an expansion tank is called a diaphragm, or bladder, expansion tank. Thus, a closed expansion tank is either a plain closed expansion tank, which does not have a diaphragm to separate air and water, or a diaphragm tank.

For a water system with only one air-filled space, the junction between the closed expansion tank and the water system is a point of fixed pressure. At this point, water pressure remains constant whether or not the pump is operating because the filled air pressure depends only on the volume of water in the system. The pressure at this point can be determined according to the ideal gas law, as given by Eq. (2.1): $pv = RT_R$. The pressure in a closed expansion tank during the initial filling process or at the minimum operating pressure is called the *fill pressure* p_{fil}, in psia. The fill pressure is often used as the reference pressure to determine the pressure characteristics of a water system.

Appropriate Size of Diaphragm Expansion Tank

If a closed expansion tank with its filled volume of air is too small, the system pressure will easily exceed the maximum allowable pressure and cause water to discharge from the pressure relief valve, thus wasting water. If the closed tank is too large, when the water temperature drops, the system pressure may decrease to a level below the minimum allowable value and cause trouble in the air vent. Therefore, accurate sizing of a closed expansion tank is essential.

For diaphragm expansion tanks, the minimum volume of the water tank, V_t, in gal, can be calculated by the following formula recommended by *ASHRAE Handbook 1992, HVAC Systems and Equipment:*

$$V_t = V_s \frac{[(v_2/v_1) - 1] - 3\alpha(T_2 - T_1)}{1 - (P_1/P_2)} \qquad (20.2)$$

where T_1 = lower temperature, °F
T_2 = higher temperature, °F
V_s = volume of water in system, gal
p_1 = absolute pressure at lower temperature, psia
p_2 = absolute pressure at higher temperature, psia
v_1, v_2 = specific volume of water at lower and higher temperature respectively, ft³/lb
α = linear coefficient of thermal expansion; for steel, $\alpha = 6.5 \times 10^{-6}$ in/in. °F; for copper, $\alpha = 9.5 \times 10^{-6}$ in/in. °F.

In a chilled water system, the higher temperature T_2,

sure at any point in the water system. The minimum tank volume should be at least 6 percent of the volume of water in the system, V_s.

An open expansion tank is simple, more stable in terms of system pressure characteristics, and low in cost. If it is installed indoors, it often needs a high ceiling. If it is installed outdoors, water must be prevented from freezing in the tank, air vent, or pipes connected to the tank when the outdoor temperature is below 32°F. Because the water surface in the tank is exposed to the atmosphere, oxygen is more easily absorbed into the water, which makes the tank less resistant to corrosion than a diaphragm tank (to be described later). Because of these disadvantages, an open expansion tank has only limited applications.

Closed Expansion Tanks

A closed expansion tank is an airtight tank filled with air or other gases, as shown in Fig. 20.8. When the temperature of the water increases, the water volume expands. Excess water then enters the tank. The air in the tank is then compressed, which raises the system pressure. When the water temperature drops, the water volume contracts, resulting in a reduction in system pressure.

In order to reduce the amount of air dissolved in the water so as to prevent corrosion and prevent air noise, a diaphragm, or bladder, is often installed in

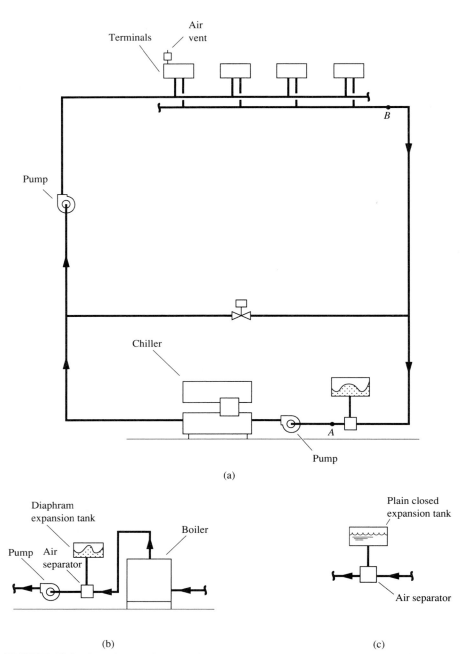

FIGURE 20.8 Closed expansion tank for a water system: (a) diaphragm expansion tank in a chilled water system, (b) diaphragm expansion tank in a hot water system, and (c) plain closed expansion tank.

in °F, is the highest anticipated ambient temperature when the chilled water system shuts down during summer.

Pump Location

The location of the pump in a water system that uses a diaphragm expansion tank should be arranged so that the pressure at any point in the water system is greater than the atmospheric pressure. In such an arrangement, air does not leak into the system and the required net positive suction head (NPSH) can be maintained at the suction inlet of the water pump.

A water pump location commonly used for hot water systems with diaphragm expansion tanks is just after the expansion tank and the boiler, as shown in Fig. 20.8b. In this arrangement, the pressure at the pump suction is the sum of the water pressure plus the fill pressure. In another often-used arrangement, the diaphragm expansion tank is moved to the highest point of the water system and the pump is still located after the boiler.

In a chilled water system, the location of the chilled water pump is usually before the water chiller and the diaphragm expansion tank is usually connected to the suction side of the water pump.

Air in Water Systems

In a closed recirculating water system, air and nitrogen are present in the following forms: dissolved in water, free air or gas bubbles, or pockets of air or gas.

The behavior of air or gas dissolved in liquids is governed and described by Henry's equation. Henry's equation states that the amount of gas dissolved in a liquid at constant temperature is directly proportional to the absolute pressure of that gas acting on the liquid, or

$$x = \frac{p}{H} \qquad (20.3)$$

where x = amount of dissolved gas in solution, percent by volume
p = partial pressure of that gas, psia
H = Henry's constant; changes with temperature

The lower the water temperature and the higher the total pressure of water and dissolved gas, the greater the maximum amount of dissolved gas at that pressure and temperature.

When the dissolved air or gas in water reaches its maximum amount at that pressure and temperature, the water becomes saturated. Any excess air or gas, as well as the coexisting water vapor, can exist only in the form of free bubbles or air pockets. A water velocity greater than 1.5 ft/s can carry air bubbles along with the water.

When water is in contact with air at an air–water interface, such as the filled air space in a plain closed expansion tank, the concentration gradient causes air to diffuse into the water until the water is saturated at that pressure and temperature. An equilibrium forms between air and water within a certain time period. At specific conditions, 24 hours may be required to reach equilibrium.

The oxygen in air that is dissolved in water is unstable. It reacts with steel pipes to form oxides and corrosion. Therefore, after air has been dissolved in water for a long enough period of time, only nitrogen remains as a dissolved gas circulating with the water.

Problems due to the Presence of Air and Gas

The presence of air and gas in a water system causes the following problems for a closed water system with a plain closed expansion tank:

- Presence of air in the terminal and heat exchanger, which reduces the heat transfer surface
- Corrosion due to oxygen reacting with the pipes
- Water logging in plain closed expansion tanks
- Unstable system pressure
- Poor pump performance due to gas bubbles
- Noise problems

There are two sources of air and gas in a water system. One is the air–water interface in a plain closed expansion tank or in an open expansion tank, and the other is the dissolved air in a city water supply.

Oxidation and Water Logging

Consider a chilled water system that uses a plain closed expansion tank without a diaphragm, as shown in Fig. 20.8. This expansion tank is located in a basement, with a water pressure of 90 psig and a temperature of 60°F at point A. At such a temperature and pressure, the solubility of air in water is about 14.2 percent. The chilled water flows through the water pump, the chiller, and the riser and is supplied to the upper-level terminals. During this transport process, part of the oxygen dissolved in the water reacts with the steel pipes to form oxides and corrosion. At upper-level point B, the water pressure is only 10 psig at a chilled water temperature of about 60°F. At this point, the solubility of air in water is only about 3.3 percent. The difference in solubility between points A and B is $14.2 - 3.3 = 10.9$ percent. This portion of air, containing a higher percentage of nitrogen because of the formation of oxides, is no longer dissolved in the chilled water, but is released from the water and forms free air, gas bubbles, or pockets. Some of the air pockets are vented through air vents at the terminals or high points of the water system. The chilled water returns to point A again and absorbs air from the air-water interface in the plain closed expansion tank, creating an air solubility in water of about 14.2 percent. Of course, the actual process is more complicated because of the formation of oxides and the presence of water vapor.

Such a chilled water recirculating process causes the following problems:

- Oxidation occurs because of the reaction between dissolved oxygen and steel pipes, causing corrosion during the chilled water transport and recirculating process.
- The air pockets vented at high levels originally come from the filled air in the plain closed expan-

sion tank; that is, after a period of recirculation of the chilled water, part of the air charge is removed to the upper levels and vented. The tank finally waterlogs and must be charged with compressed air again. Waterlogging also results in an unstable system pressure because the amount of filled air in the plain closed expansion tank does not remain constant. Oxidation and waterlogging also exist in hot water systems, but the problems are not as pronounced as in a chilled water system.

Oxidation and waterlogging can be prevented or reduced by installing a diaphragm expansion tank instead of a plain tank. Air vents, either manual or automatic, should be installed at the highest point of the water system, and on coils and terminals at higher levels if a water velocity of 2 ft/s is maintained in the pipes, in order to transport the entrained air bubbles to these air vents.

In a closed chilled water system using a diaphragm expansion tank, there is no air–water interface in the tank. The 3.3 percent of dissolved air, or about 2.6 percent of dissolved nitrogen, in water returning from point B to A cannot absorb more air again from the diaphragm tank. If there is no fresh city water supply to the water system, then after a period of water recirculation the only dissolved air in water will be the 2.6 percent nitrogen. No further oxidation occurs after the initial dissolved oxygen has reacted with the steel pipes. Waterlogging does not occur, either.

Because of the above concerns, a closed water system should have a diaphragm or bladder expansion tank. An open expansion tank at high levels causes fewer problems than a plain closed expansion tank. A diaphragm tank may be smaller than an equivalent plain tank. An air eliminator or air separator is usually required for large water systems using a diaphragm tank to separate dissolved air from water when the water system is charged with a considerable amount of city water.

20.5   CORROSION AND DEPOSITS IN WATER SYSTEMS

Corrosion

Corrosion is a destructive process that acts on a metal or alloy. It is caused by a chemical or electrochemical reaction of a metal. Galvanic corrosion is the result of contact between two dissimilar metals in an electrolyte.

The corrosion process involves a flow of electricity between two areas of a metal surface in a solution that conducts the electric current. One area acts as the anode and releases electrons, whereas the other area acts as the cathode, which accepts electrons and

forms negative ions. Corrosion, or the formation of metal ions by means of oxidation and disintegration of metal, occurs only at the anodes. In iron and steel, ferrous ions react with the oxygen to form ferric hydroxide (rust).

Moisture encourages the formation of an electrolyte, which is one of the basic elements that give rise to corrosion. Oxygen accelerates the corrosion of ferrous metals by means of a reaction with hydrogen produced at the cathode. This creates the reaction at the anode. Some alloys, such as stainless steel and aluminum, develop protective oxide films to prevent further corrosion.

For iron and steel, solutions such as those containing mineral acids accelerate the corrosion, and solutions such as those containing alkalies retard it. Because the corrosion reaction at the cathode depends on the concentration of hydrogen ions, the more acidic the solution, the higher the concentration of hydrogen ions and the greater the corrosion reaction. Alkaline solutions have a much higher concentration of hydroxyl ions than hydrogen, and as such they decrease the corrosion rate.

Water Impurities

In hot and chilled water systems, the problems associated with water mainly concern water's dissolved impurities, which cause corrosion and scale, and the control of algae, bacteria, and fungi. Typical samples of dissolved impurities in public water supplies are listed in Table 20.7.

Calcium hardness, sulfates, and silica all contribute to the formation of scale. *Scale* is the deposit formed by the precipitation of water-insoluble constituents on a metal surface. Chlorides cause corrosion. Iron may form deposits on a surface through precipitation. All of these increase the fouling factor of water.

In addition to dissolved solids, unpurified water may contain suspended solids, which may be either organic or inorganic. Organic constituents may be in the form of colloidal solutions. At high water velocities, hard suspended solids may abrade pipes and equipment. Particles that settle at the bottom may accelerate corrosion.

In open water systems, bacteria, algae, and fungi cause many operating problems. As described in Section 11.11, the possibility of bacteria existing in the cooling tower and causing Legionnaires' disease necessitates microbiological control.

Water Treatments

SCALE AND CORROSION CONTROL. One effective corrosion control method is to reduce the oxygen composition in water systems. In past years, acids and

TABLE 20.7
Analyses of typical public water supplies

Substance	Unit	(1)	(2)	(3)	(4)	(5)	(6)	(7)	(8)	(9)
					Location or Area*,†					
Silica	SiO_2	2	6	12	37	10	9	22	14	—
Iron	Fe_2	0	0	0	1	0	0	0	2	—
Calcium	Ca	6	5	36	62	92	96	3	155	400
Magnesium	Mg	1	2	8	18	34	27	2	46	1300
Sodium	Na	2	6	7	44	8	183	215	78	11,000
Potassium	K	1	1	1	—	1	18	10	3	400
Bicarbonate	HCO_3	14	13	119	202	339	334	549	210	150
Sulfate	SO_4	10	2	22	135	84	121	11	389	2700
Chloride	Cl	2	10	13	13	10	280	22	117	19,000
Nitrate	NO_3	1	—	0	2	13	0	1	3	—
Dissolved solids		31	66	165	426	434	983	564	948	35,000
Carbonate hardness	$CaCO_3$	12	11	98	165	287	274	8	172	125
Noncarbonate hardness	$CaSO_4$	5	7	18	40	58	54	0	295	5900

*All values are ppm of the unit cited to nearest whole number.

†Numbers indicate location or area as follows:
(1) Catskill supply, New York City
(2) Swamp water (colored), Black Creek, Middleburg, FL
(3) Niagara River (filtered), Niagara Falls, NY
(4) Missouri River (untreated), average
(5) Well waters, public supply, Dayton, OH, 30 to 60 ft (9 to 18 m)
(6) Well water, Maywood, IL, 2090 ft
(7) Well water, Smithfield, VA, 330 ft
(8) Well water, Roswell, NM
(9) Ocean water, average

Source: ASHRAE Handbook 1987, HVAC Systems and Applications. Reprinted with permission.

chromates were the chemical compounds commonly used to eliminate or to reduce scale and corrosion. On January 3, 1990, however, the "Proposed Prohibition of Hexavalent Chromium Chemicals in Comfort Cooling Towers" was posted by the Environmental Protection Agency because chromates are suspected carcinogens and disposal problems are associated with these chemicals.

There has been a significant improvement in water treatment chemistry in recent years. Currently used chemical compounds include crystal modifiers and sequestering chemicals. Crystal modifiers cause a change in the crystal formation of scale. As a result, scale ions cannot interlace with ions of other scale-forming elements. Another important characteristic of these crystal modifiers and sequestering chemicals is that they can be applied to water systems that have a wide range of pH values. (The pH value indicates the acidity or alkalinity of a solution. It is the negative logarithm of the hydrogen ion concentration of a solution.) Even if the chemical is over- or underfed, it will not cause operating problems. Crystal modifiers and sequestering chemicals create fewer environmental problems.

MICROBIOLOGICAL CONTROL. The growth of bacteria, algae, and fungi are usually treated by biocides to prevent the formation of an insulating layer on the heat transfer surface, which would promote corrosion and restrict water flow. Chlorine and chlorine compounds have been effectively and widely used. Bromine has the same oxidizing power of chlorine and is effective over a wide pH range. Biocide chemicals are detrimental to the environment if they are used in excess, however.

Biostat is a new type of chemical used in algae growth control. It prevents algae spores from maturing, which is an approach different from that of a biocide.

As described in Section 11.11, blow-down or bleed-off operation is an effective water treatment. It should be considered as important as treatments that use chemicals.

CHEMICAL FEEDING. Improper chemical feeding causes operating problems. A water treatment program with underfed chemicals results in an ineffective treatment, whereas an overfed program not only increases the operating cost but also may cause environmental problems. Generally, a continuous feeding of very small amounts of chemicals often provides effective and economical water treatment.

20.6 WATER SYSTEM CHARACTERISTICS

System Characteristics

Closed water systems, including hot, chilled, and dual-temperature systems, can be categorized as follows:

CONSTANT-FLOW OR VARIABLE-FLOW. A constant-flow water system is a system for which the volume flow at any cross-sectional plane in the supply or return mains remains constant during the operating period. Three-way mixing valves are used to modulate the water flow rates to the coils. In a variable-flow system, all or part of the volume flow varies when the system load changes during the operating period. Two-way valves are used to modulate the water flow rates to the coils or terminals.

DIRECT-RETURN OR REVERSE-RETURN. In a direct-return water system, the various branch piping circuits, such as *ABGHA* and *ABCFGHA*, are not equal in length (see Fig. 20.9a). Careful balance by means of throttling valves is often required to establish the design flow rates for a building loop when a direct-return distribution loop is used, as described

in later sections. In a reverse-return system, as described in Section 14.7, the piping lengths for each branch circuit, including the main and branch pipes, are almost equal (see Fig. 20.9b).

TWO-PIPE OR FOUR-PIPE. In a dual-temperature water system, the water piping from the boiler or chiller to the coils and the terminals, or to various zones in a building, can be either a two-pipe system, with a supply main and a return main, as shown in Fig. 20.10a; or a four-pipe system, with a hot water supply main, a hot water return main, a chilled water supply main, and a chilled water return main, as shown in Fig. 20.10b. For a two-pipe system, it is impossible to heat and cool two different coils or terminals in the same zone simultaneously. Changeover from summer cooling-mode operation to winter heating-mode operation is required. A four-pipe system doesn't need changeover operation. Chilled and hot water can be supplied to the coils or terminals simultaneously. However, a four-pipe system requires a greater installation cost.

Changeover

In a dual-temperature two-pipe system, *changeover* refers to when the operation of one zone or the entire water system in a building changes from heating mode to cooling mode, or vice versa. During changeover, the water supplied to the terminals changes from hot water to chilled water, or vice versa. The changeover temperature T_{co}, in °F, is the outdoor temperature at

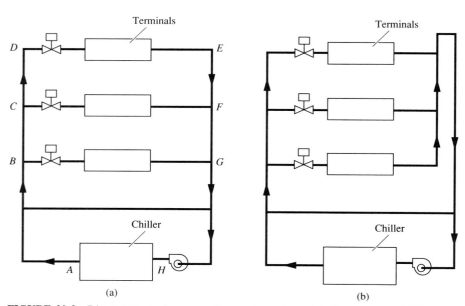

FIGURE 20.9 Direct-return and reverse-return water systems: (a) direct return and (b) reverse return.

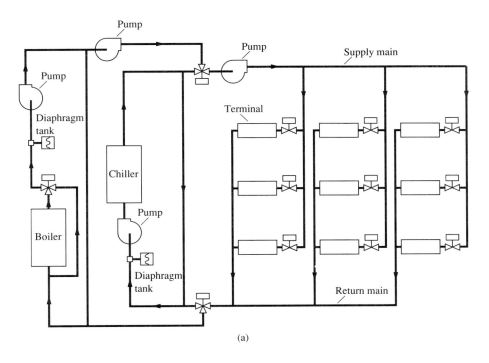

(a)

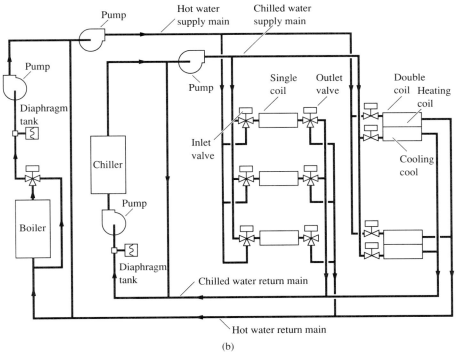

(b)

FIGURE 20.10 Multiple-zone, dual-temperature water systems: (a) two-pipe system and (b) four-pipe system.

which the space sensible cooling load can be absorbed and removed by the combined effects of the conditioned outdoor air, the primary air, and the space transmission and infiltration loss. Such a relationship can be expressed as

$$T_{co} = T_r - \frac{\sum Q_{ris} + \sum Q_{res} - K \dot{V}_{so}(T_r - T_{so})}{q_{tl}}$$

$$K = 60 \rho_{so} c_{pa} \qquad (20.4)$$

where T_r = space temperature, °F
$\sum Q_{ris}$ = sum of internal sensible loads from electric lights, occupants, appliances, and solar loads, Btu/h
$\sum Q_{res}$ = sum of solar loads through building shell, Btu/h
$\dot{V}_{so}, \rho_{so}$ = volume flow rate and density of conditioned outdoor air, cfm
c_{pa} = specific heat of air, Btu/lb · °F
T_{so} = supply temperature of outdoor air or primary air, °F
q_{tl} = transmission and infiltration losses per 1 °F of outdoor-indoor temperature difference, Btu/h · °F

Changeover usually takes from three to eight hours to complete. The greater the size of the water system, the longer the changeover period. In order to prevent more than one changeover per day, the changeover temperature T_{co} may have a tolerance of ±2°F.

Changeover may cause a sudden flow of a large amount of hot water into the chiller or of chilled water into the boiler. Such a rapid change of temperature imposes a thermal shock on the chiller or boiler and may damage the equipment. For chillers, the temperature of water entering the chiller should be no higher than 80°F to prevent excessive refrigerant pressure in the evaporator. For boilers, a temperature control system bypasses most of the low-temperature water until the water temperature can gradually be increased.

Changeover may be performed either manually or automatically. Manual changeover is simple but may be inconvenient during periods when daily changeover is required. With sufficient safety controls, automatic changeover reduces the operating duties significantly. A compromise is a semiautomatic changeover system in which the changeover temperature is set by a manual switch.

Outdoor reset control is often used to vary the supply water temperature T_{ws}, in °F, in response to the outdoor temperature T_o for a hot water system. Typically, T_{ws} is 130°F at the winter design temperature and drops linearly to 80°F at the changeover temperature.

20.7 PUMP–PIPING SYSTEMS

System Curve

When a pump is connected with a pipe system, it forms a pump–piping system, similar to fan–duct systems. A water system may consist of one pump–piping system or a combination of several pump–piping systems.

As is the case with duct static pressure control in a variable-air-volume system, the speed of a variable-flow water system using a variable-speed pump is controlled by a pressure differential transmitter installed at the end of the supply main, with a set point normally between 20 and 30 ft WG. This represents the head loss resulting from the control valve, pipe fittings, and pipe friction between the supply and return mains at the branch circuit farthest from the variable-speed pump.

Therefore, the head loss shown on a water system curve can be divided into two parts:

- *Constant,* or *fixed, head* H_{fix}, which remains constant as the water flow varies. Its magnitude is equal to the set point of the pressure differential transmitter, H_{set}, or the static head due to the height difference between the suction and discharge levels of the pump in open systems, $H_{s.d}$, in ft WG.

- *Variable head* H_{var}, which varies as the water flow changes. Its magnitude is the sum of the head losses caused by pipe friction, H_{pipe}, pipe fittings, H_{fit}, equipment, H_{eq} (such as the pressure drop through the evaporator, condenser, and coils), and components, H_{cp}, all in ft WG, that is,

$$H_{var} = H_{pipe} + H_{fit} + H_{eq} + H_{cp} \qquad (20.5)$$

Head losses H_{fix} and H_{var} are shown in Fig. 20.11. As with Eq. (19.13) for fan–duct systems, the relationship between the flow head H_{var}, flow resistance of the water system R_{var}, in ft WG/(gpm)^2, and water volume flow rate $\dot{V}_w$, in gpm, can be expressed as

$$H_{var} = R_{var} \dot{V}_w^2 = K \dot{V}_w^2 \qquad (20.6)$$

where K is the flow coefficient, in ft WG/(gpm)^2.

System Operating Point

The intersection of the pump performance curve and the water system curve is the system operating point of this variable-flow water system, as shown by point P in Fig. 20.11. Its volume flow rate is represented by $\dot{V}_P$, in gpm, and its total head is $H_P = (H_{fix} + H_{var})$, in ft WG.

Usually, the calculated system head loss is overestimated and the selected pump is oversized with a

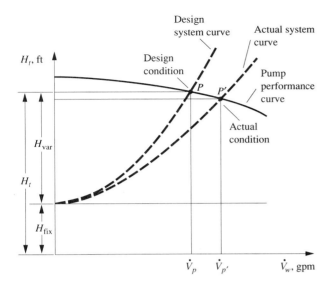

FIGURE 20.11 Water system curve and system operating point.

higher pump head, so that the actual system operating point is at point P'. Therefore, for a variable-flow water system installed with a constant-speed pump, the design system operating point is preferably located to the left of the region of pump maximum efficiency because the system operating point of an oversized pump moves into or nearer to the region of pump maximum efficiency.

Combinations of Pump–Piping Systems

As with a combination of fan-duct systems, when two pump–piping systems 1 and 2 are connected in series as shown in Fig. 20.12a, the volume flow rate of the combined pump–piping system, $\dot{V}_{com}$, in gpm, is

$$\dot{V}_{com} = \dot{V}_1 = \dot{V}_2 \tag{20.7}$$

where $\dot{V}_1$ and $\dot{V}_2$ are the volume flow rates of pump–piping systems 1 and 2, in gpm. The total head increase of the combined system, H_{com}, in ft WG, is

$$H_{com} = H_1 + H_2 \tag{20.8}$$

where H_1 and H_2 are the head of pump–piping systems 1 and 2, in ft WG.

It is simpler to use one system curve to represent the whole system, that is, to use a combined system curve. The system operating point of the combined pump–piping system is illustrated by point P with a volume flow of $\dot{V}_P$ and head of H_P. The purpose of connecting pump–piping systems in series is to increase the system's head.

When a pump–piping system has parallel-connected water pumps, its volume flow rate $\dot{V}$, in gpm, is the sum of the volume flow rates of the constituent pumps, $\dot{V}_1$, $\dot{V}_2$, etc. The head of each constituent pump and of the combined pump–piping system are equal. It is more convenient to draw a combined pump curve and one system curve to deter-

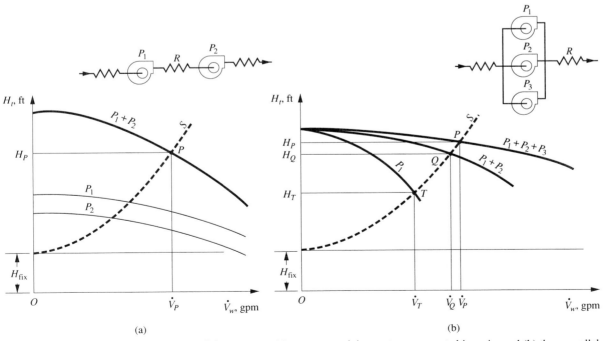

FIGURE 20.12 Combinations of pump–piping systems: (a) two pump–piping systems connected in series and (b) three parallel-connected pumps.

mine their intersection, the system operating point P, as shown in Fig. 20.12b. The purpose of equipping a water system with parallel-connected water pumps is to increase its volume flow rate.

Modulation of Pump–Piping Systems

Modulation of the volume flow rate of a pump–piping system can be done by means of the following:

- Throttle the volume flow by using a valve. As the valve closes its opening, the flow resistance of the pump–piping system increases. A new system curve is formed, which results in having a new system operating point that moves along the pump curve to the left-hand side of the original curve, with a lower volume flow rate and higher total head, as shown in Fig. 20.13a. As with fan-duct systems, such behavior is known as *riding on the curve*. Using the valve to modulate the volume flow rate of a pump–piping system always wastes energy because of the head loss across the valves (ΔH_{val} in Fig. 20.13a).

- Turn water pumps on or off in sequence for pump–piping systems that have multiple pumps in a parallel connection. Modulation of the volume flow rate by means of turning water pumps on and off often results in a sudden drop or increase in volume flow rate and head, as shown by system operating points P, Q, and T in Figure 20.12b.

- Vary the pump speed to modulate the volume flow and the head of a pump–piping system. When the speed of the pump is varied from n_1 to n_2

and then to n_3, new pump curves P_2 and P_3 are formed, as shown in Fig. 20.13b. The system operating point will move from point P to Q and then to T along the system curve, with a lower volume flow rate, head, and input pump power. The system curve becomes the modulating curve and approaches $H_{fix} = H_{set}$ when the volume flow rate is zero. Here H_{set} is the set point of the pressure differential transmitter in ft WG. Varying the pump speed requires the lowest pump power input in comparison with other modulation methods.

Pump Laws

As with the fan laws described in chapter 19, the performance of geometrically and dynamically similar pump–piping systems can be expressed as follows:

$$\frac{\dot{V}_2}{\dot{V}_1} = \frac{D_2^3 n_2}{D_1^3 n_1}$$

$$\frac{H_{t2}}{H_{t1}} = \frac{n_2^2}{n_1^2} \qquad (20.9)$$

$$\frac{P_2}{P_1} = \frac{n_2^3}{n_1^3}$$

where $\dot{V}$ = volume flow rate of the pump–piping system, gpm

H_t = total head lift, ft WG

P = pump power input at the shaft, hp

D = outside diameter of the pump impeller, ft

n = speed of the pump impeller, rpm

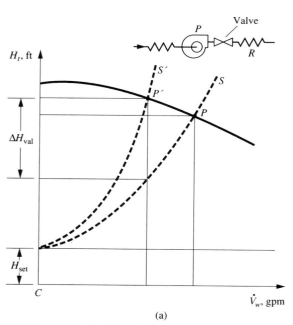

(a)

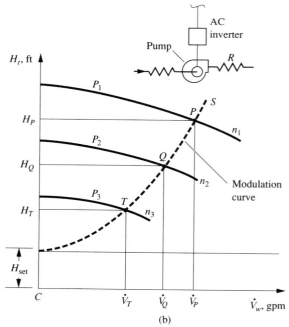

(b)

FIGURE 20.13 Modulation of pump–piping systems: (a) using a valve and (b) varying the pump speed.

Wire-to-Water Efficiency

A pump may be directly driven by a motor, or it may be driven by a motor and belts. When the energy cost of a water system is evaluated, the pump total efficiency η_p, the motor efficiency η_{mot}, and the efficiency of the variable-speed drives η_{dr}, should all be considered.

The *wire-to-water efficiency* of a water system, η_{ww}, expressed either in dimensionless terms or as a percentage, is defined as the ratio of energy output from water to the energy input to the electric wire connected to the motor. It can be calculated as

$$\eta_{ww} = \frac{\dot{V}H_t g_s}{3960 P \eta_{dr} \eta_{mot}} \qquad (20.10)$$
$$= \eta_p \eta_{dr} \eta_{mot}$$

where g_s is the specific gravity (for water, $g_s = 1$).

Total efficiency of the centrifugal pump, η_p, can be obtained from the pump manufacturer or calculated from Eq. (10.36). Pump efficiency η_p depends on the type and size of the pump, as well as the percentage of design volume flow rate during operation. Pump efficiency usually varies from 0.7 to 0.85 at the design volume flow rate. Drive efficiency η_{dr} indicates the efficiency of a direct drive, belt drive, and various types of variable-speed drives. For a direct drive, $\eta_{dr} = 1$. Among variable-speed drives, an adjustable-frequency AC drive has the highest drive efficiency. For a 25-hp motor, η_{dr} often varies from 0.96 at design flow to 0.94 at 30 percent design flow to 0.80 at 20 percent design flow. Motor efficiency η_{mot} depends on the type and size of motor. It normally varies from 0.91 for a 10-hp high-efficiency motor to 0.96 for a 250-hp motor, as listed in Table 7.8.

20.8 PLANT–BUILDING LOOP COMBINATION

System Description

Many chilled and hot water systems used in commercial central hydronic air conditioning systems often have their central plant located in the basement, rooftop, or equipment floors of the building. The hot/chilled water from the boiler/chiller in the central plant is then supplied to the coils and terminals of various zones in one building or in adjacent buildings by means of supply main pipes. Water returns from the coils and terminals to the central plant via the return mains.

Because equal-percentage-contour two-way valves installed on coils have better control characteristics and are lower in cost than three-way valves, and also because having variable water flow at the supply and return mains saves considerable pump power for part-load reduced flows, variable-flow water systems that use two-way valves on coils have become more and more popular since the 1960s.

ASHRAE/IES Standard 90.1–1989 specifies that water systems using control valves to modulate coil or terminal load must be designed for variable flow for energy savings. Water systems should be able to reduce system flow to 50 percent design flow or less, except to maintain proper operation of the equipment or energy efficiency.

Usually, a fairly constant volume flow is required in the evaporator of the water chiller to prevent water from freezing at a reduced flow and to avoid an extremely high temperature drop in the chiller during part-load operation.

If a single-loop water system, such as that shown in Fig. 20.14a, is used, it should be either (1) a constant-flow system throughout the loop or (2) a system that uses a variable-speed pump and sophisticated DDC to provide freezing protection at the evaporator as well as to reset the chilled water's exit temperature to balance the required refrigeration load, if necessary.

Neither of these arrangements can provide reliable operation and minimize the pump power consumption. Therefore, in current practice a chilled or dual-temperature water system usually consists of two piping loops:

- *Plant loop*. In a plant loop, there are chiller(s)/boiler(s), circulating water pump(s), diaphragm expansion tank(s), corresponding pipes and fittings, and control systems, as shown by loop *ABFG* in Fig. 20.14b. A constant volume flow rate is maintained along the plant loop.

- *Building loop*. In a building loop, there are coils, terminals, probably variable-speed water pumps, two-way control valves and control systems, and corresponding pipes, fittings, and accessories, as shown by loop *BCDEF* in Fig. 20.14b. The piping and control systems in a building loop are often divided into various zones. A variable flow is maintained along a building loop.

A short common pipe connects these two loops and combines them into a plant–building loop combination. Such a combination is also called a *primary-secondary circuit,* in which the primary circuit is the plant loop and the secondary circuit is the building loop.

System Characteristics

Consider a chilled water system that uses a plant–building loop combination, as shown in Fig. 20.14b.

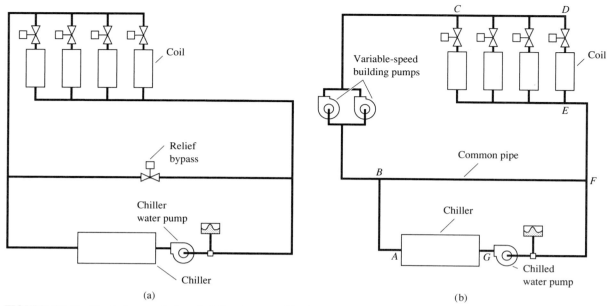

FIGURE 20.14 Single-loop and plant–building loop combination water systems: (a) single-loop water system and (b) plant–building loop combination.

When the volume flow rate of the chilled water in the building loop is at its design value, the volume flow rate in the plant loop is equal to that in the building loop. Chilled water leaving the chiller at point A flows through the junction of the common pipe, plant loop, and building loop (point B), is extracted by the variable-speed building pump, and is supplied to the coils. From the coils, chilled water returns through another junction of the building loop, common pipe, and plant loop (point F) and is then extracted by the chiller pump and enters the chiller for cooling again. No water flows in the common pipe in either direction.

When the coil load drops and the water flow rate reduces in the building loop because the control valves have been partially closed, the volume flow rate in the plant loop will be greater than that in the building loop. Chilled water then divides into two flows at junction B: water at the reduced volume flow rate is extracted by the variable-speed building pump in the building loop and is supplied to the coils; the remaining water bypasses the building loop by flowing through the common pipe, is extracted by the chiller pump, and returns to the chiller.

In actual practice, the water flow rate in the plant loop at design conditions is always slightly higher (about 5 percent) than in the building loop, to guarantee a sufficient chilled water supply to the building loop.

For a water system that includes a plant–building loop combination with a common pipe between the two loops, Carlson (1968) states the following rule: One pumped circuit affects the operation of the other to a degree dependent on the flow and pressure

drop in piping common to both circuits. The lower the pressure drop in the common pipe, the greater the degree of isolation between the plant and building loops. The head-volume flow characteristics of these loops act like two separate systems.

Compared with a single-loop chilled water system such as that shown in Fig. 20.14a, a plant–building loop combination has the following advantages:

- It provides variable flow at the building loop and constant flow at the plant loop and thus saves pumping power during periods of reduced flow in the building loop. According to Rishel (1983), the annual pump energy consumption of a plant–building loop combination with variable flow in a building loop that uses a variable-speed building pump is about 35 percent that of a single-loop constant-flow system.

- It separates the plant and building loops and makes the design, operation, performance, and control of both loops simpler and more stable.

Although the head-volume performances of the plant loop and building loop can be isolated from each other in a plant–building combination, the volume flow, temperature, and pressure of these two loops are still related to each other at junctions B and F.

According to the principles of continuity of mass and energy balance, if differences in the density of chilled water are ignored at junctions B and F, the sum of the volume flow rates of chilled water entering the junction must be equal to the sum of volume flow rates leaving that junction. Also, for an adiabatic mixing process, the total enthalpy of chilled water entering the junction must be equal to the total enthalpy of water leaving the junction. That is,

$$\dot{V}_{e.pt} + \dot{V}_{e.bg} + \dot{V}_{e.cn} = \dot{V}_{l.pt} + \dot{V}_{l.bg} + \dot{V}_{l.cn} \quad (20.11)$$

$$\dot{V}_{e.pt}h_{e.pt} + \dot{V}_{e.bg}h_{e.bg} + \dot{V}_{e.cn}h_{e.cn}$$
$$= \dot{V}_{l.pt}h_{l.pt} + \dot{V}_{l.bg}h_{l.bg} + \dot{V}_{l.cn}h_{l.cn} \quad (20.12)$$

where $\dot{V}_{e.pt}, \dot{V}_{e.bg}, \dot{V}_{e.cn}$ = volume flow rates of chilled water entering the junction from the plant loop, building loop, and common pipe, respectively, gpm

$\dot{V}_{l.pt}, \dot{V}_{l.bg}, \dot{V}_{l.cn}$ = volume flow rates of chilled water leaving the junction en route to the plant loop, building loop, and common pipe, respectively, gpm

$h_{e.pt}, h_{e.bg}, h_{e.cn}$ = enthalpies of chilled water entering the junction from the plant loop, building loop, and common pipe, respectively, Btu/lb

$h_{l.pt}, h_{l.bg}, h_{l.cn}$ = enthalpies of chilled water leaving the junction en route to the plant loop, building loop, and common pipe respectively, Btu/lb

At junctions B and F, the water pressure and temperature are the same for both the plant and building loops.

Coil's Load and Volume Flow

In AHUs or fan coils, two-way control valves are currently widely used to modulate the water volume flow rate so as to maintain a predetermined discharge temperature or space temperature at reduced loads. Coils, especially oversized coils, operate at the design load usually less than 5 percent of their total operating time. For a typical coil, nearly 60 percent of the operating time may correspond to a coil load between 35 and 65 percent of the design value.

During part-load, the required fraction of design volume flow rate of chilled water flowing through a coil $\dot{V}_w$ is not equal to the fraction of design sensible coil load q_{cs}, in Btu/h, which is the sensible heat transfer from the coil to conditioned air, as shown in Fig. 20.15a. In Fig. 20.15, $\dot{V}_{w.d}$ indicates the design chilled water volume flow rate, in gpm, and $q_{cs.d}$ the design sensible coil load, in Btu/h. This is because of the characteristics of sensible heat transfer described by Eq. (12.2), $q_{cs} = A_o U_o \Delta T_m$. When the volume flow rate of chilled water $\dot{V}_w$ is reduced, the decrease in the product of $A_o U_o \Delta T_m$ is not the same as the reduction in the chilled water volume flow rate $\dot{V}_w$. When $\dot{V}_w$ drops, the outer surface area A_o remains the same and U_o is slightly reduced. Only a considerable rise in chilled water temperature across the coil, $\Delta T_{w.c} = (T_{wl} - T_{we})$, as shown in Fig. 20.15b, can reduce ΔT_m sufficiently to match the reduction of q_{cs}. Figure 20.15 is obtained for entering water and entering air temperatures that remain constant at various fractions of the design flow.

Theoretically, when the sensible coil load q_{cs} is reduced to 0.6 of the design value, the chilled water volume flow rate should be decreased to about 0.25 of the design volume flow rate to match the

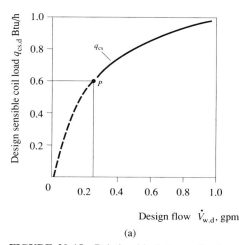
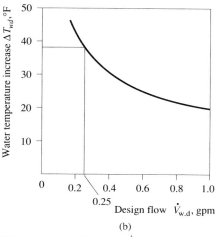

FIGURE 20.15 Relationship between fraction of design volume flow rate $\dot{V}_{w.d}$, coil load q_{cs}, and water temperature rise ΔT_w: (a) q_{cs} vs. $\dot{V}_w$ and (b) $\Delta T_{w.c}$ vs. $\dot{V}_w$.

reduction of q_{cs}. Meanwhile, the power input at the shaft of the variable-speed pump is only about 8 percent of its design brake horsepower. There is a tremendous savings in pump power for a variable-flow system in comparison with a constant-flow system.

A two-way control valve for the coil must be carefully selected. First, an equal-percentage contour valve should be used. As described in Section 17.6 and shown in Figs. 17.13 and 20.15a, when a coil is equipped with an equal-percentage valve, the sensible coil load is directly related to the valve stem travel or control output signal through the percentage water flow rate and thus provides better control quality.

Second, as described in Section 17.6, the pressure drop across the two-way control valve at the design flow should be greater than the sum of the pressure drops of the coil, the pipe friction, and the pipe fittings between the supply and return mains.

Third, if a two-way control valve is selected based on 100 percent of the design flow with a pressure drop of 15 ft WG and the actual pressure drop is only 7.5 ft WG, for a valve with a rangeability of 10 or less, it is possible that this valve will provide a 50 percent coil capacity when the valve is only 5 percent open. Such a control valve would function as a two-position on-off valve.

Fourth, the control valve closes its opening not only to provide the pressure drop for the modulated required flow; it also often further closes its opening to provide an additional pressure drop for the coils nearer to the pump for water flow balance if it is a direct-return piping system.

Plant Loop

For chilled water systems, a central plant is often installed with multiple chillers, typically two to four chillers. Multiple chillers are usually connected in parallel. Each chiller is installed with a chilled water pump that has the same volume flow rate as the water chiller. In such an arrangement, it is more convenient to turn the chillers on or off in sequence. Chillers are turned on or off depending on the sensible coil load or the required system cooling capacity. The required system cooling capacity can be found by measuring the product of temperature differential across the supply and return mains and the water flow rate. The volume flow rate in the common pipe cannot be taken as the control signal to turn the chillers on or off in sequence because the output capacity of the chillers must meet the required coil's load in the building loop. Details of optimized control of multiple chillers will be discussed in Chapter 23.

Transmitters and Variable-Speed Pumps in Building Loops

DIFFERENTIAL PRESSURE TRANSMITTERS.
These are used to maintain the minimum required pressure differential between the supply and return mains at a specific location. If only one differential pressure transmitter is installed for chilled or hot water supply mains, it is usually located at the end of the supply main furthest from the building pump discharge, as shown in Fig. 20.16. If multiple differential pressure transmitters are installed, they are of-

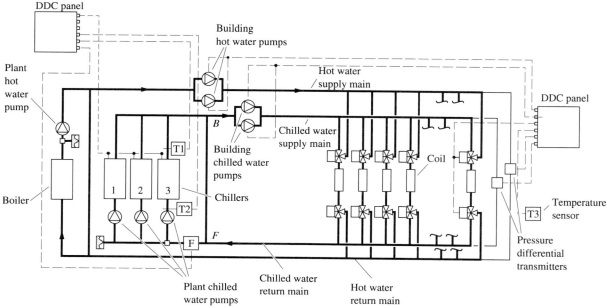

FIGURE 20.16   Dual-temperature water system using plant–building loop combination.

ten located at places remote from the building pump discharge, with a low signal selector to ensure that any coil in the building loop has an adequate pressure differential between the supply and return mains.

The setpoint of the differential pressure transmitter should be equal to or slightly greater than the sum of the pressure drops of the control valve, coil, pipe fittings, and pipe friction of the branch circuit between the supply and return mains. A low setpoint cannot ensure adequate water flow through the coils. A high setpoint consumes more pump power at a reduced flow. A setpoint between 15 and 30 ft of head loss may be suitable.

VARIABLE-SPEED BUILDING PUMPS. For identical variable-speed pumps connected in parallel, the best overall efficiency is often obtained if the pumps are operated at identical speeds. Pumps should be reduced or increased to approximately the same speed. For building pumps, a standby is usually equipped to prevent system shutdown in case a repair is needed. It is different from the plant pump, which is often included in multiple chiller/pump combined units.

System Performance

Consider a chilled water system in a dual-temperature water system that is in a plant–building loop combination, as shown in Fig. 20.16. There are three chillers in the plant loop, each of which is equipped with a constant-speed chilled water pump. In the building loop, there are two variable-speed chilled water pumps connected in parallel. One of them is a standby pump. Chilled water is forced through the water cooling coils in AHUs that serve various zones in the building. For simplicity, assume that the latent coil's load remains constant when the coil's load varies.

For such a chilled water system, the following apply:

- At a specific sensible coil's load, expressed in terms of a fraction of the design load, $q_{cs}/q_{cs.d}$, there is a corresponding water volume flow rate that offsets the coil load, expressed as a fraction of the design flow, $\dot{V}_{bg}/\dot{V}_{bg.d}$. The building variable-speed pump should operate at this volume flow rate of $\dot{V}_{bg}$, with a head sufficient to overcome the head loss in building loops.

- The supply and return temperature differential of the building loop, or the water temperature rise across the coils, $\Delta T_{w.c}$, in °F, depends on the fraction of the design sensible coil's load, $q_{cs}/q_{cs.d}$, and fraction of the design volume flow rate through the coils, $\dot{V}_{bg}/\dot{V}_{bg.d}$. The smaller the value of $\dot{V}_{bg}/\dot{V}_{bg.d}$, the greater the value of $\Delta T_{w.c}$. The water temp-

erature increase $\Delta T_{w.c}$ is always greater than the design value at a reduced $q_{cs}/q_{cs.d}$, as shown in Fig. 20.17c.

- The water volume flow rate in the plant loop, $\dot{V}_{pt}$, depends on the number of operating chillers. Whether a chiller and its associated pump are turned on or off is determined by the coil's sensible load and the chiller capacity. At design conditions, $\dot{V}_{pt}$ should be slightly greater than $\dot{V}_{bg}$.

- The water temperature drop across the evaporator, $\Delta T_{w.ev} = (T_{ee} - T_{el})$, in °F, depends on the sensible coil's load and the water volume flow rate at the plant loop, $\dot{V}_{pt}$. Here, T_{ee} and T_{el} indicate the water temperature entering and leaving the evaporator. T_{ee} and $\Delta T_{w.ev}$ are normally less than the design value during part-load. This is due to the mixing of the water flow from the common pipe at a temperature T_{el} at junction F (see Fig. 20.16).

- The water volume flow rate in the common pipe is given by

$$\dot{V}_{cn} = \dot{V}_{pt} - \dot{V}_{bg} \qquad (20.13)$$

Normally, $\dot{V}_{pt} \geq \dot{V}_{bg}$, and at design conditions, $\dot{V}_{pt} \approx \dot{V}_{bg}$. The flow direction of water in the common pipe should normally be from the discharge side of the plant loop to its return side.

- Because the water flow rate flowing through the coil is controlled according to the discharge air temperature or space temperature, only the sensible load is taken into account during analysis. The coil capacity to remove the latent load usually drops at a reduced water volume flow rate. Checking the space relative humidity should be considered at part-load operation if space relative humidity control is essential. This will be discussed in Chapters 26 and 27.

- Comparing Fig. 20.16 with Fig. 20.10a, there are additional supply and return mains in Fig. 20.16. This piping arrangement can provide zone changeover from chilled water supply to hot water supply and vice-versa rather than building changeover illustrated in Fig. 20.10a.

Example 20.1. Consider a chilled water system using a plant–building loop combination. The design chilled water flow rate is 1000 gpm with a chilled water temperature rise across the coils of $\Delta T_{w.c} = 20$°F. There are three chillers in the central plant, each equipped with a constant-speed chiller pump that provides 350 gpm at 40 ft total head. In the building loop, there are two variable-speed building pumps, each with a volume flow rate of 1000 gpm at 60 ft head. One of these building pumps is a standby pump.

At design conditions, chilled water leaves the chiller at a temperature $T_{el} = 40$°F and returns to the chiller at

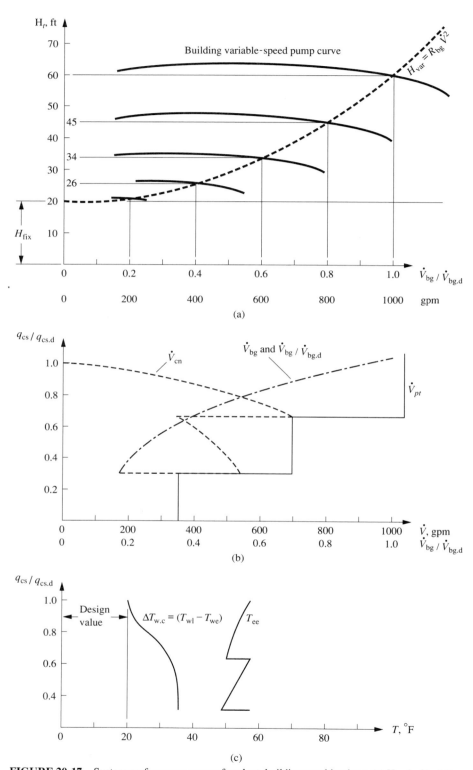

FIGURE 20.17 System performance curves for plant–building combinations: (a) Head of building variable-speed pump at various volume flow rates, (b) $\dot{V}_{bg}$ vs. $q_{cs}/q_{cs.d}$ and $\dot{V}_{bg}/\dot{V}_{bg.d}$ vs. $q_{cs}/q_{cs.d}$, and (c) $\Delta T_{w.c}$ vs. $q_{cs}/q_{cs.d}$.

60°F. Chilled water leaving the chiller is controlled at 40°F for both design and part-load operation. Once the fouling and inefficiency of the coils have been taken into account, the fractions of the design volume flow rate, $\dot{V}_{bg}/\dot{V}_{bg.d}$, required to absorb the coil load at various fractions of the sensible load, $q_{cs}/q_{cs.d}$, are listed in Table 20.8.

When the system load drops, plant chiller No. 1 will turn off when $q_{cs}/q_{cs.d}$ equals 0.65 and chiller No. 2 will turn off when $q_{cs}/q_{cs.d}$ equals 0.30. When the system load increases, chiller No. 2 turns on at $q_{cs}/q_{cs.d} = 0.35$, and chiller No. 1 turns on at $q_{cs}/q_{cs.d} = 0.7$. Plant chiller No. 3 operates continuously.

Calculate the following based on the chillers' on-off schedule at various fractions of the sensible coil's load:
1. Chilled water temperature rise across the coil
2. Water flow in the common pipe
3. Temperature of water returning to the water chiller

Solution
1. From the given information and Eq. (12.3), the water temperature rise across the coil for $q_{sc}/q_{sc.d} = 0.9$ is

$$\Delta T_{w.c} = 20\frac{q_{cs}\dot{V}_{bg.d}}{q_{cs.d}\dot{V}_{bg}} = 20\frac{0.9}{0.8} = 22.5°F$$

Values of $\Delta T_{w.c}$ at other values of $q_{sc}/q_{sc.d}$ can be similarly calculated and are listed in Table 20.8.

2. For $q_{cs}/q_{cs.d} = 0.9$, because all three chillers are operating, the water volume flow rate in the plant loop $\dot{V}_{pt} = 3(350) = 1050$ gpm. The water volume flow rate in the building loop is

$$\dot{V}_{bg} = 1000\frac{\dot{V}_{bg}}{\dot{V}_{bg.d}} = 1000(0.8) = 800 \text{ gpm}$$

Then, the water volume flow rate in the common pipe is

$$\dot{V}_{cn} = \dot{V}_{pt} - \dot{V}_{bg} = 1050 - 800 = 250 \text{ gpm}$$

3. When $q_{cs}/q_{cs.d} = 0.65$, chiller No.1 is turned off. Just before the chiller is turned off, the volume flow

rate in the common pipe is

$$\dot{V}_{cn} = 1050 - 0.38(1000) = 670 \text{ gpm}$$

Immediately after chiller No. 1 is turned off,

$$\dot{V}_{cn} = 2(350) - 0.38(1000) = 320 \text{ gpm}$$

Values of $\dot{V}_{cn}$ for other values of $q_{cs}/q_{cs.d}$ can be similarly calculated and are listed in Table 20.8.

4. For $q_{cs}/q_{cs.d} = 0.9$, after the adiabatic mixing of water from the building loop and common pipe, the temperature of chilled water returning to the water chiller can be calculated as

$$T_{ee} = \left(\frac{250}{1050}\right)40 + \frac{[(0.8)(1000)(40 + 22.5)]}{(1050)}$$
$$= 57.1°F$$

5. For $q_{cs}/q_{cs.d} = 0.65$, just before chiller No.1 is turned off,

$$T_{ee} = \left(\frac{700}{1050}\right)40 + (0.35)(1000)(40 + 34.2)$$
$$= 51.4°F$$

Immediately after chiller No.1 is turned off,

$$T_{ee} = \left(\frac{350}{700}\right)40 + (0.35)(1000)(40 + 37.1)$$
$$= 57.1°F$$

Other chilled water temperatures upon entering the chiller can be similarly calculated and are listed in Table 20.8.

Balance Valves and Direct Return Piping Arrangements

In a variable-flow water system, the imbalance in the pressure differentials between the supply and return mains of two or more branch circuits that results from the use of direct return piping arrangements may be able to be offset by opening or closing the control valves on the coils to a position that does not affect

TABLE 20.8
Magnitudes of $\dot{V}_{bg}/\dot{V}_{bg.d}$ and calculated results for Example 20.1

	$q_{cs}/q_{cs.d}$										
	0.3		0.4	0.5	0.6	0.65		0.7	0.8	0.9	1.0
	1 chiller	2 chillers				2 chillers	3 chillers				
$\dot{V}_{bg}/\dot{V}_{bg.d}$	0.17	0.17	0.22	0.275	0.34	0.38	0.38	0.45	0.60	0.8	1.0
$\Delta T_{w.c}$, °F	36	36	36	36	35.2	34.2	34.2	31.1	26.7	22.5	20
$\dot{V}_{pt}$, gpm	350	700	700	700	700	700	1050	1050	1050	1050	1050
$\dot{V}_{cn}$, gpm	180	530	480	425	360	320	670	600	450	250	0
T_{ee}, °F	57.5	48.8	51.3	54.2	57.1	58.6	52.4	53.3	55.3	57.1	60

proper system operation. If so, no balancing valve is required. Most of the imbalance between branch circuits in a building loop can thus be counteracted.

However, the designer still needs to carefully design various branch circuits as approximately balanced so as to provide better water distribution and control quality. Commissioning, including test and balancing, is still required to find the reason for the imbalance as well as any possible errors in the flow circuits and try to correct them.

If small coils are equipped with on-off two-position control valves, they should also be installed with a flow control valve to limit the maximum volume flow rate through the coil to the design value.

20.9 PLANT–DISTRIBUTION-BUILDING LOOP COMBINATION

System Description

Chilled water or chilled and hot water are often supplied to many buildings separated from a central plant in universities, medical centers, and airports. The benefits of using a campus-type central plant chilled water system instead of individual building installations are cost savings, minimal environmental impact (for example, from cooling towers), effective operation and maintenance, and reliability.

Many recently developed campus-type central plant chilled water systems use a plant–distribution-building loop combination, as shown in Fig. 20.18a. As in a plant–building loop combination, constant flow is used in the plant loop. Each chiller in the central plant has its own constant-speed chiller pump. Chilled water leaves the chiller at a temperature of 40 to 42°F. It is then extracted by the distribution pumps and forced to the supply main of the distribution loop.

Chilled water in the supply and return mains of the distribution loop operates under variable flow. Multiple variable-speed pumps are often used to transport chilled water at a volume flow rate slightly higher than the sum of the volume flow rates required in the building loops. At each building entrance, the variable-speed building pump extracts the chilled water and supplies it to the coils in AHUs and terminals in various zones by means of building supply mains. Chilled water is then returned to the water chillers through building return mains, a distribution-loop return main, and chiller pumps.

The system performances of the plant loop and building loop are similar to those in a plant–building loop combination.

Pressure Gradient of Distribution Loop

A campus-type chilled water central plant must transport several thousand gallons of water per minute to the farthest building at a distance that may be several thousand feet away from the plant. The pressure gradient of the distribution supply and return mains due to the pipe friction and fitting losses cause uneven pressure differentials among the supply and return mains ($\Delta H_{s.r}$, in ft WG) of buildings along the distribution loop, as shown in Figure 20.18b. Buildings nearer to the central plant have a greater $\Delta H_{s.r}$ than the buildings farther from the plant. Along the distribution loop, a smaller pressure gradient results in a lower pumping power but a larger diameter of chilled water pipes. A more even $\Delta H_{s.r}$ does not impose excessive pressure drops across the control valves of coils.

Using a lower pressure drop ΔH_f is an effective means of reducing the pressure gradient and pressure differential $\Delta H_{s.r}$ and of saving energy. For a distribution loop, a value of ΔH_f between 0.5 and 1 ft per 100 ft pipe length, sometimes even lower, may be used. Low values of $\Delta H_{s.r}$ can be offset at the coil control valve without affecting the coil's proper operation. A life-cycle cost analysis should be conducted to determine the optimum ΔH_f.

Using two-way distribution from the central plant, with two supply and return distribution loops, may reduce the pipe distance and the diameter of the supply and return mains. Such a distribution loop layout depends on the location of the central plant and the air conditioned buildings as well as the cost analyses of various alternatives.

Using a reverse return piping arrangement instead of a direct return one does even the value of $\Delta H_{s.r}$ along the supply and return mains of the distribution loop. However, having an additional pipe length equal to that of the return main significantly increases the piping investment. A simpler and cheaper way is to install a pressure throttling valve at each building entrance to offset the excess $\Delta H_{s.r}$. Usually, direct return is used for a distribution loop.

Variable-Speed Building Pumps

The function of a variable-speed building pump is (1) to provide variable flow and corresponding head to overcome the pressure drop of the building loop at design and reduced coil loads, and (2) to provide different magnitudes of head for the building loop according to the needs of various types of buildings.

When only a variable-speed distribution pump is used instead of both a variable-speed distribution pump and a variable-speed building pump, it is possible to have sufficient pump head to overcome the pressure drop of the building loop. However, the pressure characteristics of the supply and return mains of the distribution loop at reduced flows make it diffi-

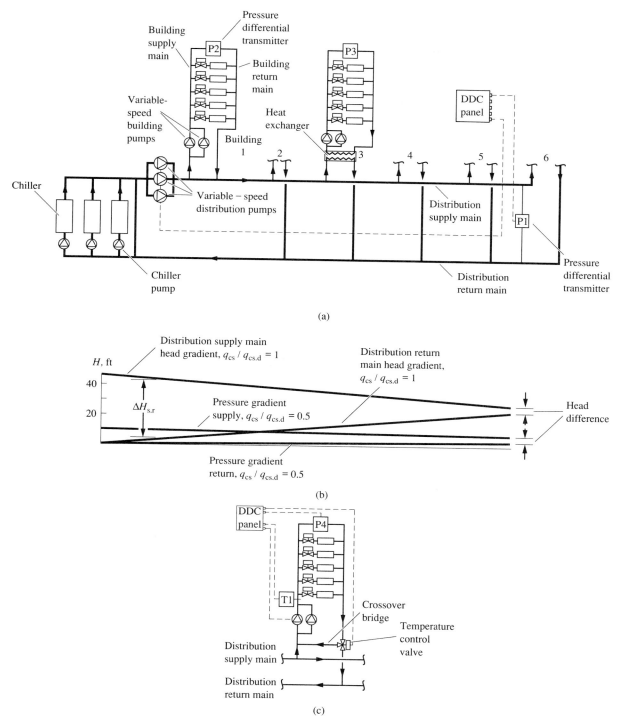

FIGURE 20.18 Chilled water system using plant-distribution-building loop combination: (a) schematic diagram, (b) pressure gradient for distribution loop, and (c) crossover bridge with temperature control valve.

cult to satisfy various load profiles in different buildings during part-load operation. Having a variable-speed building pump for each building also saves more pump energy. Therefore, the use of variable-speed pumps for both distribution and building loops is preferable.

Control of Variable-Speed Distribution Pump

Two types of controls can be used to modulate a variable-speed distribution pump to transport the required chilled water volume flow to various buildings:

- A differential pressure transmitter may be located near the farthest end of the distribution supply main, as shown in Fig. 20.18a. Theoretically, the head of the building pump should extract the exact required amount of chilled water corresponding to the coil load from the distribution supply main, force it through the coils, and discharge it to the distribution return main. Therefore, a setpoint for the pressure differential of about 5 ft may be appropriate.

- A stand-alone DDC panel measures the total water flow that returns from each building by means of flow meters and modulates the variable-speed distribution pump to supply exactly the required amount to various building loops. This type of control is more precise but more expensive.

Building Entrance

Chilled water is usually supplied directly from the distribution supply main to the building supply main. A pressure throttling valve may be used to offset the excess pressure differential $\Delta H_{s.r}$ along the distribution loop.

Although using a heat exchanger at the building entrance entirely isolates the chilled water in the distribution loop from the chilled water in the building loop, a temperature increase of about 3 to 7°F is required for a chilled water heat exchanger. Because chilled water has a supply and return temperature differential of only about 15 to 20°F, a heat exchanger is seldom used at a building or zone entrance for a chilled water system. Because a hot water system has a greater

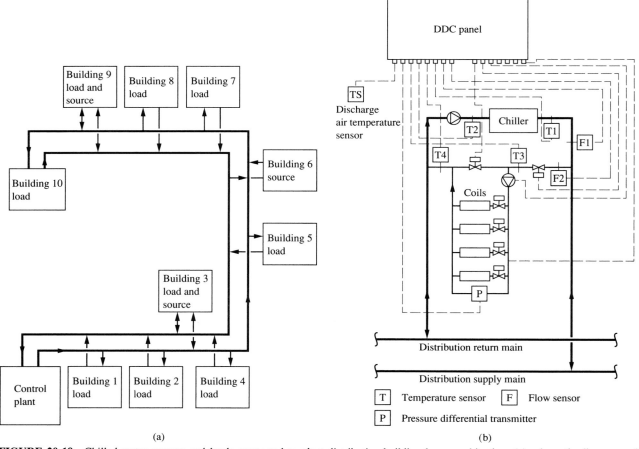

FIGURE 20.19 Chilled water sources and loads connected to plant-distribution-building loop combination: (a) schematic diagram and (b) building with sources and loads.

supply and return temperature differential, a heat exchanger is sometimes used at the building entrance for a hot water system.

A chilled water building loop may be divided into various zones based on different height levels within the building. In such an arrangement, the coils in the lower floors of the building loop will not suffer a high static pressure because the low-level water loops is often isolated from the high-level water loops by means of a heat exchanger at the zone entrance.

If a building requires a chilled water supply temperature higher than that given by the distribution supply main, a crossover bridge with a temperature control valve can be arranged for this purpose, as shown in Fig. 20.18c.

The return temperature from the coils in a building loop is affected by the cleanliness of the coil, including air-side cleanliness, and the control system in the building loop. The building's variable-speed pump is often controlled by the pressure differential transmitter located at the end of the building supply main, as shown in Fig. 20.18 a and c.

Combination of Multiple Sources and Multiple Loads

Many conversions and retrofits of campus-type chilled water systems require existing chilled water plants in addition to the developed central plant. In these cases, there are buildings with chilled water sources (chillers); buildings with chilled water coils and loads; and buildings with chilled water sources and loads connected to the same plant–distribution-building loop combination, as shown in Figure 20.19a. A standalone DDC panel is used for each type of building. There is a central microprocessor for the whole chilled water system. The optimization program may proceed as follows:

1. Measure the chilled water temperature across each load and source, as well as its rate of water flow.
2. Add all the loads together.
3. Turn on the most efficient sources first, including the demand and downtime, according to the available sources. A chiller's running capacity should match the coil load. There should be a time delay to start or stop the chiller.
4. Use predictive control and expert system control strategies to predict load changes according to past experience and outdoor conditions.

Chilled and Hot Water Distribution Pipes

Chilled and hot water distribution pipes are large pipes mounted in underground accessible tunnels or trenches. They are well insulated, although underground return chilled water mains for which $T_{ret} > 55°F$ may not be insulated, depending on a detailed cost analysis. Factory-made conduits consist of inner steel pipe, insulation, air space, and outer conduit; or steel pipe, with and without insulation, and outer casing. Expansion loops or couplings should be included, and a good drainage system is important to protect the insulating quality. Please refer to *ASHRAE Handbook 1991, HVAC Systems and Equipment* for details on design and installation.

REFERENCES AND FURTHER READING

ASHRAE, *ASHRAE Handbook 1987, HVAC Systems and Applications,* ASHRAE, Inc., Atlanta, GA, 1987.

ASHRAE, *ASHRAE Handbook 1992, HVAC Systems and Equipment,* ASHRAE, Inc., Atlanta, GA, 1992.

ASHRAE, *ASHRAE Handbook 1989, Fundamentals,* ASHRAE, Inc., Atlanta, GA, 1989.

ASHRAE, *ASHRAE/IES Standard 90.1–1989, Energy Efficient Design of New Buildings Except Low-Rise Residential Buildings,* ASHRAE, Inc., Atlanta, GA, 1989.

ASTM, *Standard Specification for Pipe, Steel, Black and Hot-Dipped, Zinc-Coated Welded and Seamless, ASTM Standard AS3-REV-B-90,* 1990.

ASTM, *Standard Specification for Seamless Copper Water Tube, ASTM Standard B88-89,* 1989.

Ahmed, O., "Life-Cycle Cost Analysis of Variable-Speed Pumping for Coil Application," *ASHRAE Transactions,* Part I, pp. 194–211, 1988.

Avery, G., "Microprocessor Control for Large Chilled Water Distribution Systems," *Heating/Piping/Air Conditioning,* pp. 59–61, October 1987.

Ball, E. F., and Webster, C. J. D., "Some Measurements of Water Flow Noise in Copper and ABS Pipes with Various Flow Velocities," *The Building Services Engineer,* pp. 33–40, May 1976.

Binkowski, R. O., "Water Treatment for HVAC Systems," *Heating/Piping/Air Conditioning,* pp. 131–133, October 1989.

Braun, J. E., Klein, S. A., Mitchell, J. W., and Beckman, W. A., "Applications of Optimal Control to Chilled Water Systems Without Storage," *ASHRAE Transactions,* Part I, pp. 663–675, 1989.

Burr, G. C., and Pate, M. E., "Conversion of Campus Central Plant from Constant Flow to Variable Flow at University of West Florida," *ASHRAE Transactions,* Part I B, pp. 891–901, 1984.

Carlson, G. F., "Central Plant Chilled Water Systems—Pumping and Flow Balance Part I," *ASHRAE Journal,* pp. 27–34, February 1972.

Carlson, G. F., "Hydronic Systems: Analysis and Evaluation—Part I," *ASHRAE Journal,* pp. 2–11, October 1968.

Carrier, *Handbook of Air Conditioning System Design,* McGraw-Hill, New York, 1965.

Griffith, D., "Distribution Problems in Central Plant Systems," *Heating/Piping/Air Conditioning*, pp. 59–76, November 1987.

Hull, R. F., "Effect of Air on Hydraulic Performance of the HVAC System," *ASHRAE Transactions*, Part I, pp. 1301–1325, 1981.

MacDonald, K. T., "Valves: An Introduction," *Heating/Piping/Air Conditioning*, pp. 109–117, October 1988

Mannion, G. F., "High Temperature Rise Piping Design for Variable Volume Systems: Key to Chiller Energy Management," *ASHRAE Transactions*, Part II, pp. 1427–1443, 1988.

Miller, R. H., "Valves: Selection, Specification, and Application," *Heating/Piping/Air Conditioning*, pp. 99–118, October 1983.

Ocejo, J., "Program Estimates Expansion Tank Requirements," *Heating/Piping/Air Conditioning*, pp. 89–93, November 1986.

Peterson, P. A., "Medical Center Expands Utilities Distribution," *Heating/Piping/Air Conditioning*, pp. 84–96, May 1985.

Pompei, F., "Air in Hydronic Systems: How Henry's Law Tells Us What Happens," *ASHRAE Transactions*, Part I, pp. 1326–1342, 1981.

Prescher, R., "Hydronic System Design Guidelines," *Heating/Piping/Air Conditioning*, pp. 132–134, May 1986.

Rishel, J. B., "Twenty Years' Experience with Variable Speed Pumps on Hot and Chilled Water Systems," *ASHRAE Transactions*, Part I, pp. 1444-1457, 1988.

Rishel, J. B., "Energy Conservation in Hot and Chilled Water Systems," *ASHRAE Transactions*, Part II B, pp. 352–367, 1983.

Solden, H. M., and Siegel, E. J., "The Trend Toward Increased Velocities in Central Station Steam and Water Piping," *Proceedings of the American Power Conference*, Vol. XXVI, 1964.

Stewart, W. E., Jr., and Dona, C. L., "Water Flow Rate Limitations," *ASHRAE Transactions*, Part II, pp. 811–825, 1987.

Uglietto, S. R., "District Heating and Cooling Conversion of Buildings," *ASHRAE Transactions*, Part II, pp. 2096–2106, 1987.

Utesch, A. L., "Variable Speed CW Booster Pumping," *Heating/Piping/Air Conditioning*, pp. 49–58, May 1989.

Waller, B., "Piping—From the Beginning," *Heating/Piping/Air Conditioning*, pp. 51–71, October 1990.

Wang, S. K., *Principles of Refrigeration Engineering*, Vol. 4, Hong Kong Polytechnic, Hong Kong, 1984.

Zell, B. P., "Design and Evaluation of Variable Speed Pumping Systems," *ASHRAE Transactions*, Part I B, pp. 214–223, 1985.

REFRIGERATION SYSTEMS: RECIPROCATING AND SCREW COMPRESSION

21.1 RECIPROCATING VAPOR COMPRESSION REFRIGERATION SYSTEMS

Types of Vapor Compression Refrigeration Systems

Most refrigeration systems used for air conditioning are vapor compression refrigeration systems. Of these vapor compression refrigeration systems, reciprocating vapor compression systems are the most widely used in medium-size systems. Medium systems usually range from 10 tons to 150 tons, sizes for which reciprocating, screw, scroll, and rotary vapor compression systems are available. According to the structure, refrigerant feed, and the purpose of cooling, reciprocating vapor compression refrigeration systems can be classified in the following catagories:

- *Direct-Expansion (DX) Refrigeration Systems or Simply DX-Systems.* Figure 21.1 shows a DX-system. In a DX-system, a direct-expansion coil is used as the evaporator and HCFC–22 and HFC–134a are the primary refrigerants used in new and retrofit projects. The air-cooled condenser may be combined with the reciprocating compressor(s) to form a unit called a *condensing unit.* The condensing unit is generally located outdoors. The DX-coil is either installed in a rooftop packaged unit consisting of fans, filters, a condensing unit, and controls, mounted on the roof of the building, or it is a

system component in a split or indoor packaged unit located indoors. In supermarkets and many industrial applications, DX-coils are also installed in refrigerated display cases and food processing and food storage facilities. Air is forced through the DX-coil by a fan for better heat transfer. DX-systems are most widely used in reciprocating, scroll, and rotary refrigerating systems.

- *Air-Cooled Reciprocating Chiller Systems.* This kind of reciprocating refrigeration system consists of a direct-expansion shell-and-tube liquid cooler used as an evaporator (as described in Section 12.6), an air-cooled condenser, single or multiple reciprocating compressor(s), accessories, and controls. The entire package is often called an *air-cooled reciprocating chiller,* and is shown in Fig. 21.2. As in DX-systems, HCFC–22 and HFC–134a are the primary refrigerants used in new and retrofit projects. In the shell-and-tube liquid cooler, the refrigerant evaporates inside the metal tubes and the chilled water fills the shell. All components and accessories of an air-cooled reciprocating chiller are enclosed in a casing made of steel sheets coated with corrosion-resistant paint.

In large air-cooled reciprocating chillers, air-cooled condenser coils are installed on both sides of the unit and the subcooling coil is at the bottom, connected to the condensing coil through a liquid accumulator. The air-cooled reciprocating chiller is usually located on the roof of the building.

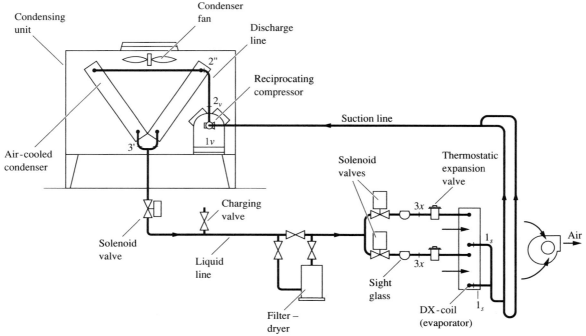

FIGURE 21.1 Direct-expansion DX reciprocating refrigeration system.

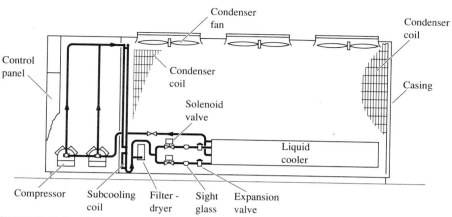

FIGURE 21.2 Air-cooled reciprocating chiller.

In locations where water is scarce and expensive, if the cooling capacity is less than 250 tons, air-cooled reciprocating chillers are often used in central hydronic air conditioning systems or for equipment cooling purposes in industrial applications.

- *Liquid Overfeed Reciprocating Refrigeration Systems.* Liquid overfeed reciprocating refrigeration systems use a liquid overfeed cooler as an evaporator, as described in Section 12.6. Figure 21.3 shows a liquid overfeed reciprocating system. Ammonia R–717 and HCFC–22 are often used as the refrigerants. An additional refrigerant pump is required to circulate the refrigerant to each evaporator at a flow rate several times greater than the actual evaporating rate. After the expansion valve, the refrigerant is throttled to the evaporating pressure. Although liquid refrigerant directly expands into vapor inside the evaporator's tubes and accumulates in the low-pressure receiver, it is not a dry-expansion evaporator. In a liquid overfeed system, a water-cooled condenser is often used.

  Liquid overfeed refrigeration systems have the advantages of a higher heat transfer coefficient of the wetted inner surface and simpler refrigerant flow control. However, they are large and heavy, and are generally used for industrial refrigeration, food storage, and sometimes in ice-storage systems.

- *Flooded Coil Reciprocating Refrigeration Systems.* In these systems, the flooded coil, or the evaporator, is full of liquid refrigerant, which wets the inner surface of the coil and provides better heat transfer. The liquid level inside the coil is maintained by a float valve, as shown in Fig. 21.3. Liquid refrigerant is circulated through the evaporator by gravity or a refrigerant pump.

  Flooded coil systems are often used for low-temperature industrial applications where there is a small temperature differential between the refrigerant and the surrounding medium.

- *Multistage Reciprocating Refrigeration Systems.* As described in Chapter 4, a multistage reciprocating system can be a compound system or a cascade system. Multistage reciprocating refrigeration systems are used for industrial applications to provide refrigeration at low temperatures (below 0°F).

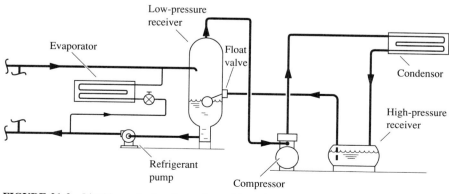

FIGURE 21.3 Liquid overfeed reciprocating refrigeration system.

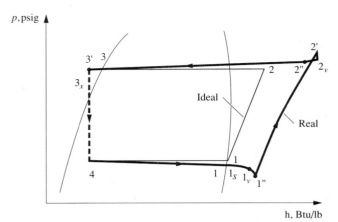

FIGURE 21.4 Real reciprocating vapor compression refrigeration cycle on *p–h* diagram.

Real Cycle of a Single-Stage Reciprocating Refrigeration System

In a reciprocating refrigeration system, the real refrigeration cycle differs from the ideal cycle because of the pressure drop of the refrigerant when it flows through the pipelines, valves, compressor passages, evaporator, and condenser. Also, the compression and expansion processes are not isentropic.

Figure 21.4 shows the real cycle of a single-stage reciprocating vapor compression refrigeration system on a *p–h* diagram. In a single-stage reciprocating refrigeration system, the pipeline from the outlet of the evaporator to the inlet of the reciprocating compressor is called the *suction line*. The suction line is shown by the $1_s 1_v$ in Figs. 21.1 and 21.4. The pipeline from the outlet of the reciprocating compressor to the inlet of the air-cooled condenser is called the *discharge line*, or *hot-gas line*. This is shown as $2_v 2''$ in both figures. The pipeline from the outlet of the condenser to the inlet of thermostatic expansion valve is called the *liquid line*. This is shown as $3' 3_x$ in both figures.

In Fig. 21.4, the thick lines represent the real refrigeration cycle, and the thin lines the ideal cycle. Various line segments indicate the following refrigeration processes:

11_s	superheat in the evaporator
$1_s 1_v$	pressure drop in the suction line
$1_v 1''$	pressure drop from the suction valve and the passage to the cylinder
$1''2'$	compression process in the cylinder
$2'2_v$	pressure drop from the discharge valve and the passage
$2_v 2''$	pressure drop in the discharge line
$2''3'$	condensing and subcooling processes in the condenser
$3'3_x$	pressure drop in the liquid line
$3_x 4$	throttling process in the expansion valve
$4 1_s$	evaporating process in the evaporator

21.2 EXPANSION VALVES AND SERVICE VALVES

Expansion Valves

In a refrigeration system, the *expansion valve* is an adjustable throttling device through which the refrigerant at condensing pressure is throttled to evaporated pressure or interstage pressure. At the same time, the expansion valve regulates its opening to feed the required amount of refrigerant to the evaporator to meet the refrigeration load at the evaporator.

Expansion valves used for reciprocating refrigeration systems can be divided into the following groups:

- Thermostatic expansion valves
- Constant-pressure expansion valves
- Float valves

Thermostatic expansion valves are the most widely used throttling or expansion devices in direct-expansion reciprocating refrigeration systems.

Thermostatic Expansion Valves

OPERATING CHARACTERISTICS. A thermostatic expansion valve regulates its refrigerant flow rate to the evaporator according to the degree of superheat of the vapor refrigerant leaving the evaporator.

Figure 21.5 shows a thermostatic expansion valve connected to a DX-coil, the evaporator. A thermostatic expansion valve consists of a valve body, a valve pin, a spring, a diaphragm, and a sensing bulb at the outlet of the evaporator. The sensing bulb is connected to the upper part of the diaphragm by means of a capillary tube. The outlet of the thermostatic expansion valve is then connected to various refrigerant circuits of the DX-coil through the nozzle and tubes of a refrigerant distributor.

When the liquid refrigerant passes through the small opening around the valve pin, its pressure is reduced to evaporating pressure. Liquid refrigerant flows through the refrigerant distributor and tubes, and vaporizes gradually as it flows inside the copper tubes. At position *x*, as shown in Fig. 21.5, all the liquid has been vaporized. By the time the vapor refrigerant reaches the outlet *o* of the evaporator, it is superheated to a few degrees higher than its saturated temperature.

If the coil's load of this DX-coil increases, more refrigerant vaporizes. Consequently, point *x* tends to move toward the inlet of the evaporator. This causes an increase in the degree of superheat at the outlet *o* as well as an increase in the temperature of the sensing bulb. Because the sensing bulb may be par-

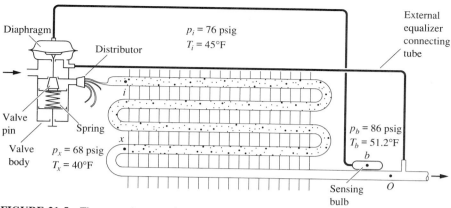

FIGURE 21.5 Thermostatic expansion valve with external equalizer.

tially filled with the same type of liquid refrigerant as in the evaporator, higher temperature exerts a higher saturated pressure on the top of the diaphragm. This lowers the valve pin and widens the valve opening. More liquid refrigerant is allowed to enter the evaporator to match the increase in refrigeration load.

If the coil's load drops, the degree of superheat of the vapor refrigerant at the outlet of the evaporator becomes smaller, resulting in a narrower valve opening. The liquid refrigerant feed to the evaporator drops accordingly.

The degree of superheat of the vapor refrigerant at the outlet can be adjusted by varying the tension of the spring in the thermostatic expansion valve.

CAPACITY–SUPERHEAT CURVE. Figure 21.6 shows the capacity–superheat curve of a thermostatic expansion valve. In Fig. 21.6, ΔT_{ss} represents the static superheat when the valve is closed at no-load condition. Static superheat ΔT_{ss} does not have sufficient pressure to open the valve from its closed position. Static superheat usually has a value of 4 to 8°F.

Term ΔT_o indicates the opening superheat, that is, the superheat required to open the expansion valve from a closed position to a fully open position corresponding to the full refrigeration load or design coil's load. Opening superheat usually has a value between 6 and 8°F.

Term ΔT_{op} indicates the operating superheat at a specific refrigeration load q_{op}, in Btu/h. Term ΔT_{rl} is the sum of ΔT_{ss} and ΔT_o, and represents the superheat setting of the thermostatic expansion valve at full-load. The recommended ΔT_{rl} is between 10 and 20°F. Superheat in excess of 20°F means that too much of the coil's surface is producing superheat, which reduces the coil capacity. Too little superheat is also risky because it may cause the liquid refrigerant to flood the compressor.

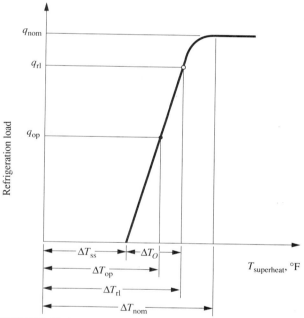

FIGURE 21.6 Capacity–superheat curve of a thermostatic expansion valve.

Term T_{nom} indicates nominal superheat. For safety, T_{nom} is generally equal to 1.1 to 1.4 T_{rl}.

EXTERNAL EQUALIZER. A thermostatic expansion valve with an external equalizer has a connecting tube that links the bottom diaphragm chamber to the evaporator outlet o, as shown in Fig. 21.5, so that the pressure under the diaphragm is the same as that at the outlet o. Because currently used DX-coils have a considerable pressure drop at the refrigerant distributor and tubes before entering the coil, thermostatic expansion valves with external equalizer tubes are often used in DX-coils.

Consider a reciprocating refrigeration system using HCFC–22 as the refrigerant. The evaporating temperature at the inlet of the evaporator is 45°F and the saturated pressure p_i is 76 psig. If the spring pressure at full load p_s is 18 psig and the pressure drop of the refrigerant flowing through the evaporator Δp_{ev} is 8 psi, in a thermostatic expansion valve with an external equalizer, the saturated pressure of the refrigerant at point x p_x is therefore

$$p_x = p_i - \Delta p_{ev} = 76 - 8 = 68 \text{ psig} \quad (21.1)$$

The corresponding saturated temperature at point x, $T_x = 40°F$.

In a thermostatic expansion valve with an external equalizer, the pressure balance at the diaphragm is

$$p_b = p_x + p_s = 68 + 18 = 86 \text{ psig} \quad (21.2)$$

The corresponding saturated temperature $T_b = 51.2°F$. Because the temperature of the vapor refrigerant at the outlet T_o can be considered equal to the bulb temperature T_b, the degree of superheat is calculated as

$$\Delta T_{rl} = T_o - T_x = T_b - T_x = 51.2 - 40 = 11.2°F$$
$$(21.3)$$

Hand-operated expansion valves are operated manually. Instead of the sensing bulb and diaphragm-operated valve stem in a thermostatic expansion valve, a hand wheel, a threaded valve stem, and threaded sockets are used to lift the valve pin or close the valve port.

LIQUID CHARGE IN THE SENSING BULB

Straight and limited liquid charge. When the sensing bulb is charged with an amount of liquid with the same properties as the refrigerant in the evaporator, such a charge is known as the *straight charge*. If there is only a limited amount of liquid charge, it is known as the *limited liquid charge*. All the liquid vaporizes at a predetermined temperature, as shown in Fig. 21.7a. Any increase above this temperature results in only a small increase in pressure on the diaphragm because of the superheated vapor.

A sensing bulb filled with limited liquid charge limits the maximum operating temperature in the evaporator and, therefore, prevents the compressor motor from overloading.

Crosscharge. The temperature versus pressure difference $\Delta T - \Delta p$ of a sensing bulb with straight charge is shown in Fig. 21.7b. At a different evaporating temperature T_{ev}, $\Delta T_{bx} = T_b - T_x$ varies accordingly.

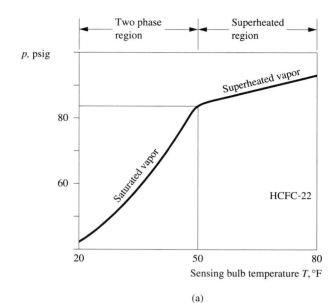

(a)

FIGURE 21.7 Thermostatic characteristics of sensing bulb: (a) limited liquid charge; (b) straight charge; and (c) cross charge.

If the liquid charge in the sensing bulb is different from the refrigerant used in the evaporator, it is known as *crosscharge*. The primary purpose of crosscharge is to provide a nearly constant degree of superheat at various evaporating temperatures, as shown in Fig. 21.7c. For a constant value of $p_b - p_x$, the degree of superheat at various evaporating temperatures is approximately equal. This prevents the refrigerant from flooding back to the compressor at higher evaporating temperatures.

HUNTING OF THE THERMOSTATIC EXPANSION VALVE.

Hunting is the quick cycling of maximum and minimum output of a controlled device. For thermostatic expansion valve, hunting results from the over- or underfeeding of refrigerant to the evaporator. It causes temperature and pressure fluctuations and may reduce the capacity of the refrigeration system.

The time lag of refrigerant flow through the evaporator to the sensing bulb may cause a continuous overshooting of the valve position. The overfed refrigerant may form unevaporated slugs. These slugs may chill the sensing bulb and suddenly reduce the refrigerant feed. The overshooting and sudden reduction cause hunting.

Hunting of the thermostatic expansion valve is determined by the following factors:

- *Valve size*. An oversized valve aggravates hunting.
- *Degree of superheat*. The smaller the degree of superheat, the greater the chance of hunting.

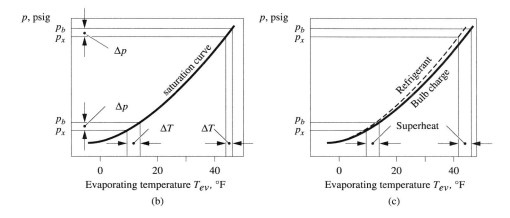

- *Sensing bulb charge*. Crosscharge tends to prevent hunting.
- *Location of sensing bulb*. A properly selected sensing bulb location often minimizes hunting.

The sensing bulb should be located at the side of the horizontal section of the suction line immediately after the evaporator outlet so that it does not sense the oil temperature or liquid refrigerant at the bottom of the tube.

Constant-Pressure Expansion Valves

Constant-pressure expansion valves, sometimes called *automatic expansion valves*, regulate the refrigerant feed to the evaporator according to the evaporating pressure at the outlet of the expansion valve. Evaporating pressure is maintained at an almost constant value.

Figure 21.8 shows a typical constant-pressure expansion valve. The valve has an adjustable spring at the top of the diaphragm to open the valve. Under the

FIGURE 21.8 Constant-pressure expansion valve.

valve seat, there is a spring that closes the valve during the shutdown period of the refrigeration plant. Evaporating pressure is exerted against the bottom of the diaphragm. An increase in evaporating pressure presses the diaphragm upward, closes the valve opening, and reduces the amount of refrigerant fed to the evaporator. When the evaporating pressure drops, the top adjustable spring presses the diaphragm down. The valve pin moves downward accordingly and creates a larger opening.

Constant-pressure expansion valves cannot control the liquid refrigerant feed to the evaporator to accommodate the variation of refrigeration load at the evaporator. Their application is therefore limited to evaporators with a fairly constant refrigeration load during operation.

Float Valves

Float valves control the refrigerant feed to the evaporator to maintain a specific liquid level. The two types of float valves used in refrigerant control are *high-side float valves* and *low-side float valves*.

A *high-side float valve* is located at the high-pressure side of the refrigeration system, between the compressor discharge outlet and the inlet of the thermostatic expansion valve. High-side float valves control the flow rate of the refrigerant feed to the evaporator indirectly by maintaining a constant liquid level in the high-pressure float chamber.

Figure 21.9a shows a high-side float valve. The valve pin is linked with a float ball in such a way that any rise in the liquid level in the float chamber opens the valve and allows the liquid refrigerant to pass through the valve to the evaporator. A bypass tube provides a passage for noncondensable gas to escape from the float chamber to the evaporator.

The high-side float valve may be located above the condenser. In a sealed refrigerant circuit, the difference in refrigerant pressure before and after the high-side float valve forces the liquid refrigerant through the valve and into the evaporator.

Because the high-pressure float chamber only holds a small and fixed amount of liquid refrigerant, most of the liquid refrigerant in a reciprocating refrigeration system is stored in the evaporator. Therefore, control

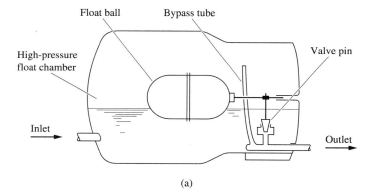

(a)

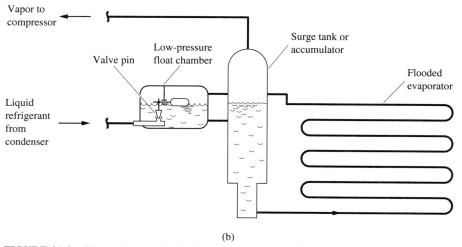

(b)

FIGURE 21.9 Float valves: (a) high-side float valve; and (b) low-side float valve.

of refrigerant charge is essential. Excessive refrigerant charge may cause floodback. Insufficient refrigerant charge may lower system capacity.

A high-side float valve can be used in direct-expansion or flooded evaporators.

The *low-side float valve* is located at the low-pressure side of the refrigeration system. The low-side float valve controls the liquid refrigerant feed to maintain a constant liquid level in the evaporator. Figure 21.9b shows a low-side float valve. As the liquid level in the evaporator drops, the float ball moves downward and opens the float valve wider so that more liquid refrigerant is fed to the evaporator. In small refrigeration systems, the float chamber is often placed directly inside the evaporator or in an accumulator, instead of in a separate low-side float chamber. Operation of the low-side float valve may be continuous or intermittent. Control of the refrigerant charge is not as important for a low-side float valve as for a high-side float valve.

In evaporators with high evaporating rates or limited vapor passages, foaming refrigerant often raises the refrigerant level during operation. Therefore, the float ball should be placed in an appropriate position to determine liquid levels accurately.

Low-side float valves are usually used in flooded coils or evaporators.

Service Valves

SOLENOID VALVES. *Solenoid valves* are widely used in the suction, discharge, and liquid lines, and in places where equipment or components are isolated for maintenance and repair. They are often automatically controlled by the DDC and other controls.

A solenoid valve in refrigerant flow control is shown in Fig. 21.10a. When the electromagnetic or solenoid coil is energized, the plunger and the attached valve pin are lifted upward to open the valve port fully. The valve is closed by gravity or spring action. For medium-size valves, a pilot-operated solenoid valve is often used. In this type of valve, the solenoid coil opens a pilot port, which produces a pressure imbalance across the diaphragm, forces it upward, and opens the main port. When the solenoid coil deenergizes, the plunger closes the pilot port. The pressure across the diaphragm is balanced through an equalizer hole.

MANUAL SHUTOFF VALVES. These valves are used to shut off or fully open the refrigerant flow manually. Figure 21.10b shows a manual shutoff angle valve.

Most manual shutoff valves have packing around the stem. They also include back seat construction: the valve seat closes against a second seat to prevent refrigerant leakage through the valve stem and back seat port.

PRESSURE RELIEF VALVES. In refrigeration systems, refrigerant pressure may rise to levels that could result in an explosion in the event of fire, malfunction, or shutdown. As in water systems, *pressure relief valves* in refrigeration systems are used to re-

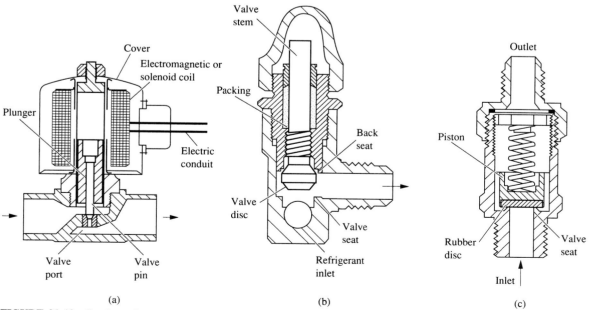

FIGURE 21.10 Service valves: (a) solenoid valve, normally closed; (b) manual shutoff valve (angle type); and (c) pressure relief valve.

lieve the refrigerant pressure when it exceeds a predetermined limit.

A typical spring-loaded pressure relief valve is shown in Fig. 21.10c. It is usually mounted on the liquid line or on the liquid receiver. The pressure required to open the relief valve is adjusted by the spring. Once it is set, the spring is sealed for protection. When the pressure of the refrigerant exerted on the rubber disc is greater than the spring load, the valve port opens, and the refrigerant pressure is relieved through the upper outlet.

REFRIGERANT CHARGE VALVE. A *refrigerant charge valve* is a manually operated shutoff valve used only during refrigerant charge. It is often mounted on the liquid line at the outlet of the condenser or on the outlet side of the receiver.

Check valves for refrigerant flow control are similar in construction and operation to the water valves described in Section 20.3.

21.3 RECIPROCATING REFRIGERATION SYSTEM COMPONENTS AND ACCESSORIES

Compressor Components and Lubrication

The housing of a semihermetic reciprocating compressor, shown in Fig. 10.25, is the enclosure that contains the cylinder, pistons, crankshaft, main bearing, oil sump, crankcase, and hermetic motor, including both the rotor and the stator. The crankcase is the bottom part of the housing of the reciprocating compressor in which the crankshaft is housed.

CYLINDER BLOCK AND PISTON. In most compressors, the cylinder block is integrated with the crankcase and forms a single casting. The crankcase and the cylinder block are usually made of high-grade cast iron. In medium and large reciprocating compressors, premachined cylinder liners or sleeves are often inserted in the crankcase, and can be replaced when they are worn. It costs far less to replace cylinder liners or sleeves than to replace the cylinder block.

The pistons are generally made of aluminum, aluminum alloy, or cast iron. For aluminum pistons, piston rings are often used to prevent leakage of gas refrigerant from the cylinder and to prevent oil from entering the cylinder. The upper piston rings are called *compression rings* and the lower rings are called *oil rings*. For cast-iron pistons, a running clearance of 0.0004 in./in. cylinder diameter provides an effective seal without oil rings.

SUCTION AND DISCHARGE VALVES. The *suction valve* controls the vapor refrigerant entering the cylinder and the *discharge valve* controls the hot gas discharging from the cylinder. Suction and discharge valves are usually made of high-carbon alloy steel or stainless steel. Spring-action ring valves are most extensively used in medium and large compressors. Ring valves are usually heat-treated to the resilience of spring steel, and must be precisely ground to a perfectly flat surface. A defect of 0.001 in. may cause leakage.

In addition, reed valves, either free-floating or clamped at one end, are also used. A reed valve opens under discharged pressure and is closed by spring action. Reed valves are usually limited to small valve ports.

A sufficient valve opening area must be provided to reduce pressure loss. A valve may open up to 0.01 in. during suction or discharge. The maximum velocity of HCFC–22 flowing through valves may reach 150 fps.

OIL LUBRICATION. Oil lubrication is necessary to form a fluid film separating the moving surfaces to protect them from wear and corrosion. Oil is also used as a coolant to carry heat away and cool the refrigerant. It also provides an oil seal between the piston and cylinder and between valve and valve plates. In refrigeration systems, mineral oils are used for lubrication. Halocarbon refrigerants are usually oil miscible so that the refrigerant–oil mixture can return to the compressor. Oil must be free of solids such as wax, must be chemically stable, and must have a suitable viscosity. In reciprocating compressors using HCFC–22 and halogenated refrigerants, a Saybolt Seconds Universal (SSU) viscosity between 150 and 300 is recommended by *ASHRAE Handbook Refrigeration 1990, Systems and Applications*. Kinematic viscosity of 1 ft^2/s is approximately equal to 4.3×10^5 SSU.

In medium and large reciprocating compressors, forced-feed lubrication through a positive displacement vane or gear oil pump is usually used. Lubricant is fed to the main bearing surfaces and crankshaft through oil passages. An oil mist formed by the crankshaft rotation lubricates the cylinder wall and piston pins. The oil pump is often mounted on one end of the crankshaft. An oil strainer submerged directly in the oil sump is connected to the suction intake of the oil pump. Because the vapor refrigerant in the crankcase is always at the suction pressure, the pressure lift of the oil pump is equal to the difference of the discharge pressure at the oil pump and the suction pressure.

In small reciprocating compressors, splash lubrication may be used. Splash lubrication uses the rotation of the crankshaft and the connecting rod to splash oil onto the bearing surfaces, cylinder walls, and other moving parts. Often, oil passages, oil rings, and other devices are used to increase oil splashing. An adequate oil level in the crankcase is necessary for satisfactory operation. Splash lubrication is simple. Its disadvantage is that bearing clearance must be larger than normal, which increases noise during operation.

For HCFC–22 and other halocarbon refrigerants, lubricating oil for the suction and discharge valves must be miscible with the refrigerant. Refrigerant piping design must accommodate the return of oil from evaporators and condensers. Oil should be separated from the suction vapor at the compressor and returned to the crankcase.

CRANKCASE HEATER. When the reciprocating compressor is installed outdoors, refrigerants may migrate to the compressor from the indoor evaporator during the shutdown period of the refrigeration system. This is mainly because the temperature and refrigerant pressure in the indoor evaporator are higher than in the outdoor compressor. If a considerable amount of refrigerant accumulates in the crankcase, it will dilute the lubrication oil. When the compressor starts to operate again, the foaming oil–refrigerant mixture may form a slug resulting in a serious loss of oil from the crankcase. Slugs entering the compression chamber may also damage the valves. For medium and large compressors, a built-in *crankcase heater* is often installed to keep the oil temperature in the crankcase 15 to 25°F higher than in the rest of the system and thus prevent refrigerant migration.

Accessories

LIQUID RECEIVER. If all the refrigerant in the system cannot be condensed and stored in the condenser during the shutdown period, a high-pressure side *liquid receiver* is needed to provide auxiliary refrigerant storage space, as shown in Fig. 21.11a. Consequently, if a shell-and-tube water-cooled condenser is equipped in a refrigeration system, a liquid receiver is not necessary. When an evaporative condenser is used, a liquid receiver is often installed to supplement its small storage capacity.

The liquid receiver is usually a cylindrical tank made of steel sheets. The top of the tank is connected to the liquid line from the condenser. It often has a liquid discharge line connected to the evaporator by a shutoff valve. A vent pipe combined with a fusible plug is mounted on the receiver to prevent the liquid

receiver from bursting when exposed to high temperatures (as in a fire). In most installations, it is common to charge the refrigerant in the receiver 80 percent full during shutdown period, and the remaining 20 percent is used to allow for liquid expansion.

To provide space for refrigerant storage during shutdown and for liquid expansion, saturated pressure exists at the interface of the vapor and liquid refrigerant inside a through-type liquid receiver. No subcooling occurs at the outlet of the liquid receiver unless a subcooling coil or other subcooling device is provided after the receiver, such as in an evaporative consenser. Because the subcooled liquid is not exposed to gaseous refrigerant in a purge-type liquid receiver, the liquid remains subcooled after flowing through it.

LIQUID-SUCTION HEAT EXCHANGER. A *liquid-suction heat exchanger* is usually mounted across the suction and liquid lines when the hot-gas discharge temperature of the refrigerant (such as HFC–134a) does not damage the lubrication oil or discharge valves. For ammonia and usually HCFC–22, the superheat of the suction vapor in the heat exchanger may raise the discharge temperature of the hot gas from the compressor to an unacceptable level, so liquid-suction heat exchangers are not used for these refrigerants.

The purposes of a liquid-suction heat exchanger are as follows:

- To subcool the liquid refrigerant and increase the refrigerant effect
- To boil liquid refrigerant in the suction line and prevent liquid refrigerant from entering the compressor
- To reduce the flash gas in the liquid line and ensure maximum capacity for the thermostatic expansion valve

Liquid-suction heat exchangers can be constructed as shell-and-coil, shell-and-tube, parallel combined of liquid and suction lines, or tube-in-tube heat exchangers. In a typical shell-and-tube heat exchanger, as shown in Fig. 21.11, the hot liquid runs through the shell. The cold suction gas flows into the inner tube and penetrates the perforations at high velocity. Liquid refrigerant is subcooled between 7 and 15°F in the heat exchanger where the heat is absorbed by the cold vapor. The pressure drop in the suction vapor should not exceed 2 psi when it flows through the heat exchanger.

FILTER–DRYER AND STRAINER. Moisture may freeze in the expansion valve, especially in a low-temperature system, and impair expansion valve operation. Foreign matter in the refrigeration system may

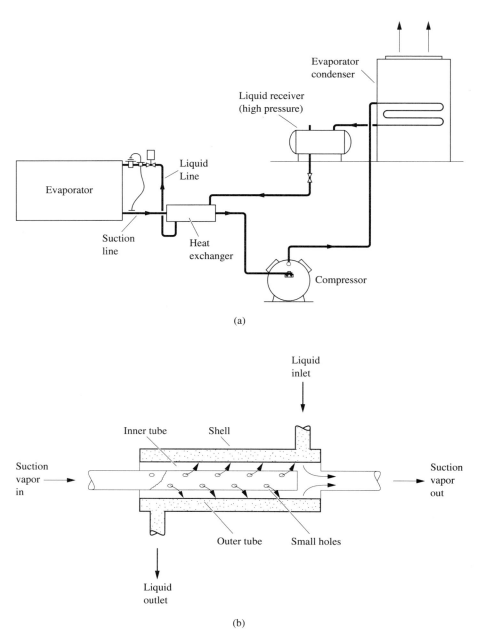

(a)

(b)

FIGURE 21.11 Liquid-suction heat exchanger: (a) schematic diagram; and (b) heat exchanger.

also damage the expansion valve, control valves, and compressor. Moisture and foreign matter must be removed from the refrigeration system by means of a filter–dryer and strainer.

Figure 21.12a shows a *filter–dryer*. It contains a molded, porous core made of material with high moisture affinity, such as activated alumina, and acid-neutralizing agents to remove moisture and foreign matter from the refrigerant. The core should be replaced after a certain period of operation. The filter–dryer is usually installed at the liquid line immediately before the expansion valve and shutoff valve.

The *strainer* contains a fine-mesh screen to remove copper filings, dirt, and other debris that may be in the system at startup. The construction of a strainer is similar to that of a filter–dryer. Instead of a porous core, there is a fine-mesh wire screen. The strainer is often installed in the suction line just before the compressor.

SIGHT GLASS. A *sight glass* is a small glass port used to observe the condition of refrigerant flow in the liquid line. Sight glasses are available with one or two ports. A double-port sight glass, as shown in

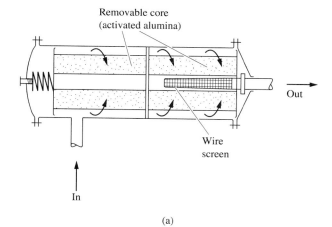

(a)

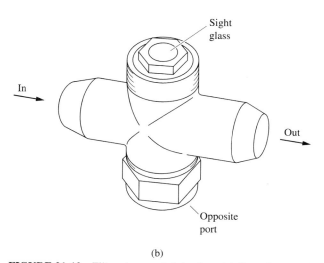

(b)

FIGURE 21.12   Filter–dryer and sight glass: (a) filter–dryer; and (b) sight glass.

Fig. 21.12b, is recommended because light enters the opposite port and makes observation easier. The sight glass is located just before the expansion valve. Bubbles seen through the sight glass indicate the presence of flash gas instead of liquid refrigerant. The presence of flash gas always indicates that evaporator capacity is reduced because of a shortage of refrigerant or insufficient subcooling.

A moisture-detecting chemical is built into the sight chamber behind the sight glass. If excessive moisture is present in HCFC–22, the chemical turns pink. If the moisture is below the safe limit, the color is green.

In addition to these accessories, there are also pressure gauges to indicate suction pressure, discharge pressure, and oil pressure. A liquid level indicator with gauge glasses is used to observe the liquid level in the receiver. It is often installed on the receiver or

on the liquid line. A hot gas muffler may be used to smooth the pulsation of hot gas discharged from the compressor. The muffler should be located as close to the compressor as possible.

Most modern reciprocating compressors minimize oil pumping. Except in systems where oil removal is necessary, as in low-temperature systems, or for refrigerant that is not oil miscible, oil separators are seldom used.

21.4   REFRIGERANT PIPING

Copper Tubing

Refrigerant pipes transport refrigerant through the compressor, condenser, expansion valve, and evaporator to provide the required refrigeration effect. For halocarbon refrigerants, the refrigerant pipes are usually made of copper. If ammonia is used as the refrigerant, the pipes are always made of steel.

As listed in Table 20.2, three types of copper tubing are used for refrigerant piping. Type K is heavy-duty and Type M is lightweight. Type L is the standard copper tubing most widely used in refrigeration systems.

Copper tubing installed in refrigeration systems should be entirely free of dirt, scale, and oxides. The open ends of clean new tubes should be capped to keep contaminants out.

Piping Design for Reciprocating Refrigeration Systems

Packaged units with factory-assembled and -integrated refrigeration systems have become more popular, and air-cooled reciprocating packaged units or chillers contain a reciprocating compressor, liquid coolers, air-cooled condensers, expansion valves, and controls in one package. However, many field built-up or split systems need proper refrigeration piping design to provide adequate cooling for air conditioning and industrial applications.

A proper refrigeration piping design must satisfy the following requirements:

- Transport the required amount of refrigerant to the evaporator, compressors, condenser, and throttling devices.
- Provide optimum pressure drop in the refrigerant lines, that is, the most economical maximum pressure drop in the suction, discharge, and liquid lines. Current practice limits the maximum pressure drop in suction, discharge, and liquid lines

corresponding to a change of saturated temperature ΔT_{suc} as follows:

	ΔT_{suc}, °F	HCFC–22 pressure drop, ψ
Suction line	2	2.91 at T_{suc} = 40°F
Discharge line	1	3.05 at T_{con} = 105°F
Liquid line	1	3.05 at T_{con} = 105°F

- For halocarbon refrigeration systems, oil should be miscible with liquid refrigerant, and oil entrained in the refrigerant should be returned to the crankcase of the compressor to maintain an adequate level for proper lubrication. Refrigerant piping design must accommodate the flow of both refrigerant and oil. Oil does not mix with gaseous refrigerant. Transportation of entrained oil in hot gas or vapor refrigerant is accomplished by ensuring a minimum refrigerant velocity not less than 500 fpm in horizontal refrigerant lines and not less than 1000 fpm in vertical refrigerant risers. Horizontal refrigerant lines must be pitched in the direction of refrigerant flow.

- Flashing of liquid refrigerant must not occur before the expansion valve. Adequate subcooling should be provided to offset the pressure drop of the liquid line and its accessories.

- Piping should be configured to prevent *liquid slugging*, in which liquid refrigerant, oil, or a combination of both floods back to the compressor.

- Maintain a clean and dry refrigeration piping system.

Size of Copper Tube, Refrigeration Load, and Pressure Drop

During the sizing of refrigerant piping, the system refrigeration load q_{rl}, in tons of refrigeration, is a known value.

Because the refrigerant velocity $v = \dot{m}_r/(\rho A_i)$, the refrigerant mass flow rate $\dot{m}_r = q_{rl}\Delta h$ and the inner surface area of the copper tubing $A_i = \pi D_i^2/4$. Here, ρ represents the density of the refrigerant and Δh is the enthalpy difference of the refrigerant leaving and entering the evaporator. For a specific size of copper tubing, the relationship between the outside diameter D and inner diameter D_i is a fixed value. From Eq. (8.35), the pressure drop in the refrigerant piping can be expressed as

$$D = K \left\{ \frac{[8/(\pi^2 g_c)](f L q_{rl}^2)}{\rho(\Delta h)^2 \Delta p} \right\}^{1/5}$$

$$= C_{pipe} \left(\frac{f L q_{rl}^2}{\rho(\Delta h)^2 \Delta p} \right)^{1/5} \qquad (21.4)$$

In Eq. (21.4), constant $C_{pipe} = K[8/(\pi^2 g_c)]^{1/5}$. The friction factor of refrigerant f can be calculated by the Colebrook equation, as shown in Eq. (8.39). If the absolute roughness of the copper tube is taken as 0.000005 ft, f has only a minor influence on outside diameter of the copper tubing D. Constant K takes into account the difference between D and D_i for a specific size of copper tubing.

The size or outside diameter of copper tubing D is determined by the following relationships:

- Directly proportional to the length of the refrigeration pipe L to the $\frac{1}{5}$ power
- Directly proportional to refrigeration load q_{rl} to the $\frac{2}{5}$ power
- Inversely proportional to the density of the suction vapor or hot gas ρ, which is affected by suction and condensing temperature, respectively
- Inversely proportional to the enthalpy difference Δh (which is closely related to the suction temperature T_{suc} and condensing temperature T_{con}) to the $\frac{2}{5}$ power
- Inversely proportional to the maximum allowable pressure drop Δp to the $\frac{1}{5}$ power

Sizing Procedure

Refrigerant piping design and size should be determined as follows:

1. Make an optimum refrigerant piping layout and measure the length of the piping.
2. Find the correction factor of the refrigeration load from Table 21.1, based on the actual suction and condensing temperatures.

TABLE 21.1
Correction factors for suction, discharge, and liquid lines for HCFC–22 or T_{suc} other than 40°F and T_{con} other than 105°F.

	Correction factors for T_{suc} other than 40°F							
T_{suc}, °F	50	40	30	20	10	0	−10	−20
$F_{suc.s}$	1.20	1.00	0.84	0.70	0.58	0.47	0.38	0.30
$F_{dis.s}$	1.01	1.00	0.99	0.97	0.96	0.95	0.93	0.91

	Correction factors for T_{con} other than 105°F							
T_{con}, °F	80	90	100	105	110	120	130	140
$F_{suc.c}$	1.11	1.07	1.03	1.00	0.97	0.90	0.86	0.80
$F_{dis.c}$	0.79	0.88	0.95	1.00	1.04	1.10	1.18	1.26
$F_{liq.c}$	1.12	1.07	1.02	1.00	0.98	0.93	0.88	0.83

Note: $F_{liq.s} = F_{dis.s}$

3. Estimate the equivalent length of refrigerant piping including the pipe fittings and accessories L_{eq}, in ft. It is usually 1.5 to 5 times the measured straight length of the piping, depending on the number of fittings and accessories.

4. Based on the corrected refrigeration load and L_{eq}, determine the tentative diameter of the copper tubing from refrigeration load vs equivalent length $q_{rl}-L_{eq}$ charts (see Figs. 21.14–21.16). All suction, discharge, and liquid line $q_{rl}-L_{eq}$ charts in this chapter are plotted based on the data given in *ASHRAE Handbook 1990, Refrigeration Systems and Applications*.

5. Find the equivalent length of the pipe fittings and accessories listed in Tables 21.2 and 21.3, based on the tentative diameter of the copper tubing. Recalculate the equivalent length of the refrigeration piping including the pipe fittings and accessories.

6. Check the size of the copper tubing from the $q_{rl}-L_{eq}$ charts based on the calculated L_{eq}. If it is equal to the tentative value, the tentative diameter is the required diameter. If not, adjust the tentative diameter.

7. Determine whether the size of the refrigeration pipe meets the oil entrainment requirement when the refrigeration load is reduced to its minimum load at part-load operation. If it does not meet the requirement, add a double riser.

Suction Lines

The suction line is the most critical refrigerant line in the piping design. Vapor refrigerant and entrained oil in the evaporator should return to the compressor free of liquid slugs.

OIL TRAP AND PIPING PITCH. In the suction line, an oil trap should be installed at the bottom of a single riser, as shown in Fig. 21.13a. The sensing bulb of the thermostatic expansion valve is often installed between the coil outlet and the oil trap. This oil trap helps to prevent liquid accumulation under the sensing bulb while the compressor is turned off. The accumulated liquid under the sensing bulb causes abnormal operation of the expansion valve when the compressor is started again. An oil trap consists of two elbows connected with a short stub of pipe.

TABLE 21.2
Pressure drop of valves expressed in equivalent length of straight pipe, in feet, for refrigeration systems.

Nominal pipe or tube size, in.	Globe*	60°-Y	45°-Y	Angle*	Gate[λ]	Swing check[$\lambda\lambda$]	Lift check
$\frac{3}{8}$	17	8	6	6	0.6	5	
$\frac{1}{2}$	18	9	7	7	0.7	6	Globe &
$\frac{3}{4}$	22	11	9	9	0.9	8	vertical lift
1	29	15	12	12	1.0	10	same as
$1\frac{1}{4}$	38	20	15	15	1.5	14	globe
$1\frac{1}{2}$	43	24	18	18	1.8	16	valves[§]
2	55	30	24	24	2.3	20	
$2\frac{1}{2}$	69	35	29	29	2.8	25	
3	84	43	35	35	3.2	30	
$3\frac{1}{2}$	100	50	41	41	4.0	35	
4	120	58	47	47	4.5	40	
5	140	71	58	58	6	50	
6	170	88	70	70	7	60	
8	220	115	85	85	9	80	
10	280	145	105	105	12	100	Angle lift
12	320	165	130	130	13	120	same as
14	360	185	155	155	15	135	angle
16	410	210	180	180	17	150	valve
18	460	240	200	200	19	165	
20	520	275	235	235	22	200	
24	610	320	265	265	25	240	

Screwed, welded, flanged, and flared connections
Losses are for all valves in fully open position.
*These losses do not apply to valves with needle point type seats.
[λ] Regular and short pattern plug cock valves, when fully open, have same loss as gate valve.
For valve losses of short pattern plug cocks above 6 in. check manufacturer.
[$\lambda\lambda$] Losses also apply to the in-line, ball type check valve.
[§]For Y pattern globe lift check valve with seat approximately equal to the nominal pipe diameter, use values of 60°. Y valve for loss.

Source: ASHRAE Handbook 1990, Refrigeration Systems and Applications. Reprinted with permission.

TABLE 21.3
Pressure drop of pipe fittings expressed in equivalent length of straight pipe, in feet, for refrigeration systems.

Nominal pipe or tube size, in.	Smooth bend elbows						Smooth bend tees			
							Flow through branch	Straight-through flow		
	90° Std*	90°Long rad.†	90° Street*	45° Std*	45° Street*	180° Std*		No reduction	Reduced $\frac{1}{4}$	Reduced $\frac{1}{2}$
$\frac{3}{8}$	1.4	0.9	2.3	0.7	1.1	2.3	2.7	0.9	1.2	1.4
$\frac{1}{2}$	1.6	1.0	2.5	0.8	1.3	2.5	3.0	1.0	1.4	1.6
$\frac{3}{4}$	2.0	1.4	3.2	0.9	1.6	3.2	4.0	1.4	1.9	2.0
1	2.6	1.7	4.1	1.3	2.1	4.1	5.0	1.7	2.2	2.6
$1\frac{1}{4}$	3.3	2.3	5.6	1.7	3.0	5.6	7.0	2.3	3.1	3.3
$1\frac{1}{2}$	4.0	2.6	6.3	2.1	3.4	6.3	8.0	2.6	3.7	4.0
2	5.0	3.3	8.2	2.6	4.5	8.2	10.0	3.3	4.7	5.0
$2\frac{1}{2}$	6.0	4.1	10.0	3.2	5.2	10.0	12.0	4.1	5.6	6.0
3	7.5	5.0	12.0	4.0	6.4	12.0	15.0	5.0	7.0	7.5
$3\frac{1}{2}$	9.0	5.9	15.0	4.7	7.3	15.0	18.0	5.9	8.0	9.0
4	10.0	6.7	17.0	5.2	8.5	17.0	21.0	6.7	9.0	10.0
5	13.0	8.2	21.0	6.5	11.0	21.0	25.0	8.2	12.0	13.0
6	16.0	10.0	25.0	7.9	13.0	25.0	30.0	10.0	14.0	16.0
8	10.0	13.0	–	10.0	–	33.0	40.0	13.0	18.0	20.0
10	25.0	16.0	–	13.0	–	42.0	50.0	16.0	23.0	25.0
12	30.0	19.0	–	16.0	–	50.0	60.0	19.0	26.0	30.0
14	34.0	23.0	–	18.0	–	55.0	68.0	23.0	30.0	34.0
16	38.0	26.0	–	20.0	–	62.0	78.0	26.0	35.0	38.0
18	42.0	29.0	–	23.0	–	70.0	85.0	29.0	40.0	42.0
20	50.0	33.0	–	26.0	–	81.0	100.0	33.0	44.0	50.0
24	60.0	40.0	–	30.0	–	94.0	115.0	40.0	50.0	60.0

Screwed, welded, flanged, flared, and brazed connections
*R/D approximately equal to 1.
†R/D approximately equal to 1.5.
Source: ASHRAE Handbook 1990, Refrigeration Systems and Applications. Reprinted with permission.

If the evaporator is located above the compressor, the horizontal section of suction line should be pitched toward the compressor at least $\frac{1}{2}$ in. per each 10 ft of pipe length.

DOUBLE RISER. When the compressor is located above the evaporator, if the compressor is equipped with a capacity control or unloading mechanism to reduce the refrigeration load q_{rl} so that the minimum q_{rl} in the vertical riser is less than the values listed in Tables 21.4 and 21.5, a double-riser is needed instead of a single vertical riser, as shown in Fig. 21.13b.

A *double riser* consists of a small riser, a large riser, and an oil trap that connects the lower ends of the risers. During full-load operation, the vapor velocity in both risers is high enough to entrain oil with the gaseous refrigerant, and there is no oil in the oil trap. However, when the q_{rl} and vapor velocity drop so that oil is no longer entrained in the large riser at part-load operation because of cylinder unloading or on-off control, oil accumulates in the oil trap. This oil blocks the vapor passage to the large riser. The vapor velocity in the small riser therefore increases to a velocity high enough to carry the entrained oil to the compressor.

If the vertical riser has a rise greater than 25 ft, an additional oil trap for each 25-ft rise is required. Thus, the upper and lower sections are drained separately, and oil leaves the traps in quantities that will not damage the compressor.

MAXIMUM PRESSURE DROP. In a suction line using HCFC–22 as the refrigerant, a pressure drop equivalent to a 2°F change in saturation temperature

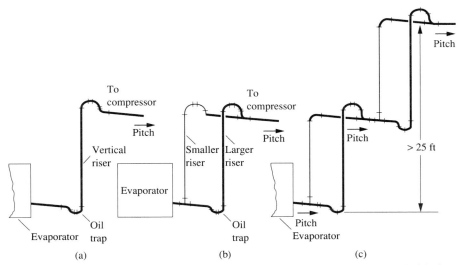

FIGURE 21.13 Single riser, double riser, and additional oil trap: (a) single riser; (b) double riser; and (c) additional oil trap.

TABLE 21.4
Minimum refrigeration load for oil entrainment up suction risers of type L copper tubing, in tons.

Refrigerant	Saturated temperature, °F	Suction gas temperature, °F	Pipe OD, in.											
			0.500	0.625	0.750	0.875	1.123	1.375	1.625	2.125	2.625	3.125	3.625	4.125
			Area, in²											
			0.146	0.233	0.348	0.484	0.825	1.256	1.780	3.094	4.770	6.812	9.213	11.970
HCFC–22	−40.0	−30.0	0.067	0.119	0.197	0.298	0.580	0.981	1.52	3.03	5.20	8.12	11.8	16.4
		−10.0	0.065	0.117	0.194	0.292	0.570	0.963	1.49	2.97	5.11	7.97	11.6	16.1
		10.0	0.066	0.118	0.195	0.295	0.575	0.972	1.50	3.00	5.15	8.04	11.7	16.3
	−20.0	−10.0	0.087	0.156	0.258	0.389	0.758	1.28	1.98	3.96	6.80	10.6	15.5	21.5
		10.0	0.085	0.153	0.253	0.362	0.744	1.26	1.95	3.88	6.67	10.4	15.2	21.1
		30.0	0.086	0.154	0.254	0.383	0.747	1.26	1.95	3.90	6.69	10.4	15.2	21.1
	0.0	10.0	0.111	0.199	0.328	0.496	0.986	1.63	2.53	5.04	8.66	13.5	19.7	27.4
		30.0	0.108	0.194	0.320	0.484	0.942	1.59	2.46	4.92	8.45	13.2	19.2	26.7
		50.0	0.109	0.195	0.322	0.486	0.946	1.60	2.47	4.94	8.48	13.2	19.3	26.8
	20.0	30.0	0.136	0.244	0.403	0.608	1.18	2.00	3.10	6.18	10.6	16.6	24.2	33.5
		50.0	0.135	0.242	0.399	0.603	1.17	1.99	3.07	6.13	10.5	16.4	24.0	33.3
		70.0	0.135	0.242	0.400	0.605	1.18	1.99	3.08	6.15	10.6	16.5	24.0	33.3
	40.0	50.0	0.167	0.300	0.495	0.748	1.46	2.46	3.81	7.60	13.1	20.4	29.7	41.3
		70.0	0.165	0.296	0.488	0.737	1.44	2.43	3.75	7.49	12.9	20.1	29.3	40.7
		90.0	0.165	0.296	0.488	0.738	1.44	2.43	3.76	7.50	12.9	20.1	29.3	40.7

Note: The capacity in tons is based on 90°F liquid temperature and superheat as indicated by the listed temperatures. For other liquid line temperatures, use correction factors in the table below.

	Liquid temperature, °F								
Refrigerant	50	60	70	80	100	110	120	130	140
HCFC–22	1.17	1.14	1.10	1.06	0.98	0.94	0.89	0.85	0.80

Source: ASHRAE Handbook 1990, Refrigeration Systems and Applications. Reprinted with permission.

TABLE 21.5
Minimum refrigeration load for oil entrainment up hot-gas risers of type L copper tubing, in tons.

			Pipe OD, in.										
			0.500	0.625	0.750	0.875	1.123	1.375	1.625	2.125	2.625	3.125	3.625	4.125
		Suction gas temperature, °F	Area, in²											
Refrigerant	Saturated temperature, °F		0.146	0.233	0.348	0.484	0.825	1.256	1.780	3.094	4.770	6.812	9.213	11.970
HCFC–22	80.0	110.0	0.235	0.421	0.695	1.05	2.03	3.46	5.35	10.7	18.3	28.6	41.8	57.9
		140.0	0.223	0.399	0.659	0.996	1.94	3.28	5.07	10.1	17.4	27.1	39.6	54.9
		170.0	0.215	0.385	0.635	0.960	1.87	3.16	4.89	9.76	16.8	26.2	38.2	52.9
	90.0	120.0	0.242	0.433	0.716	1.06	2.11	3.56	5.50	11.0	18.9	29.5	43.0	59.6
		150.0	0.226	0.406	0.671	1.01	1.97	3.34	5.16	10.3	17.7	27.6	40.3	55.9
		180.0	0.216	0.387	0.540	0.956	1.88	3.18	4.92	9.82	16.9	26.3	38.4	53.3
	100.0	130.0	0.247	0.442	0.730	1.10	2.15	3.83	5.62	11.2	19.3	30.1	43.9	60.8
		160.0	0.231	0.414	0.884	1.03	2.01	3.40	5.26	10.5	18.0	28.2	41.1	57.0
		190.0	0.220	0.394	0.650	0.982	1.91	3.24	3.00	9.96	17.2	26.8	39.1	54.2
	110.0	140.0	0.251	0.451	0.744	1.12	2.19	3.70	5.73	11.4	19.6	30.6	44.7	62.0
		170.0	0.235	0.421	0.693	1.05	2.05	3.46	3.35	10.7	18.3	28.6	41.8	57.9
		200.0	0.222	0.399	0.658	0.994	1.94	3.28	5.06	10.1	17.4	27.1	39.5	54.8
	120.0	150.0	0.257	0.460	0.760	1.15	2.24	3.78	5.85	11.7	20.0	31.3	45.7	63.3
		180.0	0.239	0.428	0.707	1.07	2.08	3.51	5.44	10.8	18.6	29.1	42.4	58.9
		210.0	0.225	0.404	0.666	1.01	1.96	3.31	5.12	10.2	17.6	27.4	40.0	55.5

Notes: The capacity in tons is based on a saturated suction temperature of 20°F with 15°F superheat at the indicated saturated condensing temperature with 15°F subcooling. For other saturated suction temperatures with 15°F superheat, use the following correction factors:

Saturated suction temperature, °F	−40	−20	0	40
Correction factor	0.88	0.95	0.96	1.04

Source: ASHRAE Handbook 1990, Refrigeration Systems and Applications. Reprinted with permission.

between 40 and 50°F is about 3 psi. When HFC–134a is used as the refrigerant, a 2°F equivalent is about 2 psi, and if ammonia R–717 is used as the refrigerant, a 1°F equivalent is about 1.5 psi.

SUCTION LINE SIZING CHART. Figure 21.14 shows the q_{rl}–L_{eq} suction line sizing chart. The abscissa of this chart is the refrigeration load q_{rl} that the evaporated refrigerant in the evaporator can provide at a corresponding mass flow rate, in tons of refrigeration, at 40°F suction and 105°F condensing temperature. The ordinate is the equivalent pipe length L_{eq}, in ft. Table 21.4 indicates the minimum refrigeration load in the suction riser so that the suction vapor velocity v_{suc} inside the copper tubing can still carry the oil.

Suction temperature is the saturation temperature of the refrigerant at the inlet of the compressor. It is usually 2°F lower than the T_x in the evaporator or the DX-coil, depending on the pressure drop of the suction line. When suction temperature $T_{suc} \neq 40°F$, and the condensing temperature $T_{con} \neq 105°F$, the corrected refrigeration load $q_{rl \cdot c}$, in tons, for sizing the suction line can be found by dividing q_{rl} by correction

factors $F_{suc \cdot s}$ and $F_{suc \cdot c}$, as listed in Table 21.1, that is,

$$q_{rl.c} = \frac{q_{rl}}{F_{suc.s}F_{suc.c}} \qquad (21.5a)$$

where $F_{suc.s}$ = correction factor for the refrigeration load in the suction line when $T_{suc} \neq 40°F$

$F_{suc.c}$ = correction factor for the refrigeration load in the suction line when $T_{con} \neq 105°F$

From Eq. (4.24), refrigeration load q_{rl}, in tons, is given as

$$q_{rl} = \dot{m}_r(h_{lv} - h_{en}) = \dot{V}_p \rho \eta_v (h_{lv} - h_{en})$$

where ρ = density of suction vapor, lb/ft³
η_v = volumetric efficiency

Here, $\dot{V}_p$ represents the piston displacement, which will be discussed later in this chapter. It is a fixed value for a specific compressor at a certain speed. Volumetric efficiency η_v is affected slightly by the compression ratio p_{dis}/p_{suc}. Therefore, correction factor $F_{suc.s}$ for the suction line is a combined index

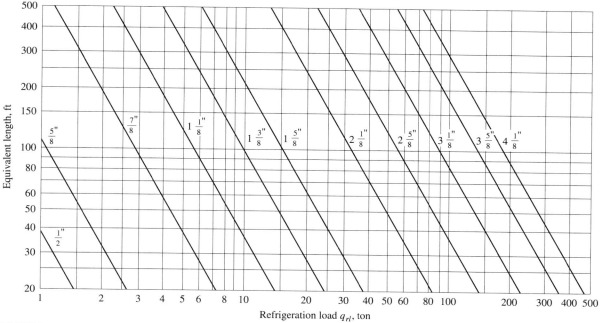

FIGURE 21.14 Suction line sizing q_{rl}–L_{eq} chart for HCFC–22 at 40°F suction temperature and 105°F condensing temperature (type L copper tube, outside diameter, 2°F change in saturated temperature).

that takes account the effect of $\rho(h_{lv} - h_{en})$ when $T_{suc} \neq 40°F$. If $T_{suc} < 40°F$, $F_{suc.s} < 1$ and $q_{rl.c} > q_{rl}$.

The correction factor $F_{suc.c}$ for the suction line takes into account the effect of $(h_{lv} - h_{en})$ when $T_{con} \neq 105°F$. If $T_{con} < 105°F$, $F_{suc.c} > 1$ and $q_{rl.c} < q_{rl}$.

SUCTION LINE SIZING. Consider the size of the suction line of a reciprocating refrigeration system in a rooftop packaged unit with a refrigeration load of 40 tons at 45°F suction temperature and 115°F condensing temperature with a subcooling of 10°F. There are two refrigeration circuits and two compressors, each with a capacity of 20 tons. Each compressor can be unloaded to 50 percent of its full-load capacity.

The suction line has a total pipe length (including horizontal lines and vertical risers of 3 ft) of 30 measured ft. There are also a manual shutoff angle valve and 10 short-radius elbows.

1. From Table 21.1, for a suction temperature of 45°F and a condensing temperature of 115°F, $F_{suc.s}$ is 1.10 and $F_{suc.c} = 0.88$. The corrected refrigeration load q_{rl} for the suction line is therefore $20/(1.1 \times 0.88) = 20.7$ tons.

2. Assume an equivalent length of the suction line L_{eq} of 90 ft. From Fig. 21.14, for a refrigeration load of 20.7 tons and $L_{eq} = 90$ ft, the next higher value of the tentative outside diameter of the suction line is $2\frac{1}{8}$ in (nominal diameter of 2 in.).

3. Recalculate the equivalent length, based on a copper tubing outside diameter of $2\frac{1}{8}$ in. The pressure drops of the refrigerant piping fittings and the accessories expressed as equivalent length of pipe are listed in Table 21.2. The calculated total equivalent length L_{eq} is:

10 short radius elbows	$10 \times 5 = 50$ ft
1 angle valve	$1 \times 24 = 24$ ft
Measured length of pipe	30 ft
Total	104 ft

4. From the capacity–L_{eq} chart, at a $q_{rl} = 20.7$ tons and an $L_{eq} = 104$ ft, the diameter of copper tubing is the same as the tentative diameter $2\frac{1}{8}$ in. Consequently, $2\frac{1}{8}$ in. should be the size of the suction line.

5. The minimum capacity of the compressor in each refrigeration circuit is $0.5 \times 20.7 = 10.4$ tons. From Table 21.4, when the liquid temperature is $115 - 10 = 105°F$, the correction factor at 105°F is 0.96. The allowable minimum refrigeration load for a suction riser with entrained oil is $7.6 \times 0.96 = 7.3$ tons, which is smaller than 10.4 tons. Therefore, a double riser is not needed.

Discharge Line (Hot-Gas Line)

The design and sizing of the discharge line (hot-gas line) is similar to that of the suction line. Oil return

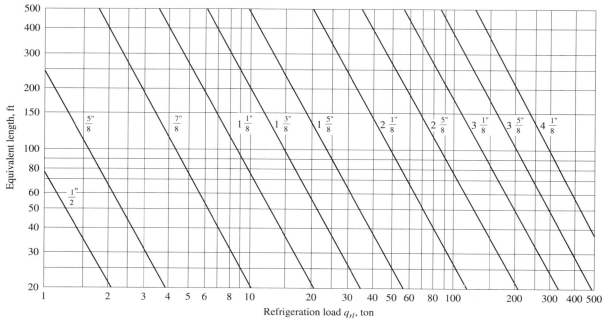

FIGURE 21.15   Discharge line sizing q_{rl}–L_{eq} chart for HCFC–22 at 40°F suction temperature and 105°F condensing temperature (type L copper tube, outside diameter, 1°F change in saturated temperature).

and optimum pressure drop are the primary considerations.

In the discharge line, oil does not mix with gaseous refrigerant. It is entrained in the gaseous refrigerant at the discharge outlet of the compressor as oil mist. The horizontal pipe run should be pitched away from the compressor at a pitch not less than $\frac{1}{2}$ in. for each 10 ft of pipe run. The minimum refrigeration loads for hot-gas risers to carry entrained oil to the condenser are listed in Table 21.5. There should be an oil trap at the bottom of the vertical riser. When the compressor is unloaded to part-load operation, if the minimum q_{rl} is less than that listed in Table 21.5, a double riser should be used. An additional oil trap should be added for each 25 ft of pipe rise.

In the discharge line, the saturated temperature difference corresponding to the maximum pressure drop is 1°F for halocarbon refrigerants. The maximum pressure drop is about 3 psi for both HCFC–22 and HFC–134a. For ammonia R–717, a 1°F saturated temperature difference is equivalent to about 4 psi.

The procedure for sizing the discharge line is similar to that for the suction line. Fig. 21.15 shows the discharge line q_{rl}–L_{eq} chart for HCFC–22 at 40°F suction temperature and 105°F condensing temperature with a pressure drop corresponding to a 1°F change in saturated temperature difference.

If the suction temperature T_{suc} is not 40°F and the condensing temperature T_{con} is not 105°F, the corrected refrigeration load is

$$q_{rl.c} = \frac{q_{rl}}{F_{dis.s}F_{dis.c}} \qquad (21.5b)$$

Correction factor $F_{dis.s}$ takes into account the effect of $(h_{lv} - h_{en})$ on $\dot{m}_r$ in the discharge line when $T_{suc} \neq$ 40°F. If $T_{suc} <$ 40°F, $F_{dis.s} <$ 1 and $q_{rl.c} > q_{rl}$.

Correction factor $F_{dis.c}$ takes into account the combined effect of density of the hot gas ρ and enthalpy difference $(h_{lv} - h_{en})$ on $\dot{m}_r$ when $T_{con} \neq$ 105°F. If $T_{con} <$ 105°F, $F_{dis.c} <$ 1 and $q_{rl.c} > q_{rl}$.

Table 21.1 lists the correction factors $F_{dis.s}$ and $F_{dis.c}$ for HCFC–22 for the discharge line.

Liquid Line

Because oil is partially miscible with liquid refrigerant such as HCFC–22, HFC–134a is miscible with synthetic oil, and oil is separated from ammonia after the compressor, oil return is not a problem in the liquid line. Therefore, the primary considerations in liquid line design and sizing are the optimum pressure drop and the prevention of flashing of liquid refrigerant into vapor before the expansion valve.

Current practice is to limit the maximum pressure drop in the liquid line corresponding to a change of 1°F saturated temperature. For HCFC–22, it is about 3 psi, for HFC–134a, it is about 2.5 psi, and for ammonia R–717, it is about 3.5 psi.

In order to prevent the flashing of liquid to vapor, liquid refrigerant must be subcooled before it is discharged to the liquid line. Recently designed air-cooled condensers for reciprocating refrigeration systems can provide a subcooling of 10 to 20°F. The liquid refrigerant should be in a subcooled state when it enters the expansion valve.

If a receiver is installed after the condenser, the liquid line between the condenser and receiver should be sized so that liquid flows from the condenser to the receiver and the gas flows in opposite direction. This section of liquid line is usually sized based on a liquid velocity of 100 fpm. According to *ASHRAE Handbook 1990, Refrigeration Systems and Applications,* the relationship between q_{rl} and the outside diameter OD of the copper tubing for HCFC–22, based on a liquid velocity of 100 fpm, is as follows:

OD Type L, in. $\frac{1}{2}$ $\frac{5}{8}$ $1\frac{1}{8}$ $1\frac{5}{8}$ $2\frac{1}{8}$ $2\frac{5}{8}$ $3\frac{1}{8}$
q_{r1}, ton 2.3 3.7 13.2 28.5 49.6 76.5 109.2

If $T_{suc} \neq 40°F$ and $T_{con} \neq 105°F$, the corrected refrigeration load $q_{rl.c} = q_{rl}/(F_{liq.s} F_{liq.c})$

In the design and planning of the liquid line layout, the following factors must be considered:

- If a receiver is placed after the subcooling section of the condenser, subcooling will occur at the outlet of the receiver only when a purge-type liquid receiver is used or if an additional subcooling setup is added after the receiver.

- Static pressure because of the liquid refrigerant column must be taken into account. At 105°F, the density of liquid HCFC–22 is 70.3 lb/ft³. For each ft of liquid column, the increase in static pres-

sure is therefore $\Delta p_s = 70.3/144 = 0.488$ psi, or roughly 0.5 psi for each ft of liquid column. At the top of a 10-ft riser, the pressure is about $10 \times 0.5 = 5$ psi less than at the bottom of the riser.

Fig. 21.16 is a q_{rl}–L_{eq} chart for a liquid line with HCFC–22 at 40°F suction temperature and 105°F condensing temperature with a pressure drop corresponding to a change of 1°F saturation temperature.

For suction temperatures other than 40°F and condensing temperatures other than 105°F, the corrected refrigeration load is calculated as

$$q_{rl.c} = \frac{q_{rl}}{F_{liq.s} \, F_{liq.c}} \qquad (21.5c)$$

Correction factor $F_{liq.s}$ takes into account the effect of enthalpy difference ($h_{lv} - h_{en}$) on $\dot{m}_r$, so $F_{liq.s} = F_{dis.s}$.

Correction factor $F_{liq.c}$ also takes into account the effect of enthalpy difference on $\dot{m}_r$. Its values are listed in Table 21.1. When $T_{con} > 105°F$, $F_{liq.c} < 1$ and $q_{rl.c} > q_{rl}$.

Consider the reciprocating refrigeration system described in the discussion of suction line sizing. The subcooling at the condenser outlet is 10°F. There are 3 manual shut-off angle valves, 1 solenoid valve (using the pressure drop as a globe valve), 1 sight glass (5 ft equivalent length), and 1 filter–dryer (150 ft equivalent length, according to the manufacturer's data) in the liquid line. Also, a 5-ft vertical riser is connected to the outlet of the condenser and the measured length

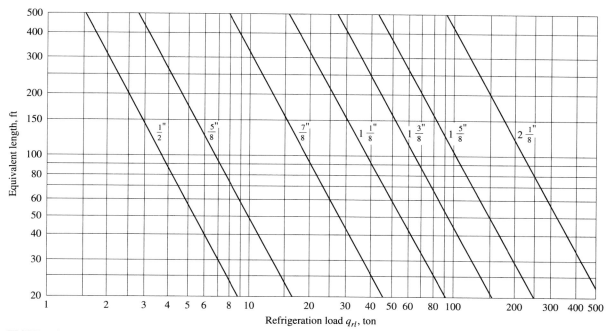

FIGURE 21.16 Liquid line sizing q_{rl}–L_{eq} chart for HCFC–22 at 40°F suction temperature and 105°F condensing temperature (type L copper tube, outside diameter, 1°F change in saturated temperature.

of the liquid line is 40 ft. Sizing of the liquid line can be accomplished as follows:

1. From Table 21.1, the correction factor for the refrigeration load in the liquid line at a suction temperature of 45°F $F_{\text{liq.s}} = F_{\text{dis.s}} = 1.01$. Condensing temperature is 115°F and a subcooling of 10°F $F_{\text{liq.c}} = 1.0$. The corrected refrigeration load is then

$$q_{\text{rl.c}} = \frac{q_{\text{rl}}}{F_{\text{liq.s}}F_{\text{liq.c}}} = \frac{20}{1.01 \times 1.0} = 19.8 \text{ tons}$$

2. Assume an equivalent length $L_{\text{eq}} = 250$ ft.
3. From Fig. 21.16, the q_{rl}–L_{eq} chart for the liquid line, the next higher tentative diameter of copper tubing for the liquid line at a $q_{\text{rl.c}} = 19.8$ tons and $L_{\text{eq}} = 250$ ft is $1\frac{1}{8}$ in.
4. Calculation of equivalent length is as follows:

3 angle valves	$3 \times 12 =$	36 ft
1 solenoid valve	$1 \times 29 =$	29 ft
1 sight glass	$1 \times 5 =$	5 ft
1 filter–dryer	$1 \times 150 =$	150 ft
Measured pipe length		40 ft
Total		260 ft

5. At $q_{\text{rl}} = 19.8$ tons and $L_{\text{eq}} = 260$ ft, the next higher diameter of the liquid line is still $1\frac{1}{8}$ in., the same as the tentative diameter. This is the required diameter of the liquid line.
6. For a condensing temperature of 115°F and a subcooling temperature of 10°F, the saturated pressure at $115 - 10 = 105$°F is 210.8 psig. When the pressure of the refrigerant in the liquid line is greater than 210.8 psig, flashing does not occur.

The lowest pressure occurs at the inlet of the expansion valve. This pressure can be calculated as follows:

Condensing pressure corresponding to 115°F	242.7 psig
Pressure drop of the liquid line	−3 psig
Pressure drop from riser $5 \times 0.5 = 2.5$	−2.5 psig
Liquid pressure at inlet of expansion valve	237.2 psig

The pressure at the inlet of the expansion value is greater than 210.8 psig, so liquid refrigerant will not be flashed in the liquid line.

PARALLEL CONNECTIONS. When two or three compressors are connected in parallel, their suction lines should be arranged so that the suction vapor is evenly distributed to the compressors. A gas equalizer should be connected to the crankcase of each compressor to equalize suction pressure, and an oil equalizer should be used to equalize the oil level in all compressors.

When two or three condensers are connected in parallel, hot gas should be evenly distributed to the condensers. If a liquid receiver is used, all condensers should be connected to a single receiver with equalizer between the hot gas supply and the top of the receiver. Refer to ASHRAE Handbooks and manufacturers' manuals for details.

21.5   CAPACITY AND SAFETY CONTROLS

Capacity Control

Refrigeration systems for comfort air conditioning or industrial applications need capacity control to meet refrigeration load variations during operation. In vapor compression refrigeration systems, control of the refrigerant flowing through the compressor is widely used as the primary refrigeration capacity control. For reciprocating refrigeration systems using DX-coils as evaporators, row or intertwined face control of DX-coil capacity is often used in conjunction with compressor capacity control for better performance. In medium and large systems, two separate refrigerant circuits are often used so that refrigerant flow in one of them can be cut off when refrigeration load drops below 50 percent of design-load.

There are four basic methods of controlling the capacity of reciprocating refrigeration systems by reducing the refrigerant flow rate during part-load operation.

ON-OFF CONTROL. *On-off* or *start-stop control* is widely used to control the refrigerant flow and the capacity of the reciprocating compressors in residential air conditioners. For reciprocating refrigeration systems equipped with multiple compressors, as used in supermarkets, on-off control is often the most suitable choice. On-off control is simple and inexpensive, but it produces a greater variation in the space air temperature and relative humidity maintained by the refrigeration system than other types of control do. In addition, cyclic operation may lead to a start-stop cyclic loss for the evaporator and condenser, which are located separately. Cyclic loss will be discussed in Chapter 22. On-off control also results in increased wear and tear on compressor components and shortens service life.

CYLINDER UNLOADER. For medium and large multiple-cylinder reciprocating compressors, a cylinder unloader is the most widely used method of capacity control. *Cylinder unloaders* reduce the capacity of the compressor by bypassing the compressed gas to the suction chamber, blocking the suction or discharge valve, or closing the suction valve late or early. Recently developed cylinder unloaders include a solenoid valve that unloads the cylinder(s) in responses to any one of the following:

- Variation of discharge air temperature in a packaged unit
- Changes in return chilled water temperature in a water chiller
- Changes in suction pressure in the evaporator

Fig. 21.17 shows a typical cylinder unloader. When a temperature sensor in a packaged unit detects a drop in discharge air temperature, the controller energizes the solenoid, draws the plunger away from the connecting port, allows the compressed gas to flow through a side passage, pushes the piston and valve stem of the unloader away from the solenoid, and opens the unloader valve. The compressed gas from the cylinder then flows to the suction side of the compressor. The two cylinders under the same head space are unloaded.

When discharge air temperature rises, the plunger of the solenoid covers the connecting port. The pressure of the compressed gas exerted on the disc closes the unloader valve. Consequently, the increased pressure in the space above the cylinder opens the check valve and permits the discharge gas to enter the discharge line. The two cylinders are now loaded.

A cylinder unloader is a form of step control. Typically, the capacity of an eight-cylinder reciprocating compressor can be controlled in four steps: 100 percent, 75 percent, 50 percent, and 25 percent.

SPEED MODULATION. Recent developments in solid-state electronics may allow adjustable-frequency AC-inverter-driven variable-speed motors to be used for capacity control in reciprocating compressors. Commercial variable-speed air conditioners with a capacity of a few tons of refrigeration were available in Japan in the 1980s.

Inverter-driven variable-speed reciprocating compressors have the advantages of lower start-stop or cycling losses, stable temperature control, and better system performance. They also have several drawbacks, such as cooling of the hermetic motor at low refrigerant flow rates, noise at high rotation speeds, and vibration at low speeds. In addition, they require a specially designed expansion valve for low-speed operation because a condensing pressure below a certain limit may affect the normal operation of the expansion valve.

HOT GAS BYPASS CONTROL. In *hot-gas bypass control*, according to the sensed suction pressure, a certain amount of hot gas that balances the reduction in refrigeration load bypasses the condenser and the expansion valve, and enters the low-pressure side directly. In a DX reciprocating refrigeration system, bypassed hot gas is often introduced between the expansion valve and the liquid refrigerant distributor, as shown in Fig. 21.18.

Although hot-gas bypass control permits modulation of system capacity to a very low percentage, this method does not save energy at reduced refrigeration load. Sometimes, it may be used as a safety device to protect the evaporator or DX-coil from frosting or excessive cycling of the compressor during on-off control when the refrigeration load is less than 20 to 25 percent of design capacity, or for industrial applications where strict temperature/humidity control is essential.

ASHRAE/IES Standard 90.1–1989 discourages the use of hot-gas bypass at reduced refrigeration loads. DOE Standard 1989 prohibits the use of hot-gas bypass in federal buildings. For capacity control of DX reciprocating refrigeration systems at low loads, *ASHRAE/IES Standard 90.1–1989 User's Manual* recommends adopting as many steps of unloading as possible by using a DDC, as described in Section 21.8, or through the installation of large and small uneven-sized compressors.

At part-load operation, because the volume flow of vapor refrigerant extracted by compressor is greater than the evaporated refrigerant in the evaporator, evaporating pressure is always lower than design value if the compressor capacity is not reduced accordingly. Greater power input to the compressor is required due to lower T_{ev}.

Safety Controls

Safety controls are used to shut down the refrigeration system during a malfunction and also to prevent damage of the system components. In reciprocating refrigeration systems, safety controls include low-pressure, high-pressure, low-temperature, frost control, oil pressure failure, and motor overload controls.

LOW-PRESSURE AND HIGH-PRESSURE CONTROLS. The purpose of *low-pressure control* is to stop the compressor when the suction pressure drops

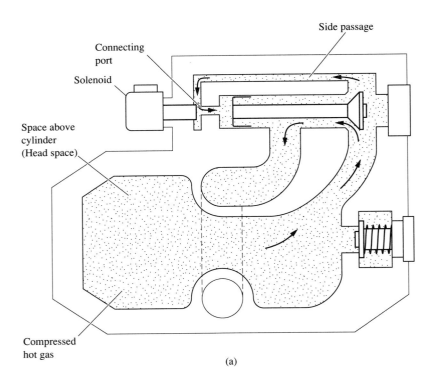

(a)

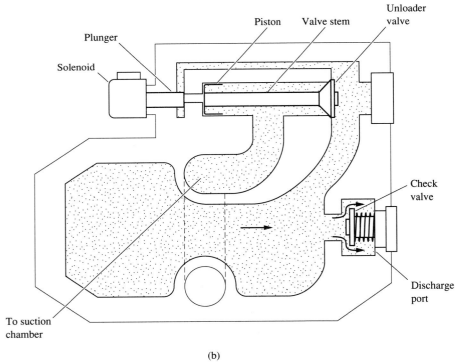

(b)

FIGURE 21.17 Cylinder unloader: (a) unloaded cylinders; and (b) loaded cylinders.

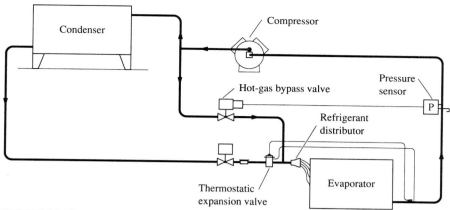

FIGURE 21.18 Hot-gas bypass control for a DX reciprocating refrigeration system.

below a preset value or when the refrigerant flow rate is too low to cool the compressor motor. Fig. 21.19a shows a typical low-pressure control mechanism. When the suction pressure falls below a certain limit, the spring pushes the blade downward, opens the motor circuit, and stops the compressor. When the suction pressure increases, the bellows expand, thus closing the contact of the motor circuit and restarting the compressor. The two adjusting screws are used to set the cut-out and cut-in pressures. *Cut-out pressure* is the pressure at which the compressor stops, and *cut-in pressure* is the pressure at which the compressor starts again.

The purpose of *high-pressure control* is to stop the compressor when the discharge pressure of the hot gas approaches a danger level. Fig. 21.19b shows a typical high-pressure control mechanism. If the discharge pressure reaches a certain limit, the bellows expand so that the blade opens the motor circuit contact and

the compressor stops. When the discharge pressure drops to a safe level, the bellows contracts and closes the contact, and the compressor starts again. As in a low-pressure control, two adjusting screws are used to set the cut-out and cut-in pressures.

In small refrigeration systems, low-pressure and high-pressure controls are often combined to form a dual-pressure control.

LOW TEMPERATURE CONTROL. The purpose of a *low-temperature control* is to prevent the temperature of chilled water in the liquid cooler from falling below a certain limit in order to protect the water from freezing. Freezing water damages the liquid cooler. Figure 21.20 shows a typical low-temperature control system. The sensing bulb senses the chilled water leaving temperature. When the temperature falls below a limit, contraction of the bellows opens the motor circuit and stops the compressor. When the chilled

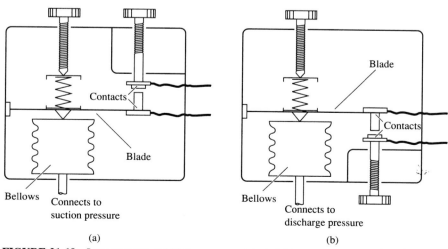

FIGURE 21.19 Low-pressure and high-pressure control: (a) low-pressure control; and (b) high-pressure control.

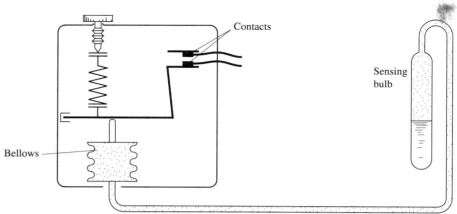

FIGURE 21.20 Low-temperature control.

water temperature rises above the limit, the expansion of the bellows closes the circuit and restarts the compressor.

FROSTING CONTROL. The purpose of frosting control is to prevent the formation of frost on the outer surface of the DX-coil. Usually, a temperature sensor is used to sense the outer surface temperature of the DX-coil. When the temperature drops to 32°F, the controller actuates a relay, which opens the circuit and stops the compressor.

OIL PRESSURE FAILURE CONTROL. The purpose of the *oil pressure failure control* is to stop the compressor when the oil pressure drops below a certain limit and fails to lubricate the main bearings and other components.

Figure 21.21 shows a typical oil pressure failure control mechanism. When the compressor starts, points a and b of the differential pressure switch and points c and d of the timer switch are in contact with each other. The differential pressure switch is affected by the differential pressure between oil pump discharge pressure and crank-case suction pressure. If the pressure of oil discharged from the oil pump does not reach a predetermined level within a certain time interval, the low oil pressure keeps the differential pressure switch a and b in a closed contact position. The heater is energized, which causes the bimetal plate to deviate from the vertical position. The bimetal plate no longer keeps the timer switch d in contact with c, and the compressor stops.

If the oil pressure reaches the preset value within the predetermined time interval, a does not make contact with b. The breaking in contact of a and b deenergizes the heater before the bimetal plate starts to bend. The bimetal plate still holds the timer switch

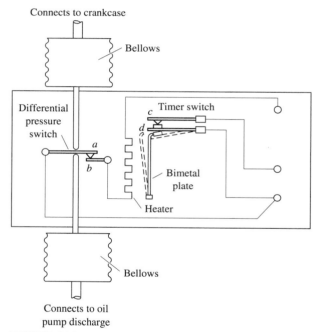

FIGURE 21.21 Oil pressure failure control.

d in contact with c, and the compressor continues to operate.

If the oil pressure drops below the preset value and does not rise again within a certain time interval, the energized heater causes the bimetal plate to bend, opens the contact of the timer switch, and stops the compressor.

MOTOR OVERLOAD CONTROL. The purpose of *motor overload control* is to protect the motor from dangerous overheating caused by long periods of overloading or failure of the compressor to start. Thermal protectors with bimetal or thermistor sensors are installed inside the motor or mounted on the outer surface of the motor shell. These sensors sense the temperature or both the temperature and electric current supplied to the motor windings. If dangerous overheating and overloading occur, a controller breaks the circuit.

Recently developed microprocessor-operated DDC panels combine all safety control mechanisms with other control systems of a reciprocating refrigeration system in one package. The DDC panel senses, analyzes, and operates the safety control systems in coordination with other systems with far more functions and more sophiscated comprehensive controls.

21.6 PERFORMANCE OF THE RECIPROCATING COMPRESSOR, EVAPORATOR, AND CONDENSER

Performance of the Reciprocating Compressor

There are two important parameters in the performance analysis of a reciprocating compressor: refrigerating capacity q_{rc}, in Btu/h or tons of refrigeration (as described in Section 4.9), and power input P_{in} or energy efficiency ratio EER, in Btu/h·W (as defined in Section 13.3). Because the q_{rc} and P_{in} are always affected by suction pressure p_{suc} and discharge pressure p_{con} (both in psig), it is convenient to relate the performance of the reciprocating compressor to suction temperature T_{suc} (in °F) and to the condensing temperature T_{con}. The saturated pressure at T_{ev} is equal to p_{suc}. Suction pressure is approximately equal to the saturated pressure of $(T_{suc} + 2)$ and discharge pressure is approximately equal to the saturated pressure of $(T_{con} + 2)$.

Consider an accessible hermetic compressor with the following operating and constructional characteristics:

Refrigerant	HCFC–22
Number of cylinders	8
Bore of the cylinder	2.65 in.
Stroke of the piston	2.25 in.
Revolutions per minute	1750
Suction temperature	40°F
Condensing temperature	115°F
Subcooling	10°F
Superheating	10°F

REFRIGERATION CAPACITY. In reciprocating compressors, the piston only performs compression work during the upward stroke in a complete rotation of the crankshaft. Piston displacement $\dot{V}_p$, in cfm, can be calculated as

$$\dot{V}_p = \pi D_{cy}{}^2 L_{st} N n \qquad (21.6)$$

where D_{cy} = diameter of the cylinder, ft
L_{st} = stroke of the cylinder, ft
N = number of cylinders
n = rotating speed of the compressor, rpm

From Eq. (10.30), the mass flow rate of the refrigerant $\dot{m}_r = \dot{V}_p \eta_v \rho_{suc}$, so from Eq. (4.25), the refrigerating capacity q_{rc}, in Btu/h, of a reciprocating compressor can be calculated as

$$q_{rc} = 60 \dot{V}_p \eta_v \rho_{suc} (h_{1'} - h_4) \qquad (21.7)$$

where ρ_{suc} = density of the suction vapor, lb/ft^3
$h_{1'}$, = enthalpy of superheated vapor leaving the evaporator, Btu/lb
h_4 = enthalpy of subcooled liquid entering the evaporator, Btu/lb

In the eight-cylinder reciprocating compressor using HCFC–22 as refrigerant described above, volumetric efficiency η_v depends mainly on compression ratio p_{dis}/p_{suc}. The smaller the p_{dis}/p_{suc} and the lower the reexpansion of clearance volume, cylinder heating, and leakage through valves and pistons, the greater the η_v. Figure 21.22a shows the η_v–p_{dis}/p_{suc} curve for this compressor. For a suction temperature of 40°F, the absolute saturated pressure is $68.5 + 14.7 = 83.2$ psia. For a condensing temperature of 115°F, the absolute condensing pressure is $243 + 14.7 = 257.7$ psia. The compression ratio is therefore: $p_{dis}/p_{suc} = (257.7 + 3)/83.2 = 3.13$. From Fig. 21.22a, $\eta_v = 0.81$.

The upper part of Fig. 21.22b shows the refrigeration capacity versus suction temperature q_{rc}–T_{suc} curves. If T_{con} remains constant and T_{suc} increases, refrigeration capacity rises accordingly for the following reasons:

- For a specific compressor rotating at a specific speed, piston displacement $\dot{V}_p$ is constant.
- Volumetric efficiency increases as T_{suc} rises
- Density of the suction vapor refrigerant ρ_{suc} increases as T_{suc} increases because of higher saturated pressure.
- For a constant T_{con} and a fixed amount of subcooling, enthalpy h_4 is also constant. Enthalpy of superheated refrigerant leaving evaporator at a fixed degree of superheat $h_{1'}$, increases nearly linearly between 40°F and 80°F as T_{suc} rises. This also causes an increase of refrigeration effect $q_{rf} = (h_{1'} - h_4)$.

These factors result in a very slight upward bend in the q_{rc}–T_{suc} curve.

If the suction temperature T_{suc} is at a constant value, an increase in condensing temperature T_{con}

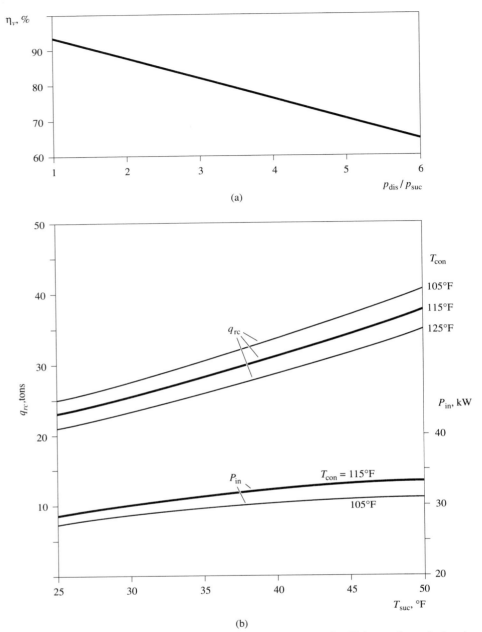

(a)

(b)

FIGURE 21.22 Refrigeration capacity, power input, and volumetric efficiency of a typical reciprocating compressor: (a) $\eta_v - p_{dis}/p_{suc}$ curve; and (b) $q_{rc} - T_{suc}$ and $P_{in} - T_{suc}$ curves.

results in a decrease in η_v and a decrease in q_{rf} because of a smaller h_4. Consequently, q_{rc} decreases as T_{con} rises and q_{rc} increases as T_{con} falls, as shown in Fig. 21.22.

POWER INPUT. The refrigeration compressor consumes most of the energy input to a refrigeration system. In a single-stage reciprocating refrigeration system, power input P, in kW, to the compressor can be calculated from Eq. (10.27) as

$$P_{in} = \dot{V}_p \eta_v \rho_{suc} \frac{(h_2 - h_{1'})}{56.85 \eta_{com}} \qquad (21.8)$$

where h_2 = enthalpy of hot gas discharge from the compressor if the compression process is isentropic, Btu/lb

η_{com} = compressor efficiency, as defined in Chapter 10

If T_{con} remains constant and suction temperature T_{suc} increases, both η_v and ρ_{suc} rise and $(h_2 - h_{1'})$ drops as T_{suc} increases. Compressor efficiency η_{com} also increases within a rather narrow range (0.75 to 0.80) when T_{suc} changes from 5°F to 50°F, and the $\eta_{com}-T_{suc}$ curve becomes nearly horizontal as T_{suc} exceeds 35°F. Because the increase in η_v and ρ_{suc} has a greater effect on P_{in} than the drop in $(h_2 - h_{1'})$ and the change of η_{com}, the $P_{in}-T_{suc}$ curve is slightly convex, indicating, P_{in} increases as T_{suc} rises, as shown in Fig. 21.22b. The $P_{in}-T_{suc}$ curves are significantly flatter than $q_{rc}-T_{suc}$ curves.

If T_{suc} remains constant and T_{con} increases, η_v falls and $(h_2 - h_{1'})$ becomes larger. The effect of the increase in $(h_2 - h_{1'})$ is greater than that of the fall of η_v. The result is a higher P_{in}.

Performance of the Evaporator

Refrigeration capacity q_{rc} is the most important parameter of an evaporator. From Eq. (12.2), q_{rc} can be calculated as

$$q_{rc} = AU\Delta T_m$$

where A = evaporating surface area, ft^2
  U = overall heat transfer coefficient, Btu/h · ft^2 · °F
  ΔT_m = log-mean temperature difference, °F

For a DX-coil or a flooded liquid cooler, surface area A is a constant, and is always based on the total outer surface area A_o including the tubes and the fins. For a liquid cooler with direct-expansion refrigerant feed, the surface area is the inner surface area of tubes plus inserts or enhanced surfaces.

Overall heat transfer coefficient U depends mainly on fluid velocity, heat flux at the evaporating surface, and the configuration of the evaporating surface of the DX-coil or liquid cooler. In DX-coils, air velocity changes only when the system load varies. In direct-expansion liquid coolers, chilled water velocity remains constant during operation.

The log-mean temperature difference ΔT_m is the temperature difference between the air temperature and the temperature of evaporating refrigerant inside the DX-coil or between the temperature of the chilled water in the shell and the refrigerant in the liquid cooler. If the air temperature and its enthalpy entering the DX-coil or the temperature of chilled water entering the liquid cooler both remain constant, an increase of suction temperature T_{suc} reduces ΔT_m, and a fall of T_{suc} increases ΔT_m. The refrigeration capacity of a DX-coil or liquid cooler is then inversely proportional to T_{suc}.

Medium or large reciprocating refrigeration systems usually have two DX-coils or two liquid cool-

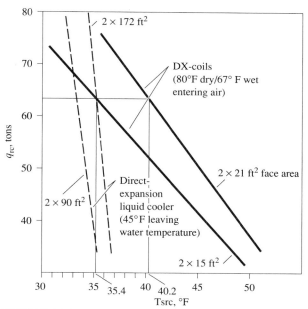

FIGURE 21.23 Performance curves $q_{rc}-T_{suc}$ for DX-coils and direct-expansion liquid coolers.

ers. Figure 21.23 compares the $q_{rc}-T_{suc}$ curves of two selected dual DX-coils and two selected dual direct-expansion liquid coolers. As q_{rc} is linearly proportional to T_{suc}, the $q_{rc}-T_{suc}$ curves are straight lines. The selected larger DX-coil has a total face area of 2×21 ft^2 and the smaller coil has a total face area of 2×15 ft^2. The selected larger direct-expansion liquid cooler has a total tube area of 172 ft^2 and the smaller one has a total tube area of 90 ft^2.

When the entering air is at 80°F dry bulb and 67°F wet bulb, each ft^2 of face area of the DX-coils has a total cooling capacity of 18,000 Btu/h at 40°F suction temperature, 115°F condensing temperature, and 600 fpm of air velocity.

Because air is directly cooled in the DX-coil, whereas chilled water is cooled before the air in a liquid cooler, the suction and evaporating temperatures in liquid coolers are lower than those in DX-coils. Because the ΔT_m of a liquid cooler is far smaller than that of a DX-coil, the $q_{rc}-T_{suc}$ curves of direct-expansion liquid coolers have a much steeper negative slope than those of DX-coils. For two evaporators at the same operating condition and q_{rc}, the larger evaporator, which has a greater evaporating surface area, has a higher T_{suc} than the smaller one, as shown in Fig. 21.23.

Performance of the Condenser

As for an evaporator, the performance of a condenser can be expressed by the following formula:

$$Q_{rej} = AU\Delta T_m$$

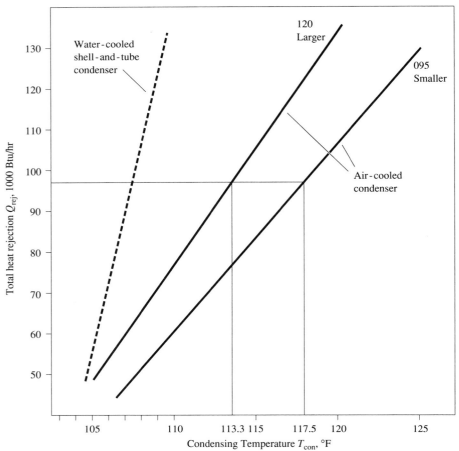

FIGURE 21.24   Performance curves Q_{rej}–T_{con} for condensers.

where Q_{rej} = total heat rejection at the condenser, Btu/h.

As described in Section 12.7, $Q_{reg} = q_{rc}$ (HRF). Here, HRF is the heat rejection factor.

In two condensers of the same type with different condensing surface areas, if their operating conditions are the same, the condensing temperature T_{con}, in °F, of the condenser with a larger condensing surface is lower than that of the smaller condenser. For a specific condenser, although the heat flux at the condensing surface and fluid velocity affect the overall heat transfer coefficient U value, the log-mean temperature difference ΔT_m is still the dominating factor that influences the performance of the condenser.

Figure 21.24 shows the Q_{rej}–T_{con} curves for condensers. Because Q_{rej} is linearly proportional to T_{con}, Q_{rej}–T_{con} curves are straight lines. Air-cooled condensers have a greater ΔT_m than water-cooled condensers, so the slope of the Q_{rej}–T_{con} curves for air-cooled condensers is smaller than that of water-cooled condensers.

21.7   SYSTEM BALANCE

Because the refrigeration cycle of a reciprocating refrigerating system is a closed cycle, if the system is operated in a continuous and steady state (that is, in an equilibrium state), according to the principle of continuity of mass and energy balance, the mass flow rates of refrigerant flowing through the evaporator, compressor, condenser, and expansion or float valve must all be equal. Also, the total amount of energy supplied to the refrigeration system must be approximately equal to the total energy released from the system.

A continuous and steady state means that the flow is continuous, and the properties of the refrigerant at any point in the refrigeration system do not vary over time. Therefore, during the design of a refrigeration system, the system components selected should have equal or approximately equal mass flow rates of refrigerant at stable operation conditions.

Of the four main system components (compressor, evaporator, condenser, and expansion valve) in a reciprocating refrigeration system, the expansion valve adjusts its mass flow rate of refrigerant according to the degree of superheat of the vapor leaving the evaporator in DX-coil and the DX liquid cooler.

Although the displacement of a selected compressor $\dot{V}_p$ is a constant, the compressor can adjust its mass flow rates within a certain range because of the increase or decrease of the density of suction gas ρ_{suc} and hot gas ρ_{dis} as T_{ev} and T_{con} vary.

There are two kinds of imbalance problems: imbalance of capacities of selected system components and imbalance caused by system load deviations from design conditions at part-load operation. Part-load operation will be discussed later in this chapter.

Balance of Capacities of Selected Components

If the imbalances in required capacities of the selected system components are small, the refrigeration system adjusts its evaporating and condensing temperatures to reach a system balance slightly different from the required design operating parameters. If the capacities of the system components are seriously imbalanced, either the refrigeration system will operate at parameters very different from the required conditions, or the system will fail to operate or even be damaged by unacceptable operating parameters such as heavy overloading, low evaporating temperature, or high pressure cut-out.

Consider a built-up rooftop reciprocating packaged system. The design refrigeration load q_{rc} is 60 tons at a suction temperature $T_{suc} = 40°F$ and a condensing temperature $T_{con} = 115°F$. If two compressors are selected, each of them has a q_{rc} of 31.6 tons. Each compressor has 8 cylinders, and the capacity of the

unit can be controlled at 100, 75, 50, 25, and 0 percent. The $q_{rc}-T_{suc}$ curve of a single compressor is shown in Fig. 21.22.

Two DX-coils are selected to match with the compressors, each with a face area of 21 ft^2 with two expansion valves and two refrigerant distributors for intertwined face control. Their $q_{rc}-T_{suc}$ curves are shown in Fig. 21.23.

When these two compressors are connected with two 21-ft^2 DX-coils, the operating point of each compressor–DX-coil combination must be the intersecting point Y of their $q_{rc}-T_{suc}$ curves, as shown in Fig. 21.25. At the intersecting point Y representing design condition, $q_{rc} = 63$ tons and $T_{suc} = 40.2°F$. The energy efficiency ratio EER is 11.8.

If two 15-ft^2 DX-coils are selected to form the compressor–DX-coil combination, the intersecting point then moves to Y', with a total refrigeration capacity about 59.5 tons at $T_{suc} = 38°F$. The drop in T_{suc} necessitates a greater energy input because of the higher compression ratio. The EER drops to 11.5. It is preferable to select the DX-coils with a face area of 21 ft^2 to reduce energy input.

Because the T_{con} of an air-cooled condenser affects its total heat rejection Q_{rej}, T_{con} is often estimated first when an air-cooled condenser is selected to determine the Q_{rej} for the compressor–DX-coil combination.

If T_{con} is estimated at 115°F, from Fig. 21.25, for a power input of 64 kW to two compressors, the required total heat rejection of the selected air-cooled condenser is

$$Q_{rej} = q_{rc} + 3413 P_{in}$$
$$= (63 \times 12,000) + (64 \times 3413)$$
$$= 974,400 \text{ Btu/h}$$

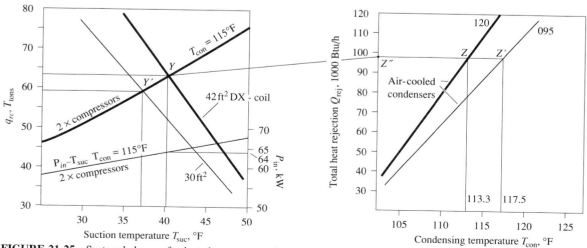

FIGURE 21.25 System balance of selected components for a reciprocating refrigeration system.

In Fig. 21.25, draw a line from point Y to point Z'' whose $Q_{rej} = 974,000$ Btu/h. Draw horizontal line Z'' Z intersecting the Q_{rej}–T_{con} curve of condenser size 120 at Z. At point Z, $T_{con} = 113.3°F$. If a smaller condenser 095 is selected, the intersecting point is Z', and the corresponding T_{con} is 117.5°F, which necessitates a higher P_{in} to the compressor.

Using a larger condenser, the final selected capacity of this direct-expansion reciprocating refrigeration system may be as follows:

Refrigeration load	64 tons
Suction temperature	40°F
Condensing temperature	114°F
Power input to the compressor motor	64 kW
Total heat rejection Q_{rej}	986,000 Btu/h

For thermostatic expansion valves, the selected capacity must be greater than the refrigerating capacity of the compressor at a pressure drop across the expansion valve of 75 to 120 psi for HCFC–22 at design conditions.

21.8 OPERATING CHARACTERISTICS OF AIR-COOLED, DIRECT-EXPANSION RECIPROCATING REFRIGERATION SYSTEMS

Operating Balance at Design Conditions

Consider an air-cooled direct-expansion reciprocating refrigeration system in a built-up rooftop packaged system with the same characteristics as described in the last section and shown in Fig. 21.25.

If the design refrigeration load from the evaporation of HCFC–22 inside the two DX-coils is only 60 tons (720,000 Btu/h) instead of the selected capacity of 64 tons, the volume flow rate of the piston displacement $\dot{V}_p$ to extract suction vapor is greater than that of the vaporized refrigerant in the evaporator. This imbalance lowers the evaporating pressure in the evaporator and, therefore, increases the specific volume v_{suc} of the suction vapor, until $\dot{V}_p$ is balanced with the volume flow rate of refrigerant vaporized in the evaporator $\dot{V}_{ev}$, in cfm. A new equilibrium between the compressor and evaporator is reached, as

$$\dot{m}_{com} = \dot{V}_p \rho_{suc} = \dot{m}_{ev} = \frac{\dot{V}_{cv}}{v_{suc}} = \frac{200 q_{rc}}{q_{rf}} \quad (21.9)$$

where q_{rc} = refrigeration load at the evaporator, tons

q_{rf} = refrigeration effect, Btu/lb

The refrigerant discharged from the compressor then enters the air-cooled condenser at mass flow rate $\dot{m}_{com}$. Because the selected condenser can condense a mass flow rate of refrigerant equal to a Q_{rej} of about 986,400 Btu/h, a mass flow rate of refrigerant greater than $\dot{m}_{com}$ is condensed in the air-cooled condenser and causes a drop in condensing pressure p_{con} and condensing temperature T_{con}. As T_{con} drops, ΔT_m and Q_{rej} both fall accordingly until the condensed mass flow of refrigerant equals the mass flow rate of refrigerant discharged from the compressor $\dot{m}_{com}$ and a new balance is again formed in the condenser.

Pressure Characteristics

Figure 21.26 shows the pressure characteristics of a typical air-cooled direct-expansion reciprocating refrigeration system using DX-coils and HCFC–22 as refrigerant.

At design conditions, subcooled liquid refrigerant at a temperature of 105°F and a pressure of 238 psig is fed to the thermostatic expansion valve. Because the saturated pressure of HCFC–22 at 105°F is 210 psig, which is smaller than 238 psig, no flashing will occur before the inlet of the expansion valve.

The pressure drop across the expansion valve is 100 psi, and the pressure drop in the refrigerant distributor and tube is 59 psi. A fraction of liquid refrigerant is vaporized in the expansion valve, distributor, and tubes. Refrigerant with a quality of about 0.15 enters the evaporator, the DX-coil, at a temperature of 47°F and a pressure of 79 psig.

In the DX-coil, all the liquid refrigerant is vaporized. The pressure drop across the refrigerant circuits of the DX-coil is 8 psi. At point x, the evaporating temperature is 42°F, and the evaporating pressure is 71 psig. At the outlet of the evaporator, vapor refrigerant is superheated to 54°F, with a pressure of 71 psig. The degree of superheat is 12°F.

The pressure drop at the suction line is 3 psi corresponding to a 2°F change in saturated temperature. Suction vapor enters the reciprocating compressor at a suction temperature saturated of 40°F and a pressure of 68.5 psig. In the compressor, the pressure is raised to the hot gas discharge pressure of 250 psig with a temperature of about 185°F.

In the discharge line, the pressure drop is 3 psi, corresponding to a 1°F change in saturated temperature. At the inlet of the air-cooled condenser, the pressure of the hot gas is 247 psig.

Hot gas is desuperheated, condensed, and subcooled in the air-cooled condenser at a condensing temperature of 115°F and a pressure of 243 psig. Liquid refrigerant is discharged from the subcooling coil of the air-cooled condenser at a temperature of 105°F and a pressure of 241 psig. The degree of subcooling is 10°F.

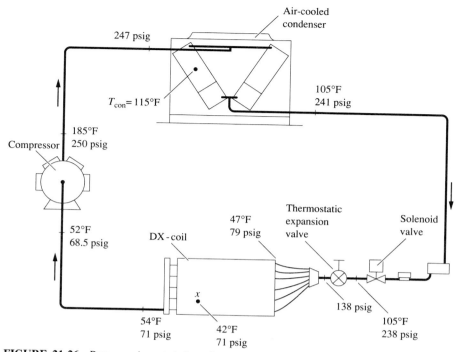

FIGURE 21.26 Pressure characteristics of a typical air-cooled direct-expansion reciprocating refrigeration system using HCFC–22 as a refrigerant.

In the liquid line, the pressure drop is 3 psi, corresponding to 1°F change in saturated temperature. Liquid refrigerant may enter the thermostatic expansion valve at a temperature of 105°F and a pressure of 238 psig.

Of the total system pressure drop of $250 - 68.5 = 181.5$ psi, a pressure drop of 159 psi, or 88 percent of the total system drop occurs in the expansion valve, distributor nozzle, and tubes. During part-load operation, when outdoor temperature is lower, this pressure drop may decrease considerably.

Part-Load Operation Using a Cylinder Unloader

The cylinders of a modern DX refrigeration system in a rooftop packaged unit using a microprocessor-based DDC panel for capacity control can be loaded or unloaded in four successive steps when discharge temperature T_{dis} rises and falls between 50 and 56°F and T_{dis} is maintained at 53°F by means of PI control mode, as shown in the lower right-hand bottom corner of Fig. 21.27. Cylinders are loaded in successive steps when T_{dis} is raised 1°F higher than the upper limit of the control band 56°F, and cylinders are unloaded when T_{dis} is 1°F lower than the lower limit of 51°F. Details of T_{dis} control for a rooftop packaged unit will be discussed in Section 29.8.

In the morning of a summer day, when the rooftop unit starts with a refrigeration load of $q_{rc} > 60$ tons from the cool-down load, all 16 cylinders are operating. The operating point may be at A_o in Fig. 21.27 with a suction temperature $T_{suc} = 42°F$. As the refrigeration load drops to a value of 58 tons, the discharge temperature T_{dis} gradually falls until it is equal to 50°F. At the same time, less refrigerant is vaporized in the DX-coil because of the drop of refrigeration load. As the 16 cylinders of the two compressors have a greater extracted volume than the vapor refrigerant evaporated in the DX-coil, the suction pressure and temperature T_{suc} drops to point A' on the 16-cylinder curve in Fig. 21.27. The DDC panel unloads four cylinders and closes one of the four liquid line solenoid valves of the two evaporators. The operating point immediately shifts to point B on the 12-cylinder curve.

As the refrigeration capacity of the 12 operating cylinders is less than 58 tons refrigeration load, more vapor refrigerant is evaporated in the DX-coils than can be extracted by the 12 cylinders, and the suction pressure and temperature of the refrigerant at operating point B tends to move upward along the 12-cylinder curve to point B'' with a T_{suc} equal to about 44°F. At the same time, because the refrigeration capacity of 12 cylinders q_{rc} is smaller than the 58-ton required refrigeration load, T_{dis} rises to 56°F. The

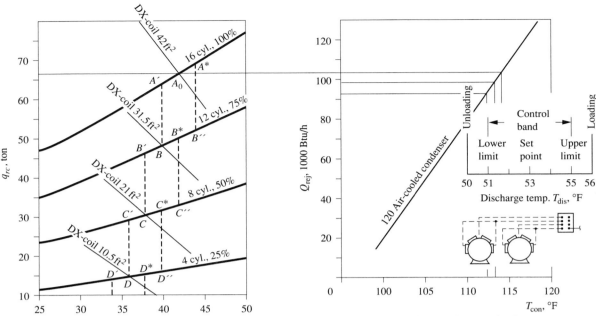

FIGURE 21.27   Part-load capacity control of a DX reciprocating refrigeration system using a cylinder unloader.

DDC panel again loads four cylinders and opens the shut-off solenoid valve in the evaporator. The operating point suddenly moves to point A^* on the 16-cylinder curve. When 16 cylinders are operating, the refrigeration capacity provided is greater than 58 tons, so point A^* again moves downward along the 16-cylinder curve. At the same time, T_{dis} drops to 50°F and unloads four cylinders and opens the solenoid valve again. The refrigeration system cycles between points A^* and A' with 16 cylinders and 100 percent of evaporator's face area operating, and B and B'' with 12 cylinders and 75 percent of the face area operating. Operating points A^*, A', B, and B'' form a cycle. Because of the variation of percentages of operating time on 16 cylinders and on 12 cylinders, the refrigeration load provided by the compressor is balanced with the required 58-ton refrigeration load.

If the required refrigeration load drops to only 44 tons, the operation point B on 12-cylinder curve drops to B'. Because of the fall of refrigeration load, T_{dis} again drops to 50°F and the DDC panel then unloads to point C on the 8-cylinder curve and shuts off one of the two compressors, and two of the solenoid valves in the two evaporators. The compressors cycle between B', C, C'', and B^* until the refrigeration capacity provided is balanced with the required refrigeration load of 44 tons by varying the time duration of 12-cylinder operation and 8-cylinder operation.

Therefore, any refrigeration load from design-load to a very low part-load can be balanced by unloading the cylinders, shutting off the compressors, and deenergizing the face area of the evaporator.

When the required refrigeration load is greater than the refrigeration provided by the compressor and the DX-coils, the cylinders are loaded and the DX-coils are energized successively in the similar manner.

Pump-Down Control

Pump-down control is an effective means of preventing migration of refrigerant from the evaporator to the crankcase of the compressor during the shutdown period. When a compressor starts, migrated refrigerant tends to mix with the oil and form slugs, which may damage the compressor.

When the discharge air temperature T_{dis} or suction pressure drops below a preset limit and is sensed by a sensor, the DDC panel deenergizes the liquid line solenoid valve, as shown in Fig. 21.26, and the valve closes. The compressor pumps all gaseous refrigerant to the condenser, where it condenses into liquid. When the vapor pressure in the evaporator falls below a certain value, the low-pressure control breaks the electric circuit and stops the compressor.

If cooling is again required, a temperature sensor senses the increase of T_{dis} or a pressure sensor senses the rise of suction pressure over a preset value, and the DDC panel energizes the liquid line solenoid valve. Liquid refrigerant then flows into the evaporator where it evaporates. Vapor pressure builds up. As the pressure exceeds the cut-in pressure of the low-pressure control, the compressor starts again.

Pump-down control acts as a capacity control by controlling T_{dis} or suction pressure. It also acts as a safety control to protect the compressor from damage from liquid slugs. Pump-down control is widely used in DX reciprocating refrigeration systems.

When outdoor air temperature is low, in order to maintain proper operation of the thermostatic expansion valve, low ambient air control should be used to modulate the dampers or cycling, or even shut off the condenser fans, as described in Section 12.8.

Main Problems in Direct-Expansion Reciprocating Refrigeration Systems

LIQUID SLUGGING. Liquid slugging is a mixture of refrigerant and oil. It is formed under the following conditions:

- Liquid refrigerant floods back from the evaporator to the crankcase of the compressor because of either an insufficient degree of superheating or hunting of the thermostatic expansion valve.
- Liquid refrigerant migrates from the warmer indoor evaporator to the colder outdoor compressor during a shutdown period in split packaged units.
- A considerable amount of oil returns to the crankcase from the oil traps.

Liquid refrigerant dilutes the lubrication oil. If a large amount of refrigerant and oil accumulates in the crankcase of the compressor, the foaming oil–refrigerant mixture forms a liquid slug, causing serious loss of oil in the crankcase when the compressor starts again after a shutdown period. A liquid slug is incompressible. When a liquid slug enters the compression chamber, it may damage the valves, pistons, and other components.

Pump-down control and crankcase heaters are effective means of preventing refrigerant migration in split packaged units. A proper degree of superheat is essential to prevent liquid refrigerant floodback to the compressor. Proper sizing of the thermostatic expansion valve and proper location of the sensing bulb prevent hunting of the thermostatic expansion valve.

OIL RETURN. For oil-miscible halocarbon refrigerants, the system design must include a means of returning oil from the condenser and evaporator to the compressor.

COMPRESSOR SHORT CYCLING. Short cycling is on-and-off control of the compressor or loading or unloading cylinders within a too-short time period (less than three minutes). It is mainly a result of close tolerance control or improper design of on-off capacity control. If the compressor is short-cycled, the cy-

cling action may pump oil away from the compressor, causing insufficient oil lubrication. It may also damage system components. Also, short cycling does not allow sufficient time to restabilize the system.

Compressor short cycling can be solved by using a time delay for restart or by overlapping unloading and loading schedules in step capacity control.

DEFROSTING. In a direct-expansion reciprocating refrigeration system, if the surface temperature of DX-coil is 32°F, frost accumulates on the coil surface. Because frost impedes air passage and reduces the rate of heat transfer of the coil, it must be removed periodically. The process of removing frost from the evaporator is called *defrosting*.

If air enters the coil at a temperature above 36°F, the ambient air flowing through the DX-coil can be used for defrosting. However, in low-temperature refrigeration systems, if the ambient air temperature is also below 32°F, an electric heating element may be used as a simple and effective way to defrost the coil.

Hot-gas defrosting is also effective, but must be planned during the design stage to ensure an adequate quantity of hot gas at a pressure high enough to flow through the coil. After defrosting, the high-pressure condensate should be returned to the liquid line, and the pressure of the desuperheated gas should be reduced gradually and released to the suction line.

The defrosting cycle can be controlled by sensing the pressure or temperature difference of air entering and leaving the DX-coil over a fixed time interval.

Minimum Performance and Equipment Selection

ASHRAE Standard 90.1–1989 specifies the minimum performance of air-cooled, direct-expansion reciprocating systems (packaged units) in terms of energy efficiency ratio (EER) and integrated part-load value (IPLV), as listed in Table 15.3. The selected air-cooled packaged unit should meet the minimum performance guidelines at both design and part-load operation.

21.9 SCREW COMPRESSION REFRIGERATION SYSTEMS

A *screw compression refrigeration system* has five main components. In addition to a compressor, evaporator, condenser, and throttling device, an oil cooler is used to cool and supply oil to the compressor. Figure 21.28 shows a typical twin-screw compressor refrigeration system.

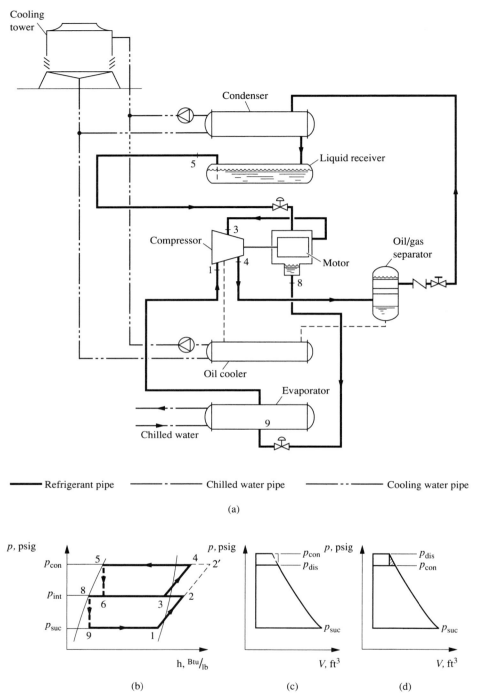

FIGURE 21.28 A typical twin-screw compressor refrigeration system: (a) schematic diagram; (b) refrigeration cycle; (c) undercompression; and (d) overcompression.

Oil Cooling

As shown in Fig. 21.28, the hot gas and oil mixture discharged from the screw compressor enters an oil–gas separator, in which oil is separated from the gaseous refrigerant. The oil is cooled in the oil cooler, flows through a filter, and divides into three circuits to

lubricate the bearings, provide hydraulic power to the capacity control system, and flow into the compressor. Oil in the compressor provides a rotor-to-rotor and rotor-to-housing seal, lubricates the rotor-driving mechanism, and cools the refrigerant during compression.

In an oil cooler, oil is usually cooled by cooling water from the cooling tower or evaporative cooler. It is sometimes cooled by river, lake, or well water, or by direct-expansion of liquid refrigerant.

Capacity Control

A sliding valve moving between the discharge end of the male rotor and the discharge port is shown in Fig. 10.26a. It is driven by oil pressure. When the sliding valve moves toward the discharge port, it creates a recirculating passage on the rotor housing, as described in Section 10.10. A portion of the trapped gaseous refrigerant returns to the suction port through this bypass passage, so the mass flow rate of refrigerant, and thus the refrigeration capacity, are reduced.

Variable Volume Ratio

There are no discharge valves in a twin-screw compressor. A screw compressor using built-in volume ratio V_i may cause over- or undercompression when the discharge pressure from the compressor p_{dis} is different from the condensing pressure p_{con}, both in psig. Fig. 21.28c shows undercompression. Undercompression causes a volume of gas at condensing pressure to reenter the trapped volume at the beginning of the discharge process. Figure 21.28d shows overcompression. The discharged gas reexpands to match the condensing pressure.

In the late 1980s, twin-screw compressors with variable volume ratios were developed. The common arrangement includes a split sliding valve, which consists of a sliding valve and a movable slide. When the sliding valve and the slide move toward the inlet end, the built-in volume ratio V_i becomes smaller. As the slide moves back and forth, the radial discharge port is relocated and thus varies the volume ratio to meet the optimum requirements, without over- and undercompression. In a screw compression refrigeration system in which suction or condensing pressure change frequently, efficiency can be improved by up to 20 percent using a variable volume ratio twin-screw compressor instead of a built-in volume ratio.

Economizer

In Fig. 21.28b, if a small portion of liquid refrigerant at point 5 is flashed at an intermediate pressure p_{int} so that the remaining refrigerant is cooled to saturated condition at intermediate pressure, point 8, such an arrangement significantly increases the refrigeration effect of the cycle. At the same time, the flashed vapor refrigerant is introduced to the partially compressed gas at p_{int}. The compression work needed for the flashed vapor at p_{int} is less than the compres-

sion work needed if the liquid refrigerant is flashed at the evaporator at p_{suc}.

Figure 21.28a is a schematic diagram with an economizer. Liquid refrigerant from the receiver at point 5 flows through the throttling device and enters the motor shell. In the motor, a portion of liquid refrigerant at p_{int} is flashed into vapor to cool the motor windings and the remaining liquid refrigerant to saturated condition point 8. The flashed vapor is then introduced to the compressed gaseous refrigerant at p_{int} in the compressor. The remaining liquid refrigerant flows through another throttling device. Its pressure drops to the evaporating pressure p_{ev} and it enters the evaporator at point 9. In the evaporator, all liquid refrigerant is vaporized. Vapor refrigerant from the evaporator is extracted by the compressor at suction pressure p_{suc}, denoted by point 1.

Most twin-screw compressors available have a capacity between 50 and 500 tons. They have efficiencies higher than reciprocating compressors, equivalent to the full-load efficiency of a single-stage centrifugal compressor. The part-load performance of twin-screw compressors is clearly superior to that of a single-stage centrifugal compressor. Also, screw compressors are quiet and compact in size, and function as positive-displacement compressors. Screw compressors are also critical in oil recirculating to provide lubrication and oil seal and cool the refrigerant. They have been used in many applications in recent years.

REFERENCES AND FURTHER READING

ASHRAE, *ASHRAE Handbook 1988, Equipment,* ASHRAE Inc., Atlanta, GA, 1988.

ASHRAE, *ASHRAE Handbook 1990, Refrigeration Systems and Applications,* ASHRAE Inc., Atlanta, GA, 1990.

ASHRAE/IES, *Standard 90.1–1989: Energy Efficient Design of New Buildings Except New Low-Rise Residential Buildings,* ASHRAE Inc., Atlanta, GA, 1989.

ASHRAE/IES, *Standard 90.1–1989 User's Manual,* ASHRAE Inc., Atlanta, GA, 1992.

Coad, W.J., "Refrigeration Piping for DX: Part III—Liquid Line Design," *Heating/Piping/Air Conditioning,* pp. 88–90, April 1990.

Coad, W.J., "Refrigeration Piping for DX: Part IV—Suction Line Design," *Heating/Piping/Air Conditioning,* pp. 93–95, May 1990.

Cole, R.A., "Reduced Head Pressure Operation of Industrial Refrigeration Systems," *Heating/Piping/Air Conditioning,* pp. 119–132, May 1985.

DOE, *Code of Federal Regulations, Title 10, Part 435, Subpart A, Performance Standard for New Commercial and Multi-Family High-Rise Residential Buildings,* DOE, Washington D.C., 1989.

Ernst, S., "Advantages of Screw Compressors," *Heating/Piping/Air Conditioning*, pp. 85–104, November 1987.

Grim, J.H., "Basic Compressor Application," *ASHRAE Transactions*, Part I, pp. 837–841, 1981.

Grimm, N.R., and Rosaler, R.C., *Handbook of HVAC Design*, McGraw–Hill, New York, 1990.

Hicks, P., and Adams, P., Jr., "Saving Older Refrigeration Systems," *Heating/Piping/Air Conditioning*, pp. 82–92, November 1989.

Kruse, H.H., and Schroeder, M., "Fundamentals of Lubrication in Refrigerating Systems and Heat Pumps," *ASHRAE Transactions*, Part II B, pp. 763–783, 1984.

Patterson, N.R., "Reciprocating Refrigeration Piping and Control," *Heating/Piping/Air Conditioning*, pp. 78–86, April 1981.

Pillis, J.W., "Advancement in Refrigeration Screw Compressor Design," *ASHRAE Transactions*, Part I B, pp. 219–224, 1986.

Schoen, A., "Resolving TEV Hunting Problems," *Heating/Piping/Air Conditioning*, pp. 69–72, July 1990.

Stamm, R.H., "Industrial Refrigeration Compressors," *Heating/ Piping/Air Conditioning*, pp. 93–111, July 1984.

Strong, A.P., "Hot Gas Defrost for Industrial Refrigeration," *Heating/Piping/Air Conditioning*, pp. 71–83, July 1988.

The Trane Company, *Helical Rotary Water Chillers*, American Standard Inc., La Crosse, WI, 1988.

The Trane Company, *Refrigeration Accessories and Controls*, The Trane Company, La Crosse, WI, 1985.

The Trane Company, *Refrigeration Compressors*, The Trane Company, La Crosse, WI, 1985.

The Trane Company, *Reciprocating Refrigeration*, The Trane Company, La Crosse, WI, 1977.

The Trane Company, *Refrigeration System Piping*, The Trane Company, La Crosse, WI, 1978.

Uekusa, T., Nakao, M., and Ohshima, K., "Control Method of a Cooling Apparatus in Low Outdoor Air Temperatures," *ASHRAE Transactions*, Part I, pp. 200–204, 1990.

Wang, S.K., *Principles of Refrigeration Engineering*, Vol. 3, Hong Kong Polytechnic, Hong Kong, 1984.

Yencho, J., "Purging Noncondensable Gases," *Heating/Piping/Air Conditioning*, pp. 75–79, February 1989.

Yun, K.W., "Supermarket Refrigeration Compressors," *ASHRAE Transactions*, Part I B, pp. 596–605, 1983.

Zimmern, B., and Sweetser, R.S., "The Centrifugal Economizer: Its Match with Screw Compressors," *ASHRAE Transactions*, Part I B, pp. 225–233, 1986.

Zubair, S.M., and Bahel, V., "Compressor Capacity Modulation Scheme," *Heating/Piping/Air Conditioning*, pp. 135–143, January 1989.

CHAPTER 22

HEAT PUMPS AND HEAT RECOVERY SYSTEMS

22.1 BASICS OF HEAT PUMPS AND HEAT RECOVERY

Heat Pumps

A heat pump extracts heat from a heat source and rejects heat to air or water at a higher temperature. During summer, the heat extraction, or refrigeration effect, is the useful effect for cooling, whereas in winter the rejected heat alone, or rejected heat plus supplementary heating from a heater form the useful effect for heating.

A heat pump is a packaged air conditioner or a packaged unit with a reversing valve or other changeover setup. A heat pump has all the main

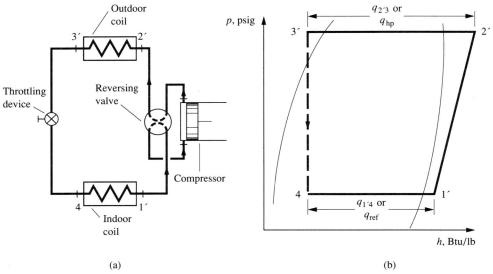

FIGURE 22.1 Heat pump cycle: (*a*) schematic diagram; and (*b*) cycle on *p–h* diagram.

components of an air conditioner or packaged unit: fan, filters, compressor, evaporator, condenser, and short capillary tube (a throttling device). The apparatus for changing from cooling to heating or vice versa is often a reversing valve, in which the refrigerant flow to the condenser is changed to the evaporator. Alternatively, air passage through the evaporator may be changed over to passage through the condenser. A supplementary heater is often provided when the heat pump capacity does not meet the required output during low outdoor temperatures.

A heat pump system consists of heat pumps and piping work; system components include heat exchangers, heat source, heat sink, and controls to provide effective and energy-efficient heating and cooling operations. HCFC–22 and HFC–134a are the most widely used halocarbon refrigerants in new heat pumps. In recent years, the use of heat pumps for heating and cooling has steadily increased. Based on the data in the EIA's *Commercial Building Characteristics*, heat pumps were used in about 15 percent of commercial buildings (by floor space) for heating and cooling at the end of 1989.

Heat Pump Cycle

A heat pump cycle comprises the same processes and sequencing order as a refrigeration cycle except that both the refrigeration effect $q_{1'4}$ or q_{rf}, Btu/lb, and the heat pump effect $q_{2'3'}$ or q_{hp}, in Btu/lb, are the useful effects, as shown in Fig. 22.1. As defined in Eqs. (4.8) and (4.9), the coefficient of performance of a refrigeration system COP_{ref} is

$$COP_{ref} = \frac{h_{1'} - h_4}{W_{in}} = \frac{q_{1'4}}{W_{in}} \quad (22.1)$$

where h_4, $h_{1'}$ = enthalpy of refrigerant entering and leaving the evaporator, respectively, Btu/lb

W_{in} = work input, Btu/lb

The coefficient of performance of the heating effect in a heat pump system COP_{hp} is

$$COP_{hp} = \frac{q_{2'3'}}{W_{in}} \quad (22.2)$$

and the useful heating effect $q_{2'3'}$ can be calculated as

$$q_{2'3'} = (h_{2'} - h_{3'}) = (h_{1'} - h_4) + (h_{2'} - h_{1'}) \quad (22.3)$$

where $h_{2'}$ = enthalpy of the hot gas discharged from the compressor, Btu/lb

$h_{3'}$ = enthalpy of the subcooled liquid leaving the condenser, Btu/lb

Here, polytropic compression is a real and irreversible process. Both the subcooling of the liquid refrigerant in the condenser and the superheating of the vapor refrigerant after the evaporator increase the useful heating effect $q_{2'3'}$. Excessive superheating, which must be avoided, leads to a hot gas discharge temperature that is too high and to a lower refrigeration capacity in the evaporator. According to the types of heat sources from which heat is absorbed by the refrigerant, currently used heat pump systems can be mainly classified into three categories: air-source, water-source, and ground-coupled.

Heat pump systems are energy-efficient cooling/heating systems. Many new technologies currently being developed, such as engine-driven heat pumps, may significantly increase the system performance factor of the heat pump system. Ground-coupled heat pumps with direct-expansion ground-coils provide another opportunity to increase the COP of heat pump systems.

HVAC&R Heat Recovery Systems

An HVAC&R heat recovery system converts waste heat or cooling from any HVAC&R process into useful heat or cooling. Here, heat recovery is meant in a broad sense. It includes both waste heat and cooling recovery. An HVAC&R heat recovery system includes the following:

- The recovery of internal heat loads—such as heat energy from lights, occupants, appliances, and equipment inside the buildings—by reclaiming the heat rejected at the condenser and absorber of the refrigeration systems
- The recovery of heat from the flue gas of the boiler
- The recovery of heat from the exhaust gas and water jacket of the engine that drives the HVAC&R equipment, especially engine-driven reciprocating vapor compression systems
- The recovery of heat or cooling from the exhaust air from air conditioning systems

Although heat pump systems sometimes are used to recover waste heat and convert it into a useful effect, a heat recovery system is different from a heat pump system in two ways:

- In a heat pump system, there is only one useful effect at a time, such as, the cooling effect in summer or the heating effect during winter. In a heat recovery system, both its cooling and heating effects may be used simultaneously.

  From Eq. (22.2), the coefficient of performance of the useful heating effect in a heat pump is $COP_{hp} = q_{2'3'}/W_{in}$, whereas for a heat recovery system, the coefficient of performance COP_{hr} is always higher if both cooling and heating are simultaneously used and can be calculated as

$$COP_{hr} = \frac{(q_{1'4} + q_{2'3'})}{W_{in}} \qquad (22.4)$$

- A heat pump is an independent unit. It can operate on its own schedule, whereas a heat recovery system in HVAC&R is usually subordinate to a refrigeration system or to some other system that uses the waste heat or waste cooling.

  Heating or cooling produced by a heat recovery system is a by-product. It depends on the operation of the primary system. A heat recovery system can use waste heat from condenser water for winter

heating only if the centrifugal refrigeration system is operating.

A centrifugal chiller that extracts heat from a surface-water source (for example, a lake) through its evaporator and uses condensing heat for winter heating is a heat pump, not a heat recovery system. Heat recovery systems that are subordinate to a centrifugal or absorption refrigeration system will be discussed in Chapters 23 and 24, respectively.

Recovery of waste heat from industrial manufacturing processes to provide heating shows great potential for saving energy. These heat recovery systems must be closely related to the specific requirements of corresponding manufacturing processes and will not be discussed here.

Heat Balance and Building Load Analysis

The building load consists of transmission gain or loss, solar radiation, ventilation load and infiltration load, people, electric lights, appliances, and heat gains from fans and pumps. Building load is actually the load of the cooling or heating coil, or DX-coil in an air-handling unit, packaged unit, or air conditioner. On hot summer days, solar radiation and the latent ventilation load must be included in the building load. However, both sunny and cloudy days may occur in cold weather; therefore, building load is calculated and analyzed with and without solar radiation in cold weather especially for the control zones facing south in a building in northern latitudes, or facing north in southern latitudes, where solar radiation is often a primary cooling load on sunny days.

In Fig. 22.2 is shown the building load analysis of a typical floor of a multistory building without solar radiation in winter. In this figure, line *ABCDE* represents the building load curve at various outdoor temperatures T_o, in °F. Point *A* on this curve represents the summer design refrigeration load q_{rl}, in MBtu/h. During summer design load, all the space cooling loads are offset by the cold supply air, and the condensing heat of the refrigeration system is rejected to the cooling tower.

When the outdoor temperature T_o drops below 75°F, on the left side of point *B* on the building load curve, then the following occur:

- The perimeter zone may suffer a transmission loss.
- Solar-radiation is excluded.
- The latent load of the outdoor ventilation air is no longer included in the building load because the outdoor air is often drier than the space air.

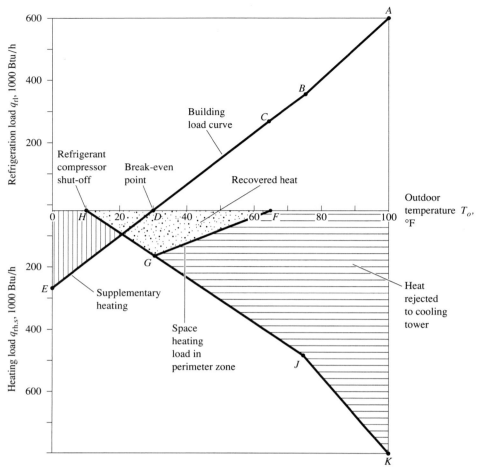

FIGURE 22.2   Building load analysis of a typical floor in a multistory building without solar radiation in winter.

If the outdoor temperature $T_o < T_F$, hot condenser water is supplied to the heating coils in the perimeter zone to satisfy the heating load. Here T_F indicates the outdoor temperature at point F.

When the outdoor temperature drops to T_D, the heat recovered from the interior zone plus the power input to the compressor is exactly equal to the heating load of the perimeter zone. No supplementary heating is needed. Point D is called the break-even point of the building. When $T_o < T_D$, supplementary heating is necessary to maintain a desirable space temperature.

As the outdoor temperature falls to T_H, the load on the cooling coil in the interior zone becomes zero. The refrigeration compressors are turned off. No recovery of condensing heat is possible. When $T_o \le T_H$, all of the heat needed for the perimeter zone will be provided by the supplementary heating.

The area of the triangle FGH indicates the heat energy recovered from the internal loads in the interior zone, which is used to offset the heat losses in the

perimeter zone by means of hot condenser cooling water.

Similar building load curves with solar radiation for entire buildings and for south-facing zones in the building should be calculated and analyzed in winter. For building loads with solar radiation, break-even point D will move to a lower T_o. Recovered heat will be greater, and supplementary heating will be less. For south-facing zones in buildings in northern climates, cold supply air may be required during sunny days in the perimeter zone even in winter.

22.2   AIR-SOURCE HEAT PUMP SYSTEMS

In an air-source heat pump system, outdoor air acts as a heat source from which heat is extracted during heating and as a heat sink to which heat is rejected during cooling. Since air is readily available everywhere, air-source heat pumps are widely used in residential and many commercial buildings.

Air-source heat pumps can be classified as individual room heat pumps and packaged heat pumps. Individual room heat pumps serve only one or two rooms without ductwork. Packaged heat pumps can be subdivided into rooftop heat pumps and split heat pumps.

According to the data in EIA *Characteristics of Commercial Buildings* (1986), air-source heat pumps were used for space heating in about 9 percent of commercial buildings (by floor space) at the end of 1986. For commercial buildings built from 1984 to 1986, 15 percent of the floor space was served by air-source heat pumps.

System Components

Most air-source heat pumps consist of single or multiple compressors, indoor coils through which air is conditioned, outdoor coils where heat is extracted from or rejected to the outdoor air, capillary tubes, reversing valves that change the heating operation into a cooling operation and vice versa, an accumulator to store liquid refrigerant, and other accessories.

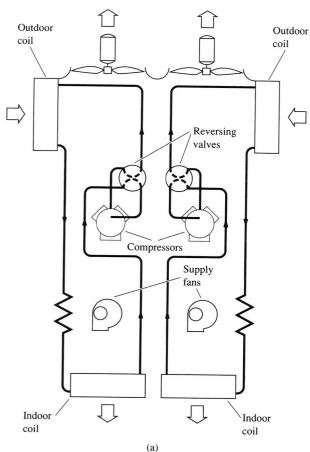

FIGURE 22.3a A typical rooftop heat pump: (*a*) schematic diagram.

A typical rooftop packaged heat pump that uses HCFC–22 as refrigerant is shown in Fig. 22.3. In this heat pump, dual circuits, consisting of two compressors, two indoor coils, two outdoor coils, two throttling devices, and two reversing valves, are often used for greater capacity and better defrosting control.

COMPRESSOR. Reciprocating compressors are the most widely used type of compressor in heat pumps. For large packaged heat pumps, dual circuits are used. In Bucher et al. (1990) the median service life of the compressor in a heat pump is 14.5 years, and the median service life of a heat pump system is 19 years, depending on the conditions of operation and maintenance. Median service life is the age at which 50 percent of the units have been removed from service and 50 percent remain in service.

INDOOR COIL. In an air-source heat pump, the indoor coil is not necessarily located inside the building. The indoor coil in a rooftop packaged heat pump is mounted on the roof. However, an indoor coil always heats and cools the indoor supply air. During cooling operation, the indoor coil acts as an evaporator. It provides the refrigeration effect to cool the mixture of outdoor and recirculating air when the heat pump is operating in the recirculating mode. During heating operation, the indoor coil acts as a condenser. The heat rejected from the condenser raises the temperature of the conditioned supply air. For heat pumps using halocarbon refrigerants, the indoor coil is usually made from copper tubing and corrugated aluminum fins.

OUTDOOR COIL. The outdoor coil acts as a condenser during cooling and as an evaporator to extract heat from the outdoor atmosphere during heating. When an outdoor coil is used as a condenser, a series-connected subcooling coil often subcools the refrigerant for better system performance. An outdoor coil always deals with outdoor air, whether it acts as a condenser or an evaporator. Like the indoor coil, an outdoor coil is usually made of copper tubing and aluminum fins for halocarbon refrigerants. Plate or spine fins are often used instead of corrugated fins to avoid clogging by dust and foreign matter.

REVERSING VALVE. Reversing valves are used to guide the direction of refrigerant flow when cooling operation is changed over to heating operation or vice versa. The rearrangement of the connections between four ways of flow—compressor suction, compressor

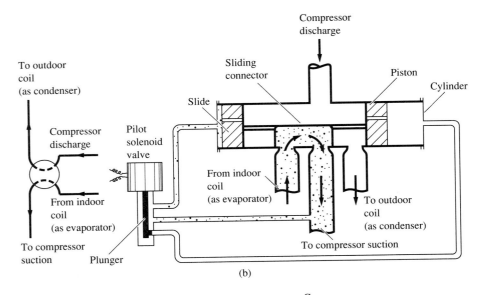

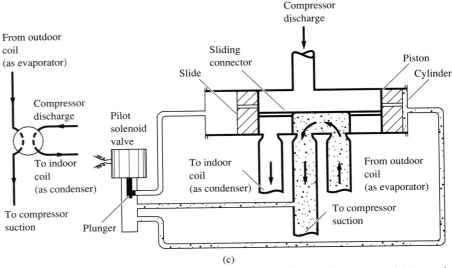

FIGURE 22.3 A typical rooftop heat pump: (b) reversing valve, cooling mode; and (c) reversing valve, heating mode.

discharge, evaporator outlet, and condenser inlet—causes the function of the indoor and outdoor coils to reverse. It is therefore called a *four-way reversing valve*.

A typical four-way reversing valve consists of a hollow cylinder with an internal slide to position the flow paths of the refrigerant. The slide is driven by the differential between the discharge and suction pressure. Both of them are introduced to opposite ends of the cylinder by a pilot valve that is energized by a solenoid coil. The operation of this typical four-way reversing valve is shown in Fig. 22.3b and 22.3c.

A reversing valve is a highly reliable system component. The efficiency losses altogether including leakage, heat transfer, and the pressure drop across the reversing valve, cause a decrease of 4 to 7 percent in heat pump performance.

Capillary Tube

In most heat pumps, a capillary tube is used as the throttling device. A *capillary tube*, also sometimes called a *restrictor tube*, is a fixed length of small-diameter tubing lying between the outlet of the condenser and the inlet of the evaporator. It reduces the pressure of the refrigerant from the high-pressure side to the low-pressure side of the system and controls the flow of refrigerant to the evaporator. The inside diameter of the capillary tube D_{in} is between 0.05 and 0.06 in. with a length L from an inch to several feet.

It is usually made of copper. In recent years, there has been a tendency to use short capillary tubes having L/D_{in} between 3 and 20 in heat pumps. These short capillary tubes are often made of brass or copper alloy. A strainer is generally located before the capillary tube to prevent clogging of the tube caused by impurities and foreign matter contained in the refrigerant.

In Aaron and Domanski (1990), tests and experiments for short capillary tubes showed that when the downstream pressure was lower than the saturated pressure of the entering refrigerant, the fluid inside the short tube had a metastable inner core of liquid with a surrounding two-phase annular ring. The flow showed a very weak dependence upon the downstream pressure and could be considered a nonideal choked flow. The diameter of the short capillary tubes was the parameter that most strongly affected the mass flow rate of the refrigerant. A greater differential between the upstream and downstream pressure also had a significant effect on the mass flow rate. Chamfering the inlet of the short capillary showed only a minor influence on the rate of refrigerant flow. By using a short capillary tube as a throttling device, one can operate the cooling mode even at an outdoor temperature of 35°F if necessary.

A capillary tube has a certain capability to balance the refrigerant flow when the refrigeration load changes. At a given condensing pressure, if the evaporating pressure is lower than normal due to a lower refrigeration load, the greater pressure difference causes a higher mass flow rate of the refrigerant through the capillary tube. Because the refrigerant flow handled by the compressor is lower due to the lower suction pressure, a decrease in the condensing pressure, and liquid level, and seal occurs in the condenser. Some vapor then passes through the capillary tube into the evaporator. All of these reduce the mass flow rate of refrigerant flowing through the capillary tube and form a new balance.

When the evaporating pressure becomes higher because of a greater refrigeration load, more liquid refrigerant accumulates in the condenser. Meanwhile, the liquid charge in the evaporator is insufficient. These effects tend to reduce the heat transfer area in the condenser and the wetted surface area in the evaporator. Consequently, the condensing pressure is raised and the evaporating pressure is lowered. A greater high- and low-side pressure difference results in a higher refrigerant flow and forms a new balance.

An accumulator is usually installed on the suction line ahead of the inlet to the compressor to store the liquid refrigerant before the defrosting process.

Other accessories, such as a filter dryer, sight glass, strainer, liquid level indicator, solenoid valves, and manual shut-off valves are the same as described in Chapter 21.

Operating Modes

The operation of an air-source heat pump can be divided into cooling mode and heating mode.

COOLING MODE. When the discharge air temperature sensor detects an increase in the air temperature above a predetermined limit at the exit of the indoor coil, cooling is required in the air-source heat pump. The DDC controller deenergizes the pilot solenoid valve, as shown in Fig. 22.3b. The low-pressure suction vapor is now connected to the lefthand end of the slide, and the high-pressure hot gas between the pistons of the slide then pushes the piston toward the left end of the cylinder. The indoor coil now acts as an evaporator and extracts heat from the conditioned air flowing through the indoor coil. After evaporation, vapor refrigerant from the indoor coil passes through the sliding connector of the slide and flows to the suction line. Hot gas discharged from the compressor is led to the outdoor coil, which now acts as a condenser.

An economizer cycle can be used when an outdoor air sensor detects the outdoor temperature dropping below a specific limit during cooling mode. The DDC controller opens and modulates the outdoor damper to admit cold outdoor air to maintain a preset discharge air temperature.

HEATING MODE. When the discharge air sensor detects a drop of air temperature below a predetermined limit at the exit of the indoor coil, heating is required. The DDC controller energizes the pilot solenoid valve as shown in Fig. 22.3c. The plunger of the solenoid valve moves upward and connects the right-hand end of the slide to the low-pressure suction line. High-pressure hot gas then pushes the pistons toward the right-hand end of the cylinder. Consequently, the hot gas from the compressor is discharged to the indoor coil, which now acts as a condenser. Heat is rejected to the recirculating air and the hot gas is then condensed into liquid form. Liquid refrigerant flows through the capillary tube and then vaporizes in the outdoor coil, which extracts heat from outdoor air. The outdoor coil now acts as an evaporator.

Electric heating can be used for supplementary heating. When the discharge air temperature sensor detects a drop in air temperature further below preset limits, the electric heater can be energized in steps to maintain the required discharge temperature. Supplementary heating is energized only when the space

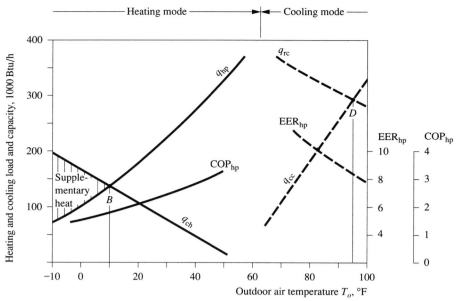

FIGURE 22.4 System performance of a rooftop heat pump.

heating load cannot be offset by the heating effect of the heat pump.

System Performance

System performance of an air-source heat pump can be illustrated by plotting the heating coils load q_{ch}, the cooling coil's load q_{rc}, the heating capacity of the heat pump q_{hp}, the cooling capacity of the heat pump q_{rc}, the coefficient of performance of the heat pump during heating COP_{hp}, and the energy efficiency ratio of the heat pump during cooling EER_{hp} against the outdoor temperature, as shown in Fig. 22.4. Coil's loads and heat pump capacities all are expressed in Btu/h.

The heating capacity of an air-source heat pump using a reciprocating compressor can be calculated as

$$
\begin{aligned}
q_{hp} &= \dot{V}_p \rho_{suc} \eta_v (q_{1'4} + q_{2'3'}) \\
&= \dot{V}_p \rho_{suc} \eta_v [(h_{1'} - h_4) + (h_{2'} - h_{3'})]
\end{aligned}
\tag{22.5}
$$

where $\dot{V}_p$ = piston displacement of reciprocating compressor, fpm

ρ_{suc} = density of suction vapor, lb/ft³

η_v = volumetric efficiency

The cooling capacity of an air-source heat pump q_{rc} can be calculated from Eq.(4.25) as described in Chapter 4. Both the cooling coil's load q_{cc} and the heating coil's load q_{ch} should be offset by the cooling capacity q_{rc} and the heating capacity q_{hp} provided by the heat pump.

When an air-source heat pump is operated in cooling mode during summer, the condensing temperature T_{con} and the condensing pressure p_{con} drop as the outdoor temperature T_o falls. These decreases result in higher η_v and, therefore, an increase in the cooling capacity of the heat pump q_{rc} as well as the energy efficiency ratio of heat pump during cooling EER_{hp}. The fall of T_o also causes a decrease in the space cooling load q_{rc} and the cooling coil load q_{cc}. The intersection of the q_{rc} and q_{cc} curves, point D, indicates the design cooling coil's load and the cooling capacity of the selected heat pump.

When the heat pump is operated in heating mode, a fall of T_o causes a decrease in the evaporating temperature T_{ev}. A lower T_{ev} results in a lower η_v, a smaller refrigeration effect $q_{1'4}$, and a lower density of suction vapor ρ_{suc}. All of these effects result in a smaller heating capacity of the heat pump q_{hp}. Although the work input W_{in} increases as T_o and T_{ev} decrease, the effect of the increase in W_{in} on the increase of q_{hp} is small.

The fall of T_o also causes a rise in space heating load q_{rh} and the heating coil's load q_{ch}. The intersection of the heating capacity q_{hp} curve and the heating coil's load q_{ch} curve, point B, is the balance point at which the heating capacity of the heat pump is equal to the required heating coil's load. When the outdoor temperature T_o drops below this balance point (such as $T_o < 10°F$, as shown in Fig. 22.4), supplementary heating from the electric heater or other heat source is required to maintain a preset discharge air temperature. The coefficient of performance of the heat pump during heating COP_{hp} also drops as T_o falls.

The heat pump and supplementary heating operate simultaneously until the COP_{hp} drops below a certain value, such as below 1 when the use of an electric heater is more cost-effective to operate the heat pump. The heat pump should be turned off.

Cycling Losses and Degradation Factor

For split packaged air-source heat pumps and individual room heat pumps, indoor coils are located inside the building and outdoor coils are mounted outdoors. When an on-off control is used for the compressor, during the off cycle refrigerant tends to migrate from the warmer outdoor coil to the cooler indoor coil in summer and from the warmer indoor coil to the cooler outdoor coil during winter. When the compressor starts again, the transient state performance shows that a 2- to 5-minute operating period of reduced capacity is required before the heat pump can operate at full capacity. Such a loss due to cycling of the compressor is called *cycling loss*.

Cycling losses are illustrated by the following indexes: cycling loss factor F_{cyc} and degradation coefficient C_d. Cycling loss factor can be calculated as

$$F_{cyc} = \frac{COP_{cyc}}{COP_{ss}} \qquad (22.6)$$

In this expression, COP_{cyc} represents the coefficient of performance of the air-source heat pump during a whole cycle. It is the ratio of useful output during the cycle to the energy input to the compressor and fan during the cycle, both in Btu/h. COP_{ss} indicates the steady-state coefficient of performance, or the COP when compressors and fans are operating continuously.

The degradation coefficient C_d can be calculated as

$$C_d = \frac{1 - F_{cyc}}{1 - LF} \qquad (22.7)$$

Here, LF represents the load factor and is defined as

$$LF = \frac{t_{on}}{t_{cyc}} \qquad (22.8)$$

where t_{on} = time in a cycle for which the compressor is turned on, min
t_{cyc} = time of a complete cycle, min

Cycling losses depend on (1) the cycling rate (whether it is 2, 3, 4, or 5 cycles per hour), (2) the indoor-outdoor temperature difference, and (3) the fraction of on-time per cycle. At design conditions, theoretically, $F_{cyc} = 1$. In Baxter and Moyers (1985), field tests of a typical heat pump in an unoccupied single-family house in Knoxville, TN between 1981

and 1983 showed that the degradation coefficient C_d in heating season is 0.26, and C_d in cooling season is 0.11.

Minimum Performance

To encourage the use of energy-efficient air conditioners and heat pumps, ASHRAE Standard 90.1–1989 specifies the minimum performance as of January 1, 1992 (see Table 15.3). The standard ratings are as follows:

Cooling. Indoor entering air temperature 80°F dry bulb and 67°F wet bulb; outdoor unit (cooling air) entering air temperature 95°F

High-temperature heating. Indoor entering air temperature 70°F dry bulb; outdoor unit entering air, 47°F dry bulb and 43°F wet bulb

Low-temperature heating. Indoor entering air temperature 70°F dry bulb; outdoor unit entering air 17°F wet bulb.

The following is a comparison of the performance of an air-cooled, electrically operated rooftop heat pump with a gross nominal cooling capacity of 121,000 Btu/h and net nominal heating capacity of 115,000 Btu/h at standard rating conditions produced in 1992 by a U.S. manufacturer with the specified minimum performance listed in ASHRAE Standard 90.1–1989 before and after January 1, 1992:

	Actual product, 1992	Standard 90–1989 beginning January 1, 1992
Cooling (EER)	9.1	8.9
Heating, COP (47°F)	3.0	3.0
Heating, COP (17°F)	2.2	2.0

Defrosting

Most air-source heat pumps use the reverse cycle to melt the frost that formed on the outdoor coil during heating mode operation in cold weather. The reverse cycle defrost switches the heating mode operation, in which the outdoor coil acts as an evaporator, to cooling mode operation, where the outdoor coil acts as a condenser. Hot gas is forced into the outdoor coil to melt the frost that accumulated on the outdoor coil. After the frost is melted, the heat pump is switched back to normal heating mode operation.

O'Neal and Peterson (1990) described the defrosting process of an air-source heat pump using HCFC-22 as refrigerant and having a short capillary tube of diameter of 0.059 in. as the throttling device. When the reversing valve was energized, the outdoor fan stopped and the defrosting cycle began. The suction and discharge pressures equalized. The sudden decrease in pressure in the indoor coil caused the liquid refrigerant to vaporize. The compressor became temporarily starved and pulled the pressure in the indoor coil down to 23 psia absolute about one minute after defrosting started. Once the suction pressure had fallen low enough, refrigerant flow began to increase. In the interval from 1.0 to 3.5 minutes after defrosting, the refrigerant changed from vapor to saturated liquid upstream of the short capillary tube. This change caused a substantial increase in refrigerant flow, and frost was melted at the outdoor coil. After 6 minutes, the refrigerant flow fell 30 percent because of the subcooling in outdoor coil. Defrosting usually lasts a few minutes up to about ten minutes, depending on the frost accumulation.

During defrost, a cold supply air from the indoor coil may cause a low space temperature and a draft. Supplementary electric heating should be considered to maintain an acceptable indoor temperature. ASHRAE/IES Standard 90.1–1989 permits supplementary heating of less than 15 minutes to be used during defrost cycles.

Defrosting only takes place on the outdoor coil. The initiation and control of defrosting can be performed by a clock or an intelligent or adaptive timer. It can also be controlled by measuring the capacity of the unit or by measuring the temperature differential between the refrigerant inside the outdoor coil and the ambient air. Defrosting terminates when the temperature of liquid refrigerant leaving the outdoor coil (or the coil temperature) rises above 60°F.

Controls

Because reciprocating compressors are used extensively for heat pumps, capacity controls using on-off, speed modulation, and cylinder unloading (as described in Section 21.5) are generally used.

Either discharge air temperature or return temperature can be used as the criterion to change automatically from cooling mode to heating mode and vice versa. A dead band of 2 to 3°F and a time delay are always required between cooling and heating mode operations to prevent short cycling.

Most of the packaged heat pumps provide controls of high pressure, low pressure, head pressure, or low ambient control; freezing protection of the indoor coil; protection from overloading, and supplementary heating. The principle and operation of these controls are the same as described in Section 21.5. A microprocessor-based DDC panel may be used to integrate all the controls in one package and to add time delay, compressor lockout, loss of refrigerant charge, and short-cycling protection controls to the sequence control of heat pump and gas furnace in heating mode operation, and air economizer and refrigeration capacity control in cooling mode operation.

Capacity and Selection

Air-source heat pump capacity is selected according to its cooling capacity because supplementary heating may be required under winter design conditions. Also, the rated cooling capacity at summer design conditions is often greater than the rated heating capacity at winter design conditions.

When an air-source heat pump is installed directly inside or above the conditioned space, the noise generated by the heat pump must be taken into consideration. Attenuation remedies should be provided if necessary to maintain an NC curve at an acceptable level in the conditioned space.

In 1992, air-source heat pump products were available ranging in cooling capacity from a fraction of a ton to about 40 tons with an indoor air flow of 16,000 cfm.

22.3 WATER-LOOP HEAT PUMP SYSTEMS

Water-Source Heat Pump Systems

Water-source heat pump systems are gaining more and more applications in residential and commercial buildings. Water-source heat pumps extract heat from the cooling water from evaporative water coolers, cooling towers, groundwater (well water), or other surface water during heating mode operation, and reject heat to cooling water during cooling mode operation. A boiler or an electric heater is usually added to provide supplementary heating.

There are three main types of water-source heat pump systems: water-loop, groundwater, and surface-water heat pump systems.

OPERATING CHARACTERISTICS OF WATER-LOOP HEAT PUMP SYSTEMS. A water-loop heat pump system conserves more energy than other heat pump systems only when simultaneous heating and cooling occur in a building. Consequently, heat rejected from the core or sunny side of the building can be transferred to the perimeter zone or shady side of the building for heating purposes during winter or intermediate seasons. A water-loop heat pump system

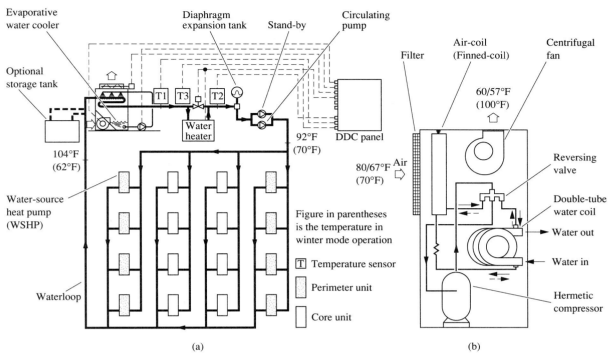

FIGURE 22.5 A typical water-loop heat pump system: (*a*) schematic diagram; and (*b*) water-source heat pump.

is essentially a combination of heat pump and heat recovery system.

A water-loop heat pump system consists of water-source heat pumps, evaporative coolers or a cooling tower with plate-and-frame heat exchangers, a boiler or electric heaters, two water-circulating pumps, an expansion tank, piping, and necessary accessories. One of the water-circulating pumps is the lead pump, the other is a standby. A storage tank is optional.

A typical water-loop heat pump system is illustrated in Fig. 22.5. During hot weather, when the outdoor wet bulb is 78°F, all the water-source heat pumps are operating in cooling mode. The water leaves the closed-circuit evaporative water cooler at 92°F and enters the water loop. After all of the heat is rejected from water-source heat pumps, the water returns to the evaporative water cooler at a temperature of 104°F and is cooled to 92°F again.

During moderate weather, water-source heat pumps serving the shady side of the building are in heating mode, whereas those serving the sunny side of the building are in cooling mode. The rise or fall in water temperature depends on the ratio of the number of cooling units to the total number of units or, more exactly, the ratio of heat rejection to heat extraction. If heat rejected from the cooling units is greater than heat extracted by the heating units, the average temperature of the water loop tends to rise, and vice versa. When the water temperature entering the evaporative

water cooler $T_{e.ev}$ rises to 90°F, the cooler starts to operate and brings the water temperature down to a lower value (85°F or lower).

During cold weather, some water-source heat pumps serving the core of the building still operate in cooling mode. If the heat extracted by the units operating in heating mode is greater than the heat rejected, the average temperature of the water loop drops. When the temperature $T_{e.ev}$ falls below 60°F, the boiler is energized. Heat is added to the water loop to raise $T_{e.ev}$ to 68°F or higher.

A water-loop heat pump system recovers the heat rejected to the water loop by the core units or by the units on the sunny side and supplies it to the units serving the perimeter zone in which heating is required to offset heat losses.

Water-Source Heat Pumps

A water-source heat pump (WSHP) usually consists of an air-coil that is a finned coil to condition the air; a double-tube water-coil, to reject or extract heat from water, as described in Section 12.7; a forward-curved centrifugal fan, which is often located downstream of the air coil; single or twin hermetic compressors; a short capillary tube; a reversing valve; an outer casing; controls; and accessories. A two-speed fan motor is often used for better capacity control. A typical water-source heat pump is shown in Fig. 22.5b. In a vertical water-source heat pump, the centrifugal fan is usually located at the top outlet, and the hermetic compressor is often mounted in the bottom of the unit. In a horizontal packaged unit, the centrifugal fan is usually located at the end of the unit.

During cooling mode operation, the air-coil acts as an evaporator and the water-coil as a condenser. Air in a typical water-source heat pump is cooled and dehumidified at the air-coil from an entering dry bulb of 80°F and wet bulb of 67°F to a leaving condition of 60°F dry bulb and 57°F wet bulb. The suction temperature is about 43°F. On the other hand, in heating mode operation, the air-coil acts as a condenser and the water-coil as an evaporator. During cold weather air is heated at the air-coil from 70°F to an off-coil temperature of 100°F. The suction temperature in winter is about 45°F. Sometimes, electrical resistance heating, in steps, is added between the air-coil and the centrifugal fan instead of using a centralized water heater.

Water-source heat pumps are available in cooling capacities of 0.5 to 30 tons and heating capacities of 8 to 200 MBtu/h. Both vertical and horizontal units can be connected with ductwork to distribute the conditioned air.

Water-source heat pumps that are used in perimeter zones and installed under windowsills are called *perimeter WSHPs*. Perimeter WSHPs usually have relatively smaller cooling and heating capacities. WSHPs that are used in the interior or core zone are called *core WSHPs*. These units often have a relatively greater cooling and heating capacity, are in either horizontal or vertical units, are connected with ductwork, and are located in special fan rooms.

Closed-Circuit Evaporative Water Coolers

A *closed-circuit evaporative water cooler*, sometimes called a *closed-circuit cooling tower*, resembles a cooling tower. However, cooling water in a cooling tower is an open-circuit system, whereas in an evaporative water cooler the cooling water is a closed-circuit system (see Fig. 22.5a). Cooling water flows through a coil on which recirculating water is sprayed. Heat is rejected from the cooling water through the tube wall and is absorbed by the vaporized liquid at the outer surface of the coil. A closed-circuit evaporative water cooler is usually located outdoors, most probably on the rooftop.

Both centrifugal fans and propeller fans can be used in closed-circuit evaporative water coolers. The capacity of a centrifugal fan is more easily controlled with a damper and makes less noise; therefore, it is more frequently used.

The capacity of a closed-circuit evaporative water cooler can be modulated by the following methods:

- *Convective cooling.* Opening of the top damper only
- *Natural draft evaporative cooling.* Spraying of recirculating water and opening of the top damper
- *Forced draft evaporative cooling.* Operating the fan and water spraying
- *Forced draft evaporative cooling with damper modulation*

If a cooling tower is used, the cooling water discharged from the cooling tower should be connected to a plate-and-frame heat exchanger, as shown in Fig. 13.6. Water from the plate-and-frame heat exchanger forms a closed loop with water coils in the water-loop heat pump system. The reason for a closed-loop water system is to prevent fouling of the water coils in water-source heat pumps.

Freezing Protection

If a water-loop heat pump system is installed where winter outdoor temperatures are always above 32°F, no freezing damage occurs to the outdoor portion of the water-loop, including the closed-circuit evaporative cooler, water pumps, and outdoor water pipework. For locations where outdoor temperatures drop below 32°F, freezing protection of the outdoor portion of the water-loop must be considered. There are three methods to provide freezing protection:

- Use pipe and equipment insulation and a sump heater in the evaporative water cooler, and circulate the water-loop continuously. Draining and isolating the evaporative cooler from the water-loop make it difficult to meet the requirement for occasional use of the evaporative water cooler in mild outdoor weather during winter.

  According to Kush (1990), field experience in an installation in Stamford, Connecticut, showed that the heat and pumping energy losses from the outdoor evaporative water cooler and piping work were about half of the energy input to the electric boiler in January 1988. Therefore, continuous operation of the evaporator water cooler produced substantial heat losses and is not suitable in locations where outdoor temperatures often drop below freezing in winter.

- Add a certain percentage of ethylene glycol solution to the water-loop to provide freezing protection. Glycol is expensive. Inhibitors must be added at the same time to resist corrosion. Adding ethylene glycol decreases the heat transfer coefficient as well as the heating and cooling capacities of the equipment.

- Isolate the closed-circuit evaporative water cooler and the outdoor portion of the water-loop from the indoor portion by means of a plate-and-frame heat exchanger. Doing this necessitates the addition of far less ethylene glycol to the outdoor water circuit than in the preceding method. In many applications, this method has proved to be the most economical.

Storage Tanks

Whether a storage tank should be used in a water-loop heat pump system depends on the local electricity rate structures and a careful analysis of initial costs and operating costs. The principle of thermal storage will be discussed in Chapter 25. In many applications, the provision of a storage tank in a water-loop heat pump system reduces the demand for electricity at peak load and results in energy savings and lower costs.

Heat rejected to the water-loop at a temperature between 60 and 90°F can be stored in the storage tank to offset nighttime heat losses if the water-loop operates at night with a setback space temperature.

Many utilities offer lower electricity rates at off-peak times. An evaporative cooler can be used to cool the water to a lower temperature. It can then be stored in the tank for use during the cooling peak load in summer to reduce costs and the demand for electricity. An electric boiler also may be energized at night to raise the water temperature to 180°F or higher. High-temperature stored water can be used in winter during the morning pick-up load to reduce electricity demand and energy costs.

Controls

WATER-LOOP TEMPERATURE CONTROL. The water temperature leaving the closed-circuit evaporative water cooler is controlled between 60 and 90°F when the water-loop heat pump system is in operation. For a typical water-loop heat pump system, a temperature sensor T1 may be located at the exit of the evaporative cooler, as shown in Fig. 22.5a. When the microprocessor-based DDC panel receives a signal from T1, it actuates the following:

At 57°F	Alert by alarm. WSHP, evaporative cooler, water heater, and pumps all shut down.
At 60 to 68°F	Energize the water heater.
At 86°F	Open the damper of the evaporative cooler.
At 88°F	Start the water-spraying pump.
At 90°F	Turn on the first-stage fan(s).
At 92°F	Turn on the second-stage fan(s).
At 106°F	Alert by alarm. System shuts down.

An outdoor air sensor T3 resets the water heater energizing temperature T2 in such a manner that when the outdoor temperature T_o drops from 60°F to 0°F, T2 increases linearly from 60°F to 68°F.

In order to maintain space air at 75°F, 50 percent relative humidity and a dew point of 57°F in summer, it is critical to maintain a minimum water-loop temperature $T_{wl} = 57°F$. If $T_{wl} < 57°F$, condensate may form on the outer surface of uninsulated pipes in the conditioned space and cause damage.

The actuating temperatures for turning off the second-stage and first-stage (fans), shutting off the water spraying pump, and closing the damper in the evaporative water cooler as water temperature decreases should be successively 2°F lower in order to prevent short cycling (For example, turn off the second stage fan at 90°F, turn off the first-stage fan at 88°F, etc.).

CAPACITY CONTROL OF WATER HEATER. If the water heater is a gas-fired hot water boiler or a hot water heat exchanger, a temperature sensor T2 senses the water temperature leaving the water heater. The heating capacity of the heater can be controlled by modulating the flow rate of gas or hot water by means of a DDC controller. Because most hot water boilers contain a limited amount of water, if the temperature sensor is located at the exit side of the boiler, the temperature response to modulation is fast. Therefore, a broadband DDC controller should be used to prevent short cycling.

If an electric boiler is used, step control in appropriate stages can be performed through a sensor located at the inlet to sense the temperature of the entering water. Step control often causes temperature fluctuation at the leaving side.

SAFETY CONTROL. In addition to the high- and low-limit control of the water-loop temperature, a flow switch or pressure differential sensor should be installed across the water pump. In case of pump failure, the entire system, including the water-source heat pumps, evaporative cooler, or water heater shuts down. After a delay of 10 to 15 seconds, the standby pump is energized. At the same time, an alarm sounds, along with an indicating light. When water flow is resumed, the evaporative cooler or the water heater and water-source heat pumps are restarted, with a delay between each stage to limit any sudden increase in starting current.

WATER-SOURCE HEAT PUMP CONTROL. When the push button is engaged, the electric circuit is energized and the outdoor air damper is opened. When the space temperature rises above the cooling mode set point, the temperature sensor T4 calls for cooling. The DDC controller then starts the fan at either the HI or LO position, which is set manually. After a 30- to 50-second delay, if the high-pressure and low-pressure safety control circuits are closed, the compressor is started. WSHP is now in cooling mode operation. When the zone temperature drops below the cooling mode set point and is sensed by T4, the DDC controller stops the compressor. The fan should be operated continuously when the space is occupied to provide outdoor ventilation air to the space.

When the zone temperature drops below the heating mode set point and is sensed by the temperature sensor T5, the DDC controller starts the compressor. WSHP is operating in heating mode. When the zone temperature rises above the heating mode set point, the DDC controller stops the compressor. There is always a dead band between the zone temperature set points of cooling and heating mode operation.

In either heating or cooling mode operation, if the discharge pressure rises above the high-pressure limit or the suction pressure drops below the low-pressure limit, the high- or low-pressure control opens the electric circuit, lights an alarm lamp, and stops the compressor.

Actual Performance of a Water-Loop Heat Pump System

For a modern office building in Stamford, Connecticut, Kush (1990) analyzed and reported the actual performance of a water-loop heat pump system that served a total area of 72,000 ft² office space on three floors. This system has 140 1-ton perimeter WSHPs, 12 10-ton core WSHPs, a closed-circuit evaporative cooler, and a 300 kW electric boiler. Stamford has a 97.5 percent winter design temperature of 9°F and 5617 heating degree-days. In summer, a 2.5 percent design dry bulb is 84°F with a mean coincident wet bulb of 71°F.

According to the results measured between November 1987 and October 1988, the breakdown of annual HVAC&R energy use was as follows:

Electric boiler	32 percent
Evaporative water cooler	4 percent
Water-loop pumps	11 percent
Perimeter WSHPs	18 percent
Core WSHPs: fan	8 percent
cooling	15 percent
heating	2 percent
Other	9 percent

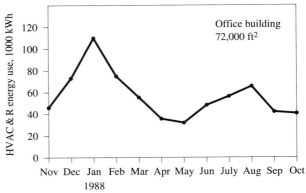

FIGURE 22.6 HVAC&R monthly energy use of a water-loop heat pump system.

The energy used for heating in core WSHPs is mainly caused by excessive outdoor ventilation air.

The monthly HVAC&R energy use of this building between November 1987 and October 1988 is shown in Fig. 22.6. The highest HVAC&R energy use, 110,050 kWh, occurred in January 1988. Of this, 66 percent was consumed by the electric boiler. The highest in summer months, 65,715 kWh, occurred in August 1988. About half of August energy use was used by core WSHPs. Year-round, the lowest energy use, 32,330 kWh, occurred in May 1988, when the core WSHPs consumed 40 percent of the energy input.

As of January 1, 1992, ASHRAE Standard 90.1–1989 specifies the minimum performance of water-source heat pumps as follows:

Indoor air	Entering water	Minimum performance
Water-source heat pump $q_{rc} < 65,000$ Btu/h		
80°F dry/67°F wet	85°F	9.3 EER
80°F dry/67°F wet	75°F	10.2 EER
Water-source heat pump $65,000$ Btu/h $\leq q_{rc} < 135,000$ Btu/h		
80°F dry/67°F wet	85°F	10.5 EER

Design Considerations

A water flow rate of 2.5 to 3.5 gpm, typically 3 gpm, per cooling ton is appropriate for water-loop heat pump systems. For most closed-circuit evaporative water coolers, a 12°F approach at 80°F outdoor wet bulb and a 15°F at 75°F are suitable.

Outdoor ventilation air rates should be adequate and should not be greater than that specified in ANSI/ASHRAE Standard 62–1989, as listed in

Table 5.9. The outdoor air damper should be closed when the conditioned space is not occupied.

Fan power in water-source heat pumps has a certain influence on energy use. Both supply volume flow rate and total pressure loss should be carefully calculated in order to select the proper unit.

As in air-source heat pumps, the noise generated by WSHPs located under the windowsill or installed directly in the conditioned space must be taken into consideration. Necessary attenuation remedies should be provided.

22.4 GROUNDWATER HEAT PUMP SYSTEMS

Groundwater heat pump (GWHP) systems use well water as a heat source during heating and as a heat sink during cooling. When the groundwater is more than 30 ft deep, its year-round temperature is fairly constant. Groundwater heat pump systems are usually open-loop systems. They are mainly used in low-rise residences in northern climates such as New York or North Dakota. Sometimes they are used for low-rise commercial buildings where groundwater is readily available and local codes permit such usage.

Groundwater Systems

For commercial buildings, the design engineer must perform a survey and study the site and surroundings to define the available groundwater sources. The design engineer should be fully aware of the legalities of water rights.

A test well should be drilled to ensure the availability of groundwater. If water is corrosive, a plate-and-frame heat exchanger may be installed to separate the groundwater and the water entering the water coil in the water-source heat pump.

Usually, two wells are drilled. One is the supply well, from which groundwater is extracted by submersible pump impellers and supplied to the WSHPs. The other well is a recharge or injection well. Groundwater discharged from the WSHP is recharged to this well. The recharge well should be at least 100 ft away from the supply well. Using a recharge well provides for resupply to the groundwater and also prevents the collapse of the building foundation near the supply well due to subsidence. If the quality of groundwater meets the requirement and if the local codes permit, groundwater discharged from the WSHP can be used as service water or can be drained to the nearby river, lake, or canal. Groundwater intake water screens of the supply well may be located in several levels where water can be extracted. For small supply wells for res-

idences, the pump motor is directly connected to the submersible pump underneath the impellers, whereas for large wells, the motor is usually located at the top of the supply well. Information regarding groundwater use regulations and guidelines for well separation and for the construction of supply and recharge wells are included in Donald (1985) *Water Source Heat Pump Handbook* published by the National Water Well Association (NWWA).

If the temperature of the groundwater is below 50°F, direct cooling of the air in the WSHP should be considered. If the groundwater temperature exceeds 55°F and is lower than 70°F, precooling of recirculating air or make-up air may be economical.

Because the groundwater system is an open-loop system, it is important to minimize the vertical head to save pump power. Standard 325–85 ARI recommends that the groundwater pump power not exceed 60 W per gpm. If the pump efficiency is 0.7, the allowable head for the well pump, including static head, head loss across the water coil or heat exchanger, valves, and piping work losses, should preferably not exceed 220 ft of water.

If a recharge well is used and the discharge pipe is submerged under the water table level in the recharge well, as shown in Fig. 22.8, the groundwater system is most probably a closed-loop system, depending on whether the underground water passage between the supply and recharge well is connected or broken.

Groundwater Heat Pump System for a Hospital

A groundwater heat pump system described in Knipe (1983) was a 1980 retrofit project for a hospital in Albany, NY (see Fig. 22.7). Altogether, 540 gpm of groundwater was supplied from a 12-in. diameter, 500-ft deep well. During summer, the groundwater temperature was 58°F, and in winter, it dropped to 46°F. About one third of the groundwater was supplied to a make-up air-handling unit. The other two thirds was sent to a heat pump which had the following operating characteristics:

Summer:
- Well $\longrightarrow$ condenser $\longrightarrow$ discharge to nearby river
- Evaporator $\longrightarrow$ chilled water to AHU and terminals $\longrightarrow$ evaporator

Winter:
- Well $\longrightarrow$ evaporator $\longrightarrow$ chilled water to AHU and terminals $\longrightarrow$ discharge to nearby river
- Condenser $\longrightarrow$ domestic hot water preheat $\longrightarrow$ perimeter heating $\longrightarrow$ condenser

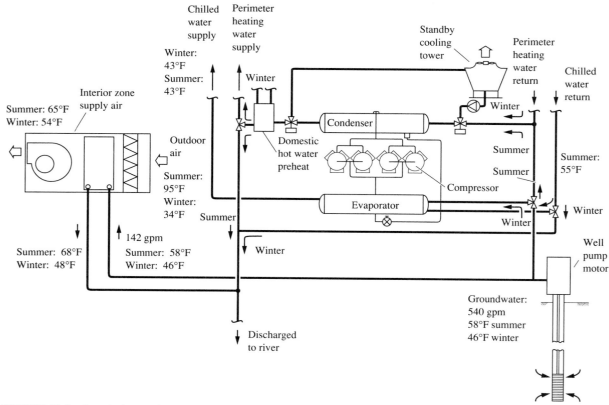

FIGURE 22.7 A typical groundwater heat pump system for a hospital.

In summer, groundwater entered the condenser at 58°F and left at 68°F. In winter, groundwater entered the evaporator at 46°F and left at 43°F.

The original installation had two 200-ton absorption chillers. Winter heating was supplied from gas-fired boilers. After retrofit, this groundwater heat pump system, including precooling and preheating, saved 30 percent of the energy used compared with the previous year's expense.

Groundwater Heat Pump Systems for Residences

A typical groundwater heat pump system for residences is shown in Fig. 22.8. Such a heat pump system usually has a rated heating capacity from 24,000 to 60,000 Btu/h. Groundwater is extracted from a supply well by means of a submersible well pump and is forced through a precooler or direct cooler. It then enters the water coil in the water-source heat pump. After that, groundwater is discharged to a recharge well. If the recharge pipe is submerged underneath the water table, as described in the previous section, such a groundwater system is most probably a closed-loop system. In the recharge well, the water level is raised

in order to overcome the head loss required to force the groundwater discharging from the perforated pipe wall through the water passage underground. The vertical head required is the difference between the water levels in the supply and recharge wells, as shown in Fig. 22.8.

The water-source heat pump for residential GWHP systems has a structure similar to the WSHP in the water-loop heat pump system, as shown in Fig. 22.5b, except that a precooler or direct cooler may be added.

Operating parameters and characteristics for a groundwater heat pump system are as follows:

- The groundwater flow rate for a water-source heat pump should vary from 2 to 3 gpm per 12,000 Btu/h (heating capacity). The well pump must be properly sized. A greater flow rate and an oversized well pump are not economical.
- The pressure drop of the groundwater system should be minimized. The pressure drop per 100 ft of pipe should be less than 5 ft/100 ft of length. Unnecessary valves should not be installed. Gate valves or ball valves should be used instead of globe valves to reduce pressure loss. A water tank is not necessary.

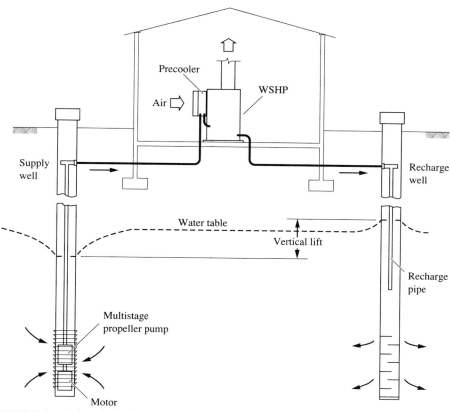

FIGURE 22.8   A typical residential groundwater heat pump system.

- Direct cooling or precooling of recirculating air by means of groundwater increases significantly the EER of the GWHP system.
- Water containing excessive concentrations of minerals causes deposits on heat pump water coils that reduce the heat pump performance.
- In locations where the annual heating degree days exceed 7000, more than 80 percent of the operating hours of the GWHP systems are for space heating.
- An electric heater may be used for supplementary heating in cold climates or in other locations where it is necessary. Operation of the electric heater when the heating capacity of GWHP is equal to or even greater than the heating load must be avoided.
- Water-source heat pumps should be properly sized. They should not be operated for short-cycle durations (cycles less than five minutes) in order to prevent cycling losses and excessive wear and tear on the refrigeration system components. Cycle durations between 10 and 30 minutes are considered appropriate.
- Extraction and discharge of groundwater must comply with local codes and regulations.
- The temperature of groundwater tends to increase with its use.

A parameter called *seasonal performance factor* SPF is often used to assess the performance of a groundwater heat pump system. SPF is defined as

$$\text{SPF} = \frac{q_{\text{sup}}}{q_{\text{cons}}} \qquad (22.9)$$

where q_{sup} = all heat supplied by the GWHP during the heating season, Btu

q_{cons} = energy consumed during the heating season, Btu

For systems with a cooling capacity $q_{\text{rc}} < 135,000$ Btu/h, ASHRAE Standard 90.1–1989 specifies the following minimum performance for water-source heat pumps using groundwater as of January 1, 1992:

Entering water temperature, °F	EER
70	11.0
50	11.5

A groundwater heat pump system has a fairly constant COP even if the outdoor air temperature varies. According to Rackliffe and Schabel (1986), the SPF and EER for 15 single-family houses in New York state during 1982 to 1984 were as follows:

System SPF (heating)   1.9 to 3   Average 2.3
System average EER
  (cooling)   5.6 to 14   Average 9.2

In many locations, groundwater heat pump systems usually have a higher SPF and seasonal EER than air-source heat pumps. WSHP system capacity remains fairly constant at very low and very high outdoor temperatures.

The main disadvantage of a groundwater heat pump system is its higher initial cost. More maintenance is required for systems using water with high mineral content. If the water table is 200 ft or more below ground level, the residential groundwater heat pump system is no longer energy-efficient compared with high-efficiency air-source heat pumps.

22.5   GROUND-COUPLED AND SURFACE WATER HEAT PUMP SYSTEMS

Ground-coupled heat pump systems can be categorized as ground-coil heat pump systems and direct-expansion ground-coupled heat pump systems. Of the ground-coil heat pump systems, both horizontal and vertical coils are used. Many types of direct-expansion ground-coupled heat pump systems are still being developed. In fact, horizontal ground-coil heat pump systems are the most widely used ground-coupled heat pump systems.

A horizontal ground-coil heat pump system is shown in Fig. 22.9. It is actually a closed-loop water-source heat pump system. The water-source heat

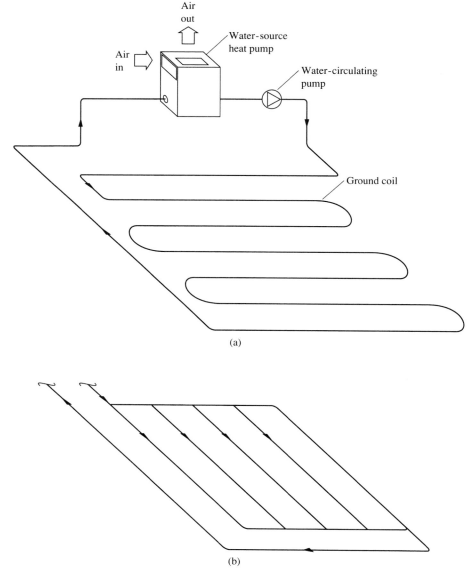

FIGURE 22.9   Ground-coupled heat pump system: (*a*) S-type ground coil; and (*b*) grid-type ground coil.

pump is similar to that used in a water-loop heat pump system that has a heating capacity between 24,000 and 60,000 Btu/h.

A horizontal ground coil is often made from polyethylene or polybutylene tubes of 1 to 2 in. external diameter in serpentine arrangement. In Ball et al. (1983), a recent advance in methodology is to bury the horizontal ground coil at two layers in a trench, typically 4 ft and 6 ft deep. Spacing between the tubes varies from 2 to 8 ft. A general guideline is to use 215 to 430 ft of copper/steel tubing for each ton of capacity for heating or cooling. When a horizontal ground-coil heat pump used in cooling mode was subjected to a heat rejection temperature that exceeded 100°F, some cracking of polybutylene pipe occurred.

A vertical ground coil is buried from 30 to 200 ft deep in drilled holes. About 200 to 250 ft length of pipe is needed for each 12,000 Btu/h heating capacity. A vertical ground coil requires less land area than a horizontal ground coil.

If the temperature of the fluid circulated in the water coil and the horizontal ground coil may drop below 32°F, aqueous solutions of ethylene, propylene glycol, or calcium chloride should be used to protect the system from freezing. A water flow rate between 2 and 3.5 gpm per 12,000 Btu/h is usually used.

In northern climates, horizontal ground-coil heat pump systems have a heating SPF of 2.5 to 3 and a cooling seasonal EER of 10.5 to 13.5. There is no significant difference in SPF between horizontal and vertical ground coils. According to Hughes et al. (1985), for newly constructed projects in upstate New York, a ground-coil heat pump system has a simple payback period of 5 to 10 years.

For water-source heat pumps that use surface water such as lake water as a heat source and sink, a plastic or copper coil, or sometimes a plate-and-frame heat exchanger, is often used to form a closed-loop system to prevent fouling of the water coil.

REFERENCES AND FURTHER READING

Aaron, D.A., and Domanski, P.A., "Experimentation, Analysis, and Correlation of Refrigerant–22 Flow through Short Tube Restrictors," *ASHRAE Transactions*, Part I, pp. 729–742, 1990.

ARI, *ARI Standard 325–85, Standard for Water Source Heat Pumps*, ARI, Arlington, VA, 1985.

ASHRAE, *ASHRAE Handbook 1987, HVAC Systems and Applications*, ASHRAE, Inc., Atlanta, GA.

ASHRAE/IES, *Standard 90.1–1989, Energy-Efficient Design of New Buildings Except New Low-Rise Residential Buildings*, ASHRAE, Inc., Atlanta, GA, 1989.

Ayres, J.M., and Lau, H., "Comparison of Residential Air-to-Air Heat Pump and Air-Conditioner/Gas Furnace Systems in 16 California Climatic Zones," *ASHRAE Transactions*, Part II, pp. 525–561, 1987.

Ball, D.A., Fischer, R.D., and Hodgett, D. L., "Design Methods for Ground-Source Heat Pumps," *ASHRAE Transactions*, Part II B, pp. 416–440, 1983.

Baxter, V.D., and Moyers, J.C., "Field-Measured Cycling, Frosting, and Defrosting Losses for a High-Efficiency Air Source Heat Pump," *ASHRAE Transations*, Part II B, pp. 537–554, 1985.

Black, G.D., "An Overview of the Four-Way Refrigerant Reversing Valve," *ASHRAE Transactions*, Part I, pp. 1147–1151, 1987.

Brown, M.J., Hesse, B.J., and O'Neil, R.A., "Performance Monitoring Results for an Office Building Groundwater Heat Pump System," *ASHRAE Transactions*, Part I, pp. 1691–1707, 1988.

Bucher, M.E., Grastataro, C.M., and Coleman, W.R., "Heat Pump Life and Compressor Longevity in Diverse Climates," *ASHRAE Transactions*, Part I, pp. 1567–1571, 1990.

Carrier Corporation, *Packaged Rooftop Heat Pumps*, Carrier Corporation, Syracuse, NY, 1980.

Donald, D.R., *Water Source Heat Pump Handbook*, National Water Well Assocation (NWWA), Worthington, OH, 1985.

Eckman, R.L., "Heat Pump Defrost Controls: A Review of Past, Present, and Future Technology," *ASHRAE Transactions*, Part I, pp. 1152–1156, 1987.

EIA, *Characteristics of Commercial Buildings*, EIA, Washington DC, 1986.

EIA, *Commercial Building Characteristics*, EIA, Washington DC, 1989.

Friberg, E.E., "Case History—Low-Rise Office Building Using Water-Source Heat Pump," *ASHRAE Transactions*, Part I, pp. 1708–1725, 1988.

Goldschmidt, V.W., "Effect of Cyclic Response of Residential Air Conditioners on Seasonal Performance," *ASHRAE Transactions*, Part II, pp. 757–770, 1981.

Howell, R.H., and Zaidi, J.H., "Analysis of Heat Recovery in Water-Loop Heat Pump Systems," *ASHRAE Transactions*, Part I, pp. 1039–1047, 1990.

Hughes, P.J., "Survey of Water Source Heat Pump System Configurations in Current Practice," *ASHRAE Transactions*, Part I, pp. 1021–1028, 1990.

Hughes, P.J., Loomis, L., O'Neil, R.A., and Rizzuto, J., "Result of the Residential Earth-Coupled Heat Pump Demonstration in Upstate New York," *ASHRAE Transactions*, Part II B, pp. 1307–1325, 1985.

Johnson, W.S., McGraw, B.A., Conlin, F., Wix, S.D., and Baugh, R.N., "Annual Performance of a Horizontal-Coil Ground-Coupled Heat Pump," *ASHRAE Transactions*, Part I A, pp. 173–185, 1986.

Kavanaugh, S.P., "Groundwater Heat Pump Performance Enhancement with Precoolers and Water Pump Optimization," *ASHRAE Transactions*, Part II, pp. 1205–1218, 1987.

Knipe, E.C., "Applications of Heat Pumps Using Groundwater Resources," *ASHRAE Transactions*, Part II B, pp. 441–451, 1983.

Kush, E.A., "Detailed Field Study of a Water-Loop Heat Pump System," *ASHRAE Transactions,* Part I, pp. 1048–1063, 1990.

Mathen, D.V., "Performance Monitoring of Select Groundwater Heat Pump Installations in North Dakota," *ASHRAE Transactions,* Part I B, pp. 290–303, 1984.

Meckler, M., "Integrating Water Source Heat Pumps with Thermal Storage," *Heating/Piping/Air Conditioning,* pp. 49–64, July 1988.

Mohammad-zadeh, Y., Johnson, R.R., Edwards, J.A., and Safemazandarani, P., "Model Validation for Three Ground-Coupled Heat Pumps," *ASHRAE Transactions,* Part II, pp. 215–221, 1989.

Mulroy, W.J., "The Effect of Short Cycling and Fan Delay on the Efficiency of a Modified Residential Heat Pump," *ASHRAE Transactions,* Part I B, pp. 813–826, 1986.

O'Neal, D.L., and Peterson, K., "A Comparison of Orifice and TXV Control Characteristics during the Reverse-Cycle Defrost," *ASHRAE Transactions,* Part I, pp. 337–343, 1990.

Niess, R.C., "Applied Heat Pump Opportunities in Commercial Buildings," *ASHRAE Transactions,* Part II, pp. 493–498, 1989.

Rackliffe, G.B., and Schabel, K.B., "Groundwater Heat Pump Demonstration Results for Residential Applications in New York State," *ASHRAE Transactions,* Part II A, pp. 3–19, 1986.

Rasmussen, R.W., MacArthur, J.W., Grald, E.W., and Nowakowski, G.A., "Performance of Engine-Driven Heat Pumps under Cycling Conditions," *ASHRAE Transactions,* Part II, pp. 1078–1090, 1987.

The Singer Company, *Electro Hydronic Systems,* The Singer Co., Carteret, NJ, 1977.

The Trane Company, *Water Source Heat Pump System Design,* The Trane Company, LaCrosse, WI, 1981.

Virgin, D.G., and Blanchard, W.B., "Cary School—25 Years of Successful Heat Pump/Heat Reclaim System Operation," *ASHRAE Transactions,* Part I A, pp. 40–45, 1985.

Weinstein, A., Eisenhower, L.D., and Jones, N.S., "Water-Source Heat Pump System for Mount Vernon Unitarian Church," *ASHRAE Transactions,* Part I B, pp. 304–312, 1984.

Wurm, J., and Kinast, J.A., "History and Status of Engine-Driven Heat Pump Developments in the U.S.," *ASHRAE Transactions,* Part II, pp. 997–1005, 1987.

CHAPTER 23

REFRIGERATION SYSTEMS: CENTRIFUGAL

23.1 CENTRIFUGAL VAPOR COMPRESSION REFRIGERATION SYSTEMS

A centrifugal vapor compression refrigeration system uses a centrifugal compressor to compress the refrigerant and provide chilled water to cool the air in air-handling units and terminals. A centrifugal refrigeration machine usually comes as a factory-assembled packaged unit and is often called a centrifugal chiller, as shown in Fig. 23.1. Centrifugal chillers are widely used for large central hydronic air conditioning systems in commercial and industrial applications.

Most centrifugal chillers are driven by open and hermetic motors. Occasionally, they may be driven by internal combustion engines. Sometimes, large centrifugal chillers are driven by steam and gas turbines.

Refrigerants

According to Hummel et al.(1991), in 1988 there were about 73,000 centrifugal chillers in the United States. Of these, 80 percent used CFC–11, 10

percent used CFC–12, and the remaining 10 percent used CFC–114, HCFC–22, or others.

In February 1992, the Bush administration announced the termination of use of CFCs as refrigerants at the end of 1995 except for equipment already in service. All CFCs used in the 70,000 existing centrifugal chillers at the beginning of the 1990s are expected to be replaced by HCFCs and HFCs before the end of 1995.

HCFC–123 will replace CFC–11. HCFC–123 is nonflammable and low in toxicity. According to Calm (1992), efficiency could drop about 5 percent in retrofits when CFC–11 is replaced by HCFC–123. A capacity reduction of 15 to 20 percent is possible.

HCFC–123 is compatible with most of the materials now used with CFC–11. Most compatibility problems due to the solvent action of HCFC–123 have been resolved. Another important difference is that the DuPont Company, a manufacturer of these refrigerants, in 1991 specified a limit of 10 ppm in air for HCFC–123 rather than 1000 ppm for CFC–11. Therefore, air monitors that incorporate an alarm device should be installed both in the plant room and in places that may have leaks to detect refrigerant concentrations. Exhaust systems should be designed

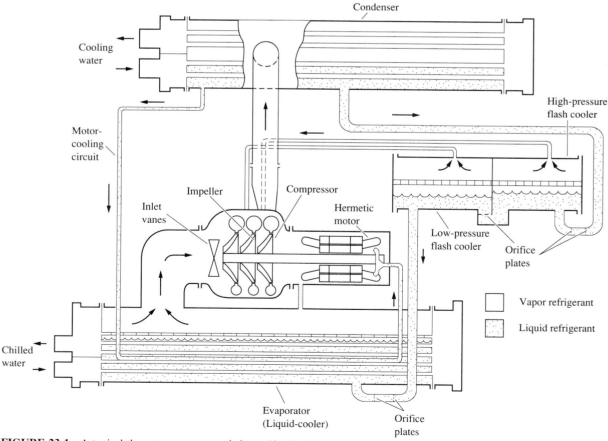

FIGURE 23.1 A typical three-stage, water-cooled centrifugal chiller.

for the refrigerant plant room. Many chillers have been successfully converted from using CFC–11 to HCFC–123.

HFC–134a will replace CFC–12. Corr et al. (1993) have shown that when HFC–134a replaces CFC–12 in chillers, HFC–134a has a power consumption and capacity similar to CFC–12, and it shows good compatibility with the polymers and metals used with CFC–12 in centrifugal chillers. However, mineral oils used with CFC–12 have a low mutual solubility with HFC–134a. Synthetic oils have been developed that are expected to be used with HFC–134a.

The 1990 Clean Air Act made the known venting of CFCs and HCFCs illegal as of July 1, 1992. Recovery/recycling equipment must be used to prevent venting during installation, operation, and maintenance.

System Components

A centrifugal chiller consists of a centrifugal compressor, an evaporator or liquid cooler, a condenser, a throttling device, piping connections, controls, and possibly a purge unit and a flash cooler.

COMPRESSOR. A single-stage, two-stage, or three-stage compressor—according to the number of internally connected impellers—may be used, as described in Section 10.10. For a refrigeration system with an evaporating temperature of 40°F and a condensing temperature of 100°F, HCFC–123 has a compression ratio of 3.59. A single-stage centrifugal compressor needs a higher peripheral velocity and, therefore, a higher speed and larger impeller to meet such a requirement than two-stage or three stage compressors.

Many centrifugal chillers with a refrigeration capacity less than 1200 tons are installed with hermetic compressors. Very large centrifugal chillers often employ open compressors in which an additional motor cooling system is used.

Centrifugal compressors need high peripheral velocity to produce head lift. They can be driven either directly by motor or through a gear train. A direct-drive compressor often has a larger diameter impeller because of the limitation on the rotating speed of a two-pole motor. On the other hand, direct-drive reduces energy loss because of the gear-train driving mechanism. Such a loss is usually 2 to 3 percent of the total power input to the compressor.

EVAPORATOR. A flooded shell-and-tube evaporator, as described in Section 12.6 is usually used for a centrifugal chiller because of its compact size and high heat transfer characteristics.

CONDENSER. Water-cooled, horizontal shell-and-tube condensers, as described in Section 12.7, are widely used for their lower condensing pressure and easier capacity control. Only in locations where local codes prohibit the use of city water as the make-up water in cooling towers are air-cooled centrifugal chillers used.

FLASH COOLER. A flash cooler is sometimes called an economizer. It flashes a small portion of refrigerant at the intermediate pressure to cool the remaining liquid refrigerant to the saturated condition in order to provide a greater refrigeration effect, as described in Chapter 4. For a two-stage compressor, a single-stage flash cooler is used. For a three-stage compressor, a two-stage flash cooler is used.

ORIFICE PLATES OR FLOAT VALVES. Orifice plates (arranged in series) or float valves are used as throttling devices in centrifugal chillers that employ flooded refrigerant feed.

The use of multiple orifices as a throttling device is shown in Fig. 23.2. This device controls the amount of refrigerant feed to the liquid cooler according to the pressure of the liquid refrigerant in the condenser.

During full-load operation, a certain liquid pressure is maintained before the first orifice plate. When liquid refrigerant flows through the first orifice plate, it encounters a pressure drop. However, the fluid pressure between the two orifice plates is still higher than the saturated pressure of the liquid. No flash of vapor occurs in the region between the orifice plates. The second orifice plate meters the maximum refrigerant flow.

As the refrigeration load in the centrifugal chiller drops, less refrigerant is delivered to the condenser

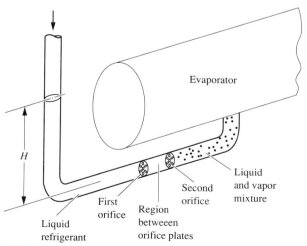

FIGURE 23.2 Orifice plates.

and evaporator, resulting in a lower condensing pressure and a higher evaporating pressure. This reduces the pressure exerted on the liquid refrigerant at the condenser due to lower condensing pressure and hydrostatic head. Because of the pressure drop across the first orifice plate, the fluid pressure between the orifice plates is less than the saturated pressure of the liquid refrigerant. Flashing of vapor occurs. Both liquid and vapor flow through the second orifice plate. Consequently, the mass flow rate of the refrigerant flow is reduced.

There are no moving parts in an orifice plate throttling device. It is simple and reliable in operation.

Purge Unit

At an evaporating temperature $T_{ev} = 40°F$, HCFC–123 has a saturated pressure of 5.8 psia which is far lower than atmospheric pressure. Noncondensable gases such as air and carbon dioxide may diffuse and leak into the evaporator through cracks and gaps. Because a condenser is always located at a higher level in a centrifugal chiller, any leakage of noncondensable gases into the system usually accumulates in the upper part of the condenser.

The presence of noncondensable gases reduces the refrigerant flow as well as the rate of heat transfer in the evaporator and the condenser. It also increases the pressure in the condenser. All of these result in a loss of refrigeration capacity and an increase in power consumption. Wanniarachchi and Webb (1982) showed that at 60 percent load, the presence of 3 percent noncondensable gas causes a 2.6 percent increase in the power input to the compressor. The purpose of the purge unit is to purge out periodically any noncondensable gases that have accumulated in the condenser.

A typical purge unit is shown in Fig. 23.3. A mixture of noncondensable gases and gaseous refrigerant is extracted by a positive displacement purge compressor from the upper part of the condenser. Oil is separated from the refrigerant in an oil separator. Noncondensable gases and gaseous refrigerant are then delivered to a finned cupronickel water-cooling coil in a brass purge drum. When the latent heat of condensation is absorbed by the chilled water in the cooling coil, the gaseous refrigerant condenses into liquid form and sinks to the bottom of the drum. Water contained in the refrigerant is separated and forms a middle layer. All of the noncondensable gases accumulate at the top of the drum as a result of their lower densities.

As the gas pressure in the drum exceeds a certain limit, the relief valve opens and the noncondensable

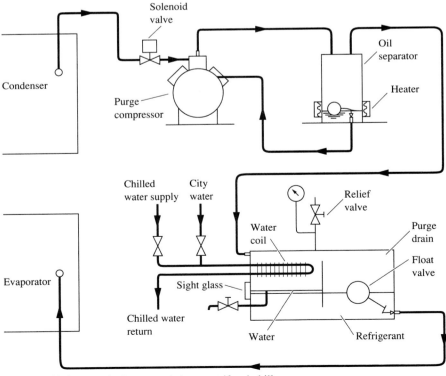

FIGURE 23.3   A typical purge unit in a centrifugal chiller.

gases are purged out from the drum. Liquid refrigerant is returned to the evaporator through a float valve, and water is drained off when it rises above a specific level, as observed through the sight glass.

Types of Centrifugal Chillers

Centrifugal chillers may be classified according to the configuration of their main components as follows:

- Single-stage or multistage
- Air-cooled, water-cooled, or double-bundle condenser (A double-bundle condenser has two bundles of condensing tubes for the purpose of heat recovery.)
- Open or hermetic
- Direct-drive or gear-drive
- Capacity control provided by inlet vanes or variable-speed drive

As mentioned in Section 10.10, because a centrifugal compressor needs a high peripheral velocity, it is not economical to fabricate small centrifugal machines. Their refrigeration capacities range from 100 to 10,000 tons. Because of their large capacity, DX-coils are not used as evaporators to condition air at such a large size and capacity, with their unacceptable air transporting and distribution losses.

23.2 WATER-COOLED CENTRIFUGAL CHILLERS

Refrigerant Flow

A typical water-cooled centrifugal chiller that uses a three-stage, hermetic, and direct-drive compressor was illustrated in Fig. 23.1. As described in Section 4.14 and shown in Fig. 4.9a and 4.9b, the flow process of HCFC–123 is as follows:

Vapor refrigerant is extracted from liquid cooler as a result of the suction of the compressor:

- It enters the first-stage impeller through inlet vanes.
- Hot gas from the first-stage impeller mixes with flashed vapor from the low-pressure (second-stage) flash cooler.
- The mixture enters the second-stage impeller.
- Hot gas from the second-stage impeller mixes with flashed vapor from the high-pressure (first-stage) flash cooler.
- The mixture enters the third-stage impeller.
- Hot gas discharged from the third-stage impeller enters the water-cooled condenser.

- Hot gas is desuperheated, condensed, and subcooled to liquid form in the condenser.
- A small portion of the liquid refrigerant is used to cool the hermetic motor.
- Most of the liquid refrigerant enters the high-pressure flash cooler through a multiple-orifice throttling device, and a small portion flashes into vapor and mixes with the hot gas discharged from the second-stage impeller.
- Most of the liquid refrigerant from the high-pressure flash cooler enters the low-pressure flash cooler through a multiple-orifice throttling device, and a small portion flashes into vapor and mixes with the hot gas discharged from the first-stage impeller.
- Most of the liquid refrigerant from the low-pressure flash cooler enters the evaporator through a multiple-orifice device.
- Liquid refrigerant vaporizes in the evaporator and produces its refrigeration effect there.

Performance Rating Conditions

According to ARI Standard 550–88, the refrigeration capacity of a centrifugal chiller is rated under the following conditions:

Water temperature leaving evaporator:		
	100 percent load	44°F
	0 percent load	44°F
Chilled water flow rate:		2.4 gpm/ton
Entering condenser water temperature:		
	100 percent load	85°F
	0 percent load	60°F
Condenser water flow rate:		3.0 gpm/ton
Fouling factor in evaporator and condenser:		0.00025 h · ft² · °F/Btu

The integrated part-load value (IPLV) (described in Section 15.8) of the centrifugal chiller at standard rating conditions can be calculated by the following formula, as specified by ARI 550–88:

$$\text{IPLV} = 0.1 \, \frac{A + B}{2} + 0.5 \, \frac{B + C}{2} + 0.3 \, C + D + 0.1 \, D \tag{23.1}$$

where A, B, C, D = kW/ton or COP at 100, 75, 50, and 25 percent load, respectively. If the operating conditions are different from the standard rating conditions for the same centrifugal chiller, a general rule for purposes of estimation can be taken as follows: When the entering temperature of chilled water T_{ee} is between 40 and 50°F, for each °F increase or decrease of T_{ee}, there is roughly a 1.5 percent difference in refrigeration capacity and energy use.

When the entering temperature of condenser cooling water T_{ce} is between 80 and 90°F for each °F increase or decrease of T_{ce}, there is roughly a 1 percent difference in refrigeration capacity; there is roughly a 0.6 percent difference in energy use.

For water-cooled centrifugal chillers with hermetic compressors, manufacturer's products that are currently available have a refrigeration capacity between 150 and 2000 tons. For water-cooled centrifugal chillers with open compressors, the refrigeration capacity varies from 150 to 10,000 tons.

For hermetic compressors using vaporized liquid refrigerant to cool the motor, no additional motor-cooling system is required. Leakage of refrigerant from the bearing seals is also less than in the open compressor. Its main drawback is that 2 to 4 percent of the liquid refrigerant must be used to cool the motor. Its refrigeration capacity is reduced, and its energy use is increased.

23.3 CENTRIFUGAL CHILLERS INCORPORATING HEAT RECOVERY

System Description

Many large commercial and industrial applications often require heating in the perimeter zone but cooling in the interior zone during the winter. If the building is considered as a whole, a heat recovery system can be used to transfer the internal heat from the interior zone to the perimeter zone to offset the winter heating load there.

A typical centrifugal chiller incorporating a heat recovery system using a double-bundle condenser is presented in Fig. 23.4. In a double-bundle condenser, water tubes are classified as tower bundles and heating bundles. Heat rejected in the condenser may be either discharged into the atmosphere through the tower bundle and cooling tower or used for heating in the coils of the perimeter zones through the heating bundle. The tower bundle may be enclosed in the same shell with the heating bundle, as shown in Fig. 23.4b, but separate plates are needed to divide the water circuits.

A storage tank stores the hot water from the heating bundle when the building is occupied. Hot water is then used for heating in the perimeter zone when the building is either occupied or unoccupied. A supplementary heater using, for example, electric power may be used to raise the temperature of the hot water from the heating bundle, if necessary. A tower bypass line is always used to maintain the cooling water entering the condenser at a required value.

Instead of a double-bundle condenser, an auxiliary condenser plus a standard condenser can be used, as shown in Fig. 23.5. Hot gaseous refrigerant always tends to be drawn to either the standard or the auxiliary condenser by the condenser's lower temperature. When heating is required in the perimeter zone, the auxiliary condenser pump is energized and the tower bypass three-way valve bypasses part or all of the cooling water returning to the standard condenser. Because the water returning from the perimeter zone and entering the auxiliary condenser is colder than that entering the standard condenser due to the heating load in the perimeter zone, the auxiliary condenser tube bundle has a lower temperature and attracts the hot gas from the compressor. The auxiliary condenser recovers as much heat as it can allow. The remaining rejected heat goes to the standard condenser and is rejected to the atmosphere through the cooling tower. A balance of heat rejection between these two condensers automatically holds.

A hot water bypass should be included if an auxiliary condenser is used, as shown in Fig. 23.5. Such a bypass ensures that water will not circulate through the auxiliary condenser when the chiller is deenergized. In Fig. 23.4a, the hot water pump P2 is energized only when the compressor of the centrifugal chiller is in operation.

Operating Modes

Depending on whether heating is required in the perimeter zone, the operating modes of a centrifugal chiller that is incorporated with a heat recovery system in a building with perimeter and interior zones can be divided into cooling mode and heating mode operation. Both the cooling and heating mode operations can be further divided into occupied and unoccupied periods, based on whether the conditioned space is occupied. The typical operation of a centrifugal chiller that is incorporated with a heat recovery system is described below.

OCCUPIED COOLING MODE. Cooling mode operation takes place when the space temperature in the perimeter zone $T_{r.p} \geq 75°F$. The occupied cooling mode starts when the chilled water pump P1 is turned on (see Fig. 23.4a.) Meanwhile, the hot water pump P2 remains inactive; tower pump P3 forces all of the cooling water to flow through the tower bundle, the modulation valve V3, and back to the cooling tower. Chilled water is then supplied to the perimeter zone through zone pump P6 or standby pump P7 and to the interior zone through zone pump P4 or standby pump P5. The chilled water supply temperature T_{ws} is usually between 42 and 45°F.

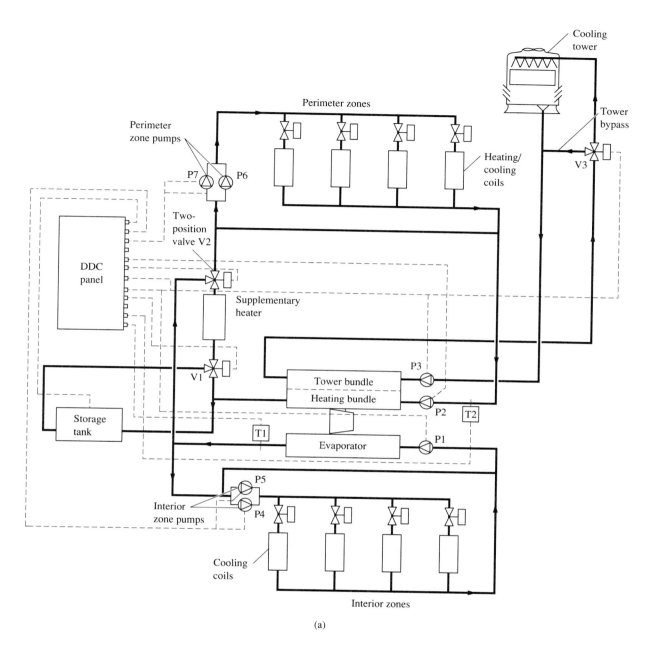

(a)

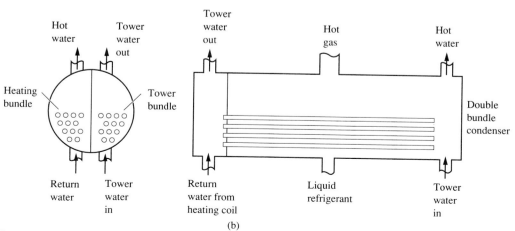

(b)

FIGURE 23.4 Centrifugal chiller incorporating a heat recovery system: (a) schematic diagram; and (b) double-bundle condenser.

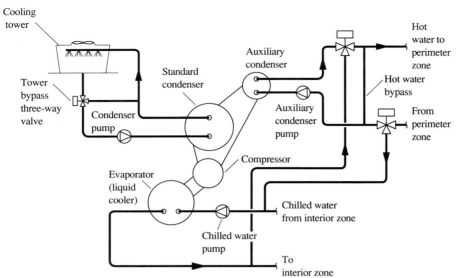

FIGURE 23.5 Heat recovery system using an auxiliary condenser.

UNOCCUPIED COOLING MODE. When $T_{r.p} \geq 75°F$ and the space is unoccupied, as determined by a timer, the DDC panel shuts down the entire centrifugal chiller and heat recovery system.

OCCUPIED HEATING MODE. When $T_{r.p} < 70°F$ and the conditioned space is occupied, the system is under occupied heating mode operation. The centrifugal chiller is energized and pump P1 supplies chilled water to the interior zone through zone pumps P4 and P5. At the same time, the DDC panel actuates the two-position valve V2 to block passage of chilled water to the perimeter zone. The hot water pump P2 is energized so that hot water from the heating bundle in the condenser is forced through the deenergized supplementary heater and supplied to the heating coils in the perimeter zone through zone pump P6 or P7 at a temperature between 105 and 110°F. If the rejected heat in the heating bundle is more than that required in the heating coils in the perimeter zone, the return water temperature T_{ret} increases before the inlet of the heating bundle and pump P2. When sensor T2 detects such a rise in temperature, the DDC panel modulates valve V1, which permits some hot water to flow to the storage tank. When the storage tank is full of hot water and sensor T2 detects a further increase of water temperature, the modulation valve is positioned so as to allow all of the hot water to bypass the storage tank. Tower pump P3 starts, and valve V3 is modulated to control the flow rate of cooling water to the cooling tower.

If the outdoor temperature T_o is so low that the heat rejection at the heating bundle is not sufficient to meet the heating requirement in the perimeter zone,

the hot water temperature detected by sensor T2 is less than a preset value. The tower pump is then stopped. Hot water then bypasses the storage tank, and the supplementary heater is gradually energized so as to maintain the water temperature sensed by T2 within a predetermined range.

UNOCCUPIED HEATING MODE. When $T_o < 70°F$ and the conditioned space is unoccupied, the DDC panel commands the following actions:

1. The hot water temperature detected by sensor T2 is reset to a value according to outdoor temperature T_o to maintain a night setback indoor temperature T_r between 55 and 60°F.
2. Only hot water pump P2 and zone pump P6 or P7 are in operation; all other equipment ceases to operate.
3. Valve 1 is modulated so that hot water from the storage tank alone is supplied to the heating coils in the perimeter zone.

If T2 senses a water temperature lower than the preset value, the supplementary heater is then gradually energized. The modulating valve V1 allows the hot water to bypass the storage tank and then enter the heater.

System Characteristics

First, compared with a cooling-only centrifugal chiller, the refrigeration cycle of a centrifugal chiller incorporating a heat recovery system has a greater pressure differential between condensing and evaporating pressure $\Delta p_{e.c}$, in psi. This is because a cooling-only chiller usually has water leaving the condenser at a temperature at 90 to 95°F during summer design conditions, whereas a chiller incorporating heat recovery needs a higher condenser-leaving temperature T_{cl} of 105 to 110°F for the hot water supply in winter heating. Although the evaporating temperature can be slightly higher in winter than in summer, an increase of 10 to 12°F in temperature difference between condensing and evaporating temperature necessitates roughly 10 to 15 percent more power input to the compressor.

Second, in order to be more energy efficient, T_{cl} should be reduced in part-load operation during winter heating. Cooling water return temperature T_{ce} should be adjusted to provide an optimum supply temperature T_{cl} at part-load.

Third, an air economizer reduces the heat recovery effect. This result is due to the reduction of the refrigeration load, which also reduces the heat recovery from condensing heat.

Finally, for the same building with equal areas of conditioned space, the heating load is smaller than the cooling load in many locations in the United States. For a refrigeration plant with multiple centrifugal chillers, usually only one or two of them include heat recovery. The remainder are cooling-only centrifugal chillers, which are more efficient.

23.4 AUTOMATIC BRUSH CLEANING FOR CONDENSERS

Principles and Operation

The cooling water from the cooling tower is usually in an open-circuit system. As described in Section 12.6, although many well-maintained cooling towers may have a long-term fouling factor for condenser tube surfaces of less than 0.0002 h · ft² · °F/Btu, the fouling factor for condenser tubes may exceed 0.0005 h · ft² · °F/Btu when significant amounts of suspended solid are contained in cooling water. High fouling factors result in a high condensing pressure and higher

kW/ton. ARI Standard 550–88 specifies a field fouling allowance of 0.00025 h · ft² · °F/Btu for condensers.

The purpose of an automatic brush cleaning system is to remove deposits and dirt from the water-side tube surface of a condenser for which the fouling factor of the condenser tubes is higher than 0.00025 h · ft² · °F/Btu. Cleaning brushes in a condenser tube of a horizontal, two-pass, and shell-and-tube condenser are illustrated in Fig. 23.6. Special nylon brushes are inserted inside the tube and catch baskets are attached to each end of the tube.

When condenser water flows through the condenser tubes, the difference in the pressure across the brush, approximately 1 psi, causes the brush to move in the direction of the water flow. As the brushes in the condenser move, they remove deposits from the inner tube surfaces. The deposits are carried away by the condenser water. The brushes are then caught by the baskets at one end. The time taken for the brushes to travel the whole length of the tube is about 1 to 2 minutes.

A special four-way valve is mounted at the cooling water entrance, as shown in Fig. 23.7a. When the four-way valve is in normal operation, the upper half of the brushes are caught by the baskets at the right-hand end of the condenser tube, and the lower half by the lefthand end baskets. During normal operation, the condenser water enters the condenser through the lower entrance so as to provide better subcooling for the liquid refrigerant accumulated at the bottom of the condenser.

After a predetermined time, a signal is given to the control system to unload the compressor by closing the inlet guide vanes or reducing the speed of the compressor. The four-way valve reverses the water direction, which causes the brushes to move in the opposite direction, from right to left for the upper half brushes and from left to right for the bottom half, until they are caught by the baskets at the ends. As soon as the brushes are caught, a signal is given to the control system to reverse the water flow again so that the four-way valve is in normal operation. The brushes stay in

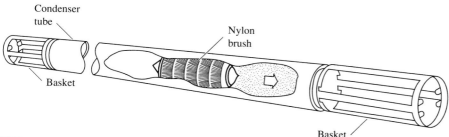

FIGURE 23.6 Cleaning brushes in a condenser tube. (*Source*: Water Services of America. Reprinted with permission.)

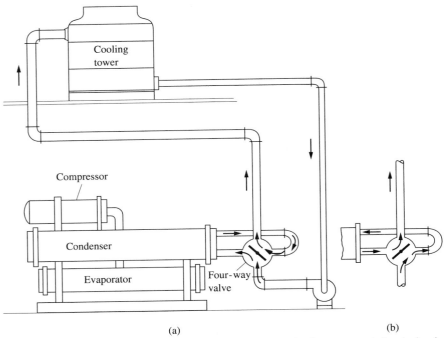

FIGURE 23.7 Normal and reverse operation of a four-way valve in an automatic brush cleaning system: (a) four-way valve at normal operation; and (b) four-way valve at reverse operation.

the end baskets during normal operation for about one to four hours until a signal reverses the water flow to repeat the process. The frequency of the cleaning cycle depends on the contaminants contained in the condenser water. The nylon brushes usually have a service life of four to five years.

Effect of Brush Cleaning Systems

Automatic brush cleaning systems have been found to accomplish the following results:

- The fouling factor drops to 0.00025 h · ft^2 · °F/Btu or below. Consequently, energy costs are lower.
- The maintenance cost can be reduced because the interval for cleaning the condenser tubes is extended from 6 months to 12 months or even longer.
- Chemicals for water treatment are still required, but their quantity can be reduced appropriately according to the actual operating conditions.

23.5 CENTRIFUGAL COMPRESSOR PERFORMANCE MAP

The performance of a centrifugal compressor can be illustrated by a compressor performance map or, simply, a compressor map. Like a centrifugal fan, a compressor map includes the volume flow, specific work or power input, efficiency, surge region, and opening of the inlet vanes. Volume flow and system head are expressed by the compressor performance curve or, simply, the compressor curve. The centrifugal compressor can be operated either at constant speed or at variable speed.

Surge of the Centrifugal Compressor

Surge is defined as the unstable operation of a centrifugal compressor or fan with vibration and noise. Surge occurs when the compressor is unable to develop a discharge pressure that is sufficient to satisfy the requirement at the condenser. During compressor surge, the pressure-flow characteristics of the compressor fluctuate up and down along the compressor curve, resulting in pressure and flow pulsation and noise.

Surge first develops at the point where the slope of the pressure lift versus the volume flow (Δp–$\dot{V}$) curve on the compressor performance map is zero, as shown in Fig. 23.8. Pressure lift is the total pressure rise of a centrifugal compressor, in psi. For a variable-speed centrifugal compressor, surge first develops at a point of zero slope along the curve that is characteristic of compressor speed n versus refrigeration load q_{ref} in tons, on the n–q_{ref} curve. This point is determined by assuming a rotating speed, in rpm, that is lower than the minimum possible speed at a required pressure lift and volume flow of the compressor.

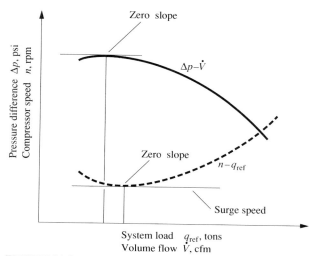

FIGURE 23.8   Compressor surge.

Surging may damage the components of the compressor. A centrifugal compressor should never be operated within the surge region on a compressor map.

Centrifugal Compressor Map at Constant Speed

A compressor map of a single-stage centrifugal compressor operated at constant speed is presented in Fig. 23.9a. The abscissa is the percentage of design volume flow of refrigerant at compressor suction $100\,\dot{V}_{suc}/\dot{V}_{suc.d}$, or load ratio

$$\text{LR} = \frac{q_{ref}}{q_{ref.d}} \tag{23.2}$$

where $q_{ref.d}$ = refrigeration load at design condition, in tons. The ordinate is the ratio of isentropic work W_{isen} to the isentropic work at design condition $W_{is.d}$, or $W_{isen}/W_{is.d}$. Isentropic work is defined as the work required for isentropic compression per unit mass of refrigerant, as described in Section 10.9. For the same compressor operated at the same conditions, the ratio $W_{isen}/W_{is.d}$ is approximately equal to the ratio of temperature lifts $R_{T.1}$ and the ratio of system heads R_{head}, that is,

$$R_{T.1} = \frac{T_{con} - T_{ev}}{T_{con.d} - T_{ev.d}} \tag{23.3}$$

and

$$R_{head} = \frac{p_{con} - p_{ev}}{p_{con.d} - p_{ev.d}} \tag{23.4}$$

where $T_{con}, T_{con.d}$ = condensing temperature and condensing temperature at design condition, respectively, °F

$T_{ev}, T_{ev.d}$ = evaporating temperature and evaporating temperature at design condition, respectively °F

$p_{con}, p_{con.d}$ = condensing pressure and condensing pressure at design condition, respectively, psig

$p_{ev}, p_{ev.d}$ = evaporating pressure and evaporating pressure at design condition, respectively, psig

System head, or *head,* is the total pressure rise of the compressor, in psi. In a centrifugal refrigeration system, system head is mainly used to raise ρ_{ev} to ρ_{con}.

The relationship between the power input to the compressor P_{in}, in hp, and the isentropic work W_{isen} is

$$P_{in} = \frac{\dot{V}_{suc}\rho_{suc}W_{isen}}{42.41\,\eta_{com}} \tag{23.5}$$

where $\dot{V}_{suc}$ = volume flow of suction vapor, cfm
ρ_{suc} = density of suction vapor, lb/ft^3
η_{com} = compressor efficiency

In Fig. 23.9a, compression efficiency η_{cp}, as defined in Section 10.9, can be calculated as

$$\eta_{cp} = \frac{\eta_{com}}{\eta_{mec}} \tag{23.6}$$

where η_{mec} = mechanical efficiency [see Eq. (10.24)]. Compression efficiency usually drops when the load ratio falls below 1, as shown by the constant η_{cp} contours.

In Fig. 23.9a, "open, 0.75 open, . . . , closed" indicates the fraction of opening of the inlet vanes, to be discussed in Section 23.6. For each fraction of opening of inlet vanes, there is a corresponding compressor performance curve that relates the load ratio and R_{head}.

In Fig. 23.9a, there are also three required system head ($p_{con} - p_{ev}$) curves, illustrated by straight lines *OS, OT,* and *OW.* The system head drops when load ratio falls below 1.

In the top left corner of Fig. 23.9a, a surge line separates the surging region of this single-stage centrifugal compressor from the normal operating region. For two- or three-stage centrifugal compressors, the surge region is smaller.

Centrifugal Compressor Map at Variable Speed

A compressor map of a single-stage centrifugal compressor operating at variable speed is illustrated in Fig. 23.9b. The abscissa, ordinate, constant compression efficiency contours, and required system head curves are the same as in the compressor map at constant speed (see Fig. 23.9a). The inlet vanes are fully open

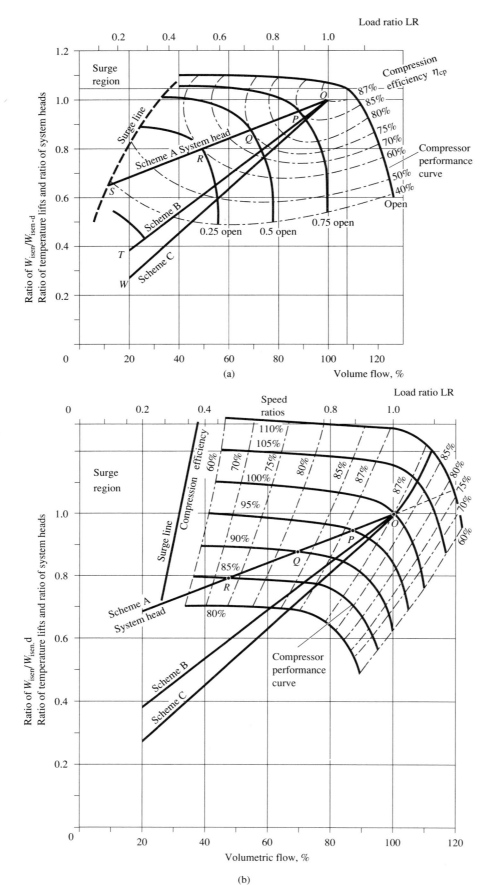

FIGURE 23.9 Compressor map of a single-stage centrifugal compressor: (a) constant speed, inlet vanes modulation; and (b) variable speed. (Adapted with permission from *ASHRAE Handbook 1988, Equipment.*)

at both design load and part-load operations. The main differences are these:

- Compressor curves are curves at various rotating speeds, such as 105 percent, 100 percent, . . . , 80 percent of speed at design condition instead of curves at various inlet vane openings on a constant speed map.
- At lower heads, the surge region is enlarged.

23.6 CAPACITY CONTROL OF CENTRIFUGAL CHILLERS

Capacity control for a centrifugal chiller is achieved mainly by modulating the volume flow of the refrigerant at the centrifugal compressor.

Differences between Centrifugal Compressors and Centrifugal Fans

During part-load operation, the following means are used to modulate the capacity of a centrifugal compressor:

- Opening the inlet vanes
- Using an adjustable frequency AC inverter to vary the compressor rotating speed
- Combining the opening of inlet vanes and variable speed control by means of an inverter
- Varying the speed of the steam- or gas-turbine-driven large centrifugal chillers

A hot gas bypass provides no energy savings at reduced system loads. ASHRAE/IES Standard 1989 discourages the use of hot-gas bypass. It should be replaced by other capacity controls.

Although centrifugal compressors and centrifugal fans are both centrifugal turbomachinery, and both centrifugal chillers and centrifugal fan–duct systems use inlet vanes and variable-speed drives to provide capacity control during part-load operation, there is an important difference between the centrifugal chiller and centrifugal fan–duct system. In a fan-duct system, the fan total pressure of the centrifugal fan is used mainly to overcome the frictional and dynamic losses in air flow. This required fan total pressure drops considerably when the air volume flow is reduced. In the centrifugal chiller, on the other hand, the system head is used mainly to lift the evaporating pressure to condensing pressure to produce the refrigerant effect. The pressure drop due to frictional and dynamic losses of refrigerant flow

is only a small part of the required head. Reduced volume flow of refrigerant at part-load has a minor influence on required system head.

The primary factor that affects the system head at part-load operation is the difference between condensing and evaporating pressures or temperatures, $(p_{con} - p_{ev})$ or $(T_{con} - T_{ev})$. For a water-cooled condenser using cooling water from a cooling tower, the outdoor wet bulb directly influences the entering temperature of cooling water and, therefore, the condensing temperature and pressure.

Required System Head at Part-Load Operation

During part-load operation, because the condensing and evaporating surface areas at reduced volume flow of refrigerant are greater than the required area, there is a drop in condensing temperature T_{con} and an increase in evaporating temperature T_{ev}.

Most probably, the outdoor wet bulb temperature also drops at reduced load. Therefore, the profiles for required system head at part-load operation for a water-cooled centrifugal chiller usually fall into the following three schemes:

- *Scheme A.* Constant 85°F condenser water entering temperature T_{ce} and constant 44°F chilled water leaving temperature T_{el}: The temperature differential $(T_{con} - T_{ev})$ at a load ratio equal to 1 is 66°F. At a load ratio of 0.2, $(T_{con} - T_{ev})$ is about 45°F.
- *Scheme B.* Constant T_{el} and a drop of 2.5°F in T_{ce} for each 0.1 increment of reduction of load ratio: the quantity $(T_{con} - T_{ev})$ at design load is still 66°F, whereas at load ratio = 0.2, $(T_{con} - T_{ev})$ is about 26°F.
- *Scheme C.* A reset of T_{el} of 1°F increase and a drop of 2.5°F of T_{ce} for each 0.1 increment of reduction of load ratio: Under this scheme, $(T_{con} - T_{ev})$ at design load is still 66°F, and at 0.2 load ratio, $(T_{con} - T_{ev})$ is about 19°F.

Required system head curves for Schemes A, B, and C at various load ratios are shown in Fig. 23.9.

On a compressor map (e.g., Fig. 23.9a), the intersections of the compressor performance curve and the required system head curve at that load ratio—points P, Q, R, . . . —are the operating points of the centrifugal chiller. The required system head curve joins the operating points together to form an operating curve of the centrifugal chiller at various load ratios.

Capacity Control Using Inlet Vanes

As in Section 10.4, when the opening of the inlet vanes at the inlet of the centrifugal compressor is reduced, it throttles the volume flow of vapor refrigerant and imparts a rotation on the refrigerant flow. This effect produces a new performance curve at a lower head and volume flow at each position of reduced vane opening. At high load ratios with larger openings, compression efficiency is only slightly affected. At low load ratios, throttling at the inlet vanes causes considerable reduction of compression efficiency.

For a single-stage centrifugal compressor using inlet vanes to modulate its capacity (see Fig. 23.9a), the following can be noted:

- If the cooling water entering temperature T_{ce} remains at the design condition, the operating curve hits the surge line when the load ratio drops to about 0.12.
- When the cooling water temperature drops 2.5°F for every 0.1 increment of reduction of load ratio, at a load ratio of 0.5 the compression efficiency of the compressor then drops below 50 percent.

Capacity Control by Variable Speed

When the rotating speed of a centrifugal compressor varies, then, just as for a centrifugal fan, a family of compressor curves is produced (see Fig. 23.9b.) A lower rotating speed results in a performance curve of lower head and volume flow of refrigerant.

One can see the following in Fig. 23.9b:

- When the load ratio ≥ 0.5, the compression efficiency of a centrifugal compressor using variable-speed control is greater than 75 percent.
- If the cooling water entering temperature T_{ce} remains at the design value, the operating curve hits the surge line at a load ratio of about 0.27.
- If T_{ce} drops and T_{el} is reset according to scheme C, the operating curve intersects the surge line at a load ratio of about 0.18.

When capacity control is to be achieved by means of varying the compressor rotating speed, a reset of chilled water leaving temperature T_{el} should be used.

Combination of Inlet Vanes and Variable-Speed Control

If a combination of inlet vanes and variable-speed modulation is used for capacity control in a centrifu-

gal chiller, when the load ratio is greater than 0.5 variable-speed control may be used for a greater η_{cp}. When the load ratio is low, inlet vane control may be advantageous for a smaller surge region. A DDC panel should be used to implement a more complicated computer program. More operator experience is required to combine inlet vane and variable-speed control successfully.

Comparison between Inlet Vanes and Variable Speed

Inlet vanes modulation is simple and has a low initial cost. At lower load ratios, it functions as a throttling device and produces considerable dynamic and frictional losses and, consequently, a lower compression efficiency. Erth (1987) compared the power consumption of centrifugal chillers, in kW/ton, of constant-speed centrifugal compressors using inlet vanes and variable-speed compressors with or without a chilled water leaving temperature T_{el} reset, as shown in Fig. 23.10. At a load ratio of 0.25, variable speed control with T_{el} reset has a power consumption of only about half its design value.

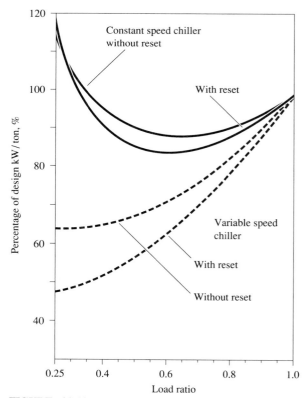

FIGURE 23.10 Comparison between constant-speed and variable-speed centrifugal chiller. (*Source*: R. A. Erth, *Heating/Piping/Air Conditioning*, May 1987. Reprinted with permission.)

Although a variable-speed compressor using an adjustable-frequency AC inverter is expensive, its energy savings may result in a simple payback of several years.

23.7 CENTRIFUGAL CHILLER PLANT CONTROLS

Functional Controls and Optimizing Controls

A large modern centrifugal chiller plant usually includes multiple chillers to prevent the entire shutdown of the refrigeration system when one chiller fails to operate. Control of a centrifugal chiller plant, therefore, includes functional controls and optimizing controls. Only a microprocessor-based DDC system can possibly perform all these duties. A proportional-integral-derivative (PID) control mode is often used for modulation control of chilled water leaving temperature for a better performance.

For functional controls, a DDC panel monitors and controls the following:

- Chilled water leaving temperature T_{el}
- Electric demand limit
- Air purge
- Safety features, including oil pressure, low temperature freezing protection at the evaporator, high condensing-pressure control, motor overheating, and time delaying
- Status features, including cooling water flow, chilled water flow, electric current, and refrigerant and chilled water temperatures and pressures at key points

To optimize the controls of a plant with multiple centrifugal chillers one should do the following:

- Optimize the on-or-off stages of the multiple chillers
- Control the condenser water temperature
- Optimize start and stop

Optimizing the on-or-off stages of multiple chillers will be discussed in a later section.

Chilled Water Leaving Temperature Control

Continuous modulation control is extensively used for chilled water leaving temperature T_{el} control in centrifugal chillers. A temperature sensor located in the chilled water pipes at the exit of the evaporator is used to sense T_{el}. The DDC panel either varies the posi-

tion of the opening of the inlet vanes or varies the speed of the compressor motor through an adjustable-frequency AC inverter. The volumetric flow of refrigerant then decreases or increases, as does the capacity of the centrifugal chiller.

The temperature of the chilled water leaving the evaporator T_{el} is often reset to a higher value at reduced load according to either the outdoor temperature T_o or the load ratio. The load ratio can be indicated by the chilled water return temperature. An increase in T_{el} may reduce the dehumidifying capability at the cooling coil. It is important to realize that at very low load ratios, such as less than 0.3, a constant-speed chiller that has a T_{el} reset uses more power (in kW/ton) than a chiller without reset because a lower system head lowers compression efficiency.

A better scheme for T_{el} reset is this: When the outdoor temperature is high (80 to 100°F), there is no reset. Only when T_o is less than 80°F, is T_{el} reset for an increase of 0.2 to 0.35°F for each °F drop in T_o. When T_o falls below a limit (45°F), T_{el} remains constant.

For energy-efficient design and operation, ASHRAE/IES Standard 90.1–1989 requires that comfort air conditioning systems include automatic controls to reset the chilled water leaving temperature T_{el} at least 25 percent of the design supply-to-return water temperature difference.

Functional Controls

ELECTRIC DEMAND LIMIT CONTROL. Electric limit or electric demand control is used to prevent the electric current from exceeding a predetermined value. A motor current is measured at the current transformers and a proportional signal is sent to the DDC panel. If the current exceeds a predetermined limit, the panel directs the inlet vanes to close or reduces the rotating speed of the compressor through the adjustable-frequency AC inverter. The limit could be from 40 to 100 percent of the electrical demand.

MOTOR PROTECTION. The compressor motor assembly is the most expensive part of the centrifugal chiller. For a typical control that protects a motor against high temperatures, if the resistance temperature detector in the motor stator winding senses a temperature under 165°F, the DDC panel introduces a time delay, normally of four minutes, after the chiller switch is turned on. If the sensed temperature exceeds 165°F, a start delay of 15 minutes is required. If the stator winding temperature exceeds a dangerous upper limit, such as 265°F, the DDC panel then prohibits the operation of the motor because of its high temperature.

There are also protection controls against current overload, temporary power loss, low voltage, and phase imbalance.

LOW TEMPERATURE AND HIGH-PRESSURE CUTOUTS. Refrigerant temperature at the evaporator is detected by a sensor. This information is sent to the DDC panel so that if the predetermined low-temperature cutout is approached, inlet vanes hold or close to provide freezing protection for the evaporator.

A pressure sensor detects the condensing pressure of the refrigerant. The DDC panel outputs override the inlet vane position instruction to close the inlet vanes if the condensing pressure reaches the predetermined limit.

HIGH BEARING TEMPERATURE. If the bearing temperature detected by a temperature sensor exceeds a preset value, the chiller is stopped immediately.

SHORT-CYCLING PROTECTION. Usually a time delay, typically of 30 minutes, is needed to restart a centrifugal chiller after it has been turned off to prevent short-cycling.

SURGE PROTECTION. A surge detection device is used to send a signal to the DDC panel. A head relief scheme is implemented by lowering the condenser water temperature. If the surge continues beyond a predetermined time period, the compressor is then shut down.

AIR PURGE. An automatic purge operation includes the operation of an air purge unit within a specific time interval, for instance, five minutes of air purge every two hours. Both noncondensable gas and condensable vapor, mainly water, leaking into the centrifugal chiller raise the purge drum pressure. If an increase in drum pressure is detected by a pressure sensor, the period of operation of the air purge unit is lengthened.

Condenser Water Control

Interaction between the centrifugal chiller and the cooling tower must be considered when one optimizes the condenser cooling water entering temperature T_{ce} so as to minimize power consumption. The control strategy is to minimize the total power consumption of the chiller and condenser tower fans.

According to Cascia (1988), the cooling water temperature should be *floated*, that is, it should rise and fall with the outdoor wet bulb temperature in agreement with the following scheme:

- T_{ce} must be maintained above the lower limit (generally around 60°F) specified by the centrifugal chiller manufacturer to maintain a necessary head pressure for proper operation. If the lower limit is reached, tower fans must be turned off first; then, the tower bypass valve should be modulated to maintain T_{ce} equal to or above the specified limit.
- If T_{ce} reaches the high limit, or if the chiller's current limit is reached, another tower fan should be added to the operating fans.

A temperature sensor should be used to sense T_{ce}. The DDC panel of the centrifugal chiller plant actuates the cooling tower fans and the cooling tower bypass valve to maintain the predetermined T_{ce}.

For optimum start-and-stop control, an adaptive control strategy that minimizes the energy required by air, water, and refrigeration systems should be considered.

Operator Interface

All of the inputs (such as switches and set points), output displays (such as operating mode indicators, key parameter monitors, and status lights), and diagnostic displays should be visible at a glance on the control display panel. They should be easy to understand and to use.

23.8 OPERATING CHARACTERISTICS OF CENTRIFUGAL CHILLERS

Sequence of Operation of a Typical Centrifugal Chiller at Startup

Many centrifugal chillers in commercial buildings are shut down at night in summer. Usually, one of the chillers is operated to serve areas needing 24-hour service, such as computer areas. The temperature of the chilled water in the shut-down system rises during the shut-down period. Chillers are often started earlier than air handling units (AHU) or terminals to pull down the chilled water temperature so that its T_{el} is maintained within acceptable limits.

When the switch of a typical centrifugal chiller is turned on, the sequence of operations is as follows:

1. The sensed chilled water leaving temperature T_{el} is higher than the set point, indicating a demand for refrigeration.

2. The interlocked chilled water pump switch starts the chilled water pump, and the chilled water flow is sensed by the water flow sensor.

3. A signal is issued to start the condenser water pump and the tower fan; the condenser water flow is sensed by the flow switch.

4. A signal is sent to start the oil pump, and the oil pressure is sensed by the pressure sensor.

5. The short-cycling protection timer is checked to see whether the time period after the previous shutdown is greater than the predetermined limit.

6. A command is sent to close the inlet vanes and is confirmed.

7. After the oil pressure has been established for 10 to 20 seconds, a start command is sent to the compressor motor starter.

8. A successful start is followed by a display of "run normal," failure to perform a successful start shuts down the compressor and the diagnostic code is displayed for correcting the fault.

After correction and reset, the start sequence may be repeated to attempt a successful start.

System Balance at Full-Load

When a centrifugal chiller is operating at full design load under the standard rating conditions, that is, $T_{el} = 44°F$, and chilled water volume flow $\dot{V}_{ev} = 2.4$ gpm/ton, $T_{ce} = 85°F$ and condenser water volume flow $\dot{V}_{con} = 3$ gpm/ton.

If the capacity of the selected evaporator is greater than the compressor at design conditions, more liquid refrigerant is then vaporized in the evaporator than the compressor can extract. The evaporating pressure gradually rise until the increase in the mass flow rate due to a higher density of suction vapor extracted by the computer is just equal to the mass flow rate of the refrigerant vaporized in the evaporator. A new system balance forms between the evaporator and compressor at a higher evaporating temperature T_{ev}.

If the capacity of the selected condenser is greater than the compressor at design conditions, more hot gas can be condensed into liquid form in the condenser than the compressor can supply. The result is a drop of condensing pressure p_{con} and temperature T_{con}. Because the condenser water entering temperature remains unchanged, a drop of T_{con} results in a lower log-mean temperature and, therefore, a reduction in the condensing capacity. A new system balance is then formed between the condenser and compressor at a lower T_{con}.

23.9 PART-LOAD OPERATION OF CENTRIFUGAL CHILLER
Part-Load Operating Characteristics

If we assume that a centrifugal chiller is operated initially at full design load and is in a condition of system balance, then chilled water leaving the evaporator is controlled at a temperature of 44°F, and the return temperature is 60°F.

If there is a reduction in the coil load in the AHUs or in the terminals, the chilled water temperature entering the evaporator T_{ee} falls. Because the rate of heat transfer and the amount of refrigerant evaporated are only slightly affected, the T_{el} is lower than 44°F, the setpoint.

When the temperature sensor at the exit of the evaporator senses a drop in T_{el} and signals the DDC panel, the positioner of the inlet vanes is actuated to close the opening. The centrifugal compressor is now operating at less than the design head, and the mass flow rate of the refrigerant extracted by the compressor is reduced from $\dot{m}_r$ to $\dot{m}_r'$, until the mass flow rate of refrigerant vaporized in the evaporator is greater than the mass flow rate of the refrigerant extracted by the centrifugal compressor. Both the evaporating pressure p_{ev} and temperature T_{ev} tend to increase. Such an effect further reduces the ΔT_m, the boiling heat transfer coefficient h_b, and the overall heat transfer coefficient U_o. The rate of heat transfer and the rate of evaporation are, therefore, decreased until the mass flow rate of evaporated refrigerant is equal to the mass flow rate of the refrigerant extracted by the centrifugal compressor.

Because the rate of condensation of the refrigerant in the condenser is greater than the reduced mass flow rate of hot gas discharged from the compressor $\dot{m}_r'$, the condensing pressure p_{con} and temperature T_{con} tend to drop until the rate of condensation is equal to $\dot{m}_r'$. Consequently, there is a reduction of total heat rejected in the condenser.

A higher evaporating pressure and a lower condensing pressure result in a small pressure difference ($p_{con} - p_{ev}$) as well as a lower hydrostatic liquid column in the condenser. As a result, the liquid refrigerant flow through the float valve or orifice plates is reduced.

When the compressor operates with the inlet vanes closed at a small angle, that is, at a load ratio between 0.8 and 1, the compression efficiency decreases only slightly. However, because a lower head and a lower volume flow have a great effect on the work input and the power consumption of the compressor, the result is a reduction in the power input. The energy performance index, in kW/ton, of the compressor is reduced less than power input because of the reduction of q_{ref}.

The reduction of the mass flow rate of the refrigerant flowing through the centrifugal compressor and the orifice plates and the reduction of the rates of evaporation and condensation create a new system balance during part-load operation.

For a constant-speed centrifugal chiller, power consumption, in kW/ton, rises when load ratio drops below 0.55 (see Fig. 23.10). This increase occurs because the fall of compression efficiency exceeds the savings in power consumption at a lower head and flow during low load ratios when inlet vanes are closed to a smaller opening.

Evaporating and Condensing Temperature at Part-Load

If other conditions remain the same, the greater the heat transfer surface area A_o in evaporator, the higher the evaporating temperature T_{ev}. Also, the greater the A_o in condenser, the lower the condensing temperature T_{con}. Consequently, greater A_o results in a lower temperature lift between T_{con} and T_{ev}.

From Eq. (12.32), if the thermal resistance of the metal is ignored, the rate of heat transfer at the shell-and-tube evaporator and condenser can be calculated as

$$q = U_o A_o \Delta T_m \qquad (23.7)$$

$$U_o = \left[\frac{1}{\eta_f h_o} + R_f \left(\frac{A_o}{A_i} \right) + \left(\frac{A_o}{A_i} \right)\left(\frac{1}{h_i} \right) \right]^{-1} \qquad (23.8)$$

where η_f = fin efficiency; for integrated fins on copper tubes, its value is between 0.9 and 0.95. From Eq. (12.35), the boiling coefficient is given as

$$h_b = C \left(\frac{q_{ev}}{A_o} \right)^n \qquad (23.9)$$

Meanwhile, from Eq. (12.41), the condensing coefficient can be calculated as

$$h_{con} = C_{con} \left(\frac{1}{Q_{rej}/A_o} \right)^{1/3} \qquad (23.10)$$

The log-mean temperature difference across the evaporating and condensing surfaces can be expressed as

$$\Delta T_m = \frac{(T_{ee} - T_{ev}) - (T_{el} - T_{ev})}{\ln[(T_{ee} - T_{ev})/(T_{el} - T_{ev})]}$$
$$= \frac{(T_{con} - T_{cl}) - (T_{con} - T_{ce})}{\ln[(T_{con} - T_{cl})/(T_{con} - T_{ce})]} \qquad (23.11)$$

Then the evaporating temperature T_{ev} can be calculated as

$$T_{ev} = \frac{e^B T_{el} - T_{ee}}{e^B - 1} \qquad (23.12)$$

where

$$B = \frac{T_{ee} - T_{el}}{q_{ev}/(A_{ev} U_o)} \qquad (23.13)$$

Similarly, the condensing temperature

$$T_{con} = \frac{e^C T_{cl} - T_{ce}}{e^C - 1} \qquad (23.14)$$

where

$$C = \frac{T_{cl} - T_{ce}}{Q_{rej}/(A_{con} U_{oc})} \qquad (23.15)$$

where A_{ev}, A_{con} = evaporating and condensing surface area, ft^2

U_{oc} = overall heat transfer coefficient based on condensing surface area, Btu/h · ft^2· F

Example 23.1. A centrifugal chiller using HCFC-123 as the refrigerant has the following operating parameters for its shell-and-tube liquid cooler at design conditions:

Chilled water entering temperature: T_{en}	54°F
Chilled water leaving temperature: T_{el}	44°F
Evaporating temperature: T_{ev}	36°F
Ratio of outer to inner surfaces A_o/A_i	3.5
Fouling factor on water side	0.00025 h · ft^2 · °F/Btu
Water-side heat transfer coefficient h_i	1700 Btu/h · ft^2 · °F

1. If the boiling coefficient of HCFC-123 can be calculated by $h_b = 2.5(q_{ev}/A_o)^{0.7}$, calculate the outer evaporating surface area required in the shell-and-tube evaporator for 1 ton of refrigeration capacity.

2. Calculate also the evaporating temperature for the same evaporator at a load ratio of 0.5.

Solution

1. From Eqs. (23.8), (23.9), and the given data, the overall heat transfer coefficient at design condition is

$$U_o = \left[\frac{1}{h_b} + \frac{R_f A_o}{A_i} + \left(\frac{A_o}{A_i} \right)\left(\frac{1}{h_i} \right) \right]^{-1}$$
$$= \left[\frac{1}{2.5(q_{ev}/A_o)^{0.7}} + 0.00025 \times 3.5 + \frac{3.5}{1700} \right]^{-1}$$
$$= 1 + 0.00734(q_{ev}/A_o)^{0.7}/2.5(q_{ev}/A_o)$$

2. From Eq. (23.11), the log-mean temperature difference at the evaporator is

$$\Delta T_m = \frac{(T_{ee} - T_{ev}) - (T_{el} - T_{ev})}{\ln[(T_{ee} - T_{ev})/(T_{el} - T_{ev})]}$$
$$= \frac{(54 - 36) - (44 - 36)}{\ln[(54 - 36)/(44 - 36)]} = 12.33°F$$

3. Then, from Eq. (23.7),

$$\frac{q_{ev}}{A_o} = U_o \Delta T_m = 12.33 \left[\frac{2.5(q_{ev}/A_o)^{0.7}}{1 + 0.00734(q_{ev}/A_o)^{0.7}} \right]$$

Rearranging, we find

$$0.00734\left(\frac{q_{ev}}{A_o}\right) + \left(\frac{q_{ev}}{A_o}\right)^{0.3} = 30.83$$

Solving by iteration, we have

$$\frac{q_{ev}}{A_o} \approx 2750 \; \text{Btu/h} \cdot \text{ft}^2$$

For $q_{ev} = 1$ ton $= 12,000$ Btu/h, then,

$$A_o = \frac{12,000}{2750} = 4.36 \; \text{ft}^2$$

That is, for each ton of refrigeration, there should be 4.36 ft^2 of outer surface area A_o for the tubes at the evaporator.

4. At 50 percent part-load, $q_{ev} = 0.5 \times 12,000 = 6000$ Btu/h. Then

$$\frac{q_{ev}}{A_o} = \frac{6000}{4.36} = 1375 \; \text{Btu/h} \cdot \text{ft}^2$$

and

$$h_b = 2.5\left(\frac{q_{ev}}{A_o}\right)^{0.7}$$
$$= 2.5(1375)^{0.7} = 393 \; \text{Btu/h} \cdot \text{ft}^2 \cdot {}^\circ\text{F}$$

Also,

$$U_o = \left(\frac{1}{393}\right) + (0.00025 \times 3.5) + \left(\frac{3.5}{1700}\right)^{-1}$$
$$= 182.6 \; \text{Btu/h} \cdot \text{ft}^2 \cdot {}^\circ\text{F}$$

5. From Eq. (23.7), at 50 percent part-load

$$\Delta T_m = \frac{q_{ev}}{U_o A_o} = \frac{6000}{182.6 \times 4.36} = 7.54{}^\circ\text{F}$$

Since the chilled water flow rate at the plant loop is constant, then at 50 percent part-load, $(T_{ee} - T_{el}) = 0.5(54 - 44) = 5{}^\circ\text{F}$. Also, from Eq. (23.13),

$$B = \frac{T_{ee} - T_{el}}{q_{ev}/(U_o A_o)} = \frac{5}{6000/(182.6 \times 4.36)} = 0.663$$

Therefore, from Eq. (23.12) the evaporating temperature at 50 percent part-load is

$$T_{ev} = \frac{e^B T_{el} - T_{ee}}{e^B - 1}$$
$$= \frac{e^{0.663} \times 44 - 49}{e^{0.663} - 1} = 38.69{}^\circ\text{F}$$

For part-load operation at a load ratio of 0.5, T_{ev} rises 2.69°F, that is, (38.69 − 36).

Temperature Lift at Part-Load for Centrifugal Chillers

Figure 23.11 is a diagram of temperature lift at various load ratios for a centrifugal chiller using HCFC–123 as refrigerant. The chilled water temperature leaving the chiller is maintained at 44°F at all load ratios, and the outdoor wet bulb also remains constant at

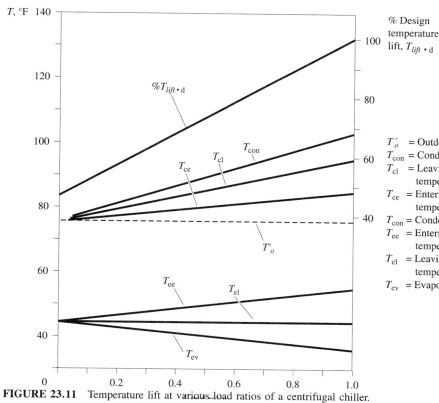

FIGURE 23.11   Temperature lift at various load ratios of a centrifugal chiller.

75°F when the load ratio varies from 0 to 1. Fouling factors for both evaporator and condenser are 0.00025 $h \cdot ft^2 \cdot °F$.

Under design conditions, if the cooling tower approach is 9°F, the evaporating temperature T_{ev} is 8°F lower than T_{el}, the condensing temperature T_{con} is 8°F higher than T_{cl}, and T_{cl} is 10°F higher than T_{ce}. Therefore, the condensing temperature is

$$T_{con} = 75 + 9 + 10 + 8 = 102°F$$

and the evaporating temperature is

$$T_{ev} = 44 - 8 = 36°F$$

The temperature lift at design conditions is, therefore,

$$\Delta T_{lift.d} = (T_{con} - T_{ev}) = (102 - 36) = 66°F$$

If the load ratio = 0, theoretically T_{con} drops to 75°F and T_{ev} rises to 44°F. Therefore, the temperature lift ΔT_{lift} drops to

$$\Delta T_{lift} = 75 - 44 = 31°F$$

This drop is about $31/66 = 0.47$, or 47 percent of the design temperature lift $\Delta T_{lift.d}$.

The drop of temperature lift is caused by the following:

- A decrease in condenser water entering temperature T_{ce} because of the smaller value of approach in the cooling tower at reduced load ratios
- A drop in condenser water temperature rise (entering and leaving temperature difference) because of the reduction in total heat rejection
- A decrease in the difference between the condensing temperature and the condenser water leaving temperature $(T_{con} - T_{cl})$
- A decrease in the difference between the chilled water leaving temperature and the evaporating temperature $(T_{el} - T_{ev})$

The actual percentage of design power input may be higher or slightly lower than the load ratios, depending on:

- whether the outdoor wet bulb is constant or drops at part-load
- whether there is a T_{el} reset
- the decrease of compression efficiency at part-load

Figure 23.12 is a plot of actual percentage design power inputs of a water-cooled centrifugal chiller operated in the summer of 1983. The calculated power input almost always exceeded the actual power input in this system because the outdoor wet bulb temperature T_o' drops at low load ratios.

23.10 FREE REFRIGERATION

Principle of Operation

Free refrigeration means the production of a refrigeration effect without the operation of a compressor. This effect can be achieved because refrigerant tends to migrate to the area of lowest temperature in the system when the compressor ceases to operate.

A free refrigeration operation for a single-stage, water-cooled centrifugal chiller is illustrated in Fig. 23.13. When the free refrigeration is turned on, the compressor ceases to operate, the inlet vanes are fully opened, and the valve in the passage that connects the condenser and evaporator is opened.

When the temperature of the cooling water from the cooling tower or other source is lower than the temperature of the chilled water leaving the evaporator, the higher saturated pressure of the refrigerant in the evaporator forces the vaporized refrigerant to migrate to the condenser, where the saturated pressure of the refrigerant is lower. The absorption of latent heat of vaporization from the chilled water in the evaporator's tube bundle causes the temperature of the chilled water to drop. The condensation of vaporized refrigerant into liquid results in the rejection of the latent heat of condensation to the cooling water through the tube bundle in the condenser, raising the temperature of the cooling water a few degrees. Liquid refrigerant drains to the evaporator as a result of gravity. Free refrigeration is thus achieved.

Automatic free refrigeration control can be provided by the DDC system. When the cooling water temperature is a specified number of degrees lower than the required chilled water temperature, the DDC panel starts the free refrigeration cycle. The shut-off valve is opened, and an interlocked circuit shuts the compressor off. The free refrigeration cycle lasts as long as there is a temperature difference between the chilled water and cooling water, and as long as the free refrigeration capacity meets the refrigeration load required.

For a two-stage centrifugal chiller with a flash cooler, or a three-stage chiller with a two-stage flash cooler, there are more connecting passages and shut-off valves. The principle of operation of free refrigeration is still the same.

Once the refrigeration required is greater than the free refrigeration capacity, as determined from the temperture of the chilled water returned to the evaporator, the DDC panel closes the vapor and liquid shut-off valves and starts the compressor. The vaporized refrigerant from the evaporator is extracted by the compressor, discharged to the condenser, and condensed into liquid refrigerant as in normal operation.

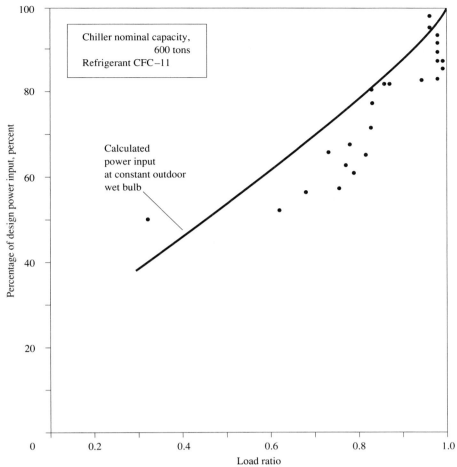

FIGURE 23.12   Actual percentage of design power inputs of a centrifugal chiller.

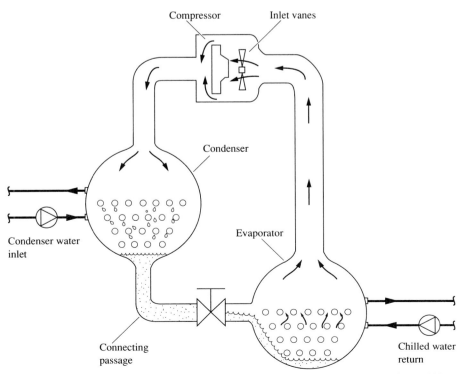

FIGURE 23.13   Free refrigeration operation for a single-stage, water-cooled centrifugal chiller.

Free Refrigeration Capacity

Free refrigeration capacity depends on the following parameters:

- The temperature difference between the chilled water leaving the evaporator and the cooling water entering the condenser
- The size of the evaporator
- Condenser water flow
- Fouling factors at the evaporator and the condenser. An allowable fouling factor is usually 0.00025 $h \cdot ft^2 \cdot °F/Btu$. A higher fouling factor means a lower free refrigeration capacity.

Free refrigeration capacity is always expressed as a percentage of the base capacity of the centrifugal chiller. For a standard size evaporator in a typical centrifugal chiller that is running in a free refrigeration cycle, if the chilled water leaves the evaporator at 48°F and the condenser water enters the condenser at 35°F, the free refrigeration capacity $q_{ref \cdot f}$ can be 45 percent of the base capacity. When $T_{en.c}$ is 40°F, $q_{ref.f}$ is 25 percent, and when $T_{en.c}$ is 44°F, $q_{ref.f}$ may be zero.

Free refrigeration saves energy input to the compressor and can be applied to locations where cooling water temperature can be at least 5°F lower than the temperature of the chilled water leaving the evaporator. A free refrigeration arrangement is optimal.

23.11 OPERATION OF MULTIPLE CHILLERS

Parallel and Series Piping

Multiple chillers can be piped in parallel, as shown in Fig. 23.14a. Each chiller usually has its own chilled water pump. If a butterfly valve is installed for each chiller–pump combination, both the refrigeration capacity and the chilled water flow are turned on and off in steps controlled by the DDC panel. Flexibility and reliability of the operation of parallel chillers are excellent. Chillers can be withdrawn from operation for maintenance without affecting the others.

When multiple chillers are piped in series, as shown in Fig. 23.14b, there are usually only two chillers in series, and they have the following characteristics:

- Chilled water coils are sized for a relatively higher temperature rise and lower water flow rate.
- The pressure drop of chilled water through two series-connected evaporators is additive.
- Control is more complicated in part-load operation.

In practice, many multiple chiller plants are often piped in parallel.

Chiller Staging

In a multiple chiller plant using a plant-building loop combination, the principle of staging the chillers, on or off, is based on these considerations: (1) The chilled water flow rate in the plant loop is always slightly higher than the building loop; (2) staging on the chiller(s) must meet the requirements of the increased system load; (3) power input to the chillers and chilled water pumps should be minimum.

Cascia (1988) recommended the following algorithm for staging the chillers on or off:

1. Calculate the required refrigeration load.
2. Determine the load ratio of each chiller and the chiller plant.

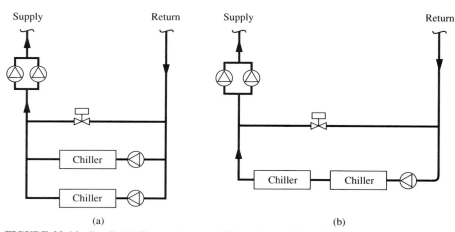

(a)   (b)

FIGURE 23.14 Parallel chillers and series chillers: (a) parallel chillers; and (b) series chillers.

3. Calculate the part-load power consumption for each chiller, and the combination of the chiller and water pumps, including the chilled water and cooling water pumps.

4. Choose the chiller combination for staging on or off for times when the load ratio varies between 30 and 100 percent:
   - the plant-loop chilled water flow rate exceeds the building-loop flow rate.
   - Select the chiller and water pump combination that results in minimum power consumption.

5. Periodically (every 30 minutes), monitor operating chillers to determine whether they are energy efficient.

23.12 AIR-COOLED CENTRIFUGAL CHILLERS

Air-cooled centrifugal chillers have a far lower minimum performance COP than water-cooled centrifugal chillers. They should be used in locations where water is scarce and the use of city water as make-up water for cooling towers is prohibited by local codes.

A typical air-cooled centrifugal chiller is illustrated in Fig. 23.15. It consists mainly of a two-stage centrifugal compressor, an evaporator, orifice plates, an air-cooled condenser, a control panel, and an outer casing. HFC–134a is often used as the refrigerant in this type of chiller for its higher density. Because the evaporating pressure for HFC–134a is higher than atmospheric, no purge unit is required.

In an air-cooled centrifugal chiller, a shell-and-tube flooded evaporator is often used. Water flow arrangements from two-pass up to six-pass are available. Seamless copper tubes with aluminum fins bonded tightly on the tube are used in the condenser. The condenser fans are propeller fans with vertical discharge. They can be cycled in response to low ambient head. A DDC panel similar to the one for a water-cooled centrifugal chiller is used for functional and optimizing control.

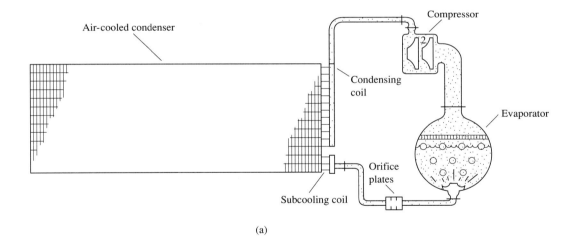

(a)

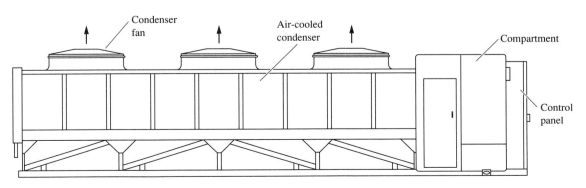

(b)

FIGURE 23.15 Air-cooled centrifugal chiller: (a) schematic diagram; and (b) outer profile.

Air-cooled centrifugal chillers are manufactured at a capacity of 130 to 340 tons. A heat recovery condenser can be added to provide both hot and chilled water simultaneously. A free refrigeration arrangement is optional. A standard unit can operate at an outdoor temperature of 115°F, whereas a high ambient unit can even run up to 125°F.

Air-cooled centrifugal chillers are usually located on the roof outdoors. The key point for efficient operation is to have sufficient clearance between the condensing coils and the surroundings, and also between units, according to the manufacturer's specification. If the chiller is shut down during winter, water should be drained from the condenser and evaporator in locations where outdoor temperatures fall below 32°F. Otherwise, glycol should be added to the chilled water for freezing protection.

23.13 MULTIPLE-CHILLER PLANT DESIGN

Chiller Minimum Performance

ASHRAE Standard 90.1–1989 and ARI Standard 550–88 specify the minimum performance for water chillers (including centrifugal chillers) as of January 1, 1992, as follows:

	COP	IPLV
Water cooled		
$\geq$ 300 tons	5.2	5.3
$\geq$ 150 tons < 300 tons	4.2	4.5
< 150 tons	3.8	3.9
Air-cooled with condenser		
$\geq$ 150 tons	2.5	2.5
< 150 tons	2.7	2.8

For HCFC–22 or other refrigerants whose ozone depletion factor is less than or equal to HCFC–22, water-cooled water chillers of capacity $\geq$ 300 tons should have a minimum performance COP = 4.7 and an integrated part-load value IPLV = 4.8.

Design Considerations

POWER CONSUMPTION. The following means are effective to reduce the power consumption, in kW/ton, as well as to enhance the minimum performance of centrifugal chillers:

- Raise the temperature of the chilled water leaving the evaporator T_{el} by using a T_{el} reset at reduced loads under the condition that space relative humidity does not exceed 65 percent.

- Lower the cooling water entering temperature T_{ce} as low as the manufacturer permits.
- If the fouling factor on the water side of the condensing surface exceeds 0.00025 h · ft² ° F/Btu, reduce the fouling factor by using a brush cleaning system or some other means, especially in locations where the outdoor air is heavily contaminated.
- Increase the surface area of the evaporator and condenser if they are cost effective.

Many manufacturers of centrifugal chillers now offer different sizes of evaporators and condensers to match a specific size of compressor. The designer can thus select the centrifugal chiller that meets the designated minimum performance.

EQUIPMENT SIZING. Although a single chiller is less costly than multiple chillers, there is a total loss of service when the chiller fails to operate or needs maintenance. Multiple chillers are usually the best choice for a large chiller plant. A plant installed with three chillers ensures 66 percent of capacity even if one of the chillers fails.

If continuity of service is critical, a standby chiller, which has a capacity equal to the largest chiller in the refrigeration plant, should be provided. A smaller chiller used during light-load conditions, especially in after-hours operation, in addition to several equally large chillers is often beneficial. The refrigeration load always increases as building usage grows. The possibility of future growth should be considered.

PLANT LOCATION AND LAYOUT. Generally, the central plant should be located near the system load as well as near the sources of utilities. For a high-rise building, the ideal location is often the hub of the building, near shopping and public areas. If space at the hub is not available, the chiller plant may be located in the basement, on an intermediate floor, or on the top floor, depending on the type of building and the architectural design.

Equipment should be laid out in an orderly arrangement for efficient piping. Vertical and lateral clearances must be maintained. Equipment may be 10 to 12 ft in height. Provision of access platforms and ladders to valves and piping should be considered in the design stage. A code-specified clearance of 3 to 5 ft from electrical panels and devices must be maintained.

Maintenance and service facilities should also be planned in the design stage. These include removal and replacement of components, storage space, washrooms, and janitorial facilities.

REFERENCES AND FURTHER READING

Alvine, R. G., "Prepurchasing Chillers," *Heating/Piping/Air Conditioning*, pp. 67–71, January 1990.

ARI, *ARI Standard 550–88, Centrifugal or Rotary Water Chilling Packages*, ARI, Arlington, VA, 1988.

ASHRAE, *ASHRAE Handbook, 1987, HVAC Systems and Applications*, ASHRAE, Inc., Atlanta, GA.

ASHRAE, *ASHRAE Handbook, 1988, Equipment*, ASHRAE, Inc., Atlanta, GA.

ASHRAE, *ASHRAE/IES Standard 90.1–1989, Energy Efficient Design of New Buildings Except New Low-Rise Residential Buildings*, ASHRAE, Inc., Atlanta, GA, 1989.

ASHRAE, *"Waterside Family Resistance Inside Condenser Tubes, (RP106),"* *ASHRAE Journal*, June 1982.

Bjorklund, A. E., "Heat Recovery and Thermal Storage at a State Office Building," *ASHRAE Transactions*, Part II, pp. 832–849, 1987.

Braun, J. E., Mitchell, J. W., Klein, S. A., and Beckman, W. A., "Models for Variable-Speed Centrifugal Chillers," *ASHRAE Transactions*, Part I, pp. 1794–1813, 1987.

Calm, J. M., "Alternative Refrigerants: Challenges, Opportunities," *Heating/Piping/Air Conditioning*, pp. 38–49, May, 1992.

Cascia, M. A., "Optimizing Chiller Plant Energy Savings Using Adaptive DDC Algorithms," *ASHRAE Transactions*, Part II, pp. 1937–1946, 1988.

Clark, E. M., Anderson, G. G., Wells, W. D., and Bates, R. L., "Retrofitting Existing Chillers with Alternative Refrigerants," *ASHRAE Journal*, pp. 38–41, April 1991.

Corr, S., Dekleva, T. W., and Savage, A. L., "Retrofitting Large Refrigeration Systems with R–134a," *ASHRAE Journal* pp. 29–33, February 1993.

Erth, R. A., "Power Inverters and Off-Design Chiller Performance," *Heating/Piping/Air Conditioning*, pp. 63–67, November 1980.

Erth, R. A., "Chilled Water Reset: Variable Speed Solution," *Heating/Piping/Air Conditioning*, pp. 79–81, May 1987.

Hummel, K. E., Nelson, T. P., and Thompson, "Survey of the Use and Emissions of Chlorofluorocarbons from Large Chillers," *ASHRAE Transactions*, Part II, pp. 416–421, 1991.

Landman, W. J., "The Search for Chiller Efficiency," *Heating/Piping/Air Conditioning*, pp. 77–81, July 1983.

Lau, A. S., Beckman, W. A., and Mitchell, J. W., "Development of Computerized Control Strategies for a Large Chilled Water Plant," *ASHRAE Transactions*, Part I B, pp. 766–780, 1985.

Leitner, G. F., "Automatic Brush Cleaning for Condenser Tubes," *Heating/Piping/Air Conditioning*, pp. 68–70, October 1981.

Lewis, M. A., "Microprocessor Control of Centrifugal Chillers—New Choices," *ASHRAE Transactions*, Part II, pp. 800–805, 1990.

Newton, E. W., and Beekman, D. M., "Compressor Capacity Control for System Part-Load Operation," *ASHRAE Transactions*, Part I, pp. 493–503, 1980.

Sauer, H. J., and Howell, R. H., "Design Guidelines for Use of an Economizer with Heat Recovery," *ASHRAE Transactions*, Part II, pp. 1877–1894, 1988.

Spethmann, D. H., "Optimizing Control of Multiple Chillers," *ASHRAE Transactions*, Part II B, pp. 848–856, 1985.

Statt, T. G., "An Overview of Ozone-Safe Refrigerants for Centrifugal Chillers," *ASHRAE Transactions*, Part I, pp. 1424–1428, 1990.

The Trane Company, *Applications Engineering Manual: Multiple Chiller System Design and Control*, The Trane Company, La Crosse, WI, 1987.

The Trane Company, *Applications Engineering Manual: Model CVHE CenTraVacs with Microprocessor Based Control*, The Trane Company, La Crosse, WI, 1987.

The Trane Company, *Engineering Bulletin: Automatic Condensing Cleaning System*, American Standard Inc., La Crosse, WI, 1989.

The Trane Company, *Air Conditioning Clinic: Centrifugal Water Chillers*, The Trane Company, La Crosse, WI, 1983.

Tao, W., and Janis, R. R., "Modern Cooling Plant Design," *Heating/Piping/Air Conditioning*, pp. 57–81, May 1985.

Utesch, A. L., "Direct Digital Control of a Large Centrifugal Chiller," *ASHRAE Transactions*, Part II, pp. 797–799, 1990.

Wang, S. K., *Principles of Refrigeration Engineering*, Vol. 3, Hong Kong Polytechnic, Hong Kong, 1984.

Wanniarachchi, A. S., and Webb, R. L., "Noncondensable Gases in Shell-Side Refrigerant Condensers," *ASHRAE Transactions*, Part II, pp. 170–184, 1982.

Water Services of America, *On-Load Automatic Brush Cleaning System*, Water Services of America, Inc., Milwaukee, WI, 1979.

Wong, S. P. W., and Wang, S. K., "System Simulation of the Performance of a Centrifugal Chiller Using a Shell-and-Tube Type Water-Cooled Condenser and R–11 as Refrigerant," *ASHRAE Transactions*, Part I, pp. 445–454, 1989.

REFRIGERATION SYSTEMS: ABSORPTION

24.1 ABSORPTION SYSTEMS

Types of Absorption Systems

Absorption systems use heat energy to produce refrigeration or heating and sometimes to elevate the temperature of the waste heat. Aqueous lithium bromide LiBr is often used to absorb the refrigerant and water vapor, and provides a higher coefficient of performance.

Current absorption systems can be divided into following categories:

- *Absorption chillers* use heat energy to provide refrigeration.
- *Absorption heat pumps* use heat energy to provide heating during winter and cooling in summer.
- *Chiller-heaters* use heat energy to provide cooling and heating simultaneously.
- *Absorption heat transformers* elevate the temperature of the waste heat to a value higher than any other input fluid stream supplied to the absorption heat transformer.

Historical Development

In the 1950s and 1960s, both absorption chillers using steam as heat input to provide summer cooling and centrifugal chillers driven by electric motors were widely used in central refrigeration plants. Steam was widely used because excess steam was available in summer in many central plants that used steam to provide winter heating, and because energy costs were of little concern.

After the energy crisis in 1973, the price of the natural gas and oil used to fuel steam boilers drastically increased. The earliest single-stage indirect-fired steam absorption chillers had a coefficient of performance (COP) of only 0.6 to 0.7. They required more energy and could not compete with electric centrifugal chillers. Many absorption chillers were replaced by centrifugal chillers in the late 1970s and 1980s.

Because of the high investment required to build new power plants, electric utility companies modified their rate structures, added high demand charges, and raised cost-per-unit charges during peak usage periods. In recent years two-stage direct-fired absorption chillers have been developed in both Japan and the United States with a COP approximately equal to 1. Although absorption chillers have a higher initial cost than centrifugal chillers, in many regions in the United States where the cost ratio of electricity to natural gas is favorable, installation of absorption chillers for use during peak or even normal operating hours is economically beneficial, and the chiller may have a simple payback period of several years.

Cost Analysis

Cost analysis is performed to determine the most economical design for a specific project. Sun (1991) compares two designs for a project with a 1200-ton design refrigeration load. One design is three 400-ton electric centrifugal chillers, and the other is two electric centrifugal chillers plus one absorption chiller.

The cost of gas for this project is $0.46 per therm in summer, from April through November, and $0.50 per therm in winter. A therm is equal to 100,000 Btu/h of heat energy input. The cost of electricity is $0.126 per kWh during peak periods, and $0.0433 for the off-peak periods. The peak period is the eight-hour weekday daytime determined by the electric utility. The additional electricity demand charge is $3.60 per kW.

If the centrifugal chiller uses 0.7 kW/ton and the two-stage direct-fired absorption chiller uses 0.12 therm/ton, the comparison cost (in dollars) of these two alternatives for the third 400-ton chiller operated 8 hours per day, 5 days per week, 8 months for summer, and 4 months for winter is as follows:

	Absorption chiller	Centrifugal chiller
Initial cost		
Chiller	200,000	80,000
Cooling tower	53,000	48,000
Demand	484	11,491
Energy		
Electricity		
Summer	1957	36,700
Winter	978	18,337
Gas		
Summer	22,969	
Winter	12,474	
Total energy cost before tax	$38,862	$66,528
Total energy cost after tax	$43,720	$74,844

The absorption chiller has an annual energy cost saving of $31,124, and a simple payback period of about 4 years.

Because the solution and refrigerant pumps for absorption chillers use a small amount of electric energy, there are demand and electric energy costs for absorption chillers.

24.2 PROPERTIES OF AQUEOUS LITHIUM BROMIDE SOLUTION

Mass Balance in Solution

The composition of a solution is generally expressed by the mass fraction of its components. In a solution containing lithium bromide (LiBr) and water,

X is used to indicate the mass fraction of lithium bromide, that is,

$$X = \frac{m_1}{m_1 + m_w} \qquad (24.1)$$

where m_1 = mass of lithium bromide in solution, lb
$\quad m_w$ = mass of water in solution, lb

The mass fraction of water in solution is $1 - X$.

If the mass fraction of LiBr in a solution is expressed as a percentage, it is known as the *concentration* of LiBr.

Vapor Pressure

In general, the total pressure of a solution is equal to the sum of the vapor pressure of the solute and the solvent. However, in the case of a lithium bromide and water solution, the vapor pressure of pure LiBr can be ignored because its value is much smaller than that of water.

When LiBr is dissolved in water, the boiling point of the solution at a given pressure is raised. However, if the temperature of the solution remains constant, the dissolved LiBr reduces the vapor pressure of the solution.

When a solution is saturated, equilibrium is established. The number of molecules across the interface from liquid into vapor per unit time is equal to the number of molecules from vapor into liquid. If the number of liquid molecules per unit volume is reduced due to the presence of a solute, the number of vapor molecules per unit volume is then also reduced. Consequently, the vapor pressure of the solution is decreased.

Equilibrium Chart

The properties of an aqueous lithium bromide solution, including vapor pressure, temperature, and the mass fraction at equilibrium, may be illustrated on an equilibrium chart based on the Dühring plot, as shown in Fig. 24.1.

The ordinate of the equilibrium chart is the saturated vapor pressure of water in log-scale mm Hg abs, and the corresponding saturated temperature, °F. The scale is plotted on an inclined line. The abscissa of the chart is the temperature of the solution, °F. Mass fraction or concentration lines are inclined lines and are not parallel to each other.

At the bottom of the concentration lines, there is a *crystallization line* or *saturation line*. If the temper-

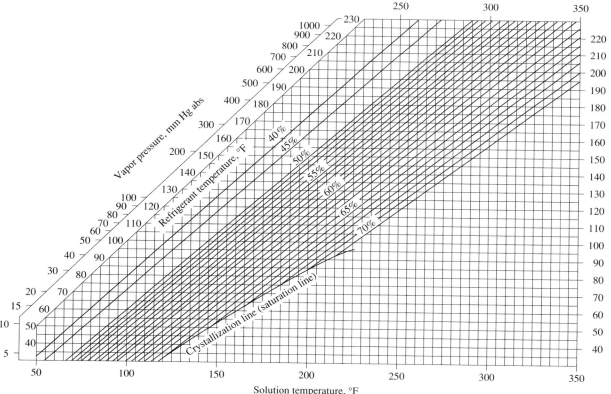

FIGURE 24.1 Equilibrium chart for aqueous lithium bromide LiBr solution. (*Source:* Carrier Corporation. Reprinted with permission.)

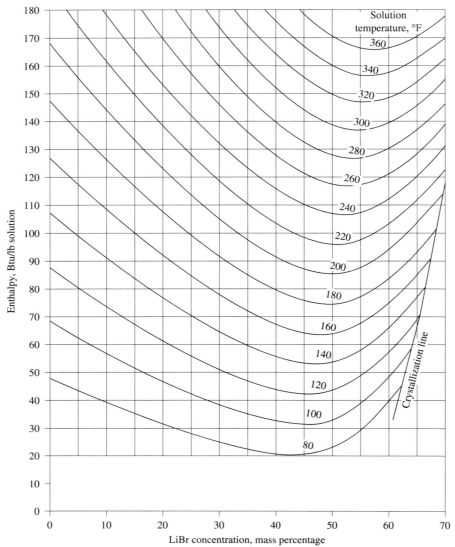

FIGURE 24.2 Enthalpy–concentration diagram for aqueous LiBr solution.
(*Source: ASHRAE Handbook 1989, Fundamentals*. Reprinted with permission.)

ature of a solution of constant mass fraction drops below this line, or the mass fraction of a solution of constant temperature is higher than the saturated condition, the part of lithium bromide LiBr salt exceeding the saturation condition tends to form solid crystals.

Enthalpy–Concentration Diagram

When water is mixed with anhydrous lithium bromide at the same temperature to form a solution adiabatically, there is a significant increase in the temperature of the solution. If the mixing process is to be an isothermal process, that is, the temperature of the process is to be kept constant, then heat must be removed from the solution. Such a heat transfer per unit mass

of solution is called the *integral heat* of the solution Δh_i, or *heat of absorption*, in Btu/lb. Based on the common rule of thermodynamics, Δh_i is negative.

If an aqueous LiBr solution is formed by an isothermal process, its specific enthalpy h, in Btu/lb, may be expressed as

$$h = (1 - X)h_w + Xh_l + \Delta h_i \qquad (24.2)$$

where h_w, h_l specific enthalpy of water and LiBr respectively, Btu/lb.

Figure 24.2 shows the enthalpy–concentration h–X diagram of an aqueous LiBr solution. The ordinate of the diagram is the specific enthalpy of the solution h, in Btu/lb, and the abscissa is the percent concentration of the aqueous LiBr solution. The curved lines represent the specific enthalpy of the solution at various concentrations of LiBr mixed with water at a constant temperature. At the right-hand side of the diagram, there is also a crystallization line. To the right of this line, excess solid LiBr may crystallize.

24.3 TWO-STAGE DIRECT-FIRED ABSORPTION CHILLERS

System Description

In an absorption chiller, water is the refrigerant and aqueous lithium bromide LiBr is the absorbant. Two-stage direct-fired absorption chillers are most commonly used today because of their higher coefficient of performance. *Two-stage* means that there are two generators. *Direct-fired* means that the heat input is supplied to a directly fired generator. Figure 24.3a is a schematic diagram of a two-stage direct-fired absorption chiller. Figure 24.3b is a flow diagram, simplified for clear illustration. ASHRAE/IES Standard 90.1–1989 requires that double-effect (two-stage) heat-operated water chilling packages (absorption chillers) should be used in lieu of single-effect equipment because of their higher efficiency, except when heat energy impact is from low-temperature waste-heat sources.

A two-stage direct-fired absorption chiller consists of eleven main components: an evaporator, an absorber, a first-stage generator, a second-stage generator, a condenser, two heat exchangers, a solution pump, a refrigerant pump, a refrigerant throttling device, and a solution throttling device. Two-stage absorption chillers are also sometimes called *double-effect absorption chillers*.

EVAPORATOR AND REFRIGERANT PUMP. An *evaporator* is composed of a tube bundle, an outer shell, spray nozzles, and a water trough. The chilled water flows inside the tubes. A refrigerant pump sprays the liquid refrigerant through the spray nozzles over the outer surface of the tube bundle to increase the rate of evaporation. A water trough is located at the bottom of the evaporator to maintain a certain liquid level for circulation. A high vacuum, typically at a saturated vapor pressure of 6 mm Hg abs, is maintained to provide an evaporating temperature of about 39°F.

ABSORBER AND SOLUTION PUMP. *Absorbers* are water-cooled tube bundles with cooling water flowing inside them. In the absorber, the vaporized refrigerant, the water vapor, is extracted and absorbed by the concentrated LiBr solution sprayed from the nozzles so that a high vacuum, 5.8 mm Hg abs, can be maintained in the evaporator. The solution pump raises the pressure of concentrated solution from slightly below 6 mm Hg abs to about 560 to 610 mm Hg abs in the first-stage generator. This LiBr solution pressure is sufficient to overcome the pressure drop in the heat exchanger, pipes, and pipe fittings.

HEAT EXCHANGERS. Heat exchangers are sometimes called economizers. Shell-and-tube, or plate-and-frame heat exchangers are most commonly used for higher effectiveness. There are two heat exchangers: a low-temperature heat exchanger and a high-temperature heat exchanger. In the low-temperature heat exchanger, heat is transferred from the hot concentrated solution discharged from the second-stage generator to the cold diluted solution pumped from the absorber. In the high-temperature heat exchanger, heat is transferred from the hot concentrated solution after the first-stage generator to the cold diluted solution from the absorber. Both heat exchangers conserve heat energy input and increase the coefficient of performance of the absorption chiller.

GENERATORS. In the first-stage generator, heat is supplied from the direct-fired burner or other waste heat source to the diluted solution and the water is then vaporized from the solution. Vapor refrigerant flows to the second-stage generator and condenses into liquid. The latent heat released during condensation is then used to heat the diluted solution. Water is again vaporized at the second-stage generator. Many types of fuel can be used in the direct-fired burner, such as natural gas, oil, or propane. Natural gas is the most widely used.

CONDENSER. A condenser is a water-cooled tube bundle in which cooling water flows. In the condenser, water is often supplied at a temperature between 85° and 95°F. Water vapor boiled off from the second-stage generator is condensed into liquid in the condenser, and the liquid water condensed inside the tubes of the second-stage generator is cooled to a lower temperature in the condenser.

THROTTLING DEVICES. Orifices, traps, and valves are often used as throttling devices to reduce the pressure of refrigerant and solution. An orifice plate on the liquid refrigerant line acts as an expansion valve between condenser and evaporator. There is also a valve on the liquid refrigerant line to regulate the pressure between the first-stage generator and condenser.

Instead of heat input from a direct-fired furnace, absorption chillers can use other heat sources such as industrial waste heat or exhaust heat from engines or gas turbines.

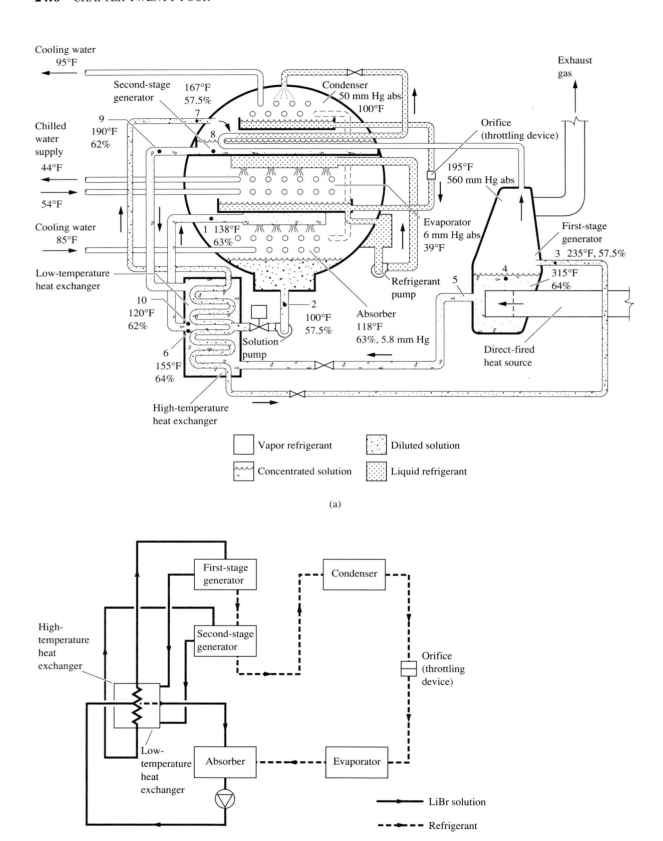

(a)

(b)

FIGURE 24.3 A typical two-stage direct-fired parallel-flow absorption chiller: (*a*) schematic diagram and (*b*) simplified flow diagram.

A single-stage absorption chiller has only one generator, whereas a two-stage absorption chiller uses part of the latent heat of condensation of the vapor refrigerant in the second-stage generator to enhance the COP. If steam is used as the input heat source in the indirect-fired absorption chiller, the combined efficiency of the boiler, loss in steam transport, and the heat transfer at the first-stage generator is significantly lower than that of the direct-fired furnace.

Air Purge Unit

Because the pressures in the evaporator, absorber, and condenser are far below atmospheric pressure, air and other noncondensable gases leak into a LiBr absorption refrigeration system. A purge unit should be installed to remove these noncondensable gases and to maintain the required pressure, temperature, and concentration in the absorption chiller.

Figure 24.4 shows a typical purge unit in a LiBr absorption refrigeration system. Such a unit usually consists of a purge chamber, a pick-up tube, a vacuum pump, a solenoid valve, and a manual shut-off valve.

When noncondensable gases leak into the system during operation, they tend to migrate to the absorber, where the pressure is lowest. Through the pick-up tube, noncondensable gases and a certain amount of water vapor flow to the purge chamber from the absorber. Water vapor is absorbed by the LiBr solution sprayed over the tube, and the heat of absorption is removed by the cooling water inside a bank of tubes.

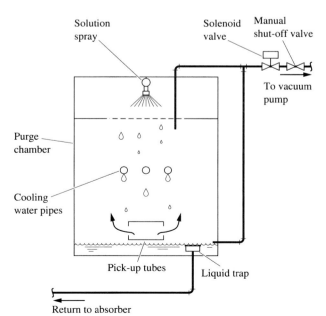

FIGURE 24.4 Purge unit in a lithium bromide LiBr absorption chiller.

The solution absorbed with water vapor is returned to the absorber through a liquid trap. Noncondensable gases are evacuated from the chamber periodically by a vacuum pump which raises the gas pressure and discharges the gases to the outdoor atmosphere.

Parallel-Flow and Series-Flow

Absorption Chillers

In a two-stage, direct-fired absorption chiller, the flow of the solution from the absorber to the generators can be either parallel flow or series flow.

In a parallel-flow arrangement, solution from the absorber is pumped to both the first- and second-stage generators separately, as shown in Fig. 24.3b. Concentrated solution from both generators is recombined and returned to the absorber. In a series-flow arrangement, the solution from the absorber is first pumped to the first-stage generator. Partly concentrated solution then flows to the second-stage generator, as shown in Fig. 24.5.

A parallel-flow arrangement has two primary advantages over a series flow arrangement:

• The flow rate of the solution entering the first-stage generator is nearly one-half of that in the series-flow arrangement. Less heat is required to raise the temperature of the solution to the saturated condition to boil off water vapor. Therefore, the COP is higher.

• The maximum pressure required for parallel flow is lower than for series flow because a lower pressure drop is needed to flow through a single heat exchanger. Parallel-flow systems operate further away from the crystallization region than series-flow systems do.

Flow of Solution and Refrigerant

For a two-stage direct-fired parallel-flow absorption chiller, the absorption cycle shows the properties of the solution at various concentrations, temperatures, and pressures as shown in Fig. 24.6. Sometimes the refrigerant is combined with the solution, and sometimes it is separate.

Two kinds of LiBr solution are used during operation: concentrated solution and diluted solution. Concentrated solution has a higher concentration of LiBr, and is created after the water vapor has been boiled off from the solution. Concentrated solution is used to absorb the water vaporized in the evaporator. Diluted solution has a lower concentration of LiBr, and is the combination of aqueous LiBr solution and absorbed water vapor.

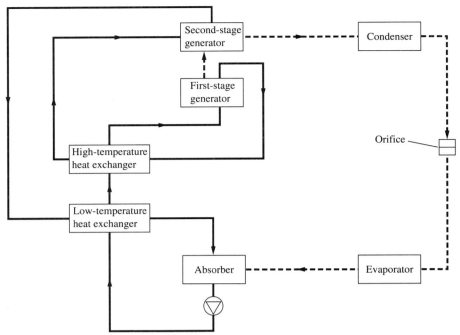

FIGURE 24.5 Two-stage direct-fired series flow absorption chiller.

An absorption refrigeration system can also be divided into two regions according to the pressure of the refrigerant and solution. The high-pressure side includes generators and condensers, and the low-pressure side comprises absorbers and evaporators. Because the evaporator and absorber (and also the second-stage generator and the condenser) are connected by vapor passages of adequate size, the pressure difference between the evaporator and absorber (and between the second-stage generator and the condenser) is small.

In a typical two-stage direct-fired parallel-flow absorption chiller operating at design load, water usually evaporates at about 39°F and at a saturated pressure of 6 mm Hg abs in the evaporator. Chilled water enters the evaporator at 54°F and leaves at 44°F. Heat

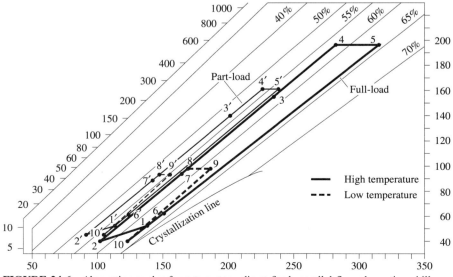

FIGURE 24.6 Absorption cycles for a two-stage direct-fired, parallel-flow absorption chiller.

is transferred from the chilled water to the vaporized refrigerant at an amount equal to the latent heat of vaporization.

Vaporized water in the evaporator is then extracted to the absorber because of its lower vapor pressure, and is absorbed by the concentrated LiBr solution. In the absorber, concentrated solution is supplied at about 138°F. It is cooled by the cooling water flowing inside the tube bundle at an entering temperature of 85°F. If the sprayed LiBr solution is at a concentration of 63 percent at 118°F during absorption, the vapor pressure of the solution is about 5.8 mm Hg abs, which is slightly lower than the evaporating pressure (6 mm Hg abs). As the water vapor from the evaporator is absorbed, the solution is diluted to a concentration of about 57.5 percent and its temperature drops to 100°F. The heat of absorption is removed by the cooling water. Diluted solution is then pumped to the generators through a solution pump.

At the outlet of the solution pump, diluted LiBr solution is divided into two streams: one that flows through the high-temperature heat exchanger and one that flows through the low-temperature heat exchanger.

In the high-temperature heat exchanger, diluted solution enters at 100°F and leaves at about 235°F. After flowing through the high-temperature heat exchanger, diluted solution enters the first-stage generator. In the first-stage generator, the solution is heated to a temperature of 315°F and the water vapor boils off at a pressure of about 560 Hg mm abs. As the water vapor is boiled off, the LiBr solution reaches a concentration of 64 percent in the first-stage generator. The concentrated solution returns to the high-temperature heat exchanger and cools to about 155°F as it leaves the heat exchanger.

The other stream of diluted LiBr solution enters the low-temperature heat exchanger at 100°F and a concentration of 57.5 percent. In the heat exchanger, its temperature is raised to about 167°F, and it then enters the trough of the second-stage generator. As the water vapor from the LiBr solution in the first-stage generator flows through the tubes submerged in the diluted solution in the second-stage generator, the water vapor condenses into liquid form. Latent heat of condensation released from the liquid refrigerant heats the diluted solution from 167°F to 190°F and boils off the water vapor from the diluted solution at a saturated pressure of about 52 mm Hg abs. Finally, in the second-stage generator, the diluted solution reaches a concentration of 62 percent and flows to the low-temperature heat exchanger, where it is cooled from 190°F to 120°F. The condensed liquid refri-

gerant flowing inside the tubes in the second-stage generator is discharged through the top inlet to the condenser.

In the condenser, liquid refrigerant is cooled to about 100°F by the cooling water. Water vapor boiled off in the second-stage generator is extracted by the condenser's low saturated pressure of 50 mm Hg abs. Water vapor condenses into liquid as the cooling water removes the latent heat of condensation. Liquid refrigerant from the second-stage generator combines with the liquid refrigerant condensed in the condenser. The refrigerant is forced through an orifice, throttled to a pressure of about 6 mm Hg abs, and returned to the evaporator.

The concentrated solution from the high-temperature heat exchanger at a concentration of 64 percent and a temperature of about 155°F is mixed with the concentrated solution from the low-temperature heat exchanger at a concentration of 62 percent and a temperature of 120°F. The mixture, at a concentration of 63 percent and a temperature of about 138°F, enters the absorber. It is then cooled to about 118°F by cooling water during absorption.

24.4 PERFORMANCE OF TWO-STAGE DIRECT-FIRED ABSORPTION CHILLER

Mass Flow Rate of Refrigerant and Solution

For a throttling process, the enthalpy of water entering the evaporator h_{le} is equal to the enthalpy of saturated liquid leaving the condenser h_{lc}, both in Btu/lb. The mass flow rate of refrigerant required for each ton of refrigeration produced in the absorption chiller $\dot{m}_r$, in lb/h, is calculated as

$$\dot{m}_r = \frac{12,000}{h_{ve} - h_{lc}} \qquad (24.3)$$

where h_{ve} = enthalpy of saturated water vapor leaving evaporator, Btu/lb.

According to the mass balance of LiBr in the diluted and concentrated solution,

$$X_a \dot{m}_{sa} = X_g \dot{m}_{sg}$$
$$\dot{m}_{sa} = \dot{m}_{sg} + \dot{m}_r \qquad (24.4)$$

Also,

$$\dot{m}_{sg} = \frac{X_a}{X_g - X_a} \dot{m}_r$$

$$C_{cir} = \frac{X_a}{X_g - X_a} \qquad (24.5)$$

where X_a, X_g = mass fraction of LiBr in diluted and concentrated solution

$\dot{m}_{sa}$ = mass flow rate of diluted solution per ton of refrigeration, lb/h

$\dot{m}_{sg}$ = mass flow rate of concentrated solution from both the first- and second-stage generator per ton of refrigeration, lb/h

C_{cir} = circulating factor

Thermal Analysis

To help determine the status of the solution in the absorption cycle of a two-stage direct-fired parallel-flow absorption chiller, as shown in Fig. 24.6, the state points of the cycle are defined as follows:

1. Mean condition of the concentrated solution entering the absorber

2. Diluted solution leaving the absorber

3. Diluted solution leaving the high-temperature heat exchanger

4. Water vapor boils off in the first-stage generator

5. Concentrated solution leaving the first-stage generator

6. Concentrated solution leaving the high-temperature heat exchanger

7. Diluted solution leaving the low-temperature heat exchanger

8. Water vapor boils off in the second-stage generator

9. Concentrated solution leaving the second-stage generator

10. Concentrated solution leaving the low-temperature heat exchanger

In Fig. 24.6, points 1, 2, 3, 4, 5, and 6 represent the high-temperature stream of solution in the absorption cycle, and points 1, 2, 7, 8, 9, and 10 represent the low-temperature stream of solution in the absorption cycle.

When the absorption chiller operates at steady-state condition, the heat balance analyses for the main system components, in Btu/h · ton, are shown below.

In the evaporator, for one ton of refrigeration load q_{rl},

$$\begin{array}{ccc} \text{Heat of leaving} & \text{Heat of entering} & \text{Refrigeration} \\ \text{water vapor} & - \text{liquid} & = \text{load} \end{array}$$

$$\dot{m}_r h_{ve} - \dot{m}_r h_{lc} = \dot{m}_r (h_{ve} - h_{lc}) = q_{rl} \qquad (24.6)$$

In the absorber,

$$\begin{array}{cccc} \text{Heat of} & \text{Heat of} & \text{Heat of} & \text{Heat to be} \\ \text{entering} & + \text{entering} & - \text{leaving} & = \text{removed from} \\ \text{water vapor} & \text{solution} & \text{solution} & \text{absorber} \end{array}$$

$$\dot{m}_r h_{ve} + \dot{m}_{sg} h_1 - \dot{m}_{sa} h_2 = q_{ab} \qquad (24.7)$$

where h_1, h_2 = enthalpy of solution entering and leaving the absorber, Btu/lb.

The enthalpy of concentrated solution entering the absorber is the mixture of the enthalpies of concentrated solutions discharged from the high-temperature heat exchanger h_6 and the low-temperature heat exchanger h_{10}, both in Btu/lb, that is,

$$h_1 = X_6 h_6 + X_{10} h_{10} \qquad (24.8)$$

where h_6, h_{10} = enthalpy of concentrated solution leaving the high- and low-temperature heat exchangers, Btu/lb.

In Eq. (24.8), X_6 and X_{10} represent the mass fractions or percentages of mass flow rate of concentrated solution per ton of refrigeration discharged from the high- and low-temperature heat exchangers.

In the first-stage generator,

$$\begin{array}{cccc} \text{Heat of} & \text{Heat of} & \text{Heat of} & \text{Heat to be} \\ \text{leaving} & + \text{leaving} & - \text{entering} & = \text{supplied to first-} \\ \text{water vapor} & \text{solution} & \text{solution} & \text{stage generator} \end{array}$$

$$\dot{m}_{r1g} h_{v1g} + \dot{m}_{1g} h_5 - \dot{m}_{a1g} h_3 = q_{1g} \qquad (24.9)$$

where $\dot{m}_{r1g}$ = mass flow rate of water vapor boiled off in first-stage generator per ton of refrigeration, lb/h · ton

h_{v1g} = enthalpy of superheated water vapor leaving first-stage generator, Btu/lb

$\dot{m}_{1g}$ = mass flow rate of concentrated solution leaving first-stage generator per ton of refrigeration, lb/h · ton

$\dot{m}_{a1g}$ = mass flow rate of diluted solution entering first-stage generator per ton of refrigeration, lb/h · ton

h_3, h_5 = enthalpy of solution entering and leaving first-stage generator, Btu/lb

q_{1g} = heat input to the first-stage generator per ton of refrigeration, Btu/h · ton

In the second-stage generator,

$$\begin{array}{cc} \text{Heat of entering} & \text{Heat of} \\ \text{water vapor} & - \text{leaving} \\ & \text{water} \end{array}$$

$$\begin{array}{ccc} & \text{Heat of} & \text{Heat of} & \text{Water vapor} \\ = & \text{leaving} & - \text{entering} & + \text{boiled off in} \\ & \text{solution} & \text{solution} & \text{second-stage} \\ & & & \text{generator} \end{array}$$

$$\dot{m}_{r1g}(h_{v1g} - h_{12g}) = \dot{m}_{2g}h_9 - \dot{m}_{a2g}h_8 + \dot{m}_{r2g}h_{2fg}$$
$$(24.10)$$

where h_{12g} = enthalpy of liquid water leaving second-stage generator, Btu/lb

$\dot{m}_{2g}$ = mass flow rate of concentrated solution leaving second-stage generator per ton of refrigeration, lb/h · ton

$\dot{m}_{a2g}$ = mass flow rate of dilution solution entering second-stage generator per ton of refrigeration, lb/h · ton

h_8, h_9 = enthalpy of solution entering and leaving second-stage generator, Btu/lb

$\dot{m}_{r2g}$ = mass flow rate of water vapor boiled off in second-stage generator per ton of refrigeration, lb/h · ton

h_{2fg} = latent heat of vaporization in second-stage generator, Btu/lb

and

$$\dot{m}_r = \dot{m}_{r1g} + \dot{m}_{r2g}$$
$$\dot{m}_{sa} = \dot{m}_{a1g} + \dot{m}_{a2g} \qquad (24.11)$$
$$\dot{m}_{sg} = \dot{m}_{1g} + \dot{m}_{2g}$$

In the condenser,

Heat of entering liquid water	+	Heat of entering water vapor	−	Heat of leaving liquid water	=	Heat to be removed from condenser

$$\dot{m}_{r1g}h_{12g} + \dot{m}_{r2g}h_{v2g} - \dot{m}_r h_{lc} = q_{con} \qquad (24.12)$$

where h_{v2g} = enthalpy of water vapor leaving second-stage generator, Btu/lb.

In the high-temperature heat exchanger, if the difference of the specific heat of solution is ignored the effectiveness ϵ_h is given as

$$\epsilon_h = \frac{\dot{m}_{a1g}(h_3 - h_2)}{\dot{m}_{1g}(h_5 - h_2)}$$
$$= \frac{h_5 - h_6}{h_5 - h_2} \qquad (24.13)$$

Similarly, the effectiveness of the low-temperature heat exchanger is

$$\epsilon_l = \frac{\dot{m}_{a2g}(h_7 - h_2)}{\dot{m}_{2g}(h_9 - h_2)}$$
$$= \frac{h_9 - h_{10}}{h_9 - h_2} \qquad (24.14)$$

Because the pump power is usually small compared with the heat supplied to the generator, it can often be ignored in thermal analysis.

If the heat loss from the system components is ignored, heat input to the absorption system must be equal to the heat removed from the system (in Btu/h · ton), that is,

$$q_{1g} + q_{rl} = q_{ab} + q_{con} \qquad (24.15)$$

Coefficient of Performance

If the pump power of the solution and refrigerant pumps are excluded, the coefficient of performance of an absorption chiller COP_c can be calculated as

$$COP_c = \frac{\text{refrigeration output}}{\text{heat input}} = 12,000/q_{1g}$$
$$(24.16)$$

where q_{1g} heat input to the first-stage generator per ton of refrigeration, Btu/h · ton.

Example 24.1. A typical two-stage direct-fired parallel-flow absorption chiller described in Section 24.3 operates under the following conditions:

1. The effectiveness of its heat exchangers is 0.65.
2. The mass flow rates of diluted solution to the high- and low-temperature heat exchangers and to the first- and second-stage generators are all equal.
3. The amounts of water vapor boiled off from the first- and second-stage generators are equal.

Calculate the coefficient of performance of this absorption chiller. Also calculate the mass flow rate of cooling water required for a temperature increase of 10°F.

Solution
1. From psychrometric table, at a condensing temperature of 100°F, the enthalpy of saturated liquid water leaving condenser h_{lc} = 68.03 Btu/lb and the enthalpy of saturated water vapor leaving the evaporator h_{ve} at an evaporating temperature of 39°F is 1080.4 Btu/lb. From Eq. (24.3), the mass flow rate of refrigerant per ton of refrigeration is therefore

$$\dot{m}_r = \frac{12,000}{h_{ve} - h_{lc}} = \frac{12,000}{1080.4 - 68.0}$$
$$= 11.85 \text{lb/h} \cdot \text{ton}$$

2. As described in the previous section, the concentration of diluted solution from absorber is 57.5 percent, and the mean concentration of solution entering the absorber is 63 percent. From Eq. (24.5), the mass flow rate of concentrated solution for each ton of refrigeration produced is therefore

$$\dot{m}_{sg} = \frac{X_a}{X_g - X_a}\dot{m}_r = \frac{0.575}{0.63 - 0.575} \, 11.85$$
$$= 123.9 \text{lb/h.ton}$$

From Eq. (24.4), the mass flow rate of diluted solution per ton of refrigeration is

$$\dot{m}_{sa} = \dot{m}_{sg} + \dot{m}_r = 123.9 + 11.85$$
$$= 135.7 \text{ lb/h} \cdot \text{ton}$$

3. From the given data, in the first-stage generator, the concentrated solution leaving the generator h_5 is at a temperature of 315°F and a concentration of 64 percent. From the enthalpy-concentration h-X diagram, $h_5 = 151.5$ Btu/lb.

Also, from the given data, the diluted solution leaving the absorber is at a temperature of 100°F and a concentration of 57.5 percent. From the h–X diagram, $h_2 = 43.5$ Btu/lb.

Then, from Eq. (24.13), in the high-temperature heat exchanger, for an effectiveness of 0.65,

$$\epsilon_h = 0.65 = \frac{h_5 - h_6}{h_5 - h_2} = \frac{151.5 - h_6}{151.5 - 43.5}$$

$$h_6 = 81.3 \text{ Btu/lb}$$

From h–X diagram, the temperature of concentrated solution leaving the high-temperature heat exchanger $T_6 = 155°F$.

From the given data, $\dot{m}_{alg} = 0.5\dot{m}_{sa} = 0.5 \times 135.7 = 67.85$ lb/h · ton. Also, $\dot{m}_{1g} = \dot{m}_{alg} - 0.5\dot{m}_r = 67.85 - (0.5 \times 11.85) = 61.95$ lb/h · ton. Then,

$$\epsilon_h = 0.65 = \dot{m}_{alg}\frac{h_3 - h_2}{\dot{m}_{1g}(h_5 - h_2)}$$

$$= 67.85\frac{(h_3 - 43.5)}{61.95(151.5 - 43.5)}$$

$$h_3 = 107.6 \text{ Btu/lb.}$$

From the h–X diagram, the temperature of diluted solution leaving the high-temperature heat exchanger $T_3 = 235°F$.

4. If the concentrated solution leaving the second-stage generator is at 190°F and a concentration of 62 percent, from the h–X diagram, its enthalpy $h_9 = 94$ Btu/lb.

From Eq. (24.14), the effectiveness of the low-temperature heat exchanger is therefore

$$\epsilon_l = 0.65 = \frac{h_9 - h_{10}}{h_9 - h_2} = \frac{94 - h_{10}}{94 - 43.5}$$

$$h_{10} = 61.2 \text{ Btu/lb}$$

From the h–X diagram, the temperature of concentrated solution leaving the low-temperature heat exchanger T_{10} is 120°F.

Similarly, in the low-temperature heat exchanger,

$$\epsilon_l = 0.65 = \frac{\dot{m}_{a2g}(h_7 - h_2)}{\dot{m}_{2g}(h_9 - h_2)}$$

$$= \frac{67.85(h_7 - 43.5)}{61.95(94 - 43.5)}$$

$$h_7 = 73.5 \text{ Btu/lb}$$

From the h–X diagram, the temperature of diluted solution leaving the low-temperature heat exchanger T_7 is 167°F.

5. Because $\dot{m}_{r1g} = \dot{m}_{r2g} = 0.5\dot{m}_r = 0.5 \times 11.85 = 5.93$ lb/h·ton, from the psychrometric table, at 315°F, $h_{v1g} = 1184.5$ Btu/lb, from Eq. (24.9), the heat input to the first-stage generator per ton of refrigeration produced q_{1g} (in Btu/lb · ton) is therefore

$$q_{1g} = \dot{m}_{r1g}h_{v1g} + \dot{m}_{1g}h_5 - \dot{m}_{alg}h_3$$

$$= 5.93 \times 1184.5 + 61.95 \times 151.5$$

$$- 67.85 \times 107.6$$

$$= 9108 \text{ Btu/h} \cdot \text{ton}$$

From Eq. (24.16), the coefficient of performance of this absorption chiller is

$$\text{COP}_c = \frac{12,000}{q_{1g}} = \frac{12,000}{9108} = 1.32$$

6. From Eq. (24.8), the enthalpy of the concentrated solution entering the absorber is

$$h_1 = X_6h_6 + X_{10}h_{10}$$

$$= 0.5 \times 81.3 + 0.5 \times 61.2$$

$$= 71.3 \text{ Btu/lb}$$

Thus, the heat removed from the absorber per ton of refrigeration q_{ab} is

$$q_{ab} = \dot{m}_r h_{ve} + \dot{m}_{sg}h_1 - \dot{m}_{sa}h_2$$

$$= 11.85 \times 1080.4 + 123.9$$

$$\times 71.3 - 135.7 \times 43.5$$

$$= 15,734 \text{ Btu/h} \cdot \text{ton}$$

From the psychrometric chart, the enthalpy of liquid refrigerant entering the condenser from the second-stage generator at 195°F, $h_{12g} = 163.1$ Btu/lb, and the enthalpy of vapor refrigerant vaporized from the second-stage generator entering the condenser $h_{v2g} = 1141.8$ Btu/lb. The heat to be removed from the condenser per ton of refrigeration q_{con}, in Btu/h · ton, is therefore

$$q_{con} = \dot{m}_{r1g}h_{12g} + \dot{m}_{r2g}h_{v2g} - \dot{m}_r h_{lc}$$

$$= 5.93 \times 163.1 + 5.93 \times 1141.8$$

$$- 11.85 \times 68$$

$$= 6932 \text{ Btu/h} \cdot \text{ton}$$

The volume flow rate of cooling water required per ton of refrigeration produced, in gpm, is then

$$\dot{V}_{gal} = \frac{q_{ab} + q_{con}}{500 \times \text{temperature increase of cooling water}}$$

$$= \frac{15734 + 6932}{500 \times 10} = 4.53 \text{ gpm/ton}$$

Hufford (1991) reported a COP based on net heat input equal to 1.29, which is quite near to the calculated value. The actual COP may be lower because of heat losses and other effects.

24.5 CAPACITY CONTROL AND PART-LOAD PERFORMANCE

Capacity Control

A two-stage direct-fired parallel-flow absorption chiller adjusts its capacity during part-load operation by modulation of the heat input to the first-stage generator through the direct-fired burner or other waste heat supply.

A sensor is often located at the outlet of the evaporator to monitor the temperature of the chilled water leaving the evaporator T_{el}. As the system refrigeration load falls, T_{el} decreases accordingly. Once a drop of T_{el} below a predetermined set point is sensed, the DDC panel or controller reduces the gas flow or other fuel supply to the direct-fired burner, and less water vapor is boiled off from the solution in the generator. Consequently, the concentration of the solution entering the absorber drops, less water vapor is extracted to the absorber and, therefore, the rate of evaporation and refrigeration effect in the evaporator are both reduced until they are balanced with the reduction of refrigeration load so that T_{el} is maintained within acceptable limits. Meanwhile, because less water vapor has been extracted to the absorber from the evaporator, both evaporative pressure and evaporating temperature T_{ev} increase.

Although less water vapor is boiled off in the generators, the rate of heat transfer at the condensing surface and the amount of water vapor to be condensed into liquid water in the condenser remain the same. The condensing pressure p_{con} and temperature T_{con} decrease until a new balance is formed.

If T_{el} rises above a limit, on the other hand, more heat is provided to the generator, and the concentration of solution and the refrigeration capacity increase and T_{el} again falls within preset limits. Of course, the increase in solution concentration should not exceed the saturation limit.

Absorption Cycle During Part-Load Operation

Part-load performance of the absorption chiller can be improved by modulating the amount of diluted solution pumped from the absorber to the first- and second-stage generators through the heat exchangers. When a drop in refrigeration load is indicated by a fall in T_{el}, the rate of solution flow is reduced in direct proportion to the reduction of refrigeration load. As a result, less diluted solution is transported to the first- and second-stage generators and less heat energy is needed to boil off the refrigerant in the generators.

The left-hand side of Fig. 24.6 shows the part-load absorption cycle of the two-stage direct-fired parallel-flow absorption chiller. During part-load operation, the following conditions exist:

- The mass flow rate of refrigerant is directly proportional to the load ratio.
- The evaporating pressure p_{ev} and temperature T_{ev} rise.
- condensing pressure p_{con} and temperature T_{con} will drop.
- The boiled-off temperatures in the first- and second-stage generators decreases.
- Heat input to the first-stage generator is reduced.
- The drop in temperature of cooling water at a lower outdoor wet bulb lowers the condensing temperature T_{con} and, therefore, the heat input.
- At a constant entering temperature of cooling water, COP at part-load is approximately the same as at design full-load. If the temperature of cooling water also drops at part-load, COP is slightly lower at part-load than at full-load.

During part-load operation, evaporating and condensing temperatures, concentrations of diluted and concentrated solution, and other operating parameters can be determined from the equations based on the heat balance analyses described in Section 24.4.

24.6 ABSORPTION CHILLER CONTROLS

Crystallization and Controls

In an absorption chiller using aqueous LiBr as the absorbant, the LiBr in a solution of constant concentration starts to crystallize when the solution temperature drops below the crystallization line on the equilibrium chart. The crystals formed are pure LiBr salt. Crystallization does not harm the absorption chiller, but it eventually decreases the concentration of the remaining solution. It is also a symptom of malfunction, and the cause of crystallization must be determined before normal operation can resume.

Absorption chillers are now designed to operate in the region away from the crystallization line shown in Fig. 24.6. There are also devices available that prevent crystallization and dissolve crystals if crystallization occurs. Crystallization is no longer a serious problem for currently designed absorption chillers, as it was before the energy crisis.

Causes of crystallization include the following:

- Air leaks into the system raise the evaporating temperature and the chilled water leaving temperature T_{el}. A higher T_{el} increases the heat input and the solution concentration to the point of crystallization.

- When the system is operated at full-load, if the temperature of the cooling water is too low, the diluted solution temperature may fall low enough to reduce the temperature of the concentrated solution to the point of crystallization.
- If the electric power is interrupted, the system ceases to operate. The temperature of the concentrated solution in the heat exchanger starts to drop, and may fall below the crystallization line.

Manufacturers have developed several devices to minimize the possibility of crystallization. One such device uses a bypass valve to permit refrigerant to flow to the concentrated solution line when conditions that can cause crystallization are detected by the sensor. A newly developed microprocessor-based DDC panel uses measured temperatures and pressures at key points to calculate the concentration of the solution to prevent crystallization.

An overflow pipe is often used to carry the concentrated solution from the generator to the absorber in case of crystallization or other failures.

Cooling Water Temperature Control

Earlier absorption chiller designs required close control of the cooling water temperature to prevent crystallization. Current absorption chiller designs include control devices to prevent crystallization, so many manufacturers now allow the temperature of the cooling water to fall to 60°F at part-load, as when the load ratio is reduced to 0.6. A three-way mixing bypass valve (as shown in Fig. 18.11) should be installed to mix bypass recirculating water from the condenser and maintain the temperature of cooling water entering the absorber at or above a predetermined value at a certain load ratio. At high load ratios, lower cooling water entering temperatures may cause crystallization. Cooling tower fans can be cycled to supplement the bypass control.

Safety and Interlocking Controls

LOW TEMPERATURE CUT-OUT. If the temperature of refrigerant in the evaporator falls below a preset value, the DDC panel shuts down the absorption chiller to protect the evaporator from freezing. As soon as the refrigerant temperature rises above the limit the DDC panel starts the chiller again.

CHILLED-WATER FLOW SWITCH. When the mass flow rate of chilled water falls below a limit, a pressure-sensitive or flow-sensitive sensor alerts the DDC panel, which stops the absorption chiller.

COOLING-WATER FLOW SWITCH. When a drop in cooling water supply is detected by the pressure- or flow-sensitive sensor, the DDC panel shuts down the absorption chiller. The chiller is started again only when the cooling water supply is reestablished.

HIGH-PRESSURE RELIEF. A high-pressure relief valve or similar device is often installed on the shell of the first-stage generator to prevent the maximum pressure in the system from exceeding a preset value.

DIRECT-FIRED SAFETY CONTROLS. As described in Section 14.3, a direct-fired absorption chiller needs controls of high-pressure and low-pressure switches, flame ignition, and monitoring for its burner and generator.

INTERLOCKED CONTROLS. Absorption chillers should be interlocked with chilled-water pumps, cooling-water pumps, and cooling-tower fans so that the absorption chiller starts only when these devices are in normal operation.

OPTIMIZING ON-OFF CONTROL. If multiple absorption chillers are installed in a chiller plant, as in a plant with multiple centrifugal chillers, optimizing controls can be used to turn the absorption chillers on and off.

24.7 OPERATING CHARACTERISTICS AND DESIGN CONSIDERATIONS

Difference Between Absorption and Centrifugal Chillers

In a vapor compression refrigeration system, compressor power input is primarily affected by head ($p_{con} - p_{ev}$) or temperature lift ($T_{con} - T_{ev}$) of the refrigeration cycle. In an absorption chiller, ($T_{con} - T_{ev}$) is also a factor that influences the heat input to the first-stage generator, but its influence is significantly smaller than that of centrifugal chillers.

For the two-stage direct-fired parallel-flow absorption chiller described in Example 24.1, the heat input to first-stage generator q_{1g} is distributed as follows:

Heat input q_{1g}	9108 Btu/h · ton	100 percent
Latent heat of vaporization h_{fg}	5422 Btu/h · ton	60 percent
Heating of diluted solution	3686 Btu/h · ton	40 percent

Latent heat of vaporization is the primary factor that affects the heat input in an absorption chiller, and temperature difference ($T_{con} - T_{ev}$) has only a minor effect on h_{fg}. This is the primary difference between an absorption chiller and a centrifugal chiller.

Evaporating Temperature

As in the centrifugal chiller, the evaporating temperature T_{ev} and pressure p_{ev} in an absorption chiller depend mainly on the chilled water temperature leaving the evaporator T_{el}. The difference $(T_{el} - T_{ev})$ in current absorption chiller design is about 5°F. A smaller $(T_{el} - T_{ev})$ means a higher COP and a large heat transfer surface area in the evaporator. It is actually a compromise between energy cost and investment.

For a $T_{el} = 44°F$ and a $(T_{el} - T_{ev}) = 5°F$, the evaporating pressure is around 6 mm Hg abs (0.238 in. Hg abs).

The vapor pressure of the concentrated solution during absorption should be slightly less than the evaporating pressure p_{ev} so that water vaporized in the evaporator can be extracted to the absorber.

Both T_{ev} and T_{el} affect the refrigeration or cooling capacity of the absorption chiller as well as the heat input to the first-stage generator.

Cooling Water Entering Temperature

The temperature of cooling water entering the absorber T_{ca}, in °F, has the following effects on the performance of an absorption chiller:

- Lower T_{ca} means a higher cooling capacity.
- Lower T_{ca} results in a lower T_{con}.
- Lower T_{ca} means a lower heat input to the first-stage generator per ton of refrigeration produced and, therefore, a high COP_c.
- At high load ratios, a too low T_{ca} may cause crystallization.

Manufacturer recommendations should be followed. For example, one manufacturer recommends that at design load, T_{ca} can be lowered to 75°F. At a load ratio $0.6 \leq LR < 0.85$, $T_{ca} \geq 65°F$. When $0.1 < LR < 0.6$, $T_{ca} \geq 55°F$.

Heat Removed from Absorber and Condenser

The total amount of heat to be removed from the absorber and condenser in a typical two-stage direct-fired parallel-flow absorption chiller is about 1.5 times the heat released from the condenser in a centrifugal chiller. As indicated in Example 24.1, the heat to be removed in the absorber is about 70 percent of the total heat to be removed. Heat removed from the condenser is about 30 percent of total heat removal.

Usually, a cooling water temperature increase of 10 to 15°F is used. For a cooling water temperature increase of 10°F when the temperature entering the ab-

sorber is 85°F, the temperature of cooling water is about 92°F entering the condenser and 95°F leaving the condenser.

Condensing Temperature

Condensing temperature T_{con} depends mainly on the temperature of cooling water available and the heat transfer surface area. If cooling water is at 85°F, allowing for a temperature increase of 7°F in the absorber and 3°F in the condenser and a difference between the condensing temperature and cooling water temperature leaving the condenser $(T_{con} - T_{cl}) = 5°F$, condensing temperature T_{con} can then be calculated as

$$T_{con} = 85 + 7 + 3 + 5 = 100°F$$

Condensing pressure p_{con} should be slightly lower than the boiled-off vapor pressure in the second-stage generator in order to extract the vapor and be condensed in the condenser.

Corrosion Control

Lithium bromide is very corrosive. It attacks steel, copper, and copper alloys in the presence of air at temperatures above 300°F. Corrosion inhibitors should be used to protect the internal components against corrosive attacks, as specified by the manufacturer.

Selection of Absorption Chiller

When selecting a two-stage direct-fired parallel-flow absorption chiller, the chiller's refrigeration capacity must meet the required system refrigeration load.

Absorption chillers are rated under the following conditions:

Leaving chilled water temperature	44°F
Chilled water temperature increase	10°F
Entering cooling water temperature	85°F
Fouling factor	0.00025 h · ft^2 · °F/Btu

At design load, under operating conditions other than the standard rating conditions, both the refrigeration capacity and the heat input should be modified.

24.8 ABSORPTION HEAT PUMPS

Functions of Absorption Heat Pump

An absorption heat pump extracts heat from a low-temperature heat source, such as waste heat or surface water, and delivers its heat output at a higher temperature for winter heating or other applications at a coefficient of performance greater than 1.

In Japan and Sweden, absorption heat pumps have been installed in industrial and district heating plants using industrial waste heat to provide hot water, typically at 165°F, for winter heating or other purposes at a COP between 1.4 and 1.7.

Absorption heat pumps can be used either for winter heating or for cooling in summer and heating in winter.

The coefficient of performance for cooling COP_c for an absorption heat pump can be calculated as described in Eq. (24.16). The coefficient of performance for heating for a two-stage absorption heat pump COP_{hp} can be calculated as

$$COP_{hp} = \frac{Q_{ab} + Q_{con}}{Q_{1g}} \qquad (24.17)$$

where Q_{ab} = heat removed from the absorber, Btu/h
Q_{con} = heat removed from the condenser, Btu/h
Q_{1g} = heat input to first-stage generator, Btu/h

Several absorbants, or working fluids, other than aqueous LiBr solution are being developed, such as $LiBr/ZnCl_2$ and $LiBr/ZnBr_2/CH_3OH$. $LiBr/H_2O$ is still the most widely used solution in absorption heat pumps.

Comparison Between Absorption and Vapor Compression Heat Pumps

Although the coefficient of performance for heating COP_{hp} for a centrifugal heat pump has a value between 4 and 4.5, and for an absorption heat pump it is only 1.3 to 1.7, electric energy used by a centrifugal machine is far more expensive than heat energy used by an absorption machine.

A life-cycle cost analysis should be performed during selection. When the ratio of cost per unit of electricity to natural gas $R_{e.g}$ is considered especially when demand charge, and higher electricity rates during peak hours are taken into account, absorption heat pumps may be more cost-effecitve in many locations.

Two-Stage Parallel-Flow Absorption Heat Pump

There are many varieties of absorption heat pumps designed to operate with a high COP_{hp}. Generally, two-stage generators are used instead of single-stage. Here two types of absorption heat pumps will be discussed: two-stage parallel-flow absorption heat pumps, and series-connected absorption heat pumps.

A two-stage parallel-flow absorption heat pump has a design similar to that of the absorption chiller shown in Fig. 24.3. Their differences are:

- An absorption chiller is used to provide refrigeration or cooling, whereas an absorption heat pump is used to provide heating in winter and sometimes also cooling in summer. Hot water for winter heating is provided by the condenser water after it absorbs heat from the solution in the absorber and the condensed refrigerant in the condenser.

- In an absorption heat pump, the temperature of hot water leaving the heat pump during winter heating is usually between 110 and 150°F and the hot water returned to the heat pump is between 100 and 130°F. Therefore, there is a higher condensing pressure p_{con} and condensing temperature T_{con} in an absorption heat pump than in an absorption chiller.

- In an absorption chiller, the evaporating temperature T_{ev} is determined by the temperature of the chilled water leaving the evaporator T_{el}. Therefore, T_{ev} is mainly determined according to the temperature of conditioned air leaving the cooling coil in air-handling units. In an absorption heat pump, T_{ev} depends on the low-temperature heat source available, such as industrial waste heat or river or lake water.

- In an absorption heat pump, when outdoor temperature is very low and the low-temperature heat source is no longer available, hot water for winter heating can be directly provided from the hot water heat exchanger connected to the direct-fired generator.

In an absorption heat pump, two heat sources are used during winter heating: a high-temperature heat source to boil off the water vapor from the diluted solution in the first-stage generator and a low-temperature heat source to vaporize water vapor from the liquid water in the evaporator.

The coefficient of performance for heating COP_{hp} for a two-stage parallel-flow absorption heat pump is between 1.2 and 1.3.

Series-Connected Absorption Heat Pump

Figure 24.7 is a schematic diagram of a 50 MW series-connected absorption heat pump installed in an incineration plant in Uppsala City, Sweden, as described by Astrand (1988).

This series-connected absorption heat pump consists of two single-stage absorption heat pumps, each with an evaporator, absorber, generator, condenser, heat exchanger, and solution pump.

Liquid water refrigerant evaporates in the evaporator. Water vapor is extracted by the concentrated

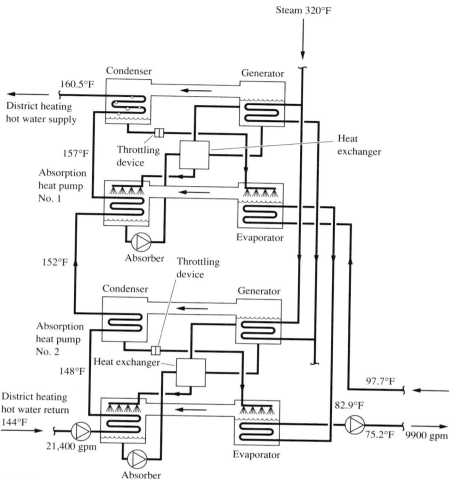

FIGURE 24.7 Schematic diagram of series-connected absorption heat pump.

solution in the absorber. The heat of absorption transferred to the hot water in the absorber is then used for district heating. The diluted solution is pumped from the absorber to the generator through the heat exchanger. In the generator, steam from the incineration plant boils off the water vapor from the diluted solution. The boiled-off water vapor is extracted to the condenser and condensed into liquid form. Latent heat of condensation is again transferred to the district heating hot water.

Concentrated solution from the generator flows to the absorber through the heat exchanger. Condensed liquid water enters the evaporator via a throttling orifice and is sprayed over the tube bundle in which flue gas cooling water flows from the incineration plant. After absorbing the latent heat of vaporization from the flue gas cooling water, liquid water evaporates into water vapor in the evaporator.

In this series-connected absorption heat pump, absorption heat pump No. 1 is operated at higher temperatures and heat pump No. 2 is operated at lower temperatures.

During operation in 1987, the return hot water from district heating was heated from 144°F to 152°F in the absorber and condenser of the No. 2 absorption heat pump, and from 152°F to 160.5°F in the absorber and condenser of the No. 1 absorption heat pump. In the evaporator, heat was extracted from the low-temperature heat source, the flue gas cooling water, which entered the absorption heat pump at a temperature of 97.7°F and left at a temperature of 75.2°F. The high-temperature heat source, steam, was supplied at 320°F at a flow rate of 66,000 lb/h from the incineration plant. The average COP_{hp} for this series-connected absorption heat pump from July 15 through August 31, 1987, was 1.61.

Absorption Chiller-Heater

An absorption chiller–heater provides cooling and heating simultaneously and, therefore, needs a low evaporating temperature and high condensing temperature. A higher condensing temperature results in a higher hot water supply temperature for winter heating. An absorption chiller–heater requires a greater pressure differential and temperature lift than an absorption chiller.

Because of the higher condensing temperature in an absorption chiller–heater, it is best to subcool the concentrated solution before it enters the absorber for better performance.

Because the chiller–heater has a higher condensing temperature, its coefficient of cooling performance COP_c is lower than that of an absorption chiller during summer cooling. A chiller-heater has a higher COP_{hr} in winter than an absorption heat pump, as described in Section 4.8.

Condensing temperature has less effect on the COP of an absorption chiller–heater than on that of a centrifugal chiller incorporating heat recovery that also provides heating and cooling simultaneously during winter. Therefore, at a higher condensing temperature, it is sometimes better to use an absorption chiller–heater than a centrifugal chiller–heater, that is, a centrifugal chiller incorporating heat recovery. Detailed cost analysis is required for an optimum decision.

A two-stage direct-fired parallel-flow absorption chiller–heater has approximately the same system components, structure, and flow processes of refrigerant and solution as an absorption chiller (shown in Fig. 24.3). The primary difference is a higher condensing temperature in the absorption chiller–heater.

24.9 ABSORPTION HEAT TRANSFORMERS

Large amounts of low-temperature waste heat are released daily from many industrial plants. *Absorption heat transformers* are systems that operate in a cycle opposite to that of absorption chillers to boost the temperature of input waste heat 40 to 100°F higher for industrial applications.

Most absorption heat transformers used today are single-stage systems using lithium bromide LiBr as working fluid for industries in Japan and Europe. Two-stage absorption heat transformers for greater temperature boosting and use with working fluids other than LiBr are being developed.

System Description

An absorption heat transformer consists of an evaporator, absorber, generator, condenser, heat exchanger, solution pump, refrigerant pump, piping, and accessories. Figure 24.8 shows the schematics and the absorption cycle on the equilibrium chart of a typical single-stage absorption heat transformer using LiBr as the absorbant.

In a heat transformer, because the heat is supplied from the waste heat fluid stream entering at low tem-

perature T_1 and leaving at lower temperature T_2, water is vaporized from the evaporator and extracted by the absorbant sprayed in the absorber at point 11. Heat of absorption is then released from the absorbant to the hot water stream entering at a temperature T_3 and raised to a temperature T_4. After absorbing the water vapor, the diluted solution at point 12 enters the heat exchanger, which is also called the recuperator. Diluted solution leaves the heat exchanger at a lower temperature, flows through a regulating valve, and enters the generator, or desorber, at point 13. Because of the heat input from the waste heat fluid streams, water vapor boils off from the diluted solution and is condensed in the condenser. The concentrated solution leaves the generator at point 14, and is pumped to a higher pressure equal to the pressure in the absorber. Liquid water in the condenser is pumped to the evaporator for evaporation again.

Operating Characteristics

The absorption heat transformer operates at two pressure levels: high pressure, including the evaporator and absorber, and low pressure, including the generator and condenser.

There are three temperature levels of input and output fluid streams:

- The fluid stream carrying the heat output from the absorber is at the highest temperature level.
- The heat source (the waste heat input to the evaporator and generator) is at the intermediate temperature level.
- The condenser cooling water in the condenser is at the lowest temperature level.

The purpose of an absorption heat transformer is to boost the temperature of the input waste heat fluid stream, and the function of an absorption heat pump is to attain a higher COP from the lower temperature heat source.

Coefficient of Performance

The coefficient of performance of an absorption heat transformer $COP_{h.t}$ can be calculated as

$$COP_{h.t} = \frac{Q_{ab}}{Q_{ev} + Q_g} \qquad (24.18)$$

where Q_{ev}, Q_g = waste heat input to the evaporator and generator, Btu/h.

$COP_{h.t}$ is mainly affected by the amount of temperature boost. The higher the temperature boost, the lower the $COP_{h.t}$. $COP_{h.t}$ usually varies between 0.4 and 0.5.

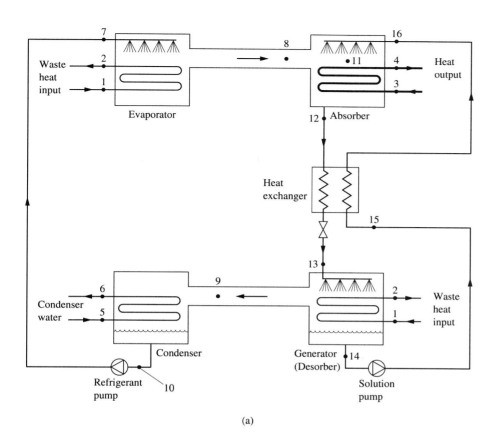

(a)

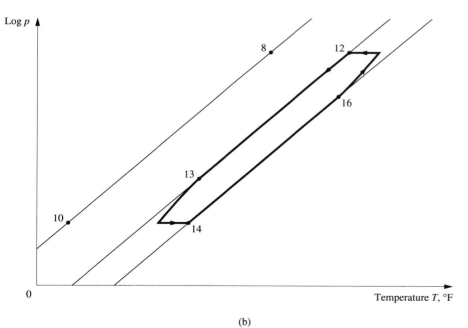

(b)

FIGURE 24.8   Single-stage absorption heat transformer: (*a*) schematic diagram and (*b*) absorption cycle.

REFERENCES AND FURTHER READING

Alefeld, G., and Ziegler, F., "Advanced Heat Pump and Air-Conditioning Cycles for the Working Pair H_2O/LiBr: Domestic and Commercial Applications," *ASHRAE Transactions*, Part II B, pp. 2062-2071, 1985.

ASHRAE, *ASHRAE Handbook Equipment 1988*, ASHRAE Inc., Atlanta, GA, 1988.

ASHRAE, *ASHRAE/IES Standard 90.1–1989, Energy Efficient Design of New Buildings Except New Low-Rise Residential Buildings*, ASHRAE Inc., Atlanta, GA, 1989.

Astrand, L. E., "Operating Experience with a 50 MW Absorption Heat Pump," *ASHRAE Transactions,* Part I, pp. 716-722, 1988.

Chuang, C. C., and Ishida, M., "Comparison of Three Types of Absorption Heat Pumps Based on Energy Utilization Diagrams," *ASHRAE Transactions*, Part II, pp. 275-281, 1990.

Davidson, K., and Brattin, H. D., " Gas Cooling for Large Commercial Buildings," *ASHRAE Transactions*, Part I B, pp. 910-920, 1986.

Davidson, W. F., and Erickson, D. C., "Absorption Heat Pumping for District Heating Now Practical," *ASHRAE Transactions*, Part I, pp. 707-715, 1988.

Fallek, M., "Parallel Flow Chiller-Heater," *ASHRAE Transactions,* Part II B, pp. 2095-2102, 1985.

Gommed, K., and Grossman, G., "Performance Analysis of Staged Absorption Heat Pumps: Water–Lithium Bromide Systems," *ASHRAE Transactions,* Part I, pp. 1590-1598, 1990.

Grossman, G., "Adiabatic Absorption and Desorption for Improvement of Temperature-Boosting Absorption Heat Pumps," *ASHRAE Transactions,* Part II, pp. 359-367, 1982.

Grossman, G., "Multistage Absorption Heat Transformers for Industrial Applications," *ASHRAE Transactions*, Part II B, pp. 2047-2061, 1985.

Holmberg, P., and Berntsson, T., "Alternative Working Fluids in Heat Transformers," *ASHRAE Transactions*, Part I, pp. 1582-1589, 1990.

Hufford, P. E., "Absorption Chillers Maximize Cogeneration Value," *ASHRAE Transactions*, Part I, pp. 428-432, 1991.

Kumar, P., and Devotta, S., "Study of an Absorption Refrigeration System for Simultaneous Cooling and Heating," *ASHRAE Transactions*, Part II, pp. 291-298, 1990.

Kurosawa, S., and Fujimaki, S., "Development of Air-Cooled Double-Effect Gas-Fired Absorption Water Chiller-Heater," *ASHRAE Transactions*, Part I, pp. 318-325, 1989.

Kurosawa, S., and Nakamura, S., "Development of Intelligent Gas-Fired Absorption Water Chiller–Heaters," *ASHRAE Transactions*, Part II, pp. 850-865, 1987.

McLinden, M. O., and Klein, S. A., "Steady-State Modeling of Absorption Heat Pumps with a Comparison to Experiments," *ASHRAE Transactions*, Part II B, pp. 1793-1807, 1985.

Sun, T. Y., "Application of Gas-Fired Absorption Chillers," *Heating/Piping/Air Conditioning*, pp. 55-58, March 1991.

The Trane Company, *Absorption Refrigeration*, The Trane Company, La Crosse, WI, 1985.

Wang, S. K., *Principles of Refrigeration*, Vol. 3, Hong Kong Polytechnic, Hong Kong, 1984.

Watanabe, K., Akimoto, M., and Ohtsuka, S., "An Approach to Support the Development of Absorption System Technology in Japan," *ASHRAE Transactions*, Part II, pp. 492-506, 1982.

Wilkinson, W. H., "What Are the Performance Limits for Double-Effect Absorption Cycles?" *ASHRAE Transactions*, Part II, pp. 2429-2441, 1987.

CHAPTER 25

THERMAL STORAGE AND DESICCANT COOLING

25.1 THERMAL STORAGE— OFF-PEAK AIR CONDITIONING

Thermal Storage System

In a *thermal storage system*, the electricity-driven refrigeration compressors are operated during off-peak hours. Stored chilled water or stored ice in tanks is used to provide cooling in buildings during on-peak hours when higher electric demand charges and electric energy rates are in effect. A thermal storage system partially or fully shifts the higher electricity demand for HVAC&R from on-peak hours to off-peak hours. An air conditioning system using such a thermal storage system is called *off-peak air conditioning*.

It is important to realize that thermal storage systems do not necessarily save energy. Thermal storage systems mainly lower the utility's monthly demand charge and electricity bill by shifting the electricity demand to off-peak hours.

In a 24-hour daily operating cycle of a power plant, on-peak hours are the hours during which the user's electricity demand supplied by the utility is at or nearly at the peak load of the utility's power plant. Off-peak hours are the hours during which the utility's power plant is not operating at peak load.

Many electric utilities' on-peak hours are between noon and 8 P.M. during summer weekdays, which includes the peak-load hours of air conditioning. Because the capital cost of a new power plant is so high, often between $1200 and $4000 per kW, electric utilities tend to increase their total output by means of customer-owned thermal storage systems, which are much less expensive. In addition, by shifting the electricity demand from on-peak hours to off-peak hours, electric utilities shift the daytime peak operation of inefficient plants such as diesel and gas turbine plants to nighttime base-load high efficiency coal and nuclear plants.

Many electric utilities use time-of-day rate structures, including demand charges and higher rates for peak hours to encourage such a shift. They also offer initial cost subsidies, typically $100 to $200 per kW, to encourage construction of thermal storage systems.

Thermal storage systems are more cost-effective because they incur lower electricity demand charges (or none at all) and lower per-unit energy costs. Additional benefits include longer operating hours of the compressors and pumps at greater load ratios and higher efficiency, operation of chillers and cooling towers at lower outdoor temperatures at night, chilled water storage for use as a fire water reservoir, and a backup source for cooling during emergencies. Drawbacks of thermal storage systems include high initial cost and complicated operation, maintenance, and control. In many cases, thermal storage systems are economically feasible only when utilities offer lower electricity rates during off-peak hours.

Electricity Rate Structures

According to Townsend (1984), the electricity rate structures in the United States can be grouped into three categories:

- *Time-of-day energy charge and demand charge.* A typical example has the following structure:

Demand charge

June through August	$9.74/kW of monthly peak demand
September through May	$7.61/kW of monthly peak demand

Energy charge (year-round)

Peak (9 A.M. to 10 P.M. weekdays)	$0.05466/kWh
Off-peak (10 P.M. to 9 A.M. weekdays plus weekends and holidays)	$0.02786/kWh

Instead of dividing the 24-hour daily operating cycle into on-peak and off-peak hours, other rate structures under this category may divide the energy charge into three periods: on-peak, semi-peak, and off-peak, each with different electricity rates. Demand charges may also differ during on-peak, semi-peak, and off-peak hours.

Time-of-day electricity rates are used by many major electric utilities in the United States.

- *Declining-block energy charge.* A typical example is:

Demand charge
$0.35 per kW of billing demand (BD)
Energy charge
July through September
First 250 kWh per kW BD @ $0.0648/kWh
> 250 kWh per kW BD @ $0.0289/kWh
October through June
First 175 kWh per kW BD @ $0.0648/kWh
> 175 kWh per kW BD @ $0.0289/kWh

Billing demand is the larger of either the month's peak demand or 100 percent of the maximum demand during the previous 11 months.

A thermal storage system saves money under this type of rate structure by qualifying for the tail-block rate of $0.0289/kWh if monthly consumption exceeds 250 kWh/kW BD in summer and 175 kWh/kW BD in winter. This is possible when the compressor is running at night, during off-peak hours.

This electricity rate structure is characterized by its small demand charge. Energy charges with different block widths are variations of this type of rate structure.

- *Simple demand charge.* The following is a typical example:

Demand charge
  $6.58 per kW BD
Energy charge
  First 6000 kWh @ $0.045/kWh
  Next 24,000 kWh @ $0.0428/kWh
  > 30,000 kWh @ $0.0402/kWh

Billing demand is the larger of either monthly peak demand or 85 percent of the maximum demand during the previous 11 June through September months.

This type of electricity rate is applied to commercial customers whose annual peak demand exceeds 1000 kW. Some utilities have a demand charge several times greater than $6.58/kW BD. For example, in Brooklyn, New York, the demand charge from June through September in the early 1990s was $21.14/k**W**. Others utilities may offer a declining-block demand charge structure.

For a thermal storage system, electricity charge savings result from lower monthly demand.

Full Storage and Partial Storage

Ton-hours, or *ton · h*, is the unit of stored refrigeration. One ton-hour is the refrigeration or heat absorption of 12,000 Btu performed by a refrigeration system with a capacity of one ton during a one-hour period.

The aim of thermal storage strategy is to incur the lowest possible demand charge, energy cost, and initial investment so that life-cycle costs are minimized. The economic benefit of a thermal storage system can also be assessed by calculating the simple payback period of its cost. A simple payback or life-cycle cost analysis of the building load profile, utility electricity rate structure, and system characteristics is always necessary.

Determination of the optimum size of a thermal storage system is based on the utility's electricity rate structure (demand charge and difference between peak and off-peak unit charges).

Direct cooling denotes the process in which compressors produce chilled water to cool the building directly. When the cost of direct cooling by a refrigeration system is lower than the cost of stored energy, the operation of the thermal storage system

is said to be at *chiller priority*. On the other hand, if the cost of direct cooling is higher than the cost of stored energy, the operation is said to be at *storage priority*. Construction costs and the utility's incentive payments should be considered. If the demand charge is low during off-peak hours and the energy cost difference between peak and off-peak hours is great, the full use of stored energy during peak hours may be most economical.

There are two kinds of thermal storage: full storage and partial storage. Figure 25.1 shows the load-time diagrams for full storage and partial storage. For a full-storage, or *load-shift*, thermal storage system, all refrigerating compressors cease to operate during peak hours and the building refrigeration load during that period is entirely offset by the chilled water supplied from the thermal storage system, as shown in Fig. 25.1a.

Partial storage, or *load leveling*, can be operated in *full leveling mode*, in which refrigerating compressors are operated at full capacity during peak hours (as shown in Fig. 25.1b), or in *demand-limited mode*, in which the building's electricity demand is limited to a certain value (as shown in Fig. 25.1c).

A utility's demand charge is the building total demand charge, which is the sum of the demand charges for HVAC&R and other uses. Other uses include electricity for lighting, escalators, computers, and electric appliances. Operating refrigerating compressors in demand-limited mode during peak hours may level the other demand charges and result in a minimized building demand charge for a partial-storage system.

The following is a comparison of different storage strategies of a thermal storage system with heat recovery and cold air distribution for an office building in Dallas, TX as described by Tackett (1989).

	Full storage	Partial storage	Demand limited
Utility incentive	$288,750	$132,500	$215,000
Net incremental cost	$291,750	$21,750	$124,250
Demand limit savings, kW	1200	400	800
Total savings	$124,250	$64,000	$92,750
Simple payback, years	2.35	0.34	1.34

The larger the thermal storage system and the capacity of the refrigeration compressors, the greater the demand charge savings during peak hours. Actually, selection of the size of the storage system is not necessarily determined by the shortest simple payback period. Life-cycle cost and system reliability should also be considered.

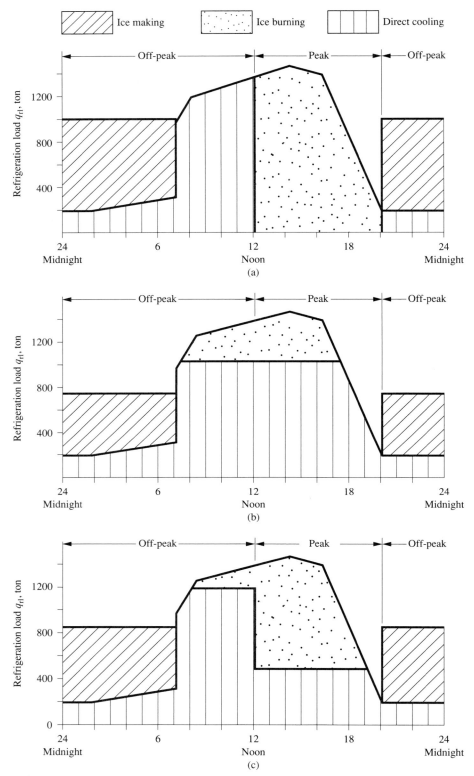

FIGURE 25.1 Full and partial storage: (a) full storage; (b) partial storage, all compressors operating; and (c) partial storage, 50 percent compressors operating.

Ice Storage and Chilled Water Storage

Two thermal storage media are widely used for air conditioning systems: ice storage and chilled water storage. At a temperature difference of 18°F, a pound of chilled water can store $18 \times 1 = 18$ Btu of thermal energy, whereas a pound of ice can store $(1 \times 144) + (60 - 35) = 169$ Btu. If the density of water is 62.3 lb/ft^3 and the density of ice is 57.5 lb/ft^3 for the same stored cooling capacity, the storage volume for ice is only about 0.12 that of the chilled water. In addition, ice storage systems generally provide chilled water at a temperature of 34 to 35°F to produce cold supply air between 42 and 49°F. Ice storage systems incorporating cold air distribution significantly reduce the volume flow rate of supply air, so air-side fan energy consumption and initial investment drop accordingly.

Generally, ice storage systems have a rather low incremental capital cost compared with conventional air conditioning systems without thermal storage. When utility incentives are taken into account, ice storage systems with cold air distribution can have a simple payback of less than 1 or 2 years. A low incremental capital cost is the main benefit of the ice storage systems.

Currently, ice storage systems can be classified in the following categories:

- Brine-coil systems
- Ice-on-coil systems
- Ice harvester systems

An *encapsulated water ice storage system* is a brine-based system with encapsulated water balls or flat encapsulated containers or bottles submerged in the brine within a storage tank. It has similar design characteristics to the brine-coil system, but often has a higher initial cost.

In an ice storage system, *ice making* or *charging* is the process in which compressors are used to produce ice. *Ice burning* (*ice melting* or *discharging*) means that ice in the storage system is melted in order to produce chilled water to offset the required refrigeration load.

A chilled water storage system is simpler. It consists of a storage tank, pumps, controls, and the piping modification necessary to add to the existing chilled water system. In most cases, existing chillers can be reused. If chilled water storage can be combined with fire protection water storage, it may be more cost effective than an ice storage system for some retrofit projects.

In newly installed chilled water storage systems, stratified tank systems are the most widely used systems.

25.2 ICE STORAGE: BRINE-COIL SYSTEMS

System Description

A brine-coil ice storage system is a brine-based system using brine flowing inside coils to make ice and to melt ice in the water that surrounds the coil. A brine-coil ice storage system consists of the following components: chillers, brine-coil ice tanks, chiller pumps, building pumps, chilled water cooling coils, controls, piping, and fittings.

Centrifugal, screw, and reciprocating chillers are usually used in ice storage brine-coil systems depending on the size of the plant and the type of condenser (water-cooled, air-cooled, or evaporatively-cooled) used. In locations where the air temperature during nighttime off-peak hours drops 20°F lower than the daytime maximum temperature, air-cooled chillers may sometimes be more efficient than water-cooled chillers.

Figure 25.2 is a schematic diagram of a typical brine-coil ice storage system for an office building near Dallas, TX.

Brine and Glycol Solution

Brine is a salt solution or an aqueous glycol solution used as a heat transfer medium. Its freezing point is lower than that of water, and depends on the concentration of salt or glycol in solution. Brine is also used as a liquid coolant to absorb heat energy in refrigeration and thermal storage systems.

Ethylene glycol and propylene glycol are brines. They are colorless, nearly odorless liquids. They are often mixed with water at various concentrations and used as freezing-point depressants to lower the freezing point of water. Inhibitors must be added to ethylene and propylene glycols to prevent metal corrosion.

The freezing point of an aqueous ethylene glycol solution with a concentration of 25 percent by mass drops to 10°F, and its rate of heat transfer is about 5 percent less than that of water. The freezing point of a propylene glycol solution with a concentration of 25 percent by mass drops to 15°F.

The physical properties of aqueous ethylene glycol solution are more appropriate for thermal storage systems than those of aqueous propylene glycol solution. In certain applications, toxicity considerations may necessitate the use of propylene glycol, and it may be required by federal EPA requirements and local codes and regulations.

Brine-Coil Ice Storage Tank

In a brine-coil ice storage system, ice is produced, or charged, in multiple storage tanks where closely

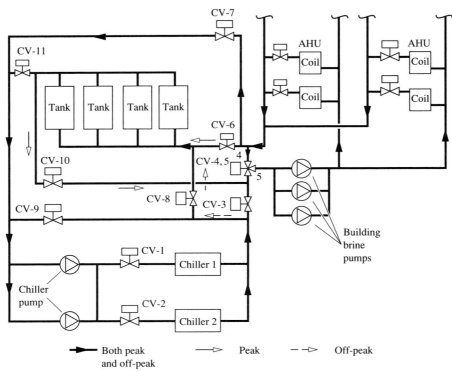

FIGURE 25.2 Schematic diagram of a brine-coil ice storage system.

spaced multicircuited polyethylene or plastic tubes are surrounded by water, as shown in Fig. 25.3. Brine, an aqueous ethylene glycol solution with 25 to 30 percent ethylene glycol and 70 to 75 percent water, circulates inside the tubes at about 24°F. The water surrounding the tubes freezes into ice up to a thickness of about 0.5 in. Tubes of glycol solution entering and leaving the tank are arranged side-by-side alternately to provide more uniform heat transfer.

Brine typically leaves the storage tank at 30°F. Plastic tubes occupy about one tenth of the tank volume, and another tenth is left empty to accommodate the expansion of ice during ice making. Multiple brine-coil tanks are always connected in parallel.

During ice burning or ice melting, brine returns from the cooling coils in the air-handling units at a temperature of 46°F or higher. It melts the ice on the outer surface of the tubes, and is thus cooled to 34 to 36°F. The brine is then pumped to the air-handling units to cool the air again.

In the storage tank, ice is stored and the high-pressure brine inside the tubes is separated from the water, usually at atmospheric pressure surrounding the tubes in the storage tank.

Case Study: Operating Modes of a Brine-Coil Ice Storage System

In a typical ice storage system using brine-coil storage tanks in a 550,000 ft² office building near Dallas, TX, as described by Tackett (1989), there are two centrifugal chillers. Ethylene glycol is used as the coolant. Each chiller has a refrigeration capacity of 568 tons when it produces 34°F brine at a power consumption of 0.77 kW/ton. If the brine leaves the chiller at 24°F, the refrigeration capacity then drops to 425 tons, with a power consumption of 0.83 kW/ton.

A demand-limited partial-storage strategy is used, that is, one chiller is operated during on-peak hours, as shown in Fig. 25.1c. Meanwhile, ice is also burned during peak hours to reduce the demand charge. Ice is charged during off-peak hours in order to reduce energy costs. The system uses 90 brine-coil storage tanks.

For summer cooling, the daily 24-hour operating cycle can be divided into three periods: off-peak, direct-cooling, and on-peak.

OFF-PEAK. This is the period from 8 P.M. until the air-handling systems start the next morning. Dur-

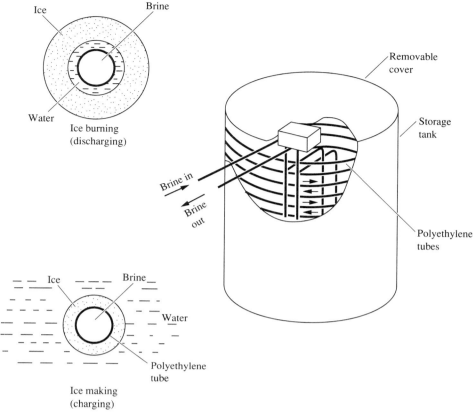

FIGURE 25.3 Brine-coil ice storage tank.

ing this period, the primary operating mode is ice making, at a maximum capacity of 650 tons. The chillers also provide direct cooling at a capacity less than 200 tons for refrigeration loads that operate 24 hours.

In this operating mode, the brine-coil storage tanks are charged. At the same time, a 34°F ethylene glycol solution is supplied to the air-handling units for nighttime cooling.

The DDC panel controls the ice storage system in the following operating sequence, as shown in Fig. 25.2:

1. Open control valves CV-1, 2, 5, 8, and 11, and close control valves CV-3, 4, 6, 7, 9, and 10.
2. Reset the temperature of glycol solution leaving the chiller T_{el} to 22°F.
3. Reset the load limit of both chillers to 100 percent.
4. Start the chiller and condenser pumps. Chiller pumps operate at high speeds during ice making to provide a higher flow rate and a high rate of heat transfer in the storage tanks as well as a greater head to overcome the pressure drop for

both the evaporator and the brine-coils in the ice storage tank. Chiller pumps operate at low speeds in direct-cooling mode.
5. Start chillers 1 and 2 following the lead/lag sequence.
6. After chillers are started, open control valves CV-3 and 7.
7. Start the building chilled water circulating pumps in sequence.
8. Modulate control valves 4 and 5, and maintain a 34°F glycol solution supply temperature to the air-handling units.

If the brine-coil ice storage tanks are all 100 percent charged, the ice storage capacity is 7500 ton · h.

When the sensors detect that the ice storage tanks are 100 percent charged, the ice-making mode is terminated. If nighttime after-hours cooling is not needed, the ice storage system shuts down. If the ice inventory (the amount of stored ice in the tanks) falls below 90 percent, ice making starts again.

There are two additional operating modes during this period:

- Ice making without direct cooling for after-hours use.
- Ice burning for after-hours use with all chillers shut off.

DIRECT COOLING. Direct-cooling operation lasts from the start of the air-handling systems until noon on weekdays. This period has two operating modes:

- *Direct cooling mode.* In this operating mode, chillers are operating, and are reset to 34°F.
- *Direct cooling with ice-burning or melting mode.* In this mode, both chillers are turned on. When the required refrigeration load exceeds both chillers' capacity, some storage will be discharged to supplement the chiller.

ON-PEAK. On-peak hours are from noon until 8 P.M. weekdays. Ice-burning mode, with or without chiller operation, is used in this period. In ice-burning or melting mode, one chiller is operated at demand-limit. This is the primary operating mode during summer cooling. The operating sequence is as follows:

1. Open control valves CV-3, 6, 7, and 10, and close CV-8, 9, and 11.
2. Open control valve CV-1 and close CV-2 if chiller 1 is required to operate. Open CV-2 and close CV-1 if chiller 2 is required to operate.
3. Modulate control valves 4 and 5 at normal open positions.
4. Reset chilled water temperature leaving the chiller to 32°F.
5. Set the load-limit of the operating chiller to 400 kW.
6. Start one condenser pump.
7. Start chiller pumps 1 and 2 at low speed. Both pumps will operate during ice burning.
8. Start one chiller according to the lead/lag sequence.
9. Modulate control valves 4, 5, 6, and 7 to maintain a 34°F chilled water supply temperature to the air-handling units.
10. Start the brine circulating pumps in sequence. During peak hours, the building's brine circulating pump needs a greater head to overcome the pressure drop of the coil in the AHU as well as the pressure drop of the brine coil in the ice-storage tank.

System Characteristics

- A brine-coil storage tank is a modulized ice storage system consisting of several closely packed storage tanks. Such an ice storage system is more flexible during the installation of storage tanks, especially for retrofit projects.
- In brine-coil ice storage systems, off-peak cooling of the building can be provided by direct cooling from the chiller(s).
- During ice burning, melted water separates the tube and ice. Water has a much lower thermal conductivity (0.35 Btu · ft/h · ft^2 · °F) than that of ice (1.3 Btu · ft/h · ft^2 · °F), so the capacity of the brine-coil ice storage system is dominated by rate of ice burning or melting.

25.3 ICE-STORAGE: ICE-ON-COIL SYSTEMS

System Description

An *ice-on-coil system* is a refrigerant-based system. Ice builds up on the outer surface of coils or tube banks, which are submerged in water in a storage tank. The refrigerant flows and evaporates inside the tubes. When the ice melts, it cools the water to a temperature between 34 and to 38°F for cooling in the AHUs.

An ice-on-coil system consists of compressors, evaporating coils, storage tanks, condenser, heat exchanger, refrigerant pumps, chilled water pumps, controls, piping, and fittings. A typical ice-on-coil ice storage system is shown in Fig. 25.4.

Screw and centrifugal compressors are often used because of their higher efficiency. For an ice-on-coil system with a capacity less than 2400 ton-hours, a reciprocating compressor may also be used. Evaporatively cooled condensers have a higher system energy efficiency ratio (EER), and are often used in many new projects.

Ice Builders

Ice builders are large, well-insulated steel tanks containing many serpentine coils, usually made of steel pipes of 1 to 1.25 in. diameter. HCFC–22 is often used as the refrigerant. The refrigerant-filled serpentine coils are submerged in water in the ice builder and function as evaporators. The ice built up on the coil is between 1 and 2.5 in. thick. When ice builds up on the coil, the suction temperature of the compressor falls to 22 to 24°F.

Ice is melted by the water circulating over it. The steel tubes of the serpentine coil should be spaced so that the built-up ice cylinders do not bridge each other. If the cylinders are bridged, the paths of water circulation are blocked. Baffle plates are sometimes added

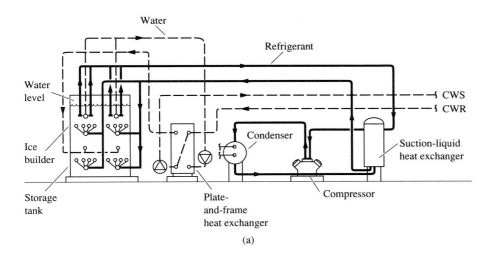

Water

Refrigerant

CWS

CWR

Water level

Ice builder

Storage tank

Condenser

Suction-liquid heat exchanger

Plate-and-frame heat exchanger

Compressor

(a)

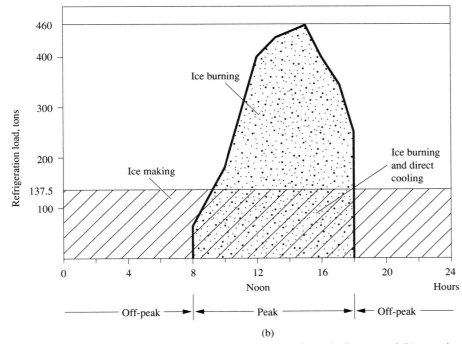

Ice burning

Ice making

Ice burning and direct cooling

Refrigeration load, tons

460

400

300

200

137.5

100

0   4   8   12   16   20   24

Noon   Hours

Off-peak ——→ ←—— Peak ——→ ←— Off-peak ——

(b)

FIGURE 25.4 A typical ice-on-coil ice storage system: (a) schematic diagram and (b) operating diagram.

to guide the water flow and provide a secondary heat transfer surface between the refrigerant and water.

The storage tanks containing refrigerant coils are usually located at a lower level or on a grade because of their weight. Because the chilled water system in a multistory building is always under a static head at lower levels, a heat exchanger, usually plate-and-frame, is used to isolate the storage tank brine system from the chilled water system connected to the AHUs. An alternative is to supply chilled water directly to the storage tanks and pressurize the tanks. This arrangement obviates the use of a heat exchanger and

a corresponding increase in brine temperature of about 3°F.

Refrigerant Feed

Two kinds of refrigerant feeds are widely used in ice-on-coil systems: direct expansion and liquid overfeed.

Direct expansion (DX) uses the pressure difference between the receiver at the high-pressure side and the suction pressure to force the refrigerant to flow through the ice builder. A suction-liquid heat exchanger can be used to cool the liquid refrigerant from the condenser for better efficiency. Direct expansion is simple and no refrigeration pump is required. Its main drawback is that 15 to 20 percent of the coil surface is used for superheat and is not available for ice build-up.

As described in Section 12.6, liquid overfeed uses a refrigerant pump to feed ice-building coils about 3 times the evaporation rate they need. Because the liquid refrigerant wets the inner surface of the ice-building coils, it has a higher heat transfer coefficient than direct expansion.

Ice-Charging Control

The thicker the ice built up on the coils, the greater the amount of ice stored in the tank. The thickness of the ice on the coil should be measured to meet ice-burning requirements during on-peak operating hours or the direct cooling period.

Because ice has a higher volume than water, as the ice builder is charged (that is, as ice builds up on the coil), the water level rises. An electric probe can sense the water level in the tank and thereby, determine the amount of ice stored in the tank.

System Characteristics

Ice-on-coil is a refrigerant-based system. It may be costly and complex. In the storage tank, stored ice occupies only about half the volume of the tank, so the ice builder must be larger and heavier.

Case Study: An Ice-On-Coil Ice Storage System

Gilbertson and Jandu (1984) presented an ice-on-coil ice storage system for a 24-story office building with an area of 265,000 ft^2 in San Francisco, California. This project was completed in the early 1980s. A schematic diagram of this system is shown in Fig. 25.4a.

This office tower had a peak refrigeration load of 460 tons and an ice storage capacity of 3300 ton-hours of refrigeration over 24 hours. It required a 137.5-ton refrigeration capacity for the ice storage systems for a hot summer day at summer outdoor design conditions. For redundancy, two identical refrigeration systems using HCFC–22 as refrigerant were installed. Each had a 70-nominal-tons compressor, an ice builder of 960 ton-hours, a water-cooled condenser, and other accessories. (Nominal tons is the refrigeration capacity, in tons, at rating conditions.) Direct-expansion refrigerant coils with a suction-liquid heat exchanger were used. When the thickness of built-up ice was

about 2 in., the ice storage capacity reached 960 ton-hours.

A plate-and-frame heat exchanger was used to isolate the storage tank from the tower chilled water system. The 34°F water from the ice builder entering the heat exchanger with an 8°F temperature increase cooled the chilled water from 54°F to 38°F.

This was a partial-storage ice storage system. The operating modes during a 24-hour cycle for this 24-story office tower are shown in Fig. 25.4b, as follows:

Time period	Operating mode
6:00 P.M. —8:00 A.M.	Off-peak, ice making
8:00 A.M. —9:00 A.M.	On-peak, ice making and direct cooling
9:00 A.M. —6:00 P.M.	On-peak, ice melting and direct cooling

The HVAC&R and plumbing system, including ice storage, for this office building cost $2.4 million, which was slightly less expensive than a conventional central system using centrifugal chillers.

25.4   ICE STORAGE: ICE-HARVESTER SYSTEMS

Ice Harvester

An *ice-harvester ice storage system* is a refrigerant-based system. It consists of the following components: chiller(s), ice harvester, storage tank, chilled water pumps, cooling coils, controls, piping, and fittings, as shown in Fig. 25.5.

Ice is produced in a harvester, which is separate from the storage tank where ice is stored. The evaporator of the chiller is a vertical plate heat exchanger mounted above a water/ice storage tank. Low-pressure liquid refrigerant is forced through the hollow inner part of the plate heat exchanger in which liquid refrigerant is vaporized and produces a refrigeration effect.

Ice Making or Charging

During ice-making or charging mode, a chilled aqueous ethylene glycol solution with a concentration between 25 and 30 percent is pumped from the storage tank and distributed over the outer surface of the evaporator plates at a temperature equal to or slightly above 32°F. It then flows downward along the outer surface of the plate in a thin film. Water is cooled and then frozen into ice sheets approximately 0.2 to 0.3 in. thick. Periodically, hot gas is introduced into one fourth of the evaporator plates by reversing the refrigerant flow. Ice is *harvested*, or released from the

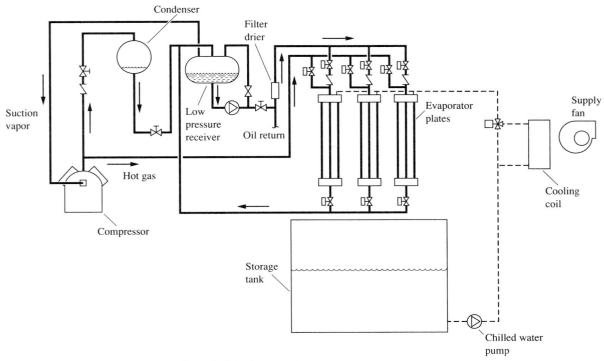

FIGURE 25.5   Schematic diagram of a typical ice harvester ice storage system.

outer surface of the plates in the form of flakes or chunks, and falls into the storage tank below. Ice is formed in 20 to 30 minutes and is harvested within 20 to 40 seconds. During harvesting, this section of the plate evaporator acts as a condenser.

Ice accumulates in the storage tank to occupy slightly less than 60 percent of the volume of the tank. Because the ice flakes are usually smaller than 6 in. × 6 in. × 0.25 in., there is a large contact area between the return brine from the cooling coils and the ice. The time required to melt the ice in the storage tank is less than one tenth of the time needed in ice making or charging.

For a reciprocating compressor using an evaporative condenser, the power consumption of the chiller during ice making is about 0.95 to 1.1 kW/ton.

Because the evaporator plates must be located above the storage tank, ice-harvesting systems need more headroom than other ice storage systems.

Chiller Operation

During off-peak hours, an ice-harvesting system can also be used to lower temperature of the brine returning from the air-handling units. Brine at a temperature of 34°F is supplied to the air-handling units to cool the air down to a supply temperature of 42 to 45°F

during direct cooling. It is then returned to the ice harvester at 50 to 60°F and distributed over the evaporator plates directly. After falling from the evaporator plates, brine is again cooled to a temperature of 34°F before it leaves the storage tank.

Because of the higher temperature of return brine distributed over the evaporator plates, the capacity of the ice harvester increases and its power consumption decreases during chiller operation. These changes in capacity and power consumption depend mainly on the temperature of return brine distributed over the evaporator plates. During chiller operation, power consumption usually varies between 0.75 and 0.85 kW/ton.

Ice harvester ice storage systems are effective in ice-making and ice-burning operations. The temperature of brine from the storage tank of the ice harvester can be lowered to 34°F, which is 2°F lower than in the brine-coil ice storage system. Ice harvester systems have been successfully used in load shifting and load leveling to reduce electricity demand and energy cost. However, melting of the ice during the harvesting process not only decreases the amount of ice harvested, but also adds an incremental refrigeration load to the system, which increases power consumption during the ice-making process. An ice harvester system using brine is an open system. More water treatment is required than in a brine-coil closed system.

25.5   CHILLED WATER STORAGE: STRATIFIED TANK SYSTEMS

Basics of Chilled Water Storage

CHARGING AND DISCHARGING. *Charging* is the process of filling the storage tank with chilled water from the chiller, usually at a temperature between 40 and 45°F. Meanwhile, the warmer return chiller water from the air-handling units or terminals, usually at a temperature between 55 and 60°F, is extracted from the storage tank and pumped to the chiller to be cooled.

Discharging is the process of discharging the chilled water, at a temperature between 42 and 45°F, from the storage tank to the air-handling units and terminals. At the same time, the warmer return chilled water from the coils fills the tank by means of storage water pumps.

LOSS OF COOLING CAPACITY DURING STORAGE. During the storage of chilled water, the following processes result in losses in cooling capacity:

- Stored chilled water is warmed by direct mixing of warmer return chilled water and stored colder chilled water.
- Heat from previously stored warmer return chilled water is transferred from the warmer tank wall to the stored chilled water.
- Heat is transferred through the tank wall from the warmer ambient air.

FIGURE OF MERIT. A more easily measured, enthalpy-based *figure of merit (FOM)* is often used to indicate the loss of cooling capacity of the stored chilled water during the charging and discharging processes in a complete storage cycle. FOM is defined as

$$\text{FOM} = \frac{q_{\text{dis}}}{q_{\text{ch}}} = \frac{\Sigma \dot{m}_w c_{\text{pw}}(T_{\text{rc}} - T_o)}{\dot{m}_w c_{\text{pw}}(T_{\text{rc.m}} - T_i)} \qquad (25.1)$$

where q_{dis} = cooling capacity available during the discharge process, Btu/h

q_{ch} = theoretical cooling capacity available during charging process, Btu/h

$\dot{m}_w, \Sigma\dot{m}_w$ = mass of flow rate and summation of mass flow rate of water, lb/h

c_{pw} = specific heat of water, Btu/lb · °F

T_{rc} = warmer return chilled water temperature filling the storage tank during discharge process, °F

T_o = outlet temperature of stored chilled water, °F

$T_{\text{rc.m}}$ = mass-weighted average temperature of return chilled water at inlet during discharge process, °F

T_i = inlet temperature of stored chilled water during charging process, °F

The smaller the losses of cooling capacity during chilled water storage, the greater the value of FOM.

Storage Tanks

Chilled water storage tanks are usually vertical cylinders. They are sometimes rectangular. Large cylindrical tanks typically have a height-to-diameter ratio of 0.25 to 0.35. Steel is the most commonly used material for above-grade tanks, and concrete is widely used for underground tanks. In certain projects, precast, prestressed, cylindrical concrete tanks with enclaved watertight steel diaphragms are used for large chilled water storage facilities with a volume over 2.5 million gallons.

All outdoor above-grade structures should have a 2-in. thick external insulation layer spray-on polyurethane foam, a vapor barrier, and a highly reflective top coating.

Stratified Tanks

Stratified tanks rely on the buoyancy of warmer return chilled water, which is lighter than colder chilled water, to separate these two chilled waters during charging and discharging. Diffusers are used to lower entering and leaving water velocity to prevent mixing. In a stratified tank, colder stored chilled water is always charged from bottom diffusers arranged concentrically, as shown in Fig. 25.6a. It is also discharged from the same bottom diffusers. The warmer return chilled water is introduced to and withdrawn from the tank through the top lateral diffusers.

According to field measurements, stratified tanks have a figure of merit between 0.85 and 0.92. Tran et al. (1989) showed that there is no significant difference in FOM between stratified tanks and membrane tanks or empty tanks. A *membrane tank* is a storage tank in which a membrane separates the colder stored chilled water and warmer return water. An *empty tank* is a storage tank in which walls are used to separate the colder and warmer chilled water.

Compared with membrane tanks and empty tanks, stratified tanks have the advantages of simpler construction and control, greater storage capacity, and lower cost. Stratified tanks are widely used in chilled water storage installations.

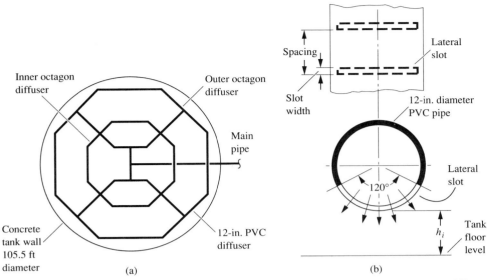

FIGURE 25.6   Double-octagon diffuser for a cylindrical stratified tank: (a) plan view of bottom diffusers and (b) 12-in. diameter PVC pipe diffuser.

Temperature Gradient and Thermocline

Vertical temperature profiles are formed during charging or discharging in stratified tanks at various time intervals. Temperature profiles may be illustrated on a height–temperature $H-T$ diagram at the beginning, the middle, and near the end of the charging process, as shown in Fig. 25.7. In the middle of the charg-

ing process along the vertical height of the storage tank, chilled water is divided into three regions: bottom colder-and-heavier water, thermocline, and top warmer-and-lighter water.

A *thermocline* is a stratified region in which there is a steep temperature gradient. Water temperature often varies from 42 to 60°F. The thermocline separates the colder chilled water from the warmer chilled water.

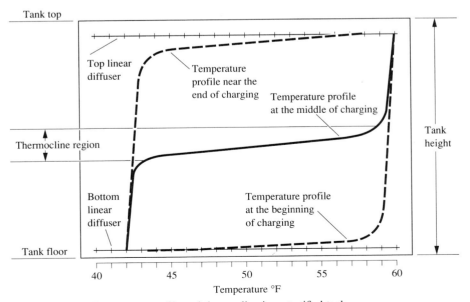

FIGURE 25.7   Temperature profile and thermocline in a stratified tank.

The thinner the thermocline, the smaller the mixing loss.

Diffusers

The layout and configuration of diffusers in a stratified tank have a significant effect upon the mixing of the colder and warmer chilled water, as well as the formation of the thermocline. The purpose of the diffusers and their connecting piping is to distribute the incoming chilled water evenly, so that it flows through the inlet openings with sufficiently low velocity (usually lower than 0.9 ft/s) to minimize mixing of colder and warmer chilled water. Wildin (1990) recommended that diffuser design and installation should take into account the following considerations:

- Warmer return chilled water should be introduced as close as possible to the top water level in the stratified tank. Colder stored chilled water should be introduced just above the bottom floor of the tank.

- The inlet temperature of chilled water should be controlled within a narrow band ($\pm 2°F$) during charging to avoid additional mixing.

- Obstructions in the flow crossing the tank, other than diffusers and the connecting piping, should be minimized.

- The primary function of diffusers is to reduce mixing. Mixing can occur at two points: at the start of the charging and discharging processes during the formation and reformation of the thermocline, and at the inlet side of the thermocline after the thermocline has been formed.

- Mixing near the inlet diffuser can be minimized if the incoming chilled water initially forms a thin layer of gravity current that travels across the tank because of the gravity difference, rather than inertia. *Gravity current* slowly pushes the chilled water originally in the tank out of the way so that mixing only occurs at the front of the gravity current when it first crosses the tank.

- There are two kinds of diffusers: linear diffusers and radial disk diffusers. Large stratified tanks usually incorporate linear diffusers.

- Inlet flow from the top diffusers should be upward or horizontal. Bottom diffusers should flow downward and have slots spreading at 120°. The cross-sectional inlet area of the branch pipe leading to the diffuser should be at least equal the total area of the diffuser openings in that branch.

- Mixing on the inlet side of thermocline depends on the inlet Reynolds number Re_i, and Froude number Fr_i. The inlet Reynolds number is closely related to the inlet velocity and is defined as

$$\text{Re}_i = \frac{\dot{V}_w}{l_{\text{dif}} \nu_w} = \frac{0.00223 \dot{V}_{\text{gal}}}{l_{\text{dif}} \nu_w} \tag{25.2}$$

where $\dot{V}_w$ = volume flow rate of chilled water, ft³/s
V_{gal} = volume flow rate of chilled water, gpm
l_{dif} = linear length of diffuser, ft
ν_w = kinematic viscosity of water, ft²/s

According to Wildin (1990), when $\text{Re}_i < 850$, loss due to mixing and loss of cooling capacity during discharge can be significantly reduced. If Re_i is determined, the length of the linear diffuser can be calculated from Eq. (25.2).

The inlet Froude number Fr_i is defined as

$$\text{Fr}_i = \frac{\dot{V}_w}{l_{\text{dif}} \left[\frac{gh_i{}^3(\rho_i - \rho_a)}{\rho_i} \right]^{0.5}} \tag{25.3}$$

where g = acceleration of gravity, ft/s²
ρ_i = density of inlet water, lb/ft³
ρ_a = density of ambient water stored in the tank at diffuser level, lb/ft³

In Eq. (25.3), h_i indicates the inlet opening height, in ft. It is the vertical distance occupied by the incoming flow when it leaves the diffuser and forms the gravity current. For the bottom diffusers, inlet opening height h_i indicates the vertical distance between the tank floor and the top of the opening of the diffuser.

SELF-BALANCING. Self-balancing means that the water flow introduced to or extracted from the tank should be self-balanced according to requirements, that is, evenly distributed at all flow conditions. This includes the following requirements:

- The piping design should be symmetric.
- Branch pipes should be equal in length.
- Flow splitters should be added at the appropriate points.
- Pipe diameter reduction should be combined with the flow splitter.
- Long radius elbows should be used.

Charging and Discharging Temperature versus Tank Volume

The FOM of a stratified tank is closely related to the temperature difference of the outlet temperature of stored chilled water during discharging T_o and

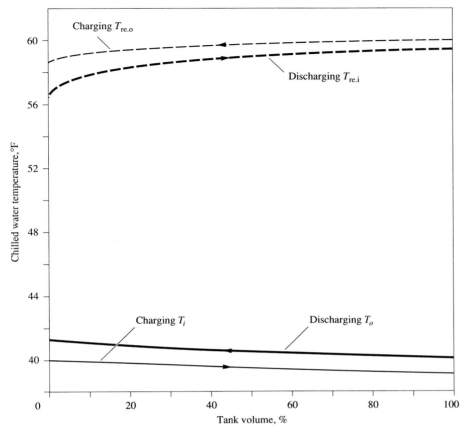

FIGURE 25.8 Chilled water temperature versus tank volume curves during charging and discharging processes.

the inlet temperature of stored chilled water during charging T_i. For a complete charging and discharging cycle, the average T_o during discharging is always greater than the average T_i because of the mixing loss and heat gains, provided that the water flow rate is constant. The smaller the $(T_o - T_i)$, the higher the FOM.

Figure 25.8 shows chilled water temperature versus tank volume curves during the charging and discharging processes of a complete chilled water storage cycle in a large stratified tank. Inlet and outlet temperatures are measured at the openings of the top and bottom diffusers. During the charging process, return chilled water is extracted from the stratified tank, cooled in the chiller, and charged into the stratified tank again through the bottom diffusers. The inlet temperature of the stored chilled water T_i gradually decreases as the stored volume increases. This is due to a comparatively greater rate of heat transfer to the inlet water from the warmer ambient water, piping, and tank wall at the beginning of the charging process.

During the discharging process, stored chilled water is extracted from the stratified tank and supplied to the cooling coils in the air-handling units and terminals. The return chilled water is introduced to the stratified tank through the top diffusers. The outlet temperature of the stored chilled water T_o gradually increases as the stored volume decreases. Because the stored chilled water has a temperature gradient below the thermocline, T_o gradually rises during the discharging process.

Both the outlet temperature of return chilled water during charging $T_{rc.o}$ and inlet temperature of return chilled water during discharging $T_{rc.i}$ should be controlled between 55 and 60°F so that stratification can be maintained in the storage tank.

25.6 CASE STUDY: A STRATIFIED TANK CHILLED WATER STORAGE SYSTEM

A full-storage stratified tank chilled water storage system was completed in August 1990 to serve a 1.142-million-ft^2 electronics manufacturing facility in

Dallas, TX. The following are the details of this project as described by Fiorino (1991).

System Description

The 2.68-million-gallon stratified tank is a precast, prestressed, cylindrical concrete water tank. The design parameters of this stratified tank chilled-water storage system are as follows:

Storage cooling capacity	24,500 ton · h
Maximum refrigeration load	3200 tons
Charge process duration	16 h
Discharge process duration	8 h
Inlet temperature during charging T_i	40°F
Limiting outlet temperature during discharging T_o	42°F
Inlet temperature during discharging $T_{re.i}$	56 °F
Maximum volume flow rate	5120 gpm
Tank diameter	105.5 ft
Tank height	41 ft
Tank volume	2.7 million gal
Usable tank volume	90 percent

There are two 1200-ton and two 900-ton centrifugal chillers to serve zones 1 and 2, as shown in Fig. 25.9. There are also five chiller pumps, one of which is a standby. Two variable-speed building pumps are installed for each zone, one of which is a standby pump. Three storage pumps are used for chilled water storage, one of which is a standby.

Concentric Double-Octagon Diffusers

Concentric double-octagon diffusers are used at both the top and bottom of the stratified tank, as shown in Fig. 25.6a. Each octagon introduces 50 percent of the total volume flow during charging process. Over time, a double-octagon arrangement provides nearly twice the effective linear length for diffusers of a single-octagon arrangement.

The total effective length of the 8 diffusers in the outer octagon is 559 ft. For a total flow of $0.5 \times 5120 = 2560$ gpm through the outer octagon, the volume flow rate per linear ft of diffuser is

$$\frac{\dot{V}_{dif}}{l_{dif}} = \frac{2560 \times 0.1337}{559 \times 60} = 0.0102 \text{ ft}^3/\text{s}$$

The kinematic viscosity of water at 42°F is 1.66×10^{-5} ft²/s, so the inlet Reynolds number is calculated as:

$$\text{Re}_i = \frac{\dot{V}_{dif}}{l_{dif}\, \nu_w} = \frac{0.0102}{1.66 \times 10^{-5}} = 615$$

Similarly, the inlet Reynolds number for inner octagon is 1068. Both of them are close to the upper limit of 805.

The acceleration of gravity $g = 32.2$ ft/s², and

$$\frac{\rho_i - \rho_a}{\rho_i} = \frac{62.4263 - 62.3864}{62.3864} = 0.00064$$

For a common inlet opening height $h_i = 5.64$ in., or 0.47 ft, the the inlet Froude number for diffusers in outer octagon is calculated as

$$\text{Fr}_i = \frac{0.0102}{[32.2 \times (0.47)^3 \times 0.00064]^{0.5}} = 0.22$$

Similarly, the inlet Froude number for diffusers in the inner octagon is 0.38.

At a lateral slot spacing of 0.5 ft in the inner octagon diffusers and 0.87 ft in the outer octagon diffusers, each linear diffuser has 32 lateral slots. If the maximum inlet velocity is 0.9 ft/s, the opening area for each lateral slot is

$$\frac{5120 \times 0.1337}{60 \times 0.9 \times 32 \times (8 + 8)} = 0.025 \text{ ft}^2$$

Because the lateral slot is spread at an angle of 120° downward, if the length of the lateral slot is about 1 ft, the width of the slot is $0.025/1 = 0.025$ ft, or about 0.3 in.

If the cross-sectional area of each linear diffuser is equal to the slot openings, the diameter of the diffuser is therefore

$$D = \left(\frac{4A}{\pi}\right)^{0.5} = \left(\frac{4 \times 0.025 \times 32}{\pi}\right)^{0.5} = 1.01 \text{ ft}$$

To provide even distribution of incoming water flow, flow splitters are used to divide the water flow evenly to the split mains and branches.

Charging Process

During full-load operation in hot weather, charging is performed from 8 P.M. until noon the next day. All four chillers and all water pumps except the standby pumps are operated simultaneously to provide direct cooling during off-peak hours as well as the required stored cooling capacity during peak hours the next day.

The direct cooling refrigeration load during off-peak hours varies from 1980 to 2600 tons, with a total of 34,800 ton · h. The required stored cooling capacity to meet the refrigeration load during peak hours was 21,300 ton·h on July 17, 1989. For a charging

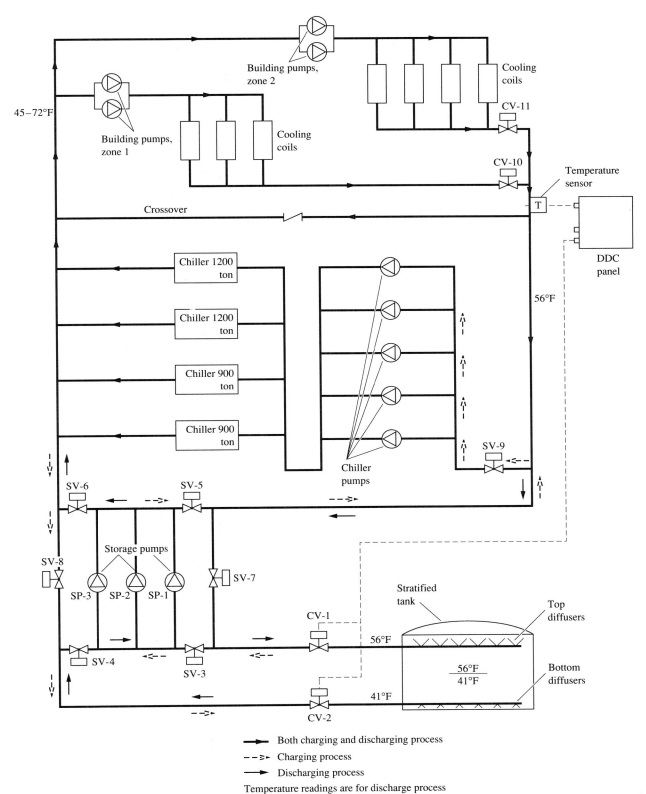

FIGURE 25.9 Schematic diagram of a stratified tank chilled water storage system for an electronics manufacturing facility in Dallas, TX.

process of 16 hours, the four chillers are operated at an average refrigeration load of about 3600 tons. The set point of the chilled water temperature leaving the chiller was 39.5°F, and the outlet return chilled water temperature is around 56°F.

Before charging, the chillers and water pumps are started. During the charging process, the control valves CV-1 and 2 and solenoid valves SV-3, 5, 8, and 9 are opened, as shown in Figure 25.9; and solenoid valves SV-4, 6, and 7 are closed. This provides both the charging of the storage tank and the supply of chilled water to building pumps during off-peak hours.

Return chilled water at around 56°F is extracted from the stratified tank through the top diffusers and the control valve CV-1 and solenoid valve SV-3 by the storage pumps SP-1 and SP-2. It flows through solenoid valves SV-5 and SV-9 and the chiller pumps, cools in the chiller, and leaves the chiller at 39.5°F. After that, the chilled water is divided into two streams. One of the streams flows through solenoid valve SV-8 and central valve CV-2 and is charged to the stratified tank through the bottom diffusers at 40°F. The other stream is extracted by the building pump and supplied to the air-handling units and terminals for direct cooling.

Discharging Process

Before the shutoff of the chillers and chiller pumps, the control valves and solenoid valves should be switched over to the following: control valves CV-1 and 2 and solenoid valves SV-1, 2, 4, 6, and 7 should be open, and solenoid valves SV-3, 5, 8, and 9 should be closed.

Stored chilled water at 41°F is extracted by storage pumps SP-2 and SP-3 via bottom diffusers and control valve CV-2. It then flows through SV-6, is extracted again by the building pumps, and is supplied to the cooling coils in the air-handling units and terminals. Return chilled water at a temperature $T_{rc} = 56°F$ is then forced through solenoid valve SV-7 and control valve CV-1, and introduced to the stratified tank through the top diffusers.

T_{rc} is maintained by changing the chilled water temperature supplied to the air-handling units from 45 to 52°F instead of 41°F. As soon as temperature sensor T senses a return chilled water temperature T_{rc} drop below 56°F, the DDC panel closes control valve CV-1 slightly. Less stored chilled water is extracted by the building pumps. The required amount of return chilled water at 56°F bypasses the crossover and mixes with the 41°F stored chilled water, which re-

sults in a higher supply temperature to the cooling coils.

During peak hours, the discharging process requires a refrigeration load from 2500 to 2800 tons, with a total of 21,250 ton · h for the 8-hour peak period.

Part-Load Operation

This electronics manufacturing facility includes clean-rooms, computer rooms, compressed air after-coolers, and manufacturing equipment, all of which needs 24-hour continuous cooling. Its daytime refrigeration load averages about 1240 tons.

When the refrigeration load is reduced or the temperature of cooling water entering the condenser drops because of a lower outdoor wet bulb temperature, the following adjustments are made during part-load operation:

- The temperature of inlet water entering the stratified tank during charging process T_i is raised from 39.5°F to 42.5°F, which increases the chillers' capacity and lowers their power consumption. All four chillers and their auxiliary equipment do not have to operate simultaneously.
- Instead of 8 P.M. daily, the start of the charging process can be delayed until nearly all the stored chilled water in the stratified tank is discharged. The discharging process might last 10 to 14 hours during part-load operation instead of 8 hours in full-load.

System Performance

According to the operating cycle from August 24 to August 26, 1990, the electricity demand dropped about 2.5 MW during peak hours as intended because four chillers and their corresponding auxiliary equipment were shut down during peak hours, as shown in Fig. 25.10.

The maximum storage capacity of the stratified tanks was 27,643 ton · h. The difference in temperature between the average outlet temperature during discharging and the average inlet temperature during charging $(T_o - T_i)$ was 1.1°F and the figure of merit was 92.2 percent.

25.7 DESICCANT COOLING SYSTEMS

Desiccant Cooling

A *desiccant cooling process* is a combination of desiccant dehumidification, evaporative cooling, supplementary vapor compression or absorption refriger-

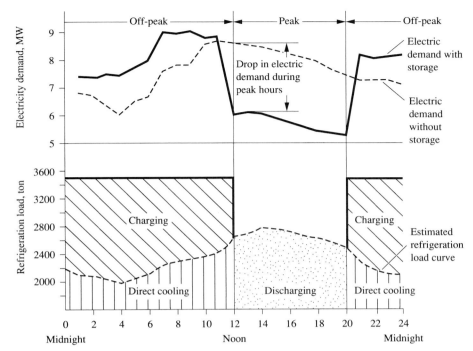

FIGURE 25.10 Electricity demand curves and storage cycle of the stratified tank chilled water storaged system for the electronic manufacturing facility in Dallas, TX.

ation, and the regeneration of the desiccant by means of waste heat or gas heating.

As described in Section 3.5, when air flows over liquid or solid desiccants, the absorption of moisture in the desiccant dehumidification process causes the heat of sorption to be released and raises the air temperature. Therefore, evaporative cooling or refrigeration should be used to cool the air to a required temperature. At the same time, a process called *regeneration* or *reactivation* is needed to force the absorbed moisture to vaporize from the desiccant—a drying process using high-temperature air to increase the temperature of the desiccant.

In an air conditioning system, the coil's load can be divided into sensible load and latent load. For comfort systems in commercial buildings, the latent load varies from 20 to 40 percent of the total coil's load. In applications such as supermarkets, it may amount to as much as 65 percent of the total coil's load.

If desiccant dehumidification can be used to offset the latent load, cheaper and more energy-efficient evaporative cooling can replace costly refrigeration. In addition, if waste heat from heat recovery is used as the regenerative heat required to reactivate the desiccant, the high initial cost of a desiccant cooling system is then cost-effective compared with vapor compression refrigeration systems in many commercial and industrial applications where latent load dominates, where cold air supply at a temperature of 40 to 45°F is required, or where electricity demand charges are high.

Desiccant Cooling System

A desiccant cooling system is a system in which latent cooling is performed by desiccant dehumidification and sensible cooling by evaporative cooling or refrigeration. Thus, a considerable part of the expensive vapor compression refrigeration is replaced by inexpensive evaporative cooling. A desiccant cooling system is usually a hybrid system of dehumidification, evaporative cooling, refrigeration, and regeneration of desiccant.

There are two airstreams in a desiccant cooling system: a process airstream and a regenerative airstream. Process air can be all outdoor air or a mixture of outdoor and recirculating air. Process air is also the conditioning air supplied directly to the conditioned space or enclosed manufacturing process, or to the air-handling unit, packaged unit, or terminal for further treatment. A regenerative airstream is a high-temperature airstream used to reactivate the desiccant.

A desiccant cooling system consists of the following components: rotary desiccant dehumidifier(s), rotary heat exchanger(s), heat pipe heat exchanger(s), direct- and indirect-evaporative coolers, absorption chiller, engine-driven vapor compression unit, fans, pumps, filters, controls, ducts, and piping. Because the others have been discussed in previous chapters, only the rotary desiccant dehumidifier, rotary heat exchanger, and heat pipe heat exchanger will be discussed here.

Desiccants

Either solid or liquid desiccant absorbs or releases moisture because of the difference in vapor pressure between the surface of the desiccant p_{des} and the surrounding air p_{sur}. In a desiccant dehumidification process, when $p_{des} < p_{sur}$, the desiccant absorbs moisture from the ambient air. In a regenerative process, $p_{des} > p_{sur}$, so moisture is released from the desiccant to the surrounding air.

Desiccants can be classified as adsorbants, which absorb moisture without accompanying physical and chemical changes, and absorbants, which absorb moisture accompanied by physical or chemical changes.

Three kinds of desiccants are widely used in desiccant cooling systems: silica gel, lithium chloride, and molecular sieves.

SILICA GELS. These are solid desiccants and adsorbants. Structurally, they contain numerous pores and capillaries in which water is condensed and contained. Silica gel has a high capacity to absorb moisture and releases it at a higher temperature. They are low in cost and available in sizes from $\frac{3}{16}$-in. beads to powder-like grains.

LITHIUM CHLORIDE (LiCl). This is an absorbant. It is in dry form when each LiCl molecule holds two water molecules. If each LiCl molecule holds more than two water molecules, it becomes a liquid and continues to absorb moisture. LiCl has a high capacity to absorb and to hold moisture. Lithium chloride is widely used in rotary wheel dehumidifiers.

MOLECULAR SIEVES. These are actually synthetic zeolites, a solid desiccant and an adsorbant in the form of crystalline aluminosilicates produced by a thermal process. Molecular sieves show physical stability and high moisture-releasing capacity at high regenerating temperatures of 248 to 428°F, and are recommended in direct gas-fired applications.

As defined in Section 6.3, sorption isotherm is a constant temperature curve that indicates the relationship between the moisture content of the desiccant X_{des}, or moisture absorbed as a percentage of its dry mass, and the relative humidity of the surrounding air φ_{sur}.

Figure 25.11 shows the sorption isotherms of silica gels, lithium chloride, and molecular sieves. LiCl has a much higher water-holding capacity than silica gels and molecular sieves.

Rotary Desiccant Dehumidifier

A *rotary desiccant dehumidifier* is a rotary wheel with a depth of 4 to 8 in., as shown in Fig. 25.12. The wheel has support spokes, typically 16, and its front surface is cut into segments. A rotary dehumidifier is often equally divided into two separately sealed sections: a dehumidifying section and a regeneration section. Solid desiccants such as silica gels and molecular sieves are in granular form lined within substrate sheets at a center-to-center spacing of about 30 mil. The blockage of the face area by desiccant, substrates, and support spokes is about 28 percent. For a face velocity of about 500 fpm, the pressure drop of air flowing through the rotary dehumidifier is about 0.5 to 0.6 in. WG.

For liquid absorbants such as lithium chloride, a porous fiberglass matrix impregnated with dessicant is used instead of particles attached to substrates by adhesive. Wheel design is similar to that of a solid desiccant wheel. To prevent the dripping of LiCl from the fiberglass matrix, the ratio of absorbed water to the mass of desiccant should be less than 10 lb H_2O/lb LiCl. Each lb of fiberglass matrix holds about 0.10 lb of LiCl. The rotating speed of rotary desiccant dehumidifiers varies from 0.5 to 6 rev/h. Both the desiccant dehumidification process and regeneration process last from 5 to 60 minutes.

The dehumidifying capacity of a rotary desiccant dehumidifier depends on the desiccant material, wheel structure, and inlet and outlet conditions of process and regeneration airstreams. Parsons et al. (1989) compared the performance of three desiccant materials on a rotary dehumidifier with a volume flow rate of 1120 cfm with a diameter of 3.3 ft. The inlet conditions of the process airstream was a temperature of 95°F and a humidity ratio of 0.015 lb/lb; the inlet temperature of the regeneration airstream was 185°F. Silica gel and LiCl were found to have similar dehumidifying capacities, a humidity ratio drop of $(0.015 - 0.006) = 0.009$ lb/lb during the dehumidification process.

The process airstream and the regeneration airstream in the rotary dehumidifier flow in parallel but opposite directions.

Rotary desiccant dehumidifiers are available in single units up to a volume flow rate of about

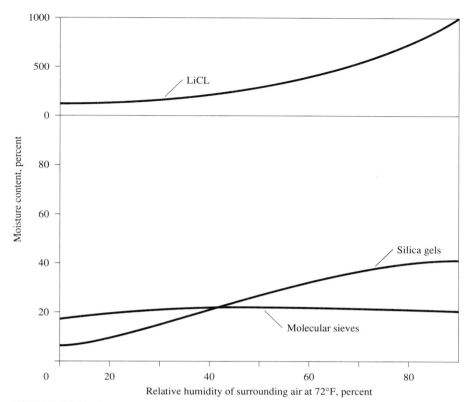

FIGURE 25.11 Sorption isotherms of some desiccants. (*Source: ASHRAE Handbook 1989, Fundamentals*. Reprinted with permission.)

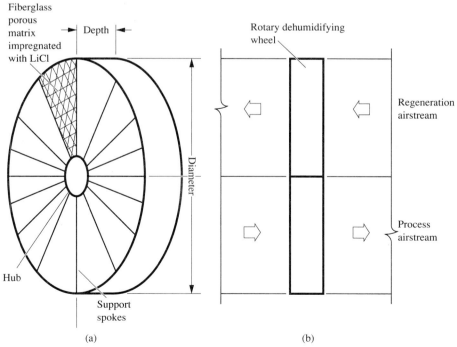

FIGURE 25.12 Rotary desiccant dehumidifier: (a) isometric view and (b) process and regeneration airstreams.

65,000 cfm with a maximum diameter of 14 ft. A wheel larger than that would be too difficult to ship and install. A horizontally installed unit with a diameter greater than 8 ft requires structural support. Manufacturers provide data on the dehumidifying capacity of their products and the required air temperature during regeneration. If a dehumidifier is to be used in conditions other than the rated conditions, the catalog data should be adjusted accordingly to ensure proper equipment selection.

Desiccant Dehumidification Process

As described in Section 3.5, desiccant dehumidification is a dehumidification and heating process on the psychrometric chart, and is illustrated by a straight line inclining downward deviating slightly from the thermodynamic wet bulb line, as shown in Fig. 3.6. It has a flatter negative slope.

The acute angle between the desiccant dehumidification process and the thermodynamic wet bulb depends mainly on the desiccant material and the water-holding capacity of the desiccant. When the desiccant approaches saturation, angle α is smaller than that when a new desiccant first absorbs moisture from the surrounding air. For a rotary desiccant dehumidifier impregnated with lithium chloride, if manufacturer's data are not available, angle α can be estimated at 5°.

The regeneration process removes the absorbed moisture from the desiccant. For the regeneration airstream, it is also a humidifying and cooling process. Molecular sieves require a higher regeneration temperature than silica gels and lithium chloride. Figure 25.13a shows the desiccant dehumidification and regeneration processes in a rotary desiccant dehumidifier.

Rotary Heat Exchangers

Rotary heat exchangers can be classified into two categories: *rotary sensible heat exchangers*, which recover sensible heat only, and *rotary enthalpy exchangers*, which recover and transfer both sensible and latent heat. Rotary heat exchangers are similar in shape and structure to rotary desiccant dehumidifiers. In an enthalpy heat exchanger, lithium chloride or other desiccant is also impregnated in the porous fiberglass matrix and used as the total heat transfer medium, as in the rotary desiccant dehumidifier. There are two major differences:

- The wheel of a rotary sensible heat exchanger is filled with an air-penetrable medium with a large internal surface area such as aluminum, monel metal, and stainless steel corrugated wire mesh, at a density of about 4 lb/ft^3.
- There are two airstreams in a rotary heat exchanger: the cold airstream and a hot airstream. In a sensible heat exchanger, the cold airstream may be the supply airstream and the hot airstream may be the exhaust airstream, as shown in Fig. 25.13b. In an enthalpy heat exchanger in summer, the hot airstream

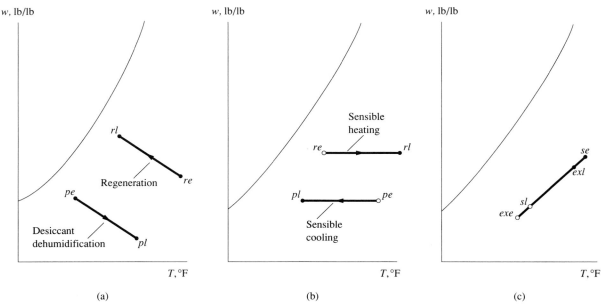

FIGURE 25.13 Conditioning processes in rotary dehumidifiers and heat exchangers: (a) rotary desiccant dehumidifier; (b) rotary sensible heat exchanger; and (c) rotary enthalpy exchanger.

may be the supply airstream, and the cold airstream may be the relief or exhaust airstream, as shown in Fig. 25.13c.

The face velocity of rotary heat exchangers varies from 500 to 700 fpm. At a face velocity of 500 fpm, its pressure drop varies from 0.4 to 0.65 in. WG, depending mainly on the structure of the heat transfer medium.

The performance of a rotary heat exchanger is indicated by its effectiveness, as defined in Eq. (12.13):

$$\epsilon = \frac{\dot{V}_c \rho_c (T_{cl} - T_{ce})}{\dot{V}_{min} \rho_{min} (T_{he} - T_{ce})}$$

$$= \frac{\dot{V}_h \rho_h (h_{he} - h_{hl})}{\dot{V}_{min} \rho_{min} (h_{he} - h_{ce})} \quad (25.4)$$

where $\dot{V}_c, \dot{V}_h$ = volume flow rate of the cold and the hot airstream, cfm

$\dot{V}_{min}$ = volume flow rate of the lower of the two airstreams, cfm

ρ_c, ρ_h = density of the cold and the hot airstream, lb/ft³

ρ_{min} = density of the lower volume flow rate of the two airstreams, lb/ft³

T_{ce}, T_{cl} = entering and leaving temperature of the cold airstream, °F

T_{he} = entering temperature of the hot airstream, °F

h_{ce}, h_{cl} = entering and leaving enthalpy of the cold airstream, Btu/lb

h_{he}, h_{hl} = entering and leaving enthalpy of the hot airstream, Btu/lb

At a face velocity of 500 fpm, the effectiveness of rotary heat exchangers is usually between 0.7 and 0.85. A higher face velocity means a higher rate of heat transfer, but the greater effect of a larger volume flow rate results in a lower effectiveness. A lower face velocity results in a lower pressure drop, smaller volume flow rate, and a higher effectiveness.

As for rotary dehumidifiers, the maximum diameter of rotary heat exchangers is also 14 ft. The limit of the volume flow rate for a single unit is about 65,000 cfm.

Heat Pipe Heat Exchangers

A *heat pipe heat exchanger* is often used as a sensible indirect heat exchanger. It consists of many heat pipes arranged in rows along the direction of air flow, as shown in Fig. 25.14a. Each sealed heat pipe con-

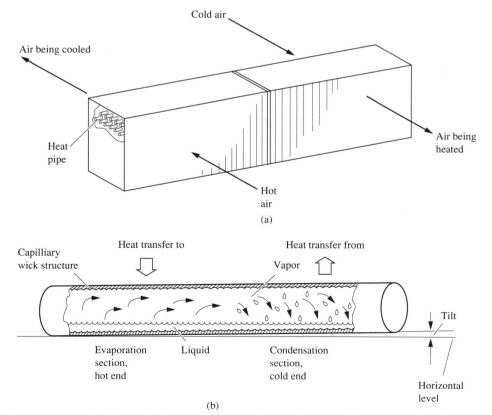

FIGURE 25.14 A heat pipe heat exchanger: (a) heat exchanger and air streams and (b) heat pipe.

tains a volatile fluid, as shown in Fig. 25.14b. When one end of the pipe, the *hot end* or *evaporation section*, absorbs heat from the airstream flowing over the pipe, the volatile fluid inside the pipe vaporizes. The vapor then moves to the other end (the *cold end* or *condensation section*) because of the higher saturated pressure in the evaporation section. After condensing heat is released to another airstream that flows over the other end of the pipe, the vapor inside the pipe condenses into liquid form and is drawn back to the evaporation section by gravity. It has then completed an evaporation/condensation cycle. The heat pipes are often slightly tilted to enable the condensed liquid to flow back to the evaporation section.

Heat pipes have an inner capillary wick structure. The outer tube is often made of aluminum, with fins of the same material for a larger heat transfer surface. The volatile fluid inside the heat pipe is usually a halocarbon compound refrigerant.

When two separate airstreams flow over the heat pipe heat exchanger, the hot airstream flows over the evaporation section and the cold airstream flows over the condensation section. These two airstreams flow in a counterflow arrangement for greater effectiveness. The airstreams are separated by a sealed partition to prevent cross-contamination.

The performance of a heat pipe heat exchanger is indicated by its effectiveness ϵ, which can be calculated by Eq. (25.4). Both the effectiveness and the pressure drop Δp of a heat pipe heat exchanger depend mainly on its face velocity v_{face}, fin spacing, and the number of rows of heat pipes in the direction of air flow. If two heat pipe heat exchangers are connected in series, the number of rows of such a series is the total number of rows of the two heat exchangers.

The greater the total number of rows of heat pipe and fins/in., the higher the ϵ and the Δp. The lower the face velocity v_{face} of the heat pipe heat exchanger, the higher the ϵ and the lower the Δp. However, the volume flow rate and the capacity of the heat pipe heat exchanger will also be smaller.

The design face velocity of a heat pipe heat exchanger is between 400 and 700 fpm. The total number of heat pipe rows usually varies from 6 to 10 rows. For a heat pipe with 14 fins/in. and a total of 8 rows with a face velocity $v_{\text{face}} = 500$ fpm, its effectiveness $\epsilon = 0.65$ and its pressure drop Δp is about 0.6 in. WG. Refer to the manufacturer's data for detailed information.

The capacity and rate of heat transfer of a heat pipe heat exchanger can be controlled by varying the slope or the tilt of the heat pipe. This adjustment increases or decreases the liquid flow inside the heat pipe and, therefore, its capacity.

25.8 SINGLE-STAGE DESICCANT COOLING SYSTEMS

Single-Stage Systems

A *single-stage desiccant cooling system* is a system in which only one desiccant dehumidifier is used to dehumidify the process air. Equipment such as direct evaporative coolers, indirect evaporative coolers, and rotary sensible heat exchangers are not considered additional stages because they are not dehumidifiers.

When the difference in humidity ratio $(w_i - w_o)$ during dehumidification in a desiccant cooling system is equal to or less than 0.009 lb/lb (63 grains of water vapor per lb of dry air), such as cooling systems in supermarkets and recreational skating rinks, a single-stage desiccant dehumidifier is usually used.

A desiccant cooling system is operated at an *open cycle* when the process air is entirely outdoor air. If the process air at the inlet to the rotary desiccant dehumidifier is a mixture of recirculating air and outdoor air, the desiccant cooling system is said to be operated at a *closed cycle*.

Loads in Supermarkets

Many early supermarket air conditioning systems are equipped with conventional vapor compression refrigeration systems to maintain a design space condition at 75°F and a relative humidity of 55 percent in the sales area where the refrigerated display cases are located. For supermarkets with large freezer volumes, there are three types of loads:

- *Refrigeration load.* The refrigeration load of the refrigerated cases depends mainly on the temperature of the foods in the display cases and the design ambient conditions or dew point temperature of the space air. The higher the space temperature and space relative humidity, and therefore the dew point, the greater the amount of moisture that may condense on the frozen surfaces, the larger the accompanying latent heat of condensation, and the greater the refrigeration load.
- *Space cooling load.* A considerable portion of the sensible space cooling load is removed by the refrigerated cases through their openings, which results in a space cooling load with a heavy latent load that may vary from 50 to 65 percent.
- *Coil's load.* The coil load or refrigeration load of the air conditioning system to maintain a required space condition in the sales area is the sum of the space cooling load, outdoor ventilation air load, and system heat gain. During summer, the outdoor

ventilation load is again mainly latent load, and is affected by the space temperature and relative humidity.

In supermarkets, typically 50 percent of the electric energy used annually is consumed by the refrigerated cases and another 15 percent is used for the air conditioning system to maintain the required space condition.

Case Study: A Desiccant Cooling System for a Supermarket

Consider a supermarket with an area of 30,000 ft² and an outdoor ventilation air requirement of 3000 cfm. The condition of the outdoor air is dry bulb 86°F, relative humidity 60 percent, and humidity ratio of 0.016 lb/lb (112 grains/lb).

If a desiccant cooling system with an impregnated lithium chloride dehumidifier is used instead of a conventional vapor compression refrigeration system, it is possible to maintain an indoor space temperature of 75°F and a relative humidity of 45 percent (0.0085). Meanwhile, the supply air volume flow rate can be reduced from the 1 cfm/ft² of a conventional system to 0.5 cfm/ft² of a desiccant cooling system. Thus, there is a significant reduction of refrigeration load of the refrigerated cases because of lower space relative humidity as well as the fan power consumption in the air conditioning system.

Figure 25.15a is schematic diagram of a typical desiccant cooling system for a 30,000 ft² supermarket. The process airstream of the desiccant cooling cycle is indicated on the psychrometric chart by the bold line in Fig. 25.15b. The static points of the process air at the exits of various components are:

Outdoor air	o
Filter	o
Rotary desiccand dehumidifier	dl
Rotary sensible heat exchanger	pl
Mixing with recirculating air	m
Indirect evaporative cooler or refrigeration	s
Supply air after supply fan	s
Space air	r

The state points at the exits of various components of the regeneration airstream are:

Outdoor air	o
Filter	o
Rotary sensible heat exchanger	xl

Gas heater	rg
Rotary disiccant dehumdifier	ro
Exhaust fan to atmosphere	ro

Calculation of Desiccant Cooling Processes

SPACE CONDITIONING LINE. For a supermarket of 30,000 ft², the total space cooling load q_{rc} is about 83,000 Btu/h. If 50 percent of this is latent load q_{rl}, then

$$q_{rl} = 0.5 \times 83,000 = 41,500 \text{ Btu/h}$$

If the supply air is at a temperature of 72.5°F, a relative humidity of 48 percent, and a humidity ratio of 0.0079 lb/lb, from Eq. (3.52), the supply air volume flow rate of the mixture of the process air and the recirculating air is

$$
\begin{aligned}
\dot{V}_s &= \frac{q_{rl}}{60\rho_s(w_r - w_s)h_{\text{fg.32}}} \\
&= \frac{41,500}{60 \times 0.075\,(0.0085 - 0.0079) \times 1061} \\
&= 14,487 \text{ cfm}
\end{aligned}
$$

It is approximately equal to $0.5 \times 30,000 = 15,000$ cfm.

Similarly, from Eq. (3.48b), for a sensible cooling load of about $0.5 \times 83,000 = 41,500$ Btu/h, the supply volume flow rate to maintain a space temperature of 75°F is

$$
\begin{aligned}
\dot{V}_s &= \frac{q_{rs}}{60\rho_s c_{\text{pa}}(T_r - T_s)} \\
&= \frac{41,500}{60 \times 0.075 \times 0.243\,(75 - 72.5)} \\
&= 15,180 \text{ cfm}
\end{aligned}
$$

ROTARY DESICCANT DEHUMIDIFIER. Assuming that the humidity difference during the dehumidification process is 0.009 lb/lb, the process air therefore leaves the rotary dehumidifier at a humidity ratio of

$$w_{dl} = w_o - 0.009 = 0.016 - 0.009 = 0.007 \text{ lb/lb}$$

For the process airstream, the outdoor air (point o) enters the rotary dehumidifier at 86°F, and a humidity ratio of 0.016 lb/lb. On the psychrometric chart, draw a line $od\,l$ from point o with an acute angle of 5° between $od\,l$ and the thermodynamic line. This line intersects the 0.007 humidity ratio line at point $d\,l$, which is the state point of process air leaving the rotary desiccant dehumidifier. At point $d\,l$, temperature $T_{dl} = 140°F$.

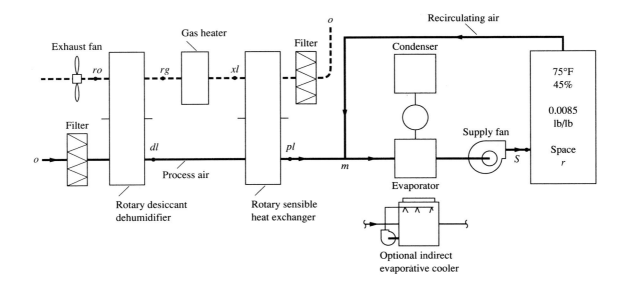

(a)

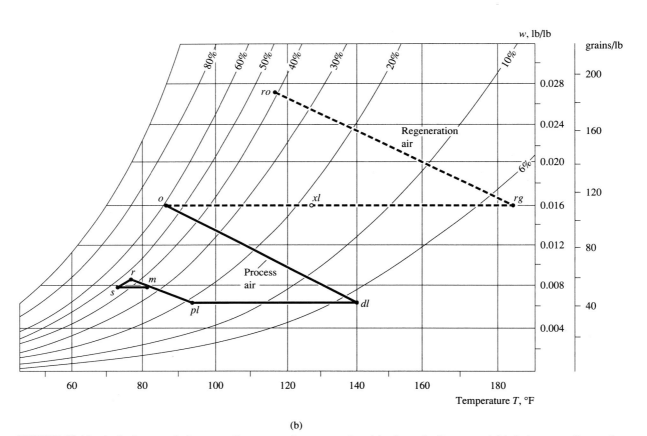

(b)

FIGURE 25.15 A single-stage desiccant cooling system for supermarket: (a) schematic diagram and (b) desiccant cooling cycle.

For process air at volume flow rate of 3000 cfm, the dehumidifying capacity of the rotary dehumidifier is

$$\dot{m}_{\text{deh}} = 3000 \times 0.075 \times 0.009 \times 60 = 121.5 \text{ lb/h}$$

If the volume flow rate of the regeneration air is also 3000 cfm and the temperature of the regeneration air entering the rotary desiccant dehumidifier is $T_{\text{rg}} = 185°F$ with a humidity ratio of 0.016 lb/lb, regeneration air leaves the rotary dehumidifier at $T_{\text{ro}} = 117°F$ and a humidity ratio of 0.027 lb/lb.

ROTARY SENSIBLE HEAT EXCHANGER. Assume that the effectiveness of the rotary sensible heat exchanger is 0.8. If the difference between the specific heat of the process air and regeneration air is ignored, and if the mean density of the process air flowing through the rotary sensible heat exchanger is $\rho_p = 0.0685$ lb/ft^3, and for the regeneration stream it is $\rho_{\text{rg}} = 0.070$ lb/ft^3, then, from Eq. (12.13),

$$\epsilon = \frac{\rho_p(T_{\text{dl}} - T_{\text{pl}})}{\rho_{\text{rg}}(T_{\text{dl}} - T_o)} = 0.8$$

The temperature of the process air after the rotary sensible heat exchanger is

$$T_{\text{pl}} = 140 - \frac{0.8 \times 0.070 \ (140 - 86)}{0.0685} = 95.8°F$$

The temperature of regeneration air leaving the rotary sensible heat exchanger T_{xl} can then be calculated as

$$T_{\text{xl}} = 86 + \frac{0.0685(140 - 95.8)}{0.070} = 129.2°F$$

MIXING OF PROCESS AIR AND RECIRCULATING AIR. Process air at a volume flow rate of 3000 cfm is mixed with recirculating air with a volume flow rate of 12,000 cfm. The condition of the mixture point m can be determined from the psychrometric chart by drawing a line that connects the space air and the process air after the rotary sensible heat exchanger, r_{pl}, so that

$$\frac{rm}{r\,pl} = \frac{3000}{15,000} = 0.2$$

From the psychrometric chart, the temperature of the mixture T_m is 81°F and the humidity ratio w_m is 0.0079 lb/lb.

INDIRECT EVAPORATIVE COOLER OR REFRIGERATION. If the desiccant cooling system is installed in a location in which the summer 97.5 percent outdoor design wet bulb is lower than 67°F, an indirect evaporative cooler can be used. Otherwise, a reciprocating vapor compressor refrigeration system should be used. For a supply volume flow rate of 15,000 cfm, the sensible cooling capacity can be calculated as

$$\begin{aligned} q_{\text{cs}} &= 60 \dot{V}_s \rho_s c_{\text{pa}}(T_m - T_s) \\ &= 60 \times 15,000 \times 0.075 \times 0.243 \ (81 - 72.5) \\ &= 139,241 \text{ Btu/h} \end{aligned}$$

The refrigeration system is mainly used for sensible cooling. This is only possible when the evaporating temperature T_{ev} in the DX-coil is higher than 50°F. If T_{ev} is lower than 50°F, a certain degree of dehumidification exists in sensible cooling process ms. The required refrigeration load is therefore greater than q_{cs}.

GAS HEATER. The temperature of regeneration air required to reactivate the desiccant LiCl is 185°F. A gas heater is used to heat the air from $T_{\text{xl}} = 129.2°F$ to 185°F. The heating capacity of the gas heater is

$$\begin{aligned} q_h &= 60 \dot{V}_{\text{rg}} \rho_{\text{rg}} c_{\text{pa}}(T_{\text{rg}} - T_{\text{xl}}) \\ &= 60 \times 3000 \times 0.243 \times 0.0685 \ (185 - 129.2) \\ &= 167,187 \text{ Btu/h} \end{aligned}$$

Operating Parameters of the Desiccant Cooling Cycle

From the above calculations, the operating parameters of the desiccant cooling cycle are shown as follows:

Point		Temperature, °F	Humidity ratio, lb/lb
Process air			
o	Outdoor air	86	0.016
dl	After dehumidifier	140	0.007
pl	After sensible exchanger	95.8	0.007
m	Mixture	81	0.0079
s	Supply air	72.5	0.0079
r	Space air	75°F, 45 percent RH	0.0085
Regeneration air			
xl	After sensible exchanger	129.2	0.016
rg	After gas heater	185	0.016
ro	After dehumidifier	117	0.027

Part-Load Operation and Controls

When the humidity ratio of the outdoor air drops or the space latent load falls during part-load operation, there are three methods of maintaining the space humidity ratio w_r and the space relative humidity φ_r if the space temperature remains constant:

- Modulation of the gas heating capacity and the temperature of the regeneration air varies the dehumidifying capacity of the rotary dehumidifier.

- In *bypass control*, a portion of outdoor air bypasses the rotary desiccant dehumidifier so that the supply air has a higher humidity ratio.

- The rotational speed of the rotary dehumidifier can be modulated by using an adjustable-frequency AC inverter. A lower rotating speed of the dehumidifier means a smaller dehumidifying capacity of the rotary dehumidifier.

The space temperature is controlled by the vapor compression refrigeration system. If the sensible cooling load drops at part-load operation, a cylinder unloader or on-off control can be used to maintain the space temperature within predetermined limits.

System Characteristics

Compared with a conventional vapor compression refrigeration system, the supermarket desiccant cooling system has the following benefits and savings:

- Because of the low humidity ratio of the supply air, the supply volume flow rate of a desiccant cooling system can be reduced to 0.5 cfm per ft^2 of conditioned area instead of 1 cfm per ft^2 for vapor compression refrigeration systems.

- Because of the lower space relative humidity in a desiccant cooling system, the refrigeration load of the refrigerated cases may drop by 15 percent.

- As the space latent load is dehumidified by desiccant dehumidification, the refrigeration load for sensible cooling in a desiccant cooling system may be 20 to 50 percent less than the refrigeration load in the conventional system.

In desiccant cooling systems, the two airstreams in the rotary desiccant dehumidifiers and rotary heat exchangers must be properly sealed to reduce leakage and cross-contamination. Rotary dehumidifiers and heat exchangers should be cleaned periodically according to the manufacturer's instructions.

25.9 TWO-STAGE DESICCANT COOLING SYSTEMS

Two-Stage Systems

A *two-stage desiccant cooling system* has two rotary desiccant dehumidifiers connected in series, or one rotary desiccant dehumidifier and an additional rotary enthalpy exchanger connected in series to dehumidify the process air. Direct or indirect evaporative coolers; rotary sensible heat exchangers; necessary vapor compression refrigeration systems; waste heat recovery from engine jackets, exhaust gas, condensers, and solar collectors; fans; filters; and gas heaters may be used for both process air and regeneration airstreams.

Normally, when the drop in humidity ratio Δw during the dehumidification of process air in a desiccant cooling system is greater than 0.012 lb/lb (84 grains/lb), a two-stage desiccant cooling system is justified.

Case Study: A Two-Stage Desiccant Cooling System for a Fast-Food Restaurant

Meckler (1991) described a gas-energized two-stage desiccant cooling system for a fast-food restaurant. This restaurant, with a conditioned area of 2000 ft^2, was built in Brooklyn, New York. Outdoor air of 95°F dry bulb, 78°F wet bulb, and a humidity ratio of 0.0168 lb/lb is dehumidified and sensibly cooled in two successive stages at a volume flow rate of 4000 cfm. It is then evaporatively cooled in a direct evaporative cooler before being supplied to the conditioned space to maintain a space condition of 80°F and 50 percent relative humidity. No vapor compression refrigeration system is used for cooling in this system.

Regeneration is separated into two airstreams. Each is heated by a rotary sensible heat exchanger, a heat pipe heat exchanger, and a gas heater connected in series before it enters the rotary desiccant dehumidifier for reactivation.

Figure 25.16 shows the schematic diagram and the desiccant cooling cycle of this system. The operating parameters of the process air and regeneration air at the exits of various components are shown below:

Point	Temperature, °F	Humidity ratio, lb/lb	(grains/lb)
Process air			
o	95	0.0168	(118)
1d	137	0.0086	(60)
1x	86	0.0086	(60)
2d	121	0.0017	(12)
2x	86	0.0017	(12)
s	55	0.0086	(60)
r	80°F, 50 percent RH	0.011	(77)
Regeneration air			
1	80	0.0203	(142)
2	131	0.0203	(142)
3	139	0.0203	(142)
4	185	0.0203	(142)
5	143	0.0283	(198)
6	135	0.0283	(198)
7	115	0.0203	(142)
8	140	0.0203	(142)
9	150	0.0271	(190)
10	125	0.0271	(190)

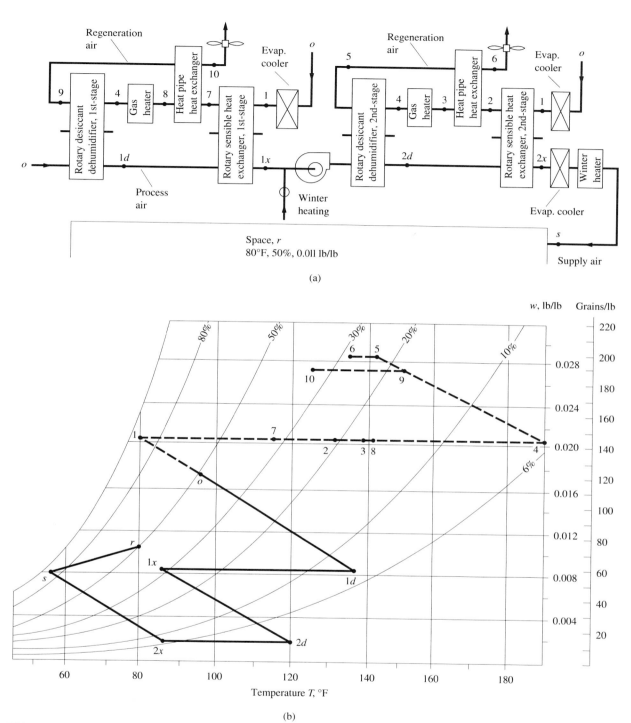

FIGURE 25.16 A two-stage desiccant cooling system for a fast-food restaurant: (a) schematic diagram and (b) desiccant cooling cycle.

System Performance

The rotary desiccant dehumidifier reduced the humidity ratio of the outdoor air up to a difference $\Delta w = 0.0082$ lb/lb.

The effectiveness of the rotary sensible heat ex-

changer varied from 0.85 to 0.89. The effectiveness of the heat pipe heat exchanger was between 0.66 and 0.7.

The coefficient of performance of this desiccant cooling system was 0.85. Although this value is slightly less than that of the two-stage direct-fired absorption chiller, using an absorption chiller for a refrigeration system of a capacity around 32 tons is not practical.

Utility rates in Brooklyn, New York in the early 1990s were as follows:

Electricity demand	June–September	$21.14/kW
	October–May	$18.84/kW
Energy rates	$0.0713 kWh at all times	
Gas rates	June–September	$5.00/MM Btu
	October–May	$6.00/MM Btu

Compared with a conventional rooftop packaged air-to-air heat pump system, the two-stage cooling system had an annual energy cost savings of about $7800.

If the two heat pipe heat exchangers had not been installed, the initial cost of the desiccant cooling system would have been lower, but the gas heating capacity and gas consumption of the system would have been greater. The decision to use the heat exchangers was determined by optimum cost analysis.

REFERENCES AND FURTHER READING

ASHRAE, *ASHRAE Handbook 1988, Equipment,* ASHRAE Inc., Atlanta, GA, 1988.

ASHRAE, *ASHRAE Handbook 1989, Fundamentals,* ASHRAE Inc., Atlanta, GA, 1989.

Banks, N.J., "Desiccant Dehumidifiers in Ice Arenas," *ASHRAE Transactions,* Part I, pp. 1269–1272, 1990.

Burns, P.R., Mitchell, J.W., and Beckman, W.A., "Hybrid Desiccant Cooling Systems in Supermarket Applications," *ASHRAE Transactions,* Part I B, pp. 457–468, 1985.

Denkmann, J.L., "Performance Analysis of a Brine-Based Ice Storage System," *ASHRAE Transactions,* Part I B, pp. 876–891, 1985.

Denkmann, J.L., "Cool Storage Retrofit of Rooftop Units and Direct Expansion Systems," *ASHRAE Transactions,* Part I, pp. 1067–1079, 1990.

Dorgan, C.E., and Elleson, J.S., "Design of Cold Air Distribution Systems with Ice Storage," *ASHRAE Transactions,* Part I, pp. 1317–1322, 1989.

Fields, W.G., and Knebel, D.E., "Cost Effective Thermal Energy Storage," *Heating/Piping/Air Conditioning,* pp. 59–72, July 1991.

Fiorino, D.P., "Case Study of a Large, Naturally Stratified, Chilled-Water Thermal Storage System," *ASHRAE Transactions,* Part II, pp. 1161–1169, 1991.

Gatley, D.P., "Successful Thermal Storage," *ASHRAE Transactions,* Part I B, pp. 843–855, 1985.

Gilberston, T.A., "Ice Cools Office-Hotel Complex," *Heating/Piping/Air Conditioning,* pp. 47–52, August 1989.

Gilbertson, T.A., and Jandu, R.S., "24-Story Office Tower Air Conditioning System Employing Ice Storage—A Case History," *ASHRAE Transactions,* Part I B, pp. 387–398, 1984.

Harmon, J.J., and Yu, H.C., "Design Consideration for Low-Temperature Air Distribution Systems," *ASHRAE Transactions,* Part I, pp. 1295–1299, 1989.

MacCracken, C.D., "Off-Peak Air Conditioning: A Major Energy Saver," *ASHRAE Journal,* pp. 12–22, December 1991.

Manley, D.L., Bowlen, K.L., and Cohen, B.M., "Evaluation of Gas-Fired Desiccant-Based Space Conditioning for Supermarkets," *ASHRAE Transactions,* Part I B, pp. 447–456, 1985.

Marciniak, T.J., Koopman, R.N., and Kosar, D.R., "Gas-Fired Desiccant Dehumidification System in a Quick-Service Restaurant," *ASHRAE Transactions,* Part I, pp. 657–666, 1991.

Meckler, G., "Use of Desiccant to Produce Cold Air in a Gas-Energized Cold Air HVAC System," *ASHRAE Transactions,* Part I, pp. 1257–1261, 1990.

Meckler, G., "Comparative Energy Analysis of a Gas-Energized Desiccant Cold-Air Unit," *ASHRAE Transactions,* Part I, pp. 637–640, 1991.

Parsons, B.K., Pesaran, A.A., Bharathan, D., and Shelpuk, B., "Improving Gas-Fired Heat Pump Capacity and Performance by Adding a Desiccant Dehumidification Subsystem," *ASHRAE Transactions,* Part I, pp. 835–844, 1989.

Pearson, F.J., "Ice Storage Can Reduce the Construction Cost of Office Buildings," *ASHRAE Transactions,* Part I, pp. 1308–1316, 1989.

Sohn, C.W., and Tomlinson, J.J., "Diurnal Ice Storage Cooling Systems," *ASHRAE Transactions,* Part I, pp. 1079–1085, 1989.

Spethmann, D.H., "Optimal Control for Cool Storage," *ASHRAE Transactions,* Part I, pp. 1189–1193, 1989.

Stamm, R.H., "Thermal Storage Systems," *Heating/Piping/Air Conditioning,* pp. 133–151, January 1985.

Tackett, R.K., "Case Study: Office Building Uses Ice Storage, Heat Recovery, and Cold Air Distribution," *ASHRAE Transactions,* Part I, pp. 1113–1121, 1989.

Townsend, S.B., and Asbury, J.G., "Cooling with Off-Peak Energy: Design Implications of Different Rate Schedules," *ASHRAE Transactions,* Part I B, pp. 360–373, 1984.

The Trane Company, "Ice Storage Systems 1987," *Applications Engineering Manual,* American Standard, Inc., La Crosse, WI, 1987.

Tran, N., Kreider, J.E., and Brothers, P., "Field Measurement of Chilled Water Storage Thermal Performance," *ASHRAE Transactions,* Part I, pp. 1106–1112, 1989.

Trueman, C.S., "Operating Experience with a Large Thermally Stratified Chilled-Water Storage Tank," *ASHRAE Transactions,* Part I, pp. 697–707, 1987.

Wildin, M.W., "Diffuser Design for Naturally Stratified Thermal Storage," *ASHRAE Transactions,* Part I, pp. 1094–1101, 1990.

CHAPTER 26

AIR SYSTEMS: CONSTANT-VOLUME AND FAN–COIL

26.1 AIR SYSTEMS

Air Systems

An *air system*, sometimes called an *air-handling system*, is a primary subsystem of an air conditioning system. Its main functions include:

- Conditioning the supply air: heating or cooling, humidification or dehumidification, cleaning and

purifying, and attenuation of objectionable noise produced by fans, compressors, and pumps

- Distributing the conditioned supply air with adequate outdoor air to the conditioned space, extracting the space air for recirculating, and exhausting or relieving unwanted space air to the outdoors

- Providing pressurization and smoke control for occupants' safety during building fires

- Controlling and maintaining required space temperature, humidity, cleanliness, air movement, sound level, and pressure differential within predetermined limits at optimum energy consumption and minimum cost

An air system consists of fans, heat exchangers (including coils, evaporative coolers, and dehumidifiers), a direct-fired gas heater, filters, a mixing box, dampers, ductwork, terminals, space diffusion devices, and controls.

In a central hydronic system, the air system includes an air-handling unit, ducts, terminals, space diffusion devices, a relief system, and controls. In a unitary packaged system, the air system includes an air handler (in which air is cooled and dehumidified by a DX-coil), ducts, terminals, space diffusion devices, a relief system, and controls. In a ducted core water-source heat pump, the air system includes a supply fan, a DX-coil, a filter, ductwork, space diffusion devices, and controls.

Mechanical ventilation systems for garages, exhaust air systems such as smoke exhaust systems, and mechanical exhaust systems for laboratories are also air systems or mechanical ventilation systems that transport air for cooling, dilution of space air contamination, or exhaust of toxic gas.

It is more convenient to define air systems as only for unitary packaged systems that include ductwork and space diffusion devices. In the room air conditioner of an individual air conditioning system, the air system, heating system, and refrigeration system are integrated into one packaged air conditioner. It is not necessary to distinguish them separately.

Terminals

A *terminal* is a factory-made device installed inside a ceiling plenum above the conditioned space or sometimes directly mounted adjacent to the external wall of the conditioned space.

Conditioned air or outdoor air is supplied to a terminal from the supply main duct through branch take-offs. It may be mixed with recirculated air, and is supplied to the control zone served by the terminal in a multizone air system.

A terminal may perform any of the following functions or a combination thereof:

- Heating/cooling or filtration
- Mixing of outdoor ventilation air or conditioned air with recirculating air, or mixing of cold supply air with warm supply air

- Controlling and maintaining a predetermined zone temperature by modulating the volume flow rate of supply air, the temperature of supply air, or the ratio of the volume flow of cold and warm air supplies

VAV boxes, reheating VAV boxes, mixing VAV boxes, fan-powered VAV boxes, and fan–coil units are all terminals, and will be discussed in this chapter and Chapter 27.

Primary Air, Secondary Air, and Transfer Air

Primary air is conditioned make-up outdoor air or a conditioned mixture of outdoor and recirculating air. Primary air is normally conditioned and supplied from a separate make-up AHU or PU to the terminals, and sometimes is supplied directly to the space.

Primary air is usually mixed with recirculating air, which is often called *secondary air*. The mixing most often takes place in the terminals or another AHU or PU, or sometimes directly in the conditioned space. The mixture of primary air and recirculating air may be conditioned again in the terminals. It is then supplied to the conditioned space to offset the space load. The ratio of the volume flow rate of primary air to the mixture of zone supply air varies from 0.15 to 0.5.

The purpose of mixing primary air with recirculating air in the terminal is to save fan energy, as well as to save money by using smaller supply and return ducts, which occupy less space.

Transfer air is indoor air that moves or supplies to a conditioned space from an adjacent area. Transfer air is entirely composed of recirculating air or secondary air. It is supply air that often has less unused outdoor air than a mixture of outdoor and recirculating air. Unused outdoor air is outdoor air whose capability of dilution of air contaminants has not been expended. Recirculating air still may contain a certain amount of unused outdoor air. Restrooms are normally supplied by transfer air.

Classification of Air Systems

Air systems can be classified into the following categories according to their system characteristics:

- All-air, air–water, or all-water systems. In an *all-air system*, space conditioning is provided by supply air only. In an *air–water system*, space conditioning is provided by the conditioned air and the chilled or hot water supplied to the terminal units installed above or in the conditioned space. In an *all-water system*, space conditioning is performed by supplying chilled or hot water to the terminal units.

- Single-zone or multizone systems. In a *single-zone system*, supply air is supplied to a conditioned space of a single control zone, in which only one zone control system is used to control the temperature, dew point, volume flow rate, or cleanliness of supply air to maintain the required parameters. In a *multizone system,* conditioned air is supplied to a space of multiple control zones. Terminal units and individual zone control systems are used to control the temperature, dew point, volume flow rate, or cleanliness of supply air to meet the variations of load and operating characteristics in each specific control zone.

- Single-duct or multiple-duct systems. In a *single-duct system*, conditioned air is transported to the conditioned space through a single supply duct. In a *dual-duct system,* conditioned air is transported to the conditioned space in the perimeter zone through two supply ducts: a warm air duct and a cold air duct. In the interior zone, only a cold air duct is needed.

- Constant-volume (CV) or variable-air-volume (VAV) systems. In a *constant-volume system*, the temperature of supply air is modulated to match the variation of space load during the part-load operation. In a *VAV system,* the supply air volume flow rate is modulated to maintain the required space temperature as space load varies.

In the late 1970s, after the energy crisis, VAV systems were widely used in commercial buildings in the United States. It is more appropriate to divide the currently used air systems into three main categories: constant-volume, variable-air-volume, and fan–coil systems. Fan–coil systems mainly use modulation of the flow rate of chilled water supplied to the coil, and thereby vary the supply temperature to match part-load operation. Fan speed can also be changed to HI-LO or HI-MEDIUM-LO modes manually or automatically by a DDC. Therefore, a fan–coil system should be considered an air system with both variable-volume and variable-supply-temperature (VVVT) characteristics.

If constant-volume and variable-air-volume air systems are again categorized by their operating characteristics, air systems can be classified mainly as follows:

- *Constant-volume.* Basic air systems and multizone, constant-volume systems with or without reheat

- *Fan–coil systems*
- *Variable-air-volume.* Single-zone VAV systems, perimeter-heating VAV systems, VAV reheat systems, dual-duct VAV systems, and fan-powered VAV systems

According to EIA *Characteristics of Commercial Buildings* (1986) and *Commercial Building Characteristics* (1991), during the late 1980s, the distribution of airsystems in commercial buildings in the United States, based on the area of floor space that they serve are approximately as follows:

Constant-volume systems	57%
Fan–coil systems	20%
Variable-air-volume systems	23%

Air-water induction unit systems were first developed in the 1930s and were widely used in multizone commercial buildings. However, because 25 to 30 percent primary air is needed to produce sufficient induced air from the conditioned space, induction units were replaced by fan–coil units after the energy crisis in new systems and in retrofit projects from the 1960s.

Performance of Air Systems

The performance of an air system can be analyzed and evaluated by the following:

- Space parameters that it maintains
- Psychrometric analysis of its air conditioning cycle
- Supply volume flow rate and coil's load
- Control and operating characteristics of the air-handling unit and air handler at full-and part-load
- Performance during emergencies such as building fires
- Evaluation of energy consumption
- Cost analysis

As described in Section 8.2, ASHRAE/IES Standard 90.1–1989 specifies that the power required by the fan motors in a constant-volume air system P_{cfm} must not exceed 0.8 W/cfm of supply air. In a VAV system, P_{cfm} must not exceed 1.25 W/cfm of supply air at design conditions. From Eq. (8.25), for a constant-volume system, the energy per unit volume flow is therefore

$$P_{cfm} = \frac{0.1175\Delta p_{sy}}{\eta_f \eta_m} \leq 0.8 \, \text{W/cfm} \qquad (26.1)$$

For VAV systems at design conditions,

$$P_{cfm} = \frac{0.1175\Delta p_{sy}}{\eta_f \eta_m} \le 1.25 \text{W/cfm} \qquad (26.2)$$

where Δp_{sy} = pressure drop of air system including coils and filters in the AHU or PU, terminal, and supply and return duct systems, in. WG

η_f, η_m = fan and motor efficiency

Air Transport Factor

Air transport factor $F_{a.t}$ is an index used to assess the energy effectiveness of an air system. $F_{a.t}$ is defined as the ratio of sensible heat removal to the energy input to the supply fan, return fan, and fans in terminal units. For an air density $\rho_a = 0.075$ lb/ft^3 and a specific heat $c_{pa} = 0.243$ Btu/lb·°F, $F_{a.t}$ can be calculated as

$$F_{a.t} = \frac{2.73\eta_f \eta_m \Delta T_s}{\Delta p_{sy}} \qquad (26.3)$$

where $\Delta T_s = (T_r - T_s)$, temperature difference between the space and supply air, °F

Although $F_{a.t}$ includes only sensible heat removal, the air transport factor is still a comprehensive factor that includes most of the primary parameters (Δp_{sy}, ΔT_s, η_f, and η_m) that affect the energy consumption of an air system.

26.2 BASIC AIR SYSTEMS

System Description

As described in Section 3.7, a basic air system is a single-zone, single-duct, all-air, constant-volume system.

In a basic air system, space air has only one state point on the psychrometric chart that must be maintained by its control system. The schematic diagram of a typical basic air system for a shopping mall and the corresponding air conditioning cycle during summer and winter operation are shown in Fig. 3.8. Usually, a packaged unit using a DX-coil for cooling and an electric heating coil for winter heating is used.

In summer, outdoor air is mixed with recirculating air, filtered and cooled in the packaged unit, and supplied to the mall. During spring and fall, 100 percent outdoor air may be used for air economizer operation. In winter, the mixture of outdoor and recirculating air is filtered and electrically heated in the packaged unit. In comfort air conditioning systems used in shopping malls, humidification is usually not provided in winter.

Supply Volume Flow Rate and Coil's Load

From Eq. (3.48b), the supply volume flow rate of this basic air system can be calculated as

$$\dot{V}_s = \frac{q_{rs}}{60\rho_s c_{pa}(T_r - T_s)}$$

From Fig. 3.8, supply temperature differential $(T_r - T_s)$ varies from 15 to 20°F depending on the required space, relative humidity, and the magnitude of supply system heat gain $(T_s - T_{cc})$.

At summer design conditions, the DX-coil's load, or refrigeration load, can be calculated using Eq. (3.21) as

$$q_{cc} = 60\dot{V}_s \rho_s (h_m - h_{cc})$$

where h_m, h_{cc} = enthalpy of mixture at point m and conditioned air off the coil, Btu/lb

If a water cooling coil is installed in the air-handling unit of a basic air system, its coil's load can be similarly calculated.

In a basic air system with warm air supply without space humidity control, the heating coil's load q_{ch} (in Btu/h) at winter design conditions can be calculated from Eq. (3.65) as

$$q_{ch} = 60\dot{V}_s \rho_s c_{pa}(T_{ch} - T_m)$$

where T_m, T_{ch} = temperature of mixture and the heated air off the heating coil, °F

In basic air systems in winter heating mode operation, the winter mode air conditioning cycle should be constructed on the psychrometric chart using the procedure described in Section 3.11. The heating coil's load and humidifying load can then be calculated based on the corresponding equations. In basic air systems used in shopping malls in locations with mild winters, heating is only required during the warm-up period. Cold-air supply is needed when the shopping mall is fully occupied.

Part-Load Operation and Controls

During cooling mode part-load operation, three types of controls are used to maintain the required parameters in the conditioned space:

- *Two-position control.* On-off control of the refrigeration compressor is used to adjust the cooling capacity of the air system in an operating cycle of a few minutes to 20 minutes. Both the air temperature off the coil and the space temperature fluctuate. Because of the storage capacity of the air system and the building structure in the conditioned space, the space temperature may fluctuate by 1 to 2°F, as shown in Fig. 17.4. Two-position control is usually used for air systems equipped with small compressors.

- *Cylinder unloader control.* When an air system includes medium and large multiple-size reciprocating compressors, cylinder unloader control is most widely used during part-load operation, as described in Section 21.5.

- *Flow rate modulation control.* For a water cooling coil in an air-handling unit, as described in Section 18.6, the reduction of space sensible cooling load causes a drop in space temperature T_r. When a drop in T_r is sensed, the DDC panel closes the two-way valve of the chilled water supply to maintain the required space temperature. The reduction of chilled water flowing through the coil raises the air temperature and humidity ratio off the coil. These effects result in a higher space relative humidity φ_{rp} at part-load than at design load, when space temperature T_r is maintained at a nearly constant value.

The part-load performance of a water cooling coil, as calculated in Example 12.3, is shown below:

	Design load	Part-load
Load ratio	1.0	0.8
Entering air	80°F dry/	79°F dry/
	67°F wet	68°F wet
Leaving air	57.5°F dry/	61.3°F dry/
	56.5°F wet	59.5°F wet
Water velocity	4 fps	2.5 fps
Water temperature		
increase	10°F	16°F
Space temperature	75°F	75°F
Space relative		
humidity	55 percent	59 percent

If the load ratio drops to 0.5, then space relative humidity at part load φ_{rp} may increase to 65 percent.

During winter mode part-load operation, when an increase in space temperature is detected, the DDC panel modulates the hot water flow in the water heating coil, modulates the gas supply to the gas-fired heater, or reduces the electric heating in steps to maintain the space temperature within predetermined limits.

Performance and Applications

Consider a basic air system, using a rooftop packaged unit, with the following operational parameters:

System pressure drop Δp_{sy}	3.75 in. WG
Supply temperature differential	20°F
Fan total efficiency η_f	0.75
Motor efficiency η_m	0.90

From Eq. (26.1), energy consumption per volume flow P_{cfm} for this basic system can be calculated as

$$P_{cfm} = \frac{0.1175 \Delta p_{sy}}{\eta_f \eta_m} = \frac{0.1175 \times 4}{0.75 \times 0.90} = 0.70 \text{ W/cfm}$$

Note that this is less than 0.8 W/cfm

From Eq. (26.3), the air transport factor is calculated as

$$F_{a.t} = \frac{2.73 \eta_f \eta_m \Delta T_s}{\Delta p_{sy}}$$
$$= \frac{2.73 \times 0.75 \times 0.90 \times 20}{3.75} = 9.83$$

Basic air systems are widely used in many commercial buildings using packaged units to serve large rooms or single-zone conditioned spaces. Basic air systems are also widely used in industrial applications such as textile mills, cleanrooms, and precision manufacturing.

The indoor air handler of a split packaged unit that is most widely used in residences to provide year-round heating and cooling is actually a single-zone basic air system. Although this air system serves living, dining, family, and bedrooms in a single-family residence, it has only one control system to maintain a representative room temperature. Manually adjusted volume dampers in the supply grilles may be used to vary the volume flow rate of supply air to various rooms through supply ducts to meet the average room loads. The volume flow rate of supply air in such an air system is usually less than 1500 cfm. Return air returns to the air handler from various rooms either by return ducts or through door undercuts, as shown in Fig. 14.5.

26.3 CONSTANT-VOLUME SYSTEMS WITH REHEAT

Reheating

Reheating is a process in which air is reheated after it has been cooled. Reheating is a simultaneous cooling and heating process in an air conditioning cycle. Figure 26.1 shows a multizone, all-air, constant-volume system with reheat.

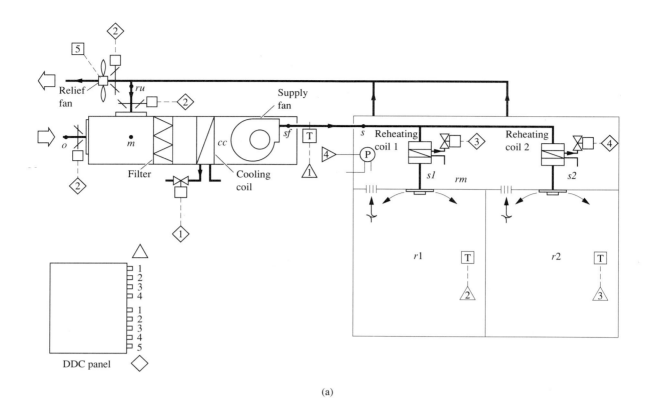

(a)

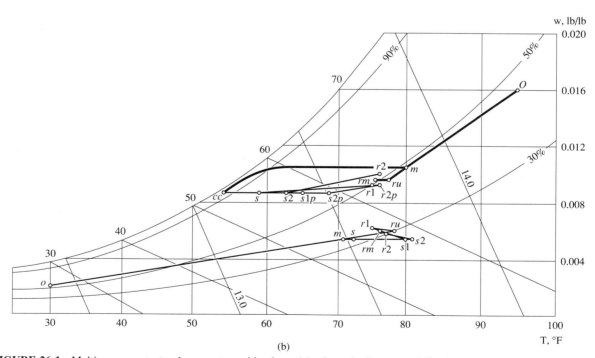

(b)

FIGURE 26.1 Multizone, constant-volume system with reheat: (*a*) schematic diagram and (*b*) air conditioning cycle.

In order to provide an energy-efficient constant-volume system design for new buildings (except low-rise residential buildings and buildings primarily for manufacturing and processing), ASHRAE/IES Standard 90.1–1989 specifies the following:

For zone controls, reheating control is prohibited except under the following conditions:

- VAV systems at minimum supply volume flow rate; this will be discussed in Chapter 27.
- Zones requiring pressurization to prevent cross-contamination in which VAV systems are impractical, such as in hospitals and laboratories
- At least 75 percent of the energy for reheating is provided from site-recovered or site-solar energy sources
- Zones where specified humidity levels are required, such as computer rooms and museums
- Zones with a supply air quantity at peak load of 300 cfm or less

Reheating is a simple and effective means of controlling space temperature and relative humidity at part-load. Reheating is used in VAV systems at minimum volume flow settings for comfort air conditioning, and also in hospitals, laboratories, computer rooms, museums, and precision manufacturing environments (such as semiconductor manufacturing plants) in order to maintain a close tolerance in space temperature ($\pm 1°F$) or space relative humidity for processing air conditioning.

Multizone, Constant-Volume System with Reheat

During summer mode operation, as shown in Figure 26.1, outdoor air at point *o* is mixed with recirculating air at point *ru*. The mixture *m* flows through the filter. It is then cooled and dehumidified to point *cc* at the cooling coil and discharged from the air-handling unit or packaged unit at the supply fan outlet *sf*. After absorbing the duct heat gain, the conditioned air enters the reheating coils RC1 and RC2 at point *s*.

During summer design load operation, RC1 is not energized. Air is supplied to room 1 at condition *s1*. Conditioned air is reheated at RC2 to maintain the required room condition in room 2. After absorbing the space cooling load, the supplied air becomes room air *r*1 and *r*2. Return air from room 1 and 2 is mixed together and forms mixture *rm* in the ceiling plenum. The return air is then returned to the AHU or PU, and the recirculating air enters the AHU at point *ru*. A portion of return air is exhausted through the relief fan to the outdoors.

During winter operation, outdoor air is mixed with recirculating air in a ratio such that its amount is always greater than the minimum outdoor air requirement. The mixture *m* flows through the filter, the deenergized cooling coil, and the supply duct, and is reheated at the reheating coils RC1 and RC2. The heated air is then supplied to rooms 1 and 2 to offset the heating load there. Both hot water and electric reheating coils can be used in multizone systems. When electric reheating coils are used, the design and installation requirements described in Section 14.4 and the requirements of the National Electric Code and local codes must be followed.

Control Systems

As described in Sections 18.8 and 18.9, the control subsystems for the air-handling unit or packaged unit in a multizone, constant-volume system with reheat may include the following:

- Discharge air temperature control
- Minimum outdoor air control
- Space pressurization and relief fan control
- Air economizer control
- Zone smoke control, which will be covered in Chapter 28

If there are only a few terminals, both the AHU and the reheating terminals can be controlled by a single DDC panel. If there are many reheating coils, a DDC panel controls the AHU and each reheating coil is controlled by a DDC controller.

When room temperature sensors T2 and T3 sense a drop in either room temperature below a limit, a signal is sent to the DDC panel. The reheating coil is then energized. For a hot water reheating coil, the two-way valve is modulated so that room temperature is maintained within predetermined limits. For an electric reheating coil, heating is added or decreased in steps. To prevent fires, electric reheating is energized only when air flow is sensed by a flow switch.

Operating Parameters and Calculations

During summer operation, the operating parameters and the corresponding calculations for a multizone, constant-volume system with reheat may be determined as follows:

1. Zone supply air condition should be determined so that the space temperature and relative humidity in each zone or room can be maintained at design-load operation by using the reheating processes.

2. The zone n supply mass flow rate $\dot{m}_{an}$ (in lb/min) and volume flow rate $\dot{V}_{an}$ (in cfm) at summer design-load can be calculated as

$$\dot{m}_{an} = \frac{q_{rsn}}{60 c_{pa}(T_{rn} - T_{rs})} \quad (26.4)$$

and

$$\dot{V}_{an} = v_s \dot{m}_{an} \quad (26.5)$$

where q_{rsn} = sensible cooling load for zone n, Btu/h

T_{rn} = temperature of zone n,°F

T_{rs} = supply temperature,°F

$v_s = 1/\rho_s$, moist volume of supply air, ft³/lb

3. The zone n supply temperature at minimum part-load T_{snp}, in °F, can be calculated as

$$T_{snp} = T_{rnp} - \frac{q_{rsnp}}{60 \dot{m}_{an} c_{pa}} \quad (26.6)$$

where T_{rnp} = temperature of zone n at part-load, °F.

4. The zone n reheating coil's load q_{hn}, in Btu/h, at minimum part-load can be calculated as

$$q_{hnp} = 60 \dot{m}_{an} c_{pa}(T_{snp} - T_{sn}) \quad (26.7)$$

where T_s = system supply air temperature entering reheating coil, °F.

System supply temperature T_s varies along the supply main duct. For simplicity, T_s can be taken as an average value.

5. The weighted mean value of zone temperature T_{rm}, in °F, and mean humidity ratio w_{rm}, in lb/lb, can be calculated as

$$T_{rm} = \frac{\dot{m}_{a1} T_{r1} + \dot{m}_{a2} T_{r2} + \cdots + \dot{m}_{an} T_{rn}}{\dot{m}_{a1} + \dot{m}_{a2} + \cdots + \dot{m}_{an}} \quad (26.8)$$

and

$$w_{rm} = \frac{\dot{m}_{a1} w_{r1} + \dot{m}_{a2} w_{r2} + \cdots + \dot{m}_{an} w_{rn}}{\dot{m}_{a1} + \dot{m}_{a2} + \cdots + \dot{m}_{an}} \quad (26.9)$$

where $\dot{m}_{a1}, \dot{m}_{a2}, \ldots, \dot{m}_{an}$ = mass flow rate of supply air for zones $1, 2, \ldots, n$, lb/min

$T_{r1}, T_{r2}, \ldots, T_{rn}$ = air temperature in zones $1, 2, \ldots, n$, °F

$w_{r1}, w_{r2}, \ldots, w_{rn}$ = air humidity ratio in zones $1, 2, \ldots, n$, lb/lb

6. The system supply volume flow rate $\dot{V}_s$ (in cfm) of the AHU or PU is calculated as

$$\dot{V}_s = \dot{V}_{s1} + \dot{V}_{s2} + \cdots + \dot{V}_{sn} \quad (26.10)$$

where $\dot{V}_{s1}, \dot{V}_{s2}, \ldots, \dot{V}_{sn}$ = supply volume flow rate for zones $1, 2, \ldots, n$, cfm

7. The condition of recirculating air entering the AHU or PU, point ru, and the mixture of outdoor and recirculating air, point m, can be determined graphically on the psychrometric chart. The cooling coil's load q_{cc}, in Btu/h, can therefore be calculated as

$$q_{cc} = 60(\dot{m}_{a1} + \dot{m}_{a2} + \cdots + \dot{m}_{an})(h_m - h_{cc}) \quad (26.11)$$

where h_m, h_{cc} = enthalpy of the mixture and the conditioned air off the coil, Btu/lb

8. At part-load, the condition of zone supply air point snp can be determined because snp must lie on horizontal line $cc\,snp$. Zone supply air temperature at part-load T_{snp} can be calculated from Eq. (26.6).

Draw a line from snp with the known sensible heat ratio for zone n at part-load, SHR$_{np}$. This line intersects zone temperature line T_{rnp} at point rnp, which is the state point of zone air at part-load. Zone relative humidity φ_{rmp} can then be determined.

The operating parameters during winter mode operation at design-load can then be determined as follows:

1. At winter design-load, $w_{rm} = w_{ru}$ and $w_s = w_m$. Here, w_{ru} indicates the humidity ratio of the recirculating air, w_s the humidity ratio of system supply air before entering the reheating coil, and w_m the humidity ratio of the mixture of the outdoor air and recirculating air, all in lb/lb. Then,

$$w_{ru} = w_s + \frac{q_{rl}}{60(\dot{m}_{a1} + \dot{m}_{a2} + \cdots + \dot{m}_{an}) h_{fg.o}} \quad (26.12)$$

Also,

$$\frac{w_{ru} - w_m}{w_{ru} - w_o} = \frac{\dot{V}_o}{\dot{V}_s} \quad (26.13)$$

where $w_o, \dot{V}_o$ = humidity ratio and volume flow rate of outdoor air, lb/lb and cfm.

2. Zone supply temperature T_{sn}, in °F, can be calculated as

$$T_{sn} = T_{rn} + \frac{q_{rhn}}{60 \dot{V}_{sn} \rho_s c_{pa}} \quad (26.14)$$

where q_{rhn} = space heating load for zone n, Btu/h.

3. The reheating coil load for zone n during winter mode operation q_{hn}, in Btu/h, is

$$q_{hn} = 60\dot{m}_{an}c_{pa}(T_{sn} - T_s) \qquad (26.15)$$

Usually, the reheating coil's load during winter mode operation is often greater than in summer mode operation, and should be taken as the design capacity of the reheating coil. However, it is necessary to determine which value is greater.

System Performance

During summer cooling mode operation, the space relative humidity of a multizone, constant-volume system with reheat at part-load is always lower than at design-load.

The energy consumption per unit volume flow rate and the air transport factor for a multizone, constant-volume system with reheat are greater than those in basic air systems because of the addition of reheating coil load, both at design- and part-load.

> **Example 26.1.** A multizone, constant-volume system with reheat for a precision manufacturing factory is operated under following conditions:

	Zone 1	Zone 2
Zone temperature, °F	75	76
Summer relative humidity, percent	50	53
Summer space cooling load, Btu/h	10,000	12,000
Summer space latent load, Btu/h	9000	8400
Summer minimum part-load, Btu/h	6000	7600
Winter heating load, Btu/h	3000	3800
Winter latent load, Btu/h	2200	2000
Summer outdoor design condition	95°F dry/ 77°F wet	
Winter outdoor design condition	30°F dry and 0.002 lb/lb	
Summer supply system heat gain, °F	5	5
Summer return system heat gain, °F	2	2
Winter supply system heat gain, °F	2	2
Winter return system heat gain, °F	1	1

The minimum outdoor air requirement is 15 percent of the total supply air. Assume that the relative humidity of the conditioned air leaving the cooling coil at summer design conditions is 95 percent. Also, the sensible ratios of the space conditioning lines SHR$_s$ of zones 1 and 2 at minimum part-load are the same as at summer design conditions.

Calculate:

1. The supply flow rate for zones 1 and 2
2. The reheating coil's load at minimum part-load during cooling mode operation
3. The cooling coil's load at summer design conditions
4. The reheating coil's load at winter design conditions

Solution

1. Because the winter heating loads are smaller than the summer cooling loads, the supply volume flow rate for zones 1 and 2 is determined according to the requirement to offset the summer cooling load.

From the given data, the sensible heat ratio of the space conditioning line for zone 1 is

$$\text{SHR}_{r1} = \frac{q_{rs}}{q_{rc}} = \frac{9000}{10,000} = 0.9$$

For zone 2, it is

$$\text{SHR}_{r2} = \frac{8400}{12,000} = 0.7$$

Draw space conditioning lines with their corresponding SHR$_s$ from zone state points $r1$ and $r2$. From Fig. 26.1b, the required zone temperatures and relative humidities can be provided if the supply air condition is determined according to line $s\,r1$.

Because $(T_s - T_{cc}) = 5°F$ and $cc\,s$ is a horizontal line, $T_{cc} = 54.5°F$, $\varphi_{cc} = 95$ percent, and $h_{cc} = 22.5$ Btu/lb. System supply temperature $T_s = 59.5°F$ and $\varphi_s = 81$ percent.

2. The condition of supply air to zone 2, point $s2$, must lie on horizontal line $cc\,s2$. It intersects space conditioning line $s2\,r2$ at point $s2$. From the psychrometric chart, $T_{s2} = 63.5°F$ and $\varphi_{s2} = 69$ percent.

The mass flow rate of supply air for zone 1 is therefore

$$\dot{m}_{a1} = \frac{q_{rs1}}{60c_{pa}(T_{r1} - T_s)}$$

$$= \frac{10,000}{60 \times 0.243(75 - 59.5)} = 44.25 \text{ lb/min}$$

For a moist volume $v_s = 1/\rho_s = 1/0.075 = 13.33$ ft³/lb, the supply volume flow rate for zone 1 is

$$\dot{V}_{s1} = 13.33 \times 44.25 = 590 \text{ cfm}$$

Similarly,

$$\dot{m}_{a2} = \frac{12,000}{60 \times 0.243 (76 - 63.5)} = 65.84 \text{ lb/min}$$

$$\dot{V}_{s2} = 13.33 \times 65.84 = 878 \text{ cfm}$$

3. From Eq.(26.6), the supply air temperature at minimum part-load for zone 1 is

$$T_{s1p} = T_{r1p} - \frac{q_{rs1p}}{60\dot{m}_{a1}c_{pa}}$$

$$= 75 - \frac{6000}{60 \times 44.25 \times 0.243} = 65.7°F$$

Supply air temperature at minimum part-load for zone 2 is

$$T_{s2p} = 76 - \frac{7000}{60 \times 65.84 \times 0.243} = 68.7°F$$

From Eq.(26.7), the reheating coil load at summer minimum part-load operation for zone 1 is

$$q_{h1p} = 60\dot{m}_{an}c_{pa}(T_{snp} - T_{sn})$$

$$= 60 \times 44.25 \times 0.243\,(65.7 - 59.5)$$

$$= 4000 \text{ Btu/h}$$

The reheating coil load at summer minimum part-load operation for zone 2 is

$$q_{h2p} = 60 \times 65.84 \times 0.243\,(68.7 - 59.5)$$

$$= 8832 \text{ Btu/h}$$

4. From Eq. (26.8), the weighted mean temperature of conditioned space at summer design-load is

$$T_{rm} = \frac{\dot{m}_{a1}T_{r1} + \dot{m}_{a2}T_{r2}}{\dot{m}_{a1} + \dot{m}_{a2}}$$

$$= \frac{(44.25 \times 75) + (65.84 \times 76)}{44.25 + 65.84}$$

$$= 75.6°\text{F}$$

From Eq. (26.9), the mean space humidity ratio is

$$w_{rm} = \frac{\dot{m}_{a1}w_{r1} + \dot{m}_{a2}w_{r2}}{\dot{m}_{a1} + \dot{m}_{a2}}$$

$$= \frac{(44.25 \times 0.0092) + (65.84 \times 0.010)}{44.25 + 65.84}$$

$$= 0.0097 \text{ lb/lb}$$

Because $T_{ru} = T_{rm} + 2 = 75.6 + 2 = 77.6°\text{F}$, and $w_{ru} = w_{rm} = 0.0097$ lb/lb, point ru can be plotted on the psychrometric chart. Draw line ru o. As ru m/ru $o = 0.15$, point m can be determined. From the psychrometric chart, $T_m = 80°\text{F}$ and $h_m = 30.9$ Btu/lb.

Therefore, the cooling coil's load at summer design condition is

$$q_{cc} = 60(\dot{m}_{a1} + \dot{m}_{a2})(h_m - h_{cc})$$

$$= 60(44.25 + 65.84)(30.9 - 22.5)$$

$$= 55,485 \text{ Btu/h}$$

5. From the given data, at winter mode operation, $w_o = 0.0022$ lb/lb and

$$T_{ru} = T_{rm} + 1 = 75.6 + 1 = 76.6°\text{F}$$

Also, because $w_{ru} = w_{rm}$ and $w_s = w_m$, from Eq. (26.12),

$$w_{rm} - w_s = \frac{q_{rl}}{[60(\dot{m}_{a1} + \dot{m}_{a2})h_{fg.o}]}$$

$$= \frac{2200 + 2000}{60(44.25 + 65.84)1061}$$

$$= 0.0006 \text{ lb/lb}$$

$$w_{ru} = w_o + \frac{w_{rm} - w_s}{\dot{V}_o/\dot{V}_s} = 0.0022 + \frac{0.0006}{0.15}$$

$$= 0.0062 \text{ lb/lb}$$

Point ru can be determined on the psychrometric chart.

Also,

$$w_m = w_s = w_{ru} - 0.15\,(w_{ru} - w_o)$$

$$= 0.0062 - 0.15\,(0.0062 - 0.0022)$$

$$= 0.0056 \text{ lb/lb}$$

and

$$T_m = T_{ru} - 0.15\,(T_{ru} - T_o)$$

$$= 76.6 - 0.15\,(76.6 - 30) = 69.6°\text{F}$$

6. If the difference in supply air densities is ignored, from Eq. (26.14), the supply air temperature for zone 1 is

$$T_{s1} = T_{r1} + \frac{q_{rhn}}{60\dot{m}_{a1}c_{pa}}$$

$$= 75 + \frac{3000}{60 \times 44.25 \times 0.243} = 79.7°\text{F}$$

Similarly, the supply temperature for zone 2 is

$$T_{s2} = 76 + \frac{3800}{60 \times 65.84 \times 0.243} = 80.0°\text{F}$$

The temperature of air entering the reheating coil's is

$$T_s = T_m + 2 = 69.6 + 2 = 71.6°\text{F}$$

Therefore, the reheating coil's load at winter mode operation for zone 1 is

$$q_{h1} = 60\dot{m}_{a1}c_{pa}\,(T_{s1} - T_s)$$

$$= 60 \times 44.25 \times 0.243(79.7 - 71.6)$$

$$= 5226 \text{ Btu/h}$$

Similarly, the reheating coil's load for zone 2 is

$$q_{h2} = 60 \times 65.84 \times 0.243\,(80 - 71.6)$$

$$= 8064 \text{ Btu/h}$$

The required reheating coil capacity for zone 1 is 5226 Btu/h. Because the summer reheating coil load at minimum part-load for zone 2 is greater than at winter load, the required capacity of the reheating coil for zone 2 is 8832 Btu/h.

26.4   FAN–COIL SYSTEMS

System Description

A *fan–coil unit system* or simply a fan–coil system uses chilled or hot water supplied to the fan–coil units to cool, dehumidify, filter, or heat the air. Conditioned air is then discharged to the conditioned space to offset the space cooling or heating load so that space temperature can be maintained within preset limits, as shown in Fig. 26.2.

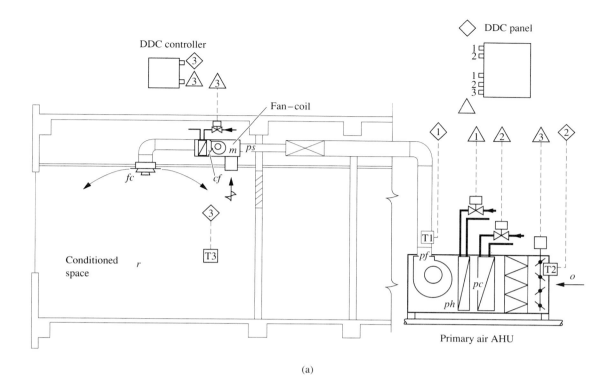

(a)

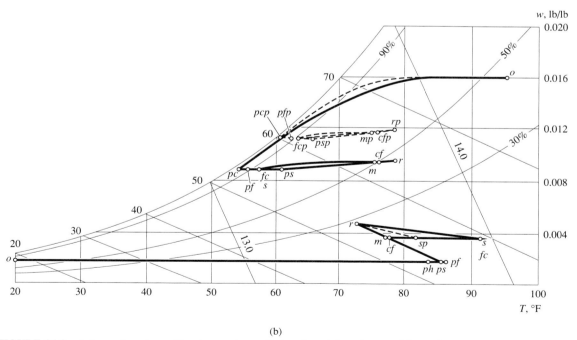

(b)

FIGURE 26.2 A fan–coil system with primary air supplied to the mixing plenum of fan–coil unit: (*a*) schematic diagram and (*b*) air conditioning cycle.

A fan–coil system consists of fan–coils, a primary air supply system, a chilled or hot water supply system, an air filter, supply ducts, diffuser(s), exhaust systems, controls, and accessories.

Outdoor air is often cooled and dehumidified, heated, or sometimes even humidified in a separate air-handling unit called a *primary air AHU*. The primary air is then transported to the fan–coil units or directly to the conditioned space via primary air supply ducts. Primary air is occasionally a mixture of outdoor air and recirculating air. Primary air is then mixed with the secondary air (the recirculating air) either in the fan–coil unit or in the conditioned space.

There are also fan–coil units in which untreated outdoor air is extracted directly through the external wall openings connected to the fan–coil. The amount of outdoor air intake in such a setup fluctuates because of wind speed and direction. Too much or too little outdoor air intake in such a setup is undesirable.

Primary air containing recirculating air is not economical because recirculating air must be transported back to the primary air AHU and conditioned there instead of in the fan–coil. Normally, there are no return fans or ducts in fan–coil systems.

Primary air supplied to the conditioned space may be balanced through the exhaust air from the restrooms or through the exhaust system to the outdoors.

Chilled water or hot water is supplied to the fan–coil through a two-pipe or four-pipe water system. In locations with a moderate winter, or where electricity rates are favorable in winter, an electric heating coil in each fan–coil unit is sometimes used for winter heating. This greatly simplifies the operation of fan–coil systems.

Air Flow Process and Types of Primary Air Intake

During summer cooling mode operation, outdoor air at point *o* is cooled and filtered in the primary air AHU. The primary air leaves the cooling coil at point *pc*, flows through the supply fan outlet at point *pf*, and is transported by the supply duct to the fan–coil or directly to the conditioned space at point *ps*. The primary air can be mixed with recirculating air, or secondary air, in three ways:

- Primary air at point *ps* is supplied to the mixing plenum of the fan–coil. It is then mixed with recirculating air, which is extracted by the fan in the fan–coil through the filter and forms mixture *m*. Air pressure at the mixing plenum is slightly less than the space pressure so that space air can be extracted into the plenum.

  The mixture then flows through the fan and coil in the fan–coil unit, cools and dehumidifies at the coil, and leaves the coil at point *fc*. Air that has been conditioned in the fan–coil is supplied to the conditioned space to offset the space load. State point f_c may or may not be on line *pc ps*, but will be near point *pc*.

- Primary air at point *ps* is supplied to the conditioned space directly. In this case, space air is filtered, and cooled and dehumidified (or heated) first. The conditioned space air discharged from the fan–coil at point *fc* is mixed with the primary air in the space. It is assumed that the mixture at point *m* offsets the space load. In such an arrangement, the recirculating airstream recirculates along its own circuit.

- Primary air at point *ps* is supplied just before the supply outlet of the fan–coil. It mixes with the conditioned air from the fan–coil at point *fc*. The psychrometric cycle of this arrangement is approximately the same as that of primary air supplied directly to the conditioned space.

Primary air supplied to the mixing plenum or before the supply outlet of the fan–coil has a more even distribution of outdoor air in the conditioned space. On the other hand, primary air supplied directly to the conditioned space has a shorter primary air supply duct and it is better to turn off one of the fan–coil units during capacity control at part-load when multiple fan–coil units are installed in a large room.

The temperature increase from fan power heat gain in the fan–coil unit is about 0.5°F for a permanent-slit capacitor fan motor and about 0.8°F for a shaded-pole fan motor. The supply duct in the ceiling plenum after the fan–coil is usually very short, so such a duct heat gain or loss is negligible.

During winter mode operation, in extremely cold weather, primary air may be preheated to point *ph* by the heating coil in the primary air AHU. After absorbing the supply fan power to point *pf* and releasing the primary air supply duct heat loss to point *ps*, the primary air is then mixed with recirculating air at point *r* and forms a mixture *m*, as shown by line *ps m r* in Fig. 26.2b. The mixture is heated to point *cf* and then point *s* by the fan power heat gain and the heat released from the coil, and is supplied to the conditioned space.

Primary air can also be mixed with heated air from the fan–coil when primary air is supplied directly to the conditioned space.

In locations with a mild winter, only primary air is heated in primary air AHU. This will be discussed later in this chapter.

Fan–Coil Unit

A *fan–coil unit*, or a fan–coil, is a terminal unit installed directly inside the conditioned space or in the

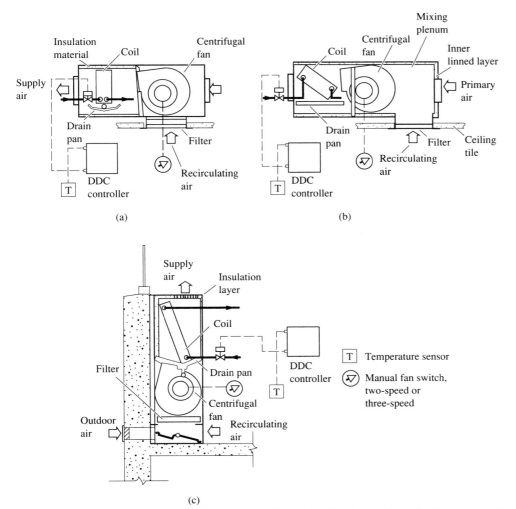

FIGURE 26.3 Fan–coil units: (*a*) horizontal unit, (*b*) horizontal unit with inner lined plenum, and (*c*) vertical unit.

ceiling plenum just above the conditioned space. A fan–coil unit includes a small motor-driven centrifugal fan or two small centrifugal fans connected in parallel, a finned coil, a filter, an outer casing, and controls. Sometimes, a cooling coil and a heating coil may be connected in series along the air flow.

A fan–coil unit can be a horizontal unit installed inside the ceiling plenum or a vertical unit mounted on the floor under the windowsill. Vertical units are usually used to offset the cold draft on the inner surface of window glass or on the external wall during cold weather. Cold drafts often flow downward along the glass because they are heavier than the surrounding air. Warm air discharged from a fan–coil during winter heating raises the inner surface temperature of window glass. Horizontal and vertical fan–coil units are shown in Fig. 26.3.

Fan–coil units are available in standard sizes 02, 03, 04, 06, 08, 10, and 12. Size 02 means a nominal flow rate of 200 cfm, 04 means 400 cfm, and so on.

FAN. Double-width, double-inlet forward-curved centrifugal fans are usually used because of their compact size and lower noise level. The fan wheels are usually made of aluminum or galvanized steel, with a diameter less than 10 inches in most cases. Fan housings are die-formed with integral scrolls and inlets.

FAN MOTOR. Permanent-split capacitor motors or shaded-pole motors are normally used. A fan–coil unit with a nominal volume flow rate of 200 cfm requires a power input of 0.125 hp for a permanent-split capacitor motor, or 0.25 hp for a shaded-pole motor.

Two-speed HI-LO switches or three-speed HI-MED-LO switches are used to vary fan speed manually or automatically (by a DDC controller). Fan motors are generally protected by a thermal overload protector. Periodic oiling of the bearing (twice a year) is often required.

COILS. Coils are usually made from copper tubes and aluminum fins. Cooling/heating coils usually have 2, 3, or 4 rows of fins, depending on the coil's cooling capacity and the sensible heat ratio of the cooling and dehumidifying process SHR_c. Two-row coils or three-row coils are widely used. Four-row coils have a greater dehumidifying capacity.

Usually, there is only one coil for both heating and cooling. A separate electric heating coil is sometimes used with two-stage step control in locations where the heating season is short or electricity demand and energy rates are low in winter.

Manual air vents are installed to prevent the formation of air pockets inside the water circuit. A galvanized steel pan with an insulating liner is often used to drain the condensate during dehumidification and to prevent outer surface condensation on the fan–coil unit.

To reduce pressure loss, the face velocity of the air flowing through the coil is usually from 200 fpm to 300 fpm.

FILTERS. Usually, low-efficiency, low-pressure-drop, permanent filters are used. They are easy to clean and replace periodically. Sometimes, disposable, low-efficiency fiberglass filters are also used.

CASING. The external cabinet is usually made of 18-gauge galvanized steel sheet with a corrosion-resistant surface coating. The cabinet is lined with insulation to prevent outer surface condensation.

Although there are fans, water coils, and filters in both fan–coil units and air-handling units, a fan–coil unit is distinguished from an AHU by the following characteristics:

	Fan–coil	**AHU**
Classification of equipment	Terminal unit	Basic equipment in air system
Location	Under window sill or in ceiling plenum	Fan room
Volume flow, cfm	≤ 1200	1200–50,000
Fan total pressure, in. WG	< 0.6	< 6
Sound power level, dB	Lower	Higher
Filter efficiency	Low	Low, medium, high
Coil row depth	2, 3, or 4 rows	2, 3, 4, 6, 8, or 10 rows
Fin spacing	Fixed	Custom made
Size of unit	Based on cooling capacity	Based on volume flow rate

Volume Flow Rate

The volume flow rate of a fan–coil unit is affected by the following factors:

- The external pressure drop from the supply duct and the supply outlet after the fan–coil
- The position of the fan switch, whether it is HI-LO or HI-MED-LO
- The elevation of the fan–coil above sea level

The *nominal flow rate* is the flow rate of a fan–coil whose external pressure drop is at a specific value when the fan switch is in the HI position and the fan-coil is at sea level. The higher the external pressure drop, the lower the volume flow rate. If the external pressure drop increases from 0.06 in. WG to 0.3 in. WG, the volume flow rate of the fan–coil may decrease to 55 percent of its nominal value.

For a typical fan–coil, the volume flow rate of fan switch position MED is about 80 percent of the HI value, and LO is only 70 percent of the HI value.

Cooling and Dehumidifying

Because the face velocity of a specific fan–coil unit at nominal volume flow rate is nearly the same for various sizes and the outer surface area of a water cooling coil is divided into dry and wet parts during cooling and dehumidifying, the coil's capacity and the sensible heat ratio of the cooling dehumidifying process SHR_c depends on the following factors:

- Dry and wet bulb of entering air
- Entering water temperature
- Water temperature rise in the coil
- The surface area and number of rows, including both pipe surface and fin area, of the coil

For a 04 fan–coil unit with an entering air temperature of 80°F dry bulb/67°F wet bulb and an entering water temperature of 45°F, if the water temperature rise is 10°F, its cooling and dehumidifying capacity q_{cc} varies from 11 to 14 MBtu/h and SHR_c varies from 0.65 to 0.80.

If the volume flow rate of a fan–coil unit deviates from the nominal value because of a greater external pressure drop or a higher altitude, its total cooling capacity should be corrected as follows:

$$q_{c.c} = C_p C_a q_{c.r} \qquad (26.16)$$

where $q_{c.c}$ = corrected total cooling capacity, MBtu/h

$q_{c.r}$ = manufacturer's catalog listed cooling capacity at a specific nominal volume flow rate, in MBtu/h

C_p, C_a = total cooling capacity correction factor for excessive external pressure drop and high altitude, respectively

The corrected sensible heating capacity of the fan–coil $q_{cs.c}$, in Mbtu/h, is

$$q_{cs.c} = C_{ps}C_{as}q_{cs.r} \qquad (26.17)$$

where $q_{cs.r}$ = manufacturer's catalog listed sensible cooling capacity at nominal volume flow rate, MBtu/h

C_{ps}, C_{as} = sensible cooling capacity correction factor for excessive external pressure and high altitude, respectively

Correction factors C_p, C_{ps}, C_a, and C_{as} can be found in the manufacturer's catalog. If these data are not available, the following values can then be estimated as C_p and C_{ps}:

External pressure drop, in. WG	C_p	C_{ps}
0.06	1	1
0.1	0.96	0.96
0.2	0.84	0.83
0.25	0.76	0.73

If the sensible heat factor of the cooling and dehumidification process $0.5 \le \text{SHR}_c \le 0.95$, for each 1000 ft higher than sea level up to an altitude of 10,000 ft,

$$C_a = 1 - \frac{0.01(\text{altitude, ft})}{1000}$$
$$C_{as} = 1 - \frac{0.03(\text{altitude, ft})}{1000} \qquad (26.18)$$

Part-Load Operation

During cooling mode part-load operation, as described in Section 18.6, when a drop in space temperature T_r is detected, the DDC controller closes the two-way valve and the chilled water flow rate decreases. If the space sensible load is reduced to 50 percent of design load, from Fig. 26.2b, the supply air from the fan–coil point *fcp* is raised to about 63°F, so that:

- Space temperature T_r is maintained at the same value as at design-load.
- The space sensible load at part-load is offset by both the fan–coil and the primary air.

- The supply temperature differential at part-load $(T_r - T_{sp})$ is greater than half of the design supply temperature differential $(T_r - T_s)$ because the fan speed is often in the LO position.
- The space relative humidity at summer mode part-load may increase to about 65 percent.
- For a room or conditioned space in the perimeter zone, the sensible heat ratio of space conditioning line SHR$_s$ is usually 0.9 to 0.97 at design-load. When sensible cooling load drops to 50 percent of the design-load value, SHR$_s$ usually falls to 0.75 to 0.85.
- When the chilled water temperature increase ΔT_w increases from 8 or 10°F to 20°F or even higher during part-load operation, the sensible heat ratio of the cooling and dehumidifying process SHR$_c$ in the fan–coil increases accordingly.
- Psychrometric analysis indicates that the chilled water supplied to the primary air AHU and the fan–coil units should be reset at a higher temperature during cooling mode part-load operation so that SHR$_c$ in the fan–coil unit increases at part-load and power input to the compressor is reduced.

When the chilled water supply temperature to the fan–coil $T_{w.f}$ and to the primary air AHU $T_{w.p}$ are both 45°F and $\Delta T_w = 10$°F at design conditions, for each 10 percent reduction of space sensible cooling load, $T_{w.f}$ and $T_{w.p}$ should be reset 1°F higher.

During heating mode part-load operation, when the space sensor detects an increase in space temperature T_r, the mass flow rate of the hot water supply to the fan–coil is modulated to maintain a predetermined space temperature, as in winter design-load.

Heating Capacity

The heating capacity of the selected fan–coil can be found in the manufacturer's catalog. Usually, for the same fan–coil, a greater heating capacity can be provided at winter design conditions because the temperature difference between the hot water and heated air is higher than between water and cooled air.

Sound Power Level of Fan–Coil Units

Because the fan–coil unit is usually located inside the ceiling plenum or directly under the windowsill, the room effect and the short supply duct are often not sufficient to attenuate fan noise in the fan–coil unit. Therefore, sound power level is often an important factor to consider during the selection of a fan–coil unit.

For a typical fan–coil unit of size 02 to 12, the sound power level L_p rating, measured in a reverberant room according to ARI Standard 443–70 for an octave band with a mid-frequency of 1000 Hz, varies from 45.5 to 52 dB. For a fan–coil unit of size 04, because $L_p = 47.5$ dB for an octave band with a mid-frequency of 1000 Hz, if the room effect is 7.5, the space NC level is about 40. The greater the size of the fan–coil unit, the higher the NC level. Refer to the manufacturer's catalog for details.

26.5   FAN–COIL SYSTEM CHARACTERISTICS

Two-Pipe and Four-Pipe Systems

The water systems used for fan–coils can be classified into two categories: two-pipe systems and four-pipe systems. The three-pipe system was discontinued because of energy loss in the common return pipe. For convenient operation and lower initial cost, current fan–coil systems often use the same temperature of chilled water supply to both the fan–coils and the water cooling coil in primary air AHUs.

A two-pipe system equipped with a supply pipe and a return pipe is shown in Fig. 26.4. In such a system, chilled water is supplied to the coil to cool and dehumidify the air during cooling mode operation. In heating mode operation, chilled water is changed over to hot water and is then supplied to the coil to heat the air.

A two-way valve is usually installed before the coil inlet because it costs less, is easier to install, and saves pump power at part-load operation when the water flow rate is reduced. A DDC controller is often used to modulate the water flow at part-load, as described in Section 18.6.

A four-pipe system is equipped with two supply mains: a chilled water supply, and a hot water supply. There are also two return mains: a chilled water return and a hot water return, as shown in Fig. 26.5.

The finned coil may be a *common coil,* as shown in Fig. 26.5a, or two separate coils, a cooling coil of 2 or 3 rows, and a heating coil of one row, as shown in Fig. 26.5b. In a common coil, the DDC controller admits the chilled water and hot water in sequence to two three-way valves and modulates their water flow at part-load operation to maintain a preset space temperature. The chilled water stream never mixes with the hot water in these three-way valves.

When two separate coils are used, hot water, steam, or electric energy can be used as the heat source.

A four-pipe system is more flexible, easier to operate, and lower in operating cost than a two-pipe system. On the other hand, it has a higher initial cost, and if reverse return pipe is used, many pipes must be squeezed into the ceiling plenum.

A drainage pipe for condensate in the fan–coil unit is needed, especially where conditioned space may have hot and humid infiltrated outdoor air.

Nonchangeover Two-Pipe Systems

When a two-pipe fan–coil system is used to serve a perimeter zone in a building, changeover from chilled water to hot water or vice versa, as described in Section 20.6, is a troublesome process and may take several hours. Therefore, in locations where winter weather is moderate, a nonchangeover two-pipe fan–coil system may be used.

In a *nonchangeover two-pipe system,* chilled water is supplied to the fan–coil throughout the year when

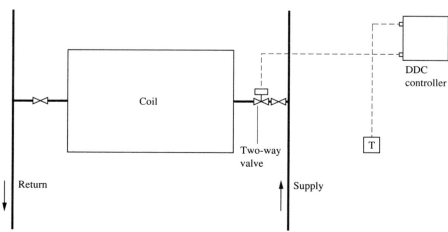

FIGURE 26.4   Supply and return mains in a two-pipe fan–coil system.

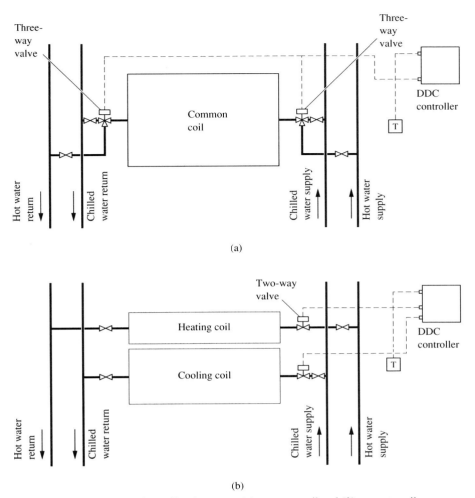

FIGURE 26.5 Four-pipe fan–coil unit system: (*a*) common coil and (*b*) separate coil.

the space is occupied. Warm primary air is supplied to the fan–coil in winter to offset the space heating load in the perimeter zone, and also in spring and fall when space heating is required. In such an arrangement, various zones at different orientations in the building that need cooling and heating simultaneously during spring and fall can be served, as shown in Fig. 26.6*a*.

In a typical nonchangeover two-pipe system, as shown in Fig. 26.6*a*, chilled water T_{el} enters the coil at a temperature of about 45°F at summer design conditions, whereas the temperature of the primary air supply T_{ps} may be maintained at 62°F. At cooling mode part-load operation, T_{el} supply for both fan–coil units and the primary air AHU can be reset gradually, up to 52°F, when the space sensible cooling load drops to 30 percent of the design value. When outdoor temperature T_o drops below 70°F, the primary air supply temperature T_{ps} begins to rise. The lower the T_o, the higher the T_{ps}.

If the space heating load of any room or control zone in the perimeter zone is offset entirely by the primary air in a nonchangeover two-pipe system, the required volume flow rate of primary air $\dot{V}_p$, in cfm, can be calculated as

$$\dot{V}_p = \frac{q_{rh}}{60\rho_{ps}c_{pa}(T_{ps} - T_r)}$$

$$= \frac{(A_{ex}U_m + 60\dot{V}_{if}\rho_o c_{pa})(T_r - T_o)}{60\rho_{ps}c_{pa}(T_{ps} - T_r)}$$

(26.19)

where q_{rh} = room heating load, Btu/h
A_{ex} = total area of the building shell in that room, ft^2
U_m = weighted average of the overall heat transfer coefficient of the building shell in that room, Btu/h · ft^2 · °F
$\dot{V}_{if}$ = volume flow rate of infiltrated air, cfm
ρ_o, ρ_{ps} = density of outdoor air and primary air supply, lb/ft^3
T_r = space air temperature, °F
T_{ps} = supply temperature of primary air, °F

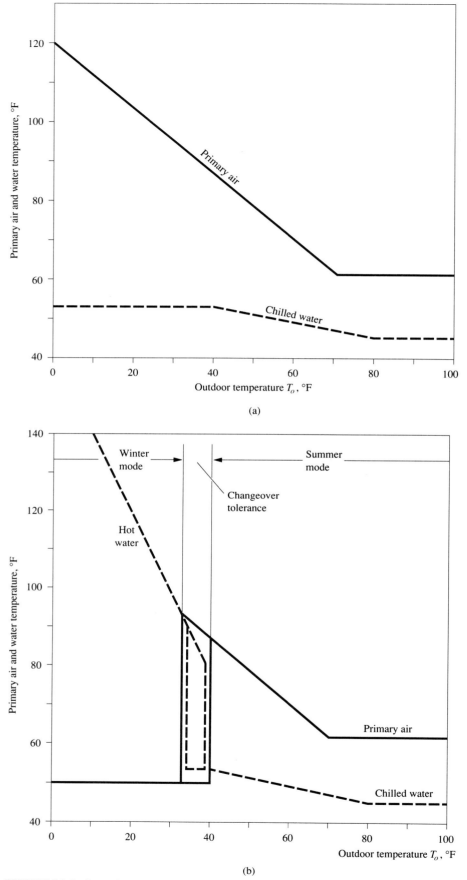

FIGURE 26.6 Operating parameters of air and water in a two-pipe fan–coil system: (a) non-changeover and (b) changeover.

The air transmission ratio is defined as the ratio of primary air volume flow rate supplied to a room in the perimeter zone $\dot{V}_p$, in cfm, to the transmission loss per degree of outdoor-indoor temperature difference T_{tran}, Btu/h · °F. According to Eq. (26.19), it can be calculated as

$$\frac{\dot{V}_p}{T_{\text{tran}}} = \frac{\dot{V}_p}{A_{\text{ex}}U_m + 60\dot{V}_{\text{if}}\rho_o c_{\text{pa}}}$$

$$= \frac{T_r - T_o}{60\rho_{\text{ps}}c_{\text{pa}}(T_{\text{ps}} - T_r)} \qquad (26.20)$$

For all the rooms in the perimeter zone supplied by the same primary air system in a fan–coil system, the temperature difference between room air and outdoor air $(T_r - T_o)$ is a constant at any time during operation. If the temperature drop due to the heat loss of the primary air supply duct is ignored, for the same primary air system, the supply air temperature difference $(T_{\text{ps}} - T_r)$ is again a constant. Because air density ρ_{ps} and specific heat of supply air c_{pa} can also be considered constant, and if the room heating load in the perimeter zone is offset entirely by the primary air, the air transmission ratio $\dot{V}_p/T_{\text{tran}}$ for all the rooms in perimeter zone in the same primary air system for a two-pipe nonchangeover fan–coil system must be a constant.

The current trend is to have an adequate amount of outdoor air supply for a better indoor air quality. Because the primary air supply must be higher than the outdoor air requirement, this often means that a nonchangeover two-pipe system has sufficient primary air for winter heating. In massively constructed buildings, the air transmission ratio can be smaller, such as $0.7 \, \dot{V}_p/T_{\text{tran}}$, because of the heat storage effect of the building shell.

One of the primary drawbacks of a nonchangeover two-pipe fan–coil system is the waste of energy during simultaneous heating and cooling in some rooms during winter heating. In a cold winter in the Northern Hemisphere, for a room facing south, warm primary air heating, solar radiation, and even some internal loads often combine to result in a higher room temperature than a room facing north in the same primary air system. Chilled water is then admitted to the fan–coil to cool the room air simultaneously. ASHRAE/IES Standard 90.1–1989 requires that simultaneous heating and cooling shall be avoided.

Changeover Two-Pipe Systems

During the summer mode operation of a changeover two-pipe fan–coil system, the space cooling load is offset by the combined cooling effect of the cooled primary air and chilled water supplied to the fan–coil, as shown in Fig. 26.6b.

During the fall season, before the changeover from summer mode operation to winter mode operation, when outdoor temperature T_o is higher than changeover temperature T_{co}, the primary air temperature increases according to T_o when T_o falls below 70°F. The heating effect of the primary air is able to offset the individual space heating load in the perimeter zone. Chilled water is used to provide cooling and dehumidifying when needed. Meanwhile, the chilled water supply temperature T_{el} should be reset between 45 and 52°F at part-load operation before changeover.

During changeover, the chilled water may be at 52°F and is changed over to hot water at a temperature higher than 80°F. Meanwhile, the warm air supply is changed over to cold air supply, typically at a temperature of 50°F. The changeover process was discussed in Section 20.6, and the outdoor changeover temperature T_{co} can be calculated from Eq. (20.6). Changeover may require several hours. In order to prevent two or three changeovers within the same day, a tolerance of about ±2°F is often used. There are many factors that influence the changeover temperature T_{co} in actual practice. Therefore, calculated T_{co} should be modified according to actual operating experience.

When $T_o < T_{\text{co}}$, a two-pipe changeover fan–coil system is in winter mode operation. The hot water temperature is reset according to outdoor temperature T_o. Cold primary air is supplied, typically at 50°F. The space cooling load of any individual control zone in the perimeter zone can be offset by the combined effect of the cold primary supply air and the transmission and infiltration loss. Zone heating loads are offset by the hot water supplied to the fan–coils. During winter mode operation, the refrigeration system that serves the perimeter zone is usually shut down. However, the changeover two-pipe system still operates with a cold source of primary air and a heat source of hot water simultaneously.

In spring, winter mode operation changes over to summer mode operation at outdoor temperature T_{co} only when the combined effect of primary air cooling and the transmission and infiltration loss does not offset the cooling load of every individual control zone in the perimeter zone.

Controls

ZONE TEMPERATURE CONTROL. As described in Section 18.6, zone temperature control in a fan–coil system is often provided by a DDC controller, which modulates the flow rate of chilled water or hot water to the fan–coils in response to the drop or increase in zone temperature sensed by a sensor.

A better type of control is modulation of the water control valves and fan speed to maintain required space temperature and relative humidity. When the fan–coil's cooling load is less than 60 percent of the maximum capacity, the fan operates at low speed. When the load is greater than 83 percent, the fan operates at high speed. At each fan speed, the two-way valve opens from the closed position to the fully open position in 12 steps, each with 8 percent of coil output.

A chilled water supply temperature T_{el} reset is preferable during cooling mode part-load operation. Return chilled water temperature from the fan–coils T_{ee} is sensed so that the reduction of space cooling load is estimated and the reset value of T_{el} can be determined accordingly. In winter mode operation, hot water is reset according to outdoor temperature T_o.

ASHRAE/IES Standard 90.1–1989 requires that water systems supplying heated or chilled water for comfort air conditioning should include controls that automatically reset supply water temperature by reduction of system load or by outdoor temperature change. Supply temperature should be reset by at least 25 percent of the design supply and return water temperature difference. Water systems of design capacity less than 600,000 Btu/h are exempt from this requirement.

In a nonchangeover two-pipe fan–coil system, chilled water flow is modulated by adjusting the opening of the two-way valve. In a two-pipe changeover system, in addition to modulation of the two-way valve, a changeover signal from the central computer or sensed changeover in supply water temperature reverses the modulating action.

In a four-pipe system with a common coil for heating and cooling, two three-way valves are modulated in sequence during part-load control, as described in previous paragraphs. A dead band of at least 5°F should be provided between the cooling and heating mode operation, as specified by ASHRAE/IES Standard 90.1–1989. In a four-pipe system with separate or split coils, modulation of two 2-way valves in sequence is often used to provide cooling and heating with reverse action during changeover. A dead band between cooling and heating is also provided.

PRIMARY AIR AHU. The controls provided by a DDC panel for a primary air AHU include discharge air temperature control, outdoor air temperature reset when T_o drops below 70°F, low-limit controls, and changeover from warm primary air to cold primary air and vice versa. The outdoor air intake damper is normally closed unless the fan is turned on and the air flow is sensed by a flow sensor.

CHILLED WATER SYSTEM. Usually, a DDC panel is used to control the temperature of chilled water that supplies to the fan–coils and primary air AHUs. In centrifugal chillers, control of leaving chilled water temperature T_{el} is provided by either modulating the inlet vanes or varying the speed of compressor motor by means of adjustable-frequency AC drives. In screw chillers, T_{el} is maintained by modulating the position of the sliding valve. In reciprocating chillers, chilled water temperature entering the chiller T_{ee} is usually maintained within predetermined limits by turning the reciprocating compressors or their cylinders on and off.

HOT WATER SYSTEM. Control of hot water temperature leaving gas-fired and oil-fired boilers T_{wb} is performed by modulating the gas or oil flow to the burner. Temperature T_{wb} is often reset according to outdoor temperature T_o.

CHANGEOVER. As described in Section 20.5, changeover can be started manually (by positioning a switch) or automatically. When changeover is automatic, a representative outdoor temperature T_o or a zone temperature that has a direct relationship with T_o can be used as a sensing parameter. When it receives the temperature signal, the central computer signals all fan–coils, chillers, pumps, and primary air AHUs to begin the changeover process.

Energy Consumption

Fan–coil systems have a smaller energy per unit volume flow W/cfm and air transport factor $F_{a.t}$ than constant-volume air systems. Consider a four-pipe fan–coil system with a total pressure drop of about 0.6 in. WG and a total pressure drop of 4.5 inches in the primary air AHU, operating under the following conditions:

- The ratio of the volume flow rate of primary air to the supply volume flow rate in fan–coil is 0.2.
- The fan total efficiency of the fan–coil is 0.45.
- The fan total efficiency of the primary air AHU is 0.65.
- The motor efficiency of the fan coil is 0.55.
- The motor efficiency of the primary air AHU is 0.8.

From Eq. (26.1), energy per unit volume flow P_{cfm} is

$$P_{cfm} = 0.1175 \frac{\Delta p_{sy}}{\eta_f \eta_m}$$

$$= 0.1175 \left[\frac{0.6}{0.45 \times 0.55} + \frac{0.2(4.5)}{0.65 \times 0.8} \right]$$

$$= 0.1175 \times 4.15 = 0.488 \text{ W/cfm}$$

For a supply air temperature differential $\Delta T_s = 15°F$, the air transport factor from Eq. (26.3) is

$$F_{a.t} = \frac{2.73 \Delta T_s \eta_f \eta_m}{\Delta p_{sy}} = 2.73 \times 15 \left(\frac{0.45 \times 0.55}{0.6} \right.$$

$$\left. + \frac{0.65 \times 0.8}{0.2 \times 4.5} \right)$$

$$= 40.5$$

Design Considerations

1. Two-pipe changeover and four-pipe fan–coil systems are used only for the perimeter zone in buildings where winter heating is required.

2. For a two-pipe non-changeover fan–coil system, the amount of primary air needed should be determined according to both the outdoor air requirements and the air transmission ratio V_p / T_{tran} when it is used for the perimeter zone. Proper zoning of the primary air system is designed so that each zone serves the area facing in the same direction, and the warm primary air supply temperature should be reset by the combined effect of outdoor temperature and solar radiation. These may help to prevent simultaneous heating and cooling during heating mode operation. A two-pipe non-changeover fan–coil may be used for the interior zone where year-round cold air supply is needed during occupied hours. Warm primary air supply is only used for morning warm-up periods in winter.

3. In locations with short, moderate winters, a careful cost analysis between a fan–coil system using an electric heating coil, a nonchangeover two-pipe system, and a four-pipe system should be performed in order to select the optimum fan–coil system.

4. A reverse return water pipe system provides a better water flow balance when a number of fan–coil units are connected to a main pipe. Shut-off valves for each fan–coil should be installed for convenient maintenance and repair.

Selection of a Fan–Coil Unit

The size of a fan–coil unit is selected mainly to meet the required total cooling capacity and sensible cooling capacity, both in Btu/h or MBtu/h.

In a fan–coil system, the temperature and enthalpy of primary air are always lower than those of space air during cooling mode operation. Therefore, the estimated total cooling capacity $q_{c.fc}$ and sensible cooling capacity $q_{s.fc}$ of fan–coils for a specific control zone are always less than the space cooling load q_{rc} and space sensible load q_{rs}, both in Btu/h. Their relationship can be expressed as

$$q_{c.fc} = q_{rc} - \dot{V}_p \rho_{ps}(h_r - h_{ps})$$

$$q_{s.fc} = q_{rs} - \dot{V}_p \rho_{ps}(T_r - T_{ps}) \qquad (26.21)$$

where h_r, h_{ps} = enthalpy of space air and primary supply air, Btu/lb

T_r, T_{ps} = temperature of space air and primary supply air, °F

The relationship between the humidity ratio of primary supply air w_{ps}, the humidity ratio of fan–coil supply air w_{fc}, and the humidity ratio of space air w_r depend on the temperature of the chilled water that supplies to the fan–coils $T_{w.f}$ and to the primary air AHUs $T_{w.p}$. It also has a direct influence on the capacity of the fan–coil units.

If $T_{w.p} < T_{w.f}$, then $w_{ps} < w_{fc}$ and $w_{ps} < w_r$. A considerable amount of both sensible and latent load is undertaken by the primary air during cooling mode operation. The cooling capacity of the fan–coil units is considerably less than the space cooling load.

If $T_{w.p} = T_{w.f}$, $w_{ps} \approx w_{fc}$ and $w_{ps} < w_r$, then a certain amount of sensible load and only a small amount of latent load are undertaken by the primary air during cooling mode operation. The sensible cooling capacity of the fan–coil is smaller than the space sensible cooling load.

As described previously, the current trend is to have $T_{w.p} = T_{w.f}$. T_{ps} and w_{ps} are influenced by $T_{w.p}$, and T_{fc} and w_{fc} are influenced by $T_{w.f}$; all of these are also affected by the cooling capacity of the coil in the primary air AHU and the fan–coil, and the condition of the entering air. For this reason, w_{ps} may be higher or lower than w_{fc}, and sometimes $w_{ps} \approx w_{fc}$ as shown in Fig. 26.7.

When the air conditioning cycle of the fan–coil unit system is at steady state equilibrium, the sensible heat ratio of the space conditioning line SHR_s is always approximately equal to the sensible heat ratio of the cooling and dehumidifying process of the selected fan–coil SHR_c at design conditions.

At summer mode design conditions, when $T_{w.p} = T_{w.f}$ and $w_{ps} \approx w_{fc}$, for fan–coil units serving an interior zone, typically $SHR_s \approx SHR_c \approx 0.8$, as shown in Fig. 26.7a. For fan–coil units serving a perimeter zone, $SHR_s \approx SHR_c \approx 0.95$, as shown in Fig. 26.7b.

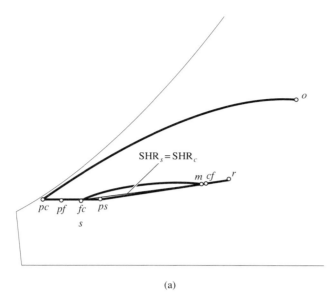

(a)

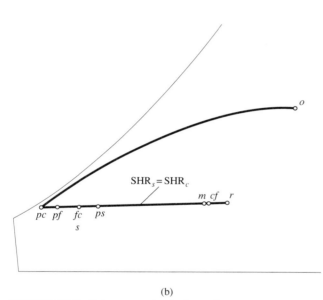

(b)

FIGURE 26.7 Primary supply air, fan–coil supply air, and sensible heat ratio of space conditioning line SHR$_s$: (a) SHR$_s$ = 0.8 and (b) SHR$_s$ = 0.95.

At summer mode part-load operation, as the chilled water flow rate decreases and the temperature rise in the fan–coil ΔT_w increases, air temperature leaving the fan–coil T_{fcp} and the humidity ratio w_{fcp} rise accordingly. If SHR$_{sp}$ = 0.8, because $w_{psp} < w_{rp}$, the cooling and dehumidifying capacity of the primary air offsets most of the space latent load. The space conditioning line *fcp rp*, the cooling and dehumidifying process in the fan–coil *cfp fcp*, and the mixing process *psp mp rp* form a cycle on the psychrometric chart, as shown in Fig. 26.2b.

An undersized fan–coil unit cannot provide the required space conditions during summer and winter design conditions. On the other hand, an oversized fan–coil unit raises the space relative humidity to an unacceptable level at part-load operation at a very low load ratio. Field surveys show that an oversized fan–coil unit can raise the space relative humidity to 75 percent on a hot summer morning just after the cool-down period.

Case Study—A Four-Pipe Fan–Coil System

Consider a large office in an office building using a four-pipe fan–coil system. The primary air is supplied to the mixing plenum of the fan–coil unit. At summer and winter design-load, the operating parameters for this fan–coil system in this office are as follows:

	Summer	Winter
Indoor space temperature, °F	78	72
Space relative humidity and wet bulb	45 percent (63.5 °F)	
Space cooling or heating load, Btu/h	29,000	12,150
Space sensible cooling load, Btu/h	26,200	
Space latent load, Btu/h		2700
Part-load cooling load, Btu/h	15,900	
Part-load sensible load, Btu/h	13,100	
Outdoor temperature, °F	90	20
Outdoor relative humidity, percent	50	80
Outdoor air, cfm	200	200

1. Assume that the temperature of primary supply air T_{ps} is 61°F and relative humidity is 81 percent. From the psychrometric chart, h_{ps} = 24.8 Btu/lb.

   At summer design conditions, the enthalpy of space air is 28.8 Btu/lb and space air density ρ_{ps} = 0.075 lb/ft³, from Eq. (26.21) the estimated fan–coil cooling capacity is

   $$q_{c.fc} = q_{rc} - 60\dot{V}_p\rho_{ps}(h_r - h_{ps})$$
   $$= 29,000 - 60 \times 200$$
   $$\times 0.075\,(28.8 - 24.8)$$
   $$= 25,400 \text{ Btu/h}$$

   The estimated fan–coil sensible cooling capacity is

   $$q_{s.fc} = q_{rs} - 60\dot{V}_p\rho_{ps}c_{pa}(T_r - T_{ps})$$
   $$= 26,200 - 60 \times 200 \times 0.075$$
   $$\times 0.243\,(78 - 61)$$
   $$= 22,482 \text{ Btu/h}$$

   If chilled water is supplied to the fan–coils and the primary air AHU both at 45°F during summer design conditions, for entering air of 78°F dry bulb and 63.5°F wet bulb and a chilled water temperature rise of 10°F, one manufacturer's catalog gives the following capacities for a size 03 fan–coil unit:

Total cooling capacity   7.2 MBtu/h
Sensible cooling capacity   6.3 MBtu/h

Because the external pressure drop in the supply duct after the fan–coil is 0.06 in.WG and the fan–coils are installed at sea level, no volume flow corrections are needed.

Four 03 fan–coil units are selected, so the total cooling capacity is

$$q_{c.fc} = 4 \times 7200 = 28{,}800 \text{ Btu/h}$$

which is greater than the estimated cooling capacity 25,400 Btu/h at summer design load. The sensible cooling capacity of fan–coils is

$$q_{s.fc} = 4 \times 6300 = 25{,}200 \text{ Btu/h}$$

which is also greater than the estimated sensible capacity of 22,482 Btu/h at summer design load.

2. Draw line $r\,fc$ from the conditioned space point r with known $\text{SHR}_s = 26{,}200/29{,}000 = 0.9$ and a $\rho_{fc} = 1/13.7 = 0.73$ lb/ft³. The temperature of supply air leaving the fan–coil unit T_{fc}, in °F, can be calculated as

$$T_{fc} = T_r - \frac{q_{rs}}{60 \dot{V}_{fc} \rho_{fc} c_{pa}}$$

$$= 78 - \frac{26{,}200}{60 \times 4 \times 300 \times 0.073 \times 0.243}$$

$$= 57.5°F \qquad (26.22)$$

where $\dot{V}_{fc}$ = corrected volume flow rate of the fan–coil, cfm

ρ_{fc} = air density at the fan outlet in the fan–coil, lb/ft³

Point fc can then be determined. From the psychrometric chart, $h_{fc} = 23.3$ Btu/lb.

3. Let ΔT_{wp} be the water temperature rise in the cooling coil of the primary air AHU. It can be assumed that the temperature of air leaving the cooling coil of the primary air AHU T_{pc}, in °F, is

$$T_{pc} = T_{w.p} + \Delta T_{wp} \qquad (26.23)$$

For a $\Delta T_{wp} = 10°F$, $T_{pc} = 45 + 10 = 55°F$

Generally, the cooling coil may have 8 to 10 rows. The relative humidity of air leaving the cooling coil $\varphi_{pc} = 98$ percent, and the humidity

ratio w_{pc} can thus be determined. For a $T_{pc} = 55°F$ and $\varphi_{pc} = 98$ percent, $w_{pc} = 0.0092$ lb/lb and $h_{pc} = 23.2$ lb/lb.

For simplicity, the fan power heat gain for the primary air AHU can be estimated at 2°F and the duct heat gain can be estimated at 4°F, so the supply temperature of the primary air AHU T_{ps}, in °F, is

$$T_{ps} = T_{pc} + 2 + 4 = 61°F \qquad (26.24)$$

Because $w_{ps} = w_{pc}$, state point ps can also be determined.

4. For a fan–coil whose primary air is mixed with recirculating air in the mixing plenum of fan–coil, draw line $r\,ps$. From Fig. 26.2b,

$$\frac{m\,r}{ps\,r} = \frac{\dot{V}_p}{\dot{V}_{fc}} = \frac{200}{4 \times 300} = 0.17 \qquad (26.25)$$

Point m can thus be determined. In this case, $T_m = 75.0°F$ and $w_m = 0.0092$ lb/lb. Because $T_{cf} = T_m + 0.5 = 75 + 0.5 = 75.5°F$ and $w_{cf} = w_m = 0.0092$ lb/lb, point cf can also be determined. From the psychrometric chart, $h_{cf} = 28.2$ Btu/lb.

5. If $\rho_{cf} = 1/13.7 = 0.073$ lb/ft³, the cooling coil's load q_{cfc} and sensible coil's load q_{sfc} of fan–coil unit, both in Btu/h, can be calculated as

$$q_{cfc} = 60 \dot{V}_{fc} \rho_{cf} (h_{cf} - h_{fc})$$

$$= 60 \times 4 \times 300 \times 0.073(28.2 - 23.3)$$

$$= 25{,}754 \text{ Btu/h} \qquad (26.26)$$

$$q_{sfc} = 60 \dot{V}_{fc} \rho_{cf} c_{pa} (T_{cf} - T_{fc})$$

$$= 60 \times 4 \times 300 \times 0.073$$

$$\times 0.243(75.5 - 57.5)$$

$$= 22{,}990 \text{ Btu/h} \qquad (26.27)$$

where h_{cf}, h_{fc} = enthalpy of air at fan outlet and supply outlet, Btu/lb

T_{cf}, T_{fc} = temperature of air at fan outlet and supply outlet, °F

Check the calculated cooling coil's load and sensible load against the selected fan–coil's cooling and sensible capacity. The calculated coil's load q_{cfc} of 25,754 Btu/h is very near to the estimated capacity of 25,400 Btu/h, and both are less than the selected capacity $q_{c.fc}$ of 28,800 Btu/h. Similarly, the calculated sensible coil's load q_{sfc} is 22,990 Btu/h. It is slightly greater than the estimated 22,482 Btu/h because of the fan power heat gain. Both are less than the selected capacity $q_{s.fc}$ of 24,000 Btu/h.

6. Calculate the cooling coil's load in the primary air AHU q_{cp}, in Btu/h. From the psychrometric chart, $T_o = 90°F$, $\varphi_o = 50$ percent, and $h_o = 38.5$ Btu/lb, so

$$
\begin{aligned}
q_{cp} &= 60\dot{V}_p \rho_{ps}(h_o - h_{ps}) \\
&= 60 \times 200 \times 0.075(38.5 - 23.2) \\
&= 13,770 \text{ Btu/h} \qquad (26.28)
\end{aligned}
$$

The cooling coil load of the primary air AHU q_{cp} is about half of the fan–coil's cooling coil load. This includes the cooling and dehumidifying load of outdoor air, system heat gain of the primary air system, and a portion of the space cooling load.

7. During summer mode part-load operation, the space temperature is still maintained at 78°F, that is, $T_{rp} = T_r$. The fan in the fan–coil can be switched from high speed to low speed while chilled water is reset to a higher temperature. If the space sensible load ratio is reduced to 50 percent of full-load and the volume flow rate of the fan at low speed is only 70 percent of the high speed, then the temperature of air leaving the fan–coil at part-load T_{fcp}, in °F, can be calculated as

$$
\begin{aligned}
T_{fcp} &= T_{rp} - \frac{q_{rsp}}{60 \times 0.7\dot{V}_{fc}\rho_{cf}c_{pa}} \qquad (26.29) \\
&= 78 - \frac{0.5(26,200)}{60 \times 4 \times 300 \times 0.7 \times 0.073 \times 0.243} \\
&= 63.3°F
\end{aligned}
$$

where q_{rsp} = space sensible cooling load at part-load, Btu/h

T_{rp} = space air temperature at part-load, °F

At part-load operation, the relative humidity of supply air leaving the fan–coil φ_{fcp} may vary between 80 and 90 percent. If $\varphi_{fcp} = 88$ percent, point fcp can be plotted on the psychrometric chart as shown in Fig. 26.2b. From the chart, $h_{fcp} = 27.2$ Btu/lb.

8. During summer mode part-load operation, the sensible heat ratio of space conditioning line $SHR_s = 13,100/15,900 = 0.82$. Draw line fcp rp from fcp with a sensible heat ratio of the space conditioning line at part-load $SHR_{sp} = 0.82$. Line fcp rp intersects the space temperature line T_{rp} at rp. Therefore, point rp can be determined. From the psychrometric chart, $\varphi_{rp} = 57.5$ percent.

9. When the space sensible load is reduced to 50 percent of full-load, if the temperature of chilled water supplied to fan–coils $T_{w.f}$ and to the primary air AHU $T_{w.p}$ are reset to 50°F at part-load operation, and the chilled water temperature increase in the cooling coil of the primary air AHU is still 10°F, the condition of supply air from the primary air AHU at part-load point psp can be determined as in full load operation, $T_{psp} = 60 + 2 + 4 = 66°F$, and $w_{psp} = 0.0110$ lb/lb.

Draw line psp rp. The condition of the mixture at part-load, point mp, can be determined from the following relationship:

$$
\frac{rp\ mp}{rp\ ps} = \dot{V}_p/\dot{V}_{fc} = \frac{200}{4 \times 300 \times 0.7} = 0.24
$$

$$(26.30)$$

From the psychrometric chart, $T_{mp} = 75.3°F$ and $w_{mp} = 0.0116$ lb/lb.

Because $T_{cfp} = T_{mp} + 0.5°F = 75.3 + 0.5 = 75.8°F$ and $w_{cfp} = w_{mp} = 0.0116$ lb/lb, from the psychrometric chart, $h_{cfp} = 31.0$ Btu/lb.

10. The fan–coil cooling coil's load at part-load q_{cfcp} and sensible coil's load at part-load q_{sfcp}, both in Btu/h, can be calculated as

$$
\begin{aligned}
q_{cfcp} &= 60\dot{V}_{fc}\rho_{cf}(h_{cfp} - h_{fcp}) \\
&= 60 \times 4 \times .07 \times 300 \\
&\quad \times 0.073(31.0 - 27.2) \\
&= 13,980 \text{ Btu/h} \qquad (26.31)
\end{aligned}
$$

$$
\begin{aligned}
q_{sfcp} &= 60\dot{V}_{fc}\rho_{cf}c_{pa}(T_{cfp} - T_{fcp}) \\
&= 60 \times 4 \times 300 \times 0.7 \times 0.073 \\
&\quad \times 0.243(75.8 - 63.3) \\
&= 11,176 \text{ Btu/h} \qquad (26.32)
\end{aligned}
$$

11. At winter heating mode design load operation, if only two fan–coil units are operated and $h_{fg.o} = 1061$ Btu/lb, the humidity ratio difference between space air and the fan–coil supply air is

$$
\begin{aligned}
w_r - w_s &= \frac{q_{rl}}{60\dot{V}_{fc}\rho_{hf}h_{fg.o}} \qquad (26.33) \\
&= \frac{2700}{60 \times 2 \times 300 \times 0.073 \times 1061} \\
&= 0.00097 \text{ lb/lb}
\end{aligned}
$$

where w_s = humidity ratio of fan–coil supply air, lb/lb.

As shown in Fig. 26.2b, $w_s = w_m = w_{fc}$, and from the psychrometric chart, the humidity ratio of outdoor air $w_o = 0.0017$ lb/lb. Therefore,

$$\frac{w_r - w_s}{w_r - w_o} = \frac{\dot{V}_p}{\dot{V}_{fc}} = \frac{200}{600} = 0.33 \qquad (26.34)$$

So

$$w_r = w_o + \frac{w_r - w_s}{0.33} = 0.0017 + \frac{0.00097}{0.33}$$
$$= 0.0046 \text{ lb/lb}$$

and $w_s = w_m = w_r - 0.0009 = 0.0046 - 0.00097 = 0.0036$ lb/lb.

12. Because at winter heating mode operation $T_r = 72°F$, from the psychrometric chart, the relative humidity at winter design condition $\varphi_r = 28$ percent.

13. If air leaving the preheating coil of the primary air AHU at a $T_{ph} = 83°F$ and $T_{pf} = T_{ph} +$ fan temperature rise $= 83 + 2 = 85°F$, because $w_o = w_{pf} = 0.0017$ lb/lb, from the psychrometric chart, air density at the fan supply outlet in the primary air AHU $\rho_{pf} = (1/13.75) = 0.073$ lb/ft^3. The preheating coil's load q_{cph}, in Btu/h, can be calculated as

$$q_{cph} = 60\dot{V}_p \rho_{pf} c_{pa}(T_{ph} - T_o) \qquad (26.35)$$
$$= 60 \times 200 \times 0.073 \times 0.243(83 - 20)$$
$$= 13,411 \text{ Btu/h}$$

14. If the heat loss of primary air supply duct is 0.5°F, $T_{ps} = 85 - 0.5 = 84.5°F$, and $w_{ps} = w_{pf} = 0.0017$ lb/lb, then point ps can be determined. Draw line $ps\ r$. Because $w_m = 0.0036$ lb/lb, point m can thus be plotted on the psychrometric chart. From the chart, $T_m = 76.2°F$.

15. The temperature of the fan–coil supply air at winter design load T_s, in °F, can be calculated as

$$T_s = T_r + \frac{q_{rh}}{60\dot{V}_{fc}\rho_{cf}c_{pa}} \qquad (26.36)$$
$$= 72 + \frac{12,150}{(60 \times 2 \times 300 \times 0.073 \times 0.243)}$$
$$= 91.0°F$$

Because $w_s = w_m = 0.0036$ lb/lb, point s can be determined and lines rs and ms can be drawn.

16. The air temperature at the fan outlet in the fan–coil $T_{cf} = T_m + 0.5 = 76.2 + 0.5 = 76.7°F$. The heating coil's load in the fan–coil unit q_{ch}, in Btu/h, can be calculated as

$$q_{ch} = 60\dot{V}_{fc}\rho_{cf}c_{pa}(T_s - T_{cf})$$
$$= 60 \times 2 \times 300 \times 0.073$$
$$\times 0.243 (91.0 - 76.7)$$
$$= 9132 \text{ Btu/h} \qquad (26.37)$$

17. At winter heating mode part-load operation, the warm supply air from the fan–coil, point sp, moves along the horizontal line $m\ sp$, and is supplied at a lower temperature T_{sp} in order to maintain the required space temperature T_{rp}.

System Characteristics and Applications

Fan–coil systems with primary air supply have the following advantages over constant-volume and variable-air-volume systems:

- Capacity control by both variable supply temperature and two or three steps of variable-air-volume control
- Less cross-contamination between rooms and zones
- A guaranteed positive supply of outdoor air
- A lower fan energy consumption than in constant-volume systems and variable-air-volume systems using inlet vanes during reduced volume flow
- Less headroom required for the primary air duct in the ceiling plenum

Fan–coil systems also have the following disadvantages:

- Higher space relative humidity at part-load than VAV systems
- More site maintenance work, including regular cleaning of the filters and coils
- Possible leaks of water and condensate above the conditioned space
- Higher air contaminated levels caused by use of low-efficiency filters
- Higher noise levels

Fan–coil systems using primary air supply are widely used in hotels and motels where the exhaust air system for bathrooms can easily balance with the primary air supply in individual rooms or specified control zones. They are also used in hospitals, schools, and offices.

Because of the increase of the minimum outdoor air (primary air) and the guaranteed positive outdoor supply to each control zone, fan–coil systems with primary supply may gain more applications in the future.

REFERENCES AND FURTHER READING

Avery, G., "No More Reheat: Banking Center Retrofit," *Heating/Piping/Air Conditioning,* pp. 98–99, March 1984.

ASHRAE/IES *Standard 90.1–1989, Energy Efficient Design of New Buildings Except New Low-Rise Residential Buildings,* ASHRAE Inc., Atlanta, GA, 1989.

Carrier Corporation, *The 42M Air Conditioning Unit,* Carrier Corporation, Syracuse, NY, 1982.

Coad, W. J., "The Air System in Perspective," *Heating/Piping/Air Conditioning,* pp. 124–125, October 1989.

Desmone, C. L., and Frank, P. L., "Air Conditioning for Precision Manufacturing," *Heating/Piping/Air Conditioning,* pp. 35–44, February 1992.

EIA, *Characteristics of Commercial Buildings,* EIA, Washington, D.C., 1986.

EIA, *Commercial Building Characteristics,* EIA, Washington, D.C., 1991.

Haines, R. W., "Temperature and Humidity Control for Electronic Manufacturing," *Heating/Piping/Air Conditioning,* pp. 153–155, November 1984.

Naughton, P., "HVAC Systems for Semiconductor Cleanrooms—Part I: System Components," *ASHRAE Transactions,* Part II, pp. 620–625, 1990.

Tao, W., "CSP: Cooling System Performance," *Heating/Piping/Air Conditioning,* pp. 77–82, May 1984.

The Trane Company, *Fan–Coil Units,* The Trane Company, La Crosse, WI, 1990.

The Trane Company, *Unitrane: A Complete Line of Fan–Coil Room Air Conditioners,* The Trane Company, La Crosse, WI, 1984.

Wang, S. K., *Air Conditioning,* Vol. 4, Hong Kong Polytechnic, Hong Kong, 1987.

Wolfert, J. E., "HVAC Systems in Shopping Centers as a Utility Business," *ASHRAE Transactions,* Part II B, pp. 680–684, 1985.

CHAPTER 27

AIR SYSTEMS: VARIABLE-AIR-VOLUME (VAV)

27.1 SINGLE-ZONE VAV SYSTEMS

A *variable-air-volume (VAV) system* is an air system that varies its supply air volume flow rate to match the reduction of space load during part-load operation to maintain a predetermined space parameter, usually air temperature, and conserve fan power at reduced volume flow.

A *single-zone VAV system* is a VAV system that varies the air volume flow rates supplied to and returned from a single-zone conditioned space to maintain a predetermined space parameter at reduced load and conserve fan power. Single-zone VAV systems are widely used in indoor stadiums, arenas, convention centers, and sometimes retail stores and shopping malls.

System Description

Figure 27.1 shows the schematic diagram and the air conditioning cycles of a typical single-zone VAV system.

During summer mode operation, the mixture of outdoor air and recirculating air at point m is filtered at the filter and cooled and dehumidified at the cooling coil. The conditioned air leaves the coil at point cc. It then flows through the supply fan of the air-handling unit (AHU) or packaged unit (PU) and the supply duct before it is discharged to the single-zone conditioned space at point s. After absorbing the space cooling load, the supply air becomes the space air r and returns to the AHU or PU through the return air system.

When air economizer cycles are used for free cooling during spring and fall, outdoor air intake may vary from minimal to 100 percent. A relief fan should be installed in combination with the supply fan, as described in Section 19.12, if the pressure drop of the return duct system is small. If the pressure drop of the return duct system is greater than 1 in. WG, a return fan in series with a supply fan should be considered.

During summer mode part-load operation, the supply and return volume flow rates are reduced to match the lower space cooling load. The temperature increase from fan power heat gain is reduced accordingly. At the same time, duct heat gain remains approximately the same, which results in a higher temperature rise for supply and return ducts during summer mode part-load operation.

During winter mode operation, if solar heat and the internal load is greater than the space transmission and infiltration loss, then a cold air supply is still needed, such as in fully occupied indoor stadiums and arenas. Outdoor air is often mixed with the recirculating air to form a mixture at a temperature around 60°F during the air economizer cycle for free cooling. Warm air may be needed only during the warm-up period to raise the space temperature to an acceptable value before the space is to be occupied.

If solar heat and the internal load is less than the transmission and infiltration loss, the mixture of outdoor and recirculating air is heated at the heating coil to point hc. It is extracted by the supply fan and forced through the supply duct. Warm air is then supplied to the conditioned space at point s and the recirculating air is returned to the AHU or PU through the ceiling plenum and connecting return ducts.

For comfort air conditioning without humidification in winter, the space relative humidity depends mainly on the space latent load and the volume flow ratio of outdoor air to supply air.

In many VAV systems in commercial buildings, the ceiling plenum is often used as the return plenum. The main source of supply fan noise from the AHU or PU is often duct-borne noise through the return system. A sound trap or attenuator should be added between the return plenum and the fan room. Radiated noise from the supply main duct in the ceiling plenum

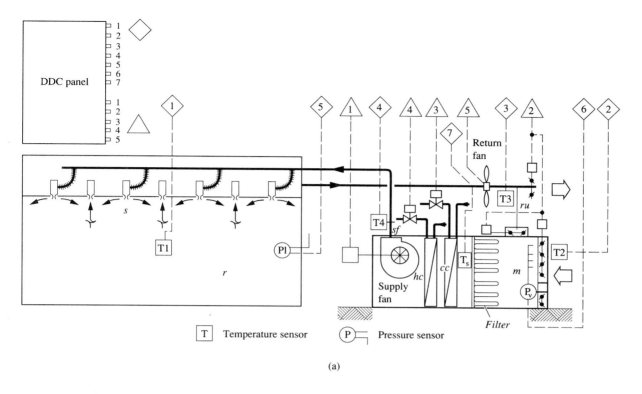

(a)

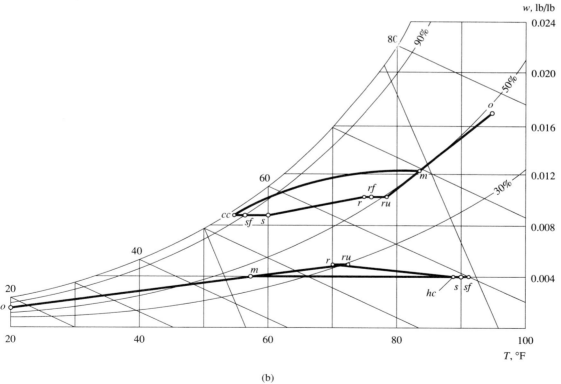

(b)

FIGURE 27.1   A single-zone VAV system: (a) schematic diagram and (b) air conditioning cycle.

immediately after the fan room should also be carefully checked.

Air Conditioning Cycle and System Calculations

The air conditioning cycle of a single-zone VAV system during summer and winter mode design-load operation can be plotted on the psychrometric chart in a procedure similar to that of a basic air system described in Chapters 3 and 26. This cycle is shown in Fig. 27.1b.

During summer mode part-load operation, the state point of outdoor air o varies, the sensible heat ratio of the space conditioning line SHR_{sp} usually becomes smaller than design-load, and the supply and return system heat gains may change. The space temperature T_r is maintained within the same predetermined limits as in design-load. The space relative humidity is slightly greater than that in design-load if SHR_{sp} becomes smaller.

The design supply volume flow rate $\dot{V}_s$, in cfm, is normally determined according to the summer mode design load, and can be calculated as

$$\dot{V}_s = \frac{q_{rs}}{60\rho_s c_{pa}(T_r - T_s)} \qquad (27.1)$$

where q_{rs} = space sensible cooling load, Btu/h
ρ_s = supply air density, lb/ft^3
c_{pa} = specific heat of moist air, Btu/lb · °F
T_r, T_s = temperature of space and supply air, °F

The design cooling coil's load q_{cc}, in Btu/h, can be calculated as

$$q_{cc} = 60\dot{V}_s \rho_s (h_m - h_{cc}) \qquad (27.2)$$

where h_m = enthalpy of the mixture of outdoor and recirculating air, Btu/lb
h_{cc} = enthalpy of conditioned air leaving the cooling coil, Btu/lb

The design heating coil's load q_{ch}, in Btu/h, which offsets the space heating load or sometimes maintains a setback temperature during unoccupied periods to reduce the warm-up load and prevent freezing, can be calculated as

$$q_{ch} = 60\dot{V}_s \rho_s c_{pa}(T_{hc} - T_{en}) \qquad (27.3)$$

where T_{hc}, T_{en} = air temperature leaving and entering heating coil, °F.

Control Systems

ZONE TEMPERATURE CONTROL. Zone temperature T_r is sensed by a temperature sensor T1. It sends an input signal to the DDC panel. If T_r drops below or rises above a predetermined set point, an output from the DDC panel actuates a motorized operator to modulate the supply fan speed through an adjustable-frequency AC drive or to vary the opening of the inlet vanes of the supply fan, as shown in Fig. 27.1a. For an AHU or PU serving a large conditioned space, several temperature sensors can be located in different areas. Either the mean value or the highest or lowest temperature can be selected to modulate the fan speed or the position of inlet vanes.

A reverse-acting relay is installed so that the control action is reversed when the cold air supply is changed over to the warm air supply and vice versa. Normally, a dead band of at least 5°F is set between cooling and heating mode operation.

AIR ECONOMIZER CONTROL. Usually, two temperature sensors T2 and T3 are installed to sense the outdoor air temperature T_o and the recirculating air temperature T_{ru}, as described in Section 18.8. Their values are compared. If $T_o \leq T_{ru}$ and $T_o \geq T_{cc}$, the outdoor damper is fully opened and the recirculating damper is closed. Here, T_{cc} represents the conditioned air temperature leaving the cooling coil, in °F. If $T_o > T_{ru}$, the outdoor damper is closed to provide the minimum outdoor air requirement, or when $T_o < T_{cc}$, the outdoor damper and recirculating damper are modulated to maintain a required mixing temperature of outdoor air and recirculating air T_m until the outdoor damper is closed to provide minimum outdoor air requirements.

DISCHARGE AIR TEMPERATURE CONTROL. Discharge temperature sensor T4 senses the discharge temperature from the AHU or PU and the DDC panel modulates the two-way valve of the cooling coil, the outdoor, recirculating, and relief dampers, and the two-way valve of the heating coil in sequence to maintain a required discharge temperature T_{dis}, in °F, as described in Section 18.9.

MINIMUM OUTDOOR AIR CONTROL. A velocity sensor P_v is installed at the outdoor air intake to measure its volume flow rate. The DDC panel modulates the outdoor damper and recirculating damper openings so that outdoor air does not drop below the minimum required amount at any time. Minimum outdoor air control overrides the control signal from discharge air temperature control.

LOW TEMPERATURE LIMIT CONTROL. A temperature sensor T5 senses the low limit of the mixing temperature to maintain a specified limit, such as 3 to 5°F below the mixing temperature, to prevent coil freezing.

WARM-UP AND COOL-DOWN CONTROL. During warm-up and cool-down control, the outdoor and exhaust dampers should be closed and the recirculating damper should be fully open as discussed in Section 18.10. Temperature sensors measure the zone temperature, outdoor temperature, and surface temperature of the building structure. The optimum starting time can thus be determined by computer software or by operator experience.

SPACE PRESSURIZATION CONTROL. If a relief fan is used, a pressure sensor P1 is used to measure the static pressure of the conditioned space or the ceiling plenum, and the DDC panel modulates the fan speed or position of inlet vanes to maintain the required space pressure. If a return fan is installed, a similar space pressure control should be used.

27.2 PERIMETER-HEATING VAV SYSTEMS

System Description

A *perimeter-heating VAV system* is a multizone air and water system. It consists of a hot water perimeter-heating system to offset the winter transmission and infiltration loss in the perimeter zone and a VAV cooling system to provide cooling for both perimeter and interior zones, as shown in Fig. 27.2. This VAV cooling system also provides the cold outdoor air for the perimeter zone during winter heating mode operation. Perimeter-heating VAV systems are suitable for multizone commercial buildings, especially in locations with long, cold winters.

During the early development of VAV systems, air skin VAV systems were used in office buildings. An

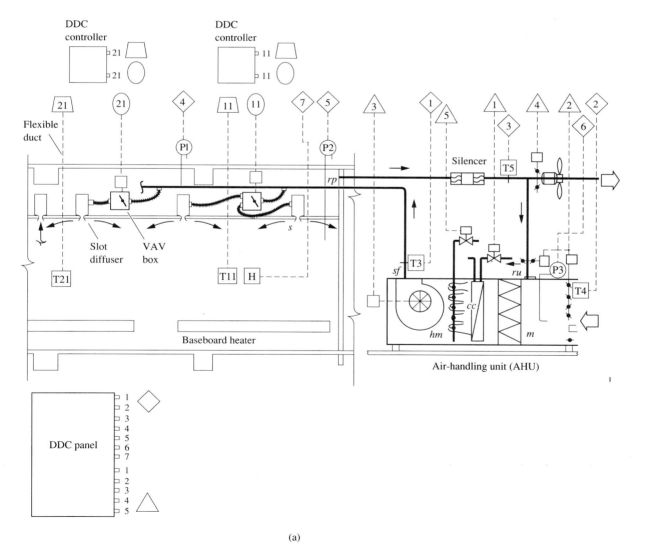

(a)

FIGURE 27.2 A perimeter-heating VAV system (including VAV cooling system): (a) schematic diagram and (b) air conditioning cycle.

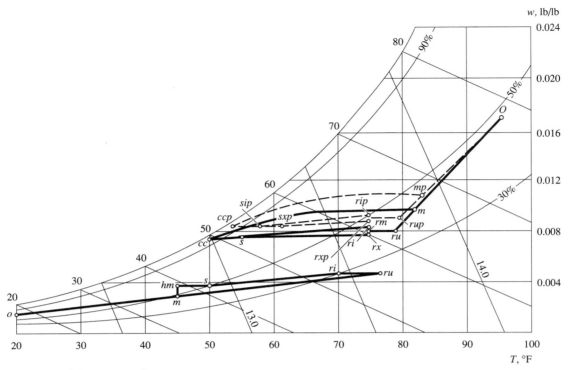

FIGURE 27.2 b   (*continued*)

air skin VAV system uses a constant-volume variable supply temperature air skin system to offset the transmission loss or transmission gain in the perimeter zone, and also has a VAV cooling system to provide cooling and outdoor air for the perimeter and interior zones. The supply temperature of the air skin system is reset according to the outdoor temperature. If several air skin systems are not used for various zones of different sides of a building, field surveys show that during a cold day in winter, rooms facing south are sufficiently warmed by the incoming solar radiation. The additional warm air supply from the constant-volume air skin system raises the zone temperature to an unacceptably high level. Simultaneous cooling from the VAV cooling system must be provided to offset the warm air supply, which results in a waste of energy.

The most widely used perimeter heating system is the hot water finned-tube baseboard heater with zone controls, as described in Section 14.7. Sometimes, an electric baseboard heater or ceiling radiating panel is also used. In the perimeter zone, the baseboard heating system must be carefully zoned in coordination with the VAV system. Each side of the building should have a separate control zone to control the capacity of the hot water baseboard heater. Heating capacity can thus correspond to the variation of the

solar load and the infiltration loss as wind direction shifts. Shading from adjacent buildings also should be taken into account in zoning control.

A VAV cooling system is a multizone, single-duct, air system that provides cooling only. It supplies cooled and dehumidified air, containing outdoor air, to both the perimeter and interior zone to offset the space cooling load (including solar, transmission, and internal loads) and maintains the required indoor temperature while the space is occupied.

The air that is conditioned in the AHU or PU is similar to that in the single-zone VAV system. In a VAV cooling system, the volume flow rate and fan total pressure of the supply fan should be modulated according to the preset supply duct static pressure between the last branch takeoff and $0.3L_{main}$ from the last branch takeoff, as described in Section 18.10. Conditioned air is distributed to various control zones through the main duct, runouts, VAV boxes, flexible ducts, and diffusers. Space air is often returned to the fan room through the ceiling plenum and its connecting return duct.

The difference between a single-zone VAV system and a VAV cooling system is that the conditioned air in a VAV cooling system discharges to various zones through VAV terminal boxes (or simply *VAV* boxes), flexible ducts, and diffusers, as shown in

Fig. 27.2a. Each VAV box serves an individual control zone. Zone supply volume flow rate is modulated at the VAV box by a DDC controller according to the zone temperature sensed by the temperature sensor, as described in Section 18.6. The modulated air flow is then supplied to the conditioned space by means of one or several slot diffusers, ceiling diffusers, or a troffer–slot diffuser combination.

A VAV cooling system is the basic VAV system used for interior zones in various multizone VAV systems, such as perimeter heating VAV, VAV reheat, dual-duct VAV, and fan-powered VAV systems. The various types of VAV systems differ mainly in the type of air system that serves the perimeter zone.

VAV Boxes

A *variable-air-volume (VAV) box* is a terminal device in which the supply volume flow rate is modulated by varying the opening of the air passage by means of a single-blade butterfly damper, a multiblade damper, or an air valve, as shown in Fig. 27.3.

A VAV box may have a single outlet or multiple round outlets.

A *single-blade damper VAV box*, as shown in Fig. 27.3a and Fig. 27.3b, has a simple construction and is simple to operate. A typical damper closes at an angle 30° from vertical and rotates in counterclockwise direction to an angle of 60 degrees in the fully open position.

An *air valve*, as shown in Fig. 27.3c is a piston damper moving horizontally inside a hollow cylinder. The opening of the air passage can be adjusted. The main advantage of an air valve is its almost linear relationship between the modulated air volume and the displacement of the piston damper.

CONTROL MECHANISMS. The control mechanism of VAV box can be classified as pneumatic or direct digital control.

In a *pneumatic control VAV box*, zone temperature is sensed by a pneumatic thermostat and the rotation of the single-blade damper is actuated by pneumatic power. When the thermostat senses an increase

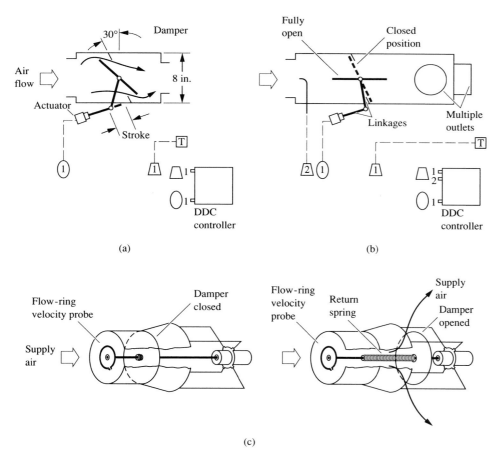

FIGURE 27.3 VAV boxes: (a) single-blade, pressure-dependent; (b) single-blade, pressure-independent; and (c) air valve.

in zone temperature, the increase of compressed air pressure on a diaphragm moves the actuator at a stroke of 1 to 1.5 in. and rotates the damper through the linkages, as shown in Fig. 27.3b. Therefore, the air passage opens wider. The spring pushes the actuator back when the air pressure decreases.

In a *DDC VAV box*, the temperature sensor sends a signal to a DDC controller. It actuates the motorized operator, moves the actuator to a certain displacement, and opens the single-blade damper wider. The damper is closed by either spring force or the reverse rotation of the motorized operator.

The pneumatic or direct digital control of an air valve is similar to that of a single-blade damper. A pneumatic control VAV box has a faster response actuator and a lower initial cost than a DDC VAV box. However, pneumatic systems require more maintenance to provide clean, dry compressed air. More importantly, DDC VAV boxes provide many sophisticated control functions, and can communicate with the central computer.

There is also a type of self-powered or system-powered VAV box control. It requires a higher supply air pressure to actuate the single-blade damper or air valve according to the input signal from the temperature sensor. A self-powered VAV box also has complicated control mechanisms, so its applications are very limited.

INFLUENCE OF DUCT STATIC PRESSURE. VAV boxes can be classified as pressure-dependent or pressure-independent.

In a *pressure-dependent VAV box*, the variation of the duct static pressure at the inlet of the VAV box caused by the opening and closing of the dampers connected to the same main supply duct influences the modulation of its supply volume flow rate. A maximum air flow limit is often used to keep the air flow below a maximum limit.

A *pressure-independent VAV box* modulates its supply volume flow rate regardless of the variation of duct static pressure at its inlet. A typical pressure-independent VAV box is shown in Fig. 27.3c. The temperature sensor resets the sensed velocity signal and the controller actuates the damper to a wider or narrower position based on the reset signal from the velocity probe. Even if the static pressure at the VAV box inlet varies from 0.5 to 3 in.WG, the volume flow is maintained according to the required value called for by the temperature sensor and the controller.

The sizes of the VAV boxes made by one manufacturer range from 04 to 20, with a corresponding volume flow rate from 225 cfm to about 4000 cfm. The pressure drop of these VAV boxes at nominal maximum volume flow rate when the damper is fully open usually varies from 0.2 to 0.5 in. WG.

In order to provide the required amount of outdoor air to the conditioned space, in a VAV cooling system, a VAV box usually reduces its volume flow to a minimum setting, typically 30 percent while the space is occupied.

Stability of VAV Box Control

The zone temperature control and sequence of operation of VAV boxes were discussed in Section 18.6. Recently designed VAV boxes using DDCs can provide the following functional controls:

- Zone temperature control by modulating the damper or air valve
- Zone temperature reset based on zone or outdoor temperature
- Maximum and minimum settings of air flow for each zone
- Morning warm-up and cool-down control

A normally closed DDC VAV box can shut off its supply volume flow when the served control zone is not occupied. This is especially useful for partially occupied floors after normal working hours.

Air flowing through VAV boxes is self-balanced along the main supply duct. However, air flow imbalance does affect the modulation range of VAV boxes. A well-designed supply duct system with minimized imbalance is preferable.

One characteristic of VAV boxes that directly affects their performance is control stability. Dean and Ratzenberger (1985) described the primary differences between an efficient and inefficient design of a VAV box and its control. An excellent design minimizes the open-loop gain and improves the overall system stability. *Gain* is defined as the real part of the logarithm of the transfer function of the control system. The primary factors that affect the open-loop gain, and thus the stability of a VAV box, are as follows:

- *Actuator Size.* In an efficient design, the stroke of the actuator is small (typically 1.0 in.), and in a pneumatic system, the active area of the actuator is typically only 2 in.^{2} instead of 8 in.^{2} A larger actuator needs more compressed air to drive a full-range stroke, so gain is greater.
- *Damper Size.* Gain is directly proportional to the size of the damper as well as the size of the VAV box.

- *Duct Pressure.* A lower excess pressure Δp_{exc}, in inches WG, during maximum and minimum air flow affects open-loop gain. Excess pressure is defined as the amount of pressure above that required to drive a nominal flow through a fully open damper.

According to Dean and Ratzenberger (1985), the typical measured performance of 26 efficient B-size VAV boxes and 27 inefficient C-size VAV boxes are shown below:

	B-size	C-size
Maximum flow, cfm	442	448
Corresponding pressure $p_{s.max}$, in. WG	1.07	1.40
Minimum flow, cfm	64	66
Corresponding pressure $p_{s.min}$, in. WG	1.24	1.59
Excess pressure, in. WG	0.08	0.62

For a step function response to a step change in internal loads, such as turning lights on or off, the deviation from set point in a pneumatic control system with a large actuator lasts about 40 minutes. In a DDC system, it lasts only about 15 minutes.

Based on their analysis, Dean and Ratzenberger (1985) recommended the following:

- A VAV system using pneumatic control VAV boxes should have relatively low main duct pressures and relatively small boxes and actuators.

- VAV boxes using DDC systems have excellent operating characteristics, although their actuators are a little slower than pneumatic controls.

- VAV boxes using direct digital control and well-designed VAV boxes using pneumatic control do not require pressure-independent values or velocity resetting. Pressure-independent VAV boxes are expensive and require additional maintenance.

- Pressure-independent VAV boxes and valves with velocity resetting improve the control stability of poorly designed pneumatic controls, make an unstable system more stable, and improve the performance of a poorly designed supply duct system.

Sound Power Level of a VAV Box

The sound power level of a VAV box depends mainly on the following:

- Volume flow of supply air for a specific size box, cfm

- Difference in static pressure $\triangle p_s$ across the VAV box, in inches WG, to provide a specific volume flow, in. WG

- The configuration of the VAV box, flexible duct(s), and diffuser(s)

The greater the Δp_s, the larger the static pressure at the box inlet. The smaller the damper opening, the higher the sound power level of the VAV box.

Noise generated by the VAV box can be transmitted to occupants in the conditioned space by duct-borne path through flexible ducts and diffusers, or a radiated path from its casing through the ceiling or plenum.

Evaluating noise levels from VAV boxes in an occupied zone has always been difficult because of the many variables that affect the results. The ADC and the ARI have jointly developed *Industry Standard 885: Acoustical Level Estimation Procedure*, as described in Chapter 16, to provide a reliable method to predict NC and RC in an occupied space.

In Smith (1989), the sound pressure levels of two terminal units in a 29-ft $\times$ 20-ft room with an 8 ft 7 in. ceiling were estimated and measured. One terminal unit (a VAV box) was operated at 500 cfm at 4.0 in. WG inlet static pressure and supplied 250 cfm to each of two diffusers though flexible ducts.

The actual sound pressure measured in the occupied space was NC-41dB. Another terminal was operated at 600 cfm with 3.0 in.WG inlet static pressure and supplied 300 cfm to each of two diffusers in the same room. The actual sound pressure measured was about NC41-dB. The measured sound pressure levels were about 2 to 3 dB higher or lower than the estimated duct-borne and radiated paths according to Industry Standard 885. If the sound power level at mid-frequency 2K and 4K Hz can be further attenuated by flexible duct and the slot diffuser, sound pressure level in the occupied space may drop to NC-35. The sound power level of a VAV box listed in a manufacturer's catalog must be tested in a certified laboratory.

Controls for AHU or PU

Controls for the air-handling unit (AHU) or packaged unit (PU) for a perimeter-heating VAV system may include the following:

- Discharge temperature control with outdoor air temperature or space air temperature reset

- Minimum outdoor air control

- Low-limit control

- Air-side economizer control

- Duct static pressure control
- Supply, relief, or return fan control
- Warm-up and cool down control
- Space humidity control
- Space pressurization control

All these controls, except space humidity control, were discussed in Sections 18.9 and 18.10. In a *space humidity control system*, a space humidity sensor measures the space relative humidity and sends a signal to the DDC panel to modulate the steam supply to the steam humidifier to maintain the space relative humidity φ_r between 24 and 30 percent (or any specified value) during winter heating mode. The steam humidifier is energized only when $\varphi_r < 24$ percent.

27.3 CASE STUDY: A VAV COOLING SYSTEM

Operating Parameters of a VAV Cooling System

Consider a multizone VAV cooling system using an AHU to serve a typical floor of 20,000 ft^2 of conditioned area in a high-rise office building that is divided nearly equally into perimeter and interior zones. A separate hot water baseboard heating system is used for winter heating in the perimeter zones. The perimeter and interior zones may each be divided into several to more than 20 control zones.

The summer and winter mode design conditions are shown below:

	Perimeter	Interior
Summer indoor temperature, °F	75	75
Space relative humidity, percent	43	45
Summer outdoor temperature, °F	95	95
Summer outdoor wet bulb, °F	78	78
Winter indoor temperature, °F	70	70
Winter outdoor temperature, °F	20	20
Winter indoor relative humidity, percent	30	30
Winter outdoor humidity ratio, lb/lb	0.0017	0.0017
Summer block cooling load, Btu/h	300,000	160,700
Winter interior zone latent load, Btu/h	—	13,300
Summer block sensible load, Btu/h	285,500	135,000
Outdoor air required, cfm	1920	1920
Summer supply system heat gain, °F	5	5
Summer return system heat gain, °F	3	3

Only summer cooling mode and the cold air supply for the interior zone during winter heating mode will be discussed in this case study.

For the perimeter zone in a typical floor of this office building, the average space cooling load den-

sity at summer design conditions is 30 Btu/h · ft^2, or 8.8 W/ft^2. The sensible heat ratio of the space conditioning line SHR$_s$ for the perimeter zone is usually about 0.95.

For the interior zone of a typical floor of this office building, the summer design load density is 15.5 Btu/h · ft^2 (4.5 W/ft^2) depending on the equipment load density. Occupant density is assumed to be 150 ft^2/person. The sensible heat ratio of the space conditioning line SHR$_s$ for the interior zone may vary from 0.85 to 0.90.

Conditioned Air Off-Coil and Supply Temperature Differential

For a supply temperature differential of 20°F, if the zone supply temperature $T_s = 55°F$, for a typical AHU with a supply fan power temperature rise of 2°F and a supply duct temperature rise of 3°F at summer design conditions, the temperature of conditioned air leaving the cooling coil T_{cc} should be $55 - (2 + 3) = 50°F$. If the relative humidity of conditioned air off the coil φ_{cc} is 95 percent and $h_{cc} = 20$ Btu/lb, then, from the psychrometric chart shown in Fig. 27.2b, the relative humidity of the supply air φ_s is 81 percent.

If a space conditioning line r_i s for the interior zone at a SHR$_{si} = 0.84$ is drawn from the state point of conditioned space r_i of 75°F and 50 percent relative humidity, line r_i s intersects the $T_s = 55°F$ line at a relative humidity of about 91 percent.

To maintain a zone temperature of 75°F and a space relative humidity of 50 percent, psychrometric analysis shows that it is impossible to have the following conditions:

- An off-coil temperature of $T_{cc} = 50°F$ and $\varphi_{cc} = 95$ percent
- A supply temperature $T_s = 55°F$
- A sensible heat ratio of space conditioning in the perimeter zone SHR$_{sx}$ of 0.95, or of interior zone SHR$_{si}$ of 0.84.

Figure 27.2b shows that when $T_{cc} = 50°F$, $\varphi_{cc} = 95$ percent, $T_{dis} = 52°F$, $T_s = 55°F$, a SHR$_{sx} = 0.95$, SHR$_{si} = 0.84$, and supply temperature differential $\Delta T_s = 20°F$, these prerequisites result in a perimeter and interior zone temperature of 75°F, and a $\varphi_{rx} = 43$ percent and a $\varphi_{ri} = 45$ percent. Here, T_{dis} indicates discharge temperature from the AHU or PU.

Supply Volume Flow Rate and Coil's Load

In any control zone in the perimeter and interior zones, the VAV box, flexible ducts, and slot diffusers should be sized according to the zone peak supply volume flow rate. The zone peak supply volume flow rate of any control zone served by a VAV box $\dot{V}_{sn}$, in cfm, at summer design conditions can be calculated as

$$\dot{V}_{sn} = \frac{q_{rsn}}{60\rho_s c_{pa}(T_r - T_s)} \qquad (27.4)$$

where q_{rsn} = zone sensible cooling load at summer design conditions, Btu/h.

In a multizone VAV system, the total supply volume flow rate the perimeter zone $\dot{V}_{sx}$, in cfm, at summer design load is the maximum possible coincident total supply volume flow rates of various zones at the same instant, or the block supply volume flow rate, which corresponds to the block load of the perimeter zone. The block supply volume flow rate in the perimeter zone at summer design conditions can be calculated as

$$\begin{aligned}
\dot{V}_{sx} &= \frac{q_{rs}}{60\rho_s c_{pa}(T_{rx} - T_s)} \\
&= \frac{285,500}{60 \times 0.075 \times 0.243(75 - 55)} \\
&= 13,054 \text{ cfm} \qquad (27.5)
\end{aligned}$$

Similarly, the total block supply volume flow rate of the interior zone at summer design conditions V_{si}, in cfm, is

$$\begin{aligned}
\dot{V}_{si} &= \frac{q_{rs}}{60\rho_s c_{pa}(T_{ri} - T_s)} \\
&= \frac{135,000}{60 \times 0.075 \times 0.243(75 - 55)} \\
&= 6173 \text{ cfm} \qquad (27.6)
\end{aligned}$$

The total supply volume flow rate for the AHU at summer design conditions is calculated as

$$\dot{V}_s = \dot{V}_{sx} + \dot{V}_{si} = 13,054 + 6173 = 19,227 \text{ cfm} \qquad (27.7)$$

Trunks, or main ducts that serve multiplezones, should accommodate the maximum possible air flow through these ducts at a given time. A diversity factor can be multiplied by the sum of individual peak volume flows to estimate the possible maximum flow.

From the psychrometric chart, the weighted mean space relative humidity for perimeter and interior zones φ_{rm} = 43.5 percent and the weighted mean humidity ratio w_{rm} = 0.0079 lb/lb.

Because the temperature of recirculating air entering the AHU T_{ru} = 75 + 3 = 78°F and $w_{ru} = w_{rm}$ = 0.0079 lb/lb, point ru can be plotted on the psychrometric chart. Draw line $ru\ o$.

$$\frac{ru\ m}{ru\ o} = \frac{3840}{19,227} = 0.20$$

so point m can then be determined: T_m = 80.5°F and h_m = 30.2 Btu/lb. The cooling coil's load q_{cc} in the AHU which is also a block load, in Btu/h, is

$$\begin{aligned}
q_{cc} &= 60\dot{V}_s\rho_s(h_m - h_{cc}) \\
&= 60 \times 19,227 \times 0.075(30.2 - 20) \\
&= 882,519 \text{ Btu/h} \qquad (27.8)
\end{aligned}$$

Summer Mode Part-Load Operation

During summer mode part-load operation, if the sensible cooling load in the perimeter zone is reduced to about 50 percent of its design load and the sensible cooling load of the interior zone is reduced to 85 percent of its design load, the total volume flow rate of the supply fan may decrease to about $\frac{2}{3}$ of its design value. Assume that the supply fan power heat gain and the volume flow are both reduced to 0.65 of the design value. The temperature rise at the supply fan at part-load is still about 2°F. If the chilled water temperature entering the coil T_{we} is reset from 45°F at design-load to 48°F at about $\frac{2}{3}$ part-load, the discharge temperature T_{dis} increases to T_{disp} = 50+2+(48−45) = 55°F. The temperature rise from duct heat gain increases because of the reduction of the supply volume flow rate. Because the increase of this temperature rise is proportional to the reduction of supply volume flow rate, the supply air temperature for the perimeter zone at part-load can be calculated as

$$T_{sxp} = (50 + 2) + 3 + \frac{3}{0.5} = 61°F$$

and the supply temperature for the interior zone at part-load is

$$T_{sip} = (50 + 2) + \frac{3}{0.85} = 58.5°F$$

The supply flow rate for the perimeter zone at part-load is

$$\dot{V}_{sxp} = \frac{0.5 \times 285,500}{60 \times 0.075 \times 0.243(75 - 58)} = 9325 \text{ cfm}$$

The supply flow rate for the interior zone at part-load is

$$\dot{V}_{sip} = \frac{0.85 \times 135,000}{60 \times 0.075 \times 0.243(75 - 55.5)} = 6359 \text{ cfm}$$

At summer mode part-load operation, the total supply volume flow is

$$\dot{V}_{sp} = \dot{V}_{sxp} + \dot{V}_{sip} = 9325 + 6359 = 15,684 \text{ cfm}$$

Assume the following conditions:

- The sensible ratio of the perimeter zone at part-load SHR_{sxp} drops to 0.91
- The SHR_s for the interior zone at part-load remains the same
- The temperature rise of the return system heat gain from the heat released by the light troffer in the ceiling plenum ΔT_{retp}, in °F, is proportional to the air volume flow, that is, $\Delta T_{retp} = 3/(15,684/19,227) = 3.7$°F

From the psychrometric chart, $T_{rup} = 75 + 3.7 = 78.7$°F.

If at summer mode part-load operation, the required amount of outdoor air remains the same as at design-load, the ratio of outdoor air to the supply air at summer mode part-load is therefore

$$\frac{rup\ mp}{rup\ o} = \frac{3840}{15,684} = 0.25$$

If the dry and wet bulb of outdoor air are the same as at design conditions, then, from the psychrometric chart, $T_{mp} = 83$°F and $h_{mp} = 31.9$ Btu/lb.

As T_{cc} is raised to 53°F because of the 3°F increase in I_{we}, the enthalpy of conditioned air off the coil at 53°F, 95 percent relative humidity is 21.6 Btu/lb and the coil's load at part-load is

$$
\begin{aligned}
q_{ccp} &= 60\dot{V}_{sp}\rho_s(h_{mp} - h_{cc}) \\
&= 60 \times 15,684 \times 0.075(31.9 - 21.6) \\
&= 726,953 \text{ Btu/h} \quad (27.9)
\end{aligned}
$$

Winter Mode Operation

At winter design conditions, the outdoor humidity ratio $w_o = 0.0017$ lb/lb. The difference of humidity ratios for the interior zone is therefore

$$
\begin{aligned}
w_{ri} - w_s &= \frac{q_{rl}}{(60\dot{V}_s\rho_s h_{fg.o})} \\
&= \frac{13,300}{60 \times 6173 \times 0.075 \times 1061} \\
&= 0.00045 \text{ lb/lb}
\end{aligned}
$$

where w_{ri} = air humidity ratio in the interior zone, lb/lb
w_s = humidity ratio of supply air, lb/lb

During winter design conditions, the space sensible cooling load in the interior zone may still equal that of summer design conditions. Therefore, the supply temperature differential for the interior zone is $(T_r - T_s) = 20$°F, or $T_s = 70 - 20 = 50$°F. The sup-

ply volume flow rate for the interior zone is still $\dot{V}_{si} = 6173$ cfm.

If the minimum supply volume flow rate to the perimeter zone for the outdoor air requirement during winter heating mode $\dot{V}_{sx.w}$ is equal to 0.30 of the design maximum flow, that is,

$$\dot{V}_{sx.w} = 0.30 \times 13,054 = 3916 \text{ cfm}$$

the temperature rise from the return system heat gain at winter design conditions is then

$$\Delta T_{ret.w} = \frac{3}{\frac{6713+3916}{19,227}} = 5.4 \text{°F}$$

and $T_{ru} = T_r + \Delta T_{ret.w} = 70 + 5.4 = 75.4$°F.

If the temperature rise from the supply fan power is still about 2°F and the temperature rise from supply duct heat gain is 3°F for the interior zone, then

$$T_m = T_s - (2 + 3) = 50 - (2 + 3) = 45 \text{°F}$$

and

$$
\begin{aligned}
\frac{(T_{ru} - T_m)}{(T_{ru} - T_o)} &= \frac{(75.4 - 45)}{(75.4 - 20)} \\
&= \frac{\dot{V}_o}{\dot{V}_s} = 0.55
\end{aligned}
$$

From the psychrometric chart, at a temperature $T_r = 70$°F and $\varphi = 30$ percent, $w_r = 0.0046$ lb/lb. Draw line rs from point r with $SHR_s = 0.84$, which intersects the $70 - 20 = 50$°F constant temperature line at point s, and $w_s = 0.0038$ lb/lb. Because $T_{ru} = 75.4$°F and $w_{ru} = w_r = 0.0046$ lb/lb, draw line $o\ ru$. Because $om/o\ ru = 0.55$, $w_m = 0.0034$ lb/lb.

In order to maintain a space relative humidity of 30 percent at winter design conditions, a steam humidifier is required with a humidifying capacity $\dot{m}_s$, in lb/h, calculated as follows:

$$
\begin{aligned}
\dot{m}_s &= 60\dot{V}_s\rho_s(w_s - w_m) \\
&= 60(6713 + 3916)0.075(0.0038 - 0.0034) \\
&= 19.1 \text{ lb/h} \quad (27.10)
\end{aligned}
$$

From the above analysis, when a perimeter-heating VAV system is used in an office building, two primary problems may be involved:

- Outdoor air supply during part-load operation. This will be discussed later in this chapter.
- If an air economizer cycle is used for free cooling in the interior zone during winter mode operation, for locations where the outdoor humidity ratio w_o drops below 0.003 lb/lb, the space relative humidity may fall below 25 percent, or even 20 percent. A

humidifier may be required. If the outdoor humidity ratio w_o is around 0.003 lb/lb, or there is a higher space latent load, space relative humidity φ_r may be around 25 percent at winter design conditions. In comfort air conditioning systems other than in settings such as hospitals and nurseries, a humidifier is not required.

Pressure Characteristics and Volume Flow Control

In a VAV cooling system with a system total pressure drop of 4.75 in. WG, as shown in Fig. 27.4, the total pressure drop, in inches WG, between various sections of the system at peak supply volume flow rate may have values as follows:

	Flow, cfm	
	Peak flow	0.5 Peak flow
Return system	0.25	0.75
AHU	2.25	0.563
Supply main duct	1.25	0.313
VAV box	0.50	0.8
Flexible ducts and slot diffuser	0.40	0.10

The purpose of duct static pressure control is to achieve the following:

- Limit the maximum static pressure in the supply main duct and maintain a preset static pressure at the point where the static pressure sensor is located.
- Provide the required volume flow rate of supply air for any branch takeoff at both design load and part-load, and prevent starve.
- Minimize fan energy use at part-load.

A *starve VAV box* is a VAV box with a supply volume flow rate lower than the amount required to offset the zone load. Starved VAV boxes are those boxes that require a duct static pressure at their branch takeoff inlets higher than that required at either design or part-load. This is because the static pressure sensor is not properly located and the supply duct and branch takeoffs are not properly designed and sized.

If the static pressure sensor is located at the end of the main supply duct just before the last branch takeoff at point D, as shown in Fig. 27.4, at peak supply volume flow rate, if the total pressure loss of the VAV box, flexible ducts, and slot diffuser $\Delta p_t = 0.9$ in. WG, the static pressure at point D is 0.9 in. WG. At peak supply volume flow rate, the branch take-offs at point C require a static pressure of 1.2 in. WG because of the pressure drop between points C and D in the supply main duct.

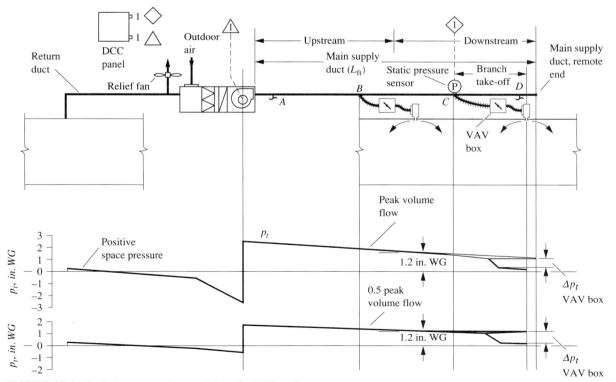

FIGURE 27.4 Typical pressure characteristics of a VAV cooling system.

During part-load operation, when the volume flow has been reduced to 0.5 peak supply volume flow, the static pressure at point D is maintained at 0.9 in. WG. The static pressure at point C is now about 0.98 in. WG. Although the system pressure and the fan energy use are minimized at part-load if the static pressure sensor is located at point D, if the branch takeoff at point C needs a volume flow nearly equal to design volume flow, the VAV box connected to point C will starve. Such a result is not desirable. Therefore, as described in Section 18.10 the following statements are true:

- If the volume flow rates of the branch takeoff connected to the upstream part of the main supply duct may remain at design volume flow while the volume flow rates connected to the downstream part of the main duct may be reduced to 50 percent of the design flow at part-load, then the static pressure sensor should be located 0.3L from the remote end of the main duct (point C in Fig. 27.4).
- If all the branch takeoffs will be reduced at part-load, or the upstream branch takeoffs have a greater reduction than downstream ducts during part-load operation, then the static pressure sensor should be located near the remote end of the main duct.
- A 0.1 in. WG should be added to the set point value as a safety factor.
- If there are two or three main or trunk ducts, two or three static pressure sensors can be installed. Each may be located 0.3L from the remote end or near the end of each main duct depending on the reduction of volume flow in its branch takeoffs. The comparator in the DDC panel can pick the lowest static pressure as the feedback value to modulate the adjustable-frequency AC drive to maintain a duct static pressure higher than preset value in all the main ducts.

As discussed in Section 18.10, microprocessor-based terminal-regulated air volume (TRAV) control modulates the volume flow of the supply fan based on the sum of measured supply volume flow rates at the VAV boxes. Such a control strategy can save more fan energy during part-load operation than conventional duct static pressure controls.

27.4 VAV REHEAT SYSTEMS

System Description

A *VAV reheat system* is similar to a VAV cooling system except that reheating VAV boxes are used as the terminal units for the perimeter zone, as shown in Fig. 27.5a. In a reheating VAV box, a reheating coil is added to a VAV box after the single-blade damper

or air valve. Reheating VAV boxes installed inside the ceiling plenum. VAV boxes are still used for the interior zone in VAV reheat systems.

When the space sensible load in the perimeter zone drops during summer mode part-load operation, the VAV box closes its opening until its minimum setting is reached. If the zone temperature drops further and falls below a certain limit, the reheating coil is energized in steps and reheats the supply air to maintain a predetermined zone temperature.

During winter mode operation, when zone temperature in the perimeter zone falls below a predetermined limit, the reheating coil is also energized to heat the air to offset the space heating load in the perimeter zone. Zone temperature can be reset at different values during summer mode and winter mode, with a dead band between them.

For a control zone in the interior zone, the zone sensible load seldom drops below 30 percent of its peak load while the space is occupied except during the evenings and weekends. Therefore, reheating VAV boxes are not used in the interior zone.

VAV reheating is a simple and effective means of maintaining preset space temperature with a lower relative humidity at summer part-load operation. Actually, the reheating of outdoor air during the air economizer cycle for winter heating in the perimeter zone to offset transmission and infiltration loss is not a simultaneous cooling and heating process. Also, the cooling capacity during simultaneous cooling and heating is considerably less when the volume flow of reheating VAV box is reduced to its minimum setting before reheating.

In order to minimize energy consumption, ASHRAE/IES Standard 90.1–1989 specifies the following criteria for VAV reheat systems to minimize simultaneous cooling and heating:

> Variable air volume (VAV) systems . . . are designed to reduce the air supply to each zone to a minimum before reheating, recooling or mixing takes place. This minimum volume shall be no greater than the larger of the following:
>
> 1. 30 percent of peak supply volume
> 2. The minimum required to meet ventilation requirement
> 3. 0.4 cfm/ft^2 of zone conditioned floor area (ASHRAE, 1989)

Reheating VAV Box

A *reheating VAV box* is a VAV box with a reheating coil. The reheating coil is usually a 2- or 3-row hot water heating coil. If an electric coil is installed, it should be a duct-mounted coil at least 4 ft down-

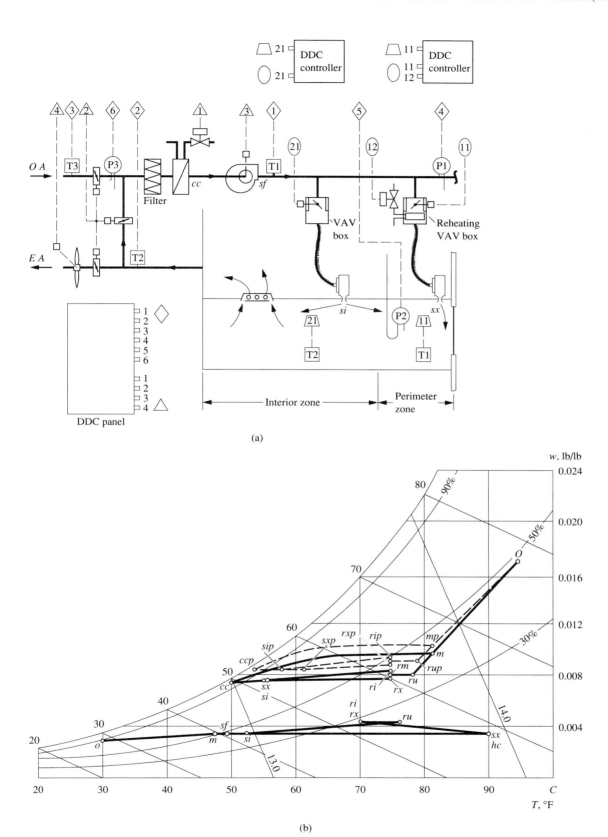

FIGURE 27.5 A VAV reheat system: (a) schematic diagram and (b) air conditioning cycle.

stream from the VAV box. Federal and local safety codes must be followed if an electric coil is installed.

To prevent excessive pressure drop at the reheating coil, the face velocity of the reheating coil is usually between 200 and 400 fpm. For each row of the reheating coil, the additional pressure drop is usually between 0.1 and 0.2 in. WG.

In the perimeter zone, a two-way slot diffuser should be located about one foot from the window. Air should be discharged downward from one of the slots in a direction toward the window, as described in Section 9.5. Air is discharged horizontally away from the window from another slot.

Characteristics of the Reheating Process

If there is only one AHU or PU to serve both the perimeter and interior zones of a typical floor in a VAV reheat system, the air discharged from the AHU or PU at winter design conditions using an air economizer cycle must fulfill the cold air supply requirement of the interior zone, typically at 52°, as shown in Fig. 27-5b. If the temperature increase from supply duct heat gain is 2° F, the temperature of air entering the reheating coil in VAV box T_{en} is 54°F.

The maximum reheating coil load for any zone or enclosed space in the perimeter zone to offset the space heating load at winter design conditions q_{chn}, in Btu/h, is calculated as

$$q_{chn} = q_{rhn} + 60\dot{V}_{sn}\rho_s c_{pa}(T_{rn} - T_{en.n}) \quad (27.11)$$

where q_{rhn} = space heating load for zone n, Btu/h
$\dot{V}_{sn}$ = volume flow rate of cold outdoor air discharged to the control zone from the VAV cooling system, cfm

If the depth of the perimeter zone is 15 ft, for a zone heating load linear density $q_{h.ft}$ = 150 Btu/h·ft, with winter indoor temperature maintained at 70°F, the supply air from the reheating VAV box is required to offset a zone heating load density of

$$\frac{q_{rh}}{A} = \frac{150}{15} = 10.0 \text{ Btu/h} \cdot \text{ft}^2$$

For a summer peak volume flow rate of 1.5 cfm/ft² floor area in the perimeter zone, at 30 percent minimum setting means $0.3 \times 1.5 = 0.45$ cfm/ft² of supply air.

The reheating coil's load for 1 ft² of floor area at winter design conditions can be calculated as

$$q_{chn}/A = 10 + 60 \times 0.45 \times 0.075$$
$$\times 0.243(70 - 54)$$
$$= 18 \text{ Btu/h} \cdot \text{ft}^2$$

One of the primary considerations in the design of a VAV reheat system at winter design conditions is the control of stratification when warm air is supplied from overhead slot and ceiling diffusers in perimeter zone. As discussed in Section 9.5, Straub and Cooper (1991) recommended that supply air temperature differential $(T_r - T_s)$ should not exceed 20°F to prevent excessive buoyancy effects. A terminal velocity of 150 fpm is required at the 5-ft level to offset the cold draft. An even distribution of air volume flow between the warm airstream blowing down toward the window glass and the warm airstream horizontally discharged toward the partition wall across from the window is preferable.

At a 30 percent minimum setting with a peak volume flow rate of 1.5 cfm/ft², the space head load per ft² that can be offset by warm air supply in perimeter zone during winter design condition is

$$q_{rh}/A = 0.3 \times 1.5 \times 60 \times 0.075 \times 0.243 \times 20$$
$$= 9.84 \text{ Btu/h} \cdot \text{ft}^2$$

or

$$q_{h.ft} = 148 \text{ Btu/h} \cdot \text{ft}^2$$

In order to control stratification, if q_{rh}/A increases to 15 Btu/h · ft, or $q_{h.ft}$ increases to 250 Btu/h · ft at winter design conditions, the minimum setting of the warm air supply must be increased to about 50 percent.

Controls

The sequence of operation of zone temperature control by means of a reheating VAV box using a DDC controller is similar to that described in Section 18.6 and shown in Fig. 18.7. The zone temperature–volume flow diagram and output diagram are both based on a proportional-integral (PI) control mode. In addition, in a DDC-controlled reheating VAV box, it is possible to provide minimum outdoor air control when supply volume flow is reduced during part-load operation. Measurements of outdoor air and supply volume flow rates in the AHU determine their ratio. A signal is sent to the central computer to reset the minimum limit of each reheating VAV box to provide adequate outdoor air. The AHU or PU control systems for a VAV reheat system are similar to those of a VAV cooling system.

Summer and Winter Mode Operation

The summer cooling mode operation, air conditioning cycle, supply volume flow and coil load calculations, and winter mode operation for the interior zone of a VAV reheat system are the same as those of a VAV cooling system.

In locations where the outdoor air humidity ratio w_o at winter design conditions is about 0.003 lb/lb or greater, space relative humidity can be maintained at around 25 percent. In comfort air conditioning systems in commercial buildings, a humidifier is usually not installed in the AHU or PU.

Outdoor air supplied to both the perimeter and interior zones must be carefully checked both at design and at minimum part-load operation. This will be discussed in later sections.

27.5 DUAL-DUCT VAV SYSTEMS

System Description

A *dual-duct VAV system* uses two supply air ducts, a warm air duct and a cold air duct, to supply both warm air and cold air to various control zones in the perimeter zone. Warm and cold air are mixed in the mixing VAV box in order to maintain a predetermined zone temperature, as shown in Figure 27.6a. For control zones in the interior zone, only a cold air duct is used to supply cold air to offset the year-round space cooling load.

At summer design conditions, minimum outdoor air is mixed with recirculating air in the AHU. The mixture then flows through the filter, the energized cooling coil, and the cold air supply fan in the cold deck of the AHU. After that, conditioned air discharges to the mixing VAV box, typically at 52°F, through the cold air duct. In the hot deck of the AHU, the outdoor and recirculating air mixture flows through the filters, deenergized heating coil, and warm air supply fan, and then discharges to the mixing box, typically at 80°F. The cold deck is the part of the AHU where air is cooled, and the hot deck is the part of the AHU where air is heated.

At summer design conditions, only cold air and a small amount of warm air leaking from the closed warm air damper are mixed together in a mixing VAV box. The mixture is then supplied to the control zone in the perimeter zone. Cold air is supplied to the interior zones via cold duct and VAV boxes. Space air is usually returned to the fan room through the ceiling plenum and the connecting return ducts.

In spring and fall, a mixture of minimum outdoor air and recirculating air may flow through the deenergized heating coil in the hot deck and discharge to the mixing VAV boxes at a temperature between 76°F and 80°F depending upon the outdoor temperature. In the cold deck, when an air economizer cycle is used, either a mixture of outdoor and recirculating air or 100 percent outdoor air is cooled and dehumidified in the cooling coil and then supplied to the mixing VAV box at a temperature around 52°F. The warm and cold air is mixed and modulated in the mixing VAV box to maintain a zone temperature between 70 and 75°F. As in summer design conditions, cold air is supplied to the interior zone through the cold duct in spring and fall to maintain a required zone temperature by modulating the volume flow in the VAV boxes.

At winter design conditions, the mixture of minimum outdoor air and recirculating air is heated at the heating coil in the hot deck and supplied to the mixing VAV boxes, usually at a temperature $T_s \le 90°$ F. In the cold deck, adequate outdoor air is mixed with the recirculating air to maintain a required mixing temperature at 52°F, if possible. The mixture flows through a deenergized cooling coil and supplies to the mixing VAV boxes in the perimeter zone and in the interior zones.

At winter design conditions, warm air or a mixture of warm and cold air at a required ratio is supplied to various control zones in the perimeter zone from the mixing VAV boxes to maintain a predetermined zone temperature. Cold air is modulated at the VAV boxes and then supplied to the interior zone to maintain required space temperature.

Number of Supply Fans

If an air economizer cycle is used, a relief fan or a return fan must be installed to relieve the space air when 100 percent outdoor air is used for free cooling. If a ceiling plenum is used as the return plenum, and if each typical floor in a multistory building is served by one or two air-handling units, the pressure drop of the return system is often less than 0.6 in. WG, and a relief fan is a better choice than a return fan, as described in Section 19.12.

A dual-duct VAV system may be installed with a single supply fan that delivers to both warm and cold ducts, or two supply fans, one for the warm duct and one for the cold duct. Although a single supply fan is simpler and lower in initial cost, and the air-handling unit can be a factory-made multizone unit rather than a field-assembled built-up AHU, a two-fan design has the following advantages:

- It allows the use of different ratios of outdoor air to supply air for the cold duct and the warm duct.
- The warm air supply fan is operated only when it is required, such as during warm-up period or in spring, fall, and winter when the transmission and infiltration loss is greater than the internal and solar loads in any control zone in the perimeter zone.
- A two-fan design simplifies control and saves energy. Different control schemes are possible for cold and warm air temperature resets, as well as for cold and warm air duct static pressure controls.

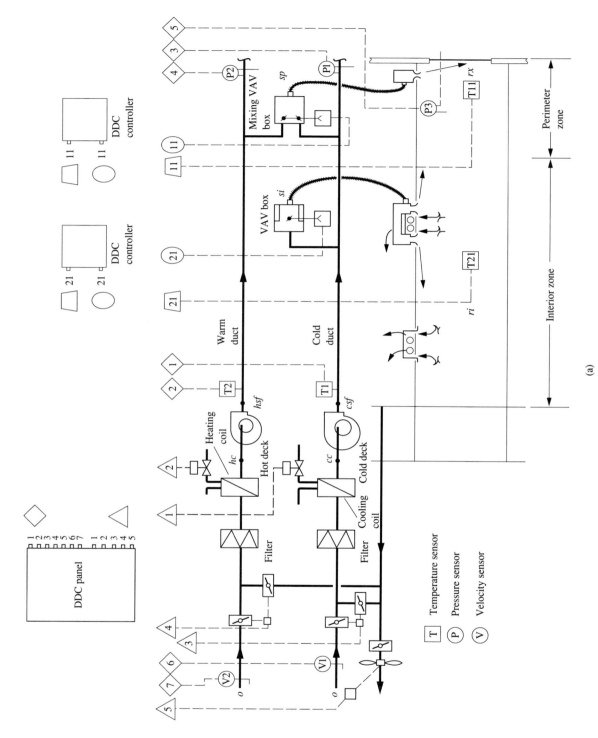

FIGURE 27.6 A dual-duct VAV system: (a) schematic diagram and (b) air conditioning cycle.

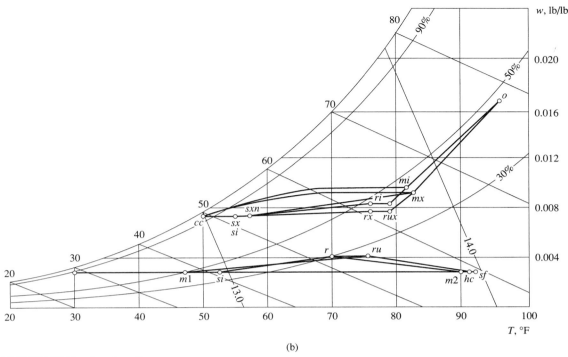

(b)

FIGURE 27.6b   *(continued)*

Mixing VAV Box

Figure 27.7 shows a pressure-dependent mixing VAV box using a specific terminal DDC controller. This box consists of two separate equal-sized air passages arranged in parallel—one for warm air and another for cold air. Each has a single-blade damper. These two air passages are then combined together and the mixture of warm and cold air is discharged through single or multiple outlets to the diffusers through a flexible duct. Two temperature sensors are connected to the DDC controller to modulate the dampers in the warm air and cold air passages separately, using two actuators.

If the mixing VAV box is pressure-independent, two additional velocity probes are added to the inlets of cold and warm air so that each damper is modulated separately by the DDC controller according to the signal from the velocity probe reset by the zone temperature sensor. As in the VAV box, a pressure-independent mixing VAV box is not necessary, although it may improve system performance if the duct and distribution system is not well-designed.

As in VAV boxes and reheating VAV boxes, currently available DDC mixing boxes have the following controls: zone temperature and reset, minimum and maximum limit, minimum outdoor air, and warm-up and cool-down controls.

In a dual-duct VAV system using a mixture of cold and warm air, as in a reheating VAV system, the min-

imum setting of zone supply volume flow from the mixing VAV box should follow the guidelines for VAV systems specified in ASHRAE/IES Standard 90.1–1989, that is, 30 percent of peak supply volume.

Operating Characteristics of Mixing VAV Box

Consider a dual-duct VAV box serving a control zone in the perimeter zone that has a summer indoor design temperature of 75°F and a winter indoor design temperature of 70°F. Figure 27.7b shows the zone temperature–supply volume $T_r - \dot{V}_{sn}$ diagram of such a DDC controlled mixing VAV box with PI control. Here, $\dot{V}_{sn}$ indicates the zone supply volume flow rate, in cfm. The operation of this mixing VAV box can be divided into cooling mode, dead band mode, mixing mode, and heating mode.

When the zone sensible cooling load is at design peak load, the mixing VAV box operates in cooling mode. Zone supply volume flow rate $\dot{V}_{sn}$ is at its peak volume flow of cold air supply $\dot{V}_{sn.d}$, that is, 100 percent peak volume flow. Zone temperature T_r is maintained at 75.2°F.

During summer cooling mode part-load operation, if a drop in the zone temperature T_r is sensed by the zone temperature sensor, the DDC controller closes the damper blade of the cold air supply. The supply volume flow rate $\dot{V}_{sn}$ is then reduced gradually in order to maintain a preset zone temperature of 75.2° F until

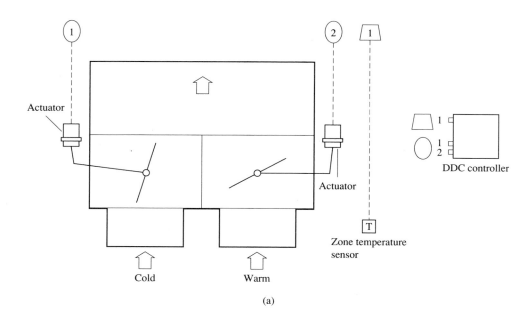

(a)

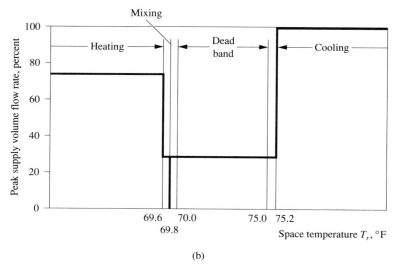

(b)

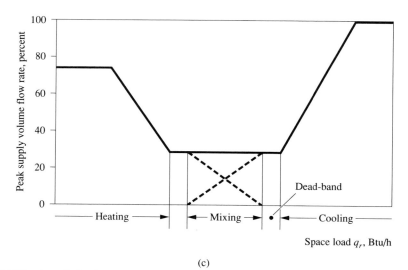

(c)

FIGURE 27.7 A typical mixing VAV box: (a) schematic diagram; (b) zone temperature–supply volume diagram, PI control mode; and (c) space load–supply volume diagram.

$\dot{V}_{sn}$ is reduced to its minimum setting, typically 0.3 $\dot{V}_{sn.d}$.

If the zone sensible cooling load still drops, then zone temperature T_r falls to 75°F or below. The mixing VAV box is now operated in dead-band mode. During dead-band mode operation a 0.3 minimum setting of cooling air supply is used to mainly provide outdoor air for the occupants and to maintain minimum zone air movements. Zone temperature is allowed to vary between 70.0 and 75.0°F.

If zone temperature T_r drops to 69.8°F, the mixing VAV box is operated in mixing mode. The cold air supply may drop from 0.3 $\dot{V}_{sn.d}$ to zero. At the same time, warm air is supplied at a flow rate such that the supply volume flow rate of the mixture of the cold and warm air supply is still 0.3 $\dot{V}_{sn.d}$ in order to maintain a zone temperature of 69.8°F, provide required outdoor air, and produce space minimum air movement.

When cold air supply has dropped to zero and warm air supply has increased to 0.3 $\dot{V}_{sn.d}$, the zone temperature may still fall to 69.6°F because of the increase in zone heating load in the perimeter zone. The mixing VAV box is then operated in winter heating mode. Warm air supply is increased to a volume flow rate $\dot{V}_{sn} > 0.3\dot{V}_{sn.d}$ until it reaches an upper limit at winter design condition $\dot{V}_{s.w}$, which may be between 0.5 and 1.0 $\dot{V}_{sn.d}$, to maintain a zone temperature of 69.6°F.

Volume flow rate $\dot{V}_{s.w}$ depends mainly on the space heating load and the supply temperature differential at winter design load. The supply volume flow rate in winter heating mode $\dot{V}_{s.w}$ should be high enough that the maximum warm air supply temperature does not exceed 90°F at winter design conditions.

Figure 27.7c shows a space load–supply volume diagram for a mixing VAV box. In this diagram, the variation of cold and warm air supply is clearly illustrated. It is assumed that the warm air supply temperature is 76°F when cold air supply is zero during mixing mode operation.

Discharge Air Temperature Control

In general, discharge air temperature from the cold deck must meet the year-round cooling requirements in the interior zone. Air economizer cycle free cooling should be applied to the cold deck whenever possible to meet the outdoor air requirements and to minimize refrigeration in order to save energy. In the hot deck, internal heat gains carried by the recirculating air should be fully utilized.

Cold deck discharge temperature $T_{c.dis}$ is usually reset according to zone demands at part-load by means of a DDC panel. Hot deck discharge temperature $T_{h.dis}$ is often reset by outdoor temperature during winter heating mode.

Figure 27.8 shows the cold and warm discharge air temperatures from the AHU of a typical dual-duct VAV system at various outdoor temperatures. The cold deck discharge air temperature $T_{c.dis}$ is maintained at 52°F at all outdoor temperatures in order to meet high zone cooling loads in the interior zone. When outdoor air $T_o > 78$°F, the cold deck is operated in minimum outdoor air operating mode. If 50°F $\leq T_o \leq 78$°F, the cold deck is operated in air economizer cycle. As soon as $T_o < 50$°F, refrigeration is turned off. Outdoor air is mixed with the recirculating air at various ratios to maintain the preset discharge temperature.

When outdoor air temperature $T_o \geq 70$°F, mixing of minimum outdoor air and recirculating air lowers warm discharge air temperature $T_{h.dis}$ from 80°F to 76°F. If $T_o < 68$°F, heating may be required in the perimeter zone, so $T_{h.dis}$ increases as T_o falls. If $T_o = 20$°F, $T_{h.dis}$ may increase to 90°F. It will stay at 90°F even if T_o falls below 20°F in order to avoid a supply differential $(T_r - T_{sn}) > 20$°F at winter design conditions and to avoid stratification.

Duct static pressure or supply volume flow control, air economizer control for the cold deck, and space pressurization controls for the AHU in a dual-duct VAV system are the same as for a VAV reheat system or a VAV cooling system.

Zone Supply Volume Flow Rate

For each control zone in the perimeter zone served by a mixing VAV box, an air leakage of 0.03 to 0.07 of its peak supply volume flow should be considered for the shut-off damper. If the difference in air density between warm and cold air is ignored, the additional supply volume flow rate $\dot{V}_{lk}$ required to compensate for the air leakage when the warm air damper is shut, in cfm, can be calculated as

$$\dot{V}_{lk} = \frac{0.05\dot{V}_{sn}(T_{s.h} - T_r)}{T_r - T_{s.c}} \qquad (27.12)$$

where $T_{s.h}$, $T_{s.c}$ = temperature of warm and cold air supply at the mixing VAX box, °F

T_r = zone temperature, °F

As in a VAV cooling system, for any control zone in either the perimeter or interior zones, the mixing VAV box, VAV box, flexible ducts, and slot or ceiling diffusers should be sized according to the peak supply

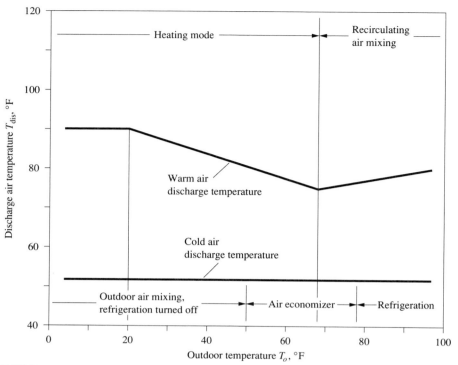

FIGURE 27.8 Warm and cold air discharge temperature in a dual-duct VAV system.

volume flow. If there is no air leakage, the zone peak supply volume flow $\dot{V}_{sn}$, in cfm, is

$$\dot{V}_{sn} = \frac{q_{rsn}}{60\rho_s c_{pa}(T_r - T_{s.c})} \qquad (27.13)$$

where q_{rsn} = zone maximum sensible colling load, Btu/h. The zone peak supply volume flow rate for cold air in the perimeter zone at summer design conditions is

$$\dot{V}_{cxn} = \dot{V}_{sn} + \dot{V}_{lk} \qquad (27.14)$$

For any control zone in the perimeter zone, the peak supply volume flow rate V_{xn} delivered to the conditioned space from the mixing VAV box, including 5 percent air leakage from the warm air damper at summer design conditions, in cfm, is therefore

$$\dot{V}_{xn} = \dot{V}_{cxn}/0.95 \qquad (27.15)$$

Because only cold air from the VAV box is supplied to the interior zone, for any control zone in the interior zone, the peak supply volume flow $\dot{V}_{in}$ at summer design conditions, in cfm, is then

$$\dot{V}_{in} = \dot{V}_{sn} \qquad (27.16)$$

Case Study: A Dual-Duct VAV System for a Typical Floor

Consider a dual-duct VAV system serving a typical 20,000-ft² floor in an office building with the same

operating parameters as the VAV cooling system described in Section 27.3.

As in the VAV cooling system, the maximum supply volume flow rate for the interior zone at summer design conditions based on block load is

$$\begin{aligned} \dot{V}_{si} &= \frac{q_{rs}}{60\rho_s c_{pa}(T_r - T_{s.c})} \\ &= \frac{135,000}{60 \times 0.075 \times 0.243(75 - 55)} = 6173 \text{ cfm} \end{aligned}$$

For a control zone in perimeter zone, if the temperature of the mixture of outdoor and recirculating air $T_{mx} = 80.5°F$, humidity ratio $w_{mx} = 0.0093$ lb/lb, $T_{sx} = T_{s.c} = 55°F$, and $\varphi_{sx} = \varphi_{s.c} = 80$ percent, draw line $sx\ mx$ as shown in Fig. 27.6b.

For a warm air leakage of $0.05\dot{V}_{sx}$ from the damper, the state point of the mixture of cold air supply and warm air leakage sxn can be determined. From the psychrometric chart, $T_{sxn} = 56.5°F$, and $w_{sxn} = 0.0083$ lb/lb.

The volume flow rate of cold air supply to the perimeter zone at summer design conditions based on the block load and including 5 percent air leakage from the warm air damper is therefore

$$\begin{aligned} \dot{V}_{scx} &= \frac{285,500}{60 \times 0.075 \times 0.243(75 - 56.5)} \\ &= 14,113 \text{ cfm} \end{aligned}$$

The total cold air supply volume flow rate to both the perimeter zone and the interior zone at summer design conditions is

$$\dot{V}_{s.c} = \dot{V}_{sx} + \dot{V}_{si} \qquad (27.17)$$
$$= 14{,}113 + 6713 = 20{,}826 \text{ cfm}$$

The volume flow rate of warm air supply is usually expressed as a percentage of peak supply volume of cold air, usually between $0.5\dot{V}_{s.c}$ and $1.0\dot{V}_{s.c}$. It depends mainly on the space heating load linear density $q_{h.ft}$, Btu/h · ft, and the supply temperature differential $(T_s - T_r)$. If the volume flow rate of the warm air supply at winter design conditions $\dot{V}_{s.h}$, in cfm, is $0.75 \, \dot{V}_{s.c}$, then

$$\dot{V}_{s.h} = 0.75 \times 20{,}826 = 15{,}620 \text{ cfm}$$

However, it is possible to use the warm air duct as a cold air duct in summer cooling mode operation so that, there is a warm/cold duct and a cold duct. Such an approach may reduce the size of the cold duct and the warm air leakages. However, changeover from cold air supply to warm air supply often produces many problems if the warm/cold duct serves zones with different orientations. This approach is not advisable.

Other parameters, such as cooling and heating coil load and space relative humidity at winter design conditions, can be determined by the same methods as for VAV cooling systems and VAV reheat systems.

Part-Load Operation

Consider a typical control zone in the perimeter zone of this dual-duct VAV system whose operating parameters at summer design conditions and summer part-load are as follows:

	Full-load	Part-load
Zone sensible cooling load, Btu/h	14,200	4260
Leaving cooling-coil temperature, °F	50	55
relative humidity, percent	95	95
Supply system heat gain, °F	5	10
Warm air supply temperature, °F	80	80

If the supply air temperature discharged from the mixing VAV box $T_{nx} = T_{m2} = 56.5°\text{F}$, the volume flow rate of conditioned air supplied to this control zone $\dot{V}_{nx}$, cfm, including 5 percent warm air leakage from the damper at summer design conditions, is therefore

$$\dot{V}_{nx} = \frac{14{,}200}{60 \times 0.075 \times 0.243(75 - 56.5)} = 702 \text{ cfm}$$

During part-load operation, as the zone sensible cooling load is reduced to 30 percent of the design load, $T_{sxp} = 50 + 10 = 60°\text{F}$. Assuming that the percentage of warm air leakage and the temperature increase due to this warm air leakage are the same at design condition, the supply temperature of the

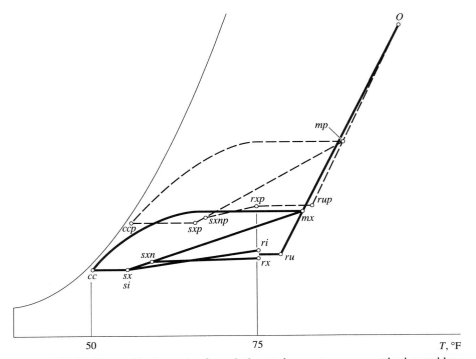

FIGURE 27.9 Air conditioning cycle of a typical control zone at summer part-load served by a mixing VAV box of a dual-duct VAV system.

mixture of cold and warm air of zone n at part-load T_{sxnp}, in °F, is now equal to $55 + 10 + 1.5 = 66.5$°F.

The supply volume flow rate from the mixing VAV box can be calculated as

$$
\begin{aligned}
\dot{V}_{\text{sxnp}} &= \frac{q_{\text{rsnp}}}{60\rho c_{\text{pa}}(T_{\text{rnp}} - T_{\text{sxnp}})} \\
&= \frac{4260}{60 \times 0.075 \times 0.243(75 - 66.5)} \\
&= 458\text{cfm}
\end{aligned}
\tag{27.18}
$$

From Eq. (27.15), The volume flow rate of cold air supply at part-load can be calculated as

$$
V_{\text{s.c}} = 0.95 \times 458 = 435 \text{ cfm}
$$

Also, the warm air supply is

$$
\dot{V}_{\text{s.h}} = 458 - 435 = 23 \text{ cfm}
$$

Figure 27.9 (p. 27.23) shows the air conditioning cycle of a typical control zone in the perimeter zone served by a dual-duct VAV system at summer mode part-load operation. From the psychrometric chart, zone relative humidity at summer part-load operation is higher than at full load condition. This is mainly because of the chilled water reset and cold discharge air temperature reset at part-load.

At winter mode part-load operation, the state point of supply air will move along the horizontal line $m1$ hc as shown in Fig. 27.6b to maintain a predetermined zone temperature.

27.6 FAN-POWERED VAV SYSTEMS

System Description

A *fan-powered VAV system,* as shown in Fig. 27.10 is a combination of VAV cooling system and fan-powered VAV units installed in each control zone of the perimeter zone in a building. The VAV cooling system provides cooling for both the perimeter and interior zones. The function of a fan-powered unit includes the following:

- It extracts warm recirculating air from the ceiling plenum to maintain a preset zone temperature when the volume flow of the cold primary air supplied from the VAV box has dropped to minimum setting at summer part-load.
- It provides the ceiling plenum warm recirculating air as the first step to offset the transmission and infiltration loss during winter heating mode operation. If the zone temperature drops further, a reheating coil in the fan-powered unit is energized to maintain predetermined zone temperature.

- It increases the air movements in the conditioned space when primary air supply falls to minimum.
- It mixes the low temperature primary air with warm plenum air to prevent the dump of a cold air jet and possible surface condensation.

If the fan-powered unit is used for mixing low-temperature primary air with warm plenum air in cold air distribution, fan-powered units are also installed in the control zones of the interior zones.

Fan-Powered Units

A *fan-powered VAV terminal unit,* or simply a fan-powered unit, consists of a small forward-curved centrifugal fan, a VAV box, a heating coil, dampers, an outer casing, and corresponding controls, as shown in Fig. 27.11 (p. 27.27). Fan-powered units can be classified as series and parallel units according to their construction and operating characteristics.

SERIES FAN-POWERED UNIT. In a *series fan-powered unit,* a small centrifugal fan is connected in series with the cold primary air stream flowing through the VAV box, as shown in Fig. 27.11a. The volume flow of the cold primary air is varied at part-load, but after this air mixes with varied induced plenum recirculating air, the fan delivers a continuous, nearly constant volume flow rate of the mixture of primary air and plenum recirculating air. The mixture is then supplied to the control zones, either heated at the heating coil or without heating, depending upon the zone load.

Because the air flow from the series fan-powered unit is constant, it is easier to select ceiling diffusers to achieve good space air diffusion. In a series fan-powered unit, cold primary air is thoroughly mixed with plenum recirculating air. The main disadvantage of a series fan-powered unit is its larger fan, which consumes more energy than a parallel fan-powered unit.

PARALLEL FAN-POWERED UNIT. In a *parallel fan-powered unit,* the induced recirculating plenum air stream discharged from the centrifugal fan is connected in parallel with the cold primary air stream discharged from the VAV box. These two airstreams are mixed either before or after the heating coil, and are then supplied to the control zones.

A parallel fan-powered unit with a heating coil located upstream from the fan is called a *draw-through parallel fan-powered unit,* shown in Fig. 27.11b. When the heating coil is located downstream from the fan, the unit is called a *blow-through parallel*

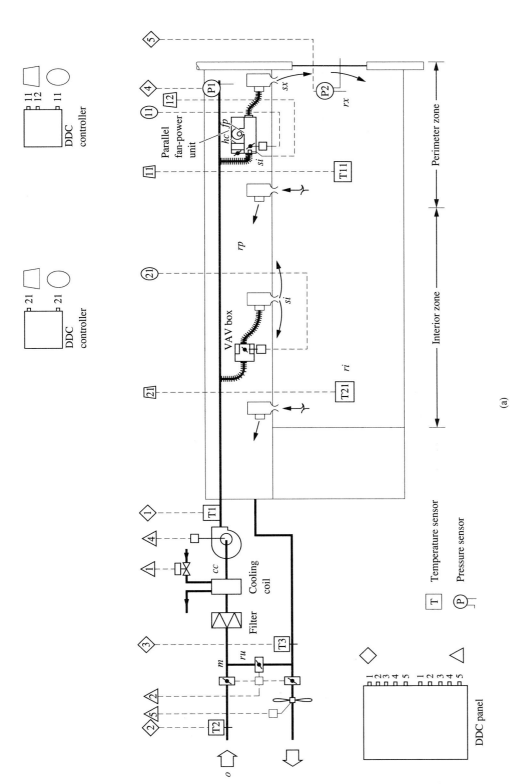

FIGURE 27.10 A parallel fan-powered VAV system: (a) schematic diagram; (b) air conditioning cycle; and (c) details of winter mode air conditioning cycle.

(a)

27.25

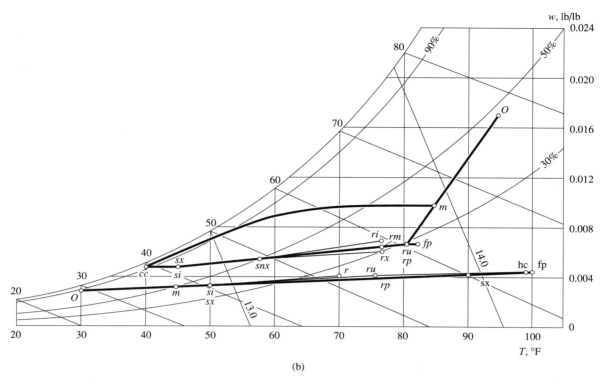

(b)

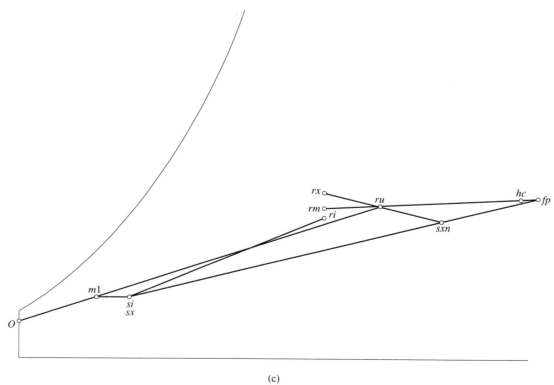

(c)

FIGURE 27.10b and c

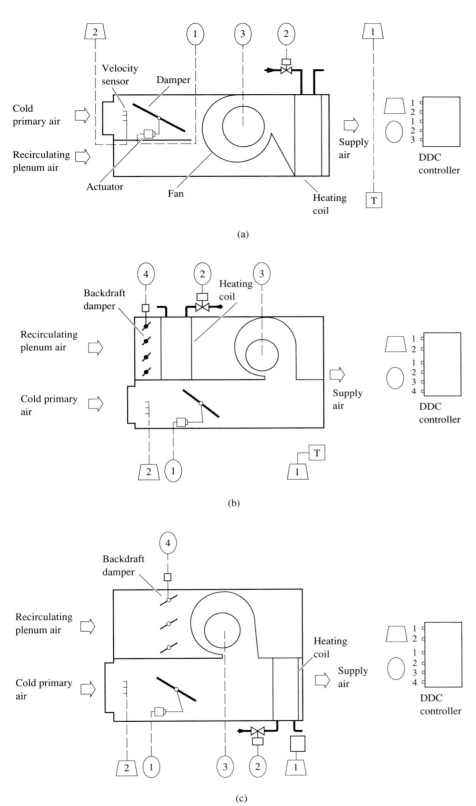

FIGURE 27.11 Fan-powered units: (a) series unit; (b) parallel unit, draw-through; and (c) parallel unit, blow-through.

fan-power unit, shown in Fig. 27.11c. The air temperature entering the heating coil of a draw-through unit is the temperature of recirculating plenum air, which is higher than the mixture of cold primary air and recirculating plenum air in a blow-through unit.

A parallel fan-powered unit operates only when there is a need to mix the warm plenum air with the cold primary air at summer part-load when the volume flow of the cold primary air is at its minimum setting, or when the mixture of cold primary air and warm plenum air is heated at the heating coil to offset the zone winter heating load. When a parallel fan-powered unit is used to mix 40 to 50°F low-temperature primary air with warm plenum air in a cold air distribution system, it is also operated during summer cooling mode.

In a parallel fan-powered unit when the cold primary air stream is reduced to its minimum setting during cooling mode part-load operation, the volume flow rate of the extracted plenum recirculating air plus the primary air is ususally equal to 0.75 to 0.85 $\dot{V}_{sn.d}$. This still results in a reduced volume flow rate delivered to the control zone. The fan in a parallel fan-powered unit extracts only recirculating plenum air, and therefore has a smaller volume flow rate and a lower fan total pressure than that in a series fan-powered unit. Because of its smaller fan size, it consumes less fan power in cooling mode operation.

Current parallel fan-powered unit designs also provide a good mixture of cold primary air and extracted plenum air. They are often used in cold air distribution because of their lower initial cost and lower fan energy consumption.

In a parallel fan-powered unit, a backdraft damper should be installed upstream from the fan. This damper is normally closed to prevent the backward flow of cold primary air in case the fan is deenergized. When the fan is turned on, the backdraft damper is fully open.

The drawbacks of a parallel fan-powered unit include the need for a careful balance of flow and total pressure of two airstreams, the difficulty in selection of the size of the unit, and its more complicated controls.

The capacity of a fan-powered unit is determined by the capacity of the fan, the VAV box, and the heating coil. The size of a fan-powered unit is determined by the size of the VAV box and the fan. The heating coil in a fan-powered unit can be a hot water coil or an electric duct heater located downstream from the fan-powered unit.

Fan-powered units are usually built in sizes corresponding to VAV box primary-air volume flow rates from 300 cfm to 3200 cfm and fan supply volume flow rates of 250 cfm to 4000 cfm. Most fan-powered units have a fan total pressure between 0.5 and 0.8 in. WG, and an external pressure of about 0.25 in.WG to offset the pressure drop of downstream flexible ducts and diffusers. The pressure drop of the VAV box and the corresponding pressure drop of flexible ducts and diffusers due to cold primary air should be offset by the supply fan in the primary-air AHU or PU.

The speed of centrifugal fan can be adjusted manually to high, medium, or low. Both radiated and duct-borne sound paths should be checked to determine whether they can meet the space required NC criteria. Because the fan total pressure is higher in a fan-powered unit than in a fan–coil unit, a temperature rise due to fan power of 0.75° F should be considered in psychrometric analysis.

The VAV box in a fan-powered unit can be either pressure-dependent or pressure-independent.

Fan Characteristics in Parallel Fan-Powered Units

In a parallel fan-powered unit used for low-temperature primary air mixing, two fan-duct systems are connected in parallel: the cold primary air stream discharged from the supply fan F1 in the AHU, and the recirculating plenum air stream extracted by fan F2, as shown in Fig. 27.12a. At the combining point A, the pressure of these two airstreams must be equal, and the volume flow rate is the sum of the volume flow rates of the cold primary air $\dot{V}_{c.p}$ and the recirculated plenum air $\dot{V}_{rec}$.

At summer design conditions, when the mixture of cold primary air and warm plenum air supplied from the parallel fan-powered unit is at its zone peak supply volume $\dot{V}_{sn.d}$, the volume flow rate of recirculating plenum air $\dot{V}_{rec} = 0.4 \dot{V}_{sn.d}$ the fan total pressure of F2 is about 0.5 in. WG, and the pressure drop across the flexible duct and the slot diffuser may be around 0.25 in. WG.

When the zone sensible cooling load is reduced to about half of the design load, the volume flow rate of the cold primary air may be reduced from 0.6 $\dot{V}_{sn.d}$ to 0.3 $\dot{V}_{sn.d}$. As the supply volume flow through the flexible duct and slot diffuser is reduced accordingly, the pressure drop between points A and o also decreases. The system pressure drop of the recirculating plenum air passage becomes smaller and a new system curve S2 is formed, as shown in Fig. 27.12b. The operating point of the fan-duct system F2 and S2 moves from point P to point Q. The result is a greater volume flow rate of the recirculating plenum air $\dot{V}_{rec}$ than that at design conditions and a smaller zone supply volume flow rate $\dot{V}_{sn}$ flowing through the flexible duct and slot diffusers.

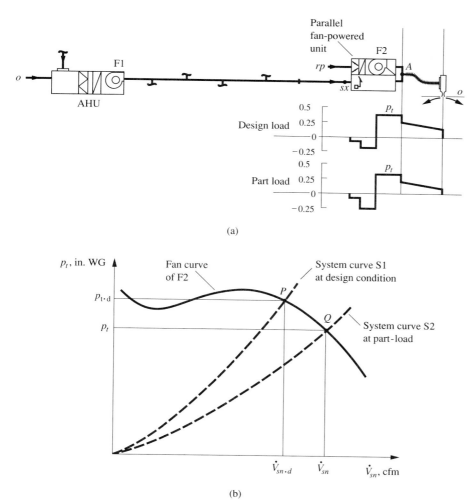

FIGURE 27.12 Fan operating characteristics of a parallel fan-powered unit system: (a) pressure characteristics and (b) fan-duct system characteristics.

Operating Characteristics of a Parallel Fan-Powered Unit

Consider a control zone in the perimeter zone of a building served by a draw-through parallel fan-powered unit that mixes low-temperature cold primary air with the warm plenum air during summer cooling mode operations, as shown in Fig. 27.10b. Because the temperature of cold primary air $T_{c.p} = 44°F$, the summer indoor space relative humidity drops to about 35 percent. Therefore, the summer indoor design temperature can be raised to 78°F. At winter design conditions, the zone temperature is maintained at 70°F with a relative humidity between 20 and 30 percent. The temperature of cold primary air is reset to 50°F. An air economizer cycle is used to provide free cooling when $T_o < T_{ru} = 81°F$.

In summer cooling mode, as shown in Fig. 27.10b, the recirculating plenum air temperature T_{rec} varies from 81°F at zone peak sensible cooling load $q_{rsn.d}$ to 82°F at about $0.2q_{rsn.d}$. Cold primary air is supplied to the inlet of the parallel fan-powered unit at point si or sx. At the same time, the centrifugal fan in the fan-powered unit extracts the recirculating plenum air and discharges at point fp from fan-powered unit's outlet. The recirculating plenum air is then mixed with the cold primary air from the VAV box. The mixture flows through the flexible duct and the slot diffuser, and supplies to the control zone at point sxn with a temperature typically about 59°F. Zone air returns to the ceiling plenum through return slots. It absorbs part of the heat released by the light troffer and mixes with return air from other control zones, which results in a mean plenum air temperature $T_{rp} = T_{rec}$. Part of

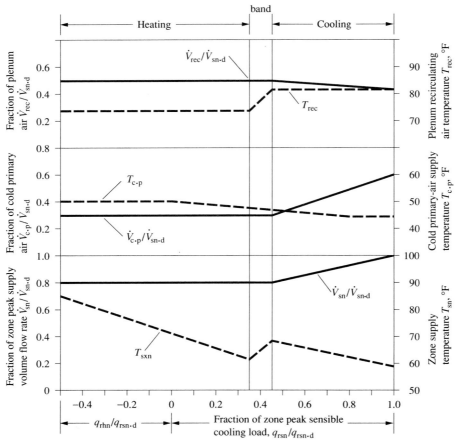

FIGURE 27.13 Operating characteristics of a draw-through parallel fan-powered unit serving a control zone in perimeter zone of a building.

the plenum air at point rp is recirculated at the fan-powered units, and the remaining part is returned to the fan room through the return duct system.

In winter heating mode in the perimeter zone, the recirculating plenum air extracted by the fan in the fan-powered unit is first heated at the heating coil to point hc. It is then discharged from the fan at point fp and mixed with the minimum cold primary air. The mixture is then supplied to the control zone to offset the zone heating load at point sxn.

Figure 27.13 shows the operating parameters of a typical draw-through parallel fan-powered unit that serves a control zone in the perimeter zone of a building.

Zone supply temperature T_{sxn}, in °F, can be estimated based on the energy balance of a mixing process

$$T_{sxn} = \frac{\dot{V}_{c.p} T_{c.p} + \dot{V}_{rec} T_{rec}}{\dot{V}_{c.p} + \dot{V}_{rec}} \qquad (27.19)$$

where $\dot{V}_{c.p}$ = volume flow rate of cold primary air, cfm, and zone sensible cooling load q_{rsn} and zone

heating load q_{rhn}, both in Btu/h, can be calculated as

$$q_{rsn} = 60 \dot{V}_{sn} \rho_s c_{pa} (T_r - T_{sxn})$$
$$q_{rhn} = 60 \dot{V}_{sn} \rho_s c_{pa} (T_{sxn} - T_r) \qquad (27.20)$$

If the calculated zone heating load q_{rhn} based on the T_{sxn} calculated from Eq. (27.19) and the volume flow rate of the selected fan-powered unit are smaller than the required zone heating load at winter design conditions, then select a larger fan and heating coil.

At summer cooling mode design conditions in the perimeter zone:

- Fraction of zone peak sensible cooling load $q_{rsn}/q_{rsn.d} = 1$. Here $q_{rsn.d}$ represents zone peak sensible cooling load, in Btu/h.
- Fraction of zone peak supply volume flow rate $\dot{V}_{sn}/\dot{V}_{sn.d} = 1$.
- Temperature of zone supply air $T_{sn} \approx 59$°F.
- Fraction of cold primary air $\dot{V}_{c.p}/\dot{V}_{sn.d} = 0.60$.
- Temperature of cold primary air $T_{c.p} = 44$°F.

- Fraction of recirculating plenum air $\dot{V}_{rec}/\dot{V}_{sn.d} = 0.40$.
- Temperature of recirculating plenum air $T_{rec} = 81°F$.

In summer cooling mode part-load when the zone sensible load in the perimeter zone has dropped to $0.45\ q_{rsn.d}$, any of the following may occur:

- $\dot{V}_{sn}/\dot{V}_{sn.d}$ may drop to 0.80 and T_{sn} may increase to 66°F.
- $\dot{V}_{c.p}/\dot{V}_{sn.d}$ falls to 0.3 and $T_{c.p}$ is reset according to zone demand.
- $\dot{V}_{rec}/\dot{V}_{sn.d}$ increases to 0.50 and T_{rec} may increase to 81.5°F.

In the perimeter zone, at winter design conditions when the zone heating load $-q_{rhn}$ is approximately 75 percent of the space peak sensible cooling load, $|q_{rhn}/q_{rsn.d}| \approx 0.75$, the cold primary air should remain at minimum setting and $\dot{V}_{sn}/\dot{V}_{sn.d}$ should remain constant. Its value may be 0.80.

The following condtions exist in winter heating design conditions:

- $\dot{V}_{sn}/\dot{V}_{sn.d}$ is about 0.80 and T_{sxn} increases to 90°F.
- $\dot{V}_{c.p}/\dot{V}_{sn.d}$ is still 0.3 and $T_{c.p}$ increases to 50°F.
- $\dot{V}_{rec}/\dot{V}_{sn.d}$ increases to 0.50 and T_{rec} drops to 73°F because of the lower zone temperature of 70°F.

When space heating load is at part-load operation, the flow rate of hot water supply to the heating coil is modulated to maintain a zone temperature of 70°F.

Fan-Powered Unit Controls

ZONE TEMPERATURE CONTROL. Zone temperature control of a parallel fan-powered unit in the perimeter zone can be divided into three operating modes: summer cooling mode, dead-band mode, and winter heating mode.

During summer cooling mode, zone temperature is controlled at a predetermined value, based on the signal sensed by a temperature sensor and the modulation of the damper blade opening in the cold primary air stream by means of a DDC controller, as shown in Fig. 27.11.

When the cold primary air supply has been reduced to minimum setting and zone temperature 70.2°F $> T_r > 77.8°F$, then the fan-powered unit is in dead-band mode operation. In dead-band mode, cold primary air is mixed with warm plenum air. Zone temperature T_r varies from 70.2 to 77.8°F. The fan-powered unit is still operating.

When zone temperature $T_r \leq 70°F$, the fan-powered unit is operated in winter heating mode operation. During winter heating mode, the cold primary air supply is at minimum setting to provide outdoor air requirement for the control zone. The fan is operating at a nearly constant supply volume flow rate. A reversing relay is actuated by the sensed temperature signal so that when zone temperature T_r falls below 70°F, the DDC controller opens the two-way valve of the hot water supply wider and admits more hot water to the heating coil.

SAFETY CONTROLS. In addition to zone temperature control, there are also safety controls. For example, the heating coil is energized and the backdraft damper is opened only when the fan is operating.

CALCULATION OF VOLUME FLOW RATE OF COLD PRIMARY AIR. In a fan-powered VAV system, if the cold primary air is modulated by a VAV box and supplied directly to the interior control zones through high-induction ceiling diffusers, then its operating modes and air conditioning cycles at summer and winter design conditions and the volume flow rate and coil load calculations are similar to those of a VAV cooling system.

In the perimeter zone at summer design conditions, cold primary air at a temperature of 42° to 46°F is first mixed with the plenum recirculating air at 81° or 82°F in the parallel fan-powered unit and is then supplied to various control zones, as shown in Fig. 27.10b. From the psychrometric chart, draw line $sx\ fp$. The ratio of the volume flow of cold primary air to volume flow of zone supply air at summer design load $\dot{V}_{c.p}/\dot{V}_{sn.d}$ often varies from 0.50 to 0.75. After selecting an appropriate $\dot{V}_{c.p}/\dot{V}_{sn.d}$, point sx and the zone supply temperature T_{sxn} can be determined.

The total volume flow rate of cold primary air supplied to the perimeter zone $\dot{V}_{sx}$, in cfm, can be calculated based on the perimeter block sensible load q_{rsx}, in Btu/h, as

$$\dot{V}_{sx} = \frac{q_{rsx}}{60\rho_s c_{pa}(T_r - T_{sx})} \qquad (27.21)$$

The coil's load of a fan-powered VAV system can be calculated in the same way as that of a VAV cooling or VAV reheat system.

The controls for the AHU or PU in a fan-powered VAV system are similar to those in VAV cooling and VAV reheat systems. It is recommended that the cold primary air supply temperature reset is based on zone demands at part-load operation for greater energy savings.

Design Considerations

Because there are two mixings (the mixing of outdoor air and recirculating air in the AHU or PU and the mixing of primary air and plenum air in the fan-powered unit), adequate supply of outdoor air is critical in a fan-powered VAV system. This will be discussed in the next section.

In a VAV system using cold air distribution, fan-powered units can be installed either in both the perimeter and interior zone, or in the perimeter zone only for lower initial cost. Cold primary air is supplied directly to the interior zone by high-induction ceiling diffusers for cold air distribution, as shown in Fig. 27.10b and 27.10c.

The mixing ratio of cold primary air to supply air and the reset of the cold primary air supply temperature for the perimeter zone are different from those of the interior zone. Different AHUs or PUs are preferable for the perimeter zone and the interior zone. An analysis of energy savings and investment costs is often necessary.

In a series fan-powered unit, the supply fan in the AHU or PU offsets the system pressure drop up to the inlet of the VAV box. The fan in the fan-powered unit offsets the pressure drop of the VAV box, elbows (if any), downstream heating coil, flexible duct, and diffuser.

In a draw-through parallel fan-powered unit, the supply fan in the AHU or PU offsets the pressure drop of the VAV box, flexible duct, diffuser, and straight-through mixing loss in the fan-powered unit when of the cold primary air stream only.

At summer design conditions, the ratio of the volume flow rate of cold primary air supplied to the fan-powered unit to the volume flow rate of the fan-powered unit $\dot{V}_{c.p}/\dot{V}_{sn.d}$ depends on the following factors:

- Temperature of cold primary air $T_{c.p}$ and the zone supply temperature differential at summer design conditions $(T_r - T_{sn.d})$
- Use of the warm plenum air at part-load and winter heating to save energy
- Outdoor air requirements

As mentioned previously, $\dot{V}_{c.p}/\dot{V}_{sn.d}$ varies between 0.5 and 0.75. During the selection of an optimum $\dot{V}_{c.p}/\dot{V}_{sn.d}$, all these factors should be analyzed carefully.

The supply ducts, branch ducts, fan-powered unit or VAV box, flexible ducts, and diffusers must be well-insulated, especially for cold air distribution. Cold primary air and recirculated plenum air should

be mixed in a perpendicular flow direction to prevent downstream temperature stratification.

A low space relative humidity and space dew point are desirable to prevent possible surface condensation during cold air distribution. When cold primary air at 44°F is supplied directly to the conditioned space through high-induction diffusers, the surface of the induction diffusers can be assumed to be 3°F higher than the supply air, or the dew point of space air should be lower than $44 + 3 = 47$°F, that is, a space temperature of 78°F and a space relative humidity $\varphi_r < 33$ percent.

At the start of the cool-down period in a hot and humid summer, direct supply of 40 to 50°F cold primary air to the fan terminals may cause condensation because of the higher dew point of stagnant air in the ceiling plenum before the cool-down period. The AHU or PU should supply 55°F air to the fan-powered unit first to reduce moisture and lower the dew point.

If humidification is required to maintain an acceptable space relative humidity in winter heating mode, a humidifier can be installed in the AHU or PU.

During unoccupied periods in winter, the AHU or PU can be shut off while the fan-powered units provide the necessary heating.

Fan-powered units installed inside the ceiling plenum must meet local safety code requirements.

27.7 MINIMUM OUTDOOR AIR SUPPLY

Basic Approach

Adequate outdoor air supply is essential for acceptable indoor air quality (IAQ) for the health and comfort of the occupants. A VAV system often reduces its outdoor air supply when the supply volume flow rate drops at part-load operation. A thorough analysis of outdoor air supply, both at design load and part-load operation, is an important part of VAV system design. An oversupply of outdoor ventilation air, other than that in an economizer cycle, is always a waste of energy.

During the early stages of development, inadequate outdoor air supply, insufficient space air motion at low flow rates, and the dumping of cold air jets at reduced supply volume flow rates were three primary problems in VAV systems. Today, slot diffusers and high-induction ceiling diffusers have eliminated the dumping of cold air jets at low supply volume flow rates. A minimum setting of the VAV box, the installation of fan-powered units, and an increase in the minimum outdoor air from 5 cfm to 15 cfm significantly improves the quality of stagnant air. The emphasis on IAQ in recent years and in future design makes outdoor air supply a critical factor in VAV system design and operation. IAQ will be discussed in Chapter 29.

Design Average, Zone Variation, and Daily and Annual Fluctuation

Traditionally, in an air system design, the minimum outdoor air supply $\dot{V}_{o.dm}$, in cfm, is often the product of the outdoor air requirement per person multiplied by the number of occupants in the conditioned space served by the air system. The calculated $\dot{V}_{o.dm}$ is often expressed as a fraction of $\dot{V}_{s.d}$ which is the volume flow rate of supply air of the air system at summer design conditions, in cfm. Based on $\dot{V}_{s.d}$ and $\dot{V}_{o.dm}$, the point m that represents the state point of the mixture of outdoor and recirculating air can be determined on the psychrometric chart. The coil's load or refrigeration load required by the air system can thus be calculated.

The design average outdoor air supply intensity, $\dot{V}_{o.dm}/A_{fl}$, in cfm/ft^2, is calculated as $\dot{V}_{o.dm}$ divided by the floor area served by the air system A_{fl}. $\dot{V}_{o.dm}$ is the outdoor air supply that can be provided by the air system at the discharge outlet of an AHU or PU with a supply volume flow of $\dot{V}_{s.d}$ for both constant-volume and VAV systems at summer design conditions. In an office building, if 20 cfm/person is supplied to the conditioned space and the occupied density is 150 ft^2/person, the design average outdoor air supply intensity is

$$\frac{\dot{V}_{o.dm}}{A_{fl}} = \frac{20}{150} = 0.13 \text{ cfm/ft}^2$$

Actual outdoor air supply density delivered to each control zone at a specific time $\dot{V}_{osn}/A_{fl}$ may be quite different from the design average value for the following reasons:

- Zone outdoor air supply intensity $\dot{V}_{osn}/A_{fl}$ is directly proportional to zone supply air intensity $\dot{V}_{sn}/A_{fl}$, and, therefore, to zone sensible load density q_{rsn}/A_{fl}.

  For an office floor in a high-rise building, if the supply temperature differential $\Delta T_s = 20°F$, typical values of zone supply air intensity calculated in Example 7.1, in cfm/ft^2, may be summarized as follows:

  Perimeter zone
facing west	1.6
facing east	1.6
facing south	1.2
facing north	1.0

  Interior zone 0.6

- For a VAV system, at operating periods other than summer design peak load, the volume flow rate of zone supply air $\dot{V}_{sn}$ may drop to a minimum setting of $0.3 \dot{V}_{sn.d}$ because of lower outdoor temperature and cloudy weather or lower internal loads. In interior control zones, $\dot{V}_{sn}$ may drop to $0.3 \dot{V}_{sn.d}$ immediately after the cool-down period, to $0.5 \dot{V}_{sn.d}$ at lunchtime, and again to $0.3 \dot{V}_{sn.d}$ in the evening.

  The minimum setting depends on system and load characteristics, outdoor air supply, and space air movement. It usually varies between 0.2 and 0.3 $\dot{V}_{sn.d}$, typically $0.3 \dot{V}_{sn.d}$, and should be carefully selected.

- Air economizer cycles are widely used in VAV systems. When a VAV system is operated in a temperature-controlled air economizer cycle, 100 percent outdoor air is used for free cooling when outdoor air $50°F \leq T_o < 78°F$ in many locations.

Critical Zone

For control zones in VAV systems in buildings in which the occupant density is high and the number of occupants remains nearly the same during part-load operation as the supply volume flow rate is reduced, the outdoor air supply in one or several control zones may be inadequate. These zones are known as *critical zones*.

In a VAV system using a temperature-based air economizer cycle for an office building, possible minimum or critical zone outdoor air supply $\dot{V}_{o.cr}$ may occur when outdoor air temperature T_o is higher than 78°F or T_o is very low in winter when the VAV system is operated at minimum outdoor air recirculating mode.

In many commercial buildings, there are control zones in which zone sensible cooling load q_{rsn}, zone supply air $\dot{V}_{sn}$, and zone outdoor air supply $\dot{V}_{osn}$ are directly proportional to the number of occupants, that is, people's load dominated. When q_{rsn} and $\dot{V}_{sn}$ drop, the occupant density O_{den} falls accordingly. The outdoor air supply for each occupant will not drop below minimum outdoor air requirements under these circumstances. There may be no critical zones. Examples include retail stores, schools, indoor stadiums, movie theaters, and ballrooms.

Recirculation of Unused Outdoor Air in Multizone Systems

In a multizone constant-volume or VAV system, if the proportion of outdoor air contained in zone supply air is determined by the requirements of the critical zone, then zones other than the critical zone may be oversupplied with outdoor air. The recirculated air will contain unused outdoor air, as defined in Section 26.1, which can be supplied to the critical zone or other less critical zones that still require outdorr air supply. Outdoor air intake can thus be reduced.

Methods of calculating outdoor air requirements for occupants and using the unused outdoor air contained in recirculating air were developed in Australia and are included in ANSI/ASHRAE Standard 62–1989. The required fraction of outdoor air supply in system supply air, considering the outdoor air requirements of the critical zone and the unused outdoor air contained in the recirculating air Y, can be calculated as follows

$$Y = \frac{X}{1 + X - Z} \qquad (27.22)$$

and

$$Y = \frac{\dot{V}_{o.cor}}{\dot{V}_s}$$

$$X = \frac{\dot{V}_{o.sys}}{\dot{V}_s} \qquad (27.23)$$

$$Z = \frac{\dot{V}_{o.cr}}{\dot{V}_{s.cr}}$$

where $\dot{V}_s$ = supply volume flow rate of the air system
$\dot{V}_{o.cor}$ = corrected outdoor air supply volume flow rate, considering critical zones and the unused outdoor air in the recirculating air, cfm
$\dot{V}_{o.sys}$ = calculated system outdoor air volume flow rate, cfm
$\dot{V}_{s.cr}$ = critical zone supply volume flow rate, cfm
X = uncorrected fraction of outdoor air supply in system supply volume flow rate
Z = fraction of outdoor air supply in the critical zone

The system outdoor air volume flow rate can be calculated as

$$\dot{V}_{o.sys} = N_{oc}\,\dot{V}_{o.req} \qquad (27.24)$$

where N_{oc} = number of occupants in the conditioned area served by the air system
$\dot{V}_{o.req}$ = outdoor air requirements as specified by ANSI/ASHRAE Standard 62–1989, cfm/person

Outdoor air volume flow rates $\dot{V}_{o.cor}$, $\dot{V}_{o.sys}$, and $\dot{V}_{o.cr}$, supply volume flow rate $\dot{V}_s$, and $\dot{V}_{s.cr}$ all must occur at the same time.

Methods of Increasing Outdoor Air Supply

- Minimum outdoor air control is often used to increase the outdoor air supply at reduced supply volume flow rates during part-load operation in VAV systems. As described in Section 18.9, when a decrease in the volume flow of the outdoor air is sensed by a velocity sensor, the DDC panel opens the outdoor damper wider and closes the recirculating damper to a smaller opening until a nearly constant volume flow of outdoor air intake is maintained.

  When outdoor temperature is very low, a higher percentage of outdoor air intake necessitates the use of more heating energy. In order to meet the outdoor air requirements of the occupants, an increase in heating energy is necessary.

  Minimum outdoor air control requires both a minimum outdoor damper and an additional outdoor damper designed to admit 75 to 85 percent of outdoor air when an air economizer cycle is used. A properly designed air passage with flow straighteners to eliminate eddies before the velocity sensor (which may significantly influence the accuracy of sensor) and a periodically calibrated sensor are of prime importance in minimum outdoor air control.

- The fraction of uncorrected outdoor air in system supply air X should be increased to an optimum value. However, if X is set above the optimum level, energy is wasted.

- In locations with hot and humid summers, if an air economizer cycle is not used, a make-up unit may precondition the outdoor air before it is supplied to the fan rooms on various floors in a multistory building. This method provides a fixed amount of outdoor air to each AHU at both design and part-load. This method has a higher initial cost and operating cost, however.

Single-Zone Constant-Volume or VAV Systems

In a single-zone constant-volume system, outdoor air volume flow rate $\dot{V}_{o.sys}$ is calculated by Eq. (27.24). Because of the diffusion of air contaminants and the dilution effect of outdoor air, an acceptable indoor air quality can be maintained in the conditioned space.

Operating conditions of a single-zone VAV system are the same as those of a single-zone constant-volume system during summer design conditions. When minimum outdoor air control is used and the supply volume flow rate is reduced to its minimum setting, say $0.3\,\dot{V}_{s.d}$, the required outdoor air supply can still be maintained.

Use of Transfer Air

ANSI/ASHRAE Standard 62–1989 allows the use of transfer air for restrooms and smoking lounges. Both restrooms and smoking lounges have large exhaust ventilation rates, 50 to 60 cfm/person, and the staying time of occupants seldom exceeds one hour. It is often difficult to determine whether the ceiling plenum air or the zone air contains more unused air. Transfer air should not be used as part of outdoor air other than for restrooms or smoking lounges. Many questions remain about transfer air and the amount of unused air. More research is needed to solve these problems.

Case Study: Outdoor Air Supply for a Multizone VAV System

Consider a multizone VAV cooling system with the operating parameters described in Section 27.3. This office floor has a conditioned area of 20,000 ft^2. There are about 140 occupants. The calculated system outdoor air volume flow rate is

$$\dot{V}_{o.sys} = N_{oc} \dot{V}_{o.req} = 20 \times 140 = 2800 \text{ cfm}$$

Because the system supply volume flow rate is 19,227 cfm, the uncorrected fraction of outdoor air supply volume flow in system supply volume flow X can be calculated as

$$X = \frac{\dot{V}_{o.sys}}{\dot{V}_s} = \frac{2800}{19,227} = 0.15$$

In the interior zone of this office floor, there is a control zone in which there are 9 occupants and a zone supply volume flow rate of 600 cfm at summer design conditions. The fraction of outdoor air in supply air in this critical zone Z can be calculated as

$$Z = \frac{\dot{V}_{o.cr}}{\dot{V}_{s.cr}} = \frac{9 \times 20}{600} = 0.3$$

At summer design conditions, the fraction of outdoor air in the system supply air considering the requirements of the critical zone and the unused outdoor air Y, which is the fraction that should be provided at the AHU or PU, can be calculated as

$$Y = \frac{X}{1 + X - Z} = \frac{0.15}{1 + 0.15 - 0.3} = 0.18$$

The system outdoor air intake is therefore

$$\dot{V}_{o.cor} = 0.18 \times 19,227 \approx 3460 \text{ cfm}$$

During summer mode part-load operation, if the system supply volume flow rate $\dot{V}_s$ drops to $0.4 \times 19,227 = 7691$ cfm and the system outdoor air intake $\dot{V}_{o.sys}$ is still 3460 cfm because of the instal-lation of minimum outdoor air control, then $X = 3460/7691 = 0.45$.

For the same critical zone in the interior zone, the zone supply volume flow rate at summer design conditions is 600 cfm and the zone supply volume flow at summer part-load is $0.5 \times 600 = 300$ cfm. This control zone has 6 occupants at design load and at part-load. The fraction of zone outdoor air supply in zone supply air $Z = (6 \times 20)/300 = 0.4$

At summer mode part-load operation, the fraction of outdoor air in supply air, considering the critical zone and unused outdoor air Y, can be calculated as

$$Y = \frac{0.45}{1 + 0.45 - 0.4} = 0.43$$

Because $Y < X$, outdoor air supply to the critical zone can be satisfied at summer mode part-load operation.

When a VAV system is installed with minimum outdoor air control, because the system outdoor air supply remains the same at summer part-load as at design-load and the number of occupants is not higher than at summer design conditions, the outdoor air requirements can be met.

27.8   FAN CONTROLS AND VAV SYSTEM PRESSURE CHARACTERISTICS

Comparison Between Adjustable-Frequency Drives and Inlet Vanes

In VAV systems, both adjustable-frequency AC drives and inlet vanes are used to control the volume flow and fan total pressure of the supply, return, relief, and exhaust fans, as described in Sections 10.4 and 19.5. Generally, fans using adjustable-frequency AC drive to vary fan speed are substantially more energy-efficient at reduced volume flow and fan total pressure than fan using inlet vanes. However, adjustable-frequency AC drive is more expensive than inlet vanes.

Englander and Norford (1992) showed that adjustable-frequency AC drive consumed 35 percent less fans energy based on the actual performance of a VAV system in a retrofit project using backward-curved centrifugal fans of a nominal volume flow rate of about 40,000 cfm. The simple payback years for the installation of adjustable-frequency AC drive was about 4 years for the specific project. From the results of energy simulation programs, other researchers also found that adjustable-frequency drives have an energy savings between 20 and 50 percent depending on the size of fan, characteristics of the building, and building location.

Field measurements also showed that many fans installed with inlet vanes saved only 20 to 30 percent, or as little as 10 percent when the

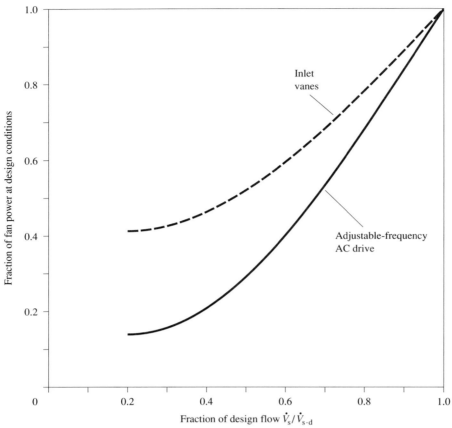

FIGURE 27.14 Comparison of fan power consumption between adjustable-frequency AC drives and inlet vanes.

volume flow rate was reduced to 0.3 of its design value. Theoretically, the energy savings should be greater than 50 percent.

Figure 27.14 compares the fan power consumption at fractions of design volume flow rate when adjustable-frequency AC drive and inlet vanes are used.

For a vane axial supply fan, blade pitch controls are often used at reduced volume flow for fan energy savings. Controllable blade pitch vane axial fans have energy savings characteristics comparable to those of adjustable-frequency AC drives, but they are also expensive.

Pressure Characteristics of a VAV Cooling System Using a Supply–Relief Fan Combination

Figure 27.15 shows the pressure characteristics of a VAV cooling system using a supply–relief fan combination. This VAV cooling system serves the fifth floor of a high-rise public library with a conditioned floor area of about 16,500 ft^2. There are two AHUs and two fan rooms, each with a volume flow rate of 12,650 cfm and a fan total pressure of 4.25 in. WG, to serve this floor. Each fan room has three relief fans and three relief dampers. The volume flow rate of relief air is modulated by relief dampers, which are controlled by an electronic controller actuated by a signal from a pressure sensor in the ceiling plenum. In the perimeter zone, winter heating is provided by the electric heating coil located downstream from the VAV box.

On September 12, 1983, the pressure characteristics of the VAV cooling system were measured. Control of relief air dampers was deliberately deenergized. The supply volume flow rate of AHU 2 was about 12,000 cfm. The pressure differences between various points in Fan Room 2 in the VAV cooling system during recirculating mode with minimum outdoor air in an air economizer cycle were as follows:

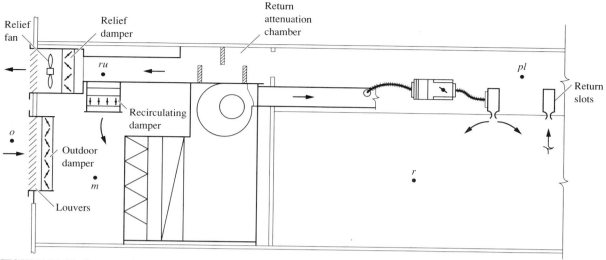

FIGURE 27.15 Pressure characteristics of a VAV cooling system using supply–relief fan combination.

	Recirculating mode	Air economizer
Outdoor damper	10% open	100% open
Outdoor air, cfm	3330	
Recirculating damper	100% open	100% closed
Relief fans operated	1 fan	All 3 fans
Relief damper	15% open	100% open
$\Delta p_{\text{pl-o}}$ in. WG	+0.06	+0.018
$\Delta p_{\text{pl-m}}$ in. WG	+0.20	
$\Delta p_{\text{o-m}}$ in. WG	+0.14	
$\Delta p_{\text{r-pl}}$ in. WG	+0.004	+0.004

During the air economizer cycle, when the outdoor damper was fully open and the recirculating damper was closed, the pressure difference between the ceiling plenum and outdoors $\Delta p_{\text{pl-o}}$ was +0.038 in. WG when only two relief fans were operating, and $\Delta p_{\text{pl-o}}$ was +0.04 in. WG when only one relief fan was operating.

The pressure difference between the conditioned space and the air in the ceiling plenum was so small because of gaps between the ceiling tiles. During recirculating mode, the space pressure was about +0.064 in. WG. Space pressure was lower when more relief fans were operating during the air-side economizer cycle.

27.9 CONTROL STRATEGIES AND INTERACTION BETWEEN CONTROLS

Control Strategies

Using a computer-simulated year-round energy performance of a prototypical building in five different locations in the United States, Mutammara and

Hittle (1990) found the following results for VAV systems:

- Proportional-integral (PI) control is a much more effective energy saving scheme than proportional control in most climates.
- The use of an air economizer cycle is recommended except in hot and humid climates. An air economizer cycle is energy-efficient and provides adequate outdoor air for occupants.
- Cold supply temperature reset for the perimeter zone is energy-efficient when PI control mode is used.

Interaction Between Controls

In a typical multizone VAV system, the following controls are used:

- Zone temperature controls
- AHU or PU discharge temperature control
- Duct static pressure control
- Minimum outdoor air control
- Air economizer control
- Space pressurization control
- Low temperature limit control

If a closed control loop contains a sensor, a controller, and a controlled device, the interaction between control loops in these controls can be summarized as follows:

- Control loops are deliberately combined to form a sequential operation. For example, the discharge temperature control in an AHU or PU is a combination of cooling coil control loop, mixed air temperature control loop, and heating coil control loop, used to control the same variable in sequence so that the actions of the controlled device of two different control loops do not occur simultaneously.

Often, the discharge air temperature is reset according to zone demand, so an additional control loop is added to the DDC panel.

- The control action of one control loop may affect the controlled variable of another control loop. For example, the closing of the damper in a VAV box of a zone temperature control loop affects the controlled variable of duct static pressure control loop—the duct static pressure.

- Part of the control actions of a control loop interact with another control loop. For example, the mixing air temperature control loop in air economizer control combines with other control loops in discharge air temperature control and forms a sequential operation.

- Two control loops operate on the same controlled device. In this case, one control loop must override the control action of other control loop. For example, minimum outdoor air control and the discharge air temperature control both modulate the outdoor and recirculating dampers. The minimum outdoor air control loop should override the discharge air temperature control loop to maintain a required mixing temperature for energy savings.

- Two control loops operate independently of each other. For example, the cooling coil control loop that modulates the two-way valve of the chilled water supply is independent from the space pressurization control loop that modulates the speed of the relief fan.

Interaction of the control loops in a VAV system should be carefully analyzed and planned. If two control loops are independent of each other without interaction, their performance will not be interrupted accidentally. Control sequence and override control action should be clearly specified.

If one control loop affects the controlled variable of another control loop, the speed of the actuator and the time constant of various system components should be analyzed to improve system stability.

Discharge Air Temperature Reset

ASHRAE/IES Standard 90.1–1989 requires that in VAV systems that supply cooled air or heated air to multiple zones, supply temperature must include au-

tomatic reset at reduced volume flow and system load, either by zone demand or by outdoor temperature.

Discharge air temperature or supply temperature reset save energy in the following ways:

- Reduce the simultaneous cooling and heating if VAV reheating and dual-duct VAV systems are used
- Lower the duct heat gain or loss
- Raise the suction temperature of the refrigeration system
- Increase the number of hours that the refrigeration system can be shut off during an air or water economizer cycle
- Increase the zone supply volume flow rate and air circulation at low zone loads

Discharge air temperature reset also has the following drawbacks:

- Lower fan energy savings because less volume flow can be reduced
- A higher space relative humidity during cooling mode operation

For a conditioned space that requires a strict relative humidity control, discharge air temperature reset is not recommended. In VAV systems using inlet vanes for duct static pressure and system supply volume flow control, discharge air temperature reset is preferable because it may save more energy. In VAV systems using adjustable-frequency AC drives for duct static pressure control, an analysis should be performed to determine whether discharge air temperature reset saves more energy.

Off-Hour Controls and Isolation

For energy efficient operation, ASHRAE/IES Standard 90.1–1989 requires that controls be provided to set back the space temperature or shut down the air system during unoccupied periods.

Systems serving zones that are not expected to be occupied simultaneously for 750 hours or more per year should be equipped with isolation devices and controls so that each zone can be shut off or set back individually.

For zones with a floor area of 25,000 ft² or more, a separate system may be more economical.

27.10 COMPARISON OF VARIOUS VAV SYSTEMS

VAV systems are used in applications where space load is varied in order to maintain a required indoor environment and to save fan energy.

Single-zone VAV systems are mainly used for single-zone conditioned spaces such as arenas, indoor stadiums, assembly halls, and retail stores.

Perimeter heating VAV systems are used for multizone conditioned spaces that require winter heating in locations with cold winters.

VAV cooling systems are used for conditioned spaces and zones that require cooling only, such as the interior zone of a building.

VAV reheat systems are simple and effective for perimeter heating in winter. Simultaneous cooling and heating may occur at low zone cooling loads. They may not be appropriate for locations where the winter heating load is very high.

Dual-duct VAV systems provide satisfactory indoor environment control for multizone high-rise buildings with flexible operating characteristics and satisfactory indoor environmental control. If the shutoff air leakage in the mixing VAV box can be minimized, and if hot water heating coils are used in the reheating VAV box, life-cycle cost analysis is required to determine whether a dual-duct VAV or VAV reheat system is more cost-effective.

Fan-powered VAV systems provide winter heating from the intake plenum air and heating coils, stimulate space air movement by maintaining a minimum supply air density from the small centrifugal fan, and provide mixing of cold primary air with the plenum air. Fan-powered VAV systems are widely used in buildings where cold air distribution is needed, such as office buildings and schools.

REFERENCES AND FURTHER READING

ADC/ARI, *Industry Standard 885: Acoustic Level Estimation Procedure*, ADC, Chicago, IL, 1989.

ASHRAE, *ASHRAE Handbook 1987, HVAC Systems and Applications*, ASHRAE Inc., Atlanta, GA, 1987.

ASHRAE, *ASHRAE Handbook 1991, HVAC Applications*, ASHRAE Inc., Atlanta, GA, 1991.

ASHRAE, *ANSI/ASHRAE Standard 62–1989, Ventilation for Acceptable Indoor Air Quality*, ASHRAE Inc., Atlanta, GA, 1989.

ASHRAE/IES, *Standard 90.1–1989, Energy Efficient Design of New Buildings Except Low-Rise Residential Buildings*, ASHRAE Inc., Atlanta, GA, 1989.

Avery, G., "The Instability of VAV Systems," *Heating/Piping/Air Conditioning*, pp. 47–50, February 1992.

Brothers, P. W., and Warren, M. L., "Fan Energy Use in Variable Air Volume Systems," *ASHRAE Transactions*, Part II B, pp. 19–29, 1986.

Chen, S. Y. S., Yu, H. C., and Hwang, D. D. W., "Ventilation Analysis for a VAV System," *Heating/Piping/Air Conditioning*, pp. 36–41, April 1992.

Dean, R. H., and Ratzenberger, J., "Stability of VAV Terminal Unit Controls," *Heating/Piping/Air Conditioning*, pp. 79–90, October 1985.

Dorgan, C. E., and Elleson, J. S., "Cold Air Distribution," *ASHRAE Transactions*, Part I, pp. 2008–2025, 1988.

Englander, S. L., and Norford, L. K., "Saving Fan Energy in VAV Systems—Part 1: Analysis of a Variable-Speed-Drive Retrofit," *ASHRAE Transactions*, Part I, 1992.

Gardener, T. F., "Part Load Ventilation Efficiencies in VAV Systems," *Heating/Piping/Air Conditioning*, pp. 89–100, February 1988.

Haines, R. W., "Supply Fan Volume Control in a VAV System," *Heating/Piping/Air Conditioning*, pp. 107–111, August 1983.

Int-Hout III, D., "Analysis of Three Perimeter Heating Systems by Air Diffusion Methods," *ASHRAE Transactions*, Part I B, pp. 101–112, 1983.

Int-Hout III, D., "Stand-Alone Microprocessor Control of Dual-Duct Terminals," *ASHRAE Transactions*, Part II, pp. 1722–1733, 1987.

Janssen, J. E., "Ventilation for Acceptable Indoor Air Quality," *ASHRAE Journal*, pp. 40–46, October 1989.

Kettler, J. P., "Efficient Design and Control of Dual-Duct Variable-Volume Systems," *ASHRAE Transactions*, Part II, pp. 1734–1741, 1987.

Krajnovich, L., and Hittle, D. C., "Measured Performance of Variable Air Volume Boxes," *ASHRAE Transactions*, Part II A, pp. 203–214, 1986.

Linford, R. G., "Dual-Duct Variable Air Volume—Design/Build Viewpoint," *ASHRAE Transactions*, Part II, pp. 1742–1748, 1987.

Lo, L., "VAV System with Invertor-Driven AHU for High-Rise Office Building in Tropical Climates—A Case Study," *ASHRAE Transactions*, Part I, pp. 1209–1217, 1990.

Lynn, M., "Balancing DDC-Controlled Boxes," *Heating/Piping/Air Conditioning*, pp. 79–84, July 1989.

Mutammara, A. W., and Hittle, D. C., "Energy Effects of Various Control Strategies for Variable-Air-Volume Systems," *ASHRAE Transactions*, Part I, pp. 98–102, 1990.

Roberts, J. W., "Outdoor Air and VAV Systems," *ASHRAE Journal*, pp. 26–30, September 1991.

Smith, M. C., *Industry Standard 885 Acoustic Level Estimation Procedure Compared to Actual Acoustic Levels in an Air Distribution Mock-up*, ASHRAE Transactions, Part I, pp. 543–548, 1989.

Shavit, G., "Retrofit of Double-Duct Fan System to a VAV System," *ASHRAE Transactions*, Part I, pp. 635–641, 1989.

Spitler, J. D., Pedersen, C. O., Hittle, D. C., and Johnson, D. L., "Fan Electricity Consumption For Variable Air Volume," *ASHRAE Transactions*, Part II B, pp. 5–18, 1986.

Straub, H. E., and Cooper, J. G., "Space Heating with Ceiling Diffusers," *Heating/Piping/Air Conditioning*, pp. 49–55, May 1991.

Tackett, R. K., "Case Study: Office Building Use Ice Storage, Heat Recovery, and Cold Air Distribution," *ASHRAE Transactions*, Part I, pp. 1113–1121, 1989.

The Trane Company, *Variable Air Volume Systems Manual*, American Standard Inc.: La Crosse, WI, 1988.

Wang, S. K., *Air Conditioning,* Vol. 4, Hong Kong Polytechnic, Hong Kong, 1987.

Wendes, H. C., "Estimating VAV Retrofit Costs," *Heating/Piping/Air Conditioning,* pp. 93–103, August 1983.

Wheeler, A. E., "Energy Conservation and Acceptable Indoor Quality in the Classroom," *ASHRAE Journal,* pp. 26–32, April 1992.

Williams, V. A., "VAV System Interactive Controls," *ASHRAE Transactions,* Part I, pp. 1493–1499, 1988.

CHAPTER
28

ENERGY ESTIMATION, GAS COOLING AND COGENERATION, AND SMOKE CONTROL

28.1 ENERGY ESTIMATION

Energy Estimation Methods

Annual, seasonal, or short-term estimation of energy use in an air conditioning or HVAC&R system is an important tool to assess energy performance and cost in both design and in daily operation. The annual energy use of various types of air, water, refrigeration, and heating systems can be predicted and analyzed. Based on the results of energy estimation, the most energy-efficient system can be selected.

For the daily operation of air conditioning systems in existing buildings, detailed simulation reveals the system's negative and positive effects on energy use and performance by comparing the estimated evaluation against the actual measured results. The energy use of many alternative improvements can be predicted and assessed. These estimates provide great potential for reducing the energy use and operating cost of existing air conditioning systems.

Three methods are widely used in energy estimation. They are the degree-day or variable degree-day method, the bin or modified bin method, and the detailed simulation method.

Degree-Day and Variable-Base Degree-Day Methods

The degree-day and variable-base degree-day methods are single-measure methods. A *degree-day* is the product of the degrees of mean daily outdoor temperature (in °F) by which the actual temperature deviates from a predetermined base value, and the number of operating days during the evaluation period. The most widely used predetermined base value for heating is 65°F. The *variable-base degree-day* method is based on the balance point outdoor temperature at which neither heating nor cooling is needed. Annual heating and cooling degree-days have been used to indicate the heating or cooling required for a specific location. They are also used as indexes to determine the thermal properties of the building shell for various locations in the United States. The degree-day method is commonly used to estimate the energy use of a heating system.

Bin Method and Modified Bin Method

The *bin method* estimates the energy use at outdoor temperature ranges called bins, and the results are weighted by multiplying the number of hours within each bin. The bin range is usually 5°F. The *modified bin method* has the advantage of using diversified off-peak load calculation, and includes the effects of air, refrigeration, and heating systems. However, the final energy estimate using the bin or modified bin method does not include most of the important factors that affect the energy use.

Detail Simulation Methods

The *detail simulation method* uses a comprehensive computer program to simulate the thermal behavior of the building and the performance of the air system, water system, and refrigeration and heating plants at various outdoor weather conditions to estimate the energy use of an air conditioning system.

Because of the powerful capability of the computer program, the detail simulation method can include most of the important factors that affect system energy use. The most widely used program is the year-round energy estimation by hour-by-hour system simulation. Short-term simulation is sometimes used to evaluate the influence of the thermal storage effect of system components on system performance.

Indexes of Annual Energy Consumption and Energy Cost

The most widely used single index of annual energy use of HVAC&R systems in buildings is the unit Btu/ft^2 · year. Building type and configuration, energy source (such as electricity, natural gas, or oil), outdoor climate, space loads, type of HVAC&R systems, and annual operating hours all affect annual energy use. Therefore, they are often used to compare the annual energy use of similar buildings located in the same area.

According to the data provided in Table 1.1 in Chapter 1, the average annual energy use for HVAC&R systems in 1983 are as follows:

Offices	204,000 Btu/ft^2 · year
Retail/service buildings	81,000 Btu/ft^2 · year

Of annual energy use for cooling, 91 percent of the energy used is electricity; for heating, 59 percent is natural gas and 16 percent is oil. The reason why retail/service buildings have a significantly lower annual energy use than offices is mainly because cooling is more common in retail/service buildings, and electicity has a far lower Btu/ft^2 · year than natural gas or oil.

Another index is annual energy cost per unit area, which is expressed in dollars/ft^2 · year. Factors that affect the annual energy use also influence the annual energy cost per unit area. In addition, annual energy cost per unit area is affected by the difference between electricity, gas, and oil prices as well as the rate structure of various utilities.

28.2 SYSTEM SIMULATION

System simulation is the representation of an actual system by a model of analogous characteristics, and is used to predict the performance and operating parameters of the system. Mathematical equations are used to describe the operating characteristics of the working substance and work and energy transfer of the system being simulated.

System Simulation Procedure

MODELING. Setting up a system model is the first step of system simulation. Modeling includes the following:

- Description of the system configuration, whether it is an air, water, refrigeration, or heating system, or whether it consists of many components or only contains a single device.
- Description of the operating characteristics of the system and the interaction between system components, whether it can be simplified to a steady-state model or it is a dynamic model. A simplification of the system model that results in an error of only 1 to 2 precent of the final result is recommended in order to simplify calculation and analysis.

DEVELOPING MATHEMATICAL EQUATIONS. Mathematical equations are developed to describe the operating characteristics of the working substance and work and energy transfer.

SOLVING FOR OUTPUTS. Computer programs are used to solve equations simultaneously, in sequence, or by iteration. The required operating parameters, that is, the outputs, can thus be obtained.

VERIFICATION. The results of predicted performance during system simulation can be verified against actual measured readings of similar models and operating conditions.

Steady-State and Dynamic Simulation

According to its operating characteristics, system simulation can be classified as steady-state or dynamic simulation.

STEADY-STATE SIMULATION. In *steady-state simulation*, the relationship between various operating parameters within a certain time increment is described by mathematical equations independent of time, such as

$$z = F(x, y, \ldots) \qquad (28.1)$$

where x, y, z = operating variables that are not a function of time

Steady-state simulation always simulates relatively long-term system characteristics, such as annual or seasonal system simulation. A time increment of one hour is typically used for analysis. Within the time increment, the operating parameters are independent of time. However, the magnitude of the same operating parameter may be different in successive hours.

In steady-state simulation, the heat capacity of the working substance is often far greater than the heat being absorbed and released from the equipment, pipes, ducts, and surroundings when the temperature of the working substance fluctuates. Therefore, heat absorbed by the equipment, pipes, and ducts, and heat transfer to or from the surroundings are often ignored. An hour-by-hour year-round energy estimation computer program is an example of steady-state system simulation.

DYNAMIC SIMULATION. In dynamic simulation, the relationship between various operating parameters is described by a mathematical equation that is a function of time, that is,

$$z = F(x, y, t, \ldots) \qquad (28.2)$$

where t = time variable

Dynamic simulation is mainly used to simulate short-term characteristics. Its purpose is to analyze the interaction between the control actions and the dynamic response, and the step change of an operating parameter on system performance. A time increment of minutes or even seconds is required to analyze the rapid response of the system during simulation. In dynamic simulation, the heat being absorbed and released from the equipment, instrument, pipes, and ducts is often taken into consideration. The variation of supply volume flow and zone temperature caused by a step change of the zone lighting load is an example of dynamic simulation.

Equation Fitting

Generally, a system component's performance with regard to one or two independent operating variables is available from the manufacturer's catalog. A polynomial expression is used to fit the catalog data into a regression equation to mathematically relate the dependent variable z with independent variable x in steady-state simulation as follows:

$$z = a_0 + a_1 x + a_2 x^2 + a_3 x^3 \qquad (28.3)$$

To relate a dependent variable with two independent variables, x and y, the polynomial expression is

$$z = b_0 + b_1 x + b_2 x^2 + b_3 y + b_4 y^2$$
$$+ b_5 xy + b_6 x^2 y + b_7 xy^2 + b_8 x^2 y^2 \quad (28.4)$$

where $a_0, a_1, \ldots, a_3, b_0, b_1, \ldots b_8 = $ coefficients

Computer programs are available to solve the coefficients according to the data from the manufacturer's catalog.

Sequential and Simultaneous Simulation

When a system comprises many components linked together by a working substance to form an open circuit, the calculation can be started at a component whose output is the input of the next linked component, and can progress in sequence to the component where the final result can be obtained. Such an approach is called *sequential simulation*.

In sequential simulation, the output of a successive component usually does not affect the output of the former component already calculated. If the output of a component does affect the output of the former component already calculated, a computing loop is used until the calculated output is equal to the assumed value.

The fuel input rate to a direct-fired gas heater is an example of sequential simulation. Calculation starts from the heating requirement at the warm air heater. The next calculation is mass flow and the temperature drop of the combustion gas, and finally the rate of fuel gas supplied to the burner is calculated.

When the working substance that links the components together forms a recirculating closed loop and many operating parameters are interrelated, all unknown operating parameters should be solved simultaneously. A simultaneous simulation is required. The power input to a compressor in a reciprocating compression refrigeration cycle is an example of closed-loop simultaneous simulation.

28.3 COMPUTER PROGRAMS FOR ENERGY USE SIMULATION

Numerous computer programs have been delevoped to estimate energy use by means of hour-by-hour detail simulation. Established programs include the following:

- DOE 2.1, developed by the U.S. Department of Energy
- Building Loads Analysis and Systems Thermodynamics (BLAST), developed by the University of Illinois in 1986
- Thermal Analysis Research Program (TARP), developed by the National Bureau of Standards in 1983

Many large manufacturers, such as the Carrier Corporation and The Trane Company, also have their own energy simulation programs.

DOE 2.1 is an hour-by-hour detail simulation computer program sponsored by the U.S. Department of Energy. It is one of the most widely used energy programs for HVAC&R systems. DOE 2.0 was developed and published in 1980. DOE 2.1A replaced DOE 2.0 in 1981, and DOE 2.1C revised the *Building Description Language (BDL) Summary* and the *Sample Run Book* of DOE 2.1A in 1984.

The DOE 2.1 energy computer program performs the following functions:

- Calculates hourly building heating and cooling loads
- Simulates the operation of various types of air systems
- Simulates the performance of the primary heating and cooling plants
- Finds the energy use of sub-systems, air, water, refrigeration, and heating systems, or air conditioning systems within a certain period and performs economic analysis

The documentation of DOE 2.1A includes a user's guide/BDL summary, sample run book, reference manual, and engineer's manual.

In the DOE 2.1A reference manual, detailed step-by-step instructions are provided to input data to load programs, air system programs, plant programs, and economic programs. Constant-volume, fan–coil, VAV systems, and many other types are listed in the air system programs with brief system descriptions and control strategies. The final outputs of the DOE 2.1A program are the output reports from load, air system, plant, and economic programs. The user's guide, reference material, and engineer's manual of DOE 2.1A are published and available.

28.4 CASE STUDY: SYSTEM SIMULATION OF A CENTRIFUGAL CHILLER

The following is an example of a system simulation to estimate the annual energy consumption of a centrifugal chiller.

System Model

The centrifugal chiller has a two-stage compressor using HCFC–123 as refrigerant. Its refrigeration cycle is the same as shown in Fig. 4.7. The centrifugal compressor is driven by a hermetic motor of an adjustable-frequency AC drive. After compression, the refrigerant is discharged to a shell-and-tube water-cooled condenser. A flash cooler or economizer is installed between the condenser and the shell-and-tube flooded evaporator. Several orifice plates are used as the throttling device between condensing pressure p_{con} and intermediate pressure p_{int}, and also between p_{int} and evaporating pressure p_{ev}. Water from the cooling tower is used as the cooling water for the condenser. A proportional-integral (PI) DDC panel is used to maintain a nearly constant temperature of chilled water as it leaves the evaporator.

During part-load operation, the DDC panel modulates the speed of the compressor to reduce both the mass flow rate of refrigerant $\dot{m}_r$ entering the compressor and the head, to create a new mass and energy balance.

The system simulation of this centrifugal chiller is considered at steady-state within a time increment of one hour. Both sequential simulation and iterations are used to calculate the final results. Many simplifications are used to reduce the calculations during simulation.

Operating Parameters Affecting Chiller Energy Performance

Operating parameters that affect the energy performance of this centrifugal chiller are as follows:

- Refrigeration load ratio R_{load}, which is defined as the ratio of the operating refrigeration load q_{ev} to the design full-load $q_{ev.d}$. R_{load} affects the rate of heat transfer at the evaporator and condenser. It also affects the efficiency of the two-stage compressor.
- Temperature of condenser cooling water entering the condenser T_{ce}, which is a function of outdoor wet bulb T'_o and the performance of the cooling tower. T_{ce} is closely related to the condensing temperature T_{con} in the condenser, and therefore the

pressure lift $\Delta p = (p_{con} - p_{ev})$, the temperature lift $\Delta T = (T_{con} - T_{ev})$, and the power input to the compressor P_{in}.

Simulation Methodology

Theoretical power input to the compressor $P_{in.t}$, in hp, can be calculated as

$$P_{in.t} = \frac{\dot{m}_r W_{in}}{2545} \qquad (28.5)$$

where $\dot{m}_r$ = mass flow rate of the refrigerant, lb/h
W_{in} = ideal work input to the compressor, Btu/lb

Work input W_{in} is directly proportional to the enthalpy difference of the refrigerant between condensing and evaporating pressure. In order to find W_{in}, it is necessary to determine the condensing pressure p_{con}, the condensing temperature T_{con}, the evaporating pressure p_{ev}, and the evaporating temperature T_{ev} first, so the following information is required:

- An evaporator model is required to determine T_{ev} at various operating load ratios R_{load}.
- A condenser model is required to determine T_{con} at various R_{load} and outdoor wet bulbs T'_o.
- Calculation of isentropic work input W_{in} to the compressor based on the two-stage refrigeration cycle at various R_{load} and T'_o is required.
- Calculation of the actual power input from the compressor model and the annual energy use is required.

Evaporator Model

As described in Section 23.9, the rate of heat transfer at the tube surface of the evaporator or the refrigeration load q_{ev}, in Btu/h, can be calculated as

$$q_{ev} = A_{ev} U_{o.ev} \Delta T_{ev} \qquad (28.6)$$

where A_{ev} = surface area of the evaporator, ft^2
ΔT_{ev} = log-mean temperature difference between the refrigerant and the chilled water at evaporator, °F

From Eq. (23.8), the overall heat transfer coefficient based on the outer surface area of the copper tubes $U_{o.ev}$, in Btu/h · ft^2 · °F, is given as

$$U_{o.ev} = \frac{1}{\frac{1}{\eta_f h_o} + \frac{A_o R_f}{A_i} + \frac{A_o}{A_i h_i}}$$

where h_o = boiling coefficient of refrigerant HCFC–123

The boiling coefficient for HCFC–123 is slightly lower than for refrigerant CFC–11. From Eq. (23.9), for a shell-and-tube flooded type evaporator, the boiling coefficient of HCFC–123 can be calculated as

$$h_o = C_b \left(\frac{q_{ev}}{A_{ev}} \right)^n$$

According to the experimental results in Webb and Pais (1991) and Jung and Radermacher (1991), for HCFC–123 in copper tubes with integrated fins of 26 fins/in., constant C_b can be taken as 2.5 and exponential index n is approximately 0.7. For enhanced surfaces, according to results of Webb and Pais, h_o is 35 percent greater.

The copper integrated-fin tubes currently used for evaporators usually have a fin spacing of 19 to 35 fins/in., typically 26 fins/in. The outside diameter of copper tubes varies from $\frac{5}{8}$ in to $\frac{3}{4}$ in. The ratio of outer surface area to inner surface area A_o/A_i is often between 3 and 4.

For a closed-circuit chilled-water system for evaporators with conventional water treatments, the fouling factor R_f can be taken as 0.00025 h $\cdot$ ft^2 $\cdot$ °F/Btu.

As described in Section 12.6, water-side heat transfer coefficient h_i, in Btu/h $\cdot$ ft^2·°F, can be calculated as

$$\text{Nu}_D = \frac{h_i D_h}{k} = 0.023 \text{Re}_D^{0.8} \text{Pr}^{0.4} \tag{28.7}$$

where D_h = hydraulic diameter, ft
k = thermal conductivity of water, Btu/h $\cdot$ ft^2 $\cdot$ °F

From Eq. (23.11), the log-mean temperature difference between refrigerant and chilled water ΔT_{ev} can be calculated as

$$\Delta T_{ev} = \frac{(T_{ee} - T_{ev}) - (T_{el} - T_{ev})}{\ln \left(\frac{T_{ee} - T_{ev}}{T_{el} - T_{ev}} \right)}$$

where T_{ee}, T_{el} = temperature of chilled water entering and leaving evaporator, °F
T_{ev} = evaporating temperature, °F

The mass flow rate of chilled water $\dot{m}_{w.ev}$, in lb/min, flowing through the copper tubes in the evaporator usually remains approximately constant during operation. Temperature T_{el} is often set and reset according to the requirements of the air system.

As in Section 23.9,

$$\frac{T_{ee} - T_{el}}{\frac{q_{ev}}{A_{ev} U_{o.ev}}} = B$$

The evaporating temperature can then be determined as

$$T_{ev} = \frac{e^B T_{el} - T_{ee}}{e^B - 1} \tag{28.8}$$

At design load, q_{ev} is equal to the design refrigeration load $q_{ev.d}$, in Btu/h. In part-load operation, however,

$$q_{ev} = R_{load} q_{ev.d} \tag{28.9}$$

and the evaporating temperature at part-load $T_{ev.p}$ can be similarly calculated.

Condenser Model

For a shell-and-tube water-cooled condenser, from Eq. (12.39) the rate of heat transfer at the condenser Q_{rej}, in Btu/h, or the total heat rejection, can be calculated as

$$Q_{rej} = q_{ev} + \frac{2545 P_{com}}{\eta_{mot}}$$
$$= A_{con} U_{o.con} \Delta T_{con} \tag{28.10}$$

where P_{com} = power input to the hermetic compressor motor, hp
η_{mot} = efficiency of hermetic motor

From Eq. (28.10) power input P_{com} is known only when condensing temperature T_{con}, work input W_{in}, and mass flow rate of the refrigerant $\dot{m}_r$ all have been calculated. Therefore, from Eq. (12.40) $Q_{rej} = F_{rej} q_{ev}$. Assume a heat rejection factor F_{rej} first. T_{con}, W_{in}, and P_{com} can then be calculated. If the calculated P_{com} and Q_{rej} do not equal to the assumed values, try another F_{rej} until the assumed and the calculated values are equal. For comfort air conditioning, F_{rej} usually varies from 1.20 to 1.35.

As in the evaporator model, the overall heat transfer coefficient based on the outer surface area of the condenser $U_{o.con}$, in Btu/h $\cdot$ ft^2 $\cdot$ °F, can be calculated as

$$U_{o.con} = \frac{1}{\frac{1}{\eta_f h_{con}} + \frac{A_{con} R_f}{A_i} + R_g + \frac{A_{con}}{A_i h_i}} \tag{28.11}$$

The log-mean temperature difference between the condensing refrigerant and the condenser water ΔT_{con}, in °F, is

$$\Delta T_{con} = \frac{(T_{con} - T_{cl}) - (T_{con} - T_{ce})}{\ln \frac{T_{con} - T_{cl}}{T_{con} - T_{ce}}} \tag{28.12}$$

where T_{ce}, T_{cl} = temperature of condenser water entering and leaving the condenser, °F
T_{con} = condensing temperature, °F

In Eq. (28.11), fin efficiency η_f, inner surface area of the copper tubes A_i, ratio of outer surface area to inner surface area of copper tubes A_{con}/A_i, and water-side heat transfer coefficient h_i can be determined as they are in the evaporator model.

As described in Section 12.7, and defined by Eq. (12.41), the condensing coefficient h_{con}, in Btu/h · ft^2 · °F, can be calculated as

$$h_{con} = C_{con}\left(\frac{1}{\frac{Q_{rej}}{A_{con}}}\right)^{1/3}$$

In a filmwise condensation shell-and-tube water-cooled condenser having $\frac{5}{8}$- or $\frac{3}{4}$-in. diameter copper tubes with integrated fins using HCFC–123 as refrigerant, constant C_{con} can be taken as 10,500.

In a condenser using a well-maintained cooling tower with proper water treatment, a fouling factor $R_f = 0.00025$ h · ft^2·°F/Btu is recommended. Industrial areas, if a brush cleaning system is installed, $R_f = 0.0002$ h · ft^2·°F/Btu.

The operating pressure of HCFC–123, like CFC–11, is lower than atmospheric pressure, so air and other noncondensable gases leak into the evaporator. The compressor transports them to a higher level and they accumulate in the condenser. Noncondensable gases reduce the condensing area and raise the condensing pressure. Their effect is similar to that of a gas-side resistance R_g, in h · ft^2 · °F/Btu, at the condenser as follows:

$$R_g = 0.00778 - 0.0173R_{load} + 0.0114R_{load}^2 \tag{28.13}$$

As in the evaporator model, C is calculated as follows:

$$\frac{T_{cl} - T_{ce}}{\frac{Q_{rej}}{A_{con}U_{o.con}}} = C$$

Condensing temperature is therefore calculated as

$$T_{con} = \frac{e^C T_{cl} - T_{ce}}{e^C - 1} \tag{28.14}$$

Cooling Tower Model

The approach of cooling water leaving the cooling tower $(T_{ce} - T_o')$ is mainly influenced by the load ratio R_{load} of the condenser, the configuration of the cooling tower, the outdoor wet bulb T_o', and the number of transfer units (NTU) of the fill.

In a counterflow induced-draft cooling tower using PVC fill or packing fill, with a cooling water flow rate of 3 gpm per ton of heat rejection and a water-air ratio of 1.2, if the outdoor wet bulb $65°F < T_o' < 78°F$,

the temperature of cooling water entering the condenser T_{ce}, in °F, can be roughly estimated as

$$T_{ce} = T_o' + \Delta T_{ap.d}K_{load}K_{wet}$$
$$K_{load} = 0.1 + 0.9R_{load} \tag{28.15}$$
$$K_{wet} = 4.8 - 0.0475T_o'$$

where $\Delta T_{ap.d}$ = approach of the cooling tower at design condition, °F
K_{load} = load factor
K_{wet} = factor considering the drop of outdoor wet bulb

Work Input and Mass Flow of Refrigerant

From the refrigeration cycle shown in Fig. 4.7b, isentropic work input W_{in}, in Btu/lb, for a two-stage compressor can be calculated as

$$W_{in} = (1 - x)(h_2 - h_1) + (h_4 - h_3) \tag{28.16}$$

The mass flow rate of the refrigerant $\dot{m}_r$ at the condenser, in lb/h, is given as

$$\dot{m}_r = \frac{q_{ev}}{(h_1 - h_9)(1 - x)} \tag{28.17}$$

and the fraction of refrigerant evaporated in the flash cooler x can be calculated as

$$x = \frac{h_{5'} - h_8}{h_7 - h_8} \tag{28.18}$$

where $h_1, h_2, h_3, h_4, h_{5'}, h_7, h_8, h_9$ = specific enthalpy of the refrigerant at state points 1, 2, 3, 4, 5', 7, 8, 9 respectively, Btu/lb

On a pressure–enthalpy p–h diagram, the specific enthalpies of saturated liquid and vapor are functions of temperature and pressure only. In centrifugal chillers used in comfort air conditioning systems, the range between evaporating temperature T_{ev} and condensing temperature T_{con} is usually between 20 and 120°F, and a simple polynomial representation can be used to calculate the specific enthalpy from the known T_{ev} and T_{con} with acceptable accuracy.

At the flash cooler, because the saturated temperature and pressure are interrelated, for simplicity, the intermediate saturated temperature between condensing and evaporating temperature can be estimated as

$$T_{int} = \frac{T_{ev} + T_{con}}{2} \tag{28.19}$$

According to the thermodynamic properties of HCFC–123 and its *p–h* diagram, the polynomial expression used to calculate the specific enthalpy of the saturated vapor of refrigerant HCFC–123 h_v between 20 and 70°F, in Btu/lb, is calculated as

$$h_v = 89.7 + 0.145(T_{ev} - 20) \qquad (28.20)$$

The polynomial expression used to calculate the specific enthalpy of the saturated liquid T_1 of HCFC–123 h_1, between 60 and 110°F, in Btu/lb, is

$$h_1 = 20.2 + 0.24(T_1 - 60) \qquad (28.21)$$

Specific enthalpy differences $(h_4 - h_3)$ and $(1 - x)(h_2 - h_1)$ to determine the isentropic work can be calculated more simply and with acceptable accuracy according to the corresponding saturated temperature of the gaseous refrigerant at condensing, interstage, and evaporating pressure along the constant entropy lines of similar profiles in the superheated region. The polynomial expression used to calculate the enthalpy of the gaseous refrigerant HCFC–123 between 60 and 120°F along the constant entropy line Δh_s, in Btu/lb, is

$$\Delta h_s = 0.135(T_{int} - T_{ev}) = 0.135(T_{con} - T_{int}) \qquad (28.22)$$

where T_{int} = temperature of the saturated gaseous refrigerant HCFC–123 at intermediate pressure, °F

Centrifugal Compressor Model

Actual power input to the hermetic motor P_{in}, in hp, can be calculated as

$$P_{in} = \frac{\dot{m}_r W_{in}}{2545 \eta_c \eta_{mec} \eta_{mot}} \qquad (28.23)$$

In Eq. (28.23), η_c is the compression efficiency of the centrifugal compressor, which is defined as the ratio of isentropic work to actual work delivered to the gaseous refrigerant during compression, as discussed in Section 10.9.

Also, mechanical efficiency η_{mec} is defined as the ratio of work delivered to the gaseous refrigerant to the work input to the compressor shaft. the difference in these two work inputs is mainly because of the loss in the bearings and gear train, and during transportion of the refrigerant in the centrifugal chiller. For centrifugal compressors operated at a certain speed, η_{mec} is considered a fixed value.

Mechanical efficiencies η_{mec} for centrifugal chillers manufactured after 1973 have the following values:

Without gear train	0.87
With gear train	0.85

Motor efficiency η_{mot} is a function of motor size and the load ratio of the chiller. Normally, for a motor size greater than 125 hp, ASHRAE Standard 90.1–1989 recommends an efficiency level greater than 92.4 percent. If the motor is to be operated more than 750 hours annually, a high-efficiency motor is most cost-effective. The efficiency of a 200-hp motor should be 96 percent or greater. If a hermetic motor is used, another 2 to 4 percent of power input is required to provide the refrigeration capacity to cool the hermetic motor.

Compression efficiency η_c is often the most influential parameter during the annual energy estimation of a centrifugal compressor. Compression efficiency η_c is a function of the volume flow rate of refrigerant V_{rf} and the system head or pressure lift Δp_t. Volume flow V_{rf} is closely related to load ratio R_{load}, and pressure lift is closely related to temperature lift $\Delta T = (T_{con} - T_{ev})$. In most centrifugal chillers, the chilled water leaving the chiller T_{el} is usually set at a constant value when the outdoor wet bulb is high. As outdoor wet bulb T'_o drops, the condenser water entering the condenser T_{ce} and, therefore, the condensing temperature T_{con} fall accordingly. It is assumed that when T'_o drops below 70°F, the reset of T_{el} offsets the fall of T'_o.

Figure 28.1 shows the compressor map of a variable-speed centrifugal compressor (shown in Fig. 23.9b). In Fig. 28.1, point A is the operating point of the centrifugal chiller at design-load conditions $(R_{load} = 1)$ and at a cooling water temperature entering the condenser $T_{ce} = 85°F$. If T_{ce} remains at 85°F, and R_{load} drops to 0, then the operating point is at point B. At point B, temperature lift ΔT is calculated as follows:

$$(T_{con} - T_{ev})_B = (T_{con} - T_{ev})_A - (T_{el} - T_{ev})_A$$
$$- (T_{con} - T_{cl})_A - (T_{cl} - T_{ce})_A \quad (28.24)$$

In a typical centrifugal chiller, if $(T_{el} - T_{ev})_A = (T_{con} - T_{cl})_A = 8°F$, $(T_{cl} - T_{ce})_A = 10°F$, and a typical $(T_{con} - T_{ev})_A = 67°F$ (similar to scheme A in Section 23.6), the operating curves, or required system head, of the compressor at other T_{ce} values can then be plotted on the compressor map shown in Fig. 28.1.

The compression efficiency η_c can therefore be calculated by a polynomial regression based on two variables, R_{load} and T_{ce}, as follows:

$$\eta_c = C_1 + C_2 R_{load} + C_3 R_{load}^2 + C_4 T_{ce}$$
$$+ C_5 T_{ce}^2 + C_6 R_{load} T_{ce} + C_7 R_{loa}^2 T_{ce}$$
$$+ C_8 R_{load} T_{ce}^2 + C_9 R_{load}^2 T_{ce}^2 \qquad (28.25)$$

where $C_1, C_2, \ldots, C_9$ = coefficients

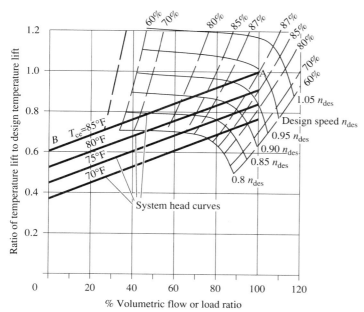

FIGURE 28.1   Compressor map of a centrifugal chiller at various load ratios R_{load} and T_{ce}.

From the manufacturer's compressor map and the added operating curves at various outdoor wet bulbs T_o', the coefficients can be determined by a computer program. In a two-stage centrifugal compressor, the mean value of compression efficiency of two individual stages is assumed to equal the overall compression efficiency η_c for simplification.

28.5   GAS COOLING AND COGENERATION

Gas Cooling

High electricity demand charges, high peak electricity rates, and the development of high-efficiency direct-fired equipment and highly reliable gas engines enable gas cooling systems to compete with electric compressors after the decline of gas cooling in the 1970s. Current gas cooling systems include the following:

- Two-stage, direct-fired, lithium bromide LiBr absorption chillers and chiller/heaters
- Desiccant cooling using direct-fired gas heaters for regeneration
- Gas engine chiller systems

Usually, a gas cooling system has a higher initial cost and a lower operating cost than an electricity-driven refrigeration system. Accurate calculation of the operating costs of a gas cooling system is important in this comparison.

LiBr absorption chillers, chiller/heaters, and desiccant cooling systems were all discussed in Chapters 24 and 25. Gas-engine chiller systems will be discussed in this section.

Cogeneration

Cogeneration is the sequential use of energy from a primary source, including natural gas, oil, and coal, to produce power and heat. Power can be electric or mechanical power or both. In a cogeneration system, the sequential use of the heat released from the flue gas and engine jacket significantly increases system efficiency and makes the cogeneration system economically attractive.

In 1978, the Public Utility Regulatory Policies Act (PURPA) permitted the interconnection of electric power lines of cogeneration systems with electric utility systems. This provides flexibility for cogeneration plants. They can either use or sell their electric power to the utility and optimize the size of the cogeneration plant by reducing or eliminating its standby generation capacity.

In 1990, hundreds of cogeneration systems were developed for internal use. *Internal use* is the production of both power and heat for use in settings such as hospitals, medical centers, university campuses, public buildings, and industrial facilities, and in their installed air conditioning systems. A successfully developed cogeneration system often relies on site technical and economical analysis, especially for local

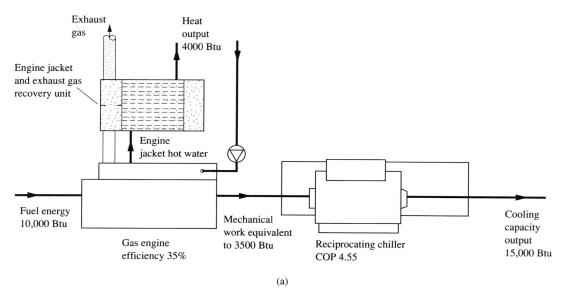

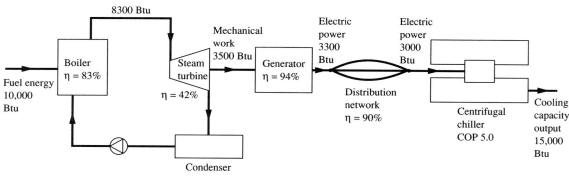

FIGURE 28.2 Energy flow in a gas-engine chiller and in an electricity-driven chiller: (a) gas-engine chiller and (b) electricity-driven chiller.

electricity demand and electricity rates. Two kinds of prime movers are widely used in these internal use cogeneration systems: gas engines and gas turbines.

Gas-Engine Chiller

A gas-engine chiller is actually a gas cooling system and a cogeneration system. Heat released from the exhaust gas and the engine jacket cooling water are all recovered to increase system efficiency. Figure 28.2a shows the energy flow of a gas-engine chiller system.

If the efficiency of the gas engine is 35 percent, the mechanical efficiency including the transmission gear train is 95 percent, and also if the chiller's COP = 4.5, for every 10,000 Btu of fuel energy input to the gas-engine chiller, there is a cooling output q_{rf} of

$$q_{rf} = 10,000 \times 0.35 \times 0.95 \times 4.5 \approx 15,000 \text{ Btu}$$

In addition, there is also a heating output of 4000 Btu from the exhaust gas and engine jacket to supply hot water for an absorption, space heating, or domestic hot water unit.

Engine-driven chillers are maintenance-intensive. The maintenance cost of a gas-engine chiller may be between 10 and 20 percent of the energy cost, and should be added to the operating cost during economic analysis.

For an electricity-driven chiller to purchase electric power, if the energy input to its power plant is 10,000 Btu, the cooling output at the chiller is also 15,000 Btu. It requires another fuel energy input of 5000 Btu to a boiler to produce 4000 Btu heat output. The simple payback period of the energy cost savings that compensate for the higher initial cost of a gas engine is usually between 2 and 5 years.

A gas-engine chiller has a prominent advantage over a motor-driven chiller because the former can vary its speeds at various operating conditions: high speed at overloads, and low speeds at part-load operation. Engine reliability is the key to user acceptance for gas-engine chillers. Several hundred engine-driven chillers were installed in the 1960s and early 1970s. According to reliability records of these systems reported in Davidson and Brattin (1986), the reliability of gas-engine chillers matches the requirements of HVAC&R systems.

Gas engines can be used to drive reciprocating, screw, or centrifugal compressors. Many manufacturers offer package units for easier field installation. The capacity of gas-engine chiller packaged units varies from 30 tons to 500 tons.

Gas-fired internal combustion engines for cooling in buildings follow the developments of gasoline and diesel engine technology. There are two kinds of gas engines: heavy-duty industrial applications and light-duty automotive engines. Industrial heavy-duty gas engines run a minimum of 30,000 hours of full-load service between major overhauls, and cost about five times as much as the automotive engine. The service life of an automobile engine is only 2000 to 5000 hours.

Recently, manufacturers have produced packaged automotive gas-engine chillers of 150-ton capacity. One manufacturer also offers one 150-ton gas engine chiller and integrated hot water absorption chillers to give a total maximum output of 180 tons. Automotive gas engines are suitable for compressors that require speeds far above 1800 rpm. This packaged gas-engine chiller is also equipped with microprocessor-based controls to coordinate and monitor the operation of the engine and chiller. An operating cost as low as half of those of similar size electricity-driven units is claimed, depending on the local utility rate structure.

Gas-engine cooling systems can also be coupled to direct-expansion DX refrigeration systems and rooftop units.

Heat Recovery from Engine Jacket and Exhaust Gas

In reciprocating engines, cooling water for engine jackets often removes about 30 percent of the heat input to the engine at the engine heads, outer housing, and exhaust manifolds. Cooling water usually reaches a maximum leaving temperature of 180°F and a maximum temperature increase of 15°F. Engines with modified gaskets and seals may permit a higher cooling water temperature. Most engines use force circulation by means of water pumps. Sometimes, gravity circulation is also used. Steam bubbles formed at the bottom of the engine jacket should be released in a separate chamber.

To prevent vapor condensation in the exhaust gas, which may form acids and damage the equipment and the exhaust gas piping, the temperature of exhaust air should not fall below 300°F. Approximately 50 to 60 percent of the exhaust heat can be recovered. In most designs, the exhaust gas heat recovery unit is also used as a silencer to reduce the exhaust gas noise. In packaged units, heat recovery from engine jacket and exhaust gas may be combined in one unit or in separate units. Hot water at a temperature of about 250°F or low-pressure steam of 5 psig can be produced by this heat recovery unit.

Cogeneration Using a Gas Turbine

Many cogeneration plants use a combustion gas turbine instead of a gas engine as prime mover. A gas turbine usually consists of a compressor section to raise the air pressure, a fuel–air mixing and combustion chamber, and an expanding turbine section. The compressor and the turbine are joined by the same shaft. Capacity may vary from several hundred brake horsepower (bhp) to more than 100,000 bhp. Gas turbines are often connected to induction generators to produce electric power through gear trains.

Steam boilers are often used as heat recovery units to produce steam at a pressure typically 15 psig from the gas turbine exhaust gas. Recovered heat can be used as process heat or to operate an absorption chiller.

28.6 SMOKE CONTROL

Fire Safety in Buildings

Fire safety is a critical design factor in tall buildings. On November 21, 1980, in Las Vegas, the MGM Grand Hotel fire took 85 lives. Smoke inhalation is the primary killer in building fires. Smoke migrates away from the fire through stairs, elevator shafts, service shafts, and other passages.

Fire protection and fire safety design in buildings includes the following measures:

- Fire compartmentation
- Fire-resistant materials and construction
- Fire alarm system
- Automatic sprinkler system
- Smoke control system including stairwell pressurization and zone smoke control systems
- Fire protection management and coordination

Smoke control systems, which include stair pressurization and zone smoke control, are features of HVAC&R system design, and will be discussed here.

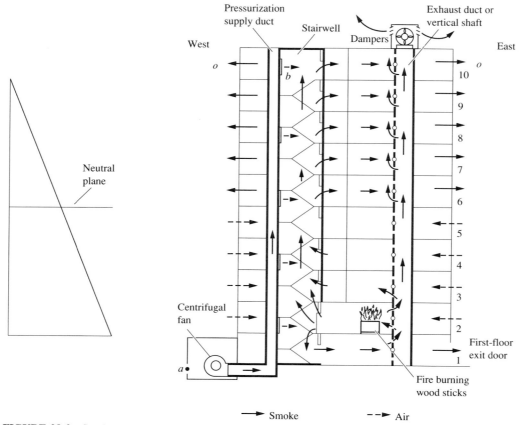

FIGURE 28.3 Smoke movements in a 10-story experimental building.

Smoke Movement in Buildings

National codes and local codes must be followed during the design stage of the smoke control system in buildings.

Figure 28.3 shows smoke movement in a 10-story experimental building.

According to *ASHRAE Handbook HVAC 1991, Applications*, the typical leakage area ratio A_{leak}/A_f, which is the ratio of leakage area A_{leak} to floor area A_{fl}, both in ft², for commercial buildings of average tightness are as follows:

External walls, including cracks around windows and doors	0.00021
Stairwell walls, construction cracks only	0.00011
Elevator shaft walls, construction cracks only	0.00084
Floors, construction cracks, and area around penetrations	0.000052

The cracks around the stairwell door can be typically taken as 0.25 ft², and the cracks around each elevator on each floor can also be taken as 0.25 ft².

A building fire can be simulated by burning wooden sticks in a second-floor corridor with a peak energy release of 900 Btu/s for a certain period while the building is under the following operating conditions:

- The stair pressurization system and smoke control system are not operated.
- All stairwell doors are closed except those on the second floor, which are opened $\frac{1}{2}$ in. by the high temperature of the hot gas.
- All windows and exit doors are closed.
- Outdoor temperature is 30°F and there is a north wind at 5 mph.

When the fire is ignited, smoke movements are mainly caused by the expansion and buoyancy forces of hot gas at a maximum temperature between 850 and 1000°F, the stack effect from outdoor–indoor temperature differences, and the wind effect.

Smoke moves from the fire on the second floor to the upper floors through stairwells, elevator shafts, service shafts, vertical risers, and floor cracks, and discharges to the outdoor atmosphere through window cracks, elevator machine-room openings, and other openings in the upper floors. Outdoor air enters the building below the neutral plane and discharges to the outdoors above the neutral plane because of the stack effect. Oxygen supply to the fire enters through second-floor window cracks, openings in the building envelope, and vertical air passages from the first and third floors.

According to Klote (1990), the CO_2 and CO levels on the highest floor of the experimental building during tests without stairwell pressurization and smoke control are 0.15 percent CO_2 and 0.015 percent CO. In tests with stairwell pressurization and smoke control, the levels are 0.002 percent CO_2 and 0.001 percent CO.

In winter, the stack effect assists the stairwell pressurization in preventing the smoke from contaminating the stairwell. This result is verified by experiments in Tamura (1990, Part II).

Effective Areas and Flow Rates

In several air flow passages connected in parallel, the effective area A_e, in ft^2, can be calculated as

$$A_e = A_1 + A_2 + \ldots + A_n \qquad (28.26)$$

where $A_1, A_2, \ldots A_n$ = air flow areas for paths 1, 2, ... and n, ft^2

In air flow paths connected in series, each with a flow area $A_1, A_2, \ldots A_n$, in ft^2, the effective area for these air flow paths connected in series A_e can be calculated as

$$A_e = \left(\frac{1}{A_1^2} + \frac{1}{A_2^2} + \cdots + \frac{1}{A_n^2}\right)^{-1/2} \qquad (28.27)$$

If the flow coefficient is taken as 0.65 and air density $\rho_a = 0.075$ lb$_f$/ft^3, the air volume flow rate $\dot{V}$, in cfm, flowing through a crack, gap, or opening can be calculated as

$$\dot{V} = 2610 A (\Delta p)^{1/2} \qquad (28.28)$$

where A = flow area or effective area, ft^2
Δp = pressure difference across the flow path or opening, in. WG

Stairwell Pressurization

A *stairwell pressurization system* uses fans to pressurize the stairwells to provide a smoke-free escape route for the occupants in case of a building fire. A stairwell pressurization system is a kind of smoke control system.

National and local codes require stairwell pressurization systems in high-rise buildings. In a stairwell

pressurization system, all interior stairwells are pressurized to a minimum of 0.15 in. WG and a maximum of 0.35 in. WG when all stairwell doors are closed.

City of New York Local Code 1979 requires a minimum air supply flow rate of 24,000 cfm plus 200 per floor for the stairwell pressurization system. The maximum allowable pressure difference between the stairwell and the floor space is 0.40 in. WG when stairwell doors are open or closed. The minimum allowable pressure difference is 0.10 in. WG when all stairwell doors are closed or 0.05 in. WG when any three stairwell doors are open. An alternative is to maintain at least 0.05 in. WG or a minimum average air velocity of 400 fpm at the stairwell door when any three stairwell doors are open. The force required to open the stairwell door must not exceed 25 lb$_f$ at the doorknob.

A stairwell pressurization system consists of centrifugal or vane axial fans, a stair pressurization supply duct with several supply air inlets, relief vents, and a control system, as shown in Fig. 28.4.

Outdoor air is extracted directly by the centrifugal fan. It is forced into the supply duct and then supplied to the stairwell through supply outlets. When the stairwell is pressurized to a pressure typically 0.10 to 0.40 in. WG higher than that of the air outside the stairwell on various floors across the stairwell doors, the smoke will not enter the stairwell, even through an open stairwell door. Air supplied to the stairwell is discharged through the open stairwell doors, leakage areas around closed stairwell doors, relief vents, or other openings to the rest of the building, and then discharged to the outdoors.

If the stairwell is an airtight enclosure or its doors have very small leakage areas, it will be overpressurized when all stairwell doors are closed. The pressure difference across the stairwell doors may be greater than 0.3 in. WG. Often, a large force is required to open the door. The total force required to open a stairwell door should not exceed 25 to 30 lb$_f$, or it will be too difficult to open stairwell doors during evacuation. Methods of overpressure relief will be discussed later in this chapter.

The air velocity at the open stairwell door on the fire floor required to prevent the backflow of smoke from the fire to the stairwell is called *critical velocity*, v_{crit}, in fpm. The outward flow air velocity from the stairwell through the open door on the fire floor should be greater than v_{crit}.

Characteristics of Stairwell Pressurization

Centrifugal or vane axial fans can be used for stairwell pressurization. Fans can be installed either at the bottom level of the building (bottom injection) or on the

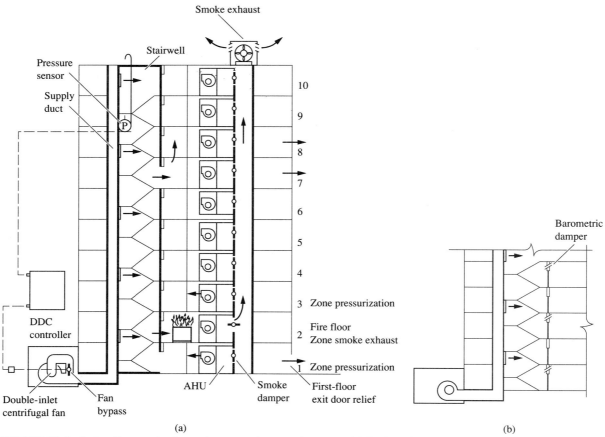

FIGURE 28.4 Stairwell pressurization and zone smoke control systems: (a) fan bypass overpressure relief and (b) overpressure relief vents (barometric damper).

rooftop (top injection). Bottom injection is preferable because it minimizes the possibility of smoke inhalation and optimizes stack effect to assist stairwell pressurization during winter. In any case, fan intake must be remote from the smoke exhaust during a building fire. If the fan room for stairwell pressurization is located on the rooftop, facilities must be provided to minimize the influence of wind pressure on fan performance.

Multiple injections, in which air is supplied from multiple inlets into the stairwell, provide a more even pressure distribution along the stairwell than a single injection from the top or bottom. Typically, each supply inlet serves two or three floors.

Open-tread stairs provide less flow resistance than closed-tread stairs. This difference becomes more prominent when occupants are walking on the stairs during evacuation.

Compartmentation of the stairwell into many sections, serving four to five floors, for example, may not provide the air flow rate necessary for stairwell pressurization when two or three stairwell doors are open at the same time.

Overpressure Relief and Feedback Control

When a stairwell pressurization system is operating and all the stairwell doors are closed, the pressure difference across the stairwell doors eventually exceeds the maximum permissible limit. Two methods are currently used to relieve the stairwell pressure. *Overpressure relief* is achieved by opening first-floor exit doors (as shown in Fig. 28.4a) or overpressure relief vents (barometric dampers, as shown in Fig. 28.4b). *Variable volume of supply air* is achieved by means of feedback control (as shown in Fig. 28.4a).

EXIT DOOR RELIEF. When a stairwell pressurization system is turned on, the interlocked control system automatically opens the first-floor exit door to relieve stairwell pressure. Overpressurized air in the stairwell is discharged to the outdoors. Because it is necessary to open first-floor exit door to evacuate the occupants during a building fire, this is a simple and effective means of overpressure relief.

Test results in Tamura (1990, Part II) showed that when stairwell pressurization was activated and the first-floor exit door was used as overpressure relief, the maximum pressure difference across the stairwell doors was lower than 0.25 in. WG in summer and winter. During nonfire conditions, pressure differences were between 0.05 and 0.1 in. WG across the stairwell doors if there was a vent in the external wall on the second floor. When a building fire was simulated on the second floor, there was no smoke backflow from the fire floor through the stairwell door when the stairwell doors on the first and second floors were open and the vent on the second floor external wall was also open. Smoke backflow occurred when stairwell doors on first, second, and third floors were open.

OVERPRESSURE RELIEF VENTS. A *relief vent* is typically an assembly in which a fire damper is connected in series with a barometric damper. The barometric damper is normally closed. If the stairwell is pressurized above a predetermined limit, the vents open and relieve air to various building floors. A counterweight in the barometric damper sets the maximum pressure limit. Fire dampers are normally closed, and open when the stairwell pressurization system is turned on in case of a building fire.

According to the experimental results of Tamura (1990, Part II), the performance of the overpressure relief vents at a supply flow rate of 28,000 cfm was slightly better than the performance of a ground-floor exit door at a supply flow rate of 17,800 cfm. If supply flow rates are the same, the difference is further reduced.

FAN BYPASS. When the stairwell is overpressurized, a pressure sensor located inside the stairwell signals a feedback control to open a fan bypass damper so that part of the supply air returns to the centrifugal fan inlet, as shown in Fig. 28.4a. The air volume flow supplied to the stairwell is reduced until the pressure at the pressure sensor drops below a preset value. If the pressure sensor is located on the rooftop or outdoors, it should be shielded from the influence of wind.

VARIABLE-SPEED OR CONTROLLABLE-PITCH FAN. When excessive pressure is detected in the stairwell, a controller actuates an adjustable-frequency AC drive to reduce the speed of a centrifugal fan or vary the blade pitch of a vane axial fan to maintain the required pressure inside the stairwell at the pressure sensor.

The test results in Tamura (1990, Part II) showed that both fan bypass and variable-speed controls require a response time of more than five minutes. Variable-speed control is slightly faster. Because reliability is the primary factor in stairwell pressuriza-

tion control, and because of its very short operating period, it may not be worthwhile to install an expensive adjustable-frequency AC inverter to provide variable-speed control.

Zone Smoke Control

In a building fire, *zone smoke control* provides smoke exhaust on the fire floor by opening the smoke damper connected to the smoke exhaust duct. At the same time, it supplies air to the floors immediately above and below the fire floor and pressurizes them, to prevent smoke contamination, by operating the air-handling units on these floors and closing the smoke dampers connected to the smoke exhaust duct, as shown in Fig. 28.4a. These mechanisms should be controlled by a DDC panel for fire protection management.

The supply and exhaust volume flow rates for zone smoke control are often matched with the HVAC&R system in buildings, especially when an air economizer cycle is used. The exhaust air from the AHU can be connected to a smoke exhaust duct, and the exhaust fan is generally located on the rooftop.

In an air system operated only in recirculating mode, a smoke exhaust system for zone smoke control should be installed. In such circumstances, the smoke exhaust volume flow rate can be determined to equal about six air changes per hour.

Design Considerations

The object of smoke control is to provide a smoke-free escape route for occupants through the stairwell to the outdoors during a building fire. This is a part of the building fire-protection scheme. The primary considerations are safety and reliability. Because the performance of stairwell pressurization and the zone smoke control are related, they should be considered an integrated smoke control system during system design.

Operation of both stairwell pressurization and zone smoke control systems must meet the requirements of national and local codes and provide a smoke-free escape route even under the following conditions:

- A fire reaches a fire temperature up to 1200°F when the sprinkler system fails to operate.
- Three or four stairwell doors, including the door on fire floor, are opened simultaneously.
- Fire floor smoke exhaust is performed mainly by a zone mechanical smoke exhaust system.
- During summer operating conditions, the stack effect does not act as an additional assistant.

Recently, it has been recommended that smoke control technology be extended to elevator shafts and lobbies used to evacuate disabled persons during a building fire. An elevator shaft and elevator lobby pressurization system should then be installed.

During a building fire, HVAC&R should operate according to fire safety and smoke control requirements. It is important to have feedback mechanisms to verify system operation and performance, such as separate alarms to signal smoke migration or burning fire, and adequate pressure levels at key points during pressurization and evacuation.

Volume Flow Rate

Volume flow rate is the primary determinant of the stairwell pressurization and zone smoke control system performance.

The volume flow rate of a zone smoke exhaust system from the fire floor is usually six air changes per hour. This volume flow rate is used by many designers and has been verified as appropriate in field tests by Tamura and Klote (1990, Part II). The volume flow rates of air supplied to the floors immediately above and below normally should be the same as the supply volume flow rate of the AHU or packaged unit for that floor.

Air discharged through the open stairwell doors and leaked through the closed stairwell doors and walls can be summarized into the following types:

- Discharged through an open stairwell door on the fire floor with a critical velocity v_{crit}
- Discharged through the open first-floor exit door
- Discharged through open stairwell doors on floors other than the fire floor and the first floor
- Leaked through cracks around the stairwell door and in the stairwell wall

The volume flow rate supplied to a stairwell pressurization system then can be calculated as follows:

$$\dot{V}_{s.p} = (v_{crit} + v_{exit})A_{door} + \dot{V}_{o.d}N_{o.d} + \dot{V}_{leak}N_{c.d}$$
$$(28.29)$$

where v_{exit} = average air velocity at the first-floor exit door, fpm

A_{door} = area of the opened stairwell door on the fire floor, ft^2

$\dot{V}_{o.d}$ = volume flow rate discharged through the open stairwell door on floors other than the fire floor and first floor, fpm

$N_{o.d}$ = number of open doors in the stairwell other than the fire floor and first floor

$\dot{V}_{leak}$ = air leakage through cracks across the stairwell wall on floors with closed stairwell doors, cfm

$N_{c.d}$ = number of closed doors in the stairwell

According to the experimental results in Tamura (1991), for a fire in a building with stairwell pressurization and smoke exhaust systems to exhaust smoke from the fire floor, the critical velocity v_{crit} can be taken as 300 fpm for a fire temperature of 1200°F and a mechanical venting of the fire floor of 5.5 ach.

In stairwell pressurization systems serving up to 10 floors, $N_{o.d} = 3$, including the door on the fire floor. In systems serving 15 or more floors, $N_{o.d} = 4$, including the door on the fire floor.

For safety, air velocity at the first-floor stairwell exit door v_{ext} can be assumed to be equal to v_{crit}.

Discharge air velocity at open stairwell doors on floors other than the fire floor $v_{o.d}$ is less than v_{crit}, in fpm. This is because the flow resistance of the cracks around doors and windows in floors other than the fire floor is greater than that of a smoke exhaust system on the fire floor.

The air volume flow rate discharged through an open stairwell door other than that on the fire floor can be calculated from Eq. (28.28) by using effective area A_e of the air flow path instead of area of the door or opening A. Air leakage through cracks in the stairwell wall can also be calculated from Eq. (28.28).

System Pressure Loss for a Stairwell Pressurization System

The frictional loss of air flow per floor inside the stairwell pressurization system $\Delta p_{f.s}$, in inches WG, can be calculated by considering the stairwell as a rectangular duct, as follows:

$$\Delta p_{f.s} = K \left(\frac{L}{D_e}\right)\left(\frac{\rho_a}{2g_c}\right)\left(\frac{\dot{V}}{A_s}\right)^2 \qquad (28.30)$$

where L = vertical height of each floor, ft

g_c = gravitational constant, 32.2 ft/s^2

ρ_a = air density, lb$_f$/ft^3

$\dot{V}$ = air supply volume flow rate, cfm

A_s = cross-sectional area of the stairwell, ft^2

In Eq. (28.30), D_e, in ft, indicates the circular equivalent of the stairwell, and can be calculated as

$$D_e = \frac{4A_s}{P_s} \qquad (28.31)$$

where P_s = perimeter of the cross-sectional area of the stairwell, ft

K is the pressure drop coefficient. It is mainly a function of the configuration of the stairwell and the occupant density on the stairs. According to the experimental results in Achakji and Tamura (1988), for a stairwell with a cross-sectional area of 134 ft^2 and a floor height of 8.5 ft, K has the following values:

Open-tread stair
  no occupants   1.6
  occupant density of 0.18 person/ft^2 2.8
Closed-tread stair
  no occupants   1.8
  occupant density of 0.18 person/ft^2 4.6

For a supply volume flow rate of 21,200 cfm to such a stairwell, the pressure drop per floor is 0.073 in. WG.

When calculating the system pressure loss of a stairwell pressurization system with multiple injection, as shown in Fig. 28.3, Δp_{sy}, in inches WG, is calculated as

$$\Delta p_{\mathrm{sy}} = \Delta p_{a-b} + \Delta p_{b-o} \qquad (28.32)$$

where $\Delta p_{a-b}, \Delta p_{b-o}$ = pressure loss between points a and b and points b and o, respectively, in. WG

Pressure loss Δp_{b-o} usually varies between 0.10 and 0.40 in. WG, and is only a small portion of the system pressure. However, it is difficult to calculate accurately because of complicated air flow paths at various operating conditions and the influence of outdoor conditions.

Pressure loss Δp_{a-b} should be calculated between points a and b according to the procedure described in Chapter 8, including the velocity pressure at the supply inlet, and Δp_{b-o} should be estimated as 0.4 in. WG. The calculated system pressure Δp_{sy} should be multiplied by a safety factor of 1.2. The actual Δp_{b-o} can be adjusted during acceptance testing.

In a stairwell pressurization system with a bottom single injection, as shown in Fig. 28.5a, the maximum pressure difference between the stairwell and outdoors often occurs at the top of the stairwell Δp_{c-o} because the cross section of the stairwell is constant. The system pressure can therefore be calculated as

$$\Delta p_{\mathrm{sy}} = \Delta p_{a-b} + \Delta p_{b-c} + \Delta p_{c-o} \qquad (28.33)$$

Here, for simplicity, Δp_{c-o} again can be taken as 0.4 in. WG. Then,

$$\Delta p_{b-c} = \Delta p_{\mathrm{f.s}} N_{\mathrm{flr}} \qquad (28.34)$$

where N_{flr} = number of floors

Actually, part of the velocity pressure discharged from the bottom inlet may be converted into static pressure, so that Δp_{b-c} is smaller. However, more data are needed before it can be calculated accurately.

In a stairwell pressurization system with top and bottom injection, as shown in Fig. 28.5b, the sys-

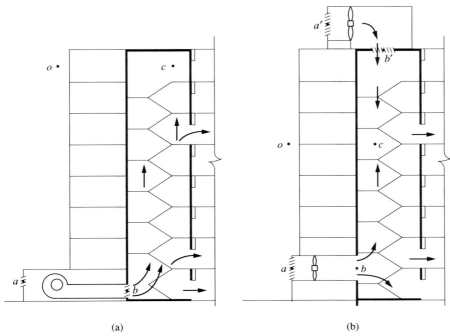

(a)   (b)

FIGURE 28.5 Bottom injection, and bottom and top injection: (a) bottom injection and (b) bottom and top injection.

tem pressure loss can be calculated as it is in bottom injection, but Δp_{b-c} may be smaller.

Example 28.1. In a smoke control system serving a 20-story high-rise building, each floor has the following constructional characteristics:

Floor area	10,000 ft²
Area of external wall	4800 ft²
Vertical distance between floor and ceiling	10 ft
Area of stairwell wall	600 ft²
Elevator shaft (3 elevators) wall area	1000 ft²
Area of stairwell door	3 × 7 = 21 ft²
Volume flow rate of supply air to conditioned space through AHU on each floor	12,000 cfm

The pressure difference across the closed stairwell door is 0.1 in. WG. The pressure differences between the stairwell and the outdoors on floors other than the fire floor whose stairwell doors remain open are also 0.1 in. WG (because the pressure drop of the open stairwell door is ignored). Determine the volume flow rates of the zone smoke control and stairwell pressurization systems.

Solution

1. The volume flow rate of zone smoke exhaust system from the fire floor $\dot{V}_{ex}$ at a rate of six air changes per hour is

$$\dot{V}_{ex} = \frac{6 \times 10,000 \times 10}{60} = 10,000 \text{ cfm}$$

The volume flow rate of air supplied to the floors immediately above and below the fire floor is 12,000 cfm.

2. The volume flow rate of air that is discharged through the open stairwell door to the fire floor $\dot{V}_{fire}$, in cfm, and is exhausted through the zone smoke control system is

$$\dot{V}_{fire} = v_{crit}A_{door} = 300 \times 3 \times 7 = 6300 \text{ cfm}$$

3. The volume flow rate discharged from the first-floor exit door is assumed to be the same as from the open stairwell door on the fire floor (6300 cfm).

4. From the given data, the leakage area on the stairwell wall, including cracks around the stairwell door, is

$$A_{leak} = 0.25 + 0.00011 \times 600 = 0.32 \text{ ft}^2$$

If the pressure difference across the closed stairwell door is 0.1 in. WG, from Eq. (28.28), the air leakage rate $\dot{V}_{leak}$ through the cracks of the stairwell wall on each floor whose stairwell door is closed can be calculated as

$$\dot{V}_{leak} = 2610A(\Delta p)^{1/2} = 2610 \times 0.32(0.1)^{1/2}$$
$$= 264 \text{ cfm}$$

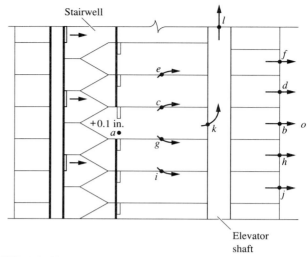

FIGURE 28.6 Air discharged from an open stairwell door through a combined air path to the outdoors.

5. It is possible that the fire floor and the pressurized floors immediately above and below are all in the lower ten floors. Because of the stack effect in this 20-story building, air mixed with the hot gas from the burning fire is discharged from the upper 10 stories.

In a 20-story building, assume that there are four open stairwell doors. One is on the fire floor, and another is the first-floor exit door. Then, the remaining two open stairwell doors should each discharge a combined air path that may be a combination of six parallel paths, as shown in Figure 28.6.

$a–b–o,$

$a–c–d–o,$

$a–c–e–f–o,$

$a–g–h–o,$

$a–g–i–j–o,$

$a–k–l–o$

6. The leakage area for the external wall for each floor is

$$A_{leak} = 4800 \times 0.00021 \approx 1.0 \text{ ft}^2$$

The leakage area for a floor area of 10,000 ft² is

$$A_{leak} = 10,000 \times 0.000052 = 0.52 \text{ ft}^2$$

The leakage area to the elevator shaft is

$$A_{leak} = 3 \times 0.25 + 1000 \times 0.00084 = 1.59 \text{ ft}^2$$

If the total opening at the top of the elevator shaft is 5 ft², assume that only 0.5 ft² will be allocated for open stairwell door. For air flow path $a–b–o$, if the pressure drop at the open door is negligible, the effective area is

$$A_e = \left(\frac{1}{A_1^2}\right)^{-1/2} = 1 \text{ ft}^2$$

For air flow path $a-c-d-o$, the effective area is

$$A_e = \left(\frac{1}{A_1^2} + \frac{1}{A_2^2}\right)^{-1/2} = \left(\frac{1}{0.52^2} + \frac{1}{1^2}\right)^{-1/2}$$

$$= 0.46 \text{ ft}^2$$

For air flow path $a-c-e-f-o$, the effective area is

$$A_e = \left(\frac{1}{A_1^2} + \frac{1}{A_2^2} + \frac{1}{A_3^2}\right)^{-1/2}$$

$$= \left(\frac{1}{0.52^2} + \frac{1}{0.52^2} + \frac{1}{1^2}\right)^{-1/2} = 0.35 \text{ ft}^2$$

For air flow path $a-k-l-o$, the effective area is

$$A_e = \left(\frac{1}{1.59^2} + \frac{1}{0.5^2}\right)^{-1/2} = 0.48 \text{ ft}^2$$

The effective area of air flow path $a-g-h-o$ is the same as that of $a-c-d-o$, that is 0.46 ft^2, and the effective area for $a-g-i-j-o$ is the same as that of $a-c-e-f-o$, that is, 0.35 ft^2. Therefore, the total effective area of this combined air flow path is

$$A_e = (2 \times 1) + (2 \times 0.46) + (2 \times 0.35) + 0.48$$

$$= 4.1 \text{ ft}^2$$

7. From Eq. (28.28), air discharged from this open stairwell door is

$$\dot{V} = 2610 A (\Delta p)^{1/2} = 2610 \times 4.1 (0.10)^{1/2}$$

$$= 3384 \text{ cfm}$$

8. From Eq. (28.29), the total supply volume flow rate of this stairwell pressurization system is then

$$\dot{V}_{s.p} = (v_{crit} + v_{exit}) A_{door} + \dot{V}_{o.d} N_{o.d} + \dot{V}_{leak} N_{c.d}$$

$$= (300 + 300)3 \times 7 + 2 \times 3384 + 264(20 - 4)$$

$$= 23{,}592 \text{ cfm}$$

REFERENCES AND FURTHER READING

Achakji, G. A., and Tamura, G. T., "Pressure Drop Characteristics of Typical Stairshafts in High-Rise Buildings," *ASHRAE Transactions*, Part I, pp. 1223–1237, 1988.

ASHRAE, *ASHRAE Handbook 1989, Fundamentals*, ASHRAE Inc., Atlanta, GA, 1989.

ASHRAE, *ASHRAE Handbook 1991, HVAC Applications*, ASHRAE Inc., Atlanta, GA, 1991.

ASHRAE, *ASHRAE Handbook 1992, HVAC Systems and Equipment*, ASHRAE Inc., Atlanta, GA, 1992.

Brodrick, J. R., and Patel, R. F., "Assessments of Gas-Fired Cooling Technologies for the Commercial Sector," *ASHRAE Transactions*, Part I, pp. 960–967, 1989.

City of New York, Local Law No. 84, 1979.

Clark, D. R., Hurley, C. W., and Hill, C. R., "Dynamic Models for HVAC System Components," *ASHRAE Transactions*, Part I, pp. 737–751, 1985.

Clark, J. A., and Harris, J. W., "Stairwell Pressurization in a Cold Climate," *ASHRAE Transactions*, Part I, pp. 847–851, 1989.

Davidson, K., and Brattin, H. D., "Gas Cooling for Large Commercial Buildings," *ASHRAE Transactions*, Part I B, pp. 910–920, 1986.

Dolan, W. H., "Gas Cooling for the Commercial Sector—Present and Future Perspective," *ASHRAE Transactions*, Part I, pp. 968–971, 1989.

Jung, D. S., and Radermacher, R., "Prediction of Heat Transfer Coefficients of Various Refrigerants During Evaporation," *ASHRAE Transactions*, Part II, pp. 48–53, 1991.

Klote, J. H., "An Overview of Smoke Control Technology," *ASHRAE Transactions*, Part I, pp. 1211–1222, 1988.

Klote, J. H., "Fire Experiments of Zoned Smoke Control at the Plaza Hotel in Washington, DC," *ASHRAE Transactions*, Part II, pp. 399–416, 1990.

Klote, J. H., and Tamura, G. T., "Design of Elevator Control Systems for Fire Evacuation," *ASHRAE Transactions*, Part II, pp. 634–642, 1991.

Lehr, V. A., "Life Safety in Tall Buildings," *Heating/Piping/Air Conditioning*, pp. 41–46, April 1990.

Lawrence Berkeley Laboratory, *DOE-2 User's Guide, Version 2.1*, National Technical Information Service, Springfield, VA, 1980.

Los Alamos National Laboratory, *DOE-2 Reference Manual, Version 2.1A*, National Technical Information Service, Springfield, VA, 1981.

Miller, D. E., "A Simulation to Study HVAC Process Dynamics," *ASHRAE Transactions*, Part II, pp. 809–825, 1982.

O'Neal, D., and Kondepudi, S. N., "Demonstrating HVAC System Performance Through System Simulation," *ASHRAE Transactions*, Part II B, pp. 116–129, 1986.

Orlando, J. A., "Packaged Cogeneration for Small Buildings," *Heating/Piping/Air Conditioning*, pp. 49–54, December 1988.

Pais, C., and Webb, R. L., "Literature Survey of Pool Boiling on Enhanced Surfaces," *ASHRAE Transactions*, Part I, pp. 79–89, 1991.

Park, C., Bushby, S. T., and Kelly, G. E., "Simulation of a Large Office Building System Using the HVACSIM+ Program," *ASHRAE Transactions*, Part I, pp. 642–651, 1989.

Schwartz, K. J., Jensen, R. H., and Antell, J., "The Role of Dampers in a Total Fire Protection System Analysis," *ASHRAE Transactions*, Part I B, pp. 566–576, 1986.

Shavit, G., "Information-Based Smoke Control Systems," *ASHRAE Transactions*, Part I, pp. 1238–1252, 1988.

Silver, S. C., Jones, J. W., Peterson, J. L., and Hunn, B. D., "CBS/ICE: A Computer Program for Simulation of Ice Storage Systems," *ASHRAE Transactions*, Part I, pp. 1206–1213, 1989.

Sinclair, J., "Cogeneration: A Success Story," *Heating/Piping/Air Conditioning*, pp. 49–54, December 1989.

Tamura, G. T., "Stair Pressurization Systems for Smoke Control: Design Considerations," *ASHRAE Transactions*, Part II, pp. 184–192, 1989.

Tamura, G. T., "Field Tests of Stair Pressurization Systems with Overpressure Relief," *ASHRAE Transactions*, Part I, pp. 951–958, 1990.

Tamura, G. T., "Fire Tower Tests of Stair Pressurization Systems with Overpressure Relief," *ASHRAE Transactions*, Part II, pp. 373–383, 1990.

Tamura, G. T., "Fire Tower Tests of Stair Pressurization Systems with Mechanical Venting of the Fire Floor," *ASHRAE Transactions*, Part II, pp. 384–392, 1990.

Tamura, G. T., "Determination of Critical Air Velocities to Prevent Smoke Backflow at a Stair Door Opening on the Fire Floor," *ASHRAE Transactions*, Part II, pp. 627–633, 1991.

VanWinckel, W. H., and Malewski, W. F., "Cogeneration and Refrigeration: A Case Study," *Heating/Piping/Air Conditioning*, pp. 59–64, December 1989.

Webb, R. L., and Pais, C., "Pool Boiling Data for Five Refrigerants on Three Tube Geometries (RP-392)," *ASHRAE Transactions*, Part I, pp. 72–78, 1991.

Wang, S. K., *Principles of Refrigeration Engineering*, Vol. 3, Hong Kong Polytechnic, Hong Kong, 1984.

Wang, S. K., *Air Conditioning*, Vol. 4, Hong Kong Polytechnic, Hong Kong, 1987.

Wong, S. P. W., and Wang, S. K., "System Simulation of the Performance of a Centrifugal Chiller Using a Shell-and-Tube Type Water-Cooled Condenser and R–11 as Refrigerant," *ASHRAE Transactions*, Part I, pp. 445–454, 1989.

CHAPTER 29

AIR CONDITIONING SYSTEMS: PACKAGED SYSTEMS

29.1 CLASSIFICATION OF AIR CONDITIONING SYSTEMS

Basic Approach

The purpose of classifying air conditioning systems is to distinguish one type from another and to provide a background for selecting the optimum air conditioning system based on building requirements. If designers cannot properly classify an air conditioning system and distinguish it from others, it is difficult for them to select an optimum system.

During the classification of air conditioning systems, the following points should be considered:

- A classification of air conditioning systems or HVAC&R systems should include the classification of air and refrigeration systems in order to define a more specific system.
- The system and the primary equipment should be compatible with each other. For example, the primary piece of equipment in a unitary packaged system is the packaged unit.
- System classification should mainly be based on practical applications, not on academic concerns. For example, if the indoor air quality is one of the primary criteria of air conditioning, an all-water system cannot provide an acceptable IAQ in current practical applications, so it should not be considered an applicable category in air conditioning system classification.
- System classification should be simple and clear, and should use current terminology.

Air Conditioning System Classification

As discussed in Chapters 1 and 15, air conditioning systems can be classified into three categories corresponding to their related equipment as follows:

INDIVIDUAL SYSTEMS. An individual-room air conditioning system, or individual system, uses a self-contained, factory-made air conditioner to serve one or occasionally two rooms.

The equipment used in individual air conditioning systems includes the following:

- Room air conditioner or window air conditioner.
- Packaged terminal air conditioner (PTAC). This is a factory-built unit mounted through the wall with a self-contained refrigeration system; a hot water, steam, or electric heating coil; fan; filter; and other accessories.
- Split indoor unit consisting of a fan, DX-coil, filter, and outdoor condensing unit.

UNITARY PACKAGED SYSTEMS. Unitary packaged systems, or packaged systems, also use self-contained, factory-made packaged units. However, they have the following characteristics:

- They serve more zones or rooms, or even more than one floor.
- They have an air system consisting of fans, coils, filters, and ductwork.
- They use a DX-coil for cooling.
- They use a gas-fired heater or an electric heater.

As described in Section 15.8, the basic piece of equipment in a packaged system is the packaged unit. Packaged units can be classified as rooftop, indoor, or split packaged units and heat pumps. A split packaged unit consists of an indoor air handler and an outdoor condensing unit. In recent models, desiccant cooling is sometimes added to a packaged unit to dehumidify and cool the air in order to reduce the energy cost.

Water-loop heat pump systems should be considered a member of the packaged system category. Although it has a centralized water system as a heat sink or heat source, air is conditioned at the DX-coil in the heat pump to provide cooling during summer and heating during winter.

CENTRAL HYDRONIC SYSTEMS. Central hydronic systems, or central systems, were described in Section 1.4. A central system consists of an air system, a water system, and a primary cooling and heating plant.

An air-handling unit (AHU) is the basic piece of equipment used in an air system. It can be either a field-assembled built-up system or a factory-made unit. Because a central system always has a chilled water system, the type of coil installed in an AHU is a water cooling coil, which is different from an air handler in a packaged system, which uses a DX-coil.

Centrifugal, screw, and reciprocating chillers and probably gas-fired hot water boilers, are the primary types of cooling and heating equipment found in a central system. In order to compensate for high electricity demands and on-peak electricity rates, thermal storage, desiccant cooling, and gas-cooling equipment and technology have been developed in recent years for use in primary cooling and heating plants.

A thermal storage system is always a part of a central system because a water cooling coil that uses chilled water or brine as the cooling medium in the air-handling unit(s) is used in central systems.

Air and Refrigeration Systems Designation

Classification of air conditioning systems also involves classification of refrigeration systems:

- An individual system uses a small, self-contained, factory-assembled refrigeration system that uses a DX-coil to cool air.
- A packaged system always has a refrigeration system that uses a DX-coil to cool air directly.
- A central system has a refrigeration system that uses chilled water as a cooling medium to cool air indirectly.

In addition, in order to designate an air conditioning system more clearly and correctly, the main characteristics of its air system and its refrigeration system may be added to the description of its basic category (that is, individual, packaged, or central).

As described in Chapters 14 and 21 through 27, more clearly specified terminology for an air conditioning system with a designated air system and primary cooling and heating plant is a combination of items from two or three of the following columns:

Air system	Refrigeration	AC system
Constant volume	Centrifugal	Rooftop packaged system
Fan–coil	Reciprocating	Indoor packaged system
Single-zone VAV	Screw	Split packaged system
Perimeter-heating VAV	Scroll	Rooftop heat pump system
VAV reheat	Absorption	Split heat pump system
Dual-duct VAV	Gas cooling	Water-loop heat pump system
Fan-powered VAV	Desiccant	Central system
	Evaporative	Ice storage system
		Chilled water (CW) storage system
		Heat recovery central system

- Some air systems are always central systems and some refrigeration systems are usually packaged systems. Fan-coil systems are always central systems and centrifugal or absorption refrigeraion systems are usually central systems. The refrigeration systems of a package system is usually a reciprocating, screw, or scroll system. A thermal storage system is always a central system.
- A VAV reheat screw central system, or simply VAV reheat central system, is a central air conditioning system that has a VAV reheat air system and a screw chiller plant to cool air in various AHUs.

A fan-powered VAV, centrifugal, ice storage central system, or simply a fan-powered VAV ice-storage system, is a central air conditioning system that has a fan-powered VAV system for cold air distribution and an ice storage system that uses a centrifugal chiller to reduce peak electricity costs. Ice storage systems are always central systems that use brine to provide cooling.

- A thermal storage system is a combination of a water system and a refrigeration system. It may be more appropriate to consider it as a designation for air conditioning (AC) systems.
- Occasionally, an air conditioning system may exist without an air system, such as a reciprocating water-loop heat pump system, or simply water-loop heat pump system, whose water-source heat pumps may be installed in individual rooms. Note that a heat pump is always a packaged system.
- If the distinctions between centrifugal, reciprocating, scroll, and screw compression are not important for an air conditioning system (for example, for a water-loop heat pump system or a single-zone VAV rooftop packaged system), just omit them.
- For central systems, most primary heating plants in commercial buildings use gas-fired, electric, and heat-recovery heating. Direct gas-fired and electric heaters are the most widely used. For simplicity, the designations of primary heating plants are not included in classification. In areas where the type of primary heating plant is important, gas heating, electric heating, and oil heating can be added after centrifugal, reciprocating, or screw compression, or the designation of refrigeration systems may be omitted.

29.2 THE GOAL OF AIR CONDITIONING SYSTEM DESIGN

As described in Section 1.8, the goal of an air conditioning system design is to achieve a high quality system that functions effectively and is energy-efficient and cost-effective.

The following are essential for a system to function effectively:

- All design criteria are fulfilled, and the requirements of the owner and the user are satisfied.
- A good indoor air quality is provided.
- The system is reliable and has adequate fire protection systems.

Energy efficiency requires that the annual energy use indexes be lower than the values specified by local codes and regulations. There is always room for improvement over prior developed projects of a similar nature in the same geographic location.

Many energy-efficient systems or pieces of equipment have a higher initial cost. A cost-effective design includes the optimum alternative by means of simple payback years or life-cycle cost analysis. Cost-effective operation involves minimizing the energy cost by choosing the appropriate energy sources for its specific location or operating during off-peak hours.

29.3 INDOOR AIR QUALITY

Acceptable IAQ

Americans are spending more and more of their time indoors. They need a comfortable and healthy indoor environment and an acceptable indoor air quality (IAQ). An acceptable IAQ is defined in ANSI/ASHRAE Standard 62–1989 as air that contains no known contaminants at harmful concentrations with which a substantial majority (80 percent or more) of the people exposed do not express dissatisfaction.

In an unhealthy building environment, uncomfortable employees do not perform well and productivity declines. Worker illness due to a poor indoor environment elevates absenteeism. A significant issue that is facing building owners, operating managers, architects, consulting engineers, and contractors today is the possibility of legal suits filed by occupants or owners who feel that their health has been damaged by poor indoor air quality. In the 1990s, indoor air quality has become one of the primary concerns in air conditioning system design, manufacturing, installation, and operation.

Achieving Acceptable IAQ

As described in Section 5.13, there are five main kinds of indoor air contaminants: particulates, nitrogen dioxide, and the volatile organic compounds formaldehyde, nicotine, and radon. Indoor air contaminants may be controlled by (1) reducing contaminant emissions, (2) capturing and exhausting at the emission source, and (3) diluting the contaminant with outdoor air. Reducing contaminant emissions requires a combined effort on the part of architects, owners, tenants, engineers, and building operating managers. For most comfort air conditioning systems, capture at the source is not practical. Therefore, dilution with outdoor air of acceptable quality is often the most practical and effective solution.

ANSI/ASHRAE Standard 62–1989 recommends two procedures for using outdoor air dilution to control indoor air contaminants. For most of the air conditioning space, the *ventilation rate procedure* involves using a prescribed quantity of outdoor air per person or per floor area, as listed in Table 5.9, to dilute the indoor air contaminants in order to provide an acceptable IAQ.

For more complicated conditions (such as abnormal indoor contaminant emissions or poor outdoor air quality), *the indoor air quality* procedure should be used and indoor air contaminants should be measured. Please refer to ANSI/ASHRAE Standard 62–1989 for indoor air quality procedure details.

Outdoor Air Intake in Office Buildings

As described in Section 5.12, more than half of all complaints regarding the sick building syndrome are due to improper use and maintenance of HVAC&R systems. The predominant factor behind such improper use of HVAC&R systems is inadequate outdoor air intake.

Persily (1989) took field measurements of the outdoor ventilation rates of 14 office buildings for approximately one year in 1983 and found that they were typically between 0.6 and 1.2 ach, with a mean value of 0.94 ach. Among these 14 buildings, 52 percent had a minimum level of outdoor air intake that was lower than the design level, 45 percent had less than 20 cfm/person, 20 percent had less than 15 cfm/person, 8 percent had less than 10 cfm/person, and 1 percent had less than 5 cfm/person.

ANSI/ASHRAE Standard 62–1989 specifies that "indoor air quality shall be considered acceptable if the required rates of acceptable outdoor air are provided for the occupied space." Minimum outdoor air requirements (abridged) are listed in Table 5.9. Strictly following the Standard 62–1989 requirements is of vital importance for acceptable indoor air quality.

The outdoor air intake of an air system depends on the following:

- Zone characteristics of a single-zone or a multizone system
- Space load density, load profile, and supply air density
- Occupant density, in person/ft^2, of floor area
- System characteristics (for both constant-volume and VAV systems)
- Operating characteristics, such as the use of an air or water economizer cycle

For a multizone VAV system:

- Properly consider the effects of unused outdoor air contained in the recirculating air, and check the amount of outdoor air being supplied to critical zones at both design load and part-load.
- Install minimum outdoor air control to increase outdoor air intake at part-load operation.

Outdoor Air Intake System Design

To provide a better IAQ, the design of the outdoor air intake and the space air diffusion system should take into consideration the following recommendations:

- The outdoor air intake system should include an airflow measuring station and minimum outdoor air control to monitor, control, and record the rate of outdoor air intake at various operating conditions.
- The outdoor air intake design should prevent the intake of exhaust contaminants, including condensation and freezing, which may provide a means of growth for microorganisms. The outdoor air intake must be located as far away from the exhaust outlet (horizontally and vertically) as possible. Codes and local regulations should be followed.
- An outdoor air intake system should be provided with air filters or even dust collectors in locations where outdoor air contaminants exceed the National Primary Ambient Air Quality Standard, as described in Section 5.13.
- Space diffusion of supply air should be flexible enough to accommodate the changing of partitions according to the tenants' needs. Improper partition placement often blocks the direct supply of outdoor air to the occupants.
- Return inlets should be located as near to the source of contaminants as possible.

Air Filtration

Physically, indoor airborne contaminants are in solid, liquid, or gaseous form. Solid particulate contaminants include tobacco smoke, paper and fibrous particles, bacteria, fungus, spores, and viruses. Their size is between 0.003 and 100 microns. Liquid contaminants include mists, water, paints, cleaning sprays, and printing inks that typically have a size between 1 and 50 microns. Gaseous contaminants such as CO,

CO_2, NO_2, and volatile organic compounds are fine particulates. Their size typically ranges from 0.003 to 0.006 microns.

As described in Chapter 15, the low- and medium-efficiency air filters (such as panel and bag filters) normally installed in AHUs and PUs are not effective against submicron-size airborne particulate contaminants. Only high-efficiency particulate air (HEPA) filters are effective in removing particles around 0.3 microns in size. To remove objectionable odors or gaseous contaminants of a molecular size between 0.003 and 0.006 microns, activated carbon filters or chemisorption media filters should be used.

Field tests have indicated that when medium-efficiency filters of 55 percent dust-spot efficiency are installed in an AHU to serve an office, the space particulate contaminant level can typically be maintained at 130 ppm. If an air filter with a dust-spot efficiency of 95 percent is installed, the space particulate contaminant level is reduced to 75 ppm.

Because HEPA, activated carbon, and chemisorption media filters are expensive, medium-efficiency air filters are installed in the AHUs or PUs of most comfort air conditioning systems. If gaseous airborne contaminants, objectionable odors, or air contaminants of submicron size must be removed to maintain acceptable IAQ, then HEPA or activated carbon filters and chemisorption media filters should be used. HEPA filters are widely used for clean rooms and clean spaces.

Operation and Maintenance of HVAC&R Systems

Effective operation and maintenance of a HVAC&R system reduces the indoor airborne contaminant emissions from the system and improves the IAQ. In case of IAQ problems, the first thing to investigate is whether the HVAC&R systems are functioning properly.

The operating conditions of an HVAC&R system that should be monitored include the following:

- Position of outdoor air dampers and minimum outdoor air controls
- Pressure drops across filters (to prevent dirty filters)
- Cleanliness of condensate drain pans in AHUs, PUs, and fan–coils
- Cleanliness and proper operations of humidifiers
- Cleanliness of coils
- Microorganism growth in such open water systems as cooling water systems, evaporating cooling systems, and cooling towers

- Minimum setting in VAV boxes
- Air balance in supply and return systems and space diffusion effective factor
- Optimum operation of exhaust or relief system

Litigation For IAQ Problems

According to Eisenstein (1992), the last decade saw the rise of litigation for IAQ problems. Both architects and engineers are vulnerable to suits filed by building occupants as well as owners for improper design and specification of materials.

Eisenstein summarized the types of suits as follows:

- Occupational/industrial exposures to single-source contaminants such as asbestos and other chemicals, and nonindustrial exposures to such chemicals as formaldehyde.
- Toxic litigation in sick building syndrome (SBS) cases. Most of the SBS cases occurred in commercial buildings with sealed windows. These air contamination cases involved toxic building materials, inadequate airflow, and improper operation and maintenance of sophisticated air conditioning systems.

The building owner has a nondelegable duty to provide occupants with a safe and healthy environment. The engineer who designs an air conditioning system is vulnerable to claims of negligence if the system is determined to be defective, so the engineer must take the following into account:

- The system must be designed to be flexible enough to account for changes in the occupancy rates and for normal changes in the use of space.
- Outdoor air intake must be carefully selected to prevent air contamination.
- Codes, regulations, and appropriate ASHRAE standards must be followed.

Engineers and contractors involved in the design and construction of an air conditioning system may be held liable under the principle of strict product liability. Under this liability, the designer, manufacturer, and installer of a defective product are considered liable regardless of whether their conduct is negligent. Recently, courts have ruled that a building ventilation system is a product, subject to the principle of strict product liability.

29.4 ENERGY CONSERVATION

Basic Considerations

As described in Section 1.11, air conditioning or HVAC&R systems in buildings consumed about one-sixth of the total annual energy use in the United States in 1990. For many commercial buildings, the utility costs for HVAC&R systems may make up the largest part of the monthly energy bill. Energy conservation has become a primary factor in the performance of HVAC&R systems as well as the economy.

Energy conservation must be achieved under the condition that a satisfactory indoor environment is maintained with an acceptable indoor air quality. A well-designed and effectively functioning DDC control system is necessary for an energy-efficient HVAC&R system. Energy conservation must be considered in every stage of HVAC&R system design and operation.

- *Design*. Various alternatives must be compared and analyzed in terms of either payback years or life-cycle cost analysis to determine which is the best and most energy-efficient.
- *Installation*. The efficiency of installed equipment, the amount of air leakage from the ducts and run-outs, the layout of ducts and pipes, and the duct and pipe insulation all affect energy use.
- *Commissioning*. The capacity of equipment, the air and water balance, and coordination between various components and control systems should be carefully measured, adjusted, and commissioned.
- *Operation*. The energy indexes for chillers, DX-systems, pumps, boilers, air systems, and heat rejection systems should be periodically checked, studied, and analyzed.

Energy Conservation Opportunities

The term *energy conservation opportunities* refers to potential energy saving measures in an HVAC&R system. For a new project, the selection of a system, subsystem, system component, or control strategy can be accomplished by (1) comparing estimated energy use from measured data from similar projects, or conducting an hour-by-hour detailed system simulation of different alternatives, as discussed in Chapter 28; and (2) justifying the optimum alternative through a simple payback-years or life-cycle cost analysis and other considerations.

ASHRAE/IES Standard 90.1-1989 recommends a procedure of compliance for speculative building by using the energy cost budget (ECB) method to achieve a most energy-efficient and cost-effective design.

The annual energy cost of a prototype or reference building (the energy cost budget) and the annual cost of the proposed design (the design energy cost (DECOS) are both calculated by hour-by-hour detailed system simulation.

If DECOS ≤ ECB, the proposed design complies with the standard and is energy efficient. During the comparison of annual energy cost breakdowns on the building envelope, load calculations, and HVAC&R system and equipment selections, energy conservation opportunities can be determined. Refer to ASHRAE/IES Standard 90.1-1989 for details.

To convert existing systems to energy-efficient operation, the energy conservation opportunities can be determined through the following steps:

1. Organize a team. One individual should be chosen to be responsible for the team.
2. Draft a plan.
3. Make a field survey by means of walking through the building; talking with the operating manager, engineers, operators, and electric company representative; simple field measurements; and tests, if necessary.
4. Proceed with an energy audit.
5. Justify the energy conservation opportunities after cost analyses.
6. Implement the plan, starting with no-cost and low-cost opportunities.
7. After energy retrofit is complete, operate and control the HVAC&R system according to the energy conservation opportunities. Check the energy costs against those from before the energy retrofit.

Energy Audits

The term *energy audit* refers to a month-by-month accounting of energy use in a building within a certain period. This energy use is checked against a budget or an energy estimate based on an hour-by-hour detailed system simulation in order to identify energy conservation opportunities. An energy audit includes the following steps:

1. Model the existing building and existing HVAC&R system according to the format of an energy estimation detailed system simulation program such as DOE 2.1. Input the required data concerning the building, weather, load, and

equipment. Calculate the month-by-month energy use of the HVAC&R system.
2. Compare the calculated monthly energy use of the HVAC&R system against the actual monitored data. Tune, or adjust, such inputs to the simulation computer program as the internal loads and fan, pump, and compressor efficiency until the difference between the calculated energy use and the actual monitored data is within the allowable tolerances.
3. Calculate the energy use of the HVAC&R system—with energy conservation opportunity implemented—by using the same tuned simulation program. Determine the amount of energy savings for each energy conservation opportunity.
4. Justify the energy conservation opportunities through payback-years or life-cycle cost analyses.

HVAC&R Energy Conservation Recommendations

The following are possible energy conservation opportunities for HVAC&R systems in buildings:

- Turn off electric lights, office appliances, and other equipment when they are not needed.
- Shut down air-handling units, packaged units, fan-coils, VAV boxes, chillers, fans, and pumps when the space or zone they serve is not occupied.
- Daily starting or stopping of the AHUs or PUs and the terminals should be optimum. This can be accomplished by means of either a custom-made program based on previous experience or a commercially available DDC computer program based on the zone temperature, the difference between indoor and outdoor temperatures, and the surface temperature of the building envelope.
- The temperature set point should be at its optimum value. Provide a dead band between summer and winter modes of operation for comfort air conditioning. The discharge air temperature from the AHU or PU, and temperature of water leaving the chiller and leaving the boiler should be reset to more energy-efficient values for part-load operation based on the space temperature, space load, or outdoor temperature.
- Reduce air leakages from ducts and dampers. Use weather strip to seal windows and external doors to reduce infiltration. Carefully design the layout of ducts and pipes to minimize the number of ducts, and pipe fittings as well as their pressure losses.
- Use more energy-efficient cooling methods such as free cooling, evaporative cooling, or groundwater

cooling instead of refrigeration. Replace refrigeration with an air economizer, water economizer, and evaporative cooler if doing so provides the same cooling results. An evaporative-cooled condenser is often more energy-efficient than a water-cooled or air-cooled condenser in many locations.

- Install cost-effective high-efficiency fans, pumps, compressors, and motors. Chillers and packaged units should have a COP or EER higher than that specified in ASHRAE Standard 90.1–1989. Equipment should be properly sized. Oversized equipment is never energy-efficient.
- Use heat recovery systems, chiller-heaters, waste heat from gas cooling engines, or heat pumps to provide winter space heating whenever applicable.
- Use variable-air-volume (VAV) systems instead of constant-volume systems and variable-flow building-loop water systems instead of constant-flow ones for comfort air conditioning systems. Large variable-speed centrifugal fans or controllable-pitch vane axial fans are often cost-effective and energy-efficient in VAV systems.
- Adopt double- and triple-pane windows with low-emissive coatings and lower U-values, and external walls and roofs with lower U-values, to reduce heat gain in the summer and heat loss in the winter in areas where the outdoor temperature is high in the summer or low in the winter.
- According to Joncich (1991,) the operating experiences of 30 active solar thermal energy systems used by the U.S. Army have shown that "solar cooling systems are not economically justifiable" and that "solar service water heating is the Army's most cost-effective application."

According to the EIA's *Commercial Building Characteristics* (1989), the popularity of energy conservation features of HVAC&R systems in commercial buildings, based on conditioned floor areas in 1989, were as follows:

Off-hours heating reduction	79%
Off-hours cooling reduction	83%
Energy management systems (EMS)	21%
EMS for conditioned space over 500,000 ft^2	57%

29.5 ENERGY COSTS AND OFF-PEAK AIR CONDITIONING

Electricity and natural gas are the two energy sources most widely used for air conditioning in buildings in the United States. Comparing the price and energy use for electricity versus natural gas for new and retrofit HVAC&R projects at a specific location is an important step in achieving lower energy costs.

It has been predicted that during the 1990s the electricity-generating capacity of the United States will fall short of demand by 60 gigawatts (1 gW = 1×10^9 W). Because of the capital-intensive nature of nuclear and coal-powered plants, the high demand for electricity will continue to push up the cost of electric energy.

Because of the high cost of electricity use during peak hours, many technologies have been developed in recent years to use electricity during off-peak hours and to adopt gas cooling equipment to replace electric chillers or compressors. These newly developed off-peak air conditioning and gas cooling technologies do not necessarily save energy, but they do reduce energy costs considerably.

These off-peak air conditioning and gas cooling technologies mainly include the following:

- Ice storage systems with cold air distribution
- Chilled water storage systems
- Direct gas-fired absorption chillers
- Desiccant cooling
- Gas engine and gas turbine cogeneration

All these have been discussed in previous chapters.

Off-peak air conditioning and cost-effective gas cooling must be taken into consideration by design engineers and owners of new and retrofit projects during the 1990s.

29.6 OZONE DEPLETION AND REFRIGERANT SAFETY

Because the production of CFCs will be terminated worldwide at the end of 1995 (except limited amounts for servicing equipment) and because knowingly venting CFCs and HCFCs has been illegal since July 1, 1992, the use of alternative refrigerants and their corresponding safety considerations has become one of the vital problems in air conditioning and refrigeration system design, retrofitting, and operation.

As far as ozone layer depletion and the safety considerations of switching to alternative refrigerants are concerned, the following actions should be considered.

Replacing CFCs and BFCs Before the End of 1995

Make a plan to replace all CFCs and BFCs with HCFC–22, HCFC–123, HFC–125, HCFC–134a, and other refrigerants with ODP ≤ 0.05 in existing refrigeration systems before the end of 1995. Use only refrigerants with ODP ≤ 0.05 for all refrigeration systems before the end of 1995.

Cost-effective conversion from CFCs to HCFCs or HFCs can be done as follows:

- Existing equipment can be used by directly replacing CFC–11 with HCFC–123, CFC–12 with HFC–134a, or other CFCs and BFCs with HCFCs or HFCs. Bear in mind that the use of HCFCs themselves will be restricted beginning in 2004. Use HFCs whenever they are cost-effective.

  For direct conversion, the capacity of a refrigeration system that uses HFC–134a is similar to one that uses CFC–12 and the capacity of a system that uses HCFC–123 may be 15 to 20 percent smaller than one that uses CFC–11.

  Because the CFCs and BFCs in most of the 70,000 existing chillers and many other refrigeration machines will be replaced by HCFC–123, HCFC–22, HFC–134a, and other refrigerants for which ODP ≤ 0.05 between 1992 and the end of 1995, making a plan beforehand will allow one to avoid the refrigerant shortage and its impact on the capacity and operation of a refrigeration plant.

- If the COP of an existing chiller or the EER of an existing condensing unit is low compared to a current model, or if the local utility offers reduced rates and bonuses for the use of electricity during off-peak hours, it may be cost-effective to replace the old chiller and condensing unit with a new one with refrigerants for which ODP ≤ 0.05.

A cost analysis should be conducted to determine the best alternative from among the following possible candidates:

- Chiller with an ice or chilled water storage system
- Direct-fired two-stage absorption system or chiller-heater
- Desiccant cooling system
- Low-power-consumption electrically driven centrifugal chiller
- Packaged system
- Refrigeration and evaporative cooling combined system
- Gas cooling system

Illegal Venting of CFCs and HCFCs

The main purpose of prohibiting any intentional venting of CFCs and HCFCs after July 1, 1992 as part of the Clean Air Act is to prevent depletion of the ozone layer.

The first step in avoiding the venting of CFCs and HCFCs is to use an ARI-certified, portable refrigerant-recovering/recycling unit to recover all the liquid and remaining vapor from a chiller or other refrigerant system. An outside recovery/reclaiming service firm may also be used to recover, recycle, and reclaim refrigerant.

A typical refrigerant recovery unit is shown in Fig. 29.1. It includes a recovery cylinder, a vacuum pump or compressor, a water-cooled condenser, a sight glass, a shutoff float switch, necessary accessories, pipes, and hoses.

Recovering refrigerant from a chiller that has been shut down involves two phases: liquid recovery and vapor recovery.

Liquid recovery is shown in Fig. 29.1a. The vacuum pump or compressor in the recovery unit creates a low pressure in the recovery cylinder. Liquid refrigerant is then extracted from the bottom of the chiller into the recovery cylinder. If the recovery cylinder is not large enough, the shutoff float switch ceases to operate the vacuum pump or compressor when the recovery cylinder is 80 percent full. Another empty recovery cylinder is used to replace the filled cylinder. If vapor enters the sight glass, which means that the liquid refrigerant is all extracted, the vacuum pump or compressor is stopped and the vapor recovery phase begins.

Vapor recovery is shown in Fig. 29.1b. The vacuum pump or compressor extracts the refrigerant vapor from the top of the chiller. Extracted refrigerant vapor is then condensed into a liquid form that flows through the water-cooled condenser and is stored in the recovery cylinder. Noncondensable gases are purged into the atmosphere from the recovery unit. Water at a temperature between 40 and 85°F is often used as condensing cooling medium. The recovered refrigerant can be recycled or reclaimed as required.

In addition to the recovery of refrigerants from the chiller or other refrigeration system, refrigerant vapor detectors should be installed at locations where refrigerant from a leak is likely to concentrate. These detectors can set off an alarm to notify the operator to seal the leak.

Classification of Safety Groups, Occupancy, and Systems

Refrigerant hazards stemming from leaks in the pipe joints, the rupture of system components, and the burning of escaping refrigerant depend on the type

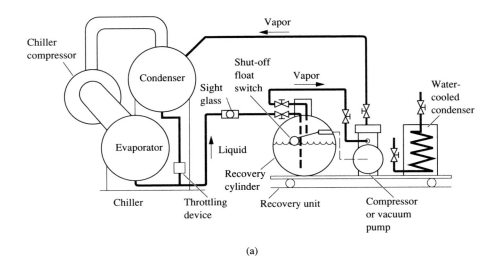

FIGURE 29.1 A typical refrigerant recovery unit: (a) liquid recovery; (b) vapor recovery.

of refrigerant, the occupancy classification, and the refrigeration system.

As described in Section 4.4, according to ANSI/ASHRAE Standard 34–1992, refrigerants can be classified into safety groups that range from lower toxicity and no flame propagation (safety group A1) to higher toxicity and higher flammability (safety group B3).

The type of occupancy may be one of five categories:

- *Institutional* or *healthcare,* such as hospitals and nursing homes
- *Public assembly,* such as auditoriums and department stores
- *Residential,* including hotels and apartments
- *Commercial,* such as offices, restaurants, and markets
- *Industrial,* such as factories and warehouses

According to ASHRAE Standard 15–1992, any refrigeration system in which a leakage of refrigerant from a failed connection, seal, or component could enter the occupied space (that is, the space normally frequented or occupied by people, except for a machinery room) is a *high-probability system.* Any refrigeration system that uses DX-coil(s) to cool the air directly or chilled water to cool the air in air washer(s) is a high-probability system.

A *low-probability system* is a refrigeration system that cannot be considered a high-probability system, such as a refrigeration system using chilled water as a cooling medium.

Application Rules for High-Probability Systems

A tightly sealed refrigeration system is always necessary to reduce refrigerant leaks that may produce refrigerant hazards. In addition, limiting the quantity of refrigerant in a refrigeration system per occupied space, thereby reducing the possible leaks from joints and seals in a high-probability system, is an effective means of reducing the hazards of refrigerants for the safety of people and property.

ASHRAE Standard 15–1992 specifies a number of rules and requirements for various refrigerants in high-probability refrigeration systems. These rules are described in the following paragraphs.

Any refrigeration system in a room air conditioner or packaged terminal air conditioner (PTAC), or any small packaged unit for which the refrigerant charge does not exceed 6.6 lb is considered to meet the system refrigerant safety application requirements.

Refrigerants that belong to the A1 safety group (HCFC–22, HFC–134a, etc.) and are used in high-probability refrigeration systems have the following restrictions:

- For institutional or healthcare occupanices (except in kitchens, laboratories, or mortuaries), the following conditions apply:
  - The quantity of refrigerant HCFC–22 in the largest refrigeration system, $m_{HCFC-22}$, is limited to 4.7 lb per 1000 ft^3 of occupied space.
  - The quantity of refrigerant HFC–134a, $m_{HFC-134a}$, is limited to 8 1b per 1000 ft^3 of occupied space.

A flame-sustaining device in an occupied space must be provided with a hood to exhaust combustion products to open air if the refrigeration system contains more than 1 lb of A1 refrigerant (except R–744 carbon dioxide).

- For commercial occupancies, the following apply:
  - The value of $m_{HCFC-22}$ in a refrigeration system is limited to 9.4 lb per 1000 ft^3 of occupied space.
  - The value of $m_{HFC-134a}$ is limited to 16 lb per 1000 ft^3 of occupied space.
- In industrial applications, the quantity of refrigerant in a refrigeration system is unlimited under the following conditions:
  - The area that contains refrigeration machines, the compressor room is separated from the area of the building that doesn't contain refrigeration machines by tight construction with tight-fitting doors.
  - The building is provided with unhampered free emergency exits
  - The number of persons in the space containing refrigeration machinery on any floor above the first floor is equal to or less than one person per

100 ft^2 of floor area, and the floor has the required number of emergency exit doors
- Detectors are installed in areas where leaked refrigerant vapor is likely to concentrate to provide an alarm if the volume of oxygen drops below 19.5 percent for systems that use HCFC–22 and HFC–134a.

Otherwise, in industrial applications, $m_{HCFC-22}$ in a refrigeration system is limited to 9.4 lb per 1000 ft^3 of occupied space and $m_{HFC-134a}$ is limited to 16 lb per 1000 ft^3 of occupied space.

Refrigerants in the B1 safety group, such as HCFC–123, that are used in high-probability refrigeration systems should be used only for process air conditioning in industrial applications. As described above for refrigerants HCFC–22 and HFC–134a in the A1 safety group, the quantity of refrigerant HCFC–123, $m_{HCFC-123}$, in a refrigeration system is unlimited under the following conditions:

- The compressor room or area containing refrigeration machinery is separated.
- Emergency exits are provided.
- The number of persons in the area containing refrigeration machine(s) is less than or equal to 100 ft^2 per person.
- Detectors are installed in areas where leaked vapor refrigerant may concentrate.

Refrigerants in the B2 safety group, such as R–717 ammonia, that are used in high-probability refrigeration systems are not allowed to be used for healthcare facilities. The following rules apply:

- When R–717 is used in high-probability refrigeration systems for process air conditioning, m_{R-717}, in lb, is limited to the following values:

	Public	Resi-dential	Commer-cial
Sealed absorption system:			
In public hallways or lobbies	0	3.3	3.3
In adjacent outdoor locations	0	22	22
Other than hallways or lobbies	6.6	6.6	22
Individual and packaged systems:			
Other than hallways or lobbies	0	6.6	22

- For industrial applications, the following hold:
  - The value of m_{R-717} is unlimited when there is a separate compressor room, emergency exits, an occupant density of less than or equal to 100 ft^2 person, and installed dectectors.
  - If m_{R-717} in a high-probability refrigeration system exceeds 0.022 lb per 1000 ft^3 of occupied

space, all refrigerant-containing parts outside the building must be installed in a machinery room.

Application Rules for Low-Probability Systems

ASHRAE Standard 15–1992 states the requirements for low-probability refrigeration systems that have complete, factory-assembled chillers tested by an approved and nationally recognized laboratory and that are used in healthcare, public assembly, residential, commercial, and industrial buildings. The following paragraphs describe these requirements.

For A1 safety group refrigerants, if $m_{\text{HCFC}-22}$ in the largest refrigeration system exceeds 9.4 lb per 1000 ft^3 of occupied space or $m_{\text{HFC}-134a}$ exceeds 16 lb per 1000 ft^3 of occupied space, all refrigerant-containing parts, except pipes and parts located outdoors, must be installed in a machinery room that meets general safety requirements.

For B1 safety group refrigerants, if $m_{\text{HCFC}-123}$ in the largest refrigeration system exceeds 0.004 lb per 1000 ft^3 of occupied space, all refrigerant-containing parts, except pipes and parts located outdoors, must be installed in a machinery room that meets general safety requirements.

For B2 safety group refrigerants such as R–717 ammonia, there are the following options:

- R–717 ammonia may not be used in any healthcare facility. In other types of buildings, $m_{\text{R}-717}$ in low-probability refrigeration systems, in lb, is limited to the same value listed earlier for high-probability systems used for process air conditioning.
- When $m_{\text{R}-717}$ in the largest refrigeration system exceeds 0.022 lb per 1000 ft^3 of occupied space, all refrigerant-containing parts, except pipes and those parts outdoors, must be installed in a machinery room that meets special requirements and $m_{\text{R}-717}$ in a refrigeration system is limited as follows:
  - 550 lbs for healthcare facilities
  - No limit for public assembly, residential, commercial, and industrial buildings

For details and precise safety requirements, refer to ASHRAE Standard 15–1992 and local codes.

Machinery Room

A machinery room is an enclosure with tightly fitted doors to safely house compressors, refrigeration components, and other types of mechanical equipment.

A machinery room must be designed so that it is easily accessible, with adequate space for proper servicing, maintenance, and operation. A machinery room must have doors that open outward and are self-closing if they open into the building; there must be an adequate number of doors to allow easy escape in case of emergency.

A mechanical room of general requirements, according to ASHRAE Standard 15–1992, must meet the following specifications:

- It must be ventilated to the outdoors by means of mechanical ventilation using power-driven fans or multiple-speed fans. The minimum volume flow rate of mechanical ventilation, $\dot{V}_{\text{m.v}}$, in cfm, required to exhaust the refrigerant from leaks or ruptured components is

$$\dot{V}_{\text{m.v}} = 100 \, m_{\text{ref}}^{0.5} \qquad (29.1)$$

where m_{ref} is the mass of refrigerant in the largest system, in lb.

When a refrigeration system is located outdoors, its location is more than 20 ft from any building opening, and it is enclosed by a penthouse or other open structure, both mechanical and natural ventilation can be used. The minimum free-aperture section $A_{\text{f.a}}$, in ft^2, required for natural ventilation of the machinery room is

$$A_{\text{f.a}} = m_{\text{ref}}^{0.5} \qquad (29.2)$$

Provisions for venting catastrophic leaks and component ruptures should be considered.

- For safety group A1 refrigerants such as HCFC–22 and HFC–134a, the machinery room must include an oxygen sensor to warn the operator when oxygen levels drop below 19.5 percent by volume. For other refrigerants, a vapor detector should be installed to actuate the alarm and the mechanical ventilation system.
- There must be no open flames that use combustion air from the machinery room except matches, lighters, leak detectors, and similar devices.

A machinery room of special requirements must meet the following specifications in addition to the general requirements:

- Inside the room, there must be no flame-producing device or hot surface continuously operated at a surface temperature exceeding 800°F.
- There must be an exit door that opens directly to the outdoors or to a similar facility.
- Mechanical ventilation for ammonia must be either continuously operating or equipped with a vapor de-

tector that actuates a mechanical ventilation system automatically at a detection level not exceeding 4 percent by volume.

- It must be provided with a remote pilot control panel immediately outside the machinery room to control and shut down the mechanical equipment in case of emergency.

Because the machinery room itself is a fire compartment, building structures and their material (including the door, wall, ceiling, and floor) should meet NFPA fire codes.

Machinery rooms are different from fan rooms, which were described in Section 15.9. In a machinery room, there are general and special requirements. The installation of mechanical ventilation and an oxygen or vapor sensor is mandatory. A central refrigeration plant is a type of machinery room.

For more details, refer to ASHRAE Standard 15–1992.

29.7 AIR CONDITIONING SYSTEM SELECTION

Selection Levels

During the design of an air conditioning system, system selection is performed mainly on three levels:

- *Air conditioning system level.* This level deals with such concerns as whether a floor-by-floor packaged system or a central system should be selected for a high-rise office building, whether it is an ice-storage system or a conventional system.
- *Air system, water system, and central plant cooling and heating system level.* As mentioned before, the different types of air systems include constant-volume, fan–coil, single-zone VAV, perimeter-heating VAV, VAV reheat, dual-duct VAV, and fan-powered VAV systems.

    The types of water systems include chilled water or dual-temperature systems, as well as constant-flow, plant–building loop variable-flow, and plant-distribution-building variable-flow systems.

    The types of central plant cooling systems include centrifugal, reciprocating, screw, scroll, absorption, gas cooling, desiccant cooling, evaporative cooling, water-source heat pumps, ice storage, and chilled water storage systems.

    The types of central plant heating systems include direct-fired gas heating, electric heating, hot water boiler, heat pump, steam from an existing facility and heat recovery systems.

- *Main system component level.* For air systems, this level involves the selection of such main system components as a supply–relief fan combination or supply–return fan combination; an air economizer or water economizer; a centrifugal or vane axial fan; an adjustable-frequency variable-speed drive, inlet vanes, or inlet-cone duct static pressure control system; or parallel or series fan-powered units.

    For water systems, this level involves the selection of such components as a variable-speed or constant-speed pump and an open expansion or diaphragm tank.

    For a refrigeration system, this level involves the selection of such components as an air-cooled, water-cooled, or evaporative-cooled condenser; a single-stage, two-stage, or three-stage compressor; direct-drive or gear-train drive; and brine-coil or ice-harvester ice storage.

    When a system or system component is selected, the corresponding control system is often also selected to provide the required performance.

Requirements During System Selection

In order to properly design an air conditioning system, the system designer and the owner should collaborate to select the system according to the following requirements:

APPLICATIONS OR OCCUPANCIES. Different applications or occupancies need different design criteria, different operating hours, and different operating characteristics. For example, a constant-volume central system will be always used for a Class 1 through Class 100 clean room used to produce semiconductor wafer. When the clean room is in operation, adequate clean air must be provided to maintain unidirectional flow to prevent the contamination of semiconductor wafer by submicron particulates. A case study of an air conditioning system for a Class 10 clean room will be discussed in Chapter 30.

A fan–coil central system is the most widely used air conditioning system for guest rooms in luxury hotels. This is because a fan–coil system can provide individual temperature and fan speed controls, as well as a positive supply of adequate outdoor air. The fan-coil can be conveniently turned off when the room is not occupied. It is isolated acoustically from adjacent rooms. The most annoying space maintenance tasks, such as changing the filters and periodically inspecting the fan–coil, can be done when the guest room is not occupied.

DESIGN CRITERIA. Specific design criteria usually dictate a specific type of air conditioning system that should be selected. For instance, a high-precision constant-temperature room that is maintained at a temperature of $22 \pm 0.1°F$ within a limited working space needs a constant-volume central system with electric terminal reheat. Electric reheat can be modulated more precisely than any other type of heating system. A central system can maintain a more uniform discharge air temperature at the AHU than a DX-coil packaged unit. Usually, such a high-precision constant-temperature room is a clean space. A constant-volume system is always preferable to a VAV system for such rooms.

SYSTEM CAPACITY. System capacity is an important factor that affects the selection of an air conditioning system. For example, for a single-story small retail shop, a constant-volume packaged system is often chosen. If the conditioned space is a large indoor stadium with a seating capacity of 70,000 spectators, a single-zone VAV central system with an air economizer that uses peripheral nozzles at different levels to supply conditioned air is often selected. A designer will not select a fan–coil central system for such a high-volume flow of supply air because fan noise and high-level maintenance would be too problematic.

SPACIAL LIMITATIONS. Spacial limitations specified by the architect also influence the selection of the air conditioning system. For instance, for a high-rise building of more than 30 stories, if rooftop space is not available for the penthouse of AHUs and other mechanical equipment, or if there is no space left for the supply and return duct shafts, a floor-by-floor DX-coil packaged system or floor-by-floor AHU central system may be the most practical choice. Because they have adequate ceiling space, low-velocity duct systems are widely used in high-ceiling industrial plants to save air transport energy.

COSTS. Initial costs and operating costs are always primary factors in the selection of an air conditioning system. For a developer who sells buildings with installed air conditioning systems, the initial cost is often the predominant factor. If the owner is also a user of the building, both the initial cost and the operating costs will be carefully considered by the owner when the system is being selected.

Energy costs are the main operating costs. As mentioned earlier, comparing the price of electricity and natural gas at a specific location at both peak and off-peak hours is always necessary to minimize energy costs.

OTHER REQUIREMENTS. Other factors such as maintainability, reliability, and flexibility affect the selection of an air conditioning system.

A dual-duct VAV centrifugal central system requires less maintenance than a floor-by-floor VAV gas-fired heating packaged system.

A VAV gas-cooling central system is probably less familiar to HVAC&R system operators and probably has more maintenance problems; it may be not as reliable as a VAV centrifugal central system.

In an office building, the tenants may change or the partitions may be remodeled, so a VAV central system with flexible duct connections often provides greater flexibility than other air conditioning systems.

PAST EXPERIENCE. Performing cost comparisons of different design options consumes a great deal of valuable design time. The time schedule of a project does not usually allow for such careful study. Because the air conditioning systems and air, water, and primary cooling and heating systems should be selected before the architectural core mechanical room and penthouse layouts are finalized, there is often a great deal of pressure to make the selection quickly. These facts compel design engineers to select systems that are most familiar to them.

29.8 PACKAGED SYSTEMS

Background

Because packaged systems are lower in initial cost and often require less space, and because factory-built packaged units are easier to install, packaged systems are used far more often than central systems in commercial buildings in the United States. As a result of the recent development of screw and scroll compressors and DDC systems, more sophisticated air systems such as fan-powered VAV systems can also be served by packaged systems. According to the *Commercial Building Characteristics* (1989) conducted by the Energy Information Administration (EIA), the amount of floor space in commercial buildings installed with packaged and central systems, expressed in millions of ft^2, were as follows:

	Up to 1989	1980 to 1989
Packaged systems	34,753	8370
Central systems	14,048	2966

The amount of floor space served by newly installed packaged units for commercial buildings between 1980 and 1989 was about 2.8 times the amount served by central systems. Among packaged systems, rooftop packaged units are the most widely used.

A typical perimeter-heating VAV rooftop packaged system, as shown in Fig. 29.2 consists of a rooftop packaged unit, in which there is an air handler

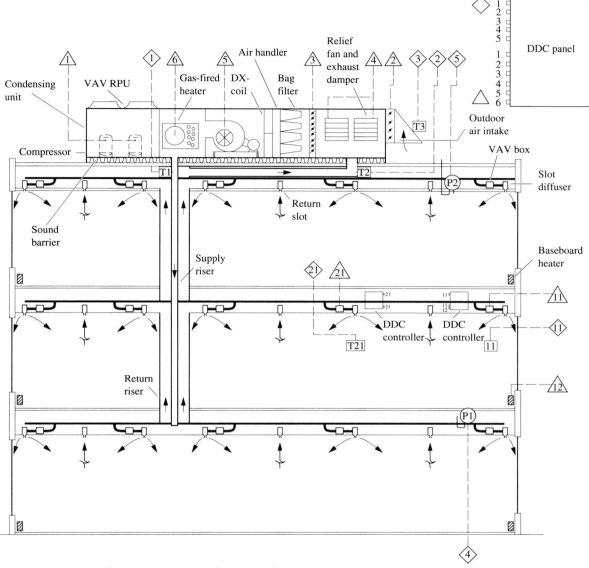

FIGURE 29.2 A typical perimeter-heating VAV reheat rooftop packaged system.

and a condensing unit; supply and return ductwork; reheating VAV boxes; ceiling diffusers; and a DDC panel. As described in Section 15.8, an air handler is composed of fans, DX-coils, a gas-fired heater (or electric heater), filters, a mixing box, DDC controls, and an outer casing. In the condensing unit, there are compressors; an air-cooled, water-cooled, or evaporative-cooled condenser; and other accessories.

Evenly Distributed Airflow At DX-Coils

In a VAV rooftop packaged system, if the air flowing through the DX-coil is not evenly distributed over the entire coil surface, then liquid slugging of the reciprocating compressor, hunting of the thermostatic expansion valve, and a decrease in the DX-coil capacity may all occur at the same time.

Refrigerant enters the various refrigerant circuits of the DX-coil and the evaporator, typically as a mixture of 75 percent liquid and 25 percent vapor, after passing through the thermostatic expansion valve and the distributor tubes. If some of the refrigerant circuits have heavy refrigeration loads and others only very light loads, the refrigerant in the circuits with heavy loads expands rapidly into vapor, resulting in a greater vapor velocity and greater pressure loss. The refrigerant in circuits with very low loads remains in a liquid state and flows to the compressor in the form of liquid slugging. Liquid slugging may cause compressor failure. Therefore, an even airflow for the DX-coil and thorough upstream air mixing must be maintained for both full and part-load operation. Liquid refrigerant lowers the temperature of the sensing bulb of the thermostatic expansion valve and causes hunting.

Discharge Air Temperature Control

For a VAV rooftop packaged unit with a capacity greater than 20 tons, a compressor is often controlled by means of three- or four-step capacity controls. Multiple compressors are usually equipped in medium and large packaged units. Each compressor may have four cylinders, so that two can be loaded and two unloaded at the same time.

AIR ECONOMIZER MODE. For typical DDC-panel-activated discharge air temperature control, as shown in Fig. 29.3, if the set point of the discharge temperature T_{dis} is set at 53°F, which gives a supply temperature at the slot diffuser of $T_s = 55$°F, there will be a dead band or control band of 4°F. When a VAV rooftop packaged system operates in an air economizer mode and $51 < T_{dis} < 55$°F, the system will float within the control band by mixing outdoor air with recirculating air.

INITIATION OF COOLING STAGES. If the outdoor temperature $T_o > 51$°F, and if $T_{dis} > 53$°F due to the fan power heat gain, and the air economizer alone can no longer balance the sensible coil load, T_{dis} then floats to the upper limit of the control band. Once

it reaches point 1, 55°F in Fig. 29.3, the first-stage cooling is energized and T_{dis} drops below 55°F. First-stage cooling is most likely provided by the cooling capacity of two cylinders in a compressor. It cycles on and off, and T_{dis} floats within the control band, with proportional-integral (PI) control mode to maintain $T_{dis} = 53$°F with the least deviation. When the first-stage cooling is turned off during cycling, it needs a time delay of at least four minutes before it can be turned on again in order to prevent hunting and possible damage to the compressor motor.

If, after the first-stage cooling has been energized, the DX-coil capacity still cannot balance the sensible coil load, T_{dis} continues to rise until it reaches point 2, which is 1°F higher than the upper limit of the control band.

If the time interval between the time the first stage turns on and the instant when T_{dis} reaches point 2 is greater than four minutes, the first-stage cooling is then locked on and the second-stage cooling is energized to cycle on and off in an attempt to maintain T_{dis} at 53°F. As in the first stage, there must be a time delay of four minutes between the time the second-stage cooling turns off and the moment it can be turned on again.

When T_{dis} floats within the control band, it may drop to point 3 because of the low sensible coil load. If the cooling capacity of having the first-stage cooling locked on and the second-stage cooling cycling still cannot offset the sensible coil load, T_{dis} will rise until it reaches a value 1°F higher than the upper limit of the control band, such as point 4. At point 4, the third-

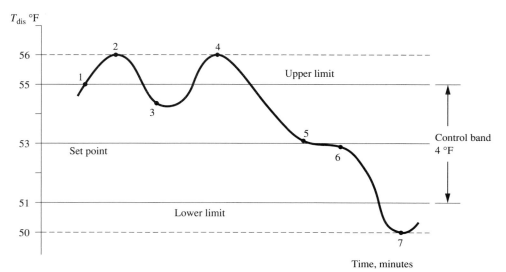

FIGURE 29.3 DDC-panel-activated discharge air temperature control for a perimeter-heating VAV rooftop packaged system.

stage cooling is cycling and the first- and second-stage cooling are locked on.

In this type of packaged unit, the greatest cooling capacity is provided when the fourth-stage cooling is cycling and the first-, second-, and third-stage cooling are locked-on.

As T_{dis} floats within the control band, any deviation from the set point of 53°F is integrated over time as part of the proportional-integral control mode in an attempt to reduce the deviation to zero, that is, to maintain a near-constant temperature at the set point of 53°F (for example, points 5 and 6 in Fig. 29.3).

If T_{dis} being controlled in air economizer and refrigeration cooling modes, the greater the deviation from the set point, the shorter the time needed for corrective action in order to provide control stability.

When a decrease in the coil's load causes T_{dis} to drop to a value 1°F lower than the lower limit of the control band, (for example, point 7), the currently cycling stage is locked off and the next lower cooling stage becomes the cycling stage.

Successive decreases in T_{dis} to 1°F below the lower limit of the control band cause repeated locking off of cooling stages until the second stage is locked off and the first stage becomes the cycling stage. When the first stage is cycled off, the liquid line solenoid valve is deenergized, but the compressor still operates to pump down the refrigerant to the condenser. When the suction pressure in the DX-coil becomes lower than the limit of low-pressure control, it opens the compressor circuit, and both the compressor and the condenser fan(s) are stopped.

A fully intertwined DX-coil will adjust its activating refrigerant circuits accordingly during step control of the compressor capacity.

At design load, the evaporating temperature inside the DX- coil is usually between 40 and 45°F. Hot gas bypass is only used to prevent the DX-coil from frosting in case T_{ev} falls below 32°F because of a sudden load change.

RESET. For a DDC-panel-controlled VAV rooftop packaged system that uses a proportional-integral control mode, T_{dis} can be reset based on either the space air temperature floating within the dead band or the outdoor temperature T_o.

When T_{dis} is reset based on the space temperature, it is preferable to have several space temperature sensors connected in series. Their average temperature T_{rm}, in °F, is used to reset T_{dis}. As T_{rm} floats within the dead band between 70 and 75°F, each °F drop in T_{rm} corresponds to a 1 to 2°F increase in T_{dis}.

During cooling mode, if T_{dis} is reset based on the outdoor temperature T_o, then T_{dis} will not be reset when $T_o \geq 70°F$. After T_o falls below 71°F, for every 5°F that T_o drops, T_{dis} will increase 1°F. For instance, at an outdoor temperature of $T_o = 50°F$, the set point of T_{dis} will be 57°F.

Fan Combination and Modulation

As described in Section 19.12, for a VAV rooftop packaged unit system that uses an air economizer with a return system pressure drop lower than 0.6 in. WG, a supply and relief fan combination is recommended because of its better performance and lower energy use. The relief fan is usually turned on only when the rooftop unit is in 100 percent outdoor air economizer mode. The volume flow rate of the relief fan is modulated based on the space pressure by either varying its fan speed or varying the opening of its damper.

Three types of supply fan modulation are widely used in large VAV rooftop packaged units to maintain a supply duct static pressure at a point between just before the most remote branch take off and about one-third from the end of the supply main duct: (1) a forward-curved centrifugal fan with inlet vanes, (2) an airfoil centrifugal fan with an adjustable-frequency AC drive, and (3) an airfoil centrifugal fan with inlet cone modulation. Because of the smaller fan inlet in an airfoil centrifugal fan, it is not recommended that inlet vanes be used in a smaller airfoil centrifugal fan (say, with an impeller diameter less than 22.5 in.) in order to prevent an extremely high inlet velocity.

Both forward-curved centrifugal fans with inlet vane modulation and airfoil centrifugal fans with inlet cone modulation have a lower cost than airfoil fans that use an adjustable-frequency AC drive. However, variable-speed adjustable-frequency AC drive fan modulation is more energy-efficient at part-load operation. A cost analysis is always helpful to determine whether having an AC inverter or having inlet vanes is optimum.

For a VAV rooftop packaged unit using inlet vanes for its forward-curved centrifugal fan modulation, consider the following:

- Because all inlet vanes are subject to a certain amount of air leakage at the completely closed position, the fan volume flow rate will still be about 10 to 30 percent of the design flow when the inlet vanes are completely closed. A modulation range from 100 percent to about 30 percent is usually sufficient for a VAV rooftop packaged system.

- A forward-curved centrifugal fan has a smaller fan surge area at a lower fan total pressure and flow rate than an airfoil centrifugal fan.

- Fan modulation resulting in an external static pressure greater than 4 in. WG in a VAV rooftop packaged unit may cause damage to the unit.
- A discharge damper should not be installed because closing the VAV boxes can provide the same shutoff function. A discharge damper may cause excessive noise due to overpressurization of the VAV boxes. It may also overpressurize the ducts.

A VAV rooftop packaged system is shut off during nighttime unoccupied hours in the summer and off-seasons. In locations where there is a cold winter, it operates to provide warm air and maintain a night setback indoor space temperature during unoccupied hours.

Minimum Outdoor Air Intake

Many VAV rooftop packaged units are installed with two outdoor air dampers: one manually operated damper that provides an outdoor intake from 0 up to 25 percent, and another one that opens when the air economizer or minimum outdoor air control is in operation. If both of these dampers are fully opened, 100 percent outdoor air can be provided.

As described in Section 27.7, during summer design conditions when a VAV rooftop packaged system is operating in the minimum outdoor air recirculating mode, the design value of the outdoor air supply (15 to 20 cfm person) is normally adequate. During part-load operation, when the system supply volume flow rate and the fan total pressure have dropped considerably due to inlet-vane or fan-speed modulation, the pressure difference between the outdoor atmospheric pressure and the negative pressure at the mixing box, Δp_{o-m}, is smaller than for the design condition. If there is no minimum outdoor air control to open the outdoor damper wider and if the number of occupants remains the same as for the design condition, the outdoor air intake becomes inadequate, and an unacceptable IAQ may result.

Some sort of minimum outdoor air control should be provided to increase the outdoor air intake for such VAV rooftop packaged systems when Δp_{o-m} becomes smaller.

Heating Requirements

For most buildings located in areas whose indoor temperature may be too low at night, heating is usually required during night setback and morning warm-up periods to maintain a desirable indoor temperature during both occupied and unoccupied periods.

During occupied hours, if a VAV rooftop packaged system is used to serve a single-zone conditioned space, heating is required if the sum of the solar heat gains and internal loads is smaller than the transmission and infiltration heat losses. Cold supply air switches over to warm air as the space temperature falls below a certain limit, such as 70°F as sensed by one or more sensors. The DDC panel energizes the heating coil in the packaged unit and starts to supply warm air.

In a VAV rooftop packaged system, heating may be provided by a gas-fired heater, an electric heater, a heat pump, or a hot water heating coil.

For a VAV rooftop packaged system serving a multi-zone office building, the winter heating required in a perimeter zone during occupied hours is often provided by a perimeter-heating system or by terminals (such as reheating VAV boxes or fan-powered units). In such circumstances, the heating capacity of the VAV rooftop packaged unit is used only for night setback and morning warm-up purposes.

If the warm air from the VAV rooftop packaged unit that is used for winter heating in a perimeter zone of an office building during occupied hours is provided by a gas heater, an electric heater, or a heat pump, then two separate rooftop package units will be installed, one for perimeter zone heating and the other for interior cooling in winter.

The reasons to supply warm air to maintain an indoor night setback temperature (such as 55°F) are (1) to prevent freezing of water pipes and water surfaces in areas where the outdoor temperature at night is below 32°F, (2) to provide an acceptable indoor temperature for emergency access, and (3) to reduce the time required to warm up to a required temperature, say 68°F or 70°F, prior to an occupied period.

During the night setback and morning warm-up periods, the following hold:

- Outdoor dampers and exhaust dampers should be completely closed.
- Inlet vanes or inlet cones of the supply fan should be fully open.
- All VAV boxes or reheating VAV boxes should be fully open.
- Both refrigeration and the relief fan should be shut off.
- The supply fan and heating coil should be turned on and cycled; that is, when the indoor temperature exceeds a certain limit, both the supply fan and heater can be turned off and deenergized. The fan and heater will be energized again when the indoor temperature drops below a certain limit.

- The termination of the night setback period is the beginning of the warm-up period. The warm air supply temperature from the VAV rooftop packaged unit during warm-up period is generally between 100 and 120°F.

Air-Cooled, Water-Cooled, and Evaporative-Cooled Condensers

In rooftop packaged units using reciprocating compressors, the energy efficiency ratio (EER) for air-cooled, water-cooled, and evaporative-cooled condensers at an outdoor dry bulb of 95°F and a wet bulb of 75°F are as follows:

Air-cooled condenser	9 to 12
Water-cooled condenser	11 to 14
Evaporative-cooled condenser	14 to 16

Air-cooled and evaporative-cooled condensers often have a higher installation cost than water-cooled condensers and cooling towers.

Although water-cooled and evaporative-cooled condensers are considerably more energy-efficient at an outdoor condition of 95°F dry bulb and 75°F wet bulb than air-cooled condensers, in many locations, the annual average dry bulb is rather low. Shaffer (1987) used a manufacturer's computer program to compare the annual energy use of air-cooled and water-cooled reciprocating chillers for mid-rise buildings using VAV reheat systems in Boston and San Diego. The annual energy use of a water-cooled condenser and cooling tower is about 10 percent higher than that of an air-cooled condenser, mainly due to the energy use of condenser pumps.

Comparing the energy use and costs of air-cooled, water-cooled, and evaporative-cooled condensers is recommended in order to make an energy-efficient and cost-effective selection.

Noise Considerations

Because of the use of large-tonnage rooftop packaged units as well as lightweight building roofs and structures in recent years, rooftop packaged units may cause objectionable noise. If noise problems are not carefully accounted for at the design stage, the result will be complaints from the owner and users. Noise problems are far more difficult to solve after the building is constructed and the rooftop packaged system has been installed.

Objectionable noise from a rooftop packaged unit usually includes direct transmission of fan and compressor noise through the building roof and ceiling, structure-borne vibration, duct-borne fan noise, duct breakout noise, and duct rumble.

The following are helpful considerations in preventing objectionable noise at the design stage of a VAV rooftop packaged system:

- Using a number of small rooftop packaged units instead of one or two large units divides the impact over a greater roof area. A rooftop packaged unit with a lower sound power level is often one size larger than the smallest unit for that capacity. The fan in such a unit is properly sized, with less restricted fan inlet and outlet conditions inside the unit. A horizontal discharge fan outlet to a plenum with 4-in. inner-lined sound absorption material and a similarly lined return plenum in the packaged unit are always advantageous for sound attenuation.

- Locate a rooftop packaged unit (RPU) at least 25 ft away horizontally from acoustically sensitive spaces. (Acoustically sensitive means areas whose NC curve is 35 dB or less, such as private offices and conference rooms.) An RPU can be located over buffer spaces. Low-use storage areas, higher-noise loading areas, and acoustically nonsensitive spaces such as rest rooms and stairwells are buffer spaces.

- Stiff structural support should be provided for the RPU. Borzym (1991) says that the best place for the RPU is directly over a column or straddling a major beam close to a column. Good locations are those near the intersection of beam with a heavy end close to a column, straddling a beam, or close to a beam. The RPU should be mounted on a spring-isolated roof curb and should float on the springs. The roof below the RPU should not vibrate.

- A supply fan that has a downshot outlet and that penetrates the roof directly produces many noise problems, such as very loud breakout sound inside the ceiling plenum. A supply fan with horizontal outlet discharge to an inner-lined plenum inside the RPU is better. Use 2-in. acoustic linings for fan, compressor, and condenser sections to reduce sound transmission directly through the roof. An even better sound-attenuation arrangement is to run the supply duct horizontally from the RPU 10 to 20 ft along the roof and then penetrate into the building. Exterior ducts must be inner-lined, externally insulated, weatherproof enclosures that are covered with noise barriers. A sufficient length of inner-lined horizontal exterior duct combined with silencers or thickly lined acoustic plenums permits duct penetration directly above acoustically sensitive spaces.

- The roof directly under the RPU within the roof curb should contain a good acoustic barrier to prevent sound transmission through the bottom panel of the RPU, the roof, and the ceiling plenum. A typical sound barrier is made of semi-rigid fiberglass insulation (at least 4 in. thick) covered by two layers of $\frac{5}{8}$-in. gypsum board. The staggered gypsum boards are used to seal the joints and seams airtight.

- The duct system should not be designed for an excessive velocity, that is, normally no greater than 3000 fpm. Design the duct fittings at minimum dynamic losses. Use a round duct or multiple round ducts instead of a rectangular duct, if possible. The important advantage of round duct is its ability to contain low-frequency noise (lower duct breakout noise). If the first run of duct must be rectangular, then use 16-gauge wall thickness instead of 22-gauge to reduce duct breakout noise. Use stiffened duct wall for a rectangular duct with a large aspect ratio to prevent rumble noise.

- Most VAV RPUs with VAV boxes or reheating VAV boxes should not be operated at a completely shut-off condition because a high total pressure at a low volume flow rate is generally a noisy operating point for fans. Fan noise may be 0 to 10 dB higher in such cases than at normal conditions.

System Controls

A VAV rooftop packaged system usually comprises the following controls:

- Zone temperature control by VAV boxes
- Discharge air temperature control
- Duct static pressure control
- Air pressure safety control to protect the duct pressure from exceeding an upper limit and damaging the duct construction as the inlet vanes completely close
- Other safety controls
- Minimum outdoor air control
- Space pressurization control provided by a modulating relief fan and a relief damper or exhaust damper
- Night setback and morning warm-up control
- Optimum start and stop control
- Smoke purge control during a fire
- Head pressure control provided by duty-cycling condenser fans for reciprocating compressors and air-cooled condensers
- Pump-down control for reciprocating compressors

The sequence of control operations for a typical VAV rooftop packaged system using a DDC panel and DDC controllers at the VAV boxes is as follows:

1. The system switch is closed to the cool position either manually or automatically by an optimum start

mechanism. The fan starter is energized, which starts the supply fan. At the same time, the DDC panel is powered to activate discharge air temperature control.

2. If $T_o > 51°F$, the air economizer cycle cannot offset the space sensible cooling load, and T_{dis} rises to the upper limit of the control band. This closes the contact of the compressor's starting circuit. If safety control circuits (such as oil pressure, high pressure, and motor overload circuits) are closed, the liquid line solenoid valve will be opened, which allows refrigerant to enter the DX-coil. The increase in suction pressure in the DX-coil closes the low-pressure limit control circuit, completes the circuit to the compressor, and starts the compressor.

3. Start the compressor and the condenser fans at the same time.

4. Based on the cooling capacity, the coil's load, and the position of T_{dis} relative to the control band, various cooling stages are locked on and off and one stage cycles in an attempt to maintain a nearly constant set point of $T_{dis} = 53°F$.

5. As soon as the supply fan is started, all the DDC controllers for the VAV boxes are energized. The position of the damper in the VAV box is modulated to maintain a near-constant preset zone temperature as described in Section 18.6.

6. When the system switch is turned off manually or is shut down automatically by the optimum stop control, pump-down conntrol will shut off the compressor and the supply fan will be turned off. All outdoor dampers, relief dampers, exhaust dampers, and dampers in the VAV boxes will be closed.

29.9   CASE STUDY: A FAN-POWERED VAV, RECIPROCATING, ROOFTOP PACKAGED SYSTEM

This case study is a retrofit project. In it, an HVAC&R system is used to serve a 48,000-ft^2 medical office building in Little Rock, Arkansas. The building was constructed in 1979. It is owned by a limited partnership of 16 physicians. The renovation of the HVAC&R system of this project won a 1992 ASHRAE Technology Award, Second Place for existing commercial buildings.

As described by Tinsley et al. (1992), prior to the 1990 renovation, the HVAC&R system was a constant-volume electric terminal reheat rooftop packaged system. There were six rooftop packaged units with a total capacity of 120 tons and 162 electric duct heaters with an approximate total capacity of 450 kW.

For the renovation, a fan-powered VAV rooftop packaged system was designed and installed. One rooftop packaged unit with a capacity of 133 tons and 100 electric-heated parallel fan-powered units were equipped to serve this medical office building.

The renovation mainly included the following:

- The air economizer and the compressors are controlled in sequence by DDC discharge air temperature control in order to maintain a supply air temperature of 55°F. When the outdoor air temperature $T_o \leq 60°F$, all compressors are locked out.

- An energy-efficient evaporative-cooled condenser is used. The energy use of the compressors, tower fans, and spray pump at full-load and at an outdoor wet bulb of 80°F is 0.79 kW/ton.

- An airfoil centrifugal fan with inlet cone modulation is used because it uses less fan energy than a forward-curved centrifugal fan with inlet vanes.

- A relief fan is used instead of a return fan. This relief fan operates only when the system is in air economizer mode.

- A lower external static pressure of 0.75 in. WG has been incorporated to overcome losses in the supply main duct, flexible ducts, VAV boxes, and slot diffusers.

- The fan-powered VAV rooftop packaged system operates only when any one of the 16 suites is occupied. When any suite is unoccupied, the primary air damper in the VAV box is completely closed. The fan and the electric heater in the parallel fan-powered unit are controlled in sequence to maintain a space temperature 10°F lower than the zone air temperature set point during unoccupied hours.

- All supply ducts have been sealed and insulated.

- The zone temperature can be maintained between 70 and 80°F in any control zone by the DDC system.

The annual energy use for the HVAC&R system in this building before renovation was 169,150 Btu/ft$^2 \cdot$ year, and the annual energy cost was \$2.92/ft$^2 \cdot$ year. After renovation, the annual energy use had dropped to 55,890 Btu/ft$^2 \cdot$ year and the energy cost was \$1.40/ft$^2 \cdot$ year, drops of 67 and 52 percent, respectively.

REFERENCES AND FURTHER READING

Ameduri, G., "Studies in Energy Retrofit: Commercial," *Heating/Piping/Air Conditioning,* pp. 81–90, March 1987.

ASHRAE Inc., *ANSI/ASHRAE Standard 62–1989, Ventilation for Acceptable Indoor Air Quality*, Atlanta, GA, 1989.

ASHRAE, *ASHRAE/IES Standard 90.1–1989, Energy Efficient Design of New Buildings Except New Low-Rise Residential Buildings*, Atlanta, GA, 1989.

ASHRAE, Inc., *ASHRAE Handbook 1991, HVAC Applications,* Atlanta, GA.

ASHRAE, Inc., *ASHRAE Handbook 1992, HVAC Systems and Equipment,* Atlanta, GA, 1992.

ASHRAE, Inc., *ASHRAE Standard 15–1992, Safety Code for Mechanical Refrigeration*, Atlanta, GA, 1992.

Borzym, J. X., "Acoustical Design Guidelines for Location of Packaged Rooftop Air Conditioners," *ASHRAE Transactions,* Part I, pp. 437–441, 1991.

Collins, P. G., "Small Hotel Guest Room HVAC Selection," *ASHRAE Transactions,* Part I, pp. 517–520, 1989.

Davidson, K. G., "Advances in HVAC Alternatives," *Heating/Piping/Air Conditioning,* pp. 59–68, September 1987.

EIA, *Commercial Building Characteristics*, EIA, Washington, D.C., 1989.

Eisenstein, H., "IAQ: Who Is Legally Responsible?" *Heating/Piping/Air Conditioning,* pp. 43–46, August 1992.

Harold, R. G., "Sound and Vibrations Consideration in Rooftop Installations," *ASHRAE Transactions,* Part I, pp. 445–453, 1991.

Hays, S. M., and Ganick, N., "How to Attack IAQ Problems," *Heating/Piping/Air Conditioning,* pp. 43–51, April 1992.

Honeywell, *Energy Conservation With Comfort*, Second edition, Minneapolis, MN, 1979.

Jardine, G. M., Robertson, W. K., and Kimsey, S. P., "HVAC Economics: A Look Back," *Heating/Piping/Air Conditioning,* pp. 73–86, September 1989.

Joncich, D. M., "Active Solar Thermal Energy Systems in the U.S. Army: Standardization for Maximum Performance and Reliability," *ASHRAE Transactions,* Part I, pp. 189–193, 1991.

Jones, J. R., and Boonyatikarn, S., "Factors Influencing Overall Building Efficiency," *ASHRAE Transactions,* Part I, pp. 1449–1458, 1990.

Jones, R. S., "Rooftop HVAC Equipment on Building Roofs," *ASHRAE Transactions,* Part I, pp. 442–444, 1991.

Lindford, R. G., and Taylor, S. T., "HVAC Systems: Central vs. Floor-By-Floor," *Heating/Piping/Air Conditioning,* pp. 43–57, July 1989.

Liu, R. T., Raber, R. R., and Yu, H. H. S., "Filter Selection on an Engineering Basis," *Heating/Piping/Air Conditioning,* pp. 37–44, May 1991.

MacCracken, C. D., "Off-Peak Air Conditioning: A Major Energy Saver," *ASHRAE Journal,* pp. 12–22, December 1991.

Nordeen, H., "Ventilation Air Control in HVAC Systems," *Heating/Piping/Air Conditioning,* pp. 71–80, November 1983.

Patterson, N. R., "Comfort and Indoor Air Quality," *Heating/Piping/Air Conditioning,* pp. 107–111, January 1991.

Persily, A., "Ventilation Rates in Office Buildings," *ASHRAE Journal,* pp. 52–54, July 1989.

Robertson, W., and Black, M., "Indoor Air Quality: Challenges for the 90s," *Heating/Piping/Air Conditioning,* pp. 26–30, February 1991.

Ross, J., "Rooftop VAV vs. Water Source Heat Pumps," *Heating/Piping/Air Conditioning,* pp. 75–79, May 1983.

Scofield, M., and Fields, G., "Joining VAV and Direct Refrigeration," *Heating/Piping/Air Conditioning,* pp. 137–152, September 1989.

Shaffer, R., "Comparison of Air and Water Cooled Reciprocating Chiller Systems," *Heating/Piping/Air Conditioning,* pp. 71–87, August 1987.

Tamblyn, R. T., "Rooftop Air Conditioning vs. Ice and Water Storage," *Heating/Piping/Air Conditioning,* pp. 117–120, March 1985.

Tinsley, W. E., Swindler, B., and Huggins, D. R., "Rooftop HVAC System Offers Optimum Energy Efficiency," *ASHRAE Journal,* pp. 24–28, March 1992.

The Trane Company, *Applications Engineering Manual: Rooftop/VAV System Design,* La Crosse, WI, 1981.

The Trane Company, *Applications Engineering Manual: Self-Contained/VAV System Design,* La Crosse, WI, 1984.

Winter, S., "Building Pressurization Control With Rooftop Air Conditioners," *Heating/Piping/Air Conditioning,* pp. 89–94, October 1982.

CHAPTER 30

AIR CONDITIONING SYSTEMS: CENTRAL SYSTEMS

30.1 COMPARISON BETWEEN FLOOR-BY-FLOOR AND CENTRALIZED SYSTEMS

Size of Systems

In a packaged system, the size of the air system, refrigeration system, and air conditioning system to serve a specific area, a typical floor, a specific zone of several floors, or the entire building, is the same. In a central system, however, the size of an air and refrigeration system is probably different.

If the air-handling units (AHU) or packaged units (PU) are installed indoors, their optimum sizes are usually between 15,000 and 25,000 cfm. Above 25,000 cfm, the headroom available in the fan room to install an AHU or PU and ducts is usually not adequate. AHUs or PUs having a supply volume flow rate below 10,000 cfm are often more expensive per cfm volume flow than larger sizes.

In order to increase the net rentable floor area, rooftop packaged units or weatherproof rooftop AHUs are often used instead of indoor units mounted in the fan room for buildings of only a few stories. In such instances, the size of rooftop PUs or AHUs is limited by (1) the products currently available (for example, the largest rooftop PUs from most manufacturers are less than 150 tons, and the largest rooftop AHUs are 60,000 cfm), and (2) the size of the air system. A system that is too large always results in a higher system pressure loss and, therefore, a greater energy cost.

Separate Air Systems

In a control zone with special requirements and a floor area greater than 1000 ft^2, it may be energy-efficient and cost effective to use a separate air system. Special requirements include the following:

- Special process temperature and humidity requirements
- Clean rooms
- Special health-care requirements
- Special operating characterisitcs, such as after-hours operation.

ASHRAE/IES Standard 90.1–1989 requires that zones with special process temperature requirements, humidity requirements, or both, must be served by air systems separate from air systems for comfort only unless the conditioned floor area of the zones is less than 1000 ft^2.

Floor-by-Floor Systems vs. Centralized Systems

In a floor-by-floor system, AHUs or PUs are installed at least one for each floor in fan rooms or mechanical rooms. For a centralized system, conditioned air is supplied or returned through vertical risers from and to the rooftop units, or from the AHUs installed in the fan rooms at the midlevel mechanical floor of a high-rise building, as shown in Fig. 1.3, or from the basement. A packaged system in a low-rise building often has only centralized rooftop PUs to supply conditioned air downward to various floors, as shown in Fig. 29.2. A high-rise building is a multistory building of four or more floors. A low-rise building has three floors or less.

Let us compare a floor-by-floor system and a centralized system, both employing conventional air distribution.

- Normally, a floor-by-floor system is smaller, or more decentralized, than a centralized system and needs a smaller duct system and less fan energy. Because the peak loads for all of the control zones in a system do not occur simultaneously, a centralized system always has the benefit of a smaller load diversity factor than a decentralized system. Therefore, its total capacity is smaller.

- Centralized systems need supply and return risers, which reduce the amount of available rental space. The air velocity in the return riser is always lower than the supply riser. Linford and Taylor (1989) recommended that a rule-of-thumb estimate for the vertical shaft area required is 1.3 ft^2 per 1000 ft^2 of conditioned area served.

- For a VAV system, it is far simpler to use a floor-by-floor system when both supply and return volume flow rates are modulated at part-load.

- Each floor in a floor-by-floor system is a separate fire compartment, which avoids duct penetration between floors for fire safety. If local codes require smoke control and purging, the supply and return duct risers for a centralized system can be easily used as part of the smoke control system during a building fire.

- The fan room or compressor(s) in a centralized system can be located far from the occupied space. Only duct-borne noise needs to be attenuated by sound traps and the longer distances of inner-lined ductwork.

- A floor-by-floor system has better redundancy, which confines any malfunction and shutdown to the individual floor.

- A floor-by-floor system has greater flexibility than a centralized system for future development, after-hours access for overtime workers, easier and more accurate tenant metering, and staged completion and rental of building space.
- A centralized system always has larger fans and compressors and, therefore, more efficient equipment. It is often able to use high-efficiency air-foil fans and adjustable-frequency variable-speed drives.
- Operation and maintenance of a centralized system are easier.
- The overall building cost of a centralized system is often less than for a floor-by-floor system for low-rise buildings of two or three floors. For high-rise buildings, a detailed life-cycle cost analysis should be undertaken to determine the optimum choice.

30.2 COMPARISON BETWEEN CENTRAL AND PACKAGED SYSTEMS

Central Systems vs. Packaged Systems

In Chapter 1, detailed comparisons between a central system and a packaged system were listed in Table 1.2. The following are additional considerations:

- In a central system, chilled water is first cooled in the evaporator and then transported to a remote AHU to cool the air, whereas in a packaged system air is directly cooled at the DX-coil by the refrigerant in the packaged unit. If both of them are using water-cooled condensers, a central system will need a 3 to 5°F lower evaporating temperature than a packaged system. Therefore, for a small- or medium-sized air conditioning system (for example, a system cooling capacity of less than 100 tons), a rooftop packaged system is often the most cost-effective choice.
- With respect to the equipment, a central system uses an AHU for its air system and a primary plant equipped with chillers and boilers for the cooling and heating sources. In a packaged system, the air handler is the air system and the condensing unit and the direct-fired gas heater or the electric heater are the heating and cooling equipment.
- In a central system, a water-cooling coil is used in the AHU for cooling. In the packaged system, a DX-coil in the air handler is always used for cooling.

- In a central system, AHUs, chillers, and boilers are all installed indoors. Packaged units may be rooftop or indoor units. Condensing units are always outdoor units.
- As described in Chapter 15, the size of an AHU is based on its volume flow rate. Its cooling and heating capacity can be varied by choosing a water coil of different row depth and fin spacing. The size of a packaged unit is based on its DX-coil and heater. Their capacities are fixed for a specific model and size.
- An AHU may be a factory-fabricated product or a field-built unit. A packaged unit is probably factory fabricated. A packaged system is simpler to design and install.
- A central system always has a larger central plant so as to provide heating and cooling. A central system has a smaller load diversity factor for its chillers and boilers than a packaged system. For a VAV system, if the size of the air-handling unit is greater than the air handler in the packaged unit, the load diversity factor of the AHU will also be smaller than the air handler.
- In Linford and Taylor (1989), central systems are compared with floor-by-floor packaged systems with cooling towers at the rooftop. Overall building costs are less for central systems if the building is 8 to 10 stories and less. For buildings above 15 stories, building costs generally favor floor-by-floor packaged systems.
- If a thermal storage system is used to provide off-peak air conditioning, it must be a central system using chilled water or brine as a cooling medium.

Capacity Control at Part-Load for Central Systems

For a VAV central system, the capacity control of a multizone air conditioning system at part-load during summer cooling mode includes an air-side control, a water-side control, and a refrigeration-side control.

For each control zone, when a drop in zone temperature T_r is detected by the temperature sensor, the DDC controller adjusts the position of the damper in the VAV box so that the volume flow rate of supply air is modulated to match the reduced sensible cooling load in that control zone.

As the dampers in the VAV boxes served by the AHU are closed to smaller openings, the duct static pressure rises. The DDC panel then closes the inlet vanes, or varies the speed of the supply fan until the supply volume flow rate of the AHU is matched by the reduced sensible cooling load of the area served by that AHU.

When the supply volume flow rate of the AHU is reduced at part-load, the discharge air temperature T_{dis} tends to drop to a lower value. As this signal is sensed by the discharge temperature sensor, the DDC panel modulates the two-way valve of the cooling coil and reduces the flow rate of the chilled water so as to maintain a nearly constant or a reset T_{dis} at part-load.

As the drop of the coil's load—represented by the product of the chilled water flow rate and the temperature difference between the entering and leaving coil $(T_{wl} - T_{we})$—is sensed by a Btu meter, the DDC panel modulates the inlet vanes of the centrifugal compressor or varies the compressor speed to reduce the refrigeration capacity to match the fall of the coil's load at the cooling coils to maintain a nearly constant or reset water temperature leaving the chiller.

The modulation of capacities between the air, water, and refrigeration sides of a central system should be equal or nearly equal to each other so that an equilibrium between the space temperature, discharge air temperature, and the leaving chilled water temperature can be maintained at a specific part-load.

Another important characteristic of the capacity control of central systems is that the control actions in all air, water, and refrigeration systems except multiple constant speed pumps used in building loops are usually stepless, continuous modulation controls.

As for packaged systems, capacity controls take place only in air and refrigeration systems. The modulation of the supply volume flow rate by means of a VAV box, to match the reduction of zone sensible cooling load, and the modulation of fan capacity, to match the reduction of system load in the air handler, are the same as in the central system. The difference is that the refrigeration capacity is modulated at the DX-coil by means of compressor capacity control.

Current microprocessor-based DDC reciprocating compressor capacity controls, using cycling of loaded and unloaded cylinders and incorporating offset free proportional-integral control modes, greatly improve the quality of discharge air temperature control of the packaged unit. The difference in capacity controls between packaged and central systems becomes insignificant.

Air and Water Temperature Differentials

Both air and water temperature differentials exist in a central air conditioning system. For a central system in cooling mode operation, the supply air temperature differential indicates the difference between the space air temperature T_r and the supply air temperature T_s. The discharge air temperature T_{dis} from the AHU after the draw-through supply fan is about 3°F lower than T_s because of the supply duct heat gain. The temperature of the conditioned air leaving the cooling coil T_{cc} is about 5°F lower than T_s as a result of the supply fan heat gain and duct heat gain.

The implications of a greater supply air temperature differential $(T_r - T_s)$, with respect to maintaining a specified space temperature and relative humidity, include the following:

- A lower supply volume flow rate $\dot{V}_s$ and, therefore, less investment for ducts, VAV boxes, flexible ducts, and diffusers
- A lower conditioned air off-coil temperature T_{cc}
- A greater risk of surface condensation due to a lower T_s
- A lower fan energy use because of the lower $\dot{V}_s$
- A higher compressor power input because of the lower T_{cc}

The fan energy saving is often greater than the increase in compressor energy input when $(T_r - T_s)$ is increased.

Because of the saving in initial cost as well as a possible saving in energy cost, the recent trend is to use a greater $(T_r - T_s)$. In current practice, the supply air temperature differential is divided into two categories:

- *Conventional air distribution*. This category has the following operating parameters:
  - A supply air temperature differential $(T_r - T_s)$ of 15 to 24°F
  - A supply air temperature of 52 to 58°F, typically 55°F
  - A lowest temperature of chilled water leaving the evaporator as low as 37°F because of the improvements in freezing protection control in evaporators
  - A space temperature of 75 to 78°F and a space relative humidity between 45 and 50 percent

  Neither ice storage systems nor glycol are used in this category.

- *Cold air distribution*. This category includes the following operating features:
  - A supply air temperature differential $(T_r - T_s)$ of 30 to 36°F
  - A supply air temperature of 42 to 47°F, typically 44°F

- A supply of water to the water cooling coils at a temperature of 34 to 38°F
- A space temperature of 78°F and a relative humidity between 35 and 45 percent

Cold air distribution is always used with an ice storage system. Fan-powered units are often used to blend the cold primary air at 44°F with the plenum air to produce a supply temperature of 55°F. Ethylene glycol or propylene glycol mixed with water, or brine, is needed for freezing protection. Adequate insulation must be provided for ducts, terminals, and diffusers to prevent surface condensation.

As described in Section 20.1, under currently accepted procedures, the chilled water temperature differential [that is, the difference between the chilled water temperatures entering and leaving the evaporator $(T_{ee} - T_{el})$] is between 12 and 24°F. A large water temperature differential saves pump power and reduces the pipe size but needs a lower $T_{e\ell}$. For a variable-flow building water loop, although the chilled water temperature return from the coils T_{wl} may be different from T_{ee} at part-load, T_{ee} is nearly equal to T_{wl} at design load. If glycol is not blended with the chilled water, the lowest chilled water temperature should not be lower than 37°F to protect against freezing.

Influence of Inlet Vanes on Small Centrifugal Fans

Inlet vanes mounted at the inlet of the centrifugal fan block a certain percentage of the air passage. For a small centrifugal fan, these vanes have a considerable effect on fan performance compared with similar types and sizes of fans without inlet vanes, even when the inlet vanes are fully open.

- The ratio of the inlet diameter D_1 to the impeller diameter D_2, D_1/D_2, is different for forward-curved and backward-curved centrifugal fans. For forward-curved fans, D_1/D_2 varies from 0.8 to 0.9, and for backward fans, it varies from 0.65 to 0.8.
- The percentage of blocked air passage for small centrifugal fans may be between 15 and 25 percent.
- Although the blocked area may be only 15 percent of the total air passage, the eddies and turbulences after the inlet vanes—even if they are fully opened—result in a greater energy loss.

The following is a comparison of fan performance between backward-curved and forward-curved centrifugal fans with inlet vanes when they are fully opened and fans without inlet vanes. These data are

taken from the manufacturer's catalog of vertical modular AHUs (1990).

	Without inlet vanes			With inlet vanes		
Blade	**BC**	**BC**	**FC**	**BC**	**BC**	**FC**
D_2, in.	20	22.25	15	20	22.25	15
$\dot{V}$, cfm	10,400	13,000	6,000	10,400	13,000	6,000
v_f, fpm	2,063	2,063	2,143	2,063	2,063	2,143
Δp_s, in. WG	3.75	3.75	3.75	2.25	3.17	3.50
Δp_t, in. WG	4.01	4.01	4.04	2.51	3.43	3.79
bhp, hp	10.93	12.66	5.79	10.38	14.45	5.95
rpm	1,841	1,645	1,296	1,839	1,646	1,291
$p_{t.i.}/p_{t.o.}$	0.63	0.86	0.94			

BC = backward-inclined or backward-curved centrifugal fan

FC = forward-curved centrifugul fan

$\dot{V}$ = volume flow rate, in cfm

v_f = air velocity at the fan outlet, fpm

Δp_s, Δp_t = fan static pressure and total pressure, respectively, in. WG

bhp = brake horsepower, hp

rpm = revolutions per minute

$p_{t.i.}$, $p_{t.o.}$ = fan total pressure with and without inlet vanes, respectively, in. WG

If the ratio D_1/D_2 for a backward-curved centrifugal fan of impeller diameter $D_2 = 20$ in. is taken as 0.75 and the percentage of blocked area at the inlet is 20 percent, the velocity at the fan inlet when the inlet vane is fully opened is

$$v_{\text{inlet}} = \frac{\dot{V}}{A_{\text{free}}} = \frac{10,400}{0.8\pi\left[(20 \times 0.75)/(12 \times 2)\right]^2}$$

$$= 10,590 \text{ fpm}$$

For such an extremely high velocity, the ratio of fan total pressure of this BC centrifugal fan of 20 in. impeller diameter when the inlet vanes are fully opened is $p_{t.i}/p_{t.o} = 0.63$. More than one third of the fan energy output is lost as a result of the installation of inlet vanes. Such a decrease in total pressure due to an extremely high velocity v_{inlet} was verified in a field installation of a VAV system using inlet vane modulation. A fan total pressure of less than 3 in. WG was available. After the inlet vanes had been dismantled, the fan total pressure increased to more than 4 in. WG.

For a BC fan with a 22.25-in. diameter impeller, the ratio $p_{t.i}/p_{t.o} = 0.86$, and for a FC fan with a 15-in. diameter impeller, the ratio $p_{t.i}/p_{t.o} = 0.94$.

During a cost comparison between inlet vanes and adjustable-frequency AC inverters, the pressure loss due to the inlet vanes when they are fully opened must be taken into consideration. Therefore, for centrifugal fans either in AHUs or PUs, installation of inlet vanes is not recommended in backward-curved centrifugal fans with impeller diameters smaller than 22.25 in. or in forward-curved centrifugal fans with impeller diameters smaller than 15 in.

30.3 CASE STUDY: A FAN-POWERED VAV, SCREW ICE STORAGE SYSTEM

A fan-powered VAV, screw compression ice storage system was designed and constructed for the 34-story office building of the Taipei World Trade Center, Taipei, Taiwan. It has a gross floor space of 1.05 million ft^2 and a total air conditioned floor space of 880,310 ft^2. The peak refrigeration load of the building is about 2530 tons. The local utility rate structure strongly favors the use of refrigeration systems during off-peak hours. This project won First Place in the 1991 ASHRAE Technology Awards for HVAC&R system designs for commercial buildings. The entrants were Hsing Chung Yu and Carl E. Claus.

Refrigeration Systems

The ice storage system was a partial storage one. Screw compressors were used for ice-making during off-peak hours. A separate flooded liquid cooler was connected to the screw compressors to provide direct cooling during peak hours as well as during after-hours operation to offset the significant nighttime load. During direct cooling, the compressors were controlled to operate at a higher suction temperature of 33°F to conserve energy.

Ice builders were used to produce ice on the coils in well-insulated storage tanks, as described in Section 25.3. A liquid overfeed system was used instead of direct expansion during the ice-making period because of its higher heat transfer. About 20 percent of the refrigerant coil surface in the ice builders could be saved. During ice melting, chilled water was supplied directly through a closed-loop water circuit to the storage tanks and the tanks were pressurized. Such an arrangement obviates the use of a heat exchanger and thus avoids the corresponding chilled water temperature rise.

Condensing heat was efficiently rejected through evaporatively cooled condensers.

Air Systems

Because of the lower chilled water temperature from the storage tank during ice melting, primary air at 45°F was introduced at the parallel fan-powered unit. It was mixed with the induced plenum air to produce a supply air of 56°F at the diffuser. Such a lower primary air temperature had the following consequences:

- Summer space relative humidity dropped to 40 to 45 percent. The thermal comfort of the occupants was greatly improved.
- Primary air volume flow rate was reduced to 40 percent of the summer full-load design value.

Two built-up centralized AHUs were installed in the basement. Each was used to supply 200,000 cfm of primary air from the bottom to the top of the entire building. Compared with the originally proposed conventional system with additional equipment floors at midlevel, this air distribution system saved considerable rental space.

The centralized air system, having floor-by-floor automatic shut-off dampers, was closely matched with the roof-mounted smoke control system. In case of a building fire, the return fan would be stopped and the rooftop exhaust fan would be started to purge the smoke from the fire floor through the opening and closing of dampers connected to the smoke exhaust system. At the same time, air would be supplied to the two floors above the fire floor as well as one floor below to pressurize these floors.

The air system also had the following features:

- Provided 15 cfm of outdoor air for each occupant
- Included cartridge-type filters and upstream prefilters with a dust spot efficiency of 85 percent
- Used controllable-pitch axial fans for supply and return air
- Included electric heaters for winter heating

A microprocessor-based DDC system was used for zone temperature; AHU operating parameters; ice charging, discharging, and direct cooling; and emergency smoke controls.

Cost and Electricity Demands

The actual bidding price of the HVAC&R system for this project was approximately $9,000,000, or $8.57 per ft^2 gross area.

Compared with the originally proposed conventional air conditioning system using centrifugal chillers, the air system power input was reduced from 1679 kW to 1136 kW, and the electricity demand dropped from 5958 kW to 4329 kW. The estimated total annual energy consumption per ft^2 gross area was 33,188 Btu/ft$^2 \cdot$ year.

Temperature Rises between Supply Air and Tank Discharge

In an ice storage system with cold air distribution, estimates of temperature increases between the chilled water temperature discharged from the storage tank $T_{dis.k}$ and the supply temperature at the ceiling diffuser T_s, both in °F, are useful in system design. Dorgan and Elleson (1989) recommended the following estimated temperature rises as a means to make a preliminary determination of T_s:

Source	Description	Temperature rise, °F	
		Range	Typical
Heat exchanger	$T_{xl} - T_{dis.k}$	2–4	2
Pump, and piping heat	$T_{we} - T_{xl}$	0.5	0.5
Cooling coil	$T_{cc} - T_{we}$	4.5–8	6
Supply fan heat gain	$T_{sf} - T_{cc}$	2–3	3
Duct heat gain	$T_s - T_{sf}$	1.5–3	2

T_{xl} = water temperature at exit of the heat exchanger, °F
T_{we} = chilled water temperature entering cooling coil, °F
T_{cc} = conditioned air leaving cooling coil, °F
T_{sf} = air temperature at supply fan outlet, °F

Using a supply fan of higher efficiency and reducing supply duct heat gain are both effective ways to decrease the fan heat and duct heat gain and to conserve energy.

30.4 CASE STUDY: CONSTANT VOLUME CENTRAL SYSTEMS FOR UNIDIRECTIONAL-FLOW CLEAN ROOMS

Indoor Requirements

The fabrication of semiconductor integrated circuits requires a highly sophisticated combination of advanced technologies. The air conditioning system for clean rooms in semiconductor wafer fabrication is one of these technologies. It must meet stringent air quality requirements for cleanliness, temperature, and humidity for the sake of successful manufacturing.

Contaminated semiconductors result in inferior products. As described in Chapter 5, clean room air cleanliness requirements for semiconductor fabrication include Class 1, 10, 100, 1000, and 10,000. Vertical unidirectional air flow with average air velocities from 60 to 90 fpm, typically at 90 fpm, are widely used.

Manufacturing an integrated circuit involves photolithography, etching, and diffusion processes. These processes may take place over many hours, even days. A stable temperature is extremely important. Because clean room production personnel wear smocks that fully cover them, clean room temperatures are controlled between 68 to 72°F with tolerances of ±0.1°F, ±0.2°F, ±0.5°F, and ±1.0°F. A closer tolerance is often maintained within the manufacturing process itself; for example, wafer reticle writing by electron beam technology needs ±0.1°F, whereas ±1.0°F is often used for the open-bay area.

In clean rooms, if the relative humidity is too low, static electricity occurs and causes defective products. If the relative humidity is too high, some chemicals will expand and may cause equipment failure. For etching and diffusing areas, the humidity should be maintained at 40 to 45 percent with a tolerance of ±5 percent, whereas in the photolithography area, a relative humidity of 35 to 40 percent with a tolerance of ±2 percent is usually maintained.

The manufacture of metal oxide semiconductors needs large quantities of conditioned outdoor air, as a component of make-up air, to replace processing exhaust air and to maintain clean room pressurization. A clean room is always maintained at a positive pressure to prevent the infiltration of contaminated air from surrounding spaces. For some clean rooms, the process exhaust air flow may be as high as 10 cfm/ft² floor area. The average process exhaust air flow for semiconductor fabrication may be between 2 and 3 cfm/ft².

Energy Use of Components

In Naughton (1990), the energy use of components in a typical vertical unidirectional flow clean room was approximated as follows:

Manufacturing equipment (75 W/ft²)	50%
Air conditioning (HVAC&R)	40%
Electric lights	6%
Building envelope and others	4%

The chiller and fan(s) each consume 45 percent of the total HVAC&R energy use. Both plant and building pumps use the remaining 10 percent.

Because of the extremely high space sensible cooling load and the very small space latent load, the sensible heat ratio of the space conditioning line SHR_s can often be taken as 0.99.

The operating cost of HVAC&R is only 5 to 20 percent of the total cost needed to produce an integrated circuit (semiconductor wafer). The air cleanliness, temperature, and relative humidity required for successful fabrication is still an extremely important goal of an HVAC&R system design. Because of high utility rates, however, a reduction in the operating cost of the air conditioning system also becomes a very influential factor in clean room design and operation.

Air Conditioning System

A typical air conditioning system for a Class 10 clean room is shown in Fig. 30.1. This system consists of the following components:

- *Recirculating air unit (RAU).* The function of a recirculating air unit is to recirculate the space air, to filter it, to cool it, to pressurize the mixture of recirculating and conditioned make-up air, and to force the mixture to flow through the ULPA filters and the clean room working area.

  An RAU comprises the following components:
  - A mixing box to mix together recirculating and conditioned make-up air
  - A prefilter with a dust spot efficiency of 30 percent
  - A chilled water sensible cooling coil
  - A recirculating fan
  - Make-up air and recirculating air dampers
  - Two sound attenuators, one located immediately before the fan inlet and the other after the fan outlet

  Usually an axial fan is used as the recirculating fan because of its higher fan total efficiency (between 75 and 82 percent) and its better operating characteristics. An unhoused centrifugal fan, often called a cabinet fan, with a fan total efficiency of 58 to 63 percent is sometimes used because of its lower sound power level.

  The chilled water entering the sensible cooling coil in the RAU is often at a temperature of 50°F.

- *Make-up air unit (MAU).* The function of an MAU is to supply outdoor air to the clean room for process exhaust and pressurization, to condition it, and to control the humidity of the clean room by cooling and dehumidifying, or heating and humidifying the make-up air.

  The system components in an MAU include the following:
  - An outdoor air damper
  - A prefilter with a dust spot efficiency of 30 percent
  - A preheating coil
  - A chilled water cooling coil
  - A make-up air centrifugal fan

- An HEPA filter of 99.97 percent DOP efficiency
- A humidifier; most often, a steam humidifier
- An MAU shut-off damper

Chilled water entering the cooling coil in the MAU for cooling and dehumidifying is at a temperature of 40°F. It is more energy efficient to have a separate chiller to provide chilled water for an MAU.

- *ULPA filters and the unidirectional-flow clean room.* The function of ULPA filters is to provide ultraclean air for the clean room. The pressurized plenum or ducted ULPA filter modules are often used for an even distribution of unidirectional downward air flow.

  For Class 1, 10, and 100 clean rooms, ULPA filters with a DOP efficiency of 99.9997 percent of 0.12 μm particles are used. For Class 1000 through Class 100,000, HEPA filters with a DOP efficiency of 99.97 percent of 0.3 μm particles and nonunidirectional flow may provide satisfactory contamination control.

  Unidirectional downward air flow produces a uniform air shower of ultraclean air. Internally generated contaminants will not move laterally against the 90 fpm air flow and will be carried away by predictable parallel airstreams. Recirculating air enters either the perforated raised floor panels or the side return inlets directly. It is then returned to the RAU to mix with make-up air again.

Operating Characteristics

The following temperature and relative humidity are to be maintained in a Class 10 clean room:

Room temperature, year-round, °F	69 ± 1.0
Space humidity, year-round, %	42.5 ± 2.5

For a Class 10 clean room with an area of 1000 ft^2, a supply volume flow rate of 90,000 cfm is required to provide an air velocity of 90 fpm in order to produce a unidirectional flow in this clean room. The outdoor air intake for process exhaust, space pressurization, and occupants is typically 6000 cfm.

Summer Mode Operation

In Mandelbaum (1991), the operating modes of clean rooms of Class 1 through 1000 are divided into summer and winter modes. When the dew-point temperature of the outdoor air T_o'' is 46°F and above, the air conditioning system is in summer mode operation.

Let us consider a hot summer day. Outdoor air at a summer design temperature of 100°F and a wet bulb of 78°F enters the MAU, as shown in Fig. 30.1a and

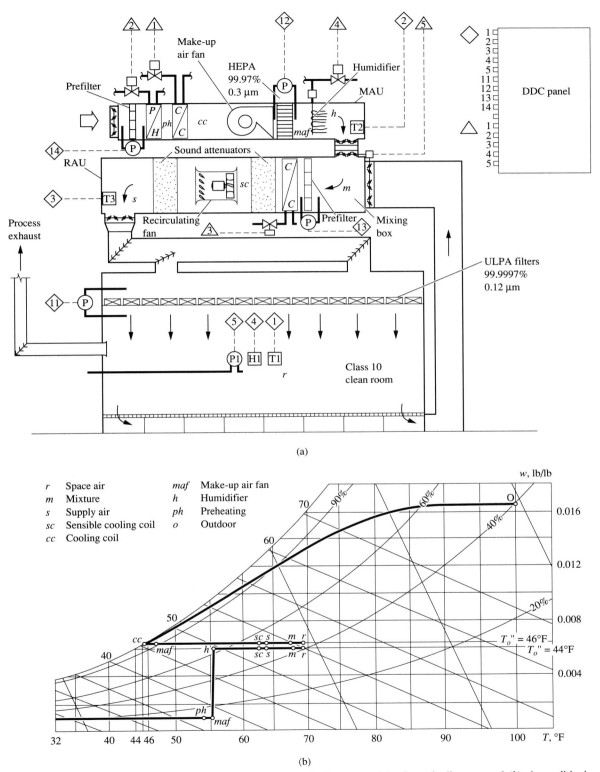

FIGURE 30.1 A typical air conditioning system for a Class 10 clean room: (*a*) schematic diagram; and (*b*) air conditioning cycle.

b. It is cooled and dehumidified at the cooling coil to a leaving coil condition of air temperature $T_{cc} = 46°F$, relative humidity $\varphi_r = 99$ percent, and dew-point temperature $\approx 46°F$. After the conditioned outdoor air absorbs the fan heat of the MAU, the air enters the RAU at a discharge temperature T_{dis} of $47.5°F$ and a relative humidity of 92 percent.

In the RAU, conditioned air from the MAU is mixed with the recirculating air from the clean room at a temperature of $69°F$, a relative humidity of 42.5 percent, and a dew-point temperature of $46°F$. The ratio of volume flow of recirculating air to make-up air is 12 to 1. The mixture m enters the sensible cooling coil at temperature $T_m = 67°F$, and a dew point of $46°F$. It is then sensibly cooled to a temperature T_{sc}.

If the maximum space sensible cooling load is 563,000 Btu/h, and if the density of the supply air $\rho_s = 0.078$ lb/ft^3, the temperature of supply air T_s can be calculated as

$$
\begin{aligned}
T_s &= T_r - \frac{q_{rs}}{60 \dot{V}_s \rho_s c_{pa}} \\
&= 69 - \frac{563,000}{60 \times 90,000 \times 0.078 \times 0.243} \\
&= 63.5°F
\end{aligned}
\tag{30.1}
$$

Usually the temperature increase due to fan heat in the RAU is $1°F$. The temperature of the air leaving the sensible cooling coil is then $T_{sc} = 63.5 - 1 = 62.5°F$. Its dew-point temperature is still $46°F$. Because of the short supply duct, large volume flow, and the fact that the surrounding space is conditioned, the duct heat gain is usually ignored.

Because the SHR$_s$ of the space conditioning line is 0.99, after the air has absorbed the space sensible cooling load and a very small amount of latent load, the space temperature is then maintained at $69°F$. The space relative humidity is 42.5 percent, and dew point is $46°F$.

In order to cool and dehumidify the make-up air at the cooling coil to a leaving dew-point temperature of $46°F$, the chilled water entering cooling coil should be provided at a temperature T_{we} of $40°F$ without using glycol for freezing protection. The chilled water leaving the chiller in the plant loop is usually $1°F$ lower, or $39°F$.

Summer Mode Part-Load Operation and Controls

Two controls maintain the required space temperature at $69°F$ and the relative humidity within acceptable limits during summer mode part-load operation: the

discharge air temperature control, incorporating dew-point control of the MAU, and the space air temperature control.

When the dry and wet bulb of the outdoor air T_o drops but its dew-point temperature is still at $46°F$ or above, the discharge air temperature sensor T2 detects the fall of T_o and the DDC panel modulates the two-way valve of the chilled water cooling coil in the MAU to reduce its water flow until a $47.5°F$ discharge air temperature and a discharge air dew point of $46°F$ are maintained.

If the space sensible cooling load falls below the design load, the space temperature drops accordingly. As the space temperature sensor T1 detects such a fall of temperature, the DDC panel modulates the two-way valve of the sensible cooling coil in the RAU, reduces its water flow, and tends to maintain a constant space temperature of $T_r = 69°F$.

In addition to the MAU discharge air temperature and space temperature controls, there are also controls for space humidity, space pressurization, filter pressure drop monitoring, and smoke detection.

Because the discharge air dew point from the MAU is $46°F$, the space dew point is also $46°F$, and the mixture of recirculating and make-up air still has a dew point of $46°F$. After sensible cooling in the RAU, the dew point of the supply air, point s, remains at $46°F$. Because the sensible heat ratio of the space conditioning line SHR$_s = 0.99$, the space humidity will be always around 42.5 percent if the space temperature T_r is maintained at $69°F$.

For a space like a clean room having only a negligible latent load, supply air dew-point control can always maintain both space temperature and relative humidity within required limits by means of sensible cooling and reheating due to the sensible cooling load.

In Fig. 30.1a, there is a pressure sensor P1 in the clean room. This sensor measures the pressure differential Δp_r between the clean room and the surrounding area. This pressure differential is generally maintained at 0.05 in. WG to prevent infiltration of contaminated air.

According to the signal from P1, the DDC panel modulates the opening of the interlocked make-up air and recirculating dampers. When there is an increase of make-up air and a decrease of recirculating air in the mixing box of the RAU, the space pressure differential tends to rise. A decrease of make-up air and an increase of recirculating air cause a drop of space pressure differential. The supply of make-up air is balanced by the process exhaust air and the exfiltrated air from door gaps due to the pressure differential.

Pressure sensors are also used for the 30 percent dust spot efficiency prefilters, HEPA filters, and

ULPA filters. If the pressure drop exceeds a predetermined limit, an indicating lamp flashes or an alarm is energized. Both signals call for replacement.

When the smoke detector is energized, a signal is sent to the building fire control system. The fire alarm and the smoke control system are energized accordingly.

Winter Mode Operation and Controls

When the dew point of the outdoor air T_o'' is 44°F or below, the air conditioning system is in winter mode operation. If $T_o'' \leq 44$°F and $T_o < 53.5$°F, the outdoor air is preheated at the preheating coil to a temperature of $T_{ph} = 53.5$°F. After absorbing the fan heat of the MAU, it enters the RAU at a discharge temperature of $T_{dis} = 55$°F. If $T_o'' \leq 44$°F and $T_o > 53.5$°F, the outdoor air is sensibly cooled to 53.5°F. Again, it is discharged to the RAU at $T_{dis} = 55$°F after absorbing fan heat.

If make-up air at a temperature of 55°F and a dew point of 44°F is mixed with winter mode recirculating air of temperature 69°F, and 40 percent relative humidity (a dew point of 44°F), the resulting mixture will be at a temperature of 67.5°F. The dew point will still be 44°F. If the mixture is cooled at the sensible cooling coil to 62.5°F according to the signal of the temperature sensor T1, and then it is supplied to the clean room at a temperature of 63.5°F and a dew point of 44°F after absorbing the fan heat of RAU, then it will maintain a space temperature $T_r = 69$°F, $\varphi_r = 40$ percent, and a dew point of 44°F at a maximum space sensible cooling load of 563,000 Btu/h. A lower space sensible cooling load results in a decrease in the space temperature T_r and, therefore, a smaller cooling capacity at the sensible cooling coil. The space temperature will still be maintained at 69°F.

If $T_o'' < 44$°F, space relative humidity φ_r will be less than 40 percent. When the space humidity sensor detects such a shortage, the DDC panel energizes the steam humidifier in the MAU to humidify the make-up air until $\varphi_r = 40$ percent. In winter mode operation, φ_r will always be maintained around 40 percent.

During winter mode operation, the chilled water temperature entering the cooling coil in the MAU is reset from 40 to 45°F for a more energy-efficient operation. Also, the preheating coil and the cooling coil in the MAU are sequentially controlled; they are not energized simultaneously.

In winter operation, outdoor air is preheated in the MAU, and the mixture of outdoor and recirculating air is again sensibly cooled in the RAU. Such simultaneous heating and cooling are mainly a result of the requirement for steam humidification. It is not appropriate to humidify the air to near-saturation. Uneven distribution may cause surface condensation.

During winter mode operation, if the dew point of the outdoor air is between 44 and 46°F, the operation mode remains in winter mode. When $T_o'' \geq 46$°F, it will change over to summer mode operation. Similarly, if 44°F $< T_o'' < 46$°F, the operating mode remains in summer mode until $T_o'' \leq 44$°F. Then it changes over to winter mode operation. Such a control strategy prevents hunting between summer and winter mode operations during intermediate seasons.

System Pressure

According to Naughton (1990) and Hunt et al. (1990), the system pressure of an MAU is about 3.8 to 4 in. WG. For an RAU, the system pressure is usually between 2.0 and 2.5 in. WG. A typical breakdown of the fan total pressure rise of the recirculating air fan in an RAU is as follows:

System Components	Pressure Drop, in. WG
RAU	
Prefilter, 30 percent efficiency	0.25
Cooling coil	0.35
Inlet sound attenuator	0.10
Discharge sound attenuator	0.15
Discharge elbow	0.10
External pressure drop	
HEPA filters, final	0.80
Distribution ductwork	0.20
Perforated raised floor panel	0.10
Return path	0.15
Total	2.20

HEPA filters of lower pressure drop, such as a final loaded pressure drop of 0.50 in. WG, are also available. However, they are expensive and need more space. This fact should be carefully analyzed and considered by the production engineer and architect.

To reduce the pressure drop across coils, the most effective method is to reduce their face velocities to 300 to 400 fpm. The result is a larger MAU and RAU. Again, the compromise is between intial and operating costs.

All duct fittings along the air flow must be carefully designed to reduce pressure losses. Square elbows should always be installed with splitter vanes.

High-efficiency fans and pumps should be used, along with high-COP and -EER chillers. According to Naughton (1990), the energy use of fans, chillers, and pumps in an air conditioning system for a Class 1 clean room may be around 50 W/ft^2 of floor area.

Effect of Filter Final and Initial Pressure Drop Difference on System Performance

During the calculation of system pressure loss of an air system for the purpose of selecting fans and AHUs or PUs, one should use the final pressure drop of the filter $\Delta p_{\mathrm{f.f}}$, in inches WG, in order to provide the required volume flow rate whenever filters are either clean or loaded to capacity. For purposes of energy estimation, the average pressure drop of the filters during the working period $\Delta p_{\mathrm{f.m}}$, in inches WG, should be employed instead of final or initial pressure drops.

For most prefilter and medium-efficiency air filter assemblies used in AHUs or PUs, the difference between the final pressure drop and the initial pressure drop of the filter assembly is between 0.4 and 0.8 in. WG. Let us consider an air system whose design volume flow rate $\dot{V}$ is 30,000 cfm, and whose system pressure loss when filters are loaded Δp_{sy} is 2.2 in. WG. At $\dot{V} = 30,000$ cfm, the difference between the final and initial pressure drops is 0.5 in. WG.

The effect of the difference in final and initial pressure drop of filters on system performance is illustrated in Fig. 30.2. In Fig. 30.2a, an airfoil-blade centrifugal fan is used. The fan and AHU or PU are selected according to these criteria: $\dot{V} = 30,000$ cfm and $\Delta p_{\mathrm{sy}} = 2.2$ in. WG. If both filters are loaded, the operating point is P. If the filters are clean, the system pressure drops to 1.7 in. WG. The operating point moves along the selected fan curve to point Q, which has a $\dot{V} = 33,000$ cfm and a $\Delta p = 1.7$ in. WG.

In Fig. 30.2b, a forward-curved centrifugal fan is used. A forward-curved fan has a slightly flatter fan curve at operating point Q, and the system volume flow rate increases to 33,500 cfm.

For Fig. 30.2c, a vane axial fan is used. A vane axial fan has a very steep fan curve. Therefore, at point Q the system volume flow rate is increased to only about 31,000 cfm.

If the air system is a constant-volume system, the fan power inputs at point Q for airfoil-blade and vane axial fans are approximately the same as at point P, whereas for forward-curved fans, their fan power input is comparatively greater.

For VAV systems, a duct static pressure or microprocessor-based supply volume flow control modulates the inlet vanes or fan speed to an operating point that approaches point P even if the filters are clean.

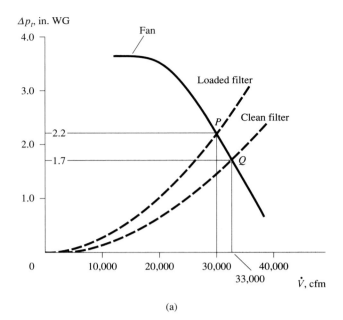

(a)

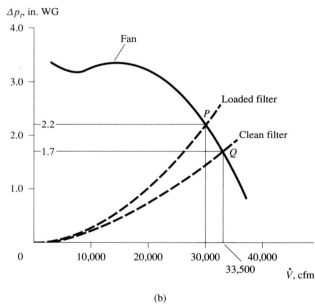

(b)

FIGURE 30.2 Effect of difference in final and initial pressure drop of filters on system performance: (*a*) airfoil-blade centrifugal fan; (*b*) forward-curved centrifugal fan; and (*c*) vane axial fan.

During the selection of AHUs or PUs, if the fan performance of the AHU or PU is expressed in volume flow $\dot{V}$ vs. fan total pressure Δp_t, the designer should ascertain that the pressure drops of the loaded filters are included in Δp_t, not those of the clean ones.

If the fan performance of PUs is expressed in volume flow $\dot{V}$ vs. external static pressure rise Δp_{ex}, most probably the PU has been tested when its filters are clean. One should check with the manufacturer's

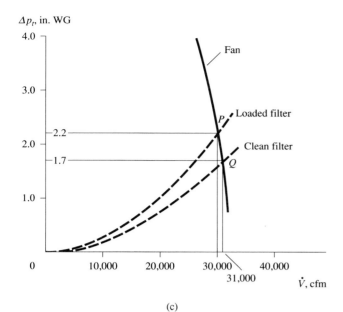

Δp_t, in. WG

FIGURE 30.2 *(continued)*

catalog whether the filter(s) are clean or loaded. If they are clean, the difference between the final and the initial pressure drop, usually from 0.4 to 0.8 in. WG, should be added to the external static pressure so as to select a PU that will provide the design volume flow when the filter(s) are loaded.

Design Considerations

A Class 1 clean room may be one of the cleanest occupied spaces on earth. The successful construction of a Class 1 clean room depends on the combined efforts of architects, mechanical engineers, production engineers, owners, contractors, HVAC&R equipment manufacturers, and many others.

One critical factor that affects the efficiency of the HEPA or ULPA filters is the sealing between the filters and the grid as well as between the grid and the building structure. As recommended in *ASHRAE Handbook 1991, HVAC Applications*, for Class 1 and 10 clean rooms, either low vapor pressure petrolatum seals or silicone dielectric gels should be used to seal the HEPA and ULPA filters into a channel or ceiling grid.

To maintain a temperature tolerance of $\pm 0.1°F$ for a manufacturing process, it is important to stabilize all internal loads during the processing period. The process should be surrounded by conditioned space for which the tolerance is $\pm 0.5°F$. Even movement of production personnel toward the processing area may cause radiant heat turbulence and temperature fluctuations.

Energy use per ft² of floor area for an air conditioning system for Class 1 through Class 100 clean rooms is 5 to 10 times higher than an air conditioning system for a commercial building. Careful analyses and improvements based on experience with similar projects will provide a satisfactory indoor environment for the manufacturing process, and reduce energy use of the air conditioning system.

Semiconductor products that are being produced now may become obsolete within a few years. As with many other HVAC&R systems, air conditioning system design for clean rooms must incorporate flexibility for change and future development.

REFERENCES AND FURTHER READING

ASHRAE, *ASHRAE Handbook 1991, HVAC Applications*, ASHRAE, Inc., Atlanta, GA.

ASHRAE, *ASHRAE Handbook 1992, HVAC Systems and Equipment*, ASHRAE, Inc., Atlanta, GA.

ASHRAE, *ASHRAE-IES Standard 90.1–1989, Energy Efficient Design of New Buildings Except New Low-Rise Residential Buildings*, Atlanta, GA, 1989.

Atkinson, G.V., and Martino, G.R., "Control of Semiconductor Manufacturing Cleanrooms," *ASHRAE Transactions*, Part I, pp. 477–482, 1989.

Dorgan, C.E., and Elleson, J.S., "Design of Cold Air Distribution Systems with Ice Storage," *ASHRAE Transactions*, Part I, pp. 1317–1322, 1989.

Hunt, E., Benson, D.E., and Hopkins, L.G., "Fan Efficiency vs. Unit Efficiency for Cleanroom Application," *ASHRAE Transactions*, Part II, pp. 616–619, 1990.

Linford, R.G., and Taylor, S.T., "HVAC Systems: Central vs. Floor-by-Floor," *Heating/Piping/Air Conditioning*, pp. 43–58, July 1989.

Mandelbaum, I., "HVAC Modifications for Semiconductor Fabrication," *Heating/Piping/Air Conditioning*, pp. 29–32, March 1991.

Naughton, P., "HVAC Systems for Semiconductor Cleanrooms—Part 1: System Components," *ASHRAE Transactions*, Part II, pp. 620–625, 1990.

Naughton, P., "HVAC Systems for Semiconductor Cleanrooms—Part 2: Total System Dynamics," *ASHRAE Transactions*, Part II, pp. 626–633, 1990.

Rose, T.H., "Noise and Vibration in Semiconductor Clean Rooms," *ASHRAE Transactions*, Part I B, pp. 289–298, 1986.

Stokes, R., "The System Choice," *Heating/Piping/Air Conditioning*, pp. 87–89, July 1983.

Tao, W., and Janis, R.R., "Modern Cooling Plant Design," *Heating/Piping/Air Conditioning*, pp. 57–81, May 1985.

Yu, H.C. and Claus, C.E., "Thermal Storage System Cools Office Building," *ASHRAE Journal*, pp. 15–17, March 1991.

COMMISSIONING OF HVAC&R SYSTEMS

31.1 INTRODUCTION

Users' Needs

Buildings and building systems must be designed, constructed, commissioned, operated, and maintained to serve the users' needs. Building types mirror every type of human endeavor: social, economic, cultural, religious, political, technical, psychological, scientific, recreational, and philosophical. Building types include the following: transportation facilities for airports, bus stations, and railroad stations; auditoriums and theaters; banks; churches, chapels, mosques, synagogues; college and university buildings; country clubs; health clubs and resorts; elementary and high schools; garages; hospitals, medical facilities, and nursing homes; hotels; industrial buildings; laboratories; libraries; office buildings; parks buildings; postal buildings; recreation facilities; restaurants; swimming pools; and stadiums. The users' needs vary according to the type of build-

ing, its overall utilization, the uses of each individual space, the expectations of the users of the building, and the characteristics of the persons utilizing each space, such as sex, age, health, education, economic status, religion, employment, physical ability, and ethnic and racial customs.

The building and all building systems must be designed, constructed, commissioned, operated, and maintained to provide the building occupants with a functional, comfortable, safe, flexible, properly illuminated, acoustically acceptable, durable, reliable, accessible and healthy environment to perform the function for which the individual building spaces are intended.

Imperfect Processes

The design, installation, operation, and maintenance of buildings, building systems, and HVAC&R systems are *imperfect* processes. The levels of performance of buildings, building systems, and HVAC&R

systems are affected by the skill, training, diligence, craftsmanship, actions, and performance of many persons including the owner, architect, structural engineer, electrical engineer, plumbing engineer, HVAC&R engineer, equipment manufacturers, contractors, installers, testing, adjusting and balancing agencies, craftspersons, operating technicians, maintenance technicians, and others. The interface and coordination of all parties are required if the building, building systems and HVAC&R systems are to perform according to the design intent and to serve the needs of the owner(s).

31.2 BUILDING SYSTEMS

For analysis purposes, building systems can be considered as being both dynamic and static. The static systems include the building envelope and structural systems. The dynamic systems include the HVAC&R systems, the plumbing systems, and the electrical systems. The dynamic systems operate and respond to the continually changing conditions imposed by the day-by-day utilization of the building and local environmental conditions. The dynamic HVAC&R systems are designed, installed, commissioned, operated, and maintained to process and control simultaneously air temperature, humidity, cleanliness, air quality, and air movements to meet the comfort requirements of the occupants and the needs of manufacturing processes in the air conditioned spaces.

The dynamic HVAC&R systems interface with the static building systems, as illustrated in Fig. 31.1. The performance of the HVAC&R systems depends on the proper design, construction, commissioning, and on-going maintenance of the static building systems. The HVAC&R engineer must monitor and verify that the interfaces of the static building systems and other dynamic systems are performing according to the design intent. HVAC&R system design is based on such items as thermal resistance of building materials; infiltration rates through doors, windows, and openings; the thermal mass of the building materials; structural system effects on the support of equipment; structural system effects on the control of pressurization; elevator, escalator, and dumbwaiter shaft effects on building pressurization; plenum effects on air-handling systems; effects of glazing on solar loads and of vapor barriers on moisture, and many other building system characteristics.

The following building system components should be monitored by the HVAC&R engineer during the design, construction, commissioning, operations, and maintenance periods for their on-going effect on the performance of the HVAC&R systems:

1. *Structural system.* Superstructure: concrete slabs, concrete joists, precast plank, concrete T beams, structural steel, open web joists, concrete floors, steel roof deck, steel beams and columns, steel siding, fireproofing for steel framing, platforms and walkways, wood joists, wood columns, wood deck, wood trusses, and other structural elements. Substructure system: footings, foundation walls, grade beams, underdrain system, piles, and waterproofing.

2. *Building envelope.* Roof system, exterior wall construction, exterior doors, exterior windows, exterior glass block, exterior plastic surfaces, exterior stairs, exterior ramps, weatherproofing, waterproofing, fire stops, louvres, with special attention to all insulating materials. Condition of cast-in-place concrete walls, precast wall panels, concrete block, brick, stone, stucco, glass block, metal siding, curtain wall wood siding, window sash, caulking, expansion joints, roof hatches, and skylights.

3. *Interior systems.* Interior partitions, ceilings, floors, doors, flooring (carpet).

4. *Vertical transportation systems.* Elevators, escalators, moving walks, dumbwaiters, cartlifts, automated cart/rail systems. (*NB*: All of these systems affect building pressurization and heating and cooling loads).

5. *Electrical systems.* (*NB*: These systems affect heating and cooling loads as well as the operation of HVAC&R equipment.) Electric service: electrical system characteristics, such as voltage, current, Hz, phase; power wiring system, control wiring system, lighting fixture types, electrical conductors, conduit and raceways, transformers, wiring devices, motor controllers, motor control centers, heat tracing systems, lightning protection systems, emergency power systems, safety devices and safety switches.

6. *Plumbing systems.* Cold water distribution system (quantity available, water pressure, temperature, and chemical analysis are all important for the water make-up to cooling towers, air washers, evaporative condensers, chillers, boilers, hot water heating system, chilled water system, and dual hot chilled water systems); domestic hot water supply and return system (important for heating load on central heating plant); storm water system (for clear water waste disposal); sanitary system (for contaminated water waste disposal); fuel gas systems (for fuel for boilers and furnaces); special water conditioning systems (for high-temperature hot water systems); and swimming/therapeutic pools (for heating load on central plant).

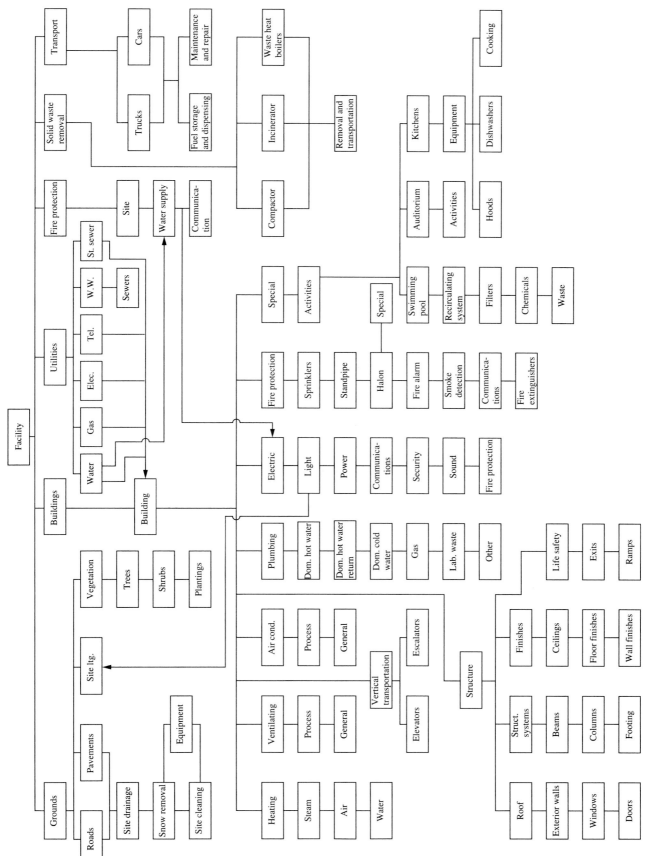

FIGURE 31.1 Static and dynamic building systems requiring commissioning.

7. *Life safety and security systems.* Smoke control system, fire standpipe system, sprinkler systems, fire suppression systems, fire alarm systems, heat detection systems, smoke detection systems, intrusion alarm systems, closed-circuit TV systems, CO_2 systems, Halon fire suppression systems, etc.

31.3 COMMISSIONING

Building Commissioning

The commissioning process begins with the owner's decision that a building is required (see Fig. 31.2). ASHRAE *Terminology of Heating, Ventilating, Air Conditioning, and Refrigeration* second edition, defines commissioning as follows: "commissioning (of an installation) process for achieving, verifying, and documenting the performance of buildings to meet the operational needs of the building within the capabilities of the design and to meet the design documentation and the owner's functional criteria, including preparation of operator personnel." Commissioning applies not only to HVAC&R systems, but also to all building systems (see Fig. 31.1). The commissioning philosophy must include all building systems as well as the building as a total system if the building is to satisfy the users' needs.

It is necessary to interface HVAC&R requirements with all other building systems, such as the plumbing, electrical, fire protection, life safety, and security systems, and to the building itself (materials, construction, building envelope, etc.). Figure 31.1 represents an overall flow sheet of those items at a typical facility that may require total facility and total building commissioning.

HVAC&R Commissioning

For the purposes of this chapter, HVAC commissioning shall be defined as the process, procedures, and methods for documenting, achieving, and verifying that the HVAC&R systems conform to the design intent and the users' needs.

Many HVAC&R engineers have generally defined HVAC&R commissioning as the advancement of an installation from the stage of static completion to full working order to specified requirements. Commissioning includes the setting to work and regulation of an installation. CIBS (Loten, 1977) used the following definitions:

1. *Setting to work.* The process of setting a static system into motion

2. *Regulation.* The process of adjusting the rates of fluid flow in a distribution system within specified tolerances

Although the definitions may differ, the implications to the HVAC&R engineers are the same, namely, that a greater involvement and a wider range of HVAC&R engineering services are required in the HVAC&R commissioning process than those normally provided under the basic engineering services. At the present time the work required by the commissioning process is covered in the *Additional Services* section of the standard engineer/owner contracts.

Usually the HVAC&R design engineer has the responsibility to formulate the management techniques and procedures to ensure that the HVAC&R systems perform to the design intent and serve the owner's needs. Typical HVAC&R engineering design drawings and project specifications should include the following: basis of design; design assumptions; description of the operational performance requirements of each HVAC&R system and subsystem; requirements for equipment manufacturers to provide (certified test data, installation instructions, and verification that the equipment is installed properly and operating properly); equipment operational and performance requirements; HVAC system configuration, metering, instrumentation, and access to provide for testing and balancing; testing, adjusting, and balancing requirements; commissioning plan; commissioning documentation; training of operations and maintenance staff; maintenance and service contracts; and maintenance and operations manuals. All the listed management procedures form part of the commissioning process. *ASHRAE Guideline 1–1989— Guideline for Commissioning of HVAC Systems* provides procedures and methods for documenting and verifying the performance of HVAC&R systems so that systems operate in accordance with the design intent.

One of many possible commissioning process arrangements is illustrated in Fig. 31.2. Design professionals management and communications procedures for the commissioning of large HVAC&R systems are presented in Fig. 31.3a, b, c, and d.

31.4 HVAC&R COMMISSIONING PROCESS

Usually the commissioning process of HVAC&R systems is part of the larger process of development and construction of buildings and facilities. For purposes of analysis, the commissioning process can be thought of as being an additional part of the standard process in the traditional design and construction process. The steps are these: project need, selection of a design team, preparation of a functional program, schematic design phase, design development phase, construction document preparation phase, bidding

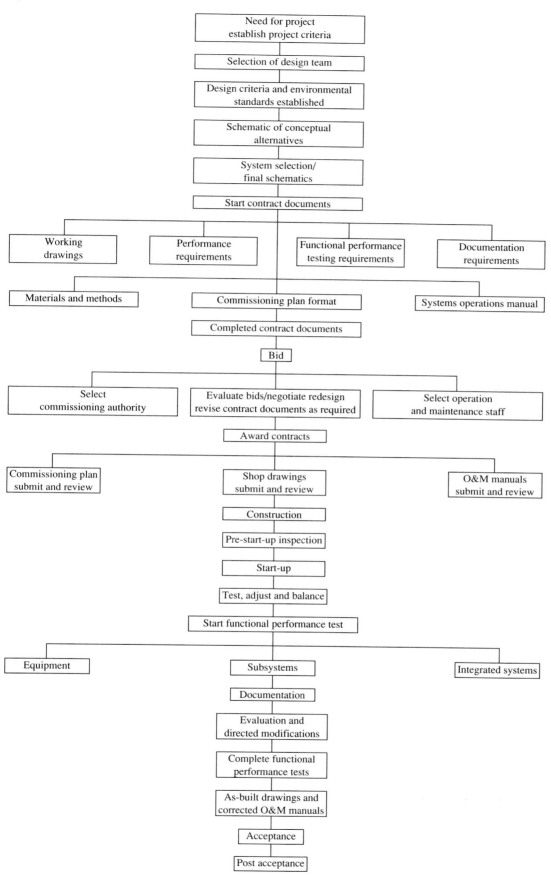

FIGURE 31.2 Typical commissioning process arrangements. *Source*: *ASHRAE Guidelines 1-1989*. Reprinted by permission.

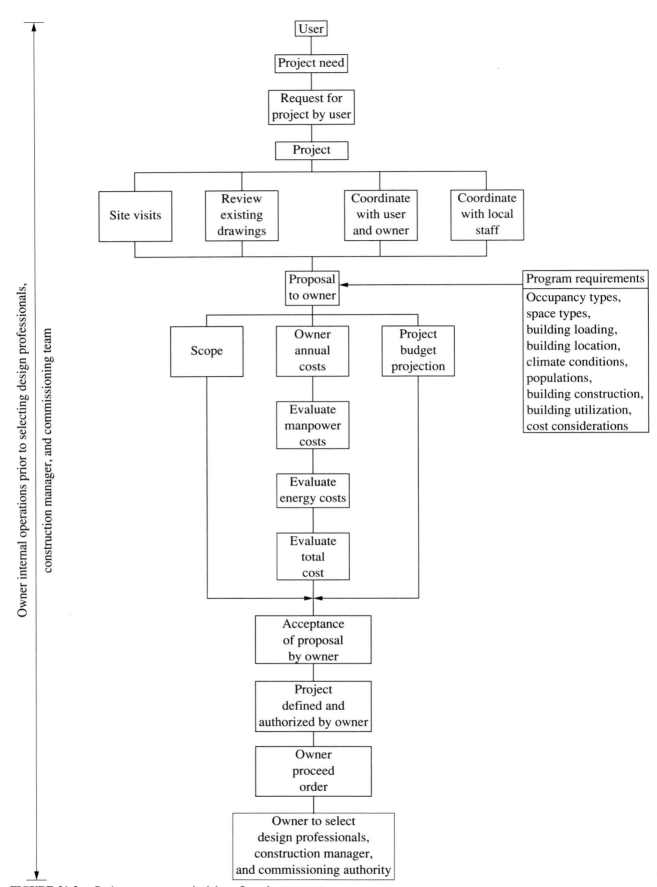

FIGURE 31.3a Project sequence methodology flow sheet.

31.6

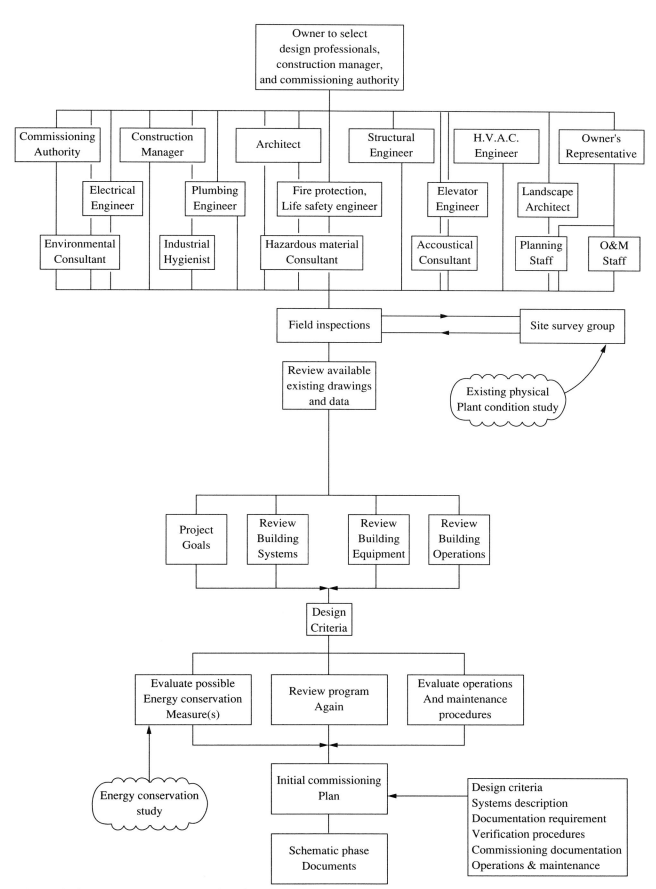

FIGURE 31.3b Project sequence methodology flow sheet.

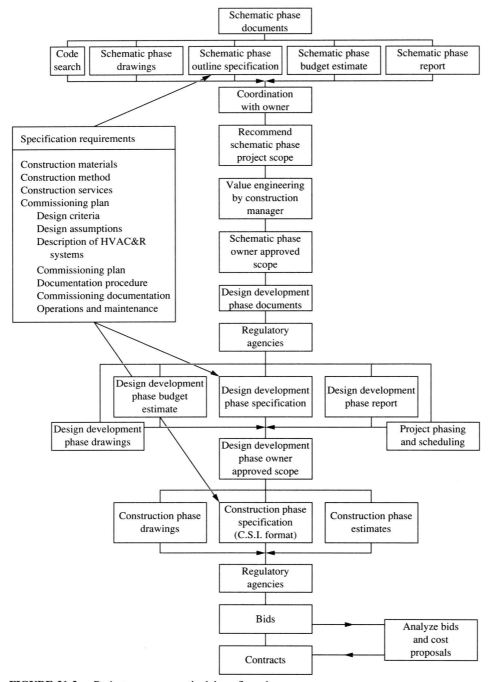

FIGURE 31.3c Project sequence methodology flow sheet.

31.8

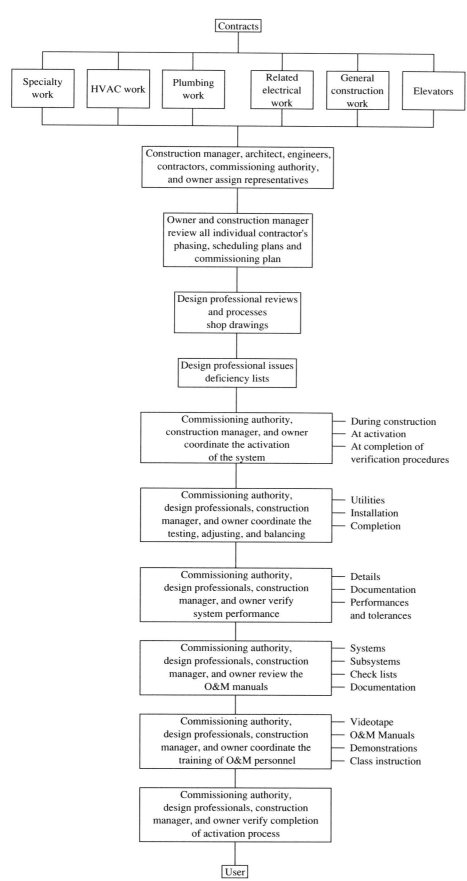

FIGURE 31.3d Project sequence methodology flow sheet.

phase, construction phase, start-up of HVAC&R systems, testing and balancing, acceptance by owner, occupancy by owner, and adjustments to suit occupants (see Fig. 31.3a, b, c, and d.)

31.5 THE HVAC&R ENGINEER'S PARTICIPATION IN THE COMMISSIONING PROCESS

The participation of the HVAC&R engineer in the commissioning process can range from the normal professional services as they presently exist in the construction industry to a *wide range* of HVAC&R commissioning services. The degree of involvement would naturally depend on the size and complexity of the HVAC&R systems, and the contractual agreements between all parties involved in the development of the building or facility.

The role of the HVAC&R engineer in the commissioning process is mainly a management function. The HVAC&R engineer designs, manages, and coordinates the procedures in the HVAC&R commissioning process, including its logical sequence. The HVAC&R engineer's design intent and the owner's needs must be communicated so that the design intent can be implemented and the owner's needs satisfied. The process of communicating and implementing the design intent by the HVAC&R engineer could consist of some and possibly all of the following activities:

1. In concert with the owner and other members of the design team, the HVAC&R engineer establishes the design intent.

2. The HVAC&R engineer prepares contract documents, drawings, specifications, commissioning requirements, etc.

3. The HVAC&R engineer prepares drawings and specifications that set forth the following:

   a. Design criteria for each and every space and for each and every system

   b. Materials, methods, equipment, and appurtenances required to meet the design criteria, including all meters, gauges, devices, and the HVAC&R system configuration required for the effective testing, adjusting, and balancing of the HVAC&R system

   c. Description of the HVAC&R systems operation and performance requirements

   d. A proposed HVAC&R commissioning plan format, including the complete specifications for the HVAC&R commissioning requirements to be accomplished by the various parties involved in the process (owners, contractors, vendors, project manager, manufacturers, owner's operations and maintenance staff, and users of the HVAC&R systems, etc.)

   e. Procedures for securing verification of the performance of HVAC&R systems at the completion of the testing, balancing, and adjusting process and for specifying check lists

   f. The required commissioning documentation to be prepared by the various contractors, subcontractors, installers, vendors, manufacturers, including the following:
      - Testing, adjusting and balancing
      - Control schematics and verification of the total HVAC system and HVAC subsystems
      - Performance of all equipment
      - Operating data, which shall include all necessary instructions to the owner's maintenance and operating staff in order to operate the system to specified performance standards
      - Maintenance data, which shall include all information required to maintain all equipment in continuous operation

   g. If necessary, identification and specification of the needed staffing, skill levels, and the means of introducing new skills to the following sectors:
      - Design team itself
      - Construction teams
      - Commissioning teams
      - Testing, adjusting, and balancing agency
      - Operation and maintenance teams

4. The HVAC&R engineer evaluates bids and makes suggestions for redesign, if budget restraints require this. The HVAC&R engineer revises the commissioning specifications and commissioning process format to satisfy design changes.

5. The HVAC&R engineer assigns a representative to work with the contractor, owner, and commissioning agency during the commissioning process. The commissioning agency or commissioning is defined by ASHRAE as follows: **commissioning authority**—qualified person, company, or agency that will plan and carry out the overall commissioning process.

6. The HVAC&R engineer reviews and processes shop drawings and submissions.

7. The HVAC&R engineer issues deficiency lists during the construction phase, at the time of systems activation, and on completion of the systems performance verification procedures.

8. The HVAC&R engineer reviews the commissioning agency's commissioning plan. ASHRAE

defines the commissioning plan as follows: **commissioning plan**—overall document, usually prepared by the commissioning authority, that outlines the organization, scheduling, allocation of resources, documentation, etc., pertaining to the overall commissioning process. The commissioning plan may be prepared by the commissioning agency or the HVAC&R engineer. It should detail how the commissioning process will be organized, scheduled, and documented. The plan shall set forth these criteria:

a. The organization of the HVAC&R contractor's commission team (installers, vendors, manufacturers; testing, balancing, and adjusting specialists; owner's operations and maintenance staff)

b. A list of activities required to commission the HVAC&R systems and their subsystems and the responsibilities of each member of the commissioning team

c. A logical sequence schedule for each commissioning activity coordinated with all members of the commissioning team

9. The HVAC&R engineer coordinates the activation of the HVAC&R systems when the installation is completed, the utility services (i.e., electrical power, water, and fuel) have been connected and are available for use, and the HVAC&R system is operational and ready for testing and balancing. The HVAC&R engineer verifies that the HVAC&R systems are ready for start-up, testing, adjusting, and balancing.

10. Testing, balancing, and adjusting work should be performed by an experienced organization subject to the approval of the HVAC&R engineer.

a. The scope of the testing, adjusting, and balancing work should be set forth in the HVAC&R specifications and on the HVAC&R drawings and should include these features:

- Details of all systems to be tested, adjusted and balanced

- Performance requirements and tolerances

- Required documentation

b. At the conclusion of the construction and installation of the HVAC&R systems and before commissioning the systems, testing, adjusting and balancing must be accomplished. HVAC&R system testing, adjusting, and balancing are the processes of checking and adjusting all building environmental systems to produce design objectives. These processes include the following:

- The balance of air and water systems
- Adjustment of the total system to provide design quantities
- Electrical measurements
- Verification of the performance of all equipment and automatic controls
- Sound and vibration measurements. These are accomplished as follows:
  - Checking for conformity of design
  - Measuring and establishing of fluid quantities of the system required to meet the design specifications
  - Recording and reporting the results

Inasmuch as the activities within buildings change over the life of the buildings, testing and balancing of HVAC&R systems may be required with each modification to the HVAC&R systems. *ANSI/ASHRAE Standard 111–1988 Practices for Measurement, Testing, Adjusting and Balancing of Building Heating, Ventilating, and Air Conditioning and Refrigeration Systems* provides these guidelines:

- Uniform and systematic procedures for making measurements in testing, adjusting, balancing, and reporting the performance of building heating, ventilation, air conditioning, and refrigeration systems in the field

- Means of evaluating the validity of collected data considering system effects

- Establishment of methods, procedures, and recommendations for providing field-collected data to designers, users, manufacturers, and installers of the systems

It is normal engineering practice to include the Standards of the AABC, NEBB and SMACNA in the HVAC&R project specifications as reference documents for use by the testing and balancing agency.

The ANSI/ASHRAE Standard 111–1988 and the ASHRAE Handbook *Fundamentals* volume recognize that accurate air flow measurements are difficult to obtain and are subject to the ability, diligence, and expertise of the testing and balancing agency, the time the testing and balancing agency is willing to expend, as well as the accuracy and applicability of equipment and air terminal manufacturers test data.

c. The HVAC&R engineering design calculations, drawings, and specifications should include provisions for implementation of the testing and balancing procedures. That is, the HVAC&R's system configuration should have been designed complete with arrangements for straight runs of piping and ductwork for testing and balancing, metering for testing and balancing, instrumentation for testing and balancing, and access for testing and balancing.

d. The HVAC&R engineering design calculations, drawings and specifications for air-handling systems should include the following:

- Make provisions to clean, test, adjust, and balance the system (primary access doors in ductwork, plenums, casings, and secondary access doors in the building walls, floors, and ceilings, for access to primary access doors, etc.)

- Provide proper balancing devices and flow control devices at the proper locations in the correctly configured ductwork system. (See the SMACNA duct standards.)

- Design, specify, and verify that the ductwork is installed to minimize air leakage (See SMACNA HVAC Duct Leakage Test Manual.) *The Architect should design, specify, and verify that plenums formed by building surfaces are air tight and all penetrations for pipes, conduits and ductwork are sealed and firestopped.*

- Design, specify, and verify that all frictional and dynamic losses and system effects are accounted for in the selection of the fan(s) and in the installation of the fan and ductwork including entries, exits, elbows, tees, transitions, obstructions, filters, coils, humidifiers, volume dampers, louvers, fire dampers, smoke dampers, fan entry effects, fan exit effects as described in Chapters 8 and 19.

- Design, specify, and verify that all Pitot tube taps in ductwork and air metering devices are installed in conformance with the manufacturer's recommendations (inasmuch as the preferred method of duct volumetric flow measurements is by Pitot tube average) and that the length of straight run before and after the device is adequate for accurate readings. If straight runs cannot be provided in adequate lengths, one should use the manufacturers' recommended straightening devices or apply cor-

rection factors. The system frictional and dynamic losses must include the pressure loss through the meter assembly and its companion ductwork.

e. The HVAC&R engineering design calculations, drawings, and specifications should generally include the following:

- Provisions to clean, test, adjust, and balance the hydronic system (access doors in ceilings, walls, and floors, drains and vents in piping, access around piping to work, clean strainers, etc.)

- Provisions for proper balancing devices at the proper locations in correctly configured piping arrangements

- Provisions and verification that the piping pressure has tested free of leaks

- Provisions for water conditioning treatment, surveillance, and verification that the water treatment is properly installed (to avoid corrosion and fouling of meters and balancing valves)

- All required metering for testing, adjusting, and balancing

11. The HVAC&R engineer should perform system performance verification in accordance with the system and subsystem performance check list that was included in the HVAC&R specifications. The verification procedures may include these steps:

   a. The engineer will witness testing procedures and verify that all HVAC&R systems and HVAC&R subsystems perform to specified requirements, including design requirements that were changed during the project construction period.

   b. The engineer will review the verified check lists submitted by the contractor.

   c. The HVAC&R engineer and contractor shall jointly make evaluations and corrections of problem conditions where an HVAC&R system, subsystem, or equipment component does not achieve specified performance standards.

   d. The engineer will review the recorded data that are submitted by the HVAC contractor for each of the seasonal operating conditions (winter, summer, and intermediate seasons) in both the occupied and unoccupied modes.

12. The HVAC&R engineer reviews the preparation of operations and maintenance manuals or in some cases may prepare operating and mainte-

nance manuals (procedures for owner/operators). (See Chapter 32 for examples of maintenance manual data.) The design professional coordinates the integration of the component maintenance manuals into a comprehensive maintenance plan and reviews and approves the maintenance plan.

13. The HVAC&R engineer coordinates the training of the owner's operations and maintenance staff. The training will be provided by contractor's forces, vendors, and manufacturers. The HVAC&R engineer generally provides the system training format to be implemented by the contractor. The design professional in concert with the commissioning team shall recommend the composition of the owner's operating and maintenance staff for the owner's consideration. The operations and maintenance staff shall be organized and scheduled for introduction to the facility over a time period that will normally be a function of the size and complexity of the project. The demonstrations, operations and maintenance instructions shall include the following:

   a. A detailed schedule of demonstration and instruction periods for specific portions of the installation with a schedule that shall indicate the parties involved, when they are to be involved, and the extent to which they are involved

   b. Instructions pertaining to format and use of the operating and maintenance manuals, with a schedule that shall set forth the parties involved, when they are to be involved and the extent to which they are involved

   c. Assignment of instruction to specialist members of the commissioning team who were involved in the installation and are familiar with the details of the equipment and HVAC systems, with a schedule that shall set forth the people involved, when they are to be involved, and the extent to which they are involved

   d. The time periods required to provide the demonstration of the systems, operations and maintenance instructions

14. The HVAC&R engineer verifies the completion of the commissioning process.

31.6 CONCLUSION

Although the commissioning process is the concerted effort of the design professionals, owners, contractors, manufacturers, vendors, and testing and bal-

ancing agencies, the responsibility for satisfying the user's needs and implementing a successful HVAC&R system rests with the HVAC&R design professional.

REFERENCES AND FURTHER READING

American Institute of Architects Service Corporation, *Masterspec Evaluations Section 15950—Testing Adjusting and Balancing,* American Institute of Architects, Washington, DC, 1986.

ANSI/ASHRAE, *ANSI/ASHRAE 111–1988—An American National Standard—Practices for Measurement, Testing, Adjusting, and Balancing of Building Heating, Ventilation, Air Conditioning, and Refrigeration Systems,* ASHRAE, Inc., Atlanta, GA, 1988.

Armstrong Pumps, *Technology of Balancing Hydronic Heating and Cooling Systems,* Tour and Anderson, Inc., North Tonnwanda, NY, 1988.

ASHRAE, "Commissioning—Construction Phase," *ASHRAE Transactions,* 1989.

ASHRAE, "Commissioning: Why We Need It—What Are the Benefits?" *ASHRAE Transactions,* 1989.

ASHRAE, "Designing Healthy Buildings—The Architect's Role in the Commissioning Process," *ASHRAE Transactions,* 1989.

ASHRAE, "The HVAC Commissioning Plan," *ASHRAE Transactions,* 1989.

ASHRAE, "An Owner's Approach for Effective Operations," *ASHRAE Transactions,* 1989.

ASHRAE, "Specifying HVAC Systems Commissioning," *ASHRAE Transactions,* 1989.

ASHRAE, *ASHRAE Guideline 1–1989—Guideline for Commissioning of HVAC Systems,* ASHRAE, Inc., Atlanta, GA, 1989.

ASHRAE, *ASHRAE Handbook and Fundamentals I-P Edition,* Chapter 13, "Measurements and Instruments," ASHRAE, Inc., Atlanta, GA, 1989.

ASHRAE, *ASHRAE Systems Handbook 1984,* Chapter 37, "Testing, Adjusting, and Balancing," ASHRAE, Inc., Atlanta, GA, 1984.

ASHRAE, *ASHRAE Terminology of Heating, Ventilating, Air Conditioning, and Refrigeration,* 2nd ed., ASHRAE, Inc., Atlanta, GA, 1991.

ASHRAE, "Building Commissioning—A Contractor's Viewpoint," *ASHRAE Transactions,* 1986.

ASHRAE, "Commissioning and Indoor Air Quality," *ASHRAE Transactions,* 1988.

ASHRAE, "Effective Methods of Building Commissioning Documentation," *ASHRAE Transactions,* 1986.

ASHRAE, "Effective O&M Requires Coordination Among the Owner, Occupant, Designer, Builder, and O&M Engineer During Design," *ASHRAE Transactions,* 1990.

ASHRAE, "Specifying Commissioning for Building HVAC Systems," *ASHRAE Transactions,* 1986.

Associated Air Balance Council, *National Standards for Testing and Balancing Heating, Ventilating, and Air Conditioning Systems,* 5th ed., AABC, Washington, DC, 1989.

Loten, A. W. "Testing and Commissioning of Building Services Installations," Symposium, An introduction to Testing and Commissioning, Chartered Institution of Building Services, United Kingdom, 1977.

National Environmental Balancing Bureau, *Procedural Standards for Testing, Adjusting and Balancing HVAC Systems,* NEBB, Chantilly, VA, 1983.

SMACNA, *HVAC Air Duct Leakage Test Manual,* 1st ed., SMACNA, Chantilly, VA, 1985.

SMACNA, *HVAC Duct Construction Standards, Metal and Flexible,* 1st ed., SMACNA, Chantilly, VA, 1985.

SMACNA, *HVAC Systems, Testing, Adjusting and Balancing,* 1st ed., SMACNA, Chantilly, VA, 1983.

CHAPTER 32

THE HVAC&R ENGINEER'S RESPONSIBILITY FOR MAINTENANCE PROGRAM DURING THE DESIGN STAGE

32.1 INTRODUCTION

HVAC&R engineering is an art based on science that includes programming; planning; calculations; analysis between alternatives; preparation of cost estimates; preparations of drawings, specifications, and contract documents; analysis of bids; and monitoring the construction, installation, and commissioning of systems.

HVAC&R engineering services are intended to result in a completed product that will satisfy the users' needs. Comfort air conditioning systems and process air conditioning systems are intended to satisfy the design requirements set forth in the owner's program for temperature, humidity, air motion, indoor air quality, energy management, life safety, fire and smoke control, condensation, surface temperature, and controls during the entire service life of the HVAC&R systems. The systems should be designed to be flexible, durable, reliable, serviceable, and maintainable.

The HVAC&R engineer usually plans and designs the HVAC&R systems to be consistent with the programmed service life of the building. If the equipment and materials included in the HVAC&R design have service lives different from the building, they should be selected and arranged so that they can be modified or replaced without major alterations or additions to the building itself. The HVAC&R design normally includes provisions for the removal and installation of future replacements.

32.2 HVAC&R SYSTEM DETERIORATION

It is the responsibility of the HVAC&R engineer to select materials and equipment that minimize the de-

terioration of the system components. The design usually includes measures to counter environmental factors such as the effects of local microclimates, topography, snow, wind, rain, ice, frost, temperature, humidity, fire, smoke, noise, vibration, radon, flood, earthquake, structural and thermal movement, chemical action and corrosion, and atmospheric pollution. Countermeasures to environmental conditions usually consist of the design and selection of durable materials and equipment to overcome the anticipated environmental conditions. Eventually, replacement, modification, addition, or major repair is required. (For example, cooling towers located in most large cities require major repairs or replacement within 15 years of original installation.) The skill, education, and experience of the HVAC&R designer affect the rate at which the HVAC&R system components deteriorate. HVAC&R designers must demonstrate their expertise in the selection and specification of HVAC&R materials, equipment, and system design and in the monitoring of the installation to minimize their deterioration. The selection, specification, and monitoring process usually involves a compromise between initial cost and life-cycle cost.

The rate of deterioration is also a function of the use or misuse of the HVAC&R systems and of the wear and tear on the system by the users and the public. In recent years, respect for property by users and the general public has declined. The effects of vandalism and the lack of concern regarding property introduce an element difficult to design into HVAC&R systems. The designer should be aware of the problem of user wear and tear and add it to the matrix of considerations to be factored into the design of the systems.

The quality of HVAC&R project construction has an ongoing effect on the rate of system deterioration. The quality of the initial construction, as well as that of subsequent repair, replacement, modification, additions, and ongoing maintenance, is a major factor in the rate of deterioration of HVAC&R systems. It requires diligence, skill, and high standards of design and construction. The standards of engineering are set forth by professional training. The standards of mechanical/electrical construction are usually defined by the unions and technical training programs. The standards of maintenance for HVAC&R system components are not as standardized as for mechanical/electrical construction, and in most cases the maintenance work covers many trades; therefore, maintenance workers cannot be expected to be as efficient as the mechanical and electrical construction workers. Materials, equipment, and designs used in the systems must be selected and interfaced within the systems so that they perform in a reliable and safe manner, under the environmental conditions imposed, with normal maintenance programs throughout their service lives.

The primary means of reducing the rate of deterioration is for the owner to formulate and institute an on-going, effective maintenance program. It is the responsibility of the HVAC&R engineer to design the means, methods, and programs for the maintenance of the systems.

32.3 MAINTENANCE DEFINITIONS

HVAC&R maintenance is the work required to preserve or restore HVAC&R systems, including materials, equipment, and appurtenances, to their original conditions such that they can be effectively used for their designated purpose, perform to the design intent, and serve in a durable, reliable, and safe manner. HVAC&R maintenance consists of routine procedures designed to ensure full service life of equipment and to prevent malfunctions. The maintenance program should be formulated in the initial design. The main goal of maintenance is to prevent deterioration and malfunction of equipment, decrease shutdowns, prevent life safety problems, reduce operating costs, enhance energy efficiency, and extend the life of the systems.

In ASHRAE Handbook 1987, HVAC Systems and Applications, Chapter 59, ASHRAE sets forth the following definitions:

"Maintenance program" defines the maintenance concept in terms of time and resource allocation. It documents the objectives and establishes the criteria for evaluation of the maintenance department to basic areas of performance such as (1) prompt response to mechanical

failure and (2) attention to planned functions that protect a capital investment and minimize the downtime or failure response.

"Failure response" classifies maintenance department resources expended or reserved for dealing with interruptions in their operation or function of a system or equipment under the maintenance program. This classification has two types of responses: repair and service.

"Repair" is defined as "to make good, or to restore to good or sound condition," with the following constraints: (1) operation must be fully restored without embellishment and (2) failure must trigger the response.

"Service" is the provision of what is necessary to effect a maintenance program short of repair. It is usually based on manufacturer's recommended procedures.

"Planned maintenance" classifies maintenance department resources that are invested in prudently selected functions at specified intervals. All functions and resources attributed to this classification must be planned, budgeted, and scheduled. It embodies two concepts: preventive and corrective maintenance. "The Preventive Maintenance" concept classifies resources allotted to ensure proper operation of a system or equipment under the maintenance program.

Durability, reliability, efficiency, and safety are the principal objectives.

"Corrective maintenance" classifies resources, expended or reserved, for predicting and correcting conditions of impending failure. Corrective action is strictly remedial and always performed before failure occurs. An identical procedure performed in response to failure is classified as a repair, providing it is optional and unrelated.

"Predictive maintenance" is a function of corrective maintenance. Statistically supported objective judgment is implied. Nondestructive testing, chemical analysis, vibration and noise monitoring, as well as routine visual inspection and logging are all classified under this function, providing that the item tested or inspected is part of the planned maintenance program.

Crisis maintenance can be defined as maintenance arising out of unforeseen and nonpredictable emergencies such as fire, flood, loss of utilities, electrical failures, vandalism, acts of terrorism, and riots. In the past, crisis maintenance was not the major concern that it currently is. Crisis maintenance can result in major HVAC&R system downtime, resulting in the inability to use the facility.

Deferred maintenance, as defined by the Association of Physical Plant Administrators (APPA) of Universities and Colleges (see Dillow, 1989), consists of "maintenance projects that were not included in the maintenance process because of a perceived lower priority status than those funded within available funding. Deferred maintenance comprises two categories of unfunded maintenance: first, the lack of which does not cause the facility to deteriorate further; and second, the lack of which does result in a progressive deterioration of the facility for the current function."

According to the *ASHRAE Handbook 1987, HVAC Systems and Applications*, "The economic merit of every system or facility is a function of its durability, reliability, and maintainability." It is difficult to focus on true cost when future qualitative values must be equated with the immediate initial cost. These qualities can be quantified as follows:

- *Durability* is the average expected service life of a system or facility. Table 32.1 shows the service life of various kinds of equipment in terms of median years. It is more specifically quantified by individual manufacturers as *design life,* which is the average number of hours of operation before failure, extrapolated from accelerated life tests and stressing critical components to economic destruction.
- *Reliability* implies that a system or facility will perform its intended function for a specified period of time without failure.
- *Maintainability* complements reliability by defining the specific time that a system or facility can operate to an economically fully restored condition.

32.4 SERVICE LIFE

Service life is a time value established by ASHRAE that reflects the expected life of a specific component. ASHRAE defines service life as the median time during which a particular system or component remains in its original service application and then is replaced. When dealing with deterioration and physical obsolescence of HVAC&R materials and equipment, service life is sometimes defined as the number of years before the total replacement of the system component is likely to be initiated. Akalin (1978) concluded that HVAC&R equipment service life is generally influenced by the following main factors:

- Replacement with an identical item becomes less costly than continued maintenance and repair
- Replacement with an identical item becomes necessary to ensure reliability and safety.

The service lives of HVAC&R equipment can be derived from ASHRAE, manufacturers, contractors, and National Bureau of Standards (NBS) records. Engineering judgment should be used only when good service life data are not available from any of these

TABLE 32.1
Estimates of service lives of various system components

Equipment Item	Median Years	Equipment Item	Median Years	Equipment Item	Median Years
Air conditioners		Air terminals		Air-cooled condensers ...	20
Window unit	10	Diffusers, grilles, and registers .	27	Evaporative condensers ..	20
Residential single or split package ..	15	Induction and fan-coil units	20	Insulation	
Commercial through-the-wall	15	VAV and double-duct boxes	20	Molded	20
Water-cooled package	15	Air washers	17	Blanket	24
Heat pumps		Ductwork	30	Pumps	
Residential air-to-air	15	Dampers	20	Base-mounted	20
Commercial air-to-air	15	Fans		Pipe-mounted	10
Commercial water-to-air	19	Centrifugal	25	Sump and well	10
Roof-top air conditioners		Axial	20	Condensate	15
Single-zone	15	Propeller	15	Reciprocating engines....	20
Multizone	15	Ventilating roof-mounted	20	Steam turbines	30
Boilers, hot water (steam)		Coils		Electric motors	18
Steel water-tube	24 (30)	DX, water, or steam	20	Motor starters	17
Steel fire-tube	25 (25)	Electric	15	Electric transformers	30
Cast iron	35 (30)	Heat exchangers		Controls	
Electric	15	Shell-and-tube	24	Pneumatic	20
Burners	21	Reciprocating compressors	20	Electric	16
Furnaces		Package chillers		Electronic	15
Gas- or oil-fired	18	Reciprocating	20	Valve actuators	
Unit heaters		Centrifugal	23	Hydraulic	15
Gas or electric	13	Absorption	23	Pneumatic	20
Hot water or steam	20	Cooling towers		Self-contained	10
Radiant heaters		Galvanized metal	20		
Electric	10	Wood	20		
Hot water or steam	25	Ceramic	34		

Source: ASHRAE Handbook 1991, Applications. Reprinted with permission. Data obtained from a national survey of the United States by ASHRAE Technical Committee TC 1.8 (Akalin, 1978). Data updated by TC 1.8 in 1986.

sources. The statistical service life should determine maintenance planning and programming for replacement, major upgrading, or modernization.

During the design process, the HVAC&R engineer should determine the service life and maintenance requirements for all system components and design the systems with consideration for the future.

32.5 CURRENT HVAC&R PROBLEMS DUE TO LACK OF MAINTENANCE

Indoor Air Quality

Many of the problems caused by inadequate performance of HVAC&R systems arise from the lack of an effective on-going maintenance program. Indoor air quality problems and sick building syndrome problems are caused, in part, by either poor maintenance programs or the lack of maintenance programs. Air contaminants are generated either within the building or external to the building and introduced into the building. The air contaminants take the form of dusts, fumes, smoke, particulates, gases, vapors, mists, fogs, radon, refrigerants, viruses, bacteria, organic vapors, and fibers. The source of the air contaminants are combustion processes (vehicles, boilers, furnaces, and domestic water heaters), building materials (insulation, wallboard, rugs, carpets, paints, fabrics), people, plants, and equipment.

The ventilation rates set forth in ASHRAE Standard 61–1989, *Ventilation for Acceptable Indoor Air Quality*, are designed to dilute the air contaminants to an acceptable level, but only an on-going effective maintenance program can ensure that the ventilation rates will be maintained at the required quantities. The maintenance program should continuously monitor and maintain the following:

- Outside air intake, recirculation and exhaust air dampers, and their minimum position
- Outside air intake air quality, return air quality, exhaust air quality, and supply air quality
- Cleanliness of intake plenum, air-handling unit casing, air-handling unit coils, air-handling unit fans, casings, and components
- Filter loading performance and condition
- Friable acoustic lining of ductwork and sound attenuation
- Cooling coil drain pan cleanliness, bacteria count, and drain line cleanliness
- Cleanliness of ductwork, sheet metal, and air distribution system components
- Water treatment and monitoring program for cooling towers and evaporative condensers
- System air quantities (on-going testing and balancing program)
- System carbon dioxide build up (CO_2 detectors)

The implementation of an on-going maintenance program to ensure an acceptable indoor air quality requires that the HVAC&R engineer include the following items as part of the system design:

- Direct and remote central console readings of minimum outdoor air quantities introduced into the building using air flow monitoring meters with notification alarms
- Direct and remote central console readings of minimum exhaust air quantities using air flow monitoring meters with notification alarms.
- Filter static pressure differential-sensing through automatic control systems and notification alarms and remote console notification
- Air-handling unit air quantity readings using air flow monitoring meters with local and remote console notification alarms
- Automated water treatment system results and notification alarms
- Volatile organic compound (VOC) concentration as CO_2 and notification alarms with local and remote console alarms
- Methodology to monitor the cooling coil bacteria count

Ineffective Automatic Controls and Automatic Control System Components

Control system characteristics and sequences of operation are described in Chapters 17 and 18. The effective and reliable functioning of an HVAC&R system depends on the proper installation, calibration, commissioning, and on-going maintenance of the automatic control systems. Automatic control system components and their sequence of operations are becoming more and more comprehensive and complex. Automatic control systems of the electric, electric-electronic, pneumatic, pneumatic-electric, and DDC types and pneumatic actuators with DDC systems and complete combinations of these are the systems in common use today. The selection of the type of system for a particular project usually goes beyond the basic technical considerations of effective performance under all modes of operation conditions, energy conservation, and reliable operation. System types and system manufacturers are often selected on the basis of the quality and responsiveness of their service organizations and their record of technical assistance. Most owners do not have maintenance staff personnel with experience in the multiple disciplines that automatic control technicians must have in order to service and maintain electronic, electric, DDC, and pneumatic equipment.

Most owners cannot afford to sustain personnel, parts inventory, tools, equipment, and on-going training programs for an effective automatic control maintenance staff. It is usually most economic to enter into a service contract with an automatic control maintenance service organization.

Safety and Health Concerns

The design of HVAC&R systems and the establishment of an effective maintenance program must include provisions to protect the health and safety of the maintenance staff and the public. The current legal system in the United States exposes HVAC&R engineers, product manufacturers, vendors, and service companies to liability for health and life safety problems arising out of the design, operation, and maintenance of HVAC&R systems. To provide HVAC&R systems that address the safety and health of maintenance workers and the public and that limit the legal liability of the HVAC&R engineer, the engineer must have a detailed, up-to-date working knowledge of all local, state, and federal laws, codes, ordinances, rules, regulations, and industry standards. The application of health and safety considerations to the design, installation, commissioning, and maintenance of HVAC&R systems involves use of industrial hygiene principles. These principles include identification and control of potential health hazards and the countermeasures to reduce, inhibit, or eliminate potential risks to health and safety.

As part of the design, construction, commissioning, and maintenance program functions, the HVAC&R engineer should carry out the following actions:

1. Identify and control contaminants, particulates, gases, vapors, aerosols, and refrigerants
2. Identify and control sound, noise, and vibration
3. Identify and control toxic and hazardous materials
4. Identify and control hazardous effluents from laboratory fume hoods, medical ventilation systems, industrial exhaust systems, chimneys, and gas engine exhausts
5. Identify and control water effluents from HVAC&R systems
6. Identify and provide proper levels of illumination for maintenance workplace safety
7. Identify and provide systems to inhibit heat stress on maintenance workers
8. Identify and include accepted ergonomic standards as part of the maintenance workplace

32.6 MAINTENANCE PROGRAM REQUIREMENTS

The *ASHRAE 1987 Applications Handbook* recommends that maintenance programs cover five areas.

Maintenance Management Policy

An effective maintenance management policy should be formulated, which defines the objectives and type of program, provides for organizing and staffing, and directs and controls its effectiveness, performance, and cost.

Detailed Inventory Records

Detailed inventories and records of the systems, system components, equipment, equipment components, and controls should be specified, collated, and available for use. These should include the following:

1. Construction drawings and specifications, including design and commissioning data
2. Records of as-built drawings
3. Microfilms of as-built drawings and specifications
4. Documentation of system changes or enhancements into as-is drawings and specifications
5. Shop drawings and equipment catalogs
6. Equipment installation, service, and maintenance instructions; troubleshooting checklists; and spare parts lists
7. Outside maintenance service organizations and spare parts sources
8. Inventory of spare parts
9. Valve charts and system flow diagrams
10. Videotapes of operations and maintenance instructions

Required Procedures and Schedules

The setting forth of the required procedures and schedules is the action part of the program, relating to operation, inspection, service, repair, and replacement. It involves the following minimum requirements:

1. Operating instructions
   a. Starting and stopping procedures, sequences, and frequency

  b. Adjustment of startup and shutdown

  c. Seasonal startup and shutdown

  d. Seasonal changeover

  e. Logging and recording

   NOTE: Log sheets should be prepared regularly (hourly in a large system) and include the following:

   1. Energy consumption and demand

   2. All air, fuel, and water temperatures, including ambient

   3. All water, fuel, and refrigerant pressures

   4. Vane openings and power limits

2. Inspection

  a. Equipment to be inspected

  b. Points to be inspected

  c. What is to be done as part of the inspection procedure.

  d. Frequency of inspection

3. Service and repair

  a. Frequency of scheduled service

  b. Exactly what is required as part of the service and maintenance procedures

  c. Scheduled service procedures

  d. Repair procedures

  e. Recording and reporting

Monitoring of Data

Data obtained by monitoring (whether obtained manually or by a computerized operations and maintenance system) can be used to predict operational and maintenance trends and uncover *temporary* system service malfunctions that require repairs, replacements, adjustments, or other modifications. Over the long term, they may predict the probable failure of an important component. Alternatively, they may discover neglect of system leaks or the need for adjustments to valves, dampers, or linkages; cleaning of coils or fan wheels; or water treatment.

Operations and Maintenance Manuals

These manuals are the central reference for organized information and instructions. Preferably, these manuals are prepared by the design engineer, but they may be prepared by others. Equipment requiring preventive maintenance for efficient operation should come with complete maintenance information. Required routine maintenance actions should be clearly printed on a permanent label and affixed in an accessible location on the equipment. The label can be provided with identifying required actions, which are described in detail in operation and maintenance manuals. When the label refers to a manual, it should identify by title and/or publication number the operation and maintenance manual for the particular model and type of product. The manufacturer should furnish copies of the manual to the owner.

32.7 DESIGN REQUIREMENTS FOR THE IMPLEMENTATION OF AN HVAC&R MAINTENANCE PROGRAM

The implementation of HVAC&R designs, including provisions for an effective maintenance program, starts with the schematic phase design and terminates with the postcommissioning of all systems. As materials and equipment are selected at each phase of the project, maintenance requirements, priorities, interfaces, alternatives, risks, and costs must be evaluated within the design context. Resources and restraints should be defined; these may include management policy, availability of skilled labor, facilities, tools, and service organizations. Maintenance criteria must be defined and may depend on the character of work and extent of work to be performed and the organization responsible (owner's employees or outside maintenance service organization).

As a part of the engineering design process, drawings, specifications, and a maintenance program must be prepared to determine the maintenance requirements of all equipment, determine the tools, diagnostic and test equipment required, set forth operational procedures and equipment functions, and provide diagnostic procedures for each HVAC&R system, subsystem, and component. It is also necessary to provide access to systems, subsystems, equipment, and components for all required inspections and service. Items to be considered include proper location of access doors, plates, and access panels and location of valves, control devices, instrumentation, thermometers, and pressure gauges so they can be readily serviced.

Identification of Maintenance Requirements and Functions To Be Included in the HVAC&R Engineering Design Process

In concert with vendors and equipment and material manufacturers, the HVAC&R engineer should evaluate the maintenance functions and requirements for each system component to ensure that they are provided for in the design. Requirements for preventive maintenance, adjusting, servicing, standby components to minimize downtime, removal of components, and access spaces should be identified as part of the engineering design process. Spare part requirements, labor, and skill levels should also be identified.

HVAC&R Maintenance Cost Analysis

The service life, replacement cost requirements, and on-going cost factors for all HVAC&R system components expected to affect the maintenance program must be identified. Typical components to be considered include the following:

1. *Maintenance organization.* Identify the maintenance functions to be performed by the owner's employees or by outside contract maintenance organizations. Indicate the extent of maintenance performed by each.
2. *HVAC&R personnel requirements.* Describe the staff and staff skills required.
3. *Equipment maintenance.* Describe characteristics of the equipment and their maintenance requirements.
4. *Maintenance support.* Identify vendor, manufacturer, spare parts supplier, outside service maintenance organizations, special equipment requirements, and special support equipment.
5. *Support facilities.* Indicate the existing and new facilities that will be required to support the maintenance program.
6. *Miscellaneous data.* Enumerate videotapes, maintenance manuals, checklists, special tools, and other data planned for each maintenance procedure.

32.8 DESIGN REQUIREMENTS CHECKLIST

General Considerations

The design HVAC&R engineer must consider the following items for each system component requiring maintenance:

1. Access openings provided
2. Access space required
3. Anticipated service life
4. Cleaning agents defined
5. Cleaning and protection procedures/times
6. Diagnostic equipment required
7. Drain pans over electric motors and under oil burners
8. Electrical facilities required
9. Electrical hazards and personnel protection required
10. Ergonomics
11. Guards and railings
12. Handling procedures required
13. Interchangeable components required
14. Ladders required
15. Long lead time spare parts required
16. Lubrication systems/materials required
17. Maintenance instructions required
18. Maintenance manuals required
19. Maintenance measurements: methods, procedures, and instruments
20. Maintenance procedures for on-going air system balancing
21. Maintenance procedures for on-going hydronic system balancing
22. Maintenance schedule required
23. Monorails and overhead structural supports required
24. Noise hazards and personnel protection required
25. Operating instructions required
26. Operating manuals required
27. Personnel protective clothing/devices required
28. Platforms required
29. Plumbing facilities required
30. Provide access doors and access plates
31. Provisions for inspection
32. Provisions for products and services: delivery, storage, and handling
33. Provisions for temporary light, power, heat, and water
34. Provisions for work area protection
35. Record drawings
36. Refrigerant safety requirements
37. Risk management data
38. Safety and health regulations and worker protection defined/posted
39. Scaffolding, rigging, hoisting, and lifting devices required
40. Shop drawings
41. Signs required
42. Skill levels required
43. Spare parts required
44. Special tools required
45. Stairs required/ramps required
46. Testing equipment required
47. Tools required
48. Tube pull space
49. Vibration hazards and personnel protection required
50. Work space required for servicing

The design engineer must consider other characteristics as well for each item to be maintained. The following checklist, using data taken from Van Cott and Kinkade's *Human Engineering Guide to Equipment Design* (Chapter 12) and modified for an HVAC&R maintenance program, is offered to serve as a guide for the preparation of the design maintenance program. The following checklist should be reviewed with all manufacturers, suppliers, and service organizations to confirm that the equipment, equipment design, installation, and commissioning include the recommendations set forth in this checklist.

HVAC&R Maintenance Safety

The HVAC&R engineer should verify the following:

1. All wiring, conduits, cables, switches, piping, valves, fittings, gauges, thermometers, meters, instruments, and control devices attached to the equipment, its control cabinets, or its components cannot be disturbed or damaged by or cause safety hazards to passing equipment, materials, or personnel.

2. Lifting devices are provided where required to service equipment.

3. All equipment, units, and components are designed to be conveniently lifted, carried, pulled, pushed, or manually turned.

4. All V-belt safety guards and protective guards are provided for all rotating equipment and exposed moving parts.

5. All access doors to hazardous areas have provisions for automatically shutting off power (such as safety interlocks on door).

6. All access covers, access doors, and access panels are self-supporting in the open position.

7. All wiring, cables, conduits, piping, ductwork, and components attached to HVAC&R units that can be partially removed (i.e., slide filter racks, coils, and tube bundles) are attached so that they are not damaged when the unit is replaced.

8. All controls that are used for maintenance or calibration only and that are not necessary to actual operation are kept distinct from the operational controls.

9. All wiring, cables, conduits, piping, ductwork, and appurtenances are routed so they cannot interfere with HVAC&R equipment, doors, or fixtures and so they cannot be stepped on or used as handholds or walking surfaces.

10. All labels identifying control devices appear on the devices as well as the control panels.

11. All access openings are free of sharp edges or projections that could constitute a safety hazard.

12. Protective clothing, gloves, and glasses to be used with system components have been recommended by the manufacturers of the components for the maintenance of their products.

13. Noise/sound personnel protective devices comply with current OSHA requirements.

14. The requirements for visual illumination and locations have been met.

15. The requirements for toxic chemical and toxic material precautions have been met, and all required material safety data sheets (MSDS) for all materials have been provided.

16. The extent of possible electrical hazards and personnel protective measures required have been identified. Verify that the electrical systems are grounded in accordance with the National Electric Code. Verify that all manufacturers provide products that comply with the National Electric Code.

17. Maintenance stands, work platforms, ladders, stairs, and walkways comply with OSHA standards.

HVAC&R Maintenance Information

The HVAC&R engineer should verify the following:

1. All mechanical identification provides complete information.

2. All mechanical identification appears on each system component that the operations and maintenance staff is required to recognize, read, or manipulate.

3. All system mechanical identifications and controls clearly indicate their function and relationship.

4. All system access points are individually labeled.

5. All system functions of each system control are clearly labeled.

6. System mechanical identifications are imprinted, embossed, or attached so that they will not be lost, mutilated, or become otherwise unreadable.

7. All system mechanical identifications and controls are labeled to correspond to notations found in system diagrams and maintenance and operations instructions.

8. All accesses have mechanical identifications indicating what auxiliary equipment is needed for service or checking.

9. All accesses have mechanical identifications that tell what can be reached through this point.

10. All appropriate mechanical identifications are used for each test point.

11. All components that require access from two or more openings are properly marked to indicate this requirement.

12. All access covers have permanent HVAC&R devices, components, and part numbers marked on the cover.

13. All component covers have mechanical identifications that provide relevant information concerning electrical, electronic, DDC, pneumatic, or hydraulic characteristics of the part.

14. All electrical components are labeled with current, voltage, frequency, impedance, and terminal information.

15. All HVAC&R displays and their associated controls are clearly labeled to show their relationship.

16. All color codes or other symbolic coding schemes used for identifying test points or tracing wire or pipe lines are easily identifiable under all illumination conditions.

17. All equipment electrical terminals are clearly marked with + or −.

18. All equipment display labels for each termination have the same code symbol as the wire or line attached to it.

HVAC&R Maintenance Equipment and HVAC&R Equipment Removal or Replacement

The HVAC&R engineer should verify the following:

1. The design and installation of all HVAC&R units, components, devices, and appurtenances that are to be removed and replaced are provided with handles or other suitable provisions for grasping, handling, and carrying, and that they are designed and located so that there is adequate clearance from obstructions.

2. All wires, cables, and conduits are routed so that they do not need to be bent or straightened when being connected or disconnected.

3. All equipment is designed and provided with maintenance facilities that provide for all environmental conditions (temperature, humidity, wind, rain, snow, ice, brightness due to solar effects, or darkness) under which the equipment is to be maintained.

4. All removable units, such as filters, coils, control devices, components, wiring, cables, and instrumentation, are easily removable.

HVAC&R System Fasteners, Fittings, Flanges, and Tools

The HVAC&R engineer should verify the following:

1. All system fasteners, fittings, and flanges are standardized.

2. All equipment units and assemblies are replaceable with common hand tools.

3. All functionally similar units are interchangeable between systems or subsystems.

4. All wires, cables, conduits, pipes, liquid and pressure lines are color coded; and both ends are tagged.

Orientation, Alignment, and Keying

The HVAC&R Engineer should verify that the manufacturers meet the following requirements:

1. All assembly parts are designed with orienting seats and pins (to save time getting parts in proper position for fastening).

2. The proper orientation of units within their respective casings are obvious, either through design of the casings or by means of appropriate labels.

3. All components and access covers are provided with alignment pins or grooves so they can be easily oriented for fastening.

4. All interchangeable units, such as DDC devices and solid-state controllers, are physically coded (keyed) so that it is impossible to insert a wrong unit.

5. All units are coded (labels or colors) to indicate the correct unit and its orientation for replacement.

6. There is some means of physical design to prevent mismating or interchanging connections.

7. The equipment design is such that it is physically impossible to reverse connections or terminals in the same, or adjacent, circuits.

8. All adjacent soldered connections are located far enough apart so work on one connection does not compromise the integrity of an adjacent connection.

9. All control linkage attachments are designed so that reversed assembly is not physically possible.

HVAC&R Maintenance Manual Control Arrangement

The HVAC&R engineer should verify that the equipment manufacturers meet the following requirements:

1. All manual controls that are intended to have a limited degree of motion have adequate mechanical stops to prevent damage.
2. All manual control devices that require only occasional resetting are provided with a cover seal or otherwise guarded against inadvertent actuation.
3. All multiposition manual controls have position control setting stabilizers to prevent leaving the setpoint between desired positions.
4. All lubrication points are accessible.
5. All manual controls are located where they can be seen and operated from the normal working position without disassembly or removal of any part of the installation.
6. All manual controls are marked to indicate which direction to operate the control.
7. All manual control devices requiring sensitive adjustments are located and guarded to prevent inadvertently displacing them out of adjustment.

Working Space Configuration

The HVAC&R engineer should verify the following:

1. Equipment is located so that awkward working positions are avoided.
2. There is sufficient space to use testing equipment instrumentation and tools.
3. All valves, dampers, switches, controls, and instrumentation can easily be reached for operation from a convenient working position.
4. All components are located so that physical interference among technicians working on the same or adjacent areas is minimized.
5. All component displays are located so that they can be observed without removing other equipment or disassembling any portion of the component installation.
6. When component maintenance activities demand the collaboration of a team, the work flow, traffic flow, and communication are analyzed and planned for maximum overall efficiency.
7. When group activity demands the use of a central visual display (computer), the lines of sight to the display are not blocked by poor arrangement of people or equipment.
8. All displays (dials, gauges, and indicators) can be easily read from a convenient working position.
9. All printed matter is displayed from a normal viewing position.
10. All wires, cables, conduits, pipes or lines, and ductwork that must be routed through walls, cas-

ings, and housings are designed for easy installation and removal without cutting or splicing of the respective items.

Accessibility

The HVAC&R engineer should verify the following:

1. All access openings are designed so that the technician can see what he or she is doing. (Clearance for a hand only obscures the thing the technician is working on.)
2. To the extent practicable, all access openings used for frequent visual inspection have transparent or quick-opening covers.
3. To the extent practicable, all HVAC&R units are placed so that they are not in recesses, behind or under ducts, pipes, conduits, furniture, or other items that are difficult to move.
4. All high-failure-rate components are easily accessible for replacement.
5. All casings, ductwork, plenums, piping, conduits, and other equipment are designed not to interfere with the removal or opening of covers of units in which work must be done.
6. All access covers that remain attached to the basic equipment are designed so that they do not have to be held open or so they do not dangle in the way.
7. All throwaway assemblies or parts are located to be accessible without removal of other components.
8. All maintenance displays are located on the side normal for maintenance tasks of inspections, checkout, troubleshooting, removal, and replacement.
9. All units that are frequently pulled out of their installed position for checking are mounted on roll-out racks, slides, or hinges.

32.9 MAINTENANCE INSTRUCTIONS AND PROCEDURES

The maintenance requirements of all systems and components should be set forth in written documents prepared and evaluated during each stage of the design development of the system. They should be collected, collated, evaluated, and formalized at each stage of the design process.

The following should be considered when preparing maintenance instructions and procedures:

1. In concert with contractors/vendors/manufacturers, instructions and procedures should be prepared in simple layperson's language using step-by-step procedures, making sure contractors/vendors/manufacturers verify that each procedure states how to start up the equipment and how to shut it down.

2. Contractors/vendors/manufacturers should provide systematic troubleshooting procedures.

3. As part of the maintenance instructions, the contractors/manufacturers/vendors should provide literature, checklists, schematics and diagrams, and videotapes.

4. The manufacturers/vendors should provide electric power, control, interlock, and interface wiring diagrams, and pneumatic control and DDC diagrams.

5. The contractors/manufacturers/vendors should provide spare parts lists and inventory data.

32.10 EQUIPMENT MAINTENANCE

The design engineer is frequently required to prepare maintenance instructions as part of the engineering services for a project. Maintenance instructions vary in format, depending on the various authorities and engineers preparing the instructions, but they have some common fundamental principles. The primary principle is that the instructions must document all of the characteristics of the equipment; how often the equipment has to be maintained and serviced; and the type of service and maintenance required. In many cases, the maintenance instructions are arranged so that the instructions also can be interfaced with the equipment inventory program, spare parts program, stock program, and work order program. The following is an example of the maintenance and operations instructions for an item of equipment: it can be extended for each item of equipment used for the specific project in order to generate a complete document:

1. Equipment Inventory Data:
   a. Equipment Name: _____
   b. Equipment Location:
       Building Name: _____ Building No.: _____
       Floor: _____ Room No.: _____
       Room or Space Designation: _____
   c. Area or Space Served:
       Building Name: _____ Floors: _____
       Areas (Spaces): _____ Rooms: _____
   d. Equipment Manufacturer: _____
       Model No.: _____ Serial No.: _____
   e. Conditions of Service: _____
       _____
       _____
       _____
       _____
   f. Installing Contractor: _____
   g. Installing Subcontractor: _____
   h. Manufacturer's Representative: _____
   i. Vendor: _____
   j. Project Name: _____
   k. Project Number: _____
   l. HVAC&R Designer: _____
   m. Equipment Provided in Accordance with: _____
       _____
       Contract Drawing No.: _____
       Specification Reference No. _____
       Construction Specificatins Institute (CSI) Specification No.: _____

n. List shop drawings, equipment cuts, catalogs, or other drawings that show this equipment: _____
_____
_____

o. Indicate all services connected to this equipment—water, drain, steam, condensate return, hot water, chilled water, condenser water, electric, etc. Give size of connections, amount used, and pressure. _
_____
_____
_____
_____

p. Indicate spare parts lists, manufacturer's maintenance and instruction manuals, or other data furnished:
_____
_____
_____
_____

q. Maintenance service organization from which parts may be obtained: _____
_____

r. Maintenance service organization from which service may be obtained: _____
_____

s. Service Agreement: Yes_____ No_____ Expires_____

t. Guarantee Number: Yes_____ No_____ Expires_____

u. Warranty Number: Yes_____ No_____ Expires_____

v. Type of refrigerant (if any): _____lbs. of refrigerant

w. Electrical Characteristics: Voltage_____ Amp_____ Phase_____

x. Electrical Circuit Data Panel Designation:
Panel Location: _____
Circuit Number: _____Fuse Size: _____Fuse Type: _____

y. Location and data on any auxiliaries: _____
_____

z. Other data: _____
_____

aa. Plant and Properties Department Equipment Designation Number: _____, _____,
_____, _____, _____

bb. Cost Accounting Control Number: _____

cc. Anticipated Service Life: _____ Basis: *ASHRAE*

2. Maintenance Schedule: (See Page _____ for a generalized example of what may be included in maintenance instructions for equipment)

3. Spare Parts List (to be provided by the equipment manufacturer)
a. _____
b. _____
c. _____
d. _____

4. Special Tools List (to be provided by the equipment manufacturer)
a. _____
b. _____
c. _____
d. _____

5. Special Instrumentation and Test Equipment Required (to be provided by equipment manufacturer)

32.11 SAMPLE MAINTENANCE MANUAL

Many variations on maintenance manuals are currently in use. Professional groups, such as ASHRAE and APPA, as well as governmental agencies (Veterans Administration, G.S.A., NavFac, and Army Corps of Engineers) have formats suitable for their particular end users.

Part I presents descriptive information, operating instructions, and inspection and maintenance data.

Part II of the manual consists of manufacturers' and fabricators' data and information and materials, including descriptive literature; operating characteristics, operating instructions, inspection instructions and procedures; parts list; spare parts; and service and dealer directory.

Part II is normally provided by the manufacturers, vendors, and fabricators, not the HVAC&R engineer; therefore, the following examples only illustrate Part I. The design engineer must include adequate information in the contract documents for the use of those preparing the maintenance instructions. The rest of this section is an example of representative maintenance instructions and is modeled after the VA format, which uses a separate operations manual and a separate maintenance manual.

Maintenance Manual (Part I) Descriptive Information

A. GENERAL DESCRIPTION OF HVAC&R SYSTEMS. A brief description of all HVAC&R systems is normally included to enable the user to understand the systems in their entirety. The following is a sample description:

Boiler plant. The boiler plant consists of two (2) built-in-place Scotch marine boiler burner units, each rated at 250 boiler horsepower, complete with field-tested firetube boilers, feed piping, double-wall oil heaters, oil piping, steam headers, steam leaders, and steam/return equalizers. The boiler burner units are capable of firing No. 2, No. 4, and No. 6 grade fuel oils and natural gas. They shall be arranged to fire natural gas as a primary fuel and No. 2 grade fuel oil as a secondary fuel.

The boiler burner units are served by natural gas (fuel) lines routed to the pilot and fuel gas trains. The gas vents from gas trains are piped to the terminal locations. The fuel oil transportation system includes two (2) storage tanks, fuel oil heaters, fuel oil transfer pumps, oil supply and return lines, suction lines, strainers, relief valves, three-way port valves, pressure gauges, thermometers, and drip pans. The oil tanks are fiberglass double-wall tanks complete with leak detection and overfill protection systems.

The boilers are served by boiler plant combustion air intake louvers with motorized dampers, ventilation louvers with motorized dampers, and control and power wiring.

The boiler plant is served by two (2) emergency breakglass stations for the emergency shutdown of all boiler plant equipment. The boiler feed system consists of one (1) condensate receiver and three (3) boiler feed pumps. Each boiler is served by its own boiler feed pump with one common standby pump piped and valved in parallel with all boiler feed pumps.

Steam distribution system. The components of the steam distribution system are as follows:

1. The steam distribution system originates at the boiler header.
2. From the steam header, valved steam mains run to a steam absorption machine and to a common steam main. The common steam main serves air-handling units, heat exchangers, and humidifiers. The condensate returns terminate at condensate pump sets. The pump discharge line from the pump sets connects to the boiler feed pump assembly and receiver.

Refrigeration plant. The components of the refrigeration plant include the following:

1. The refrigeration plant consists of three (3) 450-ton steam absorption machines.

   Each refrigeration machine is served by its own chilled water pump. A common standby pump serves all machines. The chilled water distribution system originates at the outlet of the steam absorption refrigeration machines; the chilled water supply lines connect together and run to their respective loads.

   The chilled water system serves cooling coils in air-handling units and secondary dual hot/chilled water systems.

   The return lines from the loads connect together and run to the chilled water pumps, each piped in parallel, and from the pumps to the machines. The expansion tanks are located on the suction side of the pumps. The primary chilled water system contains flow meters in the discharge lines to the refrigerant machines. Each machine has pressure gauges, thermometers, and drain valves at all inlets and outlets. In each inlet line are provided plug valves for balancing and gate valves for shutoff.

2. The cooling towers (each rated for 450 tons) are arranged, piped, wired, and controlled to serve the three refrigeration machines. The cooling towers are winterized. All exterior piping serving cooling towers are heat traced and insulated. Each tower section has an internal strainer. The lines pitch back to the building and have a drain valve and butterfly shut off lines. The lines shall tie together and run to the suction side of the circulating pumps. Each pump has the following connections: on the suction side, a gate valve, strainer, pressure gauge, reducer, and drain valve; on the discharge side, a pressure gauge test tee, drain valve, spring-loaded check valve, and gate valve. Each cooling tower pump is arranged so that each pump serves one refrigeration machine. The standby pumps are arranged to serve all machines.

At each inlet and outlet of each refrigeration machine are thermometers, pressure gauges, and drains: lubricated plugs are provided for balancing, and gate valves are provided for shutoff. The cold water make-up line to the cooling towers is provided with a master water make-up assembly consisting of a water meter assembly with a three-valve bypass, backflow preventor, separate water make-up line with make-up water control valve for each tower, and chemical feedwater treatment connections.

3. The refrigeration plant control panel contains the emergency shutdown for the entire refrigeration plant.

Secondary water system. The secondary temperature dual hot/chilled water system consists of secondary zone dual hot/chilled water pumps, steam/water heat exchanger, air removal device, dual hot/chilled water supply and return piping, drain piping, air vent piping, hand valves, balancing valves, expansion tanks, spring-loaded check valves, cold (city) water make-up line with backflow preventor, pressure-reducing and automatic feed valves, and other appurtenances.

Factory-assembled air-handling units. The factory-assembled air handling units consist of a mixing section, filter section, plenum section, steam heating coil, freeze protection rack and its plenum section, chilled water cooling coil and its plenum section, fan section, inertia base, and flexible connections.

Exhaust fans. Fans are of the following types:

1. In-line centrifugal fans
2. Roof-mounted ventilators
3. Ceiling-mounted ventilators

Additional provisions. The HVAC&R systems are provided with the following:

1. Pressure gauges and fittings on the inlets and outlets of pumps, boilers, air-handling unit coils, heat exchangers, hot water storage heaters, and refrigeration machines

2. Flow meters on all primary and secondary pumping circuits

3. Horizontal-piping hangers and supports, vertical-piping clamps, hanger-rod attachments, building attachments, saddles and shields, spring hangers and supports, and equipment supports.

4. Pipe markers, duct markers, valve tags, valve schedule frames, engraved plastic-laminate signs, plastic equipment markers and tags, and underground plastic line markers for underground pipe.

5. Pad-type isolators, plate-type isolators, equipment rails, fabricated equipment bases, inertia base frames, roof–curb isolators, isolation hangers, riser isolators, and flexible pipe connectors.

6. Mechanical insulation for low-pressure steam piping systems; low-pressure condensate piping systems; boiler feedwater piping systems; primary chilled water piping systems; cold water make-up piping systems; dual hot/chilled water piping systems; chemical treatment pumps, tanks and piping; condensate pump discharge lines; portions of steam boiler burner units; condensate receivers; drip lines to collect water condensed out at cooling coils and fan–coil units; pumps; air conditioning supply and return fans and casings; portions of water chillers and refrigeration machinery not factory insulated; cooling tower lines outside building subject to freezing; tanks (expansion, overflow, drain, blowdown, and surge); heat exchangers; all air conditioning supply and return ducts; all outside air intake ducts and plenums from outside air intake connections to air conditioning units and intake connections; breechings for boilers; all exhaust ducts and plenums beyond automatic backdraft dampers; portions of water, drip, condensate and drainage systems subject to freezing; and fuel oil suction return and pump discharge systems for fuel oil transportation systems.

Ductwork systems. These systems include the following:

1. Rectangular, round, and flat-oval metal ducts and plenums made of galvanized sheet metal.

2. Ductwork accessories include: balancing air foil dampers, low-pressure valves, control dampers, fire and smoke dampers, combination fire/smoke

dampers, turning vanes, duct hardware, flexible connections, drain pans, velocity gauges, filter gauges, access doors and access panels (in ductwork, housing, and casings), acoustical treatment (attenuators and duct lining), belt guards, drip pans over electrical equipment and panels, flexible ducts, sealants, mastics, gaskets, and tapes, air registers and diffusers, louvers, air terminals.

Automatic control systems. These include control sequences for the following:

1. Boiler plant equipment
2. Refrigeration plant equipment
3. Primary chilled water loop
4. Dual temperature water system
5. Steam/water heat exchanger
6. Fan coil units
7. Reheat coils
8. Air-handling units
9. Exhaust fans
10. Smoke purge control
11. Domestic hot water storage heater

B. SPECIFIC EQUIPMENT: DESCRIPTION, INVENTORY AND MAINTENANCE SCHEDULE. A description of the role of each piece of equipment in the total HVAC&R System, its inventory data, and its maintenance schedule is normally included.

Equipment inventory data. The following is a generalized example of what may be included in the equipment inventory for a fictitious air-handling unit.

1. Equipment Name: Air-Handling Unit No. 1
2. Equipment Location:
   Building Name: Ozona Library Building No.: A-5, Floor: Penthouse Room No.: Mechanical Room No. 1,
   Room or Space Designation: PH-1
3. Area or Space Served:
   Building Name: Ozona Floors: 4th and 5th Areas (Spaces): Stack Spaces Rooms: 400-410 & 500-510
4. Equipment Manufacturer: Oxen Fan Co.
   Model No.: OX-36B Serial No.: OX-470-92
5. Conditions of Service: _____
   _____

General	**AHU-1**
(1.) Manufacturer's model number	145
(2.) Unit type	Draw-thru
(3.) Arrangement	THD
Fan Type	
(1.) Number of fan wheels	1
(2.) Wheel diameter	27"
(3.) Wheel type	Backward-curved air foil
(4.) cfm	14,100/7050
(5.) Total static pressure	3.5"
(6.) rpm	1362
(7.) Max outlet velocity	2171 fpm
(8.) Brake horsepower	10.8 bhp
(9.) Motor horsepower	15 mhp
(10.) Static efficiency	69.15%
(11.) Operating weight w/vibration control	11,426 lbs
(12.) Vibration control	Inertia base
(13.) Motor starter data	Two (2)-speed motor See schedule
(14.) Interlocked with	RE-6, TE-1, TE-2, TF-6A, KE-1, KE-2
Reheat Steam Heating Coil	
Air	
(1.) cfm	14,100 cfm
(2.) Entering temperature, degrees (dry bulb)	38.5°F
(3.) Leaving temperature, degrees (dry bulb)	72°F

Steam
1.	Entering coil steam pressure	2 psig
2.	Capacity	511 Mbh
3.	Flow	528 lb/hr
4.	Steam control valve inlet pressure	5 psig
5.	Steam control valve (cv)	24 cv
6.	Steam traps—number	2
7.	Steam traps—size	$1\frac{1}{2}''$

Coil
1.	Coil type	Nonfreeze
2.	Number of sections	2
3.	Sections high	2
4.	Sections wide	1
5.	Casing height	25''
6.	Nominal tube length	5' 6''
7.	Rows deep	1
8.	Fins per inch	8 fpi
9.	Face area per/section	9 sq. ft
10.	Total face area	18 sq. ft
11.	Face velocity	783 fpm
12.	Maximum pressure drop	0.30'' WG
13.	Total operating weight	290 lb

Cooling coil

Air
1.	cfm	14,100 cfm
2.	Entering air temperature (dry bulb)	88°F
3.	Entering air temperature (wet bulb)	75°F
4.	Leaving air temperature (dry bulb)	58°F
5.	Leaving air temperature (wet bulb)	57°F

Water
1.	Entering temperature	45°F
2.	Leaving temperature	55°F
3.	Capacity	887 Mbh
4.	Flow	177.3 gpm
5.	Control valve (cv)	89 cv
6.	Pressure drop (ft hd.)	15 ft maximum

Coil
1.	Coil type	DC
2.	Number of sections	2
3.	Sections high	2
4.	Sections wide	1
5.	Casing height	38''
6.	Rows deep	8
7.	Fins per inch	8 fpi
8.	Nominal tube length	5' 6''
9.	Face area per section	15.03 sq. ft
10.	Total face area	30.06 sq. ft
11.	Face velocity	469 fpm
12.	Maximum pressure drop	1'' WG
13.	Total operating weight	861 lb

Preheat steam heating coil schedule

Air
1. cfm   14,100 cfm
2. Entering temperature (dry bulb)   5°F
3. Leaving temperature (dry bulb)   38.5°F

Steam
1. Entering coil steam pressure   2 psig
2. Capacity   511 Mbh
3. Flow   528 lb/hr
4. Steam control valve inlet pressure   5 psig
5. Steam control valve (cv)   24cv
6. Steam traps—number   2
7. Steam traps—size $1\frac{1}{2}''$

Coil
1. Coil type   Nonfreezing
2. Number of sections   2
3. Sections high   2
4. Sections wide   1
5. Casing height   25
6. Nominal tube length   5' 6''
7. Rows deep   1
8. Fins per inch   8 fpi
9. Face area per/section   9 sq. ft
10. Total face area   18 sq. ft
11. Face velocity   783 fpm
12. Maximum pressure drop   0.30'' WG
13. Total operating weight   290 lb

Filters
1. cfm   14,100/7050 cfm
2. Type : (flat or V)   V
3. Filter size   24'' × 24''
4. Filter depth   2''
5. Face area per filter   4.00 sq. ft
6. Number of filters   12
7. Filters high   4
8. Filters wide   3
9. Total face velocity   294 fpm
10. Total filter face area   48 sq. ft
11. Air-side static pressure drop   0.5'' WG
12. Operational weight   414 lb

Concrete bases

Fan section
1. Length   59''
2. Width   96''
3. Approximate height   9''
4. Approximate weight   4425 lb

Fan section inertia base
1. Length   59''
2. Width   96''
3. Approximate height   8''
4. Approximate weight   3933 lb

Secondary heating coil/plenum section
(1.)	Length	30″
(2.)	Width	96″
(3.)	Height	18″
(4.)	Approximate weight	4500 lb

Cooling coil section
(1.)	Length	14″
(2.)	Width	96″
(3.)	Approximate height	18″
(4.)	Approximate weight	2100 lb

Plenum section
(1.)	Length	30″
(2.)	Width	96″
(3.)	Approximate height	18″
(4.)	Approximate weight	4500 lb

Heating coil section
(1.)	Length	7″
(2.)	Width	96″
(3.)	Height	18″
(4.)	Approximate weight	1050 lb

Filter section
(1.)	Length	30″
(2.)	Width	96″
(3.)	Approximate height	18″
(4.)	Approximate weight	4500 lb

Mixing box section
(1.)	Length	36″
(2.)	Width	96″
(3.)	Height	18″
(4.)	Approximate weight	5400 lb

---

6. Installing Contractor: Gnaw HV & AC Mechanical Corp.

7. Installing Subcontractor: N/A

8. Manufacturer's Representative: UOY Suppliers, Inc.

9. Vendor: N/A

10. Project Name: Ozona University Library

11. Project Number: 91-0572

12. HVAC&R Design Professional: CPBPW Associates

13. Equipment Provided in Accordance with: Contractors Purchase Order GNAW-PO-11-92, Contract Drawing No.: HVAC-11 Specification Reference No.: (See Project Specification) Construction Specifications Institute (CSI) Specification No.: (See Project Specification)

14. List Shop Drawings, Equipment Cuts, Catalogs, or other Drawings that show this equipment: —

15. Indicate all services connected to this equipment—water, drain, steam, condensate return, hot water, chilled water, condenser water, electric, etc. Give size of connections, amount used, pressure, etc. (all items to be listed)

16. Indicate spare parts lists, manufacturer's maintenance and instruction manuals, or other data furnished:

17. Maintenance service organization from which parts may be obtained:

18. Maintenance service organization from which service may be obtained: N/A

19. Service Agreement: Yes ___ No X Expires

20. Guarantee Number: Yes X No ___ Expires 10/31/93

21. Warranty Number: Yes ___ No ___ Expires

22. Type of refrigerant (if any): N/A lbs. of refrigerant
23. Electrical Characteristics: Voltage 460 Amp 60 Hz Phase 3
24. Electrical Circuit Data Panel Designation: Panel Location: 4th Floor Circuit Number: 4-10 Fuse Size N/A Fuse Type N/A.
25. Location and data of any auxiliaries: N/A
26. Other data: N/A
27. Plant and Properties Department Equipment Designation Number: H-O___, A___, AHU___, 01___, ___
28. Cost Accounting Control Number: HOA5-AHU-1
29. Anticipated Service Life: 20 Basis: ASHRAE 33.3, 1991, Table 3.

Maintenance schedule for air-handling unit No. 1. The following is a generalized example of what may be included in the maintenance schedule for a fictitious air-handling unit.

1. General service and checks
   a. *Lubrication frequency.* As per manufacturers' instructions (for fan and motor bearings)
   b. *Filter replacement frequency.* As per filter loading but not less than every three (3) months
   c. *Belt change frequency.* As required but not less than every three (3) months

2. Daily service and checks
   a. Check and record temperatures: return air, outside air, mixed air, leaving heating coils, leaving cooling coil, and after discharge of fan
   b. Check and record steam pressure to heating coil
   c. Check and record water temperatures entering and leaving cooling coil
   d. Inspect filter sections for air bypassing filters
   e. Check that drain pans and drains are clean
   f. Check bearing temperatures
   g. Check motor rotation
   h. Check motor temperature
   i. Check motor starters
   j. Check controls for proper operation

3. Monthly service and checks
   a. Repeat all daily checks, recordings, and inspections
   b. Check, adjust, and clean outside air intake, outside air intake louvers, screens, outside air motorized dampers, and recirculating air motorized dampers

   c. Check damper blades for tight closing (outdoor air intake).
   d. Check piping appurtenances: steam traps, air vents, drains, and strainers; check piping for leaks, check for damaged insulation
   e. Clean all heating and cooling coils
   f. Check fan bearing alignments
   g. Check fan bearings for wear and tear
   h. Clean fan section and fan wheel
   i. Clean unit casings, plenums, and connecting ducts
   j. Check that motor electrical connections are tight, check for arcing of brushes, and check that motor shaft/sheave is secure
   k. Check that all motor started connections are tight, and clean all contacts
   l. Automatic temperature control systems: inspect, check, clean, adjust, repair, and calibrate all control devices, interlocks, electric-pneumatic (EP) and pneumatic-electric (PE) switches, compressors, and refrigerated air dryers

4. Yearly service and checks
   a. Repeat all daily and monthly checks, recordings, and inspections
   b. Check louvers, dampers, filter frames, coils, casings, plenums, and fans for corrosion. Clean and treat as recommended by the manufacturer
   c. Check coil fans for corrosion, wear and tear.
   d. Check fan shaft as per manufacturer's recommendations
   e. Check fan shaft for straightness, that the fan wheels are tightly fastened to the shaft and the drive sheaves are secured
   f. Clean motor frame and air passages
   g. Check motor acceleration time, voltage, current, and speed
   h. Check motor anchor bolts and vibration isolation
   i. Motor starter: clean and check contacts, clean housing, check for proper operation and overload protection
   j. Ductwork: check for dust, moisture, defective linings, defective insulation, and leaks
   k. Mechanical insulation: check and repair all damaged pipe insulation, fitting insulation, valve insulation, duct insulation, casing and housing insulation

32.12 SAMPLE MAINTENANCE PROGRAM FOR A TYPICAL HVAC&R FACILITY

Purpose of Program

The purpose of the following sample maintenance program is to prepare the owner's staff for the maintenance of a refrigeration plant.

The program herein sets forth the coordinated data and actions by the owner's staff, contractors, vendors, and designers necessary to implement an effective operations and maintenance program.

The owner's maintenance staff normally provides labor, supervision, hand tools, and equipment necessary to perform all the work and services to perform maintenance for the systems. The maintenance program presents procedures for maintenance, materials handling, materials, storage, inventory control, and related custodial services. The program is intended to be the product of the joint efforts of the owner's staff, the contractor, the various equipment manufacturers, and the designer.

The program recommends methods, materials, procedures, operations, and staffing that will endeavor to keep the refrigeration plant, maintenance, and operation costs to a practical and reasonable level.

It is the intent of the program to provide safe, uninterrupted, and efficient maintenance of the HVAC&R systems and to reduce depreciation to a minimum. It is not intended nor practical to define in detail every requirement to accomplish the intent. The overall scope of the system maintenance performed by the owner's staff includes corrective maintenance, repairs, and preventive maintenance of the systems and their components.

Maintenance Instructions

GENERAL. The following general requirements must be met:

1. Upon completion of all work, all tests, and commissioning of the systems, the HVAC&R contractor shall furnish the necessary skilled labor and helpers for operating the systems and equipment during the startup period. During this period, the contractor shall instruct the owner or his or her representatives fully in the maintenance of all equipment furnished. The contractor shall videotape the instructions for future training of new or rotated personnel.
2. The contractor shall furnish bound sets of typewritten instructions for maintaining all systems and equipment included in the facility.

RESPONSIBILITIES OF THE CONTRACTOR. The contractor shall include the following in the manual:

1. Brief description of each system, subsystem, and basic maintenance features
2. Manufacturer's name and model numbers of all components of the systems listed on the equipment schedules, drawings, control diagrams, and wiring diagrams of controllers
3. Chart of the numbers, location, position, and function of each valve and each damper
4. Step-by-step maintenance instructions, including preparation for starting summer operation, winter operation, shutdown, and draining
5. Maintenance instructions for all equipment, with each one individually described
6. Possible breakdowns and repairs
7. Manufacturer's literature describing each piece of equipment listed on the equipment schedules, drawings, control diagrams, and wiring diagrams of controllers
8. As-installed control diagrams by the control manufacturers
9. Description of sequence of operations by the design professional or control manufacturer
10. Parts list of all equipment
11. As-installed color-coded wiring diagrams of electrical motor controller connections and interlock connections
12. Manufacturer's literature describing the lubrication type, source, quantity, and schedule for each item of equipment and each component part of an item
13. The format of the maintenance manual may be the two-part format indicated on pages 59.3 and 59.4 of the *ASHRAE 1987 Systems and Applications Handbook.*
14. Include the name, address, and phone number of all outside maintenance service organizations for the following components:
    a. Automatic temperature control systems
    b. Chillers
    c. Boiler-burner units
    d. Water treatment
    e. Energy management and control system

MAINTENANCE PROCEDURES AND METHODS. Maintenance will be both preventive and corrective.

Preventive maintenance. The preventive maintenance program is formulated to retain each item at effective performance by providing systematic inspection, detection, and prevention of statistically probable failures. It consists of scheduled maintenance periods and procedures performed to extend the statistical service life by inspection, tests, adjustment, calibration, cleaning, and lubrication.

Corrective maintenance and repairs. The following features characterize the functions of maintenance and repair:

1. Corrective maintenance and repairs are performed as a result of failure to restore an item to a specified level of performance. These tasks include systems testing, diagnostics, fault isolation, and repair tasks, such as the changing of replaceable units and modules, replacement or repair of components, and equipment adjustments and alignments.

2. It is understood that before the owner's staff repairs any of the owner's equipment, the owner shall first determine from the equipment inventory data whether or not such equipment is covered by any warranties or guarantees from the manufacturer or supplier of such equipment, and in the event of any such warranties or guarantees, the owner shall then look toward the manufacturer or supplier of such equipment for said repairs.

3. All preventive and corrective maintenance on all equipment is performed in accordance with the manufacturer's recommendations for the equipment.

Maintenance. The maintenance function pertains to inspection, testing, adjusting, service, lubrication, and repair and replacement. It includes, but is not limited to, the following:

1. Initial operating instructions and training of operating staff (by contractor)
   a. Starting and stopping procedures, sequences, and frequency
   b. Adjustment and regulations
   c. Seasonal start-up and shutdown
   d. Seasonal changeover
   e. Logging and recording
2. On-going inspection (by owner's staff)
   a. Equipment to be inspected
   b. Points to be inspected
   c. Frequency of inspection
   d. Inspection methods and procedures
   e. Evaluation of observations
   f. Recording and reporting
3. Service and repair (by owner's staff after guarantee period)

a. Frequency of scheduled services
b. Schedules for service procedures
c. Repair procedures
d. Recording and reporting

Recommended Staffing

Staffing should be in accordance with the skills and time required to perform scheduled routine tasks, plus some reserve for unpredicted servicing and repair.

Staffing shall conform with industry-accepted requirements for HVAC&R staffing.

The staff, according to the size of HVAC&R system, could include an HVAC&R supervisor, HVAC&R operating technicians, HVAC&R mechanics, electricians, and pipe fitters/plumbers with backgrounds and duties normally as follows.

1. *HVAC&R supervisor.* This is an employee with experience in maintaining HVAC&R systems of the same size or larger than the owner's systems and successful supervisory-administrative experience with mechanical maintenance employees. The HVAC&R supervisor's educational background should include a minimum of a high school education, completion of trade school courses, or at least ten (10) years practical experience related to large HVAC&R systems. The supervisor shall have and maintain local or state licenses for the operation of special system equipment (chillers and boilers). The supervisor should have a working knowledge of the various skills and trades such as pipe fitting, refrigeration, electrical, sheet metal, automatic temperature controls, plumbing, and general handyperson's work. The supervisor implements the maintenance program and on-going start-up and shutdown procedures for all equipment. This person will coordinate and schedule daily maintenance assignments, inspect the quality of work, and advise management on the status of the HVAC&R system conditions.

2. *Operating technicians.* Technicians are required to staff the boiler and refrigeration plants and they shall be required to possess the regulated local and state licenses for the operation of refrigeration and boiler plants. The educational background includes a high school education, completion of a trade school course, or at least five to ten years of practical experience operating refrigeration plants and boiler plants. Their work duties include servicing equipment, observations and adjustments, corrective actions, and maintaining daily log readings and work assignments, such as preventive maintenance. These technicians are expected to be pro-

ficient in handling all work tools and equipment, and have sufficient knowledge of those pieces of auxiliary equipment necessary for repairs.

3. *Electrician.* Electricians must have a local and state electrical maintenance license, with at least 10 years of experience as a journeyperson electrician or of trade schooling and a high school education. The duties are assigned by the supervisor, and the electrician is expected to perform electrical maintenance with a minimum of supervision. The electrician's expertise is used in developing a preventive maintenance program.

4. *Pipe Fitter/Plumber.* Pipe Fitter/Plumber only performs maintenance work that does not require any licensing. The pipe fitter/plumber should have a high school education and at least 10 years of practical experience and/or trade schooling. This employee's duties are assigned by the supervisor, and this person is expected to perform pipe fitting/plumbing maintenance with a minimum of supervision.

5. *Maintenance mechanics.* The maintenance mechanic shall be encouraged to continue training to prepare for those licenses required by the operating technicians. Doing so enables the owner to promote from within the company's staff, as operating conditions dictate, and provide the flexibility of staffing the operating technicians' downtime with one of these employees. The employees should have a high school education and a minimum of 10 years of practical experience operating HVAC plants and/or trade schooling. Their daily assignments are directed by either the supervisor or the operating technicians. They perform various maintenance repair work on all HVAC&R equipment with the exception of those items included in outside service contracts. They are expected to be proficient in using tools, appliances, and any equipment required to perform maintenance functions.

The owner's personnel department evaluates qualifications, references, and background, and after selecting personnel institutes an on-going training program.

Custodial Services Related to HVAC&R Work

The owner's building custodial services staff provides all materials, labor, equipment, and services to perform custodial services for the systems.

The general overall HVAC&R-related services include the following routine cleaning functions:

1. Clean registers, grilles, and diffusers
2. Provide snow removal services (around cooling towers and air cooled condensers)
3. Provide internal refuse collection
4. Provide pest control services

Recommended Procedures and Administrative Systems

The owner's staff, with the assistance of the contractors, vendors, and designers, institutes an on-going inventory control program, and the owner's staff maintains the program.

The owner's staff, with the assistance of the contractors, vendors, and designers institutes a computerized preventive mechanical maintenance program, and the owner's staff maintains the program.

The owner's staff institutes a technical library at the site, and the owner's staff maintains the library.

The owner's staff institutes a custodial scheduling and control system, and the owner's staff maintains the custodial control.

The owner's staff develops methods and procedures for delivering and removing materials and supplies from the facility.

The owner develops space assignments for custodial materials and supplies, service rooms and service closets, shop space, and equipment storage space.

The owner designates service elevators for the use of maintenance personnel.

The owner maintains copies of the following documents for reference by the maintenance staff:

1. Architectural drawings (as-built)
2. Architectural specifications (as-installed)
3. Structural drawings (as-built)
4. Structural specifications (as-installed)
5. Mechanical (HVAC&R and plumbing) drawings (as-built)
6. Mechanical (HVAC&R and plumbing) specifications (as-installed)
7. Electrical drawings (as-built)
8. Electrical specifications (as-installed)
9. Shop drawings approved (as-built)
10. Operational instructions, both individually and as related to each system
11. Service and maintenance instructions
12. Manufacturer's installation instructions
13. Manufacturer's parts lists and sources of supply for replacements
14. Construction drawings—special equipment, boiler and refrigeration plant

15. Vendors warranties
16. Codes—federal, state, and local
17. Industry standards—underwriters laboratory standards and NFPA
18. Publications—journals
19. Vendor catalogs
20. Compliance test reports according to codes and laws
21. Acceptance certificates from inspecting agencies
22. Manufacturer's pressure test data
23. Manufacturer's performance tests on operating equipment
24. Contractor's field test reports
25. Welder's certificates and field test reports
26. Performance report on the balancing of air and water systems
27. Manufacturer's reports on motorized equipment alignment and installation
28. List of special nonstandard tools to service equipment
29. Fire protection working plans
30. Certification data
31. Performance reports for vibration isolation equipment
32. Representative and service company lists
33. Material safety data sheets

All equipment should be operated only as required and in accordance with manufacturer's recommendations and established standardized practices.

All operation and maintenance services should be performed with emphasis on safe procedures and with due regard to public and occupant safety, comfort, and convenience.

Safety and other regulations established from time to time by the owner should be adhered to by external maintenance service organizations, vendors, their agents, and their employees. The owner's staff must make every effort to ensure the maintenance of safe working conditions and clean premises in building and grounds.

All equipment should be operated and maintained as required by local, state, and national codes, underwriters' requirements, and established standardized practices.

All equipment logs, schedules, records, and preventive maintenance service schedules should be developed and maintained under the supervision and direction of the supervisor. All schedules, logs, and preventive maintenance records should be available for daily examination.

Unscheduled work, equipment modifications, installations, major repairs, and any work or services by the owner's staff should be limited to the staff's availability, skill, and licensing.

All equipment, where possible, should be operated at maximum efficiency and only when required. Operation of any equipment or systems should be monitored and controlled to meet the load demands of the building and prevent unnecessary equipment operation.

The owner's staff supervisor is responsible for the development, initiation, and implementation of all the aforementioned procedures, schedules, and services.

The owner's staff supervisor, following start-up procedures, should be assigned to this project to monitor and inspect the facility at sufficient intervals to ensure maximum performance efficiency. The supervisor must also be available for advisory services pertaining to the maintenance, improvement, or modification of the mechanical plant and its operation.

The owner's staff supervisor should make periodic inspections and/or evaluations of operating and maintenance procedures, records, preventative maintenance schedules, equipment conditions, and energy control for the purpose of maintaining operating and maintenance efficiency.

The Owner's Personnel

Inasmuch as the work to be performed requires specifically trained people, the following requirements must be met:

1. All work must be performed by skilled boiler/refrigeration plant operating technicians, pipe fitters, mechanics, electricians, and other necessary workers directly employed and supervised by the owner's supervisor.

2. The owner should provide the service of competent supervisors to supervise the work and report on same. The supervisors are responsible for maintaining work reports. The supervisors should coordinate the work of setting up an operations and maintenance (O & M) program.

3. All employees assigned by the owner for the performance of the work must be physically able to do their assigned work and must be qualified in this type of work.

Maintenance Shop Equipment

The owner must provide all necessary HVAC&R maintenance shop and diagnostic equipment needed for the performance of the maintenance work. (The following is a representative listing of shop equipment normally provided for large HVAC&R systems.)

1. Sling psychrometer
2. Set of pocket hand thermometers
3. Refrigerant leak detector kit
4. Combustion testing kit
5. Voltage and continuity tester
6. Volt-ohm-milliammeter with temperature adapters
7. Stroboscope tachometer
8. Vibration meter
9. Temperature recorders
10. Air velocity meter
11. Sound level meter
12. CO_2 monitor meter
13. CO monitor meter
14. Combustible gas detector set
15. Clamp-on volt-ohm meter
16. Inclined manometer
17. Automatic temperature control instrument testing: air regulators, air filters, and control manufacturers' test tools and instruments
18. Air balancing hoods
19. Ductwork leakage testers
20. Smoke gun/candles
21. Digital pressure gauges
22. Pressure and vacuum gauges
23. OSHA-approved oxygen/acetylene welder and arc welder
24. Grinder
25. Drills
26. Grease pump
27. Vacuum pump
28. Ultrasonic flow meters
29. Refrigerant charging and testing unit
30. OSHA-approved ladders
31. Digital thermometers
32. Infrared thermometers
33. Humidity meter
34. Voltage recorder
35. Soldering gun kit
36. Combination wet/dry vacuum cleaner
37. Barrel with hand pumpset
38. Electric short circuit locator
39. Megohmeter
40. Electrical cable testers
41. Hand oil pump
42. Punch and chisel set
43. Pipe wrenches
44. Flaring and swaging tool set
45. Tube bender
46. Inner-outer tube reamer
47. Tube cutter
48. Screwdrivers, Phillips head and Allen wrenches
49. Steel work benches
50. Inside and outside calipers
51. Bench vise
52. Claw hammer
53. Rubber mallet
54. Machinist's square
55. Dividers
56. Level
57. Tap and die sets
58. Files, cabinets, and tool kits
59. Snips—straight, left, and right
60. Wrecking bar
61. Ratchet wrench and socket set, multiple drives and depths
62. Set of open and box-ended wrenches
63. Pipe and fitting chain wrench
64. Strap wrenches
65. Heavy-duty pipe cutter
66. Locking pliers
67. Set of pipe taps
68. Pipe threader set
69. Pipe reamer
70. Pipe cutter and tri-stand
71. Gasket-cutting tool
72. High-pressure air hose and compressor set
73. High-pressure portable fan for winterizing coils

32.13 MAINTENANCE SERVICE ORGANIZATIONS

In most facilities, maintenance is traditionally divided among the operations and maintenance staff of the facility and outside firms offering specialty maintenance contracts.

Using maintenance service organizations to perform specialty maintenance work has been a part of the HVAC&R industry since its inception. In many cases owners prefer to use outside service contractors in lieu of in-house forces for maintenance work that requires special skills, multiple skills not normally found in the in-house staff, or special tools, and where economic analysis indicates that it is more beneficial to the owner to contract the work to an outside maintenance service organization.

Among the most notable examples of common maintenance service contracts are those for automatic temperature control systems, central refrigeration cycles (refrigeration machines, cooling towers, pumps, and appurtenances), central boiler plants, water treatment services, and energy management and control systems.

There are generally two types of maintenance service contracts. The more common one is for a specific period of time for specific system equipment, and can be categorized as a *general maintenance service contract*. A type currently being considered for use by some major owners of properties is the *guaranteed performance and maintenance service contract*. In the latter, a specific item of equipment is sold by a manufacturer to an owner on the basis of guaranteed performance over a specified period including the cost of the item itself and its maintenance cost over the contract period.

32.14 GENERAL MAINTENANCE SERVICE CONTRACT

The general maintenance service contract between the maintenance service organization and the facility manager (or owner) normally includes the following items: the location of the equipment to be maintained and serviced; the contract time period and the intervals of service within the contract period; the responsibility of the maintenance service organization; the responsibility of the owner; the responsibility of any other interfacing parties; the methods of ensuring and monitoring maintenance performance; laws, codes, and standards that must be adhered to; the requirements for processing and retaining maintenance records; insurance requirements, including those required by regulatory authorities; maintenance and service schedules; and provisions for the termination of the contract after the completion of the contract period.

The rest of this section is an example of how the items are normally included in a maintenance service contract.

Location

1. XYZ Building, 1050 Manning Street, Oshkosh, AZ 01234
2. Cellar north equipment room

Equipment

Two (2) centrifugal chillers—water cooled—open motor, Model 1234, Serial No. ABCD.

Contract Period and Intervals of Service

1. The maintenance firm shall provide full maintenance service for the two (2) chillers and their appurtenances, including but not limited to the furnishing of all material, labor, supervision, tools, and supplies necessary to provide full maintenance services, repair, inspections, adjustments, tests, and replacement of parts for all equipment covered under this maintenance service contract, for an initial period of two (2) years. Full maintenance service includes all chiller components, including emergency call-back service on an as-needed basis during the period of the contract, systematic examinations, adjustments, lubrication, and repair and replacement of the chiller systems and their component parts. This includes all components and accessories essential for the proper operation and functioning of each complete chiller system.

2. The monthly and yearly maintenance requirements shall be the responsibility of the contractor for the summer months starting April 15 through November 15 inclusive each year.

Responsibilities of the Maintenance Service Organization (MSO)

1. The MSO shall provide all materials, labor, equipment, and services and perform all operations in connection with the maintenance service for two (2) centrifugal chillers including controls, control panels, power wiring, control wiring, relief valves, appurtenances, and related work.

2. The MSO shall provide field engineering start-up and testing control systems along with additional services as may be required to adjust and maintain proper operation of the equipment during the contract period.

3. The MSO shall provide the services of a trained field service representative to start up, test, adjust, balance, and instruct the owner's personnel in the on-going operation of the systems.

4. The MSO shall render monthly reports to the owner's representative.

5. The MSO shall notify the owner's representative of any conditions, such as leaks or improper system operation, that may affect the maintenance of the chillers and appurtenances.

6. The MSO shall instruct the operating personnel on the usage and maintenance of the systems and equipment provided and on maintenance procedures on equipment for emergency purposes, handling of chemicals, lubricants, routine tests, routine minor equipment adjustments, and safety procedures.

7. The MSO shall instruct the owner's personnel in the safe handling of the refrigerants, lubricants, and chemicals used and provide all safety devices, safety gloves, safety glasses, safety aprons, and safety materials necessary for the owner's staff to handle the refrigerants, lubricants, and chemicals.

8. The MSO shall provide maintenance safety data sheets for all substances.

9. The MSO shall provide all work, including emergency repairs, repairs, additions, alterations and modifications required to maintain the chillers and appurtenances in compliance at all times with current laws, current codes, and current safety standards.

10. The MSO shall provide the necessary labor required to make periodic inspections of all of the systems specified and shall make all adjustments, including the replacement or repairs of the aforementioned equipment and systems, that are necessary to keep the equipment and systems in operating condition.

11. The MSO shall provide emergency call-back service by promptly responding to requests from the owner's representative by telephone (or by any other means) for emergency service at any hour on any day (including weekends and holidays).

12. The MSO shall provide all parts, materials, apparatus, and equipment. Only parts that are designed and fabricated mechanically and are structurally correct and suitable in all respects for the installation shall be used.

13. The MSO shall maintain at the site, at all times, a sufficient supply of duplicate equipment, parts, and materials, for emergency or rapid replacement.

14. All materials, design, apparatus, and construction shall be subject to the approval of the owner's insurance company and comply with all laws, codes, rules and regulations, and ordinances, both local and state, together with all standards.

   The MSO shall provide a certificate stating that the equipment complies with OSHA rules, the National Electrical Code, and the applicable Underwriters Laboratories standards with respect to motor protection, grounding, and protection against hazards.

15. The MSO shall provide the owner all tools, jigs, hoists, motors, tachometers, wiring diagrams, apparatus, equipment, materials, and instruments necessary to keep the chillers operational.

16. In the entrance and exit of all workers; in bringing in, storing, or removal of all material and equipment; and in the manner and time of performing the work, the MSO shall cooperate with those in authority on the premises so as to avoid unnecessary dust, mud, or accumulated debris and otherwise adhere to sanitation requirements; and shall prevent personal injuries and the loss of or damage to the property of the owner or the owner's employees. The MSO will make the necessary arrangements with the owner's representative for access into the building when it is required to bring in and set equipment in place. Where necessary, equipment shall be brought into the building in sections and reassembled on the job by the manufacturer's mechanics or qualified mechanics under the supervision of the manufacturer's representative. The MSO shall repair to the satisfaction of the owner, and without additional cost to the owner, any and all damages it may cause to the property of the owner.

17. Parts of the buildings designated by the owner may be used by the MSO as a shop and as storage facilities, subject to the written approval of the owner's representative.

18. The MSO shall assume all responsibilities and costs for temporary and fixed instruments and recording devices as requested by the owner for checking operating conditions.

19. The MSO shall place upon the building, equipment, and apparatus or any part thereof, only such loads as are consistent with the safety of that portion of the building, equipment, and apparatus. The MSO will check and verify that maintenance service items are not stacked on or carried over floor and roof construction such that they would stress any of the members beyond their designed life loads.

20. The MSO shall provide and install in the refrigeration plant a *Record Schedule of Operations* as required.

Owner's Responsibilities

During the contract period, the following items of work will be performed by the owner's staff:

1. The owner's staff will keep the exterior of the machinery and any other parts of the equipment clean and presentable at all times.

2. The owner's staff will perform regular adjustments, cleaning, and lubrication of the equipment and systems, as set forth in instructional data provided by the MSO.

3. The owner's staff will perform the required periodic routine minor adjustments as set forth in the instruction data provided by the MSO.

Quality Assurance and Monitoring

The MSO's authorized service representatives and each specialty technician shall be assigned to perform the specialized maintenance and shall be thoroughly qualified in all respects to perform the maintenance and repairs that may become necessary during the term of the contract. The MSO shall maintain an on-going technical and safety training program for its employees and the owner's staff. The MSO shall have and maintain backup factory-authorized service representatives who are completely qualified in all respects to assume the maintenance of the equipment in the event of sickness or other cause of absence of the assigned representative.

The service technicians of the MSO shall be required to provide a monthly written report to the owner detailing all work performed for the month. For this report the MSO shall maintain a complete written record of all emergency call-back, replacement, and repair work performed. This information shall be consolidated by the MSO into a monthly report to the owner. This monthly report shall indicate the equipment worked on, the date the work was performed, type of work (call-back, replacement, or repair), brief description of the work performed, person-hours expended, and materials used.

The quality of maintenance service shall be subject to inspection and monitoring by the owner's representative. If it is found that the quality of the maintenance service being performed is not satisfactory, and that the maintenance service requirements are not being met, the MSO will be notified of these deficiencies in writing, and it shall be the responsibility of the MSO to make the necessary corrections.

During the term of the contract, maintenance and safety inspections will be performed by the owner's representative. The MSO shall accompany the owner's representative for a review of all equipment covered under the contract. At the conclusion of this inspection, the owner shall give the MSO written notice of any deficiencies found, and it shall be the responsibility of the MSO to make the necessary corrections.

Periodic inspection of the equipment as required by ASHRAE Refrigeration Code, NFPA Standards, and the Life Safety Code shall be performed by the MSO.

The MSO shall provide personnel who are familiar with the equipment to perform tests. The tests shall be conducted in the presence of the owner's representative. It is the responsibility of the MSO to determine when these tests are due. After completion of the required tests, the MSO is required to submit a document to the Owner providing at a minimum the following information:

1. Machine(s) being tested
2. Type of test
3. Date test was performed
4. Name of organization performing test
5. Signature of individual performing tests
6. Name of the owner's representative witnessing the tests
7. Regulatory agency
8. Signature of regulatory agency representative

Insurances

The MSO shall indemnify and save harmless the owner from suits, actions, damages, and costs of every name and description relating to or in any way connected with its performance under the contract. (Normally the owner would stipulate the types of insurance required and each policy limit required.)

Codes and Standards

All maintenance service work shall be performed in strict conformance with all laws, codes, and standards.

Interfacing with the Owner's Staff and Other MSOs

All maintenance service work shall be coordinated with the work of the owner's maintenance staff.

Retention of Records

The MSO shall retain and store at the site all records including those required by regulatory authorities. All records become the property of the owner.

Maintenance Service Schedule

The following is intended only as a guide and covers only a fictitious chiller unit and its components, but based on York Corporation data for water cooled centrifugal R-11 chillers. It does not cover other related system components.

The monthly and yearly maintenance requirements shall be the responsibility of the contractor for the summer months starting April 15 through November 15 inclusive each year. All components shall be maintained according to the manufacturers' written instructions.

The seasonal start-up includes the following:

1. Inspect centrifugal chiller for any damage that may have occurred during shutdown period. Report any damage to the owner's representative.

2. Check the system 24 hours before seasonal start-up; perform the following in strict conformance with the manufacturer's instructions:

   a. Properly pressurize the units and check that each unit is properly charged and that there are no leaks in the refrigeration system. Recharge unit.

   b. Check oil sump and purge oil heaters and temperatures.

   c. Compressor lubrication circuit shall be primed with manufacturer's *Special Oil* before start-up according to the manufacturer's instructions.

   d. The compressor oil level must be maintained within the sight glass at all operating conditions. If it is necessary to add oil, do so per the manufacturer's instructions.

   e. Check chilled water pump operation. Check condenser and cooling tower operation. Check and adjust water pump flow rates and pressure drops across both the cooler and condenser as per design intent.

   f. Check control panel to see that it is free of foreign material (such as wires and metal chips).

   g. Visually inspect wiring (power and control). All electrical system components must meet National Electric Code and all other codes and regulations.

   h. Check for proper size fuses in main and control power circuits as per manufacturer's requirements.

   i. Verify that wiring matches the three-phase power requirements of the compressor, as per manufacturer's requirements.

   j. Ensure control power wiring has correct capacity and voltage as per manufacturer's instructions.

   k. Be certain all control bulbs are inserted completely in their respective wells and are coated with heat conductive compound, as per manufacturer's instructions.

   l. As per manufacturer's instructions, set the return or leaving chilled water temperature to the desired set point temperature.

   m. Check insulation. Repair/replace damaged insulation as required.

   n. As per manufacturer's instructions, ensure that all user-adjustable cutouts are properly set. Adjust them as required.

3. Testing, checking, and starting must include the following:

   a. Check compressor and motor as per manufacturer's instructions.

   b. Start machine, check all controls, including microprocessor and diagnostic display module, and calibrate as per manufacturer's instructions.

   c. Complete operating log and record settings.

   d. Check refrigerant and oil levels. Recharge/refill as required.

   e. Check solid-state starter operation and record voltage and current as per manufacturer's recommendations.

   f. Set up operating log with operator, instruct and advise operator of troubleshooting techniques, as per manufacturer's recommendations.

   g. Check condition of contacts for wear and pitting. Replace if required.

   h. Examine compressor components. Repair/replace as required.

   i. Tighten motor terminals and control panel terminals.

   j. Clean and check oil filter. Replace filter if required.

   k. Check accuracy of all temperature sensors. Calibrate if necessary (as per manufacturer's instructions).

   l. Train operators or building engineers on equipment operation as per manufacturer's recommendations.

   m. Inspect unit casing and accessories for chipping or corrosion. If damage is found, clean and repaint with a high-grade rust-resistant owner-approved paint.

   n. Check and report entering and leaving water temperature and associated pressure drop for evaporator. Record pressure difference on inlet and outlet side of chilled water pump.

   o. Check operation of the compressor lubrication system and repair/replace as required by manufacturer's instructions.

Monthly maintenance and service includes the following:

1. Adjust operating and safety controls. Record settings. With respect to operating pressures and temperatures, check to see that operating pressures and temperatures are within the **limitations recommended by the manufacturer.**

2. Complete operating log of temperatures, pressures, voltages, and amperages.

3. Check and record entering and leaving water temperatures and associated pressure drop of evaporator. With respect to operating pressures and temperatures, check to see that operating pressures and temperatures are within the **limitations recommended by the manufacturer.**

4. Check refrigerant and oil levels. Recharge/refill as required.

5. Oil pressure should be the minimum amount recommended by the manufacturer.

6. Check operation of control circuit, including functions on microprocessor and diagnostic display module. Repair/replace all defective components.

7. Check operation of the lubrication system. Repair/replace as required.

8. Record and log hours of operation and number of starts.

9. Check customer's log with operator, discuss general operation of the machine.

10. Report to operator any uncorrected deficiencies previously noted.

The following tasks are performed once each year during shutdown to evaluate the equipment status properly and to prepare each unit for the next cooling season:

1. Check the compressor motor assembly as follows:
   a. Record voltages.
   b. Megger and record motor winding resistance.
   c. Lubricate open motor system.
   d. Check the alignment on open motor assembly.
   e. Check the coupling.
   f. Check all seals and gaskets.
   g. Check inlet vane operator and linkage; lubricate and calibrate where required.

2. Check the compressor oil system as follows:
   a. Change oil, oil filter, and dryer. Oil shall be disposed in legal manner by MSO. Copy of laboratory oil analysis will be given to the owner.
   b. Submit sample of compressor oil to independent laboratory for analysis. Analyze results of test and provide recommendations for corrective action if required (e.g., change oil). Oil to be changed as per manufacturer's instructions.
   c. Check oil pump, seal, and motor.
   d. Clean the dirt leg.

e. Check heater and thermostat.
   f. Check all other oil system components, including the cooler, strainer, and solenoid valve.

3. Check the solid-state motor starter as follows:
   a. Run diagnostic check.
   b. Clean contacts or recommend replacement.
   c. Check linkage.
   d. Megger motor.
   e. Check all terminals and tighten connections.
   f. Check overloads and dash pot oil; calibrate.
   g. Clean or replace air filter where required.
   h. Dry running starter (or before start-up); check status lights.

4. Review the control panel for the following items:
   a. Run diagnostic check of microprocessor control panel.
   b. Check safety shutdown operation.
   c. Check all terminals and tighten connections.
   d. Check accuracy of display data and set points.

5. Review the purge unit as follows:
   a. Inspect the operation of the unit.
   b. Change the oil.
   c. Change the filter dryer.
   d. Clean the orifice in the liquid feedline to the coil.
   e. Clean the gas strainer.
   f. Clean the solenoid valves.
   g. Clean the purge drum; check and clean the float valve; replace gaskets.
   h. Check heater operation.
   i. Drain and flush purge shell.
   j. Check all other components for proper condition and operation.
   k. Record pressure control set point.

6. Check condenser for the following items:
   a. Check for proper water flow rate.
   b. Check flow switch operation.
   c. Remove condenser head and inspect end sheets.
   d. Mechanically brush-clean condenser water tubes.

7. Check the cooler for the following items:
   a. Check for proper water flow rate.
   b. Check flow switch operation.
   c. Check refrigerant level.
   d. Mechanically brush-clean chilled water tubes.

8. Check the system for the following items:

   a. Submit a sample of refrigerant to the laboratory for analysis. Analyze results of test and provide recommendations for corrective action if required (e.g., change oil). The refrigerant and/or oil should be changed as per manufacturer's instructions.

   b. Conduct a leak check and identify leak sources.

   c. Add refrigerant as required.

   d. Record condition of sight glasses.

   e. Check the refrigerant cycle to verify the proper operating balance.

   f. Check condenser water and chilled water heat transfer.

9. General items must also be included:

   a. Repair insulation removed for inspection and maintenance procedures.

   b. Clean equipment and surrounding area upon completion of work.

   c. Consult with operator and provide any additional training needed.

   d. Ascertain availability of factory training courses.

   e. Report deficiencies and repairs required.

Termination at End of Contract Period

If the owner does not elect to continue the services of the MSO beyond the contract period, the MSO shall remove all cartons and unused materials from the site. All equipment, tanks, feeders, pumps, valves, piping, controls, and test kits shall remain the property of the owner. The title to the maintenance equipment at the site shall be transferred to the owner at the time of the completion of the contract.

REFERENCES AND FURTHER READING

Akalin, M. T., "Equipment Life and Maintenance Cost Survey," *ASHRAE Transactions,* Volume 84, Part II, 1978.

ASHRAE, "Chapter 42, Life Cycle Costing," *ASHRAE Handbook 1984, Systems,* ASHRAE, Inc., Atlanta, GA, 1984.

ASHRAE, "Chapter 59, Mechanical Maintenance," *ASHRAE Handbook 1987, HVAC Systems and Applications,* ASHRAE, Inc., Atlanta, GA.

ASHRAE, "Chapter 33, Owning and Operating Costs," *ASHRAE Handbook 1991, Applications,* ASHRAE, Inc., Atlanta, GA.

Chen, P. T., and Chapman, R. E., "Budget Estimates for Replacement of Plant and Facility Equipment at the National Bureau of Standards," *ASHRAE Transactions,* Volume 87, Part I, 1981.

Crawford, B. M., and Altman, J. W., "Chapter 12, Designing for Maintainability," in H. P. Van Cott and R. G. Kinkade (eds.), *Human Engineering Guide to Equipment Design,* rev. ed., Sponsored by Joint Army-Navy-Air Force Steering Committee, American Institute for Research, Washington DC, McGraw-Hill, New York, 1972.

Dillow, R. O. (editor-in-chief), *Facilities Management, A Manual for Plant Administration,* 2d ed., Association of Physical Plant Administrators of Universities and Colleges, Alexandria, VA, 1989.

Dohrmann, R., and Alereza, T., "Analysis of Survey Data on HVAC Maintenance Costs," *ASHRAE Transactions,* Volume 92, Part II, 1986.

Dwyer, M. J., "Preventive Maintenance," in R. O. Dillon (editor-in-chief), *Facilities Management, A Manual for Plant Administration,* 2d ed., pp. 1072–1077, Association of Physical Plant Administrators of Universities and Colleges, Alexandria, VA, 1989.

Hansen, S. J., "Rx for Improved Air Quality, An Ounce of Operations and Maintenance Prevention," *Strategic Planning for Energy and the Environment,* Vol. 10, No. 2, pp. 7–24, 1990.

National Institute of Building Sciences, *Hospital User Manual Series Guidelines: Vol. 4. Maintenance Manual,* NIBS, Washington, DC 20006, May 20, 1988.

"Research and Development of Material, Maintainability Engineering," Department of the Army Pamphlet 705-1, June 1966.

Seboda, E. F., (ed.), *Comprehensive Maintenance and Repair Program, Guidelines and Standards for the Maintenance and Repair of State Owned Facilities,* State of Maryland, Department of General Services, Baltimore, 1978, Revised 1983.

Van Cott, H. P., and Kinkade, R. G. (ed.), *Human Engineering Guide to Equipment Design* rev. ed., Sponsored by Joint Army-Navy-Air Force Steering Committee, American Institute for Research, Washington DC, McGraw-Hill, New York, 1972.

York International Corporation, *Operations and Maintenance Data for Centrifugal R–11 Chillers—Water Cooled, Model YTC3C3B2-CGF,* York International Corporation, York, PA.

APPENDIX

I

I.1 Nomenclature

A area, ft^2

a turbulence factor

A_a face area, ft^2

A_c core area of outlet, ft^2

A_D Dubois surface area of naked body, ft^2

$A_{e.1}$ effective leakage area, ft^2

A_f area of the fins, ft^2

A_i inner surface area, ft^2

A_k net or unobstructed area of grille, ft^2

A_o outer surface area, ft^2

A_p primary surface area, ft^2

A_r cross-sectional area of the room perpendicular to air flow, ft^2

Ar Archimedes number

AS annual savings, dollars

B ratio of the outer surface area to the inner surface area A_o/A_i

Bi Biot number

C scale factor
Concentration of air contaminant, mg/m^3
local loss coefficient
cost, dollars

C_{cc} cloudy cover factor

C_d discharge coefficient
degradation coefficient

C_{di} duct installation cost, dollars

C_e energy cost, dollars

C_{cir} circulating factor

C_{iu} unit cost of duct installation, \$/ft^2

C_n clearness number of the sky

C_o local loss coefficient

c_p specific heat at constant pressure, Btu/lb·°F

c_{pa} specific heat of moist air at constant pressure, Btu/lb·°F

c_{pd} specific heat of dry air at constant pressure, Btu/lb·°F

c_{pr} specific heat of liquid refrigerant at constant pressure, Btu/lb·°F

C_{pre} pressure loss coefficient

c_{ps} specific heat of water vapor at constant pressure, Btu/lb·°F

c_{pw} specific heat of water at constant pressure, Btu/lb·°F

c_{sat} saturation specific heat per degree wet bulb, Btu/lb·°F

$C_{s.i}$ inlet system effect loss coefficient

$C_{s.o}$ outlet system effect loss coefficient

C_t scale factor for temperature lines, °F/ft

C_{to} total cost, dollars

C_v flow coefficient, a flow rate of 1 gpm at a pressure drop of 1 psi

C_w scale factor for humidity ratio lines, lb/lb·ft

$C + R$ convective and radiative heat loss, Btu/h·ft^2

CDD cooling degree-day, degree-day

COP coefficient of performance

COP_c coefficient of performance of chiller

COP_{hp} coefficient of performance of heat pump

COP_{hr} coefficient of performance of heat recovery

COP_{ref} coefficient of performance of refrigeration

CRF capital recovery factor

D diameter, in. or ft
depreciation, dollars per year

D_{aw} mass diffusivity for water vapor through air, ft^2/s

D_e equivalent diameter, in. or ft

D_h hydraulic diameter, in. or ft

D_{lv} mass diffusivity of liquid and vapor, ft^2/h

D_T mass diffusivity due to temperature gradient, ft^2/h

E evaporative heat loss, Btu/h·ft^2
electric potential, volt
efficiency

e exergy, Btu/lb

E_{dif} evaporative heat loss due to direct diffusion, Btu/h·ft^2

E_{max} maximum level of evaporative heat loss, Btu/h·ft^2

E_r unit energy cost, \$/kWh

E_{rsw} sweating due to thermoregulatory mechanism, Btu/h·ft^2

EER energy efficiency ratio, Btu/h·W

ET* effective temperature, °F

F factor
shape factor
solar heat gain factor

f friction factor
frequency, Hz

$F_{a.t}$ air transport factor

F_{block} blockage factor

F_{cl} clothing efficiency

F_{cyc} cycling loss factor

F_p performance factor

F_{rej}, HRF heat rejection factor

F_s coil core surface area parameter

F_t fin thickness, in.
fan total pressure, in. WG

Fo Fourier number

Fr Froude number

G mass velocity, lb/h·ft^2

g gravitational acceleration, ft/s^2

g_c dimensional conversion factor, 32 lb$_m$·ft/lb$_f$·s^2

Gr Grashof number

H head, ft of water column
height, ft
hour angle, degrees

h enthalpy, Btu/lb
heat transfer coefficient, Btu/h·ft^2·°F

h_b boiling heat transfer coefficient, Btu/h·ft^2·°F

h_{con} condensing heat transfer coefficient, Btu/h·ft²·°F

H_f frictional head loss, ft of water

h_{fg} latent heat of vaporization, Btu/lb

$h_{fg.r}$ latent heat of vaporization of refrigerant, Btu/lb

$h_{fg.32}$ latent heat of vaporization at 32°F, Btu/lb

h_i inner surface heat transfer coefficient, Btu/h·ft²·°F

h_m convective mass transfer coefficient, ft/h

h_o heat transfer coefficient at outer surface, Btu/h·ft²·°F
   outdoor air enthalpy, Btu/lb

h_r enthalpy of space air

h_{ru} enthalpy of recirculating air, Btu/lb

H_s static head, ft of water

h_s enthalpy of saturated air film, Btu/lb

h enthalpy of supply air

$\Delta H_{s.r}$ head difference between supply and return mains, Ft WG

$\Delta h_{s.r}$ enthalpy of saturated air film at evaporating temperature, Btu/lb

H_t total head, ft of water

H_v velocity head, ft of water

h_{wet} heat transfer coefficient of wetted surface, Btu/h·ft²·°F

HC   heating capacity, Btu/ft²·°F

HDD   heating degree-day, degree-day

I electric current, amp

i interest rate, percent

I_D direct radiation, Btu/h·ft²

I_d diffuse radiation, Btu/h·ft²

I_G global radiation on a horizontal plane, Btu/h·ft²

I_{DN} solar radiation on a surface normal to sun rays, Btu/h·ft²

i_m moisture permeability of clothing

I_o extraterrestrial solar intensity, Btu/h·ft²

I_{rad} effective radiant field, W/ft²

I_{ref} reflection of solar radiation, Btu/h·ft²

I_t total intensity of solar radiation, Btu/h·ft²

I_{tur} intensity of turbulence

I_{sc} solar constant, 434.6 Btu/h·ft²

IC   initial cost, dollars

IL   insertion loss, dB

J Joule's equivalent, 778 ft.lb$_f$/Btu

j cost escalation factor

K constant
   factor
   coefficient
   derivative gain

k thermal conductivity, Btu·in./h·ft²·°F, Btu/h·ft·°F

K' wet bulb constant

K_{cc} cloudy reduction factor

K_i integral gain

K_P power constant

K_p proportional gain
   pressure constant

K_V volume constant

L distance, thickness, ft
   sound level, dB
   latitude angle, degree

$L_{a.f}$ air flow noise, dB

L_e equivalent length, ft

L_w sound power level, dB re 1pW

L_p sound pressure level, dB re 20 μPa

L_{pA} sound pressure level in dBA

L_{pr} room sound pressure level, dB

L_w vertical distance between state points, ft

L_t horizontal distance between state points, ft

$L_{w.b}$ branch power division, dB

L_{wr} room sound power level, dB

Le   Lewis number

LHG   latent heat gain, Btu/h

LR   Lewis relation

M molecular weight
   metabolic rate, Btu/h·ft²

m mass, lb

m'' slope of air enthalpy saturation curve

$\dot{m}$ mass flow rate, lb/min, lb/h

$\dot{m}_{par}$ rate of air contaminants generated, mg/s

$\dot{m}_r$ mass flow rate of refrigerant, lb/h, lb/min

m_s surface density, lb/ft²

N number

n number of moles, mol
   number of air changes, ach
   circulation number
   amortization period, year

n' depreciation period, year

N_r number of rows

NTU   number of transfer units

Nu   Nusselt number

OC   operating cost, dollars

P power, hp or kW
   perimeter, ft
   penetration

p pressure, psi, psia, psig

P_{air} air power, hp

p_{at} atmospheric pressure, psia

P_{cfm} power per unit volume flow, W/cfm

P_{com} compressor power, hp

p_{con} condensing pressure, psig

p_{dis} discharge pressure of compressor, psig

Δp_{dy} dynamic loss, in. WG

p_{ev} evaporating pressure, psig

P_f fan power input, hp

Δp_f pressure drop due to frictional and dynamic losses, in. WG

p_{fill} fill pressure, abs psia

Δp_{fix} fixed part of system pressure loss, in. WG

$\Delta p_{f.s}$ frictional loss per floor inside the pressurized stairwell, in. WG

$\Delta p_{f.u}$ duct frictional loss per unit length, in. WG

P_H horizontal projection, ft

P_{in} power input, hp

Δp_{od} total pressure loss of damper when fully open, in. WG

P_p pump power, hp

Δp_{p-od} total pressure loss of air flow path excluding damper, in. WG

Δp_{res} residual pressure, in. WG

p_s static pressure, in. WG, psig

$\Delta p_{s.i}$ inlet system effect pressure loss, in. WG

$\Delta p_{s.o}$ outlet system effect pressure loss, in. WG

$\Delta p_{s.r}$ static regain, in. WG

p_{st} pressure due to stack effect, lb_f/ft^2

p_{suc} suction pressure of compressor, psig

Δp_{sy} system pressure loss, in. WG

p_t total pressure, in. WG

$p_{t.ex}$ external total pressure, in. WG

P_V vertical projection, ft

p_v velocity pressure, in. WG

Δp_{var} variable part of system pressure loss, in. WG

p_{vo} velocity pressure at outlet, in. WG

p_{vw} wind velocity pressure, in. WG

p_w water vapor pressure, psia

PMV predicted mean vote

Pr Prandtl number

Q rate of heat transfer, Btu/h

q rate of heat transfer, Btu/h

q_c coil's load, Btu/h

q_{cc} cooling coil's load, Btu/h

$q_{c.c}$ corrected cooling capacity, Btu/h

$q_{c.r}$ catalog listed cooling capacity, Btu/h

$q_{c.wet}$ heat and mass transfer, Btu/h

q_{ev}, q_{ref}, q_{rl} refrigeration load at evaporator, Btu/h

q_{1g} heat input to the first-stage generator, Btu/h·ton

q_{ch} heating coil's load, Btu/h

q_{int} internal load, Btu/h

q_{rc} space cooling load, Btu/h

Q_{rej}, THR total heat rejection, Btu/h

q_{rh} space heating load, Btu/h

q_{RCi} inward heat flow from the inner surface of sunlit window, Btu/h

q_{rs} space sensible cooling load, Btu/h

q_{rsp} space sensible cooling load at part-load, Btu/h

$q_{rs,t}$ sensible cooling load at time t, Btu/h

q_{sen} rate of sensible heat transfer, Btu/h

q_{tran} transmission loss, Btu/h

q_{wi} heat gain admitted into conditioned space, Btu/h

R gas constant, ft·lb_f/lb_m·°R
electric resistance, ohm
thermal resistance, h·°F/Btu
flow resistance, in. WG/(cfm)^2
ratio

R R-value, h·ft^2·°F/Btu

R_c radius of curvature, in. or ft

R_{cl} thermal resistance of clothing, h·ft^2·°F/Btu,

R_{com} compression ratio

R_{en} entrainment ratio

R_f fouling factor h·ft^2·°F/Btu

R_{fa} ratio of free area to gross area

R_g gas side thermal resistance, h·ft^2·°F/Btu

R_{load}, LR load ratio

R_o universal gas constant, ft·lb_f/lb_m·°R

$R_{T.1}$ ratio of temperature lift

Re Reynolds number

ROR rate of return

S salvage value, dollar
heat storage, Btu/h·ft^2

s specific entropy, Btu/lb·°R
dimensionless distance

S_f fin spacing, fins/in.

S_H shaded height, ft

S_W shaded width, ft

SC shading coefficient

Sc Schmidt number

SHG sensible heat gain, Btu/h

SHGF solar heat gain factors, Btu/h·ft^2

SHR sensible heat ratio

SHR_c sensible heat ratio of cooling and dehumidification process

SHR_s sensible heat ratio of space conditioning line

SHR_{sp} sensible heat ratio of space conditioning line at part-load

SP simple payback, years

St Strouhal number

T temperature, °F

T' wet bulb temperature, °F

T^* thermodynamic wet bulb temperature, °F

T_∞ bulky air unaffected by surface, °F

t_{an} annual operating hours, h

T_{co} changeover temperature, °F

T_{dew}, T'' dew point temperature, °F

T_{dis} discharge temperature, °F

ΔT_f fan temperature rise, °F

T_{en} mass temperature of building envelope, °F

T_m temperature of mixture, °F

ΔT_m log-mean temperature difference, °F

T_o operative temperature, °F
outdoor temperature, °F

T'_{os} summer mean coincident wet bulb, °F

$T_{o.ws}$ statistically determined winter design outdoor temperature, °F

T_p plenum air temperature, °F
T_R absolute temperature, °R
T_r space temperature, °F
$T_{\text{ra}}, T_{\text{rad}}$ mean radiant temperature, °F
T_{rm} average space temperature, °F
T_{rp} space temperature at part-load, °F
$T_{\text{r.p}}$ space temperature in perimeter zone, °F
T_{ru} temperature of recirculating air, °F
T_s supply temperature, °F
ΔT_{sa} temperature difference between the surface and air, °F
$\Delta T_{\text{t,r}}$ throttling range, °F
T_{ws} chilled water supply temperature, °F
TD temperature differential, °F
TL transmission loss, dB
TL_{in} breakin transmission loss, dB
TL_{out} breakout transmission loss, dB
U overall heat transfer coefficient, Btu/h·ft²·°F
u internal energy, Btu/lb
    peripheral velocity, fpm
U_i overall heat transfer coefficient based on inner surface area, Btu/h·ft²·°F
U_o overall heat transfer coefficient based on outer surface area, Btu/h·ft²·°F
UAC uniform annual cost, dollars
V volume, ft³
$\dot{V}$ volume flow rate, cfm
v velocity, fpm of ft/s
    specific volume or moist volume, ft³/lb
v_c centerline velocity, fpm
v_{con} air velocity in constricted part of damper or duct fittings, fpm
$\dot{V}_{\text{conv}}$ volume flow rate of upward convective flow, cfm
$\dot{V}_{\text{ef}}$ volume flow rate of exfiltrated air, cfm
v_{fc} face velocity, fpm
$\dot{V}_{\text{gal}}$ volume flow rate of chilled water, gpm
$\dot{V}_{\text{lk}}, \dot{V}_L$ volume flow rate of air leakage, cfm
$\dot{V}_o$ volume flow rate of outdoor air, cfm
$\dot{V}_{\text{o.dm}}$ minimum outdoor air supply volume flow at design conditions, cfm
$\dot{V}_{\text{oif}}$ volume flow rate of outdoor and infiltrated air, cfm
$\dot{V}_{\text{osn}}$ zone outdoor air supply volume flow rate, cfm
$\dot{V}_p$ piston displacement, cfm
$\dot{V}_s$ supply volume flow rate, cfm
$\dot{V}_{\text{sp}}$ supply volume flow rate at part-load, cfm
W work, Btu/lb
    mechanical work performed, Btu/h·ft²
    sound power, dB
    width, ft
w relative velocity, fpm
    humidity ratio, lb/lb
W_{in} work input, Btu/lb
W_{isen} isentropic work, Btu/lb
W_{rsw} wetted portion of human body due to sweating
w_s^* saturated humidity ratio at thermodynamic wet bulb temperature, lb/lb
X moisture content, dimensionless or percent mass fraction
x mole fraction
    quality or dryness fraction
    coordinate dimension
x_{rl} quality of refrigerant leaving overfeed cooler
y vertical drop of air jet, ft
Z compressibility factor
z elevation, ft
z_{stat} stationary level, ft

Subscripts

a air
    ambient
    absorber
ab absorber
    air at wet-dry boundary
ae, aen entering air
alv leaving air
am ambient
at atmospheric
av average
b body
    bleed
    branch
    building material
bg building
by bypass
c coil
    cooling
    cold
    convective
    condenser
    compressor
    common end
    corrected
ca cooling air
    cooled air
cc cooling coil
c.d closed door
ce entering condenser
ch heating coil
cl clothing
    cooling load
    latent cooling coil
    leaving condenser
cn common
co changeover

com compressor
con condensing
    condenser
corr correction
cr body core
cs sensible cooling coil
d duct
    design
dam damper
deh dehumidifier
des desiccant
dif diffuser
dis discharge
dl process air leaving desiccant dehumidifier
dy dynamic
e entering
ee entering evaporator
ef exfiltrated
eff effective
el elevation
    leaving evaporator
    equivalent
en entering
en.c cooling water entering condenser
ev evaporating
    evaporator
ev.c evaporative cooling
ex, exh exhaust
exf, ef exfiltration
f fan
fc fan–coil
fix fixed part
fl floor
fu furnace
g moisture gain
    gas
    globe
    ground
    generator
1g first-stage generator
2g second-stage generator
go saturated water vapor at 0°F
h higher
    heat exchanger
    heating
    hot
hg heat gain
h.t heat transformer
hu humidifier
i inlet
    input
    indoor
    interior
    intermediate
    inner surface

if infiltration
in input
    indirect
    interior
int intermediate
k conduction
l latent
    liquid
    lower
    lights
    leaving
lc liquid at condenser
le water entering evaporator
liq liquid
lk leakage
lr liquid refrigerant
lv leaving
m mixture
    mean
    maximum
    motor
mat material
max maximum
min minimum
mo, mot motor
n number
o outdoor
    output
    outlet
    outer surface
    oversaturation
o.d open door
oi inward flow from outer surface
os outer surface
o.s summer outdoor
o.sys system outdoor air
out outdoor air
p constant pressure
    people
    part-load
    pump
    process air
    primary air
par particulates in air
pd dry air at constant pressure
pl process air after sensible heat exchanger
ps water vapor at constant pressure
    primary air supply
pt plant
r space
    room
    return
    refrigerant
    radiative
    regeneration air

rc space cooling
   refrigeration capacity
rd return duct
rec recirculating
ref refrigeration
ref · f free refrigeration
rel release
res residual
   respiration
ret, rt return
rf return fan
   relief fan
   refrigeration effect
rg regeneration air entering desiccant dehumidifier
rh space heating
rl space latent
ro regeneration air leaving desiccant dehumidifier
rp return plenum
rs space sensible
r · s return system
ru recirculating air entering the AHU or PU
s supply
   steam
   saturated state
   surface
   sunlit
   straight-through end
   summer
s^* saturated at thermodynamic wet bulb temperature
sa solution at absorber
s · a saturated air film
sat saturation
sb surface at dry-wet boundary
sc subcooled
s · c cold air supply
sd supply duct
sen sensible
sf supply fan
sg solution at generator
sh shaded
s · h warm air supply
si supply air for interior zone
sil silencer
sk skin
sn supply air for zone n
sn · d zone supply air at design condition
snp supply air for zone n at part-load
sol solar
s · s supply system
s · t surface temperature
suc suction
sun sunlit
sur surroundings

sx supply air for perimeter zone
T total, overall, temperature
t time
un unconditioned
var variable part
ve saturated water vapor leaving evaporator
w water
   condensate
   winter
wb water at dry-wet boundary
   boiler hot water
we water entering
wet wet air
wl water leaving
w · o winter outdoor
ws water vapor at saturated state
xl regeneration air leaving the sensible heat
   exchanger
   exit of heat exchanger

Greek Letter Symbols

α mean temperature coefficient
   angle between the air conditioning
      process and the horizontal line on
      the psychrometric chart, degrees
   thermal diffusivity, ft^2/s
   absorptance
   spreading angle of air jet, degrees
   damper characteristic ratio
β solar altitude angle, degrees
   blade angle, degrees
γ ratio of specific heat at constant pressure
      to constant volume
   surface-solar azimuth angle, degrees
Δ difference
δ solar declination angle, degrees
ϵ emissivity
   absolute roughness, in.
   effectiveness; effectiveness factor
ϵ_{ex} air exchange efficiency
ϵ_{sat} saturation effectiveness
ϵ_{wet} wet coil effectiveness
η efficiency
η_{cb} combustion efficiency
η_{com} compressor efficiency
η_{cp} compression efficiency
η_{dr} efficiency of driving mechanism
η_f fan total efficiency
   fin efficiency
η_{fu} furnace efficiency
η_{isen} isentropic efficiency
η_{mec} mechanical efficiency
η_{mo}, η_{mot} motor efficiency
η_{ov} overall efficiency
η_p pump efficiency

η_s fan static efficiency
fin surface efficiency
η_{sat} saturation effectiveness
$\eta_{sy.h}$ system efficiency for heating
η_t fan total efficiency
η_v volumetric efficiency
η_{ww} wire-to-water efficiency
θ angle of incidence, degree
effective draft temperature, °F
μ absolute viscosity, lb/ft·s
degree of saturation
mechanical efficiency
ν kinematic viscosity, ft^2/s
ρ density, lb/ft^3
reflectance
ρ_{suc} density of suction vapor, lb/ft^3
Σ angle between tilting and horizontal surface, degree
σ Stefan-Boltzmann constant, 0.1714×10^{-8} Btu/h·ft^2·°R^4
standard deviation
τ transmittance
φ relative humidity, percent
solar azimuth angle, degrees
φ_f fin resistance number
ψ surface azimuth angle, degrees
Ω profile angle, degrees

I.2 Abbreviations

ABMA American Boiler Manufacturers Association
abs absolute
AC air conditioning, alternate current
ACEC American Consulting Engineers Council
ACI adjustable current inverter
ACP alternate component packages
ADC Air Diffusion Council
ADPI air diffusion performance index, percent
AFUE annual fuel utilization efficiency
AHU air-handling unit
AI analog input
AMCA Air Movement and Control Association
ANSI American National Standards Institute
AO analog output
ARI Air Conditioning and Refrigeration Institute
ASHRAE American Society of Heating, Refrigerating, Air Conditioning Engineers, Inc.
ASME American Society of Mechanical Engineers
ASTM American Society of Testing and Materials
AVI adjustable voltage inverter
BAS building automation system

BHP, bhp brake horse power
BI binary or digital input
backward inclined blade
BLAST building loads analysis and systems thermodynamics
BMS building management system
BO binary or pulsed output
BOCA Building Officials and Code Administrators Internationals
CABDS computer aided building design system
CADD computer aided design and drafting
CFC chlorofluorocarbons
CLF cooling load factor
CLTD cooling load temperature difference
COP coefficient of performance
CTD condenser temperature difference
DA direct acting
DC direct current
DDC direct digital control
DECOS design energy cost
DOE Department of Energy
DOP di-octyl phthalate
DSA double-strength sheet glass
DWDI double-width double-inlet
DX direct expansion, dry expansion
ECB energy cost budget
EEPROM electrically erasable, programmable, read-only memory
EIA Energy Information Administration of the Department of Energy
EMS energy management system
EPA Environmental Protection Agency
EPROM erasable programmable read-only memory
ETD equivalent temperature difference
FC fan–coil, forward-curved blade
FDA Food and Drug Administration
FOM figure of merit
GWHP ground water heat pump
GWP global warming potential
HEPA high-efficiency particulate air filters
HR heart rate
HSPF heating seasonal performance factor
HVAC&R heating, ventilating, air conditioning, and refrigeration
IAQ indoor air quality
ILD internal load density
I/O input/output
I-P inch-pound
IPLV integrated part-load value
IRS Internal Revenue Service
LiBr lithium bromide
LiCl lithium chloride
LPG liquefied petroleum gas
MAU make-up air unit

MCPC microcomputer constructed psychrometric chart
MPS manual position switch
NBC National Broadcasting Corporation
NC noise criteria curve
nomally closed
NCDC National Climatic Data Center
NFPA National Fire Protection Association
NIOSH National Institute of Occupational Safety and Health
NO normally open
NPL neutral pressure level
NPSH net positive suction head
NWWA National Water Well Association
ODP ozone depletion potentials
PC personal computer
PI proportional plus integral
PID proportional-integral-derivative
PU packaged unit
PTAC packaged terminal air conditioner
PURPA Public Utility Regulatory Policies Act
PVC polyvinyl chloride
PWM width modulated inverter
RA reverse acting
RAM read and write memory
RAU recirculating air unit
RC room criteria curve
RH relative humidity
ROM read-only memory

RTD resistance temperature detector
RTS room temperature sensor
SBS sick building syndrome
SEER seasonal energy efficiency ratio
SEUF seasonal energy utilization factor
SMACNA Sheet Metal and Air Conditioning Contractors' National Association
SI International System of Units
SPF seasonal performance factor
SSE steady-state efficiency
SSU Saybolt Seconds Univeral viscosity
SWSI single-width single-inlet
TA time averaging
TARP thermal analysis research program
TETD total equivalent temperature differentials
TFM transfer function method
TIMA Thermal Insulation Manufacturers Association
TRAV terminal regulated air volume
UL Underwriters Laboratories Inc.
ULPA ultra-low penetration air filters
VAV variable-air-volume
VDC volts direct current
VLSI very large scale integrated
VVVT variable-volume variable-temperature
WHO World Health Organization
WSHP water source heat pump
WWR window-to-wall ratio

Appendix
II

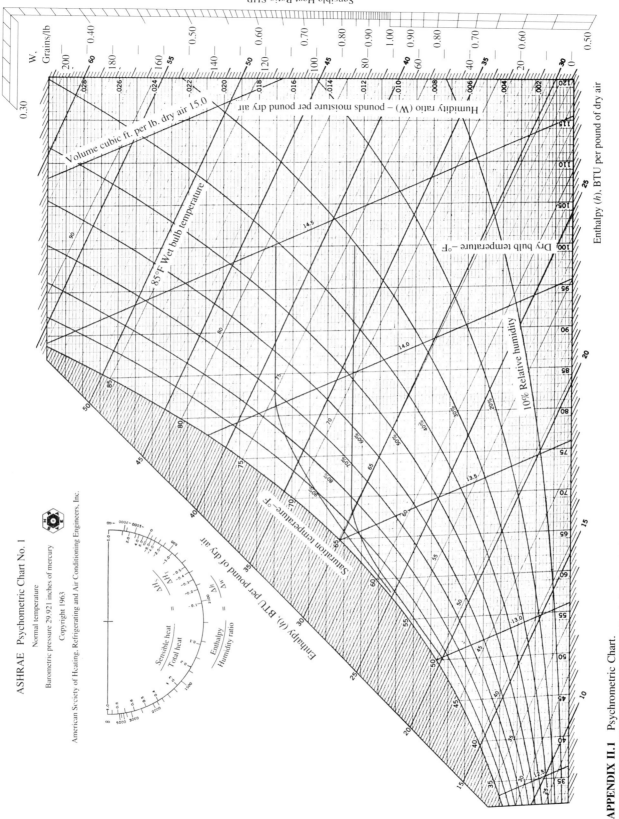

APPENDIX II.1 Psychrometric Chart.

Based on ASHRAE Psychrometric Chart No. 1. Reprinted with permission from ASHRAE Inc.

Sensible heat ratio (SHR), humidity ratio scale in grains/lb, and two cooling and dehumidifying curves were added by author.

APPENDIX II.2
Thermodynamic Properties of Moist Air (at atmospheric pressure 14.696 psia) and Water

Temp. T. °F	Humidity Ratio lb_w/lb_{da} s	Volume ft³/lb dry air			Enthalpy Btu/lb dry air			Saturated water vapor Absolute Pressure, p		Enthalpy, Btu/lb Sat. water		
		v_a	v_{as}	v_s	h_a	h_{as}	h_s	psi	in. Hg	liq. h_f	Evap. h_{ig}/h_{fg}	vapor h_g
32	0.003790	12.389	0.075	12.464	7.687	4.073	11.760	0.08865	0.18049	−0.02	1075.15	1075.14
33	0.003947	12.414	0.079	12.492	7.927	4.243	12.170	0.09229	0.18791	0.99	1074.59	1075.58
34	0.004109	12.439	0.082	12.521	8.167	4.420	12.587	0.09607	0.19559	2.00	1074.02	1076.01
35	0.004277	12.464	0.085	12.550	8.408	4.603	13.010	0.09998	0.20355	3.00	1073.45	1076.45
36	0.004452	12.490	0.089	12.579	8.648	4.793	13.441	0.10403	0.21180	4.01	1072.88	1076.89
37	0.004633	12.515	0.093	12.608	8.888	4.990	13.878	0.10822	0.22035	5.02	1072.32	1077.33
38	0.004820	12.540	0.097	12.637	9.128	5.194	14.322	0.11257	0.22919	6.02	1071.75	1077.77
39	0.005014	12.566	0.101	12.667	9.369	5.405	14.773	0.11707	0.23835	7.03	1071.18	1078.21
40	0.005216	12.591	0.105	12.696	9.609	5.624	15.233	0.12172	0.24783	8.03	1070.62	1078.65
41	0.005424	12.616	0.110	12.726	9.849	5.851	15.700	0.12654	0.25765	9.04	1070.05	1079.09
42	0.005640	12.641	0.114	12.756	10.089	6.086	16.175	0.13153	0.26780	10.04	1069.48	1079.52
43	0.005863	12.667	0.119	12.786	10.330	6.330	16.660	0.13669	0.27831	11.04	1068.92	1079.96
44	0.006094	12.692	0.124	12.816	10.570	6.582	17.152	0.14203	0.28918	12.05	1068.35	1080.40
45	0.006334	12.717	0.129	12.8946	10.810	6.843	17.653	0.14755	0.30042	13.05	1067.79	1080.84
46	0.006581	12.743	0.134	12.877	11.050	7.114	18.164	0.15326	0.31205	14.05	1067.22	1081.28
47	0.006838	12.768	0.140	12.908	11.291	7.394	18.685	0.15917	0.32407	15.06	1066.66	1081.71
48	0.007103	12.793	0.146	12.939	11.531	7.684	19.215	0.16527	0.33650	16.06	1066.09	1082.15
49	0.007378	12.818	0.152	12.970	11.771	7.984	19.756	0.17158	0.34935	17.06	1065.53	1082.59
50	0.007661	12.844	0.158	13.001	12.012	8.295	20.306	0.17811	0.36263	18.06	1064.96	1083.03
51	0.007955	12.869	0.164	13.033	12.252	8.616	20.868	0.18484	0.37635	19.06	1064.40	1083.46
52	0.008259	12.894	0.171	13.065	12.492	8.949	21.441	0.19181	0.39054	20.07	1063.83	1083.90
53	0.008573	12.920	0.178	13.097	12.732	9.293	22.025	0.19900	0.40518	21.07	1063.27	1084.34
54	0.008897	12.945	0.185	13.129	12.973	9.648	22.621	0.20643	0.42031	22.07	1062.71	1084.77
55	0.009233	12.970	0.192	13.162	13.213	10.016	23.229	0.21410	0.43592	23.07	1062.14	1085.21
56	0.009580	12.995	0.200	13.195	13.453	10.397	23.850	0.22202	0.45204	24.07	1061.58	1085.65
57	0.009938	13.021	0.207	13.228	13.694	10.790	24.484	0.23020	0.46869	25.07	1061.01	1086.08
58	0.010309	13.046	0.216	13.262	13.934	11.197	25.131	0.23864	0.48588	26.07	1060.45	1086.52
59	0.010692	13.071	0.224	13.295	14.174	11.618	25.792	0.24735	0.50362	27.07	1059.89	1086.96
60	0.011087	13.096	0.233	13.329	14.415	12.052	26.467	0.25635	0.52192	28.07	1059.32	1087.39
61	0.011496	13.122	0.242	13.364	14.655	12.502	27.157	0.26562	0.54081	29.07	1058.76	1087.83
62	0.011919	13.147	0.251	13.398	14.895	12.966	27.862	0.27519	0.56029	30.07	1058.19	1088.27
63	0.012355	13.172	0.261	13.433	15.135	13.446	28.582	0.28506	0.58039	31.07	1057.63	1088.70
64	0.012805	13.198	0.271	13.468	15.376	13.942	29.318	0.29524	0.60112	32.07	1057.07	1089.14
65	0.013270	13.223	0.281	13.504	15.616	14.454	30.071	0.30574	0.62249	33.07	1056.50	1089.57
66	0.013750	13.248	0.292	13.540	15.856	14.983	30.840	0.31656	0.64452	34.07	1055.94	1090.01
67	0.014246	13.273	0.303	13.577	16.097	15.530	31.626	0.32772	0.66724	35.07	1055.37	1090.44
68	0.014758	13.299	0.315	13.613	16.337	16.094	32.431	0.33921	0.69065	36.07	1054.81	1090.88
69	0.015286	13.324	0.326	13.650	16.577	16.677	33.254	0.35107	0.71478	37.07	1054.24	1091.31
70	0.015832	13.349	0.339	13.688	16.818	17.279	34.097	0.36328	0.73964	38.07	1053.68	1091.75
71	0.016395	13.375	0.351	13.726	17.058	17.901	34.959	0.37586	0.76526	39.07	1053.11	1092.18
72	0.16976	13.400	0.365	13.764	17.299	18.543	35.841	0.38882	0.79164	40.07	1052.55	1092.61
73	0.017575	13.425	0.378	13.803	17.539	19.204	36.743	0.40217	0.81883	41.07	1051.98	1093.05
74	0.018194	13.450	0.392	13.843	17.779	19.889	37.668	0.41592	0.84682	42.06	1051.42	1093.48
75	0.018833	13.476	0.407	13.882	18.020	20.595	38.615	0.43008	0.87564	43.06	1050.85	1093.92
76	0.019491	13.501	0.422	13.923	18.260	21.323	39.583	0.44465	0.90532	44.06	1050.29	1094.35
77	0.020170	13.526	0.437	13.963	18.500	22.075	40.576	0.45966	0.93587	45.06	1049.72	1094.78
78	0.020871	13.551	0.453	14.005	18.741	22.851	41.592	0.47510	0.96732	46.06	1049.16	1095.22
79	0.021594	13.577	0.470	14.046	18.981	23.652	42.633	0.49100	0.99968	47.06	1048.59	1095.65
80	0.022340	13.602	0.487	14.089	19.222	24.479	43.701	0.50736	1.03298	48.06	1048.03	1096.08
81	0.023109	13.627	0.505	14.132	19.462	25.332	44.794	0.52419	1.06725	49.06	1047.46	1096.51
82	0.023902	13.653	0.523	14.175	19.702	26.211	45.913	0.54150	1.10250	50.05	1046.89	1096.95
83	0.024720	13.678	0.542	14.220	19.943	27.120	47.062	0.55931	1.13877	51.05	1046.33	1097.38
84	0.025563	13.703	0.561	14.264	20.183	28.055	48.238	0.57763	1.17606	52.05	1045.76	1097.81
85	0.026433	13.728	0.581	14.310	20.424	29.021	49.445	0.59647	1.21442	53.05	1045.19	1098.24
86	0.027329	13.754	0.602	14.356	20.664	30.017	50.681	0.61584	1.25385	54.05	1044.63	1098.67
87	0.028254	13.779	0.624	14.403	20.905	31.045	51.949	0.63575	1.29440	55.05	1055.06	1099.11

APPENDIX II.2 (*continued*)

Temp. t. °F	Humidity Ratio lb$_w$/lb$_{da}$ s	Volume ft³/lb dry air			Enthalpy Btu/lb dry air			Saturated water vapor Absolute Pressure, p		Enthalpy, Btu/lb Sat. water liq. h_f	Evap. h_{ig}/h_{fg}	Sat. water vapor h_g
		v_a	v_{as}	v_s	h_a	h_{as}	h_s	psi	in. Hg			
88	0.029208	13.804	0.646	14.450	21.145	32.105	53.250	0.65622	1.33608	56.05	1043.49	1099.54
89	0.030189	13.829	0.669	14.498	21.385	33.197	54.582	0.67726	1.37892	57.04	1042.92	1099.97
90	0.031203	13.855	0.692	14.547	21.626	34.325	55.951	0.69889	1.42295	58.04	1042.36	1100.40
91	0.032247	13.880	0.717	14.597	21.866	35.489	57.355	0.72111	1.46820	59.04	1041.79	1100.83
92	0.033323	13.905	0.742	14.647	22.107	36.687	58.794	0.74394	1.51468	60.04	1041.22	1101.26
93	0.034433	13.930	0.768	14.699	22.347	37.924	60.271	0.76740	1.56244	61.04	1040.65	1101.69
94	0.035577	13.956	0.795	14.751	22.588	39.199	61.787	0.79150	1.61151	62.04	1040.08	1102.12
95	0.036757	13.981	0.823	14.804	22.828	40.515	63.343	0.81625	1.66189	63.03	1039.51	1102.55

Abridged from *ASHRAE Handbook 1989, Fundamentals*. Reprinted with permission.

APPENDIX II.3
Physical Properties of Air (at atmospheric pressure 14.696 psia)

T (°F)	ρ (lb$_m$/ft³)	c_p (Btu/lb$_m$ °F)	$\mu \times 10^5$ (lb$_m$/ft² · s)	$\nu \times 10^3$ (ft²/s)	k (Btu/hr· ft·°F)	α (ft²/h)	Pr	$\beta \times 10^3$ (1/°F)	$g\beta\rho^2/\mu^2$ (1/°F · ft³)
0	0.0862	0.240	1.09	0.126	0.0132	0.639	0.721	2.18	4.39×10^6
30	0.0810	0.240	1.15	0.142	0.0139	0.714	0.716	2.04	3.28
60	0.0764	0.240	1.21	0.159	0.0146	0.798	0.711	1.92	2.48
80	0.0735	0.240	1.24	0.169	0.0152	0.855	0.708	1.85	2.09
100	0.0710	0.240	1.28	0.181	0.0156	0.919	0.703	1.79	1.76
150	0.0651	0.241	1.36	0.209	0.0167	1.06	0.698	1.64	1.22
200	0.0602	0.241	1.45	0.241	0.0179	1.24	0.694	1.52	0.840
250	0.0559	0.242	1.53	0.274	0.0191	1.42	0.690	1.41	0.607
300	0.0523	0.243	1.60	0.306	0.0203	1.60	0.686	1.32	0.454
400	0.0462	0.245	1.74	0.377	0.0225	2.00	0.681	1.16	0.264
500	0.0413	0.247	1.87	0.453	0.0246	2.41	0.680	1.04	0.163
600	0.0374	0.251	2.00	0.535	0.0270	2.88	0.680	0.944	79.4×10^3
800	0.0315	0.257	2.24	0.711	0.0303	3.75	0.684	0.794	50.6
1000	0.0272	0.263	2.46	0.906	0.0337	4.72	0.689	0.685	27.0
1500	0.0203	0.277	2.92	1.44	0.0408	7.27	0.705	0.510	7.96

Source: Fundamentals of Momentum Heat and Mass Transfer, Welty et al., 1976. John Wiley & Sons. Reprinted with permission.

APPENDIX II.4
Physical Properties of Water (at atmospheric pressure 14.696 psia)

T (°F)	ρ (lb$_m$/ft³)	c_p (Btu/lb$_m$·°F)	$\mu \times 10^3$ (lb$_m$/ft² · s)	$\nu \times 10^5$ (ft²/s)	k (Btu/hr· ft·°F)	α (ft²/h)	Pr	$\beta \times 10^3$ (1/°F)	$g\beta\rho^2/\mu^2$ (1/°F · ft³)
32	62,4	1.01	1.20	1.93	0.319	5.06	13.7	−0.350	
60	62.3	1.00	0.760	1.22	0.340	5.45	8.07	0.800	17.2
80	62.2	0.999	0.578	0.929	0.353	5.67	5.89	1.30	48.3
100	62.1	0.999	0.458	0.736	0.364	5.87	4.51	1.80	107
150	61.3	1.00	0.290	0.474	0.383	6.26	2.72	2.80	403
200	60.1	1.01	0.206	0.342	0.392	6.46	1.91	3.70	1010
250	58.9	1.02	0.160	0.272	0.395	6.60	1.49	4.70	2045
300	57.3	1.03	0.130	0.227	0.395	6.70	1.22	5.60	3510
400	53.6	1.08	0.0930	0.174	0.382	6.58	0.950	7.80	8350
500	49.0	1.19	0.0700	0.143	0.349	5.98	0.859	11.0	17350
600	42.4	1.51	0.0579	0.137	0.293	4.58	1.07	17.5	30300

Source: Fundamentals of Momentum Heat and Mass Transfer, Welty et al., 1976. John Wiley & Sons. Reprinted with permission.

APPENDIX II.5
Units and Conversion of Units

Most of the following conversion equivalents are based on values in *ASHRAE Handbook 1989, Fundamentals* and *ASHRAE Pocket Handbook, 1987.*

Unit	Abbreviations	Equivalents
air change per hour	ach	
ampere	amp, A	
atmosphere	atm	$= 14.696$ $lb_f/in.^2$ abs., psia
		$= 33.91$ ft of water
		$= 29.92$ in. Hg
		$= 1.013$ bar
		$= 101325$ Pa
Btu per pound	Btu/lb	$= 2.326$ kJ/kg
	$Btu/lb \cdot °F$	$= 4.187$ $kJ/kg \cdot K$
Btu per hour	Btu/h	$= 8.33 \times 10^{-5}$ ton
		$= 0.293$ Watts
	$Btu/year \cdot ft^2$	$= 0.000293$ $kWh/year \cdot ft^2$
	$Btu/h \cdot ft^2$	$= 3.153$ W/m^2
	$Btu/h \cdot ft^2 \cdot °F$	$= 5.678$ $W/m^2 \cdot K$
	$Btu \cdot in./h \cdot ft^2 \cdot °F$	$= 0.1442$ $W/m \cdot K$
British thermal unit	Btu	$= 778$ $ft \cdot lb_f$
		$= 1055$ J
		$= 0.2519$ kcal
calorie	cal	$= 3.97 \times 10^{-3}$ Btu
		$= 4.187$ J
clothing insulation	clo	$= 0.88$ $h \cdot ft^2 \cdot °F/Btu$
		$= 0.155$ $m^2 \cdot °C/W$
cubic foot	ft^3	$= 0.748$ gal
		$= 0.02832$ m^3
cubic foot/pound	ft^3/lb	$= 0.0625$ m^3/kg
cubic foot/minute	ft^3/min, cfm	$= 0.748$ gpm
		$= 0.472$ L/s
cubic meter	m^3	$= 1000$ L
		$= 35.31$ ft^3
degree Celcius	°C	$= 1.8$ °F
degree Fahrenheit	°F	$= 0.556$ °C
degree Kelvin	K	$= 1.8$ °F
degree Rankine	°R	$= 1$ °F
decibel (sound power)	dB re 1 pW	
decibel (sound pressure)	dB re 20 μPa	
dollar/sq. foot	$/ft^2$	$= 10.76$ $/m^2$
foot	ft	$= 0.3048$ m
foot of water, gauge	ft WG	$= 0.4334$ $lb_f/in.^2$
		$= 2.99$ kPa
foot per minute	ft/min, fpm	$= 0.01136$ miles/h
		$= 0.00508$ m/s
foot per second	ft/s, fps	$= 0.3048$ m/s
foot-pound	$ft \cdot lb_f$	$= 1.356$ J
foot-pound/minute	$ft \cdot lb_f/min$	$= 0.226$ W
gallon (U.S.)	gal	$= 0.1337$ ft^3
		$= 8.35$ lb of water
		$= 3.785$ L

Unit	Abbreviations	Equivalents
gallon/minute	gal/min, gpm	= 0.0631 L/s
giga-joule	GJ	= 1×10^9 J
gram	g	= 0.001 kg
horse power	hp	= 33,000 ft · lb_f/min
		= 550 ft · lb_f/s
		= 0.746 kW
horse power (boiler)	hp (boiler)	= 33,476 Btu/h
hertz (cycle/second)	Hz	
inch	in.	= 0.08333 ft
		= 25.4 mm
inch of mercury	in. Hg	= 0.4912 lb_f/in^2
		= 3.3 kPa
		= 25.4 mm Hg
inch of water	in. W	= 0.3613 lb_f/in^2
		= 5.20 lb_f/ft^2
		= 249 Pa
inch of water gauge	in. WG	= 1.0 in. W + 1 atm
flow resistance	in. WG/(cfm)^2	
joule	J	= 9.48×10^{-4} Btu
kilogram	kg	= 2.2046 lb
kilometer	km	= 3281 ft
		= 0.6214 mile
kilowatt	kW	= 3413 Btu/h
		= 1.341 hp
kilowatt/ton	kW/ton	= 3.51 COP_{ref}
kilowatt-hour	kWh	= 3413 Btu
		= 3.6×10^6 J
liter	L	= 0.001 m^3
		= 0.0353 ft^3
liter per second	L/s	= 2.119 cfm
		= 15.85 gpm
metabolic rate unit	met	= 18.46 Btu/h · ft^2
		= 58.2 W/m^2
meter	m	= 1.094 yard
		= 3.281 ft
		= 39.37 in.
megawatt	mW	= 10^6 W
microgram	μg	= 1×10^{-6} g
micron	μ	= 1×10^{-6} m
		= 3.94×10^{-5} in.
mil	mil	= 0.001 in.
		= 25.4 μ
mile	mile	= 5280 ft
		= 1.61 km
mile per hour	mile/h	= 88 ft/min, fpm
		= 0.44 m/s
milligram	mg	= 0.01543 grain
millimeter	mm	= 0.03937 in.
minute (angle)	min	= 0.01667 deg
molecular weight	mol	
ounce	oz	= 0.625 lb
		= 28.35 g

Unit	Abbreviations	Equivalents
part/million (mass)	ppm	$= 1$ mg/kg
pascal	Pa	$= 0.004$ in. WG
pint	pint	$= 28.37$ in.^{3}
		$= 0.4732$ L
pound (mass)	lb	$= 7000$ grains
		$= 16$ oz
		$= 0.4536$ kg
pound (force)	lb$_f$	$= 4.45$ Newtons
pound/lb dry air	lb/lb	$= 1$ kg/kg
pound of water	lb water	$= 0.01602$ ft^3
		$= 0.12$ gal
	lb/ft $\cdot$ h	$= 0.413$ mPa $\cdot$ s
	lb$_f$/ft $\cdot$ s	$= 1490$ mPa $\cdot$ s
pound/sq. foot	lb$_f$/ft^2	$= 0.0069$ lb$_f$/in.^{2}
		$= 4.88$ kg/m^2
pound/in.^{2} absolute	psia, lb$_f$/in.^{2} abs	$= 2.307$ ft water abs
		$= 703.1$ kg/m^2 abs
		$= 6.89$ kPa abs
pound/in.^{2} gauge	psig, lb$_f$/in.^{2} g	$= 1$ lb$_f$/in.^{2} $+ 1$ atm
pound/cu. foot	lb/ft^3	$= 16$ kg/m^3
quadrillion	quad	$= 1 \times 10^{15}$ Btu
		$= 1.055$ EJ
quart	qt	$= 57.75$ in.^{3}
		$= 0.9461$ L
radian	rad	$= 57.3$ degrees
revolutions/minute	rpm	$= 1$ rev/min
square foot	ft^2	$= 144$ in.^{2}
		$= 0.0929$ m^2
square meter	m^2	$= 10.76$ ft^2
therm	therm	$= 100,000$ Btu
		$= 105.5$ MJ
ton (long)	ton (long)	$= 2240$ lb
		$= 1016$ kg
ton (short)	ton (short)	$= 2000$ lb
ton (metric)	ton (metric)	$= 1000$ kg
ton (refrigeration)	ton	$= 12,000$ Btu/h
		$= 3.516$ kW
ton-hour	ton $\cdot$ h	$= 12,000$ Btu
		$= 3.516$ kWh
torr	torr	$= 1$ mm Hg
watt	W	$= 3.413$ Btu/h

ABOUT THE AUTHOR

Shan K. Wang received his B.S. in mechanical engineering from Southwest Associated University in China in 1946. Two years later, he completed his M.S. degree in mechanical engineering at Harvard Graduate School of Engineering. In 1949, he obtained his M.S. in textile technology from the Massachusetts Institute of Technology.

From 1950 to 1974, Wang worked in the field of air conditioning and refrigeration in China. He was the first Technical Deputy Director of the Research Institute of Air Conditioning in Beijing from 1963 to 1966 and from 1973 to 1974. He helped to design space diffusion for the air conditioning system in the Capital and Worker's Indoor Stadium. He also designed many HVAC&R systems for industrial and commercial buildings. Wang published two air conditioning books and many papers in the 1950s and 1960s. He is one of the pioneers of air conditioning in China.

Wang joined Hong Kong Polytechnic as senior lecturer in 1975. He established the air conditioning and refrigeration laboratories and established courses in air conditioning and refrigeration at Hong Kong Polytechnic. Since 1975, he has been a consultant to Associated Consultant Engineers and led the design of the HVAC&R systems for Queen Elizabeth Indoor Stadium, Aberdeen Market Complex, Koshan Road Recreation Center, and South Sea Textile Mills in Hong Kong. From 1983 to 1987, Wang published *Principles of Refrigeration Engineering* and *Air Conditioning* as the teaching and learning package, and presented several papers at ASHRAE meetings.

Wang has been a member of ASHRAE since 1976. He has been a governor of the ASHRAE Hong Kong Chapter-At-Large since the Chapter was established in 1984. Wang retired from Hong Kong Polytechnic in June 1987 and immigrated to the United States in October 1987. Since then, he has joined the ASHRAE Southern California Chapter and devoted most of his time to writing.